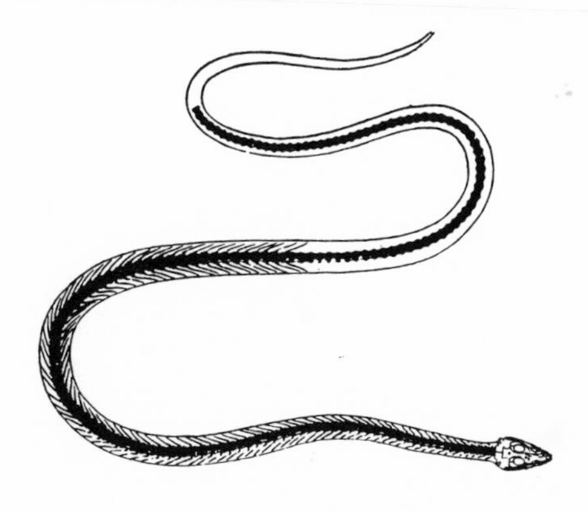

VERTEBRATE PALEONTOLOGY

VERTEBRATE

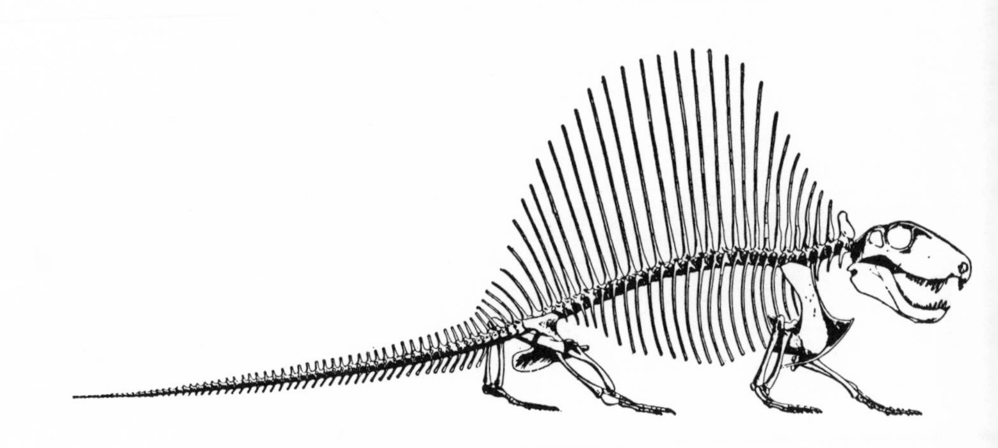

PALEONTOLOGY

ALFRED SHERWOOD ROMER

Third Edition

THE UNIVERSITY OF CHICAGO PRESS
Chicago and London

International Standard Book Number: 0-226-72488-3
Library of Congress Catalog Card Number: 66-13886

THE UNIVERSITY OF CHICAGO PRESS, CHICAGO 60637
The University of Chicago Press, Ltd., London

Third Edition published 1966. Third Impression 1971
Printed in the United States of America

Preface to the Third Edition

Although I trust that the second edition of this work, published in 1945, has served a useful purpose, the great amount of research on fossil vertebrates done during the postwar period has caused much of the text to become embarrassingly antiquated. In consequence, I have devoted much of my time during the last several years to a thorough revision. Certain general passages and the accounts of some of the mammalian and reptilian groups which have long been well known have undergone only minor modifications; but most of the sections relating particularly to fishes, amphibians, the older reptile assemblages, and the more primitive mammal types have been radically revised. About half of the text has been rewritten. Some sixty-six new illustrations have been added and about eighty others replaced. I have spent (drearily) much time in a thorough revision of the systematic listings and have included, in addition to currently accepted generic names, many synonyms which the serious student will encounter in the literature.

In the course of preparation of this new edition, I have sought and received advice and information from a host of friends and colleagues—so many, in fact, that although I have attempted to cite them all below, I feel sure that I have unintentionally overlooked some to whom I am indebted. At the Harvard Museum I am fortunate in having close at hand paleontological colleagues from whom aid has been solicited and generously given—Dr. Tilly Edinger, Professor Bryan Patterson, Dr. George G. Simpson, Dr. Ernest E. Williams. I have discussed various points in fish history with Drs. David H. Dunkle, Colin Patterson, Bobb Schaeffer, Errol I. White, and T. Stanley Westoll, and have corresponded on problems of early fish evolution with Drs. R. H. Denison, Tor Ørvig, and R. S. Miles and on lower actinopterygians with Drs. B. G. Gardiner and J. R. Nursall. Dr. Shelton P. Applegate has aided in the listing of shark occurrences and Mr. David Bardack with Cretaceous teleosts. Drs. Jean Piveteau and J.-P. Lehman kindly lent me manuscript from the unpublished fish volume of the "Traité de Paléontologie," and Dr. Colin Patterson gave me access to his unpublished work on chimaera history. Dr. Keith S. Thomson aided in the area of the Sarcopterygii. I have been given access to the forthcoming classification of teleosts by Drs. P. Humphry Greenwood, Donn E. Rosen, Stanley H. Weitzman, and George S. Myers, and have essentially followed this able attempt at resolution of the complex problem of teleost relationships and phylogeny in my text and systematic listings. Dr. R. R. Miller aided in clearing up a portion of the confusion which exists in older identifications of North American Tertiary teleosts.

In recent years workers have been active in the field of Carboniferous tetrapods. I am very grateful to Dr. Donald Baird for much valuable information and suggestions; Dr. Robert L. Carroll's current work on the Joggins fauna has proved very useful, as has Dr. Alec L. Panchen's restudy of British embolomeres. Drs. James and Margaret C. Brough have given me access to their unpublished work on Carboniferous amphibians, and I have corresponded with Dr. John W. Cosgriff regarding rhachitomes. I have followed Drs. Ernest E. Williams and Thomas S. Parsons in their interpretation of the relationships of the modern amphibian orders, and Drs. Max Hecht and Richard Estes have aided in the problem of classification of these groups. I have enjoyed various discussions of reptile problems with, among others, Drs. Edwin H. Colbert, F. R. Parrington, and Everett C. Olson. Drs. Olson and Peter P. Vaughn have discussed primitive reptile groups with me. Dr. Williams has aided me with regard to the Chelonia, Dr. Estes with the Squamata, Mr. William Sill with the Crocodilia, Dr. Dale A. Russell the mosasaurs, Dr. Joseph T. Gregory the phytosaurs, Dr. Wann Langston the dinosaurs, and Dr. C. Barry Cox the dicynodonts. I have had profitable conversations with Drs. Alick D. Walker and Alan J. Charig regarding thecodonts and Triassic dinosaurs. I have discussed problems concerning early lepidosaurs and possible euryapsids with Dr. Pamela L. Robinson, and she has kindly sent me unpublished data on her Triassic "flying reptile," *Kuehneosaurus*. Dr. James A. Hopson has furnished me with figures, as yet unpublished, of *Bienotherium*. Drs. Theodore E. White and J. Alan Holman have furnished systematic data.

I am much indebted to Dr. Pierce Brodkorb for furnishing me with data from unpublished sections of his valuable checklists of fossil birds. I have discussed various items in avian paleontology with Drs. Alexander Wetmore and Hildegarde Howard.

After a long period of quiescence, the field of Mesozoic mammals is again active and in a state of flux. I profited from discussion with colleagues such as Drs. Kenneth A. Kermack, B. Patterson, and G. G. Simpson, with whom I attended a conference, led by Professor G. Vandebroek, mainly devoted to these problems. Also in a current state of turmoil is the general field of the Insectivora (*sensu lato*) and other early placentals. I have discussed problems in this area with Drs. Malcolm C. McKenna and Leigh Van Valen. I have in great measure followed their current interpretations of insectivore relationships and have adopted Van Valen's conclusion (already suggested by earlier workers) that the deltatheridiids are ancestral to the "proper" creodonts (although preferring the available term Creodonta for the proposed order to the creation of a new term). Dr. Robert E. Sloan aided with data on the important new "Bug Creek" Cretaceous discoveries and the classification of the multituberculates. Professor Elwyn L. Simons furnished valuable information and suggestions regarding lower primates, Dr. Leonard Radinsky furnished new data on tapiroids, and Dr. Rosendo Pascual aided greatly with problems concerning South American mammals. As regards rodents, I consulted particularly Professor Albert E. Wood, and have adopted a classification of the group essentially that agreed upon (for the most part) at a recent conference held by him, Drs. Craig C. Black, Mary R. Dawson, and René Lavocat. Information on various specific points was kindly furnished by Drs. James S. Mellett, Charles A. Reed, Peter Robinson, and John A. Wilson.

There are, of course, countless problems of vertebrate phylogeny and relationships which are currently unsettled and debatable. Although I have in general attempted to mention alternative interpretations, it is necessary in an introductory work to establish a coherent working pattern by choosing one specific course in each area of conflict without, unfortunately, being able to discuss adequately the grounds upon which the decision was based. I hope, at some time in the near future, to publish a modest companion volume in which many of these problems can be treated.

Miss Nelda Wright has, as ever, aided me throughout the preparation of this revision. Miss Margo Hayes has aided in typing; Mrs. Elizabeth Pfohl has carried the main task of manuscript preparation with fortitude and intelligence.

ALFRED SHERWOOD ROMER

CAMBRIDGE, MASSACHUSETTS

Preface to the First Edition

The story of vertebrate evolution as revealed by the study of fossils is one of interest not only to the paleontologist but to those engaged in other branches of science—to the geologist and to the worker in many fields of biology. The present volume was undertaken because of the lack of any modern work in English dealing with the subject as a whole. Smith Woodward's excellent *Outlines of Vertebrate Paleontology* dates from the late nineties; and the English edition of Zittel's excellent reference work, *Grundzüge der Paläontologie,* is based on a German edition nearly as old. During recent decades our knowledge of vertebrate fossils has increased greatly, but the new facts and the new concepts based on them have been but slowly taken up in general biological or geological literature.

The most general appeal of vertebrate paleontology lies in the evolutionary story it presents. In consequence, the material is here arranged as a group-by-group treatment of the vertebrates, tracing out the various ramifications of the family tree.

Our knowledge of extinct vertebrates is based primarily upon preserved skeletal remains; and the anatomical study of these remains is the solid foundation from which alone can arise any treatment of these forms from the point of view of taxonomy, stratigraphy, ecology, or evolution. No attempt has been made here to go into minutiae, but I have not hesitated to describe at some length the major structural features of the various groups encountered.

Although the relationships of many groups are quite uncertain or in dispute, it has been necessary to adopt, tentatively at least, some one theory in each case to establish an order of treatment and set up a usable classification. The phylogenetic framework and the taxonomic scheme here established upon it are, for the most part, conservative. In general I have adopted the theories of relationships now current among workers in the various fields and have departed from common usage only where this has been necessitated by recent investigations.

To the worker in any field the history of his subject, the great figures in the science, and their individual accomplishments are matters of considerable interest. The writer feels, however, that for the beginner such a historical treatment is not particularly appropriate; it is with the history of animals, not scientists, that he is to deal. In consequence, personalities have been omitted from the text. For the more advanced student who wishes to delve into the original literature a knowledge of the workers and their achievements is of considerable interest and importance. With this in mind, brief historical notes have been introduced into the Bibliography.

The Bibliography makes no pretense of comprehensiveness and includes only a restricted list of the more important or more useful works. Particular attention has been given to monographic works dealing adequately with entire groups, to recent papers giving the latest advances in the subject, to papers containing adequate bibliographies, and especially to works which contain adequate illustrations of a group or of representative forms.

As regards illustrations, a complete skeleton and skull of a representative member of each group is shown, whenever possible, and, in addition, other significant or diagnostic skeletal details, such as the limbs or teeth. In order to facilitate comparison of forms, original figures have been reversed in numerous cases, so that all skulls and skeletons are viewed from the right, and all limb bones are of the right side; in the mammals upper dentitions are of the right side, while the lower dentitions are from the left jaw ramus, to afford easier understanding of occlusional relations. The skeletons are almost all direct copies of well-known figures. The other illustrations are, for the most part, new drawings, although the majority are adapted from preexisting figures. In contrast to the illustrations in an original scientific paper, which are generally dedicated to the faithful portrayal of often imperfect material, those in this volume aim to give the reader as complete a picture as possible of the structure of the forms portrayed. With this in mind, missing portions of skulls have been freely restored and composite restorations attempted in many cases, although the writer thereby lays himself open to possible criticism.

I am greatly indebted to many friends for assistance in the preparation of this volume. Dr. W. K. Gregory, of the American Museum of Natural History, has read nearly the entire work and has offered many helpful and stimulating criticisms. Dr. F. B. Loomis, of Amherst College, and Dr. G. G. Simpson, of the American Museum, have read the chapters dealing with mammals and have saved me from numerous sins both of commission and omission. My colleagues, Drs. R. T. Chamberlin, J Harlen Bretz, and Carey Croneis, have read portions of the manuscript; and Mr. Bryan Patterson, of the Field Museum of Natural History, has aided in the chapters on South American ungulates and edentates.

The greater part of the original illustrations, particularly the skull figures, are the work of Mr. L. I. Price, to whom I am indebted for much painstaking work as a collaborator rather than merely as an illustrator. Mr. Brandon Grove has assisted greatly with photographic work.

Finally, and most greatly, I am indebted to my wife, Ruth Hibbard Romer, for her unfailing aid throughout the course of preparation of this book.

Alfred Sherwood Romer

CONTENTS

CHAPTER 1

Introductory

The present work gives a brief account of the history of vertebrates as revealed in the fossil record. We shall find it difficult, however, to confine ourselves strictly to fossil animals, for our subject is intimately related to many other scientific fields.

Vertebrate paleontology is essentially a biological science. To the paleontologist the animals of the present constitute but a brief cross-section of the vertebrate story. To him a separation of fossils from the modern forms, which have descended from extinct types and are destined to become the fossils of the future, would seem extremely artificial. Paleontology has much to learn from many other biological fields and much to give in return. With anatomy and evolutionary studies the connections are particularly close, while not only the taxonomist and ecologist but even workers in such seemingly remote fields as physiology and medicine may benefit from the possession of a historical background.

Paleontology is also intimately connected with the history of the earth itself, the field of historical geology. Vertebrate paleontology can contribute valuable information to the stratigrapher and the paleogeographer, and, on the other hand, the geologist's study of sediments can tell the paleontologist much as to the environment in which his ancient animals lived—and died. In the study of fossil human types the field of the vertebrate paleontologist overlaps that of the anthropologist. And, finally, the study of vertebrate evolution is history of a sort—but a history written in terms of millions of years rather than decades or centuries.

Since it is probable that this book will fall into the hands of readers with varied backgrounds, this opening chapter is devoted to a number of topics which may well be omitted by those grounded in the fundamentals of biology.

Vertebrate fossils.—Few of the countless individual animals of the past have been preserved to us as fossils. When a land animal dies today, its remains are usually scattered and destroyed by flesh-eaters or disintegrated by plant roots, soil acids, and bacterial action. Similar conditions, no doubt, were present in the past.

More exceptionally and more fortunately, a dead animal may be buried deep in a deposit of mud or sand, where the soft parts will usually quickly rot away but where the skeleton will be left surrounded by a slowly hardening matrix. The bony matter may persist and become gradually altered into a complex type of mineral in which many of the original constituents of the bone remain. Meanwhile, the cavities are, in the course of time, filled by minerals brought in in solution, so that the originally porous bone becomes a heavy, solid structure. But since the mineral matter filling the cavities is usually of a different nature from that into which the substance of the bone has been transformed, fossil bones, when sectioned and studied microscopically, frequently show in perfect fashion the structural and histological details of the original.

The softer portions of the body are sometimes, although much more rarely, preserved. Certain forms recently extinct, such as the dodo or the giant moas of New Zealand, may be considered as fossils of a sort, although their bones are usually but little altered chemically; in these cases feathers, ligaments, and other soft parts have been preserved. In America gigantic ground sloths are extinct, and many of their remains are perfectly orthodox fossils; but in several instances skeletons have been found with patches of skin and hair still adherent. Extinct mammoths have been completely preserved in natural "cold storage" in the Siberian tundras, while woolly rhinoceroses have been found pickled in entirety in a waxy material from a Galician oil seep.

But even among much more ancient forms traces of soft parts are occasionally found. Specimens crushed flat in shales have in many cases preserved the outline of the skin. Certain dinosaurs appear to have been mummified by natural forces before burial, so that the surrounding material, hardening before the disintegration of the skin, has left a cast of the mummy. In Bavarian lithographic limestone deposits, impressions are sometimes preserved of such delicate structures as feathers. In some small ancient amphibians the outlines of the stomach and intestine are visible; while in specimens of a Devonian shark, remains of the muscles are so perfectly preserved that individual fibers and their cross-striations are seen under the microscope.

We may gain much information concerning fossil forms in other ways. Footprints are often found in the now solidified mud flats of former periods. Eggshells pertaining to various groups—birds, turtles, skates, dinosaurs—have been discovered. Series of growth stages are known in some forms, and in the late Paleozoic have been found numerous "branchiosaurs," which are gilled larvae of labyrinthodont amphibians. Much information can be gathered concerning the food habits of extinct animals, aside from the knowledge afforded by the teeth and skeleton. Remains of the bones of devoured animals have been found in the position of the stomach; "gizzard stones" have been discovered lying within the skeletons of several reptilian types. Tooth marks tell us of the fate of the animal whose bones bear them. Feces, the remains of digestion, known as coprolites in the fossil state, are frequently encountered. Wounds and diseased conditions may be discerned on the bones of fossils. Even though actual remains of such structures as blood vessels, nerves, brain, and muscles are almost never found, a study of the skeleton gives us much information concerning them, particularly in the cranial region.

Evolutionary theories and phenomena.—In considering many curious and widely divergent animals adapted to a great variety of modes of life, it is impossible to avoid speculation as to the processes through which their evolutionary development was accomplished. This subject cannot be adequately discussed here, but we shall mention a few of the chief problems involved and the theories brought forward for their solution.

All animals are well fitted for the lives they lead (although sometimes, it is true, the fit does not seem to be a perfect one). We also find, in almost every case where fossil ancestors are known, that these animals have descended with many modifications in structure from forms seemingly less well adapted to their present modes of life. How have these changes come about?

Occasionally we find a statement of this type: "The giraffe, descended from short-necked ancestors, has acquired a long neck because it is useful to him in browsing off the higher branches of the trees." It is true that the long neck is of use to the giraffe, but the citation of a useful result does not in itself explain the nature of the process by which the result was obtained. The statement given would suggest either that some power outside the giraffe brought about the change with a useful end in view or that the giraffe had thought the matter over and brought about the change through his own volition. The first suggestion immediately takes us outside the realm of scientific thought; the second seems obviously absurd, although certain theorists have suggested the existence of some vague "vital force," a mysterious inner "urge," which has driven animals on toward an evolutionary goal.

A theory of the inheritance of acquired characters, or the effect of the use and disuse of parts, was advocated a century and a half ago by the French naturalist Lamarck. It might be that the individual giraffe, by constantly stretching after the foliage on the higher branches, tended to increase the length of his neck during life and that this increase was transmitted to his offspring; a cumulation of such individual lengthenings was supposed to have resulted in the development of a long-necked from a short-necked form. This type of theory is plausible at first sight; but there is no evidence, despite repeated experiments, that changes taking place in the body of an individual during his lifetime produce any similar effect in his offspring.

A much sounder theory to explain evolutionary changes, and one which first brought about general belief in evolution, was that advocated by Darwin a century ago. The Darwinian theory points out that no two animals are exactly alike and that, to cite the giraffe again, a form with a neck slightly longer than the average would, especially in times of famine, have a better chance of survival than a shorter-necked form. This selective process would tend, Darwin believed, to the gradual development of the long neck.

The principle of natural selection is an exceedingly important one, and, indeed, is still the underlying concept in the general theory of evolution held by most today. But it was not, in itself, enough; many puzzles remained, primarily because in Darwin's day nothing was known of the process by which the variations upon which selection relied came into existence, and nothing was known of the mechanisms by which such variations were inherited.

A vast amount of work in the present century has resulted in the development of the science of genetics, which has given us a very concrete knowledge of the mechanisms of heredity, and much data as to the nature of variation. Inheritance is not, as once thought, a "blend" of the characteristics of the two parents, but is particulate. Each animal has, for the growth and functioning of its body and for transmittal to its offspring, thousands of hereditary units, the genes, present in the chromosomes of each cell; current studies in biochemistry are yielding valuable knowledge as to the nature and mode of operation of the genes. Each gene, although often with most notable effects on one structure or character, may influence a number of different features of the body and, on the other hand, any given feature may be influenced by a number of genes. Further, every gene is present in duplicate in every cell; every individual receives one complete set of genes from each parent and transmits a complete single set to each of its offspring. Still further, a given gene may vary, producing varied effects; among variants—alleles—one type may dominate over its partner, whose potential effects are concealed. The variations are the result of mutations, which are known to occur constantly, if relatively rarely, in every organism; their causes are incompletely known, although radiation and certain chemical agents are known to be causative agents. It must be emphasized that mutations are essentially random in nature, and are not "directed" in any way or related to Lamarckian assumptions.

Mutations, then, furnish a "pool" of variation in the genetic composition of the individuals making up a species and furnish the material upon which selection may act. Study of the potentialities of the complex hereditary mechanisms which we know today, but which were unknown to Darwin, lead to deductions as to the probable course of evolution which agree well with the known facts of the fossil story. From a combined study of genetics, paleontology, and systematics there has been drawn a synthetic theory of evolution which has been generally accepted by recent workers. The reader who is not acquainted with modern concepts would be well advised to consult some work in this field.

As our knowledge of the fossil story has gradually increased, varied patterns of evolutionary progress have become apparent. At certain times in certain groups, rapid evolutionary changes appear to have taken place. A new "ecological niche" may open up for occupancy, and an animal type may change markedly and rapidly to take maximum advantage of the opportunity presented. Sometimes a variety of major opportunities may open up for members of a group, with the result that a whole series of diversified descendants may rapidly evolve to occupy, with appropriate adaptations, a series of varied ecological niches. Such a phenomenon is termed an adaptive radiation; an outstanding example of such a radiation is that of the placental mammals at the dawn of the Tertiary, following the extinction of the dinosaurs.

Apart from such essentially explosive outbursts, the speed and pattern of evolutionary progress may show wide variation. In some types, change appears to have come essentially to a standstill over long periods of time. For example, among odd-toed ungulates, the modern horses and rhinoceroses have departed far from the structural pattern of the Eocene ancestors of the order, but their tapir cousins have survived to the present little changed, except in stouter build and minor tooth modifications, from the ancestral types.

Frequently we seem to see in evolutionary history a more or less continuous series of changes, each of modest degree, occurring at modest speed (geologically speaking). We sometimes get the impression that groups once started along an evolutionary line have kept on without deviation toward a "goal." This resulted in the building up of a theory of orthogenesis, or "straight-line" evolution, with implications of "design" by supernatural forces back of the phenomenon. However, with increasing knowledge of the fossil record, it becomes increasingly clear that there is no cause for any mystical concept in this regard. There are, it appears, really few "straight" evolutionary lines. Most lines do have side branches, and their apparent absence in some cases may be due to the fact that they were nipped in the bud; selection alone might have resulted in an advance in the single direction of greatest adaptive value.

The adaptive radiation of a number of groups tends to result in convergence in evolution—the attainment of similar adaptations by two or more unrelated types. The Tasmanian wolf is similar in many features to the true wolves of other continents; the pouched Australian wombat is similar to a woodchuck. But these resemblances do not at all imply relationships; they are merely the taking-on of similar structural features necessary for survival in similar modes of life.

Sometimes confused with convergence, but basically a very different phenomenon is that of parallelism—the tendency for two or more closely related types to undergo similar structural changes. Several types of extinct South African mammal-like reptiles had a secondary palate like that of mammals. We are sure that this feature was not present in their common ancestor but that, with a similar ancestry and similar build, there has been a tendency for related forms to acquire similar structures—to "drift" in the same evolutionary direction.

It was once assumed that most of the structural features found in modern representatives of a given group had been present in the common ancestor. But increasing knowledge of phylogenies show that this is frequently not the case; such features are often found to have developed in parallel fashion in various forms during evolutionary history. This situation makes precise definition of a group difficult; we cannot define it by the *possession* of such-and-such characters by all its members; the best we can do for a definition is to cite a trend toward the acquisition of these characteristics.

Taxonomy and classification.—Zoologists have found it necessary, in referring to living animals, to use a system of scientific names formed mainly from Greek and Latin roots, rather than variable popular terms. This necessity is even greater in the case of extinct forms for which no popular names can well exist. Every animal bears a double name: the first, that of a genus, usually including a number of related forms; the second, that of the particular species of that group to which reference is made. Thus the domestic dog, together with the wolves and jackals, constitute the genus *Canis.* When referring to the dog, the specific name is added—*Canis familiaris;* the wolf, a close relative, is *Canis lupus.* This system is not unlike our usage of given names and surnames, only the order is reversed.

Upon the genus and species as a base has been erected a system of classification originally designed as a method of "pigeonholing" animals in larger and smaller groups according to the degree of resemblance between them. Thus, for example, the dogs are obviously fairly closely related to the foxes and to other doglike tropical forms; and these are all united in a family, the Canidae. This group, in turn, is somewhat more distantly related to the cats, bears, and other flesh-eaters; and all these forms are united in an order, the Carnivora. This great group, again, has many features (such as the nursing of young and the presence of hair) which tend to bind it with an array of forms ranging from men to bats and whales; all these are considered members of a common class, the Mammalia. These mammals have a number of characters (such as the possession of an internal skeleton) also present in birds, reptiles, amphibians, and fishes; and, in consequence, all are grouped together in a phylum, the Chordata, one of the primary subdivisions of the animal kingdom. The complete list of major terms used in such a classification is thus: phylum, class, order, family, genus, and species. Further flexibility is often obtained by using prefixes to give such terms as subclass and superorder.

Originally adopted merely as a convenient method of "sorting out" animals, classification was put in a new light by the recognition of the evolutionary theory. Obviously these various taxonomic divisions represented, in a general way, the branches of the evolutionary family tree. There has been a continual attempt to make classifications "natural," to see that each group established—whether genus, family, or higher division

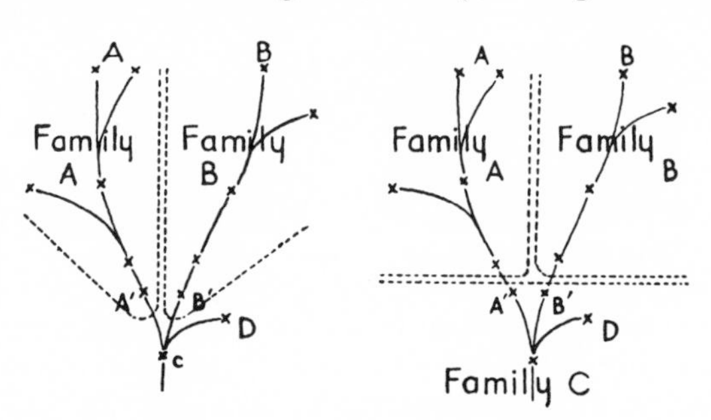

FIG. 1.—Diagram to show the contrast between "vertical" and "horizontal" classifications. A hypothetical family tree showing the descent of two living forms, *A* and *B* (horses and rhinoceroses, for example) from a common ancestral form, *C*; known forms are indicated by crosses. At the left, a vertical classification; the divisions between groups are carried as far down to the roots as possible. This makes clear the relation of early ancestors (*A′ B′*) to their descendants but separates them sharply from *C*, the common ancestor, and other side branches, such as *D*, to which they are closely related. At the right, a "horizontal" classification, which unites all the similar early forms into a common ancestral group, *C*, but obscures the relationship of *A′* and *B′* to their descendants.

—should contain only forms descended from a common ancestor.

With living types alone to consider, groups may be fairly easily established and defined. But with the continued discovery of intermediate fossil types, classification becomes increasingly difficult. The living Equidae (horses, asses, and zebras) can be easily distinguished from the related rhinoceroses by such characters as the presence of but a single toe on each foot and the absence of horns. But some of the earliest horses had three toes, as do the rhinoceroses; and many of the early rhinoceroses were hornless. We cannot always give definitions which will hold true of all members of a group; we can merely cite the characters of typical members and build the groups about them. Tendencies, rather than the arrival at specific conditions, must be our criteria. Only a few late horses are actually one-toed, but a trend in that direction was early evident; the ancestral rhinoceroses were hornless, but a tendency for the development of horns soon appeared in several independent lines.

Two types of classification are possible—"vertical" and "horizontal" (Fig. 1). Under the first system each family or other unit comprises all members of a known line from its first beginnings to its end or to modern times; the cleavage between lines is carried down to the very base of the evolutionary tree. But when, for example, forms are discovered seemingly ancestral to two distinct families or closely related to both, their inclusion in one or the other seems improper. Under such circumstances the best solution seems to be a "horizontal" cleavage, the erection of a stem group, including the base from which the long-lived later families have been derived.

"Horizontal" or "vertical" classifications of the sort just described are both acceptable. To be avoided, however, is another type of "horizontal" classification. Let us assume that of the forms shown in the diagram, *A* and *B*, although with separate pedigrees, have evolved in much the same direction. If, after setting up (as at the right of the diagram) a basal family *C*, a scientist decides, because of similarities, to unite *A* and *B* in a common family, he has created a family that is not merely "horizontal" but polyphyletic—that is, he has included forms of different origin in a common assemblage. Sometimes, because of inadequate knowledge, this has been done, inadvertently, by paleontologists, but such unnatural grouping should be avoided.

The large number of taxonomic terms with which one must become familiar in the study of fossil vertebrates tends to dishearten the beginner. But it will soon be recognized that (except in cases where names of persons or localities have been utilized) these terms are mainly compounded from but a few-score simple and easily recognizable Greek and Latin roots and are based on some real or fancied characteristic of the form described.

It is to be noted that, while considerable variation occurs in the formation of names of various larger groups, those of superfamilies usually end in *-oidea,* those of families invariably end in *-idae,* of subfamilies in *-inae,* all suffixed to the root of the name of a typical included genus, as Equoidea, horse-like forms; Equidae, horses; Equinae, modern horses; *Equus,* horse genus.

Vertebrate structure.—Before we can adequately discuss the history of the varied vertebrate types, a basic knowledge of their anatomy is a necessity. We shall here briefly review some of the more essential structures of the primitive water-living vertebrates, the fishes, especially as regards the skeletal system. (The morphological features of higher vertebrates will be discussed later.)

All vertebrates are bilaterally symmetrical animals, with the long axis of the body usually in a horizontal position and with a tendency for the concentration of organs related to the environment at the anterior end. This type of symmetry appears to be an expression of the usually active life of vertebrates, as contrasted with the comparative immobility of those invertebrates possessing radial symmetry. In land animals the development of large limbs tends somewhat to mask the primitively simple arrangement, while a radical change in the position of the body axis occurs in man and other bipeds.

Body form in fishes.—The ideal shape for a swimming vertebrate is that which engineers have approached in torpedoes and ship hulls as offering least resistance to the water—streamlined, with the maximum girth somewhat anterior to the middle. Actively swimming fish usually have this shape. Locomotion in such forms is principally accomplished by lateral undulations of the trunk and tail, a series of curves traveling backward and pushing the fish forward through the water; paired fins are usually steering organs only. Sinuosity is sometimes exaggerated, giving eel-like forms, usually degenerate end-products of evolutionary lines; while bottom-living types are usually depressed, with broad flat bodies.

Lying dorsally and ventrally along the main axis of the fish body are unpaired median fins consisting of a dorsal fin or fins projecting from the upper side of the body; a caudal fin, in the tail region; and an anal fin, lying behind the anus (Fig. 2). Dorsal fins are usually one or two in number in primitive types; a single anal fin is common. There may be great variation in the way of fusion or subdivision of the dorsal fins and fusion of either anal or dorsals with the caudal.

Two types of caudal fin are common in more primitive fish—the diphycercal and the heterocercal (Fig. 3). In the former the fleshy termination of the body containing the backbone runs straight out to the tip of the tail, and the fin is arranged symmetrically above and

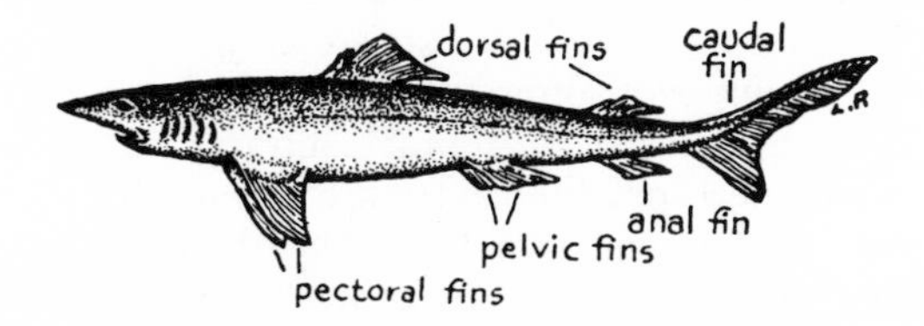

FIG. 2.—Diagram to show the position of the median and paired fins of fishes.

below it. In the heterocercal tail the body, containing the backbone, tips up posteriorly, and nearly all the development of the fin takes place beneath. In a few primitive fishes we find a reversal of this symmetry, the tip of the body tilting down and the fin erected above it—a reversed heterocercal type. The symmetrical diphycercal fin would seem logically to be the primitive type; the heterocercal, a specialized derivative. However, as we shall see, the latter is almost universal in the earliest members of most fish groups; and it is not improbable that the diphycercal type is in reality derived from this.

Notochord.—Vertebrates and their close relatives differ from the common invertebrate types in the possession of an internal skeleton for support and the facilitation of muscular movement. Among invertebrate types the skeleton, when present, is an external one, covering the surface of the body. An external skeleton was early developed by vertebrates and is present, to at least some degree, in many forms today, but internal skeletal structures are always present in addition. Earliest and most primitive of such internal structures, we believe, was the notochord—a long, slim rod usually extending from the base of the skull down the back to the tail. Composed of a soft, jelly-like material surrounded by a tough sheath, it forms in primitive vertebrates a firm but flexible supporting structure.

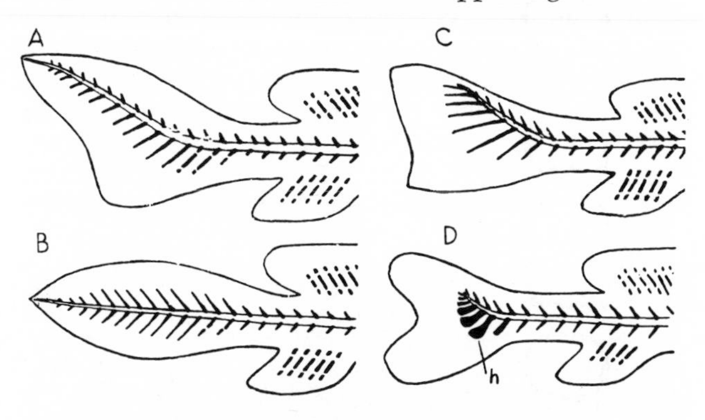

FIG. 3.—Diagram of caudal-fin types. *A*, Heterocercal type, found in most sharks, primitive lungfish, and actinopterygians and some early crossopterygians; *B*, diphycercal type, found in pleuracanth sharks and later lungfish and crossopterygians; *C*, abbreviated heterocercal, and *D*, the derived homocercal type, found in later actinopterygians. In *D* the enlarged haemal arches (*h*) are termed "hypurals." *C* and *D* are undoubtedly derived from *A*, and so probably is *B* as well. (Modified from Goodrich, *A Treatise on Zoölogy*, Part IX.)

In some water-living vertebrates it may persist throughout life; in higher types it is supplanted by the backbone but is always present in the embryo. Containing no hard parts, it is never preserved as a fossil.

Cartilage and bone.—In addition to the notochord, all vertebrates have a skeletal system composed of cartilage or bone. The former is a comparatively soft and translucent material, containing rounded cells and capable of growth by expansion. Ordinary cartilage is seldom preserved in fossil specimens, for it shrivels up and disintegrates easily upon exposure. In many fishes, however, calcium salts are laid down in the cartilage; and in this calcified condition it is a much firmer substance, capable of preservation.

Bone is, in microscopic section, easily distinguishable from cartilage by the presence of irregularly branching cell spaces, the presence of larger cavities for blood vessels which penetrate it, and its frequent arrangement in layers of deposition. It consists of a matrix rich in fibers and heavily impregnated with calcium salts; it is a much stronger supporting material than cartilage. Unlike that substance, bone is incapable of expansion and can grow only by the addition of layers on its surface. In large bones the interior often contains a large marrow cavity as well as numerous smaller canals containing blood vessels and nerves.

Cartilage is almost never found in the skin; when

present, it is confined to the deeper layers of the body. In existing lower vertebrates (cyclostomes and shark-like fishes), cartilage alone is present in the main portion of the skeleton; hence these forms lack any superficial skeletal covering except for small toothlike structures (dermal denticles) found in the skin of sharks, and spines, of similar structure, occasionally present in these fishes.

In the higher groups of vertebrates, bone is the predominating skeletal material; little cartilage remains in the adult. Two categories of bone may be distinguished. The deeper portions of the skeleton are formed in the embryo in cartilage, just as in a shark or lamprey. As development proceeds, however, this cartilage is gradually destroyed and replaced by bone. Bone so formed is called "endochondral" or "cartilage-replacement"

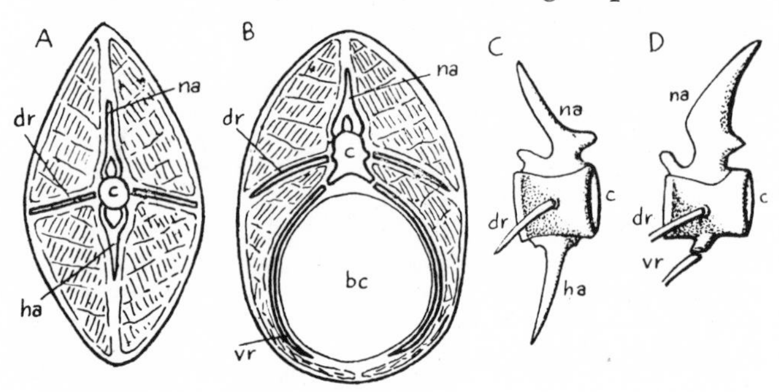

FIG. 4.—*A*, Diagrammatic section through the tail of a fish; *B*, section through the trunk, to show the relation of the vertebrae and ribs to the musculature and body cavity; *C*, diagrammatic caudal vertebra of a teleostean fish; *D*, trunk vertebra. Abbreviations: *bc*, body cavity; *c*, centrum; *dr*, dorsal ribs lying between dorsal and ventral muscle groups; *ha*, haemal arch and spine; *na*, neural arch and spine; *vr*, ventral ribs surrounding the body cavity.

bone. Quite different in origin and development are dermal or membrane bones. These are never preceded in the embryo by cartilage and are, in contrast, superficial in origin, formed in the deeper layers of the skin. In fossil lower vertebrates dermal bones often covered the body completely, commonly as large plates over the head and shoulder region and smaller scales over the trunk and tail. The dermal plates often tend to unite with replacement bones, especially in the head region; the human skull, for example, is a composite of bones having these different origins.

In dermal bones and scales (cf. Figs. 22, 68) the deeper layers usually consist of characteristic bone, arranged in laminae; the middle portion in fishes presents a spongy appearance due to the presence of numerous blood vessels. The superficial layers vary greatly. Usually, however, bone cells are absent, and the material is a hard, compact tissue comparable to the dentine which forms the main bulk of the teeth. The surface of fish scales and head plates is frequently finished by a layer of hard, shiny enamel, such as covers the teeth. Frequently the surface of the bony plates and scales of fishes shows an ornamentation of tubercles or ridges of dentine and enamel. These superficial structures seem comparable to the dermal denticles which are found in the skin of sharks and to the teeth which are present in jawed vertebrates. The spines often present on the fins of lower fish groups are characteristically formed of materials corresponding to dentine and enamel.

Since the lowest of existing vertebrates are boneless, it was once generally believed that the ancestral vertebrates were likewise cartilaginous forms and that the evolution of bone took place at a later stage in fish history. The presence of cartilage in the embryos of vertebrates which later acquired a bony skeleton was supposed to be a recapitulation of the phylogenetic history. But, as will be seen, all the oldest known fishes possessed bone; many groups show a decrease rather than increase in the degree of ossification in their skeletons as we pass from older to younger members; sharks and other cartilaginous fishes appear relatively late in the fossil record. It thus appears probable that vertebrates acquired bone at a very early stage in their history, although cartilage (with the useful power of growth by expansion) may have been present throughout as an embryonic type of structure. Cartilaginous vertebrates are thus reasonably regarded as degenerate forms which fail to mature in skeletal development and exhibit as adults an immature cartilaginous condition of the skeleton.

The internal skeletal structures of fishes may be grouped under four heads: the axial skeleton of the trunk and tail, the braincase, the branchial arch system, and the skeleton of the paired appendages.

Axial skeleton.—The main elements of the axial system are the vertebrae constituting the backbone. In some primitive forms the notochord remains large, but in most cases its growth is restricted, and in advanced types it disappears altogether in the adult. It is replaced by a series of vertebral centra of cartilage or bone. In some types the centra are formed by skeletal materials laid down as a thick sheath of tissue surrounding the notochord. In others they are formed from paired blocks of cartilage which occur on either side of the notochord and may unite to form ring-shaped centra (Fig. 4). Sometimes there are two such sets of cartilage blocks in a single body segment, and a double set of centra may result (cf. Figs. 108, 128).

The nerve cord lies directly above the notochord; a λ-shaped protecting element usually develops above the centrum in each segment. The two branches—the neural arch—meet above the nerve cord and project upward as a neural spine. Important blood vessels lie just below the notochord in the tail, and these are usually surrounded by haemal arches, or chevron bones, analogous to the neural arches above.

In many specimens of Paleozoic and Mesozoic fishes, neural and haemal arches and ribs are present as bony or calcified structures, but the position of the centra shows only a blank space (cf. Figs. 35, 52, 80, 85, 87). Presumably, centra were present but tended to remain in a cartilaginous condition; in other instances, however, much of this blank area may have been occupied by a persistently large notochord.

Ribs extending out from the sides of the vertebrae and lying between dorsal and ventral groups of muscles are usually present in the trunk region; primitively they are present in every segment from the neck to the base of the tail. In addition to these intermuscular dorsal ribs, which are the only ones present in land forms, fish usually possess a second series—ventral ribs—which lie deeper and partially surround the body cavity. In the tail the ventral ribs are continued by the haemal arches.

Accessory skeletal structures are often found in the trunk. Additional bones are sometimes placed between various muscles (the shad is all too good an example), and in most land forms there is a breastbone, or sternum, to which many of the ribs attach ventrally. The median fins are supported by a series of parallel rodlike cartilages or bones—the radials; while below them, in the fish's body, basal elements connect them with the vertebral column. (In addition they are covered superficially and stiffened distally by scales or elongated rays derived from scales or from the skin itself.)

Appendicular skeleton.—Paired appendages—the paired fins of fishes and the homologous limbs of land forms—are present in most vertebrates. There are normally two pairs (Fig. 2)—pectoral appendages, just behind the gill region or neck, and pelvic limbs, typically situated at the posterior end of the trunk in front of the anal opening. Support for the limbs is afforded by girdles contained in the body of the animal. They are composed primarily of cartilage or of replacement bone, but the pectoral girdle often has membrane bones attached to its outer and anterior margins. Running out from the side of the girdle region is the appendage proper, the skeleton of which usually consists of a complicated series of jointed cartilages or bones (cf. Figs. 27, 72). The appendages are primitively small steering and balancing organs in fishes, but they become of paramount importance in locomotion in land vertebrates. The great variations in structure in these organs will be one of the central themes in our study of vertebrate evolution.

Paired fins are poorly developed or absent in some of the lowest vertebrates. Their origin has been much debated. It seems probable, however, that they were at first merely folds which had grown out from the sides of the body just as the median fins appear to have grown out in the midline of the back and tail regions; the two types of fins often have a similar structure and similar stabilizing functions and may well have had similar origins.

There appears to have been a strong trend among the most ancient fishlike vertebrates to develop lateral outgrowths—"stabilizing keels"—on either flank; even in ostracoderms, which in general do not possess proper fins, there may be a row of spines down either flank or (as can be seen in some of the anaspids described in the next chapter) there may grow out a long stabilizing fold, presumably fixed in position but with the potentialities of producing true fins in the course of evolutionary advance.

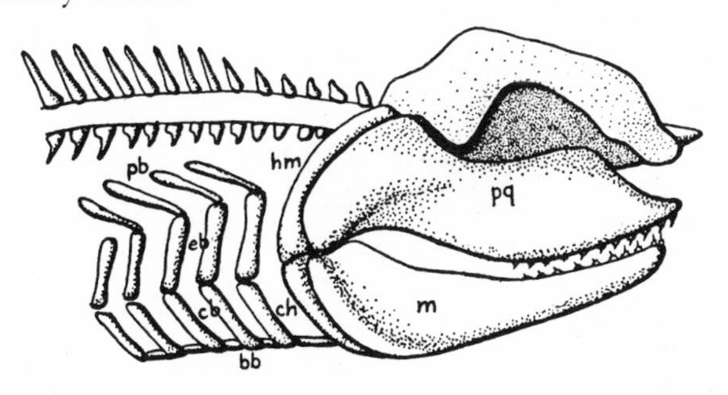

Fig. 5.—The jaws and branchial arch system of a shark, essentially the Mesozoic *Hybodus*. The braincase and spinal column are also shown. Abbreviations: *bb*, hypobranchial and basibranchial elements of the branchial arches; *cb*, ceratobranchials, the main ventral elements; *eb*, epibranchials, the principal dorsal elements; *pb*, pharyngobranchials; *ch*, ceratohyal, and *hm*, hyomandibular, elements of the hyoid arch immediately behind the jaws; *pq*, primary upper jaw (palatoquadrate); *m*, lower jaw or mandible. (After Smith Woodward.)

Branchial arch system.—A series of cartilaginous or bony bars found between the gill openings in typical vertebrates serves to stiffen the gill (or branchial) region and affords support for the muscles which open and close these slits (Figs. 5, 6). In jawless vertebrates, as far as is known, these bars form a united set of structures fused dorsally to the skeleton of the head region. In jawed forms the branchial bars are normally separate from the skull structures and are divided into upper and lower halves, bent somewhat on each other;

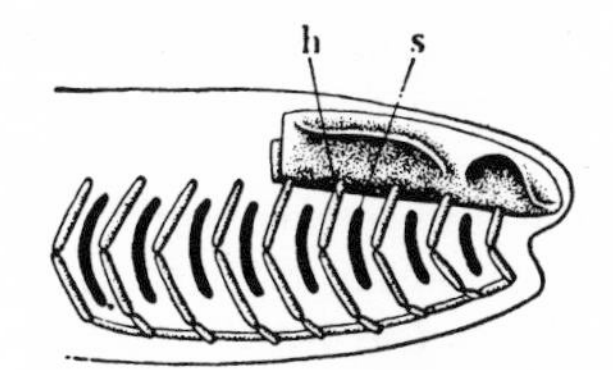

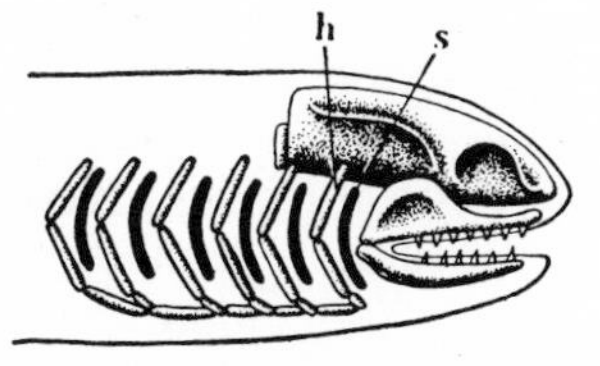

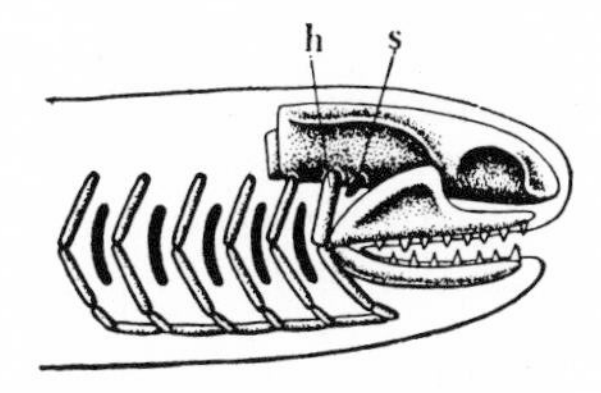

Fig. 6.—Diagrams to show stages in the evolution of jaw structures in vertebrates. *Left*, The jawless stage found in the Agnatha; the mouth was small, the gill arches similar throughout. *Center*, A hypothetical stage, possibly present in some of the older fishes, in which jaws had evolved, with the elimination of one or two of the most anterior gill bars, but the slit corresponding to the spiracle (*s*) of more advanced forms is still of normal form, and the hyoid arch (*h*) behind it is unspecialized. *Right*, The more advanced type of jaw seen in most fishes. The upper part of the hyoid arch is specialized as the hyomandibular, propping the jaw joint on the braincase; the intervening gill slit is reduced to the spiracle.

and further subdivision usually occurs. In land types, with the loss of gill breathing, there is great reduction in this originally important skeletal system.

Jaws are absent in the most primitive vertebrates. It is believed that they have been derived from an anterior pair of gill bars, the upper (palatoquadrate) corresponding to the upper half of an arch, the lower jaw (mandible) to the main ventral segment.

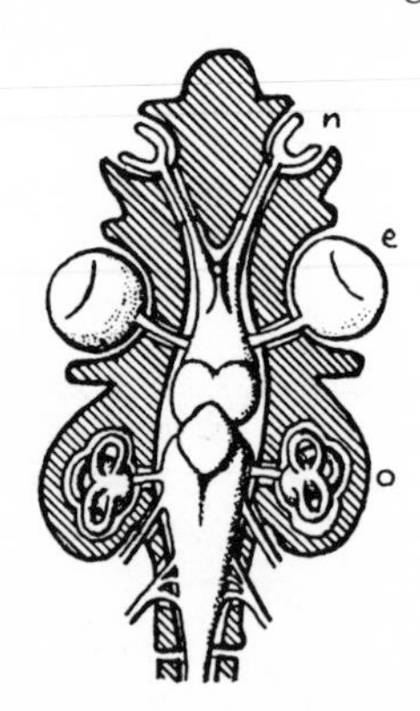

FIG. 7.—Diagram to show relation of braincase to nervous system and sense organs in fishes. The braincase is represented as sectioned in a horizontal plane. The braincase forms paired capsules enclosing the nasal sacs (*n*), protects the eyeballs (*e*), and encloses the canals and sacs of the ear (*o*).

As the jaws enlarge, they tend to encroach on the gill area behind them. In jawed fishes the gill openings are typically elongate slits. However, the jaws have pushed back so far that they crowd against the next arch (called the hyoid arch) and become connected with it in the region of the jaw joint. This blocks the space occupied by the first slit, which tends to be reduced to a small opening, termed the spiracle, in the upper part of the original slit area. As the jaw joint crowds against the hyoid arch, the main upper element of this arch—the hyomandibular—tends to assume a new function: it braces the jaw against the side of the braincase and thus aids in mastication.

Skull.—In the embryo of every vertebrate there develops a braincase, a box of cartilage enclosing the brain and articulating at its posterior end with the vertebral column (Figs. 5, 7, 63). Anteriorly, nasal capsules protect the nostrils; hollows in the sides receive the eyes; extensions from the sides of the posterior part enclose the primitive ear as otic capsules. The braincase remains cartilaginous in many fishes, but in a majority of fishes and other vertebrates it becomes more or less ossified.

In sharks the braincase and the primary upper jaws, loosely attached to it, are the only head structures. In most other fishes, however, dermal bones are present in addition. A shield of such bones covers the top and sides of the head region (Fig. 8), coming into contact with braincase and primary jaws (when present). Other dermal bones appear in the skin lining the mouth. These dermal structures tend to replace in part the deeper cranial elements and to form, with them, a unified cranial structure—the skull.

Nervous system.—The central nervous system of vertebrates consists of the brain, the spinal cord running down the body from the brain, and nerves passing out from these structures to sensory and motor end-organs. The spinal cord differs radically in position from the main nerve trunk of most invertebrates in that it lies in a dorsal position, passing down the back above the notochord and vertebral centra and protected by the neural arches between which the nerves emerge (a pair to each segment). The brain possesses an exceedingly complicated structure, of which we may note the superficial features often revealed by casts of the inside of the braincase (Fig. 9; cf. Fig. 18). A rough division may be made into three portions—forebrain, midbrain, and hindbrain. The first has paired swellings above—the cerebral hemispheres—which, primitively associated with the sense of smell, in higher forms tend to enlarge greatly and become the seat of the highest functions of the brain. Extending up behind the hemispheres is a stalked pineal body, or a similar structure termed

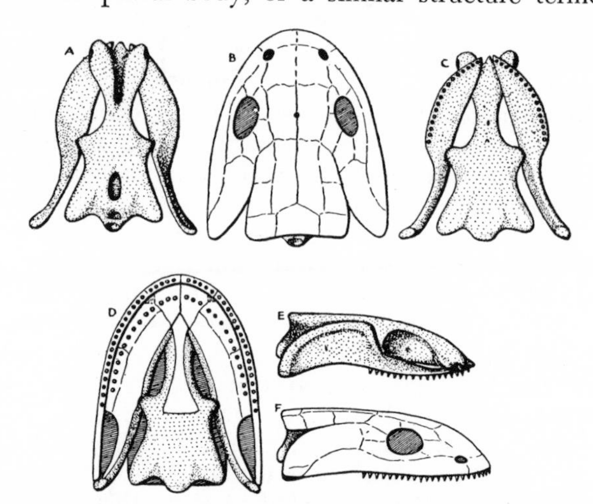

FIG. 8.—Diagrams to show comparison of the skull in forms with and without dermal bones. Cartilage or cartilage-replacement bones stippled; dermal bones white. *A*, Dorsal, *C*, ventral, and *E*, lateral views of a sharklike form, with separate braincase and "primary" jaws. *B*, *D*, and *F* are dorsal, ventral, and lateral views of the skull type found in bony fishes and higher forms. Dorsally and laterally the dermal bones cover the internal structures and unite them into a single mass. Ventrally a dermal element underlies the anterior end of the braincase; other dermal bones cover most of the "primary" upper jaws; and the lower margins of the roofing elements form tooth-bearing outer margins to the jaws.

the parapineal, or both; one or the other forms a median eye in many lower vertebrates. In most vertebrates, tissues from the infundibulum—a ventral projection from the forebrain—fuse with a pocket-like growth from the roof of the mouth—the hypophysis—to form the pituitary, an important gland of internal secretion. The midbrain roof has paired optic lobes, primarily associated with sight, but acting, in lower vertebrates, as the principal coordinating center. The main portion of the elongate hindbrain is the medulla oblongata, which has to do with touch, taste, balance, hearing, and motor responses, while above it lies the cerebellum, the "tree of life," which has control over body posture and muscular coordination.

The cranial nerves are of interest to the paleontologist, since the openings (foramina) through which they emerge are usually visible in good fossil skulls. There are always at least ten, sometimes twelve, pairs of nerves. Three go to major sense organs: I, to the nostril; II, to the eye; VIII, to the ear. Three (III, IV, VI) are small nerves which move the muscles of the eyeball. The others (V, VII, IX, X) mainly receive sensory impressions from the skin of the head and neck and lining of the mouth and convey impulses to the muscles of the jaws and throat. Each member of this series of nerves is associated with a specific gill arch. Nerve V is associated, in specialized fashion, with the jaw arch, and an anterior branch appears to have pertained to a still more anterior gill arch which has been obliterated by growth of the jaws.

Sense organs.—All vertebrates have the three major sense organs found in ourselves—nostrils, eyes, and ears. The nostrils in primitive forms do not usually connect with the interior of the mouth but are simply a pair of pockets into which water can pass and in which the sensory cells of smell are located. In some very low vertebrates (cyclostomes, some ostracoderms) there is but a single nostril. In addition to the universal paired eyes, a median eye often reaches the surface (especially in older fossil forms) through an opening in the top of the skull. The primitive ear possesses no drum or earbones, as it does in higher forms, but consists merely of a series of liquid-filled sacs and canals lying entirely inside the braincase; a fish can hear only vibrations which have passed into its body and set up vibrations in its braincase. The ear seems to have been primitively a balancing organ, and that function is still important today. This sense is located in a pair of sacs and a series of semicircular canals; the canals are usually three in number, although only two (or even one) may be present in jawless forms.

Primitive aquatic vertebrates seem to possess a sixth sense in the lateral-line organs, situated in canals or pores extending in a line along the sides of the body and forming a complicated pattern on the head (Figs. 24, 30, 37). Present generally is a main canal or groove passing forward on either side from the trunk, turning downward, and then continuing forward as an infraorbital canal; and a pair of supraorbital canals. It is believed that these organs are sensitive to water currents or pressure as an aid in swimming; and from the fact that the inner ear may communicate with the surface by a tube to the top of the head in sharks, it has been suggested that our internal ears (which also register movements and pressure in a liquid medium) may be but specialized portions of this same system.

Circulatory system.—We shall not be greatly concerned with the blood system in an elementary study of fossil forms, although the investigation of passages for the transmission of blood vessels is a topic of great importance in research on fossil skulls. We shall here note merely that, primitively, the blood is carried to the heart from the body by a series of veins; from there it is carried forward along the throat in a ventral aorta, from which branches arch up between the gill slits. After passing through capillaries on the surface of the gills, these arches reunite above, and the main current of blood passes backward to the body in the dorsal aorta. With the introduction of lungs and the reduction of gills, the system becomes, of course, highly modified. Frequently found in fossils are ventral openings into the braincase which allow passage of the carotid arteries supplying the brain.

Respiratory system.—The lower vertebrates and their relatives possess a unique breathing system of internal gills. Water passes (usually through the mouth) into the throat, thence through a series of gill chambers, where respiration occurs, to the surface of the body. In most fishes, five typical pairs of gill slits are present; originally the number may have been much higher (cf. Fig. 18). The spiracle, mentioned above, is sometimes enlarged and in skates functions instead of the mouth as an inlet for water; in many higher fish, on the other hand, it is reduced or absent. The gill slits sometimes open separately to the surface (sharks, some ostracoderms). In most fishes, however, the gills have a common posterior opening and may be covered by a structure termed the operculum. Lungs, connected with air breathing, appeared at an early stage in fish history, for they seem to have been characteristic of early bony fish and are even reported in one of the placoderms. In amphibians the gills have tended to atrophy; the spiracle becomes the eustachian tube of the middle ear; other pouches may form glands of internal secretion.

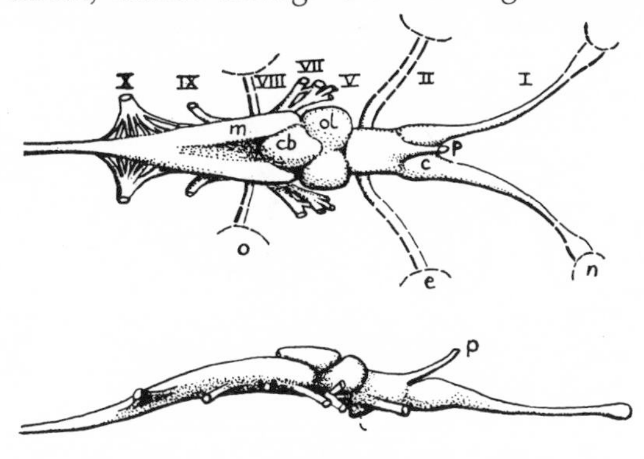

FIG. 9.—Dorsal and lateral views of restored brain and cranial nerves of *Macropetalichthys*, a Devonian placoderm. The cranial nerves are numbered (see text); the three small nerves to the eye muscles omitted. Abbreviations: *c*, cerebral hemispheres; *cb*, cerebellum; *e*, eye; *i*, infundibulum; *m*, medulla oblongata; *n*, nostril; *o*, ear; *ol*, optic lobes of midbrain; *p*, pineal organ. (After Stensiö.)

In living adult vertebrates the gill function is primarily and usually exclusively that of respiration. Very probably, however, this function was originally an incidental or secondary one. The gill system in lower chordates is mainly a feeding device; food particles are strained out from the stream of water passing through the pharynx. The gills may still have functioned in this fashion in the Paleozoic jawless fishes (ostracoderms); modern jawless types (lampreys and hagfishes) have, as adults, evolved peculiar rasping "tongues" as a substitute for jaws in the adult, but the larval lamprey is still a food-strainer.

Digestive and urinogenital systems.—We shall have little to do directly with these organ systems (jaws and teeth are discussed elsewhere), although many skeletal features are directly related to food habits and reproduction. Primitive vertebrates are, of course, egg-laying types, although even among some sharks the egg may undergo development within the mother's body. Usually the eggs are unprotected in water-living forms, although in some cases (skates and chimaeras, for example) a hard egg case, capable of preservation, is present.

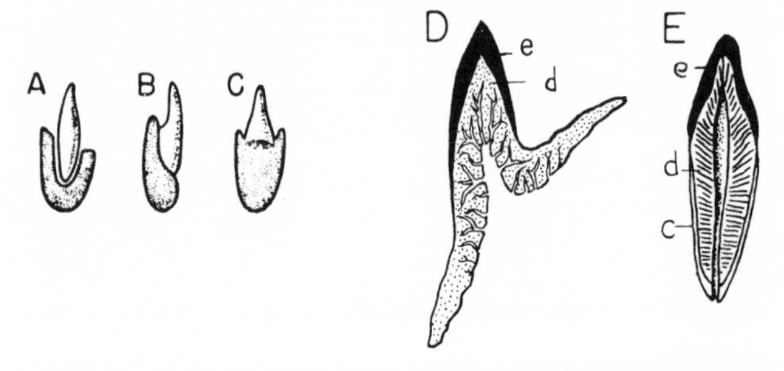

FIG. 10.—*A* through *C*, Diagrammatic sections through lower jaws to show the distinction between thecodont (*A*), pleurodont (*B*), and acrodont (*C*), tooth attachments. *D*, Section through a shark denticle, and *E*, a tooth; *c*, cement, *d*, dentine, *e*, enamel.

Muscular system.—Although muscles are practically never preserved, these organs are of importance to us since they are so intimately associated with the skeleton. Somewhat as with the skeleton, the muscle system may be divided into axial, branchial arch, and appendicular series. The axial muscles constitute the main means of locomotion and the greater part of the bulk of water-living vertebrates. They are arranged segmentally in a series of layers down the back and flanks of the fish. Between each successive segment lies a rib; opposite each segment lies a vertebra. It is probable that the segmentation of the skeleton took place primarily in relation to that of the muscles. The small muscles which move the eyeball are probably highly modified portions of this axial group. A special group —the visceral muscles—is that associated with the branchial arches; the muscles of the jaws, when these structures are formed, were derived from this group. With the development of limbs, strong sets of muscles form above and below the skeletons of these appendages.

Skin derivatives.—Not only membrane bones, previously discussed, but other structures may arise in the outer layers of the body. From the most superficial portion of the skin are formed, in various groups of higher vertebrates, horny scales, claws, true horns, feathers, or hair; these are usually incapable of fossilization. From deeper layers in the skin of many fishes are formed denticles, typically hollow cone-shaped structures, mainly composed of compact dentine and covered with a film of very hard, shiny enamel or enamel-like material (Fig. 10). In typical sharks such denticles are scattered over the surface of the skin, giving it a rough, sandpaper-like texture. In many early and primitive fishes they form the superficial portion of underlying bony plates or scales, and it is probable that shark denticles are the remnants of a bony covering originally present in the ancestors of the modern elasmobranchs.

Teeth (Fig. 10).—Present, except where secondarily lost, in all jawed vertebrates, teeth loom large in the study of fossils—notably in mammals, where molar tooth patterns are diagnostic. Presumably they were derived, originally, from tubercles on dermal plates lying at the jaw margins. In some instances they are firmly fused to the surface of dermal jaw bones—the acrodont condition. At the opposite extreme, they may be attached to the jaws only by ligaments, as in sharks. Frequently they are set in sockets, deep or shallow—the thecodont condition. They are characteristically arrayed in a series along the margins of the jaws, but they may also be present on the surface of the palate and the inner side of the lower jaws and even, in many bony fishes, on the gill bars in the floor of the mouth. In some instances, particularly where they have the form of crushing plates, individual teeth may persist throughout life; generally, however, replacement takes place. This is limited in mammals, but in lower forms there is a continued renewal. In the jaw margins this generally occurs in waves of replacement, traveling from front to back along each jaw ramus in such fashion that the dentition often has an irregular appearance.

As in denticles and the surface ornaments of dermal plates in many early fishes, teeth are composed generally of two substances, dentine and enamel; in addition, a spongy type of bone termed cement may be present around the roots, binding tooth to jaw. Enamel, shiny in appearance, forms a thin but hard surface covering which stands wear well. The main substance of the tooth is dentine, similar to bone in composition, but differing in that the cells associated with it do not lie in the substance of the enamel, but in a central pulp cavity, from which tiny canaliculi extend outward through the dentine. As suggested by transitional conditions in ostracoderms, dentine was derived from bone in early vertebrate days.

Primitive chordates (Figs. 11, 12).—The vertebrates, although a large and important group of animals, do not in themselves constitute a major division of the animal kingdom but are grouped together with a small number of other living forms of more primitive character to form the phylum Chordata, animals with a notochord. These other types are almost entirely unknown as fossils but must be considered briefly because of their bearing on the origin and early history of the vertebrates.

Amphioxus.—The Cephalochorda comprise but a few small marine forms, mainly included in the genus *Amphioxus.* This animal has the appearance of a small translucent fish; but structurally it is far inferior to any of the true vertebrates. There are no bones or cartilages, no paired fins or limbs, no jaws or teeth; there is little indication of a brain and only vestiges of sense organs. There are, however, a number of important features which tend to show that *Amphioxus* is related to the vertebrates. Although hard skeletal parts are absent, there is a well-developed notochord; a dorsally situated nerve cord runs the length of the body; there is a very highly developed series of gill slits. *Amphioxus,* although capable of activity, spends most of its time partially buried in the sand, as a "filter-feeder"; water is drawn in through the mouth and passed out through the gills; suspended food particles filtered out in the gill passages form its diet, as in the larval lamprey. There are a few specializations, such as the extension of the notochord to the tip of the snout as an aid in burrowing, but for the most part *Amphioxus* qualifies well as a vertebrate ancestor.

Tunicates.—The Urochorda include a considerable number of marine animals known as the tunicates or sea squirts; some are colonial, and some float freely in the ocean, but the central types are perhaps the solitary, sessile forms found as adults attached to rocky surfaces in shallow waters. They are motionless, essentially shapeless masses covered with a leathery tunic. Water is drawn in through an opening at the top and passed out through a second orifice at the side of the body. Superficially, they are quite unlike vertebrates or

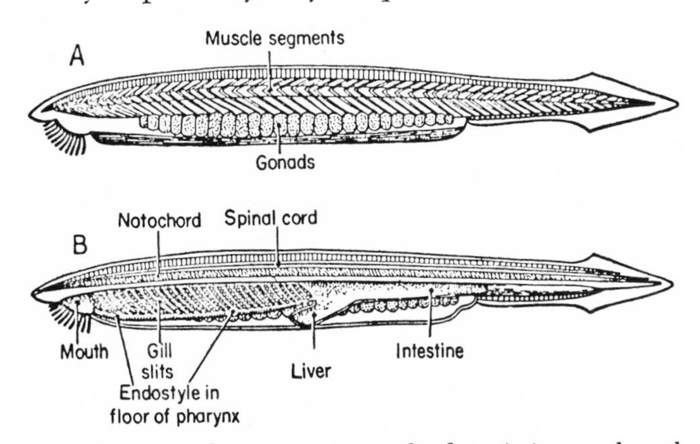

FIG. 11.—*Amphioxus,* a primitive chordate. *A,* As seen through the transparent skin; *B,* a sagittal section. (After Gregory.)

even *Amphioxus,* and internally the resemblances are, at first sight, not much greater. There is no notochord, no nerve cord. The water drawn in is strained through a barrel-like structure which occupies the greater part of the interior of the creature. Food particles collected here pass through an opening at the bottom of the barrel to the stomach and intestine, and the feces pass out through an anus into the current of water leaving the body through the lateral opening.

But this barrel which strains the food is also the breathing organ of the animal and consists of a much-elaborated set of internal gills. Confirmation of this suggestion of vertebrate relationship is furnished by the study of the larva, which, unlike the adult, is a free-swimming, tadpole-shaped animal with a long tail in which are found both a dorsal nerve cord and a well-developed notochord. After swimming about for a short time, the larval tunicate becomes attached to a rock, and the tail (and with it the notochord and most of the nerve cord) disappears. Which is the more primitive type, the sessile adult tunicate or the active larva?

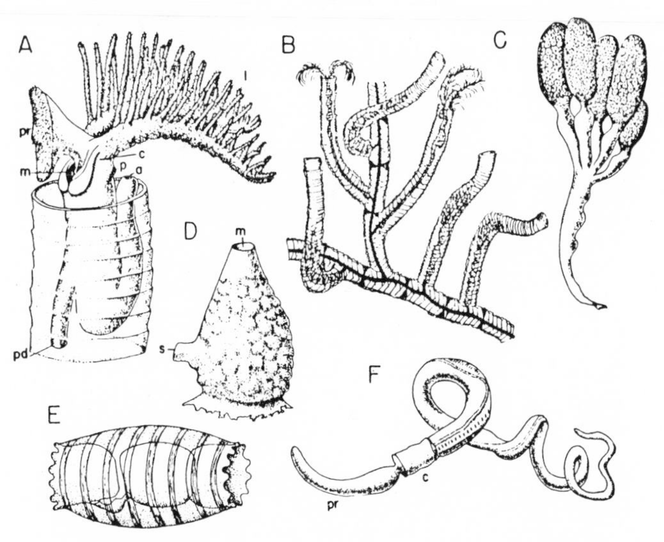

FIG. 12.—Tunicates and hemichordates. *A,* An individual of the pterobranch genus *Rhabdopleura* projecting from its enclosing tube. *B,* A part of a colony of the same. *C,* A colonial sessile tunicate; each polygonal area is a separate individual of the colony. *D,* External view of a solitary tunicate. *E,* A free-floating tunicate, or salp. *F,* An acorn worm (*Balanoglossus*). *a,* Anus; *c,* collar region; *l,* lophophore; *m,* mouth; *p,* pore or opening from coelom; *pd,* stalk (peduncle) by which individual is attached to remainder of colony; *pr,* proboscis or anterior projection of body; *s,* siphon which carries off water and body products. (Mainly after Delage and Hérouard.)

Many students, impressed with the active nature of vertebrates, have tended to the opinion that the larva represents the more primitive type, and that the sessile adult is degenerate. But careful consideration of the whole chordate picture suggests the reverse. It is more probable that the chordate ancestor of the vertebrates had, at this stage, developed a motile larva with a swimming tail furnished with muscle, notochord, and nerve cord. The primary function of this larva was to move about and "set up shop" for the adult in a favorable spot. Living tunicates have gone no further with this development. Very likely the evolution of *Amphioxus* and the vertebrates came about by a process of paedogenesis—that is, by an individual animal failing, in the course of evolution, to reach a former adult state and reproducing while still in a larval stage. Under some conditions it was presumably advantageous if the larva never "grew up" and became sessile but, rather, retained its tail and its motile habits throughout life. The ancestral larva, not the ancestral adult, opened the door to advance and freedom.

Acorn worms.—Quite different, again, are the acorn worms. These essentially sessile seashore burrowers are somewhat like the ordinary annelid worms in general appearance but have as characteristic structures a "collar," in front of which is a tough burrowing snout, or proboscis, the two sometimes resembling an acorn in its cup (hence the name). Their internal structure is quite different from that found in the annelids. Strong proof that they are, on the other hand, related to the chordates is shown by the presence of numerous and

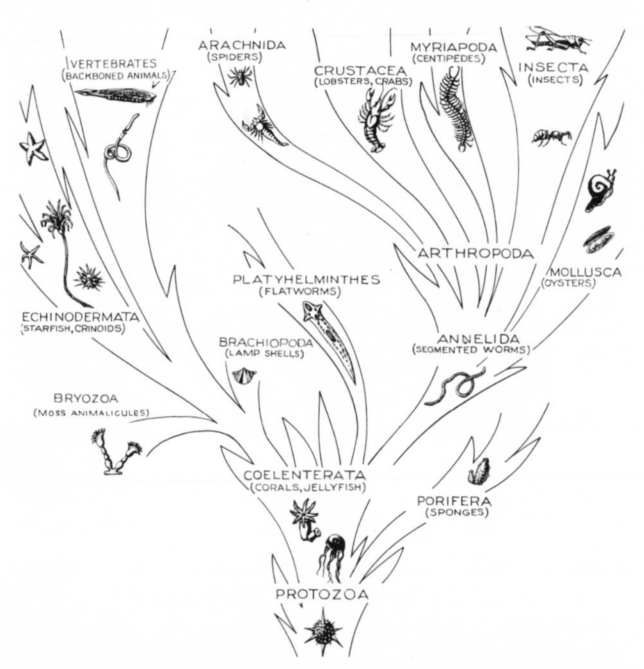

FIG. 13.—A pictorial family tree of the animal kingdom to show the probable relationships of the vertebrates.

typical gill slits. The nervous system is not highly developed, but there is a hollow dorsal nerve cord in the collar region, and a small structure in the proboscis is thought to be a rudimentary notochord. The acorn worms are assuredly far below *Amphioxus* in the level of their organization but are highly specialized and hence not on the main line of vertebrate ascent.

Pterobranchs.—In studying the acorn worms we have descended the tree of life to a point where there is little left of vertebrate characters or even of the basic chordate characters of *Amphioxus* apart from the presence of gill slits as a feeding (and breathing) device. One further, final, step brings us to forms which lack, or nearly lack, even this last remaining vertebrate resemblance and would not be suspected of relationship were it not for an arrangement of body parts comparable to that of acorn worms. These are the pterobranchs. There is no popular name for these tiny, rare, marine animals. Apart from a few doubtful types, there are but two known genera. They form little plantlike colonies, whose individuals project like small flowers at the ends of a branching series of tubes. The short body is doubled back on itself, so that the anus opens anteriorly back of the head. Proof of relationship to the acorn worms lies in the fact that there is a snoutlike anterior projection beyond the mouth, corresponding to the acorn worm proboscis, and back of this, a short collar region. But other resemblances to acorn worms—to say nothing of more highly developed chordates—are almost entirely lacking. There is little development of a nervous system, no trace of a hollow nerve cord, and not the slightest suggestion of a notochord. And the feeding mechanisms are of a very different type. True, these plantlike animals feed, as do other lower chordates, on food particles drawn in by ciliary action. But the gill mechanism, which is so important in the filter-feeding of acorn worms, tunicates, and *Amphioxus*, is almost entirely absent. One of the two pterobranchs has a single pair of small gill openings, the other none at all. Instead, there projects from the collar region large, branching, tentacle-like structures, termed "lophophores." These are richly supplied with cilia, which set up water currents, bringing food particles to the mouth.

So unchordate-like are these small creatures that one is tempted to suggest that they are degenerate forms, perhaps relatively modern in development. But it has recently been demonstrated that they are a very ancient group indeed. Paleontologists have long been familiar with a variety of small tubelike structures termed "graptolites," which were abundant in the seas long before the appearance of the oldest vertebrates. Close study shows that these tubes are similar to those which shield the modern pterobranchs. With this indication of their antiquity, it can be reasonably argued that the pterobranchs are a truly primitive group of chordates (in a broad sense), from whom, by progressive development of a gill filtering system, the more typical chordates are derived.

Vertebrate ancestry (*Fig. 13*).—At one time or another the ancestry of the vertebrates has been sought in almost every invertebrate group. Some have suggested that vertebrates have been derived directly from the coelenterates, the simplest of the metazoans. Obviously, these forms may have been the original ancestors; but they lack so many structures which are found both in vertebrates and in many other invertebrates that it seems much more reasonable to believe that the splitting-off of the group lay somewhat higher in the scale.

A theory of descent of the vertebrates from the annelid worms has been advocated. These animals are bilaterally symmetrical, as are vertebrates; they are segmented (as are vertebrates in backbone, nerves, and muscles); and they have a well-developed nerve cord. The nerve cord lies on the ventral side of the worm; but if the form be supposed to have turned over, the nerve cord would be on the dorsal side. But such diagnostic vertebrate structures as gill slits and notochord are not found in annelid worms; and since the mouth of the annelid is on the underside of the head, we should expect that the mouth of vertebrates would lie on the back of the head unless we suppose that the old mouth has closed and a new one formed.

The arthropods, most highly organized of invertebrates, have also been strongly advocated as vertebrate ancestors, especially the arachnids, a group including not only the spiders but the scorpions and a number of such water-living types as the horseshoe crab and the extinct water scorpions—the eurypterids. Arachnids have an external, not an internal, skeleton; but some of the earliest fossil vertebrates had a highly developed armor which in some cases greatly resembled that of aquatic arachnids. It has been suggested that this resemblance denotes a real relationship. According to this theory, however, it should be the undersurface of one group which should resemble the dorsal side of the other (for the nerve cord, as in annelids, lies ventrally in arachnids). Unfortunately, this is not the case; and, just as in the case of the worms, there are great difficulties involved in assuming a reversal of the top and bottom sides of the animal. In addition, it is necessary to do away with the jointed limbs and other complicated structures of the arthropod and remake the entire animal. There is no positive evidence of such a radical rebuilding. Further, the theory implies that the lower chordates are unrelated to the vertebrates, despite the many similarities in structure.

The echinoderms—starfishes, sea urchins, sea lilies, and the like—seem the most unpromising of all as potential ancestors of the vertebrates. They are radially symmetrical, in contrast to vertebrates; they have no internal skeleton, no trace of any of the three major chordate characters of notochord, nerve cord, or gill slits; in addition, they have many peculiar and complicated organs of their own. At first sight one would be inclined not even to consider them. But here embryology and biochemistry shed an unexpected gleam of light.

The early embryo of the echinoderm is a tiny, free-floating creature, almost beyond the range of vision of the naked eye. Despite the disparity between the adults, the larvae of some echinoderms are so close to the larva of the acorn worm that the worm larva was long thought to pertain to some echinoderm type. Further, in the development of the embryo, the mode of laying down the main structural elements of the body in the echinoderms is radically different from that found in other advanced invertebrate groups—but agrees well with the embryological story seen in such chordates as the acorn worm and *Amphioxus*. Still further, recent advances in biochemistry show that, in such important features as the oxygen-carrying pigments of the blood and the chemistry of muscle action, the echinoderms show significant similarities to the vertebrates and contrast in these features with most other invertebrates.

Here, at last, are some positive facts and from a quarter where, at first thought, one would least expect them. But on further reflection, this suggestion of echinoderm relationship is not as astonishing as it might seem. We think of vertebrates as active animals.

But if our reasoning in the last section has any merit, it is probable that the ancestral chordate, from which the vertebrates are descended, was, in contrast, a stalked, sessile animal making its living from tiny food particles gathered by ciliary currents along its outspread "arms." Most living echinoderms are free-moving (although their movement is of a limited and peculiar nature). But the fossil history shows clearly that the ancestral echinoderms, like the ancestral chordates, were sessile

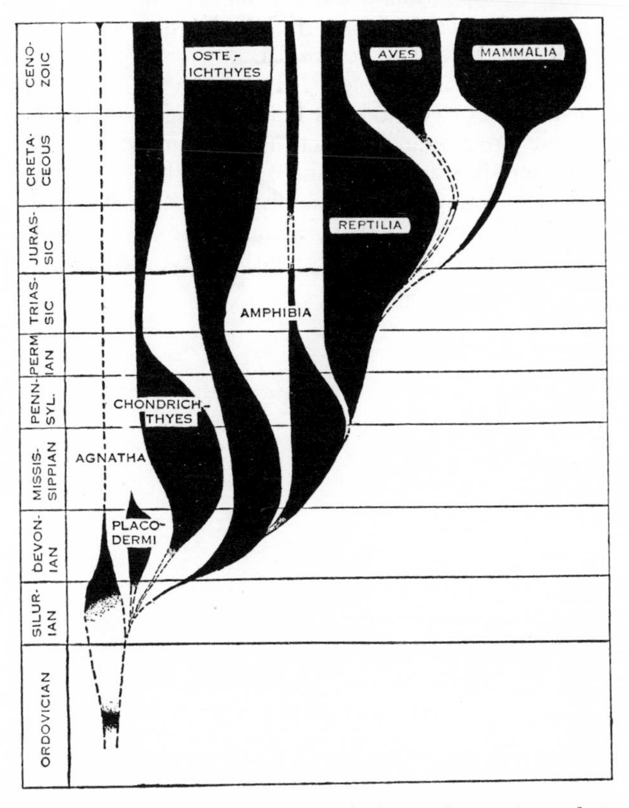

FIG. 14.—A family tree of the vertebrate classes. A rough indication of the comparative abundance of the various groups is furnished by the thickness of the various branches. In this diagram the Acanthodii are provisionally included in the Osteichthyes.

forms. The ancestral echinoderm groups are now extinct, but the living crinoids, or sea lilies, are echinoderms of similar habits. A crinoid has a complex skeleton and other specialized features. But basically it is a "stalked, sessile animal, making its living from tiny food particles gathered by ciliary currents along its outspread 'arms.' "

These are, as you may see, exactly the words used a few lines above to describe a primitive chordate. Actually, a little pterobranch seems to be of such simple and unspecialized nature that it may well be close to the humble dwellers in ancient seas from which both chordates and echinoderms, in their very different fashions, took their origins, and from which other sessile lophophore-bearing marine forms, such as brachiopods and bryozoans, may have been derived as well.

Classification of vertebrates (*Fig. 14*).—Most of the classes into which vertebrates should be divided are obvious. The higher types include the mammals (class Mammalia), birds (class Aves), reptiles (class Reptilia), and amphibians (class Amphibia). The lower, water-living types are often regarded as forming a single class—the fish, or Pisces. But some subdivision of this last group seems necessary. Superficially, the various kinds of "fish" seem quite similar in nature; they are all water-living types and in that connection have many features in common. But in structure there is a vast amount of variation. A lamprey and a codfish, for example, are in many respects as different as a frog and a man; and it is not reasonable to lump them in a single class. The lampreys and hagfishes, together with related fossil forms, may be grouped as the class Agnatha, primitive vertebrates without typical limbs or jaws. Certain primitive armored fossil fishes in which jaws and paired appendages are present but peculiarly constructed may be considered as constituting a class Placodermi. The sharks, rays, and chimaeras comprise the class Chondrichthyes; the higher bony fish, the class Osteichthyes. Although it is hardly necessary to group the various classes formally into superclasses, we may for convenience use the term Pisces ("fish") to include all the primitive water-living groups, in contrast to the Tetrapoda ("four-footed"), the land-dwelling classes. Another method of grouping is to contrast the Amniota (reptiles, birds, mammals; cf. p. 102) with the Anamnia; we may in many respects conceive of the Amphibia as merely a sort of progressive fish.

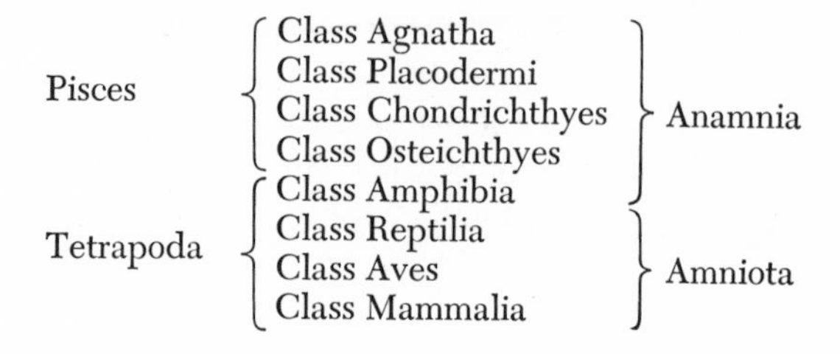

CHAPTER 2

Primitive Jawless Vertebrates

The lampreys and hagfishes, usually termed cyclostomes, are outstanding among living vertebrates in the presence of seemingly primitive characters. These peculiar forms of eel-like appearance are quite devoid of jaws such as characterize all other living vertebrates and are limbless, whereas all other living groups of fishes have (unless secondarily lost) well-developed paired fins. Turning to fossil types, we find that the most ancient of known vertebrates were forms usually grouped as the ostracoderms. These fishes were almost always encased in a heavy armor of bone or other hard material and thus seem quite unlike the soft-skinned modern cyclostomes in which the skeleton is entirely cartilaginous. But there, too, jaws appear to have been totally absent, and limbs, at the most, poorly developed. The living cyclostomes and the fossil ostracoderms are members of a common stock of primitive ancestral vertebrates which we may term the class Agnatha, jawless vertebrates.

Cyclostomes.—The living agnathous forms include the lampreys, such as *Petromyzon,* and the hagfishes and slime eels (*Myxine, Bdellostoma*) (Fig. 15). They are semiparasitic in habit; the lampreys attach themselves to living fish and suck their blood; the hags typically feed on dead and dying fishes. The body is elongate, eel-shaped; there are no scales or denticles in the slimy skin. Paired fins are absent, although median fins are developed. From six to fourteen pairs of gills are present; these differ from the slitlike structures of other living vertebrates in being spherical pockets connected with the inside and the surface by small tubes. In the lampreys the margins of the round, jawless mouth form a sucking disk for attachment to the prey; in the hagfishes the mouth is surrounded by tentacles. Jaws are absent. The cyclostomes, however, have evolved a substitute of sorts in the development of a long protrusile "tongue," armed with horny structures resembling teeth; this forms an efficient organ for rasping away the skin or flesh of the cyclostome's prey. There are paired eyes and a median pineal eye (well developed in lampreys, vestigial in hagfishes). Instead of the three semicircular ear canals of most vertebrates, lampreys have but two and hagfishes one.

The skeleton consists of uncalcified cartilage. The braincase is a rather specialized and complicated structure. An elaborate system of branchial arches is present in the lamprey; but these arches, instead of being separate elements, are fused into a peculiar basket enclosing the gills, while part of the branchial skeleton is modified into a support for the "tongue." The notochord is large and unrestricted. All forms have cartilaginous supports for the median fins, while *Petromyzon,* although lacking vertebral centra, has a row of small neural arches.

In contrast to typical vertebrates, the cyclostomes have but a single nostril. In most vertebrates the nasal openings are near the front of the head, or even slightly on the underside. In the hagfishes the nostril opens at the tip of the "snout," but in the young lampreys a large "upper lip" grows out from the front of the roof of the mouth to form the tip of the body; the originally ventral nostril (together with a second pocket-shaped structure, the hypophysis) is forced around forward and upward until in the adult it is found opening high on the top of the head just in front of the eyes—a situation unparalleled in other living vertebrates (Fig. 16).

Possessing no bones, true teeth, or other hard parts capable of preservation, the cyclostomes are unknown as fossils, although certain small Paleozoic toothlike structures—conodonts—have been thought, quite erroneously, to be cyclostome "teeth."

The lampreys and hagfishes are obviously lower in their plane of organization than any other living vertebrates. But, before concluding that they are really primitive types, we must take into account the fact that parasites are generally specialized and degenerate rather than truly primitive forms. Specialized they undoubtedly are in such respects as the peculiar "tongue." May it not be that many of the features of simplicity in their structure are really due to losses associated with parasitism rather than to true primitiveness? We may seek an answer in the study of related fossil types, including the oldest known fishes.

Ostracoderms.—In sediments of late Silurian and early Devonian age, numerous fishlike vertebrates of varied types are present, and it is obvious that a long evolutionary history had taken place before that time. But of that history we are still mainly ignorant. A restricted amount of material, mainly fragmentary in nature, is known from early phases of the Silurian. From the Ordovician only two series of finds are known. Nearly a century ago a dozen or so toothlike structures, presumably fragments of dermal armor, were found in the Russian Ordovician. Much more numerous, but for the most part equally fragmentary in nature, are remains found in a series of Ordovician sandstones extending in a line from the eastern Rockies of Colorado north through Wyoming and Montana. These tantalizing fragments tell us that vertebrates had appeared on the evolutionary scene at this early stage of the Paleozoic and (a matter of particular interest) had already acquired bone, but tell us little more. Obviously, major evolutionary events were occurring in vertebrate history during the Ordovician and Silurian, but we are still in almost complete ignorance regarding them.

This unfortunate gap in our knowledge may be due to the early evolution of vertebrates having taken place in fresh water (a theory discussed later in this chapter). There are few freshwater deposits in the preserved stratigraphic record which antedate the late Silurian. The rarity of early vertebrate remains may well be due to this situation.

As the Silurian draws to an end, and freshwater deposits tend to take the place of marine sediments, we find increasingly numerous remains of vertebrates quite unlike any existing forms in appearance. Almost all are covered with varied types of armor, the feature to which the name "ostracoderms" ("shell-skinned") is due. In some forms the internal skeleton of the head region is highly ossified; in others it is persistently cartilaginous; in none have remains of the axial skeleton as yet been found. In body outlines and in the presence of hard skeletal parts they seem remote from the cyclostomes; but we find that all these forms, like living lampreys and hagfishes, are characterized by the absence of jaws and by the absence or poor development of paired fins; and that, further, those which are best known show striking resemblances to the cyclostomes in many structural features. When we first see these ostracoderms, they already have a long history behind them and are divided into several distinct groups.

Osteostraci.—*Cephalaspis* and its relatives, constituting the order Osteostraci, are known in more detail than any other ostracoderms, and hence merit first consideration. These types are abundant in late Silurian and Lower Devonian rocks; a few forms lingered until the close of Devonian times.

Few were of great size, the length ranging generally from a few inches to a foot or two. They were completely encased in an armor of plates and scales. When sectioned, these plates are seen to be composed of true bone, while on the surface was a tuberculated, enamel-covered layer of dentine comparable to fused dermal denticles. The body scales were arranged in a series of vertical rows, each containing but a few elongate members. The body (Fig. 17) was rather fishlike in appearance, with one or two dorsal fins and a heterocercal caudal. The head was much flattened, the body higher but with a flat ventral surface and a triangular section. Structures are present which appear to be paired fins of a sort. At either lateral margin of the trunk a continuous series of scales projects outward to form a fin fold. Behind the head shield there are, in many genera, scale-covered flaps which are comparable to pectoral fins; their internal structure is unknown. The head (Figs. 18, 19)—in fact, the whole area back to the shoulder region—was covered dorsally by a nearly solid shield of bone; in all but a few of the older and presumably more primitive genera there were prominent "horns" at the posterolateral corners. The eyes, directed upward as in many bottom-living types, were situated close together near the center of the shield; between them lay a median plate with an opening for the pineal eye. Anterior to this plate was a slit comparable to that which in lampreys contains the opening for the single nostril and hypophysis. In the center of the skull, behind the eyes, was an area filled by a series of small polygonal plates, and there were similar areas near each margin.

The shield folded under onto the margins of the flattened ventral surface. Behind its front edge was a small and obviously jawless mouth, while posterior to this a series of movable plates covered the throat region. At the edge of these plates were round external openings for the numerous gill pouches.

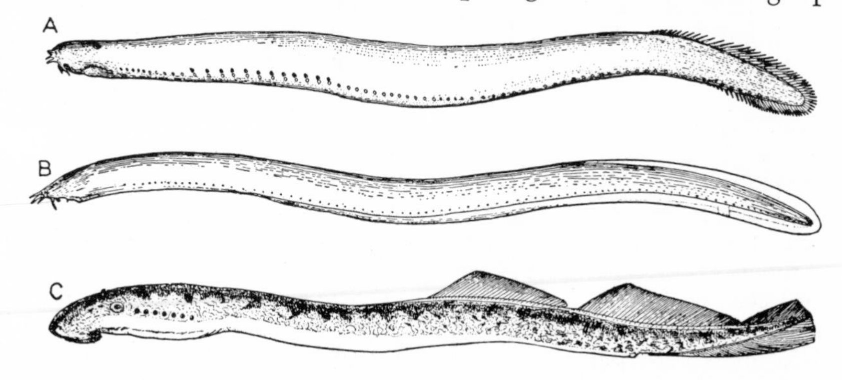

FIG. 15.—The three main types of cyclostomes. *A*, The slime hag, *Bdellostoma; B*, the hagfish, *Myxine; C*, the lamprey, *Petromyzon*. (From Dean.)

It was long supposed that, while ostracoderms had superficial armor plates, the internal skeleton was purely cartilaginous. But in the Osteostraci an ossified cranial skeleton was present. Its composition was in contrast to that of most vertebrates; for, instead of a separate braincase and jointed branchial arches, there was a single unified structure underlying the entire dermal shield. Such a fusion of braincase and gill structures, in contrast with typical fishes, seems a highly specialized feature; but it is possible that it is really a primitive undifferentiated condition and that the establishment of independent units was a later development.

Cephalaspid skulls have been carefully studied, both by reconstructions from ground serial sections and by "dissection" with fine needles. From the internal cavities most of the structure of the soft parts of the head can be made out. The brain and nerves and blood vessels are found to have been quite similar to those of lampreys and seem to show a very primitive vertebrate pattern. In the ear there were two semicircular canals, as in the lampreys; the dorsal slit referred to above is seen to lead into cavities which obviously lodged a single nostril and hypophysis in lamprey-like fashion. About ten pairs of gill cavities were usually present; their morphological relations show that it is probably the third of these gill pouches which corresponds to the spiracle with which the gill series of jawed vertebrates commences; two pouches are present here which have been crowded out of existence in higher vertebrate groups with the development of jaws and expansion of the mouth opening.

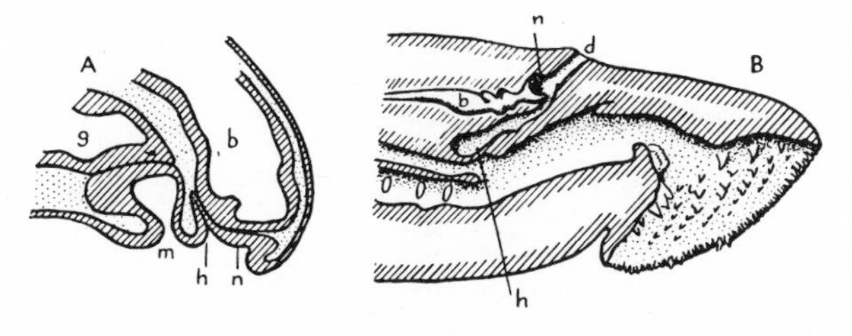

FIG. 16.—Diagrams to show the development of the dorsal nostril of lampreys. *A*, Longitudinal section through the head of a larval lamprey, showing the central position of the nostril and hypophysis; *B*, a later stage, in which hypophysis and nostril have migrated dorsally. Abbreviations: *b*, brain; *d*, dorsal common opening of hypophysis (*h*) and nostril (*n*); *g*, gut; *m*, mouth. (After Goodrich, simplified.)

Very stout tubules are found running from the ear region to the polygonal plate areas mentioned previously. The nature of these structures is uncertain. It has been advocated that the plate areas contained protective electric organs, such as are found in a number of modern teleost fishes, and that the trunks running to them were motor nerves supplying them. But there is hardly room beneath the plates to lodge an effective "battery." An alternative and more probable suggestion is that the plate areas were organs sensitive to water vibration (somewhat like the lateral-line organs) and that the trunks were filled with liquid carrying the vibrations inward to the sensitive ear region.

The Osteostraci were a homogeneous group in their fundamental structures. There is, however, considerable variation in the contours of the head (Fig. 19): there are great differences in the degree of development of the "horns"; the head may be broad or slender; a long rostral spine may be developed; the plate areas may vary in size, and the lateral ones may be subdivided. *Tremataspis* is an older, Silurian genus which may be relatively primitive in nature. Here there are no "horns" or pectoral "flippers"; instead, the head tapers smoothly back into the trunk, and a considerable portion of the trunk armor is fused with the head shield. Such an animal would have swum rather inefficiently, and with little steering ability, by a tadpole-like wriggling of the posterior part of the body. In *Cephalaspis* and other late genera a freeing of body segments from the shield would give greater mobility to the body, and the development of pectoral fin flaps would aid greatly in balance and control of direction. A similar development of paired-fin structures will be seen in other early vertebrates.

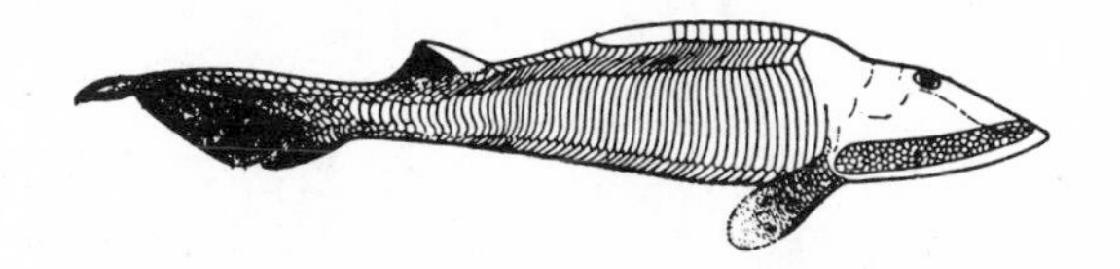

FIG. 17.—*Hemicyclaspis*, a characteristic cephalaspid of the late Silurian and early Devonian. About 1/3 natural size. (After Stensiö.)

The cephalaspids were obviously, from their depressed shape and dorsally situated eyes, bottom-dwelling forms; their small mouths and expanded gill chambers suggest that they were forms which made their living by straining food particles from the mud of the stream bottoms. *Cephalaspis* is a logical vertebrate development of the filter-feeding habits of pre-vertebrate chordates; it is essentially a large food-straining gill chamber, with a swimming tail, nerves, and sense organs.

It is obvious that these ancient types were fundamentally quite close to the modern lampreys in structure. Can the lampreys have descended from them? Many of the differences may be correlated with a change from a bottom-dwelling to a semiparasitic mode of existence. The most striking contrast lies between the well-ossified skeletal system of the ancient types and the purely cartilaginous skeleton of the lampreys. But similar degeneration is known to have occurred in other groups; and, despite their superficial dissimilarity, the osteostracans may possibly have been related to lamprey ancestors.

Anaspids (*Figs. 20, 21*).—The order Anaspida includes a number of genera, such as *Jamoytius, Birkenia, Lasanius,* and *Pterygolepis,* which were widespread in fresh waters in the late Silurian and earliest Devonian; there were late survivors (such as *Endeiolepis,* in the Upper Devonian of Canada). None was more than

10 inches in length—most about half that size. In the typical genera there was a complete covering of dermal armor, composed of a non-cellular bony material. Scales arranged in regular rows covered the body; the head region was protected by a complicated pattern of small plates, which appear to have been more or less fused into a shield on the dorsal surface. A series of spines lay along the dorsal side of the body. Unlike any normal vertebrate of later times, the tail tilted downward, rather than upward, to give a "reversed heterocercal" type of caudal fin. (So unexpected was this that for many years these forms were restored bottom side up.)

A fin of this type would tend to tilt the body upward while swimming; the anaspids (and other ostracoderms

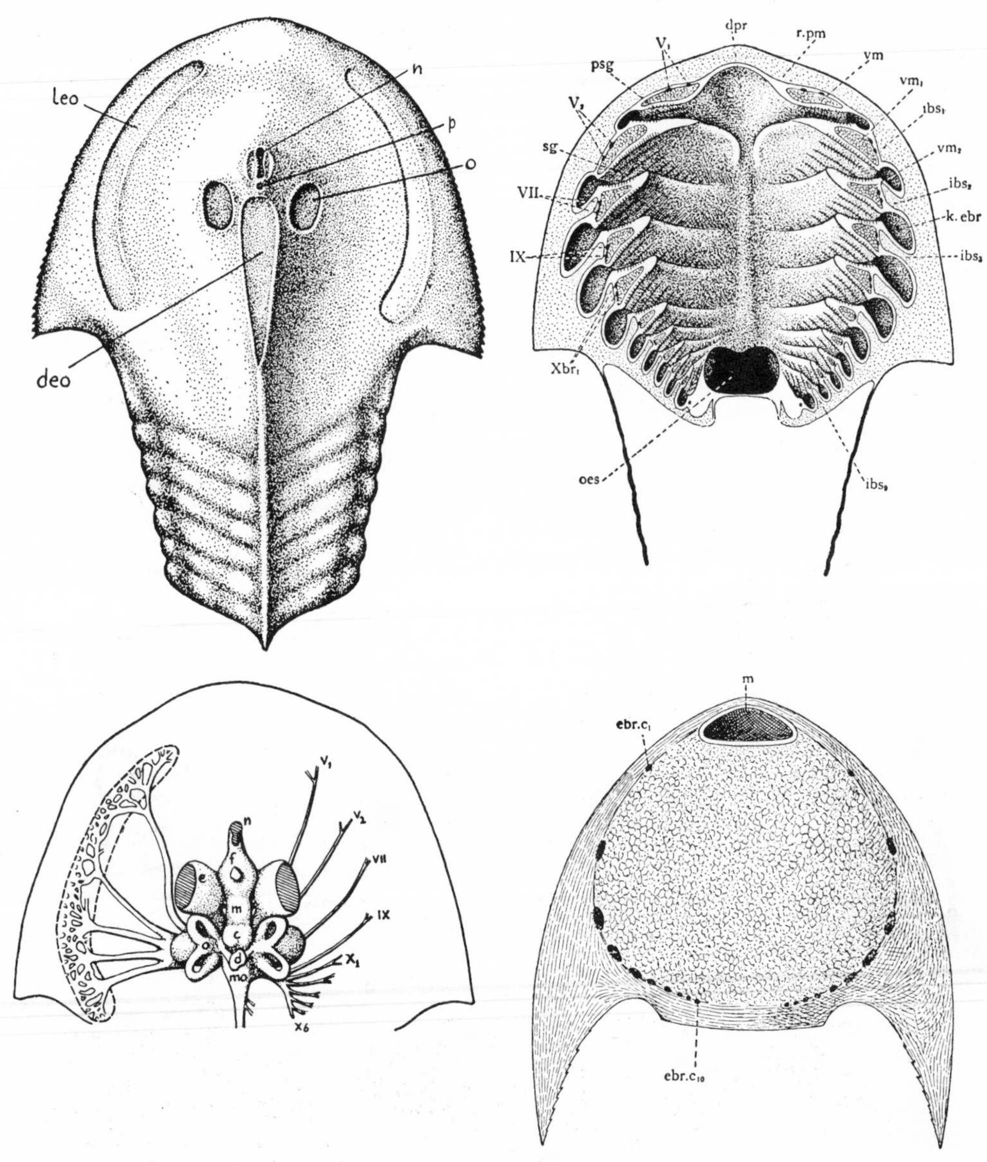

Fig. 18.—The cranial anatomy of *Kiaeraspis*, a cephalaspid. *Upper left*, Top view of head and anterior part of trunk region: *deo*, supposed dorsal electric organ; *leo*, supposed lateral electric organ; *n*, opening of single dorsal nostril and hypophysis; *o*, orbit; *p*, pineal opening. *Lower left*, Diagrammatic dissection of top of head, to show brain, nerves, and sense organs. Nerves to supposed electric organ shown on the left; other major nerves (numbered) on right, arranged in a series corresponding to the gills. *c*, Cerebellum; *d*, nerve to dorsal electric field; *e*, eye socket; *f*, forebrain; *m*, midbrain; *mo*, medulla oblongata; *n*, opening for nostril and hypophysis; *o*, ear region (two canals). In this and the next figure Roman numerals indicate cranial nerves. *Upper right*, Restoration of ventral surface of head, with the plates covering the throat removed, showing the ten gill sacs and the small, anteriorly placed mouth: *dpr*, mouth; *ibs*, partitions between gill sacs; *kebr*, ducts from gill pouches to surface; *oes*, oesophagus; *psg*, a prespiracular gill pouch lost in higher vertebrates; *rpm*, rostral region in front of gills; *sg*, gill pouch corresponding to the spiracle of higher fishes; *vm*, muscles of gill pouches. *Lower right*, Restoration of ventral surface of head, covered by small plates: *ebrc*, opening of the gills; *m*, mouth. (After Stensiö.)

with similar tails) may have been surface dwellers. Although there were no typical paired fins, there were lateral developments of a somewhat comparable nature. There was usually a stout spine in the shoulder region, and in some forms, at least, there was a lateral fin fold which sometimes ran a good part of the length of the body, roughly corresponding to the combined pectoral and pelvic fins of higher fish types. Behind the head a row of small, circular openings slanting back and down formed the exits from a series of gill pouches

No trace of internal structure has been discovered; the internal skeleton was presumably cartilaginous. The dermal armor, however, indicates that the head structure was fundamentally similar to that of the Osteostraci. The two large orbits were somewhat farther apart and faced more laterally than in cephalaspids. Between them lay a plate pierced by an opening for the pineal eye; while the nostril, just as in the last order and in the lampreys, reached the surface through an opening high on the top of the head. The mouth, as usually restored, was shaped more like that of higher vertebrates than in other agnathous types, but it is improbable that true jaw structures were connected with it.

Considerable variation in structure is found among anaspids. *Lasanius* of the Upper Silurian had lost most of its armor, except for the dorsal spines and a series of pectoral spines and their supports. Still more divergent is *Endeiolepis* of the Upper Devonian, whose body was almost completely naked, and whose dorsal spines were replaced by a long, soft dorsal fin.

That these forms were related to the Osteostraci is obvious; they were, however, less depressed and seem-

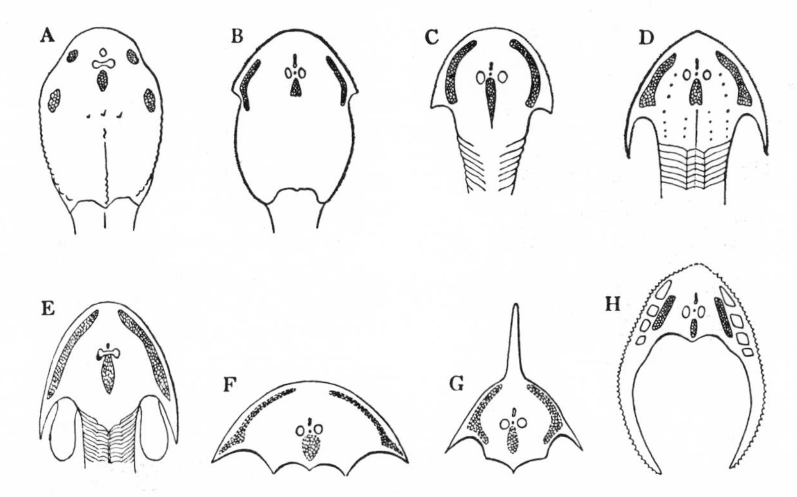

FIG. 19.—Dorsal view of the head shield in various osteostracans. *A–E*, A morphological series showing reduction in the amount of trunk region included in the cephalic shields and in the development of pectoral fins and "horns." The fins (seldom preserved) are known in this series only in *Cephalaspis*. In *A* a large proportion of the trunk is included in the shield, and pectoral fins were not developed; in *E* most of the trunk has been freed from the head shield, large sinuses for the attachment of pectoral fins are present; *B–D* show intermediate conditions: *A, Tremataspis; B, Didymaspis; C, Kiaeraspis; D, Thyestes; E, Cephalaspis; F, Benneviaspis; G, Boreaspis; H, Sclerodus.* (Mainly after Stensiö.)

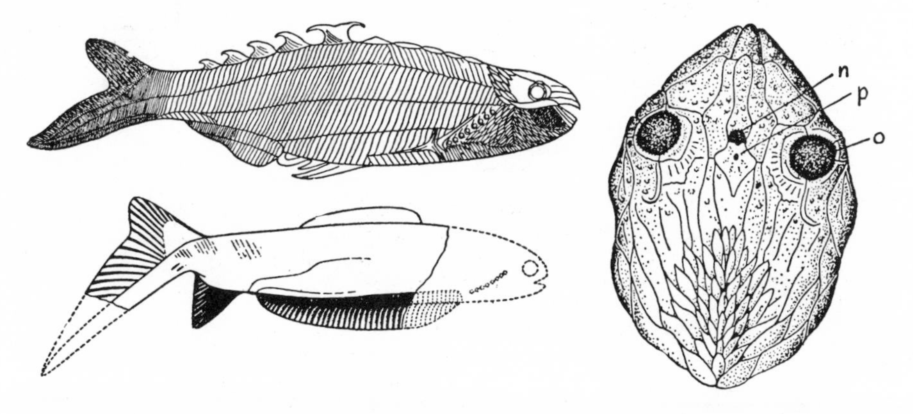

FIG. 20.—Anaspids. *Upper left, Birkenia* from the late Silurian of Scotland; original about 4 inches long. (From Stetson.) *Lower left, Endeiolepis,* from the Upper Devonian of Canada; length about 8 inches. (From Stensiö.) *Right,* Dorsal view of the head plates of *Pharyngolepis,* an early Devonian anaspid from Norway, showing the lateral (*o*) and pineal (*p*) eyes and the single nostril (*n*). (After Kiaer.)

ingly more active, surface-swimming types, suggesting a trend away from the comparatively sedentary, bottom-living existence which may have characterized the earliest jawless vertebrates. They may well have been even more closely related to the lampreys than *Cephalaspis.* It seems probable that the anaspids were undergoing reduction of armor; with the loss of the remainder of the dermal skeleton, the anaspids would be suitable lamprey ancestors.

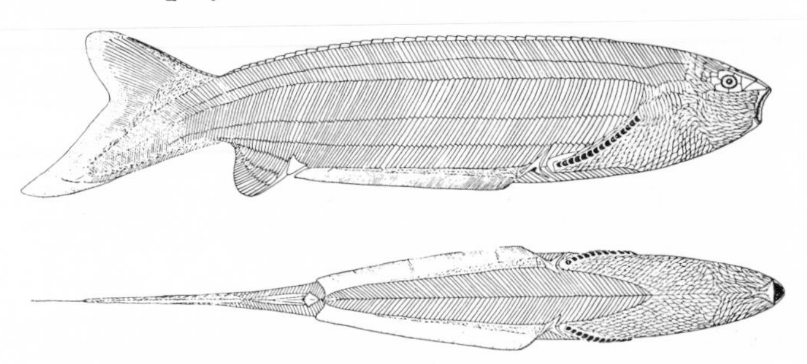

FIG. 21.—Lateral and ventral views of the anaspid *Pterygolepis,* showing particularly the long, lateral fin folds. Length of original about 4 inches. (After Ritchie.)

Heterostraci (Figs. 22–24).—A very different group of jawless vertebrates is the order Heterostraci, whose members (like the cephalaspids) were abundant in the late Silurian and early Devonian but had disappeared completely by the end of the latter period. The Heterostraci are oldest of all vertebrate orders, for dermal plates from the Ordovician show a microscopic structure of heterostracan type. *Pteraspis* is the best-known member of the group; *Poraspis, Anglaspis,* and *Cyathaspis* are other characteristic forms. Usually there was a complex armor comparable to that of *Cephalaspis,* but lacking bone cells. In contrast with the cephalaspids and anaspids, the paired eyes were far apart on the sides of the head and the pineal eye often failed to pierce the top of the skull; a striking difference lies in the fact that there was no opening for a dorsal nostril. The body was covered by scales, often diamond-shaped. In typical heterostracans the body was little flattened. The tail fin was of the reversed heterocercal type. There were no other median fins or paired fins, but there were dorsal and ventral spines on the trunk; frequently a prominent dorsal spine arose from the back of the head shield.

The anterior part of the body, including the large gill chamber, was protected by stout armor (Fig. 24). In typical Silurian types this consisted of a single oval dorsal shield, a similar ventral element, and a pair of lateral gill plates. In later types there is a tendency for the armor to break up into a number of smaller elements with the retention, however, of major dorsal and ventral plates. In *Pteraspis,* for example, there separated from the dorsal shield an anterior rostral element and smaller plates about the orbits.

A single lateral opening formed a common outlet for the gills; the position of the gill pouches is shown by a series of five to nine paired impressions on the inner surface of the dorsal shield.

On the underside of the rostrum a transverse slit indicates the position of the mouth. In some specimens, depressions in the inner surface of the rostral shield above the corners of this slit indicate that the nasal sacs were situated here and were paired structures, as in higher vertebrates. There were no jaws, but the mouth slit was bounded posteriorly by a series of parallel flexible plates, suggesting that some type of nibbling or scooping movement was possible. No internal skeletal structures have been discovered; presumably they were cartilaginous.

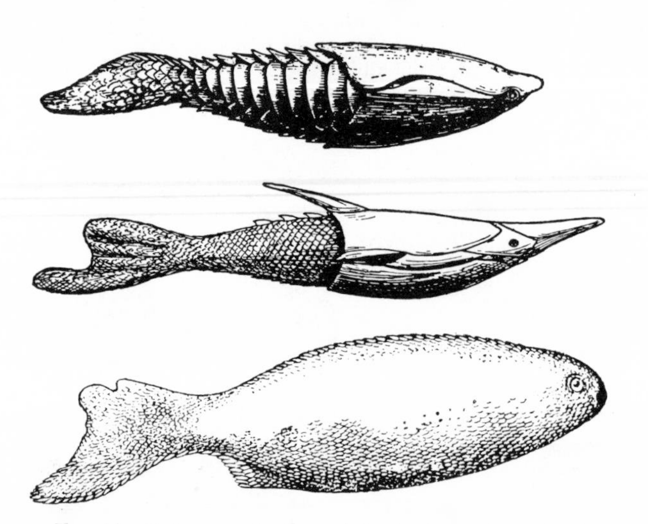

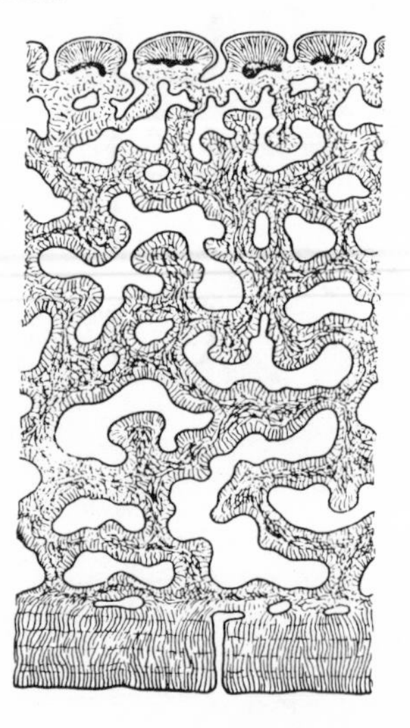

FIG. 22.—Heterostraci. *Above left, Anglaspis* from the Lower Devonian of Spitsbergen; length about 6 inches. (From Heintz.) *Center, Pteraspis* from the Lower Devonian of Europe, length about 2.25 inches. (From Heintz, after White.) *Below,* The coelolepid *Phlebolepis* of the late Silurian, about natural size. (After Obruchev.) *Right,* Enlarged section of a dermal plate of *Psammolepis,* showing an outer layer of tubercles of enamel and dentine-like materials, a thick middle layer of spongy bone, and a basal laminated layer. (From Bystrow.)

Considerable diversity is encountered among the known heterostracans. We have little knowledge of the appearance or structure of Ordovician Heterostraci. The Upper Silurian genera, such as *Poraspis* and *Cyathaspis,* are more heavily armored than the Devonian forms and with the armor concentrated in a smaller number of plates. There is great variety, too, in the shape of the "head" (Fig. 25); in some the rostrum is short and rounded, in others greatly elongated.

Flattened, bottom-dwelling relatives of the pteraspids are represented by *Drepanaspis* (Fig. 23) and other Devonian genera. Here the head and gill region, corresponding to the area included in the pteraspid shield, is very broad and flat but is covered by a series of plates comparable with those of *Pteraspis* with, in addition, a mosaic of smaller polygonal elements. The eyes were quite small; the mouth apparently was similar to that of *Pteraspis,* but in the fossils it appears at the front end of the upper rather than the undersurface.

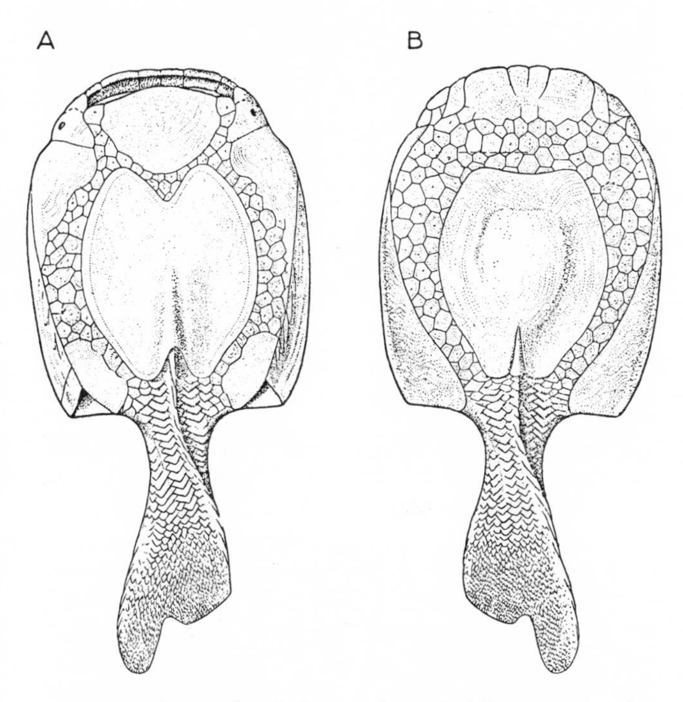

Fig. 23.—*A,* Dorsal and, *B,* ventral views of *Drepanaspis,* a flat-bodied Lower Devonian heterostracan; the general plate arrangement is comparable to that of pteraspids; original about 1 foot long. (After Heintz.)

The Heterostraci are obviously not closely related to the agnathous forms previously considered—cyclostomes, Osteostraci, Anaspida—all of which are peculiar in the possession of a single dorsal nostril. Possibly the Heterostraci are the remnants of a generalized primitive vertebrate group which might have given rise, by modification and specialization, to other jawless types. Still further, it is possible that the Heterostraci may be close to the line of ascent to jawed types, in which paired nostrils were retained. Known forms of the order, however, appear to be too specialized on their own account to be considered as such ancestors.

Coelolepids.—In the late Silurian and Lower Devonian are found a number of poorly known and tiny fishes which may be grouped in the order Coelolepida. *Coelolepis, Thelodus, Phlebolepis* (Fig. 22), and *Lanarkia* are representative genera. In many instances these forms are known only from patches of the scales which covered their bodies. These scales were minute structures, not overlapping but nevertheless forming a continuous covering for the body and hence to be considered as true scales rather than as dermal denticles of

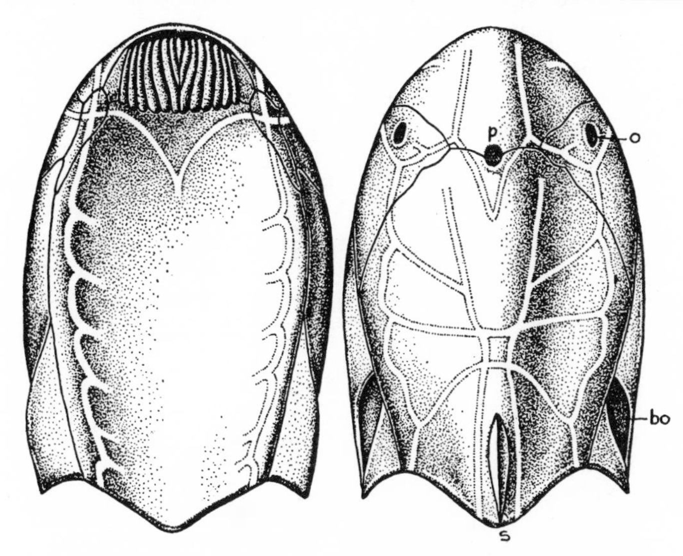

Fig. 24.—Ventral and dorsal views of the anterior portion of the body of *Anglaspis.* Lateral-line canals are indicated. The mouth lay in front of the slender series of plates near the anterior end of the ventral surface. Abbreviations: *bo,* common opening of the gill pouches; *o,* orbit; *p,* pineal opening; *s,* median dorsal spine. About twice the size of the original. (After Kiaer.)

the shark type. In a few cases the complete body outline is known, but, even so, our knowledge of these fishes is slight. We know almost nothing of internal structures and in most cases little even of superficial features. The tail is forked and apparently of the reversed heterocercal type; an anal fin may be present. The anterior part of the body (including the gill chamber) may be broad and flattened, as in most other ostracoderms. Flaps at the posterior end of this region in *Thelodus* may mark merely the region of the gill opening, as in Heterostraci, or may possibly be rudimentary pectoral fins, as in cephalaspids; in *Coelolepis* and *Phlebolepis,* however, these flaps are absent, and the head region tapers evenly into the trunk contours. In some instances pigment spots or openings show that the paired eyes were laterally placed; we have, however, no data on the pineal eye or nostrils. The mouth was small. Impressions of gill chambers have been seen in *Thelodus,* but the nature of the external opening is unknown.

The coelolepids have frequently been allied with the Heterostraci because of the similarities in body contours, but there is no proof of such association. Their scales are comparable to the superficial portion of the plates and scales of other ostracoderms. It was originally suggested that the coelolepids demonstrated a stage in the development of armor, and that typical bony plates and scales might arise by a fusion and deeper growth of small and superficial structures such as those present here. But it seems even more probable that we here have indications of a degenerative process—a reduction of armor, which, if carried a bit further, would lead to the condition of isolated dermal denticles, as seen in sharks. While it is customary to regard coelolepids as ostracoderms, even this is uncertain. It is not impossible that they may be forerunners of some group of the later jawed-fish types.

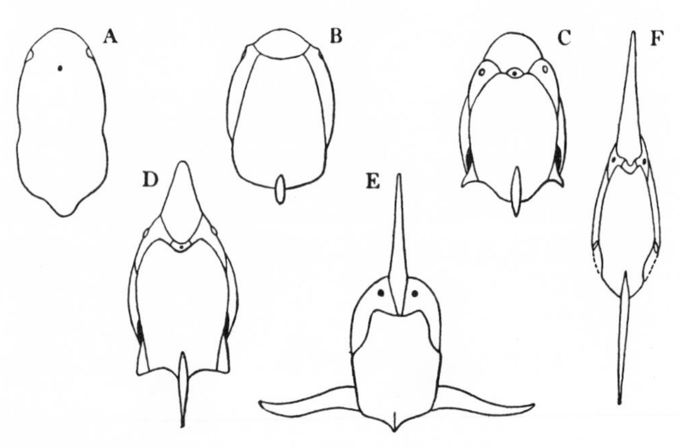

FIG. 25.—The armor in a varied series of Heterostraci: *A, Poraspis; B, Cyathaspis; C, Anglaspis; D, Pteraspis dunensis; E, Dyreaspis; F, Pteraspis rostratus. A–C* show a series in the breakdown of the originally solidly co-ossified shield into separate elements, and the development of a dorsal spine and lateral projections behind the gill opening; *D–F* show variations in the development of these structures and of the rostrum. (After Heintz.)

Primitive vertebrates.—In the previous chapter the problem of the history of skeletal materials—bone versus cartilage—was briefly discussed. Classical theory maintained that cartilage was the older material. But the paleontological data strongly suggest that bone was a primitive adult skeletal material, cartilage an essentially embryonic adaptation which is retained in the adult only as the result of degenerative processes.

The evidence from the Agnatha tends strongly to support the latter view. Modern jawless forms are boneless; all those of the Silurian and Devonian possess dermal bone, and in the cephalaspids there is, in addition, a well-developed internal bony skeleton in the cranial region. It may be argued that in the ostracoderm fauna, which first appears in abundance near the Silurian-Devonian boundary, we are dealing with forms which were just acquiring bone, and that there was a preceding stage in which vertebrates had already evolved but the skeleton was purely cartilaginous. If so, however, the time when bone was unknown must have been an exceedingly remote one, for bony tissues were present in the oldest vertebrate scraps from the Ordovician. If bone was a new invention among the older vertebrates, one would expect to find a progressive increase in ossification among the later representatives of the groups concerned. This, however, is the reverse of the true situation. In general, the later ostracoderms of the Devonian show a less substantial armor than that of Silurian forms; bone is regressive, not progressive, in its history among ostracoderms. As will be seen, a similar story of bone reduction is true of various other lower vertebrate groups.

It seems certain that, in the ostracoderms, we are dealing with forms which, despite the specializations of one group or the other, are representative of a radiation of truly primitive vertebrates. That the absence of jaws is a primitive feature need not be seriously questioned. This condition greatly limits the possible modes of life of primitive vertebrate types. The modern cyclostomes have been enabled to take up a predatory mode of existence through the development of a rasping "tongue" as a substitute for jaws, and a type of nibbling may have been present in some ostracoderms. In general, however, the older vertebrates were limited, as to food supply, to tiny organisms and bottom detritus that could be strained through the gill apparatus, much as in lower chordates today. The gills, therefore, were not only a breathing, but also a feeding, device. In relation to this, we see that in most ostracoderms the major—anterior—part of the body was much expanded to form a large set of branchial chambers, surmounted by a brain and sense organs. Posterior to this, the trunk and tail appear as a relatively small locomotor appendage, useful in transporting the feeding apparatus to favorable locations.

The modern cyclostomes lack paired fins; among the ostracoderms we see various stages in the early development of paired-fin structures. Without such fins the locomotion of the oldest vertebrates must have been of the relatively ineffective and uncontrolled type seen in a frog tadpole. Paired flaps or rudders would aid greatly in preventing rolling and pitching and, if flexibility were attained, aid in steering. The finlike developments seen among the ostracoderms are not closely comparable to the "orthodox" fin structures of sharks or of higher bony fishes but presumably represent independent evolutionary developments. Similar essays in the establishment of paired-fin systems, often of curious types, will be seen among the placoderms and acanthodians. Paired fins, it seems probable, were developed in parallel fashion among various early vertebrates in relation to their needs for more efficient locomotion.

The ostracoderms are primitive vertebrates; but if we seek among the known forms for the ancestors of higher vertebrate groups, we meet with disappointment. The cephalaspids and anaspids show very definite characteristics (as in the development of nostril and hypophysis) which seem to ally them to the cyclostomes, and the anaspids especially may well be close to the cyclostome ancestry. It is, however, very improbable (although not impossible) that the ancestors of other vertebrate groups passed through a stage with the peculiar monorhine condition seen here.

The Heterostraci and coelolepids appear, in such features as their presumably diplorhine condition, to offer better prospects of relationship to higher vertebrates. We know, however, very little about the structure of the coelolepids, and the known Heterostraci are obviously too highly specialized in many ways to be considered in themselves as ancestors of any of the higher jawed vertebrates.

Our failure to find actual gnathostome ancestors among the well-known ostracoderms of the late Silurian and Devonian is, of course, a result only to be expected from a broad consideration of the problem. The known forms are ancient vertebrates, it is true, but not the oldest of vertebrates. We know that the vertebrate stock was in existence a full period earlier, in the Ordovician; and it is possible that highly developed vertebrate types may have been in existence in the Cambrian, a hundred million years or more before the first adequately known types; a dearth, in the early Paleozoic, of freshwater deposits in which such animals may have lived may be responsible for the paucity of the pre-Silurian record. Cephalaspids, anaspids, pteraspids, and coelolepids represent not the beginning but the end of a cycle of early vertebrate evolution—end forms rather than generalized ancestral types.

The sea is generally assumed to have been the original home of life; lower chordates and lower living vertebrates are mainly marine types. It is, therefore, natural to believe that the vertebrates were originally dwellers in a saltwater environment.

But although the question is warmly debated, it is not improbable that early vertebrate evolution took place in fresh water. The active swimming characteristic of vertebrates suggests that their home lay in fresh waters where mobility was necessary to counteract the downward sweep of stream currents. Studies of kidney structures and function indicate that the primitive vertebrate kidney was one "invented" for use in a freshwater environment and that this structure was variously modified by later marine types. The fossil record is in great measure in accord with these other lines of evidence. Nearly all records of Silurian and early Devonian vertebrates are of a type that suggests that the oldest fishes lived in inland waters. Most ostracoderms were apparently stream and pond dwellers; it is not until the later part of the Devonian that sharklike fishes and placoderms become numerous in the seas; few of the higher bony fishes appear to have been ocean dwellers before the Mesozoic.

Why should primitive vertebrates have been armored? Later armored vertebrates are usually protected against their own carnivorous relatives; but we believe that the earliest vertebrates were without biting jaws and incapable of ingesting other vertebrates as food. Obviously, the enemies must have lain among invertebrate types. The inland-dwelling vertebrate would not have come in contact with cephalopods or other marine forms which one might at first think of as possible enemies. We do, however, frequently find in the same beds with the early vertebrates of the late Silurian and Devonian, members of one group of predaceous invertebrates, the eurypterids—aquatic, scorpion-like Paleozoic creatures with well-developed claws and biting mouth parts—which were, on the average, much larger than the contemporary ostracoderms. It is probable that the early vertebrates furnished a food supply for the "water scorpions" and that vertebrate armor served as a defense against these carnivorous enemies. When, in the course of the Devonian, faster-swimming fishes supplanted the comparatively sluggish ostracoderms, the eurypterids dwindled into insignificance and disappeared. The activity which characterizes the vertebrates may be related not merely to the taking-up of life in running streams but also to the necessity of escaping early eurypterid enemies.

3

CHAPTER

Archaic Gnathostomes

In the Devonian, often called the Age of Fishes, we find the ostracoderms, already discussed, and early representatives of the higher fish classes living today (Fig. 26). A considerable majority of the fish population of that period, however, belonged to groups now long extinct and peculiar in structure: the arthrodires, with heavy armor in articulated head and thoracic segments; the antiarchs, grotesque little creatures which look like a cross between a turtle and a crustacean; a series of odd forms which are armored caricatures of modern skates and rays; the acanthodians, sharklike in superficial appearance but most un-sharklike in various anatomical features.

Where to place these curious creatures has been a vexing problem. One or the other of these types has at times been thought allied to the ostracoderms, to the sharks, to the lungfish, to the "ganoids"; but in each case the supposed likenesses have been more than outweighed by the obvious differences. There are few common features uniting these groups other than the fact that they are, without exception, peculiar. All, however, are characterized by the presence of bony skeletal tissues; all have paired fins, although the fins are built on unusual plans; and all have jawlike structures of some sort or other. We shall here consider the greater part of these archaic jawed fishes as constituting a separate fish class, the Placodermi; the spiny sharks, the Acanthodii, will be here provisionally appended to the higher bony fishes (the class Osteichthyes).

Jaws.—In an earlier chapter (p. 8) we discussed in brief fashion the evolution of jaws. A series of terms is commonly used to describe different types of articulation of fish upper jaws with the braincase. Autostyly is defined as a condition in which the articulation is formed by the jaws themselves without aid from the hyoid arch behind them. Presumably in the earliest stage of jaw evolution this articulation would have been a movable one. It was once assumed that the fishes discussed in this chapter were primitive in this regard, but increased knowledge of them now shows that in the acanthodians and in at least some of the placoderms the hyomandibular bone—the major upper element of the next most posterior branchial arch—had become enlarged and aided in propping the jaws on the otic region of the braincase. This condition of joint jaw support is reasonably termed amphistyly. In some groups of fishes—notably chimaeras and lungfishes—there does exist an autostylic situation; but this is brought about by actual fusion of the upper jaws with the braincase—rather surely a specialized condition. Amphistyly appears to have been early developed in fishes, is present in early sharks as well as placoderms, and (although the situation is complicated by the presence of dermal elements) appears to have been basic in the Osteichthyes as well. When, as in modern sharks and higher actinopterygians, the hyomandibular bone bears the main burden of jaw support, the condition is termed hyostyly.

Fins.—We have seen in the ostracoderms certain early essays in the development of paired fins in the shape of spines, flaps, or folds of some sort on either side of the body—structures which presumably aided in stabilizing the fish during progression. In the "higher" fish groups, treated in later chapters, appendages (unless secondarily lost or modified) invariably consist of two pairs, the pectoral fins in the shoulder region just behind the gill slits, and the pelvic (or ventral) fins, primitively placed just anterior to the anal region; the pectorals are, in general, the more prominent of the two. Each fin articulates with a girdle of bone or cartilage, which lies inside the body wall and to which are attached muscles which move the fin. The pectoral girdle usually extends far up the flank and, in addition to internal cartilages or bones, may have, in bony fishes and land vertebrates, a covering layer of dermal bones; the pelvic girdle is less highly developed and typically consists in fishes of a small plate of bone or cartilage embedded in the undersurface of the abdomen.

The pattern of the fin itself is variable; these variations may be illustrated by a consideration of fin types seen in sharklike forms. In the past, two major theories have been advocated as to the origin of paired fins. One theory held that the primitive fin type was essentially of the sort seen in Figure 27*C*, with the skeleton including a long axis, with side branches, extending out into a leaf-shaped fin with a narrow base. Because of its supposedly ancestral nature, this fin type was named the archipterygium; it was assumed that from it might have been derived, on the one hand, broad-based fins, by such a process as suggested by Figure 27*C, B, A*, and, on the other, fins of the usual sharklike type shown in Figure 27*E*.

However, archipterygial fins are rare among fossil forms, and it is improbable that this type of fin is really primitive. An alternative idea, early advocated, was the fin-fold theory, which pointed out that paired fins could be considered as essentially comparable to median fins in structure and primary function; that they originated, to begin with, as lateral folds from the body walls which, like the dorsal median fin fold, were primarily stabilizing structures; only later, when individual paired fins acquired narrow bases, could they become effective aids in steering as well as stabilizing.

Under this type of theory, a broad-based fin, such as that of Figure 27*A*, would be the most primitive, and narrow-based fins, such as 27*C* or 27*E*, would be secondary derivatives. In its broadest development, the fin-fold theory suggested that originally there was a long fin fold down either flank of the trunk, and that pectoral and pelvic fins originated by subdivision of this original unit structure.

This fin-fold theory is in great measure plausible, and there appears to be ground for partial advocacy of it; for example, we have seen that in the *Cephalaspis*-like

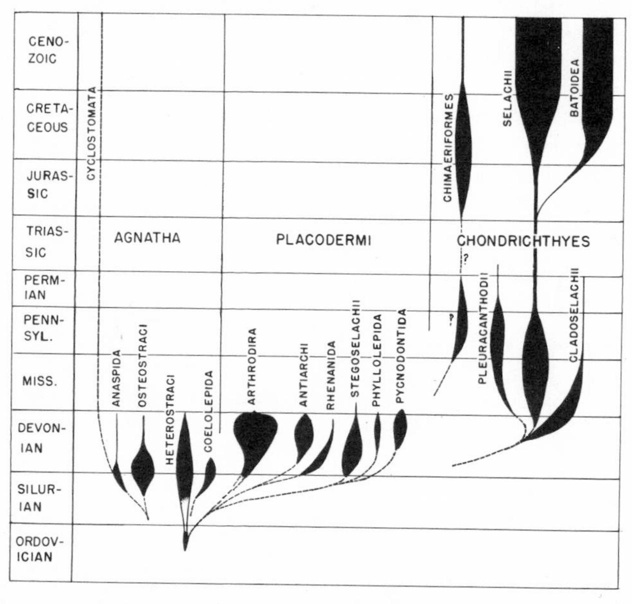

FIG. 26.—Diagram to show the evolution and distribution in time of lower fish groups.

ostracoderms there was a row of projecting scales down either flank somewhat in the fashion of a bony fin fold, and in some anaspids, at least, an actual, long, continuous fold was present; later in this chapter we shall see that in some "spiny sharks" there were, instead of simply pectoral and pelvic fins, a whole row of spine-supported finlets down each flank (Figs. 46, 47), giving almost a continuous fin fold. But one who attempts to find in ancient fishes "idealized" forms completely supporting such theories as that of the continuous fin fold is doomed to disappointment. After all, an animal cannot spend its time being a generalized ancestor; it must be fit for the environment in which it lives, and be constantly and variably adapted to it. In the specific case of appendages, it is probable that discrete paired fins arose more than once—possibly a number of times —by selective development from rudimentary stabilizing structures.

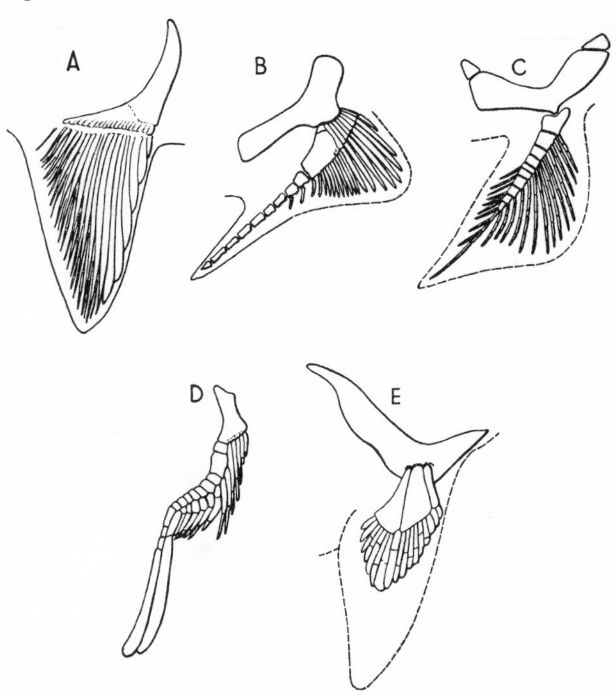

FIG. 27.—The paired fins and girdles of sharklike fishes. All except *D* are pectoral fins. *A*, *Cladoselache*, of the late Devonian, with parallel radials extending out from the girdle; *B*, "*Cladodus*" *neilsoni* (Carboniferous), with a posterior axis developed, transitional between *A* and *C* or *E*; *C*, a pleuracanth shark, with archipterygial structure; *D*, pelvic fin of male of the same, with clasping organ; *E*, *Hybodus*, a Mesozoic shark. (*A* after Dean; *B* after Traquair and Goodrich; *C* and *D* after Fritsch; *E* after Smith Woodward.)

Arthrodires.—As was said earlier, most of the archaic jawed fishes to be treated in this chapter are considered to be members of the class Placodermi, a group now extinct, which flourished greatly in the Devonian but was almost completely confined to that period. Included are a host of ancient forms termed the arthrodires, as well as forms related to them. There is great variation within the members of the Placodermi (most of which were marine), but in all cases not only were jaws and paired fins of some sort present, but all had skeletons ossified to some degree, both externally and internally. Although in some forms ossification became greatly reduced, all except the most specialized forms had armor in two main segments, one a shield for the head (and gill region), the second, a set of armor plates surrounding the thorax—the "chest region," if we may call it that. In addition, it seems probable that primitive forms possessed a scaly covering over the trunk and tail (this, however, tended to be lost in most cases). A seemingly odd but highly distinctive feature is that in all the earlier or more primitive placoderms there is invariably found a prominent bony spine projecting on either side at the shoulder region. In contrast to the ostracoderms, in which there was in general little ossification inside the armor, the placoderms frequently showed ossifications not merely in the head region but in the vertebral column as well.

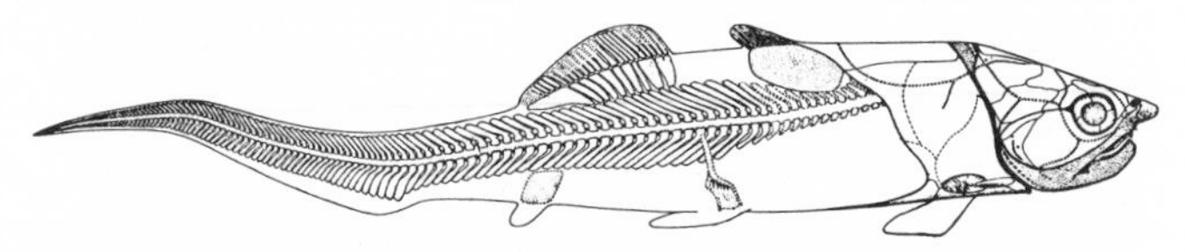

FIG. 28.—*Coccosteus*, a small arthrodire of the Middle Devonian, with reduced pectoral spines; original about 16 inches long. Rostrum and part of jaws restored. (After Stensiö.)

Outstandingly prominent among the placoderms were the Arthrodira, common in all stages of the Devonian. As the best-known members of the class they will be discussed first. We may best treat them in the reverse of chronological order, for the better-known members of the group were those of the later part of the Devonian. The arthrodires are characterized (as the name implies) by the fact that, except in the more primitive forms, the head and trunk shields were connected by a pair of well-developed ball-and-socket joints, situated well up on either side of the neck. Peculiar bony jaws were present, and there were paired pectoral and pelvic fins. These features are well shown in such typical forms as *Coccosteus* (Figs. 28, 29) and *Dunkleosteus* (often termed *Dinichthys*) (Fig. 30). The former was a Middle Devonian freshwater fish of modest size, reaching perhaps two feet in length, the latter, a giant late Devonian type; but, except for some reduction in the dermal armor in the latter, the two were basically similar in most known structures.

The body appears to have been of fairly normal fish shape, although rather broad and flat. The posterior region is adequately known in *Coccosteus.* Scales or denticles have been found in but a few arthrodires, and the skin may have been smooth and naked in most. No trace of vertebral centra has been discovered, but there were well-ossified neural and haemal arches. The column tilted up at the posterior end, suggesting the presence of a heterocercal tail. A series of plates above the backbone indicates the presence of a dorsal fin, while a ventral plate presumably supported an anal fin.

The surface armor plates were of bone, often ornamented superficially by dermal denticles, and comprised a complex series of median and paired elements, which were, in general, uniform in number and position throughout the arthrodire group. These plates cannot be homologized satisfactorily with those of higher bony fishes, and special series of names, such as those noted in Figure 30, are generally applied to them. Grooves in the surface of the plates indicate the presence of a well-developed lateral-line system. The head shield consisted of a series of plates covering the top and sides of the cranial and gill regions. Large paired orbits were present and a pineal eye as well, although this often failed to pierce the armor and reach the surface. A sclerotic ring of thin, bony plates was present in the eyeballs. This protective structure, found in many higher bony fish and tetrapods, in arthrodires consisted of four elements. In some cases notches at the anterior end of the shield mark the position of paired nostrils. These are placed close together, and there is no development of the rostrum seen in sharklike fishes. Most of the head plates were firmly united, but the plates beneath and behind the eyes and over the gill region were but loosely attached to the roofing bones in many members of the group.

The jaws were very peculiar in structure. In most cases the upper jaws are represented only by two pairs of bony elements, the front pair usually having a sharp tusklike projection, the posterior bearing a long shearing edge. Below, a single long element bearing a "tusk" in front and a shearing edge farther back was present on either side. These structures were very different from the jaws of normal fishes. The projections seen in the adult were not teeth but simply hard bony surfaces, although in some cases young forms possessed pointed toothlike structures fused to the jaw bones. There is seldom found any bony connection between the different parts of the jaw apparatus or between upper-jaw bones and the head plates. It is obvious, however, that these jaw elements are dermal ossifications which must have been attached to deeper-lying structures corresponding to the upper-jaw and lower-jaw bars—the palatoquadrate and the mandible. In most instances

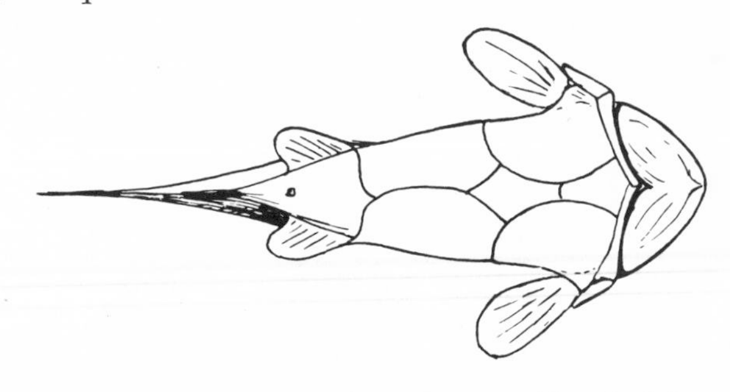

FIG. 29.—A ventral view of *Coccosteus,* with paired fins restored. (After Heintz.)

these elements appear to have remained cartilaginous; in a few, however, they have been preserved in a partially ossified condition. In many of the advanced arthrodire genera no traces of the braincase have been found, although rugose surfaces and projections for braincase attachment are found on the underside of the skull roof. Some partially ossified specimens are, however, known, and, as described later, primitive arthrodires had well-ossified braincases.

The posterolateral portions of the head shield covered the gill chambers. These areas of the shield functioned as an operculum; the gills thus had a common opening to the surface through the slit between the head shield and that portion of the trunk armor which lies in the position of a shoulder girdle; unlike the elasmobranchs and many primitive bony fishes, there was no separate spiracular opening.

The thoracic armor entirely surrounded the body. Dorsally, on either side, it articulated with the head shield by complicated peg-and-socket joints, indicating that there was a habitual movement of the head on the trunk (which, it is thought, was probably associated with the facilitation of gill-breathing, rather than feeding). The thoracic armor, both above and below, extended far back over the trunk region; below, it also extended some distance forward to protect much of the throat. The somewhat arched dorsal shield and the flattened ventral plate were composed of a number of closely connected elements with a nearly uniform arrangement in all typical members of the order. The lateral bony elements connecting the dorsal and ventral shields were situated in what is obviously the shoulder region and these elements have been compared with the dermal bones in the same situation covering the pectoral girdle in the Osteichthyes.

From the sides of the shoulder region projected a fixed bony spine, of modest length in *Coccosteus*, vestigial to absent in later and larger types. In advanced types, such as *Dunkleosteus* (*Dinichthys*), a long slit extended back from this point, nearly separating dorsal and ventral segments of the trunk armor. In a number of cases an ossified endochondral girdle element has been found lying within this slit, and it is obvious that in such forms there was a good-sized pectoral fin with a broad base (although the fin is seldom preserved). In *Coccosteus* and its relatives, somewhat more primitive types, dorsal and ventral shields are more broadly connected, and instead of a long slit in the dermal armor for the pectoral fin, there is merely an oval opening behind the pectoral spine, bounded in back as well as in front by bone. The pectoral fin was here obviously narrow based, in contrast to the advanced late Devonian types (Fig. 29). Behind the shield region a plow-shaped bone represents a pelvic girdle; in exceptional instances traces of small (and spineless) pelvic fins have been found associated with it.

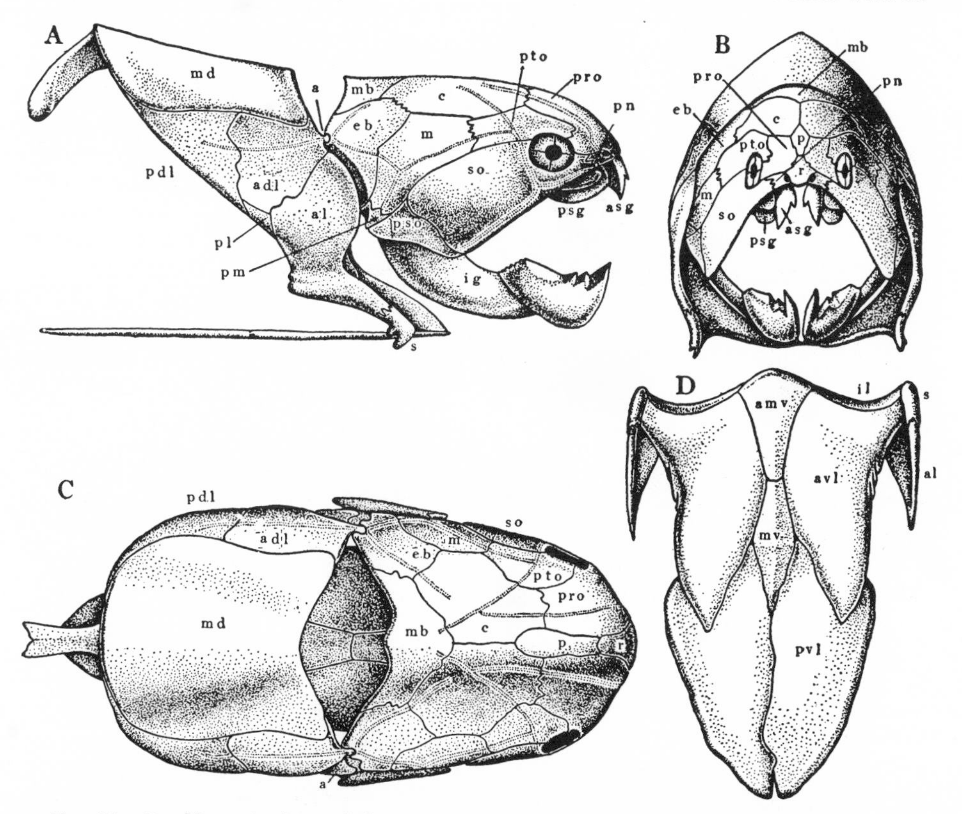

FIG. 30.—*Dunkleosteus* (*Dinichthys*). *A*, Lateral, *B*, anterior, and *C*, dorsal view of armor; *D*, ventral shield. Based on a large form in which the armored region is about 10.5 feet long. Lateral-line canals are indicated. The dermal plates of arthrodires cannot be readily homologized with those of other vertebrates and have been given a series of arbitrary names, listed below with their abbreviations: *a*, joint between head and thoracic shields; *adl*, anterior dorsolateral; *al*, anterior lateral; *amv*, anterior median ventral; *asg*, anterior supragnathal; *avl*, anterior ventrolateral; *c*, central; *eb*, paranuchal (exterobasal); *ig*, inferognathal; *il*, interlateral; *m*, marginal; *mb*, nuchal (mediobasal); *md*, median dorsal; *mv*, median ventral; *p*, pineal; *pdl*, posterior dorsolateral; *pl*, posterior lateral; *pm*, posterior marginal; *pn*, postnasal; *pro*, preorbital; *psg*, posterior supragnathal; *pso*, postsuborbital; *pto*, postorbital; *pvl*, posterior ventrolateral; *r*, rostral; *s*, spinale; *so*, suborbital. (After Heintz.)

Coccosteus and similar forms were fairly common in the Middle Devonian and seem to have ranged from fresh to brackish waters, showing a tendency toward a marine life. In the late Devonian the arthrodires, mainly inhabiting salt waters, reached a climax in size and numbers. A great assemblage of types has been recovered from the black shales of the Cleveland region and from the Wildungen region of western Germany. *Dunkleosteus* may have reached a length of 30 feet. In such giant types, in contrast with *Coccosteus*, the armor was shortened, on both the head and the trunk; the shoulder spine was tiny or absent; an unprotected gap developed between the two shields. These changes were also present in great measure in a number of contemporary forms. Some forms were smaller, but *Titanichthys* was even larger, and the average size had greatly increased since Middle Devonian days. The jaw apparatus varied considerably; *Mylostoma* had even developed crushing plates for mollusk-eating. In all, however, a reduction in the area covered by the armor and a reduction in its relative (although not absolute) thickness seem to have occurred with increase in size and tendency toward a marine life. This late Devonian climax marks the end of the group, for no arthrodires are known to have survived into the Mississippian.

Early arthrodires.—The arthrodires of the middle to late Devonian are frequently grouped as the Brachythoracici, "short-chested" forms, because of the fact that the thoracic armor was relatively restricted in development. In contrast is a large series of arthrodires almost entirely confined to the Lower Devonian, where they are most abundantly represented in freshwater deposits, notably in Spitsbergen. These earlier types are sometimes classed as the Dolichothoracici, "long-chested," because of the fact that the chest armor was much more highly developed than in later arthrodires. Arctolepida, however, is the preferable term for this suborder; *Arctolepis* (Figs. 32, 33) and *Phlyctaenaspis* are representative. Here, in contrast to many later types, the body appears to have been much flattened. The head shield was very much like that of later types, but with the eyes far forward. There is little fossil evidence as to the nature of the jaw plates, upper or lower. Whereas in later arthrodires there may be a considerable gap dorsally between head and thoracic shields, in these early forms the two sets of armor fit rather closely, but the ball-and-socket joint connection was not developed. In later arthrodires, the braincase is frequently cartilaginous, but a massively developed bony braincase (Fig. 31) has been found in the case of some arctolepids. As the alternative group name implies, the thoracic armor is here much longer in extent and more highly developed than in later arthrodires. Seldom, in these early forms, is there any indication of a division of the thoracic armor into discrete elements. Contrasting with later types is the build of the armor in the pectoral region. As we have seen, in advanced late Devonian types there was present a broad-based pectoral fin; this emerged from a long slit in the pectoral region which nearly separated dorsal and ventral divisions of the armor. In *Coccosteus*, we have noted, the slit did not extend all the way to the back and consequently the fin was constricted at the base. In the

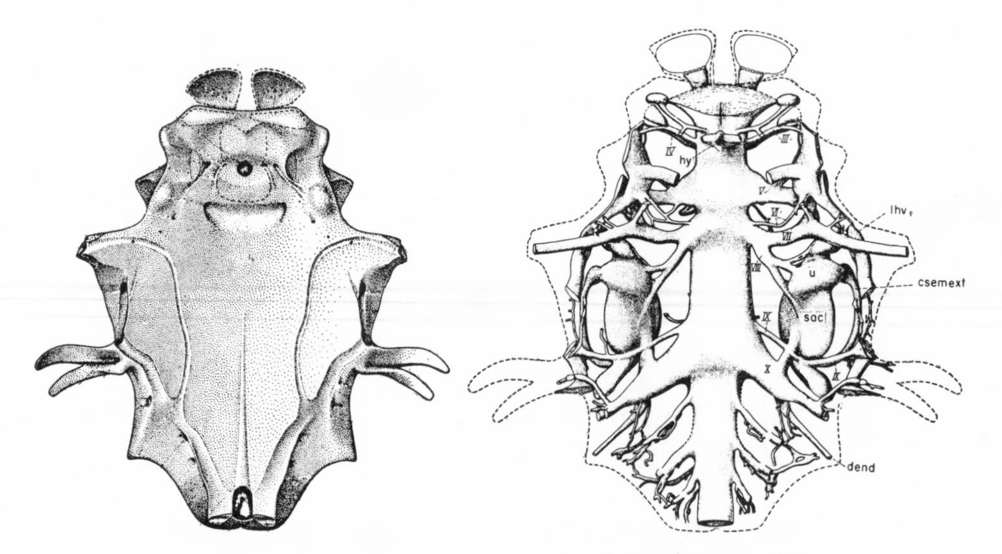

Fig. 31.—*Left,* Ventral view of the braincase, lying below the "head" shield, of the Lower Devonian arthrodire *Kudjovoniaspis.* The nasal capsules are restored. About natural size. Grooves indicate the position of blood vessels, and the foramina for nerves and blood vessels are visible. Medially, toward the front, is a ventral foramen leading into the pituitary fossa. *Right,* A model of the internal cavities of the braincase; in the center the brain cavity; on either side, canals for nerves and blood vessels and cavities for the internal-ear structures. Various nerves are indicated by Roman numerals. *csemext,* Horizontal (external) semicircular canal of ear; *dend,* endolymphatic duct from ear; *hy,* pituitary (hypophysis); *lhv,* canal for part of lateral head vein; *sac,* sacculus of internal ear; *u,* utriculus of internal ear. (After Stensiö.)

early Devonian forms, there is a still more restricted opening, essentially a circular hole, for emergence of the fin that obviously was only a small fan-shaped structure.

In contrast with this reduction of fin-base area, as we progress backward in the Devonian, is the story of the pectoral spine. As we have noted, the spine was vestigial or absent in advanced forms and modest in size in such Middle Devonian genera as *Coccosteus*. In the arctolepids, however, the spines were long, hollow, curved, cone-shaped projections of the armor. One would think that these curious structures would have been a hindrance rather than a help in swimming; the rather flat build of these fishes suggests that they were bottom-living forms, and these winglike projections may have been of use in wiggling upstream against a current. Little is known of the posterior part of the body; there is, however, evidence suggesting that, in contrast to many later arthrodires, scales were present over the back part of the trunk and tail in at least some arctolepids.

At first sight one would tend to think that the late Devonian arthrodires of the *Dunkleosteus* type are the more primitive and "generalized" members of the group, and that the grotesque arctolepids, with their long, curved shoulder spines, must be specialized end forms. This point of view has, in fact, been advocated by a distinguished student of the group. But the geological sequence strongly indicates that the reverse is the case. Further, as will be seen in later sections of this chapter, there are various groups of arthrodire relatives which differ in many regards, but which can all be traced back to early Devonian forms with well-marked shoulder spines.

The forms so far described seem to comprise the main line of arthrodire evolution. From this there were, however, many divergent branches during the Devonian. Some of these consist of forms in which the body tended to assume a depressed shape, giving rise to sluggish bottom dwellers. *Holonema* is representative of a family in which the body was moderately depressed and the trunk armor much elongated. *Homostius* (Fig. 32) and *Heterostius* are much more flattened, but with a different type of specialization of the armor. The head shield is greatly elongated posteriorly, suggesting an expansion of the gill chambers; the pectoral armor, on the other hand, is quite short dorsally, although the ventral elements of this series extend far forward beneath the head. The jaws were weak and slender; as in advanced arthrodires, the pectoral spines were lost. These forms grew to large size. They appear to have lived in brackish estuarine waters, and are frequently found in the Baltic region in late Devonian deposits, which appear to be of this origin. A still more peculiar depressed type characteristic of estuarine life is *Phyllolepis* (Fig. 34), common in the Upper Devonian and readily recognized by its plates, which are ornamented with fine concentric lines. This form was so depressed and featureless that it was long thought to be an ostracoderm allied to *Drepanaspis;* it is only recently that it has been recognized as a degenerate arthrodire relative which has lost most of its head armor except for an enlarged median element. However, the thoracic armor persisted, including, in the stout pectoral spines, the key character of primitive placoderms.

Ptyctodonts.—The name *Ptyctodus* has been applied to small bony plates, often no larger than a date pit, common in late Devonian marine deposits. Usually a hardened area is to be found on one surface, suggesting that they were grinding jaw plates of a small fish of some sort. Some Devonian ptyctodont tooth plates were much larger and more like those of arthrodires. In *Rhamphodopsis* and *Ctenurella* (Figs. 35, 36) most of the skeleton is known. The posterior part of the body was nearly naked; one, perhaps two, dorsal fins were present; a pelvic girdle and a scale-covered pelvic fin occur rather far forward on the body. There is a set of thoracic armor plates in the shoulder region, comparable in general construction to those of normal arthrodires but rather narrow. Laterally, there is in

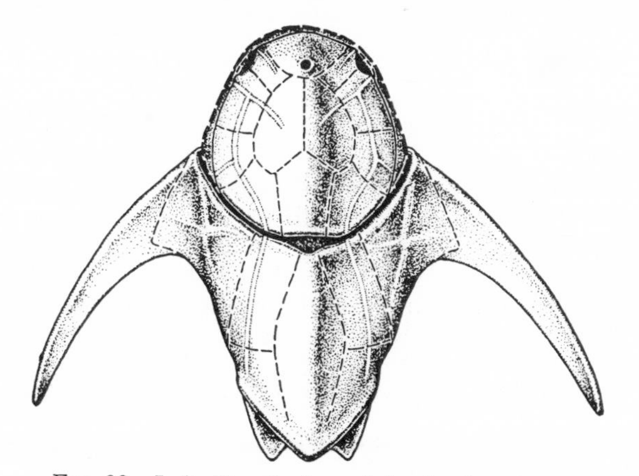

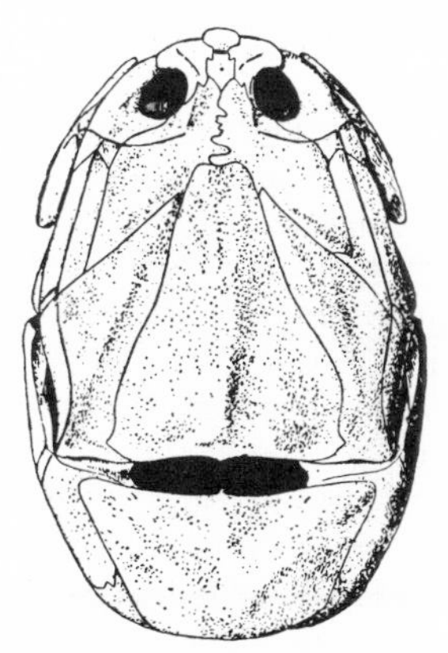

FIG. 32.—*Left,* Dorsal view of the head and trunk of *Arctolepis,* a primitive arthrodire with enormous pectoral spines. About 3/5 natural size. *Right,* Dorsal view of the armor of *Homostius,* a large flat-bodied arthrodire with long cranial and short trunk armor. About 1/15 natural size. (After Heintz.)

Rhamphodopsis a large pectoral fin spine like that of primitive arthrodires, in *Ctenurella* remains of a narrow-based pectoral fin. Dorsally, the shoulder plates in some forms support a large spine, which acted as a cut-water for an anterior dorsal fin.

The shoulder armor articulates with a head shield comparable in part to that of other placoderms, but much reduced in extent. In the mouth lay stout upper and lower dental plates. The short upper and lower jaw bars, to which the plates were attached, were partially ossified. The structures seen in these forms indicate clearly that in the ptyctodonts we are dealing with a group of arthrodire relatives which had taken to eating mollusks or some other type of hard invertebrates.

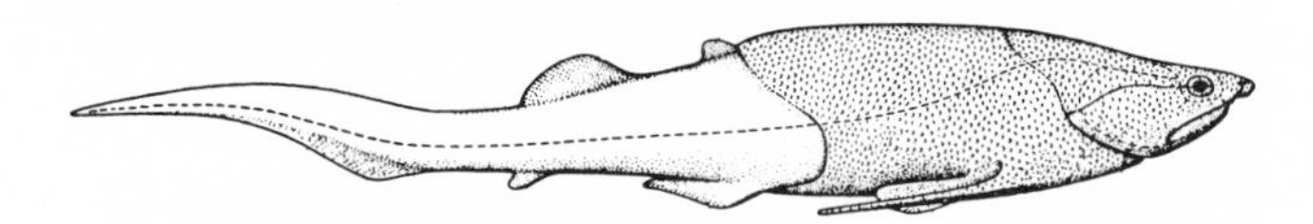

Fig. 33.—Restored side view of the Lower Devonian arthrodire *Arctolepis*. (From Stensiö.)

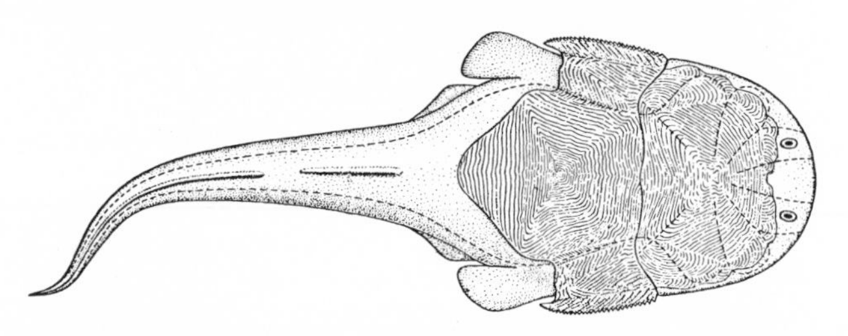

Fig. 34.—Restoration of *Phyllolepis*, a degenerate late Devonian placoderm, in dorsal view. (Anterior part of head and much of body conjectural.) About 1/4 natural size. (From Stensiö.)

Except that *Ptyctodus* may have lingered on into the very early Mississippian, there are no known post-Devonian remains of ptyctodonts. It has, however, been advocated that the ptyctodonts are the ancestors of the chimaeras, a group of rare high-seas fishes discussed in the next chapter. The limited nature of the armor in ptyctodonts suggests that we are witnessing here a stage in bone reduction that, if continued, might lead to the purely cartilaginous condition seen in chimaeras. The compact, short-jawed head structure is suggestive of that of chimaeras, and the tooth plates are similar. Quite probably chimaeras are descended from placoderms of some sort; but, as will be discussed later, it is by no means sure that the ptyctodonts are the ancestral group.

Petalichthyids.—In such forms as *Phyllolepis* and the ptyctodonts we are dealing with forms which seem reasonably associated with the arthrodires. We now come to a varied assemblage of weird fishes, mostly of early Devonian age and marine in habitat, which at first sight are not at all suggestive of arthrodire relationships. Most are incompletely known, but all appear to have well-developed fins, and are somewhat like the later sharks and rays in general proportions. But all have, in contrast with elasmobranchs, a skeleton which was ossified to at least some degree, and at least partially armored. Most significantly, the armor consists of discrete head and trunk portions, and there is always the diagnostic feature of a pair of lateral bony shoulder spines. We are surely dealing here with members of the class Placodermi which are related in some fashion to the arthrodires.

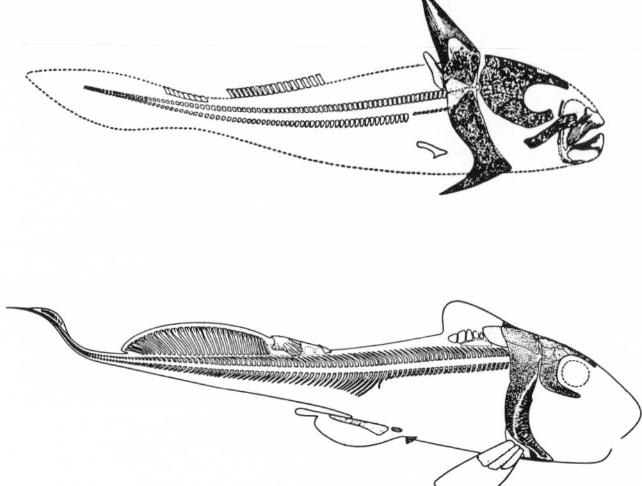

Fig. 35.—Restorations of ptyctodonts. *Above, Rhamphodopsis*, about natural size (after Watson). *Below, Ctenurella*, about 2/3 natural size. (After Ørvig.)

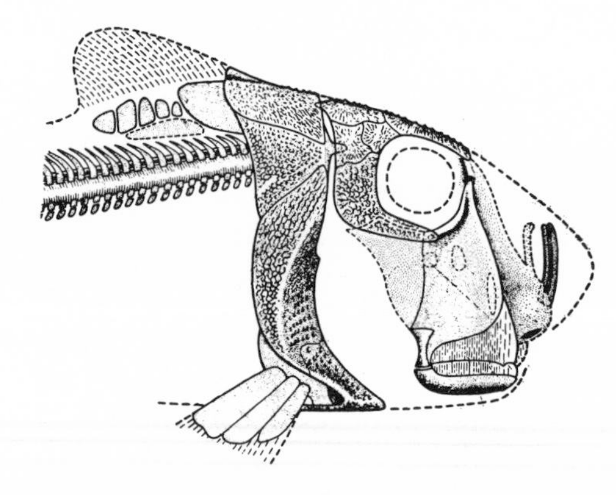

Fig. 36.—Anterior part of the body of the ptyctodont *Ctenurella*. There is a short belt of thoracic armor and, dorsally, a reduced head shield. Tooth plates and jaw remains are preserved. About 5/4 natural size. (After Ørvig.)

The greater part of these odd forms can be assembled into two groups, the Petalichthyida and Rhenanida. Of the petalichthyids, the best known are *Lunaspis* (Fig. 37) of the early Devonian, and *Macropetalichthys* (Fig. 38), widespread in Middle Devonian seas. In *Macropetalichthys* there is a greatly elongated head shield, broadly comparable to that of its arthrodiran cousins, although there are differences in the arrangement of its component plates. Beneath this shield there was present (much as in early arthrodires) a well-developed braincase. This structure lay below the anterior two-thirds of the bony head shield; posteriorly an extension, bracing the shield, ran back to connect with the front end of the vertebral column, while braincase cavities containing blood vessels show that the gills were located under the back part of the shield on either side.

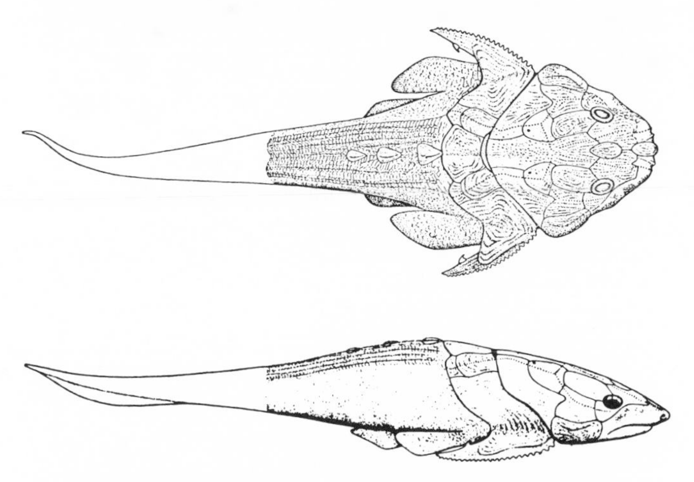

Fig. 37.—The petalichthyid *Lunaspis* of the Lower Devonian, dorsal and lateral views. Fins restored. About 1/3 natural size. (From Stensiö.)

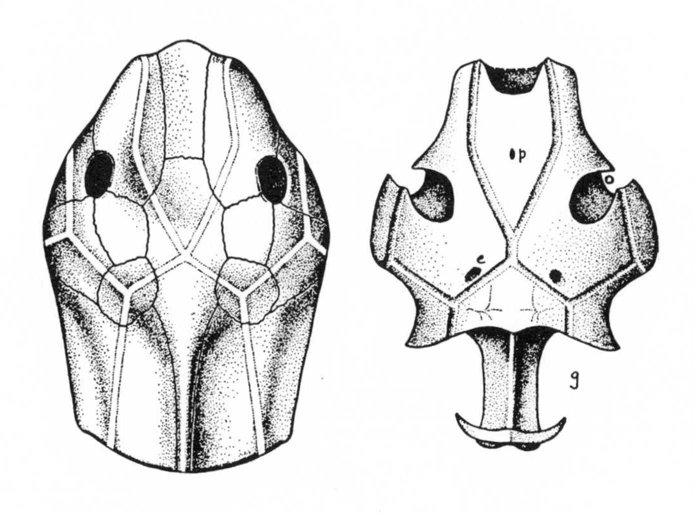

Fig. 38.—*Macropetalichthys,* an armored Devonian form related to the arthrodires. *Left,* Dorsal view of head armor; *right,* the underlying braincase. The position of the lateral-line canals is indicated. Abbreviations: *e,* opening of endolymphatic duct from inner ear; *g,* position of gill chamber beneath lateral back corners of shield; *o,* orbit; *p,* pineal opening. Original about 9 inches long. (After Stensiö.)

Until recent years, no further remains of *Macropetalichthys* had been identified. We now know, however, that behind the head shield this fish carried a set of thoracic plates with large pectoral spines very similar to those of arctolepid arthrodires. In the related Lower Devonian genus *Lunaspis,* nearly the entire fish has been recovered; the body was completely covered by large scales, as was presumably the case in the most primitive arthrodires, and there is evidence of the presence, as in those primitive forms, of narrow-based pectoral fins. It is obvious that the petalichthyids are related to the arthrodires; despite differences in armor plate proportions and bone patterns, it is not too difficult to conceive of a common ancestor for the two groups.

Probably related to the petalichthyids is a small group of forms from the early marine Devonian of the Rhineland, such as *Stensiöella* (Fig. 39) and *Pseudopetalichthys* (Fig. 40). Our knowledge of them is,

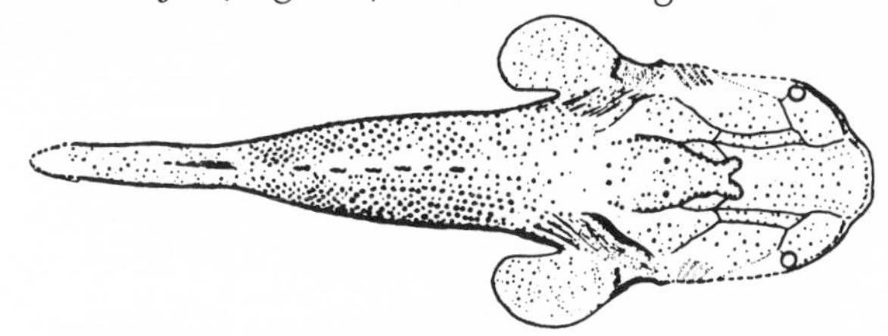

Fig. 39.—Dorsal view of the petalichthyid *Stensiöella;* about 1/3 natural size. (From Broili.)

Fig. 40.—Ventral view of *Pseudopetalichthys,* a petalichthyid; about half natural size. Beneath the head region are seen peculiar jaws and a set of branchial arches. Pectoral and pelvic girdles and remains of the associated fins are present. (From Broili.)

however, very limited, since there is only a single, rather poorly preserved slab specimen of each. Their general proportions are rather similar to those seen in *Lunaspis* and *Macropetalichthys;* and even in some features of the head, such as the long preorbital region and the arrangement of lateral-line canals, there is close similarity. In *Pseudopetalichthys* a ventral view shows short but well-developed jaws and a series of gill bars situated (as would be expected) beneath the back portion of the long "head" shield.

But there are differences from *Lunaspis* and *Macropetalichthys* as well as resemblances to these forms—notably in bone reduction. The body and tail of *Lunaspis* were well covered by scales; in *Stensiöella* there are isolated tubercles only. In the proper petalichthyids there is a well-developed (if short) bony pectoral girdle, with large shoulder spines; in the present forms girdle and spines are little developed. And in the head shield, the bony plates appear to be thin and imperfect. It is possible that these little fishes are more primitive than the true petalichthyids. But in view of the trend toward bone degeneration seen in other cases, the situation may be the reverse—there may be here a stage in armor reduction. There is a fairly general belief that sharks may be of placoderm descent, and *Stensiöella* and its relatives might well represent a transitional group. There is no definite evidence of related forms beyond the early Devonian. However, it is of interest that there has been discovered in the Mississippian a single fragmentary specimen of a fish, *Cratoselache,* which in general appears to be sharklike, but has a few

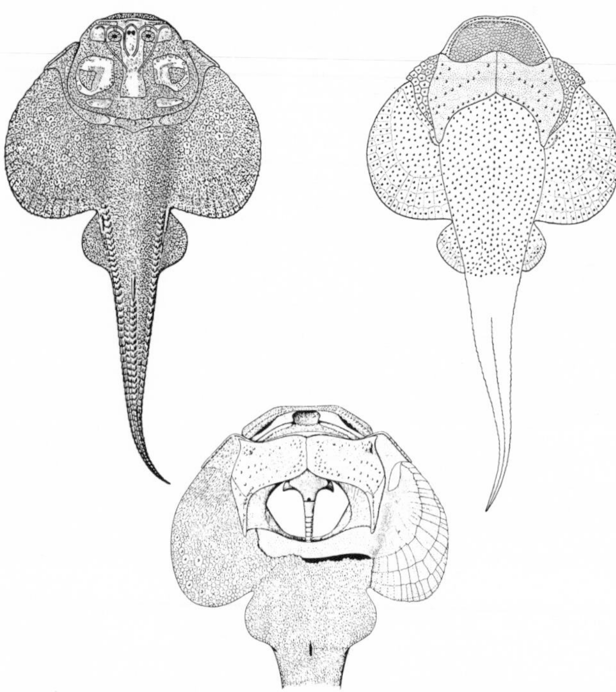

FIG. 41.—*Gemuendina,* a Lower Devonian rhenanid, about 9 inches long. *Left,* Dorsal view; *right,* ventral view; *below,* ventral view with dermal structures removed to show internal anatomy. (After Gross, Stensiö.)

vestiges of placoderm armor. The time of occurrence of *Cratoselache* is too late for it to be an actual shark ancestor; but it may well be a relict of such a transitional group.

Rhenanids.—Another, rather different series of somewhat sharklike placoderms is those which may be included in the Rhenanida. Most are from the marine Lower Devonian; *Gemuendina* and *Radotina* are typical genera. Few are known from more than fragmentary remains, but all show a skull pattern quite in contrast with that of the petalichthyids. In the latter group there was a long preorbital region, with the nostrils far forward; here the face is short and the nares placed close together between the orbits, which face upward on the flattened head. In some forms there is a complete, well-developed cranial shield; in others much of the shield is broken into a series of small plates or tubercles. *Gemuendina* (Fig. 41), known from a number of specimens from the Lower Devonian of the Rhineland, is the only form which is at all adequately known. The mouth was terminal, with short jaws transversely placed and armed with stellate pointed teeth. Gill chambers, as in other placoderms, appear to have occupied much of the area below the "head" shield. There was present a reduced thoracic armor, narrow above, broad below, comparable essentially to a dermal shoulder girdle; on either side was a stout if blunt spine —the placoderm trademark. The body was completely covered by tiny tuberculated structures which superficially resemble shark dermal denticles, but which are actually a mosaic of small bony plates. There are small pelvic fins; the pectoral fins are enormously expanded, giving the animal a superficial resemblance to the skates and rays of much later times.

Antiarchs.—Most weird, in some regards, of all the grotesque placoderms were the members of the order Antiarchi, small but abundant Devonian forms which

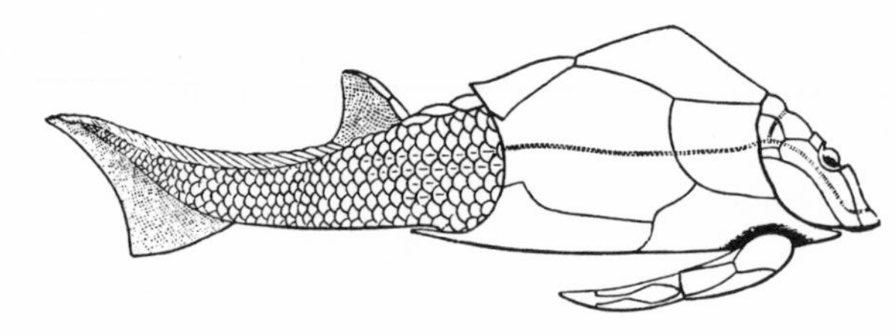

FIG. 42.—*Pterichthyodes* [*Pterichthys*], a scale-covered Middle Devonian antiarch; original about 6 inches long. (After Traquair.)

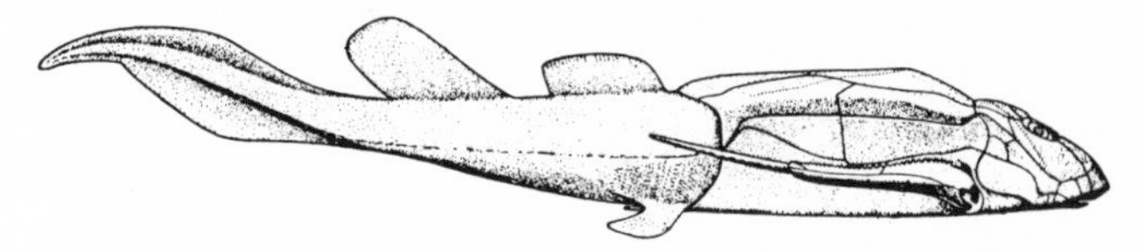

FIG. 43.—Lateral views of the Upper Devonian antiarch, *Bothryolepis,* a form with a scaleless trunk and tail. About 1/4 the size of an average specimen. (After Stensiö.)

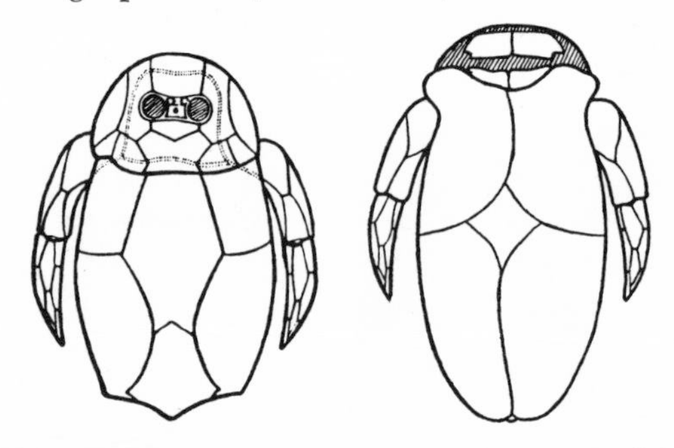

FIG. 44.—Dorsal and ventral views of the armor of *Pterichthyodes.* The elements of the thoracic armor are comparable to those of arthrodires (cf. Fig. 30). (After Traquair.)

(unlike most placoderm groups) were for the most part inhabitants of fresh water. The antiarchs are first found in Middle Devonian strata where *Pterichthyodes* [*Pterichthys*] (Figs. 42, 44) and *Astrolepis* are common. In the Upper Devonian, *Bothryolepis* (Fig. 43) is abundant and worldwide, it seems, in distribution; but the group disappeared completely by the end of the period.

The posterior portion of the trunk and tail was normally fishlike in appearance, scaled in *Pterichthyodes* but naked in *Bothryolepis,* with a heterocercal caudal and one or two dorsal fins. The anterior portion of the body was covered with an armor structurally similar to that of other placoderms and similarly divided into head and trunk portions. The underside of the solid box formed by the armor was flattened, but the thorax was arched, with a decided peak at the top. The head shield was quite short and composed of a number of plates not readily homologized with those of arthrodires. The two orbits, with the pineal eye between them, were situated close together on the top of the head and directed upward. The nostrils were placed close together on the upper surface of the shield not far in front of the eyes. (The situation in this regard is reminiscent of rhenanids, such as *Gemuendina.*) The jaws consist of small, transversely placed plates of bone even more aberrant in appearance than those of arthrodires. The gills were located under somewhat movable plates at the back corners of the head shield. The arrangement of the elements of the body shield was quite similar to that found in primitive arthrodires. Apparently this shield extended nearly the full length of the trunk.

In some specimens traces of soft internal structures have been preserved. It is certain that the intestine resembled that of sharks, dipnoans, and other primitive fish types in containing a characteristic spiral valve. Of great interest are distinct traces of a pair of large sacs connected with the floor of the pharynx. These can be interpreted only as lungs. Such structures, we believe, were present in early bony fishes of all sorts, but they are absent in sharks and have been thought to be a relatively late development in fishes. The evidence from the antiarchs suggests, however, that they may have arisen at a much lower stage in vertebrate evolution than had been suspected.

Folds of skin found at the posterior end of the trunk shield, in some *Bothryolepis* specimens, may represent pelvic fins. In the shoulder region we find, as we do in primitive arthrodires, prominent projecting bony structures. But in arthrodires these projections are fixed spines; here they are freely movable appendages composed of a large number of plates, with more the appearance of arthropod limbs than those of a normal vertebrate. Internal skeletal structures are present in the proximal part; there is a well-developed articulation with the shoulder region and (except in one instance) a second joint halfway toward the pointed tip.

These little fishes appear to have been bottom-living forms, perhaps crawling about with their peculiar flippers as much as swimming. With their small mouths and weak jaws, they presumably fed either on small invertebrates or soft vegetation.

It is obvious that in the antiarchs we are dealing, once again, with a variation on the general primitive placoderm pattern. Unlike the ptyctodonts and stegoselachians, there is here no tendency for armor reduction; but in antiarchs the development of paired appendages has gone in a direction startlingly different from that of other placoderms—indeed, startlingly different from that seen in any other type of vertebrate. Spines, as we have seen, appear to have been widespread in early fishes, as predecessors of fins, which usually develop behind the spines. Here in contrast, it is the spine itself which, in the pectoral region, becomes (paralleling the arthropods) the actual locomotor organ.

The evolutionary position of placoderms.—In many areas of paleontology the known fossil record agrees well with the evolutionary story that might have been inferred from a study of recent types alone; the ancestral forms, when found, have proved to be those which could reasonably have been expected from our knowledge of existing types. But it is quite otherwise with the story of the placoderms, as it is being gradually worked out. Except for the little acanthodian "spiny sharks" (to be discussed shortly), the placoderms are the oldest of all jawed vertebrates. They appear at a time—at about the Silurian-Devonian boundary—when we would expect the appearance of proper ancestors for the sharks and higher bony fish groups. We would expect "generalized" forms that would fit neatly into our preconceived evolutionary picture. Do we get them in the placoderms? Not at all. Instead, we find a series of wildly impossible types which do not fit into any proper pattern; which do not, at first sight, seem to come from any possible source, or to be appropriate ancestors to any later or more advanced types. In fact, one tends to feel that the presence of these placoderms, making up such an important part of the Devonian fish story, is an incongruous episode; it would have simplified the situation if they had never existed!

But they did exist; and we must attempt to fit them into the vertebrate evolutionary story. In our lack of knowledge of antecedent gnathostome types, we cannot even reasonably speculate as to their ancestry among hypothetical agnathous forms. At best, we can attempt merely to reconstruct the nature of an ancestral placoderm, as it must have existed in the Silurian, in preparation for the development of the varied types seen at the beginning of the Devonian.

In this ancestral form there surely would have been, as in nearly all placoderms, a well-developed "head" shield, covering not merely the cranium proper, but also a gill region posteriorly on either side. Back of this shield there would have been a slitlike common gill opening (no separate spiracle). Primitive jaws, including both endoskeletal and dermal structures, would have been present beneath the front part of the head shield, with, internally (as in many known forms), an extensive, ossified braincase. Posterior to the head and gill region, the trunk would have been covered (as in arctolepids and antiarchs) with a full set of armor; posteriorly, as in some known placoderms, the tail would have been completely sheathed in stout bony scales. And, finally, the placoderm trademark, in the form of a projecting spinal process on either side of the shoulder region. As to paired fins? The evidence is inadequate. Very probably small pectoral fins may have been present behind the pectoral spines, and equally small—or smaller—pelvic fins near the anus at the back end of the trunk armor. But it is not impossible that fins were primitively absent in this group and developed independently of those seen in acanthodians and higher bony fishes.

The varied grotesque types which evolved during the Devonian from such a hypothetical ancestor present, in themselves, a remarkable story. But are any of these odd creatures antecedent to the fishes of later time? At first one would be tempted to a vigorous denial of the possibility. But, as we have seen, we must consider seriously the possibility that at least the sharks and chimaeras of later days may have descended from such impossible ancestors.

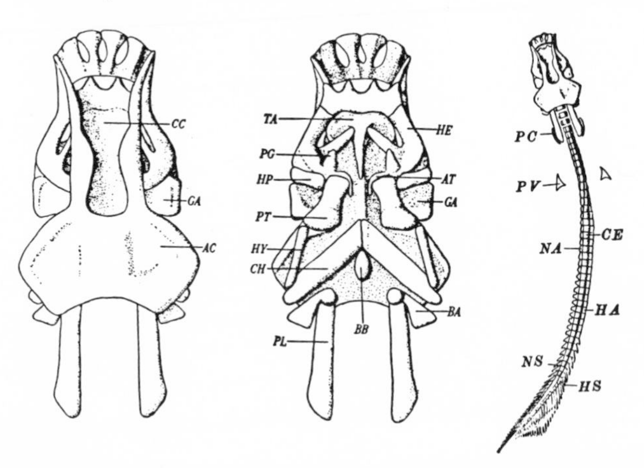

FIG. 45.—*Palaeospondylus*, a small and problematical Middle Devonian fish. *Left,* Dorsal view of the skeletal structures of the head region; *center,* ventral view of the same. Dorsally is seen a troughlike cranial cavity (*CC*), roofed posteriorly in the region of the auditory capsule (*AC*). Beneath the braincase are structures difficult to interpret; however *TA* is probably a ventral braincase structure; *HE, PG, AT, HP, GA,* and *PT* may be parts of a peculiar jaw apparatus; *HY* is presumably an unspecialized hyomandibular; *CH,* the ceratohyal; *BB,* a basibranchial; *BA* and *PL* more posterior branchial structures. *Right,* The entire fish, the original about 1.5 inches long. *PC,* Presumed pectoral girdle; *PV,* pelvic girdle; *NA,* neural arch; *CE,* centrum; *HA,* haemal arch; *NS,* neural spine; *HS,* haemal spine. (After Moy-Thomas.)

Palaeospondylus.—From a single Middle Devonian quarry in Scotland have come numerous examples of *Palaeospondylus* (Fig. 45), a tiny animal with a long name. In structure it is far removed from any other early fossil type. There is no trace of dermal armor of any sort, but a considerable series of ossified internal structures has been preserved. There are small plates which appear to have been girdles for paired fins; however, little, if any, trace has been seen of the fins themselves. There was a well-developed caudal fin. While the backbone in most early vertebrates appears to have consisted largely of ordinary cartilage at best and is hence inadequately known, this tiny form had well-preserved ring-shaped centra and neural arches. There was a well-developed braincase with otic capsules at the posterior corners, and in the center a trough for the brain, roofed at the rear; anteriorly, a series of projections stiffened the rostrum in the presumed nasal region. On the underside of the skull is a series of distinct paired elements of varied shapes. These are difficult to interpret but may represent jaws and gill arches modified for some unusual mode of life. The more anterior sets of these structures are presumably upper and lower jaws; back of them, slender bars may represent the hyoid arch.

The relationships of this little form have been the subject of endless argument. One suggestion is that it was the larval form of some contemporary fish. But usually the vertebrae of small embryos are cartilaginous, and the vertebrae of *Palaeospondylus* are better developed than those of the adults of its contemporaries. It has been considered as an agnathous type, but such features as the presumed limb girdles tend to rule out this conclusion. It is perhaps best to regard it, provisionally, as a placoderm in which reduction of armor has reached the end of the story with the complete loss of dermal bone.

Acanthodians.—Much more normal in general appearance than any of the placoderms were the Acanthodii (Figs. 46–48), frequently, but rather inappropriately termed "spiny sharks." These Paleozoic fishes were generally of small size, but a few inches in length, and usually found crushed flat in slabs of shale. Interpretation of their structure has therefore proved difficult. The acanthodians were superficially sharklike in their proportions and in the presence of a heterocercal caudal fin. This fin type, however, appears to have been the general primitive one in all fishes above the

FIG. 46.—*Climatius*, a Lower Devonian acanthodian with accessory paired fins, about 3 inches in length. (Modified from Traquair and Watson.)

ostracoderm level, at least; and in most other characteristics the acanthodians show little specific resemblances to the elasmobranchs. They are covered, not with isolated dermal denticles, as are sharks, but with a complete armor of true scales, diamond-shaped and of small size. A complete scaly covering appears to have been present in primitive ostracoderms and, as suggested earlier in this chapter, was probably present in ancestral placoderms as well; the acanthodian condition further confirms the conclusion that ancestral fishes were completely armored. It is of interest that the microscopic structure of acanthodian scales is closely comparable to that of the ganoid scales characteristic of primitive ray-finned bony fishes (Fig. 68*A*), with, externally, numerous layers of enamel-like material and, internally, a series of sheets of bonelike substance. Larger dermal plates, sometimes in a complex pattern, may be present in the shoulder region.

The head is covered by a continuation of the dermal armor of the body (Fig. 48). Here we find a series of small plates, arranged in definite patterns but not forming large bony elements or a shield (except for a prominent ring about the orbit). This is in contrast to the heavier armor found in most other Paleozoic vertebrates (sharks excepted). The gill region appears to have been covered, as in jawed fishes other than sharks, by opercular structures, presumably folds of skin reinforced by small dermal plates or scales. This covering, however, seems to be of an unusual nature. In higher bony fish groups the operculum, when present, grows backward from the region of the hyoid arch. Here each gill appears to grow a small flap, but a main operculum—stiffened by parallel bony bars or rays—grows back from the posterior margin of the jaws and covers the whole gill series.

The fins are of an unusual type. All, except the caudal, are stiffened anteriorly by dermal spines, sometimes of considerable size. In many cases only the spines are visible in the specimens; in others, however, there is evidence of a true fin structure. This consists primarily of a web of skin attached to the spine; occasionally there are traces of an internal skeleton in the fin and of tiny scales covering its surface. There are one or two dorsals and an anal. Of paired fins there are

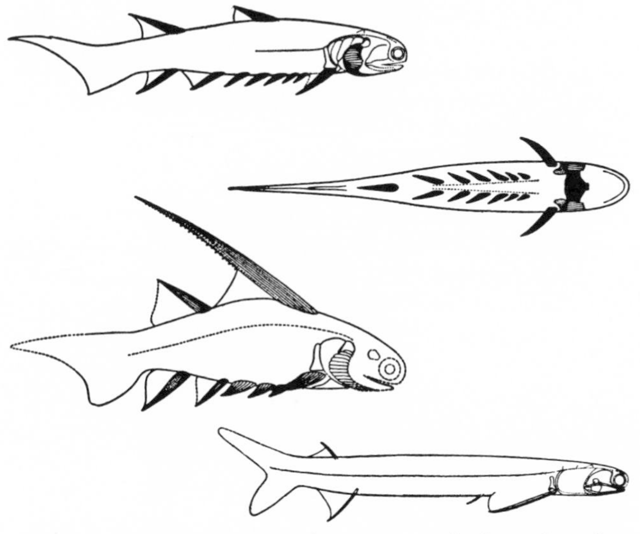

FIG. 47.—Outline drawings of various acanthodians. In order, lateral and ventral views of *Euthacanthus* of the Lower Devonian with five supernumerary pairs of fin spines, original about 6 inches long; *Parexus*, a Lower Devonian form about 5 inches long, with large dorsal fin spines; *Acanthodes* of the Permian, an elongate and apparently degenerate type reaching a foot in length. (From Watson.)

always readily distinguishable pectorals and pelvics in normal position. Most acanthodians, however, exceed this number and have additional smaller spine pairs between the major ones—in one genus as many as seven pairs of spines are present. It is obvious that there was little motion possible in the paired fins, and the series of fin spines seems obviously comparable in general to the pair of spine rows seen on the flanks of some ostracoderms.

Owing to the usual small size and crushed type of preservation of the material, we know little of the internal structure of acanthodians. The postcranial skeleton is quite unknown in most forms; in a few cases vertebral ossifications have been observed. The internal skeleton of the head and gill region is adequately known only in *Acanthodes* (Fig. 48), which is, unfortunately, a late and (we suspect) rather degenerate genus.

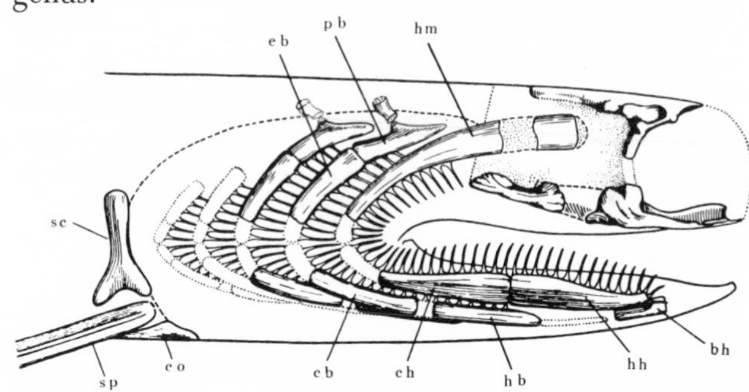

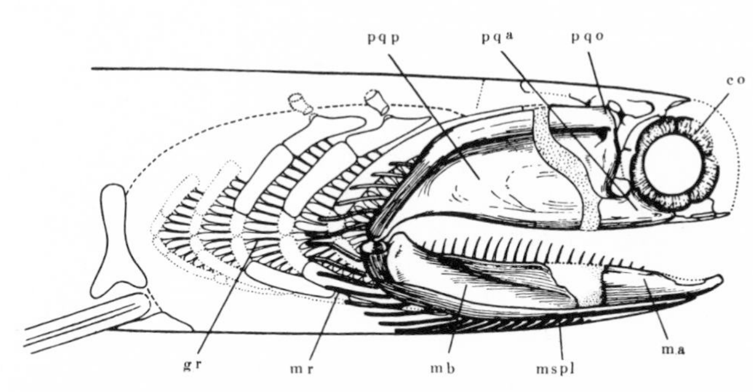

FIG. 48.—The skeleton of the head and gill region of *Acanthodes*. In the lower figure the superficial dermal plates and scales have been removed (except for a circumorbital series, *co*), but all deeper structures are shown. The upper jaw is ossified in three pieces (*pqo, pqa, pqp*), the lower jaw in two (*ma, mb*), while the latter is braced by a splintlike dermal bone (*mspl*). Rays (*mr*) extending back from the jaws presumably supported an operculum.

In the upper figure the jaws are removed to show the underlying structures. The braincase includes a number of ossified areas. Abbreviations: *hm, ch, hh, bh,* elements of hyoid arch—hyomandibular, ceratohyal, hypohyal, basihyal; *pb, eb, cb, hb,* elements of more posterior branchial arches—pharyngobranchial, epibranchial, ceratobranchial, hypobranchial; *sp, sc, co,* spine of pectoral fin and elements perhaps homologous to the scapula and coracoid of the shoulder girdle in bony fishes and tetrapods. (After Watson and Miles.)

A considerable part of these internal structures was well ossified. There appears to have been a braincase of fairly normal fish build. This was ossified to a variable degree; in *Acanthodes* there were several bony regions separated by areas in which ossification was feeble or lacking. The orbits were large, the snout very short, the nostrils close together near the tip of the snout—features quite unlike those of sharks and especially comparable to those of ray-finned bony fishes. Obviously the acanthodians were "eye fishes" rather than "nose fishes," depending in great measure on vision rather than smell for major contacts with the outside world.

Upper and lower jaw bars of replacement bone were present and closely comparable in shape to those of crossopterygians (cf. Fig. 69*D*). In some genera there is but a single ossification in each jaw element; in *Acanthodes,* however, the palatoquadrate ossifies in three portions, the lower jaw in two. The mandible is braced by a splintlike dermal bone. The palatoquadrate is movably joined to the braincase by a well-defined basal articulation with the basisphenoid region, as in crossopterygians, for example, and in strong contrast with elasmobranchs; and more posteriorly and dorsally by an articulation with a postorbital process. The teeth are variably developed; they are seldom found on the upper jaw, and some genera, including *Acanthodes,* were toothless. The replacement teeth in each series are sometimes seen to form little whorls, and there may be an especially prominent median whorl at the jaw symphysis.

Back of the jaws is found a series of five well-developed gill arches. It was at one time believed that the first of these, the hyoid arch, was unspecialized and not concerned in jaw support. Additional material, however, shows that the main upper element of this arch, the hyomandibular (ossified in two portions), is enlarged and aids in propping the jaws on the braincase.

The acanthodians are the oldest of known groups of jawed vertebrates. No complete skeletons are known from the Silurian, but bone beds of that period yield a wealth of spines and scales which appear to be acanthodian in nature. In the Lower Devonian the group had already reached its peak in numbers and variety, and acanthodians are one of the most abundant of vertebrate groups in freshwater deposits of this age. *Climatius, Euthacanthus,* and *Parexus* are representative primitive genera. The body was short and deep; the fin spines were large—particularly so in *Parexus*—and from three to five pairs of extra fins were present. In other genera of this age, however, the body tends to be rather more elongate and the fin spines more slender, and there is little development of extra fins. Beyond the Lower Devonian, acanthodians are relatively rare (perhaps because of the competition of higher bony-fish types then abundant). They are present, however, in many Carboniferous localities. A last survivor of the group is *Acanthodes* of the Lower Permian, a rather degenerate type, with an elongate, somewhat eel-like and partially scaleless body and a toothless mouth.

The more typical acanthodians were small freshwater dwellers. It is probable, however, that some of the acanthodians went to sea, as did other fish groups in the Devonian and Carboniferous, and there developed into forms of large size. In later Paleozoic marine deposits there are frequently found isolated spines, usually classified by paleontologists as "ichthyodorulites." Most were presumably carried by sharklike fishes higher in the evolutionary scale; others, however, may have pertained to acanthodians. This seems certainly true of *Gyracanthus,* a name given to a common type of large Carboniferous fin spine.

This order of early fishes shows an interesting combination of primitive and seemingly specialized features. In some respects, at least, the nature of the paired fins—particularly their variable number—is suggestive of primitive conditions. The acanthodians were at a stage when vertebrates had not, so to speak, made up their minds as to how many fins were the proper number. The prominence of spines in the paired-fin structures appears, at least at first sight, to be a specialization. But it will be recalled that cephalaspids and anaspids among jawless vertebrates show paired spines which may be the forerunners of paired fins; some placoderms, as we have seen, show even greater spiny developments; the dorsal fins, at any rate, bore spines in primitive sharks. Possibly spine outgrowth may have been an initial stage in the evolution of paired fins; the appearance of a fin membrane behind the spine and its elaboration into a typical fin may have been subsequent stages in limb evolution.

How and where do the acanthodians fit into the general picture of early fish evolution? They were surely descended from some jawless type, but show no special connections with any of the known ostracoderm orders described in the last chapter; their early development presumably took place in Ordovician-Silurian fresh waters from which no sedimentary deposits have persisted. Now that the early appearance of bone is generally accepted, the presence of a bony skeleton, internally as well as dermally, in these oldest jawed vertebrates need not astonish us. At one time there was a tendency to unite the acanthodians with the arthrodires and their kin in the class Placodermi. But, actually, apart from the fact that both groups are ancient and armored, there is little to unite them. There are, we have noted, some resemblances to the sharks, and it is not impossible (despite the claims made earlier in this chapter for the placoderms) that early acanthodians, with loss of ossification, may have given rise to elasmobranchs; the presence of fin spines in many early sharks is suggestive.

Stronger, however, are suggestions of relationships to the higher bony fishes, most particularly the ray-finned forms, the actinopterygians. The bony fin spines, nonexistent in the Osteichthyes, are, of course, an obstacle. On the other hand, the well-developed skeleton, with a good scaly covering and internal ossification, and the basal articulation of jaws and braincase, form bonds with the Osteichthyes and contrast with the sharks. And in such features as the details of scale structures, the dominance of vision, and the possession of a heterocercal tail fin without an upper (epicaudal) lobe, the Acanthodii and the older members of the Actinopterygii are very close. Not improbably the acanthodians are an early branch from the unknown ancestral stock, from which arose the varied components of the Osteichthyes; we shall here include them (albeit with some doubt) in that class of fishes.

CHAPTER 4

Sharklike Fishes

With the consideration of such forms as the sharks, skates, and chimaeras, we emerge from the fish "underworld" of grotesque ostracoderms and placoderms, and reach the level of "orthodox" fish types, built on recognizable plans, with well-developed jaws and paired fins. These higher gnathostome fishes are customarily considered as forming two classes, readily separable—the Chondrichthyes, or cartilaginous, jawed fishes, treated in the present chapter, and the Osteichthyes, the higher bony fishes.

The class Chondrichthyes, as customarily considered, includes the sharks and their close allies, the skates and rays, forming the subclass Elasmobranchii, and in addition, the subclass Holocephali, represented in modern seas by the peculiar "ratfishes," the chimaeras. As was suggested in the last chapter (and will be discussed later in the present one), it is quite possible that the sharks and chimaeras have evolved independently from placoderm ancestors, and that, in consequence, the class Chondrichthyes may not be a natural group, in that these two types are not, so to speak, brother and sister, but merely "cousins" in an evolutionary sense. However, since the case for separate origins is far from proved and the two main types are, at the least, fairly closely related, we may reasonably retain the group Chondrichthyes for the present.

Sharks and chimaeras have many common characteristics which readily distinguish them from the Osteichthyes. One obvious feature is that in modern sharks and chimaeras the males bear on their pelvic fins projecting claspers which aid in the internal fertilization appropriate to the large-yolked, shell-covered eggs characteristic of the group. Lungs are absent—a feature presumably associated with a contrast in the environmental history of the two higher fish groups. A majority of modern bony fishes are marine types, but it is generally agreed that they entered the seas at a relatively late date and that their early, lung-bearing ancestors were freshwater forms. It is quite otherwise with Chondrichthyes. Apart from one rather aberrant Paleozoic family and occasional later strays, the cartilaginous fishes appear to have been ocean dwellers since their first appearance in the Devonian. Possibly their remote ancestors may have dwelt in inland waters; but the advent of sharklike fishes into the sea was certainly early, and was accompanied by special structural and functional adaptations to a saline environment quite different from those evolved (presumably at a later date) by marine bony fishes.

Skeleton.—Most distinctive of the class, however, is the feature implied in the name "cartilaginous fishes"—the nature of the skeleton. We have seen a great development of bone among ostracoderms and placoderms and will find bone the dominant skeletal element in adult animals among the vertebrates considered in later chapters. In the Chondrichthyes, bone is utterly absent. The internal skeleton is purely cartilaginous. These cartilages are, in many cases, calcified, but they never show true bone structure. The dermal skeleton is likewise strongly reduced. Dermal denticles, corresponding to the superficial ornaments of bony plates, may be present; teeth may be abundant and varied; spines of one sort or another are not uncommon; but never are there dermal bones or bony tissues of any sort.

It will be noted that this condition of the skeleton renders the study of chondrichthyan history a difficult matter. Throughout the geological range of the group, finds of isolated teeth and spines are not uncommon; but unless these are similar to types found in modern forms, it is difficult, sometimes impossible, to deduce the nature of the animal which bore them. Calcified braincases or vertebrae are found, but found all too infrequently; ordinary cartilage disintegrates too rapidly to be readily preserved. It is, thus, only under exceptionally favorable conditions that we can gain any idea of the general organization of extinct types of sharklike fishes, and there are many dubious points in the history of the group.

In past times it was generally assumed that the absence of bone in the Chondrichthyes was a primitive condition and that the sharks represented an evolutionary stage antecedent to that of the bony fishes. This assumption appears, at the present day, to be a highly improbable one. Bone, as we have seen, appears in groups much lower down the evolutionary scale; and if we believe the sharks to be primitive in this regard, we must believe, rather improbably, that bone was evolved a number of times by the vertebrates. Nor are the sharks, as one might expect according to earlier beliefs, an early group geologically. They are, in fact, the last of major fish groups to appear in the fossil record. No member of the class is known before the latter half of the Devonian; even some Osteichthyes are present in Lower Devonian strata. It might be argued that absence of an earlier record is due to the fact that the ancestral sharks were soft-bodied and not preserved. But the first sharks were far advanced in the evolution of jaws; and it is difficult to believe that the teeth of these supposed earlier forms should have escaped our attention.

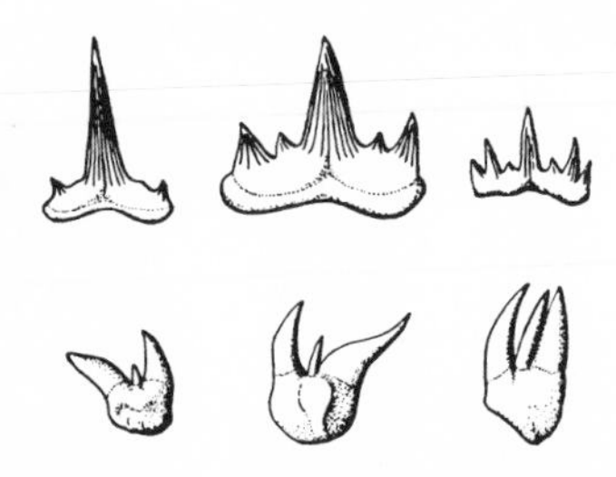

FIG. 49.—*Above, Cladodus* teeth from the late Devonian and Carboniferous; *below,* pleuracanth teeth from the Permian. Somewhat reduced.

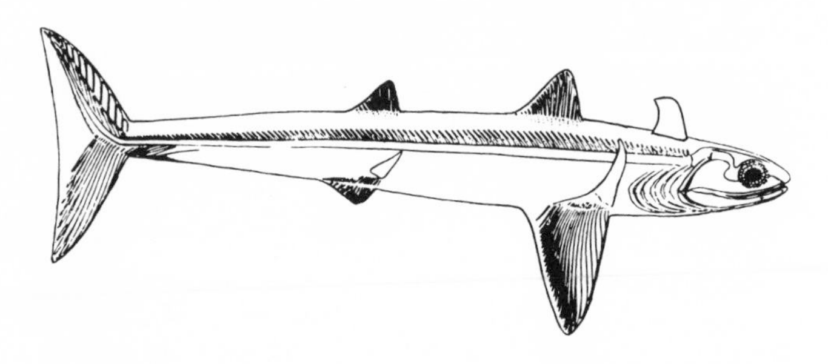

FIG. 50.—*Cladoselache,* a late Devonian sharklike fish. Original specimens range from about 1.5 to 4 feet in length. In addition to the normal paired fins, a pair of horizontal "rudders" may be noted at the base of the tail fin. Some specimens, at least, had a spine in front of the first dorsal fin; whether one was present on the second dorsal is uncertain. (After Harris and Dean.)

The record, in fact, fits in better with the opposite assumption: that the sharks are degenerate rather than primitive in their skeletal characters; that their evolution has paralleled that of various other fish types in a trend toward bone reduction; and that their ancestry is to be sought among primitive bony, jaw-bearing fishes of the general placoderm type. No well-known placoderms can be identified as the actual ancestors of the Chondrichthyes, but we have noted that some of the peculiar petalichthyids appear to show morphologically intermediate stages in skeletal reduction. Increasing knowledge of early Devonian placoderms may some day bridge the gap.

Primitive elasmobranchs.—The Chondrichthyes, living and fossil, are, as we have said, clearly separable into two subclasses, the Elasmobranchii and the Holocephali. Of the two, the elasmobranchs—the sharks and related types—are the better known and more abundant. The oldest geological record of sharks is in late Middle Devonian beds, in which are found a few specimens of the type of tooth known as *Cladodus* (Fig. 49), and characteristic of various early elasmobranchs. These teeth are found in considerable numbers in Carboniferous rocks and finally disappear in the Permian. They consist of a tall central cusp with a broad base on which are found one or more pairs of smaller lateral tubercles.

In the Cleveland shales of the late Devonian an exceptional type of preservation has given us numerous specimens of *Cladoselache,* a shark possessing this tooth type, in which much of the body structure can be made out (Figs. 50, 51). This form is found in concretions in which are preserved not only remains of the teeth and calcified cartilages of the skeleton but also impressions of the skin and body outline, and even traces of such soft tissues as muscle and kidney. There were two dorsal fins, each with basals and radials consisting of parallel rods of cartilage. A low but stout spine was present behind the head. The tail fin was very markedly heterocercal, the posterior end of the vertebral column tipping sharply upward. Well-developed pectoral and pelvic fins were present, the former much the larger, while far back at the base of the tail fin was a pair of outgrowths from the side of the body, seemingly a third pair of lateral keels or fins, absent in later

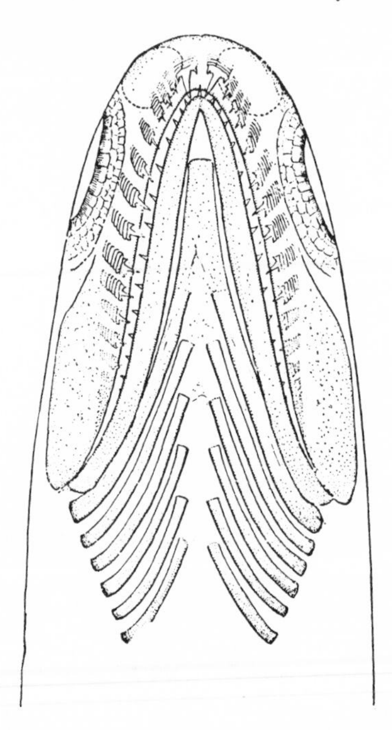

FIG. 51.—Ventral view of head and throat region of *Cladoselache* to show the ventral segments of the branchial arches and the jaws seemingly homologous with them. The position of the nasal capsules is indicated by dotted outlines; the orbits, seen laterally, are surrounded by enlarged scalelike denticles. (After Dean.)

sharks. Like the median fins, the pectorals (Fig. 48*A*) and pelvics had a skeleton in which the projecting portion—the radials—consisted of long, unjointed, parallel rods of cartilage. The basal portion of the fin is elongated anteroposteriorly, sometimes consisting of discrete blocks of cartilage forming an axis within the body wall, sometimes consisting of a single fused basal structure. The fin was a very broad-based flap, which appears to have projected nearly straight outward from the body in life, with limited possibilities of motion. It is quite probable that this is a very primitive type of fin structure from which may have arisen, by a constriction of the base, the fan-shaped fin of most later sharks and even the leaf-shaped appendage of a number of Paleozoic fish types. In *Cladoselache,* in contrast to acanthodians and placoderms, the paired fins are spineless; but it is possible that the spineless condition is a secondary one. It may be emphasized that no trace has been found in *Cladoselache* of the claspers which are characteristic of later shark types.

The braincase and jaws are fairly well known; they were very similar to those seen in pleuracanth sharks, illustrated in Figure 63. In modern sharks—and, indeed, in most fossil elasmobranchs as well—the postorbital part of the braincase is very short; in this ancient type (rather as in many placoderms) there was a long otic and occipital extension of the braincase back of the orbital region. The jaws were amphistylic. They were supported, in part, by a moderately developed hyomandibular. In addition, the upper jaws articulated with a prominent postorbital process. This type of jaw suspension is apparently the primitive one for elasmobranchs but is very uncommon in modern sharks. There were large paired eyes, with concentric rows of denticles about them, and paired nostrils. Behind the jaws (Fig. 51) lay the hyoid and five succeeding branchial arches.

The essential features of *Cladoselache* are repeated in another sharklike fish of equal antiquity—*Ctenacanthus,* present from Upper Devonian to Lower Permian; its characteristic fin spines are found abundantly in the Carboniferous. In *Ctenacanthus* very stout spines, with numerous beaded, longitudinal ridges, acted as cutwaters for the dorsal fins. The dentition—of the *Cladodus* tooth type—was identical with that of *Cladoselache;* the paired fins were of similar structure, except that the radial elements were segmented, as in modern sharks, rather than single rods, as in *Cladoselache.* It is probable, however, that there was a normal anal fin rather than the peculiar paired structures reported here in *Cladoselache.*

Cladoselache and *Ctenacanthus* may be considered as representative of a very primitive order of sharklike fishes, the Cladoselachii. The amphistylic jaw suspension, the long braincase, the broad-based fins, and the absence of claspers are diagnostic characters which are held by most to be truly primitive for elasmobranchs, and it is generally believed that the cladoselachians are a group close to the base of the entire elasmobranch assemblage. But since our knowledge of the structure of early sharks is very fragmentary, it is not

impossible that the absence of claspers is a specialization, and possibly (although not too probably) the broad-based fins of cladoselachians are derived (like those of advanced arthrodires) from smaller, narrow-based types.

Hybodontoids.—The ancient cladoselachians became extinct during the Permian; they were replaced by more progressive sharks, presumably descended from them, which appear to have come into existence at the end of the Devonian and flourished in the late Paleozoic and Mesozoic. These sharks—the hybodontoids—form an intermediate stage in selachian evolution. They were persistently primitive in that the amphistylic type of jaw support was retained, but the braincase tended to assume shortened modern proportions;

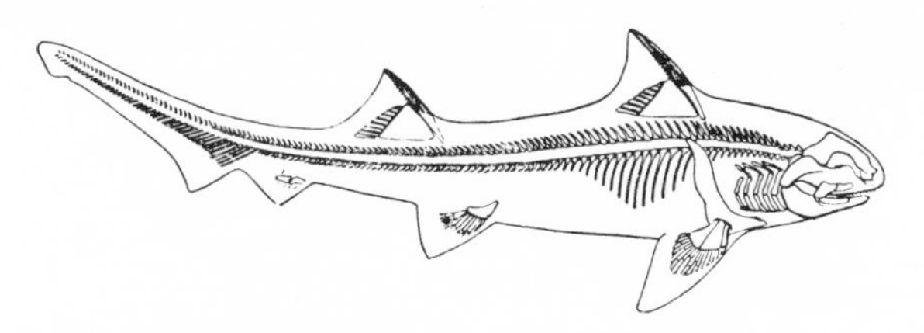

FIG. 52.—*Hybodus hauffianus,* a Mesozoic shark; original about 7.5 feet long. (From Smith Woodward.)

and they had advanced considerably as regards their paired fins. These came to be, as in living sharks, flexible, narrow-based structures rather than broad folds. It is generally held that this more flexible type of fin had evolved from that of cladoselachians by the freeing of the posterior margin from the body wall so that the basal elements project outward into the fin itself. These basal elements in the pectoral fin usually consist, in both hybodontoids and modern shark types, of three elements—propterygium, mesopterygium, and metapterygium, the last the most important of the three; the pelvic fin is similarly built but usually simpler in structure. An important feature is the fact that claspers (Fig. 27*D*) had developed on the pelvic fins of the males. These structures are present in all the sharks yet to be considered (as well as the skates and rays derived from them), and cause us to separate all these forms as an order, Selachii, from the archaic cladoselachians. Best known of primitive selachians is *Hybodus,* a common Mesozoic shark (Figs. 6, 27, 52). The general appearance is, for the most part, quite similar to that of cladoselachians, and a striking resemblance to *Ctenacanthus* lies in the fact that the dorsal fins and their spines are almost identical in nature (the striated spines can be distinguished, however, by the smooth rather than beaded ridges).

The dentition shows some variation in the hybodont group but tends to adhere to a single general pattern. The teeth (Fig. 53) are readily derivable from those of the *Cladodus* type. Those near the front of the mouth are generally sharp cusped. Farther back in the jaws, however, the cusps are usually lower, often blunt, and sometimes completely reduced, so that a low, rounded crown is produced. In all sharks replacement teeth are formed in rows inside the jaw and gradually swing up into place (Fig. 53*E*). In hybodontoids it appears probable that a number of replacement teeth in each row were exposed and functioning at the same time. This feature, together with the flattening of the back teeth, results in the development posteriorly of broad triturating surfaces, suitable for crushing mollusks; the numerous sharp anterior teeth were used for seizing prey. A wide variety of food was thus available to the hybodonts.

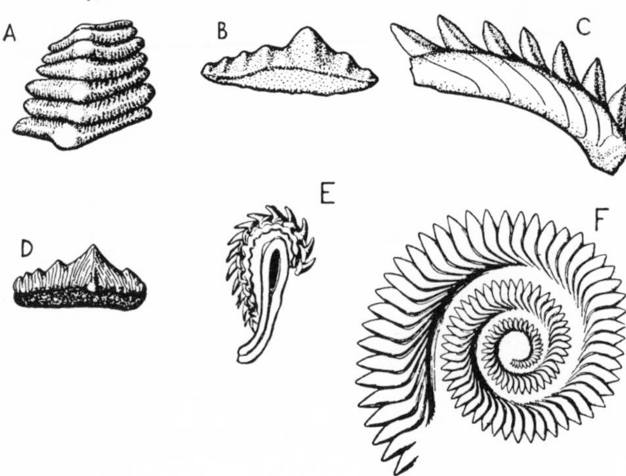

FIG. 53.—The teeth of some early selachians. *A, Orodus* (Carboniferous), crown view of a tooth series (×3/4); *B*, same, lateral view; *C, Edestus* (Carboniferous) tooth whorl, reduced; *D, Hybodus* (Triassic), natural size; *E*, section through the jaw of a Recent shark to show a crowded row of replacement teeth; *F*, a tooth whorl, presumably placed in the jaw symphysis of the Permian *Helicoprion* (reduced)—the earlier teeth were not shed as in typical sharks but were apparently retained and buried in the inner part of a complex spiral of replacing teeth. (*C* after Dean; *E, F* from Smith Woodward.)

Hybodus is a Mesozoic form; a number of sharks of this sort were present during that era and were especially common in the Jurassic. They were, however, a much older group. There are a few teeth which may be hybodontoid in nature in the Upper Devonian, suggesting that at an early date the hybodontoids had evolved, through tooth and fin modifications, from the primitive cladoselachians. In the Carboniferous, teeth and spines of *Orodus* (Fig. 53*A, B*) and other hybodont genera are not uncommon finds.

In many hybodontoids (as in such older fishes as acanthodians) there appears to have been a central series of teeth in the midline of both upper and lower jaws; and in some late Paleozoic genera, as *Edestus* and *Helicoprion* (Fig. 53*F*), these symphysial whorls may be highly developed. In *Helicoprion* the teeth, in contrast to usual shark conditions, did not drop out after they had reached the edge of the jaw, but curved downward and inward to form the core of a complex growing spiral.

Pavement-toothed sharks.—In the Carboniferous seas there abounded a wide range of invertebrates as potential food sources for early sharks. Many of these forms —mollusks and arthropods, notably—had hard shells, and a powerful crushing dentition was needed to cope with them. We noted above that the posterior part of the hybodont dentition was at least partially adapted for a diet of this sort. A number of families of Paleozoic sharks, however, went even further in this type of specialization, with the development of a crushing pavement of teeth, and, in some cases, at least, had fused the upper jaw with the braincase to give a firmer bite. Although forms of this sort appeared at the end of the Devonian and persisted into the Permian, their tooth plates are most common in the marine limestones which make up most of the deposits of the Lower Carboniferous—the Mississippian—both in Europe and in North America. Frequently these "pavement-toothed sharks" are bracketed together as an order Bradyodonti. However, there is no guarantee that the various families are at all closely related to one another; more probably they represent a number of parallel developments from the primitive shark stock.

Of these pavement-toothed families, one—the Cochliodontidae—is probably related to the chimaeras, and will be mentioned later. The others may be briefly described at this point. In the Petalodontidae, such as *Petalodus* (Fig. 54*A*), the teeth (as the name indi-

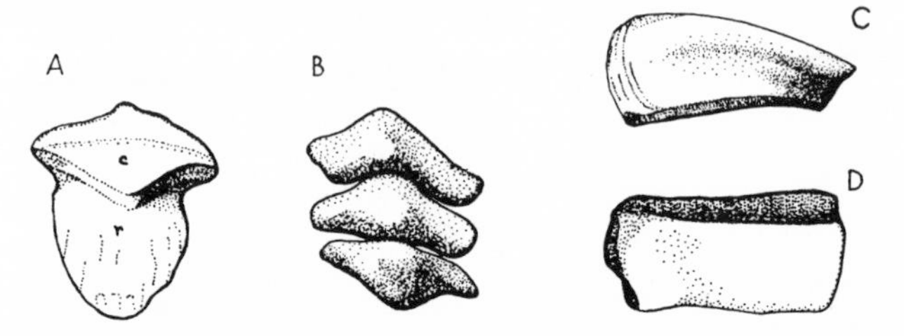

FIG. 54.—Teeth of pavement-toothed "sharks" and cochliodonts. *A, Petalodus* (×1/2); *c*, crown; *r*, root. *B, Helodus*, three teeth, crown view (×3/2). *C, Deltodus*, crown view (×1/4). *D, Psammodus* (×1/2).

cates) are not unlike flower petals in shape. There is a broad lozenge-shaped crown, which appears to have been closely applied to the adjacent teeth, a group of such teeth forming a solid crushing pavement. Each tooth has a long root, sometimes subdivided, extending downward at an angle from the surface of the crown. In two instances, specimens have been found which show something of the body contours. *Janassa* appears to have paralleled the modern rays in developing enlarged pectoral fins and a broad, depressed body (Fig. 55). Of a second family, that of the Psammodontidae (Fig. 54*D*), only tooth plates are known. The dentition consisted of a few large quadrate plates, arranged above and below in two rows which met in the midline. The origin of these plates is not known; while they were probably formed by a fusion of teeth, there is no indication of this in the specimens. In the obscure group of Copodontidae, likewise known mainly from the teeth, the dentition consisted of a single subquadrate tooth medially placed at the front end of the upper and lower jaws.

Two further Carboniferous pavement-toothed types are known from single genera in which body outlines as well as dentitions are preserved. *Helodus* (Figs. 54*B*, 56) has frequently been included among the cochliodonts (considered later), but appears to be quite distinct from that group. The jaws show rows of successional teeth, rather like the cheek teeth of hybodonts and edestids, in which there is a partial trend toward fusion. The short upper jaws are fused with the braincase. The body was somewhat depressed and the pectoral fins somewhat enlarged, again (as in *Janassa*) showing the initiation of ray-like adaptive characters; there was a large dorsal spine. Quite in contrast was little *Chondrenchelys* (Fig. 56). As in *Helodus*, upper jaw and braincase were fused, but here toothplates were present on the jaws. The body pattern was quite different from that of any other known marine elasmobranch. There were no fin spines of any sort; but in other respects the body was very similar to that of the pleuracanths, for the tail was of the diphycercal type, there was a long dorsal fin continuous with the caudal, and the paired fins were of the archipterygial type. Presumably, however, these pleuracanth resemblances are due to parallelism rather than to real relationship.

As we have seen, varied elasmobranchs were abundant in late Paleozoic seas. However, most of these types disappeared before the end of the Paleozoic. This may be correlated with a marked change at this time

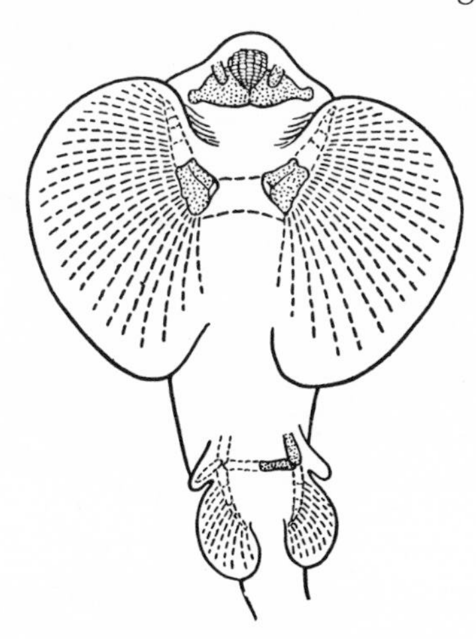

Fig. 55.—The Permian petalodont *Janassa*, seen in ventral view. About 1/5 natural size. (From Jaekel.)

in the nature of the marine invertebrates—a change to which these sharks, directly or indirectly dependent upon them for food, failed to adapt—and, further, a very considerable reduction, it would seem, in the numbers and variety of the invertebrate forms present. Only the seemingly versatile and omnivorous hybodonts survived into the Triassic. During the Mesozoic many sharklike fishes became once more adapted to invertebrate diets, and numerous bony fishes entered the sea to form an additional source of food supply. There began, from the hybodonts as a basal stock, a great deployment of more modern types of sharks and of the related skates and rays. The hybodontoids themselves produced, in the Cretaceous, a short-lived side branch, the ptychodonts (Fig. 61*B*), with flattened tooth plates for crushing invertebrates. Neither typical hybodonts nor their ptychodont offshoot survived the end of the Mesozoic.

Primitive shark survivors.—There are, however, archaic sharks which appear to be relatively primitive descendants of the old *Hybodus* group.

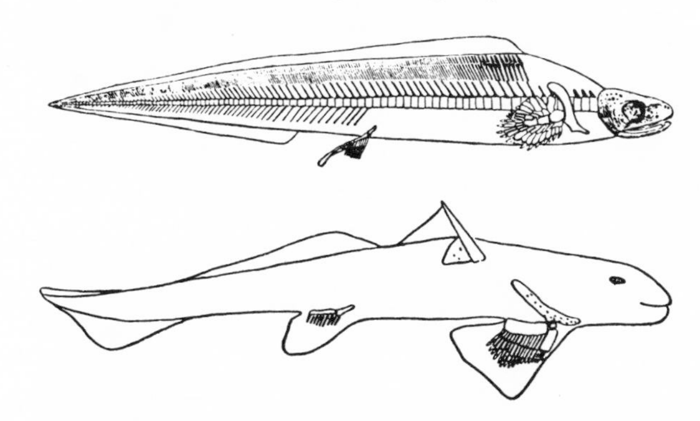

Fig. 56.—Two aberrant mollusk-eating sharks of the Carboniferous, *Chondrenchelys* (*above*) and *Helodus* (*below*). *Chondrenchelys* about 7 inches long; the *Helodus* specimen figured, about 14 inches. (From Moy-Thomas.)

The Port Jackson shark, *Heterodontus* [*Cestracion*] of the Pacific Ocean is in many respects a little-modified hybodont descendant. Like those ancient forms, this shark is, in great measure, a mollusk eater and retains the hybodont type of dentition, with sharp anterior fangs and low, rounded posterior teeth, forming a shell-crushing apparatus (Fig. 57*H*). Its general appearance and structure, including the presence of stout spines on its two dorsal fins, is comparable to that of hybodonts. In its jaws, however, there has been a structural change, for the palatoquadrate no longer articulates, as did that of the older sharks, with a postorbital process of the braincase. Nevertheless, *Heterodontus* is still fundamentally amphistylic, for the upper jaws are firmly braced against the braincase more anteriorly and ventrally, as well as receiving support from the hyomandibular. Several genera of the later Mesozoic appear to be intermediate between the hybodonts and this group, and *Heterodontus* itself appeared in the Jurassic.

Other archaic surviving families are those represented by *Hexanchus* [*Notidanus*] and *Chlamydoselache*. These are slender-bodied, fast-swimming predaceous sharks, lacking fin spines and having but a single dorsal fin—features in which they are quite different from the hybodonts. Six or seven gill slits, rather than five, are present, possibly a primitive feature but equally possibly a secondary increase. All the teeth (Fig. 57*A*, *B*) are sharp cusped but quite unlike those of the older selachians; in *Hexanchus* they are peculiar saw-toothed structures; in *Chlamydoselache,* three-pronged and superficially comparable to those of pleuracanth sharks of the Paleozoic. In their jaw structure, however, they are persistently primitive. *Hexanchus* has the postorbital connection of upper jaw and braincase developed as in the oldest sharks; in *Chlamydoselache* the jaw fails (barely) to make contact here, but the construction is still essentially amphistylic. Both sharks are rare today. *Chlamydoselache* is represented as a fossil only by occasional teeth from the Tertiary; *Hexanchus* had developed in the Jurassic.

"Modern" sharks.—In the more typical modern sharks remaining for consideration, direct articulation of jaws with braincase has been abandoned, and the jaws are hyostylic, supported by a prop formed by the large hyomandibular. Selachian development in this "modernized" phase began in the Jurassic; and a rapid deployment resulted, before the close of the Mesozoic, in the appearance of most of the dozen or more existing families. Two main lines of development were early evident, and most of these sharks can be readily grouped into two suborders, the Galeoidea and the Squaloidea.

The galeoids, including the great majority of the living sharks, are, in general, active predaceous types, with fusiform bodies which are not at all depressed, the spiracle small or even absent, and an anal fin always present. The oldest forms are from the Upper Jurassic, when there appear a number of genera closely related to modern forms. In the Cretaceous, galeoid sharks have diversified into a number of genera, most of them abundantly represented by teeth in Tertiary marine deposits and surviving to the present day (Fig. 57*C–G, K–O*). We may cite some common forms *seriatim: Carcharias* [*Odontaspis*], the sand sharks, with large, awl-shaped teeth; *Scapanorhynchus,* the long-snouted "ghost shark"; *Isurus* and *Lamna,* the mackerel or porbeagle sharks, with long, slender, flattened teeth; *Carcharodon,* the large white shark—a maneater—bearing large triangular teeth with serrate edges (one Tertiary species may have had a mouth gape of 6 or 7 feet); *Squalicorax* [*Corax*] of the Cretaceous and early Tertiary, with hook-shaped serrate teeth; the nurse shark, *Ginglymostoma; Galeocerdo,* the tiger shark; *Hemipristis,* common in the early Tertiary; *Sphyrna* [*Zygaena*], the hammerhead, with eyes placed on peculiar lateral projections of the head. Still other common types appear in the early Tertiary, including *Alopiopsis,* the long-tailed thresher shark, and *Mustelus,* the smooth dogfish or "hound."

A further series of sharks not uncommon at the present day is that of the Squaloidea. This group shows some features suggestive of an approach toward bottom-dwelling habits and the condition seen in the skates and rays. As in them, the spiracle is usually large, and the anal fin is lacking; the body, however, is well rounded in squaloids, and the pectoral fins are of normal shape. This group of sharks is as old a one as the galeoids, for *Protospinax,* a form related to the modern spiny dogfish (*Squalus*) is an Upper Jurassic shark, and in the Upper Cretaceous there appear a number of representatives of this group.

To be considered as sharks rather than as skates or rays are two additional forms—the saw shark, *Pristiophorus,* and *Squatina* [*Rhina*] (Fig. 60*A*). The former has a long-toothed rostrum, like the saw rays noticed below, and a flattened head, but is otherwise similar to the squaloids. Saw sharks are known from the Upper Cretaceous. *Squatina,* the angel shark, is still more ancient, appearing in the Jurassic, and has paralleled the skate type of organization; the body is flat, the eyes are dorsal in position, and the pectoral fins are much enlarged. However, the gills are lateral in position rather than ventral (as in skates), and the pectorals lack any attachment to the head, so that we must consider *Squatina* as a shark, although a very skatelike one.

Skates and rays.—All remaining elasmobranchs may be included in an order which we may term the Batoidea, in contrast with the order Selachii, the true

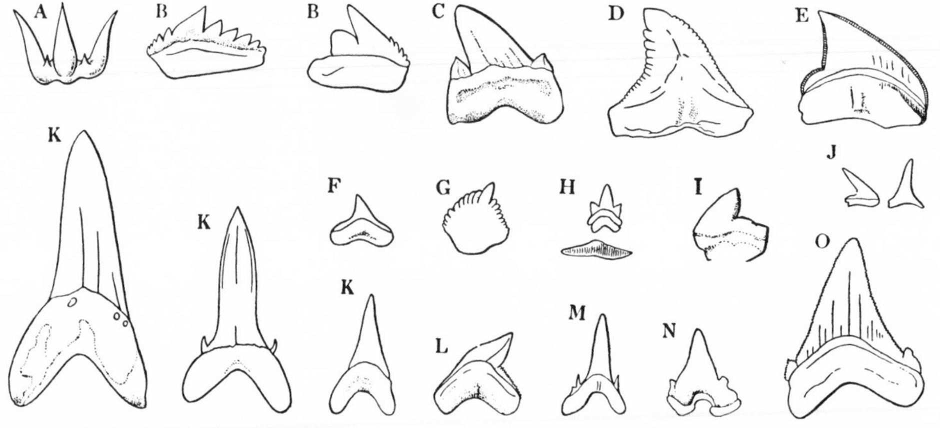

FIG. 57.—Teeth of various Cretaceous and Tertiary sharks: *A, Chlamydoselache; B, Hexanchus; C, Otodus; D, Hemipristis; E, Squalicorax* [*Corax*]; *F, Sphyrna; G. Ginglymostoma; H, Heterodontus,* anterior and posterior teeth; *I, Squalus; J, Squatina* [*Rhina*]; *K, Isurus; L. Galeocerdo; M, Scapanorhynchus; N, Scyliorhinus* [*Scyllium*]; *O, Carcharodon.* Mostly half natural size; *Squatina* and *Squalus* somewhat enlarged; *Carcharodon* ×1/4. (After Fowler and Smith Woodward.)

sharks just discussed. The batoids have become bottom-living types, often very depressed in shape. The eyes are dorsally placed; the spiracle, used for intake of water, is much enlarged; the gills are ventral in position; and the anal fin is lost, as in squaloid sharks, in correlation with bottom dwelling. The enormously expanded pectoral fins extend forward above the gills to attach to the sides of the head. Swimming is accomplished by an up-and-down motion of these large fins; the tail tends to be reduced and is often a mere whiplash; the dorsal fins are reduced or absent. Skates and rays are mostly mollusk eaters, and in correlation with this the teeth are usually flattened and often form a solid crushing pavement.

Within this group many varied types have been developed. The most primitive members, and among the oldest in their geological appearance, are types intermediate in structure between sharks and rays—forms in which the pectoral fins are enlarged but in which the body is still somewhat rounded, and well-developed dorsal and caudal fins are still present. *Rhinobatos,* the living banjo fish, is a form of this sort with ancestors, such as *Aellopos* [*Spathobatis*] (Fig. 58), already present in the Jurassic. Sawfishes (Fig. 59) were common in the Upper Cretaceous, and *Pristis* is a surviving genus of this group. They resemble the banjo fishes in their general organization but are notable for the development of a long rostrum armed with teeth on either margin. We have mentioned a similar development in *Pristiophorus;* but in that saw shark the diagnostic skate characters are absent, and the two are generally regarded as exhibiting parallel development rather than any close relationship.

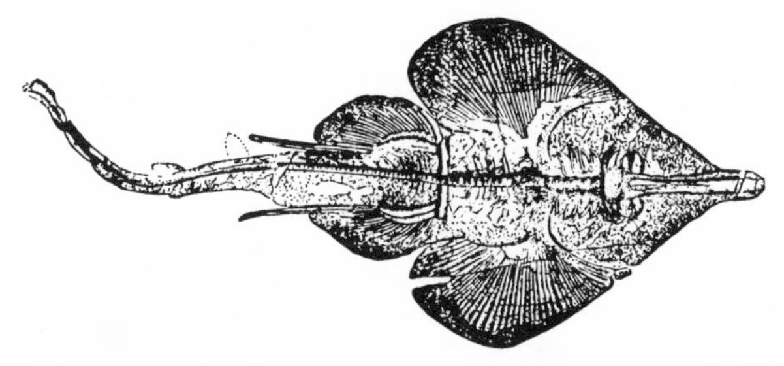

Fig. 58.—*Aellopos* [*Spathobatis*], a Jurassic "banjo fish"; about 1/20 natural size. (From Smith Woodward.)

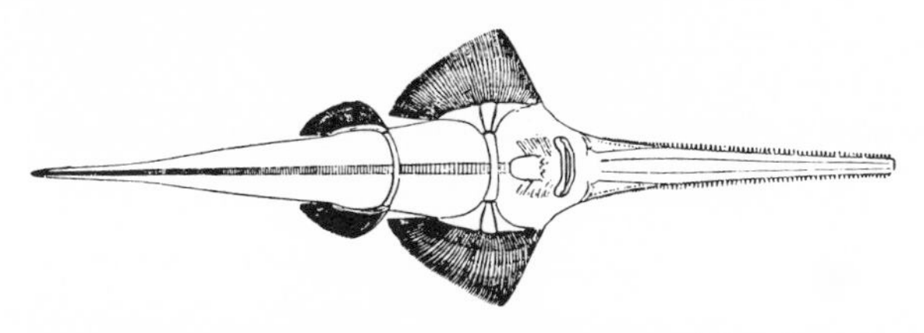

Fig. 59.—*Sclerorhynchus,* a Cretaceous sawfish; about 3 feet in length. (From Smith Woodward, *Outlines of Vertebrate Paleontology.*)

In the Cretaceous, typical skates and rays, including most of the modern types, were present. In *Raja* (Fig. 60*B*), still a common skate today, we find developed the normal characters of the group, such as a short body, very broad pectoral fins extending forward along the sides of the snout, a slender tail, and median fins reduced almost to the vanishing point. In the Cretaceous were also present the sting rays of the *Dasyatis* [*Trygon*] type (Fig. 60*C*), commonly with a poisonous "stinger" developed from a modified fin spine on the tail; the pectoral fins are so expanded laterally that the body is broader than long and so greatly produced anteriorly that they are fused in front of the snout. Still another group to develop at that time was that of the eagle rays, such as *Myliobatis* (Fig. 60*D*) and *Rhinoptera.* Here the fins are greatly expanded laterally but do not meet in front of the head, and the mouth has a highly developed pavement of flat crushing teeth (Fig. 61*A*). In the Tertiary, skates and rays are known from numerous localities, and a final addition to the roster of fossil skates is made by the appearance in the Eocene of the torpedoes (Torpedinidae), with rounded bodies in which some of the musculature is modified into powerful electric organs.

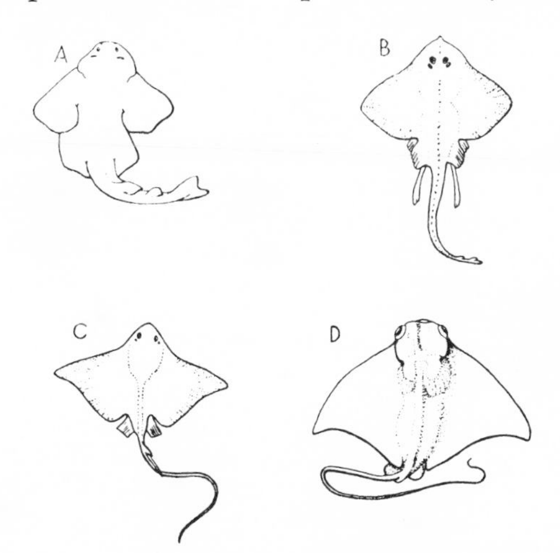

Fig. 60.—Various skates and rays: *A, Squatina* [*Rhina*]; *B, Raja; C, Trygon; D, Myliobatis.* These forms are known both from living representatives and from Tertiary fossil types. (*A* through *C* after Goodrich; *D* after Bridge.)

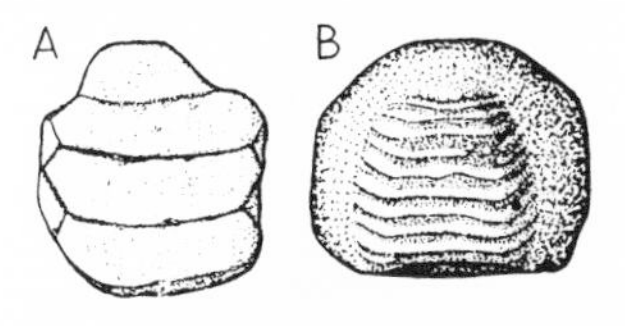

Fig. 61.—*A,* Part of a fossil dental battery of the ray *Myliobatis,* with three complete tooth plates and fragments of others. *B,* A tooth of the large Cretaceous mollusk-eating shark *Ptychodus,* 1/2 natural size.

Pleuracanths.—An early side branch from the "main line" of elasmobranch evolution is that seen in the order Pleuracanthodii. These, in contrast to most shark types, were freshwater forms. They appeared in the late Devonian, and were abundant in the Carboniferous and early Permian; a few persisted into the Triassic. The best-known genus has been frequently termed *Pleuracanthus;* however, this generic term had been early used for an animal of another sort, and *Xenacan-*

thus must be used as a substitute. The body of *Xenacanthus* (Figs. 62, 63) was long and slim, with an elongated dorsal fin stretching far down the back. The tail, unlike that of most early fishes, was diphycercal; but the tip tilted slightly upward, suggesting derivation from a heterocercal ancestor. The anal fin was a peculiar double structure. The paired fins (Fig. 27*C*, *D*) were of the archipterygial type, with a main axis and side branches. This type of fin may have been derived by the gradual development of a posterior axis from one of cladoselachian type; the branches were not fully developed on the posterior margin of the axis, suggesting a transitional condition. A clasper is present in the males.

The braincase, jaws, and branchial arches are sometimes preserved in calcified cartilage; the jaw construction is of the typical amphistylic type seen in *Cladosel-*

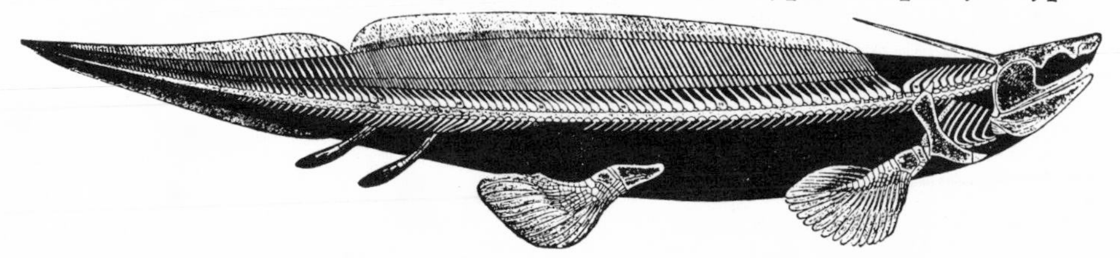

Fig. 62.—*Xenacanthus* [*Pleuracanthus*], a Carboniferous and Permian sharklike form; an average specimen perhaps 2.5 feet long. (After Fritsch, from Smith Woodward.)

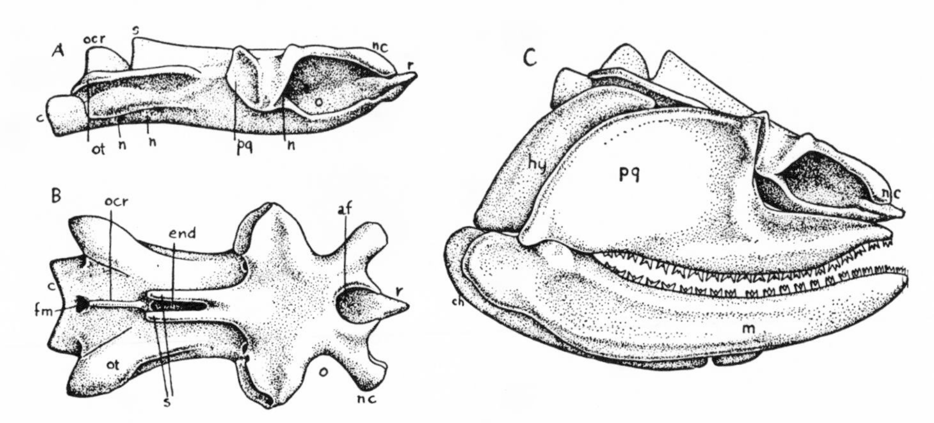

Fig. 63.—The cranial anatomy of a Paleozoic "shark," *Xenacanthus* [*Pleuracanthus*]. *A*, Lateral view of braincase; *B*, dorsal view; *C*, lateral view with jaw and hyoid arch articulated. One-third the size of a large specimen. Abbreviations: *af*, anterior opening (fontanelle) in roof of braincase; *c*, occipital condyle; *ch*, ceratohyal; *end*, common opening for endolymphatic ducts from internal ear; *fm*, foramen magnum; *hy*, hyomandibular; *m*, mandible; *n*, exits for certain cranial nerves; *nc*, nasal capsule; *o*, orbit; *ocr*, occipital crest for attachment of body muscles; *ot*, otic region; *pq*, upper jaw (palatoquadrate bar) and its point of attachment to postorbital process of braincase; *r*, rostrum; *s*, paired processes to which a large spine (peculiar to pleuracanths) is attached.

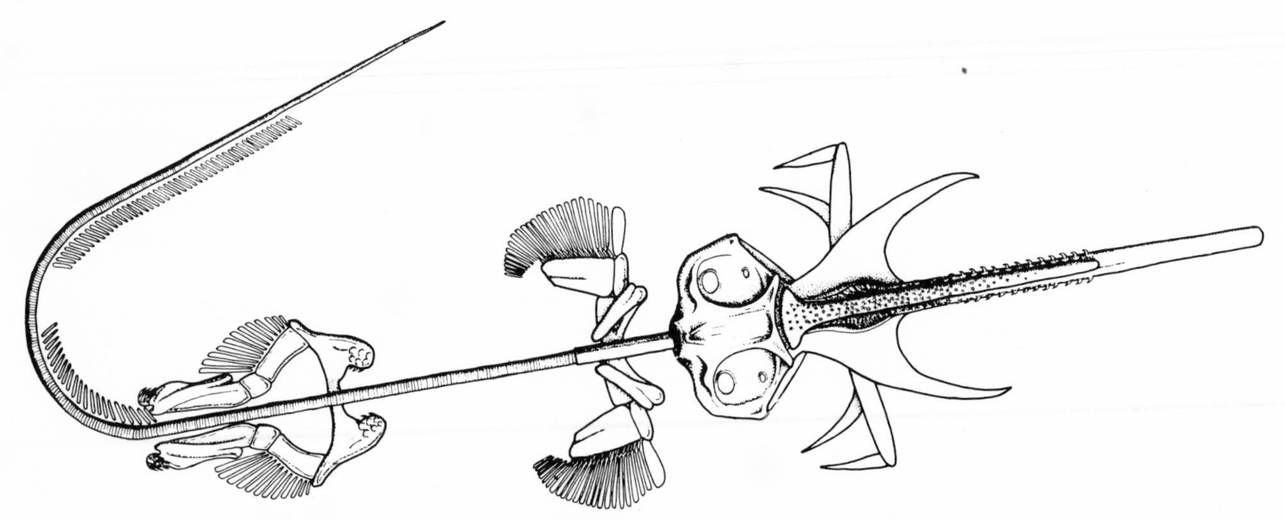

Fig. 64.—Dorsal view of the Jurassic chimaera *Squaloraja*. About 2/5 natural size. (From Patterson.)

ache. A long movable spine, bordered by two rows of small denticles, projected from the back of the head. The teeth, loosely attached to the jaws, typically consisted of two divergent prongs and a central cusp (usually small), set on a button-like base (Fig. 49). These teeth, when found separately, are frequently termed *Diplodus* or *Dittodus.*

Little is known of the ancestry of the pleuracanths. In some cases teeth are known in which the three cusps are nearly equal in size, suggesting a derivation from the *Cladodus* type by a reduction of the central cusp and emphasis on the lateral ones.

Chimaeras.—Among the cartilaginous fishes of modern times we find, besides the sharks and skates, a second group, relatively rare and quite different in many features. These are the chimaeras or ratfish, of the subclass Holocephali. They are fairly active swimming types but often with the tail fin reduced to a whiplash, with little indication of its true heterocercal nature. In many features (such as the construction of the paired fins and the presence of a clasper) they resemble the elasmobranchs. But there are some striking differences. There is often a large rostrum, and the males have an odd clasping organ on the "forehead." Dermal denticles are absent from the skin in advanced modern forms; there is commonly a large spine in front of the dorsal fin; and the pectorals are enlarged and fan shaped. The gills, placed far forward, beneath the braincase, are covered with a flap of skin, an operculum, contrasting with the open slits of the sharks; and there is no separate spiracular opening. The usual dentition consists merely of a large tooth plate on each upper and lower jaw ramus and an additional pair of plates at the front of the upper jaw. This construction appears to be related to eating habits of chimaeras; although omnivorous, "shelled" invertebrates appear to be the basic diet. The upper jaws are firmly fused with the braincase, giving a firmer leverage for the bite; this unusual "solid-headed" condition gives its name to the subclass and is probably secondary. There is, of course, no need of support for the jaws from the hyomandibular, and the simply built hyoid arch is attached to neither jaws nor braincase.

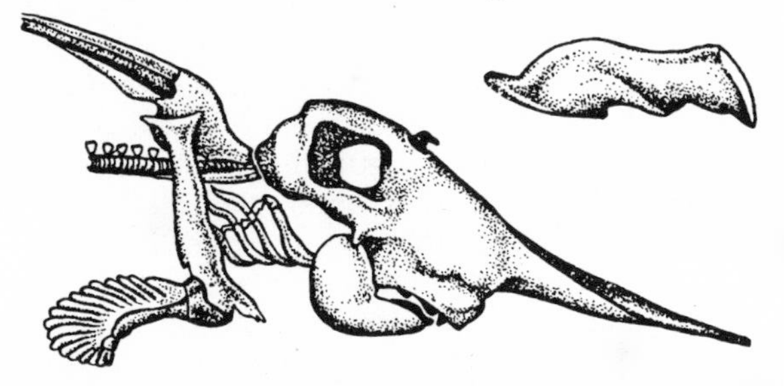

Fig. 65.—Skull and shoulder region of a Jurassic chimaera, *Acanthorhina.* (After Fraas.) About 1/5 natural size. *Upper right,* Mandibular dental plate of the Jurassic chimaera, *Myriacanthus.* (After Dean.)

These high-sea fishes are far from common today, but occasional tooth plates and spines are found in Tertiary and Cretaceous deposits, and in the Jurassic chimaeras were relatively abundant and varied. Some of that age were fairly similar to modern genera, but other Jurassic forms show considerable variation. For example, *Squaloraja* (Fig. 64) has an exceedingly long rostrum and a complex series of tentacular outgrowths from the head but lacks a dorsal fin spine and has a flattened head. In *Myriacanthus* and *Acanthorhina* (Fig. 65) there is likewise a long rostrum, but the head is not flattened, and the usual dorsal fin spine is present.

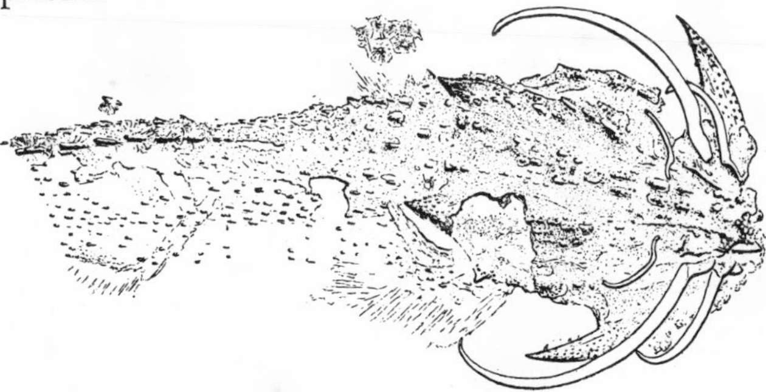

Fig. 66.—Dorsal view of a specimen of the Permian cochliodontoid *Menaspis.* About 2/3 natural size. (From Jaekel.)

No trace of forms which can be definitely included in the Holocephali are known in older geologic periods. Their earlier pedigree is none too clear. In the Carboniferous and Permian are a number of forms, mainly known from tooth plates, which constitute the family Cochliodontidae. In typical genera, such as *Cochliodus* and *Deltodus* (Fig. 54*C*), the main tooth structure consists of a long spirally curved plate in each half of each jaw, although there appear to have been smaller anterior teeth. This is a dentition closely comparable to that of Jurassic chimaeras, on the one hand, and on the other, to that of the Devonian ptyctodonts, which (we have noted earlier) have been advocated as holocephalian ancestors.

In most instances we know nothing of cochliodonts except the tooth plates. In two genera, however, *Menaspis* of the Permian (Fig. 66), and the Carboniferous *Deltoptychius,* part of the body is known. In modern holocephalians the body covering consists, at the most, of reduced dermal denticles; in *Menaspis* there are series of large tubercles over head and body, suggesting the last phase of a breakdown of an armored covering of placoderm type. Although there is no rostrum of holocephalian pattern, there are series of tentacles rather like those of *Squaloraja* and, further, stout bony spines projecting laterally at either back corner of the head region. *Deltoptychius* shows a similar pattern, but here we find, in addition, a large dermal bone, traversed by a pair of lateral-line canals, covering most of the length of the center of the skull roof. These facts point strongly toward the derivation of chimaeras from placoderms. But the structures seen in *Menaspis* and *Deltoptychius* do not point toward the ptyctodonts; rather, toward some member of the poorly known stegoselachian groups.

5

CHAPTER

Bony Fishes

The groups thus far considered, although prominent in the older geologic periods, became in later times much reduced in numbers; the higher bony fish constituting the class Osteichthyes have, on the other hand, continually increased in importance (Fig. 67). Appearing in Devonian fresh waters, they flourished greatly and by the end of the Paleozoic had almost sole possession of the lakes and streams. By that time, they also had invaded the seas, and at the present time the group includes almost all freshwater fishes and the vast majority of marine forms as well.

In an earlier chapter we tentatively assigned the acanthodians to the Osteichthyes. This problematical group apart, the bony fishes may be grouped in two subclasses—Actinopterygii and Sarcopterygii (Choanichthyes)—already quite distinct at their first appearance in the fossil record. The former, the ray-finned fishes, were at first rare, but rapidly became the dominant types; their modern representatives, the teleosts, are the common fishes of today. The Sarcopterygii, or fleshy-finned fishes, include the Crossopterygii, ancestral to land vertebrates, and the aberrant Dipnoi, the lungfishes. Flourishing at first, these groups are now close to extinction.

To give a concise definition of the Osteichthyes is impossible. In their general organization all bony fishes show variations on a common structural pattern of advanced type, but in no one specific feature can we sharply differentiate them from lower fish groups. Bone, for example, is not a new material in the Osteichthyes, although its retention here is a useful point of distinction between these fishes and the modern Chondrichthyes, which have lost all vestiges of bony structure. The presence of some sort of lung or air sac seems to have been a characteristic of all early Osteichthyes; this feature is in contrast with conditions in other surviving fish types. But, as noted previously, a lung may have developed at a still earlier stage in fish history.

The primitive members of both Actinopterygii and Sarcopterygii appear to have been active swimming types with a fusiform body and a heterocercal caudal fin. This last structure may be variously modified, and in almost all late members of the class the tail tends to become a symmetrical structure. In lungfish and crossopterygians this symmetry is attained by the evolution of a diphycercal tail; in the ray-finned types, superficial symmetry was reached by a different process. One dorsal fin was present in the earlier actinopterygians, two in the sarcopterygians; a single anal fin is found in all.

Skeleton.—Bone appears to have been a common feature in ancestral vertebrates. But in the fish classes already considered, ossification is, in general, far from complete in forms so far discovered, and there was a strong tendency toward bone reduction and the retention of the embryonic cartilaginous state. In early Osteichthyes, on the other hand, skeletal ossification was of an advanced nature; and, while there is even here a trend toward reduction, this trend is, in general, much less marked; the process of bone development is, so to speak, much more firmly embedded in the embryological pattern of these fishes. A dermal covering of thick scales and plates was present in the most primitive Osteichthyes. These structures may be much modified but are seldom lost. The internal skeleton, too, was highly ossified in early bony fishes, except for a few structures such as the vertebral centra. In most later members of the class these internal bones are preserved with little loss, and there may even be further progress in ossification of the backbone. In only a few groups are there conspicuous regressions from a highly ossified condition: the sturgeons and paddlefishes among the Actinopterygii, the later lungfishes, and, to a lesser degree, the later coelacanths among the Sarcopterygii.

Scales.—In the older bony fishes the trunk and tail were completely encased in an armor consisting of continuous and usually overlapping rows of bony scales, presumably rhomboidal in shape in the most primitive stage. Two main types of scale are found in early bony fishes—the cosmoid and the ganoid (Fig. 68). The former type was present in primitive crossopterygians and lungfish. Its base consisted of bone arranged in parallel layers. Above was a layer of spongy bone filled with spaces for blood vessels. The upper part of the scale consisted of a substance termed cosmine, similar to the dentine of teeth, arranged about numerous pulp cavities; the surface was covered with a thin layer of enamel. The whole suggests a series of fused denticles united with a basal plate of dermal bone (or rather, as was discussed earlier, shark denticles represent superficial pieces of such scales which have "floated free" with the dissolution of the underlying bony plate).

A second ancient type, found in the older actinopterygians, was the ganoid scale. The same three units are present as in the cosmoid scale, but here the enamel-like material, deposited in thick layers, is termed ganoine. Embryologically these scales arise by the deposition in onion-like fashion of concentric layers of all three materials around a central core composed of a single "cosmoid-scale" unit. It is of interest that almost the same type of scale was present in acanthodians. Possibly the ganoid scale is the primitive type, and the cosmoid has resulted from a reduction of the superficial ganoid layers; or, conversely, the ganoid type has developed by elaboration from a simple cosmoid structure.

The term "ganoid" is often used to designate certain ancient bony fish. Properly, such a term may be applied to primitive actinopterygians which have scales of the ganoid type. But frequently (and unfortunately) the term is applied to any primitive-looking fish with thick shiny scales, and to forms with cosmoid, as well as ganoid, scales, and hence should be avoided.

Both types of scale have undergone great modification and reduction; only one living fish type has a cosmoid scale; few have the ganoid type. In both subclasses there was a strong tendency toward the loss of both the enamel-like surface and the underlying cosmine, leaving a relatively thin and simple bony scale.

Skeleton of the head (*Figs. 69, 70*).—The cranial skeleton is of a double origin, consisting of dermal bony plates comparable to enlarged scales and of internal bones—endochondral bones—replacing embryonic cartilages; in skull and jaws (and even in the shoulder girdle) these two types of components are combined to form complex structures. Similar complexes were present in ostracoderms and placoderms, but our relative lack of knowledge of internal anatomy in these forms and the frequent lack of ossification have caused our account of them to be incomplete and inadequate. In the sharks the absence of dermal components has resulted in a deceptive secondary simplicity. In the Osteichthyes, however, there emerges a general pattern of cranial construction, which, although complex, merits description in detail not only because it is to be found, with variations, in all subdivisions of the group but because it is the basal plan upon which the skulls of all tetrapods are built.

We shall note in preliminary fashion the essential units of which it is constructed before considering in

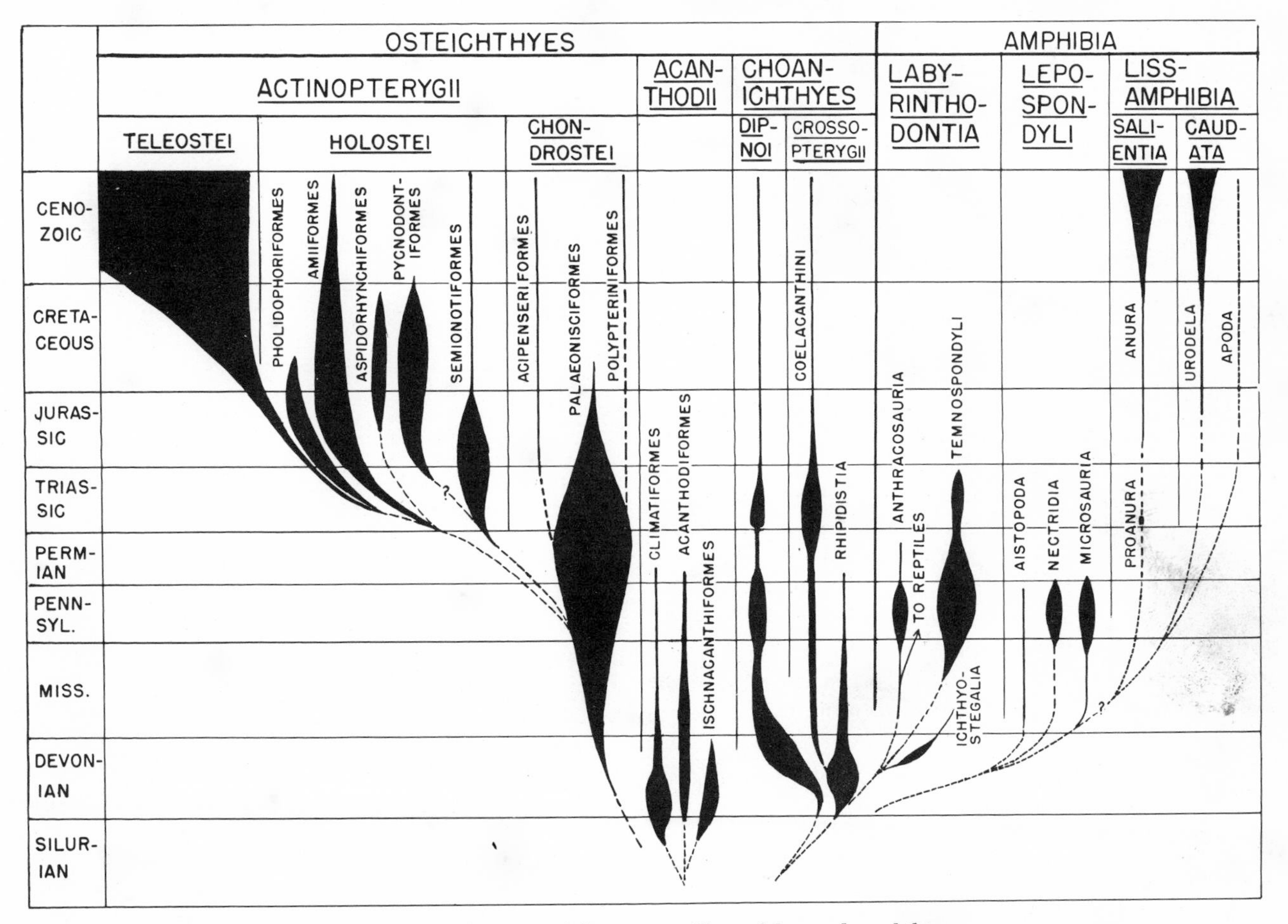

FIG. 67.—Development of the groups of bony fishes and amphibians

more detail the various dermal and cartilage-replacement bones which take part in its formation. The skull itself is made up of three units: (1) an ossified braincase, comparable to that of a shark, with a median dermal plate below it; (2) upper jaw and palatal structures, including ossified equivalents of the cartilaginous shark upper jaw and additional dermal bones of the palate; and (3) a shield of dermal bones covering the top and sides of the head and in contact with the deeper structures already noted (cf. Fig. 8). Additional skeletal features of the anterior part of the body include: (4) the lower jaw, also a combination of dermal elements and replacement bones; (5) the gill arches, including a prominent hyomandibular, all endochondral in nature; (6) a series of dermal gill coverings; and (7) the shoulder girdle, composed of both types of bones.

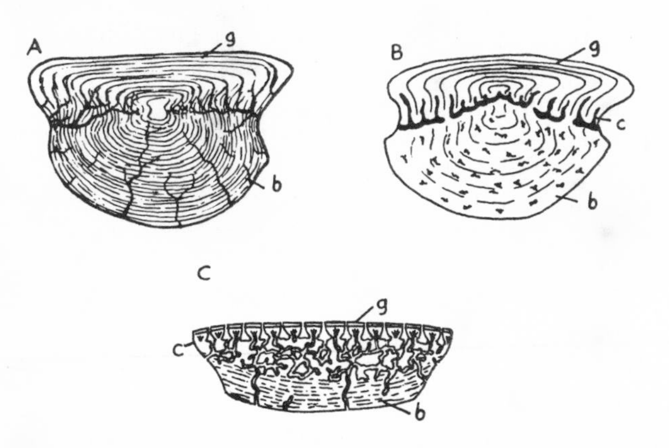

FIG. 68.—Vertical sections through the scales of, *A*, *Acanthodes*, an acanthodian; *B*, *Cheirolepis*, a Devonian actinopterygian; *C*, *Megalichthys*, a Carboniferous crossopterygian. All highly magnified. Abbreviations: *b*, basal layers of bone or bonelike material; *c*, cosmine layer; *g*, superficial region—enamel or layers of enamel-like material (ganoine). (After Goodrich.)

The names used for vertebrate bony elements are customarily derived, where possible, from those used in man or other mammals. In the early periods of the study of fish anatomy, connecting links between fish and higher vertebrates were unknown; names were arbitrarily applied to fish bones which appeared to occupy the same general position as their mammalian namesakes, although real identity could not be proved. Today our information concerning the transitional forms from fish to higher types is much greater, and in crossopterygians, closest to land forms, most of the bones can be identified with considerable confidence; lungfishes and actinopterygians, however, show such departures from the crossopterygian pattern that the nomenclature used for them is still an arbitrary one in many instances.

The canals which carry the lateral-line sense organs are distinctive features of the superficial elements of the fish cranial skeleton. They have rather constant relations to the dermal bones along their courses. The main pair of canals passes forward from the body to the lateral portions of the skull roof (there is typically a cross-commissure between them here) and continues anteriorly to the region above and behind the orbit. A suborbital canal here turns downward around the orbit and forward toward the snout. Another canal—the supraorbital—runs forward above the orbit; in many cases (but not primitively) it connects posteriorly with the main canal. A third major canal curves back and down over the cheek region and then forward along the lower jaw. In addition, there may be accessory canals or rows of pits containing individual lateral-line organs.

Dermal bones.—All early bony fishes have a shield of dermal bones forming a nearly solid covering for the top and sides of the head region. The shield is notched posteriorly at either side in the region of the spiracle; its continuity is broken only by the openings for external nares, the large orbits for the paired eyes, and the pineal opening, when present. The braincase is fused to the lower side of the shield; at its lateral margins the shield carries the principal tooth row of the upper jaw and is in contact here with the palatal structures.

It is probable that in the unknown early ancestors of the bony fishes the skull roof was composed, in the embryo, of a mosaic of small plates. During the evolution of the various groups there appears to have been initiated (more strongly at first in the back part of the skull) a sort of "struggle for existence" among these numerous and probably variable elements, and, as a result, later types have but a few relatively large skull elements. Evidence for such a belief lies in the facts that variable small elements persist in the rostral region of both ray-finned fishes and crossopterygians and that the earliest dipnoans are still largely in a mosaic stage. If this be true, it is likely that we shall never be sure of all homologies between the different bony-fish groups, since it is improbable that the "survivors" have been the same bones in all cases. Nevertheless, a common pattern and grouping of elements can generally be discerned in ray-finned fishes and crossopterygians, and, to a lesser degree in dipnoans. These groups of bones include:

1. A variable series of small elements in the rostral region of the skull in primitive forms; usually reduced or much modified in later types.
2. A paired longitudinal series on either side of the dorsal midline, typically including, from front to back, nasals, frontals, parietals (bordering the pineal opening, when present), and postparietals (dermal supraoccipitals). The anterior part of this series may primitively occur as a mosaic of small elements at the sides of the rostral region.
3. Marginal tooth-bearing elements—maxilla and, more anteriorly, a premaxilla, which in some cases appears to be a specialized rostral element.
4. A circumorbital ring of bones, in which it is sometimes possible to identify equivalents of the tetrapod elements in this region—prefrontal, postfrontal, postorbital, jugal, and lacrimal.
5. Bones bordering the posterior margin of the skull

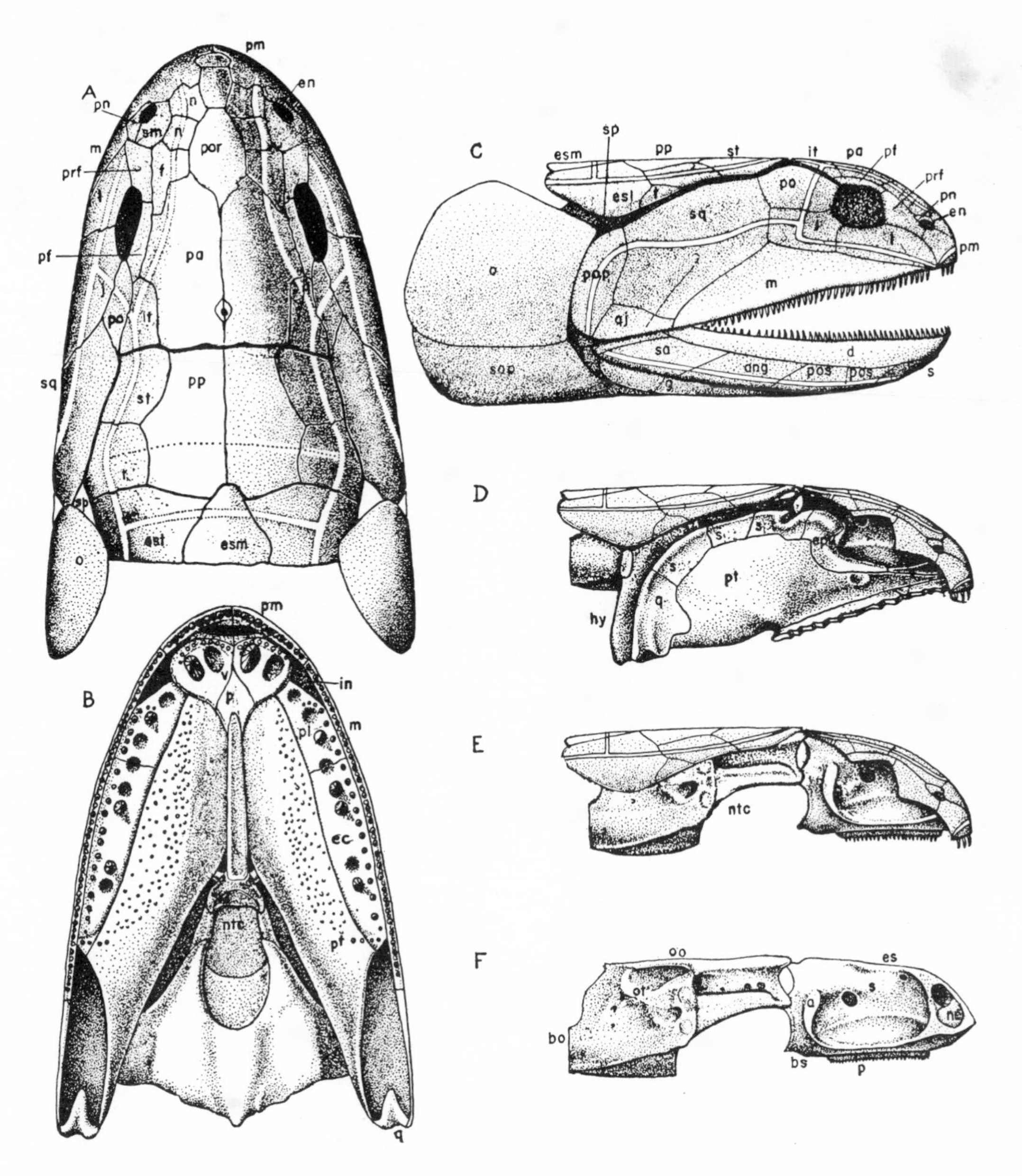

FIG. 69.—The skull in rhipidistian crossopterygians (composite). *A*, Dorsal, and *B*, palatal views of the skull of *Eusthenopteron* of the Upper Devonian of North America; *C*, lateral view of the skull and opercular region of *Osteolepis* of the Middle Devonian; *D, E, F*, a diagrammatic "dissection" of the skull of *Osteolepis* to show internal structure. In *D* the opercular plates and the bones of the cheek region removed to exhibit the primary upper jaw. In *E* the primary jaw and hyomandibular removed to show the braincase in relation to the dermal skull roof. In *F* the braincase isolated. The upper jaw and braincase of *Osteolepis* are incompletely known and are restored, in part, from other crossopterygians. Abbreviations for this and other bony-fish skulls are given below (many of the identifications are in dispute; some frequently used synonyms are given in parentheses): *a*, articular facet on braincase for upper jaw; *ang*, angular; *ar*, articular; *bo*, basioccipital region of braincase; *br*, branchiostegal elements of throat region (gulars); *bs*, basisphenoid region of braincase; *cl*, clavicle; *ct*, cleithrum; *d*, dentary; *ec*, ectopterygoid; *en*, external naris; *ep*, epipterygoid (metapterygoid); *es*, ethmosphenoid portion of braincase; *esl*, lateral extrascapular; *esm*, medial extrascapular; *f*, frontal (nasal); g, gulars; *hy*, hyomandibular; *in*, internal naris; *iop*, interopercular; *it*, intertemporal (postorbital or dermosphenotic); *j*, jugal; *l*, lacrimal (infraorbital); *m*, maxilla; *n*, nasal; *nc*, nasal capsule; *ntc*, area occupied by notochord; *o*, opercular; *oo*, otico-occipital portion of braincase; *ot*, otic region of braincase; *p*, parasphenoid; *pa*, parietal (frontal); *pf*, postfrontal (supraorbital); *pl*, palatine; *pm*, premaxilla; *pn*, postnarial; *po*, postorbital; *pop*, preopercular; *por*, postrostral or internasal; *pos*, postsplenial; *pot*, posttemporal; *pp*, postparietal (parietal); *prf*, prefrontal; *pt*, pterygoid; *pv*, vomer (prevomer); *q*, quadrate; *qj*, quadratojugal; *r*, rostral; *s*, suprapterygoid(s), splenial; *sa*, surangular; *sc*, sclerotic ring; *sct*, supracleithrum; *sm*, supramaxilla; *smp*, symplectic; *so*, suborbital(s); *soc*, supraoccipital; *sop*, subopercular; *sp*, spiracular cleft; *sq*, squamosal; *st*, supratemporal (dermal sphenotic); *t*, tabular (supratemporal-intertemporal); *v*, vomer. (Data in part after Watson, Stensiö, and Westoll.)

roof, termed extrascapulars. They appear to be enlarged scales lying beyond the limits of the skull proper and have no homologues in tetrapods.

6. Small lateral elements on the "table" formed by the posterior part of the skull roof and adjacent to the spiracular slit; variable but, in general, corresponding to the supratemporal, intertemporal, and tabular bones of land vertebrates.

7. Elements of the cheek region, sometimes with bones corresponding to the squamosal and quadratojugal of tetrapods, and always with an element, the preopercular, lost in all but the most primitive of land vertebrates.

Dermal bones are also present on the undersurface of the skull in the skin lining the roof of the mouth. A median dermal structure, the parasphenoid, is always present beneath the anterior end of the braincase. Paired groups of palatal elements supplement and in great measure replace the embryonic upper-jaw cartilage. These include a large pterygoid bone and a lateral series, frequently of three elements—vomer, palatine, and ectopterygoid, the first functioning primarily as a floor for the nasal capsule. These bones usually are tooth-bearing. This dentition presumably represents that borne on the jaw cartilage of sharks; the lateral tooth row of bony fishes is not represented, it would seem, in elasmobranchs. Behind the ectopterygoid is a fossa through which descend the powerful muscles which close the jaw.

The lower jaw (Fig. 71) consists, for the most part, of dermal bones. The dentary, carrying the marginal tooth row, is a major external element, with bones termed splenial (and postsplenial), angular, and surangular below and behind it. Internally a series of dermal elements, which often bear teeth, include a long prearticular, and above it, a series of coronoids.

The gill region is sheathed by opercular elements—laterally a large operculum and suboperculum; more ventrally, in the "throat" region, a variable series of gular plates, which in advanced actinopterygians become slender rodlike structures, the branchiostegal rays.

The external surface of the shoulder girdle is formed by the most posterior elements of the dermal bone series—a clavicle below (this is lost in higher ray-finned fishes) and a much larger fish element, the cleithrum. The girdle is joined to the table of the skull by supracleithrum and posttemporal.

Replacement bones.—A braincase corresponding rather closely in its form to that seen in early sharks is present in primitive bony fishes as a highly ossified structure. Dorsally the braincase is tightly bound to the dermal bones of the skull roof. Laterally it appears to have been movably articulated with the palatal structure by means of a basal process part way back on either side (adjacent dermal elements take part in the articulation). The braincase of the Osteichthyes is composed of a large number of ossifications; in the early types, however, these elements tended to be fused in the adult into larger units, and in some early Osteichthyes the braincase consists of two or three such units, within which there is little or no evidence of

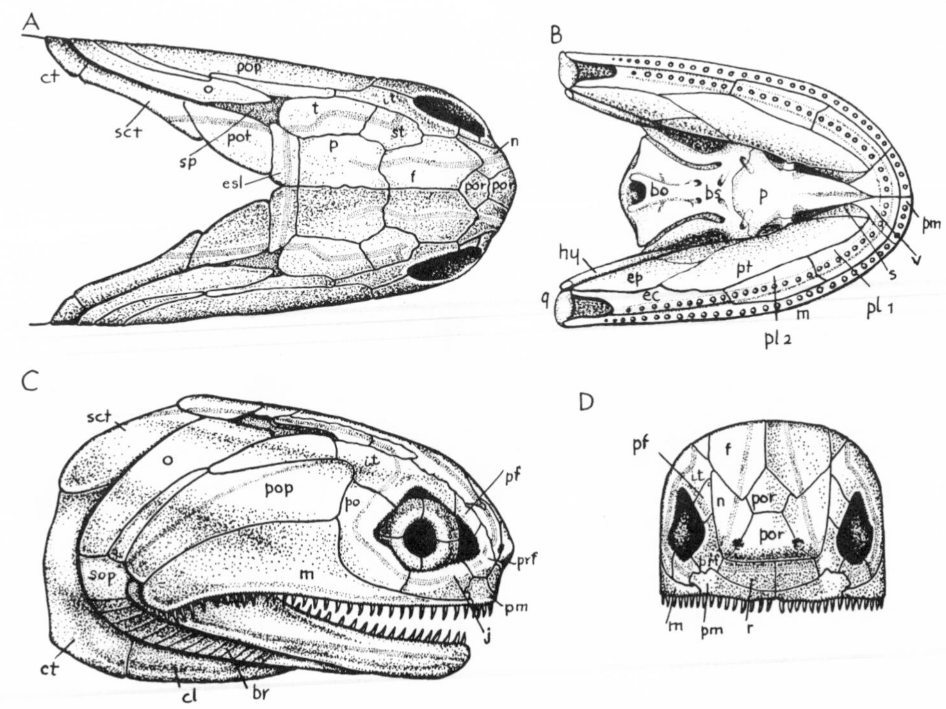

FIG. 70.—*A*, Dorsal, *C*, lateral, and *D*, front, views of the skull and shoulder region of *Cheirolepis*, a Middle Devonian palaeoniscoid; about natural size. (After Watson.) *B*, Restoration of the palate of a palaeoniscoid, based mainly upon *Haplolepis* as described by Watson. For abbreviations see Figure 69.

subdivision. In most later forms, in contrast, cartilaginous areas remain in the adult, and sutures become visible between the bony elements. These elements are rather variable but tend to include (1) a basioccipital and exoccipitals posteriorly (a supraoccipital is present in higher actinopterygians); (2) variable otic ossifications in the ear region, to which the terms proötic, opisthotic, epiotic, sphenotic, pterotic, and intercalar are commonly applied; (3) a basisphenoid in the region of the basal articulation; (4) more anteriorly a tubelike structure comparable to the sphenethmoid of early land vertebrates, or paired elements to which the mammalian terms alisphenoid, orbitosphenoid, and ethmoid are very arbitrarily assigned.

The cartilaginous upper jaw, we noted, has been partially replaced functionally by dermal palatal elements, but ossifications in the cartilage are also present, including, posteriorly, the quadrate which articulates with the lower jaw and variable "suprapterygoid" elements, one of which may articulate with the basal process of the braincase and is thus comparable to the epipterygoid of tetrapods. The lower-jaw cartilage is almost completely replaced by dermal elements and usually survives only in the form of the articular bone of the jaw articulation.

The hyomandibular, propping the jaw joint, is typically large and well ossified; the remaining branchial arches are variably ossified.

Postcranial skeleton.—Because generally obscured by the thick scaly covering, the internal skeleton of the body is seen relatively rarely in early fossil bony fishes. Of axial skeletal structures, neural and haemal arches are frequently seen to be ossified, as are ribs, in many cases, and basal and radial elements in the paired fins. The vertebral centra in most early fishes remained unossified; in later actinopterygians (Fig. 4) they ossify, generally as spool-shaped structures. In crossopterygians and a few dipnoans partial or complete ossification may occur in the Paleozoic; in some well-ossified forms the large notochord was partially confined by dorsal structures seemingly comparable to the pleurocentra of early land vertebrates and ventral elements comparable to the intercentra of tetrapods (Fig. 108), and ring-shaped centra are sometimes developed.

The shoulder girdle is a compound structure including the dermal elements described above and a relatively small "primary" girdle, variably ossified, with which the fin skeleton articulates. The pelvic girdle is a small and typically a wedge-shaped ventral plate.

Radically different types of paired fins (Fig. 72) are found in the two subclasses of bony fishes. In most actinopterygians the fins have a comparatively broad base with a skeleton consisting of a number of parallel bars of bone or cartilage. This is essentially the structure seen in the cladoselachians, but in the actinopterygians the internal skeletal elements are usually short, and there is, consequently, only a comparatively small, scale-covered, fleshy lobe; the greater part of the fin consists of a web of skin stiffened by dermal-fin rays. It is to this feature that the group owes its name.

In contrast, as the name Sarcopterygii implies, a large lobe of flesh and bone is developed in the fins of crossopterygians and lungfish, and the dermal rays merely form a fringe about its margins. This lobate fin, when most highly developed, is of the type termed an archipterygium, already seen among pleuracanth sharks. Only one stout element articulates with the limb girdle. Beyond this the crossopterygians usually exhibit an irregular branching arrangement—an abbreviated archipterygium. The dipnoans and some crossopterygians, however, have the full archipterygial structure—a leaf-shaped fin, the skeleton of which consists of a long, jointed, central axis with numerous short side branches.

The fins of sharks are stiffened by long, slim, horny

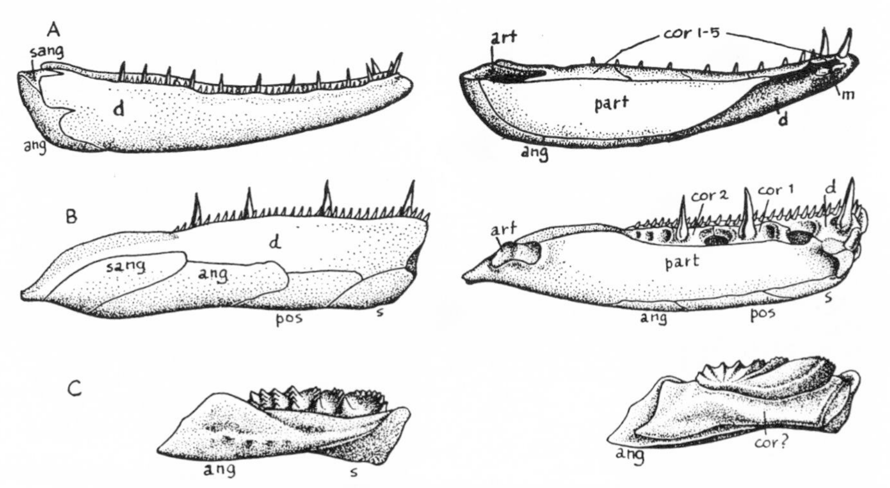

FIG. 71.—The lower jaw in bony fish. Outer view at the left, inner view at the right. *A*, *Nematoptychius*, a primitive actinopterygian; *B*, *Megalichthys*, a Carboniferous crossopterygian; *C*, *Sagenodus*, a Carboniferous lungfish (the articular unossified). Abbreviations: *ang*, angular; *art*, articular; *cor*, coronoid; *d*, dentary; *m*, Meckelian bone in anterior end of jaw; *part*, prearticular; *pos*, postsplenial; *s*, splenial; *sang*, surangular. The bone marked *cor?* in *Sagenodus* is probably the prearticular. (After Watson.)

rays. In bony fishes these structures may be present but are much reduced. The fleshy lobe of the fin is covered with scales, and in addition the remainder of the fin is stiffened by fin rays composed of long, slim, modified scales arranged in rows. Such rays are present both in paired and in median fins.

Lungs.—Gills, never more than five in number, are always present, covered by an operculum and opening by a common slit just in front of the shoulder region. In addition, a spiracle appears to have been present in primitive types but is lost in most recent bony fishes. However, lung breathing, which we naturally associate

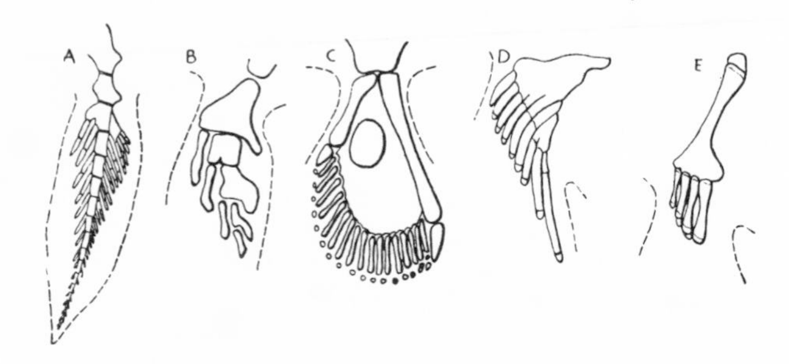

FIG. 72.—The paired fins of bony fishes. Pectoral fins of: *A, Epiceratodus,* a modern lungfish; *B, Eusthenopteron,* an Upper Devonian crossopterygian; *C, Polypterus,* a modern relative of the palaeoniscoids. Pelvic fins of: *D,* the sturgeon, *Scaphirhynchus; E, Polypterus.* (Mainly after Goodrich.)

only with land forms, had its beginning early in the history of bony fishes (Fig. 73) and, we have noted, may have developed even earlier in vertebrate history. Probably the most generalized condition of the lung is that found in the primitive living actinopterygian *Polypterus,* in which there is a simple bilobed sac opening out of the bottom of the throat and situated in the position of lungs in land animals. It is probable that a similar type of lung existed in the typical crossopterygians, and a functional lung is present in all three living lungfish.

The explanation of this unexpected development of lungs in water-living forms (cf. p. 33) is suggested by the habits of the living types mentioned above. These live in tropical regions in which there are alternate periods of rain and drought. During the wet season, lungs are generally unnecessary; but, with the drying-up of streams, with water present only in stagnant pools, ordinary fish die by the myriad, while those which have lungs may survive until the next rains. Most Devonian bony fish, as far as we can tell from the sediments, appear to have lived under quite similar conditions—in freshwater streams and lakes which were subject to periodic droughts. Lungs were a considerable asset to them under these circumstances.

But lungs had obvious disadvantages. They seem to have lain originally in the underpart of the chest and would have tended to make the fish top-heavy. In relation to this fact we find that in the lungfish the lung has shifted around to the top of the body, although its duct may still be attached to the bottom of the throat. The lungs are vestigial in the sole living species of the crossopterygian group, and in all actinopterygians above the *Polypterus* level lungs as such are absent.

While a lung is a useful adjunct under the peculiar conditions which we assume to have been present generally in the Devonian and which still exist in some tropical regions today, it is of little use under more normal climatic conditions. Except for *Polypterus* and *Calamoichthys* we find in actinopterygians not lungs but a single dorsal sac, the air bladder, probably derived from the ancestral lung. This sac may still aid in obtaining oxygen, and may take on such extra functions as that of aiding hearing by acting as a resonating chamber. Its main function, however, is that of a hydrostatic organ. By filling or emptying this sac, the specific gravity of the fish is altered, and it can float higher or sink lower in the water.

In lower vertebrates the nostrils are simply sacs opening to the outer surface of the head. In land types there are, as well, internal openings (choanae) from the nostrils into the roof of the mouth. Such internal openings are not normally present in actinopterygians, but are found in dipnoans and appear to have been present in rhipidistian crossopterygians. It is probable that in the latter group these openings were used in air breathing, as in tetrapods, permitting the intake of air into the lungs without the necessity of opening the mouth and running the risk of "shipping water." They are not, however, normally used in air breathing in lungfishes.

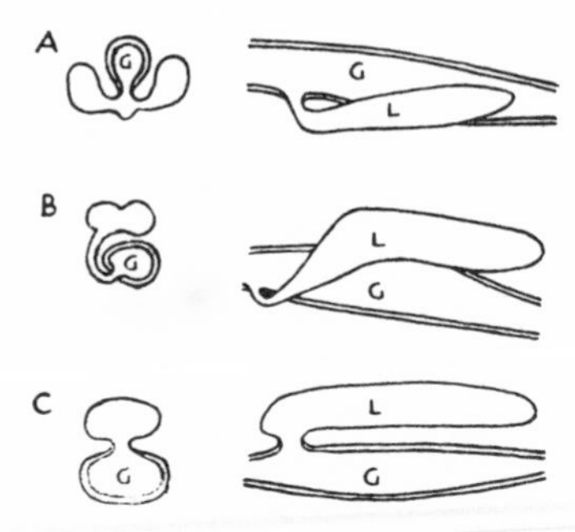

FIG. 73.—Diagram to show the development of lungs in fish and tetrapods. *Left,* Cross-sections of gut and lungs; *right,* longitudinal sections. *G,* Gut; *L,* lung. *A,* Paired ventral lungs, found in tetrapods, African and South American lungfish, and *Polypterus; B,* lung dorsal but duct ventral, as in Australian lungfish; *C,* single dorsal lung—air bladder—with dorsal duct, as in most actinopterygians.

Bony-fish ancestry.—The appearance of the typical bony fishes in the geologic record is a dramatically sudden one. Apart from their possible relatives, the acanthodians, there are no traces of the group in the Silurian and there is only a single incompletely known form of Lower Devonian age. In the Middle Devonian, however, all the major types—ray-finned forms, crossopterygians and lungfishes—appear full fledged and diversified, and at once dominate the scene. The initial stages in their development must have taken place long before, quite probably in the upper reaches of river systems from which no deposits have been preserved.

The common ancestor of the bony-fish groups is unknown. There are various features, many of them noted above, in which the two typical subclasses of bony fish are already widely divergent when we first see them—features such as fin structure, scale structure, and so on. So marked are these differences that it has been suggested that the Osteichthyes are an artificial assemblage and that ray-finned fishes and sarcopterygians represent two or three progressive lines which have arisen separately from an archaic gnathostome stock. However, this is at present mere speculation, and it may be that some common ancestor existed in the waters of Silurian times. We have seen that some major features of bony-fish organization are to be met with among the acanthodians. The known acanthodians are too specialized in various respects to be in themselves the ancestral Osteichthyes. But belief in the descent of the later bony fishes from unknown forms closely related to the acanthodians is extremely reasonable. Still farther back, one may believe, the ancestral osteichthyans and the placoderms may have arisen from a common ancestral gnathostome type. But this is, of course, pure speculation, for which at present there is not the slightest scrap of fossil evidence.

Ray-finned fishes.—Of the two major groups of bony fishes, the actinopterygians perhaps merit first consideration because of their seemingly more primitive nature and because the ray-finned fishes, although comparatively rare at the first, were destined to play the chief role in fish history. Today only four genera of sarcopterygian fishes survive, while modern representatives of the Actinopterygii are to be numbered by the tens of thousands. In the early days of the study of fishes, the living ray-finned forms were divided into three groups—Chondrostei, Holostei, and Teleostei, in ascending order. The names were based upon mistaken assumptions as to the history of bone development; and today, with a great store of knowledge of fossil forms, we realize that actinopterygian history is much more complex than was once believed. Nevertheless, these three terms still form a convenient framework on which to build a discussion of this vast group and will be used here as infraclasses within the subclass Actinopterygii.

1. The Chondrostei include mainly primitive Paleozoic and dominant Triassic types and certain specialized or degenerate survivors.

2. The Holostei were the characteristic ray-finned fishes of middle Mesozoic days and are also near extinction.

3. The Teleostei began their expansion in the late Mesozoic and have been dominant since the Cretaceous.

These terms indicate successive grades of development within the Actinopterygii rather than true phyletic units; for the holostean level of organization was very probably attained by more than one group of primitive forms, and the teleosts may possibly be similarly polyphyletic. However, this classification, even if not entirely natural, is one which it is best to preserve until our knowledge of the details of the complex evolutionary history of the ray-finned fishes is much more adequate than is the case at present.

In ray-finned fishes there is typically but a single dorsal fin, in contrast to the sarcopterygians. Above the front edge of the dorsals and caudal in earlier actinopterygians are frequently found rows of large V-shaped fulcral scales, forming a protective cutwater for the fins. These are not found in other fish groups and disappear in advanced ray-finned types.

The scales primitively had the typical ganoid structure; in early forms they were usually rhomboidal in outline. In later types they may be much modified. The shape may become cycloid with the development of a rounded posterior margin, and in many teleosts become ctenoid, with the development of numerous tiny spinelets on their surfaces. In most Mesozoic forms the scales lose the cosmine layer. The ganoine covering becomes thin in many holosteans and is lost in teleosts, which have simply-built bony scales; and in various cases the entire scaly covering may be reduced or lost.

In the earliest actinopterygians the general arrangement of the bony plates of the head was similar to the generalized pattern described above (Fig. 70). There are, however, marked contrasts with the crossopterygian type. There is rarely a pineal opening; a spiracular slit was present early, but the spiracle was lost in higher groups; there is normally no internal opening for the nostrils. The orbits are large in primitive actinopterygians, and the large eyeballs are stiffened by sclerotic plates, usually four in number. These forms appear to have relied largely upon vision for their knowledge of the outer world, while most other fish groups appear to place great reliance on the sense of smell; they are "eye-fishes" rather than "nose-fishes." Whereas the mouth is nearly terminal in crossopterygians, it is here preceded by a short terminal rostrum in primitive forms, and the external nares in typical actinopterygians, usually double, are well up on the side of the head, rather than close to the jaw margins. The cheek is mainly occupied in early genera (again in contrast with crossopterygians) by a large preoperculum and an expanded maxilla. A small quadratojugal may be present primitively, but there is no equivalent of the squamosal; there may develop, instead, variable small bones termed suborbitals. From the first, great reliance was placed on the jaw support afforded by the broad hyomandibular element, and in the later ray-finned fishes an extra bone (the symplectic) is added to bind the lower jaw and hyomandibular firmly. In all early actinopterygians the well-ossified braincase presumably developed in the embryo from a considerable number of separate ossifications; in lower members of the subclass, however, the braincase bones are welded together in the adult into three units—an occipital region at the back, paired otic structures on either side, and anteriorly a single mass of bone (sphenethmoid) extending forward to the nasal region. The three together form a very solid structure, contrasting with that described later for the Crossopterygii.

The tail was heterocercal in the oldest known actinopterygians, as was also the case in early sarcopterygians; but it differed in that there was practically no development of fin membrane above the backbone. In consequence, only exceptionally has a member of the present group been able to develop a diphycercal tail—an evolutionary process in which an upper lobe (here generally nonexistent) must be greatly developed to equalize the lower one. A similar result, however, has been attained by the development of a homocercal tail (Fig. 3*D*), a superficially symmetrical type found in the teleosts, in which only the lower lobe is concerned but in which the scaly tip of the tail is reduced and no longer projects outside the rounded posterior termination of the body. The final result is a tail fin externally somewhat like the diphycercal one but constructed in a very different way.

We have noted that, in contrast with other bony fishes, there is usually little extension of the flesh and internal skeleton into the paired fins (Fig. 72*D, E*), which are mainly supported by long dermal rays. The internal skeleton of the fins consists generally of a number of short parallel bars of bone which articulate directly with the girdle. In the most primitive actinopterygians the paired fins appear to have extended rather stiffly outward from the body as fixed planes, much as in cladoselachian sharks; in later types, with reduced heterocercal tails and consequent changes in swimming habits, the fins became narrower at the base and much more flexible in movement.

Within the actinopterygians there are, in later days, great modifications in the cranial pattern. Particularly noticeable is a series of changes in the cheek region, in the course of which, as described later, the maxilla is much reduced, freed from the cheek, and even eliminated from the mouth margins. In typical holosteans and teleosts the braincase tends to retain in the adult sutural lines between the numerous component bones of the embryonic condition.

Many of the skull elements in actinopterygians were early named in accordance with their seeming resemblance to elements in the skull of mammals or other tetrapods; additional names, frequently compounds, were added to this artificial system to take care of the numerous variations in pattern. In our illustrations here we have used the customary names for actinopterygian skull bones, although the homology of various of the elements with those in tetrapods and crossopterygians is doubtful; for example, the bones here termed "parietals" are probably the true postparietals, and the true parietals are those termed "frontals" in actinopterygians.

Like the other bony fishes, the Actinopterygii appear to have originated in fresh waters, and only a few members of the group became marine forms during the Paleozoic. By the end of the Triassic, however, the main center of actinopterygian evolution lay in salt waters. Among the modern teleosts the vast majority are marine types, and a proportion, at least, of modern freshwater dwellers are not, we may suspect, "original inhabitants" but have reinvaded the fresh waters from the salt.

Primitive actinopterygians, the Palaeonisciformes.—We shall classify as the infraclass Chondrostei the groups of ray-finned fishes which are primitive in nature or which have remained on a low structural level despite specializations of various types. Such forms include (1) the order Palaeonisciformes, primitive forms, mainly Paleozoic and Triassic; (2) Polypteriformes, living African fishes retaining much of the original palaeoniscoid structure; and (3) Acipenseriformes, degenerate modern types.

Early actinopterygians appear, together with crossopterygians and lungfish, in Middle Devonian freshwater deposits. Rare at first, they so increased in numbers that by the beginning of the Carboniferous they were already the commonest of freshwater groups, and a few appear to have already invaded the sea. Most of the more characteristic Paleozoic ray-finned fishes were once assigned to *Palaeoniscus* (properly a Permian genus) or were thought to belong to a relatively few closely related genera. In recent decades, however, our knowledge of them has greatly increased, and we now realize that these older actinopterygians include a great variety of fishes which will here be included in the order Palaeonisciformes. A majority of them were forms of modest size, which at first glance would have resembled in life their actinopterygian successors—the herring- and minnow-like teleosts of modern streams and seas. But there were fundamental differences. The scales were almost always of the thick and shiny ganoid structure, usually with a rhomboidal shape. The tail was heterocercal, and the cranial pattern was that described above as typical of primitive ray-finned forms; there were large eyes; a long mouth gape with a maxilla expanded at the back over the solid cheek, to which was bound a large preoperculum; and a shoulder girdle with clavicle as well as cleithrum.

Despite the similarity in general appearance of many of these fusiform palaeoniscoids, closer study reveals a great variety in structural detail, and a considerable array of subgroups and families have been identified. *Cheirolepis* (Figs. 70, 74, 75) of the Middle and Upper Devonian is an early form and a primitive one, notable for the tiny size of its square scales, which give it much the appearance of an acanthodian. Prompt, however, in making their appearance were more typical members of the order. *Palaeoniscum* (Fig. 75) (from which the group takes its name) is a Permian genus, but five typical palaeoniscoid genera were present (although rare) in the late Devonian, and a series of two dozen genera or more were present and abundant in the Mississippian.

In recent decades a surprising number of variants on the palaeoniscoid pattern have been discovered in late Paleozoic formations. We shall mention a few, for the most part some of the numerous Mississippian forms from Great Britain. *Canobius* (Fig. 75) has a deep, large-eyed head in which the jaw gape was shortened without bringing about the specializations in construction seen in the later holosteans. *Cornuboniscus* (Fig. 76) is a proper palaeoniscoid in all regards, although, exceptionally, there was a distinct, if short, fleshy lobe in the pectoral fin—an unusual actinopterygian character, but also present in the modern *Polypterus.* Most members of the group are properly scale covered, but *Carboveles* (Fig. 76) has scales only at the base of the tail. Still more remarkable is *Tarrasius* (Fig. 76), in

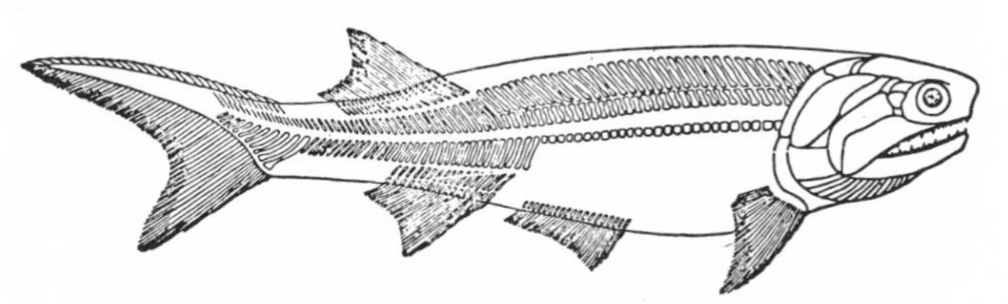

FIG. 74.—The internal skeleton of the primitive palaeoniscoid *Cheirolepis,* about 1/3 natural size. (After Smith Woodward.)

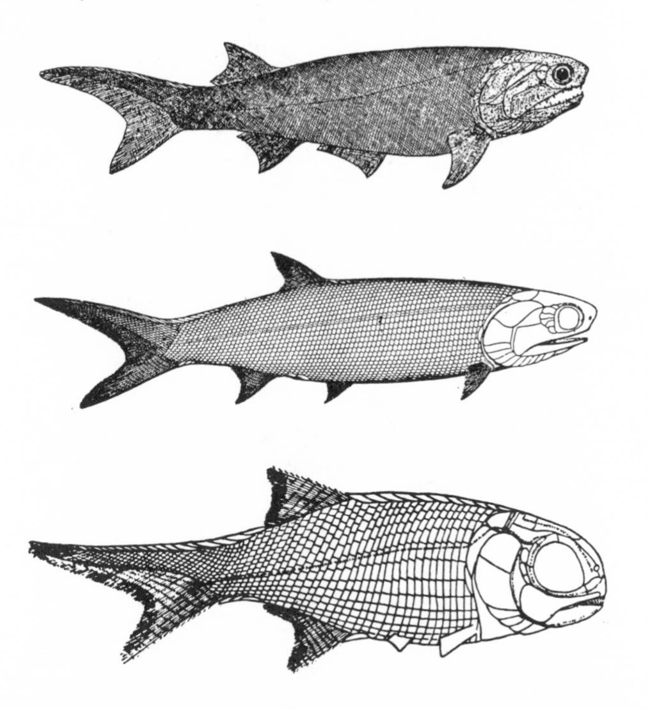

FIG. 75.—Palaeoniscoids. *Above, Cheirolepis,* a primitive Devonian genus, about 1/3 natural size. (After Traquair.) *Center, Palaeoniscum* of the Permian, about 1/3 natural size. (From Traquair.) *Below, Canobius* of the Carboniferous, about natural size. (From Moy-Thomas and Dyne.)

which scales are also reduced. This little fish is normal in build as far as the head region is concerned, but the body is elongate, and dorsal, caudal, and anal fins are fused into a continuous median fin fold analogous to that of modern lungfishes. Deep-bodied fishes, adapted for life in quiet waters, have appeared at various stages in actinopterygian history. Primitive examples are such Carboniferous genera as *Platysomus* and *Chirodus* (*Amphicentrum*) (Fig. 77).

The Carboniferous was the time of the greatest abundance and greatest variety for palaeoniscoids. In the Triassic the group dwindled rapidly in numbers and variety, as regards more primitive types, for they were being replaced by subholostean and holostean

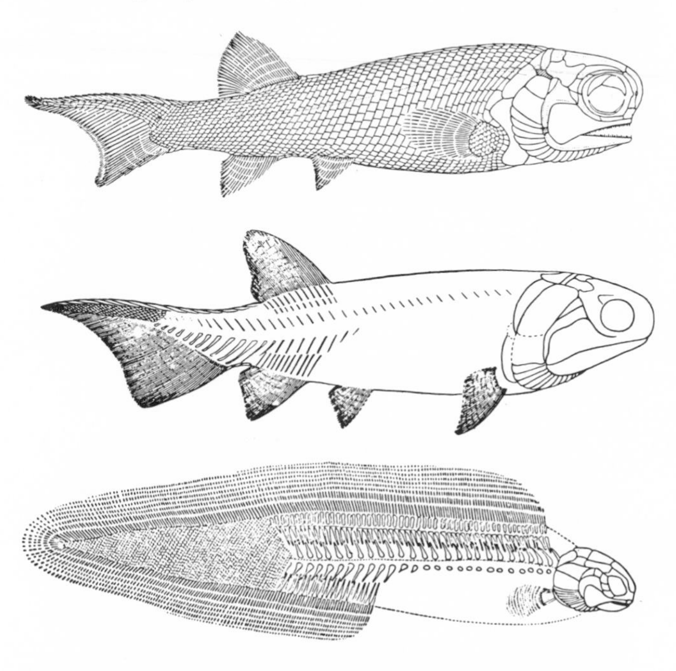

FIG. 76.—Aberrant Carboniferous palaeoniscoids. *Above, Cornuboniscus,* about 1.5 times size of original. (After Moy-Thomas.) *Center, Carboveles,* in which most of the squamation has been lost; about 3/5 size of original. (After White.) *Below, Tarrasius.* Most of the scales are lost, and the median fins are fused into a continuous structure, about 2/3 natural size. (After Moy-Thomas.)

forms evolved from them. A few rare palaeoniscoids of relatively conservative structure lingered on into the Jurassic and even the lower Cretaceous before this ancient actinopterygian order became extinct.

Surviving chondrosteans.—Although the palaeoniscoids have been long extinct, they are represented in modern times by two groups of specialized or degenerate descendants. *Polypterus* (together with an eel-like relative, *Calamoichthys*) is an inhabitant of fresh waters in tropical Africa. The body is covered by thick and shiny scales, and the internal skeleton is highly ossified. The median fins are peculiarly developed; the dorsal is represented by a series of small "sails" to which the genus owes its name; and the tail is a nearly symmetrical one. *Polypterus* is remarkable for the retention of a pair of ventral lungs; their presence may have been influential in the survival of this form in a region subject to seasonal drought.

Polypterus is also notable for the presence of a well-developed fleshy lobe in its pectoral fin. Largely for this reason the genus was long thought to be a crossopterygian. But in recent decades it has become increasingly apparent that *Polypterus* is, instead, a modified survivor of the palaeoniscoid group. The scales, when sectioned, show the ganoid structure of actinopterygians rather than the cosmoid plan found in the Crossopterygii; the skull pattern, while modified, shows no crossopterygian structures; many features of the soft anatomy are of a characteristic actinopterygian type. A somewhat lobate fin, we have noted, may occur

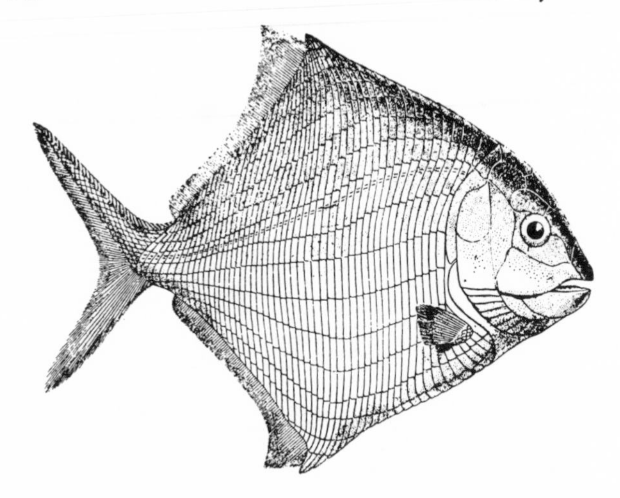

Fig. 77.—*Chirodus* (*Amphicentrum*), a deep-bodied Carboniferous palaeoniscoid, about 1/2 natural size. (From Traquair.)

in palaeoniscoids; the fin structure of *Polypterus*, with parallel radials, is closer to the ray-finned type than to that of the sarcopterygian fish. All in all, this archaic fish is surely a modified palaeoniscoid descendant. Connecting links, however, are as yet missing; we know almost nothing of the history of the genus.

More degenerate survivors of the older actinopterygians are the living sturgeons, such as *Acipenser*, and the paddlefishes (*Polyodon*, etc.) of the Mississippi and of Chinese rivers. In these forms ossification has been greatly reduced. In the sturgeon there is but little ossification in the internal skeleton, and the scales have been reduced to a few rows of large bony scutes; the paddlefish is still more degenerate, for there remain only a few small scales at the base of the tail. Head structures, too, have changed greatly, for these bottom scavengers have developed a long rostrum in front of weak jaws. On the other hand, the fins are persistently primitive, and there is a typical heterocercal tail such as is found today in no other bony fishes.

Sturgeon remains are reported from the Tertiary, and paddlefishes are represented by Eocene and Cretaceous forebears. The Pennsylvanian palaeoniscoid *Phanerorhynchus* has a sturgeon-like rostrum, but this may be merely parallelism. However, *Chondrosteus* (Fig. 78) of the Jurassic appears to show the transition from the palaeoniscoids; for this fish, while much better ossified than the modern genera, already showed considerable loss of both dermal and replacement bones, loss of much of the squamation, and rostral development.

These forms show the most extreme example of bone reduction to be found among the Osteichthyes. If their evolution had proceeded but a little further, they would have slumped to a purely cartilaginous state comparable to that of the sharks and, in default of fossil evidence, would probably have been considered as primitive cartilaginous fishes—the reverse of the true situation.

Subholosteans.—A very considerable proportion of the Triassic actinopterygian fauna consists of an array of families and genera of fishes which, in both superficial appearance and many anatomical features, resemble the palaeoniscoids and have been generally included in that group. They show, however, various progressive structural trends in which they approach the holostean stage. To such forms the name "subholostean" is commonly applied.

An obvious progressive feature is that the scale-covered lobe of the tail tends to be reduced from the full heterocercal condition; it may become a very slender structure or an abbreviated stub. The scales, although of ganoid type, frequently show the loss of the middle cosmoid layer. There is some reduction in number of the fin rays, as in holosteans (and, indeed, in some true palaeoniscoids). The jaws may be somewhat shortened, but almost never is there any indication of the specialized modifications described below as accompanying gape reduction in true holosteans.

The subholosteans, it is certain, are not a single natural group, but represent a number of types which have evolved from typical palaeoniscoids and independently acquired advanced characteristics. How to treat them in a formal classification is a worrisome

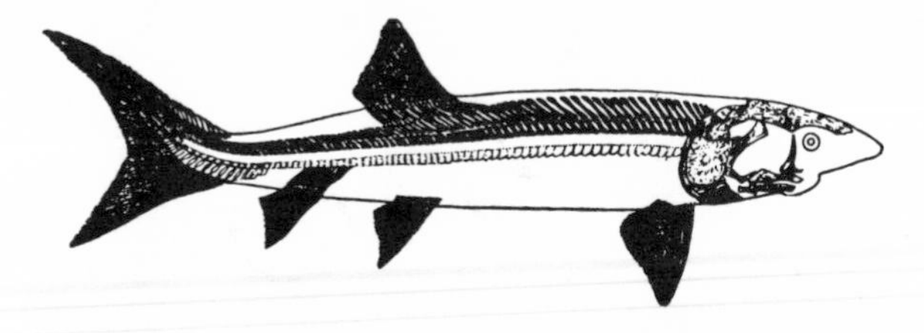

Fig. 78.—*Chondrosteus*, a Jurassic fish intermediate between palaeoniscoids and sturgeons, about 3 feet long. (From Smith Woodward.)

problem. What is done in the present volume is to expand our concept of the Palaeonisciformes to include these progressive types and consider them to constitute a series of advanced suborders of that order. Some of the subholostean groups (mainly Triassic) include only a few forms of little evolutionary significance, and need not concern us greatly. We may note, for example, deep-bodied forms of advanced nature, such as *Dorypterus* (Fig. 79) and *Bobasatrania*, which may have descended from such deep-bodied but basically primitive palaeoniscoids as *Platysomus*, or, alternatively, may have evolved independently. Again, equally specialized, but in quite a different manner, was *Saurichthys*

(Fig. 80), widespread in the Triassic and early Jurassic, a long-bodied, long-beaked, sharp-toothed predaceous fish, comparable to a modern pike in proportions and probable habits. *Saurichthys* is advanced in reduction of the heterocercal tail, specialized in sturgeon-like fashion in reduction of the scales to a few longitudinal rows of scutes, but basically primitive in other features. Despite the lack of connecting links, *Saurichthys* obviously represents an independent line of evolution from palaeoniscoid forebears.

A minor further advanced group is that of the Triassic Pholidopleuridae, which show a mixture of primitive and advanced features. In the change of the tail to an almost typical homocercal condition and the loss from the scales not only of the cosmine layer but also of the superficial covering of ganoine, these fishes are as advanced as the teleosts; however, the fins, except the caudal, show typical primitive palaeoniscoid structures, and the skull, except for a few specialized features, is of a palaeoniscoid type.

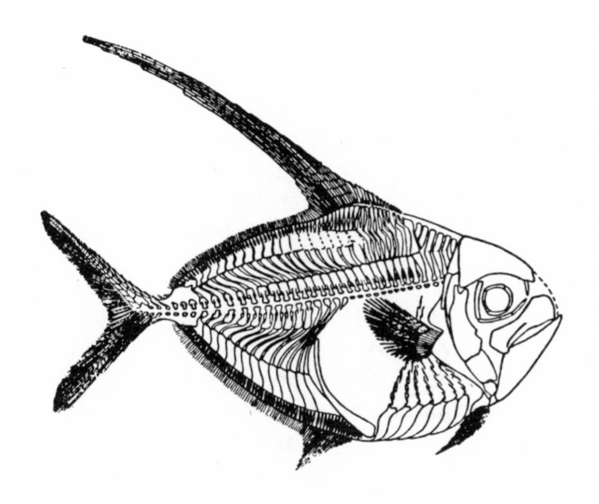

Fig. 79.—*Dorypterus*, an advanced deep-bodied palaeoniscoid, about 1/2 natural size. (From Westoll, modified after Gill.)

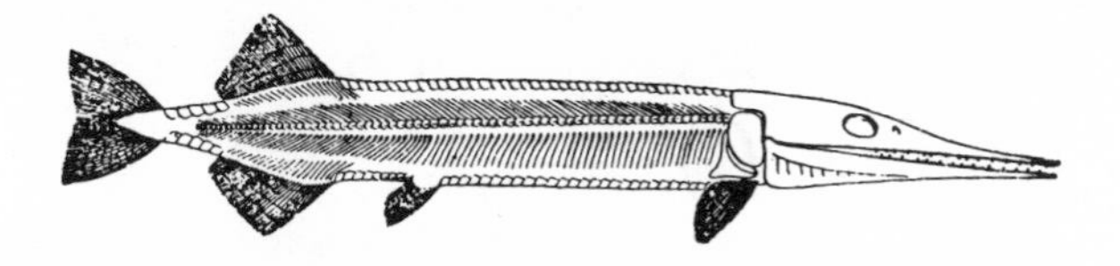

Fig. 80.—*Saurichthys* [*Belonorhynchus*], a predaceous Triassic and Lower Jurassic subholostean, about 20 inches long. (From Smith Woodward.)

More prominent in the Triassic faunal assemblages was a series of subholostean families which are here arranged in the suborders Perleidoidei and Redfieldoidei. These forms retained many palaeoniscoid features, but were advanced in a partial abbreviation of the heterocercal tail, a reduction in fin-ray numbers, and a shortening of the jaw. Typical are *Perleidus* and *Redfieldia* (Fig. 81). Members of these groups exhibit a considerable range in ecology and in structure and possible habits. For example, *Colobodus* was a large codfish-like form, several feet in length; *Habroichthys*, in contrast, was a tiny plankton-feeder barely an inch long. *Dollopterus* had huge pectoral fins like those of a modern "flying fish." *Cleithrolepis* (Fig. 82) demonstrates the potentiality of development of a deep-bodied form within this group.

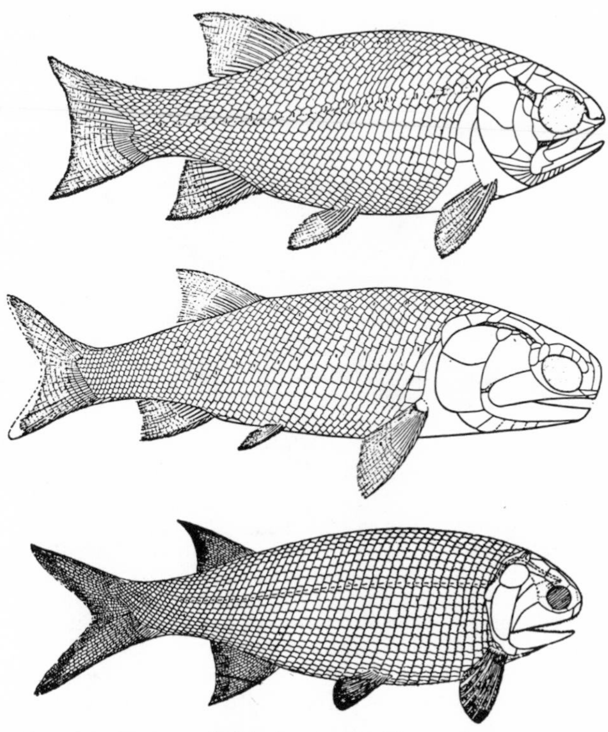

Fig. 81.—Triassic "subholosteans." *Above, Parasemionotus,* about 4 inches long; *center, Perleidus,* about 7 inches long; *below, Redfieldia* [*Catopterus*], about 5.5 inches long. (After Lehmann, Brough.)

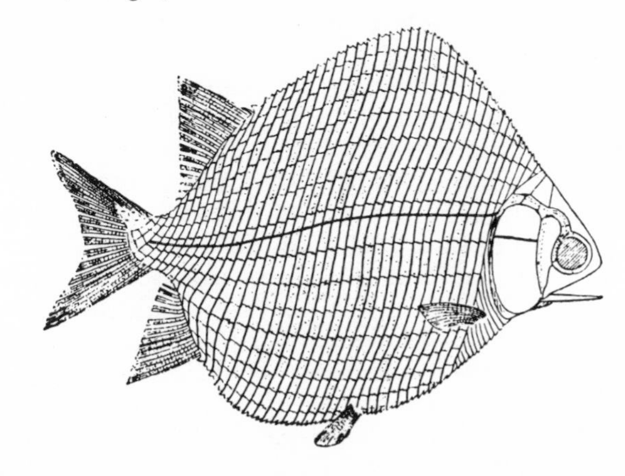

Fig. 82.—*Cleithrolepis*, a deep-bodied Triassic subholostean, about natural size. (From Brough.)

Holosteans.—We shall here group as the infraclass Holostei a large and varied assemblage of actinopterygians intermediate in structural features between Chondrostei and teleosts; forms particularly abundant in the middle Mesozoic and represented today only by two American freshwater fishes—the garpike, *Lepisos-*

teus, and *Amia*, the so-called freshwater dogfish or bowfin.

In the holostean grade of organization are seen, to begin with, the progressive features encountered in many subholosteans. The heterocercal tail is always much abbreviated; in the more progressive Holostei it gained a superficially symmetrical appearance, but there is not the specialized internal condition characteristic of the true homocercal tail. The scales have lost the middle layer of cosmine, but the superficial

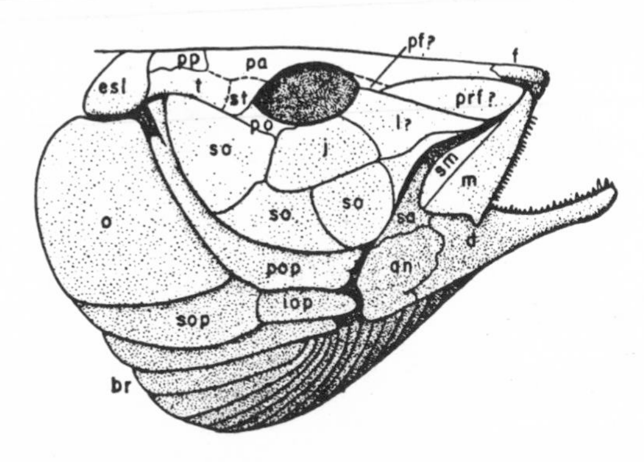

FIG. 83.—Side view of the skull of the Jurassic holostean, *Heterolepidotus*. In contrast with palaeoniscoids (Fig. 70*C*) the gape is shortened, the cheek covering is reduced, and the small maxilla (and new supermaxilla) separated from it; the lower jaw is supported by deeper structures (cf. the teleost condition, Fig. 90). *an*, Angular. For other abbreviations see Figure 69.

ganoine scale covering may persist to a variable degree. Bony fin-ray reduction in the fins is marked. These structures, originally fine, numerous, and segmented along their length, have been reduced to a relatively few stout and unjointed structures, which in the median fins usually correspond in number to the supporting structures of the internal skeleton. The single dorsal air bladder functions to some degree as a lung in living forms, although its chief use is hydrostatic. The spiracle has been lost, and in the shoulder girdle the clavicle has disappeared. Ossifications are often developed in the central region of the vertebral column.

It is in the jaw mechanism that distinctive holostean characters are best seen. In palaeoniscoids generally, the mouth opening was elongated, with the jaw joint far back; in subholosteans the jaw articulation frequently moved well forward, but in most cases the essential palaeoniscoid structural features remained unchanged. In the holosteans, however, the architecture of this region is much modified (Fig. 83; cf. Fig. 90). The maxilla is generally reduced in length, losing its original posterior expansion, and becomes freed posteriorly from the preoperculum (and internally from the ectopterygoid as well). Dorsally, too, the maxilla may be loosened from its original contact with the bones below the orbit; one or more special supramaxillary elements may develop along its upper edge. There is thus initiated a reduction of the maxilla which is to result, in many advanced teleosts, in its entire elimination from the functional margin of the mouth. The preoperculum is reduced in size to a crescent forming an anterior margin to the gill cover, and a new interopercular element farther ventrally helps fill the gap left in the throat armor by the forward retreat of the jaws. The orbit is still bordered posteriorly and ventrally by a variable ring of circumorbital elements, but the originally complete covering of the cheek has been lost; this region may be left more or less bare or be covered by a variable series of "suborbital" plates.

With the breakdown of the cheek region, the entire thrust of the lower jaw must be carried by the internal upper-jaw structures and the hyomandibular. The hyomandibular, the quadrate, and the epipterygoid are all much enlarged as jaw supports, and an extra bone—the symplectic—may unite the quadrate with the hyomandibular.

The braincase is known in but few holosteans. In the older holosteans, ossification was fairly complete, but, in contrast with typical palaeoniscoids and even most subholosteans, sutural distinctions between the numerous bones comprising the braincase are generally apparent.

While we are here considering the Holostei as a major unit group, it is far from certain that it is a truly natural assemblage. Most holosteans, as we shall see, can be reasonably interpreted as derived from a common ancestor; but some forms—notably semionotids and pycnodonts—may have evolved, independently of the "main line," from ancestral palaeoniscoids.

Semionotoids.—Within the Holostei may be distinguished a number of groups, which we shall here consider as orders. First to appear in time is the prominent order Semionotiformes, of which *Semionotus* of the Triassic and *Lepidotes* (Fig. 84) and *Heterolepidotus*

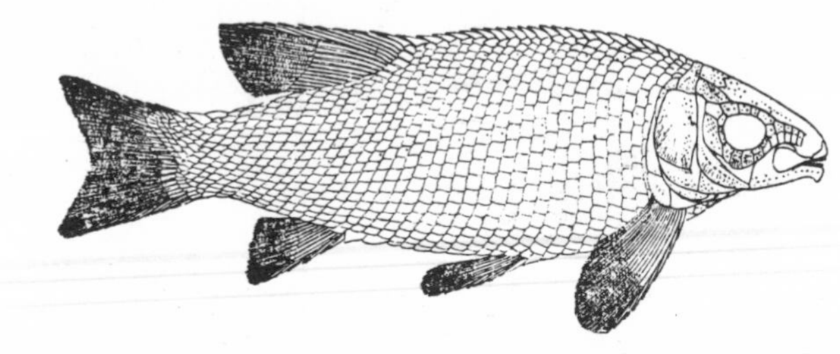

FIG. 84.—*Lepidotes minor*, a Jurassic semionotid; original about 1 foot long. (From Smith Woodward.)

of the Jurassic are typical representatives. These are rather heavily built fishes, and *Dapedium* (Fig. 85) is a deep-bodied form analogous to types already seen in earlier actinopterygians. The short mouth was armed with peglike or rounded teeth, indicating that some hard type of invertebrate material formed the food supply. The scales in this group are still thick and shiny structures, with a heavy layer of ganoine. The earliest semionotid appeared in the Upper Permian, giving rise to a strong probability that the family arose from some palaeoniscoid stock, independently of other holosteans. Abundant in mid-Mesozoic days, the semionotids appear to have failed in competition with the rapidly evolving teleosts in the Cretaceous, and none survived the end of that period.

The modern garpike (*Lepisosteus*) of American waters is usually bracketed with the Semionotidae. The gars, whose pedigree traces back only to the Eocene, have, like the semionotids, retained thick ganoid scales and exhibit basic similarities to that family in the internal construction of the skull. But they must have had a long independent history, for they are very different in adaptive features associated with predaceous habits. The teeth are sharp, and great elongation of the jaws is associated with marked modifications in dermal bone arrangement. The body, too, is markedly different —long and slim, with dorsal and anal fins far back, and an advanced, nearly homocercal tail construction.

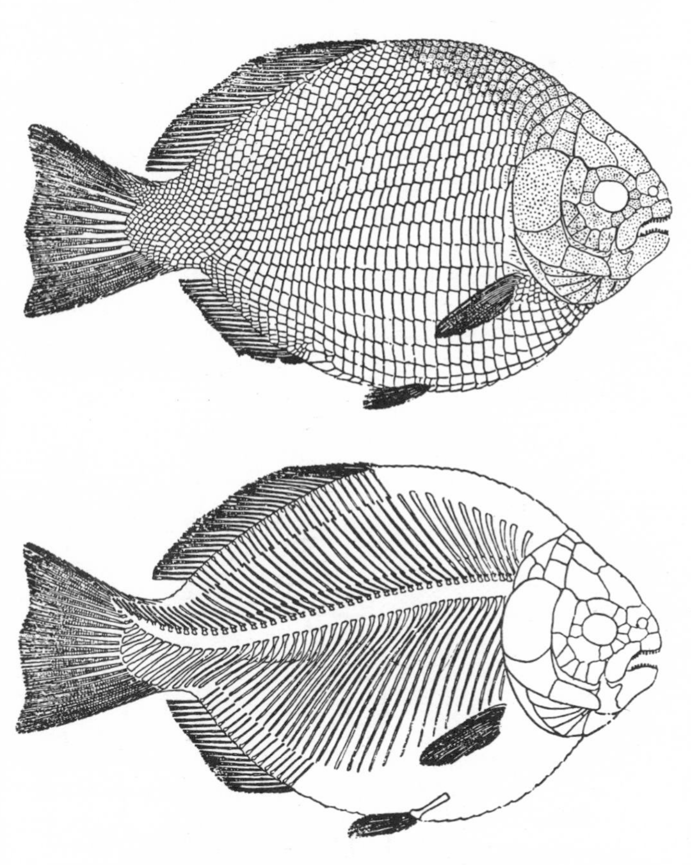

FIG. 85.—*Dapedium*, a deep-bodied Jurassic holostean (Semionotidae), about 14 inches long. (From Smith Woodward.)

There are no transitional forms known between semionotids and *Lepisosteus;* the gar body pattern was already fully established when the family first appeared in early Tertiary times.

Aspidorhynchus (Fig. 86) and *Belonostomus* of the late Jurassic and Cretaceous are holosteans which resemble the gars in the retention of thick ganoid scales and in body adaptations related to fast-swimming and presumably predaceous habits. However, they have very different cranial structures which make close relationship improbable, although these fishes, like the gars, are of unknown ancestry. The jaws are moderate-

ly elongate, but, in addition, there is a long, toothless rostrum, a structure which finds a parallel in other holosteans and teleosts.

Pycnodonts.—Still more remote from the typical holosteans in many of their structures are the Pycnodontidae, a family which appeared in the late Triassic, flourished greatly in the Jurassic and Cretaceous, and died out in the Eocene.

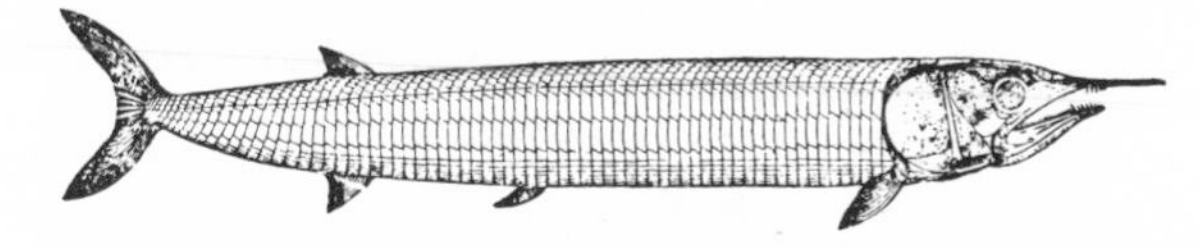

FIG. 86.—*Aspidorhynchus,* an aberrant predaceous Jurassic holostean, about 2 feet long. (From Assmann.)

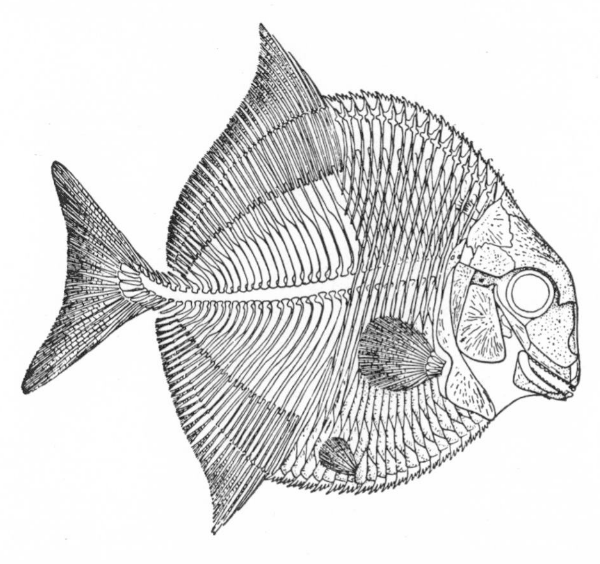

FIG. 87.—*Proscinetes* (*Microdon*), a Jurassic pycnodont, about 5 inches long. (From Smith Woodward.)

Included are such genera as *Pycnodus* and *Proscinetes* (*Microdon*) (Fig. 87). In these peculiar fishes the body was extremely deep and narrow, nearly circular in outline when seen in side view, with long dorsal and anal fins, and with the tail reduced to the abbreviated heterocercal condition characteristic of the pycnodonts' holostean contemporaries. The squamous covering of the body was lost or transformed into a latticework of jointed rods. There was a short, deep beak terminating in a tiny mouth filled with pebbly teeth, perhaps adapted for coral nibbling in the quiet reef waters which may have been a favorite environment.

The ancestry of the pycnodonts is far from certain. It has been suggested that they come from a long line of deep-bodied ancestors, beginning with such primitive, if deep-bodied, palaeoniscoids as *Platysomus*, and with such deep-bodied forms on the subholostean level as *Bobasatrania* as connecting links. However, there is no proof that this is the case; deep bodies seem to be readily assumed by actinopterygians of all sorts, and we may be dealing with parallelism. On the other hand, there is evidence suggesting, reasonably, that the pycnodonts are merely a further specialization of the semionotids, in which there was a strong trend toward a deep-body shape and in some of which there occurred a partial loss of scales, as in pycnodonts.

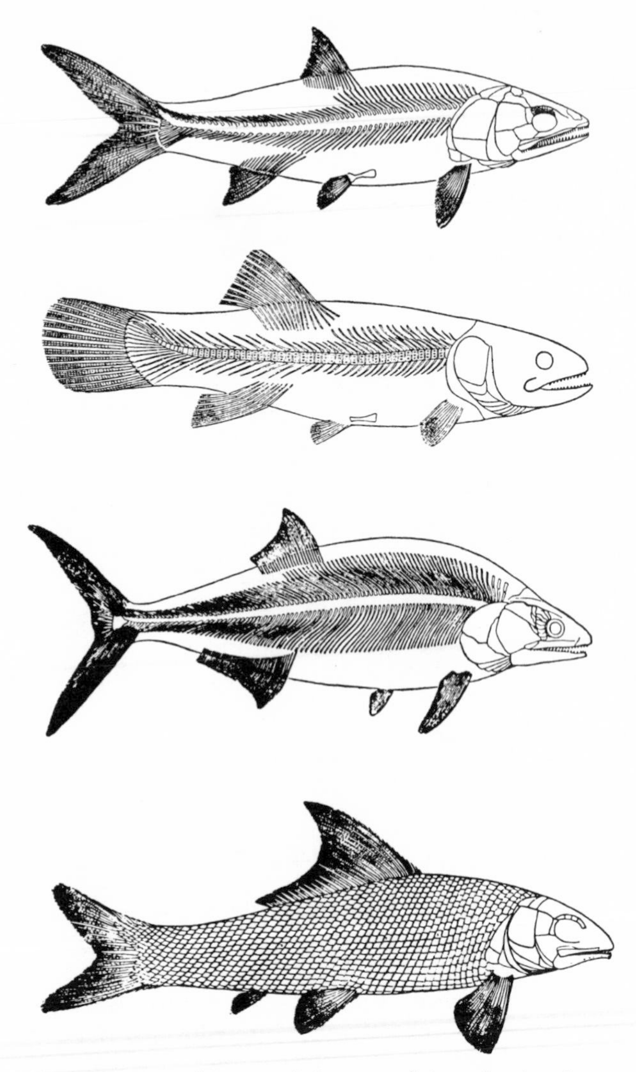

FIG. 88.—Typical Jurassic holosteans of the order Amiiformes. In order: *Caturus* (Caturidae) (this and the following much reduced in size); *Amiopsis* (Amiidae); *Hypsocormus* (Pachycormidae), about 1/20 natural size; *Ophiopsis* (Macrosemiidae), 1/3 natural size. (From Smith Woodward.)

Amiiformes (*Fig. 88*).—Apart from the problems concerning the semionotid and pycnodont groups, a major portion of the holosteans seems fairly certain to constitute a natural group, of which the living freshwater bowfin, *Amia,* is the living survivor, and which may be called the order Amiiformes. Structurally ancestral to the typical members of the group are the members of the flourishing Triassic family Parasemionotidae. These are, in general, still on the subholostean level of organization; reduction of the heterocercal caudal has begun (but not advanced far); the number of fin rays is much reduced; and the scales, although still thick and shiny, have lost the middle cosmine layer in holostean fashion. The jaws are not merely shortened, as in subholosteans generally, but show structural changes pointing definitely toward the true holostean condition.

Beyond the parasemionotids, the members of this order have reached the definitive holostean evolutionary stage. In these fishes we find typical holostean skull structures, with relatively few specializations. The body is generally of a conservative fusiform shape; the tail fin is greatly reduced but without any homocercal specialization; the scales are still covered in most cases with ganoine but are tending to become thin. A relatively primitive family, mainly late Triassic and Jurassic in age, is that containing *Caturus* and *Furo* [*Eugnathus*]. A small but distinct scaly lobe remains on the tail, and fulcral scales (lost in higher groups) persist; the scales, too, tend to be rather thicker here than in later families. The Macrosemiidae are contemporaries which tend to develop into somewhat elongate and delicate fishes with long dorsal fins. *Pachycormus* and *Hypsocormus* represent a group of mainly Jurassic fishes in which some members developed a heavy bony rostrum. In a Cretaceous genus (*Protosphyraena*) this rostrum became much elongated and comparable to that of the swordfishes among teleosts. In relation to

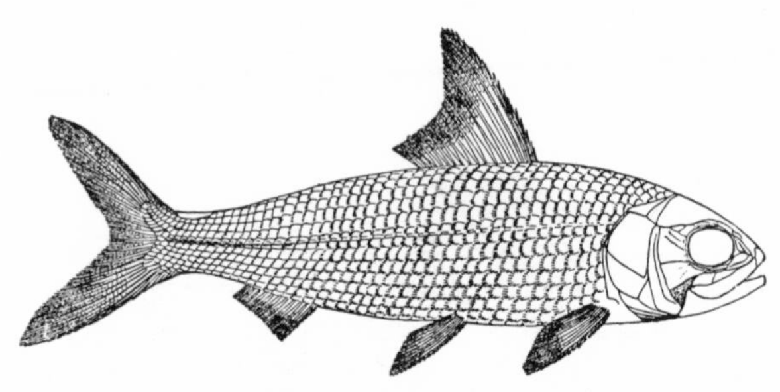

FIG. 89.—*Pholidophorus,* an advanced Jurassic holostean, about 9 inches long. (From Smith Woodward.)

rostral support the body has a rather deep, "hump-backed" shape. A fourth family is that of the Amiidae, represented by Jurassic and Cretaceous genera and by numerous remains of the living *Amia,* or its close relatives, in Tertiary freshwater deposits. This family has very probably descended from caturid ancestors, but, in addition to the diagnostic feature of a more or less elongate dorsal fin, shows such degenerate or advanced characters as an elongate body, reduction of braincase ossification, thin scales, and development of a nearly homocercal tail.

Pholidophoriformes.—A final group of holosteans is that of a series of families such as the Pholidophoridae (Fig. 89), Archaeomaenidae, and Oligopleuridae, which closely approach the teleosts in structural features and appear to be transitional to them. So close, indeed, is the resemblance that some writers include them among the teleosts; and, on the other hand, *Leptolepis* and its relatives, here considered as primitive teleosts, are often bracketed with these families as advanced holosteans. In these forms—small fishes which range from the late Triassic into the Cretaceous—the scales are quite thin, although covered with a film of ganoine, and the tail is close to the homocercal condition. Skull structure in, for example, the development of a supraoccipital element in the braincase, is becoming very teleost-like. The ancestors of part, if not all, of the Teleostei may well have been members of this advanced holostean group.

The ancestry of these little forms is none too certain. One would at first assume that they had arisen from generalized holosteans, such as the caturids. But since some pholidophorids were already present in the Triassic, these advanced types may also have arisen directly from "subholostean" ancestors among the parasemionotids.

Holostean history is, as we have seen, essentially a Mesozoic story. Except in the case of the semionotids, these forms are unknown until the Triassic. By the end of that period they had essentially replaced the subholosteans, and the known fish faunas of the Jurassic and Lower Cretaceous are composed almost exclusively of holostean genera. By Upper Cretaceous times, however, their descendants—the teleosts—had almost completely replaced them. A few pycnodonts remained in the Eocene; in the later Cenozoic only gars and *Amia* remain of this once important group.

Teleosts.—The teleosts are the successful fishes of today, the culmination of the series of phylogenetic stages seen in the more typical chondrosteans and holosteans. Primitive transitional types appeared before the end of the Triassic and were abundant in numbers, although not in variety, in the late Jurassic. In the Cretaceous a wide range of teleosts was evolved, and by Eocene times every major group of these fishes had appeared. In the Upper Cretaceous they far outnumbered the holosteans, and they have since been the dominant fishes of both fresh and salt waters.

The diagnostic features of the teleosts are merely the further expression of evolutionary tendencies already seen at work in the lower ray-finned fishes. The thin, deeply overlapping scales are so reduced that the shiny ganoid surface layer disappears, and the fulcral scales are lost. The internal skeleton is composed almost entirely of bone, including complete ossification of the vertebrae (in contrast to the condition in many holosteans). In the skull (Fig. 90) a supraoccipital bone, small or absent in lower actinopterygians, appears as a new and prominent structure which, in advanced groups, separates the parietals. The vomers are fused into a single median bar. In other groups of fishes, batteries of crushing teeth, when present, tend to form on the palatal bones above and on the inner surface of the lower jaw; in numerous teleosts in which such dentitions develop, the upper tooth battery tends to concentrate on the parasphenoid, while the opposing lower teeth are attached to the bones of the pharyngeal floor. The lower jaw has a reduced number of elements, only the bones termed dentary, angular, and articular persisting. With few exceptions, the median gular has disappeared.

The tail has become homocercal; superficially it is perfectly symmetrical, but internally there is still a trace of the uptilted end of the axial skeleton, and the fin rays are supported by enlarged haemal arches, known as hypural bones. Functionally, this tail is very similar to the diphycercal type attained in other fishes;

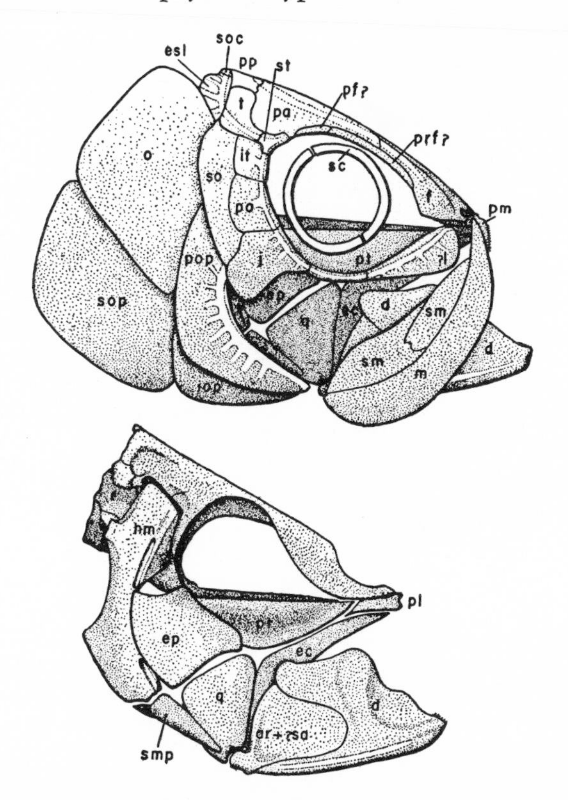

FIG. 90.—The skull of *Leptolepis*. *Upper*, Superficial lateral view of skull, jaw, and opercular apparatus. *Lower*, The dermal bones of the head removed to show the braincase and lower jaw and the support of the latter structure by the palatoquadrate and hyoid elements. *hm*, Hyomandibular. For other abbreviations see Figure 69. (After Rayner.)

but here the entire fin is developed from the lower lobe of the original heterocercal tail fin, whereas the diphycercal type is formed equally from upper and lower lobes. The primitive ray-finned forms, as we have noted, almost entirely lacked an upper lobe. The diphycercal type was thus difficult to attain, but in the homocercal tail the same functional end has been attained by other means.

Within this group there has been a tremendous amount of variation. Primitive teleosts possessed a normal fusiform body shape; but later members of the group have evolved in almost every conceivable direction—into the elongate eel type of body, into deep-bodied forms paralleling platysomids and pycnodonts, and into such weirdly shaped types as the sea horse. The scales may be lost entirely, as in typical catfish;

or the body may be armored, as in other catfish and sea horses, in bony plates, giving a superficial resemblance to some of the old ostracoderms. The fin rays may be stiffened into spines, as is the case in the higher members of the group. The median fins may coalesce; the pelvic fins in advanced groups tend to move forward and reach a position beneath the shoulder region or

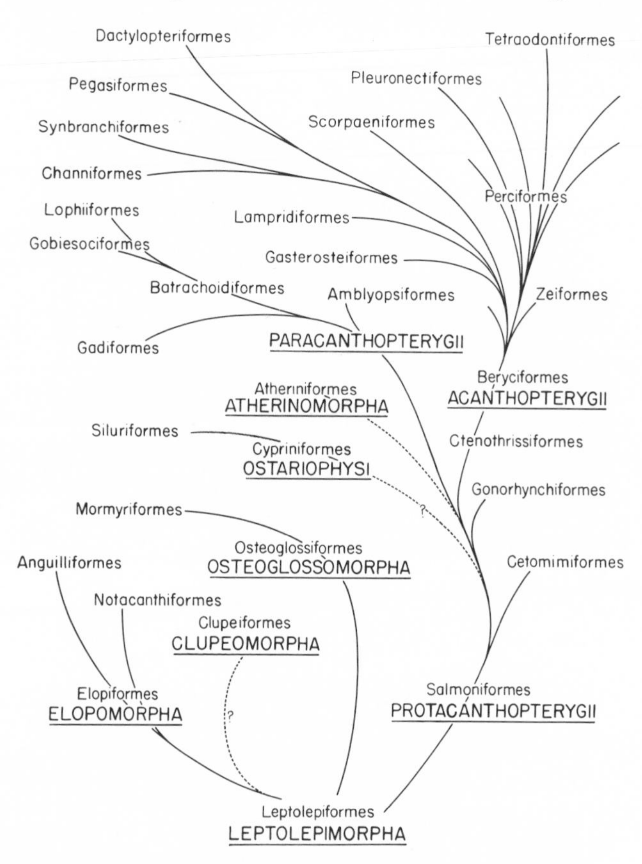

FIG. 91.—A diagrammatic "family tree" of the teleost orders

even, as an extreme, beneath the "chin." In the upper jaw the premaxilla tends to lengthen and exclude the maxilla from the margin of the jaws. The latter in advanced forms is so articulated that it aids in the protrusion of the mouth, pushing the premaxilla forward when the jaws are opened. The air bladder has lost its breathing function in most teleosts. In many marine forms its connection with the gut may be lost, the sac manufacturing gas within its own membranes; in still other marine fishes the bladder has vanished.

Modern teleosts carry in the liquid-filled sacs of their internal ears tiny calcareous concretions—otoliths, which have characteristic shapes depending upon the varied contours of the membranous sacs in which they are formed. These ear stones are not uncommon as tiny fossils in various Tertiary clays and give valuable evidence of the presence of forms unknown from ordinary skeletal evidence.

The vast majority of living teleosts are marine in habitat, and so apparently were most of the older known forms, thus suggesting that the group originated in the sea. Very probably this was the case, but it must be noted that freshwater fish-bearing deposits are rare in late Mesozoic rocks. A fraction of modern genera—notably members of the carp-catfish groups—dwell in inland waters, and a number (the salmon and eel are familiar examples) divide their time between the two habitats. At the other extreme, many teleosts (particularly, it would seem, "refugee" members of primitive groups) have become abyssal, deep-sea types.

The teleosts comprise not only some twenty or thirty thousand or more living species, but a vast array of fossil forms as well. How to arrange and classify this huge assemblage is a perplexing problem. Modern students of living fishes have tended to divide the group into as many as three dozen or so distinct orders and hundreds of families. No general agreement has ever been reached as to the evolutionary relationships of the various orders. We shall here consider them as arrayed, according to concepts currently being developed, in some nine superorders which may represent natural phylogenetic units (Fig. 91).

Leptolepomorpha.—Abundant remains of *Leptolepis* (Figs. 90, 92), a fish of herring-like appearance and of modest size, are found in Upper Jurassic marine deposits. This marks the beginning of the rise of the teleosts to a position of dominance which was to reduce the holosteans to a subordinate position in the Cretaceous and to near-extinction in Tertiary and Recent times. *Leptolepis*-like scales and fragmentary remains of this type have been found as far back as the middle Triassic, but it was not until late Jurassic times that these primitive teleosts became at all prominent. A dozen or more Jurassic and Cretaceous genera on

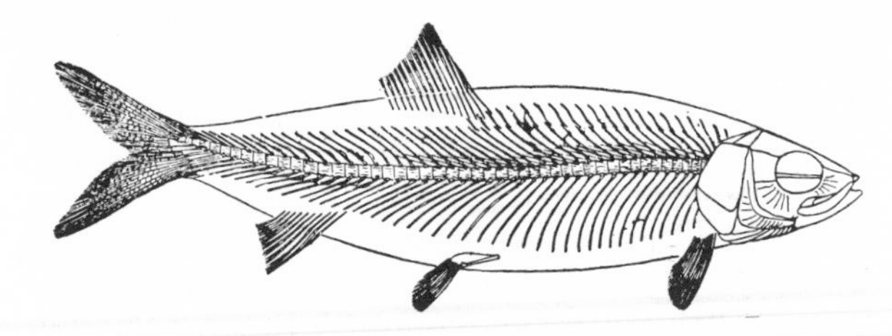

FIG. 92.—*Leptolepis dubius,* a primitive Jurassic teleost; original about 9 inches long. (From Smith Woodward.)

this primitive level of teleost organization are recognized. In nearly every regard *Leptolepis* and its relatives are quite generalized in structure, as befits potential ancestors of the highly varied teleosts of later times. They are so primitive in such features as the lack of expanded hypural bones beneath the homocercal tail and the retention of a trace of ganoine on the surface of the thin scales that many students of fishes prefer to class them as advanced holosteans, close to but just below the teleost level.

Elopomorpha.—By early Cretaceous times the evolution of teleosts was advancing rapidly. Before the period was far advanced, three, or perhaps four, stocks of varied if still primitive types of teleosts had differentiated. From these may have arisen all later teleost types. All the early representatives of these stocks had advanced over the *Leptolepis* level of organization in such features as the complete loss of ganoine from the scales and the presence of true hypural bones supporting the tail fin.

Most primitive in nature of these early differentiated groups was a series of forms directly ancestral to the living tarpons, *Elops* and *Megalops*, and to the ladyfishes, *Albula*. A considerable series of tarpon relatives, such as *Osmeroides* (Fig. 93), *Spaniodon, Pachyrhizodus,* and *Istieus,* was present in the Cretaceous seas, as were, less commonly, relatives of the ladyfishes. These forms, together with their descendants, may be considered as constituting the superorder Elopomorpha; the Cretaceous types mentioned plus later ones leading to the tarpons and ladyfishes constitute the order Elopiformes. Related to this group and descended from them are the members of the order Notacanthiformes, obscure modern deep-sea forms such as *Notacanthus* and *Halosaurus* which had Upper Cretaceous forerunners.

Successful as the elopomorphs were in the Cretaceous, typical members of the group have amounted to little in later geologic times. But there is an odd quirk to the story, for (unlikely as it would seem) the eels—the order Anguilliformes—are probably much-modified descendants of the tarpon group.

The eels are widely separated morphologically from other teleost orders. The body is very long and slender; the pelvic fins are absent in modern forms; the scales are rudimentary or absent; and the upper part of the opercular covering has been lost. These carnivorous fishes have a peculiar, specialized type of mouth structure; for example, the premaxillae appear to be absent and are replaced functionally by the vomers. The group as a whole is marine, although a few species spend their adult life in fresh waters. There are more than a score of families of living eels; with them are to be associated several deep-sea forms, such as *Saccopharynx*, the gulpers, with eel-like bodies but enormous mouths. These last forms are unknown as fossils. The fossil record of the eels is fragmentary. There are over a score of living families, of which only a fraction are known from fossils, and these are generally close to the living forms. The Cretaceous genera *Urenchelys* and *Anguillavus* show rather more primitive conditions in the median fins, and pelvic fins were still present in the latter.

There remains, nevertheless, a notable morphological gap between even primitive eels and any other teleost group—a gap which no known fossils even begin to close. We gain, however, a strong clue as to eel relationships from the developmental story. Eels are notable for the presence of a "leptocephalus" larva; the infant eel, in strong contrast to the adult shape, is a thin, translucent leaf-like object. A similar leptocephalus larva is found in the living members of the elopiform group—and in no other known fishes whatever. It thus seems probable, despite the absence of connecting links, that the eels originated from early members of the general group of tarpon-like teleosts, and hence can be included in the Elopomorpha.

Clupeomorpha.—Nearly as primitive as the members of the *Elops* group, but far more successful in modern times, is a great group of fishes of which the herring, *Clupea,* is typical, and of which there are numerous familiar relatives, such as the shad, sardines, and anchovies. The clupeoids lack a few technical features of primitive nature still present in the elopoids and the presumably primitive leptocephalus type of larva; as regards relations to other primitive teleost groups, they lack the adipose fin which is a trademark of the salmon and their relatives, and lack the peculiar weberian ossicles, a hearing device characteristic of the carp-catfish assemblage. Like the elopomorphs, the group was established early, for there are Lower as well as Upper Cretaceous herring-like forms and numerous Tertiary finds, belonging to fifty genera; *Diplomystus* (Fig. 94), widespread in the late Cretaceous and early Tertiary, is but one of many examples.

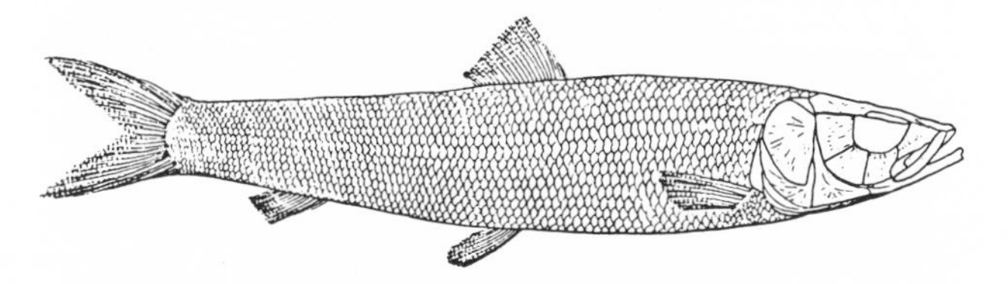

FIG. 93.—*Osmeroides* [*Holcolepis*], an Upper Cretaceous relative of the tarpons. Original about 2.5 feet long. (From Arambourg.)

The point of origin of the clupeomorphs is somewhat uncertain. In many regards they approach the tarpon group in structure. Possibly they were an early offshoot of the elopomorphs; but equally possibly they arose independently from the ancestral teleost stock.

Osteoglossomorpha.—Sharing dominance in the Cretaceous seas with the tarpon ancestors was a great series of fishes, many of large size, which are ranged in the families Ichthyodectidae, Saurocephalidae, and Thryptodontidae, and include a number of well-known genera such as *Ichthyodectes, Xiphactinus* [*Portheus*] (Fig. 94), and *Saurocephalus.* These fishes are similar in many ways to the ancestral elopomorphs (and are, in fact, often confused with them) but differ in various technical details.

These ancient types disappeared from the seas at the end of the Cretaceous. It is, however, quite probable that there descended from them a series of existing fishes which are the most primitive of freshwater teleosts. We have here ranged most of these modern forms, together with their possible marine Cretaceous ancestors, in the order Osteoglossiformes. Most of the living genera are tropical and are found in South America as well as Africa and southern Asia, and even Australia. *Osteoglossum,* for example, is a large river fish present in Brazil, in Egypt, and in the East Indies as well. Certain African members of this series of freshwater fishes, notably *Mormyrus,* the elephant-snouted fish, are so distinctive that they are sometimes placed in a separate, but related order. The curious distribution of these fishes in the tropics of both hemispheres (paralleled, we shall see, by the characins) has been used as an argument for the existence of a transatlantic land bridge in Cretaceous or early Tertiary times. But while little is known of the fossil history of these freshwater fishes, the presence in the Eocene of North America of *Phareodus,* a typical osteoglossoid, makes it reasonable to believe that the modern discontinuous distribution is due to the retreat to the modern tropics of a group once present in northern areas.

Protacanthopterygii.—The teleost asemblages so far considered have, in the main, had successful careers as representatives of more primitive levels of teleost evolution, but have either remained as relatively primitive forms or, in the case of the eels, have become specialized but not truly progressive. With the present superorder we deal with a major group which, although basically primitive in nature, appears to represent the main evolutionary line leading toward more advanced forms. The basal primitive stock is that of the Salmoniformes, of which the familiar representatives are the salmon and trout of northern temperate waters, salt and fresh (the group is represented in southern temperate-zone waters as well). A key character (if a seemingly minor one) is the presence of a fatty adipose fin back of the normal dorsal. Salmon-like fishes are found in the Upper Cretaceous, but, although primitive, there are no indications of close relationship with either ancestral osteoglossomorphs or elopomorphs, and their line of descent from teleost ancestors is somewhat obscure. Included in the salmon order are the stomiatioids—voracious, large-mouthed deep-sea fishes—whose ancestry likewise traces to the Upper Cretaceous. Related to the salmon are the "Haplomi," a small group of northern freshwater fishes of which the one familiar member is the long-jawed predaceous pike, *Esox,* which appears in fossil form in the early Tertiary.

Progressive members of the salmoniform order are the Myctophoidei (Iniomi). They include a large and varied series of deep-sea fishes, of which the lantern fishes, such as *Myctophum* [*Scopelus*], are typical (Fig. 98). The popular name refers to the fact that luminescent organs, sometimes arranged along the body like a series of portholes, are generally developed in Tertiary and living representatives. The myctophids are advanced toward the spiny-finned teleosts in such features as the exclusion of the maxilla from the mouth gape, but they are clearly of salmonid descent; even the trademark of the adipose fin is present. The group is an ancient one, for myctophoids were already abundant in the Upper Cretaceous, including such genera as *Enchodus, Benthesikyme* [*Leptotrachelus*], and *Eurypholis* (Fig. 95).

A small but probably important extinct order is that including *Ctenothrissa* (Fig. 96) and a few allies of the marine Cretaceous. The myctophoids are suggestive of relationship to the ancestry of the higher, spiny-finned fishes; the Ctenothrissiformes may be directly ancestral. *Ctenothrissa* is a small, rather deep-bodied fish, in which, as in spiny-finned forms, the pec-

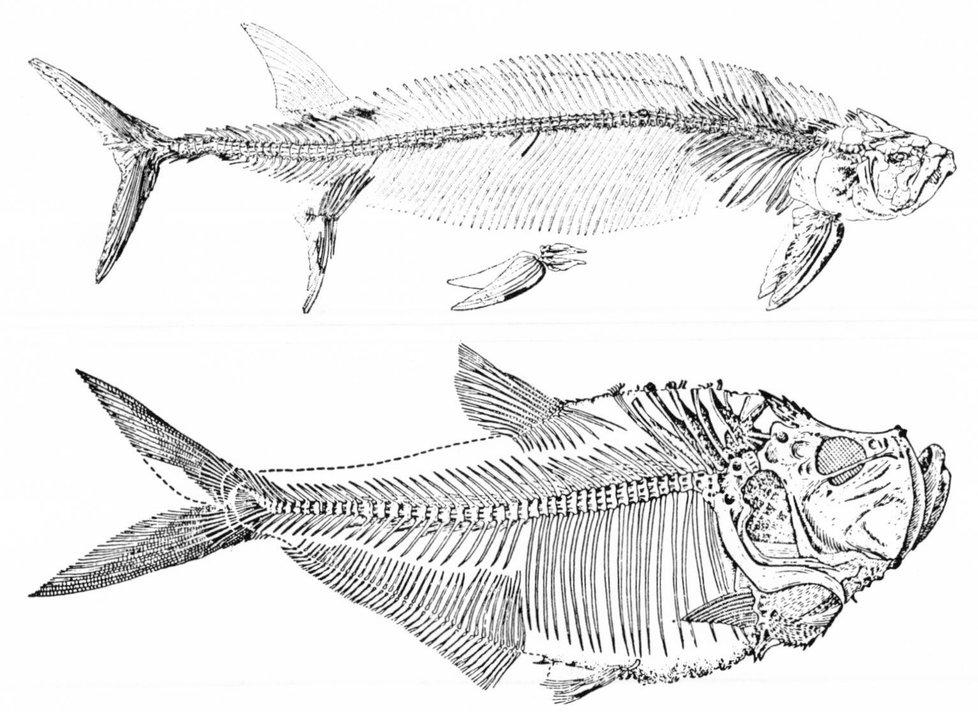

FIG. 94.—*Above, Xiphactinus* [*Portheus*], a large Upper Cretaceous teleost; original about 12 feet long. (From Osborn.) *Below,* The Eocene clupeid *Diplomystus,* about 1/5 natural size. (From Jordan.)

toral fins have moved far up the sides of the shoulder region, and the pelvic fins have migrated forward to a position between them. Again, the maxilla is here excluded from the mouth margin. Only further modification of mouth parts and spine development is needed to turn the descendants of these little fishes into primitive actinopterygians.

Of other Protacanthopterygii, the Cetomimiformes include various aberrant deep-sea fishes which may be myctophoid derivatives; we know nothing of them as fossils. More interesting because of their evolutionary implications are the Gonorhynchiformes, of which *Gonorhynchus* and *Chanos,* the milk fishes, both from the Indian Ocean are representative; there are a few African freshwater forms as well. The group, although unimportant today, has a long history, the gonorhynchids going back to the Upper Cretaceous, while the chanids even have several Lower Cretaceous representatives. They are surely related to salmonoids and myctophoids, but in their cervical and cranial regions show conditions suggestive of those to be expected in the ancestors of the Ostariophysi, now to be considered.

Ostariophysi.—In this superorder we come to the first of four major teleost groups which include all further teleosts and which all seem to have been derived from

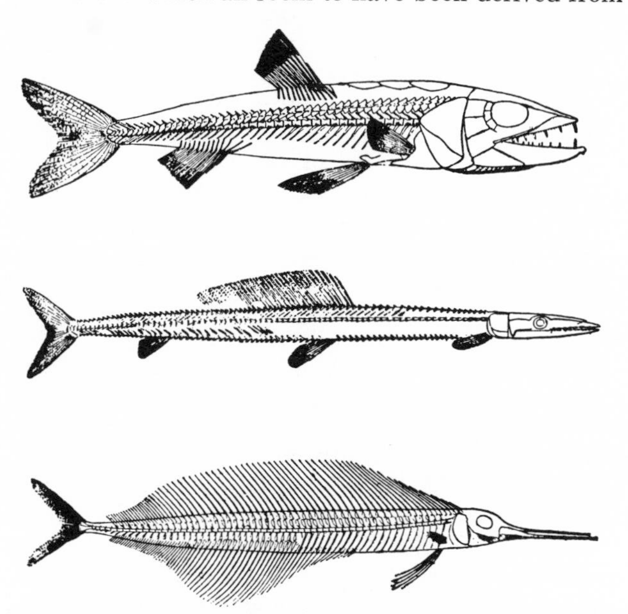

FIG. 95.—Teleosts. *Above, Eurypholis,* a Cretaceous myctophoid, about half natural size. *Center, Benthesikyme* [*Leptotrachelus*], a long-bodied Cretaceous myctophoid, about 1/4 natural size. *Below, Palaeorhynchus,* an Eocene relative of the mackerels (scombroids), about 1/6 natural size. (From Smith Woodward.)

protacanthopterygian ancestors. The members of this superorder constitute the greater part of the freshwater fish fauna of modern times. The more primitive members of the group constitute the order Cypriniformes. Except for eel-like *Gymnotus* of South America, all of the numerous members of the order can be arrayed in two suborders. The Characoidei, the characins, are small fishes extremely numerous in the streams and lakes of Africa and South America. The suborder Cypri-

noidei includes a vast number of freshwater fishes. The name derives from the carp, *Cyprinus,* but most members of this order are small minnows, various of which are known as shiners, dace, roach, and so on; there are perhaps two thousand or so species, mostly inhabiting northern temperate regions. More specialized are the cat fishes, the order Siluriformes. Strikingly different

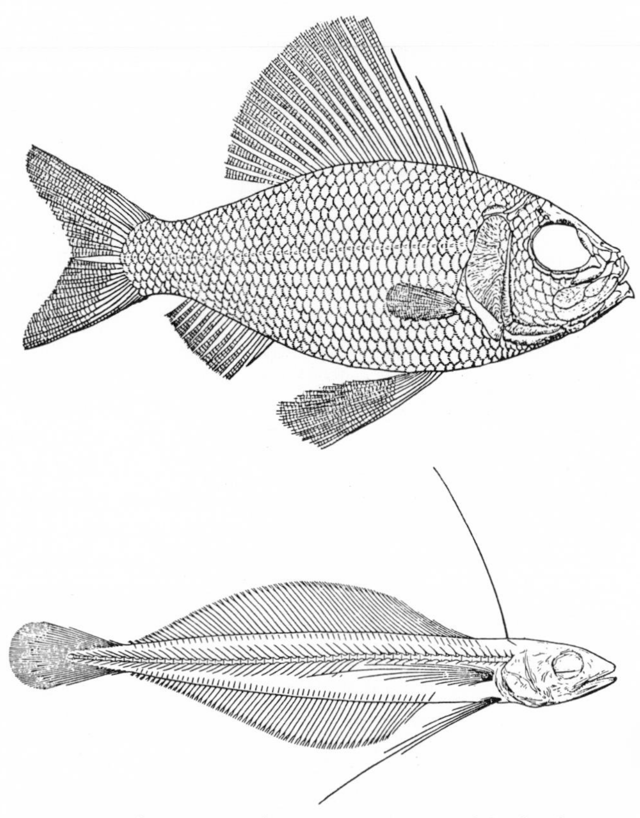

FIG. 96.—*Above, Ctenothrissa,* a Cretaceous fish closely approaching the spiny-finned teleosts in structure, about 1/4 natural size. (From Patterson.) *Below, Bregmacerina,* a Miocene codfish relative, about natural size. (From Danelchenko.)

in appearance from characins or cyprinoids, they are very abundant in all tropical regions, mainly in fresh waters but with a few venturing into coastal seas. Despite the variations in superficial characters in the members of these ostariophysan groups, they are clearly related and clearly distinguished from all other fishes by a peculiar internal set of structures termed the weberian ossicles. Hearing in these fishes is well developed. Parts of the anterior vertebrae have been transformed into a chain of small ossicles; these transmit vibrations picked up in the air bladder forward to the liquids and sensory endings in the internal ear. Apart from this one specialization, the Ostariophysi are relatively primitive members of the teleost assemblage. Their pedigree, however, has been difficult to trace and the fossil record is poor; a few Eocene forms are known, but most recorded finds are from the late Tertiary and Pleistocene.

Atherinomorpha.—In this superorder and its only component order, the Atheriniformes, are included a large number of small living fishes of varied appearance and habits. Among types placed in the order are the "flying fishes," such as *Exocoetus* and *Cypsilurus,* which can plane over the water for some distance, using the large pectoral fins; and their close relatives, the half-beaks such as *Hemiramphus,* in which the lower jaw protrudes forward like a spear; the gar fishes, *Belone,* which look like miniature marine garpike in that both jaws are elongated; the silversides, small coastal and lake fishes such as *Atherina;* and the numerous little freshwater and coastal minnow-like killifishes, such as *Cyprinodon* and *Fundulus.* Typical members of the order are small slender fishes, persistently primitive in many characters (as for example the position of the pelvic fins), but advanced in others, such as exclusion of the maxilla from the mouth gape and a tendency for fin-spine development in various members of the order. The Atheriniformes appear to represent a series of progressive side branches from the main line of preacanthopterygian evolution. There are scattered Tertiary records of the main components of the order and very few Cretaceous forms, but the group history is not well known, although we may believe them to have branched off from the protacanthopterygians—perhaps from the myctophoids.

Paracanthopterygii.—In this final superordinal group before reaching the acanthopterygians we here place a number of orders, mainly of specialized character, which have advanced toward the level of the true spiny-finned group, but (as the name suggests) have paralleled the acanthopterygians in advancing from the salmoniform level rather than being in any way ancestral. Among these parallel character developments may be cited the forward movement of the pelvic fins and a trend toward the development of fin spines and modifications of mouth parts. Most of the orders included in the Paracanthopterygii have fossil representatives dating back to the Eocene, but none are known from the Cretaceous; the group appears to have evolved at a relatively late time, presumably from myctophoids.

Most primitive members of the superorder are a few North American freshwater fishes, such as *Amblyopsis* and *Percopsis,* the pirate perch, forming the Amblyopsiformes, dating from the Eocene. These small minnow-like forms have departed relatively little from the general salmoniform type except that there is a tendency for development of spines in the dorsal fin, as well as for forward movement of the pelvic fins. A further minor component of the superorder is that of the Gobiesociformes, not positively identified in the fossil record, which consists of the clingfishes, *Gobiesox,* and a few other tiny saltwater fishes. There are no fin spines, but the pelvic fins have moved forward and form between them a sucking disc by which they cling to rocks in the tidepools where they dwell.

Important today and common as fossils throughout the Tertiary, and with few late Cretaceous representatives, are the Gadiformes (Anacanthini), including such forms as the cod (*Gadus*) and haddock (*Melanogrammus*) (Fig. 96). The body is typically rather long and tapering posteriorly. Dorsal and anal fins are elongate or may be divided into two or three parts, but, as the name Anacanthini indicates, there is but little development of fin spines.

Constituting an order of their own, but almost unknown in the fossil record are the toadfishes, the Batrachoidiformes (Haplodoci), such as *Batrachoides* and *Opsanus.* Here, as in the cods, there is a somewhat long and tapering body, long dorsal and anal fins, and forwardly moved pelvic fins; there are, however, a large head and strong teeth in these very predaceous fishes and a short spiny anterior portion of the dorsal fin.

The Lophiiformes (Pediculati), the anglers, are perhaps the most grotesque of all teleosts. In the common angler or goose fish, *Lophius,* the head region is enormous, the gape broad, so that large prey can be swallowed entire. The most distinctive feature is that the first dorsal fin spine, overlying the head, carries a dangling tassel which acts as a lure to prospective victims. The order includes more than a dozen families, nearly all of which are deep-sea fishes; two Eocene fossil anglers are known, but, as might be suspected, there is no fossil record of the deep-sea types; and although it is probable that they are related to the toadfish, there is no fossil proof of such relationship.

If we survey the various orders which we have, above, included in the Paracanthopterygii, it is obvious that while all show advanced characters paralleling the spiny-rayed teleosts in one feature or another, they do not show a common evolutionary trend, but a wide variety of patterns. Surely all of them were derived (as were the spiny-finned forms themselves) from the protacanthopterygian superorder, but there is no guarantee that they are monophyletic; quite possibly they represent a number of discrete groups which have evolved in parallel fashion from ancestors of salmonoid or myctophoid nature.

Spiny-finned teleosts.—The teleost groups so far considered are members of a radiation which had its origins well back in the Mesozoic. Now to be considered is a vast assemblage of advanced forms, representing a major upward step in teleost history, which had its beginnings only late in the Cretaceous. This assemblage is that of the spiny-finned teleosts, the superorder Acanthopterygii, which includes a great majority of modern marine fishes as well as a modest number of freshwater forms. Most of the spiny-finned teleosts are (as we shall see) included in a very comprehensive order Perciformes, but various primitive or aberrant members of the group are given independent ordinal status.

An obvious feature of typical acanthopterygians, and one to which this term refers, is the development of stiff spines—perhaps primarily for defense—replacing part of the softer fin spines present in lower teleost orders. This trend is most obvious and common in the dorsal fin, which typically consists of two parts, the anterior portion stiffened by a limited number of stout spines and often distinctly separated from the soft-rayed posterior region of the fin. Spines also develop in the anal fin. Again, the scales—usually smooth with rounded posterior edges in lower teleost orders—are here generally of the ctenoid (comblike) type, with tiny spinelets developed on their surfaces. Scales are confined to the body in lower groups; here the dermal bones tend to sink beneath the skin and scales grow forward over the posterior part of the skull in many cases. In typical spiny-finned fishes there has been a major change in mouth structure (foreshadowed in some lower groups, Fig. 97). The premaxilla is enlarged and becomes the sole marginal element in the upper jaw; the maxilla, toothless, is excluded, and functions merely as a lever to push the premaxilla forward to a protrusive position of the mouth characteristic of most actinopterygians. With mouth protrusion, the cavity of the mouth and pharynx is expanded, and food materials can be sucked in. Teeth tend to be lost from the mouth cavity proper, but generally are developed powerfully farther back, on the branchial elements lining the pharynx. But more important, it would seem, in the evolution and success of acanthopterygians than spine development or mouth changes was a radical change in methods of locomotion. In most lower teleosts steering is performed mainly by body undulation, the paired fins being little more than horizontal planes which prevent pitching. The situation is changed in typical higher teleosts. The body is generally much shortened, with a sharp decrease in the number of vertebrae, narrow from side to side, and typically somewhat deep. Greater reliance in propulsion is put on the tail, and reliance for steering (and braking) placed in the pectoral fins. These fins are almost always well developed, but have moved, for greater efficiency, high up the side of the body. As an aid toward proper balance, the relatively unimportant pelvic fins move forward, generally to a position below the pectorals, with the pelvic girdle attached to the cleithrum.

Primitive spiny-rayed forms.—As was seen earlier, *Ctenothrissa* and its allies of the Upper Cretaceous are small protacanthopterygian teleosts, which are still primitive in many ways, but have attained the basal body pattern of the acanthopterygians. While, as said above, the central stock of spiny-finned forms is that of the Perciformes, certain more primitive forms are present in the seas today, and had already developed and become moderately abundant before the end of the Cretaceous. These are the members of the Beryciformes. Both the living members of the group—the little squirrel fishes and soldier fishes—and their Upper Cretaceous forebears, such as *Hoplopteryx* (Fig. 98) and *Berycopsis* (Fig. 97), are moderately deep bodied, and exhibit almost all the major characters expected in generalized ancestors of the acanthopterygians. They differ, however, from the typical Tertiary forms in a few technical features, such as the retention of an orbitosphenoid bone in the braincase, lost in late acanthopterygians, and the presence of eighteen or nineteen rays in the tail fin, whereas more advanced forms never have more than seventeen. As regards the main body of the beryciform group, we are here surely dealing with acanthopterygian ancestors. An early berycoid side branch is that of the order Zeiformes, represented by a few living tropical shore fishes, such as the "John Dory," and by a few Tertiary fossils.

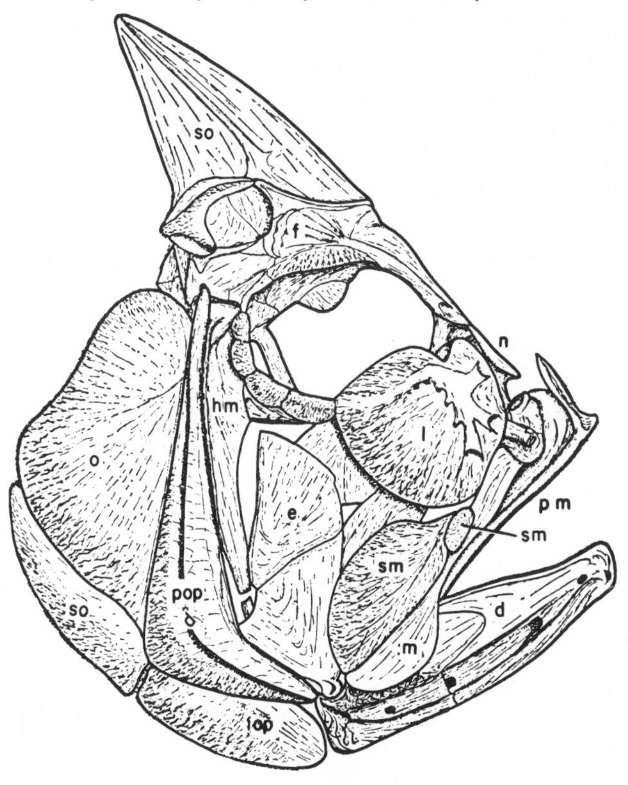

Fig. 97.—Side view of the head of the Cretaceous *Berycopsis*, which, with maxilla excluded from the jaw gape and the premaxilla enlarged, illustrates the skull pattern of primitive spiny-finned teleosts. Abbreviations as in Figure 69. (From Patterson.)

Although the main direction of acanthopterygian evolution lay in upward progress toward the Perciformes, there appear to have diverged, at an early, berycoid stage, a number of odd types, mainly deep-sea forms, which may be briefly mentioned. Some are often included, with some doubt, within the confines of the Beryciformes; others are regarded as constituting a distinct order Lampridiformes (Allotriognathi). Best known of the Lampridiformes are *Lampris,* the opak or moonfish, an exceedingly large, scaleless, and rare high-seas fish, and *Lophotus,* the likewise naked deep-sea bananafish. The two genera mentioned are known from the Tertiary, but otherwise we are almost completely ignorant of the history of these aberrant forms.

The Gasterosteiformes (including the Thoracostei) are small fishes of which the sticklebacks (*Gasteros-*

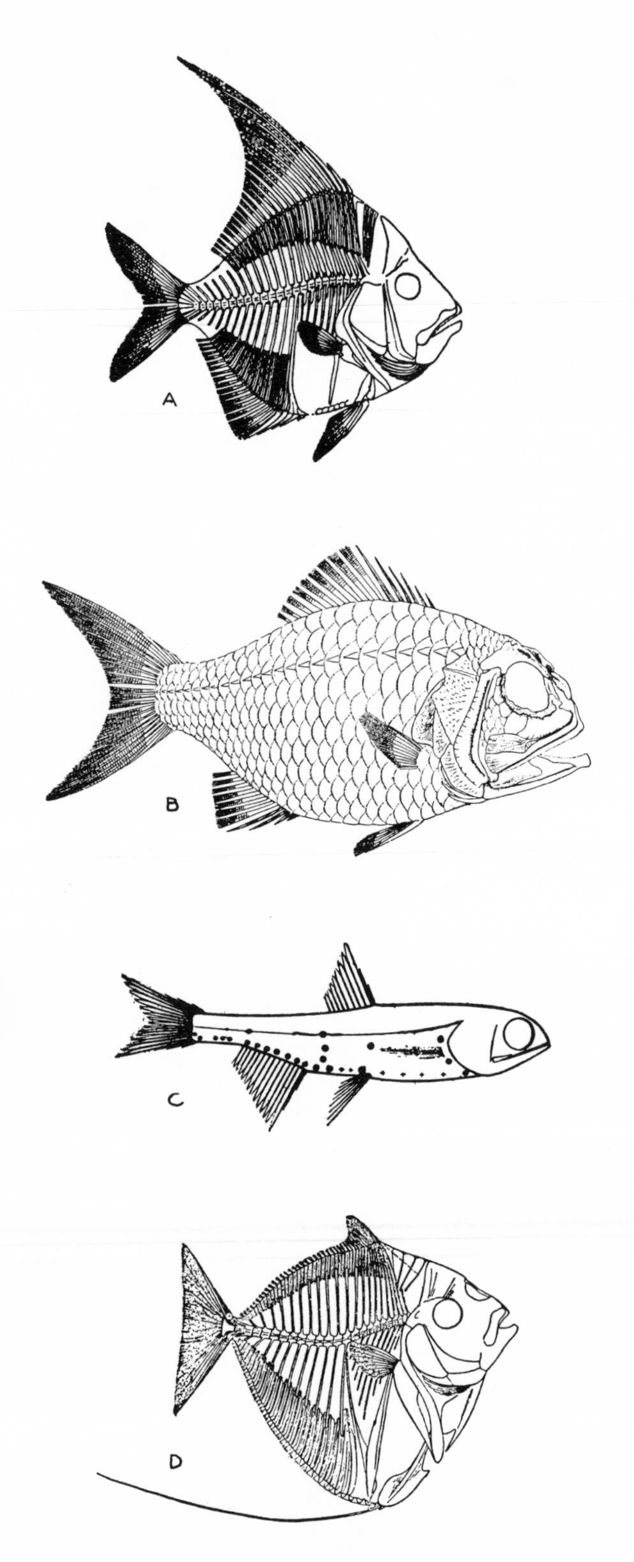

Fig. 98.—Teleosts. *A*, *Aipichthys*, about 1/2 natural size, and *B*, *Hoplopteryx*, about 1/4 natural size; both are primitive spiny-finned teleosts (order Beryciformes). *C*, *Myctophum*, a late Tertiary to Recent myctophoid, about natural size; the spots are remains of light organs. *D*, A deep-bodied percomorph (*Mene*) from the Eocene, about 1/6 natural size. (*Myctophum* after Arambourg; others from Smith Woodward.)

teus) are relatively primitive and the pipefish (*Syngnathus*) and sea horse are more advanced types. They appear to represent an early but highly specialized side branch of the acanthopterygian group. The body is elongate, typically slender, and encased in bony armor; the small mouth lies at the end of a tubular snout. Presumably because the armor makes for good preservation, fossil sticklebacks and pipefishes are moderately abundant in the Tertiary.

Almost unknown as fossils are several minor groups, all of isolated position, which may have likewise diverged from the spiny-finned stock at an early stage. These are the Channiformes, including only the "snakehead," *Channa* [*Ophiocephalus*], a small coastal fish which can long survive out of water through the use

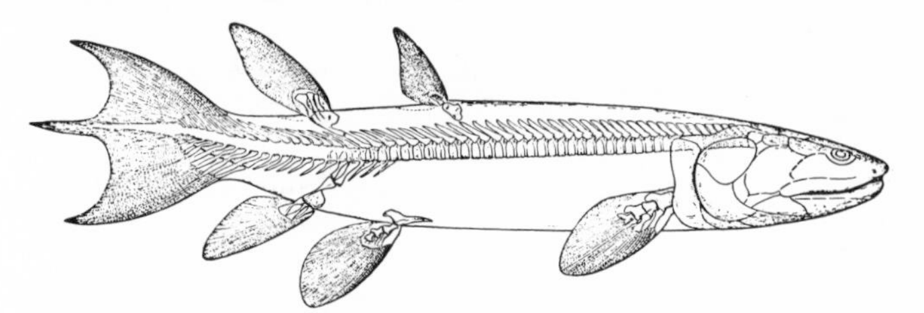

Fig. 99.—The skeleton of the Upper Devonian rhipidistian crossopterygian *Eusthenopteron;* average specimens 1 to 2 feet long. (From Gregory and Raven.)

of lunglike pockets extending back from the gill cavities; the Synbranchiformes, small coastal fishes from southern Asia and eastern Africa of rather eel-like appearance; the Dactylopteriformes, with large pectoral fins giving them some power of "flight" in a fashion paralleling the exocoetids; and Pegasiformes, slender little armored fishes of the tropical Indo-Pacific.

Striking in appearance are the mail-cheeked fishes, the Scorpaeniformes (Scleroparei), including a considerable variety of small marine forms, such as the scorpion fishes (*Scorpaena*), sculpins (Cottidae), and gurnards and sea robins (Triglidae). The pectoral fins tend to be much enlarged. The group name of "mail-cheeked fishes" is due to specialized cranial structures. The head may be secondarily armored with dermal bone external to the normal cranial elements, and projecting spines are often present on the preopercular. A characteristic feature is that the suborbital bones are much enlarged and brace the face against the preopercular. There are a fair number of Tertiary finds of members of this group, but these do not show any trend toward connecting the mail-cheeked fishes with other orders. Possibly they may have originated from advanced Paracanthopterygii; more probably, they represent still another side branch from the base of the acanthopterygian stock.

Perciform fishes.—Except for a few doubtful cases, not a single advanced spiny-finned form is known from any Cretaceous deposit. But, starting from the beryciform stock, an immense radiation of progressive actinopterygians began at the dawn of the Tertiary. This radiation must have been of a rapid, explosive nature, for we find in early stages of the Tertiary not only more generalized types, but a host of forms which had already achieved a variety of specializations. By the end of the Eocene we find representatives of almost every major type of spiny-finned teleost, including identifiable members of sixty or so distinct families.

The great majority of these spiny-rayed fishes are currently regarded as forming a vast order—the Perciformes. The boundaries of the order are not well defined; just how far a type may depart from the typical perciform pattern before being considered distinct enough to be classed in a discrete order is something that no two competent workers currently agree upon. We shall here use the term Perciformes in a very broad sense and divide it into a number of suborders.

Suborder Percoidei. Despite attempts to split the group into suborders, so many forms show so great an adherence to the typical pattern that a considerable fraction are retained in a central stem suborder, the Percoidei. We cannot, within reasonable limits, give an account of all of them, but will list and comment on a few of the more prominent families. Nearly all the families mentioned are known as early as the Eocene. Serranidae, the basses, are considered by all as typical perciforms. There are numerous modern genera, mainly tropical, but a few are found in temperate zones and even in fresh waters. About a dozen forms, including *Serranus,* have been described from the Eocene; the family was thus prominent in early times. Percidae, the perch and darters of temperate Eurasia and North America; a numerous group as regards genera and species, both as fossils and as living forms. This family and the next are two of the three major freshwater types among the spiny-finned teleosts. Centrarchidae, the little sunfishes abundant in North America. Carangidae, the pompanos, somewhat like the serranids, and equally common both in the fossil record and in modern warmer seas. Lutjanidae, the snappers or pargos, abundant warm-ocean forms and, like the pompanos, favorite food fishes. Sparidae, the deep-bodied porgies and their numerous relatives. Sciaenidae, still another abundant group of food fishes, including the croakers, roncadors, weakfish, and so on. Cichlidae, forms which have in tropical fresh waters much the position of the perch in temperate zones. These little fishes are widely spread in tropical America, southern Asia, and, especially, Africa, where they are numerous and varied. This discontinuous distribution in the Old World and New World argues for a former transatlantic land connection. As noted elsewhere, such a connection in Tertiary times is improbable. Unfortunately, the fossil record of cichlids is poor. Chaetodontidae, the butterfly fishes, and Pomacentridae, the damsel fishes, are deep-bodied tropical fishes, often brilliantly colored, and particularly abundant in tropical coral reefs.

Apart from the flatfishes (Pleuronectiformes) and plectognaths (Tetraodontiformes) we shall further include all the remaining "higher" spiny fishes in this great perciform order, arrayed in a score or so of suborders. Apart from a few groups of minor nature or without fossil representatives, these suborders will be now briefly considered *seriatim.*

Suborder Mugiloidei. This series of fishes are advanced, spiny-finned forms in most regards, but the pelvic fins are found somewhat back toward their original posterior position. This was at one time thought to indicate that the mugiloids were relatively primitive, but current belief is that the posterior position of the pelvic fins is a secondary one. There are three families, all found occasionally in fossil form from the Eocene on. The Sphyraenidae are the barracudas, slender swift swimmers of very voracious habits. The mullets, Mugilidae, are clumsier in form and, in contrast, are primarily mud grubbers. The thread-fins, such as *Polydactylus,* living along warm sandy shores, are small mullet-like forms, but are readily distinguishable by the fact that each pectoral fin is sharply divided into two portions—the upper part normal in build, the lower consisting of long, free, thread-like rays. Their fossil pedigree is almost unknown.

Suborder Labroidei. This includes the Labridae—wrasses, tautog, etc.—and the Scaridae, parrot fishes. Labroidei are moderately deep-bodied fishes, often brightly colored, with small "nibbling" mouths but extremely powerful dentitions. They are common as fossils from the Eocene onward.

Suborder Trachinoidei. A series of a score or so of small families, of which the weavers, such as *Trachinus,* and *Uranoscopus,* the "stargazer," with eyes atop the head, are representative. They are in general small fishes, with a long body tapering posteriorly, and with correspondingly long dorsal and anal fins; many inhabit southern, colder ocean waters. They are seldom found as fossils, although the two genera mentioned are reported from the Eocene.

Suborder Blennioidei. A great series of widely distributed small fishes, dwelling mainly along rocky shores, comprises this suborder; most representatives are but a few inches in length. The body is elongate and tapering, with long dorsal and anal fins; the pelvic fins, placed far forward, are small, but the pectorals are generally large. The group shows considerable variation, more than a dozen families often being recognized. They are, however, rare as fossils—perhaps in part because of their small size.

Suborder Ammodytoidei. The sand lances, such as *Ammodytes* are small, slender, silvery fishes, which have the habit of burying themselves in the sand under the surf in shallow water. They are quite aberrant—for example, there are no spines in the dorsal fin—but they presumably are an offshoot of the Perciformes, perhaps related to the last group. Fossil sand lances have been recognized in the Eocene and Miocene.

Suborder Callionymoidei. The dragonets are a modest group of small, scaleless shore fishes with flattened heads. The typical genus, *Callionymus,* has been reported from several stages of the European Tertiary.

Suborder Gobioidei. The gobies, mainly arrayed in two families (Eleotridae and Gobiidae), are an immensely abundant group of small fishes found in temperate and tropical shallow waters, and in bays, brooks, and lakes of coastal regions. A peculiar feature is that in typical gobies the two pelvic fins are placed close together beneath the breast to form an adhesive disk. Some six hundred or so living species of gobies have been described, but they are very poorly represented in the fossil record, with only half a dozen or so reported Tertiary finds. It would seem that the gobies are essentially a "modern" group of spiny-rayed fishes, evolving rapidly in relatively recent times.

Suborder Acanthuroidei. The surgeon fishes, such as *Acanthurus* and *Teuthis,* have short, deep bodies; the name is due to the presence, in typical forms, of defensive weapons in the form of sharp spines on either side of the tail base. A number of Eocene forms are known. The acanthuroids are frequently suggested as ancestors of the plectognaths, considered later.

Suborder Scombroidei. Here the main family is that of the Scombridae, including forms ranging from the relatively small mackerels (*Scomber*) to the giant tuna (*Thunnus*); placed in separate but related families are the sailfish *Istiophorus* and swordfish *Xiphias,* in which the upper jaw is prolonged into a long swordlike structure, and such relatives as *Palaeorhynchus* (Fig. 95). With a streamlined, fusiform body, and forked tail, the predaceous scombroids are the fastest swimmers among all fishes, ranging widely, as surface dwellers, over the open ocean. There are numerous Tertiary fossils from the Eocene on.

Suborder Stromateoidei. These are high-seas fishes of the warm regions of the Indo-Pacific; they form a further series of long-bodied forms, with poorly ossified skeletons. Not common today, they are represented in the Tertiary record only by two Eocene finds.

Suborder Anabantoidei. These are small fishes, including the "climbing perch," *Anabas,* which live in fresh and brackish waters along the coasts of the East Indies, southern Asia, and East Africa. They are able to exist for some time out of water because of the development of membranes for air breathing in the gill cavities. The anabantoids are almost unknown in fossil form.

As noted earlier, the division of the Perciformes into proper suborders is in great measure a matter of individual judgment rather than of clear-cut distinction. For example, the Oligocene-Recent shark suckers or remoras, the Echeneidae, here included in the Percoidei, are sometimes placed in a distinct suborder.

Flatfish.—The flounders, soles, and halibuts forming the order Pleuronectiformes (or Heterosomata) are, in a sense, teleost analogues of the skates, in that they have become flattened bottom dwellers. This flattening, however, has been accomplished in a peculiar fashion. The flatfish have been derived from thin, deep-bodied ancestors which have settled to the bottom on one side, with resulting asymmetrical developments of the new upper and lower surfaces in such features as position of the eyes and structure of the jaws. There are no hard spines in the long dorsal and anal fins, but since the major characters of the group are those of higher teleosts, a majority of students of the group believe that the absence of spines is secondary and that the flatfishes are of perciform origin. There is a good fossil record of flatfishes from the Eocene on.

Plectognaths.—In the order Tetraodontiformes or Plectognathi is a series of fishes which varies considerably in appearance but is united by common anatomical features. The mouth is tiny but armed with heavy teeth; the jaws are powerful and gain strength by a fusion of the upper jaw with the braincase (a feature to which the name plectognath refers). The body is deep and short; the spiny dorsal fin tends to be reduced or lost, as are the pelvic fins.

There are two suborders. The Balistoidei include a series of deep-bodied flat-sided forms, such as the triggerfish *Balistes* and the filefishes, in which the spiny dorsal fin is generally reduced to a single large upward-projecting spine. There are half a dozen families in this suborder, and a considerable series of forms was already present in the Eocene. Included in the same suborder are the tropical trunkfishes such as *Ostracion,* also known from the Eocene. The name refers to the fact that the body is completely enclosed in a rigid box formed by a series of polygonal bony plates. In the trunkfishes the spiny dorsal has vanished, and this is also true of members of the second suborder, Tetraodontoidei. This includes the familiar little puffers, such as *Tetraodon* and *Sphoeroides,* in which the belly can be expanded to a ridiculously large size; the porcupine fishes, such as *Diodon,* in which the rounded body is protected by spines in hedgehog-like fashion; and the gigantic oceanic sunfishes, such as *Mola.* This last form, in contrast to the puffers and porcupine fishes, is flattened from side to side, the body so shortened that the fish seems to be merely an enormous head. The sunfish is not known earlier than the Miocene; the other two families were already present in the Eocene. While there are no truly ancestral forms, it is not improbable that, as suggested earlier, the plectognaths arose from the surgeon-fish group of perciforms.

The history of many other groups of fishes is of interest in connection with the evolution of vertebrates as a whole; that of the later ray-finned fishes has no such interest, for these forms are quite unrelated to the development of further vertebrate types. But for the history of fish alone, no other group can approach them in importance. Since the Paleozoic, almost all other fish groups have tended toward extinction, while the actinopterygians have remained dominant types, and group after group of them has flourished in succession. During the Paleozoic the primitive chondrosteans were the ruling order but have now disappeared except for a few aberrant forms. In the late Triassic and Jurassic the holosteans, their descendants, had taken their place and are found in great abundance and variety. By the Cretaceous the older types had been almost terminated by the teleosts. Today among the teleosts the spiny types have attained dominance in the seas, the Ostariophysi in fresh waters.

The fleshy-finned fishes.—The remaining bony fishes, comprising the Sarcopterygii (or Choanichthyes), are, beyond the Devonian, relatively unimportant in their own right but of great evolutionary interest as the stock from which the land vertebrates were derived. The two component groups of this subclass are the Crossopterygii, often termed fringe-finned or lobe-finned fishes, and the Dipnoi, or lungfishes. The Sarcopterygii were common inhabitants of fresh waters during much of the late Paleozoic, but today there survive only three tropical lungfishes and a rare, deep-sea crossopterygian.

The living lungfish are so aberrant (and, as we now realize, degenerate) that they were long thought to occupy a position among bony fishes remote from the crossopterygians and ray fins, and the two latter groups were thought to be fairly closely related. With our present knowledge of fossil forms, however, the historical picture is an entirely different one. As we trace back actinopterygians, crossopterygians, and dipnoans to their first appearance in the Devonian, we find that ray fins and lobe-finned crossopterygians were then already remote from one another. Lungfish and crossopterygian lines, however, converge as we follow them backward. In the Middle Devonian the early representatives of both groups were not only very similar in general appearance but, apart from differences in skull structure, were also alike in many fundamental characters. These indicate that the two were members of a common bony-fish stock and had diverged from common ancestors not far remote in time or structure.

There are many striking similarities in the postcranial skeleton (cf. Figs. 100, 105). A heterocercal tail was present in primitive members of both groups of Sarcopterygii. This type of tail was also present, as we have seen, in early actinopterygians. There was, however, a seemingly small but potentially important difference. In the ray fins there is usually no epaxial (epichordal) lobe, that is, no fin development above the body axis. In consequence, when a symmetrical tail developed in later actinopterygians, it had to arise from the lower (hypaxial) lobe alone, as seen in the homocercal tail of teleosts. In the fleshy-finned fishes, on the other hand, a small but distinct epaxial lobe was present even in the adult and was presumably large in embryonic stages. When a symmetrical tail was developed (as happened in most later members of the subclass), this upper lobe was able to expand, equal the lower lobe in size, and produce a truly symmetrical, diphycercal tail.

A minor but clear-cut distinction between early actinopterygians and sarcopterygians is the fact that the older ray-finned genera almost invariably have a single dorsal fin; crossopterygians and lungfish, two. More important is the contrast in paired-fin construction already discussed. The sarcopterygians invariably have an archipterygium of some sort—a full, leaf-shaped structure in lungfish and a few crossopterygians, and an abbreviate archipterygium in other lobe-finned forms.

A fundamental point of agreement between the fleshy-finned groups and of contrast with ray-finned fishes lies in the scale structure. Many later sarcopterygians have simple bony scales, but in mid-Devonian times both the early lungfish, *Dipterus*, and its crossopterygian contemporaries possessed scales of the cosmoid type described above (Fig. 68). These differ markedly, of course, from the ganoid scales of actinopterygians.

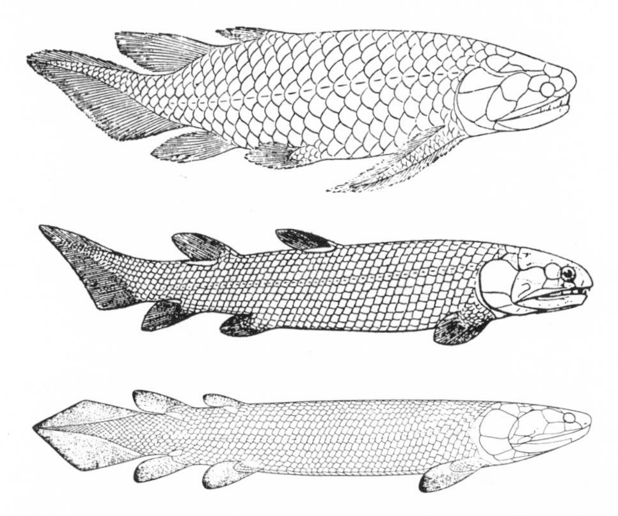

FIG. 100.—Rhipidistian crossopterygians. *Above, Holoptychus* of the late Devonian, with dipnoan-like appendages; maximum size about 2.5 feet. *Center, Osteolepis* of the Middle Devonian, with a heterocercal tail; about 9 inches long. *Below, Gyroptychius,* also Middle Devonian but with tail approaching diphycercal condition; about 9 inches long. (After Traquair.)

Little can be said of the internal bony skeleton in early fleshy-finned fishes in general because of the presence of an obscuring scaly covering in many specimens. Of axial structures, many forms show bony neural and haemal arches, ribs, and fin supports. As in actinopterygians, ossifications in the region of the vertebral centra tend to be incomplete. Centra are rarely reported in lungfish; in Devonian crossopterygians they are sometimes seen in a form comparable to temnospondylous amphibians (Figs. 99, 108*A*); in Carboniferous rhipidistians they may form complete rings, pierced for the notochord.

Certain basic features are present in the cranial structures of both groups. A foramen for a pineal eye is rarely present in actinopterygians but is common in early crossopterygians, and was present in the most archaic lungfishes, although it was lost in later representatives of both groups. The eyes are not so large as in early actinopterygians; sarcopterygians are "nose-fishes" rather than "eye fishes." Sclerotic plates, when found, are numerous, instead of limited to four, as in actinopterygians.

Both crossopterygians and lungfishes had well-ossified cranial skeletons in the earliest stages of their careers, and certain structural comparisons and homologies of bony elements may be made between them. However, the diagnostic features of the two groups lie in the skull and dentition; they had already diverged markedly in these respects at their first appearance, and their cranial structures are best considered separately.

Primitive crossopterygians.—Fragmentary crossopterygian materials, assigned to the genus *Porolepis,* are known in Lower Devonian strata, and we may be sure that the evolution of this important group of tetrapod ancestors was well under way at this time. In the "Old Red Sandstone" and other continental deposits of the Middle Devonian we find the group represented in abundance and variety. *Osteolepis* (Fig. 100) is the best-known genus of the time and is generally regarded as a very primitive form. This lobe-fin shows a general structural plan which is repeated, with variations, in a considerable number of other genera of the Devonian and later Paleozoic periods. The body was covered with rhombic scales of typical cosmoid structure. The tail was heterocercal, in presumably primitive fashion. Fully archipterygial paired fins are seen in some crossopterygians, but in the great majority, including *Osteolepis,* there is a short, broad lobe. The internal structure of the fins is not known in *Osteolepis;* in other genera this type of fin contains a much-abbreviated archipterygium. The main skeletal axis is short; side branches are irregular distally, but three prominent proximal pieces appear to be diminutive predecessors of the three major bones of the limb of later tetrapods (Fig. 72*B*).

In the head region (Fig. 69) both dermal elements and replacement bones are in a highly ossified state. The structures seen in such genera as *Osteolepis* and *Eusthenopteron* are worthy of detailed description because they are preserved with few changes of importance in primitive land vertebrates.

As has been said, the nomenclature of fish bones has been in a confused state, and names familiar in higher vertebrates have been used in an arbitrary fashion. In recent years, however, evidence has been obtained which enables us to identify with considerable confidence most of the skull elements of crossopterygians in terms of those of tetrapods.

Most early crossopterygians exhibit a skull covered by a complete shield of dermal bones. The appearance of this shield, however, differs greatly from specimen to specimen. In some the cosmine layer and its shiny, enamel-like surface is widespread over the skull and may superficially fuse large series of elements of the snout and other regions into large plates; in others the cosmine areas are restricted, and all the sutures between individual bones are clearly visible. Such differences were once thought to be of taxonomic value, but it has been reasonably suggested that they represent merely fluctuations, perhaps seasonal in nature, in the lives of the individuals. The fused condition represents periods of stability; the reduction of cosmine allows the elements to separate and growth to occur.

Many of the bones of the crossopterygian skull roof are relatively large and stable structures. Suggestions of the mosaic pattern which we believe to have been characteristic of ancestral bony fishes are, however, seen in the individual variability of the skull pattern in any series of specimens, particularly in the case of the rostral (and postrostral) bones in the center of the snout region. On either side of these structures is a longitudinal row of small elements which collectively correspond to the nasals and frontals of land forms. This row is continued by a pair of plates surrounding the median eye. Their homology was long debated, but it is now certain that they are the tetrapod parietals, which similarly surround the pineal eye. A final posterior pair of elements, occupying half the length of the skull roof, are homologues, despite marked dissimilarity in size, of the postparietals (dermal supraoccipitals) of lower land vertebrates. Back of these is a transverse row of plates, termed extrascapulars, which are absent in land forms and which are to be regarded as expanded scales rather than as part of the true skull structure.

The margins of the shield are formed by well-developed maxillary and premaxillary elements, bearing a marginal tooth row. The two are usually firmly united and separate the external nostrils from the jaw margin. The naris is sometimes seen to be bordered by one or two small bones, which may correspond in some way to a small bone—the septomaxilla—similarly situated in many land vertebrates. Circumorbital elements are commonly five in number and may be homologous with the five present here in early land forms. A row of bones lateral to parietal and postparietal is, in general, comparable to that formed by the similarly situated intertemporal, supratemporal, and tabular of lower tetrapods. In the cheek region the crossopterygians lack the expansion of the maxilla seen in the palaeoniscoids; there is, instead, a large squamosal, as well as a preopercular and a quadratojugal. In palatal view (Figs. 69*B*, 101), crossopterygians, like actinopterygians, exhibit a dermal parasphenoid bone covering the lower surface of the anterior part of the braincase; and a set of dermal bones, including pterygoid, vomer, palatine, and ectopterygoid, form most of the palatal surface. The choanae, or internal nares, lie between vomers and palatines, bounded laterally by maxilla and premaxilla. The lower-jaw dermal elements (Fig. 71) differ in some respects from those seen in palaeoniscoids; there is a large dentary externally, bounded below by a series of four elements, including splenial, postsplenial, angular, and surangular; internally there are a prearticular and three coronoid elements. Large opercular and subopercular bones cover the gill region laterally, again as in actinopterygians; the ventral gular series comprises a small unpaired median gular, large paired median gulars, and a still more lateral row of small plates completing the armor below the jaws. A well-developed system of lateral-line canals and accessory pit-line grooves is present in and on the dermal bones.

The dentition is of particular interest because of peculiar features carried over unchanged into the Amphibia. The teeth are sharp, pointed structures—the crossopterygians were dominant predators in Devonian fresh waters. Superficially, they usually show longitudinal grooves, which in section (Fig. 102) are seen to represent infoldings of the enamel surface. These folds are often of a complicated nature and cause the term "labyrinthodont" to be applied to this sort of dental structure. The lateral teeth are usually well developed; but, in addition, large fangs as well as smaller denticles are present on the palatal bones and on the corresponding bones on the inner surface of the lower jaw. The fangs are frequently seen to be arranged in pairs, which are replaced alternately so that at least one member of a pair is always functional.

In a number of respects the cartilage-replacement bones correspond to those of actinopterygians. In the lower jaw there is a single replacement element, the

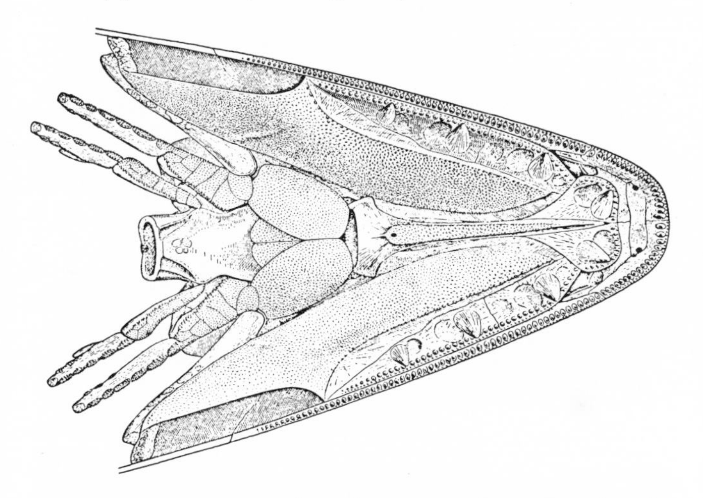

Fig. 101.—The palate of *Eusthenopteron;* the structure is mainly comparable to that seen in Figure 69*B*, but dermal bones are seen in the roof of the mouth, covering part of the posterior segment of the braincase and the upper parts of the anterior gill arches. (From Jarvik.)

articular, and a well-developed quadrate opposes it in the upper jaw. Although other upper-jaw replacement bones are not so well known, there appear to be suprapterygoid elements, one of which articulates with the basal process of the braincase and may be termed an epipterygoid. There is a stout hyomandibular, articulating by a double head with the surface of the ear region of the braincase.

The braincase is, as in early actinopterygians, well ossified throughout but is unique in certain respects. It was built in two distinct units, anterior and posterior, between which a certain amount of motion was possible, at least in many cases. This may have been an adaptation in these carnivorous fishes for taking up the jar caused by a powerful snap of the jaws (articulated to the front part of the braincase) and leaving the posterior segment of the skull (and body) unaffected. The union of the two parts is accomplished not only by articular surfaces but also by the fact that the notochord, almost unrestricted, runs forward through the

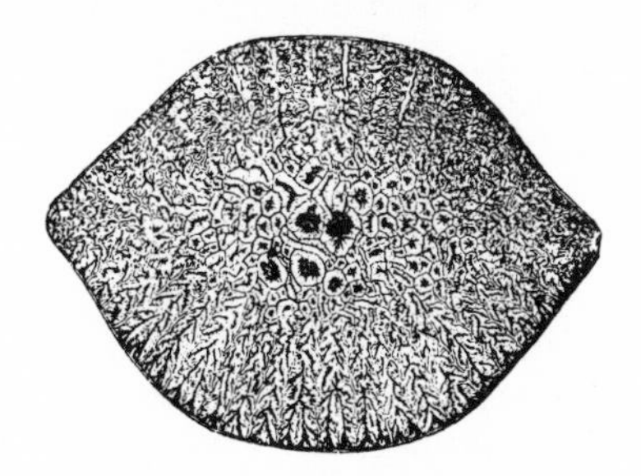

Fig. 102.—Section of tooth of *Holoptychus*, a Devonian crossopterygian, to show the labyrinthine infolding of the enamel. (After Pander.)

posterior segment in a large canal beneath the brain cavity and, in life, ended by a union with the hinder end of the front segment. In vertebrates generally, the cartilages which form the braincase are laid down in the embryo in anterior (trabecular) and posterior (parachordal) groups. Here this embryonic division persists into the adult stage. In *Osteolepis* and various other genera a straight and open suture is present between parietals and postparietals, corresponding to the line of cleavage of the two halves of the underlying braincase. In some individuals, however, this suture is not so marked, and it seems probable that, in general, the movement was not a violent one but merely a "shock absorber."

Although there is in many land vertebrates a zone of weakness in the adult between anterior and posterior portions of the braincase, we knew until recently of no tetrapod in which there was a clear distinction between the two halves, and in none was there such a structure as the large notochordal canal which runs through the floor of the posterior braincase segment. In consequence, it seemed unlikely that tetrapods could be of crossopterygian descent, despite the similarities in many other regards. However, the recent discovery that the oldest amphibians, the ichthyostegids (Fig. 119), retain a notochordal canal and show at least traces of braincase subdivision, has removed this objection.

The genera so far considered pertain to a crossopterygian suborder termed the Rhipidistia. They were, as a group, common in later Devonian and Carboniferous days but became extinct in the early Permian.

Their disappearance may perhaps be linked with geological changes which reduced the areas with a stagnant-water type of environment and made lung-breathing of less importance for survival. More important, however, may have been the fact that they were replaced as freshwater predators by the amphibians, which had arisen from them.

Three rhipidistian families can be readily distinguished. Most typical is that of which *Osteolepis* (Fig. 100) is the commonest Devonian representative; *Megalichthys* of the Carboniferous and *Ectosteorhachis* of the American Lower Permian were late survivors. These were fishes of modest size, 1 or 2 feet in length, with broad and relatively flattened heads, rhomboidal scales, and tails which, heterocercal in *Osteolepis,* tended toward a diphycercal type. The marginal teeth were small, if sharp, and their diet presumably consisted of small invertebrates inhabiting the freshwater ponds and streams in which members of this family lived. Closely related to the osteolepids were the Rhizodontidae, of which *Eusthenopteron* (Fig. 99) is a well-known Devonian representative. The rhizodonts, with a highly developed dentition of sharp-pointed teeth, would appear to have been a highly predaceous group, apparently fast swimmers, with a long and slender body, cycloidal scales, and a tail which generally assumed a characteristic trifid diphycercal pattern (repeated in the coelacanths). Members of this group tended to grow to large size, and specimens of *Rhizodus* and *Rhizodopsis,* Carboniferous representatives of the family, appear to have reached several yards in length.

Quite different in many ways from the osteolepids and rhizodonts were the Holoptychidae, which were modestly successful in the Devonian but did not survive the end of that period. *Porolepis* of the Lower Devonian was a forerunner; *Holoptychus* (Fig. 100) and *Glyptolepis* were characteristic forms of middle to late Devonian days. *Holoptychus* was a relatively short and deep-bodied fish, attaining several feet in length, with cycloidal scales on which (in contrast with osteolepids) there remains little of the cosmine layer. The tail was persistently heterocercal and—as in lungfish and in contrast with other crossopterygians—the pectoral fin was of a fully developed archipterygial type. The cranial anatomy was distinctive in various features; for example, there are a number of "extra" plates in the deep cheek region, and on the posterior part of the skull roof a broad pair of plates occupy the position of the intertemporal bones as well as that of the parietals, and carry a segment of the lateral-line canals. As a dental specialization, the holoptychids carry a pair of tooth whorls at the lower-jaw symphysis, somewhat after the fashion of edestid sharks. Most rhipidistians appear to have been freshwater forms, but tooth whorls of the holoptychid genus *Onychodus* are widely distributed in Devonian deposits which are, in part at least, marine in origin. It has been claimed that both the *Holoptychus* group and the osteolepids independently gave rise to tetrapod groups in diphyletic fashion; however, re-examination of the evidence fails to support this conclusion.

Coelacanths.—Much longer lived than the rhipidistians were the Coelacanthini, a side branch of the crossopterygians which rapidly evolved a series of specialized and rather degenerate structures. The body, generally rather stocky in many forms and covered by relatively thin scales, terminated in a three-pronged tail similar to that seen in *Eusthenopteron.* Dorsal and anal fins persisted but became fan-shaped structures with narrow bases; and, while the paired fins, as far as is known, seem to have had a bony structure of the usual crossopterygian type, the scaly lobe was reduced and the ray-supported web increased. In the head the duplex condition of the braincase persisted, but there was a rapid and considerable reduction in ossification after the Devonian, so that most coelacanths show here a series of isolated bony elements with large intervening areas, presumably formed in cartilage. The dermal elements, too, are much modified. There is no pineal opening. The gape of the jaw is shortened, the head becomes short and deep, and the typical development of a sharp angle in the skull outline where the two halves of the braincase join gives these fishes a "Roman-nosed" appearance. The dermal plates of the cheek and gill covers tend to be much reduced; on the other hand, there are numerous extra "supraorbital" elements along the sides of the snout. Very characteristic is an early reduction and loss of the lateral tooth row except for a cluster of teeth on the premaxillae and the tip of the dentaries; in connection with this, the maxilla and dentary are reduced (lungfish show a comparable history in this regard). The hyomandibular is reduced; and, although the upper jaws still articulate with the braincase, the primitive basal articulation is lost, and a new connection develops more dorsally. A notable difference from rhipidistians is the apparent absence of choanae.

Primitive coelacanths, not too far removed in many features, are present in the Devonian, with such genera as *Diplocercides* and *Dictyonosteus,* and in the genera *Rhabdoderma* of the Carboniferous and *Coelacanthus* of the Permian the characters of the group were fully established. The early members of the group were, in general, freshwater types; but in the Triassic, when coelacanths were common and varied, we find them in marine deposits as well. The group survived in this new environment; *Holophagus* [*Undina*] of the Jurassic and *Macropoma* (Fig. 103) in the Cretaceous are later Mesozoic representatives.

Beyond the Cretaceous there are no fossil records of coelacanths; and it was long customary to state that the coelacanths (and hence the Crossopterygii as a whole) became extinct at the end of the Mesozoic. But absence of a group furnishes only negative, not positive, evidence. Two decades ago a South African fishing boat, dredging deeper than usual, brought up an unfamiliar fish which proved, to the astonishment of the scientific world, to be a surviving coelacanth, *Latimeria,* looking in life very much like the restorations of its Mesozoic forebears. The coelacanths had shifted, during their evolution, from fresh waters not merely to salt but finally to the deep-sea regions little represented in fossil deposits of any age.

Lungfish.—The lungfish, of the order Dipnoi, when first found in the early Devonian, had already acquired characteristic specializations in skull and dental structures but showed, as has been noted, many fundamental resemblances to crossopterygians. In the late Paleozoic and Triassic, lungfish were moderately abundant and varied in the freshwater environment in which they arose and to which they were adapted. In later days, however, their fossil remains become rare, and they are represented today by only three tropical forms.

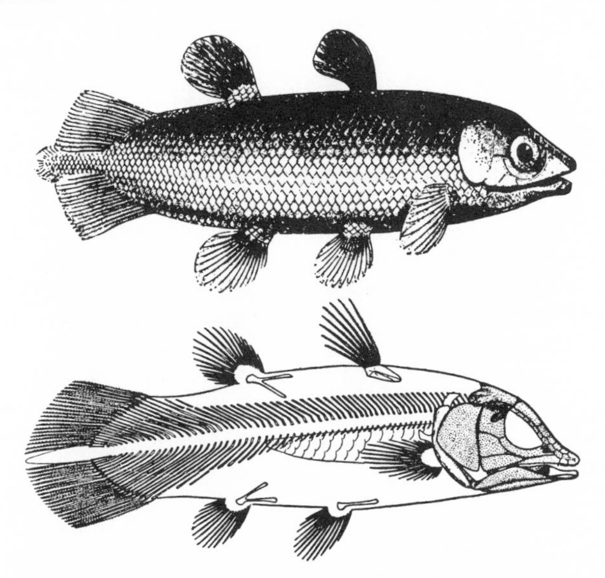

Fig. 103.—*Macropoma,* a Cretaceous coelacanth, original about 22 inches long. *Above,* Restored with squamation; *below,* with scales removed. (From Smith Woodward.)

Dipterus (Figs. 104, 105) of the Middle Devonian is one of the oldest known and most primitive lungfish. The body appears fundamentally similar in construction to that of the contemporary crossopterygians. In *Dipterus* two dorsal fins and a heterocercal tail like that of *Osteolepis* were present. In later types the median fins became much modified, the first dorsal disappearing and the second spreading back to join the caudal. Meanwhile, the tail fin had tended to straighten out and fuse with the anal, so that in the living forms there results a continuous symmetrical diphycercal fin bordering the posterior part of the body.

The paired fins in *Dipterus* and all well-known fossil forms are leaf-shaped structures similar to that in the modern Australian lungfish. The internal skeleton is unknown in fossils generally but was presumably of the typical archipterygial pattern found in the modern genus, with a central axis and numerous short side branches. The scales and head plates in *Dipterus* were of the cosmoid type seen in contemporary crossopterygians. In later types the body scales, with loss of the enamel and cosmine layers, tended to become thinner and larger.

Functional lungs, comparable in all essentials to

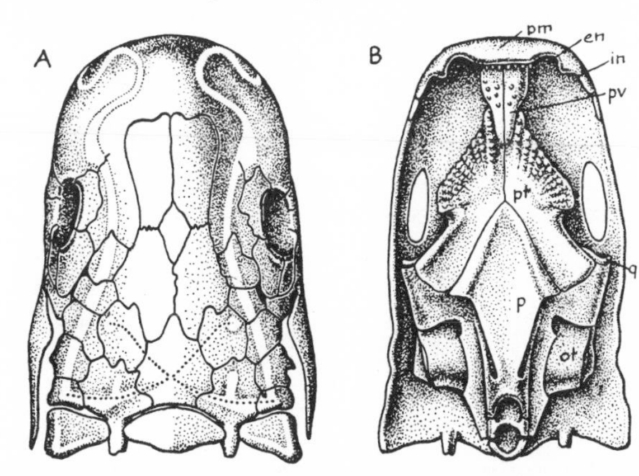

Fig. 104.—*A,* Dorsal, and *B,* ventral, views of the skull of *Dipterus,* a primitive Devonian lungfish. The homology of the dermal roofing elements is in doubt, and hence these have been left unlabeled. Anterior part of braincase, including nasal capsules, unossified. For abbreviations see Figure 69. (After Pander, Watson, and Goodrich.)

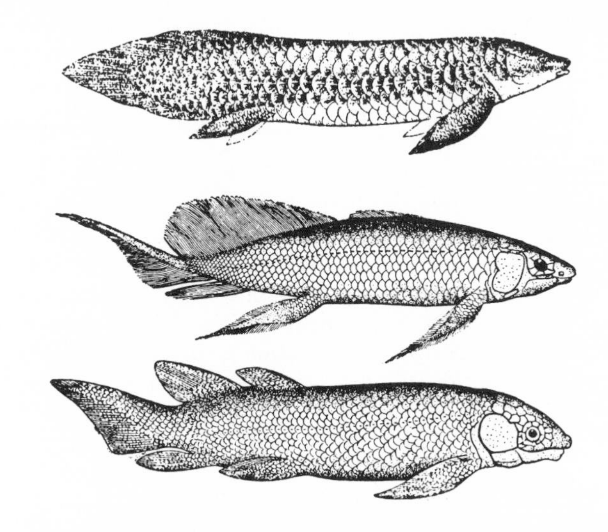

Fig. 105.—Dipnoans. *Below, Dipterus* of the Devonian, primitive genus with a heterocercal tail and median fins comparable to those of crossopterygians. *Center, Scaumenacia* of the Upper Devonian; the median fins are concentrating posteriorly. *Above,* The Recent *Neoceratodus,* with all median fins fused to form a symmetrical caudal. (After Traquair, Hussakof, Norman.)

those of primitive land animals and to those already described in the actinopterygian *Polypterus* are present in the modern lungfishes. The presence of similar lungs in the fossil Dipnoi is a reasonable inference. Modern lungfish live under conditions of seasonal aridity presumably comparable to those under which all the primitive bony fish lived. The Australian lungfish, *Neoceratodus,* cannot survive the complete drying-up of the streams, but in foul and stagnant pools may come to the surface and breathe air. The African lungfish, *Protopterus,* and *Lepidosiren* of South America are able to aestivate—to retreat into a burrow and withstand a complete drying-up of the water. The specialized eel-like shape and structure of these two forms suggest that this habit was not one characteristic of the more typical Paleozoic forms. However, vertically placed cylindrical structures of hardened clay which appear to represent mud-filled burrows have been recently discovered in the American Pennsylvanian and the Permian red beds. In some of these have been found remains of the lungfish *Gnathorhiza,* a presumed ancestor of the modern African and South American forms. Apparently the aestivating habit developed early in this special group of dipnoans.

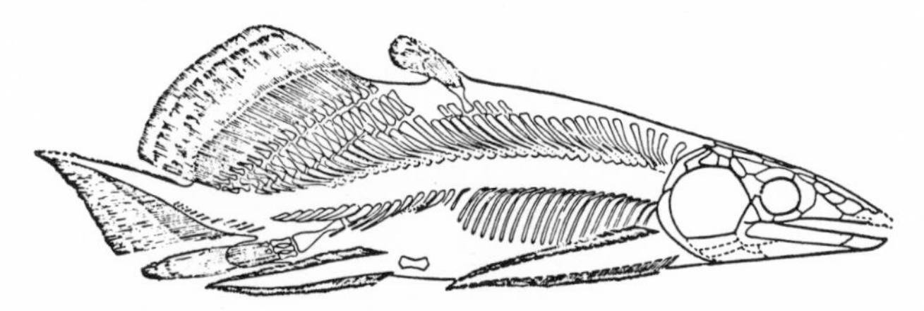

FIG. 106.—*Fleurantia,* a long-snouted Upper Devonian lungfish, about 1/3 natural size. (From Graham-Smith and Westoll.)

Despite the important diagnostic features linking lungfish to crossopterygians, even the earliest forms show striking differences in skull and dentition. The dentition, particularly, forms a characteristic trademark of the Dipnoi. Marginal teeth were abandoned in the earliest days of lungfish history, and only in a very few Devonian and Carboniferous forms are remnants of such structures reported. In some Devonian genera a partial substitute is found in thickened bony rims at the front end of upper and lower jaws, which appear to have functioned as a cropping device in food gathering. Absent, too, are teeth on the palatal margins, and evolutionary development was centered on the formation of tooth plates close to the midline of the mouth. The prevomers usually carry small tooth plates or tooth clusters, but teeth are mainly concentrated on the pterygoids above and prearticulars below. In a few cases the teeth are scattered over the palatal surface of these bones, but even in *Dipterus* there are radial rows of large teeth, and in most later genera these rows form connected ridges borne on thickened fan-shaped plates; such plates, unmistakable in origin, are the commonest remains of fossil lungfish. This type of plate for shearing and crushing, with upper and lower ridges interdigitating, appears adapted to a dipnoan food supply of small mollusks and other invertebrates.

The oldest lungfish exhibit a skull roof which in great measure is still in the primitive stage of a mosaic of small elements. These are particularly small and variable in the anterior portion of the head; more posteriorly they are larger and more constant, with a tendency for the formation of large median plates. It is obvious that in lungfish the process of mosaic reduction was occurring in a fashion rather different from that seen in crossopterygians, and hence it seems extremely difficult to identify "normal" skull elements in the lungfish pattern.

There is a strong trend toward skull-roof reduction in the lungfish series. Cosmine and enamel layers may be present in *Dipterus* and some of the other Devonian forms, and there appears to have been, as in contemporary crossopterygians, a periodic formation of a solid anterior shield. These outer layers are, however, absent in most later lungfish, both in dermal bones and body scales. In most, the anterior end of the head is bare of dermal bones, the lateral walls are gradually reduced, and in living genera there remain but a few large dorsal plates. Opercular and gular elements, except for a large principal gular, also tend to be reduced. With development of the characteristic lungfish dentition, both upper and lower jaws lose much of their dermal structure; some of the elements absent in later dipnoans are represented in *Dipterus* by small ossicles. The loss of lateral teeth has been accompanied by the loss of the premaxilla and maxilla and by reduction of the dentary and other lower-jaw elements. The lower jaw in later dipnoans includes but three bones—two outer dermal elements, which are probably a splenial and the angular, and internally that bearing the tooth plate, presumably the prearticular. The palatine and ectopterygoid have disappeared from the palate.

As in various other fish groups, the lungfish show a marked phylogenetic reduction in the degree of ossification of replacement bones—a reduction more rapid than in any other group of bony fishes and more complete than in any others except the sturgeons and paddlefishes. The braincase had lost almost all ossification, apparently, by Carboniferous times. It was still ossified, however, in a few recorded Devonian specimens. It was a single compact structure which lacked the bipartite specialization of the crossopterygians.

The cartilage elements of the jaws and gill arches show little ossification even in early forms. In *Dipterus* the articular and quadrate, forming the jaw joint, were ossified, but even these structures later slumped back to a cartilaginous state. The recent genera show us that dipnoans have changed radically in basic jaw construction from that found in other bony fishes; the primary upper jaws have fused to the braincase, with, of course, reduction of the hyomandibular, which plays a prominent part in jaw support in other groups. This specialization is parallel to that found in the chimaeras and is obviously related to a similar function—strong support for the crushing tooth plates.

Much of the description above has been based on *Dipterus,* a primitive form which appears in the Middle Devonian. Among the primitive features to be noted in this genus are the unspecialized median fins, retention of good cosmoid scales, and a complete skull shield. A contemporary, *Dipnorhynchus,* is still more primitive (it retains a "pineal" opening, lost in all other dipnoans, for example), but is imperfectly known.

From *Dipterus* a "main line" of lungfish evolution can be traced to the living Australian lungfish. In the Upper Devonian genera *Scaumenacia* (Fig. 105) and *Phaneropleuron,* common fishes of their period, the first dorsal fin is much reduced, the second much enlarged and moving backward, and the anal is likewise moving back to a position close to the caudal; in the latter form dorsal and caudal are joined. The Carboniferous genera *Sagenodus* (Fig. 71) and *Ctenodus* are represented mainly by abundant and characteristic tooth plates, and the body outline is little known. Some late Paleozoic lungfish, however, appear to have completed the fusion of the median fins into the symmetrical and deceptively simple-looking diphycercal tail seen in the living forms. *Ceratodus* is a genus common in the Triassic, with large, flat tooth plates which show little trace of the separate teeth of which they are composed. In later deposits remains are rare; the living Australian lungfish appears to be closely related to the Triassic form, although frequently placed in a separate genus, *Neoceratodus* (or *Epiceratodus*).

From this central line of dipnoan evolution there have been various side branches. Notable in the Upper Devonian are two genera which became long-snouted types, *Rhynchodipterus* of Scotland and Greenland and *Fleurantia* (Fig. 106) of America. In *Fleurantia* there are, instead of well-formed tooth plates, merely a few rows of cusps on the pterygoid and a shagreen of smaller teeth; *Rhynchodipterus* is thought to be toothless. A similar reduction of the tooth plates appears in the Carboniferous *Uronemus* and Permian *Conchopoma,* which, however, seem derived from the *Sagenodus* group and are apparently little specialized in other respects. A final side branch of the dipnoan stock is that seen in the living African and South American genera, *Protopterus* and *Lepidosiren.* In these forms the body is rather long and eel-like, the fins degenerate, and the tooth plates reduced to a few shearing ridges. *Gnathorhiza* of the early Permian has teeth of this sort, but we have no further information regarding the history of these forms.

6 CHAPTER

Amphibians

We now turn from a consideration of the fishes—primitive vertebrates for whom water is the natural habitat—to the history of land vertebrates, the tetrapods. These forms have initiated one of the most radical advances in vertebrate history in the development, from fish fins, of limbs capable of being used for progression on land. The change from water to land life, begun in the Devonian by the ancestral amphibian stock, was completed by the reptiles in late Paleozoic days; later, from reptilian types, were developed the birds and mammals. Once the water was left behind, wide and repeated radiations of tetrapod types occurred. The four-footed vertebrates have adapted themselves to almost every conceivable type of land habitat and have, moreover, taken to the air and, backed by the advances made in land life, have reinvaded the seas. Snakes, birds, men, and whales are but a few examples of the widely varied types evolved from the primitive and ancient forms which in late Paleozoic times first left the streams and pools to walk upon the land.

Amphibian life.—Most primitive and most ancient of tetrapods are the amphibians, which as a class are, without question, the basal stock from which the remaining groups of land vertebrates have been derived. Living amphibians include but three interrelated, comparatively unimportant vertebrate groups: the frogs and toads (Anura or Salientia), the salamanders and newts (Urodela or Caudata), and some rare, limbless, wormlike types (Apoda or Caecilia or Gymnophiona). All are highly specialized and have in many respects departed far from the primitive type described below.

The life history of a salamander, however, shows the essential characters of the class. The eggs are usually small and without the protective membranes or shell found in the eggs and developing embryos of reptiles and higher types. Primitively, amphibian eggs are laid in the water, whereas the reptilian egg is laid on land. The young embryo, without any great supply of yolk for nourishment, hatches out at a very immature stage as a water-dwelling and water-breathing larva, which must find its own food. Later, when the salamander approaches maturity, a metamorphosis takes place; the gills disappear, lungs develop rapidly, and the animal may become a land, instead of a water, type.

But such an amphibian is not even then freed entirely from an aquatic environment, for the eggs normally must be fertilized and laid in that element. Consequently, typical amphibians must return to the water at the breeding season, and complete adaptation to land life alone is impossible. Various devices have been developed by amphibians to help them avoid, to some extent, this necessity for a double mode of life and a double set of adaptations, but never with complete success. It is no wonder that among the salamanders, and apparently among many extinct groups as well, there have been numerous types which have, so to speak, given up the struggle and reverted, as neotenous forms, to a permanent water life and in some cases to a retention of gill breathing.

This type of reproduction and development is the best definitive feature of the class; in the frogs the "double life" cycle is emphasized further by the development of the tadpole larval type, which must undergo a radical metamorphosis in almost every regard to reach the adult condition. For living groups, certain adult morphological characters (particularly those of the skeleton) have been used to distinguish the group from the reptiles; but among the fossils practically all anatomical differences break down.

Among existing types the salamanders and newts approach most closely the body form which is probably the primitive one (the frogs, tailless and with long hind legs, are among the most highly specialized of land vertebrates). The salamander body and tail are elongate; median fins have disappeared; but the tail is often high and flattened, and may be used as a swimming organ in rather fishlike fashion. The paired fins of the ancestral fish have been transformed into tetrapod limbs, which in salamanders are usually quite small and feeble as compared with those of reptiles and other higher tetrapods, but are quite large as compared with fish appendages. While these limbs are capable of a great amount of independent motion, it is of interest that a salamander progresses more or less in fish fashion on the ground, the body being thrown into

S-shaped curves, which may advance the limbs with very little movement on their own part.

Structure of a primitive amphibian.—But while a salamander may resemble the probable tetrapod ancestors superficially, the skeletal anatomy of all modern amphibians is very much simplified and modified from primitive conditions. The structure of some primitive fossil types will be described below in some detail, since the fundamental pattern of the skeleton of those forms is basic for an understanding of the history not only of later amphibians but of the reptiles and higher tetrapods as well. Almost every skeletal element of a bird or of a man may be traced back to these primitive types, although later forms have modified the shape and relationships of parts greatly and there have been frequent losses of bones.

Our description will be based in great measure upon such primitive Carboniferous genera as *Palaeogyrinus* (Fig. 109), *Pteroplax* (Fig. 131), and *Megalocephalus* (Fig. 110), members of the Labyrinthodontia, an important group of primitive amphibians. Such types were common inhabitants of the coal swamps. Some labyrinthodonts were small, with lengths of but a few inches, but many were considerably larger, measuring several feet—or even yards—in length. The general proportions were not unlike those of a modern salamander, but the body in the older labyrinthodonts was somewhat higher and more rounded in section. Most of the life of such an archaic tetrapod was still spent in the water, and the small palaeoniscoid fish, then abundant, may have served as a major source of food supply (all ancient amphibians appear to have lived on animal food). There was still much resemblance to the crossopterygian fish type both in structural features and in a continuation of the ancestral aquatic mode of life. But the development of tetrapod limbs, rendering land locomotion possible, is an obvious and striking difference.

Bony scales, present in the fish ancestors, have tended to be reduced in land forms, although horny scales may functionally replace them in reptiles (modern amphibians have a soft, moist skin). Ancient amphibians and some modern reptiles retain bony scales in a ventral armor of V-shaped rows along the belly, a structure of use in protecting the low-slung body while traveling over rough ground. Reduced bony scales covering the flanks and back are found in some of the older Amphibia, but (except for vestiges in the Apoda) they have disappeared in living tetrapods, although later reptiles (and even mammals) may redevelop bony armor.

Axial skeleton.—In most of the older fishes the neural arches of the vertebral column were well ossified, as were the haemal arches in the tail; the centra were usually cartilaginous. We have noted, however, that centrum ossifications were present in crossopterygians; in amphibians this region is usually highly ossified in one fashion or another, for the backbone must carry the weight of the body, and cartilage could not stand the strain.

In lower tetrapods the nature of the centra is of great significance in classification. Two fundamentally different types of construction are found among the amphibians. One is the lepospondylous ("husk vertebra") type, found in numerous small Paleozoic forms (Fig. 107) and in modern amphibians. In this type the centrum is a single structure, often essentially spool-shaped and pierced lengthwise by a hole for the persistent notochord.

More important phylogenetically is a second type ("arch vertebra") found in the ancient Labyrinthodontia (Figs. 108, 128) and persisting today in modified form in all the higher vertebrate classes. This type is a direct inheritance from the crossopterygian one.

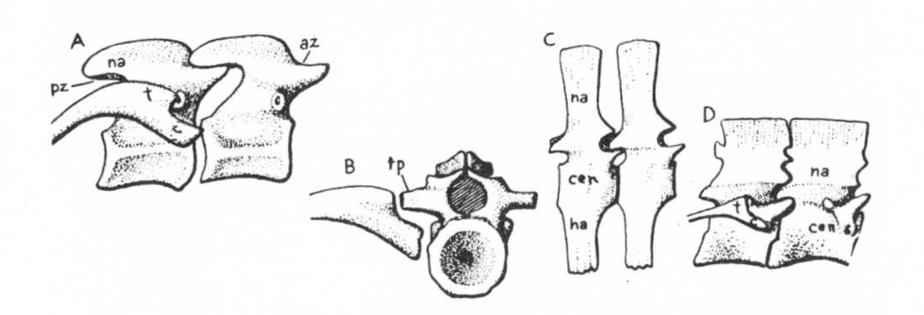

Fig. 107.—The vertebrae of Paleozoic lepospondyls. *A, B,* Lateral and anterior views of vertebrae of *Lysorophus; C, D,* caudal and dorsal vertebrae of *Crossotelos,* an elongate nectridean. For abbreviations see Figure 108. (*A, B* after Sollas; *C, D* after Williston.)

We have noted that in some members of that fish group each segment of the body contained two sets of ossified arch structures formed in the central region of the vertebrae—the anterior ones termed the intercentra, the posterior termed the pleurocentra. These structures are characteristically present, although in varied fashion, in the Labyrinthodontia. Several different subtypes of arch vertebrae are found in this group. The rhachitomous condition is apparently the most primitive one, and seemingly taken over without change from crossopterygian ancestors. Here the intercentrum in each segment is a median ventral element, wedge-shaped in side view, crescentic in outline as seen end-on; the pleurocentra, which together correspond to the true centrum of higher classes of vertebrates, are small paired blocks on either side between neural arches and intercentra. The rhachitomous type is present in the oldest known amphibians, the Ichthyostegalia, and in a majority of the "main line" of the labyrinthodonts, termed the Temnospondyli.

In many late Permian and Triassic temnospondyls the pleurocentra are reduced or absent and the intercentrum alone is present below the neural arch, sometimes growing upward to form a complete ring around the notochord—an advanced "stereospondylous" condition. In another major labyrinthodont group, the Anthracosauria, the evolutionary process has gone in a different direction. The pleurocentra increase in size and fuse to form a complete ring which is destined to become the true centrum in higher vertebrate classes. Beyond this point, anthracosaurian vertebral evolution proceeded in two different directions. In the line leading to reptiles, the intercentra were reduced to small ventral wedges between successive pleurocentra. In a second line, the intercentra grew to form complete rings

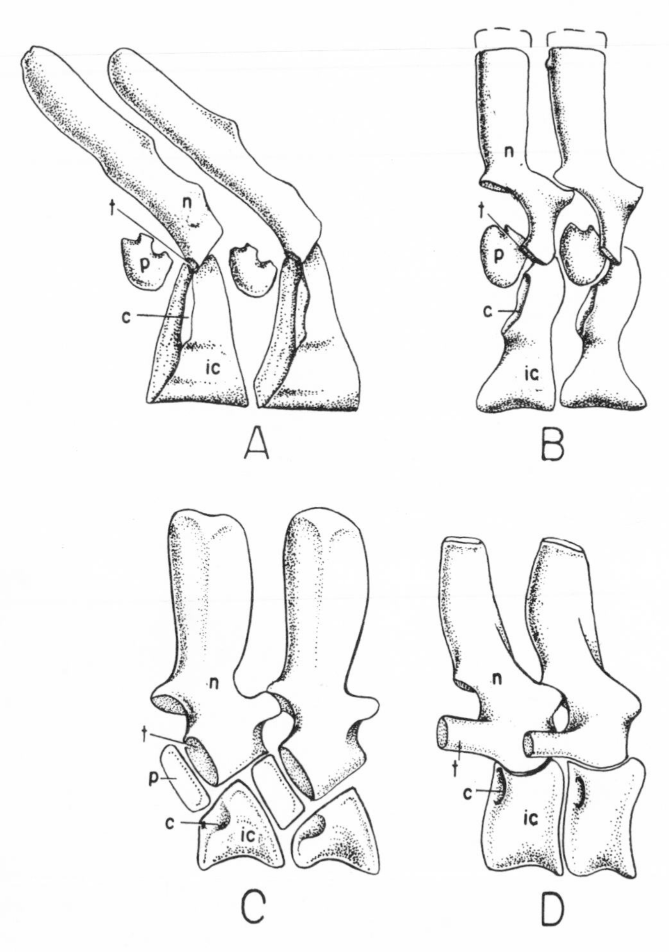

Fig. 108.—Labyrinthodont vertebrae in side view. *A*, The crossopterygian *Eusthenopteron*. *B*. *Ichthyostega* of the late Devonian. *C*, *Eryops*, a typical rhachitome, with reduced pleurocentra. *D*, *Mastodonsaurus*, an advanced stereospondyl. *c*, Articulation for capitulum of rib; *ic*, intercentrum; *n*, neural arch; *p*, pleurocentrum; *t*, attachment for tuberculum of rib.

so that there resulted the embolomerous condition, in which each segment included two complete ring-shaped central structures.

Neural arches are well developed in all land vertebrates. The segments of the column connect with each other not only by juxtaposition of the centra but also by special articular processes—zygapophyses—on the arches. At the back of each arch there is a pair of round, flat surfaces, facing down and usually somewhat outward. These posterior zygapophyses articulate with corresponding anterior zygapophyses, flat surfaces on the anterior side of the arches, facing upward and inward. In the tail are Y-shaped haemal arches (chevron bones); when intercentra are present, they are formed in common with these arches.

Of the two types of fish ribs, only the dorsal set is present in a tetrapod. Two-headed ribs appear to have been primitive, the true head or capitulum articulating with the intercentrum, and the tubercle meeting a transverse process from the neural arch. However, the two heads tend to be well separated only in the anterior part of the trunk; farther back along the column the heads tend to fuse. Phylogenetically, too, the heads may fuse throughout the column or become variously modified in other ways. Primitively ribs are found along the length of the body from the first vertebra in the neck to the proximal part of the tail. The longest ribs are in the thoracic (chest) region, but there is little sharp differentiation of a cervical (neck) or lumbar (waist) region in primitive amphibians. Opposite the pelvic girdle a specialized rib, the sacral, gives attachment for the girdle in most amphibians; in higher classes (but in almost no amphibians) there are usually two or more such sacral ribs.

Skull.—The skull of primitive labyrinthodonts, such as *Palaeogyrinus* (Fig. 109) and *Megalocephalus* (Fig. 110), is completely roofed by dermal bones; in life these apparently lay close below the surface, barely covered by the skin. The skull pattern is essentially that described for rhipidistian fishes, but there have been marked changes in proportions. In the crossopterygian there was a long skull table behind the orbit and a relatively short snout. In typical tetrapods the skull table is relatively much shortened and the snout greatly lengthened, with consequent modifications in the degree of development of the bones concerned.

The rather variable fish roofing pattern has tended to become a fixed set of bones that, with relatively few changes, can be followed throughout the labyrinthodonts and the higher, amniote classes. Nasal and frontal elements, variable in crossopterygians, are prominent and elongate structures. The parietals, enclosing the pineal opening, are placed far back on the skull. There are well-developed premaxillae and maxillae. A circumorbital series includes five bones: prefrontal, postfrontal, postorbital, jugal, and lacrimal. The last is variable in development, but sometimes extends all the way from orbit to nostril and comes to carry the tear duct (a fish eye is naturally bathed in water, but that of a land form must be kept moist and have a duct to carry off superfluous liquid). There is a single outer nasal opening, or external naris, typically bounded by premaxilla, maxilla, nasal, and lacrimal; in primitive types a small element—the septomaxilla—forms part of the margin of the opening and may exclude the lacrimal. A large squamosal and a quadratojugal are present on the cheek, both having a contact with the quadrate bone at the jaw articulation. At the side of the parietal lie one or two temporal elements—a supratemporal and an intertemporal, the latter frequently absent. At the back of the roof are postparietals (der-

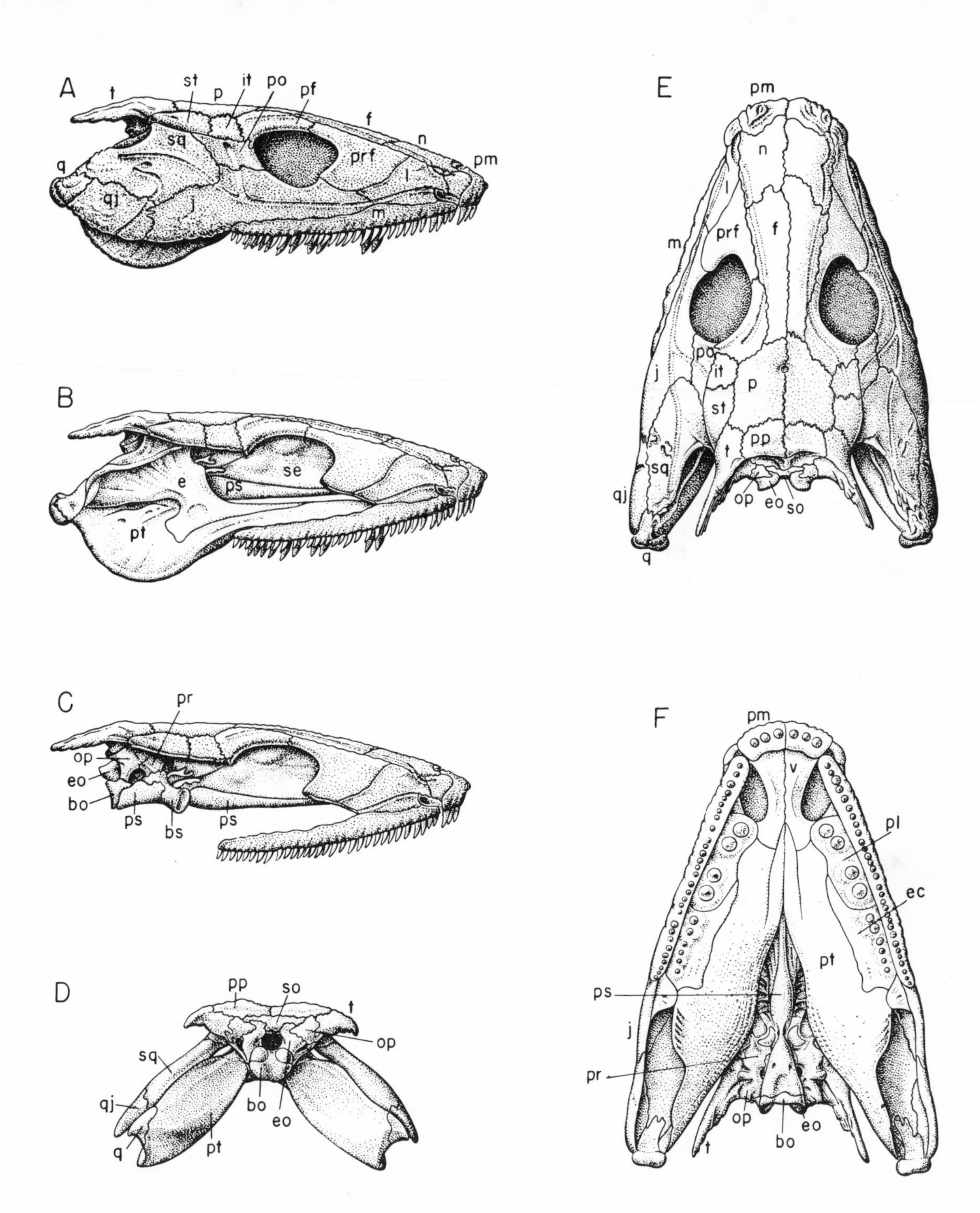

FIG. 109.—The skull of the embolomerous labyrinthodont amphibian *Palaeogyrinus*, partially restored to show the structure of a primitive tetrapod skull. Length of original about 7.5 inches. (Based on figures of Panchen.) *A*, Lateral view; *B*, the same, dermal bones of the cheek region removed to show the primary upper jaw (quadrate, epipterygoid, and associated dermal bones —pterygoid and ectopterygoid); *C*, the same, primary upper jaw removed; *D*, occiput; *E*, dorsal view; *F*, palatal view. Lateral-line canals stippled. Abbreviations for this and other amphibian and reptilian skulls: *a*, angular; *ar*, articular; *bo*, basioccipital; *bs*, basisphenoid; *c*, coronoid; *d*, dentary; *do*, postparietal (dermal supraoccipital); *e*, epipterygoid; *ec*, ectopterygoid; *en*, external nares; *eo*, exoccipital; *f*, frontal; *fm*, foramen magnum; *in*, internal nares; *ina*, internasal; *it*, intertemporal; *j*, jugal; *l*, lacrimal; *m*, maxilla; *n*, nasal; *on*, otic notch; *op*, opisthotic; *or*, orbit; *p*, parietal; *pa*, prearticular; *pap*, palpebral; *pd*, predentary; *per*, periotic; *pf*, postfrontal; *pl*, palatine; *pm*, premaxilla; *po*, postorbital; *pop*, preopercular; *pos*, postsplenial; *pp*, postparietal; *pr*, proötic; *prf*, prefrontal; *ps*, parasphenoid; *pt*, pterygoid; *pv*, vomer (prevomer); *q*, quadrate; *qj*, quadratojugal; *r*, rostral; *s*, stapes; *sa*, surangular; *se*, sphenethmoid; *sm*, septomaxillary; *so*, supraorbital; *soc*, supraoccipital; *sop*, subopercular; *sp*, splenial; *sq*, squamosal; *st*, supratemporal; *t*, tabular; *v*, vomer.

mal supraoccipitals), much reduced in extent from the condition seen in crossopterygians, and, at the corners of the skull table, tabulars.

With the loss of the gills, the operculum which covered them has disappeared, breaking the sheet of bones which originally extended unbroken from the skull to the shoulder region. The disappearance of the operculum has converted the slit in which the spiracle lay into an open otic (ear) notch, bounded by the tabular above and the squamosal below.

The primitive palate resembles that of the crossopterygians in its basic plan. The premaxilla and maxilla bear a marginal row of sharp teeth, typically with a labyrinthine structure. The teeth in early amphibians (and most reptiles as well) were replaced frequently. Waves of replacement appear to have passed from front to back of each jaw in such fashion that frequently we find that every other tooth in a row is mature and fully functional, while those between may be represented by "young" teeth or by alveoli in which replacements were presumably forming. Inside the marginal bones lie the vomers, the palatines, and the ectopterygoids, which in primitive types, just as in the rhipidistians, usually carry large teeth, often in alternately replaced pairs. The pterygoids are large and stretch back to the quadrates; toward the midline, processes from them form part of a basal articulation with the braincase, movable in primitive forms. In the primary upper jaws there ossify only the quadrate, for lower-jaw articulation, and an epipterygoid, lying farther forward above the pterygoid and entering into the braincase articulation. The front end of the braincase is bounded below by a slim parasphenoid, between which and the pterygoids are openings, primitively small—the interpterygoid vacuities. The internal opening of the nostril (choana) is between premaxilla, maxilla, vomer, and palatine.

The braincase is well ossified. In contrast to the two-part braincase of crossopterygians, it is typically a fused single structure in which few sutures are visible. At the back may be present four occipital elements—the median basioccipital and supraoccipital and the paired exoccipitals. A single rounded surface mainly on the basioccipital—the occipital condyle—joins the skull with the first vertebra in primitive amphibians and reptiles. In typical crossopterygians a large notochordal canal ran forward from this point through the floor of the braincase; this is absent in amphibians above the most primitive level. At the sides the opisthotic (paroccipital) extends upward to the tabulars; in front of this and generally fused with it is the proötic, the two enclosing the inner-ear region. Ventrally in front of the basioccipital lies a well-developed basisphenoid, which internally is excavated for the pituitary. It articulates with the epipterygoids and pterygoids laterally and is covered below by the parasphenoid. An anterior braincase element is a tubular ossification, the sphenethmoid; in contrast with crossopterygians, the nasal capsules at the front end of the braincase fail to ossify in most cases.

The lower jaw of the labyrinthodonts resembles that of the crossopterygians. On the outer surface lies the long, tooth-bearing dentary, and below this a row of further dermal elements including two splenials, angular and surangular. On the inner side, close to the dentary, are coronoids (primitively, three in number) usually bearing an inner series of teeth. There is a long prearticular, and, as in most fish, the primary jaw is represented only by the small articular.

With the loss of gill breathing in the adult, the gill arches naturally are reduced, although these elements

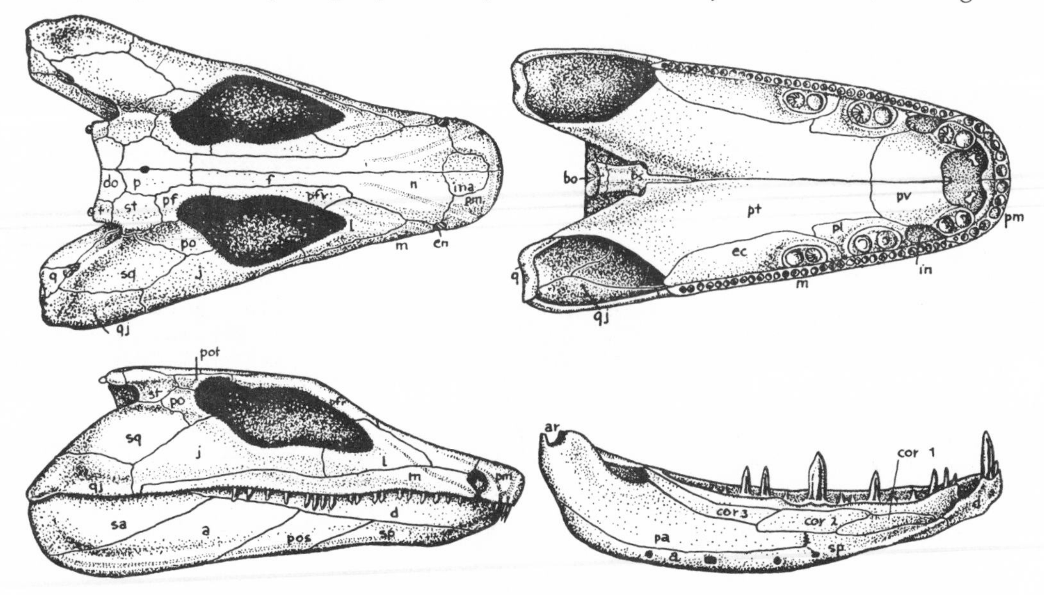

Fig. 110.—The skull of the Carboniferous amphibian *Megalocephalus* [*Orthosaurus*], dorsal, ventral, and lateral views and inner view of jaw; length of original about 12 inches. Lateral-line canals stippled. *pfr*, Prefrontal; *pof*, postfrontal. For other abbreviations see Figure 109. (After Watson.)

are still well developed in many amphibians. In amniotes there is much further reduction, although even in mammals the cartilages of the tongue and throat are remnants of this skeletal system. They are, however, seldom found in fossils and need little further consideration. The fate of the hyomandibular is discussed below.

Limb girdles.—The skeleton of the limbs and girdles is naturally highly developed. The shoulder girdle (Fig. 111) lies close behind the head, but in tetrapods the original connection with the skull has been lost, for in a land animal independent movements of head and shoulder are a necessity. The cleithrum—the upper bone of the dermal girdle—is retained in primitive land forms, as is the clavicle. In addition, there is developed a median ventral dermal element, the interclavicle, lying on the underside of the chest. Primitively the lower ends of the two clavicles and the interclavicle between them are expanded to form a broad, flat chest plate. The upper ends of the clavicles and the cleithra are, however, much reduced, while the primary shoulder girdle beneath tends to increase in size, since there had been a great increase in the musculature passing from it to the limb. The upper portion of the primary girdle is a flat scapular blade, below which is an articular surface for the humerus—the glenoid cavity. Ventrally there is a large, flat coracoid plate turning in toward the midline of the underside of the body. Primitively, the primary girdle appears to have included but a single ossification, seemingly homologous with the

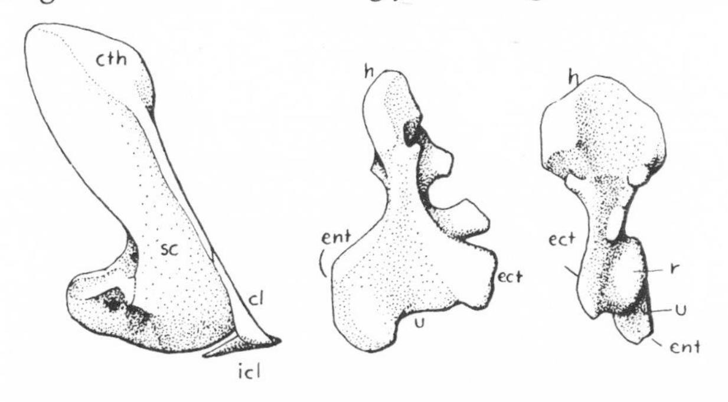

Fig. 111.—*Left,* Shoulder girdle of *Eryops: cl,* clavicle; *cth,* cleithrum; *icl,* interclavicle; *sc,* scapulocoracoid (primary girdle). *Center* and *right,* Dorsal and anterior views of the humerus of *Eryops: ect,* ectepicondyle, a muscular process on the outer (front) side of the humerus; *ent,* entepicondyle on the inner or back side; *h,* head; *r,* articular surface for radius; *u,* articular surface for ulna.

scapula of higher classes. In later tetrapods (seldom in amphibians) one or two additional coracoid ossifications are developed.

The pelvic girdle (Fig. 112) is a far larger structure than the small plate present in fishes. It has been enlarged to accommodate more muscles, and, while the fish plate is confined to the ventral side of the body, the primitive tetrapod girdle extends far dorsally and is joined to the backbone by one or more specialized sacral ribs. This extension is a functional necessity for land dwellers, for the hind legs in a primitive tetrapod must not only support much of the weight, but must

also give much of the push that propels the body forward. The dorsal extension of the girdle is usually a separate ossification, the ilium; in very primitive forms there is a long posterior process, presumably for attachment of ligaments leading to the muscles of the powerful tail. The head of the limb rests in a cavity known as the acetabulum. Below this is a broad ventral plate which joins its fellow of the opposite side in a

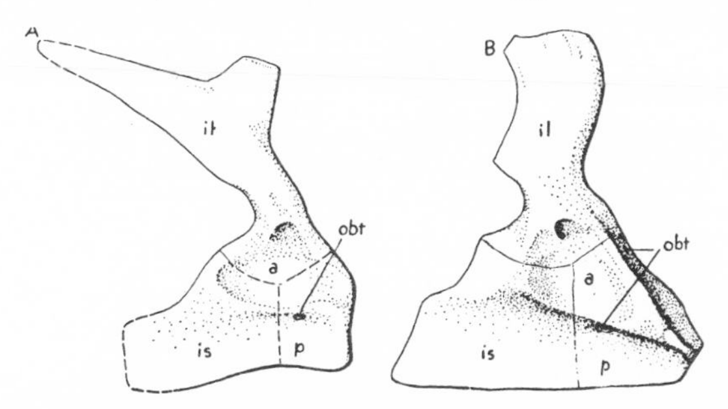

Fig. 112.—Pelvic girdles of *A,* the primitive type seen in embolomeres, etc., and *B,* the rhachitome *Eryops.* Abbreviations: *a,* acetabulum; *il,* ilium; *is,* ischium; *obt,* obturator foramen in pubis; *p,* pubis.

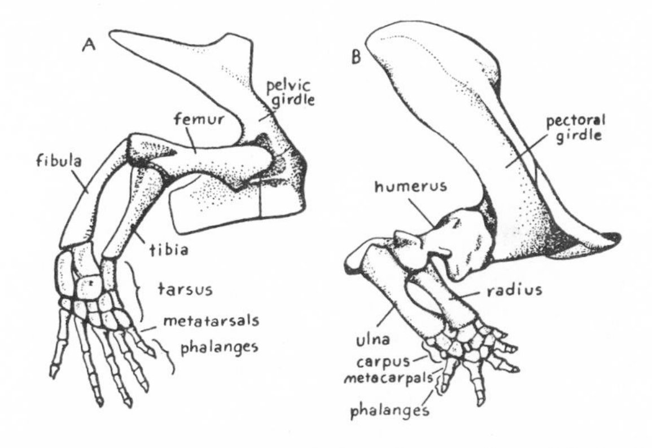

Fig. 113.—*A,* Lateral view of the pelvic girdle and limb of a primitive labyrinthodont (composite); *B,* lateral view of the pectoral girdle and limb of *Eryops.* (After Miner.)

median symphysis; this plate consists of two ossifications, the pubis (containing a nerve foramen) in front and the ischium behind; the pubis is often unossified in amphibians.

Limbs.—The limb skeleton is, of course, large in comparison with that of the paired fins of fishes. In the pectoral limb (Fig. 113*B*) the first segment is a broad and powerful element, the humerus (Fig. 111). Its head fits into the glenoid cavity of the shoulder girdle; distally it articulates with two elements which form the second segment of the limb—the radius and the ulna. The former lies on the inner side of the forearm and rests directly under the humerus, supporting much of the weight. The ulna lies outside and usually has a head, the olecranon ("funny bone"), extending above the edge of the humerus; muscles pulling on this extend the arm.

The radius and ulna articulate below with a complex

of small elements which constitute the carpus, or wrist (Fig. 114*A*). Typically there is a proximal row of three elements, the radiale lying under the radius, the ulnare beneath the ulna, and an intermedium lying between the two. Below the last is a large central element (which may articulate directly with the radius), followed by a row of three smaller centralia lying toward the radial side and, finally, a series of five distal carpals, each of which lies opposite the head of a toe.

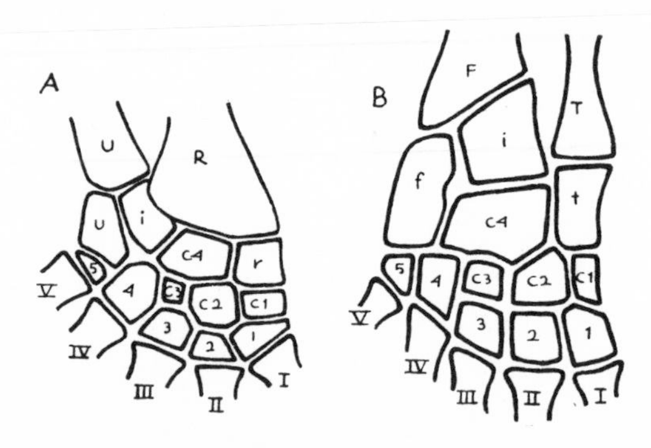

Fig. 114.—Nomenclature of the carpal and tarsal elements. *A*, Diagram of the right carpus of a primitive amphibian; *B*, diagram of the right tarsus. Abbreviations: *I–V*, metacarpals or metatarsals. 1–5, Distal carpals or tarsals; *c*1–*c*4, centralia. *F*, Fibula; *f*, fibulare; *i*, intermedium. *R*, Radius; *r*, radiale. *T*, Tibia; *t*, tibiale. *U*, Ulna; *u*, ulnare.

Each toe, or digit, of land forms (Fig. 115) consists of a first segment—a metacarpal—lying within the palm region of the front foot, and beyond this a variable number of elements—phalanges—making up the joints of the free portion of the toe. The number of toes in the truly ancestral amphibians is uncertain; quite possibly there may have been some variation in early stages of tetrapod evolution. Primitive reptiles had five in the front foot, and at least some of the anthracosaurian amphibians related to reptile ancestry had that number. However, four is a common number in living amphibians and in most early forms as well.

The inner digit, corresponding to the human thumb, receives the special name of the pollex. The number of phalanges varies, although the inner toes tend to be shorter than the outer ones. Reptiles typically have a phalangeal formula (counting the number of joints in successive toes from the pollex out) of 2–3–4–5–3 (or 4); amphibians generally have less.

The pelvic limb (Fig. 113*A*) may be compared, part for part, with the pectoral, although a different set of names is used for the bony elements. The femur, usually longer and slimmer than the humerus, constitutes the proximal segment. The next consists of the tibia within and the fibula on the outside. The ankle is the tarsus (Fig. 114*B*); and the names of the tarsal elements are similar to the carpal ones, except that, of course, the names tibiale and fibulare replace radiale and ulnare; the first toe is the hallux. Five toes, each including a metatarsal and phalanges, are commonly present in the hind foot.

The pose of the primitive land limbs is quite unlike that found in most living forms, although the straddling walk of a turtle is comparable (and, indeed, may have been retained from those early days with comparatively little modification). The proximal segments of the limbs are extended nearly straight out from the body, with the forearm and lower leg extended down at nearly a right angle. It is obvious that in such a position only short, broad strides could be taken—an assumption confirmed by numerous Carboniferous slabs showing footprints. A great deal of the animal's strength was used merely in keeping the low-slung body off the ground (footprint specimens almost never show a belly-drag). Walking must have been a slow and difficult process, especially for an animal of any size.

Other organs.—The probable condition of organs of other systems in such primitive vertebrates may be deduced from conditions still found in living forms. Gills would, of course, be retained in the water-dwelling young, and remains of feathery external gills have been found in a number of small, presumably larval, Paleozoic specimens. In the adult, gill breathing would have gone and the lungs alone been functional; internal nostrils were, of course, present. Usually in tetrapods all the gill openings disappear. In connection with this change, the circulatory system has been considerably modified; for, instead of passing to the gills, the blood must now pass to the lungs to be purified. This results in the destruction of most of the aortic arches which passed between the gills. The last one is enlarged, and forms part of the passage for blood to the lungs; more anteriorly another (the fourth) persists to carry the blood to the body. With this double flow of blood through the heart there is the beginning of a division of this organ into two separate sets of compartments; this division, however, is incomplete in amphibians.

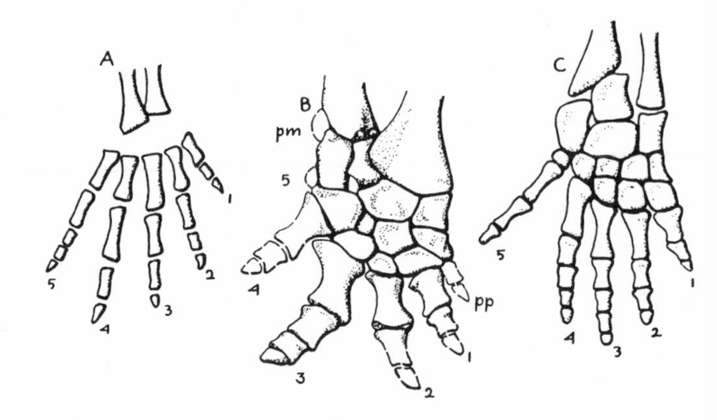

Fig. 115.—The feet of amphibians. *A*, Manus of *Diplovertebron*, a small anthracosaurian; *B*, manus of the rhachitomous amphibian *Eryops; C*, pes of the rhachitomous amphibian *Trematops*. Abbreviations: *pm*, postminimus; *pp*, prepollex. (*A* after Watson; *B* after Gregory, Miner, and Noble; *C* after Williston.)

The brain of amphibians is still small, especially the cerebral hemispheres, which are as yet almost entirely concerned with smell. The sense organs have undergone considerable modifications. The nostrils are adapted for olfaction under the changed circumstances of air instead of water for a medium. The eyes are usually large; the sclerotic plates, already mentioned as present in many fish groups, were retained in primitive tetrapods. A median eye was present in all early forms, but tended to be lost in post-Triassic tetrapods of most groups.

The ear region has undergone great change (Fig. 116*A,B*). The balancing function is, of course, retained. Hearing, however, is difficult in a land type, for delicate air vibrations cannot effectively set up pulsations which reach the ear through the thickness of the skull. We find in relation to this situation that most tetrapods have developed a new mechanism which amplifies the vibrations and transmits them to the internal ear. The spiracular gill pouch is retained, but near its outer end it is closed by a thin tight membrane—the ear drum—which picks up the sound vibrations. This drum primitively lies in the otic notch, which corresponds to the spiracular slit of the fish. Attached to the drum there is typically a small bone, the stapes, or stirrup (sometimes termed the columella), which at its inner end abuts on the otic region of the skull. Beneath this inner end is a foramen (fenestra ovalis) leading into the liquid-filled spaces of the internal ear, so that air waves caught by the drum and transmitted by the stapes set up vibrations in the internal ear; hearing thus results. The stapes is the fish hyomandibular, which was similarly propped up against the otic region of the fish skull. With the tightening of connections between jaws and skull, the hyomandibular's old function of a jaw support became superfluous, and, lying in the region just behind the spiracle, it was free to be adapted to this new function.

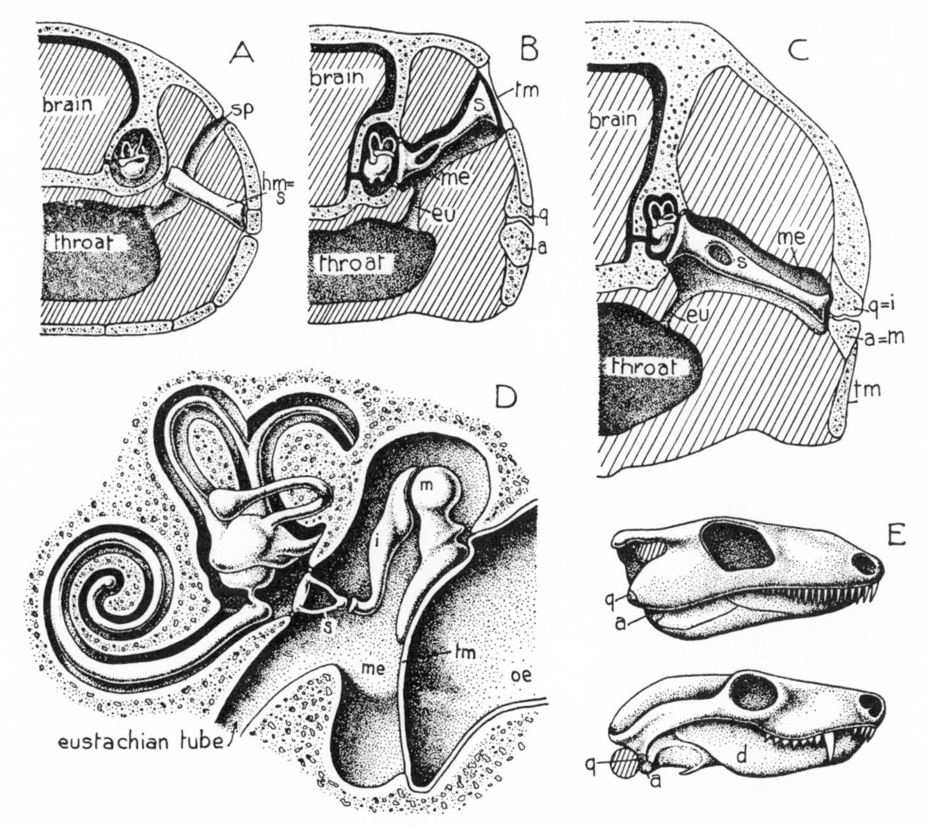

FIG. 116.—*A–D*, A series of stages in the evolution of the ear apparatus. *A*, Cross-section of half of a fish skull in the ear region. The ear structures consist only of the deep-lying sacs and semicircular canals. *B*, An amphibian. The hyomandibular bone (*hm*) of the fish is pressed into service as a sound transmitter, the stapes (*s*); the first gill slit, the spiracle (*sp*), becomes the Eustachian tube (*eu*) and the middle-ear cavity (*me*), while the outer end of the spiracle is closed by the tympanic membrane (*tm*). *C*, A mammal-like reptile. The stapes passes close to two skull bones (*q*, quadrate; *a*, articular) which form the jaw joint. *D*, Man (the ear region only, on a larger scale). The two jaw-joint bones have been pressed into service as accessory ossicles, the malleus (*m*) and the incus (*i*). *E*, A primitive land animal and a mammal-like reptile to show the relation of the eardrum to the jaw joint. At first in a notch high on the side of the skull (the otic notch) occupying the place of the fish spiracle, it shifts in mammal-like reptiles to the jaw region. In mammals the jaw comes to be formed of one bone only (*d*, the dentary), and the bones of the jaw-joint region are freed to act as accessory hearing organs; *oe*, tube of outer ear.

The lateral-line system of fishes is still retained in water-living amphibians. In bony fishes the lateral-line organs of the skull were in canals within the substance of the dermal bones. In primitive Devonian amphibians this is still the case. In later amphibians they have become superficial in position, but grooves for them can often be seen on the skull roof of fossil labyrinthodonts. In higher vertebrate classes these canals are lost entirely, and even among the amphibians, adult forms which are land dwellers lose them. Their presence or absence in a fossil is a good clue to its habits, for this sensory system is of use only in the water.

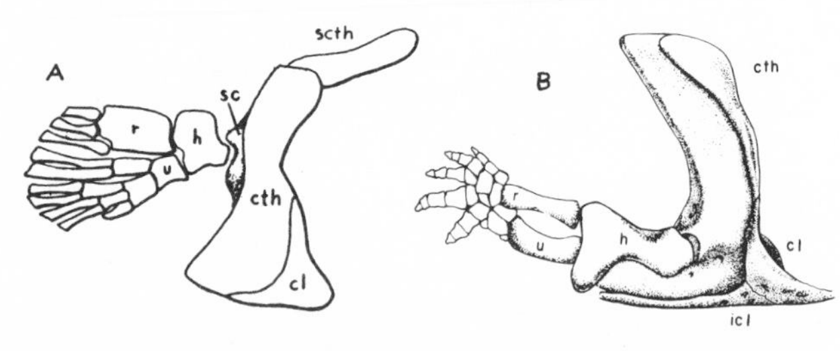

FIG. 117.—*A*, The pectoral girdle and fin of the Devonian crossopterygian *Sauripterus* (after Gregory); *B*, diagrammatic representation of a tetrapod limb placed in a comparable position. Abbreviations: *h*, humerus; *r*, radius; *u*, ulna; for other abbreviations cf. Figure 111.

Tetrapod ancestors.—From the description above, it is obvious that among the bony fishes the extinct crossopterygians were the closest to the ancestors of the tetrapods. The lungfish of today are very similar to the amphibians in their embryology and certain soft parts, but this merely indicates the retention of characters present in ancestral sarcopterygians; in such respects as teeth and skull structure the lungfish are definitely on an evolutionary side line and cannot themselves be tetrapod ancestors. As regards the actinopterygians, such features as the absence of internal nostrils and the lack of a large fleshy fin from which a land limb might develop rule this group out of consideration.

In almost every respect the rhipidistian crossopterygians fulfill the requirements for tetrapod ancestors. If a shift in skull proportions be taken into account, the skull-roof patterns in the two groups are readily comparable; the palatal and jaw structures are identical. Even the details of the dentition, with the labyrinthine structure and the peculiar pit-and-tooth arrangement on the palate, were identical in the two groups. The crossopterygian braincase, we have seen, is exceptional in its division into two units and the presence of the huge notochordal canal. These are structures not normally present in tetrapods, but primitive Devonian amphibians are transitional in braincase structure. The paired fins of crossopterygians were smaller than the limbs of tetrapods, but the structure of the skeleton was essentially similar (Fig. 117). In many crossopterygians there was a single proximal bone (corresponding to the humerus or femur of tetrapods), two bones in the next segment (as in the forearm or lower leg of land forms), and beyond this an irregular subdivision which is roughly comparable to the foot skeleton of primitive land types. The better-known crossopterygian genera are probably too specialized or too late in time to have been the actual ancestors, but the rhipidistians as a group seem unquestionably to represent the stock from which land vertebrates have sprung. A dual origin of tetrapods, from two fish groups (one or both being rhipidistians) has been advocated, but there is little positive evidence in favor of this theory (although, as discussed later, the early appearance of specialized lepospondyls presents a worrisome problem).

Origin of tetrapods.—The "why" of tetrapod origin has been often debated. Why did there develop forms capable of emerging onto land? Not to breathe air, for that could be done by merely coming to the surface of the pool. Not because ancestral amphibians were driven out in search of food—they were carnivores for whom there was, at first, little food on land. Not to escape enemies, for their fish ancestors were among the most powerful of vertebrates found in the fresh waters from which they came.

Their appearance on land seems, curiously, to have resulted from adaptations primarily useful in aiding them to survive in the water.

The earliest known amphibians appear to have led much the same sort of life as the related contemporary crossopterygians. Both lived normally in the same streams and pools; both were predaceous types, presumably feeding largely on small actinopterygian "minnows." As long as there was plenty of water, the crossopterygian probably was the better off of the two, for he was obviously the better swimmer—legs were in the way. The Devonian, during which land adaptations originated, was seemingly a time of seasonal droughts when life in fresh waters must have been difficult. Even then, if the water merely became stagnant and foul, the crossopterygian could come to the surface and breathe air as well as the amphibian. But if the water dried up altogether, the amphibian had the better of it. The fish, incapable of land locomotion, would be, literally, stuck in the mud, and, if the water did not soon return, must die. But the amphibian, with his short and clumsy but effective limbs, could crawl out of the pool and walk overland (probably very slowly and painfully at first) and reach the next pool where water still remained—and resume his normal existence as a water dweller!

Once this process had begun, it is easy to see how a land fauna might eventually have been built up. Instead of seeking water immediately, the amphibian might linger on the banks and devour stranded fish. Some types might gradually take to eating insects (primitive ones resembling cockroaches and dragon flies became abundant in the Pennsylvanian), and, finally, plants. The larger carnivores might take to eating their smaller amphibian relatives. Thus a true terrestrial fauna might eventually be established.

The problem of the proper classification of the amphibians has been—and still remains—a vexatious one, and no existing system seems completely satisfactory. We here utilize a scheme which at the moment appears best, in accord with present knowledge of the group (but which may, in turn, become obsolete with increased data regarding amphibian evolution). The two contrasting types of vertebral structures found in amphibians are here regarded as distinctive of major subdivisions of the Amphibia. The term Lepospondyli has long been used for a number of fossil groups, here considered as constituting a subclass, with the lepospondylous type of vertebral structure; the modern orders with a similar vertebral structure are here grouped as the subclass Lissamphibia. The "arch" type of centrum is found only in one group of amphibians, the subclass Labyrinthodontia. The labyrinthodonts include by far the larger number of known fossil amphibians from the older geologic periods. They flourished in great abundance and variety from the late Devonian to the end of the Triassic. The oldest types exhibit a structural transition from crossopterygian fishes to tetrapods. Other, later members of the group are clearly related to the origin of reptiles. But, as we shall see, the labyrinthodonts, despite their prominence in the early history of tetrapod evolution, give us no evidence as to the origin of the modern amphibian orders.

The earliest amphibians.—Amphibians are abundant and varied in the Carboniferous, and it is therefore certain that they originated well back in the Devonian, at a time when their crossopterygian relatives and ancestors were still flourishing. Our knowledge of Devonian amphibians is, however, meager. In the famous Upper Devonian beds at Scaumenac Bay in Canada there has been found, among the plentiful remains of

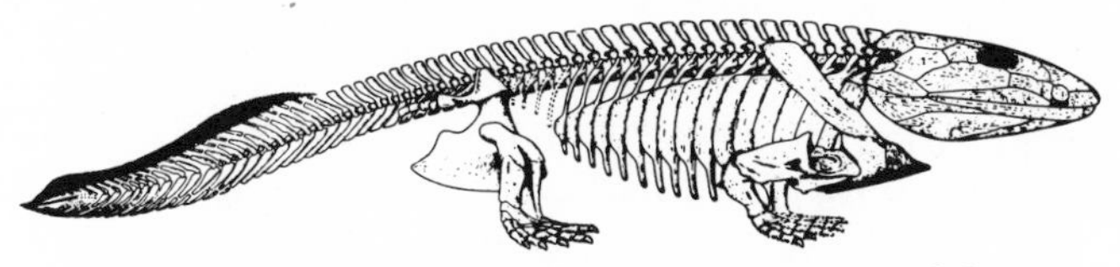

FIG. 118.—The oldest known amphibian skeleton, *Ichthyostega* of the late Devonian, about 3 feet long. (From Jarvik.)

fishes, a single small and incomplete skull roof of unusual nature, which has been named *Elpistostege* (Fig. 121*A*). As may be seen, the pattern of the bones is comparable, in general, to that of crossopterygian fishes, but appears even closer to that of the ichthyostegal amphibians discussed below. We have noted that a major difference between crossopterygian and amphibian skulls lies in the relative development of front and back portions; *Elpistostege* is intermediate in its proportions. It thus appears to represent in skull structure a transitional stage between the two groups. We know nothing at all of its postcranial skeleton—whether it bore fins or legs. Perhaps significant in attempting to reach a decision is the fact that the roof

shows no trace of the transverse break behind the parietals, which in crossopterygians is associated with the bipartite braincase. We may very tentatively assign it to the Amphibia and associate it with the more highly developed ichthyostegids next to be described.

In very late Devonian freshwater deposits of eastern Greenland has been discovered a series of skulls and postcranial skeletal materials described under the generic names *Ichthyostega* (Figs. 118, 119), *Ichthyostegopsis*, and *Acanthostega*. These genera are the oldest known forms which can be definitely classed as amphibians. They were animals of fairly good size; a skeleton of *Ichthyostega* is about 3 feet long. The skull-roof pattern is, in general, highly comparable to that of Carboniferous and Permian labyrinthodonts. It is, however, primitive in a number of respects which are suggestive of the crossopterygian condition. The snout is rather less developed than in most amphibians, and there is present a rostral element comparable to the rostrals of many bony fishes; the skull table is unusually long for an amphibian. There is no evidence of the fish operculum, but there is in *Ichthyostega* a remnant of the preopercular bone, which in bony fishes connects it with the cheek. The snout is so rounded that the nasal openings are invisible in dorsal view. In most land forms the internal and external openings are separated by a broad bar of bone; here they are separated only by a slender process of the maxilla—presumably a primitive condition. The palate is primitive in nature, with small interpterygoid vacuities. We have described earlier the braincase of crossopterygians, in

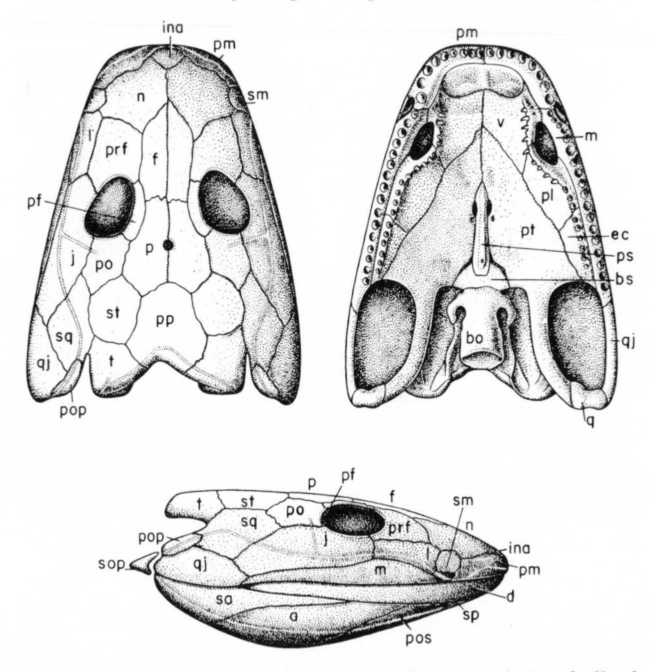

FIG. 119.—Dorsal, ventral. and lateral views of the skull of *Ichthyostega;* length of original about 8 inches. For abbreviations see Figure 109. (After Jarvik.)

which, quite in contrast to all "orthodox" animals, fish or tetrapods, this structure is built in two quite discrete parts, and with a large canal for the notochord piercing the floor of the posterior moiety. This unique braincase pattern was long thought to exclude the typical crossopterygians from tetrapod ancestry. The ichthyostegid pattern, however, removes this stumbling block, for it is intermediate in nature. The two halves of the braincase are in contact, although a distinct line of demarcation remains, but, on the other hand, the large notochordal canal still persists.

In recent years the postcranial skeleton of *Ichthyostega* has been discovered. Some piscine features are present, notably a small dorsal tail fin, in which, in contrast to typical later tetrapods, a row of bony fishlike tail supports is present above the neural spines. The stocky limbs and limb girdles, however, are already of the basic early amphibian type, although very short. Noteworthy is the centrum construction. To some degree this was persistently cartilaginous, but, as is essential for land locomotion, a fair degree of ossification was present. In each segment there is a large element, the intercentrum, crescentic in end view, wedge-shaped in side view, which surrounded the notochord ventrally and laterally, and high up on either side, a small block of bone, the pleurocentrum. This is, of course, an "arch" type of vertebra—the rhachitomous pattern, persisting in a great array of later amphibians. This type of vertebral column is a direct inheritance from the crossopterygians, in some of which an identical construction is present.

In most ways the ichthyostegids are, it would seem, ideal tetrapod ancestors. They are, however, a bit off the main line in a few regards. The intertemporal bone in the skull roof is here absent, although retained in many later labyrinthodonts, and the cheek region, still rather flexibly articulated with the skull roof in some Carboniferous forms, is here already firmly consolidated. There is no certain record of any ichthyostegids in later geologic times except for *Otocratia* of the Scottish Mississippian (Fig. 121*B*), which shows an ichthyostegid skull pattern. It has been suggested that the ichthyostegids form a point of departure for the evolution of the lepospondyls, but there is no positive evidence for this.

Temnospondyls.—Beyond this archaic ichthyostegal stage, the known labyrinthodonts appear to be divided into two major orders, Anthracosauria and Temnospondyli. The anthracosaur group, discussed later, led the way toward reptilian development; the temnospondyls were self-contained, giving rise, as far as we know, to no later tetrapod orders, but flourishing in a variety of adaptive forms not only during the late Paleozoic, but even through the whole extent of the Triassic period. A basic diagnostic difference between the two groups lies in vertebral construction. In the anthracosaurians the originally small pleurocentra, as we shall see later, tend to grow and fuse into a major structure which becomes the true centrum in reptiles and higher classes. In the temnospondyls, on the other hand, vertebral structure is more conservative. The pleurocentra remain small and vanish in advanced types, while the intercentra are persistently large and in many even form a complete ring, paralleling in appearance (although not in origin) the centrum developed in the anthracosaurians. A minor, but readily observed diagnostic feature is the fact that, almost without exception, the tabular bone of the skull roof is relatively small in temnospondyls and has no contact with the parietal, whereas in all anthracosaurians this contact is present (cf. Figs. 110, 120–22 with Figs. 109 and 133). In specimens in which the palate is preserved, temnospondyls are seen to have the internal nares widely separated by broad vomers, whereas in anthracosaurs the vomers form a relatively narrow bar between the narial openings, and usually lack the prominent tusks generally present here in temnospondyls. In these diagnostic features the ichthyostegals resemble the more typical temnospondyls.

There is easily discernible a "main line" of temnospondyl evolution from the Carboniferous upward through the Permian and Triassic. In members of the central temnospondyl stock the skull-roof pattern is a nearly uniform one, in the form of a moderately elongated oval, with the orbits on the average somewhat back of the middle. The pattern of roofing bones tends toward constancy, the only exception of note being the loss of the little intertemporal bone in all except the more primitive groups. Primitively the skull was, as in crossopterygians, fairly high, but in advanced forms it becomes progressively flattened, with the orbits facing directly upward. The palate was in early forms essentially a closed structure, with small interpterygoid vacuities; in later evolutionary stages there is a strong trend toward the development of large palatal vacuities on either side of a broad parasphenoid bar. The basal articulation of palate and braincase, originally movable, becomes a solid attachment, with a broad fusion of pterygoid and parasphenoid in many Triassic forms. The braincase was highly ossified in crossopterygians, and in the ichthyostegals as well; beyond this early stage, however, the nasal capsules are unossified, and in advanced temnospondyls much of the braincase, in the orbital and otic regions, is cartilaginous. The occipital condyle of the skull was primitively a single rounded structure; in most temnospondyls, however, the single condyle broadens and eventually becomes double (as is true of modern amphibians), a situation facilitating up-and-down movements of the head, but rendering sideways motion almost impossible.

The body primitively appears to have been more or less rounded in cross-section, in fishlike fashion. But in body, as in head, there appears in most cases to have been a progressive flattening and broadening. The limbs in many primitive forms appear to have been sturdy, if short. In a few groups (notably in the early

Permian) there is some advance in limb construction, facilitating progression on land. But in general, later types were regressive, with only small and feeble legs. These changes in body patterns appear to have been associated with changes in life habits. The early labyrinthodonts were water dwellers, living, despite their ability to emerge on occasion on to the land, in much the same fashion as their crossopterygian relatives. In a few types there was progress toward a true terrestrial habitus. But with the development and spread, in the Permian and Triassic, of reptiles better equipped for land life, this avenue of evolutionary advance was closed to the amphibians, and the later temnospondyls appear to have slumped back to a purely aquatic existence.

As regards the central temnospondyl line, we currently have fairly adequate knowledge of most features of this evolutionary history, and any attempt to split the series into successive stages is rather arbitrary. We may, however, array most of the temnospondyls in two groups, the more primitive forming the suborder Rhachitomi, the more advanced the suborder Stereospondyli. In the Rhachitomi we see the temnospondyl type of vertebra fully developed (Fig. 108*C*). The intercentra are the major elements in the construction of the centrum, but the pair of smaller pleurocentra are well ossified, and may fill completely the chinks between successive intercentra, making a column well built to support body weight when the animal emerges from the water. In the typical Stereospondyli (Fig. 108*D*) of the Triassic the intercentra have grown upward to form complete rings about the notochord, and the pleurocentra are absent; we shall, however, include in this group less advanced forms of the late Permian and early Triassic in which the intercentra have not yet become completely ring-shaped, but the pleurocentra are reduced or absent.

Rhachitomi.—Within the Rhachitomi, as here restricted, three successive stages are discernible. In the first, known at present only from the somewhat aberrant Carboniferous loxommids mentioned below, the skull was relatively high and narrow, palatal vacuities were practically nonexistent, the palate was still movably attached to the base of the braincase, the intertemporal bone still present, and the occipital condyle single. A second stage, better known, was characteristic of the Pennsylvanian, but with survivors such as *Edops* (Fig. 120), in the earliest Permian. Here some depression of the skull is present, and there are definite interpterygoid vacuities of modest size. Of a third stage, with numerous representatives in the late Carboniferous and early Permian, *Eryops* (Figs. 122, 123) of the Texas Permian red beds is best known. Here the limbs were well developed, and the animal's life, presumably centered in swampy regions, truly amphibious. There was, however, a marked trend toward skull flattening, the palatal openings were of increased size, the palate firmly connected, by a narrow bar, to the braincase floor, and the condyle partially divided.

Besides the main progressive series, numerous side

branches developed at every stage of temnospondyl evolution. In a sense, the ichthyostegals can be considered as the earliest temnospondyl side branch, since, despite a few specializations, they possessed all major diagnostic features of the group. Of rhachitomes proper, a very early group of aberrant forms include *Loxomma* and its relatives, such as *Megalocephalus* (Fig. 110), widespread in the Carboniferous. The loxommids are very primitive in almost every regard, but are characterized by a peculiarly enlarged and keyhole-shaped orbit, perhaps lodging a facial gland as well as the eye socket. At the still primitive rhachitome level of which *Edops* is representative, diverged

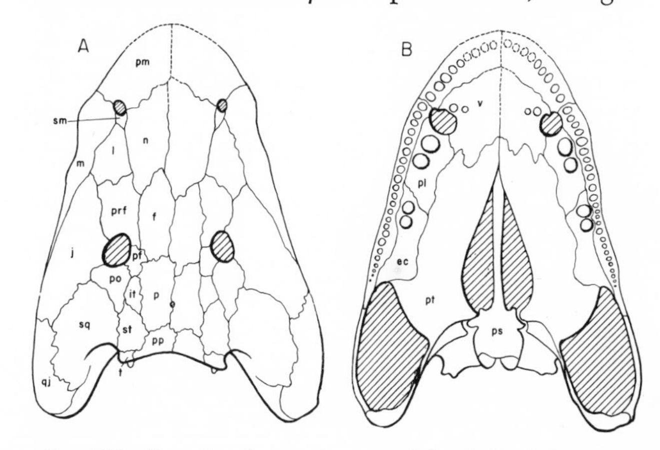

Fig. 120.—Dorsal and ventral views of the skull of the primitive rhachitome *Edops* (× 1/10).

Trimerorhachis (Fig. 121*C,D*) and its relatives, known from the late Carboniferous and early Permian, and possibly ancestral to later forms. Here the skull is much flattened, the face is short, the postorbital part of the skull elongate, and the palatal openings enlarged, but there are such primitive features as retention of the intertemporal bone and of the movable basal articulation. Somewhat parallel were *Colosteus* and *Erpetosaurus* (Fig. 121*E, F*), which had rather similar skull proportions and structures, but had lost the intertemporal bone.

From the more advanced rhachitome stage represented by *Eryops* and its relatives, side branches, some originating in the late Carboniferous, were numerous. Notable were the Dissorophidae, such as *Cacops* (Fig. 122), common in the early Permian, but already widespread in the Pennsylvanian. These were sturdily built little animals which seem to have been more truly terrestrial in habits than any other ancient amphibians. Presumably in relation to defense against the reptiles, already flourishing at that time, the advanced Permian members of the family developed a stout bony armor over the back.

We may here note the history of the "branchiosaurs" (Figs. 124, 125). In the late Pennsylvanian and early Permian deposits of Europe there have long been known many skeletons of tiny amphibians which show a number of points of resemblance to the rhachitomes, but differ not only in their small size, but in a much less ossified skeleton, a shorter skull, and other features. In many specimens clear traces of gills were present, and hence the name "*Branchiosaurus*" was appropriately given them. The vertebrae were feebly ossified and poorly preserved and were thought to differ from those of the Rhachitomi; on this account the branchiosaurs were considered to represent a separate order, the "Phyllospondyli"; a number of other Paleozoic amphibians were provisionally included in this group.

A study of growth stages, with changing proportions and increasing ossification in larger specimens, shows, however, that the branchiosaurs are merely larval labyrinthodonts, many of them gradually transforming, with increase in size, better ossification, and loss of gills, into rhachitomes of the *Eryops* type. The "order

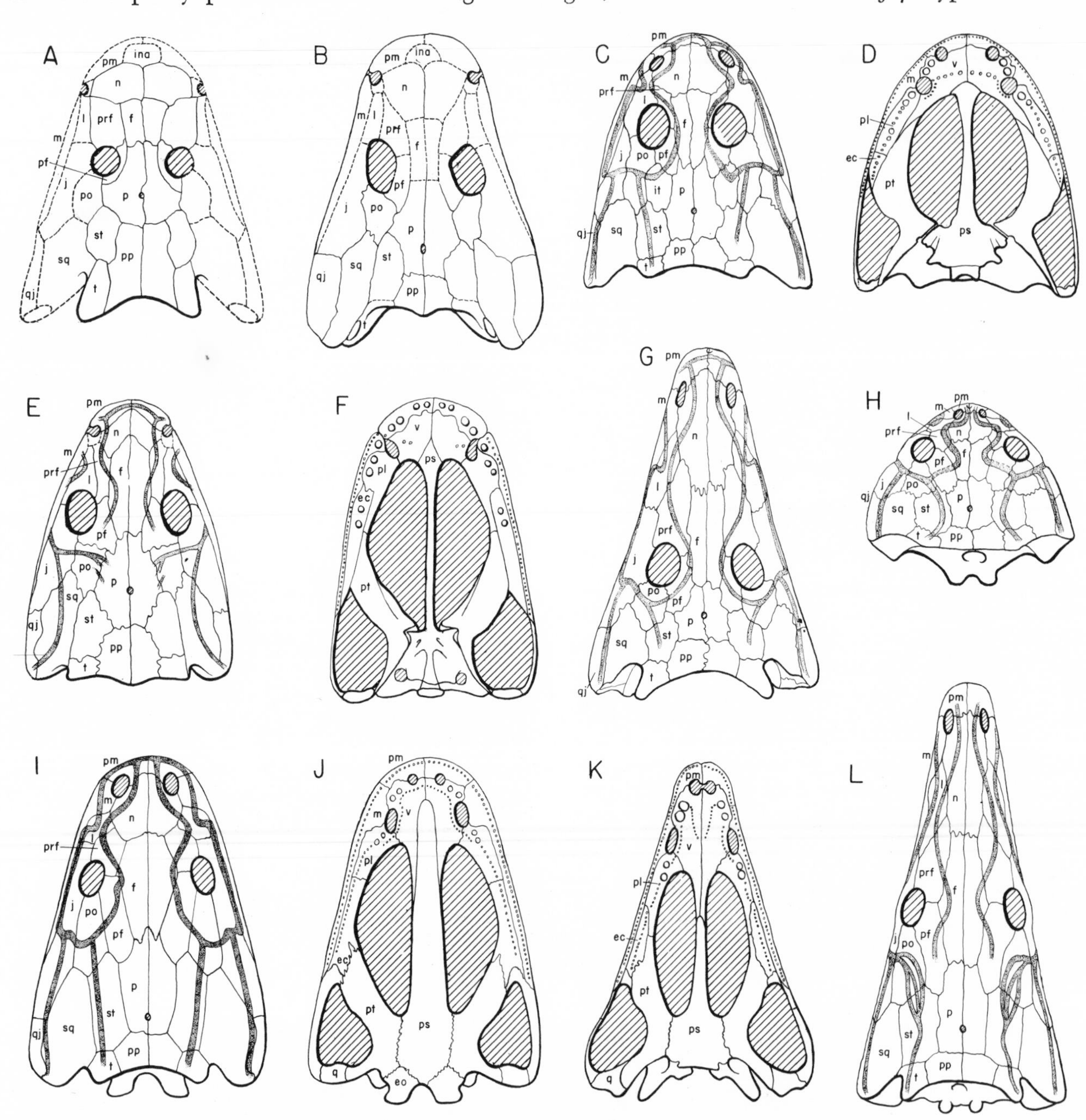

FIG. 121.—Dorsal and palatal views of the skulls of primitive and temnospondyl labyrinthodonts. *A*, *Elpistostege;* *B*, *Otocratia;* *C*, *D*, *Trimerorhachis;* *E*, *Colosteus;* *F*, *Erpetosaurus;* *G*, *Benthosuchus;* *H*, *Batrachosuchus;* *I*, *J*, *Metoposaurus;* *K*, *Benthosuchus;* *L*, *Trematosaurus.* (*A* after Westoll; *B*, *H* after Watson; *C*, *D* after Case; *G*, *K* after Efremov; *I*, *J* mainly after Fraas; *L* after Bystrow.)

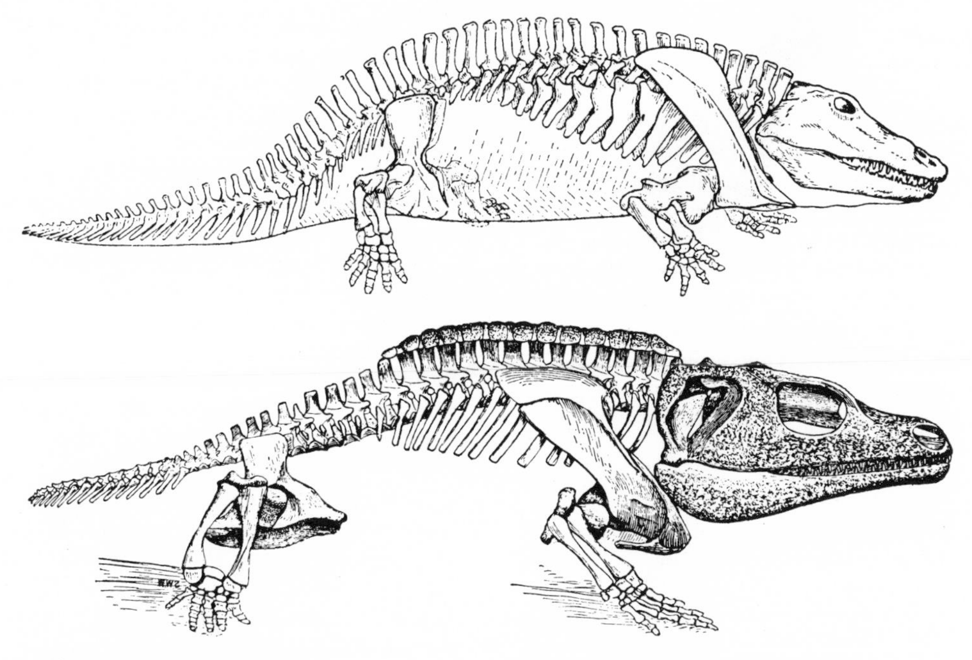

FIG. 122.—Permian rhachitomes. *Above, Eryops,* original about 5 feet in length (from Gregory); *below, Cacops,* with a peculiarly developed otic notch; original about 16 inches long (from Williston).

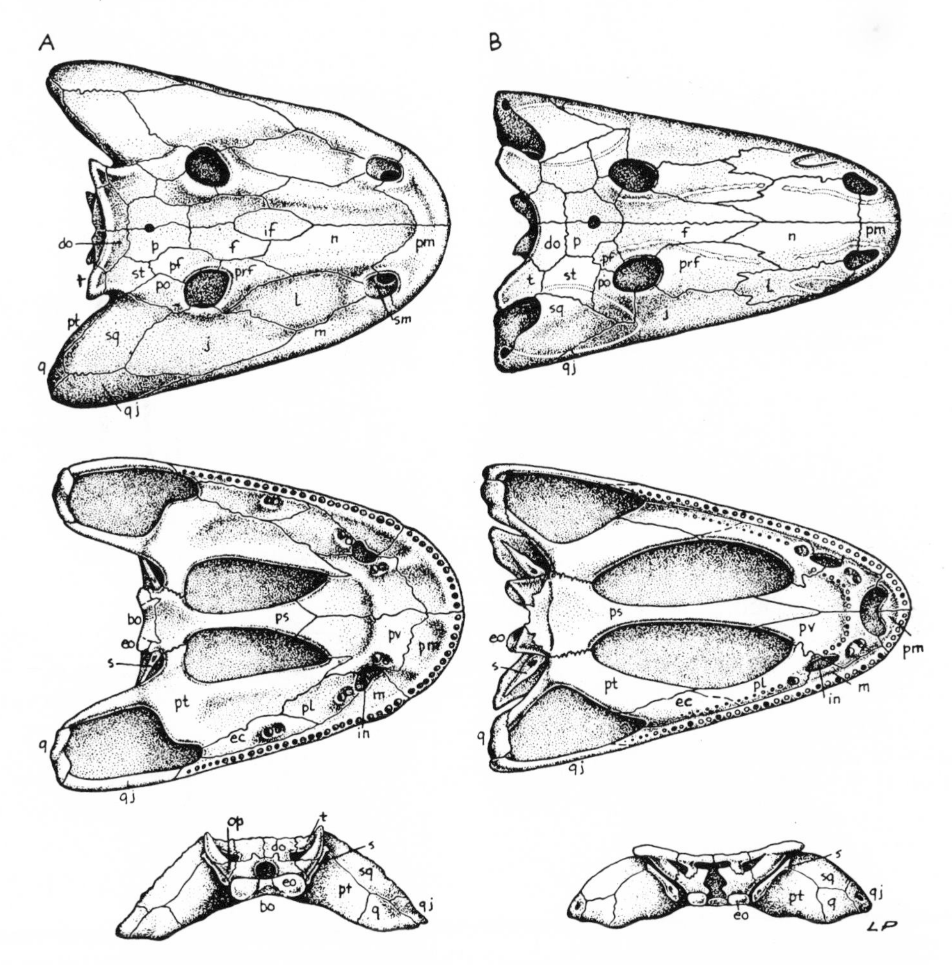

FIG. 123.—The skulls of *A,* the rhachitomous labyrinthodont *Eryops* and *B,* the stereospondyl *Capitosaurus*—dorsal, ventral, and occipital views. *if,* Interfrontal. For other abbreviations see Figure 109. (*Eryops* mainly after Broom, Watson, and Williston, length about 18 inches; *Capitosaurus* after Schroeder and Watson, length about 12 inches.)

Phyllospondyli," long included in every account of the Amphibia, is thus a purely imaginary group.

Even in early Permian times there began to develop long-snouted rhachitomes, such as *Archegosaurus,* which were presumably specialized fish eaters, and in the early Triassic there are found worldwide, aquatic long-snouted forms, such as *Trematosaurus* (Fig. 121*L*), transitional in vertebral structure between rhachitomous and stereospondylous types. These are found mainly in marine beds. Amphibians in general cannot survive in salt water. The trematosaurs are the

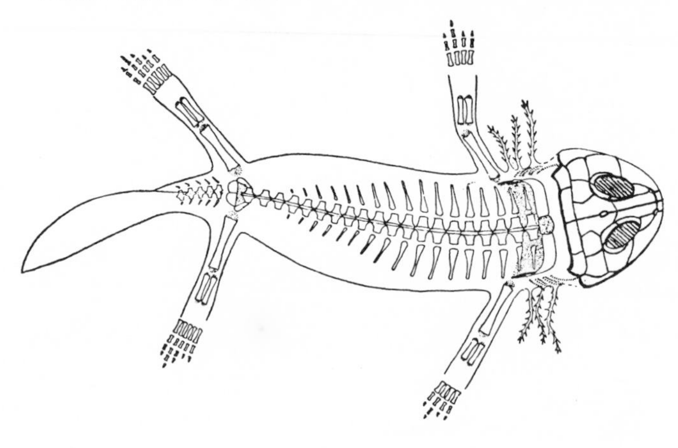

Fig. 124.—The skeleton of *"Branchiosaurus,"* a larval rhachitome from the Lower Permian, with remains of both external and internal gills. (After Bullman and Whittard.)

only major exception, apparently. Evidence suggests that there was at this stage considerable migration of ray-finned fishes into the seas. Possibly the trematosaurs followed them in this change from fresh to salt waters (perhaps, like some modern fishes, rearing their young inland).

Stereospondyls.—Resuming the temnospondyl "main line" story, beyond the evolutionary level of *Eryops*-like forms, we reach the Stereospondyli, in which further skull flattening occurs, the palatal vacuities increase further in size, the attachment of palate to braincase is a broad bar on either side of the prominent flat parasphenoid, the braincase is largely cartilaginous, the occipital condyle definitely double, and the limbs, where known, reduced. A minor but characteristic feature is that the palatal dentition, originally composed mainly of a few large pairs of tusks, tends to be supplemented or replaced by long rows of teeth of smaller size. Forms such as *Rhinesuchus,* which are clearly derived from *Eryops*-like ancestors, are common in the later Permian of South Africa and persist into the early Triassic. Although obviously ancestral to typical stereospondyls, the pleurocentra may persist although reduced, and the intercentra, though large, are not yet ring-shaped; such animals are hence sometimes reasonably termed "neorhachitomes." A stage beyond is abundantly represented in the early Triassic, from Greenland, Spitsbergen, and Russia, to Madagascar, by such genera as *Benthosuchus* (Fig. 121*G, K*) and *Wetlugasaurus,* in which pleurocentra have vanished, although the intercentral ring is still incomplete.

A final "main line" stage of stereospondyl evolution is reached in *Capitosaurus* (Fig. 123) and its allies, equally widespread, and the most common Triassic labyrinthodonts. A long series of genera has been described. Readily recognizable variations are found in the region of the otic notch. This is relatively small but deeply incised, and in *Cyclotosaurus,* for example, is completely surrounded by bone. The giant of all labyrinthodonts is *Mastodonsaurus* of the European Triassic, whose great flat skull attains 4 feet or so in length, and in which the pleurocentra are complete rings.

Among stereospondyls, as among rhachitomes, there were prominent divergent branches. From a primitive stereospondyl stage there apparently split off the brachyopids, such as *Batrachosuchus* (Fig. 121*H*), which as far as known were almost exclusively inhabitants of the southern continents (plus India), and are mainly early Triassic in age. The skull was quite broad and short—notably the facial region—but other features, such as the double condyles and palatal construction, suggest their relationship to other stereospondyls rather than to short-faced temnospondyls of more primitive nature. The vertebral centra consisted of incompletely developed intercentra of primitive stereospondyl type. A much more advanced stereospondyl

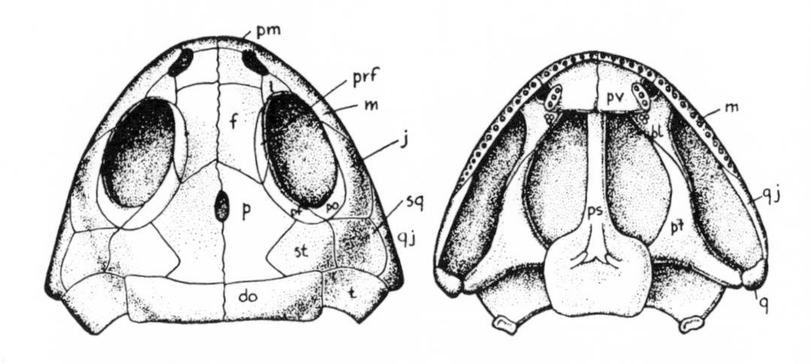

Fig. 125.—Skull of *"Branchiosaurus,"* dorsal and ventral views, enlarged. For abbreviations see Figure 109. (After Bullman and Whittard.)

variant, with the fully ring-shaped type of vertebral centrum, is *Metoposaurus* [*Eupelor, Buettneria*] of the Upper Triassic of northern continents (Fig. 126). In most regards *Metoposaurus* closely paralleled the contemporary capitosaurs. Here, however, the orbits were very far forward in the skull, the postorbital region greatly elongated. In proportions the metoposaurs rather resembled the more ancient *Trimerorhachis* types, but there is no evidence of relationship.

Plagiosaurs.—Finally to be considered among the temnospondyls is a most aberrant group, the plagiosaurs, such as *Plagiosaurus* and *Gerrothorax* (Fig. 127), not uncommon in late phases of the Triassic. Here the skull was developed in a fashion rather comparable to that of the brachyopids, but tended to be even shorter and broader. The postcranial skeleton is well known only in *Gerrothorax;* the body was broad and very flat, and armored above as well as below, the tail reduced, the limbs small. In this genus, at any rate, external gills for water breathing persisted in the adult—an example of neoteny, the persistence of larval characters in the adult, a phenomenon seen in a number of modern salamanders. At one time the plagiosaurs were thought to be allied with the brachyopids. But the vertebrae of brachyopids are relatively primitive, whereas those of the plagiosaurs are highly specialized structures, the centrum being a single imperforate cylinder of considerable length. Little is known of plagiosaur ancestry, but, in the late Permian of East Africa, *Peltobatrachus* shows a vertebral construction possibly antecedent to that of plagiosaurs, and a partially armored body. The skull of plagiosaurs, specialized as it is, shows basic temnospondyl features, but there appears to have been a long separate line of evolution leading to these grotesque Triassic end forms.

Anthracosaurs.—The temnospondyls, as we have seen, flourished in the late Paleozoic and Triassic, but despite a long and varied history, became extinct at the end of Triassic times and left no descendants. Quite in contrast in many ways is the story of the order Anthracosauria. The anthracosaurs are as yet relatively poorly known; they appear to have given rise to relatively few types, and they declined early, becoming extinct toward the close of the Permian. They are, however, of outstanding importance as including the ancestors of the reptiles, and, hence, of all advanced terrestrial vertebrates.

Although relatively few members of this order are adequately known, they can be readily distinguished from the temnospondyls by diagnostic features of the skull and—more important—of vertebral construction. We have already noted the readily observed feature of skull-roof pattern by which the two can be separated, the anthracosaur tabular invariably having a good sutural contact with the parietal; again, we have noted the contrast between the two groups in the region of the internal nares and vomers. As an item of difference in postcranial structure, it may be mentioned that as far as known, no temnospondyl has more than four digits in the front foot, whereas all anthracosaurian specimens in which the feet are known show (as in reptiles) five toes in the manus.

The basic skeletal distinction between temnospondyls and anthracosaurians lies in the direction taken in vertebral evolution. In temnospondyls, as we have seen, the intercentrum is a persistently dominant element; the pleurocentra are never conspicuous, and eventually disappear. In anthracosaurians the intercentra are generally present, but it is the pleurocentra which come to play the major role (Fig. 128). In a very primitive stage, the intercentrum retains its role as an important element, but the pleurocentra grow down on either side of the notochord to form a pair of half-rings. In a second stage of anthracosaurian development, the intercentra are persistently large, but the two pleurocentral half-rings have fused to form a complete ring centrum, analogous to, but not really comparable with the advanced stereospondyl "centrum" (formed from the intercentrum).

From this point, anthracosaurian evolution appears to have radiated in several directions. In one group, the Embolomeri, the intercentrum, as in stereospondyls, becomes, like the pleurocentrum, a complete ring, so that each segment of the backbone has two ring-shaped structures. In a second group, the Seymouriamorpha, the intercentrum may still occupy a considerable share of the central region, but tends toward a reduction in ossification dorsally. And, finally, as the reptilian stage is approached and reached, the intercentrum is reduced to a small ventral wedge, and the pleurocentral ring tends to elongate and take over nearly all the central region as the typical centrum of the higher vertebrate classes.

The two earlier stages in this evolutionary process were certainly reached in the early Carboniferous history of anthracosaurians, but because of the scarcity of Mississippian fossil remains, are poorly known. *Pholidogaster* (Fig. 129) of the Mississippian of Scotland is the oldest known anthracosaurian of which we have a skeleton, and is to date the only surely known form in which the adult shows pleurocentra in the form of paired half-rings (the young of some later anthracosaurians show this condition as a growth

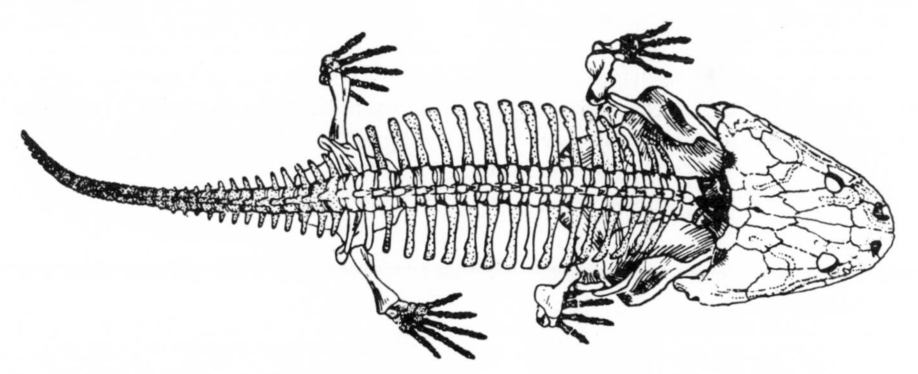

FIG. 126.—Dorsal view of the skeleton of the Triassic stereospondyl *Metoposaurus* [*Eupelor, Buettneria*], 1/20 natural size. (From Sawin.)

stage). Some obscurely known Carboniferous forms appear to represent the second stage in anthracosaurian vertebral development, but it is only adequately seen in *Diplovertebron* (Fig. 130) of the Pennsylvanian, presumably a late survivor of a primitive group.

Embolomeres.—Better known than these theoretically important but (at present) obscure ancestral groups are the members of the order Embolomeri, a major side branch of the anthracosaurians in which both intercentrum and pleurocentra form complete

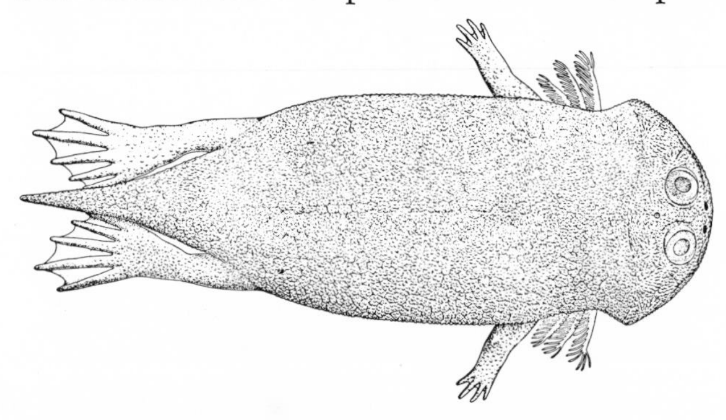

FIG. 127.—Dorsal view of the specialized late Triassic temnospondyl *Gerrothorax,* restored; original about 3 feet long. (After Nilsson.)

rings in every body segment. Embolomere remains—particularly the characteristic centra—are frequent in Carboniferous deposits, and it was once believed that the embolomeres were the truly ancestral tetrapods and that, as a corollary, all Carboniferous labyrinthodonts were embolomeres. This is not the case; but the embolomeres were a flourishing water-dwelling side branch of the anthracosaurians, in the Carboniferous and early Permian, persistently primitive in many features. *Pteroplax* [*Eogyrinus*] (Fig. 131) and *Palaeogyrinus* (Fig. 109) of the British Pennsylvanian, and *Archeria* of the early Permian of Texas are among the better known members of the order. Our foregoing account of primitive amphibian structure was based in great measure on such embolomeres. The skull, as expected in primitive forms, was relatively high and narrow and sometimes elongated; there was but a single occipital condyle. The skull-roof pattern was of a very primitive type. There was a deep otic notch; the cheek plate below it appears to have been but loosely attached to the skull roof. Both intertemporal and supratemporal elements are present in the skull table; the tabular, in anthracosaurian fashion, is in broad contact with the parietal. The palate is primitive in nature, with small vacuities, and there persists a movable articulation between braincase and palate on either side. The braincase is very well ossified. In the only forms in which the skeleton is at all well known, the body was much elongated, with a powerful tail but small limbs. Such embolomeres were, we believe, persistent water-dwelling fish eaters.

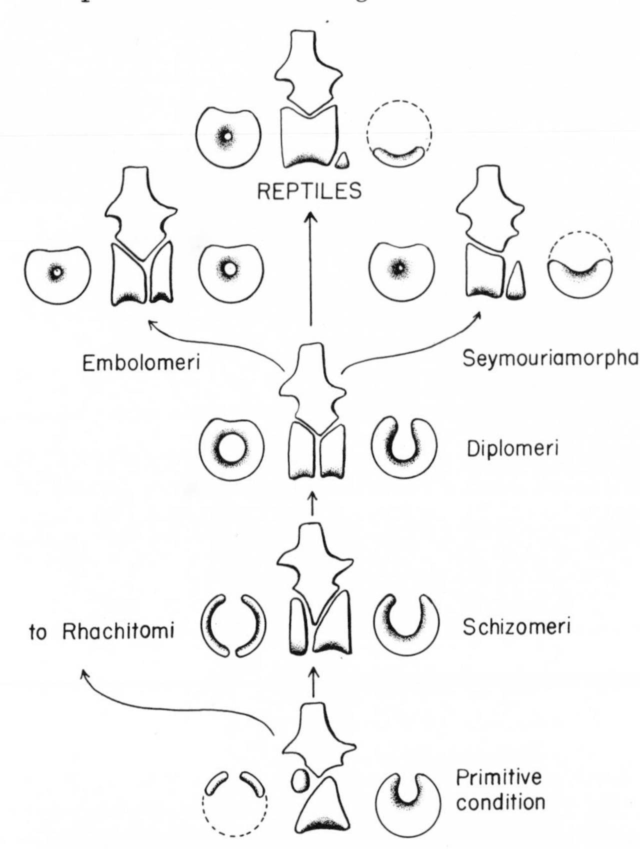

FIG. 128.—The evolution of vertebral central elements in anthracosaurian labyrinthodonts. In each stage the centra are shown from in front, the right side, and back. The pleurocentra begin as small paired elements, increase in size, and fuse to form a ring which becomes the definitive reptilian centrum. The intercentrum long persists as a large crescent, and in embolomeres forms a ring comparable to that made by the pleurocentra; in seymouriamorphs and reptiles, however, it decreases in size.

Seymouriamorphs.—*Seymouria* (Figs. 132, 133), a form of modest size from the Lower Permian redbeds of Texas, is one of the most frequently discussed of early tetrapods, for it exhibits such a combination of amphibian and reptilian characters that its proper position in the classification of vertebrates has been much disputed. Many of its structural features are those common in the skeletons of a wide variety of early tetrapods—labyrinthodonts and reptiles alike. The skull

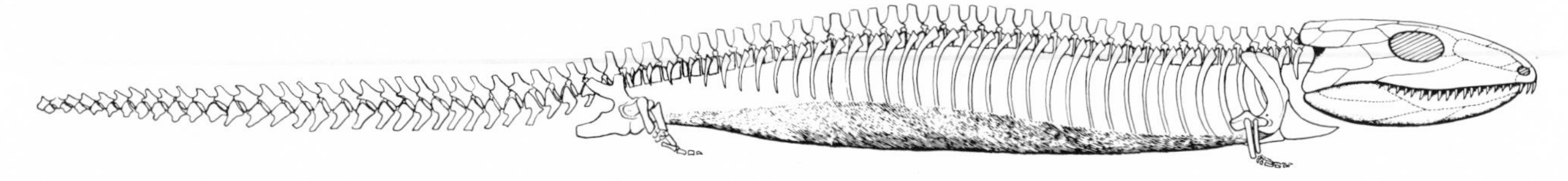

FIG. 129.—*Pholidogaster,* oldest known anthracosaurian, from the Lower Carboniferous of Scotland. Length about 44 inches.

roof is identical in pattern with that of anthracosaurs, even to the presence of an intertemporal bone, not present in reptiles. The palate is likewise primitive, with small interpterygoid vacuities. There is a single occipital condyle, as in primitive members of both classes, and the braincase shows features of both groups. Characteristic, however, of *Seymouria* and its relatives, but not of either typical anthracosaurs or of most reptiles, is the fact that the otic notch is very large and extends far forward. The limbs and girdles show many characters common to all primitive tetrapods. There are a number of features comparable to those in reptiles rather than to typical amphibians. The humerus is pierced by a foramen found in primitive reptiles but not in typical amphibians; the interclavicle has a long stem, whereas that of most amphibians is short and broad; the manus is five toed; the phalangeal formula is 2–3–4–5–3 (4), as in reptiles; the shoulder girdle, as in reptiles but not in typical amphibians, has a ventral coracoid ossification in addition to the scapula. These points suggest that *Seymouria* should be placed among the reptiles. But the humeral foramen is found in embolomeres and ichthyostegids, and is thus a truly primitive character. A stemmed interclavicle is occasionally found in other amphibian groups, and other anthracosaurians had a five-toed manus, although, as far as known, with a lower phalangeal count. The ilium, usually slender in amphibians, is here expanded somewhat, as in reptiles generally. Reptiles commonly have two or more sacral ribs; amphibians, one; *Seymouria* appears to be in the process of incorporating a second rib. The vertebrae (Fig. 128) are like those of primitive reptiles (cotylosaurs) in their broad and swollen neural arches and in the structure of the central region, with a complete disk of true centrum and a wedge-shaped intercentrum; but the intercentrum is unusually large, although not reaching as far dorsally as in primitive anthracosaurians.

Seymouria thus seems to be an anthracosaurian

which stands almost exactly on the dividing line between amphibians and reptiles; we have here a demonstration of the fact that there is no clear-cut distinction between the two classes in skeletal structures. The distinction between them is fundamenally one of modes of development. Did *Seymouria* lay its eggs, frog fashion, in the water, or produce shelled, land-laid eggs of amniote type? The probable answer as to the position of *Seymouria* is to be found not from a consideration of this one form, but from a broader study of a number of other Permian and late Carboniferous genera which are likewise included in the order Seymouriamorpha. For example, *Kotlassia* (Fig. 134), a late survivor of the group in the Upper Permian of Russia, and the early Permian *Discosauriscus* (Fig. 135) show in flattened skulls, relatively feeble ossification of the skeletons, and other features, a series of regressive changes associated in typical amphibians with a reversion to permanent water-dwelling existence. Again, there have been discovered in the early Permian of Moravia gill-bearing larvae of *Discosauriscus* similar to the "branchiosaur" larvae of rhachitomes, and indicating that in reproductive features the seymouriamorphs were definitely amphibians, although surely close to the line leading to reptiles and their shelled amniote eggs.

A further borderline form, generally regarded in the past as an archaic reptile, but perhaps better considered as an advanced anthracosaur, is *Diadectes* (Figs. 136, 137) of the early Permian. This large and clumsy animal has a specialized palate and transversely broadened cheek teeth, which appear to have had chewing potential; perhaps *Diadectes* is one of nature's first experiments in terrestrial herbivores (nearly all early tetrapods appear to have been predaceous in habits). The animal has many features seen in early reptiles, but these are all characters which it shares with *Sey-*

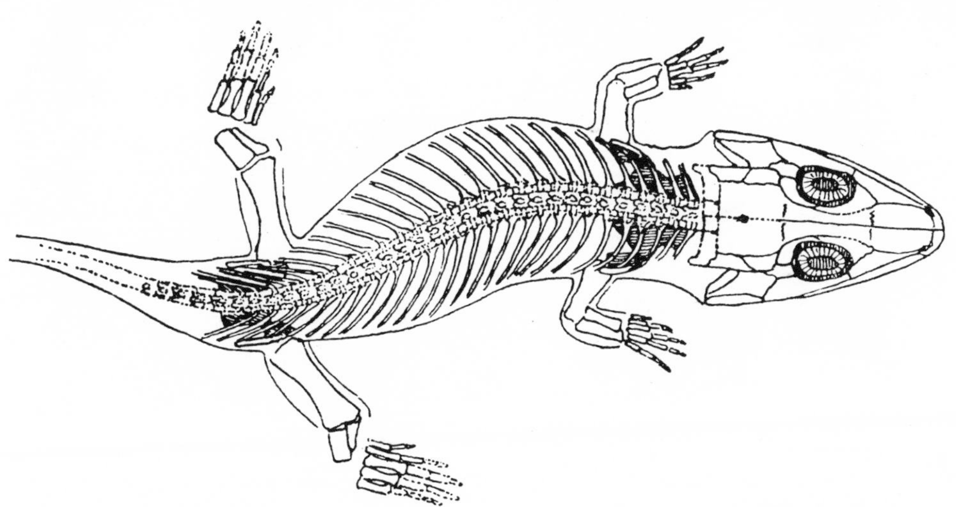

Fig. 130.—Dorsal view of the skeleton of *Diplovertebron*, a small and relatively primitive anthracosaurian from the mid-Pennsylvanian of Bohemia; about natural size. (From Watson.)

mouria. Notable is its enormously developed otic notch, in which the eardrum has been found to be ossified in some specimens. This notch can be derived by shortening of the jaw from the already large notch of *Seymouria,* and it is not improbable that *Diadectes* is an ally of the seymouriamorphs.

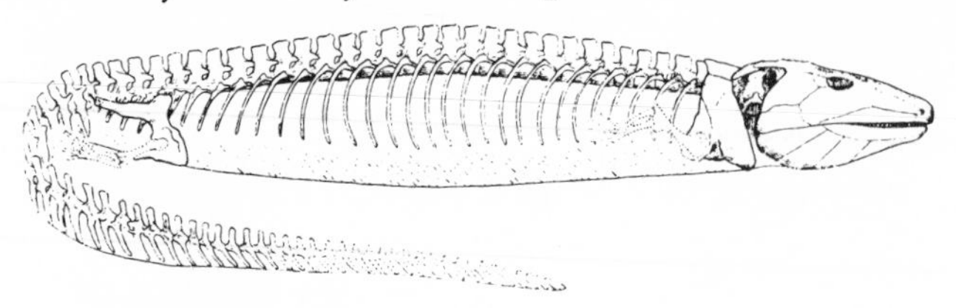

Fig. 131.—*Pteroplax* [*Eogyrinus*], a Carboniferous embolomerous amphibian; estimated length about 15 feet. (After Gregory, modified from Watson.)

Ancient lepospondyls.—We have distinguished from the "arch" type of vertebral build seen in labyrinthodonts a lepospondylous type (Fig. 107), in which the centrum forms as a single, spool-shaped bony cylinder around the notochord. Such vertebrae are found in the modern amphibian orders, here termed the subclass Lissamphibia; we shall, however, restrict the technical use of the term Lepospondyli to a subclass including numerous small forms of the late Paleozoic which appear to belong to three ordinal types—Aistopoda, Nectridea, and Microsauria. Such lepospondyls are known from rocks of early Mississippian age; all three orders flourished in the coal swamps of the Pennsylvanian; only a few forms remained in the Permian; none of the ancient groups has been found in later periods. These lepospondyls were animals of modest size. Most were but a few inches long; the largest, a foot or two at the most. It is currently assumed that the early labyrinthodonts were the true base of the tetrapod stock, but the very early appearance of lepospondyls and the great diversity which they exhibit in coal swamp days suggest that they branched off at the very beginning of amphibian evolution—possibly from the ichthyostegids.

Fig. 132.—*Seymouria,* an early Permian seymouriamorph; about 2 feet long.

Aistopods.—The order Aistopoda includes but three Pennsylvanian genera—*Ophiderpeton* (Fig. 138), *Phlegethontia* (139*A-C*), and *Dolichosoma* (Fig. 139*D*)—in which the limbs are reduced or lost and there is an elongate snakelike body, with up to two hundred vertebrae. The skull of *Ophiderpeton* differs greatly from that of the other two genera, but that the group is a natural one is shown not only by the unusual body

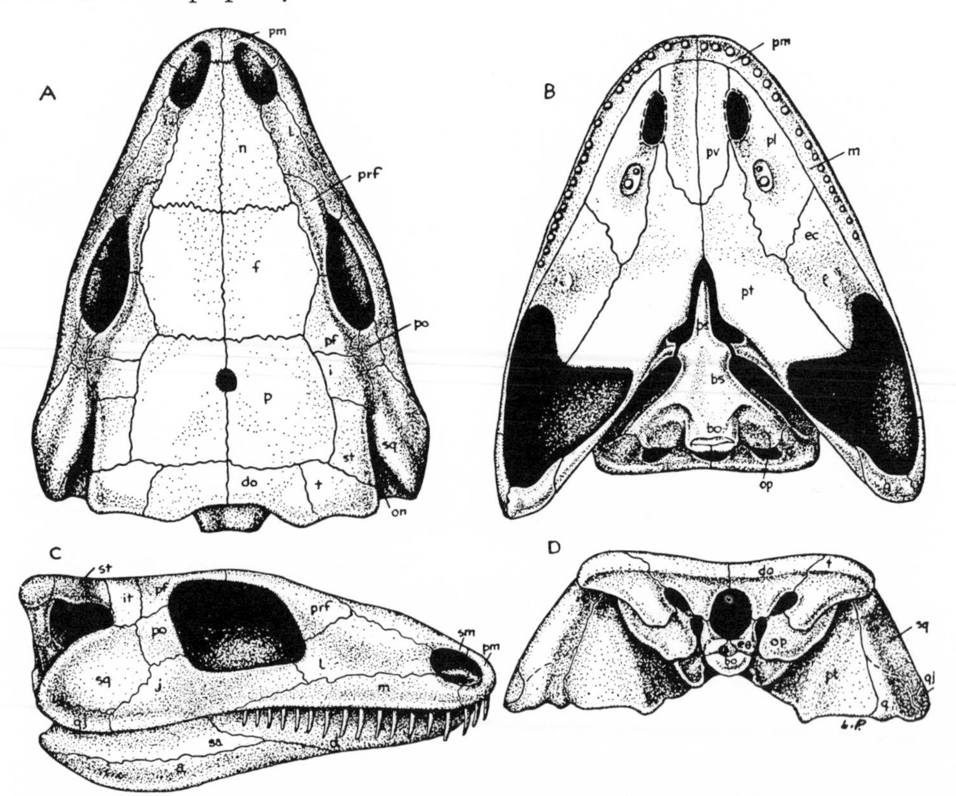

Fig. 133.—The skull of *Seymouria. A,* Dorsal; *B,* ventral; *C,* lateral; and *D,* posterior views. Length of original about 4.5 inches. For abbreviations see Figure 109. (After Williston and White.)

shape, but also by structural features, such as the presence in all of peculiarly forked single-headed ribs. One would expect that such specialized forms would represent a late development among the older amphibian groups; but material as yet undescribed proves that the aistopods had developed early in the Mississippian and hence are among the oldest of all known amphibians.

Nectrideans.—A more varied series of amphibians is that included in the Nectridea. Its members are characterized by the fact that the caudal vertebrae (Fig. 107*C*) have expanded, fan-shaped neural and haemal spines. The nectrideans of the Pennsylvanian are of two very distinct types, both of which had early Permian survivors. In such forms as *Urocordylus* and *Sauropleura* (Fig. 140) the body was very long, and, as in the aistopods, the limbs had been almost or entirely lost, while the skull was long and pointed. These eel-like types were locally very abundant in the Coal Measures swamps.

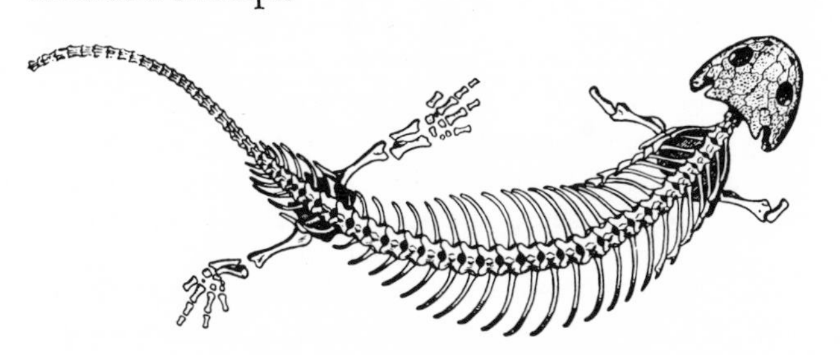

Fig. 134.—The skeleton of the Upper Permian seymouriamorph *Kotlassia,* about 1/12 natural size. The manus is unknown. (From Bystrow.)

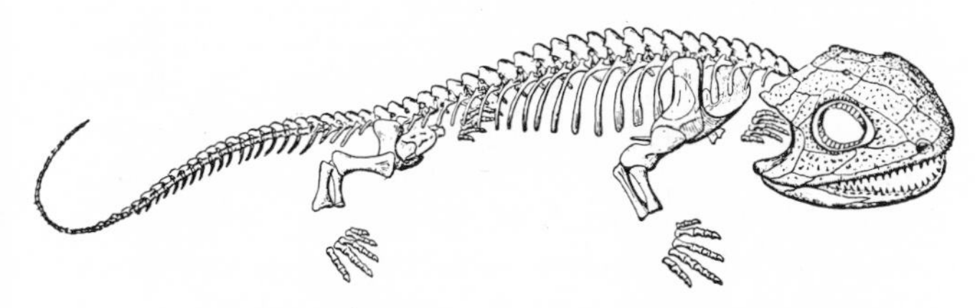

Fig. 135.—*Discosauriscus,* a small early Permian seymouriamorph, slightly less than natural size. (After Spinar.)

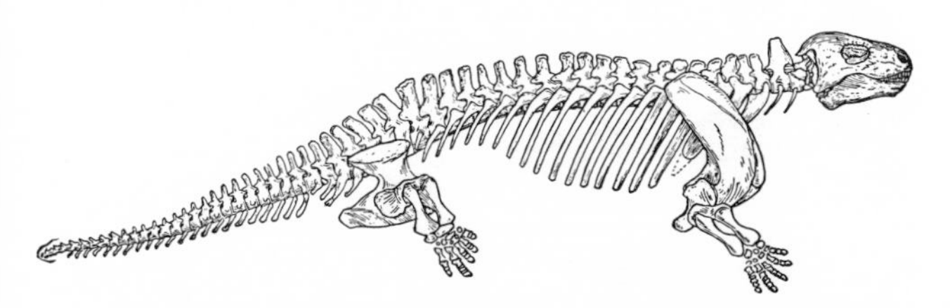

Fig. 136.—*Diadectes,* a highly specialized seymouriamorph of the early Permian. Maximum length about 10 feet. (After Gregory.)

A second group of nectrideans were forms with grotesque "horned" skulls. *Diplocaulus* and *Diploceraspis* (Fig. 141) of the Lower Permian were the last and most highly specialized. The broad flat body in some cases was 2 feet or so in length. These nectrideans exhibit a series of degenerative changes paralleling those seen in the later labyrinthodonts. As in stereospondyls, there were a large parasphenoid and huge interpterygoid vacuities; two very distinct condyles were developed; the braincase was almost unossified; and the skull as a whole was very much flattened. The greatest specialization lay in the fact that the tabulars were, so to speak, pulled out backward and laterally to form the major part of huge hornlike extensions of the skull

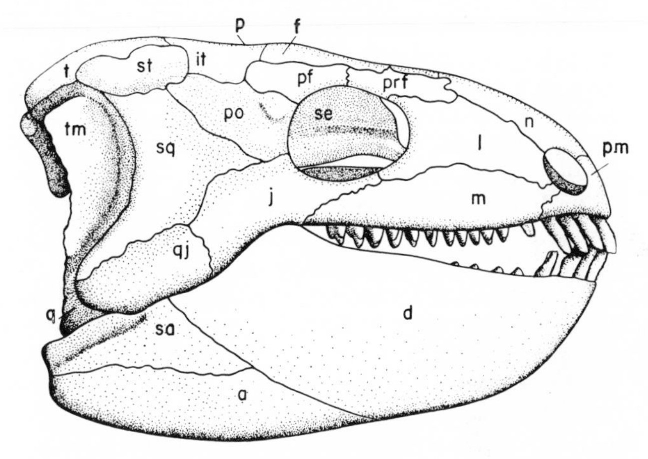

Fig. 137.—Lateral view of the skull of *Diadectes;* about 8 inches long. (After Olson and Watson.)

Fig. 138.—*Ophiderpeton,* a snakelike Pennsylvanian lepospondyl, estimated length about 2.5 feet. (After Fritsch.)

roof. The jaws, however, had not elongated, and lay far forward on the underside of the skull, while the eyes, also, had retained their anterior position, but looked directly up from the flat skull roof. The limbs were quite small, and the clavicles and interclavicle expanded into huge flat plates. Obviously, these odd creatures were pond dwellers, quite unable to successfully venture onto the land. In the Pennsylvanian are morphologically ancestral forms in which the "horns" were not so long, the skull a bit higher, the interpterygoid vacuities small, the limbs not so disproportionately small, and the body less flattened. They suggest that the descent of the nectrideans can be traced back to a form with a structure comparable (except in the vertebrae) with that found in early labyrinthodonts.

Microsaurs.—A third group of ancient lepospondyls

is that of the Microsauria. This term, "little reptiles," is hardly appropriate, for, although they paralleled the early reptiles in some regards and have sometimes been considered to be reptile relatives, the resemblances are merely parallelisms and their affinities may lie rather with the modern orders of Amphibia. As with the other lepospondyls, the peak of development of microsaurs lay in the Pennsylvanian, and there were but a few early Permian survivors. Microsaurs generally retained a rather normal type of body build, as seen in *Microbrachis* (Fig. 142), although the trunk tended to be a bit elongate and the limbs rather weak; as far as known, there were no more than three digits in the manus. There are variations in skull structure, but a basic type appears to be shown in such forms as *Euryodus* (Fig. 143) and *Pantylus*. As may be observed, the elements present are for the most part comparable with those of labyrinthodonts. However, the temporal region at the side of the skull table offers definite diagnostic features. This area in labyrinthodonts, we have seen, is occupied by a row of two or even three bones—intertemporal, supratemporal, and tabular (cf. Figs. 109, 110, 120, 121, 123); in microsaurs, in the same area, we find only a single large element, provisionally identified with the supratemporal. Often confused with the microsaurs are some small true reptiles (captorhinomorphs) present in Pennsylvanian deposits.

Sometimes included in the microsaurs, sometimes considered as a separate group, is a series of forms of which *Lysorophus* of the early Permian (Figs. 107*A*, *B*; 144) is best known. This little animal, which appears to have dwelt in or about pools of water, had paralleled aistopods and some nectrideans in limb reduction; the size and proportions are much like those of a large angleworm. The skull is much modified. In contrast to most Paleozoic amphibians, there was no pineal opening; and the skull was no longer completely roofed, for most of the circumorbital bones were gone and the orbit was open below. The parasphenoid was large, as in most degenerate amphibian types, but the occipital condyle had remained single. In the "throat" the branchial arches were large and ossified, suggesting that this form retained gill breathing throughout life—a feature probably true of various other lepospondyls. *Lysorophus* (Fig. 144) is obviously highly modified; but it is of interest that (as in the case of the equally specialized aistopods), forms from early Mississippian levels in Scotland appear to be relatives, albeit not so highly modified as *Lysorophus*. It is obvious that we still have much to learn about the earliest phases of amphibian evolution.

The modern orders.—We have now concluded the story of the two major groups of ancient amphibians, and may turn to a discussion of the three orders existing today, which we shall here consider as constituting the subclass Lissamphibia. Between them and the Paleozoic group is a broad evolutionary gap, not bridged by fossil materials. There are many major structural differences between the frogs and toads (Salientia or Anura) and the salamanders (Urodela or Caudata), and between either of these modestly successful groups and the curiously wormlike Apoda. But frogs and salamanders—and to a lesser degree the apodans—share a number of seemingly significant common characteristics, strongly suggesting a common ancestry—an ancestry, however, to which the fossil record, in its current stage of exploration, gives no certain clue. In both frogs and salamanders the skull is much flattened, and there has been much reduction in ossification of the braincase. Exoccipitals bearing the paired condyles and a sphenethmoid are almost universally present, and there is sometimes an opisthotic; but other braincase elements, such as basioccipital, supraoccipital, and basisphenoid, are practically unknown. The pineal opening is lost, and many roofing bones have disappeared; the skull is no longer solidly roofed. Almost without exception every bone of the circumorbital series has vanished, together with the temporal elements, tabulars, and dermal supraoccipitals, leaving of the dermal bones only a reduced central row and a lateral rim along the jaw. As in most of the advanced members of the older groups, there are

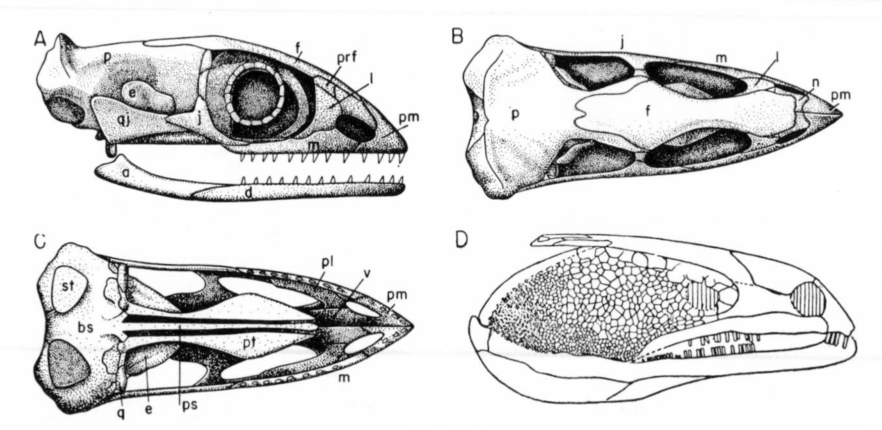

FIG. 139.—The skull in aistopods. *A–C*, Lateral, dorsal, and ventral views of the skull of *Phlegethontia*, about 5 times natural size (after Gregory and Turnbull); *D*, lateral view of the skull of *Dolichosoma*, about 1/2 natural size. (From Steen.)

in the palates of frogs and salamanders large interpterygoid vacuities separated by a parasphenoid bar. Both ectopterygoids and epipterygoids disappear, and the other elements may be much modified; the pterygoids are immovably attached to the braincase. Characteristic of many modern amphibians is the presence —in contrast to all other tetrapods—of two auditory ossicles fitting into the oval window of the ear, the stapes and a bone termed the operculum. The jaw elements are much reduced; at the most, we find a dentary, an angular, a prearticular, an articular, and a single coronoid, as compared with the ten bones present in early labyrinthodonts; and there may be still

FIG. 140.—*Sauropleura*, a snakelike Carboniferous nectridean. About 7.5 inches long. (After Steen.)

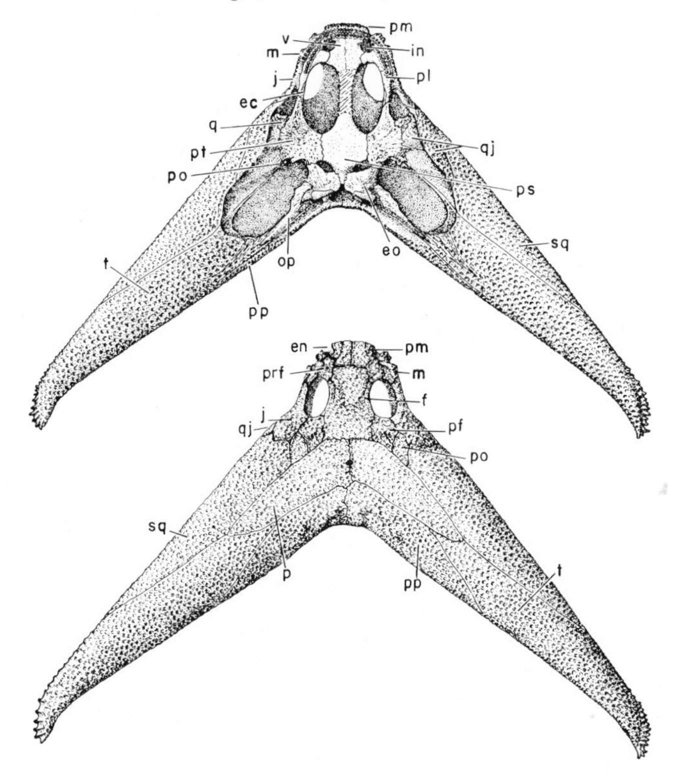

FIG. 141.—Ventral and dorsal views of the skull of *Diploceraspis*, a "horned" nectridean of the early Permian; about 2/5 natural size. (After Beerbower.)

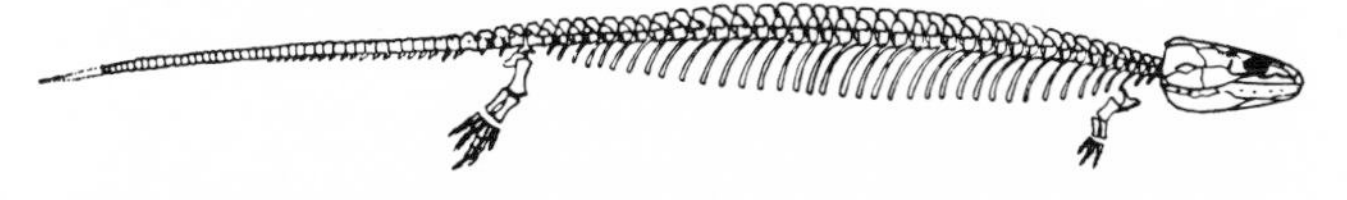

FIG. 142.—*Microbrachis*, a late Pennsylvanian microsaur, about 2/3 natural size. (From Steen.)

further reduction. The teeth of lissamphibians contrast with those of all other land vertebrates in that there is a separation of base and crown by a zone of weakness. No functional significance is known for this condition.

There is much loss of ossification in the limb skeleton; the pubis is never ossified and the carpus and tarsus may remain largely formed of cartilage. There are never more than four typical digits in the front foot, though an apparently secondary prepollex may occur in some groups. There is no trace of scales in either frogs or salamanders.

Anurans.—Most flourishing of the lissamphibians are the frogs and toads, here considered as forming the

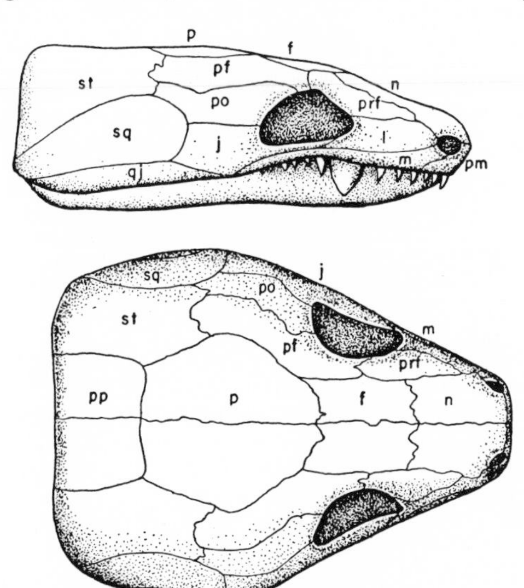

FIG. 143.—The skull of *Euryodus*, an early Permian microsaur, in dorsal and lateral views, about twice natural size. For abbreviations see Figure 109. (After Olson.)

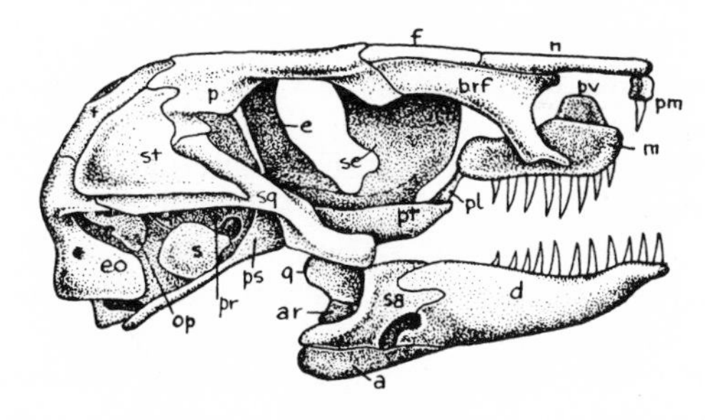

FIG. 144.—*Lysorophus*, a Permian microsaur, with a reduced skull pattern, about 3 times natural size. For abbreviations see Figure 109. (After Sollas.)

superorder Salientia. There are numerous peculiar skeletal specializations (Fig. 145). Most obvious are the long hind legs and the absence of a tail. The axial skeleton is highly modified. The vertebral centra—spool shaped, and primitively separated by cartilaginous disks—are not usually separate structures, the adult vertebra being totally ossified from centers descending from the neural arches. There are only eight to ten vertebrae in the entire backbone (usually nine), although at the posterior end of the column, beyond the sacrum, there is a peculiar long spike of bone, the urostyle, which appears to represent a dozen or so tail vertebrae fused into a solid mass. Except in a few relatively primitive forms, the ribs have been replaced by long transverse processes.

The ilium is extended into a long rod which reaches forward a considerable distance to connect with the last of the vertebrae. The hind legs are relatively long and often powerfully developed for swimming or jumping; tibia and fibula are fused into a single element, as are the radius and ulna in the front leg. In the hind leg an extra joint is formed by elongation of two proximal tarsals, which themselves occasionally fuse. The shoulder girdle retains all the primitive parts, although it is rather peculiar in form; in many anurans

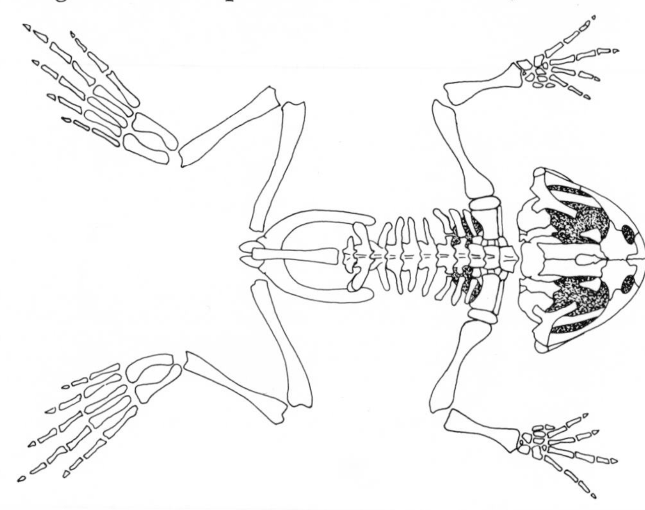

Fig. 145.—*Neobatrachus*, an early Jurassic frog, about natural size. (After Reig.)

the two sides of the girdle are braced on one another ventrally by apposition to the sternum—the firmisternal condition—as contrasted with the alternative arciferous condition, in which the two unfused halves of the girdle overlap ventrally. There is a separate coracoid ossification, an unusual feature in amphibians. Alone among living tetrapods, the frogs apparently have a vestige of the cleithrum in a partially ossified area above the scapula proper, and this bone is very well developed in certain fossil forms.

Arrangement of the fifteen or so existing anuran families in suborders is based to some degree on types of larvae. This is not altogether satisfactory when the fossil forms are considered, but is supported by some osteological evidence. Three main groups of anurans may be discerned. Generally believed to be primitive among living frogs and toads are the two genera forming the family Leiopelmatidae, in which there are nine presacral vertebrae rather than the usual eight. Free ribs are still present, and intervertebral cartilages fail to ossify. The Discoglossidae also have free ribs, and possess the same kind of tadpole as do the leiopelmatids. Together with some fossil forms, these families form a basal group—the Archaeobatrachia. The pipids are peculiar aquatic forms, and share the same kind of tadpole with the equally peculiar burrowing rhinophrynids, and may be grouped as the Aglossa. The remainder of the living frog families (with the exception of the microhylids) have similar tadpoles. Though much diversification has occurred within this group, they seem to be related and may be termed the Neobatrachia.

Records of late Cretaceous and Tertiary frogs are not uncommon; most are fragmentary and poorly preserved, and hence often difficult to assign to their proper position in the system of classification. It seems clear, however, that most of the modern families have been in existence since the early part of the Cretaceous. In recent years several good specimens of frogs have been found in the Jurassic of South America and Europe. These appear to be allied to the two more primitive groups of frogs—the Archaeobatrachia and the Aglossa. Nevertheless, neobatrachians are recorded from the late Jurassic as well.

Although these finds carry the frog story far back in time, they do not tell us much of frog evolutionary history, for even the "primitive" frog families differ only in relatively minor features from the more "advanced" ones. The basic pattern of anuran structure was already established by the early Jurassic and exemplified by the South American *Vieraella*—essentially a modern frog in its general adaptation, despite its great age.

The frog story can be carried back one stage further, into the Triassic, overlapping the history of the older groups in time, but without closing the morphological gap. In the early Triassic of Madagascar has been found the skeleton of a small animal, *Triadobatrachus* [*Protobatrachus*] (Fig. 146). The specimen displays a

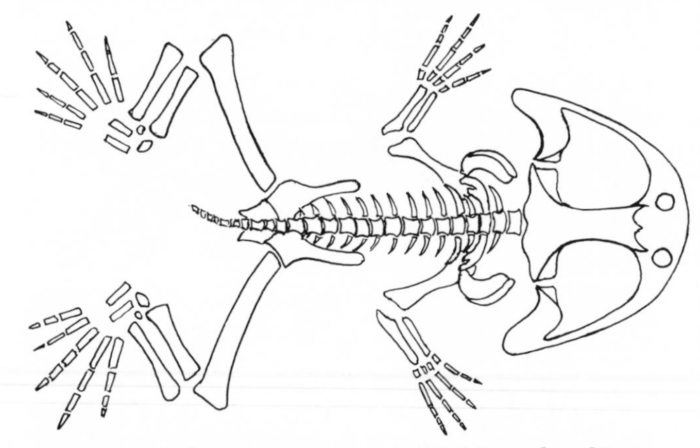

Fig. 146.—A sketch restoration of *Triadobatrachus* [*Protobatrachus*], a Triassic froglike form; data from Piveteau; feet, front of skull, etc., restored.

skull which, although incomplete, appears to be basically similar to that of modern frogs and toads. The postcranial skeleton, on the contrary, is not particularly froglike except for elongation of the ilium. The trunk is moderately shortened; ribs are present; the limbs are moderately long and slender and seem to show the tarsal elongation seen in modern anurans. Apparently skull specialization preceded major trunk and limb adaptations in anuran evolution. The cranial specializations seen in frogs are, as we have noted, similar in many ways to those present in the later temnospondyls, and an ancestry for the group among the labyrinthodonts has been advocated. No transitional types, however, are known, and one Carboniferous fossil type which was suggested as a frog ancestor proves to be a member of a typical rhachitomous family.

The salamanders and Apoda.—Although the newts and salamanders included in the order Urodela possess a long series of characters in common with the frogs, they differ markedly from the anurans in many respects. A few have paralleled certain of the ancient lepospondyls in limb reduction, but in most salamanders, limbs of normal build are present, although they are usually rather small. Externally, the salamanders, with a long body and well-developed tail, resemble primitive amphibian types. Internally, however, the skeleton shows many modifications. The limb girdles are largely cartilaginous. There is, generally, only one reduced ossification in the shoulder girdle, and the dermal girdle has disappeared completely. Many salamanders have adopted terrestrial life and often lay their eggs in moist places on land, but there is a general tendency in the group to persist in water living; a few types never leave the water, continuing, as neotenic forms, to breathe with gills throughout life. Elongation of the body and limb reduction accompany this tendency in some modern forms, and are characteristic of many of the early fossil types.

There are a number of salamanders known from the late Cretaceous and Tertiary. One large Miocene specimen (belonging to the genus *Andrias,* a species of which survives in Japan and China) was, some hundreds of years ago, regarded as the skeleton of a poor human sinner drowned in the flood, "Homo diluvii testis." The oldest known salamander is a late Jurassic genus. It is disappointing that even the older fossil salamanders show no primitive characteristics. The modern structural pattern of the urodeles was, it would seem, established by Jurassic time; there has since been little important evolutionary advance.

The Apoda (or Caecilia or Gymnophiona) include a few genera of small, tropical forms of very peculiar build. There are no limbs, and the general appearance is that of a large earthworm. Alone among living amphibians, the Apoda have retained scales, although these are small and buried in folds of skin. The skull is technically roofed, for there is a complete covering to the top and sides. But there has been a considerable loss of elements and much change in the shape and position of those remaining. The orbits are very small, for these forms are practically blind; and there are, instead, pits lodging tentacles which functionally take the place of the eyes as sensory organs. The solidly built little head is quite possibly derived from one in which much of the dermal covering had been lost; it seems likely that there has been a secondary enlargement of the remaining elements, resulting in the formation of a more effective burrowing organ.

Lissamphibian ancestry.—Whence have the modern orders been derived? Certain of the cranial characters of frogs and urodeles (large palatal vacuities, double condyles, and so forth) resemble those of advanced temnospondyls. But there are no conclusive lines of evidence indicating any direct relationship between the temnospondyls and the modern groups, and the adult vertebral pattern is quite different. We thus appear to be driven to a consideration of the ancient lepospondyls as possible ancestors. The modern orders are morphologically lepospondylous in adult vertebral construction, hence no obstacle is apparent in this regard. Of the known older groups, it is generally agreed that the elongated, snakelike aistopods are ruled out, although a few modern urodeles have evolved a somewhat comparable body shape. The nectrideans too seem improbable as ancestors, for not only do the nectridean vertebrae seem too specialized, but also all well-known members of the group became peculiar "horned" forms of the *Diplocaulus* type, or tended, as in *Sauropleura,* to specialize in a contrary direction as long-snouted, long-bodied forms not at all appropriate as ancestors of modern amphibians. We are thus left with the microsaurs and their *Lysorophus*-like associates. Here there are tendencies toward elongation of the postorbital skull region and no positive evidences of connection between microsaurs and urodeles. The differences between these early forms and the frogs and toads are broader still.

The temnospondyls show the greatest number of skull characters in common with the Lissamphibia; on the other hand, the lissamphibian vertebral construction suggests descent from Paleozoic lepospondyls. But the peculiar dentition and other specializations of the Lissamphibia are not seen in any Paleozoic group. We need to know much more about the smaller amphibian life of late Paleozoic and Triassic times before we can bridge the gap between modern and ancient amphibian groups.

7 CHAPTER

Stem Reptiles

There is no more interesting story in the fossil record than that of the rise and fall of the reptiles. Springing from labyrinthodont amphibians during the Carboniferous, they became the dominating group of vertebrates during the whole of the Mesozoic and gave rise to many curious and spectacular types. Now they are in decay; but it must not be forgotten that the birds and mammals which have triumphed over them are their descendants.

Class characteristics.—The major definitive character for the class Reptilia is the fact that, in contrast to amphibians, they have developed a type of egg which can be laid on land—the amniote egg. This type of egg (very familiar to us, since it was retained by birds) contains a large supply of nourishing yolk, so that there is no need for development of a larval form that must, as in the case of the frog tadpole, make its own living at a premature stage. A shell affords protection; membranes (notably the amnion, from which this egg type gets its name) surround the developing embryo and enable it to be bathed in liquid, with avoidance of desiccation. Other embryonic membranes form an embryonic bladder to care for the nitrogenous waste due to the active metabolism of the growing embryo and, highly important, form an embryonic lung, capable of absorbing atmospheric oxygen through the porous shell. Because of these devices, a discrete larval stage is unnecessary, and the young reptile can hatch from the egg as a well-developed little replica of the adult form.

It was at one time assumed that the amphibian ancestors of the reptiles were already tending strongly toward terrestrial life, and that the development of this egg type was a final step in the conquest of the land. However, the fossil record indicates that many of the earliest true reptiles were still water dwellers to a considerable degree. It is probable that the egg came ashore, so to speak, before the adult, and was developed as a device which avoided the vicissitudes of larval life—better protection against enemies and against drought during the developmental periods. But even if the initiation of this new developmental method came about in this fashion, it afforded the potentiality for a true conquest of the land, of which the reptiles rapidly took advantage.

In some respects the reptiles are persistently primitive. For example, typical reptiles are, unlike their avian and mammalian descendants, cold-blooded—that is, there is no effective mechanism for the regulation of body heat. In consequence, the activity of the animal varies with the external temperature, and reptiles characteristically are dwellers of warmer climatic zones. In other regards, however, reptiles show a higher type of organization than amphibians, which is apparent in every part of the body—in skeleton, in muscle, in the circulatory system. Even in the brain, the reptile is on a higher plane; for, while this structure is still small compared with that of birds or mammals, it is, nevertheless, a much better organ than the amphibian brain; in its small cerebral hemispheres appear for the first time the rudiments of the higher mental centers which have become predominant in the brains of mammals.

Reptiles living today can easily be distinguished from amphibians by means of soft parts; and among living types it is also feasible to distinguish between these classes by means of skeletal features. Reptiles, for example, have but one condyle on the skull, modern amphibians two; reptiles typically have five toes in the manus, living amphibians four or less. The sacrum in reptiles includes at least two vertebrae, whereas there is but one in amphibians.

But the fossil evidence obliterates these and other supposed differences. Primitive Paleozoic reptiles and some of the earliest amphibians were so similar in their skeletons that (as was seen in the case of *Seymouria* and *Diadectes*) it is almost impossible to tell when we have crossed the boundary between the two classes. The true test, of course, is the type of egg the animal laid; but direct evidence for this is practically impossible of attainment in the case of extinct forms.

The ties between primitive reptiles and anthracosaurian amphibians are so many and so clear that it is reasonable to believe that the reptiles stem from that group. It has been suggested from time to time that the reptiles may have originated from amphibians in polyphyletic fashion, the little microsaurs, particularly, being brought up for consideration in addition to the anthracosaurians. But the developmental pattern associated with the amniote type of egg is complex in nature and yet uniform throughout the Reptilia and their descendants; it is next to impossible that this pattern could have been "invented" separately by two different amphibian groups.

Structure of primitive reptiles.—The basal reptilian group is the order Cotylosauria, the "stem reptiles." Apart from the possible inclusion of "border" forms such as *Diadectes* and *Seymouria* and certain advanced types of the late Permian and Triassic (to be considered later), these stem forms constitute the suborder Captorhinomorpha. We will consider the structure of captorhinomorphs in some detail as introductory to the study of reptiles generally. The group is a very ancient one, for although the best-known members are of Permian age, we have in recent years come to recognize the presence of primitive reptiles in the faunas of the Coal Measures swamps. One of them, *Hylonomus* (Figs. 147, 148) found in hollow tree stumps in the Coal Measures of Nova Scotia, dates from an early stage in the Pennsylvanian period. The captorhinomorphs—and hence the reptiles as a class—are thus of ancient lineage, and must have separated from the anthracosaurs far back in the Carboniferous. Beyond the Coal Measures, the captorhinomorphs continued into the early Permian in modest numbers, but then became extinct.

The early members of the group are as yet incompletely known, and our description will be based in considerable measure on better-known Permian genera —*Limnoscelis* (Figs. 149, 150), *Romeria, Captorhinus* (Figs. 151, 152), and *Labidosaurus* (Fig. 153). The last two are end forms, somewhat specialized in dentition and other features; *Limnoscelis* and *Romeria* are more primitive. *Limnoscelis* was about 5 feet in length (including a long tail) and rather clumsy in build; the other genera were much smaller and with less massive limbs.

The skull in these reptiles, as in this class in general, has remained rather high and narrow, in contrast to the flattening prevalent among almost all amphibians. Much of the skull pattern seen in primitive amphibians is repeated in primitive reptiles, but there are noteworthy differences and evolutionary advances. One prominent change is the elimination of the characteristic otic notch of amphibians. In reptiles the eardrum has moved backward and downward to a position on the side of the head above the jaw articulation. In captorhinomorphs its presumed position is not clearly defined by bone, but in numerous advanced forms there develops a new type of otic notch—a concavity in the posterior margin of the bones covering the cheek (cf., for example, Figs. 157, 161, 165, 202). In *Limnoscelis* an anteroposterior ridge is found at the edge of the skull table, marking the place where the edges of the former amphibian type of notch have joined; in *Captorhinus* this has been obliterated. The elimination of the old otic notch is mechanically advantageous, giving to the region of the jaw articulation a vertical bracing and suspension from the skull roof and braincase region.

The bony pattern of the skull table has somewhat altered in primitive reptiles. We have noted previously the gradual reduction of the postparietals from the crossopterygian condition in which they extend perhaps a third the length of the skull (Fig. 69) to their relatively modest size in most labyrinthodonts. In early reptiles they are still further limited in extent, and in later types they will be seen to disappear completely or to abandon the roof for a position on the occipital surface. The tabulars have undergone a similar reduction. In amphibians the skull table was bordered laterally by a supratemporal and frequently an intertemporal as well. The latter is absent in reptiles; the former is present as a small structure in a number of reptiles (Figs. 149, 175, 179, 268) but has generally disappeared.

Apart from the skull table and otic-notch regions, the roofing bones of early reptiles are very similar to those of the more generalized labyrinthodonts. The pineal opening is generally present, and the lacrimal bone extends the full distance from orbit to naris.

The palate is readily comparable with that of labyrinthodonts. *Captorhinus* is specialized in its maxillary dentition, but normally there is a single row of marginal teeth. There are slender vomers separating the choanae, palatines, and small ectopterygoids. The prominent pterygoid (above which lies a rodlike epipterygoid) articulates movably with a basal process of the braincase. Narrow interpterygoid vacuities separate the pterygoids from one another and from a slender parasphenoid lying beneath the front end of the braincase. There is a large vertically placed quadrate in *Limnoscelis,* as in many later reptiles, but this bone is less prominent in other captorhinomorphs. Two palatal features, however, differ from those of amphibians. The old system of palatal fangs with accompanying

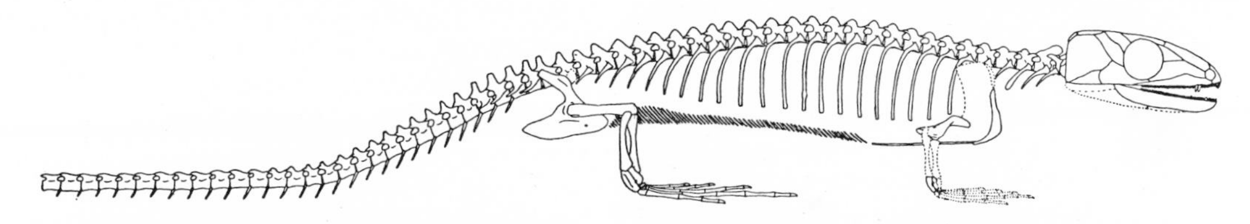

FIG. 147.—The skeleton of the oldest known reptile, the captorhinomorph *Hylonomus,* from the early Pennsylvanian; about 1/5 natural size. (From Carroll.)

pits is not repeated here, although the palatal bones are frequently tooth-bearing; and the pterygoid bears a prominent transverse flange, often tipped by a row of teeth. This type of palate is present today relatively unchanged in *Sphenodon* (Fig. 198) and many lizards.

The braincase is relatively high and narrow, in contrast with the flat condition developed in most amphibians. Posteriorly it is well ossified, containing the same elements found in primitive amphibians. A series

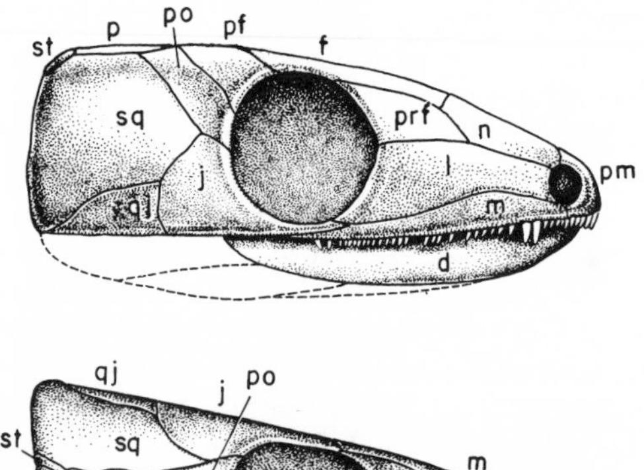

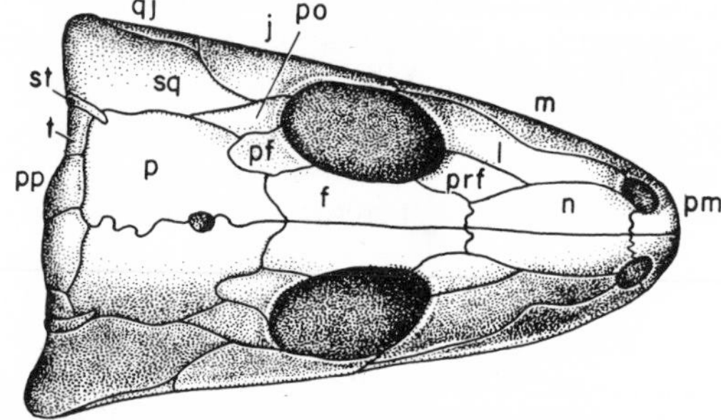

Fig. 148.—The skull of *Hylonomus* in lateral and dorsal views; about natural size. (After Carroll.)

of occipital and otic bones forms well-developed walls for the brain cavity, and anteriorly there is a sphenethmoid element. Midway of the length of the braincase, above the region of the palatal articulation, there is, however, a large unossified gap in the lateral walls. There is, as in all typical reptiles, a single occipital condyle. With the closure of the otic notch, stout "paroccipital" processes of the opisthotic bones have come to run laterally to brace the region of the jaw joint. The large openings above these processes may remain open (as in *Captorhinus*) or may be covered by flanges descending from the skull roof. There is typically a large stapes running laterally from the ear capsule to the region behind the quadrate, where the eardrum was presumably situated.

The lower jaw is generally comparable to that of amphibians, but the number of elements is reduced. There is but a single splenial, and one coronoid (rarely two).

The vertebrae of cotylosaurs are almost identical in structure with those of advanced anthracosaurians such as *Seymouria* and *Diadectes,* strongly suggesting their relationship. The centrum, a stout cylindrical structure pierced for the notochord, is the element frequently termed pleurocentrum in amphibians; the intercentrum is present as a crescent of modest size. A diagnostic feature of most cotylosaurs (but one found in a few other early reptiles and in *Diadectes* and *Seymouria* as well) is the fact that the zygapophyses are placed far apart laterally and the supporting neural arches have a pronounced swollen appearance, with a convex outline when seen in end view. There is still little regional differentiation of the vertebral column, for ribs may be present throughout the trunk and the base of the tail. As in many labyrinthodonts, the anterior ribs are double headed, the capitulum attaching to the intercentrum, the tubercle to the transverse process; more posteriorly the two heads tend to fuse, the capitular attachment transferring upward onto the centrum near the base of the neural arch.

The limbs are of the short, stubby type found in many primitive amphibians, with the proximal elements projecting at right angles to the body. The massive shoulder girdle is close behind the head, there

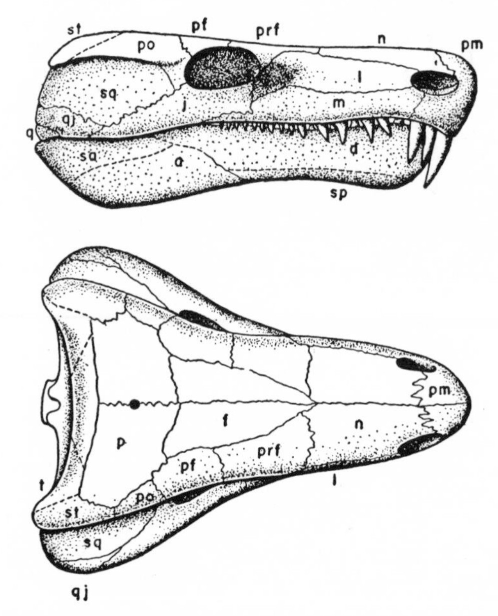

Fig. 149.—*Limnoscelis,* a large, primitive reptile; skull in dorsal and lateral views; length about 10.5 inches. For abbreviations see Figure 109. (After Williston.)

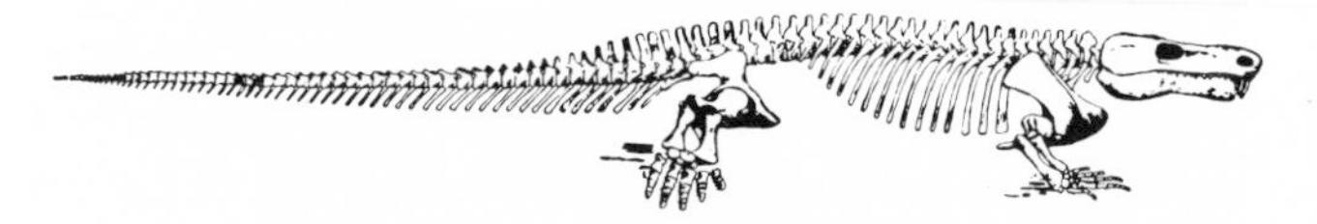

Fig. 150.—The skeleton of *Limnoscelis;* length about 5 feet. (From Williston.)

being but a short neck. It is very similar in many respects to that of such an amphibian as *Eryops.* However, in reptiles the ventral part of the primary girdle (Fig. 154*C*) is (in contrast with most amphibian groups) a separate coracoid element. The interclavicle has a long posterior stem for the attachment of powerful breast muscles. In most primitive reptiles the cleithrum is present, but reduced to a splint. The pelvic girdle (Fig. 154*D*) consists of the usual three elements; but the ilium is larger and more expanded than in amphibians, a feature presumably related to greater walking powers and increased musculature.

The limb bones are in many respects similar to those of a well-developed ancient amphibian such as *Eryops*. Most early reptiles possess an entepicondylar foramen, an opening for nerve and blood vessels on the back (or inner) side of the humerus near the distal end. This structure appears to have been present in the most ancient tetrapods, but except for the anthracosaurians, was generally lost in amphibians. In primitive reptiles the carpals (Fig. 273) are, for the most part, comparable to those of early amphibians; however, there are never more than two central elements. In the tarsus (Fig. 274) the centralia are similarly reduced. Here, however, another and more important change in foot structure has occurred. The inner proximal element—the tibiale—has been reduced or eliminated, and the tibia now articulates with the inner margin of the intermedium (the mammalian astragalus). This permits the foot to swing more readily into a forward position and gives a very free movement between shank and foot. There are primitively five toes on both front and hind feet and a phalangeal formula of 2–3–4–5–3(4). Only in anthracosaurians have we seen among amphibians a five-toed manus or as high a phalangeal formula.

From such primitive structural types as these early cotylosaurs there have been enormous variations among later orders of reptiles. The development of bipeds from quadrupeds and the evolution of flying forms and of many marine types have caused many startling changes in the general body organization. We shall here outline some of the main trends.

The reptile skull.—The skull in reptiles has tended in

general to become progressively higher and narrower in shape, in contrast with the flattening found among amphibians. Many of the dermal roofing elements may disappear. The supratemporal and, later, the tabulars and postparietals tend to become reduced and lost. The pineal may be absent, and there may be considerable variation in the bones about the orbits.

Especially to be noted is an interesting series of variations in the cheek region of the skull, frequently used in the characterization of major reptile groups. The temporal muscles closing the jaws have their attachment under the bones covering this region. It appears that when a muscle arises from a broad plate of bone, its action is facilitated if an opening (fenestra) develops through which the muscle may bulge when contracted. In correlation with this we find that in most later reptiles the cheek is fenestrated, that is, openings appear between various bones. These fenestrae vary considerably, apparently being related to different developments of the temporal muscles (Fig. 155). Frequently there is but one opening. In some cases this is high up on the skull and bounded below by the squamosal and postorbital; in others (mammal-like forms) it is low on the side, and these two bones are in contact above it. These two types of opening are termed the upper (superior) and lateral temporal vacuities. In a large number of types (the crocodiles and dinosaurs, for example) both openings may be present, with the postorbital and squamosal meeting between the two.

There are many variations in the palate. In some types the palatal bones may fuse into a nearly solid plate. In several groups the internal nostrils may be

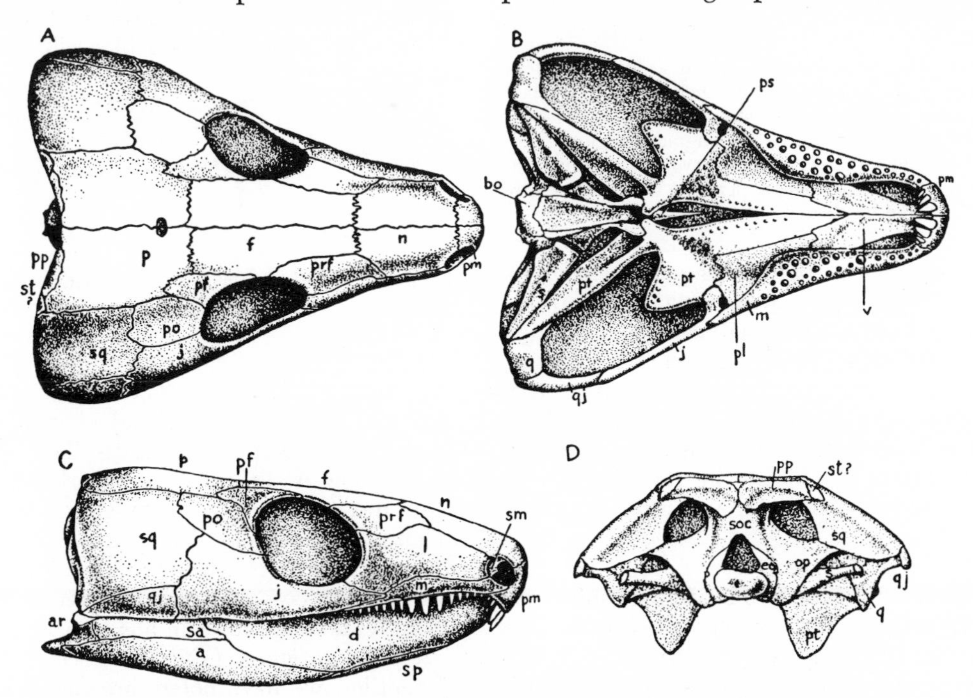

FIG. 151.—*Captorhinus*, a cotylosaur. *A*, Dorsal; *B*, ventral; *C*, lateral; *D*, occipital views of skull. Length of original about 2.75 inches. For abbreviations see Figure 109. (Data from Price.)

shifted back from their original anterior position by the construction of a shelf—a secondary palate—in the roof of the mouth.

Axial skeleton.—In the backbone there is a universal tendency for the reduction of the intercentra. Even in most Permian forms they are merely small ventral crescents, and they usually disappear completely in Mesozoic types. The true centra tend to lengthen. They

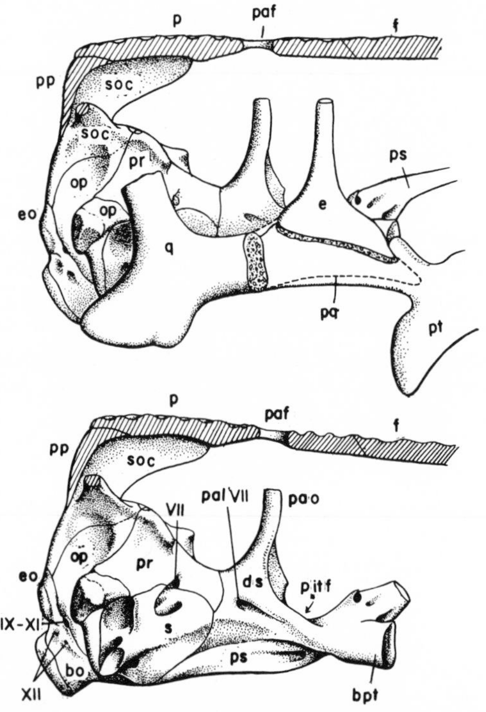

FIG. 152.—The skull of *Captorhinus*, "dissected" in lateral view. *Upper*, The dermal bones of the cheek removed to show the posterior portion of the palatal structure; the broken line indicates the probable extent of the cartilaginous region of the palatoquadrate cartilage (in which epipterygoid and quadrate are ossifications). *Lower*, The palatal elements removed to show the braincase. Length of original about 2.5 inches.

are primitively hollow at both ends (amphicoelous), but the opening for the notochord usually disappears; the ends may be flush across (platycoelous), while in many cases articulations are developed between successive centra, one end being hollowed, the other swelling out to fit into this socket. Various terms have been devised to describe this and other conditions. If, for example, a vertebra is concave in front and convex behind, it is termed procoelous; the opposite condition is opisthocoelous.

The ribs vary greatly in structure (cf. Figs. 173, 209, 262). The attachments of the two heads may shift, or a single-headed condition, with attachment either to centrum or to neural arch, may result. With the development of a proper neck in most reptiles, the ribs in this region may shorten or even disappear and a well-marked cervical region be formed. The number of sacral ribs tends to increase in large forms and particularly in bipedal types.

In addition to the ventral continuation of the ordinary ribs, reptiles often have independent structures called "abdominal ribs"—dermal, bony, splintlike elements, ranged along the belly in V-shaped rows. These structures are seemingly the last remnants of the old fish scales which were developed in amphibians in a similar position on the underside of the body. While reptiles are scaled (in contrast with living amphibians), the scales are really horny scutes (although plates of bone often develop beneath them) and are not homologous with fish scales.

Limbs.—There is, of course, much variation in the limb structure. In the shoulder girdle the cleithrum tends to disappear very early in reptile history (it is found in but a few Triassic forms and in none today), and the other dermal elements may be lost as well. In most reptiles a scapula and a single coracoid constitute the elements of the primary girdle; but in one major group (mammal-like reptiles) a second, posterior, coracoid element is added (Fig. 269). The peculiar screw-shaped glenoid of primitive type disappears by the close of Permian times. There is never any change in the number of pelvic bones, but in many cases a fenestration of the ventral plate occurs in relation to the origin of a large limb muscle, a development functionally comparable to the fenestration seen in the skull (cf. Figs. 164*C*; 186; 270*B*, *C*).

The limb bones in general tend to become much slimmer than in primitive forms, and many variations in structure occur in relation to varied locomotor needs. In the humerus the entepicondylar foramen characteristic of primitive reptiles usually disappears, but there often develops an ectepicondylar foramen on the outer side of the distal end of the bone (Fig. 271*B*, *C*). Reduction takes place in carpus and tarsus

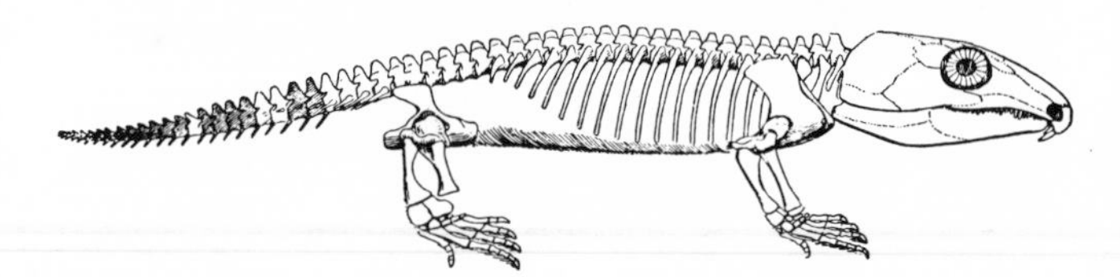

FIG. 153.—*Labidosaurus*, a somewhat specialized Lower Permian captorhinomorph cotylosaur; original about 26 inches long. (From Williston.)

(cf. Figs. 273*A*, 274*A*). There is usually but a single centrale, and the outer distal element usually disappears in both carpus and tarsus. The number of joints in the toes is, in general, remarkably uniform. The outer toe, especially in the hind foot, tends to be set off at an angle and may disappear. In bipeds among the dinosaurs there are strong tendencies for reduction in number of toes, and the limbs may, of course, be strongly modified to form fins or paddles.

Reptile radiation (*Fig. 156*).—The earliest remains of reptiles, we have noted, are found in the Pennsylvanian. Known specimens from that period are few; but, since practically all our knowledge of Carboniferous fossils is gained from coal-swamp deposits, we know nothing of the life of higher and drier regions where reptiles might well have already been numerous. Dating from the end of the Pennsylvanian and the beginning of Permian times are deposits (redbeds) of a more terrestrial type in which are found abundant reptiles, including not only primitive cotylosaurs but also more advanced types. Amphibians were still numerous at that time; but long before the close of the Permian they had been pushed into the background, and the radiation of the reptiles was underway. The Permian and earlier Triassic were notable for the prominent development of mammal-like reptiles, but by the end of the latter period almost every other major reptilian group had made its appearance. In the Jurassic and Cretaceous these groups increased in size, in abundance, and in diversity. Toward the close of the Cretaceous, however, the great reptilian orders waned; and the beginning of the Cenozoic is marked by the end of the reptilian dynasties and the radiation of the mammals. Only a few reptilian groups have survived—the lizards and snakes, the turtles, the crocodiles, and the rhynchocephalian *Sphenodon*.

These varied reptilian types have been arranged in from fifteen to twenty or more orders by different authors. It is probable that they did not all spring directly and independently from the base of the reptile stem, and various workers have attempted to assemble larger groups of presumably common parentage as subclasses or superorders. It is difficult to select a proper basis for such classification; the most obvious differences between groups lie in the varied construction of the limbs and other superficial features which are closely related to environmental adaptations and are hence untrustworthy (seagoing reptiles, for example, are all likely to develop paddle-like legs, but this means merely convergent adaptation and not real relationship). It was early suggested that the nature of the temporal openings was a better criterion, and attempts to work out reptilian phylogenies have been based mainly on this feature. It was once thought that all reptiles originally had two temporal openings, and hence forms with two openings (or two bars across the temple, which amounts to the same thing) were termed "diapsid" (two-arched); and forms with one opening (in which it was presumed that the two openings had fused into one), "synapsid" (fused-arched) reptiles. But it soon became obvious that this division was inadequate. Truly primitive forms must have possessed a completely roofed skull like that of their amphibian ancestors, and for this condition the term "anapsid" was coined. Further, the single opening found in a number of groups is obviously not homologous throughout, being sometimes a lateral one, to which the term "synapsid" is applied and sometimes a superior opening, for which, when characteristically developed, the term "euryapsid" may be used. Still fur-

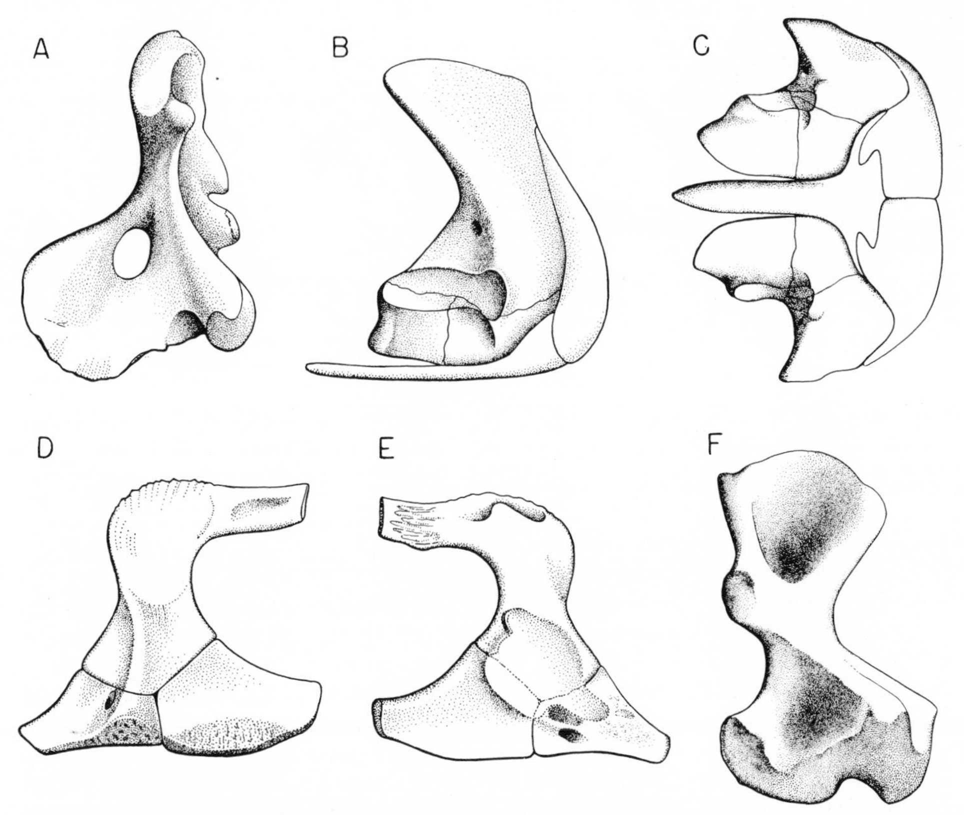

FIG. 154.—Girdles and limb bones in cotylosaurs. *A*, Humerus of *Limnoscelis* in dorsal view. *B, C,* Lateral and ventral views of the shoulder girdle of *Labidosaurus*. *D, E,* External and internal views of the pelvic girdle of *Limnoscelis*. *F,* Femur of *Limnoscelis* in ventral view.

ther, temporal fenestrae may shift in position or become secondarily closed. These structures may be used for purposes of classification—but must be used, like any other single character, with caution.

Procolophonians.—Earlier in this chapter we discussed at some length the skeletal characteristics of members of the Captorhinomorpha as truly primitive

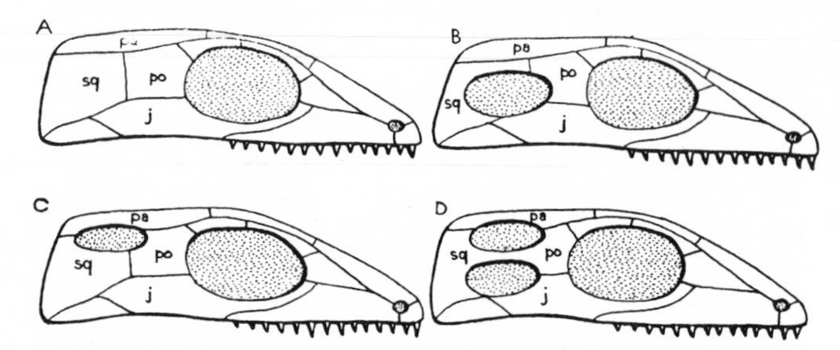

FIG. 155.—Diagrammatic side views of reptilian skulls to show various types of temporal opening. *A*, No opening—"anapsid" condition; *B*, a lower opening with postorbital and squamosal meeting above—"synapsid" condition; *C*, an upper opening with postorbital and squamosal meeting below—"euryapsid" condition; *D*, both openings present—"diapsid" condition. Abbreviations: *j*, jugal; *pa*, parietal; *po*, postorbital; *sq*, squamosal.

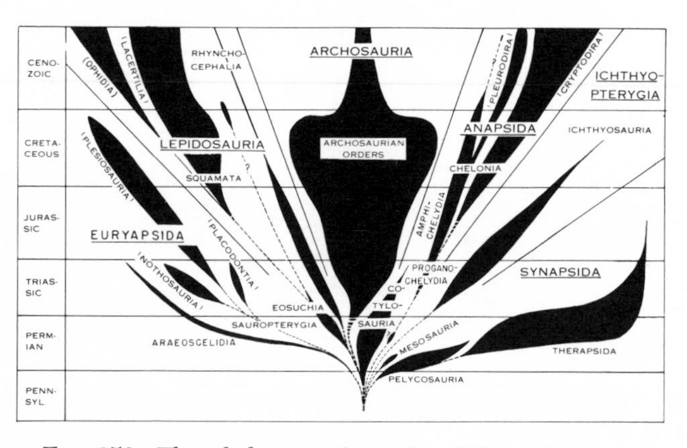

FIG. 156.—The phylogeny of reptiles. (The archosaurs are shown separately in Figure 207.)

reptiles of ancient pedigree. Certain later reptiles—notably the important synapsid group, leading toward mammals—may well have sprung directly from typical captorhinomorphs. For other lines, however, the evolutionary lineage may have been less direct, and intermediate stages may be expected. One interesting group of reptiles which may be included in the order Cotylosauria, but is suggestive of relationships to more advanced types, is that which will here be termed the suborder Procolophonia, whose members are found in the Permian and Triassic.

A seemingly primitive member of the group, although too late in time to be itself the common ancestor, is *Nyctiphruretus* (Figs. 157, 158) of the Russian Upper Permian. This small animal has a build in skull and body which is in many regards comparable to that seen in captorhinomorphs. There are, however, significant differences, most notably in the cheek region. The jaws have shortened, so that the jaw articulation lies well in front of the plane of the occiput; there is a strongly developed, vertically placed quadrate, bracing the jaw articulation on the skull-roof region. The back of the cheek, in the region of the quadrate and the squamosal bone which covers it, is somewhat concave, suggesting the initiation of an otic notch of the new, reptilian type.

It is possible that forms of this type gave rise to important later groups of reptiles, such as the turtles and the great groups of two-arched (diapsid) reptiles, in which the cheek region is constructed in a comparable manner. But within the confines of the cotylosaurs there developed from such a stock two families, the procolophonids and pareiasaurs, which played a prominant role in the Permian and Triassic faunas.

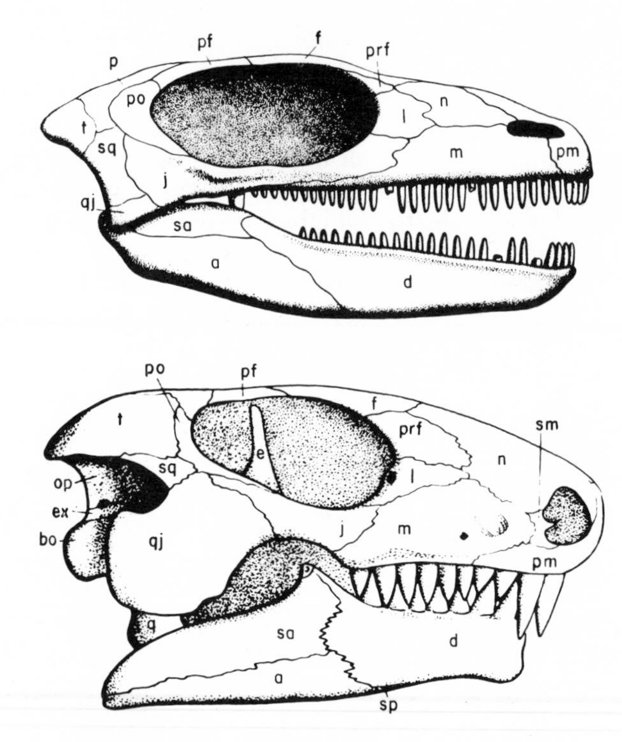

FIG. 157.—Side views of the skulls of (*upper*) *Nyctiphruretus* of the Russian Permian, a short-jawed procolophonian cotylosaur suggestive of the ancestry of many later reptile types, about 1 inch long (after Efremov); *lower*, *Procolophon* of the Lower Triassic of South Africa, about 2 inches long. (After Broili and Schroeder.)

Nyctiphruretus itself can perhaps be included in the family Procolophonidae, which takes its name from *Procolophon* (Figs. 157, 158) of the early Triassic of South Africa. These forms appear in the late Permian and are widespread if modest members of every well-known Triassic fauna, persisting, as the last surviving cotylosaurs, to the end of that period. The procolophonids were small reptiles, usually a foot or two in length, rather slim-limbed, and with transversely expanded cheek teeth resembling those of *Diadectes*.

A second, more prominent group of procolophonians were the pareiasaurs of the Middle and Upper Permian of Europe and Africa (Figs. 159, 160). These were the largest of cotylosaurs, ranging up to 10 feet or so in length. With increase in size, the problem of weight support becomes a serious one. It would have been physically impossible for such a large animal to have walked with its limbs spread out at the sides, and in this connection we find that in pareiasaurs the limbs had rotated in toward the body and bore the weight more vertically. This change in pose was accompanied by changes in the limb bones similar to those which we shall see in the reptilian ancestors of mammals. While the feet of most cotylosaurs were of a very primitive character, those of the pareiasaurs tended to be considerably modified; the fifth toe, for example, tended to be reduced or lost in many cases.

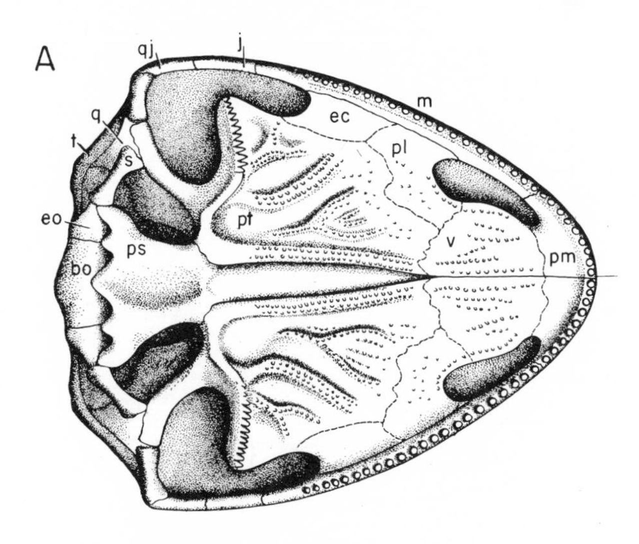

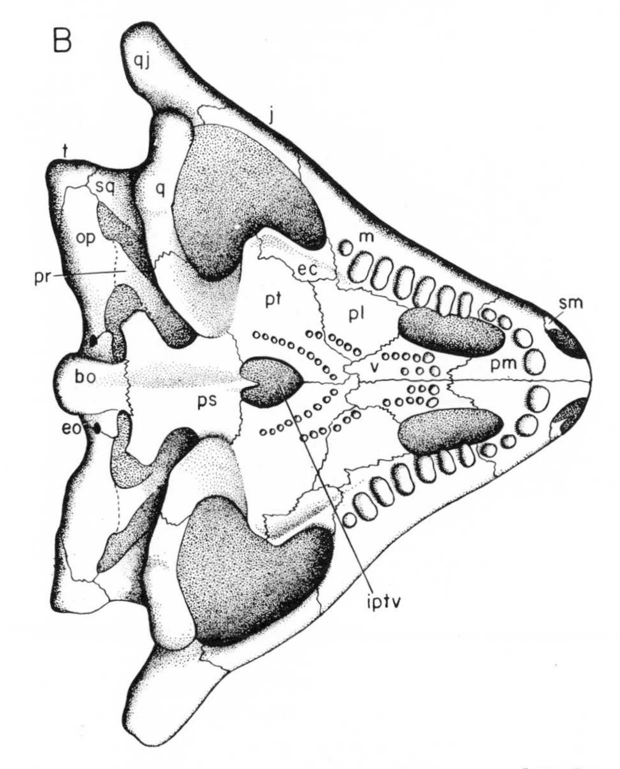

Fig. 158.—Palatal views of *A, Nyctiphruretus,* and *B, Procolophon.*

The skull was very large and grotesque, often with warty or hornlike protuberances. The teeth along the margins of the mouth were serrated and leaf-shaped, and there were rows of small teeth on the palate. No excavation for the eardrum comparable to that of the procolophonids appears to be present. In reality, however, it was well developed but covered superficially by a great expansion of the cheek plate, so that this notch faced backward rather than laterally. These forms were obviously noncarnivorous, slow, and harmless reptiles, and for defense there were usually plates of bone studded along the back.

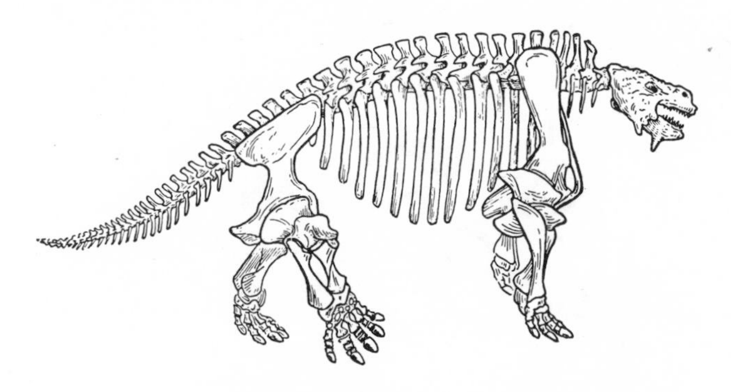

Fig. 159.—The pareiasaur *Scutosaurus* from the late Permian of Russia; original about 8 feet long. (From Gregory.)

Numerous remains of pareiasaurs have been found in the Middle Permian beds of South Africa, the skeletons often complete and right side up in the position in which they presumably became mired in their native swamps. In later Permian times they became rarer in Africa but had spread to Europe, where skulls and skeletons have been found in Russia and Scotland.

The millerettids.—As was said earlier in this chapter, the nature of the temporal region has been generally used as a key character in classifying reptiles. Obviously the primitive reptiles would have been forms in which the skull roof was solidly built. The cotylosaurs are, therefore, generally regarded as the major members of a subclass Anapsida—reptiles in which there are no temporal openings (and hence, of course, no temporal arches, to which the term "apsida" refers). Also included, customarily, in the Anapsida are the turtles, in which the cheek may be variably reduced but in which no true temporal fenestra is present.

But although the presence or absence of temporal openings gives a useful clue to reptilian relationships, and terms based on such characters may be used to designate major groups, we should not be hidebound in our thinking in this regard. It is probable that considerable variation in the development of temporal openings occurred in the early days of reptile history, and we should use common sense, rather than technicalities, in assigning one form or another to its proper place in classification and phylogeny.

We may consider, as an example, the family Millerettidae, the best known members of which are small reptiles from the Upper Permian of South Africa, such as *Milleretta* (Fig. 161) and *Millerosaurus*. In many regards the skulls of these forms resemble the captorhinomorphs of the earlier Permian. However, here, as in procolophonians, the jaw has shortened. The articulation lies well forward of the level of the occipital condyle, with the squamosal and underlying quadrate forming (as in procolophonids) a somewhat concave posterior border to the cheek above it, as an incipient otic notch of reptilian (not amphibian) type. In contrast, however, to the procolophonids, there is here a fairly broad cheek expanse back of the orbit; and here, within the family, we witness the development of a lateral temporal fenestra. In one genus the cheek is solid; in *Milleretta* there is a small gap in ossification between jugal and squamosal; in *Millerosaurus* there is a well-developed fenestra.

One might argue this would make us place the millerettids among the synapsids, the mammal-like reptiles, in which the presence of a lateral fenestra is a

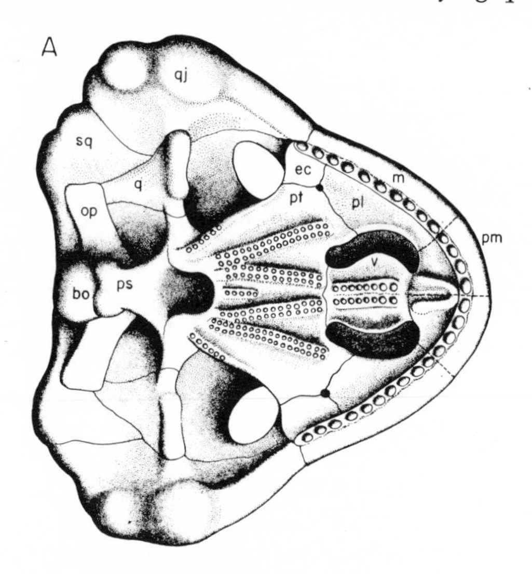

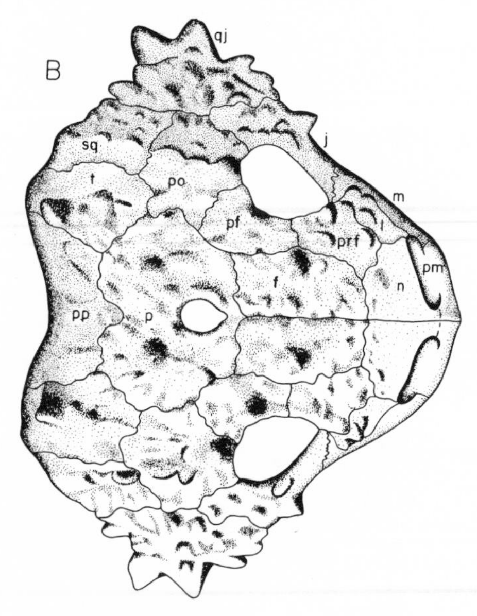

Fig. 160.—*A*, The skull of the pareiasaur *Bradysaurus* in palatal view, length about 16 inches; *B*, the skull of *Nannopareia* in dorsal view, length about 4 inches. (*A* after Boonstra and Seeley, *B* after Broom and Robinson.)

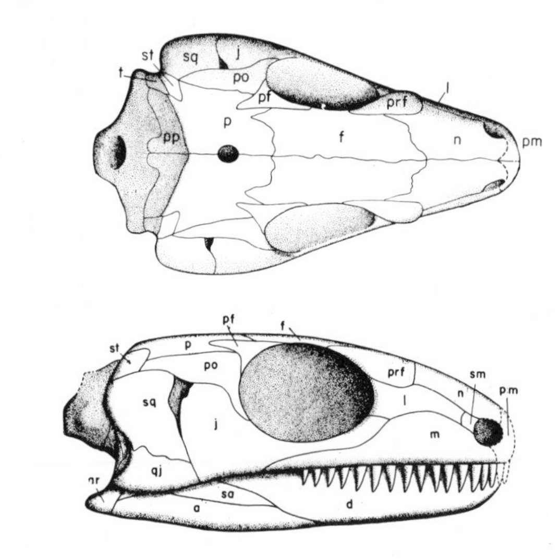

Fig. 161.—The millerettid *Milleretta*, skull in lateral and dorsal views; length about 2.75 inches. (After Watson.)

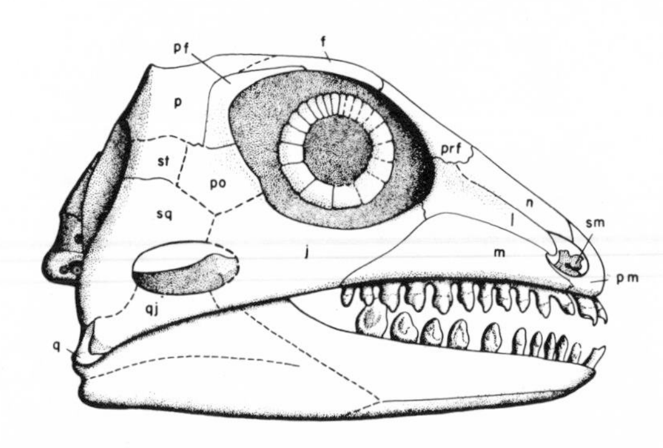

Fig. 162.—The Lower Permian reptile *Bolosaurus*, about 1.5 inches long. (After Watson.)

major diagnostic feature. But in no other regard is there any notable comparison between millerettids and synapsids, and surely the opening here has evolved in parallel fashion.

Are these millerettids merely a short-lived, sterile offshoot of the cotylosaurs? Possibly. But it has been suggested that they are ancestral to the most ancient diapsids (such as *Youngina*) which make their appearance in these same late Permian African beds. To transform a millerettid into a primitive diapsid, it is necessary only to add an upper opening to the lateral one already developed among the members of this family. This possibility is a reasonable one. For the time, however, it is perhaps preferable to retain the millerettids among the cotylosaurs, and bracket them with the procolophonians, which they resemble in their short jaws and incipient otic notch.

Here we may mention a further early form which had, like the millerettids, developed independently a lateral temporal opening, but differs from them in many other regards. *Bolosaurus* (Fig. 162), a tiny reptile from the early Permian of Texas, is peculiar in its dentition, with cheek teeth of bulbous form with a sharp cusp at one margin. Unlike the millerettids, there is no jaw shortening and no indication of an otic notch; further, the lateral opening is small and placed low down the side of the cheek, quite in contrast to that group. Surely in this case we are dealing with a short-lived side branch of the captorhinomorph stock.

8 CHAPTER

Varied Reptilian Types

The greater proportion of the Mesozoic and Tertiary reptiles belong to large and important orders which can be assembled more or less satisfactorily into major subclasses of advanced reptiles—Lepidosauria, Archosauria, Euryapsida, and Synapsida. There remain, however, several orders which occupy isolated positions and seem to represent independent stocks rising directly from the ancestral cotylosaurs. Of such a nature are the turtles, mesosaurs, and ichthyosaurs, which are treated in the present chapter.

Chelonians.—The turtles, forming the order Chelonia, may technically be included as members of the Anapsida, since the skull roof, although often emarginated and incomplete, has never developed true temporal openings. In almost every other respect, however, the turtles have departed widely from the structural pattern of their cotylosaur ancestors.

The chelonians are the most bizarre, and yet in many respects the most conservative, of reptilian groups. Because they are still living, turtles are commonplace objects to us; were they entirely extinct, their shells—the most remarkable defensive armor ever assumed by a tetrapod—would be a cause for wonder. From the Triassic the turtles have come down to present times relatively little changed; they have survived all the vicissitudes which have swamped most of the reptilian groups and are in as flourishing a condition today as at any time in the past.

The turtle shell (*Fig. 163*).—The armor plate of the ordinary modern turtle is composed of two materials—horny scutes representing the ordinary reptilian scales on the surface and bony plates underneath. The arrangement of the outlines of scutes and bones does not coincide; in general, there is an alternating arrangement, which gives greater strength to the combined structure. The scutes are not, of course, preserved in fossils, although the outlines are often indicated by grooves in the bones which lay beneath them.

The bony shell is divisible into a dorsal carapace and a ventral plastron, connected by a bridge at the sides. At the center of the carapace is a row of eight bony neural plates fused to the neural arches of the vertebrae beneath. Lateral to them on either side is a row of longer plates, also eight in number; these are fused to the eight underlying ribs and are hence known as pleural plates (or costals). Circling the edge of the carapace is a ring of small elements—the peripherals (or marginals). In addition, there may be in the central row an extra nuchal element in front of the neurals and a few pygals behind them.

A much smaller number of bony elements is present in the ventral shield—the plastron. Near the front is a median plate—the entoplastron. The other elements, which are paired, consist (from front to back) of epiplastra, hyoplastra, in some primitive types one or two pairs of mesoplastra, then hypoplastra, and xiphiplastra.

The shell is widely open in front for the withdrawal of the head and front legs; behind, for the hind legs and the stubby tail.

It would seem that the carapace, at least, has arisen through the development of new plates of dermal bone. Such plates are developed in many crocodiles and some lizards, as well as in other groups now extinct. The turtles differ from them in the consolidation of the plates into a complete, compact covering. As regards the plastron, however, it may be that no new elements were needed. The old fish scales had been retained in many primitive reptiles as the so-called abdominal ribs. It is possible that the turtle plastron for the most part arose by an enlargement and consolidation of these elements. With regard to the front of the plastron, it is interesting to note that the paired epiplastra articulate internally with the scapula, as do the clavicles of ordinary reptiles, while between them the single entoplastron lies in the position of an interclavicle. The anterior elements of the shield thus appear to represent the old dermal shoulder girdle.

Body skeleton.—The backbone is, of course, much modified in connection with the development of the shell. The tail is short and easily tucked away. There are two sacral vertebrae and only ten trunk vertebrae in front of it, all except the first immovably attached to the median plates of the shell. This represents a great shortening of the body from primitive conditions; probably nearly a dozen vertebrae have dropped out. There are eight ribs, fused to the underside of the pleural plates; this fusion and that of the neural arches with the median elements effectively brace the shell. There are eight cervicals in the neck; in modern forms, in which the head is withdrawn into the shell, these have very complicated articular arrangements.

Limbs.—The limbs and girdles are of rather peculiar construction, as might be expected from the requirements of a shell-enclosed life. The limbs spread out at the sides, giving a broad trackway and a short stride, much as was the case in the cotylosaurs; and it is not improbable that this old-fashioned method of walking was present in the group when the shell first appeared and has become permanently fixed. In detail, however, the limbs are far from primitive.

In the shoulder girdle the dermal bones have been taken over into the plastron. The primary girdle (Fig. 164*B*) is a triradiate structure with the glenoid cavity at the union of the branches; the scapula includes a slim ascending blade and a long projection which goes down to touch the clavicles (the epiplastra); while the third prong is a single coracoid. In the pelvic girdle (Fig. 164*C*) much of the ventral region, instead of being a solid plate of bone, is occupied by a large opening or fenestra which has separated the pubis and ischium almost completely from each other. The feet are quite specialized; notable is a tendency (rather unusual among reptiles) for a reduction of the number of phalanges in the toes; there is generally a phalangeal formula of 2–3–3–3–3.

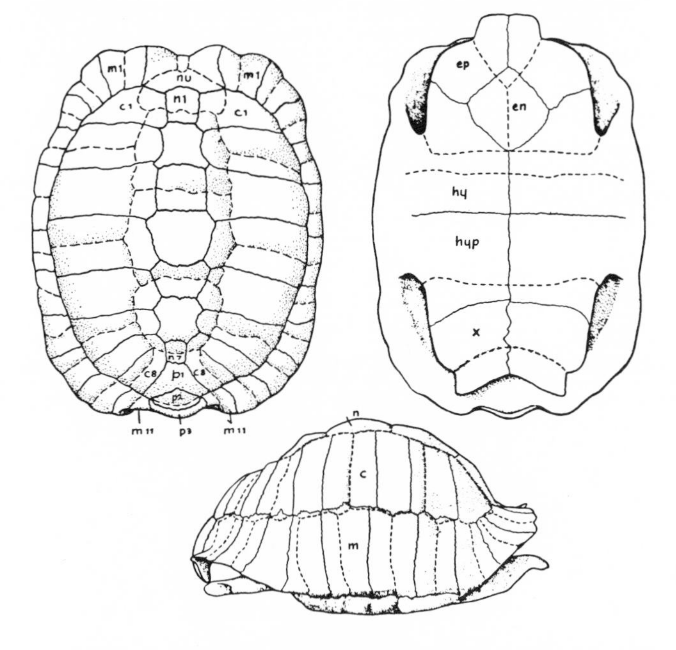

Fig. 163.—*A*, Dorsal; *B*, ventral; *C*, lateral views of the shell of a fossil *Testudo*. Broken lines are outlines of horny scutes. Abbreviations: *en*, entoplastron; *ep*, epiplastron; *hy*, hyoplastron; *hyp*, hypoplastron; *nu*, nuchal; *n*1–*n*7, neurals; *p*, pygal plates (postneurals); *pe*, peripheral plates; *p*11, most posterior (eleventh) peripheral; *pl*, pleural plates; *x*, xiphiplastron. (After Andrews.)

Skull (*Fig. 165*).—The skull is also highly specialized in a number of ways. Modern turtles entirely lack teeth, their function being taken over by a horny beak which sheaths the edges of the jaws. The palatal bones have fused and united to give a solid structure quite different from that of primitive reptiles. There is a large eardrum, placed in an otic notch, which is, in modern reptile fashion, low down on the side of the skull; the large quadrate is characteristically hook-shaped and may almost completely surround the tympanum. The otic bones of the ear capsule have fused broadly with the quadrate and squamosal, so that the region of the jaw articulation is solidly bound into the braincase. The pineal eye no longer pierces the roof of the skull, and the two external nostrils have a common bony opening. A number of elements are lost, including the postparietals, the tabulars, the supratemporals and intertemporals, and the postfrontals; a single element, generally interpreted as a prefrontal, in modern turtles fills the area originally occupied by that element and by the nasal and lacrimal as well. The cheek is rather puzzling in many respects. In many forms most of this region is bare of dermal bones. The gap so formed is not a true temporal opening, for it has no back rim of superficial dermal bone. Seemingly, there has been merely an emargination, an eating-away of the skull roof from the back, or, less commonly, from below. From this point of view the turtles are technically anapsid, and to be associated with the cotylosaurs. It has been argued that this hole was once a true temporal opening and that its posterior margin later disappeared; but some living turtles have a solidly roofed skull, and so did the oldest known fossils.

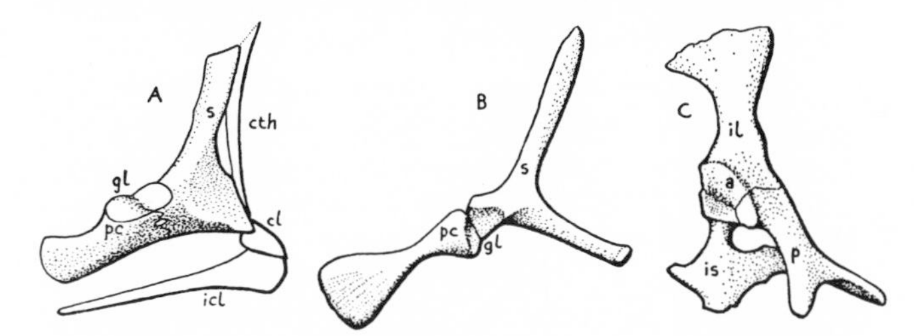

Fig. 164.—Limb girdles in chelonians. *A*, Shoulder girdle of the primitive *Proganochelys* [*Triassochelys*]. The girdle is still close to the primitive reptilian type; a cleithrum is present; the dermal elements forming part of the plastron are shown. *B*, Shoulder girdle of *Toxochelys* (Cretaceous). *C*, Pelvic girdle of the Oligocene *Testudo laticunea*. For abbreviations see Figure 154. (*A* after Jaekel; *B*, *C* after Hay.)

Primitive turtles.—The previous description has been based on the more normal turtles. The oldest forms are found in Triassic deposits; and here in such a form as *Proganochelys* [*Triassochelys*] (Figs. 166, 167) we meet with a type which was in some respects more primitive.

All the shell elements of later turtles were already present, and their arrangement was essentially that of later forms. In addition, several pairs of mesoplastra—intermediate plastral elements not found in the modern turtles of northern continents—were present. It is probable that the limbs could not be withdrawn, but some protection was afforded them by a series of scutes projecting beyond the normal marginal region. Head and tail were not capable of being retracted and were covered with bony tubera. The limb girdles were rather primitive, particularly the shoulder girdle (Fig. 164*A*). The clavicles and interclavicle were already incorporated into the plastron but show their original character clearly, and there was possibly even a small cleithrum. Here and in other primitive forms the pelvis was fused to both carapace and plastron. In the skull, nasals, lacrimals, and prefrontals appear to be present and distinct; the external nares are separated; the temporal region is completely roofed; while the most striking feature of all is that, although a horny beak is apparently already developed, there are still teeth on the palate. Although the back margin of the cheek is somewhat concave, there is not the sharp curvature of the quadrate around the eardrum seen in later forms.

These Triassic turtles may be considered to form a primitive chelonian stock, as the Proganochelydia. Dominant in the Jurassic and Cretaceous was a somewhat more advanced group, the Amphichelydia, in which teeth were lost and the skull had attained a more modern aspect, although there are some archaic skeletal features, and it is probable that the head still could not be withdrawn. There were a number of families, some of them apparently transitional to the later pleurodires and cryptodires. Two amphichelydians are known to have lingered on into the Eocene. Apart from this, the only known survivor of the group appears to be *Meiolania* of the Australian Pleistocene, a large form with a peculiarly horned skull; the remains are incomplete, but some idea of its size may be gathered from the fact that the skull was 2 feet across.

Pleurodires.—A side branch from these archaic chelonians forms the suborder Pleurodira, the "side-neck" turtles. A distinguishing feature of the Mesozoic types included in the previous suborder was the impossibility (as shown by the structure of the neck vertebrae) of drawing the head into the shell. This has been accomplished in the pleurodires; but, in contrast with most living turtles, the head is withdrawn by a sideways bending of the neck. Nasals and lacrimals may be present or absent, as in cryptodires, but in contrast is the fusion of the pelvic girdle to the shell. A number of primitive features still persist, such as the presence of mesoplastra in many forms. The oldest forms surely assignable to the pleurodires are from the Upper Cretaceous, best known in North America; beyond this time, however, remains from the northern continents are rare, and today they are confined to the three southern continents. Representative is *Podocnemis*, a genus widespread in the Cretaceous but today present only in South America and Madagascar. The present distribution of pleurodires is an example, repeated in many other groups, of the late survival in the tropics, often in discontinuous fashion, of archaic types once widely distributed.

Cryptodires.—More progressive and successful has been the suborder Cryptodira, in whose members the head is withdrawn straight back into the shell by means of an S-shaped curvature of the neck vertebrae. Mesoplastra are never present, nor is the pelvis ever

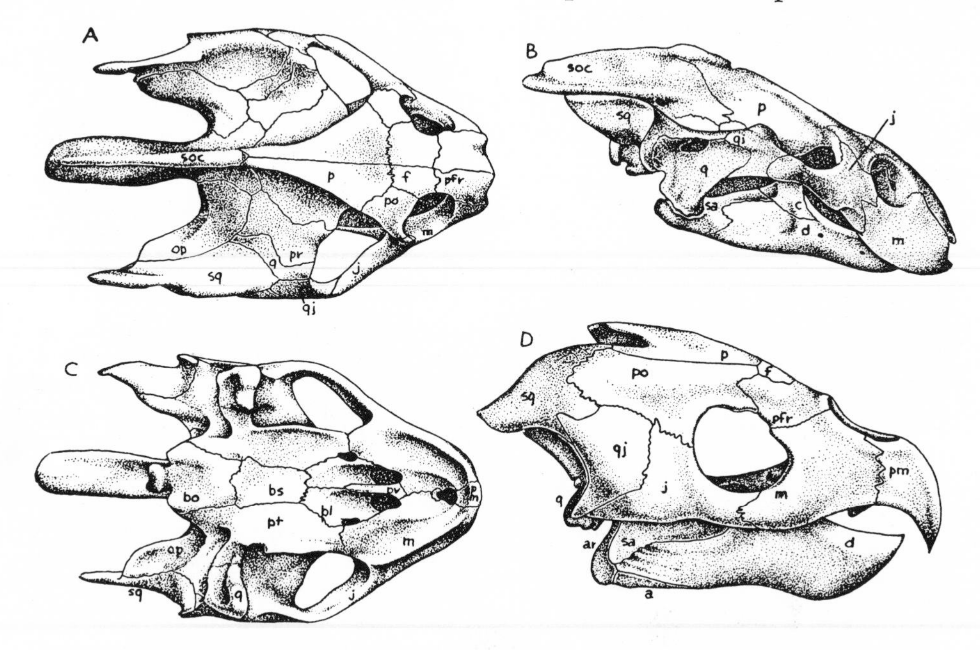

FIG. 165.—The skull in chelonians. *A*, Dorsal; *B*, lateral; and *C*, ventral, views of the skull of *Trionyx tritor*, an Eocene trionychid turtle, jaw restored from a related species; skull length about 6 inches. *D*, Lateral view of *Archelon*, a Cretaceous marine turtle, skull length about 2.5 feet. *pfr*, Prefrontal. For other abbreviations see Figure 109. (*A-C* after Hay; *D* after Wieland.)

attached to the shell. The first cryptodire appeared at the end of the Jurassic; by Upper Cretaceous times, cryptodires had become the dominant types in northern areas and invaded the seas as well.

The generalized stock of the cryptodires seems to lie among a series of marsh turtles and terrapins, of which a rare living Central American form, *Dermatemys,* is generally considered the most primitive; it has numerous Cretaceous relatives. Primitive, too, are the snapping turtles, of the family Chelydridae, and marsh dwellers of the family Testudinidae, such as *Chrysemys, Clemmys,* and *Emys,* although the pedigree of these two families is restricted to the Tertiary where their remains are common; the testudinid *Stylemys,* for example, is an extremely abundant fossil in the White River Oligocene. Presumably the turtles arose from primitive reptiles which were still

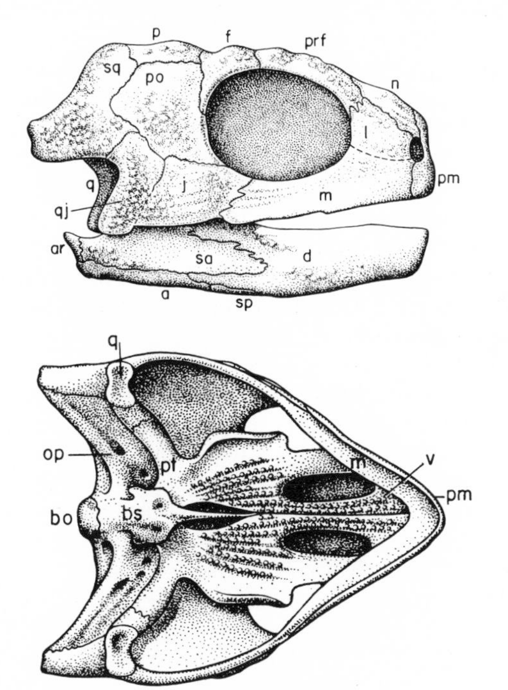

FIG. 166.—The skull of *Proganochelys* [*Triassochelys*], a primitive Triassic turtle still retaining teeth; lateral and ventral views. Length of original about 5.25 inches. For abbreviations see Figure 109. (After Jaekel and photographs.)

FIG. 167.—The carapace of *Proganochelys* showing impressions of the horny scutes. Length of original about 2 feet. (After Jaekel.)

partially water dwellers, and probably many of the older amphichelydians were water lovers, as are the modern turtles mentioned above. However, the Testudinidae includes the most terrestrial of all chelonians in the tortoises—*Testudo* and its relatives—which, apart from technical features, are readily distinguished by their highly arched and well-ossified shells. The tortoises are found throughout the Tertiary.

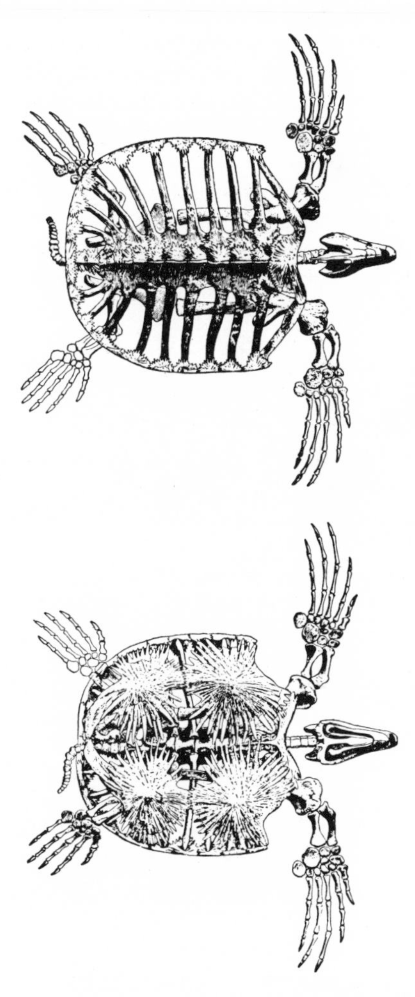

FIG. 168.—*Archelon,* a Cretaceous marine turtle, dorsal and ventral views. Original about 12 feet long. (From Wieland.)

A number of species of large size are found today on various isolated islands where the absence of land enemies has afforded them even more protection than that which is given them by the shell; some fossil tortoises were equally large, notably one from the late Pliocene of India, which is often referred to a separate genus, *Colossochelys.*

Full aquatic adaptations have been attained by a series of marine turtles. A number of such types appeared in the Cretaceous, and a few are present in the oceans today. Of the extinct forms the best known are *Protostega* and the related *Archelon* (Figs. 165*D*, 168) of the Cretaceous. These old sea turtles were large types, up to 12 feet in length. Land turtles seldom grow to any great size; but in the water, of course, many of the problems of support do not arise. The heavy armor of a turtle, however, means a much-increased specific gravity and a consequent high expenditure of energy in maintaining the body in the water; it would be advantageous if the armor were lightened and the weight consequently reduced. This was rendered possible by the fact that in taking to the sea many land enemies were left behind, and for a large form there were few or no marine enemies; it was consequently possible for a marine turtle secondarily to reduce its armor to a considerable extent, leaving a small amount of bony framework.

The compact shape of a turtle's body and the small size of the tail renders swimming by a fishlike movement out of the question. Consequently, it was necessary for the feet to develop into powerful swimming paddles, by which the turtle might row itself through the water. The proximal part of the limbs of these oceanic forms is short but powerful, while the fingers are elongated and broadened to form large, webbed flippers.

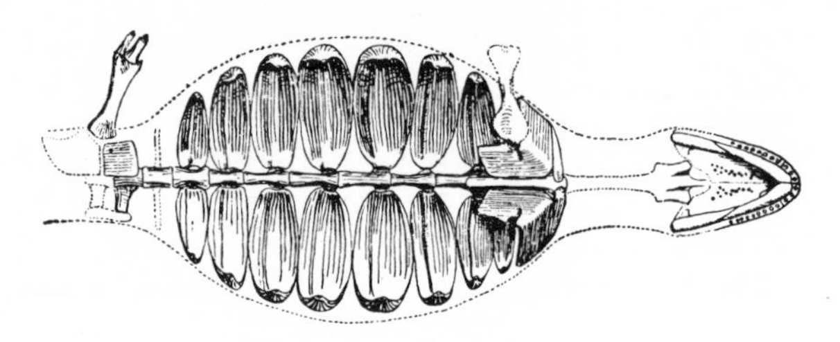

FIG. 169.—Ventral view of *Eunotosaurus*, a small Permian reptile possibly ancestral to the chelonians, about 2/3 natural size. (From Watson.)

Of a number of such turtle types which evolved in the Cretaceous, most are extinct; but *Chelonia* [*Chelone*] and a few related forms have survived to modern times. This living genus is, in general, still similar in build to the oldest forms.

A second marine type is *Dermochelys*, the leathery turtle. In this form there is almost no connected dermal skeleton at all—merely a series of small bony plates studded in the skin of the back. It has been argued that this really represents the primitive chelonian condition and that the leathery turtle is the most primitive known member of the entire order. But there is no fossil evidence to substantiate this, for the complete development of the armor had taken place by the Triassic; and, of the few imperfect fossils which appear to represent relatives of *Dermochelys*, none is earlier than the Eocene. We have just seen that other marine turtles tend to reduce their armaments somewhat; and it is probable that *Dermochelys* is merely a marine type in which reduction has gone much further than is usually the case.

These oceanic turtles are the first examples that we have encountered of reptiles returning to a marine existence. Later, however, we see further examples of this type of adaptation in other reptilian groups.

Trionyx, the living soft-shelled turtle, is a river-dwelling type with a very flat body in which the armor is somewhat reduced and on which horny scutes are entirely lacking, the plates being covered only with a leathery skin. Sometimes placed in a separate suborder, these turtles are perhaps also to be regarded merely as aberrant cryptodires. *Trionyx* [*Amyda*] (Fig. 165*A–C*) appears at the end of the Jurassic, and from the late Cretaceous on trionychids are common fossil forms.

Turtle origins.—But little light is shed on the ultimate origin of the turtles from a study of fossil members of the order. Some early Triassic forms, such as *Proganochelys*, show a structure slightly more primitive than that of most later turtles. But even at that time the armor was almost perfectly developed; we are dealing definitely with a true turtle and not with a transitional type. The ancestors of the group must be sought farther back—in Permian times. Obviously the remote ancestors lay among the cotylosaurs, most probably among the forms here grouped as the Procolophonia. The skull pattern of *Nyctiphruretus*, for example, is not too remote in pattern from that of *Proganochelys* (cf. Figs. 157 and 158 with Fig. 166). But these forms were not armored and have a "normal" postcranial structure.

A possible connecting form is *Eunotosaurus* (Fig. 169), a small reptile from the Middle Permian of South Africa. Unfortunately, the roof of the skull is unknown; we cannot settle the problem of the presence or absence of a temporal opening. Teeth were still present. The limbs, too, are poorly known, but the girdles appear to have been of a primitive character. The greatest point of interest lies in the skeleton of the trunk. There was only a small number of vertebrae—this suggesting the reduction which has taken place in turtles. Especially significant seems to be the fact that following the small first rib were eight exceedingly broad ribs which extended far laterally, almost touching one another at their edges. This may be compared with the condition in turtles in which there are eight ribs supporting the costal plates which make up most of the carapace. This creature was not, of course, a true turtle; on the other hand, it was far from the typical cotylosaurs. Perhaps we may include it provisionally in the Chelonia in a broad sense but place it in a separate suborder, the Eunotosauria.

Mesosaurs.—Although the basic reptile stock was a potentially terrestrial one, many reptiles, it would seem, refused to leave the ancestral waters, and there developed a variety of aquatic groups. Earliest were *Mesosaurus* (Figs. 170, 171) and its allies, constituting the Mesosauria. These were slimly built little reptiles about a yard in length, inhabiting bodies of fresh water. Their remains are known only from deposits in South Africa and South America which appear to be of late Carboniferous or early Permian age. This distribution has been, not unreasonably, used as an argument for the former union of these continents, for although *Mesosaurus* was obviously a competent swimmer in fresh waters, it is difficult to imagine it breasting the South Atlantic waves for 3,000 miles.

The mesosaur skull was long and slim, the neck and body elongate and there was a long, powerful, laterally compressed tail. The last seems to have been the principal swimming organ. But, while the front legs were not particularly long and presumably functioned merely as steering organs, the hind legs were long and strong, suggesting that they also were of considerable use in locomotion.

The vertebrae were of a primitive type with broad arches like those of cotylosaurs. The dorsal ribs (which attached by a single head to the transverse processes) were curiously thickened, rather banana-like; a similar structure is found in some marine sirenians among mammals, and this "pachyostosis," it is suggested, is due to some physiological change connected with aquatic life.

The shoulder girdle had the usual reptilian elements—a single coracoid and a scapula, a clavicle and an interclavicle. As was general in aquatic forms, the scapula was short and broad, the coracoid large. The pelvis was rather primitive but was poorly ossified. The limbs were not much modified for aquatic life; the long bones were little shortened. Presumably, the spreading digits of the feet were webbed. In the toes of such marine forms as the ichthyosaurs and plesiosaurs there was a tendency for increase in the number of joints. Here, however, the phalangeal formula was normal except for an extra joint or two in the fifth toe in the pes.

There was a long, slender, fish-eating type of snout; the nostrils were pushed far back—an aquatic adaptation for easier breathing; the tooth-bearing palate is derivable from the generalized type seen in most primitive reptiles. The temporal region was long unknown; recent studies suggest (with some doubt) that there were a single large lateral opening. The marginal teeth are very numerous, and exceedingly slender and delicate. It is difficult to imagine them operating as a typical biting apparatus. Tiny crustaceans, which may have served as their food, are found in the beds with *Mesosaurus*, and the teeth may have functioned as a straining apparatus.

The very early appearance of the mesosaurs suggests that they were an extremely early offshoot of the stem reptiles, the cotylosaurs. They do not appear in typical Permian deposits and hence had a very short span of life. It has been suggested that they were ancestral to the ichthyosaurs. But common features are merely such as might (and did) occur in other aquatic reptiles; the ichthyosaur hind legs are reduced, not enlarged, and the temporal region is very differently constructed. The lateral temporal opening suggests inclusion of the mesosaurs in the synapsids, or mammal-like reptiles. No other features of mesosaurs, however, show any resemblance to that group, and it is probable that the mesosaurs are an independent and early side branch from the Carboniferous stem reptiles which (as in the case of *Bolosaurus*) we can retain within the confines of the subclass Anapsida, despite the apparent presence of an odd type of temporal opening.

Ichthyosaurs.—Of all reptiles, those most highly adapted to an aquatic existence were the ichthyosaurs, which well deserve their name of "fish reptiles." They seem to have occupied the place in nature now taken by the dolphins and porpoises. They were particularly abundant in the Jurassic, but their life span seems to have covered the greater part of the Mesozoic.

Numerous marvelously preserved specimens of ichthyosaurs have been recovered from slabs of black shale of the early Jurassic of southern Germany. The typical Jurassic forms (Fig. 172) seem to have been very fishlike in their superficial appearance. There was a large, fishlike caudal fin. The body was short and deeply fusiform and apparently somewhat compressed laterally. Perfectly preserved specimens show the body outline and reveal the presence of a large dorsal fin, unsupported by bone. The limbs were reduced (especially the pelvic ones) to short steering paddles, the main swimming motion being a fishlike undulation of the body and the tail fin. The vertebral centra (Fig. 173*A*, *B*) were short amphicoelous disks to which attached both heads of the two-headed ribs. The caudal series of vertebrae appear to be broken in fossil specimens, with the tip of the tail bent downward. This was once thought to be a postmortem effect, and the tail was restored as a straight structure. But we now know that this seeming break was natural and that ichthyo-

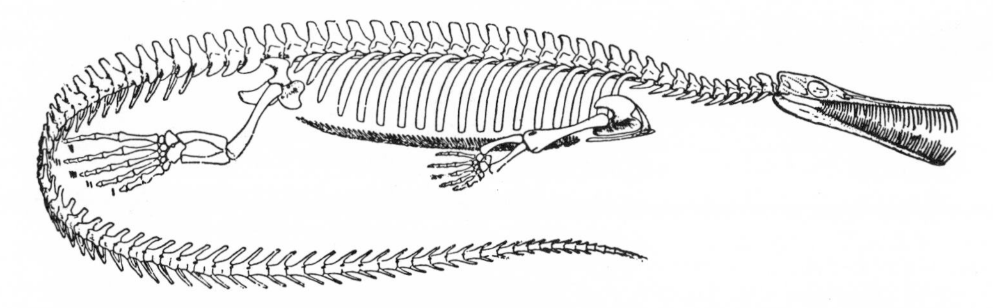

Fig. 170.—The skeleton of *Mesosaurus*, a late Carboniferous or early Permian aquatic reptile from the Southern Hemisphere; length of original about 16 inches. (After McGregor.)

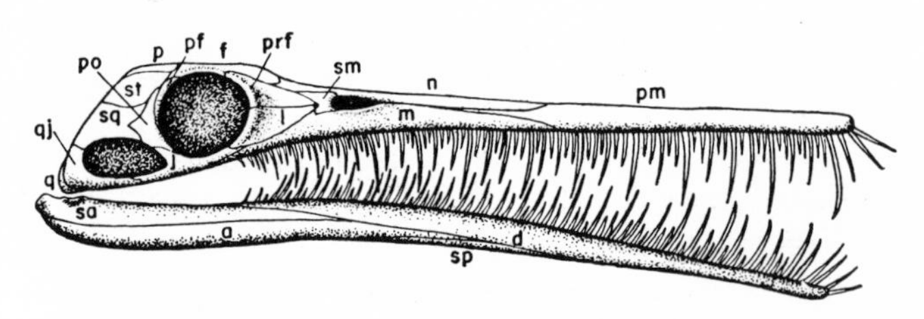

Fig. 171.—The skull of *Mesosaurus*, about natural size. For abbreviations see Figure 109. (After von Huene.)

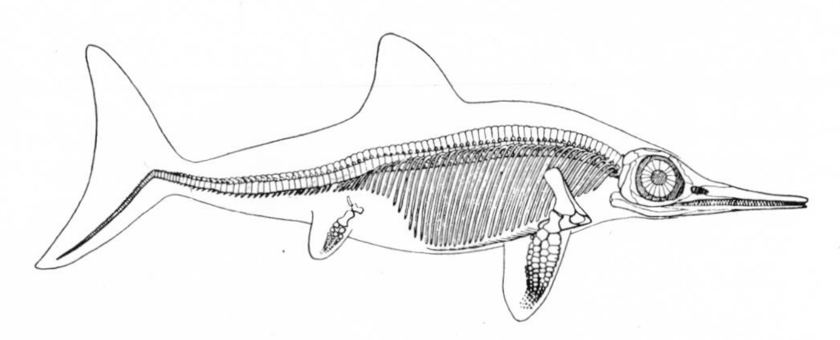

Fig. 172.—A Jurassic ichthyosaur, much reduced. (Simplified from E. von Stromer.)

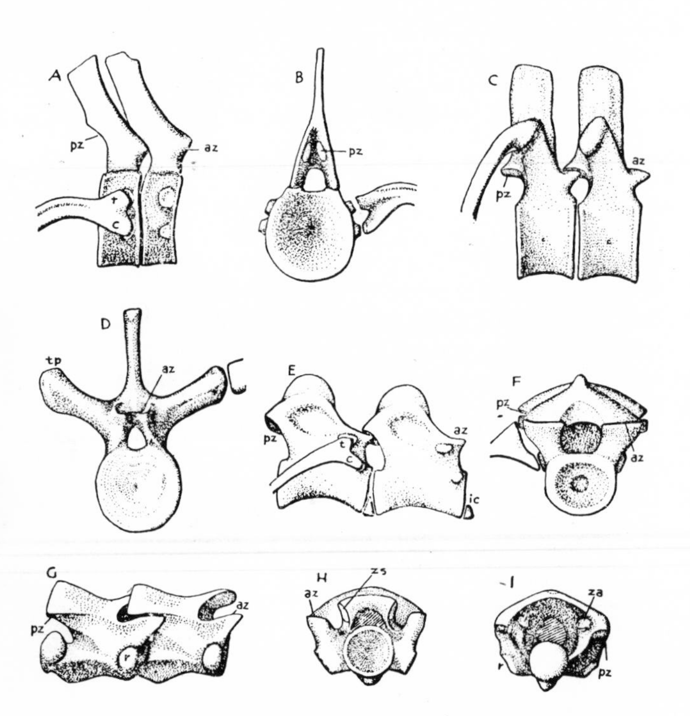

Fig. 173.—Dorsal vertebrae of various reptiles. *A, B,* Lateral and posterior views of vertebrae of the ichthyosaur *Ophthalmosaurus; C, D,* lateral and anterior views of vertebrae of the plesiosaur *Cryptocleidus; E, F,* lateral and anterior views of vertebrae of *Araeoscelis; G, H, I,* lateral, anterior, and posterior views of vertebrae of the Cretaceous snake *Coniophis*. Abbreviations: *az,* anterior zygapophysis; *c,* capitulum of rib; *ic,* intercentrum; *pz,* posterior zygapophysis; *r,* rib articulation on centrum in squamata; *t,* tubercule of rib; *tp,* transverse process of neural arch; *za, zs,* zygantrum and zygosphene, additional articular processes of neural arch. (*A–D* after Andrews; *E, F* after Williston; *G–I* after Marsh.)

saurs had developed a reversed, but otherwise shark-like, heterocercal tail fin (Fig. 176).

All the bones of the girdles were present but were none too well ossified; and the elements were small, especially the dorsal ones. The pelvis had lost its connection with the sacral vertebrae. The bones of the limbs were much shortened and, except for the humerus and femur, tended to become hexagonal or circular disks (Fig. 174*B*). The paddle seems to have moved as a unit, flexible in structure but without any great motility between individual bones. There was

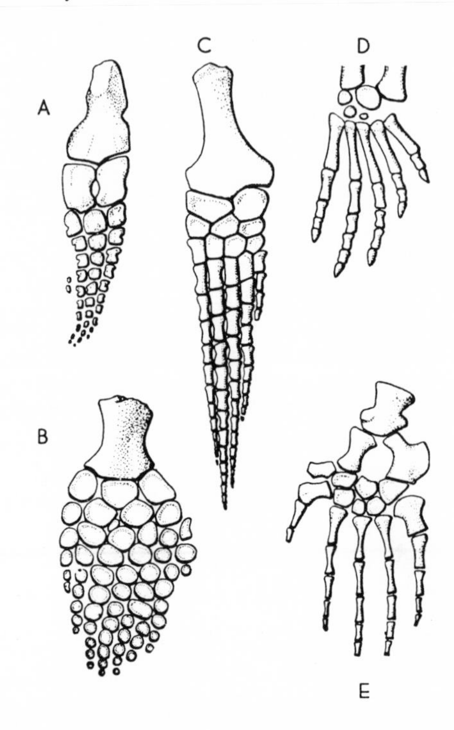

Fig. 174.—Limbs of various aquatic reptiles. *A,* Pectoral limb of the Triassic ichthyosaur *Merriamia; B,* same of the Jurassic *Ophthalmosaurus; C,* same of the Cretaceous plesiosaur *Elasmosaurus; D,* pes of the nothosaur *Lariosaurus; E,* pectoral limb of the mosasaur *Clidastes*. (*A* after Merriam; *B–E* after Williston.)

always considerable hyperphalangy, or addition of extra finger joints, as in other marine forms. In addition, the number of toes varied; in some cases there was a decrease to three, in others hyperdactyly, with as many as seven or eight digits; sometimes a single toe is observed to divide into two, part way down the paddle.

The skull (Fig. 175) was, of course, highly modified for aquatic life. The beak was long; the external nostrils had moved far back; the eyes were very large, with well-developed sclerotic plates; a pineal opening was present. With the enormous enlargement of the eyes, the temporal region had become much shortened. There was a large temporal opening, which lay high on the skull at either side of the parietals and was thus an upper temporal fenestra. But, in contrast with other reptiles, its lateral border is thought to be formed by the postfrontal and supratemporal bones, the postorbital and squamosal elements lying farther down the cheek region and not entering into its borders. The stapes, usually slim in reptiles, had a massive structure, suggesting that these types had undergone considerable change in their mode of hearing, as have the whales and other secondarily aquatic forms.

These creatures were extreme in their marine adaptations, and their limbs were obviously unfitted for use

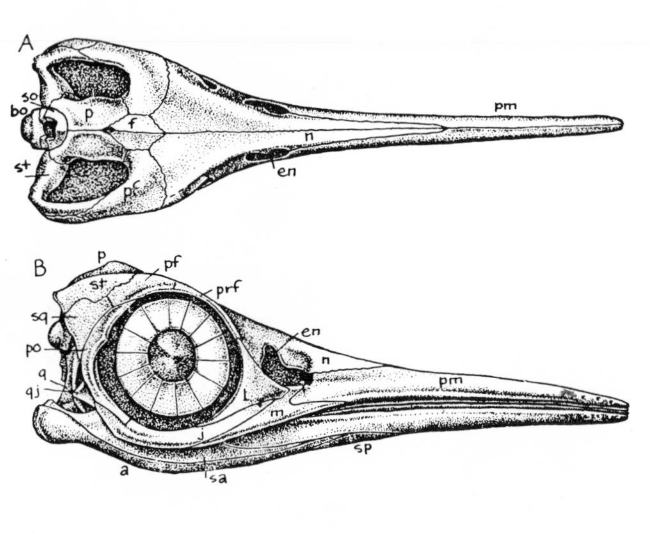

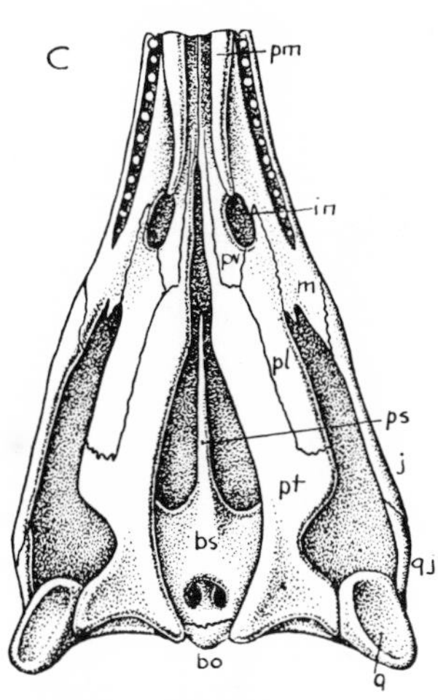

FIG. 175.—The ichthyosaur skull. *A*, *B*, Dorsal and lateral views of the skull of *Ophthalmosaurus*, skull length about 40 inches; *C*, posterior portion of palate of *Stenopterygius*. For abbreviations see Figure 109. (*A*, *B* after Gilmore; *C* from Abel after Owen and Smith Woodward.)

on land. It was long ago suggested that reproduction must have taken place in the water (in many snakes, for example, and in lizards, the eggs are retained in the mother's body until they are hatched). In agreement with the idea that the young were born alive are specimens which actually show skeletons of young ichthyosaurs inside the body of a large individual. It has been argued that these may have been youngsters which had been eaten by mistake. But several specimens show the young partially emergent from what would have been the cloacal region in life. The mother here apparently died during childbirth, or (there are human parallels) labor may have taken place after the death of the mother.

The description above is that of a typical Jurassic ichthyosaur. Triassic types, such as *Mixosaurus*, were more primitive. The tail (Fig. 176) was much straighter, with little development of the fish type of caudal fin; the limbs were somewhat less specialized (Fig. 174*A*); the face shorter. These forms are found in marine Triassic beds, mainly in Nevada and Spitsbergen; recently there has been discovered in Nevada an extensive "graveyard" of giant Triassic ichthyosaurs.

From these same deposits come fragmentary remains of *Omphalosaurus*, which represented a short-lived, shell-eating side branch of the group. The skull

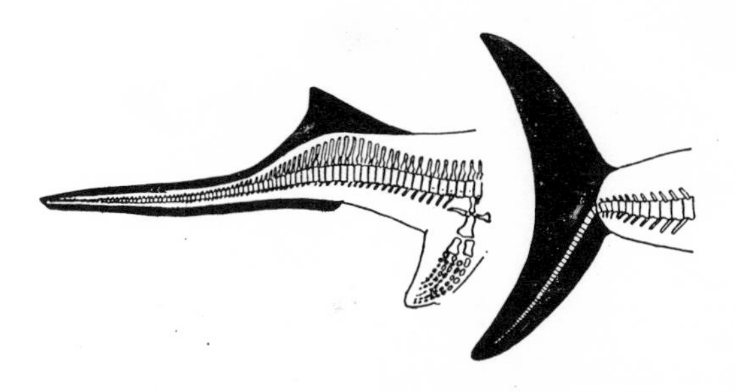

FIG. 176.—Tails of *Mixosaurus* (Triassic) and a Jurassic ichthyosaur, to show development of sharklike tail. (From Williston, after Wiman.)

was, in sharp contrast with that of ordinary ichthyosaurs, short and strongly built, with flattened, button-like teeth for crushing mollusks. These forms seem to have paralleled the contemporary placodonts discussed later; like them they did not persist beyond the Triassic.

Most of the later ichthyosaurs were once included in a single genus, *Ichthyosaurus*. There were, however, a number of distinct types—deep-bodied and slender forms; broad-finned (*Eurypterygius*) and narrow-finned (*Stenopterygius*) series; and one ichthyosaur, *Eurhinosaurus*, has a short lower and long upper jaw like those of a swordfish.

In the Cretaceous, the ichthyosaurs became rarer; they are quite unknown in the very well-known Upper Cretaceous fauna of the Kansas chalk, where plesiosaurs and mosasaurs abound, and they seem to have become extinct well before the close of the period.

We noted earlier that the ancestral reptiles appear still to have been water dwellers to a considerable degree. The ichthyosaurs may have descended directly from such forms, without there ever having been a truly terrestrial stage in their pedigree. But the aquatic adaptations of all ichthyosaurs are far more highly developed than in any of their early tetrapod ancestors. Although the Triassic ichthyosaurs were slightly more primitive than their Jurassic descendants, they were already very highly specialized marine types. No earlier forms are known. The peculiarities of ichthyosaur structure would seemingly have required a long time for their development and hence a very early origin for the group, but there are no known Permian reptiles antecedent to them. In many regards their cranial structures are comparable to those of primitive aquatic pelycosaurs of the *Ophiacodon-Varanosaurus* group. But the pelycosaurs already had a temporal region of a quite different sort. Not improbably the two groups had a not-remote common ancestor among the captorhine cotylosaurs. But the ichthyosaurs have diverged so far that we may regard them as the types of a distinct subclass termed the Ichthyopterygia.

CHAPTER 9

Euryapsid Reptiles

Treated in the present chapter is a varied series of reptilian types now entirely extinct but not uncommon in Permian and Mesozoic deposits. They were, in the main, amphibious or aquatic forms, of a highly diversified nature, but showing a common diagnostic feature in the type of temporal opening present. This opening was situated high up on the skull roof; it was essentially comparable to the superior temporal fenestra of diapsid types and was bounded below by a broad cheek plate. A similarly placed opening was, as we have seen, present in the ichthyosaurs. In the present group, however, the postorbital, squamosal, and parietal bones were the major elements forming its margins; the supratemporal, which plays a prominent role in the boundaries of the opening in ichthyosaurs, is absent. These forms are here considered as making up a subclass Euryapsida (sometimes termed Synaptosauria).

We are here dealing with a most puzzling major chapter in reptilian history. Most of the later euryapsids, from the mid-Triassic on, are members of well-known, solidly established aquatic groups—nothosaurs, plesiosaurs, placodonts—reasonably considered to be related to one another, but of unknown ancestry. Apart from them we find, starting in the early Permian, a scattered series of other animals which have (or are suspected of having) the euryapsid type of temporal opening.

Araeoscelis.—One of the very few early Permian reptiles, apart from the cotylosaurs, and the pelycosaurs to be discussed in a later chapter, is *Araeoscelis* (Figs. 173*E, F;* 177; 178) of the Texas redbeds (there is an almost identical contemporary in Europe). This was a small and lightly built creature, measuring, tail apart, but a foot or so in length. Presumably its habits were not dissimilar to those of the more agile little lizards of the modern tropics. Except for slenderness, the postcranial skeleton had not, in most regards, departed far from the primitive reptilian pattern seen in its captorhinomorph ancestors. The one noticeable specialization is a considerable elongation of the neck vertebrae. The skull, in nearly all features, is likewise closely comparable to that of captorhinomorphs (or of the more primitive pelycosaurs). Only the cheek region is distinctive. Here, high above a broad plate of bone made up chiefly of the large squamosal and the postorbital, is a well-developed upper temporal opening—a feature quite unknown in any contemporary.

What is the evolutionary significance of *Araeoscelis?* Surely it is an offshoot of the ancestral captorhinomorphs. It has been argued that it was a sterile offshoot from that group (as was *Bolosaurus*). This is quite possible, and there are no series of definitely annectant types to tie *Araeoscelis* to later forms with a comparable temporal structure. Equally possible, however, is an opposing point of view, to which we will here provisionally adhere—namely, that *Araeoscelis* is representative of a poorly known primitive radiation of small euryapsid reptiles in the Permian and Triassic, which were destined to be superseded by the lepidosaurs, but which, before this occurred, gave rise to an amphibious—and eventually marine—branch which continued to flourish in the form of nothosaurs, plesiosaurs, and placodonts. The forms which are tentatively included in this early euryapsid radiation will be considered as members of the order Araeoscelidia.

A few of these forms, possibly related to *Araeoscelis,* may be briefly mentioned. *Protorosaurus* (Fig. 179) of the European Upper Permian was a form somewhat larger than *Araeoscelis,* but with similar body proportions. It has been frequently associated with that form, and the two (with various other types) are frequently bracketed as forming an order Protorosauria. But the temporal region of *Protorosaurus* is unknown, and it has been argued that it was actually a diapsid. Under the circumstances, the use of an ordinal name based on this genus is best avoided. *Trachelosaurus* of the European Lower Triassic is known from a disarticulated skeleton, features of which suggest relationship with the sauropterygians. But here, again, the skull structure is unknown.

A further possible relative is the grotesque *Tanystropheus* of the Middle Triassic. In this shore dweller, the trunk and tail are rather normally built. But the neck, in abrupt contrast with the trunk, is extremely elongated. The contrast between neck and body is so extreme that when a complete skeleton was found for the first time, it proved that bones of the front part of the animal had been described by earlier workers as belonging to a flying reptile, while the trunk had been thought to be that of a primitive dinosaur! Stout front feet suggest digging habits; but the general mode of life of this grotesque animal is difficult to imagine. Not improbably, *Tanystropheus* (Fig. 180) is a member of a line of amphibious euryapsids and is distantly related to sauropterygian ancestors; but decisive skull structures cannot be made out in the known material, and it is equally possible that the creature is a lizard relative.

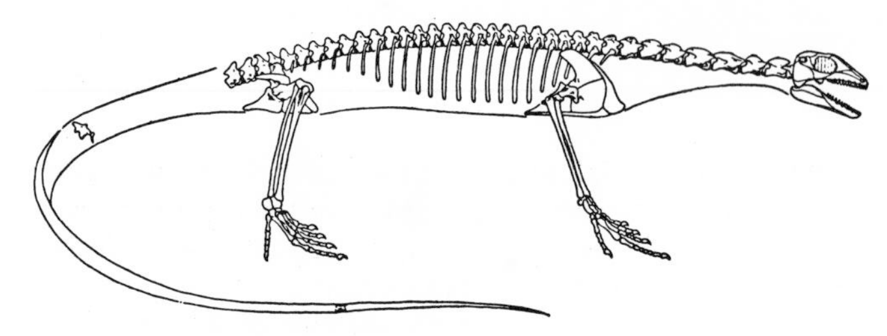

Fig. 177.—*Araeoscelis* of the Lower Permian, about 1/3 natural size. (From Vaughn.)

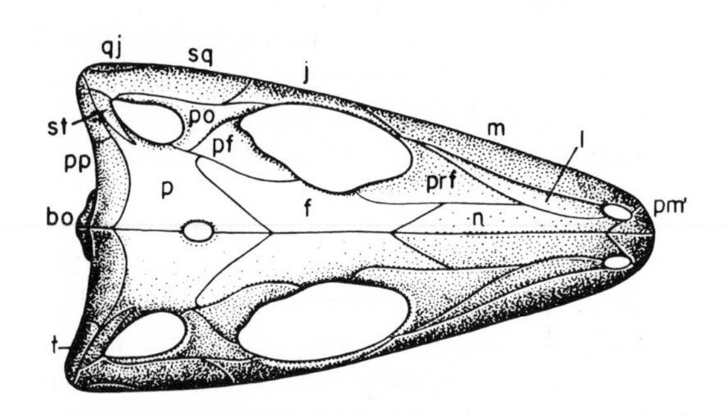

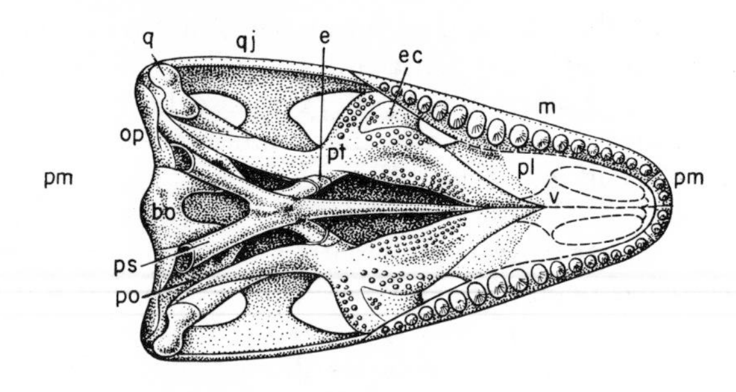

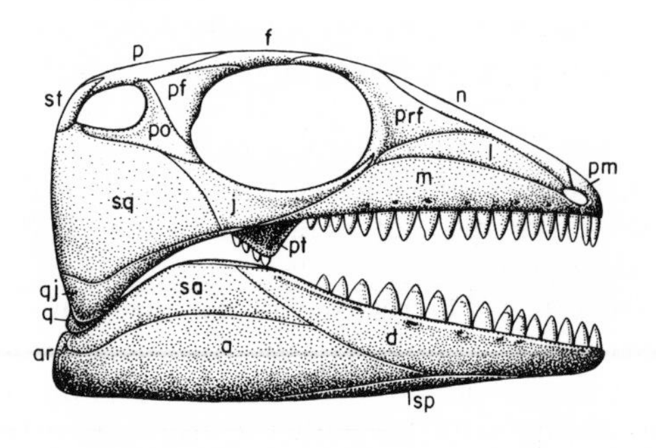

Fig. 178.—The skull of *Araeoscelis* in dorsal, ventral, and lateral views; length of original about 1.75 inches. (After Vaughn.)

Radically different from any of the forms so far mentioned, but definitely euryapsid in skull structure, is *Trilophosaurus* (Figs. 181, 182) of the late Triassic of North America. This was a stoutly built reptile of modest size, with a normal neck length, but with a skull which superficially gives the impression of a cross between an archaic diadectid and a rhynchocephalian. Short and deep, the skull has a toothless beak, presumably horn covered; the cheek teeth were transversely widened and three cusped, looking much like those of *Diadectes,* and there is, somewhat as in *Diadectes* and chelonians, a highly developed otic notch.

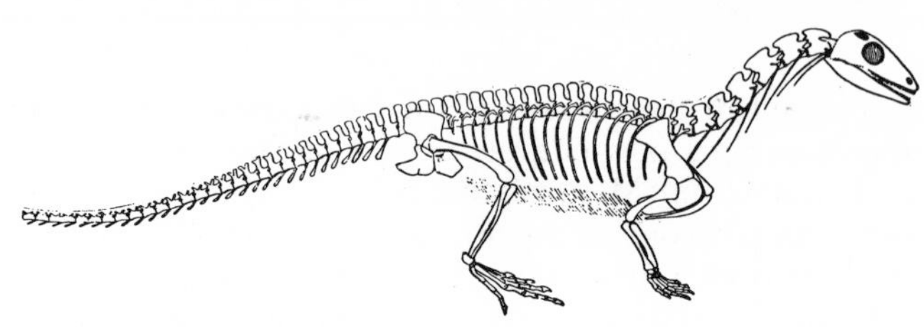

Fig. 179.—*Protorosaurus* of the late Permian, about 1/10 natural size. (After Gregory.)

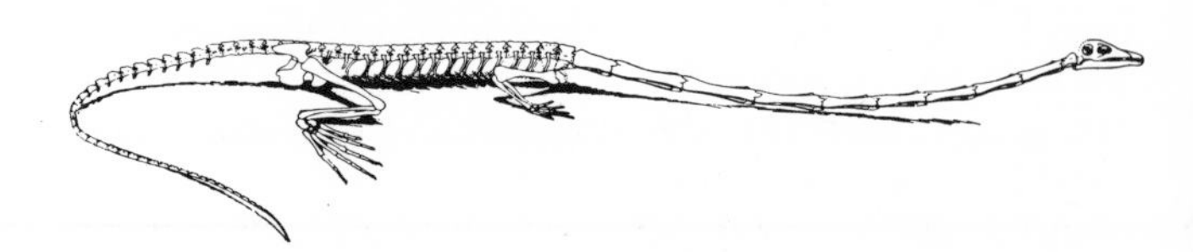

Fig. 180.—*Tanystropheus,* a grotesquely long-necked Triassic reptile; average specimens about 2.5 feet long. (From Peyer.)

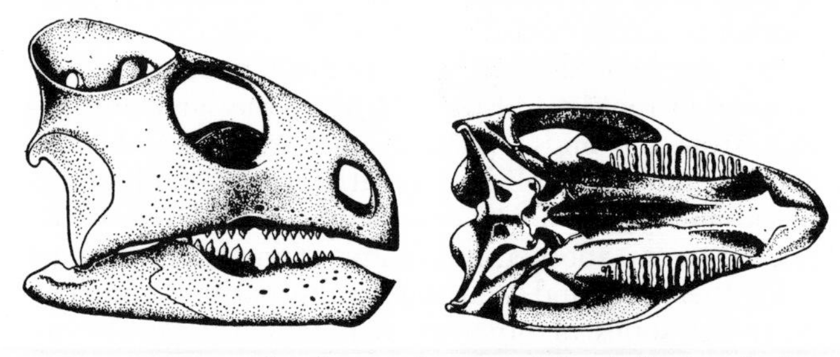

Fig. 181.—Lateral and ventral views of the skull of *Trilophosaurus,* an Upper Triassic euryapsid; length about 5 inches. (After J. T. Gregory.)

Apart from one "nest" of specimens in the Texas Triassic, we know little of the history of *Trilophosaurus* and its relatives. They were, however, widely distributed in time and space, for there are similar forms in the late Triassic of Europe and a possible relative in the North American Cretaceous.

Sauropterygians.—Leaving this problematical area, we reach firm ground with a consideration of a variety of aquatic euryapsids which flourished during the Mesozoic. The plesiosaurs, one of the most important of marine reptilian types, may be included with related Triassic forms, the nothosaurs, in the order Sauropterygia.

In these reptiles, aquatic adaptations were developed, and, in addition, there were a number of common structural features which tend to show their mutual relationship. There was a single upper temporal opening of euryapsid type, which, as in *Araeoscelis*, lay above a broad cheek plate formed mainly by the squamosal. The jaw articulation usually lies well below the level of the tooth row, and the cheek margin in front of the articulation is curved and emarginated to a variable degree, with loss of the quadratojugal. The pineal eye had been retained, while the nostrils had, as in many marine forms, moved well back along the top of the skull. There was no secondary palate, but the interpterygoid vacuities were nearly or entirely closed by a fusion of the pterygoids in the midline below the old roof of the mouth. The vertebrae (Fig. 173*C*, *D*) were rather primitive in build, amphicoelous or, at the most, flat ended. The ribs of the trunk region had but a

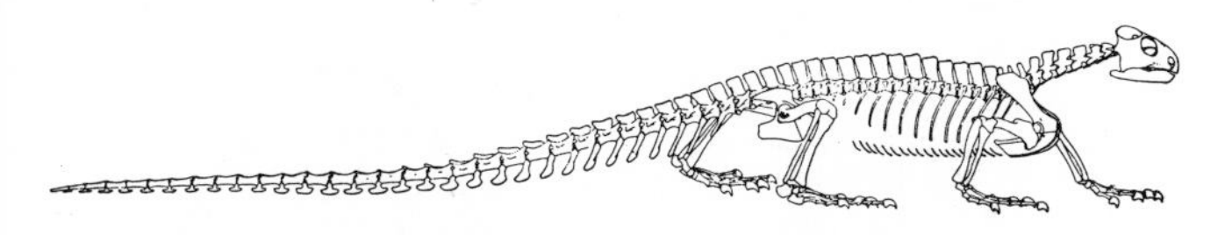

FIG. 182.—The skeleton of *Trilophosaurus*, about 1/30 natural size. (After Gregory.)

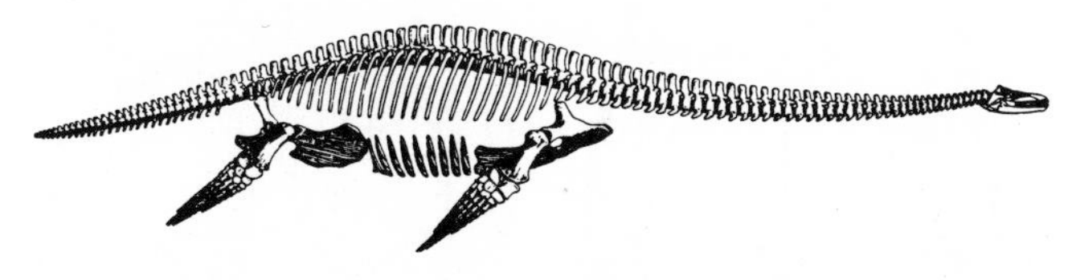

FIG. 183.—*Muraenosaurus*, a long-necked Jurassic plesiosaur, about 21 feet in length. (From Andrews.)

single head which articulated with the transverse process of the neural arch rather than the centrum. The old ventral-rib system had not only been retained but powerfully developed to form a basket-like structure along the belly, presumably in connection with the origin of powerful muscles running to the limbs.

The limbs and girdles were usually considerably modified for aquatic life. The dorsal elements of the girdles (scapula, ilium) were reduced in size; the ventral ones tended to be expanded for muscle origins. The sacral articulation was reduced in relation to the lessened need of support as terrestrial life was abandoned. The limbs of the nothosaurs were still more or

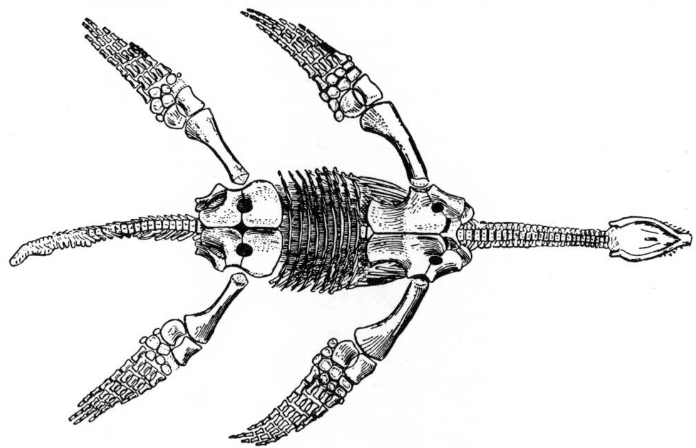

FIG. 184.—*Thaumatosaurus*, a Jurassic plesiosaur, ventral view, to show the abdominal ribs and expanded girdles, about 11 feet long. (After Williston, adapted from Fraas.)

less primitive and not suited as well to walking as to swimming; in the plesiosaurs they were long, paddle-like structures.

Plesiosaurs.—The plesiosaurs are the best-known members of the sauropterygians, as well as the last to survive, and the most specialized for aquatic life; they are common in many Jurassic and Cretaceous deposits. Many of them were of considerable size, with a maximum length of about 50 feet. Although they were obviously well fitted for marine life, their adaptations were very different from those of the rival ichthyo-

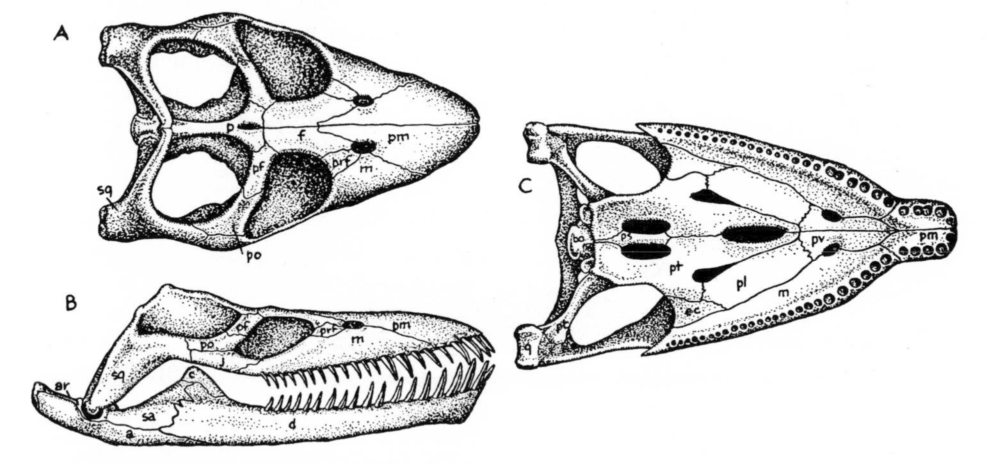

FIG. 185.—The skull in plesiosaurs. *A*, Dorsal view of skull of *Muraenosaurus*; *B*, lateral view of same, skull length about 14 inches; *C*, palatal view of *Thaumatosaurus*, skull length about 14 inches. For abbreviations see Figure 109. (*A*, *B* after Andrews; *C* after Fraas.)

saurs. The latter swam by a fishlike undulation of the body and tail; the fins were merely for steering, and the head did not move independently of the body. In the plesiosaurs, however, the body (Figs. 183, 184) was broad, flat, and inflexible, and the tail seems to have been of comparatively small importance; swimming seems to have been accomplished by rowing the body along by means of the well-developed paddles. The

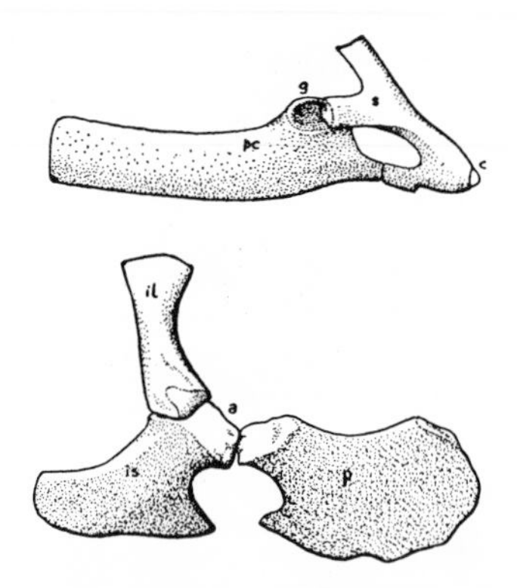

FIG. 186.—*Above,* Shoulder girdle of the plesiosaur *Microcleidus.* The scapula has grown downward and backward to gain a ventral contact with the coracoid. *Below,* Pelvis of *Muraenosaurus.* Abbreviations: *a,* acetabulum; *c,* clavicle; *g,* glenoid; *il,* ilium; *is,* ischium; *p,* pubis; *pc,* precoracoid; *s,* scapula. (After Watson and Andrews.)

head was set on the end of a flexible neck, often long. A plesiosaur has been compared by an old writer to "a snake strung through the body of a turtle."

The skull (Fig. 185) was variable in shape, most forms having a short face, while in others shortness of the neck was compensated by an elongated beak. The lower border of the cheek was typically arched and sometimes excavated to some degree, as in some chelonians, but there is no reason to believe that a lower temporal opening had ever developed in plesiosaurs or other sauropterygians. The external nares had been pushed back dorsally to a position comparable to that found in ichthyosaurs. The palate was highly modified; the two pterygoids met each other beneath the basisphenoid, separated to disclose a small interpterygoid vacuity, and then met again farther forward. The pointed teeth, set in sockets, were confined to the margin of the jaws.

The scapula (Fig. 186) had only a small dorsal blade, but ventrally the coracoid had expanded into a huge ventral plate for the attachment of strong muscles which gave a powerful down-and-back stroke to the paddle. In the pelvic girdle (Fig. 186) the ilium was very short and only loosely attached to the tip of the sacral ribs. The pubis and ischium, however, had developed to form a very large plate containing a small vacuity. Between pectoral and pelvic girdles there stretched a highly developed ventral basket of abdominal ribs.

The humerus and femur were well developed, while the bones of the distal segment of the limb were much shortened. The paddles (Fig. 174*C*) were very long but less specialized than those of the ichthyosaurs; for, although there was considerable hyperphalangy, there

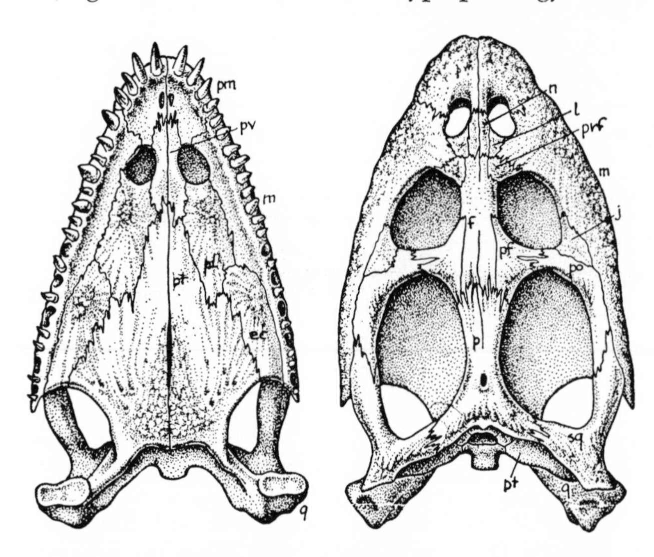

FIG. 187.—The skull of the nothosaur *Simosaurus,* dorsal and ventral views; length of original about 12 inches. For abbreviations see Figure 109. (After Jaekel.)

was never any hyperdactyly, and the phalanges retained the shape of normal finger joints.

Plesiosaurs first appear in the Rhaetic deposits, laid down at a time transitional from the Triassic to the Jurassic. In the marine Liassic deposits of the Lower Jurassic of Europe the group was already abundant and diversified, and numerous excellent specimens

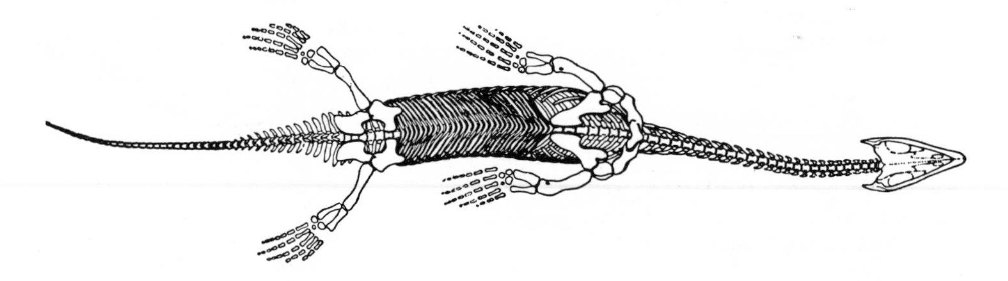

FIG. 188.—*Ceresiosaurus,* a Triassic nothosaur, ventral view; original about 3.5 feet long. (From Peyer.)

have been recovered from the Liassic of England and Germany. Two types, quite different in adaptive features, are early apparent. *Plesiosaurus, Muraenosaurus* (Figs. 183; 185*A, B*), and *Thaumatosaurus* (Figs. 184, 185*C*) belong to a group with small, short heads, a long neck, and short proximal limb bones. *Pliosaurus* and *Trinacromerum,* on the other hand, represent a series which had a relatively short neck but tended to increase the length of the head in compensatory fashion; humerus and femur tended to be more elongate than in the long-necked genera. The pliosaurs tended, even in the Jurassic, to become large forms; *Stretosaurus* of the English Upper Jurassic, and *Kronosaurus* of the Australian Lower Cretaceous had a body length of more than 40 feet.

Jurassic plesiosaurs are mainly from European deposits; a second major fauna is that of the last appearance of the group, in the late Cretaceous, mainly known from specimens from western North America. All Upper Cretaceous genera show specializations in the union of the two heads of the cervical ribs into a single structure and a shortening of the elements of the second segment of the limbs. Long-necked and short-necked groups, however, persisted. The long-necked forms are the more numerous and include such genera as *Elasmosaurus,* with seventy-six vertebrae in its flexible neck. The short-necked forms have a modest cervical count—thirteen, for example, in *Brachauchenius.*

Nothosaurs.—Much more primitive sauropterygians are present in the Triassic strata, particularly the marine beds of the Middle Triassic of Europe. *Nothosaurus, Simosaurus* (Fig. 187), *Ceresiosaurus* (Fig. 188), and a number of other genera of that age were types not yet completely adapted to aquatic life and, on the average, considerably smaller than the plesiosaurs. Except for relative slenderness, the general structure of the body and limb girdles was quite similar to that of the plesiosaurs. The limbs, however, were little specialized for aquatic life, except that the distal segments were somewhat shortened. The digits (Fig. 174*D*) were of rather normal build but were probably webbed; in some cases there was a mild increase in the number of phalanges. In the skull the nostrils were "in transit" between the normal position and that found in plesiosaurs. The palate shows much the same general type of construction as in plesiosaurs but was a completely solid structure; the two pterygoids met in the midline for their entire length to form a continuous plate. In this respect the typical nothosaurs were already a bit more specialized than the plesiosaurs and hence, despite the similarities in other respects, cannot have been plesiosaur ancestors.

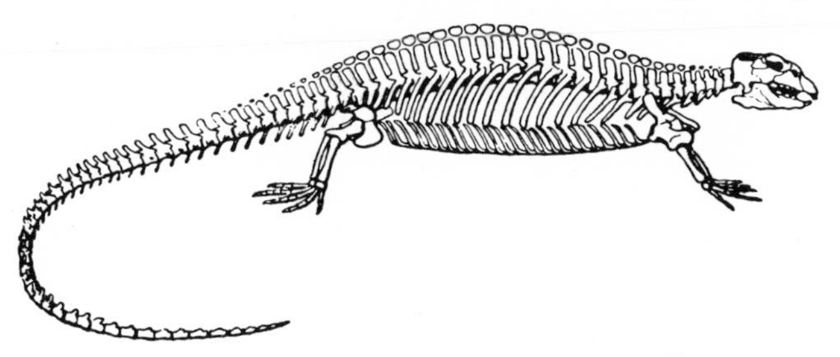

Fig. 189.—The skeleton of *Placodus,* original about 7 feet long. (From Peyer.)

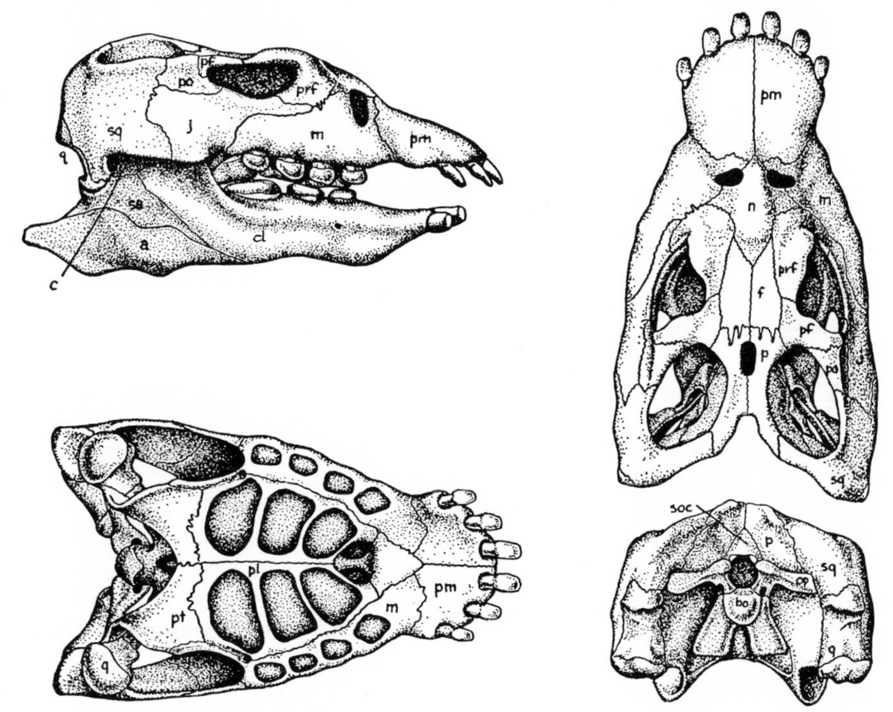

Fig. 190.—The skull of the placodont reptile *Placodus;* original about 10 inches long. For abbreviations see Figure 109. (After Broili.)

A possible plesiosaur ancestor is *Pistosaurus* of the Middle Triassic. The skull, which alone is preserved, is for the most part similar to that of nothosaurs. However, the palate is not completely closed; in this and other details of skull construction we find suggestions of plesiosaur relationships.

The exact ancestry of the nothosaurs is uncertain. Unlike the case of the ichthyosaurs, there is no reason to believe that the aquatic trend in the sauropterygians is a direct inheritance from primitive reptiles; rather surely we have here a return to the water from terrestrial forebears, and the sauropterygians may well have been derived from primitive captorhinomorphs by way of terrestrial forms of the general *Araeoscelis* type.

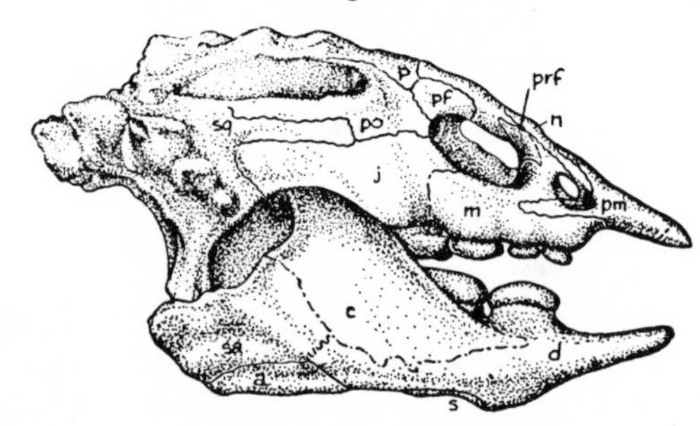

FIG. 191.—Lateral view of the skull of the placodont *Placochelys;* length of original about 6 inches. For abbreviations see Figure 109. (After Jaekel and von Huene.)

Placodonts.—Placodus is the best-known member of a small group of aberrant Triassic euryapsids which had taken up mollusk eating as a livelihood. Connected with this is the most remarkable specialization of *Placodus* (Figs. 189, 190), the development on the palate and lower jaws of enormous flat teeth capable of crushing mollusk shells. These teeth were not unlike those of some of the pavement-toothed sharks and, indeed, were long thought to pertain to fish. Presumably there were very powerful jaw muscles for this heavy work, for the coronoid bone of the jaw forms a long process, directed upwardly beneath the deep cheek, for muscular insertion. This remarkable coronoid development simulates that found in mammals and is paralleled in but few other reptilian types. The front teeth were heavy, projecting pegs across a broad muzzle in *Placodus* and may have served to root up mussels.

Apart from adaptations for this specialized diet, *Placodus* was not dissimilar to the nothosaurs in many features of limb and trunk structures; the skulls were essentially similar in such features as the position of the nostrils and the solid palate. However, the placodont neck and trunk were relatively short and the trunk rounded in cross-section; and in the *Placodus* skull the tooth-bearing palatines have pushed back at the expense of the pterygoids. *Placodus* shows a further specialization in that numerous nodules of bone were found associated with the skeleton, indicating the beginning of body armor.

The *Placodus* specializations are intensified in other placodont genera of the middle and late Triassic of Europe. The crushing teeth may become larger but fewer in number; the snout may become slim and, in *Placochelys* (Fig. 191), toothless, the sharp beak perhaps functioning as a pair of forceps in picking out mollusks; another genus, *Henodus*, became almost completely toothless and presumably had developed stout horny plates, analogous to those of a turtle, for masticatory purposes. The more specialized placodonts, such as *Placochelys* (Fig. 192) developed the armor to a higher degree; and in *Henodus* the short, broad, and flattened body was completely encased in a shell comparable to that of the turtles. However, the placodont armor was composed of a much larger number of plates than in the Chelonia, arranged in a mosaic somewhat comparable to the carapaces of ankylosaur dinosaurs or glyptodont edentates; the similarities between placodonts and turtles are surely due to convergence rather than true relationships.

One stage backward in the evolutionary history of the placodonts is seen in the Middle Triassic genus *Helveticosaurus.* The basic structure of this reptile is that of a true placodont. But the body is, in contrast with placodonts, rather elongated, much as in nothosaurs, and no armor is present. Most significantly, there

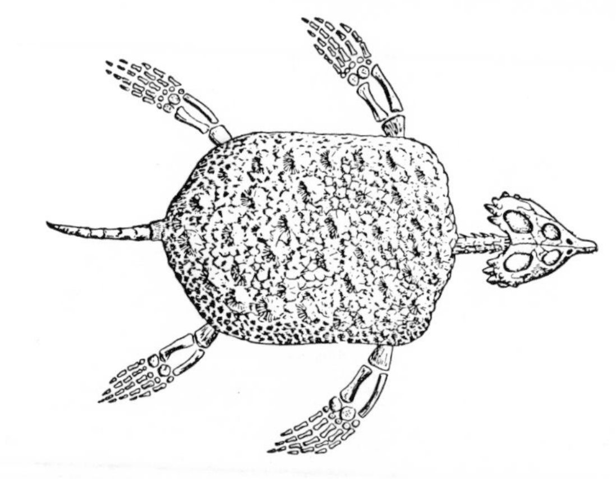

FIG. 192.—The skeleton of *Placochelys* in dorsal view; about 1/12 natural size. (From Gregory, after von Huene.)

was no development of tooth plates; instead, both jaw margins and palate bore numerous simple teeth, pointed and somewhat recurved. We have here a form which has not diverged far from a common stem with the sauropterygians.

We have no positive clues about the ancestry of the aquatic euryapsids as a whole. Ultimately, of course, the captorhinomorph cotylosaurs are the source from which they were derived. And while such a primitive euryapsid as *Araeoscelis* shows no positive indications of a trend in a sauropterygian direction, it may represent a first stage in the evolution of later euryapsids.

CHAPTER 10

Lepidosaurian Reptiles

The reptiles so far described have been either anapsids, usually with a solid skull roof, or forms with but a single temporal opening. However, except for marine types, the great bulk of Mesozoic reptiles were two-arched forms, with a temporal opening both above and below a bar formed by the postorbital and squamosal bones. This is the diapsid condition (cf. Fig. 155), and all forms with this double opening have often been grouped as members of a single subclass (the Diapsida). But many of the diapsids were advanced types showing evidences of close relationship to one another but not especially to more primitive, two-arched forms. We prefer to regard the advanced groups—the crocodiles, dinosaurs, and pterosaurs—as members of a subclass Archosauria, which will be considered in later sections. In this chapter we shall treat only the more primitive, two-arched reptiles and types derived from them; these constitute the subclass Lepidosauria—using a term frequently applied to the "scaly" reptiles (lizards, snakes, and *Sphenodon*) which are the modern representatives of the group.

Eosuchians.—No two-arched reptiles are known from the early Permian. First of such types to appear were a number of small reptiles from the Upper Permian and Lower Triassic of South Africa, of which *Youngina* (Fig. 193) is the best known. This form appears to have had a comparatively primitive postcranial skeleton, with moderately slender limbs, normal reptilian foot structure, and (as far as can be told) body proportions not unlike those of typical lizards.

In the skull were retained many primitive features, in which *Youngina* is in contrast to the archosaurians to be described in the next chapter. The palate was of a very generalized reptilian type, with pterygoids movable on the braincase and with teeth on the palatal bones, in addition to a row of teeth set in sockets on the jaw margins. Small supratemporal, postparietal, and tabular bones were still present at the back of the skull, a pineal eye had been retained, and (unlike many archosaurs) there was no opening (the antorbital fenestra, cf. Figs. 210, 212) between the bones on the side of the slim snout. As in various cotylosaurs, the posterior margin of the cheek was concave, showing the initiation of an otic notch of reptilian type.

Youngina, nevertheless, had two temporal openings —a lateral one far down the side of the cheek region and a large upper opening on the top of the broad skull table. We are dealing with a diapsid reptile, but a very primitive one; and it seems fitting that *Youngina* and its relatives should constitute a discrete order, the Eosuchia. *Youngina* is, of course, readily derivable from the cotylosaurs, and, as mentioned previously, it may well be that the ancestors were the millerettids, in which a lateral temporal opening was already developing.

It may be that primitive eosuchians gave rise to the advanced archosaurs; it seems, however, definite that the eosuchians were the direct ancestors of the lizards and snakes of the order Squamata. The lizards differ from primitive diapsids in a variety of features; most distinctive, however, is the different construction of the temporal region. In lizards there is an upper temporal opening with a bar below it; below this bar the cheek is widely open, and at the back the quadrate is freely movable. It was once thought that this one-arched condition of the lizards was primitive, but it is now generally agreed that the ancestor of the lizards was a diapsid, with two temporal openings (and hence two arches), and that the lizard condition of the cheek region was attained by the loss of the lower arch. In accordance with this is the presence, in the early Triassic of South Africa, of *Prolacerta* (Figs. 194, 202*D*), a genus similar in many ways to *Youngina,* but with the lower temporal arch incomplete.

Until recently it was believed that the lizards developed at a relatively late date, for few fossils attributable to this group had been found in rocks earlier than those of the late Cretaceous. In recent years, however, a number of lizard-like fossil forms, some of them of remarkable structure, have been found in Middle and Upper Triassic deposits. It appears that there was an early Mesozoic radiation of lepidosaurians, descended from *Youngina*-like ancestors, but with the lower temporal bar reduced.

How to classify them is somewhat of a problem; they tend, on the whole, to show a series of intermediates between ancestral eosuchian and true lizard structures. We shall here use, as a point of cleavage, the condition of the quadrate region. In true lizards the quadrate is flexibly articulated with the squamosal above it, thus giving potentialities of an increased jaw gape. This situation is brought about by the loss of the lower temporal arch, and, further, by a reduction in the development of the squamosal, which primitively lapped downward over the quadrate.

Using this criterion, we can regard *Prolacerta* as an advanced eosuchian rather than a true lizard, despite the reduction of the temporal arch and the fact that it has advanced far in the reduction of certain small roofing bone elements absent in lizards. Also in this category is, for example, *Macrocnemus* of the European Middle Triassic. This is a small animal with a pointed skull, a moderately long neck, and very long and slender legs. In default of knowledge of the temple region of the skull, its grotesque contemporary, *Tanystropheus*, mentioned in the last chapter, may belong here as well.

Apart from forms advancing toward and to the lacertilian condition, a number of aberrant types seem to have evolved from the primitive eosuchian stock. In *Thalattosaurus* of the marine Middle Triassic of North America, and *Askeptosaurus* of beds of similar age in Europe, the body and tail, with a total length of 5 feet or so, were long and slender, the limbs small and paddle-like; we are dealing with an aquatic side branch of the eosuchians. In the skull (Fig. 195) the snout was elongated, suggesting fish-eating habits; as in many aquatic forms, the nostrils had moved far back toward the orbits. In the temporal region the lateral opening is large, and the bar below it slender but complete; the upper opening is merely a narrow slit in *Askeptosaurus*, and in *Thalattosaurus* it has closed completely.

Champsosaurus (Figs. 196, 197) is a reptile found in late Cretaceous and early Tertiary deposits in both Europe and North America, which appears to have been a fish-eating dweller in fresh waters. In its general proportions it gives the appearance of a small crocodilian, and its long-snouted skull is superficially comparable to that of a gavial. The external nostrils were at the tip of the long snout; the internal nares were carried far back in a fashion somewhat paralleling the crocodilians. The temporal region, typically diapsid in nature, is considerably expanded, presumably accommodating powerful jaw muscles. *Champsosaurus* has been frequently included in the Rhynchocephalia, but it lacks the rhynchocephalian beak and the teeth are socketed rather than acrodont; hence it seems best to regard it as a specialized late survivor of the eosuchian group of primitive diapsids.

Rhynchocephalians.—Persistently primitive in many features, but with a few characteristic specializations, are the Rhynchocephalia, which first appeared in the Triassic and survive in the shape of a small reptile, *Sphenodon,* inhabiting a few islands off the coast of New Zealand. *Sphenodon* (Fig. 198) looks very much like a lizard and, indeed, resembles the lizards greatly in many features; on the other hand, in its retention of numerous primitive characters, it appears to have departed little from the structure of the more primitive Eosuchia. It is sharply marked off from the lizards, in which the cheek is open beneath the upper opening and bar; here both temporal arches (and temporal openings) are retained. It is easily distinguished from the eosuchians by two specialized features: the teeth are fused to the edge of the jaws (acrodont), rather than set in sockets, and there is a small overhanging beak on the upper jaw, a characteristic to which the group name refers.

Most of the diagnostic characters of the rhynchocephalians are primitive ones to which little reference need be made. The palate is as primitive as that of *Youngina,* the pineal eye is still present, the vertebrae

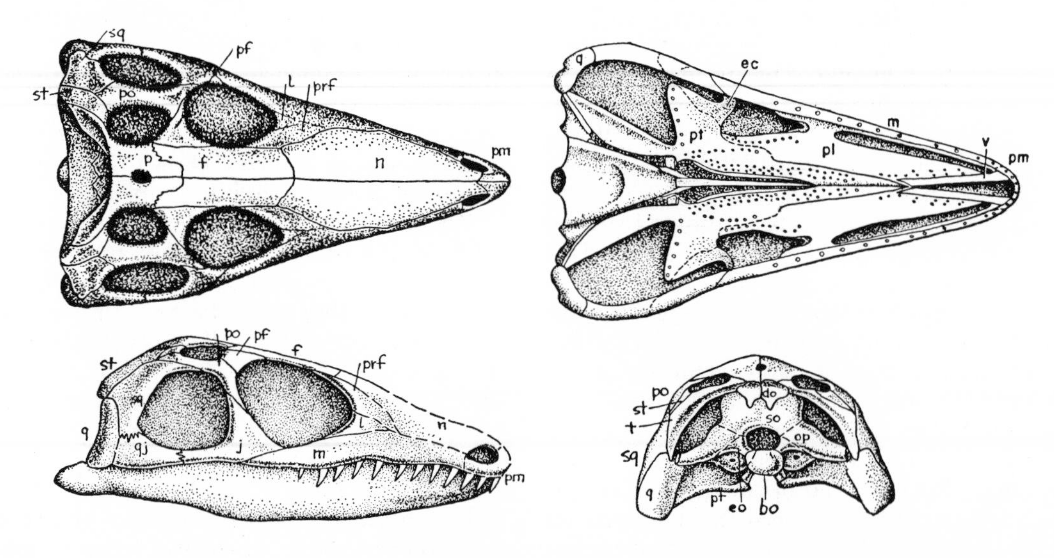

Fig. 193.—*Youngina,* a primitive, two-arched reptile from the Upper Permian of South Africa—dorsal, palatal, lateral, and occipital views; length of original about 2.5 inches. For abbreviations see Figure 109. (After Broom, Watson, Olson.)

are amphicoelous rather than having the solid structure of most land reptiles of post-Triassic times. There have been some changes from primitive conditions, however: the lacrimal, supratemporal, and tabular bones have disappeared; the ribs attach to both arch and centrum by a single broad head rather than a double one; and (as in lizards but not in eosuchians) a large opening has developed in the primitively plate-like ventral part of the pelvis.

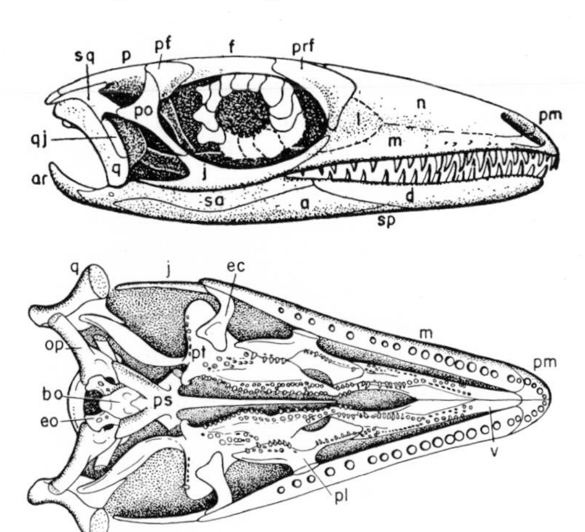

Fig. 194.—Lateral and ventral views of *Prolacerta;* length of skull about 2.5 inches. (After Camp and Parrington.)

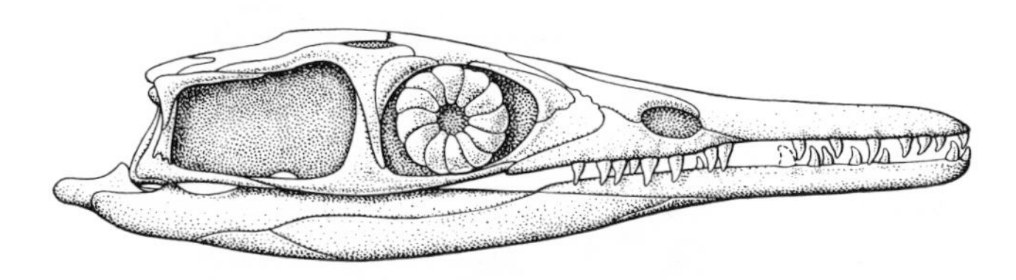

Fig. 195.—The skull of the Triassic aquatic eosuchian *Askeptosaurus;* length about 1 foot. (After Kuhn-Schnyder.)

Sphenodon appears to be a persistent survivor of an ancient group, for *Homoeosaurus* (Fig. 199) of the Jurassic appears to have been almost identical in structure, and the ancestry of this type may without doubt go back through some imperfectly known Triassic genera to such a form as *Youngina.* Very few changes are needed to make a rhynchocephalian out of an eosuchian.

For the most part, the history of the Rhynchocephalia has been an uneventful one. Occasional finds of small sphenodonts occur in European and American deposits from the Triassic to the Lower Cretaceous. Beyond this point there are no records, and it is probable that by the Tertiary the sphenodonts had become restricted to their present isolated home in New Zealand.

A few unimportant and obscure variants of the rhynchocephalian pattern are recorded in Triassic and Jurassic deposits. The one prominent development in rhynchocephalian history occurred in the Triassic, with the appearance, explosive development, and present extinction of the family Rhynchosauridae (Figs. 200, 201). These were rather heavily built quadrupeds, reaching perhaps 6 feet or so in size at a maximum. The skull was short and broad, obviously with powerful muscles to cope with vegetable food. There was a parrot-like beak. In the upper jaw a large tooth plate

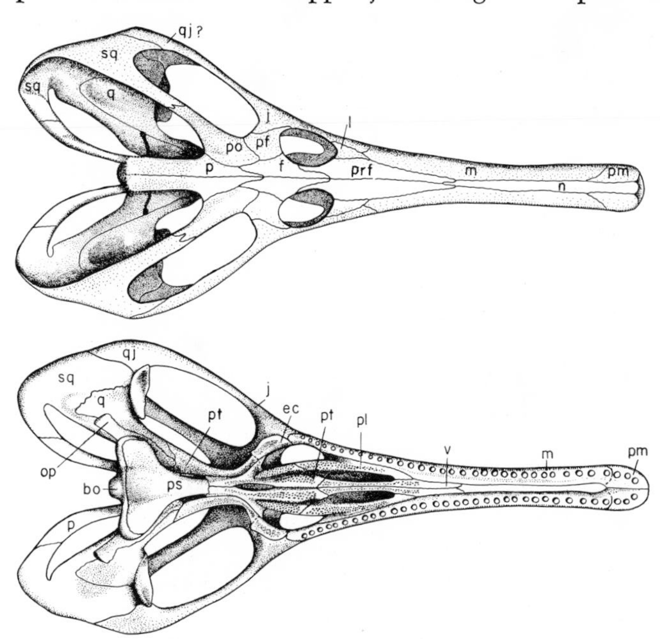

Fig. 196.—The skull of *Champsosaurus,* dorsal and ventral views; length about 13 inches. (After Brown and Williston.)

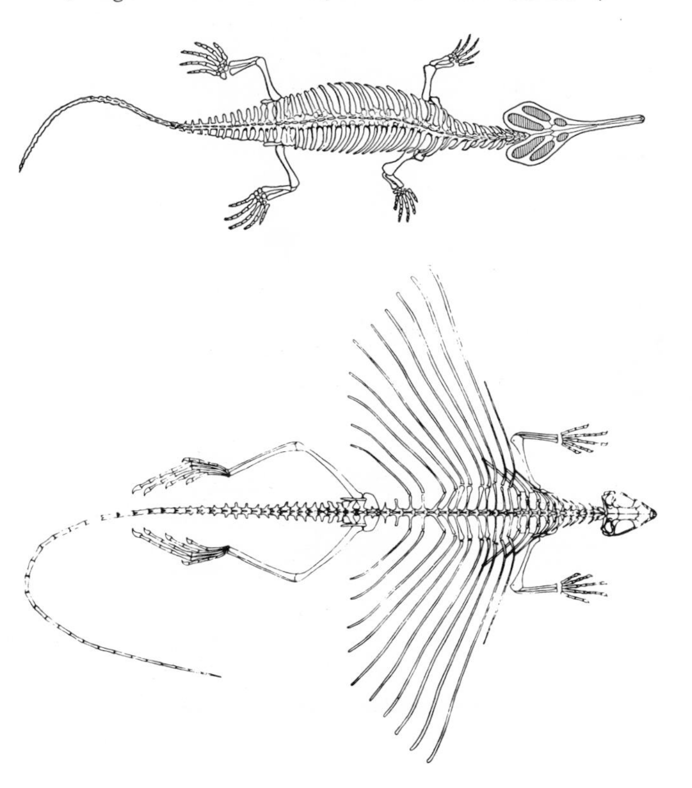

Fig. 197.—*Above,* The skeleton of *Champsosaurus,* about 1/15 natural size. (After Gregory.) *Below,* The skeleton of the Triassic "flying lizard" *Kuehneosaurus,* about 1/5 natural size. (From Robinson.)

was present on either side of the palate, with a number of longitudinal rows of teeth. The lower jaw was modified, at its upper edge, into a narrow chopping blade which fitted into a lengthwise groove in the upper tooth plate. The flora of the Triassic included several groups in which there had developed a "fruit" with a soft external pulp surrounding a hard shell which had good nutritive material within. The rhynchosaur chopping device presumably served to break this shell.

A few sparse remains of primitive rhynchosaurs are known from the early Triassic; a few late stragglers are found in late Triassic deposits. It is in the Middle Triassic that the group reached its climax. Rocks of this age in the northern continents are almost exclusively marine. But in the Middle Triassic continental formations of East Africa, Brazil, and Argentina the rhynchosaurs are extremely abundant; perhaps a third or so of all animals discovered in them are members of this briefly flourishing group.

The Squamata.—The order Squamata includes the lizards and snakes, the most successful of modern reptiles; there are, it is estimated, over six thousand species belonging to this order living today, chiefly in tropical regions. We have noted that a diagnostic feature of the lizards lies in the fact that there is but a single temporal opening—the upper one, lying above a bar formed by the postorbital and squamosal (Fig. 202). Instead of a lower opening being present, the cheek region lies open at the side behind the eye; the quadratojugal has disappeared; and the squamosal is much reduced. As a result of this loss of cheek covering, the quadrate bone is freely movable, giving greater flexibility to the jaws. The pineal eye persists in most lizards. Most of the center row of bones of the skull roof, originally paired, have fused into single median elements. The lacrimal bone is usually absent, the postparietal has gone, and so has either the tabular or the supratemporal (there is a single bone to represent the two). On the other hand, some lizards have developed a whole series of new superficial dermal bones which may cover the entire skull roof.

The palate is of the primitive type seen in many Permian reptiles and in *Sphenodon.* The teeth are not set in sockets but are fused to the edge (as in *Sphenodon*), or, more commonly, to the inner sides of the jaws (pleurodont) and are thus specialized in character; palatal teeth, present in eosuchians, have disappeared.

The primitive amphicoelous type of vertebra has persisted in a few living lizards, but in most of the Squa-

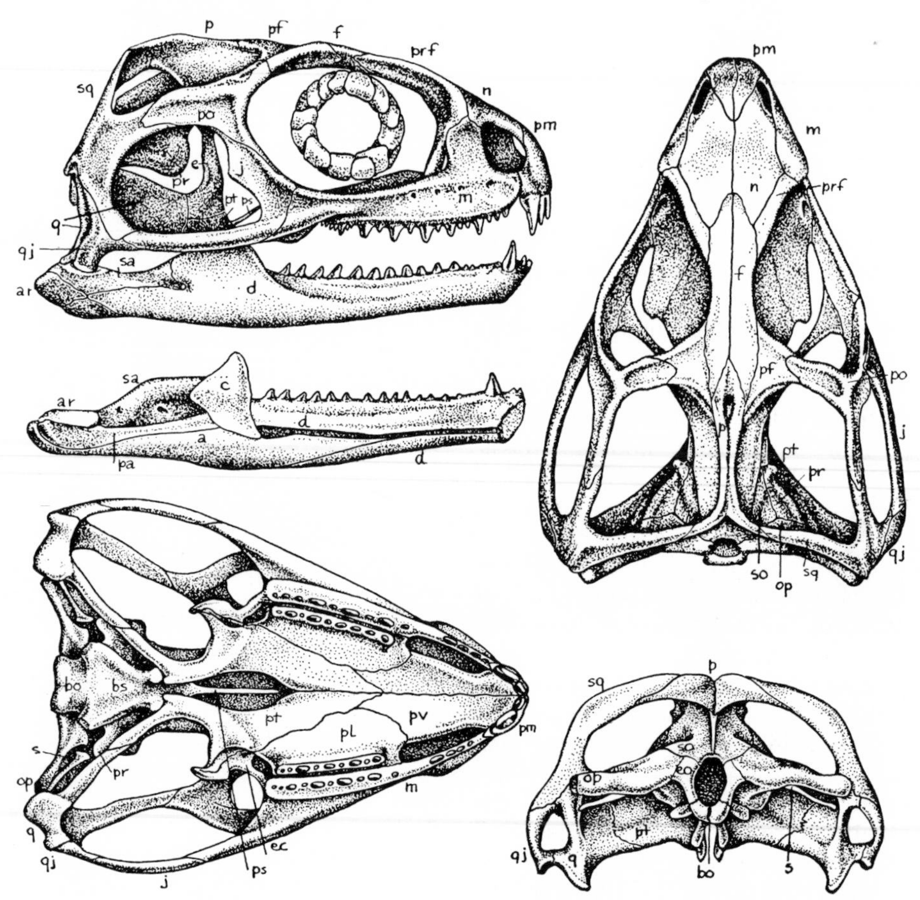

FIG. 198.—The skull of the living rhynchocephalian *Sphenodon. so,* Supraoccipital. For other abbreviations see Figure 109.

mata a good articulation between the centra has been brought about through the development of the procoelous condition in which the swollen posterior end of one vertebra fits into a socket in the one behind (Fig. 173*G–I*). The ribs have, as in *Sphenodon,* a single head; but, in contrast, this is narrow here and articulates only with the centrum. In other features, the general structure is not unlike that just described for the Rhynchocephalia and may not be far from that of the eosuchians.

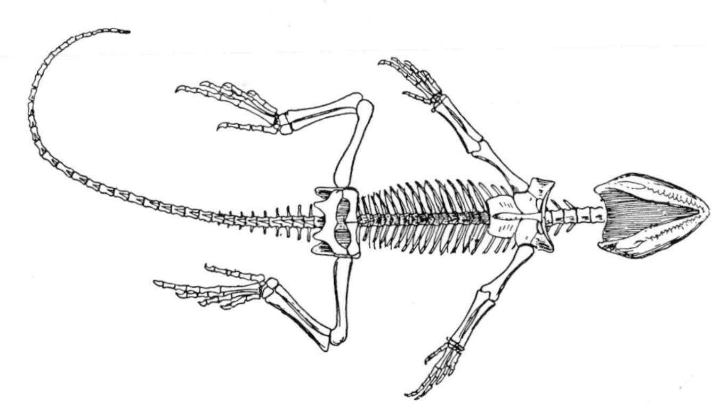

FIG. 199.—*Homoesaurus,* a Jurassic rhynchocephalian in ventral view, about 7.5 inches long. (From Williston's *The Osteology of the Reptiles,* by permission of the Harvard University Press.)

Early lizards.—The Squamata are divided into two suborders: the Lacertilia (or Sauria), the lizards; and Ophidia (or Serpentes), the snakes. The first are, of course, the more primitive of the two and the first to be represented in the fossil record. We noted above the recent discovery, mainly in the middle and late Triassic, of a highly varied series of forms which appear to lie close to the evolutionary boundary between eosuchian ancestors and lacertilian descendants. Any line of cleavage between the two must necessarily be arbitrary. We have, earlier in this chapter, suggested that the development of mobility of the quadrate be used as a dividing line. Probably, with increased knowledge, a number of late Triassic forms which have crossed this boundary will become known. One which has definitely done so is *Kuehneosaurus* (Fig. 206) of the very late Triassic of Europe, in which the quadrate has been freed but primitive characters are present in other regards—the central series of skull elements are still paired, and, for example, there is a large lacrimal bone, and teeth are still present on the palate.

One would expect at this stage that ancestral types of lizards would be generalized in nature in all regards. Not so *Kuehneosaurus.* Most of the skeleton was of normal build. But in the trunk (Fig. 197) the ribs, instead of showing a normal downward curvature, could be projected stiffly outward on either side. Surely a pair of horizontal "sails," giving gliding ability, was present. One modern lizard, *Draco,* can extend its ribs outward to form a somewhat similar planing structure; but in *Kuehneosaurus* we have, at a very early date, an even more highly specialized adaptation of the same sort. A comparable form is known from the Upper Triassic of North America. Our knowledge of small reptiles of the Mesozoic is currently very inadequate, owing in great measure to the fact that known Jurassic and Cretaceous deposits are mainly marine. Considering the high degree of specialization shown in these "flying lizards," it is not improbable that further advances in paleontological discovery will reveal a variety of early lizard specializations.

There are numerous families of later lizards, many of which need not be mentioned in detail. There are various methods of classifying them; we shall here

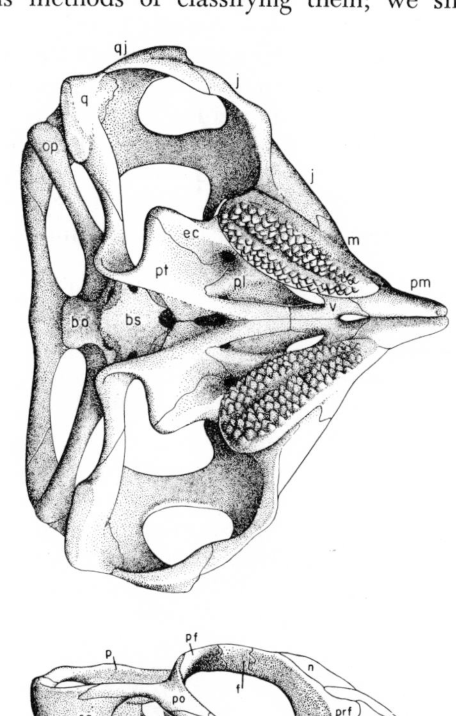

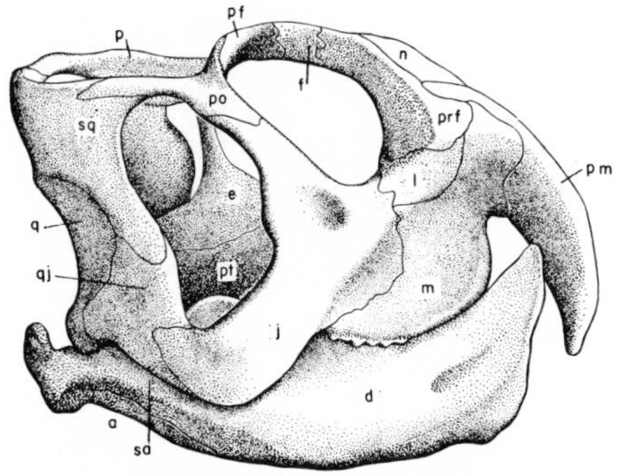

FIG. 200.—Lateral and ventral views of the skull of the rhynchosaur *Scaphonyx,* about 1/2 natural size.

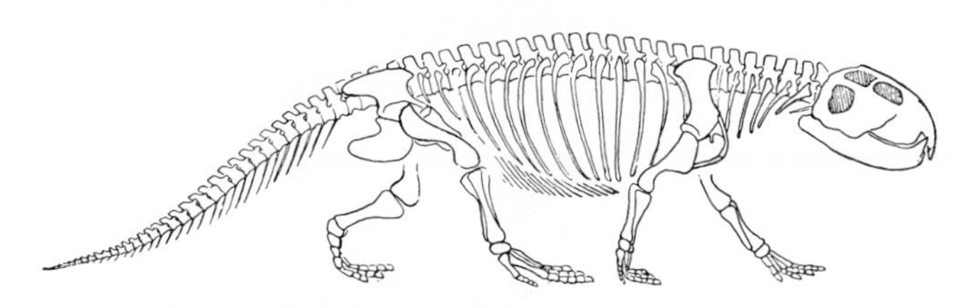

FIG. 201.—The skeleton of the rhynchosaur *Scaphonyx,* about 1/12 natural size. (From von Huene.)

consider them to make up five infraorders, which will be briefly reviewed.

Nyctisauria (*Gekkota*).—Most of the little tropical geckoes have amphicoelous vertebrae, a presumably primitive feature; they are not, however, particularly primitive in other respects. Few fossils are definitely attributable to this group, but several poorly represented Jurassic types may mark an early appearance of geckoes.

Iguania.—The iguanids are common, and often large, New World lizards; the agamids, distinguished by acrodont rather than pleurodont teeth, are Old World relatives. The two families include lizards of rather primitive character. Allied to the agamids are the chameleons, highly specialized African forms with skull specializations as well as peculiar grasping feet and a highly protrusible tongue. Agamids and chameleons are definitely known from the Upper Cretaceous, iguanids from the Eocene onward, and two specimens from the European Jurassic may be forerunners of the Iguania.

Leptoglossa (*Scincomorpha*).—Included here are a host of lizards of moderately advanced types, such as the widespread skinks (Scincidae), the "true" lizards (Lacertidae) of the Old World, and the Teiidae and Xantusiidae of the American tropics. In skinks, osteoderms are found underlying body scales, and commonly plates of this sort are present covering the normal bones of the skull roof. Although many members of the group have normal body proportions, numerous skinks and teiids have especially reduced limbs and tend toward a snakelike body form. There are possibly Upper Jurassic forerunners of the group, and numerous teiids occur in the Upper Cretaceous.

Annulata (*Amphisbaenia*).—Although perhaps of scincoid derivation, the tropical amphisbaenoids have become so highly specialized for a burrowing life as limbless, elongate, and rather wormlike forms that they merit distinction as a discrete group. Indeed, some would advocate placing them in a separate suborder of the Squamata, or even in a separate order. The skull elements are tightly bound together (somewhat as in the Apoda) to form a compact wedge-shaped burrowing organ. A number of fossils are known from the early Tertiary.

Diploglossa.—Most remaining lizards can be placed in a single further suborder which, however, includes a considerable variety of types. The Anguidae, appearing in the Cretaceous, include today but a few forms, some of which have normally developed limbs, while others have snakelike proportions, as the "glass snake" (*Ophisaurus*) and the European "slow worm" (*Anguis*). Osteoderms are highly developed in the anguids, and in the early Tertiary there was present a series of forms, such as *Placosaurus* [*Glyptosaurus*], with normal limbs and a relatively massive armor of osteoderms. The "Gila monster," *Heloderma*, dating from the Oligocene and surviving in the American Southwest, is a comparable form, but generally regarded as a bit more advanced structurally, and placed in a separate family.

Varanus, the "monitor lizard," of the Old World tropics, is the living representative of an ancient lizard family, present in the Cretaceous, which is of an advanced nature in such characters as the complete enclosure in bone of the front part of the braincase, and the development of a flexible joint in the middle of

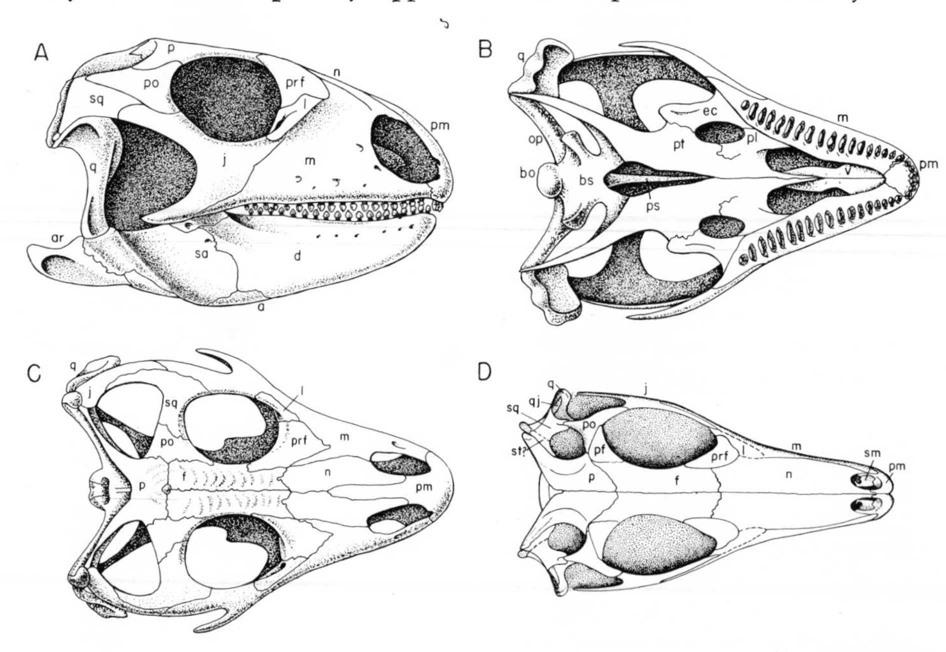

Fig. 202.—*A–C*, Lateral, ventral, and dorsal views of the skull of the Cretaceous lizard *Polyglyphanodon;* length of original about 3.5 inches. (After Gilmore.) *D*, Dorsal view of the skull of the eosuchian *Prolacerta*. (After Camp.)

each lower jaw—thus facilitating the swallowing of large prey, already aided by quadrate flexibility. In certain regards the monitors are suggestive of relationships to the snakes. More definite are clear affiliations to the extinct aquatic lizards described below, and frequently the term Platynota is used to include varanids and these fossil forms. *Varanus* includes the largest of living lizards, for an East Indian monitor reaches a length of 12 feet, and *Megalania* of the Pleistocene was perhaps double that size.

Marine lizards.—The most striking series of fossil lizards is that of the varied marine forms of the Cretaceous, which are closely related to the monitors and other platynotans. Two groups incompletely adapted to marine life appeared early in Upper Cretaceous times in Europe. The dolichosaurs, including *Acteosaurus* (Fig. 203) and *Dolichosaurus*, were small liz-

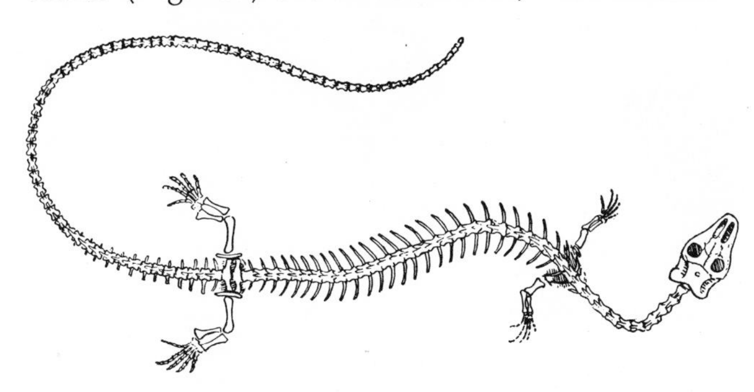

Fig. 203.—*Acteosaurus*, a Middle Cretaceous semiaquatic lizard; original about 16 inches long. (From Williston's *The Osteology of the Reptiles*, by permission of the Harvard University Press.)

ards with short limbs of fairly normal structure; the body was greatly elongated, indicating that they could not be ancestral to other aquatic Cretaceous types. The second group was that of the aigialosaurs, also fairly small and also probably only semiaquatic in their habits. Here the neck was short, the body not greatly elongate, the limbs somewhat paddle-like, and there was a powerful swimming tail. *Aigialosaurus* is the best-known genus; there was a late Jurassic ancestral form.

From the aigialosaurs, apparently, came the mosasaurs (Figs. 204, 205), Upper Cretaceous marine lizards, worldwide in distribution but especially common in the chalk rocks of Kansas; *Platecarpus, Tylosaurus, Mosasaurus,* and *Clidastes* are among the best-known types. They were of large size, about 15 to 30 feet in length. There was a long head, a short neck, and a long, slim body and tail. The tail was the main swimming organ, the limbs functioning only in steering. In the limbs (Fig. 174*E*), as usual in aquatic forms, the proximal bones were much shortened, but the toes were well developed, spreading, and presumably webbed. The normal number of digits was present, but there was a slight increase in the number of joints. The skull was much like that of the varanid lizards; and, as in varanids, there was a well-developed joint midway of the jaws, between the angular and splenial bones. The teeth were rather unlike those of most lizards in that they were set in pits rather than fused to the inner sides of the jaws.

The mosasaurs were mainly fish-eaters, it would seem, probably existing on the large variety of primitive teleosts which were their contemporaries; however, in *Globidens* the teeth had peculiar spheroidal crowns, and it would seem that this form was a mollusk-eater. They appear to have been an ecologic replacement of the declining ichthyosaurs. Despite the profusion of mosasaur types in the late Cretaceous rocks, they perished as soon as did their contemporary fellow fishermen, the plesiosaurs.

Snakes.—Ophidia or Serpentes, the snakes, are the newest of reptilian groups. They first appeared in the Cretaceous but they are not, in general, well-known as

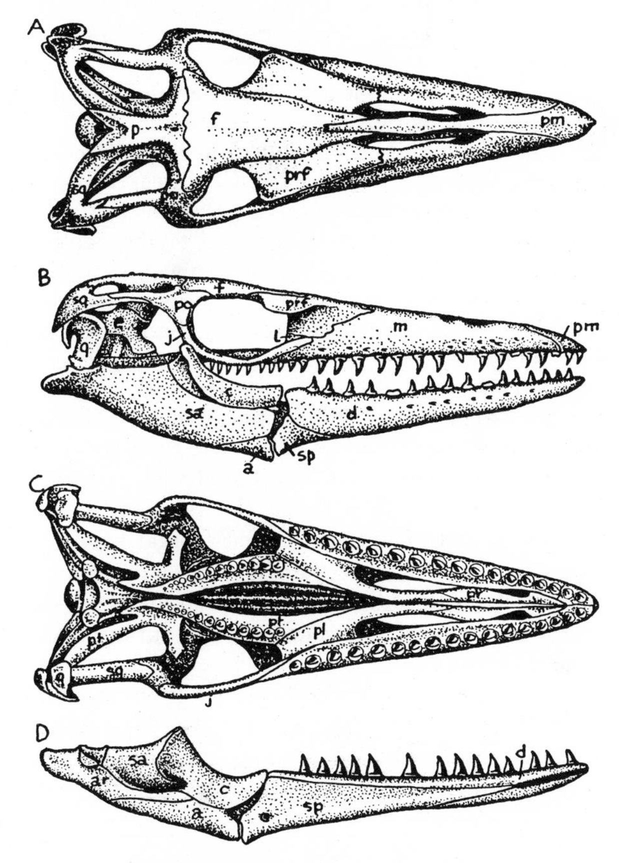

Fig. 204.—The skulls of mosasaurs. *A, Tylosaurus,* dorsal view, original about 38 inches long; *B, Clidastes,* lateral view, original about 20 inches long; *C, Platecarpus,* ventral view, original about 22 inches long; *D,* inner view of jaw of *Clidastes.* For abbreviations see Figure 109. (After Williston.)

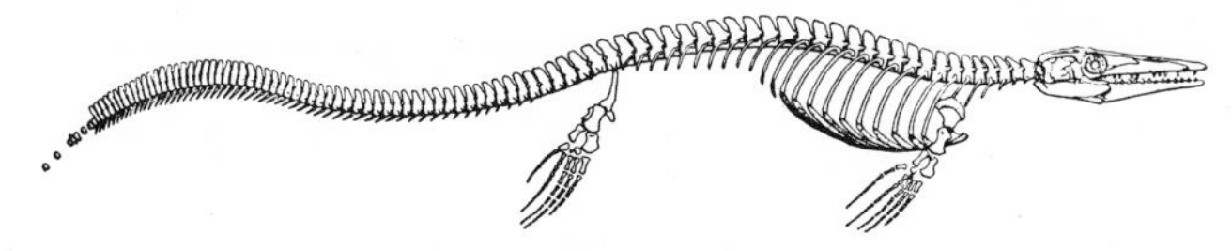

Fig. 205.—*Tylosaurus dispar,* a mosasaur from the Niobrara Chalk of Kansas; original about 26 feet long. (From Osborn.)

fossils; their lightly built skulls are seldom preserved intact, and known remains generally consist of strings of vertebrae or isolated elements.

In snakes almost all traces of limbs are lost, as in some of the true lizards. The limbless condition seems to be associated with the fact that in all slimly built reptiles, locomotion, even with limbs, utilized to some extent the sinuous motion of the body held over from the fish stage; the legs merely held the ground gained by twisting the body. With the development of a freely movable rib system to aid in sinuous movement and the addition of stout, horny scales, which could hold onto the surface and prevent a backslip, the loss of legs became a possibility, and a new type of locomotion made its appearance. In snakes the vertebrae may increase in number up to several hundred, with, of course, little regional differentiation, and extra articulations are present between the vertebrae (Fig. 173*G-I*).

It is in the head skeleton that the most distinctive specializations of the snakes are seen. The skull (Fig. 206) has been highly modified in the direction of greater motility of its components for swallowing large prey. Even the upper arch of the cheek is lost, leaving the quadrate very loosely attached to the skull. The quadrate also has but a loose connection with the palate, while the palate and the anterior portion of the skull may both move freely on the braincase region. The dentary is freed considerably from the more posterior elements of the jaws, and the two branches of the jaw are connected only by ligaments. These all make possible a wide jaw gape. The teeth are usually directed backward, so that alternate jaw movements tend to push the prey down the throat. The pineal opening is lost. The parietal and frontal have grown down so that, together with a new ossification, they completely enclose the front of the braincase, a rather necessary feature for brain protection during the swallowing of large objects.

The snakes are certainly derived from the lizards, and while there is no certainty as to the point of origin, a relationship to the varanoids is quite probable, suggested particularly by the flexible jaw structure of that group. As to the fashion in which their peculiar type of locomotion evolved, the most widely held current theory is that the ancestral snakes were burrowers. We have noted that in various lizard groups such as skinks, anguids, and amphisbaenids, there was a strong tendency toward development of snakelike body proportions, and many of these forms are burrowers. Most snakes of today have resumed life on the surface, but there are still burrowing types amongst the more primitive members of the suborder.

Although the snake pattern of anatomy, once established, clearly marks off the typical snakes from the lizards, there are both living and fossil types which are problematical as to position. *Typhlops,* for example, is a small, wormlike tropical burrower, already present in the early Tertiary, which has certain of the diagnostic characters of snakes, but is so specialized that many believe that it evolved from lizards independently of the true serpents; it is here considered (together with a few other living genera of somewhat similar pattern) to typify a separate superfamily of snakes, the Typhlopoidea. Again, in the Cretaceous and Eocene we find a number of poorly known forms, such as *Palaeophis* and *Simoliophis,* which appear to have had elongate snakelike bodies but vertebrae more on the pattern of the monitor lizards; quite possibly these, if better known, would be found to represent a group truly intermediate between the snakes and their lizard ancestors, but their marine habitat renders this improbable.

These forms apart, all snakes can be marshaled in two superfamilies, the primitive Booidea and the more advanced Colubroidea. Of the first, the typical members are those of the Boidae, the boas and pythons. The familiar forms are the giant "constrictors"; many boids, however, are small in size and fossorial in habits, and in addition to the boids proper, the superfamily includes a number of other small tropical burrowers. Booids are first definitely present in the Eocene. Possibly related are such late Cretaceous forms as the 6-foot *Dinilysia* of Patagonia and a burrower, *Coniophis,* from North America.

The remaining snakes, the superfamily Colubroidea, are characterized by an advanced type of jaw structure, with a further "loosening up" of the various elements concerned with swallowing prey; in contrast to

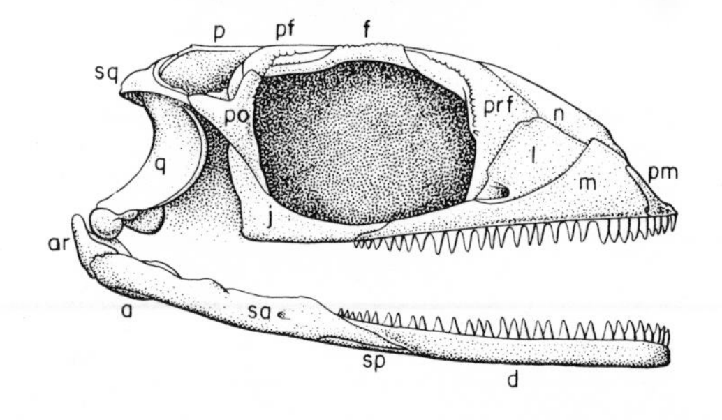

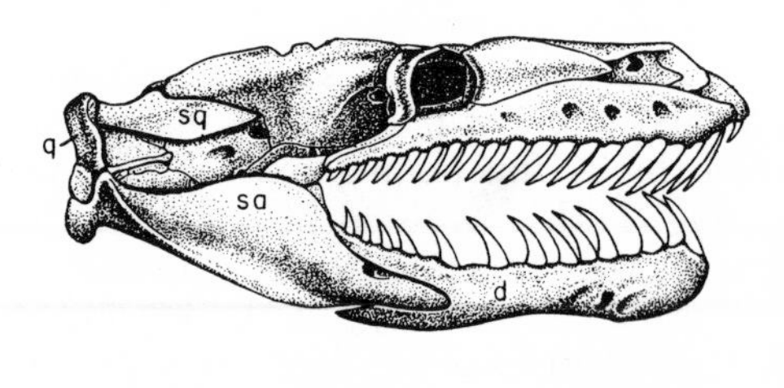

FIG. 206.—*Left,* The skull of the early lizard *Kuehneosaurus,* about 1.5 times natural size. (After Robinson.) *Right,* Skull of a primitive living snake, the python. For abbreviations see Figure 109. (After Williston.)

the boids, for example, the maxilla is but loosely connected with the other skull elements. The basic family is that of the Colubridae, with about three hundred living genera, including all the familiar non-poisonous snakes of temperate regions as well as a host of tropical forms. There is no certain record of colubrids earlier than the Miocene, although they were presumably evolving earlier in the Tertiary.

Among the higher snakes there is a strong trend toward the development of poison glands. Even among the colubrids there are numerous tropical genera with poison sacs associated with teeth in the back part of the mouth, and, probably deriving from the colubrids, there have arisen three further families of highly poisonous snakes. The Elapidae, with a modern center of distribution in the East Indies and Australia, include such forms as the cobras (*Naja*), the mambas of Africa, and, in the Western world, the little coral snakes. In this family there is a pair of fixed poison fangs in the premaxilla. The elapids are poorly known as fossils, except for a Miocene form. Related to and perhaps to be included in the elapid family are *Hydrophis* and its relatives, highly poisonous marine snakes of the Pacific and Indian oceans; they are unknown as fossils. The Viperidae are further specialized in the development of retractile fangs which are the sole remaining teeth in the upper jaw margins. The so-called true vipers are Old World forms. The "pit vipers," so called because of a heat-perceiving pit gland between nose and eye, are mainly New World forms, of which the rattlesnakes (*Crotalus*) are typical; however, the copperhead genus (*Agkistrodon*) is found in Asia as well. As in the case of other higher snake families, the vipers can be traced only to the Miocene.

Viewed broadly, snake history would appear to include the probable development of primitive burrowing forms in the late Mesozoic, a re-emergence to the surface of booid types at about the end of the Cretaceous, and in the mid-Tertiary the development and rapid rise of the colubrids and poison-bearing derivatives. In contrast to the extinction or seeming evolutionary stagnation of other reptile types, the snakes are today a group of reptiles still "on the make."

11 CHAPTER

Ruling Reptiles

As has been seen, the mutual relations of many of the reptilian types are none too clear, and the establishment of superordinal or subclass groups is a dubious matter. In contrast are the groups now to be considered—a series of orders including the crocodiles, dinosaurs, flying reptiles, and related primitive forms which may confidently be unified in the subclass Archosauria, the ruling reptiles (Fig. 207).

Diagnostic characters possessed by all these forms are few. All possess two temporal openings—the diapsid condition; but two openings are also present in rhynchocephalians and other reptiles which seemingly have little to do with the groups under consideration. Of greater importance are the comparable evolutionary trends seen in the development of the various orders and indicative of the potentialities inherent in the ancestral stock. Many similarities in structural features among end forms of different archosaurian lines have not been inherited as such from a common ancestor but have been independently acquired by members of the different groups. This, however, does not debar such characters from consideration as indications of relationship. Study of fossil forms increasingly indicates that there has been an enormous amount of parallelism in evolution; but this study also appears to demonstrate that close parallelism occurs only in closely related forms.

Bipedalism.—The most characteristic archosaurian evolutionary tendencies are to be seen in the limb-and-girdle structure and are connected with significant changes in the locomotion of most members of the group.

The clumsy style of walking of the primitive cotylosaurs was improved upon by later reptiles in various ways. Many reptilian orders tended merely to improve a four-footed style of progression; the archosaurs are unique in that many of them tended toward a bipedal gait, the animal running semi-erect on its hind legs. Some of the most primitive archosaurs had hardly begun the changes potentially leading to bipedalism, and some groups never attained this condition. Others, having once acquired, partially, at least, a bipedal gait, have slumped back to a four-footed pose; but even in such forms the group history has left structural marks. The changes have to do with the presence of a highly developed tail and, most especially, with great development and modification of structure in the hind limbs; little attention, so to speak, was given to the front legs. It has been suggested that the initiation of these changes was associated with improvements for swimming in archosaur ancestors of amphibious habits. If so, this was a fortunate "preadaptation."

In the archosaurian bipedal pose, the anterior part of the body tilts forward at a considerable angle from the pelvis. This posture could not be maintained without a tail as a balancing organ, and this structure is almost always long and powerful. In addition to its balancing function, it also sheathes powerful muscles which move the hind limbs.

This changed position has entailed many structural changes in the limb bones. The entire weight of the body in bipeds is supported from the hips; and the sacrum is strongly built, a number of ribs in addition to the primitive two usually being incorporated in it. The ilium is elongated, and the acetabulum lies high up near the backbone. The body could not, of course, be supported in a biped with the legs spread out at the side of the body. Instead, the hind legs have been turned forward, giving them a fore-and-aft motion, and are considerably straightened, so that essentially they form two long and powerful pillars extending from the pelvis to the ground.

The head of the femur now lies under the upper margin of the socket of the acetabulum and no longer pushes in against it, and the bottom of the cavity becomes open in advanced forms. The femur (Fig. 208), originally straight, develops a head at the side of its upper end to fit into the acetabulum, while, part way down, an extra process—a fourth trochanter—is often developed for the attachment of tail muscles. The tibia tends to become long, the fibula reduced in importance. An extra joint, making for additional speed, may be added to the leg by an elongation of the metatarsal elements. The tarsal bones, which might become an element of weakness in this new type of limb, tend to be reduced, leaving the astragalus and calcaneum (fibulare) as the only prominent elements. In many early types the astragalus tended to be firmly connected with the tibia and the calcaneum with the foot, the main motion between leg and foot occurring at a movable articulation between these two elements. This type of foot persists in the crocodilians. With the forward turning of the limb, the foot, originally twisted, extends straight forward from it; and the central, rather than the lateral, toes tend to be the more highly developed. The third toe often becomes the longest of the set; the second and fourth, at either side, slightly shorter. The remaining toes usually become comparatively unimportant; the fifth open disappears entirely, while the first is sometimes turned backward, apparently acting as a prop. Although the comparative length of the toes may be greatly changed, the number of joints has tended to remain constant; and in forms as far apart as carnivorous dinosaurs and birds the feet are almost identical in structure and in phalangeal formula.

A characteristic series of modifications takes place in the pubis and ischium (Fig. 208, *top row*). Instead of retaining the primitive platelike structure, both bones tend to be elongated and directed downward, thus producing, with the ilium, a triradiate type of pelvis. This structural feature is probably related to the fact that, with the femur playing back and forth close to the body, the muscles running to it from these bones must shift fore or aft to gain a longer play. Even the most primitive of archosaurians show at least the beginning of this tendency, while some forms have a still more complicated pelvic structure.

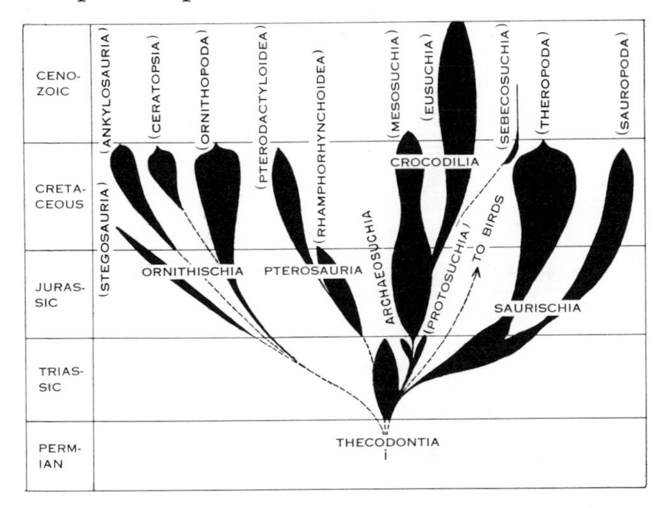

FIG. 207.—The phylogeny of the archosaurian reptiles

The front legs naturally became less important with the development of the bipedal gait. They are usually considerably shorter and much weaker than the hind legs. There is some tendency toward a loss of fingers in advanced bipeds, and but two or three of the inner ones may persist. Certain dinosaurs which have become quadrupedal again seem to have effected some secondary relengthening of the front legs, but even in most of these the anterior limbs are much shorter than the hind ones. In the shoulder (Fig. 208, *lower row*) there is only a single coracoid bone, and the scapula tends to be rather long and slim. The cleithrum is lost, as in most advanced reptiles; clavicles or interclavicle or both may drop out.

Skull.—The primitive archosaurs were carnivores, with simple, pointed teeth; in contrast with most other reptiles, teeth have disappeared from the palatal bones in all except primitive forms. In some cases there is a tendency for some or all of the teeth to be lost and to

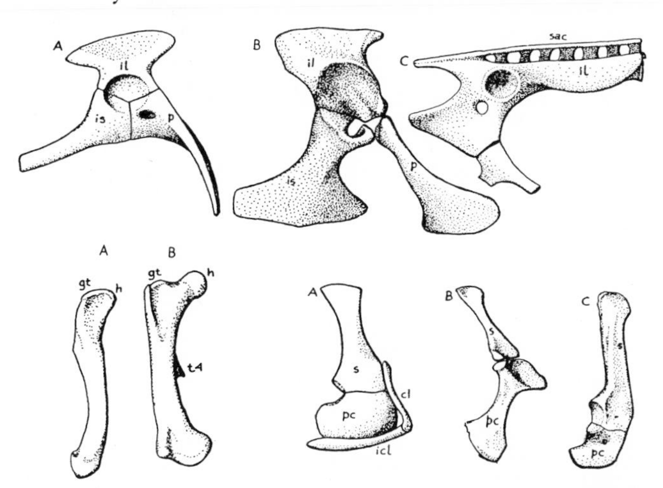

FIG. 208.—Girdle and limb elements of archosaurs. *Above,* Pelvic girdles: *A,* the carnosaur *Ornithosuchus; B,* the Jurassic crocodile *Steneosaurus; C,* the Cretaceous pterosaur *Pteranodon.* Abbreviations: *il,* ilium; *is,* ischium; *p,* pubis; *sac,* sacrum. In *C* the homologies of the ventral element are uncertain. (*A* after Broom; *B* after Andrews; *C* after Eaton.) *Lower left,* Anterior views of right femur of *A,* a crocodilian; *B,* a hadrosaurid dinosaur. Abbreviations: *gt,* greater trochanter for attachment of iliac muscles; *h,* head, fitting into acetabulum; *t4,* fourth trochanter for attachment of caudal muscles. *Lower right,* Shoulder girdles of archosaurians: *A,* the thecodont *Euparkeria; B,* the Jurassic crocodile *Steneosaurus; C,* the sauropod dinosaur *Camarasaurus.* Abbreviations: *cl,* clavicle; *icl,* interclavicle; *pc,* precoracoid; *s,* scapula. (*A* after Broom; *B* after Andrews; *C* after Marsh.)

be replaced by a horny bill, and the cheek teeth were sometimes modified for plant feeding. In contrast with most reptiles, the teeth are firmly set in deep sockets (thecodont). There is generally some loss of skull elements, particularly postfrontals, postparietals, and tabulars. A pineal opening is almost invariably absent. With the effect, perhaps, of lightening the skull, an antorbital fenestra is usually present between the orbit and the nostril; this fenestra persists in most later archosaurs. The quadrates are large and extend far up the back corners of the skull; in all except the most primitive types the back edge of the cheek may be concave in outline for attachment of the front edge of an eardrum, as in many lepidosaurs. The two pterygoids have tended to meet to form a median plate in the roof of the mouth, with palatal vacuities present at either side. In the lower jaw appears a fenestra on the outer surface between dentary, angular, and surangular.

Trunk.—The ribs almost always retain the primitive double-headed condition. But a common archosaurian feature is the tendency for the two heads to crowd closer together; in the cervical region the capitulum moves back to arise from the centrum below the transverse process, and in the trunk both heads may arise from the process itself (Fig. 209). Except for the most primitive of thecodonts, there appears to have been generally a modest armor of a pair of rows of plates down the back in early archosaurs. In many forms among thecodonts, crocodilians, and ornithischian dinosaurs there was an increase in armament; in others, such as saurischian dinosaurs and pterosaurs, this plating is lost.

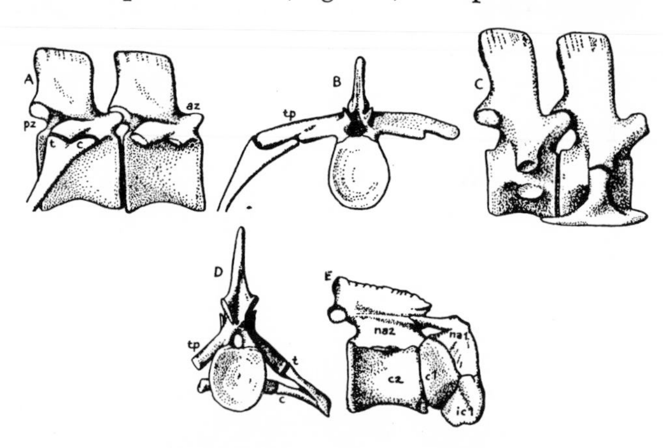

FIG. 209.—Vertebrae of the Jurassic crocodile *Steneosaurus. A,* Dorsal vertebrae; *B,* same, anterior aspect; *C,* cervicals; *D,* same, posterior aspect; *E,* atlas and axis. Abbreviations: *az,* anterior zygapophysis; *c,* capitulum of rib; *c1, c2,* centra of first and second vertebrae (atlas and axis); *ic1,* intercentrum of atlas; *na1, na2,* neural arches of atlas and axis; *pz,* posterior zygapophysis; *t,* tubercle of rib; *tp,* transverse process of neural arch. (After Andrews.)

Thecodonts.—Most of the characters and tendencies just discussed are seen to a variable degree in the various advanced archosaur types—the dinosaurs, pterosaurs, and crocodiles—and in the primitive birds (which seem clearly derivable from the archosaurs). These forms make their appearance in the late Triassic and Jurassic. We should expect to find in the strata deposited during the late Triassic the ancestors of the later ruling reptiles, a common stock in which archosaurian characters were first making their appearance.

Such a group is that which we shall here consider as comprising the order Thecodontia. In it are included a number of types of Triassic reptiles which vary considerably among themselves but which, as a whole, show the initiation of archosaurian tendencies.

The two-arched condition of the skull had already been acquired by some reptiles in late Permian times. The early diapsids which we have treated as constituting the Eosuchia are possibly ancestral to the primitive ruling reptiles but show almost no suggestion of true archosaurian characters. It is quite possible that the archosaurs arose independently from cotylosaurian ancestors; the fact that in the most primitive of known archosaurs of the early Triassic (Figs. 211, 212) there was no shortening of the jaw or trend toward a reptilian otic notch suggests the possibility that the archosaurs rose directly from the primitive captorhinomorph cotylosaurs rather than by way of procolophonoids or eosuchians, in which such modifications were already present.

Euparkeria, a "typical" thecodont.—An early representative of the Thecodontia, and one which typifies a relatively early stage in the development of the progressive characters of the Archosauria, is *Euparkeria* of the Lower Triassic of South Africa (Figs. 210, 211).

Euparkeria is in many regards an almost ideal ancestor for later archosaurian types. The skull is moderately high and slenderly built; there are well-developed antorbital vacuities and a fenestra in the outer surface of the lower jaw; there is a tall quadrate, somewhat concave posteriorly to receive a tympanum in the reptilian type of otic notch, with the squamosal projecting somewhat back above the notch. The pineal foramen has been lost, but the retention of postparietal and postfrontal bones is a primitive feature. The palate is primitive in construction. There are, in contrast to most later types, numerous small teeth on pterygoid and palatine; there are long, if narrow, interpterygoid

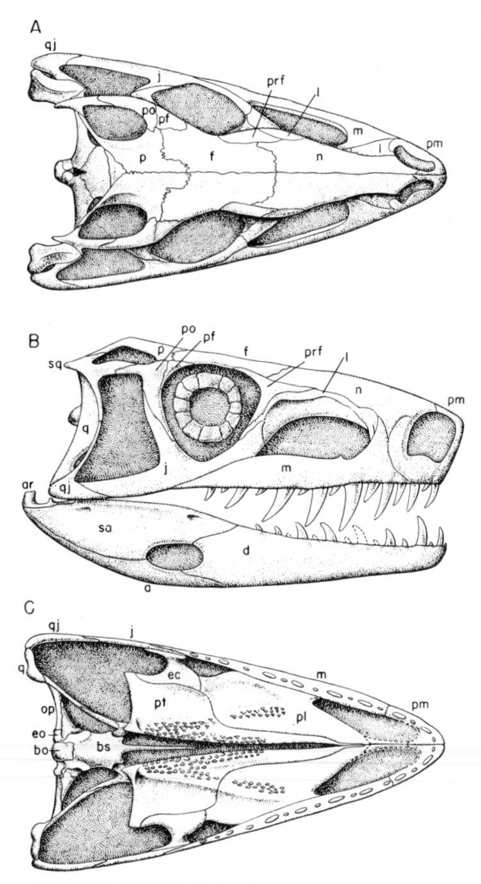

FIG. 210.—*A,* Dorsal; *B,* lateral; and *C,* ventral views of the skull of the primitive thecodont *Euparkeria;* skull length about 3.5 inches. For abbreviations see Figure 109. (After Ewer.)

vacuities on either side of a slender parasphenoid, and, in contrast with typical later archosaurs, there persists a primitive movable basal articulation between basisphenoid and pterygoid.

The general skeletal structure indicates that while *Euparkeria* could have progressed on all fours, it was well adapted for speedy bipedal progression as well. The presacral column had been shortened to twenty-two vertebrae, and, as a primitive character, there were still but two sacral vertebrae rather than the higher number developed in more advanced bipeds. Clavicles and interclavicle were present although slender. In the pelvis the iliac blade is still relatively short and the acetabulum imperforate, as in thecodonts generally (and in contrast with most dinosaurs and crocodiles). The pubes are turned sharply downward, but are still broad and with a small thyroid fenestra as well as an obturator nerve foramen. The ischium is essentially primitive, as a broad plate with little downward angulation, and there is a continuous symphysis the whole length of pubis and ischium.

The hind limbs are half again as long as the fore limbs, but in contrast with strictly bipedal animals, the tibia is no longer than the femur. The manus is incompletely known. In the hind foot, all toes are complete, but the fifth toe is shortened, while the other four toes all point forward; toe 4 is already somewhat reduced in length, indicating a trend toward the birdlike symmetry of toes 2 through 4 seen in later bipedal types. Down the back runs a paired series of small overlapping armor plates, essentially segmental in arrangement.

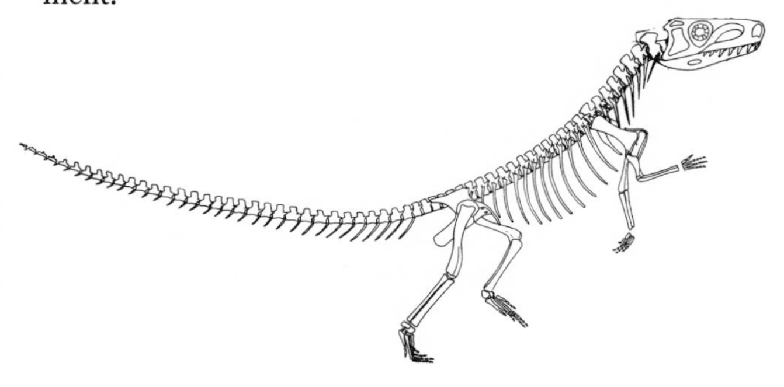

FIG. 211.—The skeleton of the Lower Triassic thecodont *Euparkeria,* about 1/10 natural size. (Modified after Ewer.)

All in all, *Euparkeria,* while showing a few primitive features in the skull, exhibits a structural pattern from which could be derived that of nearly all later archosaurs.

Proterosuchians.—In the early Triassic are found some thecodonts, such as *Chasmatosaurus* (Fig. 212) and *Erythrosuchus* (Fig. 213), which are still more primitive than *Euparkeria.* In their skulls these forms show primitive features retained in *Euparkeria* in the presence of palatal teeth in *Chasmatosaurus,* movable basal articulations, and postparietal and postfrontal bones; further, a pineal foramen, almost always absent in archosaurs, is persistent and, most notably, there is no development of a concave quadrate region for eardrum reception in an otic notch; in primitive reptilian fashion the back margin of the cheek slants downward and backward to the jaw articulation. Still further, there is little development of positively archosaurian characters in the postcranial skeleton, apart from a downturning of the pubis and, it appears, some slight disproportionate elongation of the hind legs. While the material is none too complete, it would seem that these reptiles were still primitive quadrupeds in which little, if any, trend toward bipedalism had taken place.

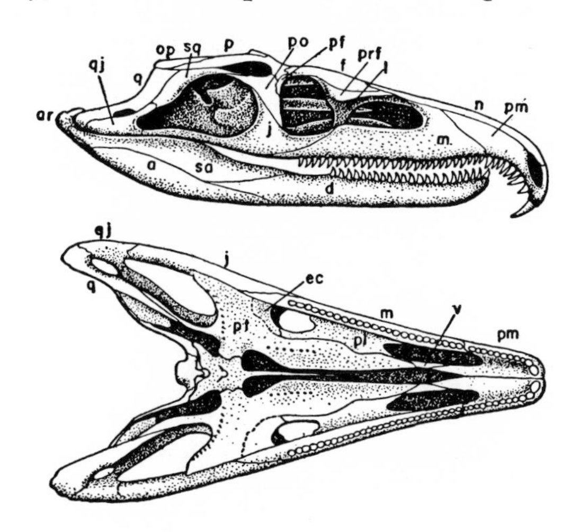

FIG. 212.—Lateral and ventral views of the skull of *Chasmatosaurus,* a Lower Triassic thecodont with primitive palatal structure; length about 18 inches. For abbreviations see Figure 109. (After Broili and Schroeder.)

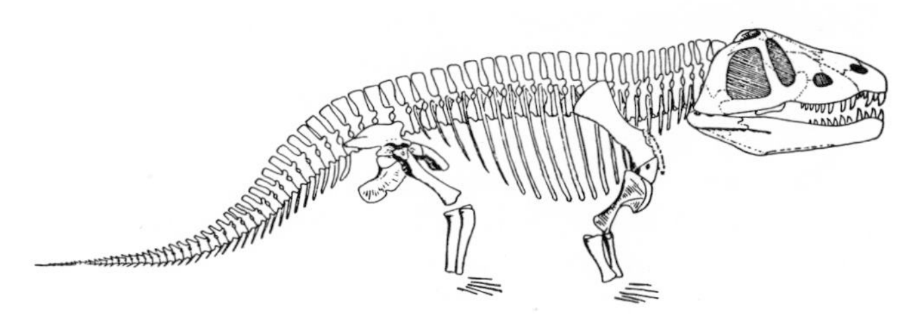

FIG. 213.—The skeleton of a large early Triassic thecodont, *Erythrosuchus,* about 1/40 natural size. (From Huene.)

Because of their primitive position, these genera and their relatives are frequently considered to form a discrete suborder of thecodonts, as the Proterosuchia. A number of other early to middle Triassic four-footed thecodonts, most incompletely known, are not improbably members of this group. In Europe, reptilian remains are rare in the earlier Triassic; there are, however, numerous five-toed footprints of quadrupeds which look much like those which might be made by human hands and hence have been termed "Chirotherium" prints. Very probably they were made by early archosaurs of this group.

Advanced bipeds.—*Euparkeria* was still primitive in certain features, but shows the beginnings of progress toward advanced members of the order, often grouped as the suborder Pseudosuchia. In the middle and late

Triassic are found a number of small bipeds, such as *Saltoposuchus* (Fig. 214), which show increased disproportion between short front legs and long hind ones; generally the double row of armor plates down the back was retained. It is obvious that it was forms of this sort from which arose the pterosaurs, birds and dinosaurs. There are no known thecodonts which show positive indications leading toward the first two groups mentioned, nor toward one of the two dinosaurian orders, the Ornithischia. More troublesome is the situation with regard to the other dinosaur group, that of the Saurischia. As we shall see in the next chapter, the saurischians became abundant and varied before the

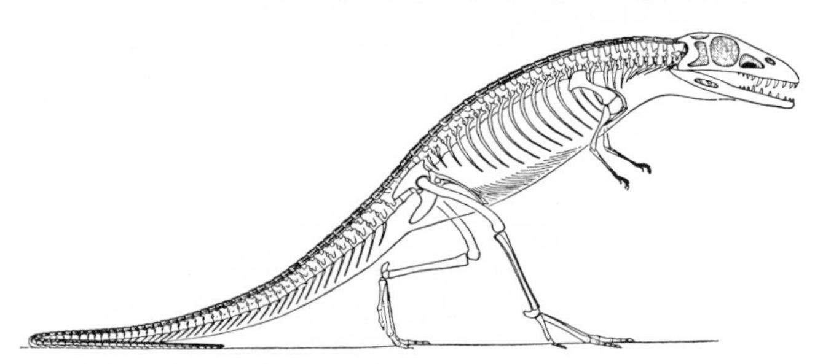

Fig. 214.—*Saltoposuchus*, a lightly built Triassic thecodont, about 3.75 feet in length. (From Huene.)

end of the Triassic and, in their more primitive members, are not far advanced structurally over the thecodonts. In consequence, it is difficult to establish a boundary between the two groups. Various Triassic forms, such as *Ornithosuchus* and *Scleromochlus*, which were long classed as advanced thecodonts, are now regarded as having "crossed the boundary" and are classed as primitive dinosaurs.

Aetosaurs.—Although some of the primitive thecodonts were, it seems, truly primitive quadrupeds, later four-footed members of the order show in various features, such as a considerable disproportion in front versus hind leg length, indications that they had regressed from ancestors which had been at least partially bipedal. Among such forms the aetosaurs of the late Triassic were notable (Figs. 215, 216). These were forms in which the bony covering, originally limited to a pair of small dorsal rows, had developed into a solid armor plating over both back and belly. *Aetosaurus* and *Stagonolepis* were small European types, *Desmatosuchus* and *Typothorax* large American representatives. The small head terminated in a blunt, apparently piglike, rooting snout.

Phytosaurs.—Among the most abundant of fossil reptiles from the later Triassic are the phytosaurs. These specialized thecodonts, heavily armored like the contemporary aetosaurs, appear to have been similar in habits and general appearance to the modern crocodiles and were ecologically the predecessors of these forms, although not in themselves directly ancestral. The skull (Fig. 217) was much elongated in correlation with probable fish-eating habits, and the jaws were armed with a powerful battery of sharp teeth. The position of the nostrils was one seen in various other water-living types, the external openings being far back, almost between the eyes. The two openings were close together near the top of the skull and were in some cases situated in a sort of crater rising above the level of the skull top; this seems to have been a device for breathing with most of the body under water.

The body (Fig. 216) was crocodilian in general shape, the pelvis primitive in structure. The limbs were short. The hind legs were considerably longer than the front ones, although, with the taking up of an aquatic life, these animals had completely given up any aspirations toward bipedalism. A variety of genera, such as *Phytosaurus* and *Rutiodon* (a profusion of names has been given to phytosaurs), are known from North America, Europe, and even India; but they appear never to have reached the southern continents. Except for one doubtful specimen, all known phytosaurs are of late Triassic age. The phytosaurs were comparatively successful for the time but apparently failed in competition with the crocodiles and disappeared at the close of the Triassic.

The history of the thecodonts was a brief one. But their development was an event of the greatest importance in the evolution of Mesozoic life—a necessary prelude to the later expansion of ruling reptile types.

Of the forms descended from this group, the dinosaurs and birds will be treated in later chapters; we shall include here an account of the other archosaurian orders—the crocodiles and flying reptiles.

Crocodiles.—The crocodiles and alligators and their relatives, the order Crocodilia, have been among the least progressive of ruling reptiles but nevertheless are the only members of the archosaurian stock which survived beyond the Age of Reptiles. Derived from the thecodonts during the course of the Triassic, they have

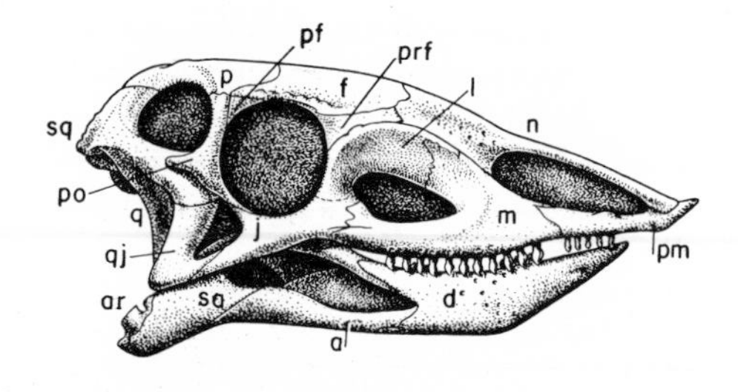

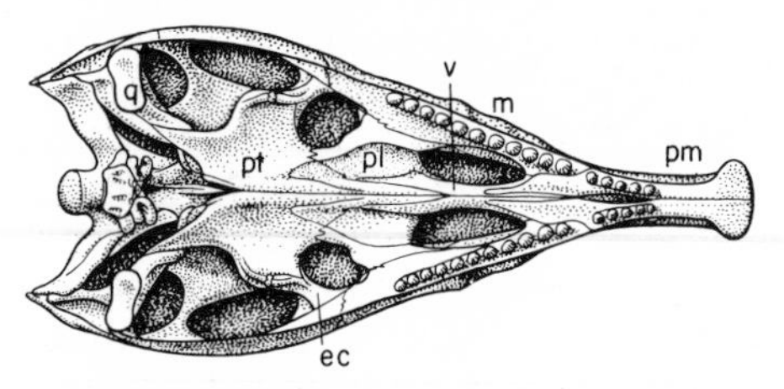

Fig. 215.—Lateral and ventral views of the skull of the armored aetosaur *Stagonolepis;* original about 10.5 inches long. For abbreviations see Figure 109. (After Walker.)

undergone comparatively little modification in later times. In many structural features they are not far from the primitive archosaurian type, and in their amphibious mode of life they resemble greatly their somewhat distantly related predecessors, the phytosaurs.

The skull (Fig. 218) in many respects is of a primitive archosaur pattern; it is always rather elongate—extremely so in fish-eating types, such as the living Indian gavial—and is generally much flattened. The typical archosaur antorbital opening is generally lost. Notable is a peculiar development in the otic-notch region, in which a process of the squamosal extends downward to transform the notch into an opening into a deep, enclosed pocket.

The palate is much modified. We have seen that in the phytosaurs breathing difficulties were overcome to

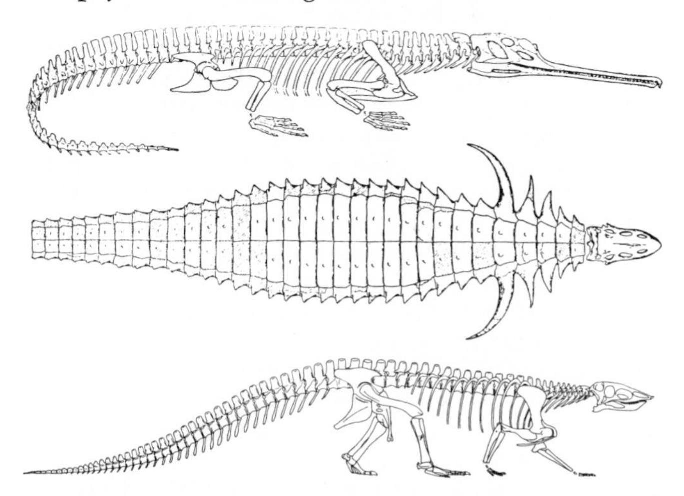

Fig. 216.—*Above, Mystriosuchus,* a Triassic phytosaur, original about 11.5 feet long. (From McGregor.) *Center,* Dorsal view of the armor of the aetosaur *Desmatosuchus,* part of body illustrated about 12 feet in length. (From Case.) *Below,* Side view of the skeleton of the aetosaur *Stagonolepis,* armor omitted; length about 9.5 feet. (From Walker.)

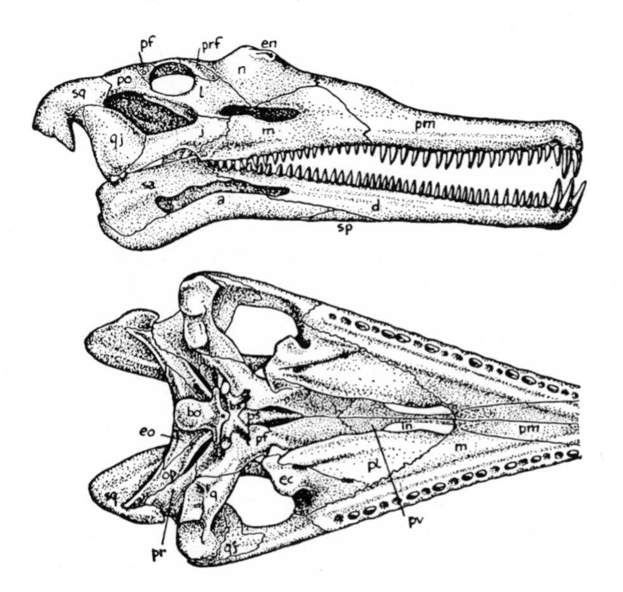

Fig. 217.—Lateral view of skull and posterior part of the palate of the phytosaur *Rutiodon;* skull length about 3.5 feet. For abbreviations see Figure 109. (After Camp.)

a considerable extent by shifting the external opening of the nostrils far back, the internal openings consequently lying in the back part of the mouth. In the crocodiles the two external nares, with a common bony opening, are at the tip of the snout, so that breathing can be accomplished even if only this part of the body is exposed. Internally the danger of shipping water

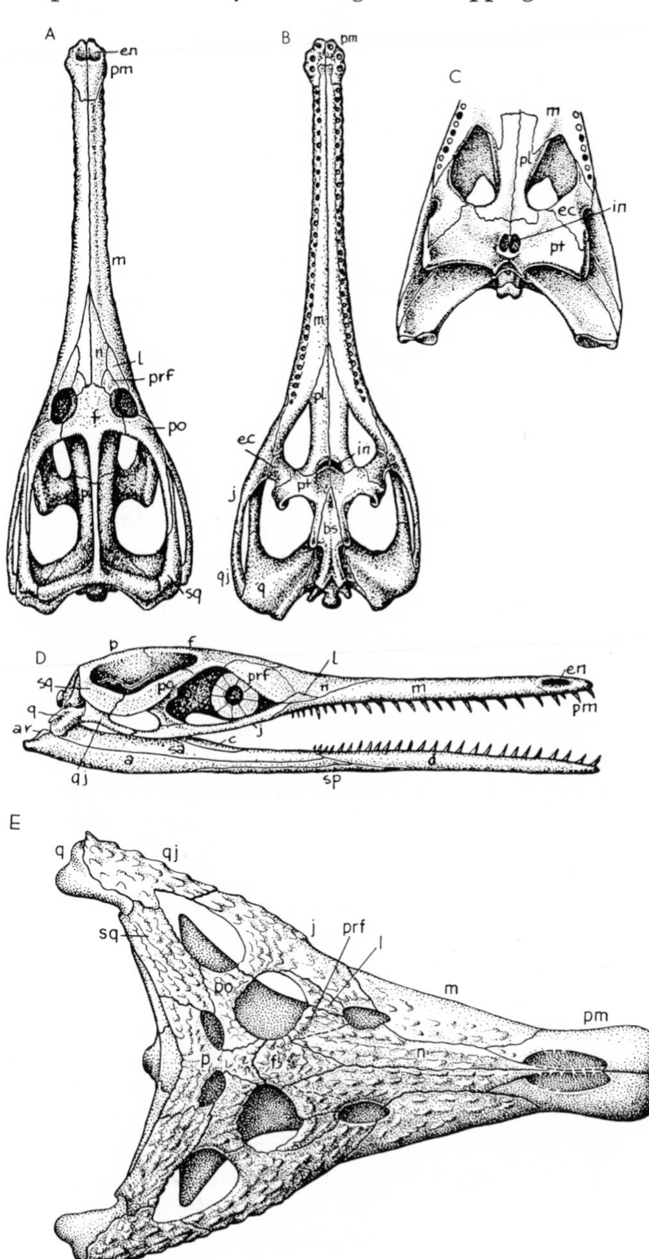

Fig. 218.—The skull in crocodilians. *A,* Dorsal, and *B,* ventral, views of the Jurassic *Steneosaurus,* skull length about 30 inches; *C,* part of palate of the Eocene *Crocodylus affinis,* to show posterior movement of internal nares; *D,* lateral view of the Jurassic marine crocodile *Geosaurus,* skull length about 15 inches; *E,* dorsal view of the primitive Triassic crocodilian *Proterochampsa,* about 12 inches long. For abbreviations see Figure 109. (*A, B* after Andrews; *C* after Mook; *D* after Fraas; *E* after Sill.)

into the air supply is prevented by the development of a secondary palate. Below the original inner opening of the nostrils, the premaxillae, maxillae, and palatines have formed a secondary shelf which extends backward parallel to the primary roof of the mouth; the air passes above this new palatal structure to the posteriorly situated internal nares. In modern forms even the pterygoids have met in the midline and formed a further prolongation. As a result, the inhaled air does not enter the mouth proper at all but passes back separately into the throat, which can be closed off by a flap of skin. A similar secondary palate has been formed in mammals but almost never developed to the extent seen in the crocodiles.

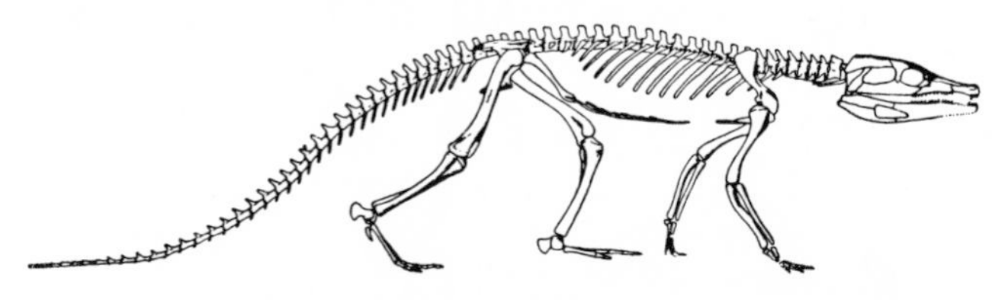

FIG. 219.—The skeleton of the primitive crocodilian *Protosuchus*, about 1/12 natural size. (After Mook and Colbert.)

The general body shape is rather lizard-like, with a long flattened tail which is the main swimming organ. In modern forms the vertebrae are procoelous. There is always a well-developed set of dermal armor plates down the back beneath the horny scales and sometimes down the ventral side as well—a feature inherited from primitive thecodonts.

The modern crocodiles are quadrupeds and, as ordinarily observed, have a slow, sprawling gait. Fast locomotion, however, is accomplished with the body high off the ground and the hind limbs straight under the body, as in all advanced archosaurians; and, while crocodiles are not bipeds, it is suggestive that the front legs are always much shorter than the hind. The contrast in length was very strongly marked in the earliest Jurassic types, and it is highly probable that some degree of bipedal locomotion had developed in the archosaurian ancestors of the group before a trend toward an aquatic life appeared. In the shoulder (Fig. 208) an interclavicle is retained, but the clavicles are lost. The two outer toes of the front foot are reduced in common archosaurian fashion, while, as in dinosaurs, the carpals are much reduced, leaving only two proximal and one distal element; the proximals are usually much elongated (Fig. 220*A*). There are but two sacral ribs, a primitive feature. The pelvis (Fig. 208) has a triradiate structure, with pubis and ischium extending down separately from the acetabular region, much as in the saurischian dinosaurs. The crocodile pelvis is peculiar, however, in that the pubis has been excluded from the acetabulum by the ischium. The hind legs are quite long; as in most archosaurians, there is some elongation of the metatarsals, and the fifth digit is reduced to a stump of the metatarsal. But the foot (Fig. 220*B*) is somewhat divergent from that of many other archosaurs in that the first toe is still in line with the others and the fourth toe is slightly reduced. A feature retained from thecodont ancestors is the presence of a mesotarsal joint between the astragalus and calcaneum, the former bone being closely attached to the lower leg elements, the calcaneum (fibulare) (with a posterior tuber for leg tendon attachment) being functionally part of the foot.

The Crocodilia may be divided into several suborders; apart from very primitive Triassic forms about to be discussed, they may be grouped as the Mesosuchia, the typical Jurassic and Cretaceous families; Eusuchia, more progressive families, particularly of the Cenozoic; and Sebecosuchia, established for peculiar South American fossil types.

Ancestral crocodiles.—Except for palatal structure and such details as the exclusion of the pubis from the acetabulum, crocodile anatomy is not remote from that of their thecodont ancestors; and, unless these features were preserved in a specimen, it would be difficult to tell an ancestral crocodile from a member of the Thecodontia.

A number of types from the middle and later part of the Triassic show sufficient crocodilian characters to merit admission to that order, although lacking enough advanced features to allow them to be placed in the mesosuchian suborder which includes the typical Mesozoic forms. For example, *Protosuchus* (Fig. 219), a well-armored form from the late Triassic of North America, shows crocodilian features in its limb girdles, although the skull, as far as known, is little advanced. Closer in skull structure are a few incompletely known forms of middle to late Triassic date, such as *Proterochampsa* (Fig. 218*E*) of South America, in which the fused nares open in crocodilian fashion on the roof of the snout and a considerable length of secondary palate has already been formed by the maxilla.

Since most sediments were laid down in water and the crocodilians are water dwellers, it is not unexpected that we have an abundant record of members of this group; in many fossil localities crocodilians are rivaled in numbers of specimens only by turtles. The story of typical crocodilians begins with Jurassic representatives. These forms, and most of those of the Cretaceous as well, show, however, a few features in which they are definitely more primitive than surviving types. The upper temporal openings were not so small as in living forms; the internal nostrils still lay between the palatine bones, for the pterygoids had not yet been pressed into service to continue the secondary palate; the vertebrae still retained flattened or even slightly hollowed ends; and the antorbital opening found in many other archosaurs is still present in some cases (it is absent in later crocodiles). These older fossils may be conveniently placed in a suborder Mesosuchia.

The Jurassic is dominantly marine in its sediments, and hence our knowledge of crocodiles of this period is mainly confined to side lines of the group; these tend to be long-snouted forms which may, in general, have been fish eaters. *Teleosaurus* and *Steneosaurus* (Fig. 218*A, B*) are representative of a family of such crocodiles. These include the oldest known mesosuchians, for they are present in the Liassic deposits of the Lower Jurassic; other crocodiles do not appear until later in the period. *Pholidosaurus* is representative of another long-snouted Upper Jurassic group which appears to have lived in fresh waters or estuaries and which may have given rise to *Dyrosaurus* and other genera which persisted into the early Tertiary.

The teleosaurs were apparently marine but retained much of the normal body build of the group; from them not improbably are descended a highly specialized marine group which flourished in the later Jurassic and persisted into the early Cretaceous. *Metriorhynchus* and *Geosaurus* (Figs. 218*D;* 220*C, D*) are here considered as mesosuchians, but they are so distinctive in marine adaptations that many writers have classed them as a separate suborder, the thalattosuchians—sea crocodiles. They were unarmored crocodiles in which the limbs were considerably modified to serve as paddles. The main swimming organ was the tail, which had redeveloped into a fishlike fin rather similar to that of early ichthyosaurs, with the backbone turned down into the lower lobe, the upper lobe presumably being supported by toughened skin. These aquatic crocodiles were not uncommon in the late Jurassic seas, but for some reason they do not seem to have been so successful as were other reptilian marine types and soon disappeared.

Still further types were present in the late Jurassic. *Alligatorellus* and *Atoposaurus* were tiny crocodiles—one less than a foot in length—with reduced armor and with a very short broad skull. In most crocodilians the external nares are united into a single tubular opening; here they are still separated by a bar of bone in primitive fashion. *Notosuchus* and other forms of the Upper Cretaceous of South America are rather larger, but armorless and with similarly built short skulls, and are possibly descended from the atoposaurs.

All the families so far mentioned are side branches of the crocodilian stock; the main line leading to later families appears to lie through *Goniopholis* and its allies of the late Jurassic and Cretaceous. The body build was normal in its construction; the skull was broad and moderately long, and the secondary palate extended farther back than in the more primitive mesosuchians.

Eusuchians.—In the typical Cenozoic crocodiles we find progressive features not present in the mesosuchians. A diagnostic feature (Fig. 218*C*) is the extension of the secondary palate far back beneath the skull, so that the pterygoids, as well as the palatines, enter extensively into secondary-palate construction. The vertebrae, earlier flat ended, are procoelous in structure. The dorsal armor is retained, but that of the belly is absent in most cases. With a few Eocene exceptions, all post-Cretaceous crocodilians belong to the Eu-

suchia. Their origins, however, lay in the Mesozoic, for *Hylaeochampsa* of the European Lower Cretaceous was already a small eusuchian, presumably derived from the goniopholids.

Apart from some early specialized genera of the late Cretaceous, all the eusuchians are considered to be of three types—gavials, alligators, or crocodiles (in a broad sense). The gavials, most remains of which are considered to pertain to the genus *Gavialis,* are readily distinguished by a long slender snout, sharply marked off posteriorly from the rest of the skull and are generally placed in a separate family Gavialidae. But there are few clear-cut distinctions between alligators and

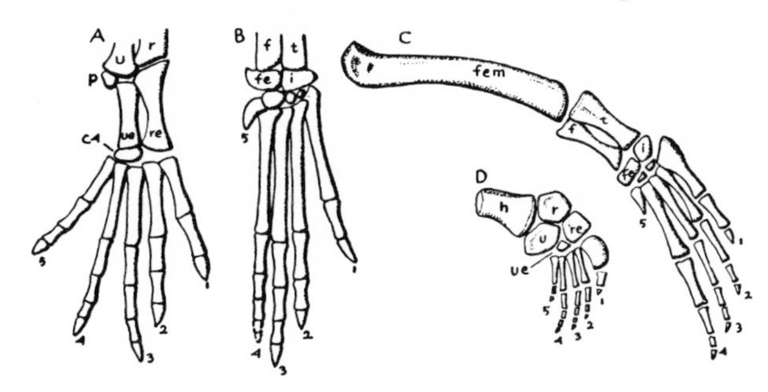

FIG. 220.—Limbs of crocodilians. *A,* Manus, and *B,* pes, of the Jurassic *Alligatorellus; C, D,* hind and front legs of the Jurassic marine crocodile *Geosaurus.* The digits are numbered. Abbreviations: *c4,* fourth distal carpal; *f,* fibula; *fe,* fibulare (calcaneum); *fem,* femur; *h,* humerus; *i,* astragalus; *p,* pisiform; *r,* radius; *re,* radiale; *t,* tibia; *u,* ulna; *ue,* ulnare. (*A, B* after Lortet; *C, D* after Fraas.)

crocodiles. A supposed key character is the fact that in alligators the first mandibular tooth (and often the fourth as well) fits into a deep pit in the palate; in the crocodiles these pits are absent, but the upper jaw may be notched for the fourth lower tooth. This, however, is hardly a feature worthy of differentiating between families; and it is perhaps better to "lump" the varied Tertiary forms in a common family Crocodylidae. The giant of the group was *Deinosuchus* [*Phobosuchus*] of the Upper Cretaceous, with a 6-foot skull and an estimated length of 40 or 50 feet. Members of the family were abundant and varied in the early Tertiary in the northern continental areas as well as in regions now in the tropics—a distribution suggesting that at that time northern climates were more equable than today.

Sebecosuchians (*Fig. 221*).—A peculiarly aberrant form is the Lower Tertiary *Sebecus,* of South America, which fits into neither mesosuchian nor eusuchian suborders. The snout, instead of being depressed, is high and narrow, giving the skull much the appearance of some of the phytosaurs. There is a secondary palate, but it is short and broadly open behind, so that there is no development of the typical crocodile tube for the air passages; and the teeth are compressed structures like those of predaceous dinosaurs. Recently discovered is a Cretaceous relative, *Baurusuchus,* with an equally high skull, but with a dentition much reduced except for a pair of enormous canines on each upper jaw. This group may have had a long independent history in South America.

Flying reptiles.—While the archosaurians did not succeed in invasion of the sea, in the air they had better success, for they not only gave rise to the birds but produced within their ranks a second aerial type—the flying reptiles of the order Pterosauria. These curious forms flourished in the Jurassic, and some large types survived to late Cretaceous times. Although there may have been pterosaurs on the continents, practically all known remains have been found in saltwater deposits, suggesting a mode of life somewhat like that of terns today.

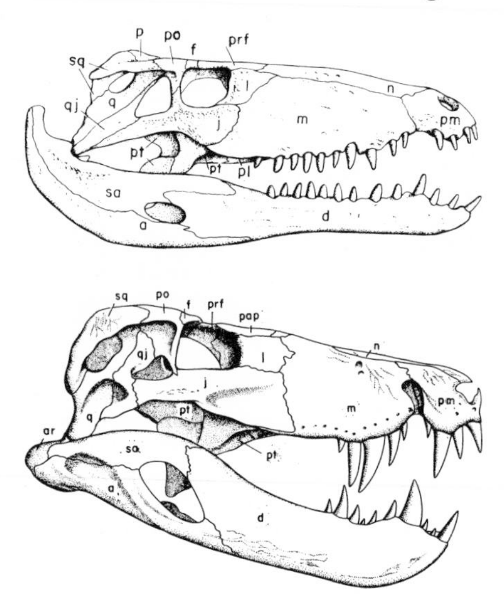

Fig. 221.—Skulls of predaceous South American crocodilians. *Upper, Sebecus,* about 20 inches long. (After Colbert.) *Lower, Baurusuchus,* about 15 inches long. (After Price.)

The structure of the pterosaurs may perhaps be best appreciated by taking as a type the fairly representative, but rather primitive, Jurassic form, *Rhamphorhynchus* (Figs. 222, 224). This little reptile was about a foot and a half in length, with a long skull, a short trunk, a long tail, and, as the most striking feature, the development of the arms into elongate wings. Specimens found in sediments which preserve many delicate structures show no trace of scales (or of feathers); small hairlike structures have been described, however, which may have given some insulation to the body surface. The wings were composed of a membrane of skin, as in the bats, but supported only by an attachment to a single elongated finger.

Skeleton.—The head (Figs. 223, 224, 226) was typically archosaurian in structure, although (as in birds) the bones tended to be fused together and the sutures obscured. There were two temporal openings, below which the quadrates slanted forward so that the jaw articulation lay below the eyes. The orbits were large; the nostrils were situated well back on the long beak; the teeth were sharp and pointed. These forms appear to have been fish eaters; *Rhamphorhynchus* was peculiar in that the teeth sloped forward, as spearing structures, rather than backward as in most carnivorous reptiles. The neck was elongated and the cervical ribs small, giving the head great freedom of motion; the trunk, on the other hand, was very short, with only ten

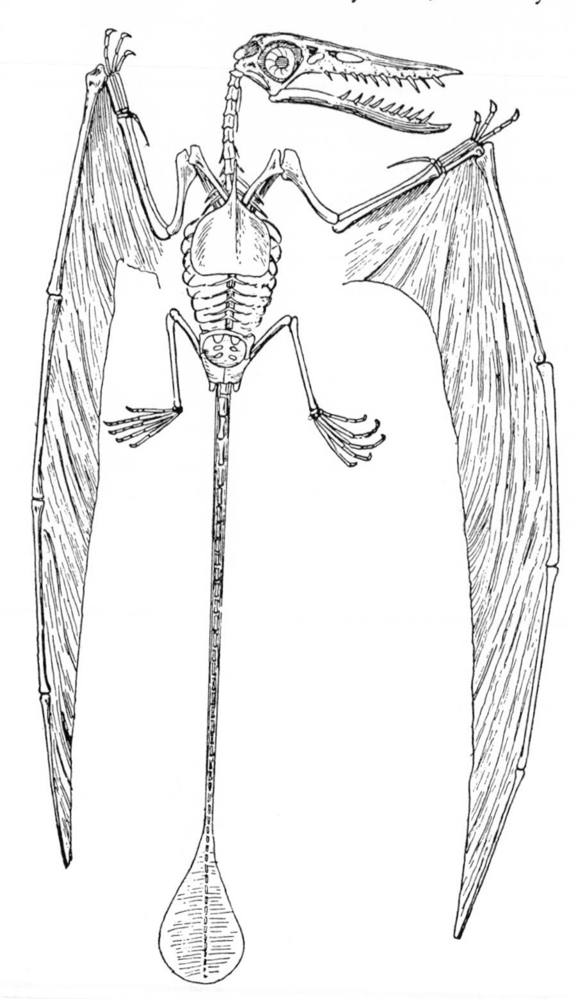

Fig. 222.—*Rhamphorhynchus,* a long-tailed Jurassic pterosaur. About 1/4 natural size. (From Williston's *The Osteology of the Reptiles,* by permission of the Harvard University Press.)

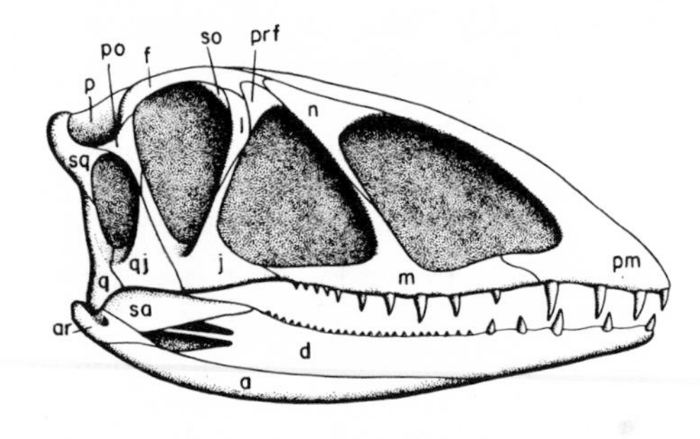

Fig. 223.—Lateral view of the skull of the primitive pterosaur *Dimorphodon;* original about 8.5 inches long. (After Owen, Huene, Arthaber.)

or so dorsal segments; there was a long sacral region of more than a half-dozen vertebrae and an elongate tail. This last structure was stiffened by strong ligaments and, as we know from impressions, bore a small steering rudder of skin.

The limbs were, of course, highly modified, the "arms" particularly so; many of the bones were hollow and air filled, as in birds, thus decreasing the weight. There was a very large sternum on the ventral surface of the chest from which arose (as in birds) the main muscles having to do with wing movements. The dermal shoulder elements were absent, but scapula and coracoid were well developed. The shoulder girdle was, of course, subject to considerable stress during flight and, in this group, tended to be better attached to the body than is usually the case; not only was the bottom of the coracoid propped against the edge of the sternum, but in some pterosaurs (cf. Fig. 225, *Pteranodon*) the upper edge of the scapula fitted into a notch on the side of a fused series of dorsal vertebrae.

The humerus was short and powerful, usually with a large process for the attachment of breast muscles, the radius and ulna long and placed close together. The carpus (Fig. 224) was much shortened and reduced and apparently without much motion in it; there was attached at its front edge a small, splintlike "pteroid" bone which seems to have supported a span of skin running up onto the neck. There is no trace of a fifth finger. The other four metacarpals ran out close together to the bases of the fingers. Three of the fingers were short but with the normal number of joints and bore small claws at their ends. The fourth finger, however, was enormously elongated with four long phalanges; this fourth finger formed the entire support for the wing membrane, which ran back to attach along the side of the body and the thigh.

The pelvic limb was also peculiar in structure. There was the usual long ilium of archosaurs, tightly attached to the numerous sacral ribs (cf. Fig. 206C, *upper row*). Below lay a large plate which is sometimes thought to represent not only the ischium but the pubis as well. However, anterior to this there was attached a T-shaped element which met its fellow in the midline. This is thought by some to have been an extra element, a prepubis; by others it is considered to be the pubis itself. The hind legs were moderately long but slim and so articulated that it would have been difficult for the animal to walk on them in any ordinary manner.

Flight.—Obviously, flight was the normal mode of locomotion; wing membranes are perfectly preserved in a number of specimens. The action of the wings, however, must have been considerably poorer mechanically than in birds. These structures were supported only by one elongate finger; and, since the membrane was probably fairly soft, no partial movements could have been accomplished—merely a flapping of the whole structure. It seems highly probable that these forms did not move the wings so much as does the average bird but relied more upon a soaring flight for their progression. An obvious fault in construction appears to lie in the fact that, without internal supports in the wing membrane, any break or tear would have ruined the whole wing structure. This is in contrast to birds, in which a few feathers may be lost without serious results, and to bats, in which several fingers are inserted in the membrane.

It is difficult to see how ptersosaurs might have got about when not in flight. Older restorations show them walking about on all fours with the wings turned back and the three small front toes on the ground, but the structure of the hind legs makes such a pose almost impossible. A bipedal mode of progression is out of the question. It seems obvious that the small front fingers were used for clutching, and the animals may have rested hanging from a limb or a rock ledge. Perhaps, as in bats, the body may have been suspended upside

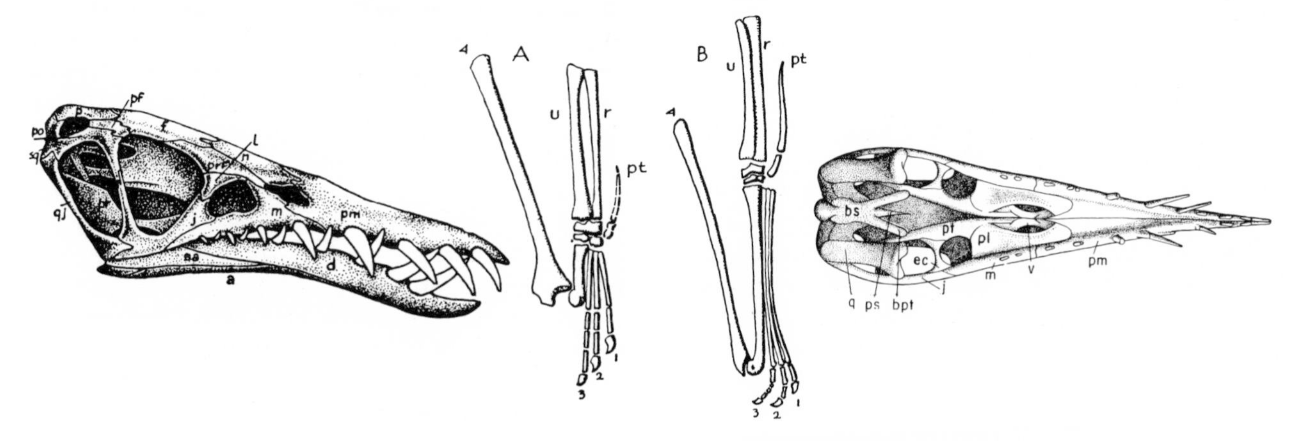

FIG. 224.—*Left and right,* The skull of the pterosaur *Rhamphorhynchus* in lateral and ventral views; length of original about 5 inches. (After Jaekel.) *Center,* Forearm and manus of pterosaurs. *A, Rhamphorhynchus; B, Pterodactylus.* The distal digits of the wing finger are omitted. Abbreviations: *pt,* elements supporting the membrane between arm and neck; *r,* radius; *u,* ulna; 1–4, first to fourth digits. (After Williston.)

down from a branch, in which case the hind legs would have made good clutching organs. But it seems difficult to believe that "emergency landings" would never have been necessary; and how the animal could get itself into the air again from level ground is difficult to understand.

The brain of the pterosaurs has been revealed by casts. The cerebral hemispheres are large for a reptile; very prominent are the regions which have to do with

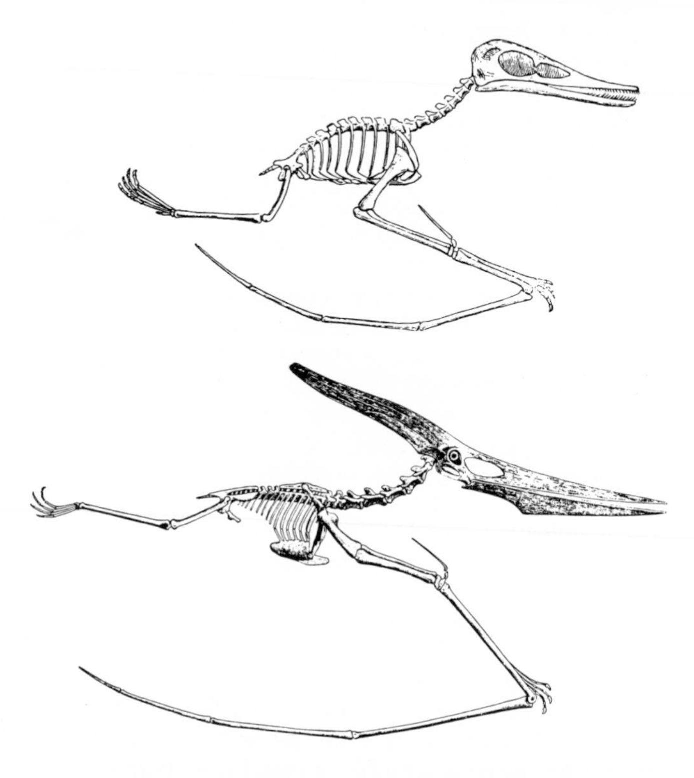

FIG. 225.—*Above, Pterodactylus,* a small, short-tailed Jurassic pterosaur. (From Williston's *The Osteology of the Reptiles,* by permission of the Harvard University Press.) *Below, Pteranodon,* a giant Cretaceous flying reptile; maximum wing spread about 27 feet. (From Eaton.)

sight, while the olfactory areas are negligible, suggesting that, as in the birds, sight was all important and smell almost lost. The many other analogies with birds and the real relationship of the two groups within the archosaurian stock suggest that perhaps other bird characters may have been present. These forms quite probably were somewhat warm blooded, for it seems very necessary that there should have been a continual supply of energy for flying; and it is not impossible that the "improvements" in the circulatory system found in birds may have been present here (they are foreshadowed in the crocodiles, the only living archosaurians).

Rhamphorhynchoids.—The pterosaurs are divided into two suborders—the Rhamphorhynchoidea and Pterodactyloidea. *Rhamphorhynchus* (Fig. 222) is typical of the Rhamphorhynchoidea; *Dimorphodon* (Fig. 223) a more primitive member. The Rhamphorhynchoidea is the more primitive group, exclusively Jurassic in age and including the only pterosaurs known from the early part of that period (there are no known Triassic forms). Its members agree in a number of primitive features. The originally long reptilian tail was still present; teeth were usually well developed; the metacarpals in the wing were comparatively short (Fig. 224*A*); the last toe in the foot was unreduced; and the fibula was present, although small.

Pterodactyloids.—But in the Jurassic there had already appeared a second, derived group—the pterodactyloids. *Pterodactylus* (Fig. 225) was a small form, some specimens being no bigger than a sparrow. The most prominent difference from the primitive types lay in the fact that there was but a stub remaining of the originally long tail. The fifth toe in the hind leg was reduced; the wing metacarpals were much elongated (Fig. 224*B*). The dentition was much modified, and the teeth more slender and delicate structures. In the genus *Ctenochasma* (Fig. 226), they are exceedingly

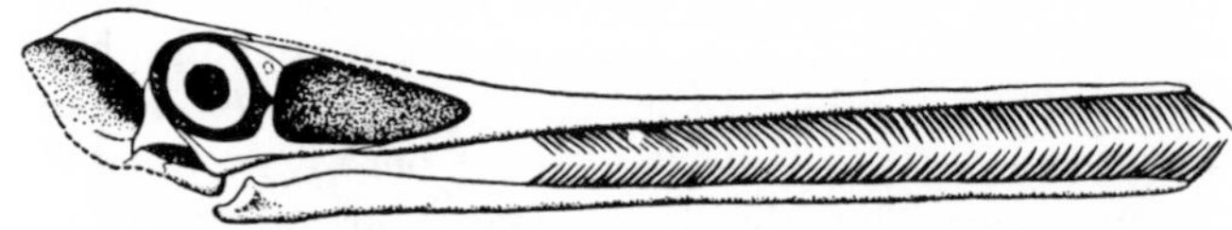

FIG. 226.—*Ctenochasma,* a Jurassic pterodactyloid pterosaur with long jaws armed with numerous slender teeth; length about 6 inches. (After Broili.)

numerous and bristle-like in appearance, suggesting the straining devices found in the bill of certain ducks. In *Pterodactylus* itself the teeth tended to be lost from the back part of the mouth at least and presumably were succeeded by a horny beak. In the Cretaceous only a few forms survived, but among them were remarkable types. For example, in *Dsungaripterus* of the Lower Cretaceous, teeth were present in part of the jaws, but anteriorly both upper and lower jaws terminated in long sharp spikes—presumably an adaptation for fish spearing. During the Cretaceous the size tended to increase, and all teeth were sometimes lost. The culmination of these tendencies is seen in *Pteranodon* (Fig. 225) of the Upper Cretaceous of Kansas, a form with a wingspread of as much as 25 feet, a long, toothless beak, and a crest nearly as long as the main part of the skull extending back from the occiput.

This was the end of the pterosaurs; the group became extinct before the close of the Cretaceous.

Pterosaur history.—The structure of these forms clearly indicates that they were derived, as were other archosaurians, from some member of the Thecodontia in which a bipedal gait had been evolved. The arms were freed from the necessity of functioning in terrestrial locomotion and were thus available to take up other employment. Possibly, the Triassic ancestors of the pterosaurs were small climbing types; and a membrane may have developed, as in a number of other tree-dwelling vertebrates, as an aid in planing from tree to tree and breaking the fall on landing. The fifth finger was often reduced in archosaurians; and the fourth—the longest of the series—was well adapted, after the loss of the claw joint, to support a wing membrane. Why, however, the loss of ordinary walking powers should have taken place is not clear. Some light may be shed on this and other problems if ever the missing connecting types, which must have been present in the Triassic, are found as fossils.

It seems obvious that these forms were not so well adapted for a flying existence as the birds. The feathered structure of the bird's wing seems much better than the pterosaur membrane. The pterosaurs appeared first in the field; for, when the first primitive birds appeared late in the Jurassic, the flying reptiles were at their peak. But, by Upper Cretaceous times, highly developed flying birds had been evolved; competition with them may have been a factor in the elimination of the last of the pterosaurs.

12 CHAPTER

Dinosaurs

Most interesting of ruling reptiles, and perhaps the most spectacular animals of any age, were the dinosaurs. Springing from the primitive thecodont stock in the Triassic, dinosaurs were already abundant by the end of that period and were the dominant land types during the remainder of the Mesozoic. Among their numbers is included almost every large terrestrial vertebrate of the Jurassic and Cretaceous.

The popular conception of the dinosaurs is that of a single group of large reptiles. This is, however, inaccurate. Most dinosaurs were large—some reaching a weight of 40 or 50 tons—but some true dinosaurs were no larger than a rooster. Further, the dinosaurs were not a single group but were already divided at their first appearance into two distinct stocks, related only in that both were archosaurs descended from the primitive ruling reptiles, the Thecodontia. The term "dinosaur" is thus one which can be used only in a popular sense; scientifically the dinosaurs are arrayed in two separate orders, defined below.

Certain of the dinosaurs were carnivores, as had been their thecodont ancestors; the majority, however, abandoned this mode of life and became herbivores. We have noted the trend toward bipedalism among the thecodont ancestors of the dinosaurs, and it is probable that the ancestral dinosaurs in all lines were bipedal to at least some degree. Some forms—the post-Triassic carnivores—attained a purely bipedal mode of progression, but other dinosaurs never reached this stage. Many herbivores reverted to a four-footed gait, although many features of their limb and girdle structure attest their descent from partly bipedal ancestors. The acetabulum is (in contrast to thecodonts) usually perforated to receive the head of the femur; except in a few primitive types, the dermal shoulder girdle has disappeared.

The dinosaurs are divided into two orders, the Saurischia and the Ornithischia, both included in the subclass Archosauria but no more closely related to one another than to the other members or descendants of the ruling reptile group—the crocodiles, pterosaurs, and birds. The distinctions between the two dinosaurian orders were clean-cut. A key character (to which the names refer) is that, while the Saurischia have the triradiate pelvic structure (Fig. 227*A*, *B*) developed in many thecodonts, the early Ornithischia had a two-pronged pubis, resulting in a tetraradiate type of pelvis (Fig. 227*C*, *E*). The saurischians were primitively carnivores, and all carnivorous dinosaurs belong to that order. Some saurischians, however, became four-footed plant-eaters and developed into great amphibious forms, among which are the largest of all reptiles. The ornithischians, on the other hand, were herbivorous from the first. Primitive ornithischians appear to have been partly bipedal, perhaps browsing about on all fours but becoming erect bipeds when speed was called for. The well-known duckbilled dinosaurs of the Cretaceous are end forms of this line. But in this order also there was reversion to a quadrupedal gait in various armored types and in the horned dinosaurs of the Cretaceous.

Saurischians.—The order Saurischia perhaps merits first consideration, since many of its early members were still exceedingly close in structure to the ancestral thecodonts; in fact, there are a number of Triassic forms which are so close to the thecodont-saurischian boundary that it is difficult to decide in which order to place them. It is not at all improbable that the order Saurischia is to some degree polyphyletic in that several different (although related) thecodonts may have given rise to the various groups within the order.

The saurischian skull is readily derivable from that in thecodonts. There are such characteristic archosaurian structures as the antorbital opening and an opening on the outer side of the jaw. The vertebrae were primitively amphicoelous or with flat ends on the centra (platycoelous); in Jurassic and Cretaceous forms, opisthocoelous vertebrae were common, particularly in the front end of the column. The backbone generally included about ten cervicals and thirteen or so trunk vertebrae. This is a somewhat lower number than the primitive reptilian count, presumably related to a better balance of the body in a bipedal pose. The tail is long and well developed—an effective balancer in bipeds, but retained in quadrupeds as well. In the shoulder, the dermal girdle is reduced and generally lost; the scapula is elongate and bladelike, the coracoid rounded. The pelvis is persistently of a typical triradiate type, and, in contrast to ancestral thecodonts, the acetabulum is open, allowing the head of the femur to be solidly socketed for body support. There is always some reduction in the digits of manus and pes (Fig. 228), although the reduction process did not progress far in most Triassic forms or in the later quadrupedal sauropods. The front limbs are almost invariably shorter than the hind, suggesting at least tendencies toward bipedalism.

We shall here consider the saurischians to be divisible into four distinct groups which may be assembled into two suborders, Sauropodomorpha and Theropoda. In the first we include the Prosauropoda, primitive, partly bipedal Triassic forms, and, descended from them, the Sauropoda, large, quadrupedal herbivorous and amphibious types of the Jurassic and Cretaceous. The Theropoda are advanced bipeds, carnivorous in habits, prominent in the same two periods but with Triassic forebears. There are two theropod groups, the lightly built Coelurosauria and the large Carnosauria.

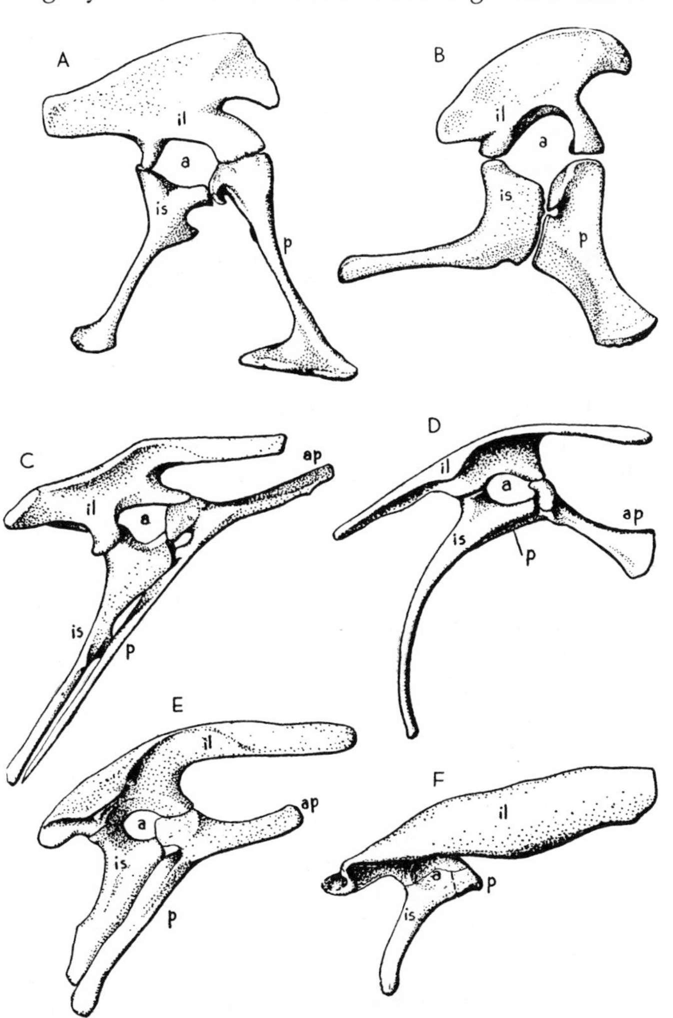

Fig. 227.—The pelvis in dinosaurs: *A*, *B*, saurischians; *C–F*, ornithischians. *A*, The theropod *Allosaurus; B*, the sauropod *Camarasaurus; C, Thescelosaurus*, a primitive ornithopod; *D, Centrosaurus*, a horned dinosaur; *E, Stegosaurus; F, Euoplocephalus*, a Cretaceous armored form. Abbreviations: *a*, acetabulum; *ap*, anterior process of pubis; *il*, ilium; *is*, ischium; *p*, pubis. (*A, E* after Gilmore; *B* after Marsh; *D, F* after Brown.)

Prosauropoda.—In contrast to the ornithischians, early records of which are scanty, the saurischians became prominent in terrestrial life before the close of the Triassic. Certain Triassic types are ancestral to the later theropods, but numerous other saurischians of that age which grew precociously, so to speak, into large, semibipeds, appear to represent an independent branch of the order. They will here be considered as constituting an infraorder Prosauropoda (or Palaeopoda).

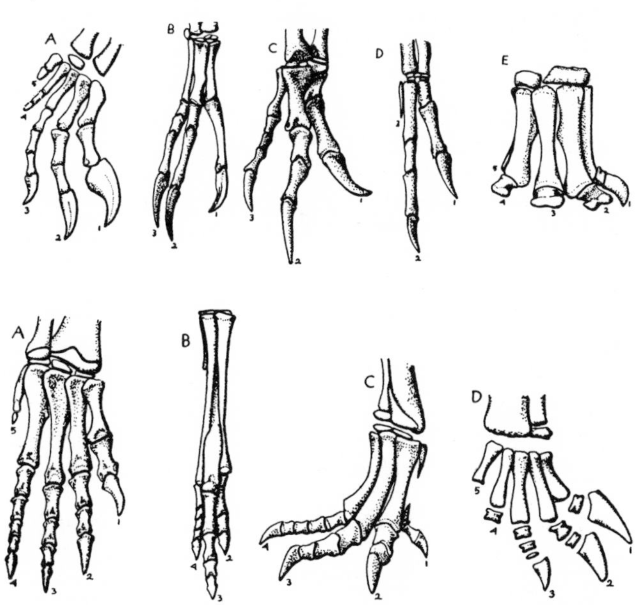

Fig. 228.—*Above*, The manus in saurischian dinosaurs: *A, Yaleosaurus* [*"Anchisaurus"*]; *B, Ornithomimus; C, Allosaurus; D, Gorgosaurus; E, Diplodocus*. (*A* after Marsh; *B, E* after Osborn; *C* after Gilmore; *D* after Lambe.) *Below*, The pes of saurischian dinosaurs: *A, Yaleosaurus; B, Ornithomimus; C, Allosaurus; D, Camarasaurus*. (*A* after Marsh; *B* after Osborn; *C, D* after Gilmore.)

A wealth of prosauropod genera are known from late Triassic deposits, notably in South Africa and Europe, but with representation in other continents as well. *Plateosaurus* (Figs. 229, 230) shows in most regards the general appearance of members of this primitive saurischian group. Many were of considerable size; *Plateosaurus*, for example, was 20 feet or more from snout to tail. The build was rather clumsy, in comparison with later theropods; the fore limbs were not greatly shortened, and it is obvious that these animals were not purely bipedal. The limb bones are here solidly built, unlike the advanced bipeds, in which they are hollowed. The ilium is short fore-and-aft, with only three sacral vertebrae anchoring it to the vertebral column. The pubis is primitive, as a broad plate, and in primitive fashion the pelvic symphysis extends the full length of the curved lower margin of pubis and ischium. The manus retains five digits, although the outer ones may be somewhat reduced; in the hind foot, the outer toe is reduced, but the other four are well developed and all point forward.

In most of the better known prosauropods, such as *Plateosaurus* and *Yaleosaurus* ("*Anchisaurus*"), the skull was relatively small, the teeth more or less flattened, spoon-shaped or leaflike—a type repeated in later sauropods, and one adapted to a herbivorous diet, not one of flesh. There are, however, a number of late Triassic dinosaurs, such as *Teratosaurus* and *Gryponyx*, in which the teeth are sharp pointed, indicative of carnivorous habits. But in other regards these dinosaurs (although poorly known) appear to be assignable to the prosauropods rather than to the predaceous theropods of later periods.

In general, this early assemblage of primitive saurischians was doomed to extinction by the end of the Triassic. The carnivores of this group are antecedent but rather surely not ancestral to the flesh eaters of the Jurassic and Cretaceous, which appear to have evolved independently from thecodonts. Even as regards *Plateosaurus* and its relatives, most, at least, seem to have left no descendants. But in the case of such a form as *Melanorosaurus* of South Africa, in which little suggestion of a trend toward bipedalism is present, we may be dealing with forms close to the ancestry of the later sauropods.

Theropods.—Quite independent in origin from thecodont ancestors, it is thought, were the purely bipedal predators of the later Mesozoic, classed as the suborder Theropoda. The front limbs were much reduced and obviously not adapted for use in locomotion. The "fingers" tended to be reduced progressively from the lateral to the medial side, so that only three, or in extreme cases two, digits remained. For bipedal posture the pelvis (Fig. 227*A*) was more strongly braced, with four or five sacral vertebrae; the pubis was narrow (with reduction of the proximal plate containing the obturator foramen) but expanded at the distal end, where the pubes met in symphysis. In theropods the femur is generally somewhat shorter than the tibia; in life this would have resulted in the development of considerable speed, a short swift stroke of the thigh being accompanied by a long swing of the tibia (in slow and ponderous tetrapods the femur is long, the tibia short). In the hind foot (Fig. 228*A*, *B*, *C*, *lower row*) the fifth toe (as in most archosaurs) had been reduced to a short metatarsal with, at the most, a vestigial nubbin or two representing the phalanges. In many theropods the first toe as well was considerably shortened and tended to be set off at an angle from the other three. In the second to fourth toes the metatarsals tended to be much elongated and closely appressed, and the central digit was somewhat longer than its neighbors, resulting in a very birdlike foot structure. The development of this type of hind foot was already under way among Triassic bipeds. So similar are such feet to those of their avian relatives that when numerous dinosaur footprints were discovered a century and a half ago in the Triassic rocks of the Connecticut Valley they were thought to belong to ancestral birds.

Coelurosaurs.—Two groups are clearly demarcated within this carnivorous assemblage, both already present in the Triassic. Generally considered as the more primitive of the two are the members of the infraorder Coelurosauria. These are relatively small forms, very lightly built, with thin-walled bones. The skull was small, the orbits large, the neck relatively long and slender. Coelurosaurs of slender build were already common in the late Triassic; *Procompsognathus*, *Saltopus*, and *Scleromochlus* were among the better-known types in Europe, *Coelophysis* (Fig. 231) and *Podokesaurus* in North America. All were small—very small for dinosaurs. *Coelophysis*, even including the long tail, was less than two yards in length, *Procomp-*

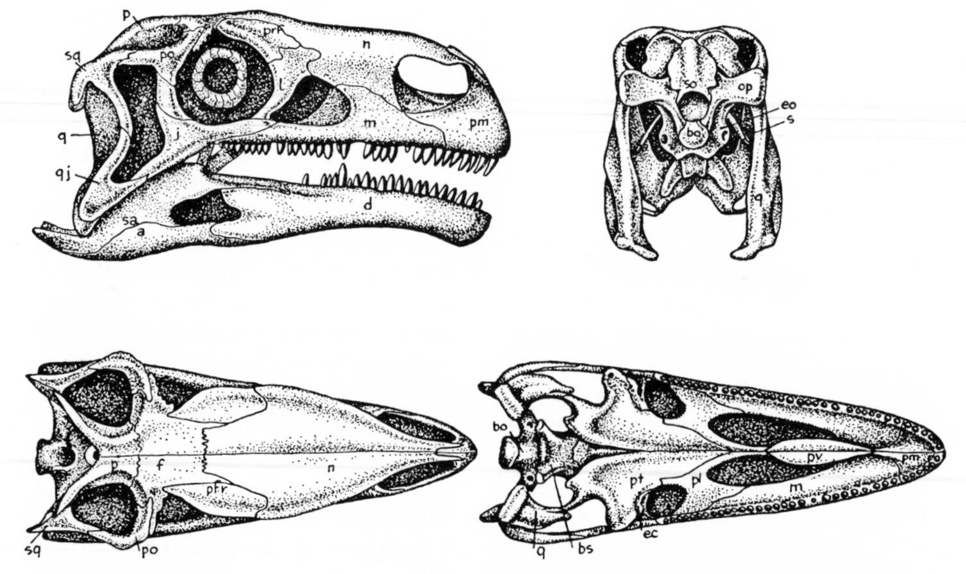

Fig. 229.—The skull of the Triassic prosauropod *Plateosaurus;* length of original about 13 inches. *prf*, Prefrontal; *so*, supraoccipital. For other abbreviations see Figure 109. (After Huene.)

sognathus and *Podokesaurus* but half that size. Small coelurosaurs of this sort continued into the Jurassic. *Compsognathus* (Fig. 235), known from a skeleton in the lithographic stone of southern Germany, was no larger than a fowl; *Coelurus* [*Ornitholestes*] (Fig. 232) of the late Jurassic Morrison beds of North America was considerably larger with a 6-foot length. In these Jurassic forms, reduction of the "hand" was far advanced; there were but three functional digits, although in *Compsognathus* there were rudiments of the other two. In *Coelurus* the short pollex was somewhat divergent, suggesting a grasping power. The obvious agility of this little dinosaur and the clutching power of the hand led to a belief that "*Ornitholestes*" might have preyed upon the contemporary primitive birds. This is improbable; presumably, small reptiles such as lizards and perhaps the small early mammals may have been the staple articles of diet of this and other tiny dinosaurs.

Fig. 230.—*Plateosaurus,* a Triassic prosauropod; length about 21 feet. (From Huene.)

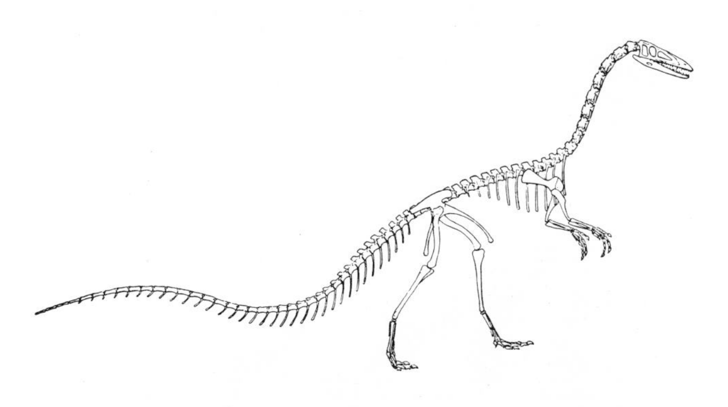

Fig. 231.—The Triassic coelurosaur *Coelophysis;* about 8 feet long. (After Colbert.)

Ostrich-like dinosaurs.—Almost no remains of such small coelurosaurs are known from Cretaceous deposits, but their descendants are to be recognized in *Ornithomimus* [*Struthiomimus*] (Figs. 228, 233, 235) of the North American Cretaceous and related Old World contemporaries. *Ornithomimus* was about the size of an ostrich and not unlike that bird in general proportions, with very long, slim hind legs, a long neck, and a small head. The foot was three toed and very birdlike. The three metatarsals were very long and closely appressed; the central one was much reduced proximally, its neighbors taking over most of its weight-supporting function. Despite the fact that these forms were obviously purely bipedal in habits, the front limbs were quite long. The three inner fingers were all well developed, and the first was seemingly opposable to the other two, forming an effective grasping organ

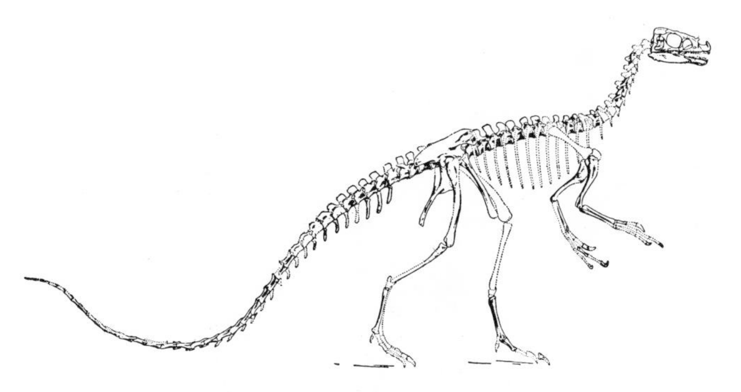

Fig. 232.—*Coelurus* [*Ornitholestes*], a small late Jurassic coelurosaur, about 6 feet long. (From Osborn.)

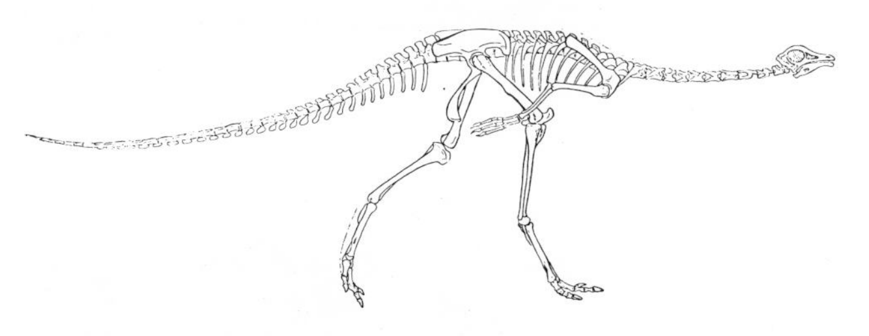

Fig. 233.—*Ornithomimus* [*Struthiomimus*], a Cretaceous coelurosaur, length about 8 feet. (From Gregory.)

(Fig. 228*B, upper row*). The skull was quite small and very lightly built, with a superficially birdlike appearance. This resemblance is greatly heightened through the fact that, almost alone among dinosaurs, *Ornithomimus* and its relatives had entirely lost their teeth and presumably had replaced them functionally with a horny bill. The jaws were short and weak.

The odd combination of specializations found in these ostrich-like forms—great swiftness, a grasping ability in the well-developed "hands," and the weak and toothless jaws—has caused considerable discussion as to their mode of life. The best idea put forward was that these forms made a living by robbing eggs from the nests of other dinosaurs. Teeth are unnecessary for egg-eating, and the shell can be effectively broken with a horny beak; the "hands" would be of great use in handling the eggs; and the speed was necessary to escape the enraged parents!

Carnosaurs.—Parallel with the evolution of the little coelurosaurs was that of flesh-eaters of larger size, the Carnosauria. These were more massively built animals; the limb bones, although hollow, had much thicker walls than in coelurosaurs. The skull, again in contrast, was large, the neck short and heavy. The hind legs were stoutly built, as is necessarily the case in a large animal.

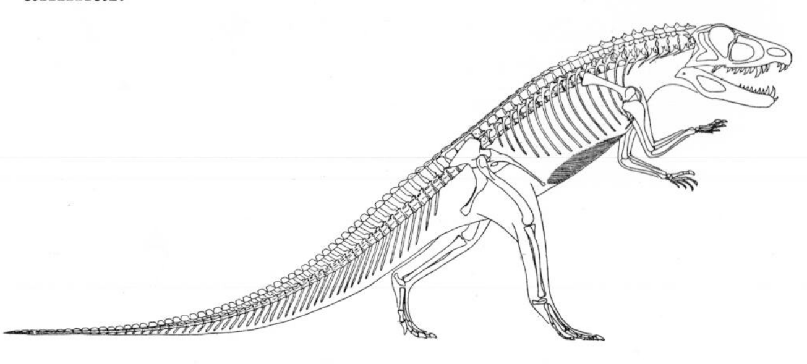

Fig. 234.—The skeleton of a Triassic ancestral carnosaur, *Ornithosuchus;* maximum size about 12 feet. (From Walker.)

Remains definitely attributable to carnosaurs are rare in the Triassic; the one form which appears to be clearly a forerunner of this group is *Ornithosuchus* (Figs. 234, 235) of the Upper Triassic of Scotland. This animal, which reached a maximum length of about 12 feet, is primitive in many regards—so primitive, in fact, that it might be (and has been) regarded as an advanced thecodont rather than a primitive dinosaur. Among primitive features, in contrast to any other known saurischian, it had retained the double row of armor plates down the back found commonly in thecodonts; a dermal shoulder girdle is still present; there are, in contrast with other theropods, but three sacral vertebrae, and the ilium is short; the pubis is broad and still retains an obturator foramen, lost in typical theropods; the acetabulum is only partly open; all five digits of the hind foot are complete, although the outer toe is short and slender. On the other hand, the skull is closely comparable, despite its smaller size,

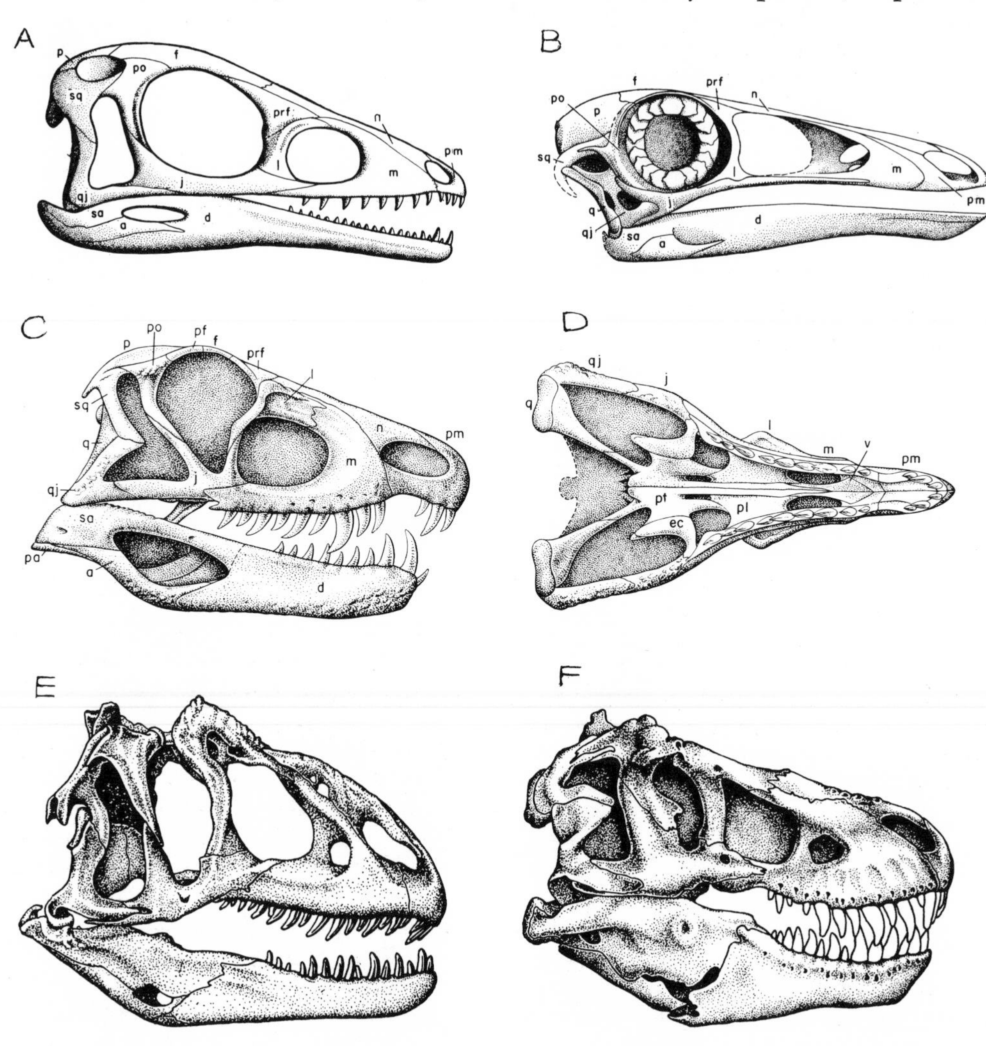

Fig. 235.—Skulls of coelurosaurs and carnosaurs. *A–B,* Coelurosaurs. *A, Compsognathus,* original about 2.75 inches long (partly after Heilmann); *B, Ornithomimus,* original about 13 inches long (after Osborn). *C–D,* Lateral and palatal views of the primitive carnosaur *Ornithosuchus,* length about 10 inches. (After Walker.) *E, Allosaurus* [*Antrodemus*], a large late Jurassic carnosaur; length about 27 inches. (After Gilmore.) *F, Tyrannosaurus,* a late Cretaceous carnosaur; skull length about 50 inches. (After Osborn.) For abbreviations see Figure 109.

to that seen in the great carnivores of the Jurassic and Cretaceous.

In the Jurassic, carnosaurs become common as the major carnivores of the day, replacing the predatory members of the Palaeopoda. There are numerous scattered finds in European deposits, representing nearly every stage of the Jurassic, to which the generic term *Megalosaurus* is usually applied. Most of these remains are, however, fragmentary and tell us little beyond the fact that large flesh-eaters were abroad in the land. Well-preserved specimens of large Jurassic carnivores are found only in the Morrison beds of North America; from these beds have been obtained nearly complete skeletons of *Allosaurus* [*Antrodemus*].

Allosaurus was an animal with a total length of about 34 feet; it was seemingly powerful enough to have attacked any of its herbivorous contemporaries, even the giant sauropods. The skull (Figs. 228, 235) was quite large in proportion to the body; the teeth were large and sharp, the jaws long and deep. The quadrate was loosely united to the other skull elements; there appears also to have been some possibility of movement between the frontal and the parietal bones; and there was some flexibility between front and back parts of the lower jaw. These appear to be adaptations for bolting large masses of flesh—a type of structural modification which we have seen carried to an extreme in the snakes and which is also present to some extent in birds.

The body was rather massively built. The neck vertebrae were opisthocoelous, and extra articulations, in addition to the normal zygapophyses, tended to add to the strength of the backbone. The hind legs, in correlation with the size of the creature, were stout and only moderately long, and the femur was somewhat longer than the tibia. The metatarsals were rather short, but otherwise the foot was quite birdlike, with three well-developed toes and a small first toe turned backward as a prop, much as in many avian types. The front limb was short but fairly stoutly built, with three spreading, claw-tipped digits; it was obviously not a supporting structure.

Little is known of the carnivores of the early Cretaceous. From Jurassic carnosaurs may have been derived *Spinosaurus* of the Cretaceous of Egypt. This large flesh-eater is poorly known but was remarkable in that the neural spines were greatly elongated, rather after the fashion of some of the Permian pelycosaurs described in a later chapter; some spines had a length of about 6 feet.

Although a few small megalosaurids survived, the characteristic carnosaurs of the late Cretaceous were the still larger and more powerful tyrannosaurids, of which *Tyrannosaurus* (Figs. 235, 236, 248*B*) and *Gorgosaurus* of western North America and eastern Asia are the best-known forms. *Tyrannosaurus* reached a length of some 47 feet and stood about 19 feet high in walking pose; *Gorgosaurus* was nearly as large. These reptiles were the largest terrestrial flesh-eaters known from any period in the history of the world. The body was heavy in construction, and the hind legs were massively built. The front legs were absurdly small and seem to have been practically useless; they were too short to reach the mouth and seemingly too weak to have been of any assistance in seizing or rending the prey. Only two small digits were present in *Gorgosaurus* (Fig. 228*D*, *upper row*), and probably the same was true in *Tyrannosaurus*.

The skull was very large and massively built, with the fenestrae of small size. The jaws were armed with powerful teeth, some half a foot in length, recurved, and with serrate edges. The cervical vertebrae, which bore the weight of the heavy head, were short and broad.

The reign of the tyrannosaurs was a brief one, for they disappeared at the end of the Cretaceous. Their extinction was obviously consequent upon the extermination of the herbivorous dinosaur types, which presumably formed their food supply.

Fig. 236.—*Tyrannosaurus*, a gigantic Cretaceous carnosaur; length about 30 feet. (From Gregory.)

Sauropods.—We have still to consider the history of the sauropods—Jurassic and Cretaceous saurischians which became herbivores, reverted to a quadrupedal pose, and reached gigantic proportions. The ancestors of these forms are unquestionably to be sought among the large Triassic forms which we have termed the Prosauropoda. Members of this group included not only the largest of all reptiles but the largest four-footed animals of any time. Among the better-known sauropods are *Apatosaurus, Brontosaurus, Diplodocus* (Figs. 237–39), and *Camarasaurus* (Figs. 227*A*, 237, 238)—forms whose remains are among the most spectacular attractions of many museums.

The most obvious contrasts between the sauropods and their carnivorous relatives lie in the quadrupedal pose of the body and in the bodily proportions. The sauropods were massively built, with powerful limbs, a long tail, and a long neck terminating in a small head.

The skull (Fig. 238) in sauropods seems absurdly small in proportion to the size of the animal. It was generally quite lightly built, with a large temporal fenestra and a large antorbital opening (there was a

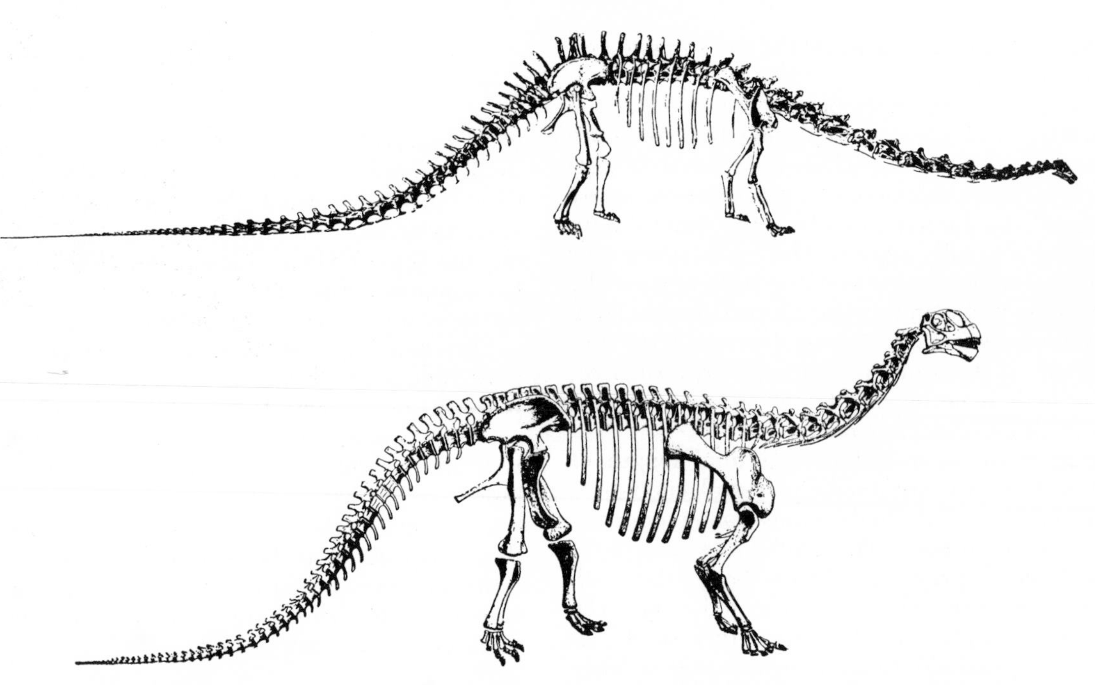

Fig. 237.—The skeleton of sauropods. *Above, Diplodocus,* longest of late Jurassic sauropod dinosaurs (87.5 feet). (From Holland.) *Below, Camarasaurus,* a comparatively small late Jurassic sauropod dinosaur; original of figure about 18.5 feet long. (From Gilmore.)

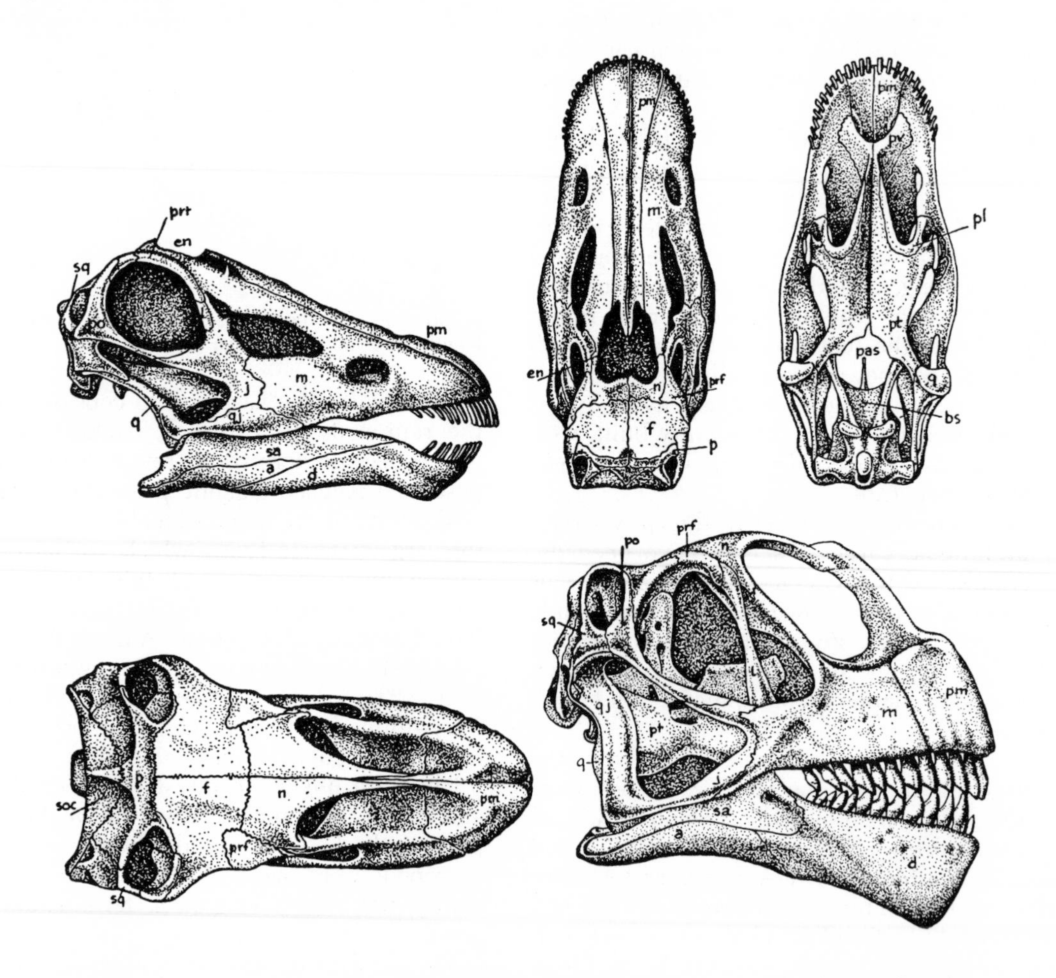

Fig. 238.—The skull of sauropods. *Above, Diplodocus,* lateral, dorsal, and palatal views; length about 2 feet. (After Holland.) *Below, Camarasaurus,* length about 1 foot. (After Gilmore.) *pas,* Parasphenoid. For other abbreviations see Figure 109.

second smaller one in *Diplodocus*). The orbits were of good size and were situated high up on the side of the head. In *Camarasaurus* the external nares were very large and reached high up on the short skull; in *Diplodocus* the two openings were fused and situated at the top of the skull between the eyes. The position of these organs suggests an amphibious mode of life for the sauropods; the animal could breathe and look about with only the top of the head exposed above the water.

The jaws were short and weak, the quadrate slanting downward and forward to the jaw articulation. The teeth were small and slim, peglike or spoonshaped; there were seldom more than a dozen or so in each jaw half. It seems almost impossible that this dental and jaw apparatus could have been capable of cropping enough fodder to supply such a huge body, although the food material may have been some soft type of water vegetation which could be eaten with little effort.

The brain was small in all dinosaurs but excessively so, proportionately, in the sauropods, in which the endocranial cavity was but a small recess in the posterior part of the small skull. Presumably the brain had few functions other than working the jaws, receiving sensory impressions, and passing the news back down the spinal cord to the pelvic region, from which originated the nerves working the hind legs. Here there was situated, in many dinosaurs, an enlargement of the spinal cord several times as big as the entire brain.

A monumental construction of the backbone was required to carry the weight of the enormous body and transfer it to the legs. The anterior vertebrae, which had only the tiny head to support, were small; but in the trunk the centra of the vertebrae become progressively massive as we pass up over the arch of the back to the hip region. The neural spines, low in front, increase rapidly in height and reach a maximum at the top of the arch just in front of the pelvis. Presumably an interlacing series of ligaments and tendons passed between these spines and helped to strengthen the back. In many sauropods the neural spines of the anterior vertebrae were cleft at their upper ends, and probably a stout longitudinal ligament lay between the two prongs.

But the dead weight of the backbone itself was a great burden; in correlation with this we find that adaptations had developed which greatly decreased the weight of the vertebrae (Fig. 239). These structures were generally cavernous; great areas were hollowed out at the sides of the centra and arches; these were presumably filled with air sacs connecting, as in birds, with the lungs. Weight was thus much reduced, but all the essential framework of the vertebrae was left intact.

The pelvis (Fig. 227*B*) was stoutly constructed; the ventral elements had more of the primitive platelike aspect than was the case in most theropods. The hind-limb bones were massive, with the femur considerably longer than the tibia; presumably the limbs extended straight down from the body in columnar, elephantine fashion. The front legs were generally much shorter than the hind, a feature suggestive of the probable bipedal ancestry of sauropods. They bore less of the weight than the hind limbs and may have been less straight, the elbows projecting somewhat. In both front and hind feet (Fig. 228) the metapodials and phalanges were short, stout, and spreading; footprints show that the limbs terminated, as in elephants, in a broad pad in which the toes were encased. The details of foot construction are poorly known in most cases, but from one to three of the digits in each foot bore a large projecting claw, which presumably prevented slipping.

The pose of sauropod limbs has been a much debated subject. It has been argued that, since in most reptiles the femur projects sideways from the body, the sauropods, too, should be mounted in this fashion. As a matter of fact, this cannot be done without doing violence to the articular surfaces of the bones. But,

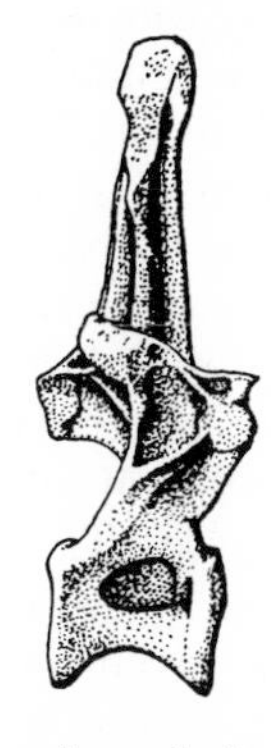

FIG. 239.—A dorsal vertebra of the sauropod dinosaur *Diplodocus*. (After Hatcher.)

more than this, it would have been impossible for the animal to have supported an enormous weight with any type of limb other than a straight column. With the limbs sprawled sideways, no mere muscles could have withstood the direct pull of the score or more of tons that the hind legs of a big sauropod had to support; the animal would have caved in.

But, even granting the columnar position of the legs, it is difficult to see how these dinosaurs ever walked on land; the elements of physics show that there are natural limits to the possible size of a four-footed vertebrate.

An elephant does not and cannot have the slim limbs of a gazelle; *Tyrannosaurus* and great sauropods could not have the slender hind legs of the little coelurosaurs. The weight of an animal varies in proportion to the cube of a linear dimension. But the strength of a leg, like that of any structural supporting element, is proportionate to its cross-section, which increases only by squares. If a reptile doubles his length, his weight is approximately eight times as great, but his legs are but four times as strong. Hence in large animals the bulk of the legs must increase out of all proportion to the rest of the body.

The legs of sauropods were large, but, even so, it seems doubtful whether they could have borne so many tons of weight. For this reason, as well as because of cranial features noted earlier, it appears probable that the sauropods were amphibious types which spent most of their lives in lowland swamps and lagoons, where they were buoyed up by the water, and problems of support and locomotion were greatly simplified.

Sauropods are unknown in the Triassic, although we have noted ancestral types, and some later Triassic South African forms appear to have been very close to the sauropod condition. In the older part of the Jurassic there are fragmentary remains, particularly in Europe, to most of which the name *Cetiosaurus* has been applied. In the Morrison beds of North America and beds of similar age in East Africa have been found a considerable number of well-preserved specimens of a variety of sauropods. *Camarasaurus* of North America was a comparatively small and unspecialized form. *Brontosaurus* was a bulkier relative, which reached a length in one specimen of some 67 feet and an estimated weight of 30 tons. *Diplodocus* was a slimly built form, which may not have been nearly so heavy as the last but which, with a whiplash at the end of his tail, reached the record for length—87½ feet. *Brachiosaurus*, known both from North America and East Africa, was the real giant of the group. The tail was comparatively short, but, even so, the length may have been close to 80 feet. The body was extremely stout, and (in contrast with almost all other sauropods) the front legs were long. Above the shoulders there extended a long neck which could place the head above the level of the roof of a three-story building; this build presumably was an adaptation for life in deep waters. A guess at the weight of this great animal would be somewhere close to 50 tons.

It is probable that sauropods were still abundant in Lower Cretaceous times, although there are relatively few fossiliferous deposits. In the Upper Cretaceous, sauropods are rare in the northern continents; for example, only two specimens have been found in all the rich dinosaur beds of that age in North America. In India and the southern continents, however, *Titanosaurus* and related genera seem still to have flourished at this time.

The decline and final disappearance of these great reptiles may perhaps be correlated with geologic events. Their environment seems to have been a restricted one and if, as seems likely, continental elevation toward the end of the Mesozoic greatly reduced the area of the lowland swampy regions in which the sauropods dwelt, their extinction was almost inevitable; changes taking place in the vegetation at this time may also have been of great importance.

Ornithischians.—Members of the second order of dinosaurs, the Ornithischia, never reached the size attained by some of their saurischian cousins but are perhaps of even greater interest because of the variety of bizarre types into which they developed. The most characteristic feature of the order, one by which the skeletons of these forms may be told at a glance, was the tetraradiate type of pelvis (Fig. 227*D*). The ilium and ischium were roughly comparable in shape and position to those elements in the saurischians, but the pubis was peculiarly constructed. Its main portion had not only rotated downward but also backward, so that it lay parallel to and close beside the ischium, while—presumably in relation to the need for support of the abdomen in bipedal pose—a new process had developed which projected forward and outward along the margin of the belly.

The limbs in ornithischians were never so efficiently developed for bipedal locomotion as those of the theropods, and the front legs were never so much reduced or shortened as in those forms. It is probable that few ornithischians were entirely bipedal in habits; the front legs were very likely used in locomotion at times in even the most lightly built types. A great number of ornithischians, apparently representing several independent phyletic lines, reverted completely to a quadrupedal gait.

The ornithischians were, even at their earliest appearance, herbivorous forms. The teeth were generally leaf-shaped, with crenulated edges. In only the most primitive genera were there teeth in the front part of the mouth; in the vast majority this region was toothless and presumably covered with a horny beak. In the lower jaw the beak capped a median predentary bone, an element not found in any other reptilian group.

The Ornithischia may be divided into the following suborders: (1) Ornithopoda, bipedal forms; (2) the Stegosauria, quadrupedal forms with a double row of protective plates and spines down the back and tail; (3) the Ankylosauria, heavily armored, rather turtle-like quadrupeds; and (4) the Ceratopsia, horned dinosaurs.

Primitive ornithopods.—*Camptosaurus* (Fig. 240), common in the late Jurassic and early Cretaceous of Europe and North America, may be used as a basis for the consideration of the ornithopods, the most primitive of ornithischian suborders; some of its contemporaries and predecessors were smaller and somewhat more primitive in structure but are less adequately known. Various specimens of this genus show a range in length from about 6 to 20 feet.

The skull was long and low and, as compared with saurischians of similar dimensions, rather heavy in its construction. An antorbital opening was present but was of small size. The external nares were oval openings much larger than those of most reptiles and were nearly completely surrounded by the enlarged premaxillae; the nasals, which bounded them above, were also elongated and stretched back along the top of the skull to the level of the orbits. An extra element, a supraorbital bone, lay in the upper anterior margin of the orbit.

As in sauropods (but in contrast with carnivorous dinosaurs), the jaws were rather short and did not reach the full length of the skull. Here (and in ornithischians generally) the quadrate extended down well below the level of the upper tooth row, so that when the jaw was closed the whole cheek battery of lower teeth tended to meet the upper series, effectively, as a unit. The lower jaw was heavily built; a powerful musculature is suggested by the fact that a coronoid process (seen elsewhere among reptiles only in placodonts and mammal-like forms) extended up beneath the edge of the cheek region of the skull for the attachment of the temporal muscles, while a process projecting back from the articular region was developed for the muscles opening the jaws. The front part of the mouth was toothless in *Camptosaurus,* as in most ornithischians, and presumably covered by a horny beak; the teeth, leaflike in shape, were confined to a single row in the cheek region. (Teeth, however, were still present in the premaxillary region in a smaller contemporary.)

The normal position of the body was presumably the bipedal one, with the neck well erect, for the occipital condyle of the skull projected downward rather than backward, indicating that the head was held at right angles to the backbone. The neck vertebrae were opisthocoelous in *Camptosaurus,* as also were the anterior trunk vertebrae in many more advanced forms. The arch of the back was stiffened in ornithopods by a latticework of tendons which (particularly in the duckbilled dinosaurs) were often ossified and thus preserved in fossil specimens.

The primitive ornithopods had the typical tetraradiate type of pelvis described above (Figs. 214, 227*C*). The ilium was considerably lengthened; both pubis and ischium were greatly elongated and of subequal length, and the former bone had a long anterior process. The hind limbs in *Camptosaurus,* as in ornithopods in general, were somewhat more massive and shorter than those of theropods of equal size; presumably in these herbivores the demand for speed was not so great as in carnivorous types. The tibia never exceeded the femur in length; and, although the ornithopods were digitigrade types, walking on their toes rather than on the flat of the foot, the metapodials were but little elongated. In primitive ornithischians, as in saurischians, the fifth toe was functionless, and the first was reduced in length (Fig. 241*D*). Claws appear to have been present but were usually rather blunt.

The front legs were not so short as in most bipedal saurischians; in *Camptosaurus* they were about two-thirds the length of the hind limbs. All five digits were present, but in even the most primitive of known ornithischians the outer two were reduced, much as in

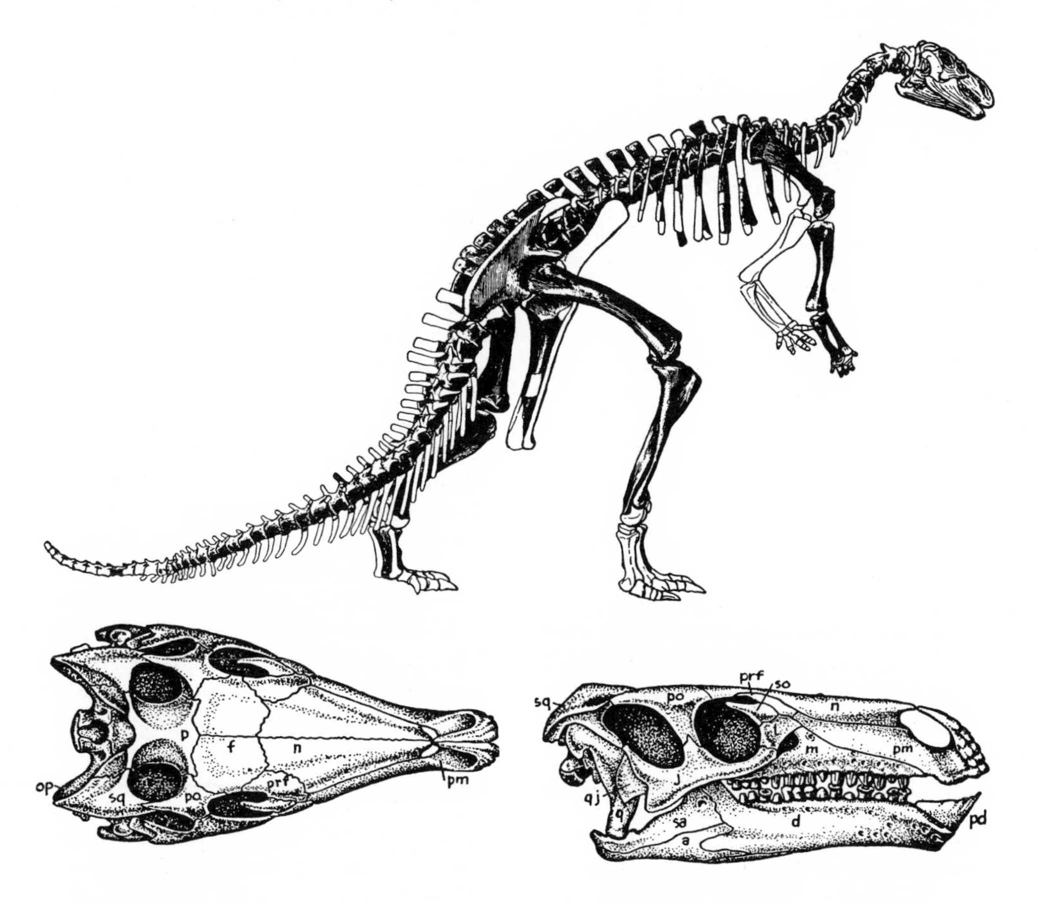

Fig. 240.—*Camptosaurus,* a Jurassic ornithopod. *Above,* The skeleton. *Below,* Dorsal and lateral views of the skull; length about 16 inches. For abbreviations see Figure 109. (After Gilmore.)

the early theropods (Fig. 241*A*). The terminal phalanges of the manus in *Camptosaurus* were clawed, but those of more advanced forms were broadened and presumably were covered by small hooflike structures rather than claws. Very probably the front legs were used for support while feeding or walking slowly.

In contrast with the saurischians, the ornithischians were rare in the early days of dinosaurian history. There are no intermediate types known which might serve to connect them with any of the better-known thecodont families. Until recently, Triassic ornithischians were known only from a few fragmentary remains of teeth and jaws. Recently, however, additional late Triassic remains have been found, notably in South Africa; one of these new forms, *Heterodontosaurus*, has already been described from a skull and partial skeleton. Quite surely a considerable radiation of the order was taking place during the Jurassic; but since most known Jurassic deposits are marine, we know almost nothing of the group until late in that period.

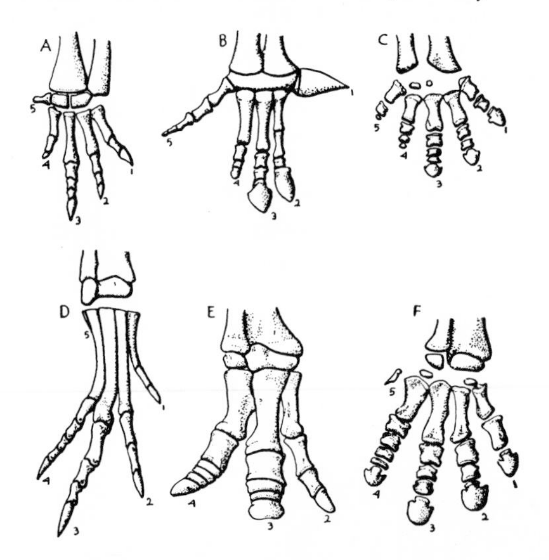

Fig. 241.—*A–C*, Manus, and *D–F*, pes, of ornithischian dinosaurs. *A, D, Hypsilophodon; B, Iguanodon; C, F, Centrosaurus; E, Anatosaurus.* (*A, D* after Hulke and Abel; *B*, after Dollo; *C, E, F* after Brown.)

Hypsilophodon of the Lower Cretaceous Wealden beds of Europe is, despite its relatively late time of appearance, the most primitive of known ornithopods. This animal was scarcely more than a yard in length, and some structural features suggest arboreal habits comparable to those of the tree kangaroo of Australia. Teeth were still present in the premaxilla in *Hypsilophodon;* and an interesting feature is the fact that there appear to have been two rows of small bony plates running down the midline of the back, much as in some thecodonts. There are several other incompletely known small forms of this general type in the Morrison beds of North America and even from the Upper Cretaceous. *Camptosaurus*, described above, was a somewhat larger and more advanced form.

Best known of European ornithopods is *Iguanodon* of the Wealden, a form twice the size of *Camptosaurus.* Numerous remains have been found both in England and on the Continent, and the genus is reported from Asia and northern Africa as well. The most striking of European dinosaur finds was the discovery of more than a score of individuals belonging to this genus during the excavation of a Belgian coal mine. Apparently, a herd of these large reptiles had fallen into a crevasse in the older Carboniferous rocks and was buried there. An interesting specialization in *Iguanodon* is that the terminal phalanx of the short "thumb" is a stout but pointed spike of bone which may have been an excellent defensive weapon (Fig. 241*B*).

Two small bipeds from the Cretaceous of Asia—*Psittacosaurus* (Fig. 243) and *Protiguanodon*—are in most respects normal ornithopods. The skull, however, has a deep and powerful beak; this and other features suggest that these genera are ancestral to the horned dinosaurs of the later Cretaceous.

A curiously aberrant group of Upper Cretaceous forms is that represented by *Stegoceras* ["*Troödon*"] (Fig. 244) and *Pachycephalosaurus.* In these forms the postcranial skeleton is of a normal ornithopod build, but the head is grotesque in its development. Above the jaws (with a feeble dentition) and the orbits, the skull expands upward into an enormous bony dome with a rugose surface and surrounded with an "ornamental" area of spikes and rugosities, presumably defensive in nature. The internal structure of the dome belies the intellectual appearance of the animal, for it is formed entirely of thick and solid bone. The dentition is primitive in the retention of premaxillary teeth. The skull is comparable in certain respects to that of the contemporary ankylosaurs, but the resemblance is apparently superficial only.

Duckbilled dinosaurs.—Most prominent of ornithopods in the Upper Cretaceous were the hadrosaurs or trachodonts, duckbilled forms which appear to have been almost universal in distribution and are represented in modern museums by numerous skeletons from western North America. The skull structure in these forms varied enormously, but the skeleton was similar in all types. The length of the body averaged about 30 feet; and, while in many features they were similar to *Camptosaurus* of the preceding period, the build was somewhat heavier. In the pelvic girdle the main body of the pubis was much shortened. The hind foot (Fig. 241*E*) had but three toes, which terminated in hoofs rather than in claws. In the manus the fifth finger had disappeared and the first was reduced; the remaining three digits ended in small hoofs.

In several instances hadrosaur mummies have been discovered in western deposits. These appear to represent specimens which had dried and mummified before burial, so that a natural cast of the skin has been preserved. This shows that the hadrosaurs were unarmored, although the skin was covered with a mosaic of small scales. A web of skin was present between the digits of the manus, and the same condition probably held true for the feet. This indicates that the hadrosaurs were amphibious in habits, probably feeding in swampy pools or about their margins. The hind limbs and the stout tail with long neural and haemal arches were probably effective swimming organs.

The skull in such types as *Anatosaurus* [“*Trachodon*”] and *Edmontosaurus* (Fig. 242A) is readily comparable with that of a camptosaur, despite some specializations. The toothless beak was flat and greatly broadened and was presumably covered by a ducklike bill. The nostrils, extending far back along the facial region, were completely surrounded by the premaxillae and nasal bones.

The teeth had preserved the primitive leaflike structure but had multiplied enormously in numbers to form a seemingly efficient grinding apparatus for tough vegetable food. In each jaw half there was not one, but several, parallel longitudinal rows of teeth closely pressed against one another, and beneath each tooth in these series its successors were already formed and ready to function. It has been estimated that in some cases there were as many as seven hundred teeth in the mouth of a single animal.

The appearance of numerous crested types was a peculiar feature of hadrosaur history (Fig. 245). In some forms, such as *Kritosaurus*, a swelling—a sort of “Roman nose”—developed above the external nares, here situated well up the forehead. In *Corythosaurus* we find a tall, domelike structure capping the skull. This was shaped like a rooster’s comb but was formed of bone, the premaxillae and nasals being the elements concerned. In *Lambeosaurus* (Fig. 242*B*) this crest was present, but there was, in addition, a backwardly directed hornlike prolongation of the same bones; while in *Parasaurolophus* the crest was absent but the backward projection was a very long, tubular structure, formed entirely by the premaxillae and nasals, which thus stretched the entire length of the skull—and beyond. These peculiar crests and prongs were composed solely of the bones originally surrounding the nostrils. In cases where the internal structure of the crest has been investigated, the air passage from the external nostrils follows a tortuous passage up through this area before descending to the choanae. Various suggestions, mainly in connection with possible underwater feeding, have been made as to the possible function of this peculiar structure—air storage, a trap to prevent water inflow, extra olfactory mucosa. None, however, is particularly satisfying.

Hadrosaurs were exceedingly numerous in the late Cretaceous but died out completely at the end of the period. Perhaps their disappearance, like that of the sauropods, may be partly accounted for by continental elevation and the consequent drying of their marshy haunts. But not improbably a main cause for extinction was the contemporary gradual replacement of

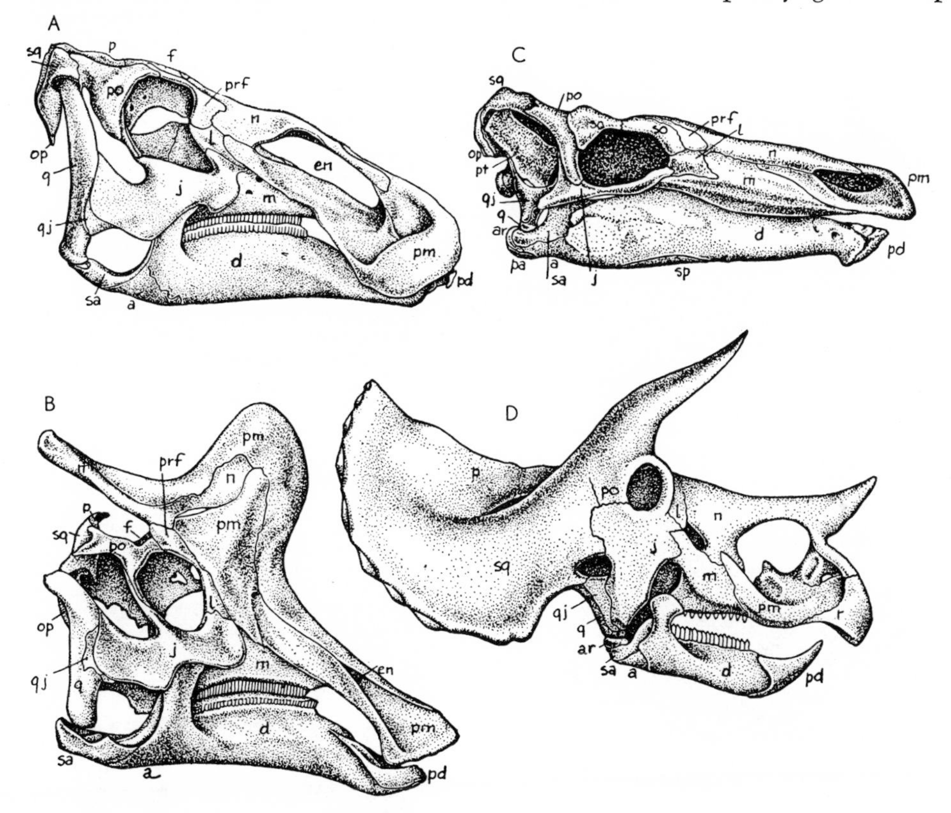

FIG. 242.—The skull in various ornithischian dinosaurs. *A*, *Edmontosaurus*, an Upper Cretaceous duckbilled ornithopod; length of skull about 3.5 feet. *B*, *Lambeosaurus*, a crested duckbill; skull length about 32 inches. *C*, *Stegosaurus;* skull length about 16 inches. *D*, *Triceratops*, an Upper Cretaceous horned dinosaur; skull length (including frill) about 5.67 feet. For abbreviations see Figure 109. (*A*, *B* after Lambe; *C* after Gilmore; *D* after Hatcher, Marsh, Lull.)

Mesozoic plants by modern types of vegetation to which they were not well adapted.

Stegosaurs.—Since the front legs were probably used to some extent in locomotion in most ornithopods, it would seem that reversion to a four-footed pose might readily occur. Such was the case in many ornithischian forms. But these slow-moving quadrupeds, left thus at the mercy of the carnivores, universally acquired some type of protective device in the way of armor or horns.

Early in appearance among quadrupedal ornithischians were members of the suborder Stegosauria, of which *Stegosaurus* (Figs. 242*C*, 246), of the late Jurassic Morrison beds, is the most familiar. This was a fairly large dinosaur, with a length of 20 feet or more. The skull was small, the front legs short, and the back arched high over the long hind limbs. The most conspicuous peculiarity of this quadruped lay in the series of plates and spines arranged in a double alternating row down the entire length of the neck, trunk, and tail. For the most part these structures were flattened plates, which projected nearly vertically above the back. They were roughly triangular in outline, with thickened bases which were presumably tied into the skeleton by tough ligaments. The largest plates lay above the hips, the size decreasing from this point fore and aft. The tip of the tail bore two pairs of long spikes. These plates and spines would seem to have afforded protection against an attack aimed from above on the backbone and spinal cord. But there is little indication of any armor over the remainder of the body, and it would seem that *Stegosaurus* might have been easily crippled by a flank attack.

The vertebrae were still somewhat amphicoelous or, at the most, had flat-ended centra. In the pelvis (Fig. 227*E*) the recurved ilium extended far forward; the pubis and ischium were both quite flat and broad, and the anterior process of the pubis was well developed. The hind legs were exceedingly long, with a columnar build, giving the creature a very high hip region. The front legs, on the contrary, were very short, seemingly an indication of the bipedal ancestry of the stegosaurs.

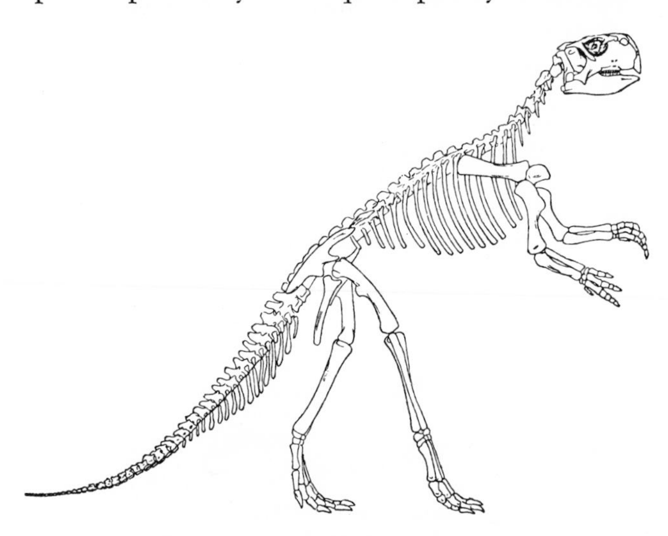

FIG. 243.—The skeleton of *Psittacosaurus*, a beaked Cretaceous ornithopod, apparently related to ceratopsian ancestry; length about 4 feet. (After Osborn.)

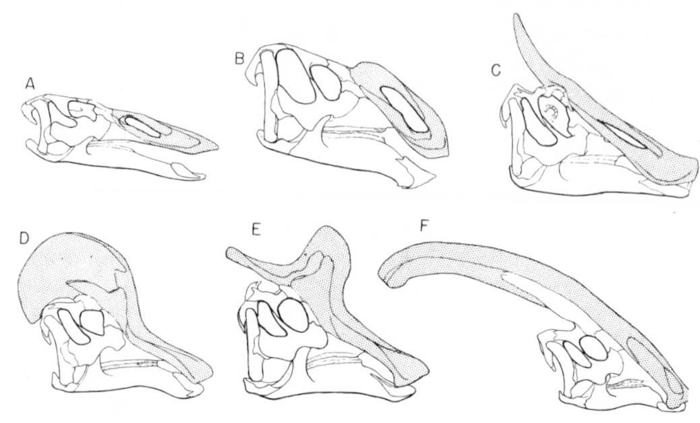

FIG. 245.—Diagrammatic views of the skulls of hadrosaurs to show variations in crest development. Premaxillary and nasal bones shaded. *A, Anatosaurus; B, Kritosaurus; C, Saurolophus; D, Corythosaurus; E, Lambeosaurus; F, Parasaurolophus.* (Data from Lull and Wright.)

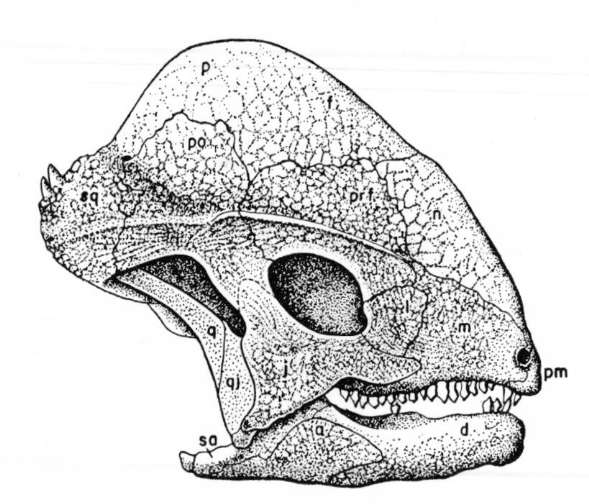

FIG. 244.—Lateral view of the skull of *Stegoceras* ["*Troödon*"], a Cretaceous ornithopod with a greatly thickened skull roof; length about 7.5 inches. For abbreviations see Figure 109. (After Brown and Schlaikjer.)

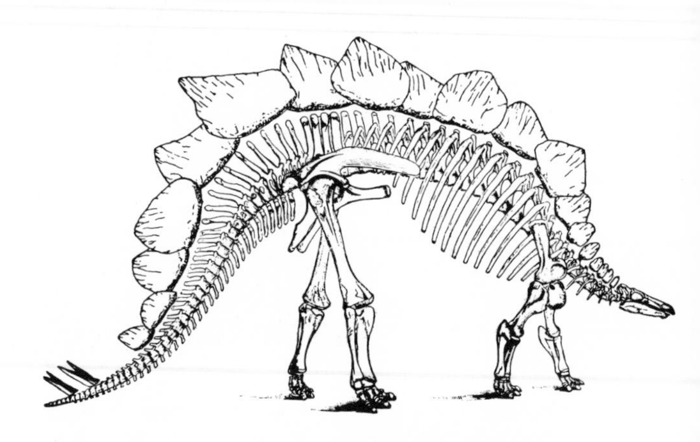

FIG. 246.—*Stegosaurus*, a Jurassic armored dinosaur about 18 feet in length. (Modified after Marsh and Gilmore.)

All five digits were present in the manus, but the outer two were reduced; the pes was three-toed. The toes terminated in flattened, hooflike structures.

The skull was very small and long but low; here, as in the sauropods, the brain was exceedingly tiny and was vastly exceeded in size by a swelling of the spinal cord in the sacral region. There was but a single row of about two dozen small teeth in each jaw half.

Stegosaurus-like forms were widespread in the late Jurassic and early Cretaceous. *Kentrurosaurus* of East Africa was somewhat smaller than *Stegosaurus*. Here plates were present only over the middle of the back and were comparatively small; spines were present not only on the tail but also over the anterior part of the body. It is thought that this animal may still have been somewhat bipedal in habits. A still earlier relative (in fact, one of the oldest known ornithischians of any sort) was *Scelidosaurus* of the Lias (Lower Jurassic) of England, a reptile with a length of about 13 feet.

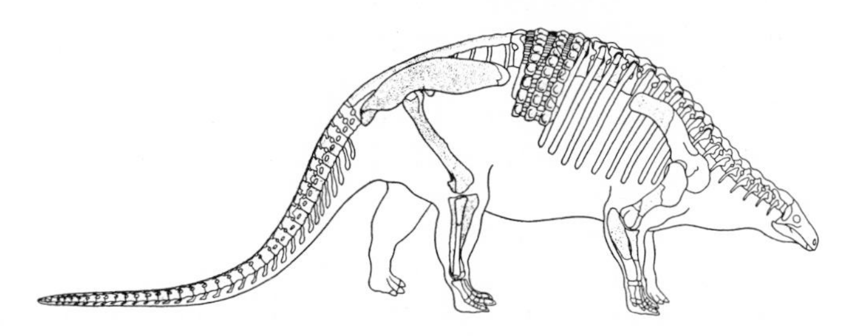

FIG. 247.—*Nodosaurus*, a Cretaceous armored dinosaur, 17.5 feet in length; a few segments of the armor are indicated. (From Lull.)

Armor plates were present, but their exact arrangement is uncertain. Probably there were two rows of oval plates with longitudinal keels, which in the shoulder region appear to have developed into longer, spiny structures. The potentiality for the development of such a type of armor was inherent, apparently, in the ancestral ornithischians; dorsal dermal plates were present in thecodonts, and we have noted their presence in the primitive ornithopod *Hypsilophodon*.

Ankylosaurs.—An entirely different group of armored forms, the Ankylosauria, succeeded the stegosaurs in the Cretaceous. These dinosaurs had an armor of quite a different nature and probably originated at a much later time from the primitive ornithopod stock. *Euoplocephalus* [*Ankylosaurus*], *Panoplosaurus* (Fig. 248*C*), *Edmontonia* (Fig. 248*D*), and *Nodosaurus* (Fig. 247) of the Upper Cretaceous are among the better-known members of the group. They have been termed, not inappropriately, "reptilian tanks," for, with the exception of the turtles, they were the best armored of any reptiles and show many analogies to the shelled glyptodonts of the Age of Mammals.

The body was broad and flattened, the proportions not unlike those of a "horned toad." The entire back was covered by a tough mosaic of larger and smaller bony plates which formed a seemingly efficient carapace. The front legs, which projected somewhat at the sides, were afforded protection by long spines extending outward from the shoulder region; the tail was encased in rings of bone and sometimes armed with long, bony spikes.

The skull was large and broad but rather short. The temporal openings had been closed over, and the flat skull roof so formed was further reinforced by an extra layer of polygonal bony plates. The dentition was weak, and teeth were seemingly entirely absent in some cases. The limbs were short but stout. In the pelvic region (Fig. 227*F*) the ilium flared out widely

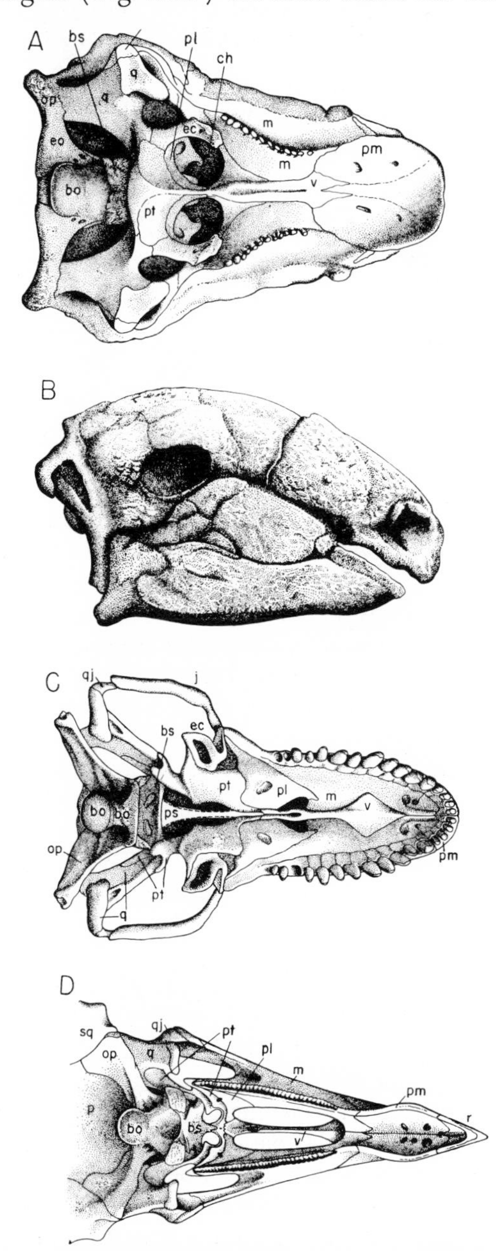

FIG. 248.—*A*, Palate of the nodosaur *Edmontonia*, length about 16 inches (after Russell); *B*, lateral view of the skull of the nodosaur *Panoplosaurus*, length about 15 inches, the skull bones covered by encrusting bony plates (after Lambe); *C*, palate of *Tyrannosaurus* (after Osborn); *D*, palate of *Triceratops*. (After Hatcher, Marsh, Lull.)

over the hip region and, reinforced by bony plates on its upper surface, effectively sheltered the thighs. The pubis had become reduced to form little more than a part of the acetabulum.

In both the Lower and the Upper Cretaceous of Europe have been found somewhat smaller armored types, such as *Acanthopholis* and *Struthiosaurus*, which seem to be rather more primitive members of the same group. There have been found, in connection with them, numerous plates and spines indicating the presence of a carapace somewhat similar to that of the

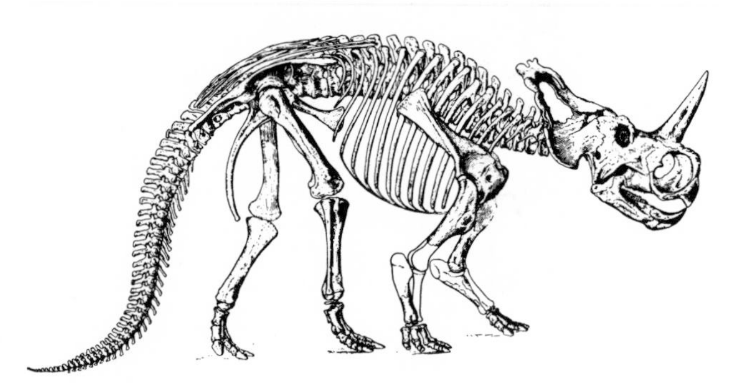

FIG. 249.—*Centrosaurus*, a ceratopsian dinosaur; about 17 feet in length. (From Brown.)

large forms described above. The skull was relatively small and primitive in nature, without the broad shape of typical ankylosaurs; and temporal openings, although small, were still present.

Horned dinosaurs.—Last of the ornithischian groups in time of appearance were the horned dinosaurs, the suborder Ceratopsia. The entire history of this group is confined, so far as known, to the Upper Cretaceous. Most forms are from North America, a few finds have been made in eastern Asia, and there is a possible South American ceratopsian. *Triceratops* (Figs. 242*D*, 248*A*) and *Centrosaurus* (Figs. 227*D*, 249) are representative forms. These were quadrupedal dinosaurs of moderate size—*Triceratops*, for example, ranging from about 16 to 20 feet in length.

The main point of interest in these forms is the cranial structure. The head appears to be exceedingly large, making up a third or so of the total length of the body. But half of this structure is really not part of the true head region but a great frill of bone formed by extensions of the parietals and squamosals which extended back over the neck nearly to the shoulders. This frill obviously must have afforded considerable protection to the neck region, a favorite point of attack by carnivores. In some, as in *Triceratops*, this frill is a solid plate of bone; in other cases there is found a probably more primitive condition in the presence of large openings on either side.

A second curious feature of the skull lay in the development of horns. The bony horn cores are not dissimilar in appearance to those of a modern bison; and, indeed, the first-discovered specimen of ceratopsian horns was ascribed to that ruminant! In *Triceratops* there was a pair of large horns on the postorbital bones over the orbits and a median horn over the nasal region. The degree of development of these horns varied considerably. In *Centrosaurus*, for example, the nasal horn is very large and the brow horns undeveloped; in other ceratopsians, such as *Torosaurus*, the brow horns, in contrast, are large and the nasal horn little developed.

The remainder of the skull was much modified in correlation with the development of these defensive structures. The temporal openings had been reduced almost to the vanishing point. The nasal region of the skull was greatly enlarged, and the powerful beak was formed by a newly developed rostral bone comparable with the predentary in the lower jaw. Only a single row of teeth was present in the cheek region.

The vertebrae were platycoelous. The neck was short. In the pelvis (Fig. 227*D*) the main body of the pubis had been reduced to a short spike, although the anterior process of that bone was well developed. The hind legs were, as in all ornithischian groups, much longer than the front and terminated in four stubby, hoof-capped toes (Fig. 241*F*). (The inner toe appears to have been reduced in some cases.) All the digits were present in the manus, but the outer two were reduced (Fig. 241*C*).

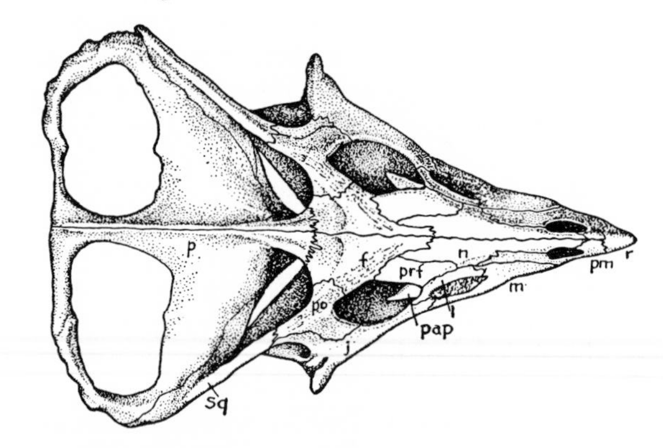

FIG. 250.—Dorsal view of skull of *Protoceratops*, a small primitive ceratopsian dinosaur from Mongolia. For abbreviations see Figure 109. (After Brown and Schlaikjer.)

Apart from variations in horns and neck frills, most ceratopsians adhere closely to a common pattern. The one really divergent form is *Pachyrhinosaurus*, in which, instead of horns, an enormous median craterlike structure—perhaps filled in life by a horny "bumper"—is present on the broad forehead.

A small primitive ceratopsian is *Protoceratops* of the Upper Cretaceous of Mongolia (Fig. 250), of which there have been found numerous specimens ranging from very young individuals to adults, as well as a number of "nests" of eggs. *Protoceratops* belies its name, for horns are not present; the nasal region, however, is elevated as a potential horn-core area, and rugosities are present in some individuals in the areas in which the paired horns later developed. In contrast, the frill was well developed, although fenestrated. The structure of the frill in this form suggests that it originated primarily as a pulling-out of the back margin of the temporal region for the better accommodation of the jaw muscles and that its function as a neck protection was secondary. *Protoceratops* exhibits a primitive feature in the persistence of teeth in the premaxilla. No earlier stages in ceratopsian history are definitely known; but, as noted before, *Psittacosaurus* and *Protiguanodon* are quite possibly ancestral types.

Numerous as were the horned dinosaurs in the late Cretaceous, they disappeared as completely at the close of the period as did the other dinosaur groups. Except for the primitive palaeopods and the stegosaurs, all of the major types of dinosaurs were still in existence in the closing phases of the Mesozoic; but by the beginning of the Cenozoic all had vanished. The reign of the dinosaurs was over; the Age of Reptiles was at an end.

13

CHAPTER

Birds

Although the birds are grouped as a separate vertebrate class—Aves—they are, apart from the power of flight and features connected with it, structurally similar to reptiles. Indeed, they are so close to the archosaurians that we are tempted to include them in that group, and we may well discuss them at this point.

Feathers.—Birds have been called by an old writer "glorified reptiles." Feathers are in reality almost the only distinctive feature of the class, for almost every other character can be matched in some archosaurian group. Large quills form the expanse of the wing, taking rise from the back of the forearm and from the reduced fingers which form the distal part of the wing support. On the tail (which, as a bony structure, is very short) is set a spreading fan of stout feathers used as a rudder. The rest of the body is covered with a thick overlapping set of smaller, softer feathers, which form a very efficient insulation for the control of the bird's body heat.

But, although the feathery covering of birds is in contrast with the horny scales which normally cover a reptile body, the difference is, in reality, not so great as it seems; feathers are comparable to such scales, although with a complex structure of barbs and barbules instead of the simple scale shape.

Flight adaptations.—Unlike ordinary reptiles, birds are warm-blooded; maintenance of a high body temperature is a necessity, for flight requires a great energy output over a long period. Also connected with this necessity, which demands a large oxygen supply and an efficient circulatory system, is the fact that the heart is divided completely into four chambers; those receiving fresh blood from the lungs and leading it to the body are completely separated from those taking used blood from the body to the lungs. This is a condition also found in mammals, but in birds the great arch of the aorta, which carries the blood from the heart to the body, passes over and down the right side of the chest, rather than the left, as in ourselves.

The nesting habit, contrasting with the usual lack of care of reptilian eggs, seems also to have been developed in connection with flight. The eggs must be kept warm for the maintenance of body temperature in the embryo. Further, while a young reptile makes its own way in the world from birth, the need for protection of the young until they have matured enough to undertake the complicated business of flight seems to have rendered essential the care and feeding of the fledglings in the nest. Brain and sense organs are much modified in relation to flight. Birds depend for their main contact with the world upon sight rather than smell (in contrast to reptiles and most mammals). The eyes are large, and sclerotic plates are commonly developed in the eyeball. The bird brain is vastly enlarged compared with that of a reptile; but this enlargement, unlike that of mammals, is not related to expansion of the gray matter of the cerebral cortex which is the major seat of "intelligence," but to the expansion of centers related to sight, the delicate muscular coordinations necessary for flight, and with complex innate behavior patterns.

Skeleton.—In the skeleton (Fig. 251) there are many modifications connected with flight. With the useful result of lessening the specific gravity, there are not only air sacs connected with the lungs within the body, but also air-filled cavities within many of the skeletal elements, including portions of the skull, neck vertebrae, humerus, and femur.

Wings.—The anterior limb as the actual support of the flying organs has been much modified. We find, on either side, a long, slim, backward-slanting scapula and a single coracoid attached to the edge of the sternum. There are slender clavicles which usually fuse in the midline to form the wishbone. The sternum is a large plate, usually with a great keel in the middle in flying birds; and to it are attached the powerful chest muscles which exert the main propulsive pull on the humerus during flight (these muscles form the white meat of the chicken).

The humerus is short and stout in most modern flying birds, with a heavy process near the head for the attachment of the chest muscles. Both bones of the lower arm are well developed and rather long; but the ulna, which carries important wing feathers at its back, is the stronger. There are four carpals, the two distal ones fused with the metacarpals. We find a reduction in the hand rather similar to that in dinosaurs, for only three fingers are represented. Despite some conflicting embryological evidence, these appear to be (as in some dinosaurs) the inner three; the fourth and fifth have vanished completely. Even the three remaining are not complete. The pollex is short and usually has but one phalanx. The three metacarpals are fused with one another; the second finger usually has two free phalanges, the third but one. Almost never do these fingers bear claws; they are buried in the flesh and function only as supports for the feathers.

Backbone.—As was presumably the case with its archosaurian ancestors, the modern bird walks in a semierect position. The free vertebrae usually have a complicated but easily movable saddle-shaped set of centrum articulations in addition to the zygapophyses. The neck includes a rather variable number of elongate and freely movable vertebrae. The dorsal region, however, is very short, with only about six to ten vertebrae; and of these, several in front are often fused, as in some pterosaurs, with the effect of giving a stronger support to the muscles running to the wings. A number of posterior dorsals and the proximal caudals as well are joined to the original sacral vertebrae to make up an elongate synsacrum. The tail is short; following a half-dozen small free vertebrae, the few remaining ones are fused in modern flying birds into the pygostyle, which forms a support for the spreading tail feathers.

Legs.—The hind legs are reminiscent of the dinosaurs in their structure. The ilium is very much elongated and firmly bound to the synsacrum. As in dinosaurs, the acetabulum is perforated. The pubis passes down and backward, as in the ornithischian dinosaurs; there is, however, no forward branch, although this may be represented by a small process. The slim ischium passes down and back, parallel to the pubis. There is usually no ventral union of the bones of the two sides, but the ischium is typically braced by an upward extension which meets the posterior end of the ilium, and pubis and ischium may join one another distally. The femur is short and stout, but the tibia is elongated, while the fibula is always reduced and attached to the tibia and may disappear except for its proximal end. As in the dinosaurs, the main joint with the foot lies in the middle of the tarsus; the proximal elements are fused with the tibia, the distal ones with the metatarsals. The foot, too, is essentially like that of many dinosaurs. The three middle toes are generally well developed; the fifth has gone without leaving a trace; and the first, often reduced or absent, has only an incomplete metatarsal and most commonly is turned to the rear, where it may aid in clutching a perch. The three principal metatarsals are firmly fused into a single element with separate distal ends for the toes. As in dinosaurs, the digits, with few exceptions, have the original count of 3, 4, and 5 phalanges, although the central toe is commonly the longest.

Skull.—The distinctly unreptilian appearance of the bird skull (Figs. 252, 255) is due principally to the large orbits and large braincase; its structure is one easily derivable from the archosaurian type. Most of the sutures between bones become obliterated early in

life, making the identification of elements difficult. The condyle, as in reptiles, is single; and there is no pineal opening. The orbits are very large and generally incompletely surrounded by bone. In front of the orbits (again as in dinosaurs) is an antorbital opening; postfrontal and postorbital are lacking. Behind the orbit is a single temporal opening, which, however, is presumably the two diapsid openings fused into one by loss of the bar separating them. Because of the incomplete nature of the circumorbital ring, the temporal opening usually communicates freely with the orbit.

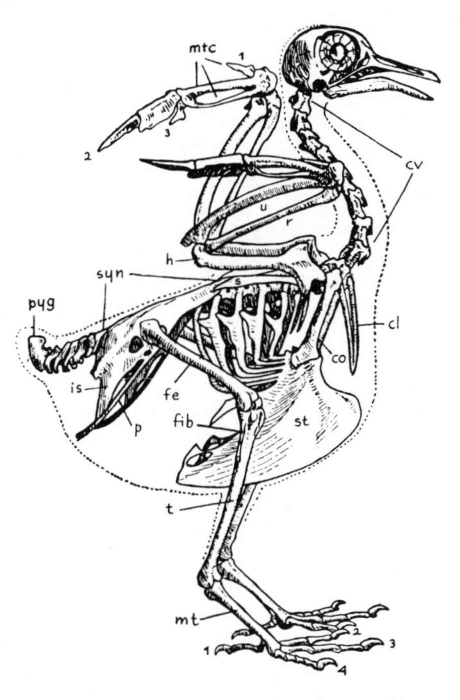

Fig. 251.—The skeleton of a modern bird. Abbreviations: *cl*, clavicle; *co*, coracoid; *cv*, cervical vertebrae; *fe*, femur; *fib*, fibula; *h*, humerus; *il*, ilium; *is*, ischium; *mt*, metatarsus (including distal tarsals); *mtc*, metacarpals; *p*, pubis; *pyg*, pygostyle; *r*, radius; *s*, scapula; *st*, sternum; *syn*, synsacrum; *t*, tibia; *u*, ulna. Digits of manus and pes numbered. (Modified from Gerhard Heilmann, *The Origin of Birds* [copyright, New York: D. Appleton & Co.], used by permission.)

The jugal and quadratojugal pass forward as a slim bar from the quadrate; this last element is freely movable on the skull, having loose joints with squamosal, pterygoid, and quadratojugal.

The palatal structure (Figs. 255, 256) is easily derivable from the archosaurian type. There is never any secondary palate, but the internal nares (and the external ones as well) are usually well back in the skull. Palatines and parasphenoid are present; but there are no ectopterygoids. In most birds the vomers (fused) are small or absent, the pterygoids short; the palatines extend back to a suture with the base of the braincase and have movable articulations with the pterygoids. In the lower jaw all six typical reptilian bony elements are present; and, as in most archosaurians, there is an external opening between angular and surangular. Modern birds are, of course, toothless, and the beak is covered with a horny bill.

Archaeopteryx (*Figs. 252, 253*).—Modern birds show many reptilian features; but even closer to the archosaurians was the earliest known bird, *Archaeopteryx,*

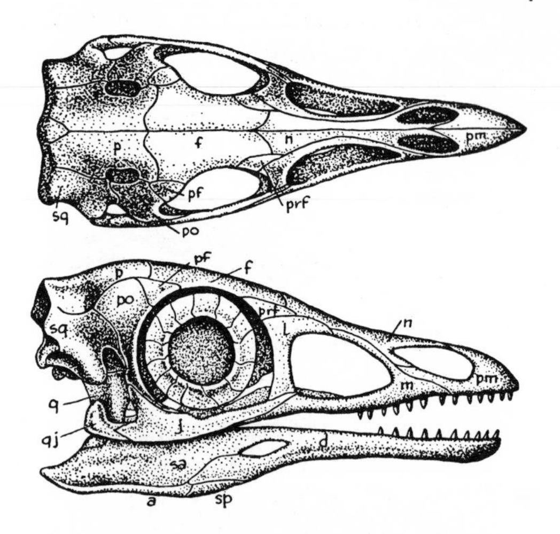

Fig. 252.—Attempted reconstruction of the skull of the Jurassic bird *Archaeopteryx;* length of original about 2 inches. For abbreviations see Figure 109. (After Heilmann, *The Origin of Birds* [copyright, New York: D. Appleton & Co.], used by permission.)

of which three skeletons have been found in the late Jurassic lithographic stone of Germany. The remains so closely resemble those of some of the smaller bipedal dinosaurs that they might well have been taken for reptiles were it not for the impressions of feathers which surround them on the stone slabs on which they are preserved. The skull, as far as it can be seen, was already rather birdlike, with an expanded braincase; and, as in modern birds, the sutures were mostly closed. However, well-developed teeth, implanted in sockets, were present in the jaws. The backbone was quite simple in structure, for the centra had the primitive amphicoelous character, and all were free as far as the sacrum, which included only five or six vertebrae. The tail still had the long structure of the typical dinosaur, and the feathers were arranged in two rows along its sides. As would be expected, the hind legs were quite similar to those of archosaurs, for they are but little changed even in modern birds. The wings had not yet completed their transformation. The sternum was not preserved but must have been small and could not have afforded much support for pectoral muscles (incidentally, ventral ribs were still present). The front limb was still a hand rather than a wing, for the three fingers which it possessed were very similar to those of the carnivorous dinosaurs; the metacarpals were separate, the number of joints was complete, and each finger was clawed. It would hardly be suspected that this appendage was used for flight, had not the impression of wing feathers been found back of the ulna and manus. The flying powers, however, must have been slight, for the spread of the wing was much less than that of rather poor fliers, such as the pheasant, among modern birds. These early forms probably were forest types and did little more than plane from tree to tree or to the ground.

Archaeopteryx was already definitely a bird, but was still very close to the archosaurian reptiles in most structures and was obviously descended from that group. The pelvis is suggestive of relationship to ornithischians, but on the other hand, the limb structure parallels that of carnivorous saurischians in many regards. Surely the birds arose, independently of either dinosaur stock, from Triassic thecodonts, and there is little in the structure of some of the small bipedal thecodonts to debar them from being avian ancestors.

There has been much discussion as to the origin of flight. The most generally held theory is that the ancestral bird was a tree dweller and that flight started as a slight parachute effect as the "proavis" jumped from

Fig. 253.—*Archaeopteryx,* an archaic Jurassic bird. *Left,* Specimen as preserved, about 1/8 natural size. *Right,* Restoration. (*Left* after Evans; *right* after Heilmann, *The Origin of Birds* [copyright, New York: D. Appleton & Co.], used by permission.)

branch to branch, the developing wings breaking the fall on landing. A second theory is that the ancestors were ground types and that the feathered arms and tail helped increase running speed by acting as planes.

Archaeopteryx was obviously sharply marked off from later types by several features which are not found in birds of later geological times. These include the absence of pneumaticity, the primitive reptilian structure of the wing bones, and especially the long reptilian tail. These place *Archaeopteryx* in such sharp contrast to all later forms that it is regarded as representing a distinct subclass, Archaeornithes.

Hesperornis.—All other birds may be included in the subclass Neornithes, characterized by such features as a reduced tail with a fan of feathers, a well-developed sternum, usually with a good keel, and the reduction and fusion of the metacarpals to the typical bird condition.

Almost nothing is known of bird life in the early Cretaceous. The chalk beds (particularly the American Niobrara chalk) give us, however, a glimpse of bird life in the later part of that period. But this glimpse is a biased one, for the remains from such marine deposits are, of course, restricted to water-loving types. Most Upper Cretaceous remains are fragmentary but are sufficiently varied to suggest that by this time bird evolution was already far advanced and that a good fraction, at least, of modern bird orders was already in existence. However, the best known of Cretaceous birds is a very distinct type, *Hesperornis* (Figs, 254, 255), which uniquely combines primitive and special-

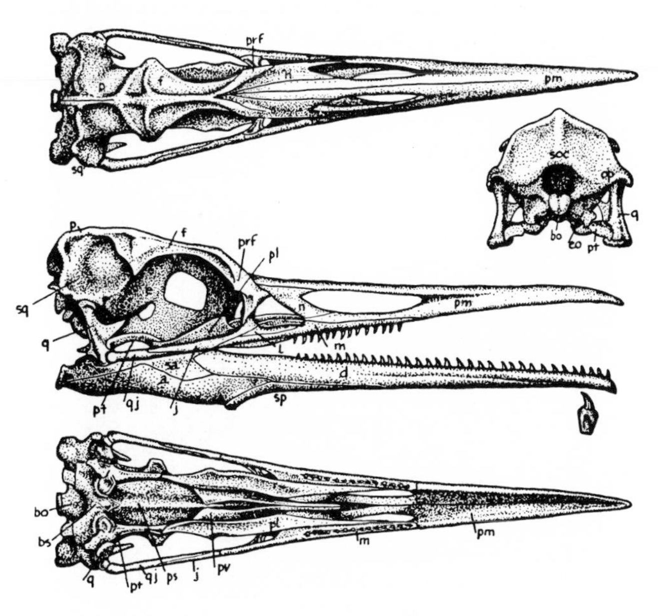

FIG. 255.—Restoration of the skull of the Cretaceous toothed bird *Hesperornis.* Dorsal, lateral, ventral, and occipital views; length of original about 10 inches. For abbreviations see Figure 109. (After Heilmann, *The Origin of Birds* [copyright, New York: D. Appleton & Co.], used by permission.)

FIG. 254.—Cretaceous birds. *Above, Ichthyornis,* a small form with well-developed wings, about 8 inches in height. It is uncertain whether the toothed jaws belong to this bird. *Below, Hesperornis,* a flightless diver, about 1/12 natural size. (The feet should be directed laterally.) (From Marsh.)

ized characters. *Hesperornis* was persistently primitive in that teeth were still present, placed in grooves in upper and lower jaws. However, there were no teeth toward the front of the mouth, and probably the horny beak characteristic of all more advanced birds was already in process of formation.

Hesperornis was probably much like the modern loon in habits, for it seems to have been a diving bird with powerful hind legs which spread sidewise at the ankles to give a vigorous swimming stroke. *Hesperornis* had degenerated in connection with its water life, for the wings had been almost completely lost, and of the arm bones there remained only a slender humerus, while the sternum was unkeeled. This is the first of the many types of birds which have taken up a water life, but there is no reason to believe that *Hesperornis* was an actual ancestor of later aquatic forms.

Palaeognathous birds.—For most of the rest of this chapter we shall review the various orders of typical "modern" birds. Before doing so, however, we may discuss the interesting but puzzling series of flightless forms commonly termed the ratites, represented today by the ostrich, rhea, cassowary, emu, and kiwi, and including a large number of interesting fossil types. These forms, here grouped together with the tinamous as a superorder, Palaeognathae, have a number of common features. Most of them, however, are associated with the fact that they have lost the power of flight. The wing skeleton in flightless forms is naturally reduced, although there are often vestiges of all three "fingers"; the sternum, with weak pectoral musculature, lacks a keel; and the hind legs are powerful running organs. Since the tail is not used for flight, there

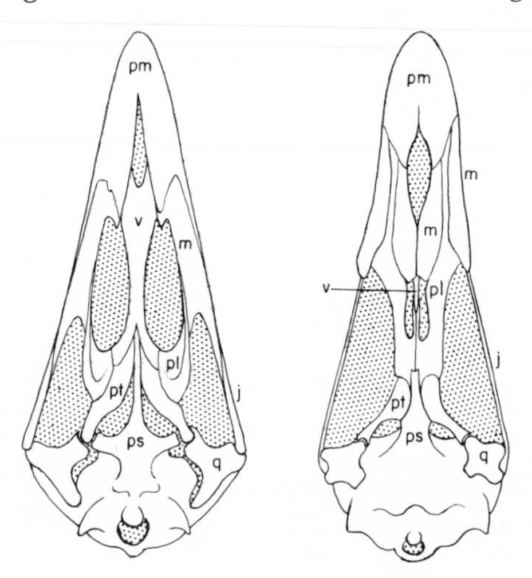

FIG. 256.—*Left,* A typical palaeognathous palate (*Dromaius*). *Right,* A "neognathous" type (*Anas*).

is usually no pygostyle for tail-feather support. The feathers, in correlation with the loss of flight and consequent need for a streamlined body surface, are usually soft and curly. But some presumably primitive features are found in the lack of the distal fusion of pubis and ischium found in most modern birds, and there is, most particularly, a special type of palatal structure termed palaeognathous (Fig. 256). Bird palatal structures vary considerably, but there is usually considerable flexibility and freedom of movement between the various units in palate and upper jaw; for example, as mentioned earlier, the palatines (attached laterally to the maxillae) usually move freely on the pterygoids, and the vomers are generally reduced or absent. In the ratites and tinamous there is, in contrast, a much firmer union of elements; the single median vomer is a long bone (the ostrich is an exception) which is bound firmly to the pterygoids posteriorly, and the palatines likewise are solidly connected with the pterygoids.

Perhaps representing the central stock from which the ostrich-like birds have arisen are the tinamous (Tinamiformes) living in South America today and almost unknown as fossils. These forms have the appearance of quail or grouse but are quite different from those game birds in their structure. They share with the ratites the palaeognathous type of palate. They are essentially ground dwellers and take to the air only in an emergency. They are exceedingly poor fliers, having little control over their flight and seldom going more than a hundred yards or so. Tail feathers are reduced or absent, but, unlike other ostrich-like birds, there is a keeled sternum.

From such a stock may have come the remaining, flightless members of the superorder. Most familiar are the ostriches (*Struthio*) of Africa and Arabia, the largest of living birds. A distinguishing feature of the ostrich, in contrast to other flightless types, is the fact that there are only two toes—the third and fourth. On the South American pampas the place of the ostrich is taken by the *Rhea,* which has a somewhat different structure (including a three-toed foot) and is smaller. In Australia and New Guinea are the emus (*Dromaius*) and cassowaries (*Casuarius*). Among their distinguishing features is the fact that the wing is more reduced than in either of the last two types, there being only one digit preserved, and this projecting but little from the body.

Still more interesting were certain fossil types which were once present in two island regions. In Madagascar lived a number of species of "elephant-birds," *Aepyornis,* some of them as large as an ostrich, with greatly reduced wings and heavy legs. There have been found numerous eggshells, some with a capacity of two gallons. It is possible that these large birds lived until quite recent times, and their extermination may have been due to man.

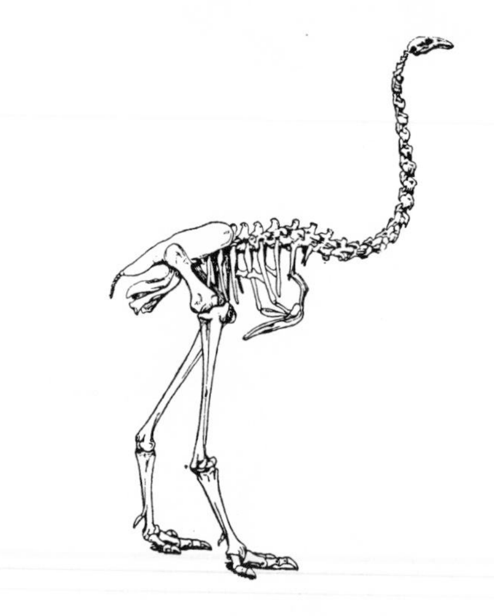

FIG. 257.—*Dinornis maximus,* one of the New Zealand moas; height about 7 feet. (After Andrews.)

Quite different from the ostrich-like forms in size and general appearance, but also palaeognathous, are the kiwis (*Apteryx*) of New Zealand. These small birds are also practically wingless ground types with very long bills and small eyes. New Zealand is remarkable today for flightless birds, for there are found not only the kiwi but numerous flightless rails and other ground dwellers belonging to higher bird groups. In the past there were much larger ratites, the moas (*Dinornis,* etc.) (Fig. 257), which ranged in size from species no larger than a turkey to creatures 10 or 11 feet in height. Numerous remains of these birds have been found, and even feathers have been preserved. There is proof from native camp sites that these forms were still alive when man reached the island and that they were used for food.

Origin of flightless types.—It has been suggested that these various groups of flightless birds are really primitive forms that have descended from types in which flight had never been developed. But upon reflection it is obvious that such a theory is highly improbable. Even in the Cretaceous, *Ichthyornis* had already developed powerful wings, and such a form as *Hesperornis* was obviously specialized rather than primitive. Further, we know that in many other higher groups of birds certain types have also lost the power of flight; the dodo, the great auk, and the flightless rail of New Zealand are all types which have reduced wings and cannot fly but yet are surely descended from flying forms.

The clue to the matter seems to be that, for many birds which seek their food on the ground, flight is necessary mainly as a protection from enemies. With freedom from carnivores, there is relatively small reason why flight should be maintained. Almost all these flightless forms are practically free from enemies. There are no carnivorous mammals in New Zealand; until the arrival of man there was no ground-living enemy of any kind for the moas and kiwi. In Australia the only carnivores before the arrival of man and his dog were comparatively harmless primitive mammalian types (marsupials). The island of Madagascar is almost devoid of carnivores. The ostrich and rhea are in a worse situation, for they live in continental regions where carnivores abound. But they inhabit open country where carnivores cannot approach without being seen, and the speed of these forms enables them to outrun their enemies. Further, no placental carnivores entered South America until late geological times, and it may be that the ostriches developed in Africa before the days of "modernized" carnivores. It would thus seem that birds are likely to return to a ground life wherever conditions permit. When flight is thus abandoned, most of the common adaptations seen in these forms, such as reduction of wings and tail and development of powerful legs, would naturally follow.

Most of the resemblances between the various ostrich-like types need not, then, really indicate relationship; and below are mentioned other large birds, now extinct, which in the past evolved along similar lines.

But the palatal structure (although sometimes questioned) does, it is generally agreed, give firm, positive evidence that these forms are really related and members of a common group. A probable suggestion is that there existed in the Tertiary a group of birds with this palatal type, which originally had complete wings and a keeled sternum but were comparatively poor fliers. The living tinamou may be a relatively unmodified relict of this primitive type. In parts of the world where favorable conditions existed, these forms might have tended to stay on the ground, lose the power of flight, and develop into ostrich-like forms; in other regions they would have lost out in competition with better flying types and disappeared.

The "modern" bird orders.—Apart from the ratites and tinamous, all existing birds are customarily grouped, in contrast to these palaeognathous forms, as the superorder Neognathae. (The names are not too well chosen, since the terms imply—highly improbably—that the specialized ratite palate is truly "ancient" and all other types "new.") The remainder of this chapter will be devoted to a discussion, *seriatim,* of the long series of a score or so orders in which the "modern" bird orders are arrayed, with mention of the more interesting fossil members of each group. The sequence followed is that usual in ornithological works. The first part of the series is mainly composed of various types which are aquatic or frequent shores or bodies of water, then various field and forest types, and finally the perching and song birds which make up a large fraction of the familiar bird world. This sequence does not necessarily imply that bird evolution progressed from water-seeking forms to inland dwellers, although it is true that the orders in the first part of the sequence do appear to show some relatively primitive traits.

Apart from *Archaeopteryx,* there are no identifiable Jurassic bird remains, and fossil birds remain rare in the Cretaceous. We have already mentioned the curious diver *Hesperornis* of that age; except for one further form, other Cretaceous bird remains are fragmentary. It was once believed that all Cretaceous birds were toothed, as is *Hesperornis.* Currently it is suspected that this is not the case, and that certain, at least, of the Cretaceous forms may be forerunners of modern groups; but the material is so poor and sparse that we can as yet make little of it.

Bird remains become more plentiful when we reach the Tertiary. But even here, it must be confessed, our knowledge is none too good. Birds are mostly small forms, and their bones and skulls are very fragile; consequently, we have but few remains, mostly of water birds. With many groups of mammals and reptiles we know more fossil forms than living ones; but among birds, against about twenty thousand living species, we have but four or five hundred fossil forms. Even these are, for the most part, known only from a fragmentary bone or so. This leads to great difficulties in attempting to work out the paleontological history of the group, for the birds of today, despite their varied plumage, songs, and habits, are very similar to one another in their structure. They are divided into many orders; but the differences, for example, between a humming bird and an albatross are much less than those between a seal and a cat, or between a stegosaur and a duckbilled dinosaur—forms which are commonly placed in a single order among mammals or reptiles. The different bird orders have, in general, no more differences between them than exist between families in other classes of vertebrates, and, anatomically, differences between bird genera are often so slight that fossils are very hard to place.

Despite these difficulties, it seems probable that we can trace many of these orders, in a broad way, well back in geologic times. Of the fossil history of two minor orders (Coliiformes, Caprimulgiformes), we know little or nothing, and the doves (Columbiformes) and parrots (Psittaciformes) are not recognizable before the Miocene. But for almost every other group, representatives have been recognized as early as the Eocene. Thanks to the fact that our Cretaceous bird remains are mainly from marine deposits, some of the fragments from that age belong to members of five water-loving orders still existing today—the loons (Gaviiformes), grebes (Podicipediformes), pelicans (Pelecaniformes), the group (Ciconiiformes) including waders such as herons and flamingos, and the Charadriiformes, including gulls, terns, and auks.

1. *Order Gaviiformes.*—The loons, *Gavia,* are northern water birds of primitive aspect which are excellent swimmers and divers and in many adaptational features recall the Cretaceous *Hesperornis.* Unlike that genus, however, the loons are good fliers with well-developed wings (used for swimming also). The feet are webbed; in relation to swimming habits, the short legs are set far back on the body in loons (and grebes as well).

2. *Order Podicipediformes.*—The grebes are smaller, cosmopolitan water birds which are, again, good swimmers as well as fliers. The feet, however, are not webbed; each toe has instead a separate broad flap of skin.

3. *Order Procellariiformes.*—The albatrosses, shearwaters, and petrels are strong-flying oceanic birds, with very long, narrow, and powerful wings; the feet, with reduced halluces, are webbed, but swimming is of secondary importance; peculiar tubular nostrils are a diagnostic feature; there is a hooked horny beak.

4. *Order Sphenisciformes.*—The penguins, although primitive in many respects, are in many others among the most highly specialized of birds. The wings are totally incapable of functioning in flight but are powerful flippers used in swimming. The wing bones are flattened and fused into a compact, powerful fin. The foot, too, is modified for swimming; it is webbed, and the metatarsal bones are but partially fused, in contrast to their intimate union in typical modern birds.

The penguins appear to have been derived from marine flying birds similar to those of the last order, which, like the diving petrels, acquired the habit, when in pursuit of food, of using the wings for "flying" under water as well as in the air. Presumably in the penguin group submarine swimming became more important than flying; flight adaptations were abandoned and the wings perfected as swimming organs. The evolution of penguins appears to have taken place in the temperate zone of the Southern Hemisphere, which is also the center of distribution of the related Procellariiformes. Some penguins are found in Antarctic waters; others have even followed a cold current northward to the Galapagos, but none has crossed the tropics to colder northern regions, where the (unrelated) auks and their allies fill a somewhat similar ecologic niche. Numerous fossil penguins, some as large as a man, have been reported from early and mid-Tertiary deposits in Patagonia and New Zealand, and on Seymour Island off the shores of the Antarctic continent.

5. *Order Pelecaniformes.*—Tropic birds, pelicans, gannets, cormorants, and frigate birds are characteristic members of this group. They are, again, water-loving—mainly marine—fish-eating birds with good powers of flight. There is a long beak; the nostrils are rudimentary, and most have a food pouch in the throat. All four toes are included in the webbed foot. Tertiary marine deposits have yielded fairly abundant remains of fossil forms related to modern members of the order and, in addition, remains of a number of types now extinct. Most interesting are rare specimens of oceanic birds which, in place of long-lost teeth, have developed curious toothlike outgrowths of bone along the jaws. A bird jaw of this unusual type, named *Odontopteryx,* has long been known from the Eocene of England; recently there have been found, in Miocene American deposits on the Pacific shore, remains of a giant bird of this sort, *Osteodontornis,* with an apparent wingspread of 5 yards or so.

6. *Order Ciconiiformes.*—This is a large group of water-loving, mainly tropical, fish-eating waders, with long legs, long necks, and (usually) long bills, and includes the herons, storks, and flamingos. The flamingos have webbed feet, but the others have so far abandoned aquatic habits for the land and for tree perching that the web is absent. Their shore-dwelling habits are perhaps responsible for the fact that the herons (Ardeidae) and storks (Ciconiidae) are common fossil types.

7. *Order Anseriformes.*—The ducks, geese, and swans and their relatives excel at both flying and swimming. The broad flat bill, short but powerful legs with webbed toes, and strong wings are characteristic features of this group, whose members are dominantly residents of northern regions.

8. *Order Falconiformes.*—This group includes the diurnal birds of prey: the vultures of the Western Hemisphere (including the condor and the "turkey buzzard"); the distinct vultures and buzzards of the Old World; the various hawks, falcons, ospreys, and eagles; and, in addition, the aberrant secretary bird of Africa. The powerful hooked beak and equally powerful claws in most members of the order are obvious specializations for seizing and tearing the prey. These forms are quite common as fossils. A Pleistocene vulture from the California tar pits, *Teratornis,* is the largest of all known flying birds.

9. *Order Galliformes.*—This order contains, besides the common fowl, many familiar game birds and related forms, such as pheasants, partridge, grouse, quail,

turkey, and peafowl. The members of this group are, for the most part, predominantly terrestrial in their habits, are not swimmers, and are capable of only short flights. The beak is short, the wings short and rounded, the feet well developed for running and scratching. An interesting member of the order is the hoactzin (*Opisthocomus*) of South America, in which claws are present on the first two digits of the nestling's wing. This adaptation, which permits the young bird to clamber about the tree, appears possibly to be a reversion toward the condition found otherwise only in *Archaeopteryx*.

10. *Order Ralliformes.*—The order includes the cranes, rails, and many presumably allied types, mainly tropical in habitat. Typical members of the group are marsh birds, waders, and heavy fliers, which resemble the herons in their long legs, but differ in technical details of skull construction. The rails (family Rallidae) are common as fossils. Among the less characteristic members of this rather heterogeneous order may be mentioned the ponderous bustards of the Old World and the seriemas or cariamas—long-legged South American ground birds with short wings and limited powers of flight. Fossil cariamas are known as early as the Oligocene in South America.

Included in the same order because of their relationship to the cariamas is a striking group of large South American fossil birds ranging from Oligocene to Pliocene in age, of which *Phororhacos* (Fig. 258) is a representative. This was a long-legged bird as tall as a man, with a powerful beaked skull as large as that of a horse; it was obviously flightless, for the wings were much reduced. The ability of these forms (as well as the rheas) to survive on the ground was rendered possible by the fact that during most of the Tertiary, the placental carnivores, those characteristic enemies of ground birds, were absent from South America.

11. *Order Diatrymiformes.*—Placed here because suspected of possible relationship to the cariamas is a further group of large, extinct flightless birds that are characteristic of the early Eocene of North America and Europe. *Diatryma* (Fig. 258) was approximately 7 feet in height, with much-reduced wings, massive legs, and (as in the phororhacids) a large head with a powerful beak. The presence of this great bird at a time when mammals were, for the most part, of very small size (the contemporary horse was the size of a fox terrier) suggests some interesting possibilities—which never materialized. The great reptiles had died off, and the surface of the earth was open for conquest. As possible successors there were the mammals and the birds. The former succeeded in the conquest, but the appearance of such a form as *Diatryma* shows that the birds were, at the beginning, rivals of the mammals.

12. *Order Charadriiformes.*—This order includes such marine types as gulls and terns, auks, and numerous shore birds, such as the long-legged and long-beaked plovers, sandpipers, and snipe; most are web footed. The auks, excellent swimmers by the use of

their wings, are northern analogues of the penguins of southern waters. The flightless great auk of the North Atlantic is now extinct, but hardly to be considered as a fossil form, having been in existence until a little over a century ago. *Mancalla,* of the Pliocene of the American Pacific coast, was an ancient parallel to the great auk. The gulls and terns, with powerful wings, rival certain of the groups previously mentioned as oceanic fliers.

Fig. 258.—*Left, Diatryma,* an Eocene flightless bird, about 7 feet in height. (From Matthew and Granger.) *Right, Phororhacos,* a giant flightless bird from the Miocene of South America; original about 5 feet in height. (From Andrews.)

13. *Order Ichthyornithiformes.*—Apart from *Hesperornis,* the only Cretaceous bird of which any adequate remains are known is *Ichthyornis* of the Kansas chalk (Fig. 254). This is very different indeed from its contemporary. *Ichthyornis* was about the size of a tern, may have had similar habits, and resembled the terns and gulls in many features. Quite in contrast to *Hesperornis,* the wings were powerfully developed and the sternum well keeled for the reception of powerful wing muscles; *Ichthyornis* was presumably a strong oceanic flying form. But despite the resemblances to the terns and gulls, *Ichthyornis* had primitive features—for example, the vertebrae (which are in modern birds highly ossified and with peculiar saddle-shaped ossifications) were still amphicoelous in primitive reptilian fashion.

The skull is poorly known. The original describer associated with it a tooth-bearing jaw. The supposed presence of teeth in both *Hesperornis* and *Ichthyornis* led to the belief that all Cretaceous birds were toothed, and that "modern" toothless birds evolved only in the Cenozoic. Recent study, however, shows that it is extremely doubtful whether this jaw belongs to *Ichthy-*

ornis; it may be that of a "baby" mosasaur! In the light of this find, our concepts have changed; the retention of teeth, as in *Hesperornis,* may have been the exception rather than the rule, and we may well look for true members of modern orders among the Cretaceous bird remains.

Ichthyornis is presumably closely related to the Charadriiformes and may well be an ancestral type; in view of its primitive features, however, it may be preferable to consider it as ordinally distinct.

14. *Order Columbiformes.*—The doves and pigeons are a cosmopolitan group of strong-flying vegetable-feeding forest dwellers. Both bills and legs are short. An interesting series of specialized forms, of which the dodo of Mauritius is the most familiar, became large flightless types inhabiting islands in the Indian Ocean; these are now extinct.

15. *Order Psittaciformes.*—The parrots are a group of tropical birds with characteristic powerful, hooked bills and with a type of foot structure termed zygodactylous, in which two of the four toes (first and fourth) are turned backward in perching.

16. *Order Cuculiformes.*—Cuckoos and road runners are apparently related to the parrots but without the hooked beak. As in the parrots, the foot is zygodactylous.

17. *Order Strigiformes.*—The owls are nocturnal birds of prey. The adaptive features, such as a short but sharp-hooked beak and powerful talons, are similar to those of the day predators (Falconiformes), but more basic structural characters show the two groups to be distinct; the large forward-turned eyes of the owl furnish one readily recognizable contrasting feature.

18. *Order Caprimulgiformes.*—The oilbirds, goatsuckers, and nightjars (including the familiar whip-poorwill) are peculiar, twilight insect feeders, thought to be related to the owls, with a compact body and a broad, short bill bordered with stiff bristles.

19. *Order Apodiformes.*—This order includes swifts and hummingbirds, two very distinct but related assemblages of small birds with rapid flight and correlated wing and breast specializations.

20. *Order Coliiformes.*—The colies, or mousebirds, are a small group of African birds which superficially resemble finches but are structurally quite distinct.

21. *Order Trogoniformes.*—The trogons are gaily colored tropical forest birds (such as the Mexican quetzal) with short, powerful bills and short, weak feet in which the second toe is turned backward with the first.

22. *Order Coraciiformes.*—This group is a rather miscellaneous assemblage, mainly tropical, and includes kingfishers, hoopoes, bee eaters, rollers, hornbills, and so forth, which have little in the way of obvious common features except arboreal habits, the usual possession of an elongate bill, and three forwardly directed toes partly fused together.

23. *Order Piciformes.*—This includes the woodpeckers and the toucans and various other tropical birds. The Piciformes are all stout billed and have the peculiar zygodactylous type of foot noted above for the parrots and cuckoos.

24. *Order Passeriformes.*—The perching birds are the "highest" of bird orders. A characteristic feature is the perching type of foot, with the very large first toe directed straight back and opposed to the other three, giving an effective grasp on the limb. Flight is usually highly developed, and arboreal life is general. There are a few relatively primitive tropical families; the others are songbirds (suborder Passeres), which include all the smaller familiar birds. The Passeriformes comprise about half the present bird population. Known fossil remains are few when considered in relation to the modern abundance of the group; however, small size and woodland habitat are features which tend to reduce the chances of adequate fossilization in this group.

CHAPTER 14

Mammal-like Reptiles

The relatively late time at which the mammals took over the world's supremacy from the reptilian dynasties would lead one to think that the stock from which they sprang must have been one developed at a comparatively late date in reptilian history. This, however, is exactly the reverse of the true situation. The mammal-like reptiles, constituting the subclass Synapsida, were among the earliest to appear of known reptilian groups and had passed the peak of their career before the first dinosaur appeared on the earth.

Primitive synapsids were already present in the Pennsylvanian—perhaps early Pennsylvanian—and took a leading role in the archaic reptilian radiation of Permian times. Although the central types were carnivores, herbivores developed as well. The majority of known Permian reptiles were members of this group; apart from the cotylosaurs, all other reptile stocks were at that time represented only by small and seemingly rare forms.

In the Triassic, however, conditions altered. The synapsids dwindled in numbers during the period and (except for a few rare or doubtful genera) vanished at its close, to be supplanted mainly by the developing archosaur stocks. Many millions of years were to pass before the mammalian descendants of the synapsid group were to rise to a position of dominance.

Since the various synapsid types cover the entire range of the vast structural gap existing between the very primitive reptiles and forms extremely mammal-like, it is hardly to be expected that there would be many diagnostic features common to all members of the subclass. The central stock of the group, as has been said, consisted of carnivorous types. There was in the dentition a tendency toward a differentiation of the tooth row, with the development of varied types of teeth to subserve various functions—a tendency which reached its climax in the mammals. Synapsids progressively improved the primitive clumsy limb structure found in the stem reptiles. Here, however, the trend was solely toward improvement in quadrupedal locomotion; there were (in contrast to the archosaurs) no bipedal tendencies.

The one diagnostic feature common to all members of the group was the presence of a single lateral opening in the temporal region—a condition found in no other major reptilian stock. This fenestra was primitively situated on the lateral, rather than the upper, surface of the skull and was of small size, with the postorbital and squamosal joining above it; it became enlarged, however, and extended upward to reach the parietal in more advanced forms. The pineal opening persisted in almost every synapsid. The eardrum appears to have lain, as in reptiles generally, at the back margin of the skull above the jaw joint, but there is seldom any notch or excavation of the skull connected with it. There was comparatively little loss of cranial elements; but the postparietals (usually fused into a single bone) were, with the tabulars, pushed down onto the back of the skull, and in advanced forms the elements surrounding the orbits were reduced in number. The vertebrae were primitively amphicoelous, and double-headed ribs were present the length of the trunk. A diagnostic feature is the presence of two coracoid elements in the shoulder (Fig. 269)—a new posterior one in addition to the anterior element which alone is present in most other reptiles. The dermal shoulder girdle was always retained and usually included a splint cleithrum in the older types.

The whole series should, perhaps, be divided into a considerable number of groups of equal rank. But it so happens that the primitive Lower Permian forms are found abundantly only in the Texas redbeds of North America (more rarely in Europe), while the later Permian and Triassic types are found mainly in the South African Karroo series, in South America, and in Russia. This cleavage of the group in time and space may be used as a basis for a somewhat illogical but fairly convenient division into two orders—the Pelycosauria, primitive forms; and Therapsida, advanced types.

Pelycosaurs (Fig. 259).—These primitive synapsids are represented, although sparsely, well back in the Upper Carboniferous, and are among the commonest animals found in the redbeds of Texas, dating from early Permian times. Scattered remains are found in early Permian beds in other areas of North America and western Europe; representatives from later times or other regions are rare and of doubtful nature. A primitive member of the order was *Varanosaurus* (Fig. 260), a small form not more than 3 feet in length even with the inclusion of a very long tail. The proportions of this form were not unlike those of many lizards. In internal structure there were many similarities to the captorhinid cotylosaurs from which these types had arisen. The limbs, while rather longer and slimmer than those of cotylosaurs, still resembled them in most features. The neural arches of pelycosaurs (Fig. 262) were (as in most reptiles) narrower than those of typical cotylosaurs and the neural spines somewhat taller; in *Varanosaurus,* however, the arches are still somewhat swollen—presumably a primitive feature. Intercentra were still present throughout the column, and the old scales were still present ventrally, as abdominal ribs, in most pelycosaurs. The snout was rather long; the skull was rather higher and narrower than in the cotylosaur types. The teeth along the jaw margins were numerous and mostly of rather uniform size in *Varanosaurus* and its close allies. A pair of teeth lying near the front end of each maxilla may, however, be considerably longer; they appear to represent the first stage in the development of the canine teeth characteristic of mammals. *Varanosaurus* was not improbably a fish-eating reptile which spent much of its time in the water—habits which here may be truly primitive in character and reminiscent of the amphibian forebears of the group. The related Permian genus *Ophiacodon* (Figs. 261, 269*A*, 272*A*, 273*A*, 274*A*) was a much larger animal, one species of which may have been nearly 12 feet in length. Small relatives are known from the Coal Measures. Some Permian forms of this group, such as *Eothyris,* developed a highly predaceous dentition, but the dominant carnivores of the early Permian are *Dimetrodon* and its relatives, described later.

A second and very different group of pelycosaurs is that represented by *Edaphosaurus* (Figs. 263, 264, 271*B*) and *Casea* (Fig. 265). These animals were apparently herbivores. The postcranial skeleton was, on the whole, rather similar to that of *Varanosaurus* but with a somewhat clumsier build and a broad barrel-shaped trunk, presumably for the reception of masses of soft vegetation. *Edaphosaurus* presents an unusual feature because of the presence of greatly elongated neural spines rising high above the back and studded with small crossbars, giving somewhat the effect of a full-rigged ship. It seems certain that the spines were connected by membranes, forming the whole into a "sail."

The skull was short and very different from that of *Varanosaurus.* The teeth were rather blunt, and there is no trace of canine differentiation; *Edaphosaurus* has, in addition, highly developed upper and lower masticatory plates, studded with teeth, lying within the mouth.

Despite its specializations, *Edaphosaurus* was an an-

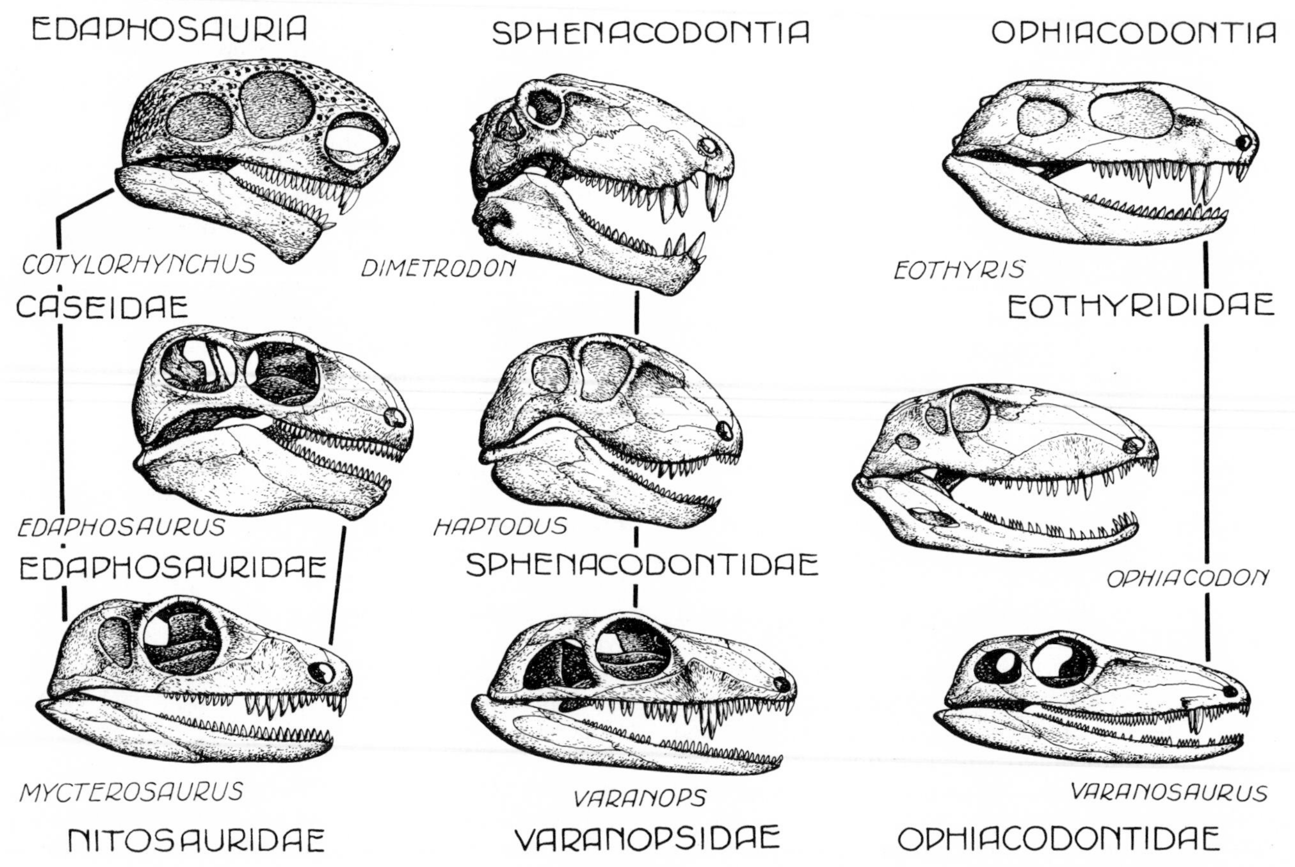

Fig. 259.—A pictorial "family tree" of the pelycosaurs

cient type, for characteristic spine fragments are recorded from the late Carboniferous. The earliest species were small; the largest Permian specimens, on the other hand, had a length of about 11 feet. A herbivorous family related to *Edaphosaurus* but lacking the spines and with a skull that was shorter, broader, and less specialized includes *Casea* (Fig. 265) and its relatives. *Casea* was a small pelycosaur, but a large relative—*Cotylorhynchus* (Fig. 266)—is estimated to have weighed about a third of a ton and is the giant among pelycosaurs. There are known smaller and much more

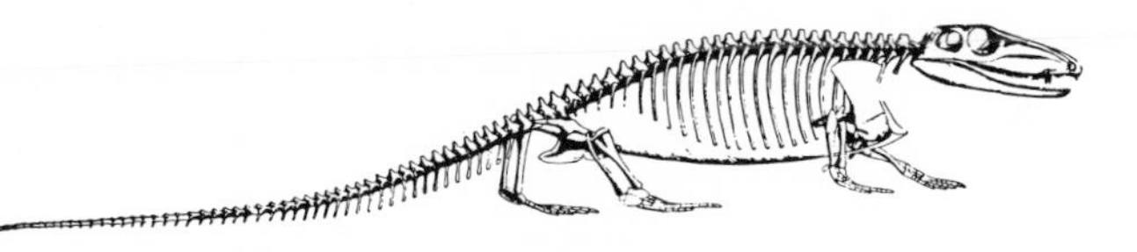

FIG. 260.—*Varanosaurus*, a primitive pelycosaur; length of original about 5 feet.

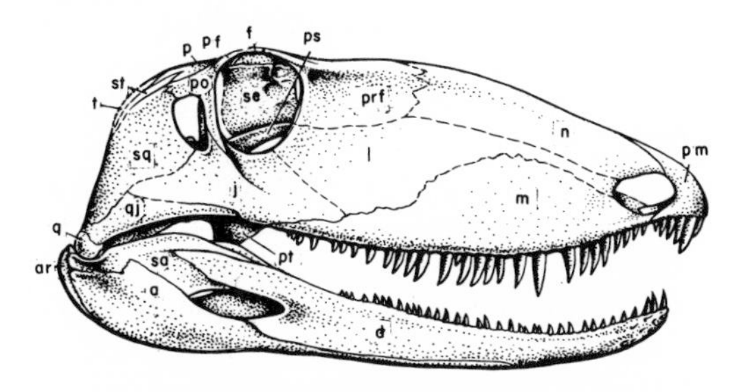

FIG. 261.—The skull of *Ophiacodon;* original about 15 inches long.

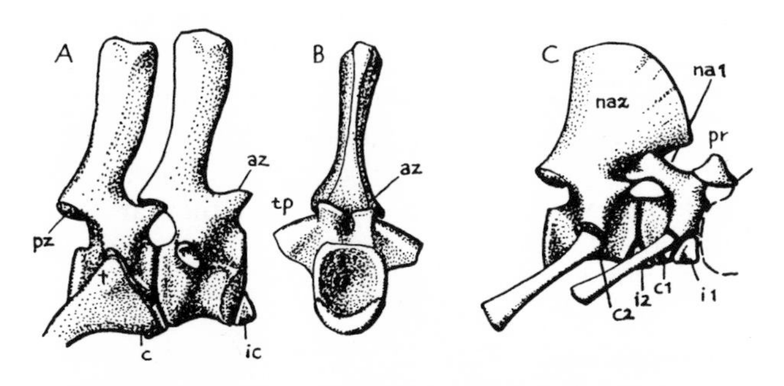

FIG. 262.—Vertebrae of a primitive pelycosaur, *Ophiacodon*. *A, B,* Posterior cervical vertebrae, lateral and anterior views. Abbreviations: *az,* anterior zygapophysis; *c,* capitulum of rib; *ic,* intercentrum; *pz,* posterior zygapophysis; *t,* tubercle of rib; *tp,* transverse process. *C,* Atlas and axis, condyle of skull in dashed line; *c1, c2,* centra of atlas and axis; *i1, i2,* intercentra of the same; *na1, na2,* neural arches; *pr,* proatlas, a neural arch whose centrum is presumably fused into the basioccipital. (After Case and Williston.)

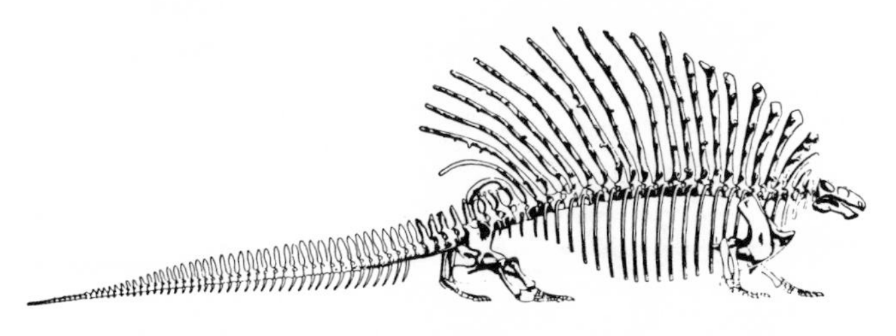

FIG. 263.—*Edaphosaurus* (*Naosaurus*), a long-spined, herbivorous pelycosaur; maximum length about 11 feet.

primitive pelycosaurs, rather like *Varanosaurus* in proportions and probable habits, which appear to be the ancestors of this herbivorous branch.

A third and final major group of pelycosaurs is that which includes *Dimetrodon* (Figs. 267, 268, 270*A*, 271*A*, 278*A*), an aggressive carnivore which is the commonest of reptiles in the Lower Permian of North America. *Dimetrodon* is specialized in that it has a great "sail" comparable to that of *Edaphosaurus* but lacking the crossbars. There were, however, other members of this group in which the sail was not developed, and the dimetrodonts are of interest as a stock

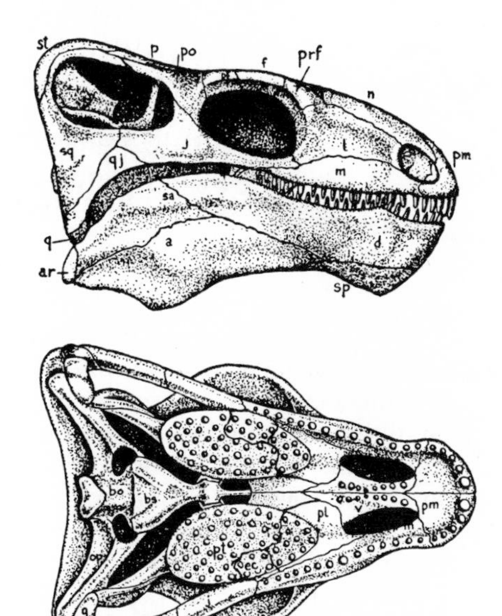

FIG. 264.—The skull of *Edaphosaurus;* length of specimen about 6 inches. For abbreviations see Figure 109.

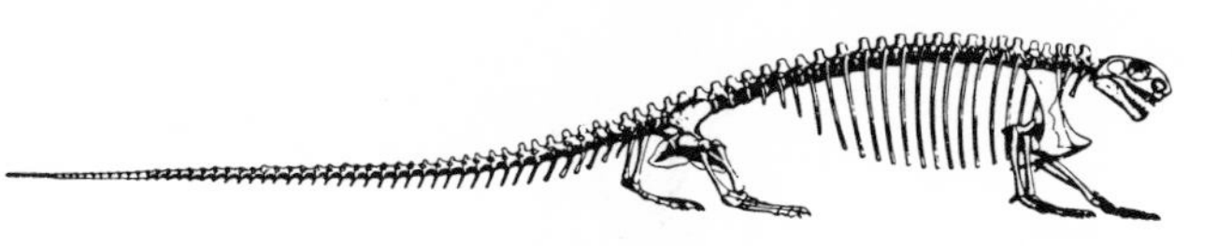

FIG. 265.—*Casea*, a short-spined herbivorous pelycosaur related to *Edaphosaurus;* length about 4 feet.

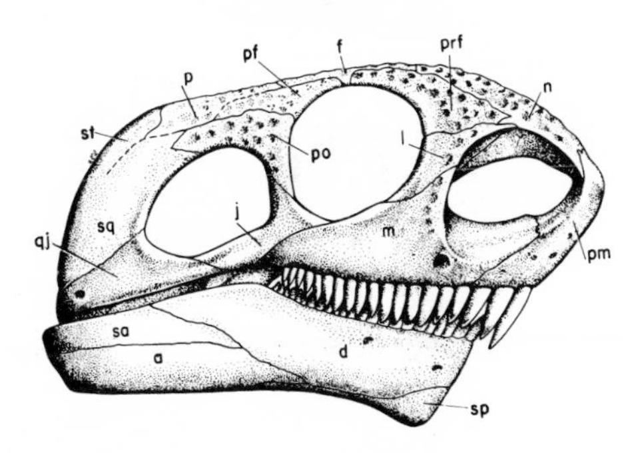

FIG. 266.—The skull of the herbivorous pelycosaur *Cotylorhynchus;* length of original about 8 inches.

from which the therapsids (and eventually the mammals) appear to have been derived. The function of the pelycosaur sail has been much debated. In *Dimetrodon* the sail area in species of different sizes is found to vary proportionately to the volume of the form concerned—hence disproportionately tall in big species. This strongly suggests that the sail was an early essay in development of a temperature-regulating organ which was able to absorb or radiate heat, foreshadowing the more effective devices present in the pelycosaur's mammalian descendants.

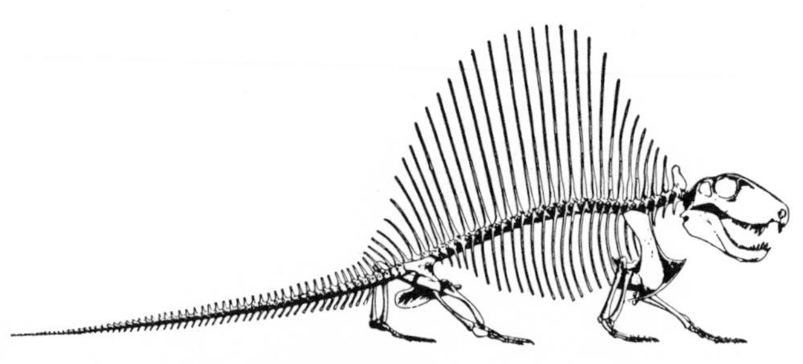

Fig. 267.—*Dimetrodon,* a long-spined predaceous pelycosaur; maximum length about 11 feet.

The postcranial skeleton of *Dimetrodon* and its relatives, forming the family Sphenacodontidae, is generally comparable to that of other pelycosaurs; but the body was rather slim, and the limbs, although primitive and sprawling in construction, are somewhat more slender than in most early Permian reptiles and suggestive of a relatively rapid gait. The skull was high and narrow, the face rather long; posteriorly the skull margins curved downward and even forward toward the jaw articulation—a feature repeated in many therapsids. The jaws were armed with compressed, sharp-edged teeth, well differentiated—particularly in the upper jaw—for predaceous habits. In the premaxilla were a limited number of teeth for biting and grasping, which are comparable to the incisors of therapsids and mammals. Separated from them by a "step" which may contain a few small teeth, we find on the maxilla a pair of large teeth, alternately replaced, which correspond to the canines of higher types, and behind these a series of smaller cheek teeth. The lower teeth are differentiated also, but to a lesser degree. The angular bone near the back of the jaw bears a curious flange, perhaps for muscle attachment, separated by a notch from the main portion of the mandible. This feature is repeated in therapsids and is one of the many diagnostic characters indicating that the sphenacodonts gave rise to the therapsids which in the later Permian and early Triassic succeeded the pelycosaurs as the dominant land vertebrates of their times. As in the case of the other two major pelycosaur groups, the sphenacodonts appear to be traceable back to small and primitive forms; these primitive types, in turn, are not far removed, except for the development of a temporal opening, from the captorhinomorph cotylosaurs, and there is general agreement that the pelycosaurs are derived from this primitive reptilian group.

Therapsids (*Fig. 275*).—Except for a few rare or doubtful forms, the pelycosaurs are known only from the late Carboniferous and, especially, the early fraction of the Permian represented by the Wichita and Clear Fork redbeds of the American Southwest and their equivalents. Derived from the pelycosaurs and succeeding them were the therapsids. As members of the order Therapsida may be grouped the numerous mammal-like types which, best known from the richly fossiliferous Karroo beds of South Africa, flourished from Middle Permian to Lower and Middle Triassic times. To this group belong all the carnivorous reptiles of the later Permian, as well as a host of herbivorous forms. So varied are these types that they are often considered to constitute a number of distinct (although closely related) orders.

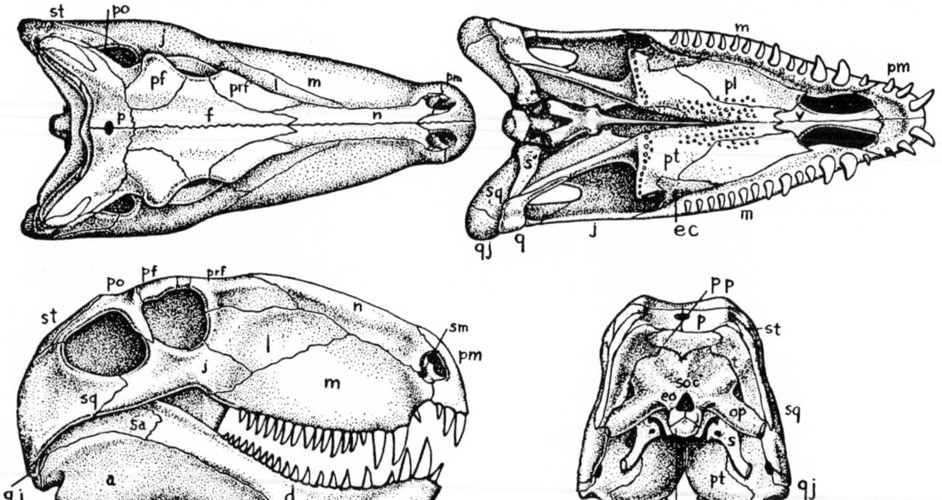

Fig. 268.—The skull of *Dimetrodon. Above,* Dorsal and palatal views; *below,* lateral and occipital views. For abbreviations see Figure 109.

Because of the wide radiation of the group, absolute diagnostic characters are rare. The evolutionary trend of the main line of therapsid development may be illustrated by a consideration of the structure of one of the later and more progressive forms.

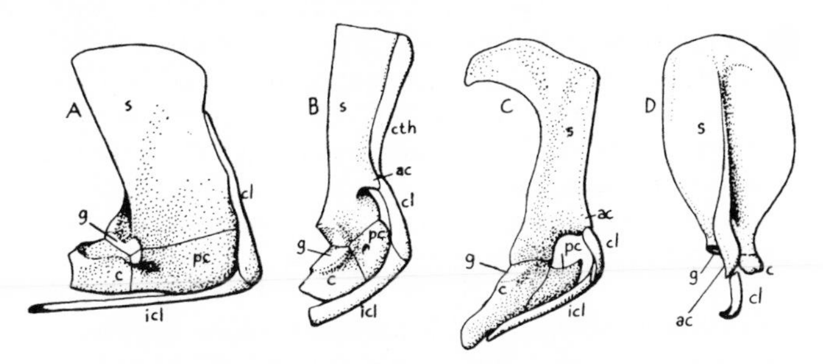

Fig. 269.—Shoulder girdles of synapsids and mammals: *A*, the pelycosaur *Ophiacodon; B*, the therapsid *Kannemeyeria; C*, the monotreme *Ornithorhynchus; D*, the marsupial *Didelphis*. Abbreviations: *ac*, acromion; *c*, coracoid; *cl*, clavicle; *cth* cleithrum; *g*, glenoid cavity; *icl*, interclavicle; *pc*, procoracoid; *s*, scapula. (*A* after Williston; *B* after Pearson.)

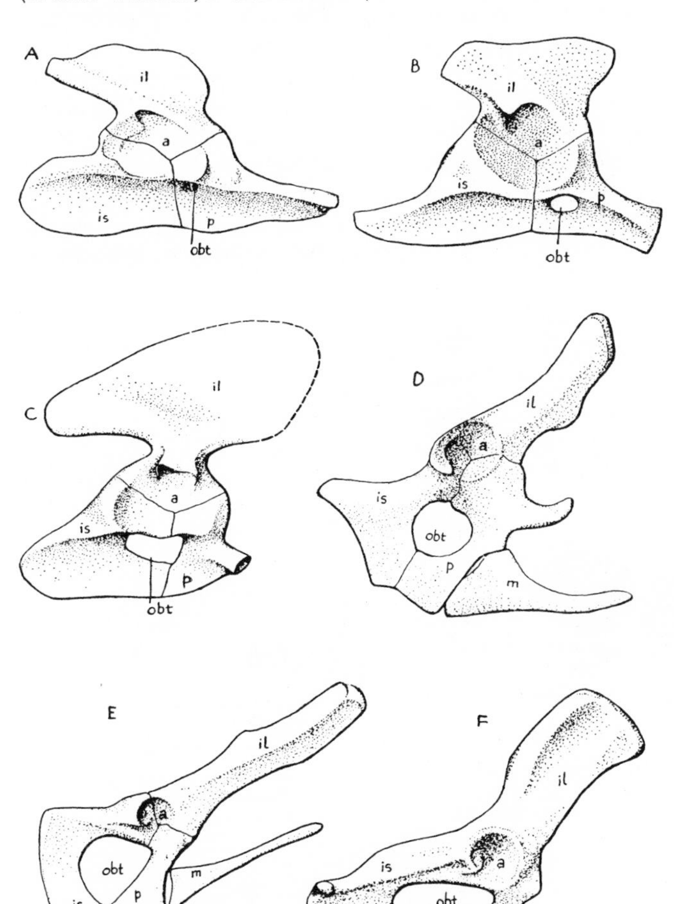

Fig. 270.—Pelves of synapsids and mammals: *A*, the pelycosaur *Dimetrodon; B*, the gorgonopsian *Lycaenops; C*, the cynodont *Cynognathus; D*, the monotreme mammal *Ornithorhynchus; E*, the marsupial *Didelphis; F*, the Miocene canid *Daphoenodon*. Abbreviations: *a*, acetabulum; *il*, ilium; *is*, ischium; *m*, marsupial bone; *obt*, obturator foramen—enlarging to form a fenestra in therapsids and mammals; *p*, pubis. (*B* after Broom; *C* after Gregory and Camp; *F* after Peterson.)

Structure of Cynognathus.—*Cynognathus* and *Thrinaxodon* (Figs. 270*C*, 271*C*, 272*B*, 276, 277, 278*C*) were Lower Triassic therapsids which we shall note later to be members of the very mammal-like cynodonts. These reptiles were rather lightly built and seemingly active carnivores.

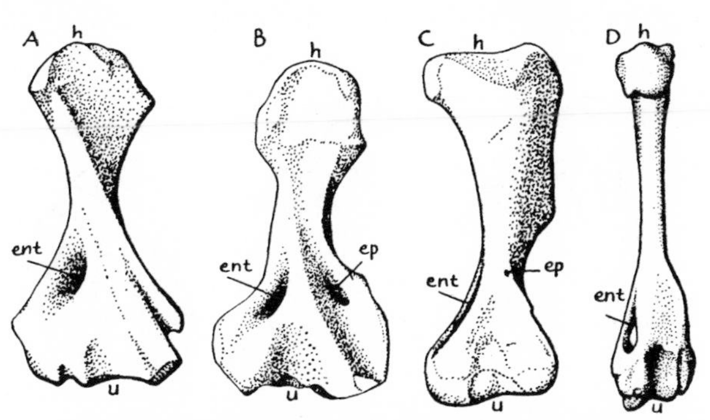

Fig. 271.—The humerus of synapsids and mammals, dorsal views: *A*, the pelycosaur *Dimetrodon; B*, the pelycosaur *Edaphosaurus; C*, the cynodont *Cynognathus; D*, the Oligocene canid *Daphoenus*. Abbreviations: *ent*, entepicondylar foramen; *ep*, ectepicondylar foramen; *h*, head; *u*, articular surface for the ulna. (*C* after Watson; *D* after Hatcher.)

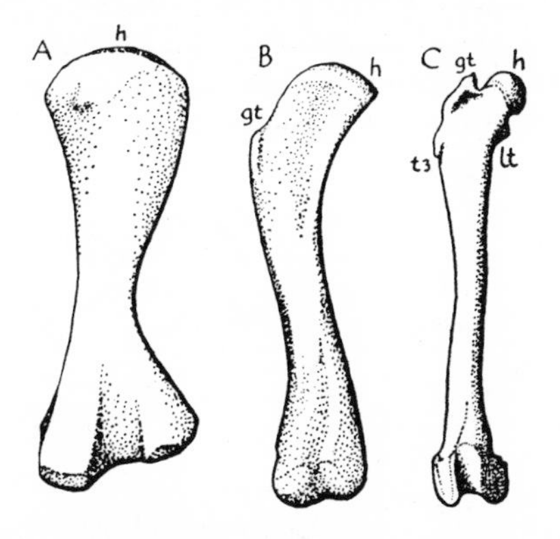

Fig. 272.—The femur in mammal-like reptiles and mammals: *A*, the pelycosaur *Ophiacodon; B*, a cynodont; *C*, the Oligocene canid *Daphoenus*. Abbreviations: *gt*, greater trochanter; *h*, head of femur; *lt*, lesser trochanter; *t3*, third trochanter. (*A* after Case and Williston; *B* after Watson; *C* after Hatcher.)

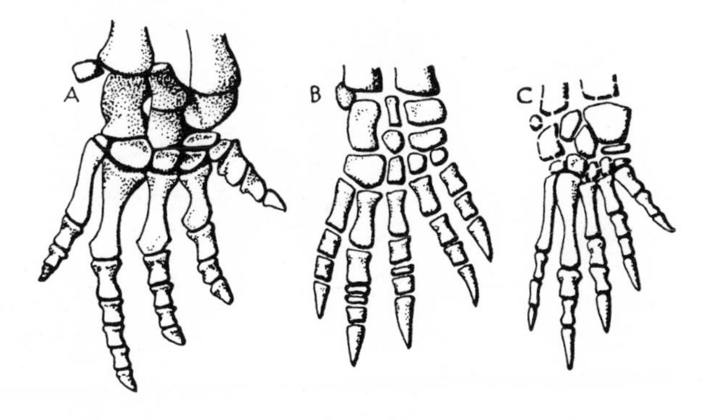

Fig. 273.—The manus in mammal-like reptiles: *A*, the pelycosaur *Ophiacodon; B*, the gorgonopsian *Lycaenops; C*, the bauriamorph *Ericiolacerta*. (*A* after Case and Williston; *B* after Broom; *C* after Watson.)

In the skull the pineal opening is small but still present. Many therapsids have a small median bony element—the preparietal—associated with it; this is, however, absent in cynodonts. The temporal opening is much expanded and reaches high up the skull; between the two openings the parietals form but a narrow longitudinal ridge over the small braincase, rather

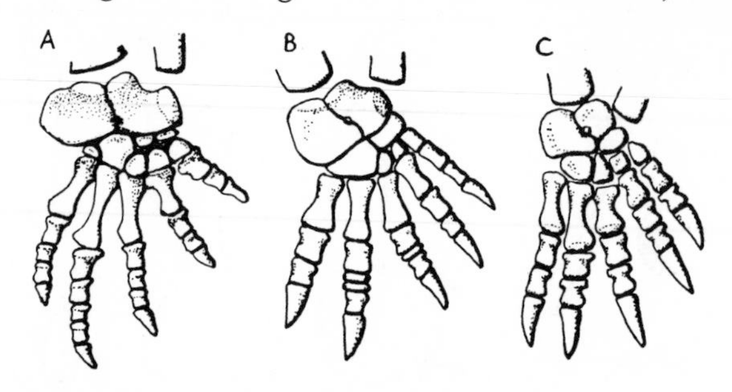

Fig. 274.—The pes in mammal-like reptiles. *A*, The pelycosaur *Ophiacodon;* *B*, the gorgonopsian *Lycaenops;* *C*, the therocephalian *Whaitsia.* (*A* after Case and Williston; *B*, *C*, after Broom.)

comparable to the sagittal crest of mammals. The postorbital and squamosal no longer meet each other above this fenestra (but have gained a secondary contact below it). The circumorbital series of bones has dwindled; the prefrontal and lacrimal are small and the postfrontal has disappeared. There is, however, a postorbital bar separating orbit and temporal opening—a structure which disappears in certain advanced therapsids. The quadratojugal is small and closely attached to the equally reduced quadrate, and these two bones are rather loosely connected with the rest of the skull. On the underside of the skull the most obvious development is the presence of a secondary palate. The vomers have fused into a single bar which sends up a process in the midline. Extensions inward on either side toward this bar from the premaxillae, maxillae, and palatines have formed a flat plate extending across below the old level of the roof of the mouth, so that a hard secondary palate is formed with the internal openings of the nostrils passing backward through the space above it. There are no longer any interpterygoid vacuities, for the two pterygoids have fused with each other and the narrow base of the braincase to form a solid, longitudinal bar, lacking the flexible basicranial articulation present in many primitive reptiles. The pterygoids have shortened and no longer extend back to the quadrates, although the epipterygoids above them may reach this far back.

The condyle has divided into a double structure, in contrast with the primitive single condition. The occipital surface of the skull is a nearly solid plate made up of the combined occipital bones and the opisthotics together with the postparietal and the tabulars above. In addition to the ordinary bones of the braincase, the epipterygoid forms a flat plate applied to the side of

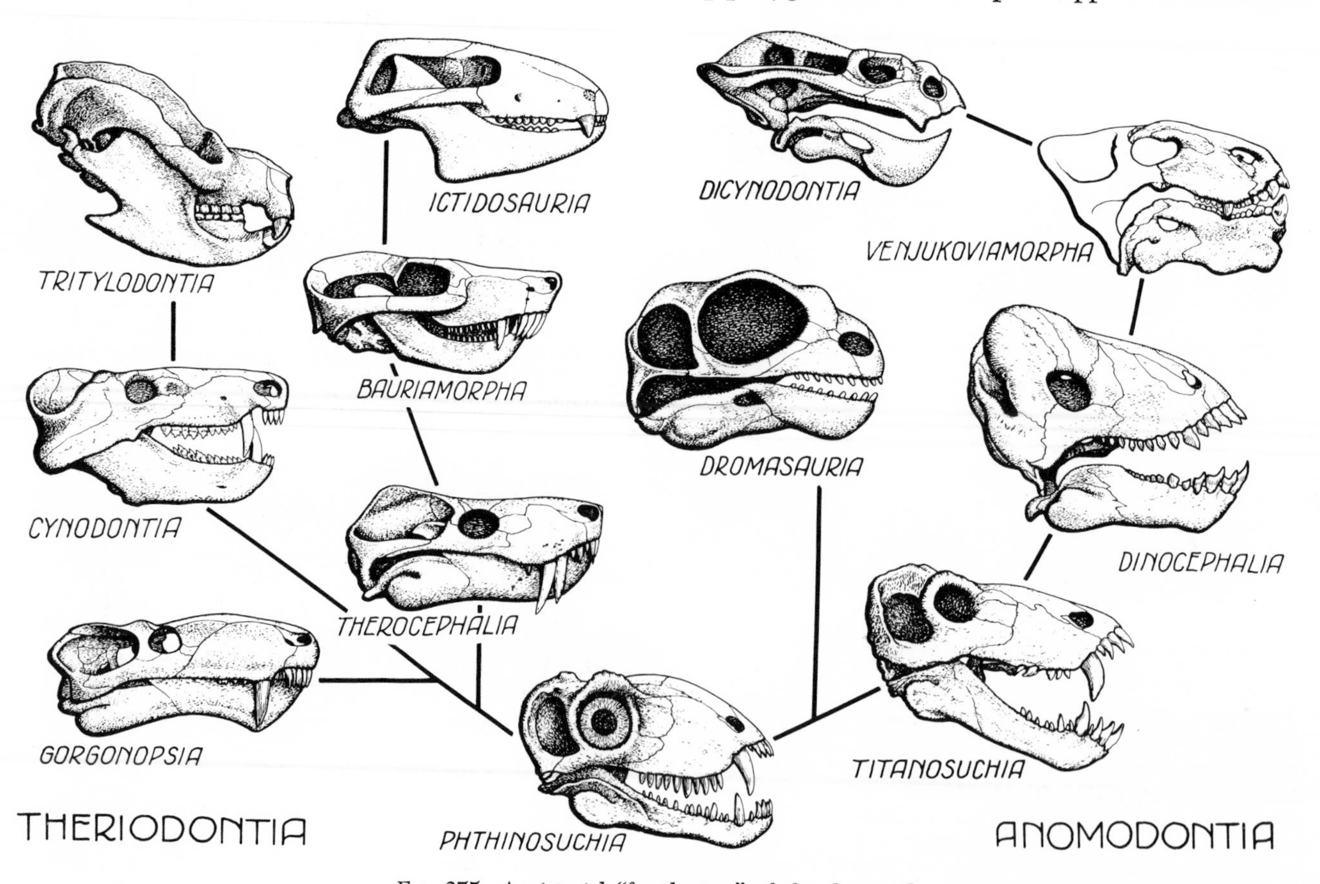

Fig. 275.—A pictorial "family tree" of the therapsids

the braincase in front of the proötic; and in the more anterior portion there appears to be an ossification corresponding to the sphenethmoid of amphibians and reptiles and the presumably homologous presphenoid of mammals. The stapes extends as a slim rod from the inner-ear opening to the quadrate.

All the bones of the primitive reptilian lower jaw are present (except that there is but one coronoid) (Fig. 278*C*), but the dentary is greatly enlarged at the expense of the other elements, sending backward and upward a large process which nearly touches the squamosal.

The dentition is differentiated into a small number of nipping teeth, comparable with the human incisors, in the premaxilla and front of the dentary; a single "canine" tusk (rather than the *Dimetrodon* pair) lies behind a step in the maxilla, and a series of cheek teeth, most of them with several cusps, is present. Teeth are absent from the palate and the inner surface of the lower jaw.

Intercentra have disappeared from the backbone, although the centra are still amphicoelous. There are still ribs all the way from the neck to the base of the tail, but there is more of a distinct neck region, with shorter ribs, than in pelycosaurs, and the ribs in front of the pelvis tend to fuse to the vertebrae in anticipation of the lumbar region developed in mammals.

The limbs in carnivorous therapsids, such as *Thrinaxodon, Cynognathus* or *Lycaenops* (Figs. 270*B, C,* 271*C*, 273*B*, 274*B*, 276), are greatly changed from the primitive sprawling position. The elbow has been moved back and the knee forward, so that the legs tend to be more underneath the body, making support easier (Fig. 276). Associated with this are many changes in the musculature and in the shape of many of the bones. The therapsid shoulder girdle (Fig. 269*B*) still consists of a dermal girdle, including a stemmed interclavicle and clavicles (the cleithrum has disappeared), and a primary girdle of scapula and two coracoid elements. The front edge of the scapula is turned out (the beginning of the mammalian spine) with the clavicle attaching to the projecting spine at the lower end of the ridge. In the pelvis (Fig. 270*B, C*) the ilium extends forward rather than back, as primitively, in relation to a changed position of the hind leg and its muscles, and is attached to an increased number of sacral ribs. The pubis and ischium have shifted toward the back (also in correlation with muscular

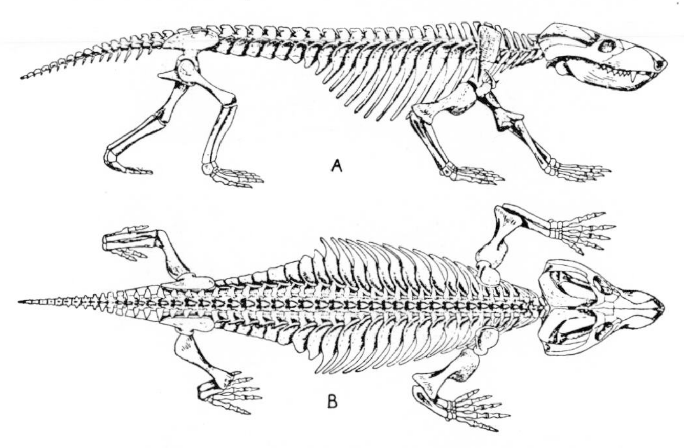

Fig. 276.—The skeleton of the cynodont *Thrinaxodon*, in lateral (*A*), and (*B*), dorsal views; length about 1.5 feet. (After Brink.)

changes), and a circular opening has appeared between, and partially separating, them. The femur (Fig. 272*B*) has developed a head at the side rather than at the end of the bone in relation to its forwardly rotated position, and on the outer side there is a large trochanter for muscles which come from the ilium and aid in its backward push. The toes are all, except the first, of nearly equal length, in correlation with the more directly fore-and-aft motion of the limbs; and each has but three functional segments (Figs. 273*B, C;* 274*B, C*). But while this foreshadows the mammalian con-

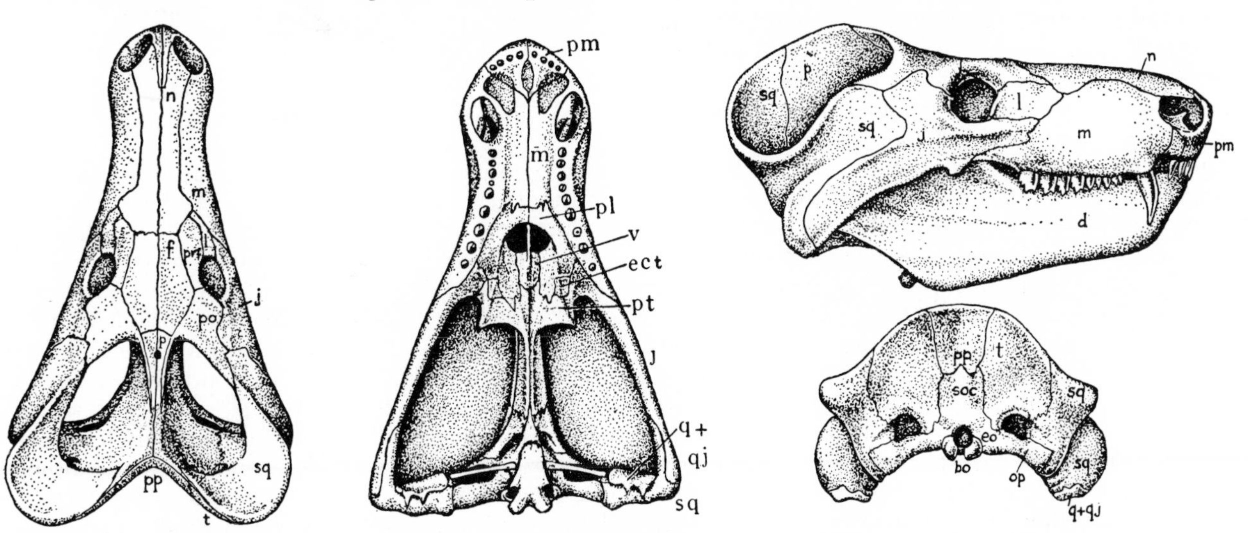

Fig. 277.—The skull of the cynodont *Cynognathus;* length of original about 18 inches. For abbreviations see Figure 109. (After Seeley, Broom, and Broili and Schroeder.)

dition, a relatively primitive phalangeal formula of 2-3-4-4-3 still persists in *Cynognathus* and related types, for vestiges of "superfluous" phalanges are still present in the third and fourth toes.

The skeleton of such a reptile approaches in many features that of the mammals. It would be of interest to know whether these forms resembled mammals as much in other features; but we know almost nothing of other organ systems, except that the brain was still small and apparently reptilian in organization. The warm-blooded condition of mammals is seemingly correlated with their active mode of life. Activity seems to have been a keynote in the development of these carnivorous therapsids, and it is not impossible that this physiological change had already begun in them. A cold-blooded reptile may cease breathing for a time without harm, but not a mammal. The secondary palate of mammals seems to be an adaptation preventing interference with breathing during mastication; the presence of this structure in advanced therapsids is suggestive. Whether hair was already beginning to replace scales is, of course, unknown. Nor have we any knowledge of reproductive processes; but, since the most primitive living mammals are still egg-laying types, this was presumably also the case in *Cynognathus*. The nursing habit seems to have been earlier in appearance, and that it was early enough to have been present in advanced therapsids is quite possible.

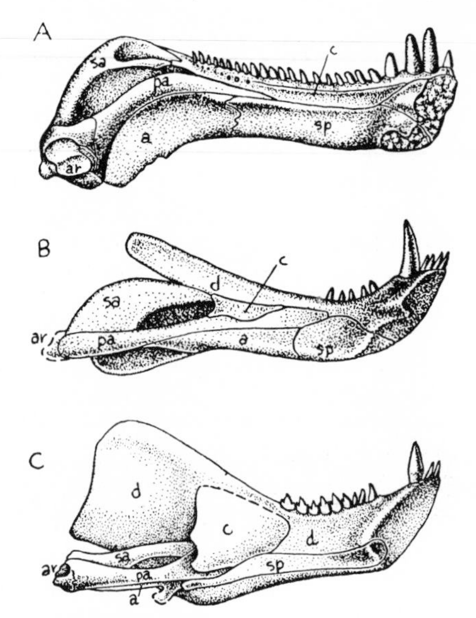

FIG. 278.—Jaws of synapsids, from the inner side. *A*, The pelycosaur *Dimetrodon; B*, the gorgonopsian *Cynarioides; C*, the cynodont *Cynognathus*. For abbreviations see Figure 109. (*A* after Williston; *B* after Broom; *C* after Broom and Watson.)

Primitive therapsids.—Most primitive of therapsids is a series of forms such as *Phthinosuchus* (Fig. 279) and a few relatives, mainly found in Russian beds which lie but a little higher in the Permian than the American redbeds containing the typical pelycosaur faunas. Of phthinosuchids we know little except the skull and jaws, but they seem to be so primitive and so close to the sphenacodont pelycosaurs in most regards that we appear justified in considering them as constituting a basic suborder of therapsids, the Phthinosuchia. As the figure shows, the appearance and general proportions of the skull and jaws are notably similar to *Dimetrodon*, even in such points as the prominent flange on the angular bone. However, the temporal opening is larger than in pelycosaurs, the jaw articulation is set at a point definitely lower and farther forward, and there is a distinct lower canine as well as a single (rather than double) upper one. The palate is for the most part sphenacodont-like, with no secondary palate, and a good supply of teeth; however, the fusion of palate with braincase and elimination of the flexible basal articulation are well advanced. The postcranial skeleton is poorly known, but it seems probable that the shift in limb posture characteristic of therapsids was already well under way; very probably the general appearance would have been much that of the form shown in Figure 283 (although this is an animal a bit more progressive).

Beyond this presumably primitive stage, there appears to have been an early cleavage of therapsids into two great groups, which between them furnish the great majority of all known reptiles of later stages of the Permian (to say nothing of a good fraction of those of the early and Middle Triassic). One group, the suborder Theriodontia, consists of forms, progressive to a variable degree, which are predominantly carnivorous in habits. The second group, the Anomodontia (using this term in a broad sense), includes a variety of therapsids which shifted to a herbivorous diet.

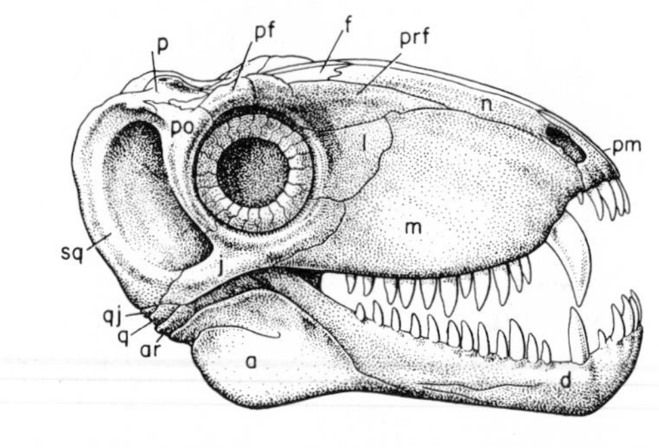

FIG. 279.—The skull of the primitive therapsid *Phthinosuchus* in lateral view; length about 8 inches. (After Efremov.)

Theriodonts.—The gorgonopsians, such as *Scymnognathus and Lycaenops* (Figs. 270*B*, 273*B*, 274*B*, 278*B*, 280), were in many features very primitive theriodonts and were very numerous in Middle and Upper Permian times. There were many primitive features in the group. The postorbital and squamosal still met above the temporal opening; a postfrontal bone was still present; there was no secondary palate, although the vomers had fused into a single bar; the palate still bore teeth; the condyle was still single; the dentary, although large and with an ascending branch, did not greatly exceed the other elements in size; the postcranial skeleton was somewhat more primitive and more heavily built than in the cynodonts; the phalangeal formula was still 2-3-4-5-3, although the "extra" joints were small.

Succeeding the gorgonopsians as the characteristic carnivores were the cynodonts, of which *Cynognathus* and *Thrinaxodon*, described earlier, were typical members. In many regards they seem like advanced descendants of the gorgonopsians, but they cannot have evolved from known members of that group because of such points as the fact that the cheek teeth of gorgonopsians are much reduced. We earlier noted various advances made by cynodonts, such as the development of a secondary palate (cf. the palates in Figs. 268, 277, 281) and growth of the dentary bone of the lower jaw. Cynodonts appear in the final stages of the Permian, and typically carnivorous members of the group flourished in the early Triassic, where *Cynognathus* and its relatives were the common carnivores. There was considerable variation in size, the smallest cynodont being not much bigger than a rat, the largest the size of a wolf.

There are a few doubtful scraps of cynodonts from the late Triassic, but it was long thought that the career of this group was essentially finished at the end of the early phases of that period. However, in recent years the discovery of Middle Triassic faunas in East Africa and South America has revealed an unexpected chapter in the history of cynodonts. Typical cynodonts have a carnivorous type of dentition; the cheek teeth, although sometimes with several cusps in a fore-and-aft series, were sharp and capable in cutting and slicing. Even in the early Triassic, however, some cynodonts, such as *Diademodon*, show cheek teeth with

broadened crowns somewhat comparable to the molar teeth of mammals; this is termed a "gomphodont" dentition. In the recently discovered Middle Triassic faunas these gomphodonts are exceedingly abundant. The nature of the diet and their mode of existence are matters for speculation.

To return again to primitive theriodonts, we find in the later phases of the Permian a group paralleling in many ways the gorgonopsians—the Therocephalia (Figs. 273*C*, 274*C*, 280, 281). These forms include a series of rather massively built carnivores which resembled their gorgonopsian cousins in such features as the absence of a secondary palate. However, they were already advanced in that their feet had undergone reduction to the mammalian phalangeal formula of 2-3-3-3-3 and in their trend toward the development of a narrow parietal crest; a useful key character lies in the fact that a pair of lateral fenestrae had developed on the palate.

Derived from the therocephalians (as indicated, for example, by the key furnished by the palatal fenestrae) was a rather advanced group, paralleling the cynodonts and termed the Bauriamorpha. As we shall use this term, it includes not only a few early Triassic forms, such as *Bauria* itself (Fig. 282), but also a number of late Permian genera of somewhat more primitive character. Here a secondary palate has developed, as in cynodonts; and even more mammal-like is the fact that the postorbital bar has disappeared. But although the dentary is quite large, the other bones of the lower jaw are much less reduced than in cynodonts. Unlike

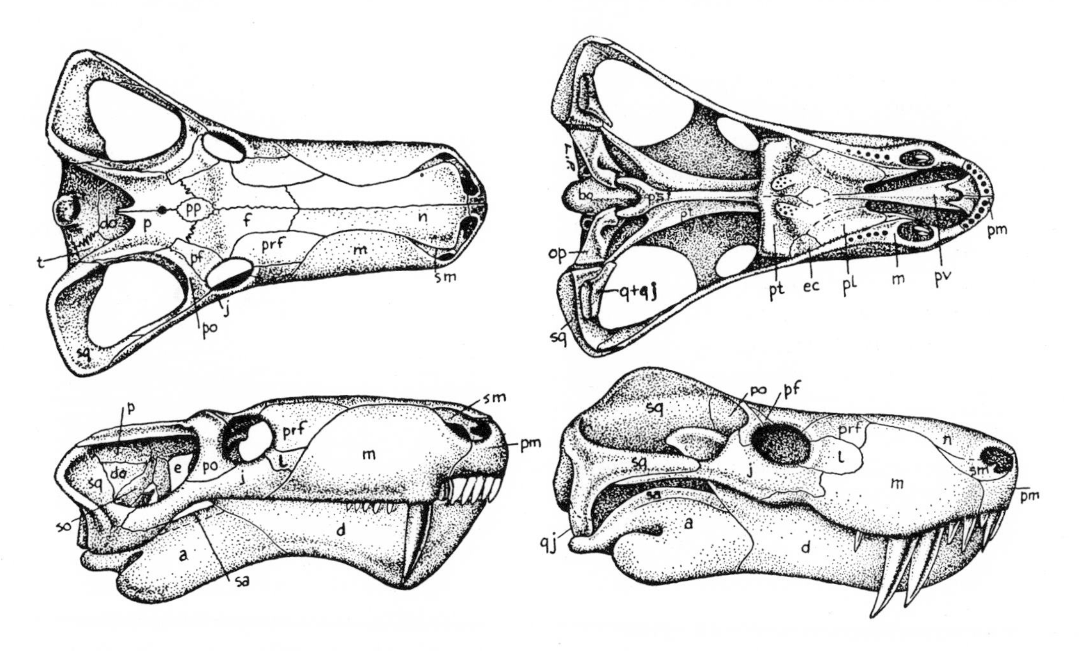

Fig. 280.—The skull in primitive theriodonts. *Upper left*, Dorsal view of the skull of the gorgonopsian *Scymnognathus*, length about 1 foot; *upper right*, palatal view; *lower left*, lateral view. *Lower right*, Lateral view of the skull of the therocephalian *Lycosuchus*, length about 11 inches; *pp*, preparietal; *so*, supraoccipital. For other abbreviations see Figure 109. (After Broom and Watson.)

the cynodonts, the bauriamorphs, as far as known, did not persist beyond the earlier phases of the Triassic.

Anomodonts.—Before discussing some of the members of the carnivorous therapsid "main line" which closely approach the mammalian condition, let us review the history of the important herbivorous side branch of the order, which we will term the Anomodontia. The forms included are highly varied, ranging

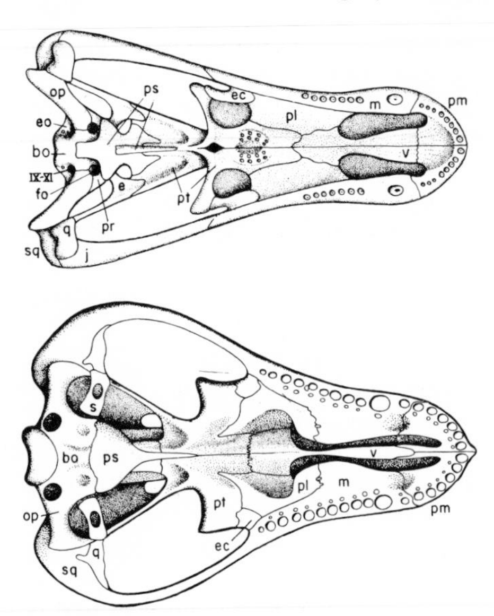

FIG. 281.—*Above,* Palate of the therocephalian *Scylacosaurus,* about 1/8 natural size (after Boonstra); *below,* palate of the primitive cynodont *Procynosuchus,* with secondary palate incompletely formed, about 1/2 natural size (after Broom). For abbreviations see Figure 109.

from primitive types not readily distinguishable from early theriodonts, to the peculiar "two-tuskers," the dicynodonts. Here the trend is for specialization, rather than advance toward a mammalian condition. As common (if relatively unimportant) features we may note that all members of the group appear to have, from the beginning, reduced the toe count to 2-3-3-3-3, and none is known to have retained a coronoid bone on the inner surface of the jaw.

The earliest anomodonts to appear are the Dinocephalia, common in the Middle Permian of Russia and South Africa, but unknown later. Except perhaps for some of the most primitive forms, none appear to be antecedent to later anomodonts. There is a strong tendency for increase to large size, and, further, in most cases, for a great thickening of the bones of the head roof. Two subgroups are apparent—a more primitive one, the superfamily Titanosuchoidea, and the more specialized Tapinocephaloidea.

The titanosuchoids, as their dentitions suggest, were still essentially carnivorous in habits, for there are usually powerful canine tusks. However, even in members of this group there is a tendency for the development of incisors which are chisel-shaped but with a pronounced "shoulder" partway down the inner side; this is a trademark of the herbivorous dinocephalians.

Titanophoneus (Figs. 283, 284) of the Russian early Middle Permian was little different in its appearance and structure from a primitive theriodont. Its size was a modest one, the length of head and trunk being only about 5 feet; a contemporary had but half these proportions. In the South African beds, however, there was a trend toward large size; *Jonkeria,* massively built, measured 12 feet from head to pelvis, and (perhaps in relation to gigantism) there is considerable thickening of the bones of the skull roof.

The herbivorous dinocephalians, such as *Ulemosaurus* (Fig. 284) and *Moschops* (Fig. 285), showed an even stronger trend toward large size and massive

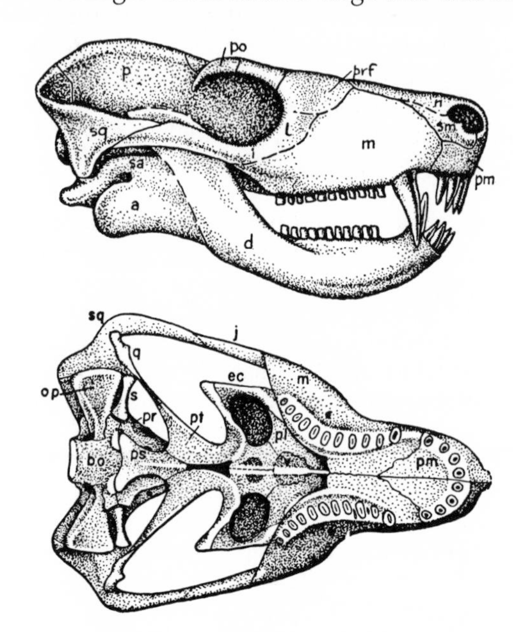

FIG. 282.—*Bauria,* an advanced theriodont, lateral and palatal views of the skull; length about 6 inches. For abbreviations see Figure 109. (After Broom and Boonstra.)

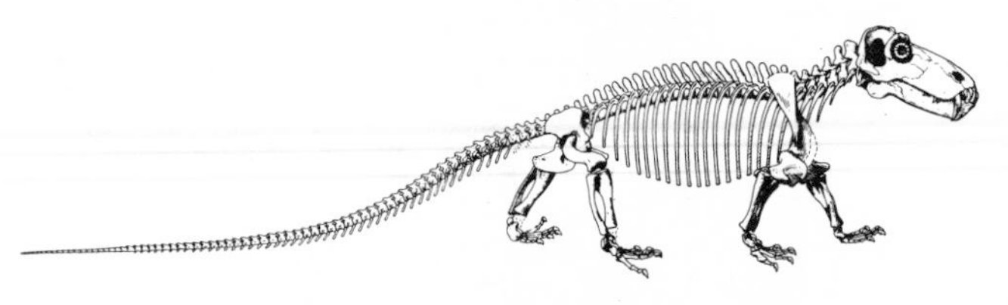

FIG. 283.—The skeleton of the primitive dinocephalian *Titanophoneus* in lateral view; length of original about 6.5 feet. (After Orlov.)

build and a massive dome of bone atop the skull; the canines are reduced to the general level of the other teeth, and nearly the entire dentition is of the chisel-shaped cropping type mentioned above.

More advanced anomodonts were not long in making their appearance. We may mention briefly the dromasaurs, such as *Galepus* (Fig. 286) of the Middle Permian of South Africa. They are tiny, lightly built animals, presumably a sterile side branch of the group. Their skulls show clearly a feature characteristically

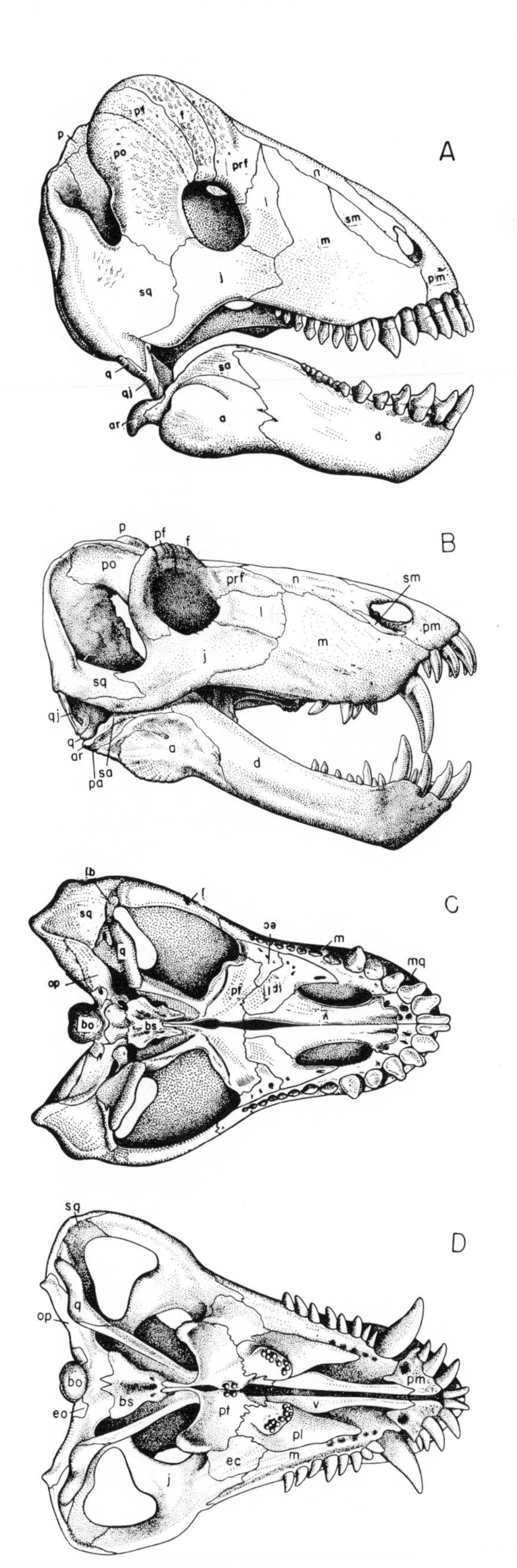

FIG. 284.—The skull in dinocephalians. *A*, Side view, and *C*, palate, of the herbivore *Ulemosaurus;* length about 16 inches (after Efremov); *B*, *D*, similar views of the carnivore *Titanophoneus;* length about 16 inches (after Orlov). For abbreviations see Figure 109.

developed in anomodonts—the jaw articulation is far below the level of the tooth row and is supported by a descending branch of the squamosal. This characteristic was already apparent in many dinocephalians and is very prominent in the dicynodonts.

The dicynodonts (Figs. 287, 288) were the most successful, temporarily at least, of all anomodonts—indeed, of all therapsids. They appear in the Middle Permian beds of South Africa, and, as apparently well-adapted herbivores, are by far the most abundant animals in the deposits of the late Permian; perhaps 90

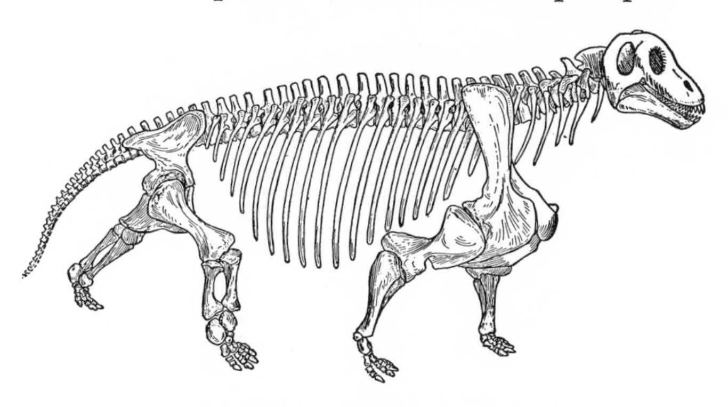

FIG. 285.—The herbivorous dinocephalian *Moschops;* about 8 feet long. (After Gregory.)

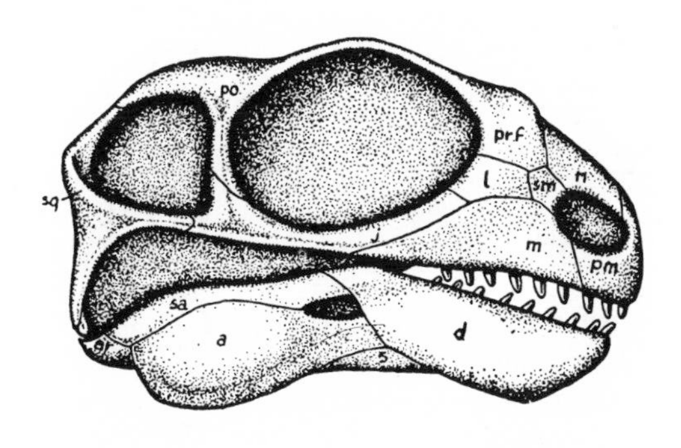

FIG. 286.—*Galepus*, a dromasaurian; length about 1.5 inches. *s*, Splenial. For other abbreviations see Figure 109.

per cent of skulls discovered in beds of this age are those of dicynodonts. In the Triassic their abundance decreases, but large forms appear to have been widespread during the middle and late Triassic, and are present in such varied areas as South America, western North America, and China.

The dicynodonts are for the most part very uniform in their general structural pattern, and in consequence we find a host of forms of various dates and areas assigned to the typical genus *Dicynodon*. The skull pattern is distinctive; there is a long postorbital region, with an enormous fenestra for jaw muscles, bounded below by a jugal bar; at the back, a highly developed process, formed by the squamosal and quadratojugal, runs far down and forward to support the jaw. The dentition is highly specialized. There is a toothless beak, obviously covered by a turtle-like horny bill, and although some genera preserve small teeth in the cheek region, it seems rather certain that horny pads were present on the jaws here as well. In most cases the only teeth remaining are a pair of large upper canine tusks (whence the name), and even these are absent in various instances (possibly a sex difference in some cases).

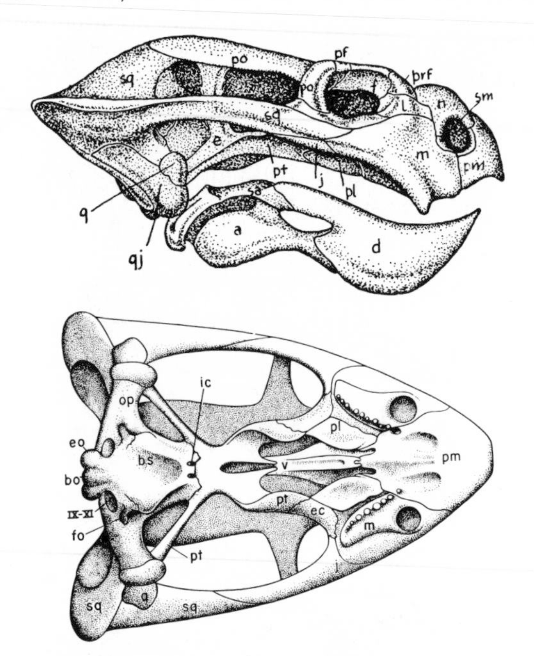

Fig. 287.—*Above,* A *Dicynodon* skull in side view; *below,* palate of *Synostocephalus,* a dicynodont retaining rudimentary cheek teeth. Abbreviations as in Figure 109. (After Broom.)

Many of the Permian forms were small, with skulls only a few inches in length. However, as mentioned above, there was a trend for size increase in the Triassic. *Kannemeyeria* (Fig. 288), the common early Triassic genus, had a skull nearly 2 feet long, and in later Triassic forms, such as *Placerias* of North America and *Ischigualastia* of Argentina, skulls may be a yard in length. Possibly the abundance of dicynodonts may have been due to marsh-loving habits, and one early Triassic genus, *Lystrosaurus* (known from Asia as well as South Africa), is believed to have been aquatic. It is clear that the dicynodonts, although presumably related to the dinocephalians, must have begun a separate career at a very early stage of anomodont evolution. *Venjukovia* of the early Middle Permian of Russia, while still retaining a complete dentition, appears to be ancestral to the dicynodonts, and shows skull proportions intermediate between those of primitive dinocephalians and dicynodonts.

Therapsid to mammal.—In the varied therapsid types, we span nearly the entire evolutionary gap between a primitive reptile and a mammal. In the skull the more primitive pelycosaurs had few features (apart from the small lateral temporal opening) not present in cotylosaurs; but in one group or another of the more advanced therapsids almost every diagnostic feature of mammals had been attained. A considerable number of bony elements had been lost, the pineal eye had been eliminated, the quadrate and quadratojugal were much reduced, the temporal opening had expanded, and the bar behind the orbit sometimes disappeared. The occipital condyle was double in higher types as in mammals, and the palate had evolved strongly in a mammalian direction. The jaw elements were never reduced to the dentary alone, as in a mammal; but the other elements had become quite small. The teeth were generally differentiated into incisors, canines, and molars, much as in mammals.

The limb skeleton, too, approached the mammalian condition (cf. Figs. 269–74). The pelycosaurs still had the primitive type of limb, with the proximal elements held out horizontally from the body. In the higher therapsids, the limbs had swung around practically into the fore-and-aft position of the mammals and had gone through a series of structural changes which were a logical consequence of this changed posture; these changes included such features as the development of a spine and acromion on the scapula, a forward growth of the ilium, the development of an opening between pubis and ischium, and the reduction of the phalanges in many types to a formula of 2-3-3-3-3. The skeleton of the advanced therapsids was almost as mammal-like as that of such a primitive mammal as the duckbill of Australia.

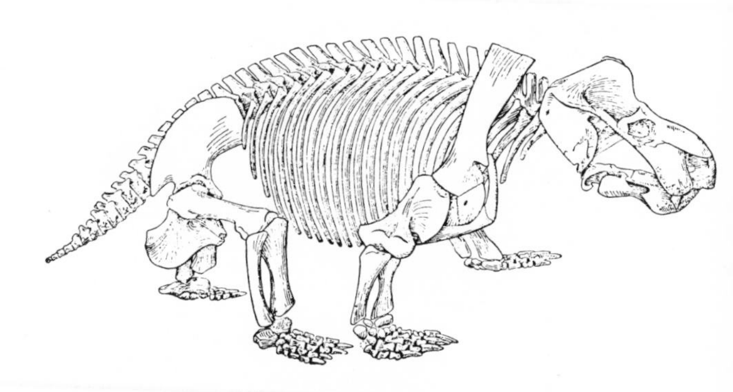

Fig. 288.—The skeleton of *Kannemeyeria,* a Lower Triassic dicynodont; length about 6 feet. (From Pearson.)

It is certain that such mammalian characters as the reduction in phalanges and the formation of a secondary palate had been independently acquired in several therapsid groups. The cynodonts are sometimes thought to be mammalian ancestors, but there are minor features which debar them, some believe, from such a position. *Bauria* in some features is closer still but may likewise prove to be somewhat off the path. But although the details of the phylogenetic history are still uncertain, the therapsid ancestry of mammals seems established.

By late Triassic times the typical therapsid groups so far described had apparently become extinct. In the mid-Jurassic there appear forms which can be definitely classified as mammals. Between, there lies an evolutionary "no-man's-land," a time when the transition from reptiles to mammals was occurring. Unfortunately, our knowledge of this transition is still poor. In a variety of continental formations which lie close to the Triassic-Jurassic boundary there is a gradually increasing number of specimens of animals morphologically astride the reptile-mammal boundary. Such specimens, however, are frequently fragmentary in nature and difficult to place in one class or the other.

The reasons for our difficulties in interpretation and classification of this fragmentary material are apparent. Many of the diagnostic features of the mammals lie in their soft anatomy and reproductive processes, and there is little or nothing to be learned of such matters from fossils. In the skeleton, we have seen that the advanced therapsids have already bridged most of the gaps between the typical reptilian organization and that of the Mammalia. It is generally agreed that the dividing line may be reasonably (if arbitrarily) established, osteologically, at the point where the "extra" jaw elements are lost or greatly reduced and changed in function and the dentary of the lower jaw articulates directly with the squamosal. But these postdentary elements are already small in advanced therapsids, and we can be sure that there was no sudden "jump" from

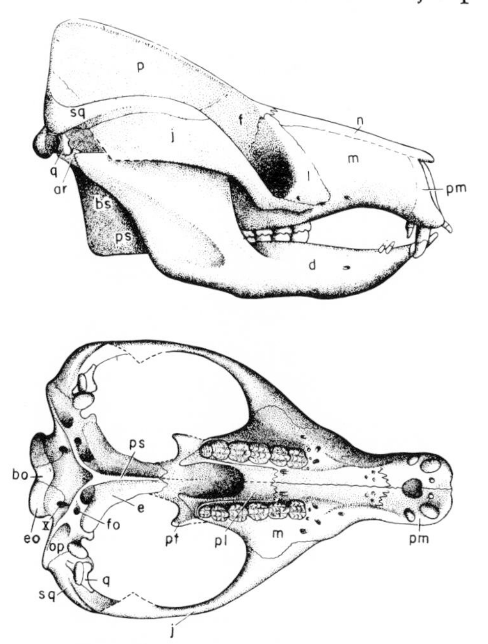

Fig. 289.—The skull of the tritylodont *Oligokyphus* in lateral and ventral views; original about 3.5 inches long. Abbreviations as in Figure 109. (After Kühne.)

one type of articulation to another. A fraction of the specimens of this age seem to be referable to the Mammalia and will be mentioned in the following chapter. Others are still obscure. Here we shall mention two groups from this border zone which can, it seems, still be classified among the reptiles, although advanced in nature.

Nearly a century ago there was found in the late Triassic of South Africa a partial skull with a peculiar dentition to which the name *Tritylodon* was given. Its nature and zoological position were obscure. In recent times, however, additional material of animals of this sort has been found in a variety of places, not only in South Africa, but in Europe (*Oligokyphus;* Figs. 289, 291), China (*Bienotherium;* Fig. 290), and North

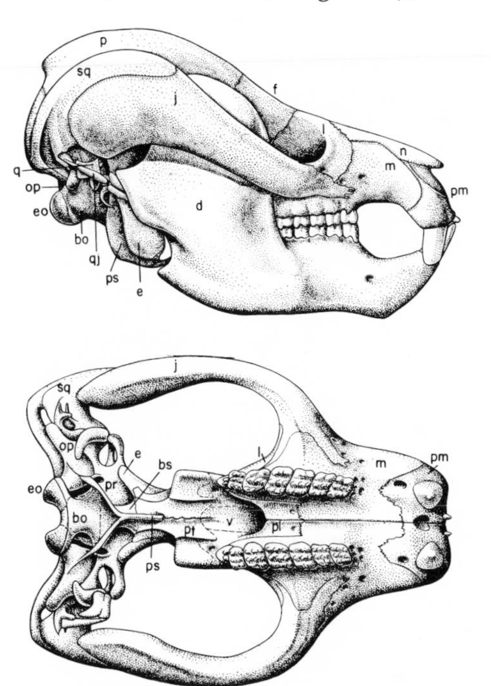

Fig. 290.—The skull of the tritylodont *Bienotherium* in lateral and ventral views; original about 5 inches long. Abbreviations as in Figure 109. (After Hopson.)

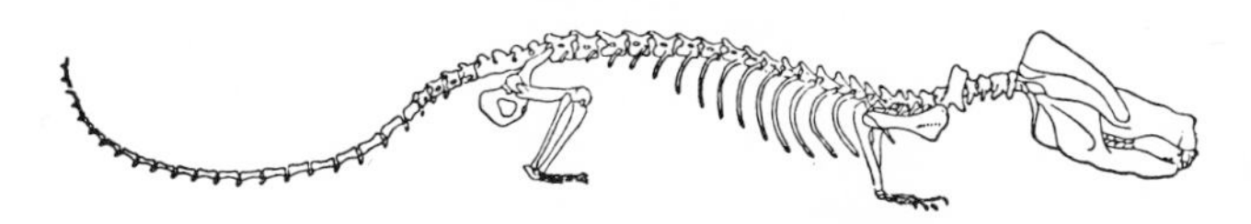

Fig. 291.—The skeleton of *Oligokyphus;* original about 20 inches long. Abbreviations as in Figure 109. (After Kühne.)

America, and much of the structure is now apparent. Most characteristic is the dentition. Anteriorly there are large incisors; then following a gap (a diastema), a battery of highly developed cheek teeth with two to four longitudinal rows of cusps appears. A dentition of this general sort has developed, in comparable fashion, in various mammals (Mesozoic multituberculates, herbivorous marsupials, rodents). The tritylodonts, which, however, are merely parallel in dental evolution to these mammals, are still to be considered therapsids, for while the "extra" bones of the lower jaw are much reduced and the very large dentary is close to the point of meeting the squamosal, the old articular-quadrate joint persists. The skull, as can be seen, is very close to the mammalian condition in many regards. There is a highly developed secondary palate;

the postorbital bar has completely disappeared, and prefrontal and postorbital bones have vanished; posteriorly there is a high and narrow sagittal crest of mammal-like appearance above the small braincase.

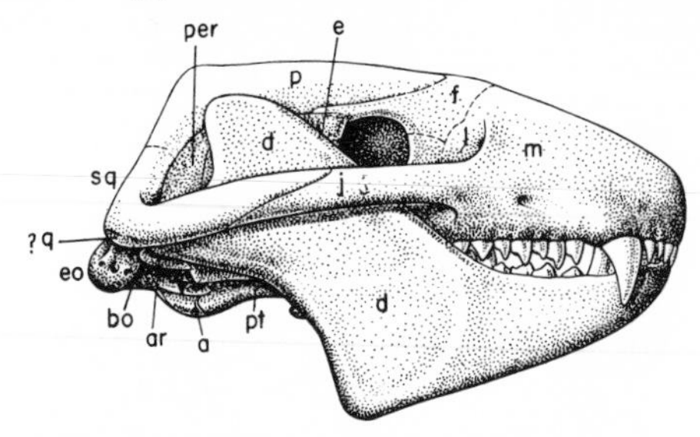

FIG. 292.—Skull of the advanced therapsid *Diarthrognathus;* original about 1.5 inches long. Abbreviations as in Figure 109. (After Crompton.)

The origin of the tritylodonts is uncertain, but derivation from cynodonts (which, alone among therapsids, tend to develop crowned molars) is reasonable. The tritylodonts were the last survivors of the Therapsida, for a last lingering genus has been found in the Middle Jurassic of England.

Less specialized than the tritylodonts and seemingly closer to mammalian ancestry are the Ictidosauria of the late Triassic, of which *Diarthrognathus* (Fig. 292) of South Africa is the only surely assignable form. This is a small animal, lacking the dental specializations of the tritylodonts but equally advanced in many cranial characters. In one regard it is a step more advanced than the tritylodonts. As described, the old quadrate-articular joint is still present; but close to and forming a minor part of the same joint is a squamosal-dentary contact. Thus *Diarthrognathus*, possibly of bauriamorph descent, appears to be almost exactly on the boundary between reptiles and mammals, and the ictidosaurs, when better known, may prove to be true ancestors for at least part of the Mammalia.

Of extremely doubtful nature are molar teeth present in considerable numbers in European Rhaetic bone beds. Formerly called *Microlestes* or *Microleptes*, but now termed *Thomasia* and *Haramiya*, they are of very small size, two-rooted, with oval crowns on which there is a central basin surrounded by a cuspidate rim. There is no real clue as to their affinities, therapsid or mammal, but the tooth pattern gives a faint suggestion of ictidosaur relationships.

CHAPTER 15

The Structure of Mammals

We now enter upon a consideration of the mammals, the most intelligent and the most successful of land vertebrates. Arising, it seems, at about the end of the Triassic from the mammal-like reptile stock, the group remained comparatively unimportant until the extinction of the great reptiles at the end of the Mesozoic. From that time on, however, mammals increased rapidly in numbers and diversification, so that now they include the greater part of the animals with which one is ordinarily familiar. They not only inhabit the surface of the earth but have invaded the air (bats) and returned to the seas (whales, seals, sirenians). In size they range from tiny shrews and mice to certain of the whales which are the largest of all known animals, exceeding even the greatest dinosaurs in size. Mammals are of particular interest to man, not only because they include many of his animal friends and enemies and much of his food supply, but also because he himself is a member of this group.

Diagnostic characters.—Many of the diagnostic features of mammals lie in their soft anatomy and hence cannot be used in paleontology. As the name implies, they alone among vertebrates suckle their young after birth. There is thus a period of infancy and "education" which, it would seem, enables mammals to make better use of their finer physical and mental equipment. The young are, except in the most primitive forms, born alive, the shell-less egg being retained within the mother's uterus. The yolk dwindles, and in most cases a connection, the placenta, is formed by which the young can be nourished by transfer between the blood streams of the mother and the young. Mammals are warm-blooded, that is, they maintain a high body temperature with a consequent potentiality of maintenance of continuous activity; the presence of hair (apparently originally developed between the scales) aids in the preservation of the body heat. Horny scales are only rarely present; superficial dermal ossifications are uncommon. There is a four-chambered heart, as in birds, with a perfect separation of the oxygenated and the nonoxygenated blood, and only one arch of the aorta; but, whereas in birds the arch is the right, here it is the left.

The brain is much enlarged (Fig. 293), the growth of the higher centers in the cerebral hemispheres being the most characteristic change; the hemispheres grow upward and backward, commonly cover the midbrain and often overlie the cerebellum, and usually further increase their area of superficial gray matter by an infolding of the surface. In the brains of the higher mammals the entire dorsal surface of the cerebrum represents a new outgrowth and infolding of nervous tissue, collectively called the neopallium or neocortex, the "new brain"; this has overgrown the old brain, forming a "supercontrol" system that dominates the old vertebrate brain beneath it. By its means, memory is greatly enriched, and the animal is enabled to make a more or less intelligent choice between conflicting sensory stimuli, guided by memory of past reactions and their results. It is of interest that in many groups of mammals (the horses, for example) there was a strong tendency for brain growth during Tertiary times; the older members of such groups had relatively much smaller brains than their late Tertiary and modern representatives.

It is often blithely assumed that brain size, or at least brain size in proportion to body bulk, is a simple index of intelligence in mammals. Not so. The structure and degree of folding of the cortex weigh heavily. Further, while in any given group the brain shows some increase in *absolute* size in large representatives (since there are more sensory and motor connections to be cared for), it is generally *relatively* small—size increase is not needed for maintenance of intelligence.

Activity guided by intelligence is the keynote of mammalian success, contrasting with the sloth and stupidity of reptiles. The warm-blooded condition has rendered possible the activity without which the many skeletal adaptations of mammals would be useless; the mammalian mode of development has rendered possible the ontogenetic elaboration of the complicated brain structure.

Among osteological characters which are usually considered as diagnostic are the following: the double condyle of the skull; the presence of epiphyses on certain, at least, of the bones; only one element (dentary) on each side of the lower jaw, this element articulating with the squamosal; the tympanic membrane supported by a tympanic bone of some sort; the presence of three auditory ossicles; the presence of only marginal teeth in the jaws; only two definite tooth generations; two or more roots in the cheek teeth; and a single bony nasal opening. Some of these characters, however, were already present or foreshadowed in the therapsids, as we have seen.

Skeleton.—The study of the skeleton in mammals presents an entirely different problem from that encountered in the reptiles. In that class there is an extremely wide variation in skeletal patterns and in the elements present. The mammalian skeleton has evolved from the type seen in the mammal-like reptiles and thus is, in a sense, highly specialized to begin with. Once established, however, it has varied but little. Almost never are new skeletal elements added, and seldom are elements lost, except in the distal portions of the limbs in relation to locomotor adaptations. But, although the bones of the skeleton are essentially stabilized, tooth structure is, as will be seen, exceedingly variable. Teeth are, from their composition, the hardest of skeletal materials and the parts most frequently preserved as fossils. In consequence, dentition and limb structure, as the two most variable and diagnostic features of the mammalian skeleton, will bulk large in our consideration of the various groups.

The mammalian skeleton contrasts with that of reptiles in its thorough ossification; exceedingly little cartilage is present in the adult. One development in this regard deserves particular mention—the epiphyses (Fig. 294). In reptiles the articular ends of bones are frequently covered by cartilage; growth in length of the bones takes place by the gradual replacement of this cartilage by bone. Such a process, however, hinders the development of a firm articulation of one element with another. In mammals this difficulty has, so to speak, been solved: the articular surface is formed of bone and finished early in life. But between the ends and the main body of the bone there is a layer of cartilage in which additional bone is laid down. Lengthening of a bone may thus take place without interference with its articulations. When the element has reached its maximum size, this zone of cartilage disappears, and the end segments, known as epiphyses, fuse with the main body of the element. After their fusion, growth is over. This is in contrast with typical reptilian conditions, in which no definite adult size is attained and growth may continue, seemingly, throughout life; epiphyses are rarely present.

Skull.—The interpretation of the mammalian skull (Fig. 295) is rendered easier if the reptilian ground plan upon which it is based is kept in mind. It is a highly complex structure, differing from that of therapsids mainly in features connected with the great expansion of the brain and with changes in the nasal apparatus and the auditory mechanism. Frequently neglected by superficial workers, the braincase, with its various foramina and canals and blood vessels, can yield important evidence as to relationships.

As seen in dorsal view, the greatest change from the reptilian condition is that caused by the growth of the brain and the consequent swelling-out of the braincase; the temporal arches stand out from the sides of the skull as if they were appendages to it rather than part of the original roof. There is no pineal eye, and the bony external nares are fused into a common anterior opening. There is the normal tetrapod series of parietals, frontals, nasals, and premaxillae. The parietals spread down over the sides of the braincase. If the brain is small, the jaw muscles meet above, causing a median sagittal crest to develop between them. This structure is most likely to be present in large forms, since the brain does not increase in proportion to the rest of the body. There is no bar behind the orbit in primitive mammals, and the reptilian postorbital bone is gone. The bar may re-form in advanced members of many groups, but this bone does not return. The pre-

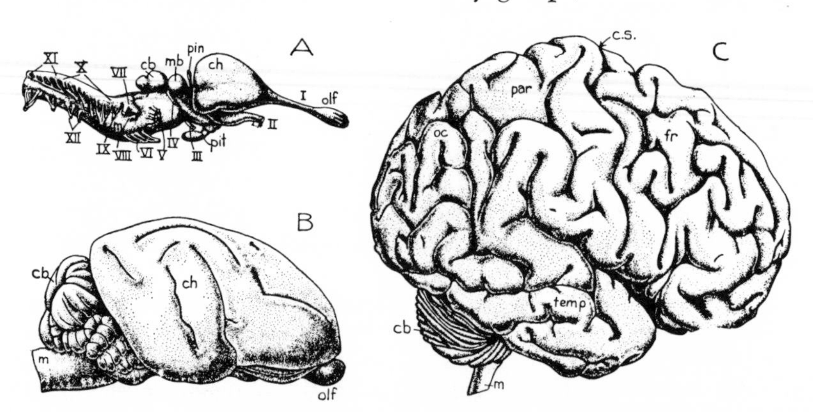

FIG. 293.—The brain in *A*, a reptile (alligator); *B*, a lemur; *C*, a human (newborn child). In all, the brain stem is drawn at about the same actual length; the enormous increase in the relative size of the cerebral hemispheres (*ch*) may be noted. In the alligator the cranial nerves are numbered. Abbreviations: *olf*, olfactory bulb; *pin*, pineal eye; *mb*, midbrain region, with optic lobes above; *pit*, pituitary; *cb*, cerebellum; *m*, medulla oblongata.

In the cerebral hemispheres of man the various lobes—frontal, parietal, occipital, and temporal—are indicated, as well as the central sulcus (*c.s.*) dividing the motor area from the sensory area behind it.

frontal and postfrontal are likewise absent; the lacrimal is small and confined to the neighborhood of the orbit and the tear duct. In front, the nasals project freely over the nares, and the premaxillae push up to meet the nasals on either side. The maxillae are large and deep, as they were in mammal-like reptiles. The jugal lies beneath the orbit and takes part in the formation of the temporal arch. The squamosal forms the hinder end of this bar and also the new jaw articulation—the glenoid cavity—as well as extending up over the lateral wall of the braincase. In contrast to reptiles, quadratojugal and quadrate are absent from their former place at the back corner of the skull.

As in reptiles, there is usually a considerable amount of fusion of the elements which make up the occipital plate. The basioccipital, lying below the foramen magnum, is fused with the exoccipitals. The latter form much of the paired condyles, extend out laterally into the paroccipital processes, and may fuse in turn with the supraoccipital. Embryology often reveals a distinct postparietal (dermal supraoccipital) which usually fuses with the supraoccipital. There is usually a transverse (lambdoidal) crest across the upper margin of the occiput for the attachment of neck muscles.

On the underside of the braincase the basisphenoid lies in front of the basioccipital, bounded posterolaterally by the auditory region and internally containing a cavity (sella turcica) in which the pituitary body is lodged. At the side of the basisphenoid, and generally fused with it, is the alisphenoid, which extends up as a wing around the side of the braincase and meets the roofing bones above. This has been derived from the epipterygoid—a bone of the upper-jaw region of lower vertebrates, here pressed into service to cover what otherwise would have been a gap in the wall of the expanding mammalian braincase. Anterior to the basisphenoid is the presphenoid, which forms the most interior part of the floor of the brain cavity. On either side is a plate of bone termed the orbitosphenoid, but which is continuous with the presphenoid and essentially a part of it. These bones are seemingly derived from the reptilian (and amphibian) sphenethmoid. Still farther forward in the braincase of many mammals there lies, at the hind end of the nasal chamber, a mesethmoid element. From this element and from the inner portions of the nasals and maxillae, scrolls of bone extend into and subdivide the nasal cavities. These are the turbinals which are covered in life with mucous membranes.

In the front part of the mouth the bony structure is not unlike that of the higher therapsids. The secondary palate is formed by extensions from the premaxillae, maxillae, and palatines back to the internal nares. Behind the palatine is the pterygoid, closely articulated to the basicranial region. It no longer reaches back to the quadrate region and ends behind in a winglike structure. (The ectopterygoid has vanished.) Below the braincase, in front of the presphenoid and partially concealed from ventral view by the secondary palate, is a long, slim median bone, of dermal origin—the vomer. This is probably derived by fusion from the paired elements to which we have applied this name in lower vertebrates (although another theory would derive the vomer from the reptilian parasphenoid).

Auditory region.—The capsule enclosing the inner-ear region in reptiles and primitive amphibians (cf. Fig. 109) was formed of two bones—proötic and opisthotic—usually fused, lying well back in the lateral wall of the braincase, and having a lateral opening (fenestra ovalis) on the outer surface, with which the stapes articulated. In mammals the otic capsule is a single compact structure—the petrosal. This is in contact posteriorly and ventrally with the occipital complex and is covered superficially in great measure by the squamosal. The petrosal is frequently exposed, as the mastoid process or region, at the posterolateral margin of the skull between these two adjacent elements.

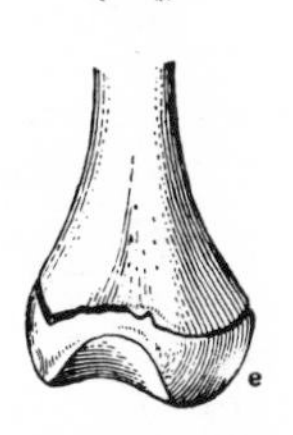

FIG. 294.—A mammalian femur at an immature stage. (For economy of space, much of the shaft has been omitted.) The two articular regions and the principal points of muscular attachment long remain as separate pieces of bone—epiphyses (*e*)—between which and the shaft there remain areas of cartilage in which new bone may form.

In mammals the eardrum tends to be surrounded and supported by a small bone termed the tympanic, primitively a ring-shaped or horseshoe-shaped structure, which is thought to be a remnant of the angular bone of the reptilian lower jaw. Between eardrum and otic capsule in mammals is the air-filled middle-ear cavity, containing the delicate auditory ossicles discussed below. There is a tendency in most mammalian groups for this cavity to be surrounded by bone, which may form an auditory bulla. This fuses with the petrosal, the fused mass being termed the periotic. The bulla is, however, formed in somewhat variable fashion. The tympanic bone may grow inward from the eardrum to form the entire structure; in many forms, however, a new element, termed the entotympanic, not present in any reptile, takes part in bulla formation.

Skull openings.—The various openings in the skull (foramina and fenestrae) are numerous, complicated, and important. They may be roughly divided into (1) openings out of the braincase for cranial nerves and (2) openings for blood vessels and various ducts.

Mammals possess twelve cranial nerves, passing directly from the brain to the sense organs and muscles of the head and throat. We shall briefly list these and their more usual relations to the skull openings (Fig. 295*E, F*). Nerve I, the olfactory, reaches the nasal chamber by perforating the anterior end of the braincase in the ethmoid region. Nerve II, the optic, passes out to the eye through the optic foramen in the orbitosphenoid. The three small nerves to the muscles of the eye (III, IV, VI) and part of the large nerve V (trigeminal) pass out through the foramen lacerum anterius (or foramen sphenorbitale) between orbitosphenoid and alisphenoid inside the cavity of the orbit. Two other branches of nerve V generally pass out through the foramen rotundum and foramen ovale, both in the alisphenoid bone. Nerve VII (facial) passes through the petrosal and emerges at the edge of the skull from the stylomastoid foramen. Nerve VIII, the auditory nerve, passes into the petrosal to the inner ear and does not, of course, reach the surface of the skull Nerves IX, X, and XI (glossopharyngeal, vagus, and accessory) pass out through the posterior lacerate or jugular foramen between the basioccipital and the ear capsule, while XII (hypoglossal) sometimes accompanies them but originally emerged through an anterior condyloid foramen in the exoccipital.

The lacrimal duct usually pierces the lacrimal bone at the anterior edge of the orbit and runs inward to reach the nose. An infraorbital canal carries blood vessels and nerves forward from the orbit through the maxilla into the face. On the palate the incisive foram-

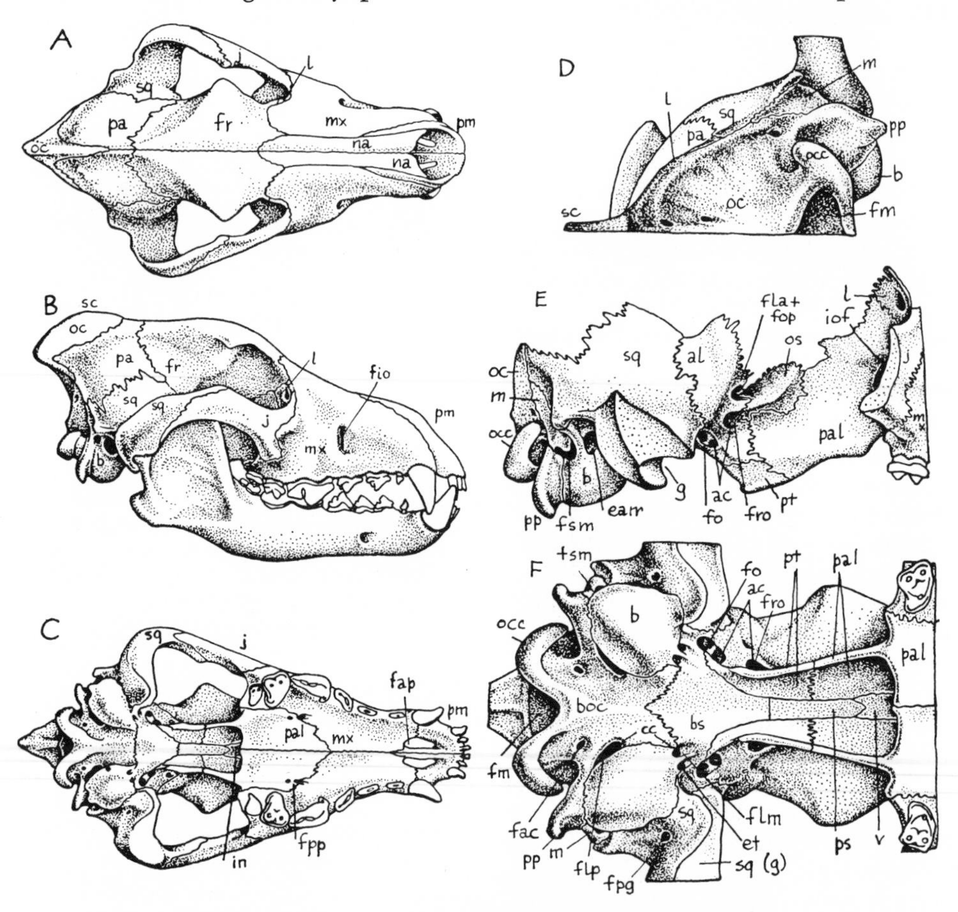

FIG. 295.—The skull of the Pleistocene wolf, *Canis dirus*, to show the structure of the mammalian skull. *A*, Dorsal, *B*, lateral, and *C*, ventral views. *D*, Occipital view of right half of skull (top at left of figure). *E*, Lateral view of posterior part of skull with zygomatic arch and roofing bones removed. *F*, Ventral view of same. Abbreviations: *ac*, alisphenoid canal; *al*, alisphenoid; *b*, auditory bulla; *boc*, basioccipital; *bs*, basisphenoid; *cc*, carotid canal; *eam*, external auditory meatus; *et*, eustachian tube; *fac*, anterior condyloid foramen; *fap*, anterior palatine foramen; *fio*, infraorbital foramen; *fla*, anterior lacerate (sphenorbital) foramen; *flm*, median lacerate foramen; *flp*, posterior lacerate foramen; *fm*, foramen magnum; *fo*, foramen ovale; *fop*, optic foramen (here fused with *fla*); *fpg*, postglenoid foramen (transmitting a nerve); *fpp*, posterior palatine foramen; *fr*, frontal; *fro*, foramen rotundum; *fsm*, stylomastoid foramen; *g*, glenoid fossa (jaw articulation); *in*, internal nares; *iof*, infraorbital foramen; *j*, jugal; *l*, lacrimal carrying tear duct, also lambdoidal (transverse occipital) crest; *m*, mastoid portion of periotic; *mx*, maxilla; *na*, nasal; *oc*, occipital (including basioccipital, exoccipital and supraoccipital); *occ*, occipital condyle; *os*, orbitosphenoid; *pa*, parietal; *pal*, palatine; *per*, periotic; *pm*, premaxilla; *pp*, paroccipital process of exoccipital; *ps*, presphenoid; *pt*, pterygoid; *sc*, sagittal crest; *sq*, squamosal; *v*, vomer.

ina between premaxillae and maxillae are openings into the mouth in many mammals for Jacobson's organ (absent in man)—an accessory nasal pouch for smelling mouth contents; farther posteriorly the palatine foramina transmit blood vessels. There is often an alisphenoid canal carrying an artery forward through the base of the alisphenoid bone. The internal carotid artery which supplies blood to the brain usually passes forward from the hind end of the skull through a carotid canal running forward under cover of the petrosal from near the posterior lacerate foramen, while at the front end of that bone the artery passes up into the braincase through the foramen lacerum medium, an opening between petrosal and alisphenoid. In this region there is an opening into the auditory bulla for the eustachian tube leading from throat to middle ear. In the jaw there is a foramen (inferior dental) on the inner side near the posterior end, through which enter nerves and blood vessels, and a mental foramen well forward on the outer side to give a blood and nerve supply to the chin and lower-lip region.

Jaw and auditory ossicles.—In the lower jaw there is but a single element on each side—the dentary. This has an ascending ramus with a coronoid process (for the temporal muscle which closes the jaw) and an articular process. At the lower back corner of the jaw there is often an angular process for muscle attachment.

Of the other elements of the reptilian jaw (cf. Fig. 278), the splenial, coronoid, and surangular seem to have disappeared in mammals. The angular has become the tympanic bone. This leaves the articular (with which the prearticular is fused) and, in the upper jaw, the quadrate (to which in mammal-like reptiles the quadratojugal was fused). The fate of these two bones has been a curious one.

In mammal-like reptiles the sole auditory ossicle—the stapes—ran out to touch the quadrate; this element, in turn, was in contact with the articular bone of the jaw (Fig. 116). In the mammalian auditory region this "stirrup bone" is the inner member of a chain of three ossicles, an outer one—the malleus or "hammer"—receiving vibrations from the eardrum and transmitting them to the stapes through the incus or "anvil." The origin of these two "new" elements was long disputed. But a study of their embryology confirms the conclusion to which a consideration of the jaw changes undergone in the evolution of mammals leads us. The malleus is the old articular; the incus, the reptilian quadrate. Once important bones, they had become useless for their original purpose and, lying close to the auditory region, were salvaged and put to a new use. This change was presumably facilitated by the position of the eardrum, close to the jaw region, in mammal-like reptiles. Very likely quadrate and articular had begun to function in sound transmission before they had lost their function as jaw elements.

The history of the auditory ossicles is one of the best examples of a change of function to be found in vertebrates. These elements, it will be recalled, were once

part of the shark jaw apparatus; and this, in turn, was derived originally from branchial bars. Accessory breathing organs were transformed into biting structures and these, finally, into part of the hearing apparatus.

Some portions of the reptilian branchial apparatus, especially the hyoid, remain in the tongue and throat of mammals, but they are paleontologically unimportant.

Dentition.—Teeth in mammals are confined to the margins of the jaws. They consist (Fig. 296) of incisors, canines, premolars, and molars. In the upper jaw the incisors are in the premaxilla, the others in the maxilla. All the teeth except the molars have, normally, a "milk" or deciduous set preceding the permanent ones. The deciduous teeth appear, in general, in rather regular order from front to back (although canines

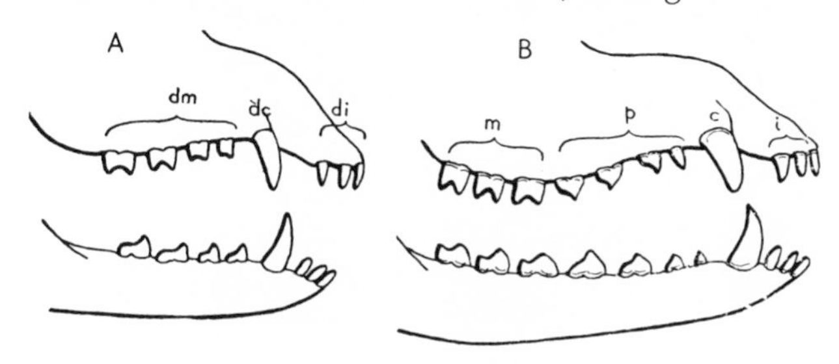

FIG. 296.—Diagrams to show the types and numbers of teeth in a primitive placental mammal. *A*, Deciduous ("milk") teeth. Abbreviations: *di*, deciduous incisors; *dc*, deciduous canine; *dm*, deciduous premolars ("milk molars"). *B*, Permanent dentition. Abbreviations: *i*, incisors; *c*, canine; *p*, premolars; *m*, molars.

with heavy roots may lag a bit in their appearance). When the premolar row has been completed, the molars appear in order, posterior to them. But meanwhile the replacement of the anterior teeth begins again in fore-to-aft sequence. The molars are "permanent" teeth but are actually posterior, unreplaced members of the same set of teeth to which belong the deciduous teeth anterior to them. This tends to explain the fact that in many cases the "milk" premolars are more like the molars than are the permanent teeth which replace them.

In a primitive placental mammal there were three incisors in each half of each jaw. No mammal has more than one pair of canines above and below. Four premolars and three molars in each half of the jaw is the number present in primitive placentals. This set of statements may be written simply as a dental formula,

$$\frac{3 \cdot 1 \cdot 4 \cdot 3}{3 \cdot 1 \cdot 4 \cdot 3} \times 2 = 44;$$

and a similar system may be used for any set of teeth. Many primitive mammals had a larger number of teeth; in placentals the formula given is usually the maximum, and there is often a considerable reduction; our own formula, for example, is

$$\frac{2 \cdot 1 \cdot 2 \cdot 3}{2 \cdot 1 \cdot 2 \cdot 3} \times 2 = 32.$$

The incisors are generally nipping teeth of moderate size. In some cases (as in rodents) a pair may be developed as strong chisels, or they become tusks, as in the elephant. The canines are primitively long, stabbing, cone-shaped teeth. They are emphasized in carnivores; in herbivores they are usually reduced or lost but are sometimes retained as defensive weapons. The cheek teeth—premolars and molars—usually have at least two roots in mammals. The four premolars are

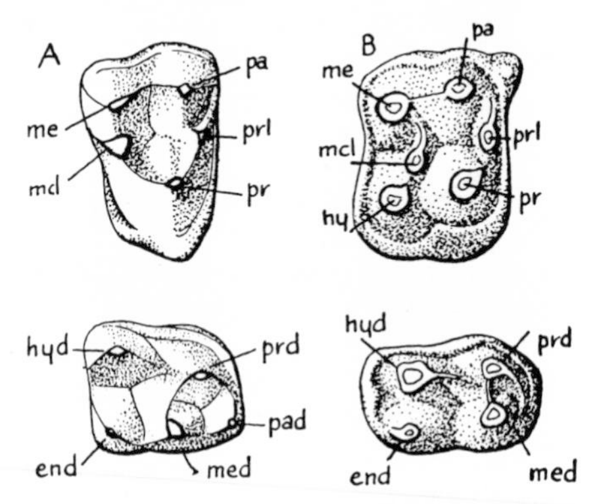

FIG. 297.—Molar teeth of primitive placentals. *Upper row,* Right upper molars; *lower row,* left lower molars. *A, Omomys,* an Eocene tarsioid, with a tritubercular pattern; *B, Hyracotherium,* a Lower Eocene horse, showing the evolved quadritubercular pattern. Abbreviations: upper molars—*hy,* hypocone; *me,* metacone; *mcl,* metaconule; *pa,* paracone; *pcl, prl,* protoconule; *pr,* protocone; lower molars—*end,* entoconid, *hld,* hypoconulid; *hyd,* hypoconid; *med,* metaconid; *pad,* paraconid; *prd,* protoconid. In addition to the cusps shown, primitive lower molars frequently bear a sixth cusp, a hypoconulid, at the posterior margin.

primitively simpler in structure than the molars. In many herbivore types they tend to become more complicated and may resemble the molars. In herbivorous mammals the molars usually take on a complicated pattern, but in carnivores they tend to become reduced.

In primitive mammals there was a closed tooth row, without pronounced gaps; but herbivores tend to develop a gap, or diastema, between the front teeth which secure the food and the cheek teeth which masticate it.

Cheek teeth were primitively low crowned (brachyodont); but in many large animals or in forms in which there is unusual wear, as in gnawing or grass eating, they become high crowned (hypsodont). The roots primitively closed, limiting the amount of crown material formed, but in some forms part or all of the teeth may have a continuous growth.

Molar structure.—Reptilian teeth are usually simple cones. Mammalian molars generally have a broad crown bearing a number of cusps or ridges arranged in a complicated pattern which varies widely from group to group. It was only natural that there should have arisen various theories to account for the origin of molar patterns and to homologize the various cusps and crests found in different types. There have been some fantastic theories based on the idea that each cusp represents an originally separate tooth; however, the tritubercular theory set forth below is generally accepted today in at least certain of its aspects.

The theory starts with the type of molar teeth found in many primitive placentals, particularly Paleocene and Eocene forms (Fig. 297*A*). In the upper jaws the molars are triangular in shape, with three major cusps. A single cusp is found at the inner apex; this was once believed to represent the original reptilian cone and hence is called the protocone. External to this is the paracone; back of the paracone is a third, the metacone. Along a ridge leading out and forward from the protocone there is often a smaller cusp, the protoconule (or preferably paraconule); along a similar posterior ridge, a metaconule. Paracone and metacone may be partially joined by ridges. On the outer margin of the tooth there was primitively a prominent shelf, termed a cingulum; and cingula also tended to be present along the margins anterior and (particularly) posterior to the protocone.

In the lower jaw there was a similar triangle, but with the base inside and the apex outside. Similar names are given to these cusps, but with the suffix *-id:* protoconid, paraconid, metaconid. Back of the triangle, or trigonid, is a low extension, the heel or talonid. This bears two cusps—an external one termed the hypoconid and an internal one, the entoconid; in addition, there is often a third, smaller cusp—the hypoconulid. A cingulum is often present along the anteroexternal margin.

This primitive type of molar tooth, often with fairly sharp cusps, has been termed "tuberculosectorial" or (preferably) "tribosphenic." The triangles are asymmetrical, the front edge of the upper tooth and the hind edge of the lower trigonid being perpendicular to the edge of the jaw. Each upper tooth centers behind its proper lower mate (Fig. 298); and if it were not for the heel below, which received the protocone in its basin, the two teeth would pass each other. A shearing action is obtained by this partial passage, while the partial opposition of the cusps gives a chopping effect.

Mammalian cheek teeth, in contrast to those of typical reptiles, tend to have divided roots. In primitive placental types the upper molars had three roots, one beneath each major cusp, and the lower molars two, one each below trigonid and talonid.

In many carnivorous mammals there has been comparatively little change from the generalized condition described above. In herbivorous or mixed feeders, with the development of a grinding surface, the pattern tends to become complicated (Fig. 297*B*). In the upper molars the tooth tends to square itself up, usually by the addition of a fourth cusp, the hypocone, at the inner back corner; the intermediate cusps may remain or drop out. The lower teeth tend also to square up, but in a more complicated way, by building up the heel to a level of the rest of the tooth and dropping the paraconid from the front end so that the four definitive cusps are protoconid, metaconid, entoconid, and hypoconid.

The cusps present may form various types of patterns. They may remain separate but low and rounded —a bunodont (hillock) type of dentition characteristic of such mixed feeders as men and pigs (Figs. 325, 326, 402*F, G*). Each cusp may tend to expand into a crescent (selenodont type) (Fig. 402*J–P*). Rows of cusps may tend to fuse into ridges (lophodont); for example, the two outer cusps of the upper molars may form a ridge called the ectoloph; an anterior row of cusps may join to form a protoloph; or the back ones may form a metaloph; in lower teeth there may be similar cross crests termed metalophid and hypolophid (Fig. 299). Vertical ridges, or styles, may form on the edge of the teeth to add to the complication of the pattern.

This tritubercular terminology affords an exceedingly useful method for the description of placental molar

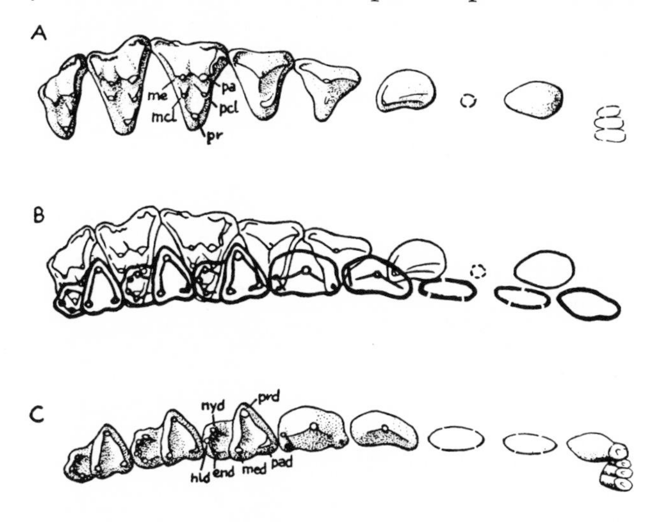

Fig. 298.—Cheek teeth and canines of *Didelphodus*, a primitive placental, to show occlusional relations. *A*, Right upper tooth row; *C*, left lower tooth row; *B*, the two superimposed to show occlusional relations. Each upper tooth lies behind and outside its homologue in the lower jaw. The two rows partially pass each other, with a shearing action along the diagonal line between each upper tooth and the lower tooth behind it. The passage of the upper tooth is stopped by the heel (talonid) on the corresponding lower molar; the protocone fits into the basin of the talonid. For abbreviations see Figure 297. (After Gregory.)

teeth. But the theory originally attempted also to explain the evolution of tritubercular placental teeth from the single-cusped tooth common in reptiles. It assumed that there had developed three cusps in an anteroposterior row, as in the triconodonts described in the next chapter (Figs. 305*A*, 306*A*, 307*B*). Then a rotation of the cusps into the triangular shape seen in the Jurassic symmetrodonts (Figs. 305*B*, 306*B*, 307*C*)

was thought to have taken place; it was assumed that from this condition through the Jurassic pantotheres, or trituberculates, arose the type of tooth found in primitive placentals (Figs. 306*F*, 307*E*).

But this part of the tritubercular theory is far from acceptable. In the first place, it is highly improbable that the triconodonts have anything to do with the evolution of the remaining mammals, and there is no evidence of the supposed rotation of the cusps. Further, in the supposed rotation of cusps it was assumed that the original reptilian cone remained at the apex

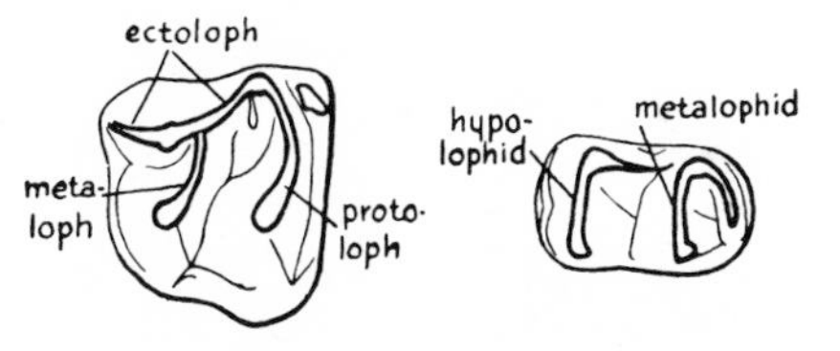

Fig. 299.—Right upper and left lower molars of a fossil rhinoceros, to illustrate a common type of lophodont tooth.

of the triangle, internal above and external below, and it was for this reason that the names protocone and protoconid were applied to the apical cusps. If this were true, it would be expected that this apical cusp, as the oldest historically, would appear first in the embryological development of teeth and that when we trace forward in the tooth row of an animal this cusp would be homologous with the principal cusp of the premolars. This is true in the lower teeth: the protoconid appears to be the primitive reptilian cusp. But in the upper molars, the paracone—not the protocone —is the first to appear in tooth development, and it is the paracone and metacone which are in line with the main cusp of the premolars (Fig. 298). Probably, then, the paracone is really the original reptile cusp; the protocone is a newer development at the internal border, while the metacone was formed by splitting off from the paracone.

But while we must divest the names of phylogenetic meaning, it remains true that, starting with a primitive placental, the tritubercular nomenclature is a very valuable tool for the description and comparison of placental molars.

Postcranial skeleton.—The skeletal features of such a generalized mammal as an opossum or certain insectivores are the logical outcome of the adaptations for rapid quadrupedal progression seen in the therapsids; the opossum is a "living fossil," with a skeleton changed but little from the ancestral mammals of the Mesozoic. The reptilian ancestors of the mammals presumably were ground dwellers. But the flexibility of the skeleton of the more generalized mammals and the fact that "thumb" and "big toe" (pollex and hallux) tend primitively to be set off to a slight extent from the other digits, as if some grasping power were early present, suggest a primitive arboreal trend.

In the axial skeleton (Fig. 300; cf. Figs. 260, 276) there is greater regional differentiation than in reptiles. Ribless lumbar vertebrae are clearly set off from the rib-bearing dorsals. The cervical region also appears to be ribless, for the ribs are short and fuse in the embryo with the vertebrae. The tail is often long but is merely an appendage, not an integral part of the body, as in most reptiles. The sacral vertebrae tend somewhat to increase in number, and the originally separate sacral ribs tend to be lost and to be replaced by the transverse processes.

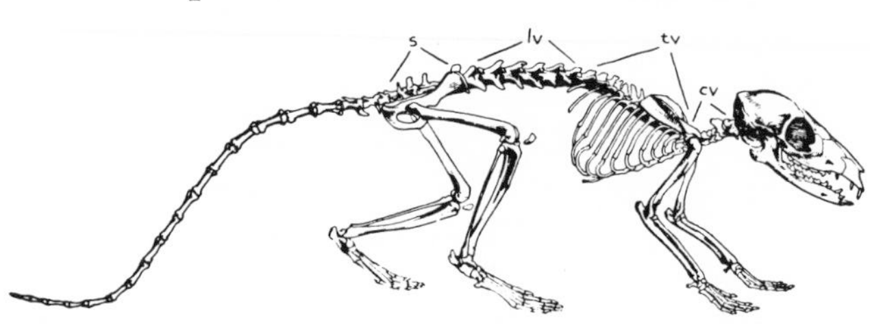

FIG. 300.—Skeleton of the living tree shrew *Tupaia*. Abbreviations: *cv*, cervical vertebrae; *lv*, lumbar vertebrae; *s*, sacrum; *tv*, thoracic vertebrae. (After Gregory.)

The number of vertebrae is much less variable than was the case in reptiles. In general, there are about twenty-five to thirty vertebrae in front of the sacrum. The number of cervicals is almost invariably seven; the long-necked giraffe and the neckless whale differ, not in number of neck vertebrae, but in their proportions. The number of rib-bearing dorsals commonly ranges from about twelve to fourteen, lumbars from five to seven, sacrals from two to five. The caudals are, of course, rather variable; there are rarely as many as two to three dozen. The chief changes from the therapsid condition are seen to be: (1) fusion of the cervical ribs with the transverse processes, (2) loss of ribs in the lumbar region, and (3) tendency for reduction of the tail.

The breastbone, or sternum, unimportant in reptiles, becomes a major feature of the skeleton in mammals. All the longer dorsal ribs attach firmly to it. A tough membrane (the diaphragm) closes off the barrel-like chest region posteriorly. Continued breathing is a necessity for warm-blooded types; this is accomplished by expansion and contraction of the ribs (held together by the sternum) and the diaphragm.

The general contour of the backbone is similar to that of the therapsid ancestors. The cervical vertebrae slant downward from the head. Above the shoulder region, at about the beginning of the dorsal vertebrae, the curve of the back rises, to descend through the lumbar region to the sacrum and tail. The neural spines of the more anterior trunk vertebrae slant backward, those of the lumbar region forward; in the posterior part of the dorsal series there is a transitional (anticlinal) vertebra, whose spine points upward.

The first two vertebrae, the atlas and axis, are of particular interest (Fig. 301). The atlas consists of a ring composed of its own neural arch plus its intercentrum. The axis has, in addition to its own proper elements, the centrum of the atlas attached to its front end as the odontoid process.

Limbs.—The limbs have continued the rotational process noted in our discussion of the therapsids, so that they lie parallel to the body and almost underneath the trunk. Humerus and femur work in a fore-and-aft plane (Fig. 302; cf. Fig. 276), functioning as a

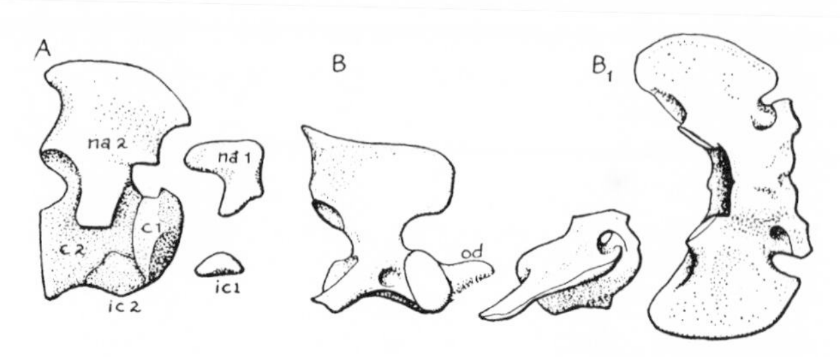

FIG. 301.—The atlas and axis in *A*, the therapsid *Moschops*; *B*, the Oligocene canid *Daphoenus*. Lateral views from right side. B_1, Dorsal view of mammalian atlas, posterior end to the left. In both *A* and *B* the two elements are separated. The atlas consists of the neural arch and the intercentrum of the first vertebra (separated in *A*); these form a ringlike structure in mammals (B_1). In mammals the centrum of the first vertebra is fused with the axis as the odontoid process, fitting into the cavity of the atlas. Abbreviations: *c1*, *c2*, centra of the first and second vertebrae; *ic1*, *ic2*, intercentra of same; *na1*, *na2*, neural arches; *od*, odontoid process of axis (*c1*).

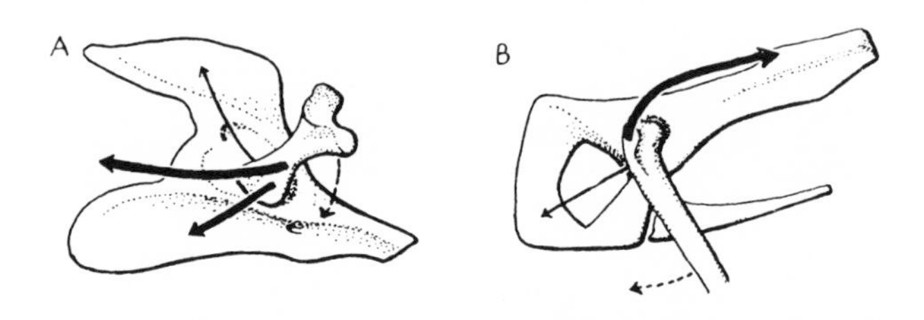

FIG. 302.—Diagram to show the differences in the musculature of the hip region between a pelycosaur (*A*) and a mammal (*B*). In *A* the principal muscles pulling the femur downward and backward (shown in heavy lines) attach to the tail and ventral part of the pelvic girdle. In *B* these muscles are reduced and replaced to a considerable extent by muscles attached to the greater trochanter and inserting on the ilium.

type of lever with the fulcrum (elbow or knee) at one end, the weight to be raised near the other end at the articulation with the girdle, and the major force (muscles) attached just beyond at the very tip (greater tuberosity of the humerus, greater trochanter of the femur). The chief muscles are, consequently, dorsal in position, and therefore the limb girdles tend to be reduced ventrally and expanded above. This is particularly marked in the shoulder girdle (Fig. 269*D*). In typical mammals the interclavicle and the two ventral coracoids have disappeared except for a nubbin of the posterior (true) coracoid element underneath the scapula. The spine of the scapula with which the clavicle articulates is in reality its old front edge; the region anterior to the spine is a new development, an addition on the front for the accommodation of muscles moving up from the coracoid. In the pelvis, too (Fig. 270), there has been a rotation of bone related to rotation of muscles; and the ilium extends far forward, not at all back, while the pubis and ischium are as much behind as below the acetabulum.

The humerus and femur have, of course, changed considerably from the primitive condition because of their rotation. The humerus (Fig. 271) has, near its proximal end, greater and lesser tuberosities for attachment of the muscles connecting it with the body; distally the primitive reptilian entepicondylar foramen is often present. The femur (Figs. 272, 302) has greater and lesser trochanters proximally for muscular attachment and often has a third trochanter more distally (a useful diagnostic character in many cases). In carpus and tarsus we lose, unfortunately, the simple and logical reptilian names for the elements and have to adopt a series of individual ones which vary considerably from author to author. In the carpus (Fig. 303*A*, *B*), instead of radiale, intermedium, and ulnare, we speak of scaphoid, lunar, and cuneiform, while an accessory bone on the outer side is called the pisiform. The centrale retains its name, while the four distal carpals are, in order, the trapezium, trapezoid, magnum, and unciform. In the tarsus (Fig. 303*C*, *D*) the intermedium and fibulare are termed the astragalus and calcaneum. The former develops a rolling joint over which the tibia moves freely; the latter projects back as a prominent heel bone to which the calf muscles attach. The distal tarsals become the internal, middle, and external cuneiforms and the cuboid. The primitive phalangeal formula is that which we have seen develop in some mammal-like reptiles, namely, 2-3-3-3-3 (Figs. 273, 274, 304). The pollex and hallux are usually somewhat shorter and thicker than the other toes and diverge somewhat from them, suggesting a grasping power in the primitive mammals. The feet in primitive forms presumably were armed with sharp claws.

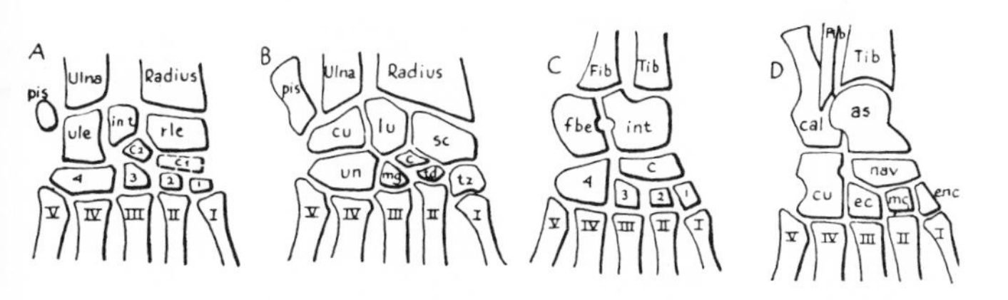

Fig. 303.—Diagrams to show the contrast in nomenclature of the carpal and tarsal elements in mammals and reptiles. *A*, Carpus of reptile; *B*, carpus of mammal; *C*, tarsus of reptile; *D*, tarsus of mammal. Metacarpals and metatarsals in Roman numerals, distal carpals and tarsals of reptiles in Arabic. Abbreviations: *as*, astragalus (talus); *c*, centrale; *cal*, calcaneum; *cu*, cuboid; *ec*, ectocuneiform; *enc*, entocuneiform; *fib*, fibula; *fbe*, fibulare; *int*, intermedium; *lu*, lunar; *mc*, mesocuneiform; *mg*, magnum; *nav*, navicular; *pis*, pisiform; *rle*, radiale; *sc*, scaphoid; *td*, trapezoid; *tib*, tibia; *tz*, trapezium; *un*, unciform; *ule*, ulnare.

Locomotor adaptations.—From the type of skeleton described above, there have been wide variations. Locomotor adaptations to different modes of life are varied and often striking. Among ground dwellers, fast running has evolved in many different groups. The beginning of such an adaptation is seen in the dog; the horse or antelope shows an extreme development. The originally flexible skeleton becomes adapted almost entirely to a fore-and-aft motion of the legs, so that other motions become awkward or impossible. The foot, which originally lay nearly flat on the surface, in a semiplantigrade (palm-walking) position, tends to be lifted from the ground; and the metacarpals and metatarsals elongate, forming an extra limb segment (as in the hind legs of some dinosaurs and birds); the divergent hallux and pollex tend to be reduced and disappear. The animal now walks on the toes (digitigrade). The clavicle disappears, thus freeing the shoulder from any bony connection with the body and breaking the jar of landing on the front feet after a bound.

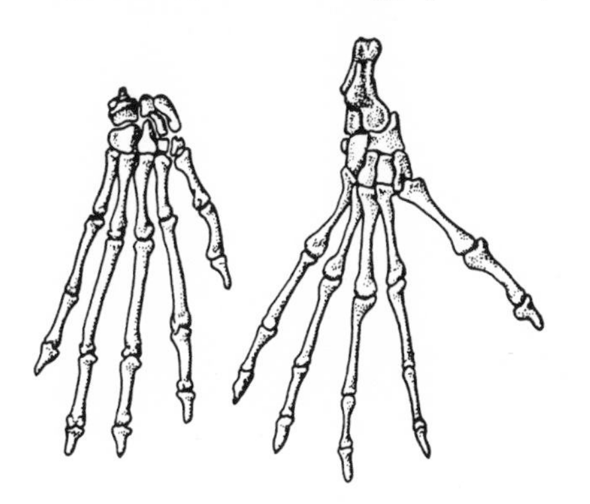

Fig. 304.—Manus and pes of *Notharctus*, an Eocene lemur. (After Gregory.)

The dog has reached a stage such as that outlined above. But many forms have progressed beyond this, with the development of hoofs and further reduction of toes, to the ungulate (hoofed) condition to be considered in a later chapter; while heavy ungulates have, in relation to the carriage of weight, developed ponderous, straight legs analogous to those of the sauropod dinosaurs—these are the graviportal (weight-bearing) types.

Still other types which have progressed considerably toward fast locomotion on the ground are leaping forms, as the kangaroos and jumping rats, in which the body is held partially or fully erect and the propulsive force is a sudden spring with the hind legs.

In a number of groups, especially among primitive

placental insect eaters, underground burrowing types have developed in which the most prominent adaptation is the development of very strong front legs for digging purposes.

In tree-living forms the skeleton remains flexible in correlation with the necessity for agility. Adaptations are present for holding onto the trees, often by an opposable thumb and big toe but sometimes merely through sharp claws for digging into the bark. The tail is usually well developed as a balancing organ and is sometimes prehensile. Some forms tend to swing more especially by the hands (brachiation), with interesting evolutionary results. These adaptations are discussed in detail in connection with the primates. A further tendency on the part of tree-living types has led in a number of groups to the development of gliding forms, in which a membrane has developed in connection with the limbs; and a final stage in this evolutionary process has resulted in the development of true flight in the bats.

Again (as in reptiles), mammals may tend to revert to an aquatic existence. A number of marine mammals, such as seals, whales, sea cows, sea otters, have developed. The limbs are retransformed into finlike organs; the fusiform body shape of the fish tends to be reassumed; a tail fin, and even a dorsal fin, may be redeveloped.

CHAPTER 16

Primitive Mammals

Living mammals are easily divisible into (1) monotremes (the duckbill and spiny anteaters of Australia), which are egg-laying types; (2) marsupials, including the opossums and a large number of Australian forms which bear the young alive but at an immature stage; (3) placentals, in which the young are retained longer within the mother's body and a highly developed placenta is present. The first are called the subclass Prototheria, or primitive mammals; the second and third together may be termed the Theria, or true mammals, with the term Metatheria applied to the marsupials and Eutheria to the placentals. Such divisions are satisfactory for living forms. But the monotremes are almost unknown as fossils, and marsupials and placentals appear only at the end of the Cretaceous. Most Mesozoic fossil forms pertain to orders not particularly close to living groups; and, since we have no knowledge of their mode of reproduction, they are difficult to classify.

The mammal-like reptiles which flourished in Permian and early Triassic times had become very rare by the end of the latter period, and only a single specialized therapsid persisted into the Jurassic. Mammals presumably came into existence toward the end of the Triassic, but we know extremely little about their history during almost the entire span of the Mesozoic. The oldest known mammals appear in "Rhaetic" beds at about the Triassic-Jurassic boundary. In the Jurassic, almost all known mammalian remains come from two English localities in the Purbeck beds and Stonesfield slate and one small bone pocket at Como Bluff, Wyoming. In the early Cretaceous, before the appearance of marsupials and placentals, our knowledge is confined to fragments from the Trinity sands of Texas and the Wealden of southern England. Further, almost all the earlier Mesozoic remains consist of isolated teeth, or, at the most, jaws; prior to the Upper Cretaceous we have not one satisfactory skeleton and very little skull material. Even in the Upper Cretaceous, mammal remains are rare, mostly fragmentary, and are found only in a few areas in North America and Mongolia. In consequence, we are still in the dark about much of the history of Mesozoic mammals. We know little except the dental anatomy of most of the forms which have been found; and the sparseness of the record suggests that groups which have escaped discovery may well have existed.

This absence of material is, presumably, not due to any great rarity of mammalian life during these times but to the small size of the Mesozoic mammals. On the average they were no bigger than a rat or a mouse; and even if present would tend to be overlooked by collectors unless the recently developed methods of careful screening were used.

Monotremes.—Before considering these early fossil types, it will be advisable to discuss the living members of the order Monotremata, a group which is paleontologically almost entirely unknown but which must have had a long independent history. The duckbill (*Ornithorhynchus*) and spiny anteater (*Echidna* or *Tachyglossus*, etc.) of the Australian region—the only known monotremes—are among the most bizarre and paradoxical of living vertebrates. They have such mammalian features as milk glands (of primitive structure), hair, and but one lower-jaw element. However, they are exceedingly primitive in that, in contrast to all other mammals, they still lay eggs, while many other reptilian characters are present in their skeletons and soft anatomy. They are highly specialized in their modes of life and in consequent adaptations, such as the loss of teeth, development of a bill, and peculiar limb structure for digging or swimming.

The duckbill has a broad horny bill; the anteater, a long slim beak suitable for ant eating. Both are toothless as adults. The anteater has no trace of teeth; in the duckbill there are a few irregularly shaped molar rudiments in the young which have been compared (but with little success) to those of the fossil multituberculates. The braincase is moderately expanded; the frontal bones are unusually small, the nasals large. There is no lacrimal, and the jugal is reduced or absent. There is a large palatine forming part of the long secondary palate; the ectopterygoid (in contrast with other mammals) has been retained; there is a large vomer and an additional pair of small problematical bones at the front of the palate. The orbitosphenoid is large; ptyerygoid and alisphenoid are present but peculiarly developed. There is no formation of an auditory bulla; the tympanic bone is a loose open ring, and the recess for the middle ear is only partially enclosed by extensions from the petrosal and alisphenoid. The jaws are reduced.

The presence of unfused cervical ribs is again a primitive feature. The shoulder girdle (Fig. 269*C*) is emphatically reptilian; for an interclavicle is still present, there are two ventral coracoid elements, and in the scapula there is no spine or supraspinous area. In the pelvis (Fig. 270*D*) there are long bones extending forward from the front of the pubis, as in marsupials. The limbs are highly specialized for digging.

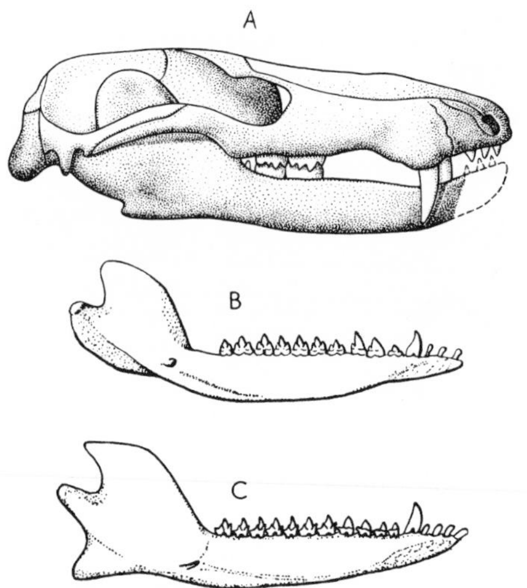

Fig. 305.—*A*, The skull of a late Triassic triconodont, *Sinoconodon; B*, inside of jaw of *Spalacotherium*, a symmetrodont, about 1.5 times natural size; *C*, inside of jaw of *Amphitherium*, a pantothere, about 2 times natural size. (*A* after Patterson and Olson; *B* and *C* after Simpson.)

There are remains, some of large size, of both monotreme types in the Pleistocene of Australia. Beyond this we know nothing about the history of this group; the absence of teeth, coupled with other marked specializations in the living monotremes, renders comparison with early fossil forms almost impossible. Certainly, they are very primitive, as well as highly specialized; and it is highly probable that they represent a line of descent from mammal-like reptiles entirely separate from that of other living forms. Their survival in the Australian region may be attributed partly to their specialized mode of life and partly to the long isolation of that region from other continents.

Mesozoic mammals.—The half dozen or so groups to be considered next include the entire known mammalian fauna of the few localities in Europe and North America in which late Triassic or Jurassic mammals are present; except for the aberrant multituberculates, these types are unknown beyond the early Cretaceous. The study and interpretation of these forms is, of course, handicapped by the generally fragmentary nature of the preserved materials. In addition, there is, for many of them, a further problem—namely, whether they are actually mammals in a proper sense of the word. Many of the diagnostic characters of the class Mammalia relate to soft anatomy and physiology and hence cannot be determined from skeletal remains—particularly fragmentary remains—and, further, certain of these characters were not improbably already present in advanced therapsids. As regards the skeleton, the most useful criteria have to do with the jaw and ear ossicles. Reptiles, we have seen, have a number of elements in the lower jaw, an articular-quadrate jaw articulation, and a single ear ossicle; proper mammals have only a dentary in the jaw, have shifted to a dentary-squamosal articulation, and have three auditory ossicles. This gives us, it would seem, a clear-cut reptile-mammal distinction. But with increasing knowledge this is becoming blurred. Obviously, this shift did not occur suddenly, as a single step. Advanced therapsids, we have seen, may be transitional in the mode of jaw articulation; and, on the other hand, there is evidence that in some Mesozoic forms which we here consider as mammals, one or more of the old reptilian jaw elements were still present and that, although the new dentary-squamosal jaw articulation is present, the old articular and quadrate may still be involved. Thus, some of the supposed Mesozoic mammals may still be in, technically, a therapsid —or semitherapsid—evolutionary stage. Further, there is no guarantee that only a single therapsid type crossed the reptile-mammal boundary; several may have done so, and thus the class Mammalia may be to some extent polyphyletic in origin.

Having now given due warning as to the nature of the materials and the dangers of drawing too positive conclusions from them, we shall review, in sequence, the known Mesozoic groups.

Triconodonts.—The order Triconodonta has long been known from a series of forms of both Middle and Upper Jurassic which grew to relatively large size; *Triconodon* of the Upper Jurassic approached the size of a cat. Very probably they were true carnivores, rather than insectivorous types. The dental formula reached as high as $4 \cdot 1 \cdot 4 \cdot 5$ for the lower jaw; incisors, large canines, relatively simple premolars, and molars were differentiated in typical mammalian fashion. The molars in typical triconodonts (Figs. 306*A, G;* 307*B;* 308*B*) had three sharp conical cusps of about equal height, arranged in a fore-and-aft row; upper and lower cheek teeth met in a longitudinal shearing device. In *Amphilestes*, a more primitive member of the group, the middle cusp was the tallest, and the three were not so discrete; and it is obvious that these teeth might easily have been derived from a type such as that seen in *Cynognathus* and certain other therapsids. The lower jaw lacks the projecting angular process characteristic of therian mammals. A natural cast shows the brain to have been very small and primitive.

Recent discoveries have extended the geological range of the triconodonts both upward and downward. A triconodont, *Astroconodon,* has recently been found in the Texas Lower Cretaceous. More important is the finding of much of the skull and postcranial remains of a triconodont, *Sinoconodon* (Fig. 305*A*), in the late Triassic of China, demonstrating (as might be expected from their seemingly primitive nature) that the triconodonts are among the oldest, as well as most primitive, of mammalian types.

The triconodonts were once believed to demonstrate an early stage in the development of the triangular, three-cusped type of molars characteristic of higher mammals. This view is no longer held, and the triconodonts appear to represent an isolated side line, without known descendants. It is, further, quite possible that they have evolved independently from the therapsid level.

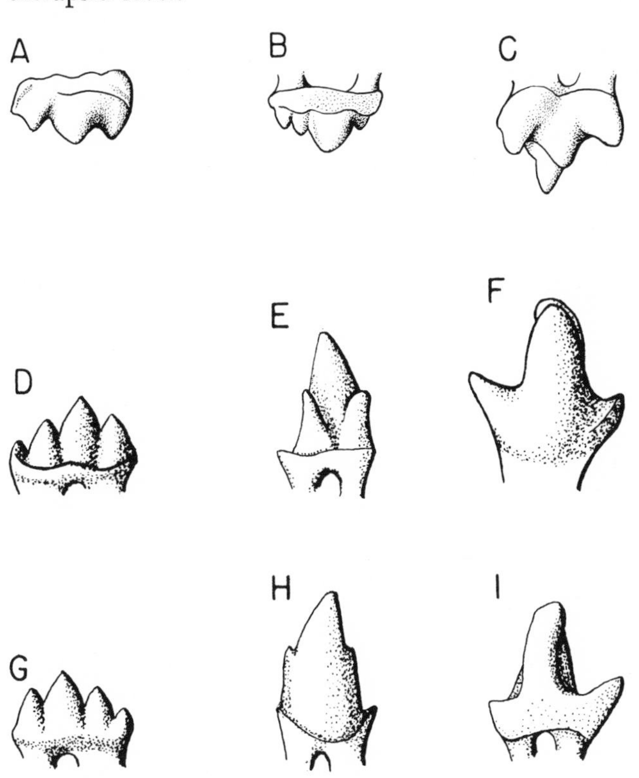

FIG. 306.—Side views of cheek teeth of early mammals. *A–C,* External views of right upper molars of *A, Priacodon,* a triconodont (×5.5); *B, Eurylambda,* a symmetrodont (×8); and *C, Melanodon,* a pantothere (×9). *D–F,* External views of left lower molars of *D, Priacodon* (×6); *E, Spalacotherium,* a symmetrodont (×8); and *F, Dryolestes,* a pantothere (×18). *G–I,* Internal views of left lower molars of *Priacodon, Spalacotherium,* and *Dryolestes.* (After Simpson.)

Morganucodonts.—Morganucodon (and the probably identical *Eozostrodon*) was first described on the basis of teeth (Figs. 307*A*, 308*A*) found in limestone fissure fillings of approximately Rhaetic age in Great Britain. In recent years considerable further material has been found in such fissures; a skull (as yet undescribed) is reported from the Rhaetic of China, and a related form is now known from South

Africa. The cheek teeth resemble those of triconodonts (except that the most anterior of the three main cusps is small). The skeleton, as far as known, appears to be of a very primitive and essentially therapsid nature; some features of braincase construction appear to be still on a reptilian plane, and the lower jaw retained part, at least, of the old reptilian elements. *Morganucodon,* again, appears to lie close to the reptilian-mammalian boundary.

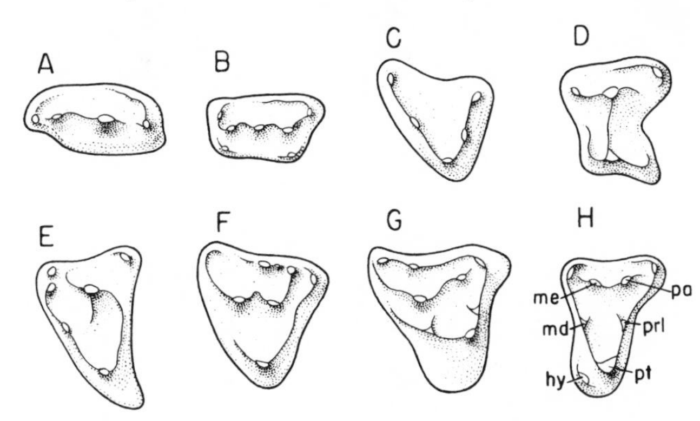

FIG. 307.—Crown views of right upper molars of primitive mammals; outer margin above, front edge to left. *A, Morganucodon; B,* a triconodont; *C,* a symmetrodont; *D, Docodon; E,* a pantothere; *F,* a Lower Cretaceous therian; *G,* a primitive marsupial; *H,* a primitive placental. In the placental the cusps are labeled as in Figure 297. (After Simpson.)

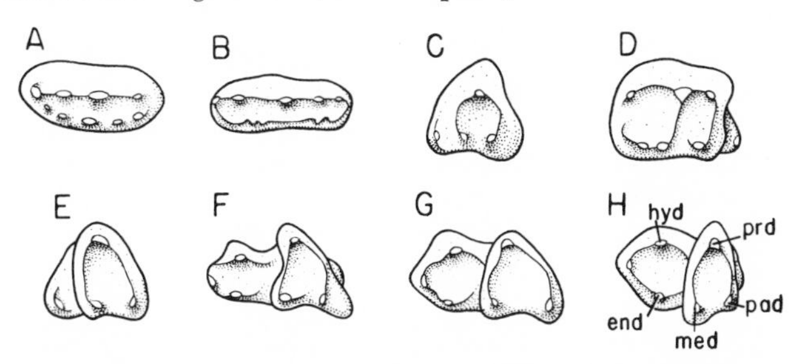

FIG. 308.—Crown views of left lower molars of the same types as those in the preceding figure; outer margin above, front to right. In the placental the cusps are labeled as in Figure 297. (After Simpson.)

Docodon.—This Upper Jurassic form is known only from the dentition (Figs. 307*D*, 308*D*). Here we find (as in groups later described) expansion of the crown surface of the molars, which are irregularly quadrate. The lower molars have a complex cusp pattern, but with a main cusp toward the buccal (outer) side; in the upper molars there are two main cusps, one on the buccal, one on the lingual (inner) side. *Docodon* was at one time considered to be a pantothere but appears to be a distinct type. Its phylogenetic position is uncertain. Relationships with monotremes or *Morganucodon* have been suggested but rest on relatively little evidence.

Multituberculates.—A specialized but highly successful and long-lived side branch of the early mammalian stock is that of the order Multituberculata (Figs. 309, 310). These forms appear to have been the first herbivorous mammals, with skull and tooth specializations somewhat analogous to those seen among the later rodents, for the treatment of vegetable food. In dental development they closely parallel some of the later marsupial types and *Tritylodon* and its allies, discussed earlier. The later multituberculates tended to grow (as do herbivores generally) to relatively large size. Contrasting with the diminutive size of most other Mesozoic mammals, some forms were as large as woodchucks; and multituberculates may have been not so far from that type of animal in superficial appearance and habits, although the skeleton suggests terrestrial to arboreal, rather than fossorial adaptations. The history of the group is a long one. They appear in the late Jurassic and, unlike contemporary orders, survived beyond the end of the Mesozoic to the true Eocene before being wiped out, presumably by the competition of advanced placentals—notably the rodents, which usurped the ecological niche long occupied by the Multituberculata. The life span of the multituberculates—on the order of 100 million years—far exceeds that of any other mammalian order.

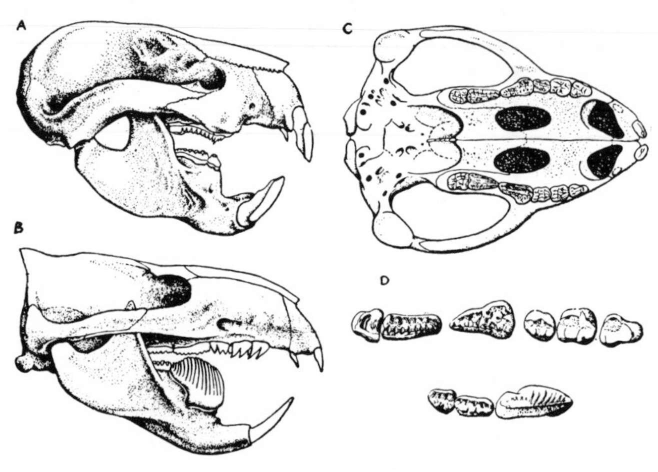

FIG. 309.—Multituberculates. *A*, Skull of *Taeniolabis*, original about 6.75 inches long; *B*, side view of *Ptilodus* skull, original about 2 inches long; *C*, palatal view of the same; *D*, upper and lower cheek teeth of *Ptilodus*, about twice natural size. (Mainly after Simpson.)

The skull is a massive one, broad posteriorly and roughly triangular in outline dorsally, although the snout, capped by very large nasal elements, is blunt and stoutly built. The great width across the arches indicates the presence of powerful jaw muscles. A curious feature of the zygomatic arches is the total loss of the jugals. In some forms (such as *Ptilodus* [Fig. 309*B*]) there are large vacuities in the palate, both anteriorly and posteriorly. The basicranial region is very broad and flat, and with structural details unlike those of any other mammalian order. A brain cast shows the olfactory bulbs (and hence presumably the sense of smell) to have been developed to a remarkable degree. The heavy jaw resembles that of triconodonts in lacking an angular process.

The dentition is of a very specialized type. Anteriorly, in rodent-like fashion, a pair of large incisors is developed above and below; smaller lateral incisors may persist in the upper jaw. Following a diastema (with loss of canines), there is a well-developed series of cheek teeth. The molars are the diagnostic multituberculate feature. The lower molars are elongated, with two parallel rows of cusps; the upper molars opposed them with two, sometimes three, similar cusp rows, the whole forming an excellent masticatory device. The premolars were rather variable in nature. In the lower jaw most multituberculates have tended to transform one or several lower premolars into sharp, striated, shearing blades.

The postcranial skeleton (Fig. 310) is for the most part definitely mammalian, but with many primitive features. The limbs appear to have been capable of sprawling more widely to the side than is usual in mammals, and, as in monotremes, the scapula is still reptilian in lacking the supraspinous area above the spine.

The first multituberculates appear in the late Jurassic of Europe and North America as the family Plagiaulacidae, including such genera as *Ctenacodon* and *Plagiaulax*. As far as known, all the essential features of multituberculate structure had already developed. These forms, however, were relatively small, had but two short rows of cusps in the upper molars, and developed a modest shearing power in several of the lower premolars. The order was a flourishing one in the late Cretaceous and Paleocene, and a few survivors persisted into the true Eocene. Two new groups had appeared to replace the plagiaulacids. *Ptilodus* is representative of one which tended to add a third row of cusps to the more elongate upper molars. The more anterior lower premolars are reduced or absent, but the last one has been greatly enlarged as a shearing or slicing structure. The molars are elongate. Among members of a third group was *Taeniolabis*, a giant of the tribe (relatively speaking), with a broad, blunt skull half a foot long, very heavy incisors, and large stout molars. Three rows of cusps are present in the upper molars, and in further con-

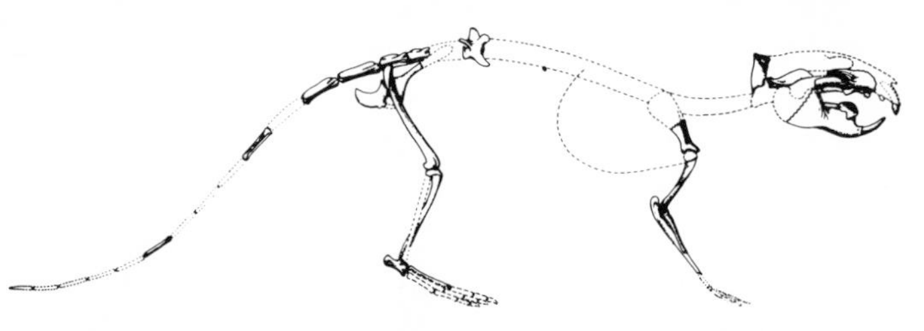

FIG. 310.—Essay at restoration of the skeleton of the Upper Cretaceous multituberculate *Mesodma*, about 2/3 natural size. (From Sloan.)

trast to other families, both upper and lower premolar series are reduced nearly to the vanishing point.

The multituberculates certainly were a sterile side branch of the early mammals; they are not ancestral to any later order, although certain marsupials paralleled them in dental development. Their ancestry is unknown; because of their isolated position they are customarily regarded as forming a distinct mammalian subclass, the Allotheria. Their dentition presents similarities to that of *Tritylodon* and its allies, once thought on this account to be multituberculates; but the two types are dissimilar in all other known features. Monotreme relationships have been suggested; but the multituberculates show no positive resemblances to them. The dentition may have been derived from such a type as that seen in the triconodonts—by a doubling-up of an anteroposterior row of molar cusps—but there is no positive evidence of phylogenetic connection.

Symmetrodonts.—In the order Symmetrodonta we reach, for the first time in our recital of Mesozoic types, a group which may be related to the ancestry of the higher, therian mammals. Best known are a number of late Jurassic genera, such as *Spalacotherium* (Figs. 305*B*, 306*E*, 307*C*, 308*C*) and *Eurylambda* (Fig. 306*B*), of comparatively good size and presumably predaceous habits (of a modest sort). Recent discoveries have extended the range of the symmetrodonts, for they are now known from the Lower Cretaceous and also from the Rhaetic—thus ranging them, with the triconodonts and *Morganucodon*, among the oldest of known mammals. These forms had three separate cusps in each molar tooth, and hence were at one time included in the triconodonts. But in the symmetrodonts the three cusps were arranged (as the name implies) in a symmetrical triangle, the base external above, internal beneath, with the major cusp at the apex of each triangle. Early advocates of the tritubercular theory believed that they had arisen from triconodonts by a rotation of the cusps. But there is no evidence of such rotation, and they probably had nothing to do with triconodonts. The symmetry of the molar teeth in this group is in contrast to the molar shape of pantotheres, and the lower molars lack the "heel" characteristic of that order and of all later mammalian groups.

Pantotheres.—Most important of Jurassic orders from an evolutionary point of view is that of the Pantotheria (Figs. 305*C*; 306*C*, *F*; 307*E*; 308*E*; 311). Nearly all known forms are from the late Jurassic; *Amphitherium*, the only Middle Jurassic genus, is a generalized form known only from the lower jaw. In *Amphitherium* this is long and slender, and, in contrast to other Jurassic orders, there is a well-developed angular process, which appears to be homologous with that in higher mammals. The pantothere tooth count is variable, but in addition to four incisors, a canine, and four premolars, there may be as many as seven or eight molar teeth. The *Amphitherium* lower molars show a three-cusped, asymmet-

rical trigonid, comparable to that of later groups; and particularly significant is the presence of a "heel" or "talonid," absent in the orders so far mentioned and small here.

In the late Jurassic mammal localities of Europe and America, pantotheres are abundant and varied. The Dryolestidae, such as *Melanodon*, appear to be closer to the generalized and primitive pattern. The lower molars are rather similar to those of *Amphitherium*. The upper ones are (as in early marsupials and placentals) asymmetrical triangular structures. At the internal apex there is a prominent cone which fits into the basin of the talonid, as does the placental protocone. More laterally and centrally there is a second prominent cusp, and along the outer margin there are further minor cusps.

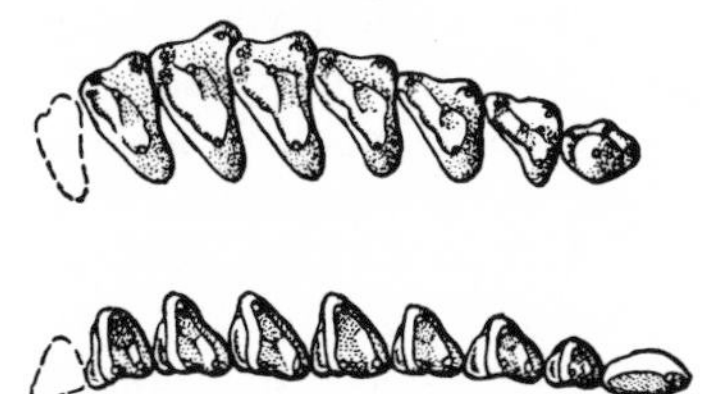

FIG. 311.—Reconstruction of the right upper and left lower rows of cheek teeth of a Jurassic pantothere. (After Gregory and Simpson.)

It is generally believed that the pantotheres include the ancestors of the higher mammalian assemblages, and hence attempts have been made to interpret the pantothere upper molar cusps in terms of those which (as described in the last chapter) are found in marsupial and placental mammals. Despite many intensive studies by competent specialists, no interpretation has been universally accepted. The majority opinion, however, is to the effect that the large, centrally placed cone of the pantothere molar represents the single tip of the primitive reptilian tooth and that this major cone is the homologue of the combined paracone and metacone of higher mammals, the inner cusp being (as it appears to be) the protocone.

Lower Cretaceous therians.—The ancestral pantotheres are typically late Jurassic; their marsupial and placental descendants appear in the late Cretaceous. In the early Cretaceous, one would, hence, expect to find therian mammals which had advanced from the pantothere condition to become forms from which the later marsupials and placentals were to arise. We know very little of Lower Cretaceous mammals. But in recent years some bits of evidence have appeared to support this theoretical conclusion. In the Lower Cretaceous of Texas have been found isolated teeth (*Pappotherium*) truly therian in nature. In beds of comparable age in Manchuria there has been found the skull of a mammal, *Endotherium*, which may possibly be of the same category.

The early evolution of mammals.—This concludes

our review of Mesozoic mammalian types. Study of the various groups gives us some ideas, although cloudy ones, as to the evolution of mammals during the reign of the dinosaurs, and consideration of these assemblages as a whole (leaving aside the specialized multituberculates) may yield some idea of the general nature and mode of life of these ancestral mammals. They were generally small creatures, averaging about the size of rats and mice, and were perhaps somewhat like these forms in their general appearance. Their food must have consisted mainly of insects; they were potential carnivores, but their size limited them to small prey. Other materials, however, such as birds, fruit, and eggs, may have made up part of their diet. The general degree of evolution of the known skeletal parts suggests that many of them were above the monotreme level of organization and that, therefore, they not only nursed their young but presumably bore them alive, although at a relatively early stage in development. Their brains, as far as can be determined, were still poor and small by modern mammalian standards, although much better than those of their reptilian ancestors. Presumably they were retiring in habits, possibly nocturnal, and to some extent arboreal. Inconspicuous and small they had to remain, for, as contemporaries of the dinosaurs, the threat of death from the great carnivorous reptiles lay constantly over them.

But this long period of "trial and tribulation" was perhaps not altogether disadvantageous. It was, it would seem, a period of training during which mammalian characteristics were being perfected, wits sharpened, developmental processes improved, and the whole organization undergoing a gradual evolutionary change from reptilian to true mammalian character. As a result, when (at the close of the Cretaceous) the great reptiles finally died out and the world was left bare for newer types of life, the mammals had evolved into a group prepared to take the leading place in the evolutionary drama.

Marsupials.—In the late Cretaceous beds we find that, although the multituberculates still flourished, the typical Jurassic orders had disappeared. Seemingly developed from the pantothere stock, we find representatives of the two great living groups of mammals—the marsupials and the placentals. The representatives of both groups were still small, still insectivorous to omnivorous in habits, and again are poorly known. Abundant were small opossum-like forms—the earliest of the marsupials, or pouched mammals.

Today the marsupials constitute a well-defined

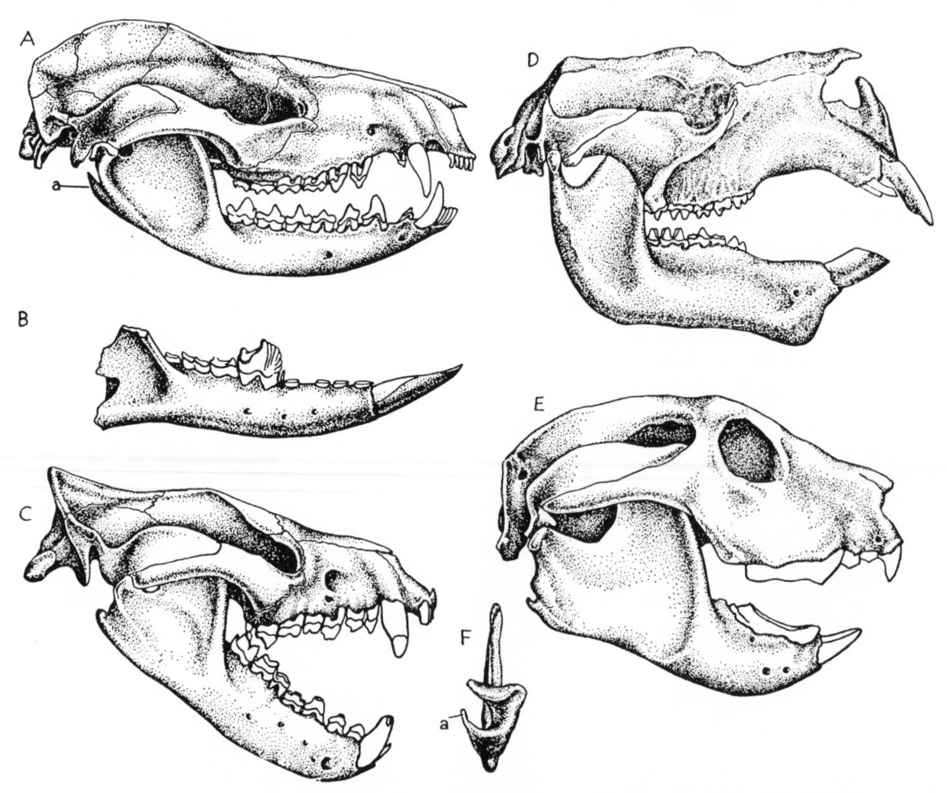

FIG. 312.—Skulls of marsupials. *A*, The living opossum, *Didelphis; B*, jaw of *Abderites,* a Miocene South American caenolestid (×7/8); *C*, *Borhyaena,* a South American Miocene carnivore, skull length about 9 inches; *D*, *Diprotodon,* a giant Australian Pleistocene diprotodont, skull length about 40 inches; *E*, *Thylacoleo,* a large Australian Pleistocene type with highly developed shearing teeth, skull length about 1 foot; *F*, opossum jaw (*right*) from the rear to show inflected angle (*a*). (*B* after Ameghino and Simpson; *C* after Sinclair; *D* after Owen; *E* after Owen and Anderson.)

group including the opossums and a few other New World mammals, together with a large array of types constituting the greater part of the fauna of the Australian region. Certain of their distinctive features are suggested by the name. Almost all have a pouch placed on the belly of the female, which contains the teats and in which the young are carried after birth. This marsupium is rendered necessary because birth takes place at a very small and immature stage, and this, in turn, is due to the fact that an efficient placenta has been evolved in but a few members of the group. Among the definite osteological characters that separate them from the monotremes is the fact that (as in higher types) the coracoids are reduced to a single nubbin of bone on the underside of the scapula, a spine and a supraspinous fossa above it are developed on the latter bone, and the interclavicle is lost. Among the more prominent key characters which separate all known forms from higher, placental mammals are the inflected (or inturned) angular process of the jaw (Fig. 312*F*), the usual presence of four molars and but three premolars, and the presence of a pair of marsupial bones articulating with the front of the pubes as in the monotremes (Fig. 270*E*).

In such a marsupial as the opossum, the brain is relatively very small and poorly developed. The braincase region of the skull (Fig. 312*A*) is consequently small compared with that of the average placental; sagittal and transverse crests are, in consequence, usually prominent. There is almost never any development of a postorbital bar behind the small orbit, nor is there ever a well-formed auditory bulla of placental type, although the alisphenoid usually sends back a process to protect the middle ear.

In the oldest known marsupials there appear to have been as many as five upper incisors, as is still the case in the opossum; there are seldom more than the conventional three in the lower jaw. Large canines were undoubtedly present in the ancestral forms as they are today in many of the more primitive carnivorous types. In the cheek series there are never more than three premolars and usually four molars. The problem, however, is complicated by the fact that there is almost no tooth replacement, the last premolar being the only one with a predecessor; the more anterior teeth are seemingly permanent milk teeth.

In the more generalized forms (Fig. 313) the teeth are tribosphenic, with triangular upper molars and a trigonid and talonid below. In the upper molars there is developed a trigon similar to that of placentals but, in addition, an outer row of "stylar" cusps, presumably a primitive feature and one present in some primitive types of placental molars, is seen. In herbivores the teeth tend to square up with addition of a hypocone and more or less development of a grinding surface.

A clavicle is almost always present in marsupials. There is seldom any reduction of toes in the front foot. In the hind foot the hallux appears to have been

primitively opposable. There is often (especially in herbivorous types) a curious sort of syndactyly (Fig. 313) in which the fourth toe is strong and elongated and the second and third toes are united and balance the fifth toe, while the first digit may be lost. Claws are always present. It has been suggested that the primitive marsupials were arboreal, and many of the structural features tend to bear this out.

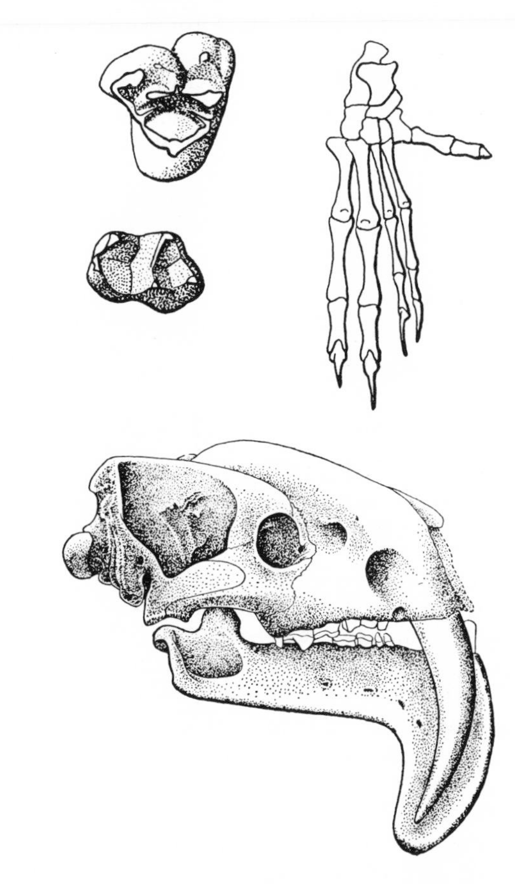

Fig. 313.—*Left,* Upper right and lower left molar teeth of an Upper Cretaceous opossum. In the upper molars there is a well-developed row of outer cusps in addition to the normal trigon. (Much enlarged; after Osborn.) *Right,* Foot of a living diprotodont marsupial to show syndactylous structure. *Lower,* The skull of *Thylacosmilus,* a Pliocene South American carnivorous marsupial comparable to the placental sabertooths; about 1/4 natural size. (After Riggs.)

The living marsupials have been frequently divided into two major suborders, the Polyprotodonta and Diprotodonta; we shall include a majority of the marsupials, living and extinct in these familiar groups, although adding two further suborders for certain living and fossil types which do not fit into this simple scheme.

American polyprotodonts.—The polyprotodonts, of which the American opossum is typical, form the basic stock of the order; insectivorous, omnivorous, and carnivorous forms are included. A primitive feature of the polyprotodonts (as the name implies) is the retention of an unreduced and unspecialized set of incisors; there may be as many as five in the upper jaw. The molars are generally triangular in shape and usually sharp cusped; there is no syndactyly in the foot. There are several families, of which the Didelphidae are the stem group.

Opossums (*Didelphis*, etc. [Figs. 312*A*, 313]) and their relatives, found today in both Western continents, are in almost every respect ideal ancestors for the whole marsupial group. There are five upper incisors and four lower ones; the upper molars are triangular in shape and rather sharp cusped (the diet is omnivorous, with a tendency toward the carnivorous side). The tail is usually prehensile, the limbs are of normal construction without any reduction of toes and with a well-developed, clawless, opposable

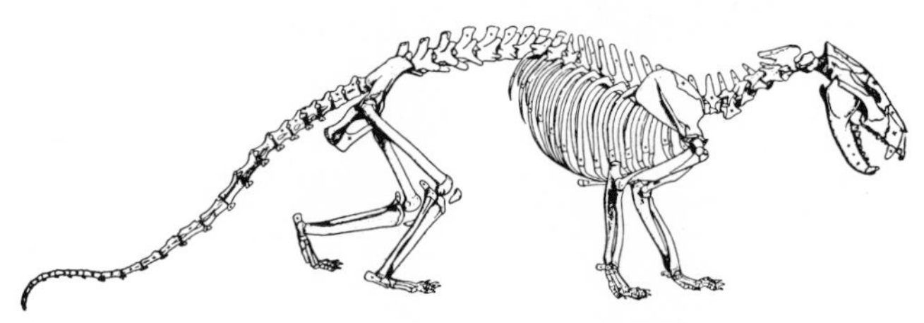

Fig. 314.—*Prothylacynus*, a Miocene carnivorous marsupial from South America, about 4 feet long. (From Sinclair.)

hallux used for grasping in the opossum's arboreal mode of locomotion. In the late Cretaceous of North America are found forms very similar to the living opossums, and in the early Tertiary these forms are known from both Americas and from Europe as well; but since the Miocene they seem to have been confined to the Western Hemisphere. The opossums have lived continuously in South America since then; but in North America there are no fossils of this group in the late Tertiary, and we cannot be sure whether or not our living opossum is really an oldest inhabitant or whether he is a recent reimmigrant from the south—probably the latter.

South America seems to have been separated from the other continents from the beginning of the Tertiary (and probably well before that time). Opossum-like marsupials, and in addition a restricted number of placental mammals, are present in the oldest, Paleocene, mammal-bearing deposits of that continent. But no placental carnivores appear to have gained admission, and thus there came about a good opportunity for the development of purely carnivorous marsupial types. These South American carnivores are separated as a distinct family, the Borhyaenidae. They were terrestrial forms, rather short-legged with large skulls. The claws were strong and compressed; and the hallux, as in running types generally, was reduced. The teeth, as in true carnivores, tended to the shearing type. These carnivores ranged in size from that of the opossum to that of a bear and were especially abundant in the Miocene, where *Prothylacynus* (Fig. 314) and *Borhyaena* (Fig. 312*C*), a puma-like form, were typical. Toward the end of the Tertiary, however, there arrived higher ungulate types, which displaced their former prey, and placental carnivores which replaced the marsupials as flesh eaters. In consequence the borhyaenids soon became extinct. A curious late survivor in the Pliocene was *Thylacosmilus* (Fig. 313), with large stabbing tusks like those of the saber-toothed tiger. Another type developed from the primitive opossum stock is seen in *Necrolestes* of the Miocene, a small form so similar to the placental insectivores that it was long mistaken for a member of that group.

Australian polyprotodonts.—It is generally believed that Australia became separated from the rest of the world at so early a time in mammalian evolution that, although primitive marsupials had reached this area in some fashion, no placentals of any sort gained admission; and, except for bats and some rodents, none have entered since except those brought in by man. We know almost nothing of early Tertiary vertebrate history in Australia, but it seems reasonable to assume that the earliest mammalian fauna was composed of primitive opossum-like marsupials, from which have descended the varied living marsupial groups which constitute the great bulk of the Pleistocene and Recent mammals of that continent. The most direct descendants of the primitive settlers appear to be carnivorous types which are called (incorrectly) "native cats," or "opossums" (*Dasyurus*), and a number of mouselike, insect-eating forms. Somewhat more advanced is the Tasmanian devil (*Sarcophilus*), a powerful carnivore with shearing molars, while a most specialized type is the Tasmanian "wolf" (*Thylacynus*), a very wolflike animal similar to some of the South American Miocene marsupials. There are some interesting side branches included in this group, such as a marsupial anteater—with typical ant-eating adaptations in the long snout, reduced teeth, and long claws—and a marsupial molelike type.

Transitional in structure between polyprotodonts and diprotodonts (and hence perhaps meriting a suborder of their own) are the bandicoots, such as *Perameles*, of the Australian Pleistocene and Recent. Rabbit-like in size and appearance except for the long pointed snout, they are omnivorous in diet. The dentition resembles that of the polyprotodonts except that the molars tend to become grinding teeth and assume a square shape with blunter cusps. In the long and slender hind limbs the toe pattern is that of diprotodonts—syndactylous, with an emphasis on the fourth toe, very much as in the kangaroos.

Diprotodonts.—A second prominent group of marsupials is that of the diprotodonts, confined to Australia. It is obvious that they are a herbivorous development from a primitive opossum-like stock. However, many structural changes have occurred. There are never more than three upper incisors, and although three are generally present below, the central pair are always developed as chisel-like teeth similar to those of rodents. The molars are quadrate in shape. The hind foot is syndactylous.

Seemingly closest to the common diprotodont ancestors are the phalangers (*Phalanger*, etc.)—arboreal types, the marsupial "squirrels," including even flying "squirrels." Somewhat more advanced is *Potorous* [*Hypsiprymnodon*], the musk "kangaroo," in which, as in kangaroos, the last lower premolar is a long, straight-edged cutting tooth similar to that of the multituberculates. The true kangaroos (*Macropus*, etc.) have specialized further, in the development of long hind legs for hopping, with a reduction of the lateral toes and emphasis on the fourth.

Two interesting living forms are characterized by the fact that in the upper jaw, as well as in the lower, there is a large pair of chisel-like teeth, the other upper incisors being small to absent. These are *Phascolarctos*, an arboreal form rather like a large teddy bear in size and appearance, and *Phascolomis*, the wombat, rather like the woodchuck in habits. Both have large Pleistocene relatives, and in addition there are a number of large extinct Pleistocene marsupials from Australia with a diprotodont dentition. *Diprotodon* (Figs. 312*D*, 315) is the largest known marsupial and was a form about the size of a large rhinoceros but with a lumbering build suggesting similarity to the living wombat. *Thylacoleo* (Fig. 312*E*) was even more odd, a form the size of a lion with reduced molars and with the last premolars developed into large shearing teeth. It has been argued that *Thylacoleo* was a carnivore, but the more generally accepted view is that the shearing teeth were used in cutting fruit.

The description, above, of the Australian marsupial fauna was based almost entirely on living types. With good reason; except for a few giants such as those just mentioned, we know of no Australian fossils not closely related to living forms. A very considerable amount of material has been recovered from the Pleistocene; but in general, the Pleistocene fauna was very close to that of modern times. Until the last few years we knew nothing of Tertiary Australian mammals, except for a phalanger skull, probably Oligocene. Recent exploration has brought to light a modest amount of fossil material of Pliocene, Miocene, and even perhaps late Oligocene age. But all forms found appear to be attributable to modern groups. No doubt an interesting radiation of marsupials took place in the still earlier Tertiary of Australia. But as yet this story is a closed book.

Caenolestoids.—The two suborders discussed above are standard divisions of the marsupials which have been recognized for a century. But certain living and fossil South American forms do not fit in well with either group; and for their reception the creation of a further suborder, Caenolestoidia is necessary. *Caenolestes* is a small, living, South American form, mouselike in size and appearance (and long thought

to be a rodent). There are four upper incisors, and the feet are not syndactylous, as is the case in the polyprotodonts. But although there are three or four lower incisors, the middle ones are strong and elongate as in diprotodonts, the molars are quadrate, and the last lower premolar in some related fossil genera is a cutting type like that of kangaroos. This group goes back in South America to the Eocene. Some of the earlier forms, such as *Polydolops* and *Abderites* (Fig. 312*B*), went far in diprotodont specializations, for there was but a single upper and lower incisor, as in the more advanced Australian forms. The upper molars were rather like those of multituberculates, with two short parallel rows of cusps; some older writers tended to group them together, but this is

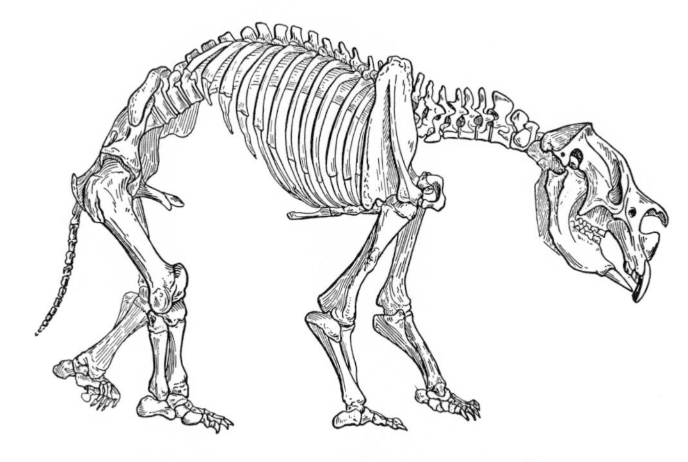

FIG. 315.—The skeleton of *Diprotodon*, a giant herbivorous marsupial from the Pleistocene of Australia, about 11 feet in length. (From Gregory.)

merely convergent evolution. Presumably in the radiation of the marsupials in the late Cretaceous or Paleocene, primitive types which were still polyprotodonts migrated both to Australia and to South America. In the former continent some developed into the true diprotodonts. The caenolestoids with the same ancestry, and presumably the same evolutionary tendencies, paralleled them closely and also paralleled the earlier mammalian herbivores, the multituberculates.

Marsupial history.—It has often been assumed that the marsupials are the ancestors of the placentals. That the ancestors of the placentals passed through a stage in which the young were born immature, as in marsupials, is theoretically probable; but that these ancestors were marsupials in the sense in which we know them (with a pouch, marsupial bones, and inflected jaw angle) is improbable. Indeed, we know primitive placentals fully as old as the earliest opossum-like forms. Surely both the metatherian marsupials and the eutherian placentals have arisen from a common ancestral stock—of which, as yet, we know little—of early Cretaceous age.

But, while not on the direct line of descent, the more primitive marsupials (such as the opossums and some of the smaller dasyurids) in their mode of life, in their insectivorous-omnivorous diet, and in many structural features give us a good picture of the Mesozoic forms from which the Tertiary mammals have come. It is of interest to note that the opossums and primitive diprotodonts are essentially arboreal types, in agreement with the suggestion that the ancestral mammals were to some degree tree-living forms.

It is probable that primitive marsupials had a worldwide distribution in the late Cretaceous. But early in the Tertiary they became almost entirely restricted to two regions—South America and Australia—presumably because they were unable to compete successfully with more highly developed placentals. In South America late Tertiary placental invasions caused the extinction of many types, and only *Caenolestes* and opossums have survived. But in Australia the marsupials, in the absence of competition, radiated out into many different modes of life. A number of forms are still arboreal, but the great majority have become ground dwellers of many sorts. From the primitive omnivorous-insectivorous diet the living marsupials have branched out into purely carnivorous and herbivorous modes of life. In Australia the marsupials, given their chance for development, have paralleled almost every type which the placentals have produced on the other continents.

CHAPTER 17

Placental Origins; Insectivores; Bats

Except for the primitive but aberrant monotremes and the marsupials just considered, all living mammals are members of a single, major, more advanced group which has been dominant since the beginning of Tertiary times—the Eutheria, commonly termed placental mammals. In these mammals we find a well-developed placenta formed from the allantois, one of the embryonic membranes; this permits of a long period of prenatal development and of birth at a more advanced stage than was possible in marsupials. The popular name is somewhat misleading, for some of the living marsupials have paralleled the true placentals to a considerable extent in the development of this structure; but there are many other points of difference between the two groups. There is no pouch in placentals, marsupial bones are absent, there is almost never the inflected angle of the jaw characteristic of marsupials, and the brain becomes considerably advanced in structure over that of the more primitive types. The dental formula was originally

$$\frac{3 \cdot 1 \cdot 4 \cdot 3}{3 \cdot 1 \cdot 4 \cdot 3}$$

and (except in whales) almost never exceeds that number, and there is typically one complete replacement of all the teeth except the molars.

Primitive placental structure.—From what we know of living and fossil types it is possible to picture a typical primitive placental at a late phase of the Age of Reptiles. Such a type would have been small in size, as had been the ancestral pantotheres of the Jurassic, and the limbs short. The clawed feet would have been applied rather fully to the ground in a semiplantigrade position, and there presumably would have been a somewhat opposable pollex and hallux. Among technical diagnostic characters in the limbs would have been the presence of a clavicle, a central bone in the wrist, an entepicondylar foramen in the humerus, and a third trochanter on the back of the femur—all characters lost in many later placentals. There would have been about twenty thoracic and lumbar vertebrae between neck and pelvis.

The brain was presumably comparatively small and poorly organized—that is, with the olfactory lobes and olfactory bulbs large and unreduced and with the neopallium but little developed. The skull, however, would have been fairly large (in correlation with the smallness of the beast), with a long facial region and large eyes, and there would have been no development of a postorbital bar. The dentition would have contained the primitive placental number of teeth and in general appearance would have resembled that of the older pantotheres. The lower molars would have been tribosphenic in character, with trigonid and talonid; the upper molars would have been triangular, with an internal protocone and, toward their outer edges, not the amphicone present in the older types, but two cusps, the paracone and the metacone, presumably derived from the division of the primary reptilian cone.

Probably these forms were inconspicuous forest dwellers and nocturnal to some degree at least. Certain characters in their descendants (such as the frequently divergent pollex and hallux and consequent grasping power) suggest an early trend toward an arboreal habitat, although in most placental types a purely terrestrial life seems to have been taken up at an early date. The diet presumably still consisted essentially of insects but was varied with other types of soft plant and animal food, such as fruits, buds, worms, and grubs.

Forms such as that described above were purely hypothetical creatures early in this century. In recent decades we have obtained, however, skull remains from the Upper Cretaceous of Mongolia which are definitely placental in nature; and, still more recently, a number of placentals have been identified in the late Cretaceous of North America. Nearly all the remains are fragmentary; we know almost nothing of the postcranial skeleton, and in most instances we have little but teeth and jaw fragments.

Small as the earliest placentals were, they were nonetheless the most efficiently organized land dwellers of the times, better in brains and in mode of development than the marsupials and multituberculates that were their mammalian contemporaries. With the disappearance of the dinosaurs, placentals spread and diversified rapidly during the Paleocene. In the true

Eocene the multituberculates made their last appearance, marsupials became rare in the Northern Hemisphere, and the placentals had already become differentiated into the ancestors of almost all the known orders of later Tertiary and Recent times.

Insectivores.—As was the case with their Jurassic ancestors, the early placentals were presumably basically insectivorous in diet. There are still in existence a number of mammals, such as the shrews, moles, and hedgehogs, which have retained similar feeding habits and which have, despite various specializations, many primitive characters. These forms, grouped with related fossil types as the order Insectivora, are regarded as the most direct modern descendants of the primitive placentals. The insectivores are in general small in size and few in number of forms. Most of them are either highly specialized or are isolated island dwellers, as is usually the case with surviving members of primitive groups.

Since the order includes the supposed stem of the later placental types, the Insectivora can be characterized for the most part only negatively as lacking the specializations of other derived groups. Many common features were noted above as characteristic of a generalized ancestral placental. The tympanic cavity lacks the ossified auditory bulla characteristic of most progressive placentals, but may be partially surounded by processes from neighboring bones, with the tympanic bone an open ring encircling the eardrum. In many instances the zygomatic arch is incomplete in modern forms. The teeth are almost always sharp-cusped, and in a majority of cases there is but little departure from primitive placental types of dentition. The incisors are somewhat variable in development, and the canines are often somewhat reduced. There is usually little modification or reduction in the limbs except in burrowers.

As we shall use the term here, the order Insectivora will serve as a category for several types of animals. We shall include primitive early placental types found in the Cretaceous and early Tertiary, the more "modern" groups of insectivores which have surviving members, and, in addition, a number of twigs of the placental evolutionary tree which have departed from the insectivore pattern but are hardly worthy of characterization as distinct orders.

Proteutherians.—Attempts to sort out phylogenetic lines among the early placentals of the late Cretaceous and Paleocene are extremely difficult, and there are currently many uncertainties. In part these are due to the generally fragmentary nature of the material; for a majority of forms we have little evidence except for that supplied by the dentition. In great measure, however, our difficulties are due to the fact that divergencies between groups which later became

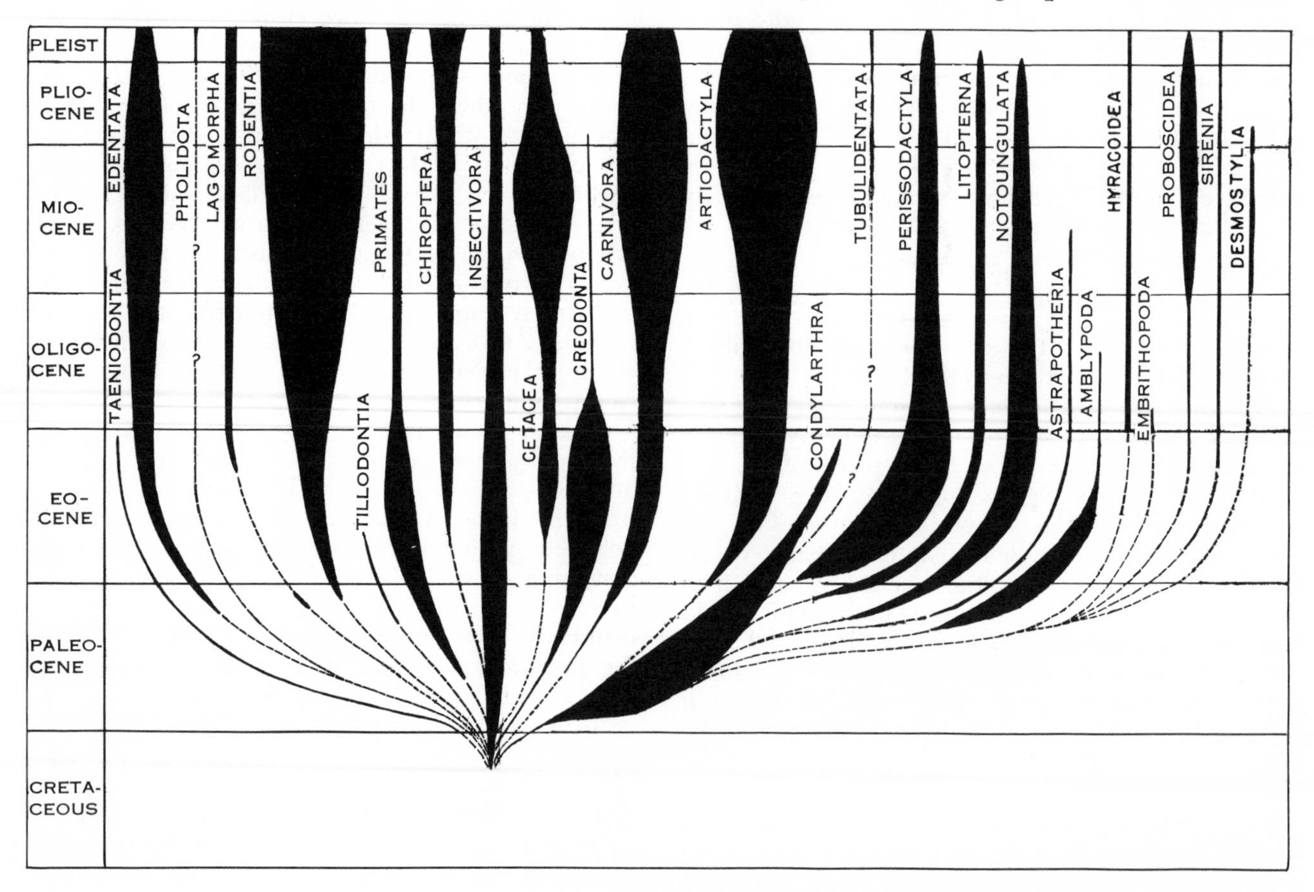

FIG. 316.—The chronologic distribution of the placental mammals

radically different were then very small. Were we living in the late Cretaceous we would probably include all placentals then existing in a single family, and even in the Paleocene most forms could be reasonably placed in a single order.

There are, hence, many unresolved questions concerning early placental evolution and classification. Currently, however, it would appear that there were, in the late Cretaceous, basal members of two distinct "insectivore" stocks, the Leptictidae and Deltatheridiidae. Both are primitive in many regards, and both have been advocated as truly basal placental types. In the two groups are seen contrasting types of upper molars (cf. Figs. 298 and 318*A*, deltatheridians, with 318*B*, a leptictid). In the deltatheridians the paracone and metacone lie close together, far in from the outer edge of the tooth; in the leptictids, in contrast, these two cusps are well separated and lie close to the outer edge of the crown. It has been argued that the deltatheridian pattern is truly primitive, paracone and metacone representing a splitting into two (so to speak) of the originally single major central cusp of the pantothere molar, while the leptictids represent a later development in which these two cusps have separated more widely and moved toward the outer margin of the tooth. The situation is currently in a state of flux. But while the deltatheridian pattern may be an ancient one, it is improbable that forms with this pattern are directly ancestral to many later placentals except, as discussed later, the creodont carnivores. We shall here accept the current theory that the more "orthodox" type of molar seen in leptictids is that from which the dentitions of most later placentals have been derived. We may, in consequence, consider the leptictids and a number of other early or primitive insectivores related to them as constituting a basal eutherian stock which we shall (reasonably) term the suborder Proteutheria.

The best known of leptictids are late Oligocene survivors, such as *Leptictis* and *Ictops,* but fragmentary remains of typical leptictids, such as *Procerberus,* are present in the Cretaceous. *Zalambdalestes* (Figs. 317*B*, 318*B*) of that age has apparently departed but little from the ancestral pattern. The dentition of leptictids is of a typical "modernized" placental type, for the paracone and metacone are widely separated and placed far toward the outer edge of the broadly triangular upper molars, with

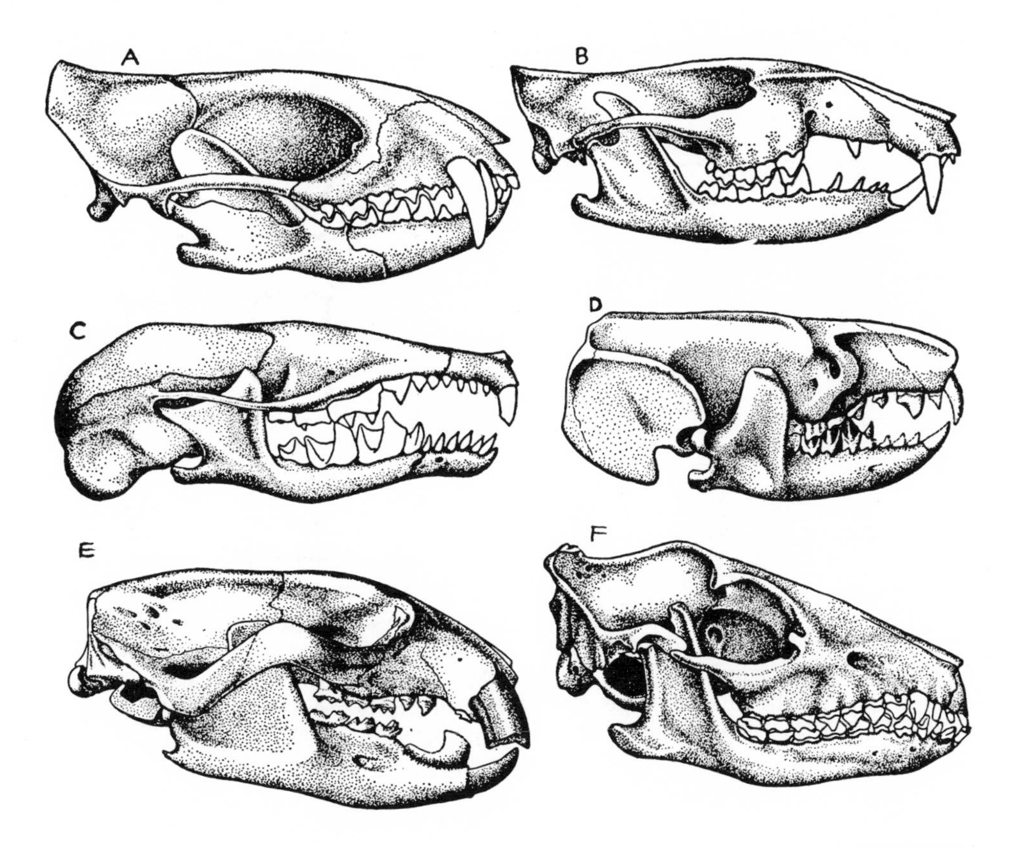

FIG. 317.—Skulls of primitive placentals (Insectivora and Creodonta). *A, Deltatheridium,* and *B, Zalambdalestes* of the late Cretaceous of Mongolia; skull lengths of about 1.75 and 2 inches, respectively; *C, Proscalops* of the Oligocene and Miocene, length about 1 inch; *D, Apternodus,* and *E, Sinclairella* of the Oligocene, with lengths of about 1.75 and 2.25 inches; *F, Anagale* of the Mongolian Oligocene, length about 2.25 inches. (*A, B* after Gregory and Simpson; *C* after Matthew; *D* after Schlaikjer and Scott; *E* after Scott and Jepsen; *F* after Simpson.)

reduction of stylar cusps; there is often, as in *Diacodon* (Figs. 318*D,* 319*B*), an incipient hypocone.

Apart from their presumed prominence as potential ancestors of major placental orders, the proteutherians appear to have given rise to a number of short-lived side branches. Prominent were *Pantolestes* of the Eocene and its Paleocene relative, *Propalaeosinopa* (Figs. 318*G,* 319*D*). *Pantolestes* was about the size of

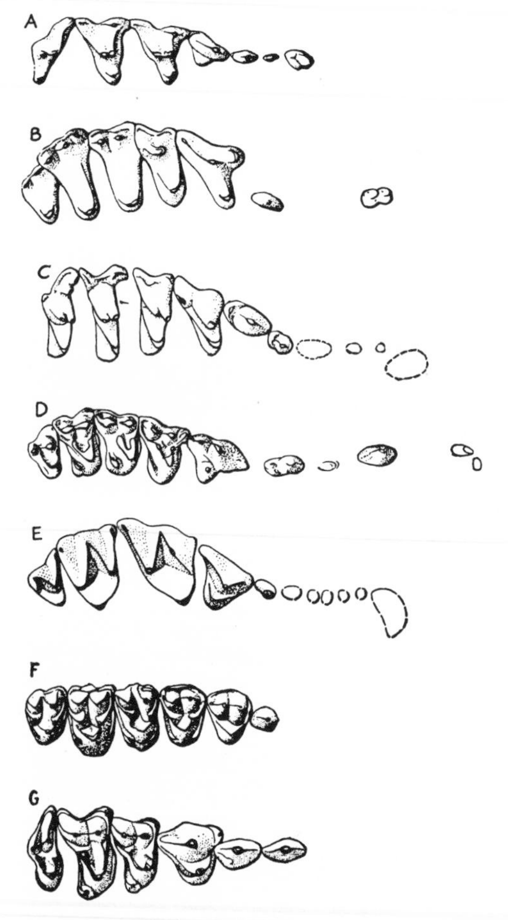

FIG. 318.—Right upper cheek teeth of primitive placentals. *A, Deltatheridium* (×2.5); *B, Zalambdalestes* (×3); *C, Palaeoryctes* (×3); *D, Diacodon* (×5/4); *E, Proscalops* (×4 approx.); *F, Elpidophorus* (×3.5 approx.); *G, Propalaeosinopa* (×3).

an otter, was apparently predaceous in habits, and started on the road toward a water-living type of existence. The braincase was elongate; the face was short and the jaw heavy, the dentition suggesting mollusks as a possible diet. Passing over several minor groups, we may mention a family containing *Apatemys, Sinclairella* (Fig. 317*E*), and their Paleocene to Oligocene relatives. There are rodent-like incisors (an upper pair very strongly developed) and, as in mixodectids (discussed later), rodent relationships have been claimed.

Of greater evolutionary interest among proteutherians is a series of forms of which the tree shrews, such as *Tupaia,* of the oriental region are the modern representatives. These living types are small arboreal animals with a squirrel-like appearance; anatomically they show a mixture of primitive features with others suggestive of relationship to primates. Features suggesting primate connections include the presence of a relatively large brain in which the olfactory region is small, the presence of a postorbital bar, a middle ear region built like that of lemurs, and considerable opposability of pollex and hallux. For these reasons many workers class them with the primates, and they may well represent the insectivore group from which the primates sprang. They are, however, persistently primitive in many regards. Claws, rather than primate nails, tip the toes; three incisors in primitive fashion rather than the two of primates are present on each jaw ramus; the cheek teeth are persistently primitive. Little is known of tree shrew ancestry. A possible member of the family is present in the Paleocene; *Anagale* (Fig. 317*F*) of the Asiatic Oligocene appears not to be (as once believed) a true tupaiid, but may well be a proteutherian not distantly related. *Anagale,* however, has powerful claws, suggesting that it dug for its food rather than having been a tree dweller.

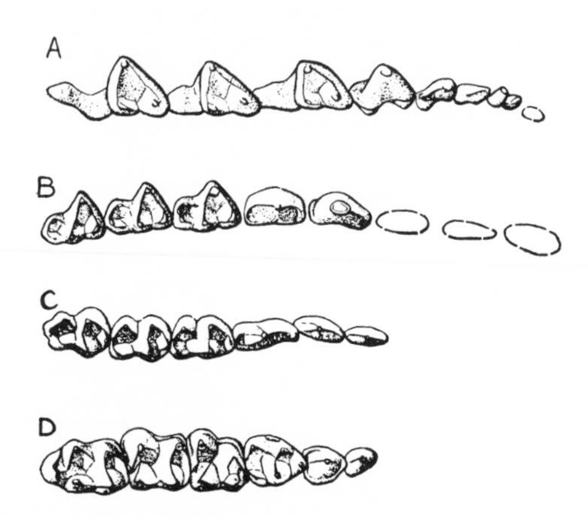

FIG. 319.—Left lower cheek teeth of primitive placentals. *A, Deltatheridium* (×2.5); *B, Diacodon* (×5/4); *C, Propalaeosinopa* (×3); *D, Elpidophorus* (×3.5 approx.). (*A* after Gregory and Simpson; *B* after Matthew; *C, D* after Simpson.)

"Modern" insectivores.—The forms so far considered were essentially truly primitive placentals or short side twigs derived from this ancestral stock. We now come to deal with members of the major "modern" insectivore assemblage, here termed the Lipotyphla, of which hedgehogs, moles, and shrews are typical. Members of this group have, despite specialization, successfully continued during the Cenozoic in the insectivorous mode of life of the Mesozoic ancestral mammals. The eyes tend to be rather poorly developed; the zygomatic arch is feeble and sometimes lost; there is a long tubular snout region.

The most primitive of living lipotyphlan families is that of the Erinaceidae, the hedgehog and its relatives. The molar cusps are simple tubercles; the upper molars have the paracone and the metacone placed close to the lateral margin and, in living members, are squared up through the development of a hypocone. There may be some enlargement of incisor teeth, but the zygomatic arch, although slender, is always retained. Living members are confined to the Old World. The spiny hedgehog, *Erinaceus,* is the best known; but more generalized ratlike or shrewlike types, such as *Gymnura,* are found in the Malay region. Members of the hedgehog family are found in the Tertiary of North America as well as Eurasia, and are moderately common from the Oligocene and Miocene on. Primitive relatives of the hedgehogs, perhaps best grouped in a separate family, Adapisoricidae, are abundantly represented in the early Tertiary, particularly Paleocene and Eocene, in such forms as *Geolabis* and *Nyctitherium,* and *Gypsonictops* of the late Cretaceous appears to be an early hedgehog ancestor. The extinct Dimylidae of the European middle Tertiary are hedgehog relatives.

The living shrews (Soricidae) and moles (Talpidae) are characterized by the paracones and metacones forming a W-shaped outer wall on the quadrate upper molars. This feature serves to distinguish them from other insectivores; but otherwise they are structurally not far from the hedgehogs and their fossil allies. Typical shrews (*Sorex,* etc.) are small, mouselike forms which in their habits and general appearance probably closely simulate the ancestral insectivores. They are, however, considerably specialized in structure; for example, the zygomatic arch is incomplete, the middle incisors are elongated, and the lower canine and most of the premolars have disappeared. The moles in a broad sense include not only the true moles (such as *Talpa*), with powerful, highly specialized limbs for digging, but also some more primitive forms with a more shrewlike appearance. The moles are in some respects more primitive than the shrews, for there are no enlarged incisors and the zygomatic arch is preserved. Members of both the shrew and the mole families are known as early as the Eocene; *Proscalops* (Figs. 317*C*, 318*E*) is a well-known American Miocene member of the latter group.

Zalambdodonts.—Living today in the West Indies is a single very distinctive insectivore, *Solenodon;* in sub-Recent deposits of Haiti and Puerto Rico are remains of a second aberrant type, *Nesophontes.* Both have developed specialized features, such as V-shaped upper molars, which are narrow fore-and-aft, with a sharp internal apex and with paracone and metacone far in from the outer border of the tooth, the two cusps partially or completely fused. In such characters they differ markedly from the soricids, but they may be descended from early North American shrew relatives. *Apternodus* (Fig. 317*D*) and several other early Tertiary forms are possible forebears.

The type of upper molar just described is often termed "zalambdodont," with a shape contrasting with the broader molars of typical insectivores in which the paracone and metacone are well toward the outer edge of the tooth and well separated from one another. A second series of zalambdodonts is from the African region, mainly found today (and in the Pleistocene) on the semi-isolated island of Madagascar. The smaller members of the group tend to parallel the shrews and hedgehogs of other regions in their adaptations; the largest form, the tenrec (*Tenrec* or *Centetes*), is a giant among insectivores, its tailless body measuring about 20 inches. Closely related to the Madagascan forms is the otter shrew, *Potamogale,* a good-sized form from West Africa, which is a fish eater with a strong swimming tail. A further modern zalambdodont is the Cape golden mole of South Africa, *Chrysochloris,* which superficially resembles ordinary moles in its adaptations. Presumably its ancestry is comparable to that of other zalambdodonts; but this "mole" differs from the *Tenrec* group and the otter shrew in many characters. In addition to Pleistocene forms, representatives of the tenrec, otter shrew, and golden mole types are known from the Miocene of East Africa. Their earlier history is unknown. Possibly they are specialized derivatives of the more typical insectivores, as is thought to be the case with the West Indian forms. But it is also possible that they are derivatives of the ancient deltatheridians described in Chapter 19.

Elephant shrews.—An odd series of small African mammals is that of the elephant shrews (*Macroscelides,* etc.), hopping types with an elongate proboscis. A peculiarity of the dentition is that the last molars have been sharply reduced and are generally absent; the cheek teeth anterior to these, however, are large and squared up in ungulate-like fashion. The diet is omnivorous, and one fossil genus has high-crowned teeth, suggesting a herbivorous trend in this form. In the past, the elephant shrews were generally believed to be allied to the tree shrews (and hence perhaps related to primate ancestry), but re-examination indicates that there is no real relationship between the two groups. Recently a series of African fossils extending back to the early Oligocene has been identified as pertaining to the elephant-shrew group. They appear to be an isolated African offshoot from some primitive insectivore stock.

The "flying lemur."—Living today in the East Indian islands is *Galeopithecus* (*Cynocephalus*), the "flying lemur," a herbivorous, tree-living form about the size of a large squirrel. The popular name cannot be called appropriate, for the animal is not a lemur and does not fly. It is, however, the most highly developed of any of the mammalian gliding forms. A very large membrane stretches from the neck to the hand and foot and onto the tail, enabling the animal to "plane" for a considerable distance. The upper molars are triangular in shape but peculiarly constructed, with a tendency for multiplication of small cusps. *Galeopithecus* is not an ancestor of the bats but may represent morphologically a stage in the development of the bat type of flying apparatus. The fingers are not greatly elongated (although webbed) and retain claws, but the hands are rather large as compared with the animal's bulk.

Through most of the Tertiary there is no record of this interesting type. But in the late Paleocene and the Lower Eocene of North America have been found inadequately known forms, such as *Plagiomene* (Fig. 320), with molars of similar structure. Possibly related to the flying lemur group, although represented by little more than dentitions, are a number of small Paleocene forms such as *Mixodectes* and *Elpidophorus*

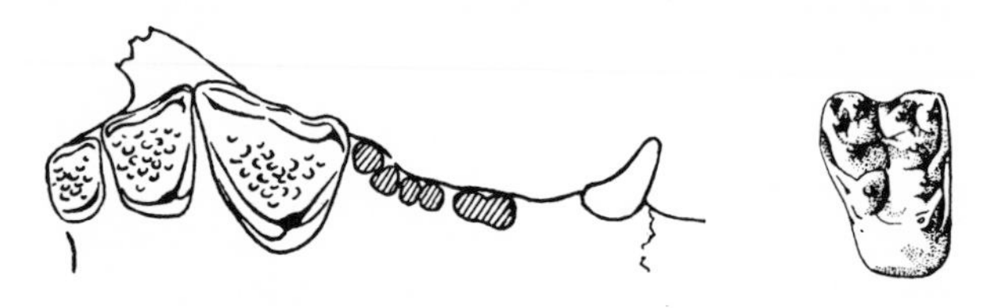

Fig. 320.—Possible relatives of the modern "flying lemur." *Left,* Right upper cheek teeth of the Paleocene *Zanycteris* (×2.5). *Right,* An upper molar of *Plagiomene,* of the Lower Eocene (×3 approx.). (After Matthew.)

(Figs. 318*F*, 319*C*). They have sharp-cusped teeth and the anterior teeth are enlarged, somewhat as in rodents and certain presumed early primates. The "flying lemur" and its fossil relatives form the Dermoptera, often considered a separate order but here included in the Insectivora, since, except for the development of a flying membrane, the structural features are not incompatible with those of the present group.

Bats.—Only in the bats—the order Chiroptera—has true flight been developed by mammals. As in the pterosaurs (and in contrast with birds), the wings are formed by webs of skin; but, instead of their being supported by a single elongate finger, nearly the whole hand is involved. The thumb, a clutching organ, is free and clawed; the other four fingers are all utilized in support of the wing membrane; claws are lost on these fingers (except the second in fruit bats); and, as would be expected, the end phalanges are reduced and may be absent. In having the wing expanse broken by the long digits, the bat has evolved a more flexible and less easily damaged wing than that of the pterosaur. In connection with the bat's habit of hanging by the hind legs, those structures, as well as the pelvis, are peculiarly developed but rather weak.

The orbit is but rarely closed behind. All the teeth (Fig. 321*B, C*) may be present except for an upper incisor and the first premolar, but the anterior teeth may be considerably and variously reduced. The greater number of the bats are insectivorous types, the lower molars being tribosphenic, while the upper molars are often of the old triangular shape or a squared type with a W-shaped ectoloph, a large protocone, and a smaller hypocone. But there is considerable variation in molar structure, for the diet in the various forms ranges from insects and fruit to fish and blood.

The sense of hearing is always highly developed (the auditory region of the skull is greatly swollen), and bats fly in the dark by "echolocation." In many cases grotesque fleshy outgrowths occur about the ears and nose, housing delicate tactile sense organs—adaptations also obviously connected with the usually nocturnal flying habits of the group.

The bats are divided into two suborders. The Megachiroptera, the fruit-eating bats or "flying foxes," are large Old World tropical forms. They are little specialized in most respects, never having the peculiar ear and nose ornaments already mentioned, and with a long, foxlike face. In connection with the fruit-eating habits, the teeth, especially the molars, are often specialized. Except for a few specimens from the Oligocene and Miocene of Europe, we have no record of their history.

Vastly larger in the number of contained types is the suborder Microchiroptera, comprising all other bats. Typically small insect eaters, as were presumably the ancestral chiropterans, there is a considerable range of

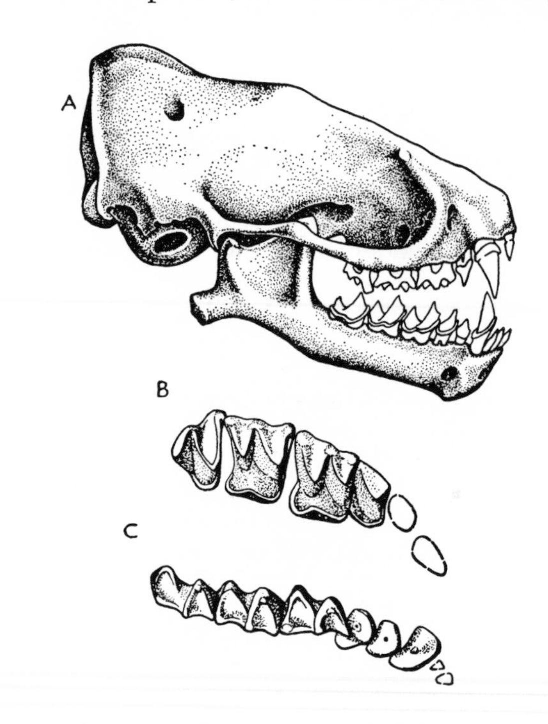

Fig. 321.—Bats. *A,* Skull of a member of the existing genus *Tadarida* from the Miocene of Europe, skull length about 5/8 inches; *B, C,* upper and lower teeth of the same, about 6 times natural size. (After Revilliod.)

diet in the various families. The incisors tend to be specialized and reduced; in connection with this and with the frequent development of nasal excrescences, the premaxilla is usually small or vestigial, the face is short, and there is a large nasal opening (Fig. 321*A*).

It is possible that some of the known insectivorous genera of the Paleocene were bats or at least types ancestral to them. But so close are some of the bats to the insectivores in dental characters that we cannot confirm or reject such suggestions in the absence of knowledge of limb and body structure. Flying adaptations must certainly have developed in the Paleocene, or at least by the earliest Eocene, because Middle Eocene deposits in both Europe (Fig. 322) and North America have yielded skeletons of microchiropterans with well-developed wings. The American form was more primitive than typical members of the suborder, for the claw on the second finger was still present. By the end of the Eocene and the beginning of the Oligocene the evolution of the varied modern families of bats was far advanced, for representatives of a number of modern groups were present in Europe. Fossil chiropterans are found here and there throughout the Tertiary, but it is only rarely that much except teeth and jaws are found of these small and delicate forms. They are common as fossils only in Pleistocene cave deposits.

It seems obvious that the bats, essentially insectivorous in their beginnings, have been derived from an arboreal insectivorous group, for in such features as are not connected with flight the bats might well be included in the order Insectivora, and certain anatomical considerations suggest relationships with the primates, likewise to be considered as intimately related to the insectivores.

Taeniodonts.—An obscure Tertiary group constituting the order Taeniodontia may be considered here for want of a better connection. The order includes but a few forms from the Paleocene and Eocene of North America; they comprise a short-lived archaic outgrowth of the primitive insectivore stock. Such an end-type as *Stylinodon* from the Middle Eocene was comparatively large for its day, the skull reaching a foot or

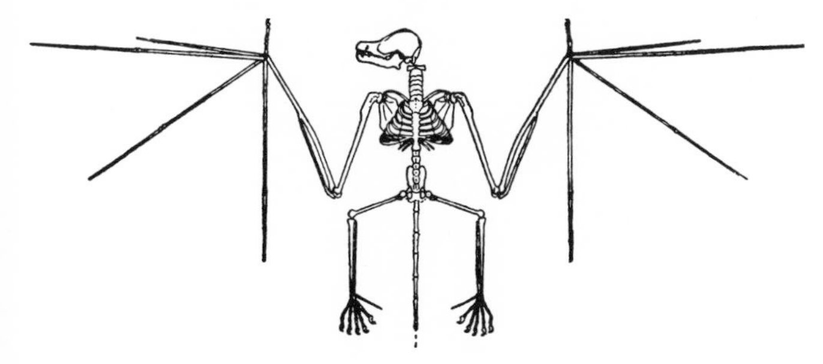

Fig. 322.—Skeleton of a fossil Eocene bat, *Palaeochiropteryx.* (From Revilliod.)

so in length. All the teeth had become high-crowned, rootless, simple pegs growing from a persistent pulp, and the enamel covering was limited to bands along the sides of the teeth. Only a single pair of rodent-like incisors (the lateral) was retained, but the canines were well developed. The skull was short, the jaws deep and powerful; a leaf-eating mode of life has been suggested for these forms. The limbs in later taeniodonts were short and stout, the toes reduced, and powerful claws developed. Somewhat earlier forms, such as *Psittacotherium* (Fig. 323) of the Torrejon and *Ectoganus* of the Wasatch, were more primitive in that the teeth (except perhaps the canine) were still rooted and the molars were enamel covered, while the Paleocene *Conoryctes* and *Onychodectes* (Fig. 324*B*) were

so close to the insectivore ancestors that it is difficult to decide in which group to place them. In the early taeniodonts the skull was still of primitive shape, long and low with long and not particularly deep jaws; the teeth were not merely rooted and enamel covered but

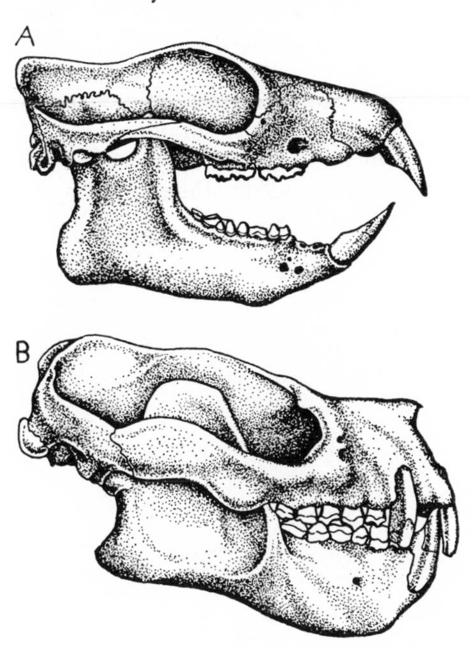

Fig. 323.—Aberrant Eocene placentals. *A, Trogosus,* a Middle Eocene tillodont, skull about 13 inches long; *B, Psittacotherium,* a Lower Paleocene taeniodont, skull about 9.5 inches in length. (*A* after Marsh; *B* after Wortman.)

Fig. 324.—*A,* Upper cheek teeth of *Esthonyx,* a primitive tillodont (×2/3 approx.); *B,* upper molars of *Onychodectes* (×3/4) to show the tritubercular origin of the teeth of taeniodonts. (*A* after Cope; *B* after Wortman.)

still retained much of the tribosphenic molar pattern of the early insectivores; and the limbs were slender and primitive in nature. Except for the loss of the central incisors, there was little in these forms to show that they were incipient taeniodonts.

Obviously, the group constitutes an early "experiment" on the part of the insectivore stock in the creation of a herbivorous form with grinding teeth. Taeniodonts, however, were seemingly never common, and soon disappeared when placed in competition with the herbivorous ungulates. Some writers have suggested that they are related to the edentates, the so-called toothless mammals; they have some resemblances to the ground sloths, but, except for a partial loss of enamel from the teeth, these resemblances are of a superficial nature. Surely they are an offshoot of lepticoid insectivores, and possibly related to the pantolestids.

Tillodonts.—Here, too, may be mentioned the order Tillodontia, another short-lived archaic offshoot from some point near the base of the placental stem. An end form, *Trogosus* (*Tillotherium*) (Fig. 323*A*), of the Middle Eocene, was the size of a brown bear. The skull, carnivore-like in general proportions, was long and low, the snout slim, the braincase small. The second pair of incisors in both jaws was enlarged and rootless, much as in rodents; the canines were tiny; and the low-crowned molars suggest a herbivorous or omnivorous diet. The beast had plantigrade, clawed, five-toed feet and presumably a rather bearlike gait. Lower Eocene forms were much more primitive; *Esthonyx* (Fig. 324*A*) was less than half as big as the Middle Eocene type, the incisors not so disproportionately large and still rooted. The tillodonts have sometimes been grouped with the rodents, but there is no reason for considering them related except for the large gnawing incisors, a feature which developed in many groups of mammals. A more reasonable comparison, based on similarity of molar teeth, is one with certain of the arctocyonids, an ancient group sometimes considered as primitive carnivores, but here ranged with the ancestral ungulates (condylarths).

CHAPTER 18

Primates

The order of Primates is of especial interest, since its members include not only lemurs, monkeys, and apes, but man himself. Unfortunately, however, fossil remains of this group are more rare than is the case in any other large group of mammals. The reasons for this paucity of material are fairly obvious. Primates are, for the most part, tree dwellers; deposits in which fossil vertebrates are to be found are not normally formed in forested regions. Again, primates are mostly dwellers in the tropics, whereas most of the known Tertiary fossil beds are in what are today zones of temperate climate. It is only in Paleocene and Eocene times, when these regions (as shown by the vegetation) had essentially tropical conditions, that primates are found to any extent in the fossiliferous deposits of Europe and North America.

Primates may be classified in various fashions. Often monkeys, apes, and men are considered as constituting a suborder Anthropoidea and all the highly varied lower forms "lumped" as a suborder Prosimii. This, however, is a rather arbitrary and unnatural grouping. Although the position of some of the early forms is far from clear, fossil and living primates are, perhaps, better arrayed in five assemblages which we shall consider as suborders: (1) Plesiadapoidea, early aberrant forms, with chisel-like incisors; (2) Lemuroidea, the lemurs, typically small, four-footed forms of rather squirrel-like appearance, found today in the Old World tropics but present in the Paleocene and Eocene in Eurasia and North America; (3) Tarsioidea, which includes *Tarsius,* a curious, small, hopping, ratlike creature from the East Indies, and its fossil Paleocene and Eocene relatives, occupying a position structurally intermediate between lemurs and higher primates; (4) Platyrrhini, South American monkeys and marmosets; and (5) Catarrhini, the more advanced monkeys of the Old World, the great manlike apes, and man.

Primates are essentially arboreal, only a few forms (such as the baboons and man) having returned to a life on the ground. (There is considerable evidence suggesting that the primitive placentals were to some extent tree dwellers to begin with.) Arboreal life is apparently responsible for much of the progressive development of primate characteristics; and, although man is not a tree dweller, this life of his ancestors has left its mark deeply upon him and is perhaps in great measure responsible for his attainment of his present estate.

Arboreal adaptations: Limbs.—Locomotion in the trees has left the postcranial skeleton of the primates in a condition much closer to that of the primitive placentals than is the case in most groups. Flexibility is necessary for climbing trees, and there is none of the restriction of limb movement to a fore-and-aft plane found in ungulate groups and in many carnivores and rodents. Very rarely is there any reduction or fusion of bones in the lower segment of the limbs. The clavicle is retained, in contrast with running forms, its connection of shoulder with trunk relieving the strain on the muscles when the animal is suspended from a limb.

In contrast with such arboreal types as the squirrels, in which climbing is accomplished by digging the claws into the bark, the primate hold is generally accomplished by grasping boughs or twigs. The claws characteristic of primitive mammals have been in most cases transformed into flat nails serving as a protection to the finger tips; claws, however, still persist in various instances in lemuroids and even marmosets.

In primitive placentals the pollex and hallux were presumably somewhat divergent. This grasping characteristic has usually been retained and emphasized in primates (Fig. 304). In most cases it is the hallux which is the most highly developed and divergent and hence more opposable to the other digits. In contrast, the thumb is sometimes reduced, or even absent, in forms which habitually progress by swinging from branch to branch, the other four fingers here hooking over the bough and the thumb not only being useless but actually in the way. Otherwise, there is seldom any reduction of digits; the primates are in this respect more conservative than most mammalian groups.

Quadrupedal locomotion is the rule among primates, bipedal tendencies becoming apparent for the most part only among the higher members of the order. The hands, however, while primarily possessing their grasping function in relation to locomotion, are well adapted (in contrast with most mammals) for seizing food and other objects; and even among the lemurs there is a tendency toward a sitting posture and the release of the front feet from supporting the body.

The primitively long tail is retained in almost all the lower primates and is apparently of considerable use in balancing. In many South American monkeys it has developed into a prehensile "fifth hand." In a number

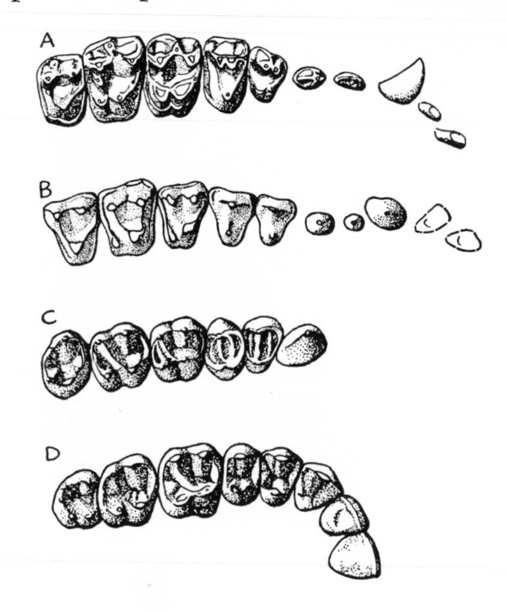

Fig. 325.—Upper right dentitions of primates. *A*, *Notharctus*, an American Eocene lemur; hypocone split off from protocone (×3/2 approx.). *B*, *Pronycticebus*, a European Eocene lemur; hypocone derived from cingulum (×5/2 approx.). *C*, *Dryopithecus*, a late Tertiary anthropoid (×2/3 approx.). *D*, Mousterian youth (×1/3 approx.). (Mostly after Gregory.)

of cases, however, the tail is reduced or absent, and there is no external tail in the manlike apes or in man.

Teeth.—An omnivorous diet appears to have been general in early primates, and similar food habits characterize many living types (although there has been a strong herbivorous trend). The dentition (Figs. 325, 326), hence, is usually less specialized than in most mammalian groups. In contrast to the three incisors present in primitive placentals, no more than two are present in primates, and these may be variously modified or reduced in lemurs. The canines are usually unreduced, projecting moderately above the other teeth, and are often larger in the males.

While four premolars are present in a few primitive fossil forms, no living primate possesses more than three; and the catarrhines have but two. Related to this is a shortening of the tooth row evident in the group as a whole, although (as in the baboons) there may be a secondary lengthening brought about by elongation of the molars. The premolars are never molarized and are typically bicuspid.

The molars of the most primitive forms (Fig. 297A) are often very similar to those of the early placentals, but the originally sharp cusps have tended to soften down toward a bunodont condition. A hypocone is usually added above. Below, as in many other groups, the paraconid disappears to give an essentially four-cusped tooth; but a hypoconulid on the back margin is persistent in many cases, and even in man this fifth cusp is often present. There is little tendency toward a hypsodont condition, since the food is mainly of the softer kinds, such as fruit and buds.

Sense organs.—Arboreal life has had a profound effect upon the sense organs. A ground-dwelling mammal is in great measure dependent upon smell for his knowledge of things about him, and his olfactory organs are generally highly developed while sight is usually comparatively poor. The reverse is true in these arboreal types. For locomotion in the trees good eyesight is

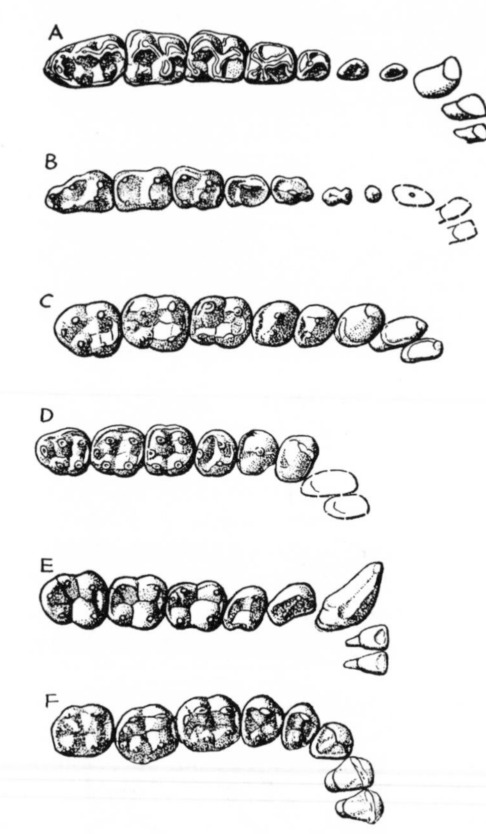

Fig. 326.—Lower left dentitions of primates. *A*, *Notharctus*, an Eocene lemur (×3/2 approx.); *B*, *Pronycticebus* (×5/2 approx.); *C*, *Parapithecus*, an Oligocene catarrhine (×5/2 approx.); *D*, *Propliopithecus*, an ancestral Oligocene anthropoid (×3/2 approx.); *E*, *Dryopithecus*, a late Tertiary anthropoid (×2/3 approx.); *F*, Mousterian youth (×1/3 approx.). (Mainly after Gregory; *C*, *D* after Schlosser.)

essential; and throughout the primate group there has been a progressive series of advances in the visual apparatus. Even in the lemurs the eyes are large, and there is a tendency to rotate them forward from their primitive lateral position. This results, in such forms as *Tarsius*, in a condition in which the two fields of vision are identical. With the overlapping of visual fields, there comes about in primates, as in many other mammals, a sorting-out of nerve fibers so that, instead of two mental pictures, one for each eye, there is a doubling-up of the impressions from the common field of sight, giving stereoscopic vision with its effect of depth. In no other group of animals is the overlap of fields so great; in all higher primates (as in ourselves) the stereoscopic effect is complete. In addition to this improvement, there is in *Tarsius* and the higher primates a special central area in each eye in which detail is much more clearly seen.

With this increase in vision goes a corresponding decrease in the sense of smell. Even in the lemurs the olfactory sense is reduced, and in the manlike apes and man it is in as rudimentary a state as in any land-dwelling placental.

Brain.—The brain, as well, has been profoundly influenced by arboreal life. Locomotion in the trees requires great agility and muscular coordination, which in itself demands development of the brain centers; and it is of interest that much of the higher mental faculties is apparently developed in the frontal area alongside the motor centers of the brain. Again, the development of good eyesight rendered possible for primates a far wider acquisition of knowledge of their environment than is possible for forms which depend upon smell. Perhaps still more important in the development of primate mentality has been the development of a grasping hand. This is not only an advantage in feeding, but undoubtedly has contributed materially to primate mental development through the potentiality of gaining knowledge by the examination of objects.

A constantly increasing brain size has been a characteristic feature of primate development. The cerebral hemispheres have grown in size and in all the higher Anthropoidea completely cover the cerebellum. Among the smaller monkeys are to be found the highest relative weights of brain to body in mammals.

Skull.—These changes in dentition, sense organs, and brain have been associated with great changes in the primate skull. In many lemurs there is present an elongate skull of primitive appearance, with a long facial region and a long, low braincase. But with the reduction of the sense of smell and the concurrent abbreviation of the tooth row, the muzzle in most primates has shortened considerably. The orbits were rather large in even the most primitive of the lemurs, and even in the Eocene were almost invariably bounded posteriorly by a postorbital bar—a feature in which they were in advance of almost all known forms of that period. In the higher lemurs the orbits tend to turn forward, their median boundaries crowding close together above the nostrils—a position in which they are found in all higher primates. The originally superficial bar separating orbit and temporal opening becomes in the higher forms a solid partition. In many forms the upper margin of the orbits is above the level of the braincase, projecting as a supraorbital ridge.

The expanding brain naturally requires an enlarged braincase; and, except for some primitive types, the primate braincase tends to be high vaulted and rounded. A sagittal crest, present in the early lemurs, is rare in later and higher types, for with expansion of the braincase the temporal muscles, between which the sagittal crest forms, fail to meet. These crests are to be found only in the larger forms and in heavily muscled, herbivorous types. In the trend toward erect posture in higher primates the braincase tends to assume a roughly globular shape and to be balanced on the upper end of the vertebral column; in consequence, the occipital condyles and foramen magnum are shifted to a position varying from the back to the underside of the skull. The two halves of the lower jaw, primitively loosely connected, become fused in higher types.

Plesiadapoids.—As noted in the last chapter, the tree shrews of the Oriental region are very commonly included among the primates, but are here, despite their seemingly transitional nature, retained amongst the Insectivora. These doubtful forms apart, the living primates below the monkey level comprise a varied series of lemurs (mainly from Madagascar) and *Tarsius* of the East Indies. The lemurs are seemingly primitive forms; *Tarsius*, although specialized, is believed to represent a more progressive stage in primate evolution, transitional to the monkeys and apes.

Almost nothing is known of the antecedents of these living groups in the later Tertiary. In the Paleocene and the Eocene, however, there are, in both North America and Europe, numerous genera of small mammals which are obviously primitive primates ancestral or related to both modern stocks. The taxonomic assignment of many of these forms has, however, been a difficult one. Most of the remains are very fragmentary and consist of teeth and partial jaws. Primate teeth are rather primitive in nature, and dental differences between lemurs and tarsioids are not marked; it is, therefore, not surprising that many of these genera have been variously and variably assigned to the insectivores, lemurs, and tarsioids.

Forms which most would agree are definitely on the primate side of the ordinal boundary first appear in the Middle Paleocene. They pertain to three families, of which *Plesiadapis* (Figs. 327, 328), *Carpolestes*, and *Phenacolemur* are typical; the three are confined to the Paleocene and Lower Eocene and are principally known from the American West. For the most part, they are represented by fragmentary remains, mainly of the dentition. The molar teeth seem to agree with those of early lemurs, and in *Plesiadapis*, in which alone the skull is preserved, the ear region has the structure seen in typical lemuroids. However, the only foot material associated with any of these forms shows the presence of claws rather than nails and, most especially, all members in which the condition is known had enlarged chisel-like incisors (a Paleocene trend already noted among the insectivores), and *Carpolestes* had enlarged lower premolars rather like those of kangaroos. These ancient families are obviously well off the line leading to higher primates. Quite possibly (as is done provisionally in the classification in this volume) they should be placed in a category distinct from the other lower primate assemblages.

Early lemurs.—More typical primates, lacking the specializations seen in these aberrant Paleocene forms and probably close to the ancestry of modern lemurs, appear, apparently as immigrants from some unknown

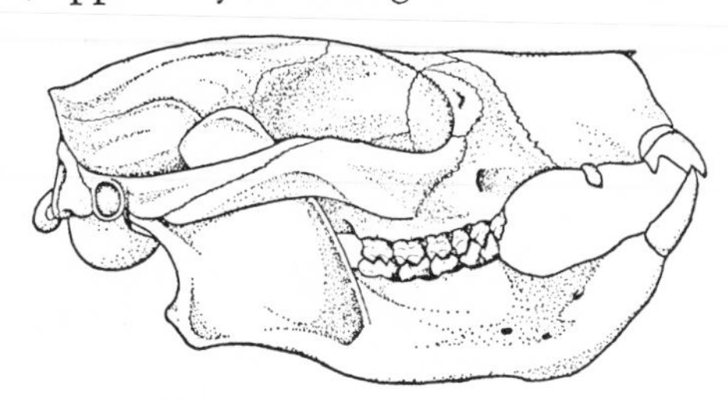

Fig. 327.—*Plesiadapis,* a late Paleocene primate with enlarged incisors; length about 2.25 inches. (From Russell.)

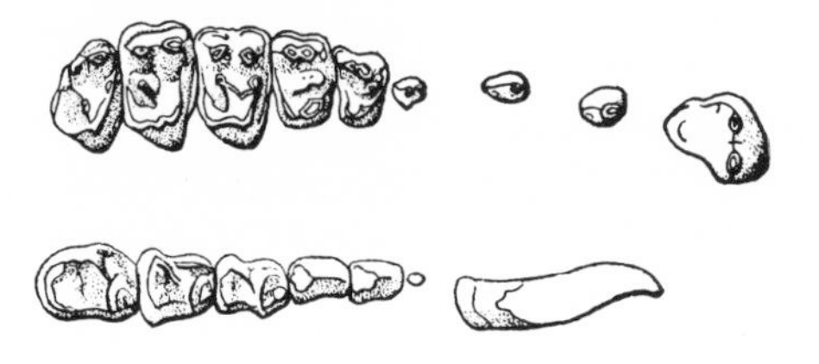

Fig. 328.—Upper and lower dentitions of *Plesiadapis* (×2 approx.). (After Simpson.)

region, in the Lower Eocene and continue to be moderately abundant in later Eocene phases in both Europe and North America. *Notharctus, Adapis,* and *Pronycticebus* are among the better-known types (Figs. 304; 325*A, B;* 326*A, B;* 329*A*). *Notharctus* was a small mammal with a skull about 2 inches long, which probably resembled the ordinary lemurs of today in general appearance. It was already quite lemur-like in its proportions and many structural features. The dentition, however, was more primitive in a number of respects. Although some members of the group had but three premolars, *Notharctus* still retained the full complement of four. Further, the upper incisors were comparatively normal; the canines were of primitive type; and the lower incisors were little, if at all, procumbent. In *Notharctus* a hypocone had already developed in the upper molars. Some of the European forms had likewise developed this additional cusp; but it is of interest that *Adapis* and *Pronycticebus* had obtained it by an upgrowth of the cingulum, *Notharctus* by a budding-off from the protocone.

Madagascar lemurs.—By the close of the Eocene the lemurs had disappeared from the fossil record in the North Temperate zone, and we know little of their further history until Pleistocene and Recent times. Here we find them present in Old World tropical regions, where conditions today are probably not dissimilar from those prevailing farther north in the Eocene. The main center of the group is on the island of Madagascar. This region presumably was cut off from the mainland during most of the Tertiary; and there are but few carnivores—a fact which may have been responsible for the survival of these primitive primates (as well as of various insectivores).

With one exception, we may include all the Madagascar lemurs in the family Lemuridae, using this term in a broad sense. They give us as near an approach as we can obtain today to the general appearance, habits, and structures of our Eocene primate ancestors. A typical Madagascar lemur of today is a fairly small arboreal animal, nocturnal in habits, with a bushy, furry covering which contrasts with the relatively sparse hairy coat of most higher primates. The limbs are moderately long, the ears pointed, and the eyes are directed more laterally than forward. The lemur face may undergo some shortening, but there is in most cases a comparatively long and foxlike muzzle. The braincase is long and low in the more primitive living

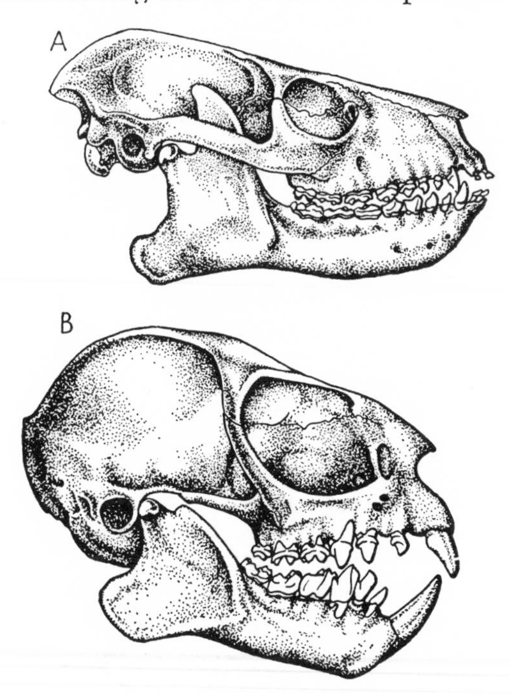

Fig. 329.—*A, Notharctus,* an Eocene lemur, length of skull about 3 inches; *B, Tetonius,* an Eocene tarsioid, length of skull about 1.8 inches. (After Gregory.)

types but somewhat expanded in the more advanced lemurs. In living lemurs the upper incisors are often small or absent, while the lower ones, together with an incisor-like canine, slant forward as a fur-combing organ. The upper molars vary from the triangular to the quadrangular pattern. The lemur tail, usually long, is never a grasping organ. Pollex and hallux are always widely separable from the other fingers; the big toe is especially well developed and has a flat nail in all forms, whereas the other digits are variable in their covering. Typical lemurs have a clawlike nail on the second toe but normal nails on the other digits.

There are numerous divergences from this general type among the living and recently extinct Madagascar lemurs. There is, for example, a small, hopping form; and *Indri* and a few other lemurs are rather large and somewhat monkey-like in superficial features. Still other variants have been discovered in Pleistocene bog deposits on the island. *Megaladapis* attained the size of a hog, with a skull well over a foot long, peculiarly long and decurved nasal bones, and small eyes; it seems, nevertheless, to have resembled the Eocene lemurs closely in many features. It may have survived in Madagascar until a few centuries ago. A second extinct group was represented by *Archaeolemur,* also large for a lemur (a 5-inch skull), but with large eyes turned remarkably far forward and a large brain, there being no sagittal crest. The face was considerably shortened. This extinct form paralleled in a number of respects the higher primates and is one of several abortive "efforts" of the lemur group to reach a monkey status.

A final member of the Madagascar primate fauna is the aye-aye, *Daubentonia* (*Cheiromys*). It is a nocturnal animal about the size of a small cat, with a short, high skull of the shape found in the higher lemurs and monkeys. All the digits except the big toe have clawlike structures rather than nails; the middle finger is very long and slim, and is said to be used to extract grubs from their holes in the wood. The single remaining pair of incisors is enlarged in rodent fashion and is used in gnawing through bark in search of insect food. This feature is reminiscent of the enlarged incisors of many early fossil primates, but has probably resulted from a parallel evolution among the Madagascar lemur group.

Higher lemurs.—All the living non-Madagascar lemurs are included in a higher group, the family Lorisidae. The lorises of the Indian region are typical; in Africa the group is represented by the potto and the "bush baby," *Galago* (of which a fossil forerunner is known from the African Miocene). The muzzle is rather short, the braincase high, the eyes turned somewhat forward, and the premolars reduced to two; the lower front teeth, however, have the typical arrangement of the Madagascar lemurs. The index finger is reduced, giving a better grasp between the thumb and the remaining fingers. The advanced skull structures of these lemurs are of the sort required to turn a lemur-like primitive primate into a more advanced type; but it is generally held that this is a development parallel to, and not on, the main evolutionary line.

Tarsius.—An exceptionally interesting primate group is that which includes a single living animal, *Tarsius,* in addition to numerous Paleocene and Eocene fossil genera. These forms have often been classed as lemurs but are quite advanced in structure and may be appropriately placed in a separate suborder, the Tarsioidea, intermediate between the true lemurs and the higher primates.

Tarsius is a small, nocturnal, tree-living form from the East Indies. The tail is very long, and the hind legs are adapted for a hopping gait. The braincase is large and rounded, and the foramen magnum is pushed somewhat forward onto the underside of the skull in contrast to the usual lemur condition. The visual apparatus is much more advanced than that of the lemurs. The eyes are exceptionally large and are turned completely forward from the primitive lateral position, with the orbits very close together above the nose. In lemurs there is only a superficial bar separating orbit and temporal fossa; in *Tarsius* the alisphenoid extends outward below the frontal and jugal and closes off the greater part of the opening between the two regions, much as in the higher primates.

While vision has thus advanced, the sense of smell has been proportionately reduced. The foxlike snout of the typical lemur has disappeared, and the nose is reduced to a small nubbin tucked beneath and between the eyes. There is little projection of the muzzle (the tooth row is rather short); and we have here the beginning of the type of face found in monkeys, apes, and man. There are some specializations of the front teeth, but the cheek teeth are of primitive pattern with fairly typical tribosphenic molars.

Specializations have developed in the living *Tarsius* in connection with the hopping gait. Alone among primates *Tarsius* has a partial fusion of tibia and fibula. A third segment is introduced into the limb by means of an elongation, not of the metatarsals but of the calcaneum and navicular (a similar development has occurred in a Madagascar lemur and in *Galago*). This seemingly odd fashion of developing a third joint is readily understandable when we consider the situation. The ordinary animal which introduces an extra segment for hopping or, indeed, for fast locomotion of any sort (as the kangaroo, jumping mouse, or any typical ungulate) has no grasping ability and is usually a ground-living type. The primate, however, is in great measure dependent for security in the trees on the powers of opposability of the great toe. If the metatarsals were to elongate as in terrestrial forms, this grasp would be lost; elongation must take place proximal to the origin of the big toe, and the tarsal bones give the only possibility for this development.

Apart from this peculiar limb specialization, *Tarsius* is close to the higher primates in most features. One of many additional indications that this form is a relic of a group transitional to the monkeys and apes lies in the structure of the placenta. In the lemurs this structure was of a type probably present in primitive eutherians. But in *Tarsius* the placental connections between mother and young are concentrated into a pancake-shaped structure which is shed at birth, as in the higher primates.

Eocene tarsioids.—Between the early Tertiary and Recent times there is a major gap in the history of the group. But in Eocene deposits we find in Europe and North America numerous remains of several series of genera suggestive of relationship to *Tarsius; Tetonius*

(Fig. 329*B*), *Omomys,* and *Necrolemur* are representative. Members of the ancient tarsioid assemblage lingered on into the Oligocene and even early Miocene in North America. Many of them are known merely by the dentition. This was often, as in *Tarsius,* close to the primitive placental type in molar construction, the hypocone appearing only rarely. In some genera the first lower premolar was still present as a small tooth; but normally there were, as in *Tarsius,* but three premolars, and in one case there were but two. The anterior teeth were quite variable in structure, and many show a trend toward the chisel-like development of incisors we have commented upon in other groups of primates and insectivores.

Although the teeth are to some degree indicative of relationship to *Tarsius,* better evidence can be obtained from skull structure. In a few forms, such as *Tetonius* and *Necrolemur,* the skull is preserved and shows an approach to the *Tarsius* condition, with a short face and huge, forwardly turned orbits with a developing partition behind them. The braincase is generally expanded and in two forms, at least, brain casts show a very considerable development of the cerebral hemispheres—far beyond the lemur stage and approaching that of monkeys and apes. Significant, too, of the relationships of these early Tertiary types not only to *Tarsius* but to higher primates as well are the technical details of the auditory bulla. As mentioned earlier, two elements are often present in this structure; an entotympanic bone which is fused to the ear capsule and in primates forms the bulk of the bulla, and an ectotympanic element, usually external to it, surrounding the ear drum. In *Tarsius* and higher primates the ectotympanic forms a tube leading outward from the bulla; in both Eocene and living lemurs it is, in contrast, a simple ring, lying inside the opening of the bulla. In the few forms now being discussed, the ectotympanic is tubular. Such facts strongly support the belief that this series of early Tertiary forms includes not only specialized types and ancestors of the living *Tarsius,* but ancestors of the higher primate groups as well.

Advanced primates.—Beyond the tarsioid stage we reach the level of higher primates—monkeys, apes, and (eventually) man. In all these forms we find further notable advances over the forms so far discussed. The eyes are large and face forward, as in *Tarsius,* and the orbits are completely separated from the temporal fossa. The sense of smell is unimportant, and the snout is greatly reduced. The monkey normally is a four-footed walker, but there is a great tendency for an upright sitting posture and the freeing of the hands for the manipulation of food and other objects. The tail is variable; in some South American types it is prehensile, but in the higher forms it is frequently reduced.

There are never more than three bicuspid premolar teeth, and the Old World forms have but two. The last molar tends toward reduction and is actually lost in one monkey family. The upper molars are still tritubercular in a few of the more primitive forms, but generally the typical quadrate pattern is developed in both jaws. The brain is comparatively large in all the members of the higher primate assemblages. The braincase is typically much expanded, and crests are only occasionally developed. The foramen magnum tends to be under, rather than at the back of, the skull, with the face turned forward almost at right angles to the backbone. The two halves of the jaw are fused.

There are geographically two distinct groups of higher primates: (1) the South American monkeys, and (2) the Old World monkeys, apes and men, none of which ever reached the Western Hemisphere in prehuman times. It has been customary in the past to unite the two groups in a common suborder Anthropoidea. This is both confusing (as a minor matter) and, from an evolutionary viewpoint, highly improper. For one thing, the term "anthropoid"—that is, manlike—is often used in a quite different sense, denoting the manlike apes, such as the chimpanzee and gorilla. More important is the fact that by including the monkeys of both South America and the Old World in a common unit, we imply that we are dealing with a natural group and that they had a common monkey ancestor. This is highly improbable. It seems certain that the ancestors of the South American monkeys reached that continent from the north. But no North American primate had evolved beyond the tarsioid stage, and hence New World and Old World higher primates represent two independent lines of advance and should be classified as distinct groups.

South American monkeys.—Of these two advanced assemblages, the less progressive is that of the suborder Platyrrhini, comprised of two families of monkeys now living and found as fossils only in South and Central America. Superficially, many are quite similar in appearance to Old World monkeys but seem to be more primitive in a number of respects. They are, on the average, much smaller than their Old World relatives. Almost never is there any reduction of the primitively long tail. The thumb is but little opposable to the other fingers. Three premolars are still present in all cases, whereas the oldest fossil representatives of the Old World apes had already lost the second premolar. The group name ("flat nose") refers to the fact that the nostril openings are usually far apart and face outward.

The smallest of the two families, both in number of forms and in the size of the individuals, is that of the marmosets, the Callithricidae. *Callithrix,* the common marmoset, is a small creature, squirrel-like in size and general appearance, with thick fur and a bushy tail. Squirrel-like, too, is the fact that the toes (except the hallux) have the nails compressed and clawlike, so that the progression is mainly by digging the claws into the bark; the hallux has lost much of its opposability. A peculiar feature in all except one marmoset is that the last molar has been lost, the only case of its complete reduction among primates (although man is approaching this condition). The incisors, and the lower canine with them, are rather procumbent and lemur-like. A late Oligocene marmoset is reported, but otherwise the early history of the family is unknown. It has been thought that the marmosets are the most primitive of monkeys, but features such as the molar loss suggest that they are specialized rather than primitive.

A more important and larger group is that of the Cebidae, the typical South American monkeys. Here the average size is somewhat larger, the hairy covering somewhat thinner (a tendency carried much further in Old World types), and the teeth are little specialized and with a persistent cheek complement of three premolars and three molars. In some cebids, alone among primates, a prehensile tail has developed.

Most primitive of living cebids is perhaps the group represented by *Aotus* [*Nyctipithecus*], the owl monkey or dourocouli, a small form with a presumably primitive thick woolly covering and a small brain. It is a nocturnal form, while most monkeys are active in the daytime. (Parenthetically, this raises the question as to whether the primitive primates were nocturnal, for this is the case in *Tarsius* and almost all of the lemurs.) The owl monkey is quite similar to the marmosets and may be close to the base of the platyrrhines. *Homunculus* of the Miocene belongs to this group.

The owl monkeys and several other fairly primitive forms lack a prehensile tail. This "fifth limb" is developed in the remaining cebids. Of these, the commonest group is that best known through the capuchin monkey, *Cebus*, usually to be found accompanying the organ grinder. Closely related are the spider monkeys, *Ateles*. Here the limbs are very long and slender, the thumb reduced or absent, the nails rather claw-like. These monkeys are clever acrobats, swinging from the branches by any of their five useful appendages. A final group is that of the howling monkeys, *Alouatta*. These are the largest of the New World forms, some as large as a good-sized dog, and are leaf eaters with heavy molar teeth. The larynx is expanded into a huge ossified shell, with a resonating chamber within, by means of which night is made mournful in the South American forests.

Little is known, unfortunately, of the fossil history of the South American monkeys. None are known before late Oligocene and early Miocene times. South America, as we have noted, seems to have been an isolated area from the dawn of the Tertiary; presumably the platyrrhines gained entry from the north by "island hopping" down islands which may have existed in the Central American and Panamanian region.

Catarrhines.—The later primates of the Old World form a well-defined group, the suborder Catarrhini, including the living monkeys of Asia and Africa, the anthropoid apes and man, and their fossil representatives. The name indicates one of the many features in which they are more advanced than the South American monkeys; the nostrils are closer together and open forward and downward, with a smaller bony opening There is a rather general tendency toward an increase in size which culminates in the chimpanzee, man, and the gorilla. Although originally arboreal, some of the higher or more specialized types, such as man and the baboons, tend toward a terrestrial life. The tail is often shortened or absent. Primitively four footed, some have tended to an arm-swinging type of locomotion—"brachiation"—in which the body is held erect; man has attained an erect posture. The thumb is usually moderately developed and opposable and the big toe universally opposable except where secondarily modified in man. The hairy covering is always thin and the face naked.

The brain is large; but since the body is, on the average, larger, the ratio of brain size to body is reduced, and sagittal and occipital crests, related to temporal muscles, are more common than in South American monkeys. The face, primitively short, tends to elongate in many instances in correlation with a tendency toward a vegetable diet and a consequent increase in tooth size. The second premolar is always absent, the tooth formula being invariably

$$\frac{2 \cdot 1 \cdot 2 \cdot 3}{2 \cdot 1 \cdot 2 \cdot 3}.$$

The molars are quadrangular and essentially four cusped.

First stages in the evolution of the catarrhines undoubtedly occurred in the late Eocene and early Oligocene of the Old World, but our knowledge of this story is very limited. We have no adequate skeleton or even a skull of the early members of the group; at best we may have jaws, but for the most part, early Tertiary monkeys and apes—and even most of those of later epochs—are represented by fragments of the tooth row or individual teeth. There are a few dubious tooth specimens from the late Eocene of Burma. Somewhat better is the situation in the early Oligocene beds of the Egyptian Fayum. A number of specimens has been collected here, and further work is being undertaken; half a dozen genera are present. Two of these, *Parapithecus* (Figs. 326*C*, 330*B*) and *Apidium*, are difficult to place with accuracy, but rather surely pertain to primitive catarrhines of some sort, and some details have suggested that they had not advanced far beyond the tarsioid level. Others, mentioned below, appear to be primitive ancestors of the great apes. One form, *Oligopithecus*, may perhaps be ancestral to Old World monkeys.

Old World monkeys.—The living Old World monkeys constituting the family Cercopithecidae include a large number of forms with a wide range of structure. They are, in terrestrial locomotion, four-footed forms, walking flat on palm and sole. The more primitive forms are arboreal, but the baboons and their relatives show a progressive tendency toward a ground-dwelling life. Toughened skin areas—the ischial callosities—are present on the buttocks. The molars are four cusped, the hypoconulid (in contrast with great apes and men) being present only on the last lower molar; and, characteristically, two cross crests are developed, forming a fairly effective mechanism for dealing with the vegetable diet toward which this family has strongly tended.

Two groups may be distinguished, the larger one being that of the subfamily Cercopithecinae. This includes many primitive African and Asiatic arboreal monkeys and a series of forms leading structurally to the baboons. There are well-developed cheek pouches for storing food, while the stomach (in contrast to the other subfamily) is simple. The genus *Cercopithecus* includes most of the African monkeys commonly found in zoological gardens (as the green monkey, vervet monkey, and Diana monkey). Typical arboreal forms with a short face, they probably stand near the base of the monkey stock. The macaques (*Macaca*) are monkeys of fairly large size and mostly Asiatic in distribution, but with a representative—the Barbary "ape"—in northern Africa and Gibraltar; there are fossil forms from the Pliocene of Eurasia. The tooth row is considerably elongated and the snout consequently projecting. The macaques are only partially arboreal, some species living in great measure on the ground. From this, several intermediate types lead to the baboons (*Papio*), which have become completely terrestrial, with a four-footed plantigrade gait. The tooth row of these exclusively herbivorous types is much elongated, and a doglike muzzle is developed. The tail is short and in some of the related types may be practically absent. Baboons of several types are found in the Pleistocene of northern Africa and Asia.

The second subfamily of the modern Old World monkeys is the Semnopithecinae, a comparatively small group including *Semnopithecus,* the langurs, and their relatives (mostly Asiatic, with one African genus). There is never any facial elongation, the body is slim, the hind legs especially long, the tail usually well developed. Instead of cheek pouches, there is a complicated stomach for food storage. Cross crests are highly developed. All the members of the group are good arboreal types.

As mentioned above, remains of cercopithecids have been found in Pleistocene deposits and to a lesser extent in the Pliocene; *Mesopithecus* of the Pliocene (Figs. 330*A*, 331), known from a nearly complete specimen, appears to be an ancestral langur. The earlier history of the group, however, is almost unknown. The living Old World monkeys furnish us with a good picture of the general nature of a primitive catarrhine; but the cercopithecids, as we know them, are not direct ancestors of the great apes or men.

Oreopithecus.—Before continuing with the story of the more advanced primates, we may briefly treat a relative of theirs, *Oreopithecus.* Fragmentary remains of this primate were found, many decades ago, in an Italian early Pliocene coal seam. *Oreopithecus* was

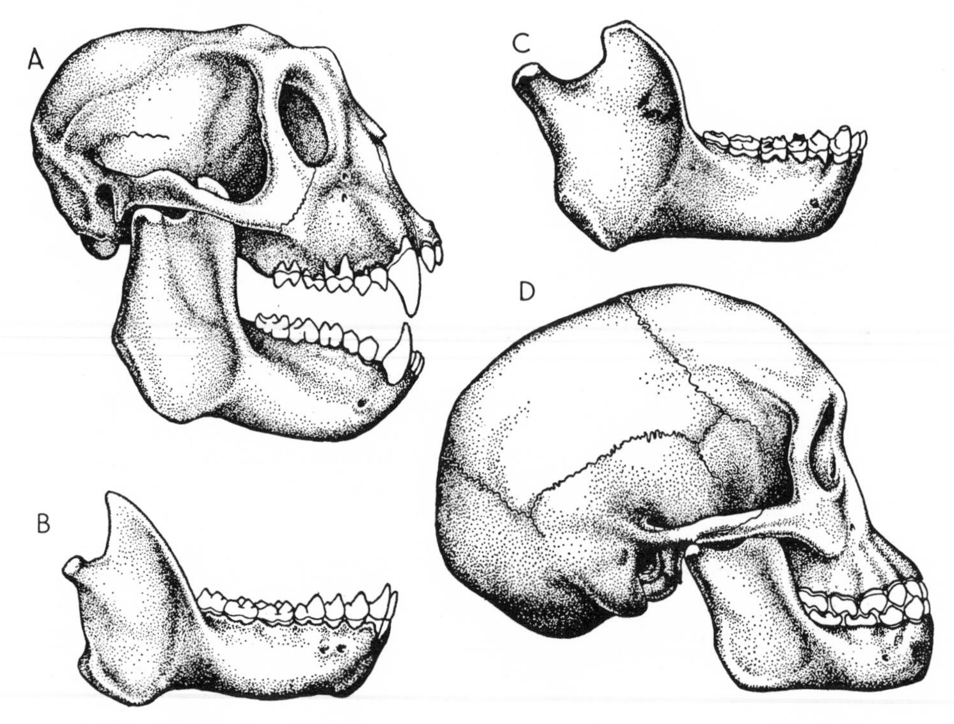

Fig. 330.—*A*, The skull of *Mesopithecus,* a Pliocene Old World langur, length of original about 3 inches; *B*, the jaw of *Parapithecus,* a catarrhine from the Lower Oligocene of Egypt, original about 1.5 inches long; *C*, the jaw of *Propliopithecus,* a primitive anthropoid, from the same deposits, original about 2.75 inches long; *D*, restored skull of a young individual of *Australopithecus.* (*A* after Gaudry; *B, C,* after Schlosser; *D* after Dart and Broom.)

long ignored as being a mere monkey of some sort or other, but recent restudy showed that it was, instead, a member of the ape-man complex in its dental characters. Exploration yielded much further material including a nearly complete skeleton of the animal. It would have had about the proportions of a chimpanzee, with long arms indicating a brachiating habit. Dental details, however, distinguish it from both men and great apes, and it seems most probable that it represents a modest independent line of ape advance.

Manlike apes.—The family Pongidae (generally termed Simiidae in the older and nontaxonomic literature) includes four living types: the gibbon, the orangutan, the chimpanzee, and the gorilla. They are of large size, ranging from the comparatively small gibbon to gorillas several times as heavy as a man. In the quadrangular molars a fifth cusp (hypoconulid) is well developed below, while in the upper teeth the hypocone is comparatively small, the old trigon still being apparent; the monkey cross crests are not developed. The molars tend to be rather long, and there is some lengthening of the face in living forms. The skeleton is rather close to the human type. The chest is broad in contrast to that of the monkeys; the ilia tend to broaden, as in man. The hands are rather similar to those of man, but the fingers are very long as compared with the thumb and the arms are long. The legs, on the contrary, are short; and the hallux, as well as the remaining toes, is much better developed and more finger-like than in man. These skeletal characteristics have seemingly developed in relation to the anthropoid-ape type of locomotion. With increasing size, normal four-footed progression in the trees has become increasingly difficult, and the members of the group often progress by brachiation—swinging from bough to bough with the body suspended from the hands. The feet are used more than ever as grasping organs and have so much lost their primitive mammalian character that most apes cannot walk flat on their soles, but must support their weight on the outer side of the foot.

It will be noted that in swinging by the arms the body is necessarily erect. Further, the front legs of all forms being much longer than the hind, the body is necessarily tilted up considerably in front in quadrupedal progression. Bipedal locomotion has thus begun among the arboreal types.

These arboreal, brachiating adaptations are characteristic of the living great apes, most notably the gibbons and, to a lesser degree, the orang. As noted below, these traits appear to have been less developed in ancestral great apes. The brain is large, especially in the gorilla. But in these large forms its relative growth, as might be expected, has hardly kept up with the increasing body weight. The upper margins of the orbits project in most forms as supraorbital ridges above the front edge of the braincase. There is, as in most mammals, no projecting chin, the jaw sloping away under the symphysis.

Possibly a few teeth from the late Eocene of Burma may pertain to the great ape family, but our earliest

definite evidence of the group comes from the early Oligocene of the Egyptian Fayum, where there are several genera, represented by remains of dentitions and jaws, which are surely small ancestral apes. *Propliopithecus* (Figs. 326*D*, 330*C*) was once believed to be an ancestral gibbon, but appears to be a rather generalized ape ancestor; *Aeolopithecus* is more gibbon-like; *Aegyptopithecus* a more advanced type.

Gibbons.—Smallest and most primitive of the living anthropoids are the gibbons (*Hylobates*) of the Malay region. The average gibbon stands somewhat less than a meter high in the erect position. Alone among the living anthropoids, he customarily walks erect when on the ground, with the arms used as balancers. The gibbon, however, is almost entirely a tree dweller and

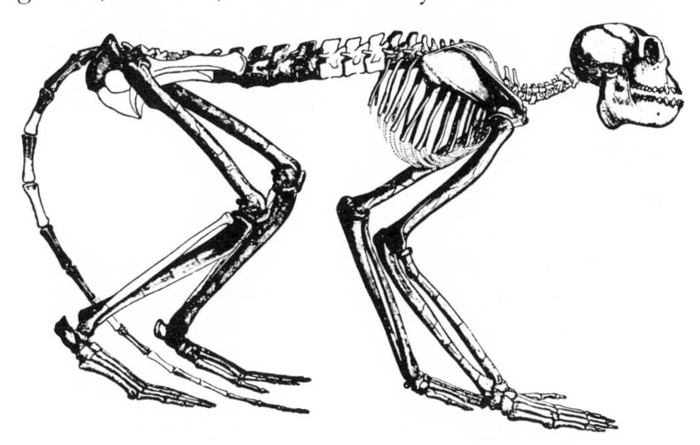

FIG. 331.—*Mesopithecus*, a Pliocene relative of the living langurs. Original about 15 inches long. (From Gaudry.)

with his long arms, is a clever acrobat. The gibbon brain averages about 90 cubic centimeters. This seems a small figure compared with that of man, but is far above the average for a mammal of that size. The braincase of the gibbon is practically smooth, and the supraorbital ridges small. Vestigial ischial callosities are present.

Pliopithecus of the Miocene and Pliocene was very similar to the living gibbon in cranial structures, but had not, the fragmentary skeletal material indicates, reached the long-armed condition of the modern form. From *Pliopithecus* it is a short step back to *Aeolopithecus* of the Fayum.

Orangutans.—The orangutan, the "man of the woods" of Borneo and Sumatra, has been generally known as *Simia*, but the taxonomic purists insist it must be called *Pongo*. This ape stands next above the gibbon in the scale of living primates. Considerably larger in size (adult males reaching nearly 5 feet), the orang is still a true arboreal type and is but a clumsy four-footed performer on the ground; the arms are somewhat shorter, reaching only to the ankles. The brain here is vastly larger than that of the gibbon, reaching 550 cubic centimeters in capacity. The braincase is expanded, crests rarely develop, and the supraorbital ridges are negligible. The tooth row, however, is long and the snout projecting, with a characteristic concave lateral profile to the face.

Almost nothing is known of the history of the orangs. Presumably their ancestors diverged from the line leading to the still higher primates and man by the early Miocene, at the latest.

Higher anthropoid apes.—The highest living members of the anthropoid group are the chimpanzee, *Pan*, and the *Gorilla*, both inhabitants of tropical Africa. These two forms are closely related and are hardly separable generically. Both are less pronounced in their arboreal adaptations than the lower apes and monkeys; the highland gorillas spend much of their time on the ground, although walking as quadrupeds rather than as bipeds. Both are large, the chimpanzee exceeding 5 feet in height and old male gorillas more than 6 feet, with a weight of 600 pounds. The arms are relatively short compared with those of the lower anthropoids. Those of the chimpanzee reach a bit below the knee and those of the gorilla not quite so far. The brain is large in both—that of the gorilla having in some cases a capacity of between 500 and 600 cubic centimeters, or at least half that found in some human races. Nevertheless, sagittal and occipital crests and supraorbital ridges tend to develop, being especially heavy in the old male gorillas. Associated with the long tooth row, there is considerable development of a muzzle. The canines are enlarged, and the dental arcade is U-shaped with the molars in two parallel rows, in contrast with the hyperbolic tooth row of man. Among the details in which these forms agree with man is the absence of the centrale in the hand, a bone otherwise almost universally present in the order.

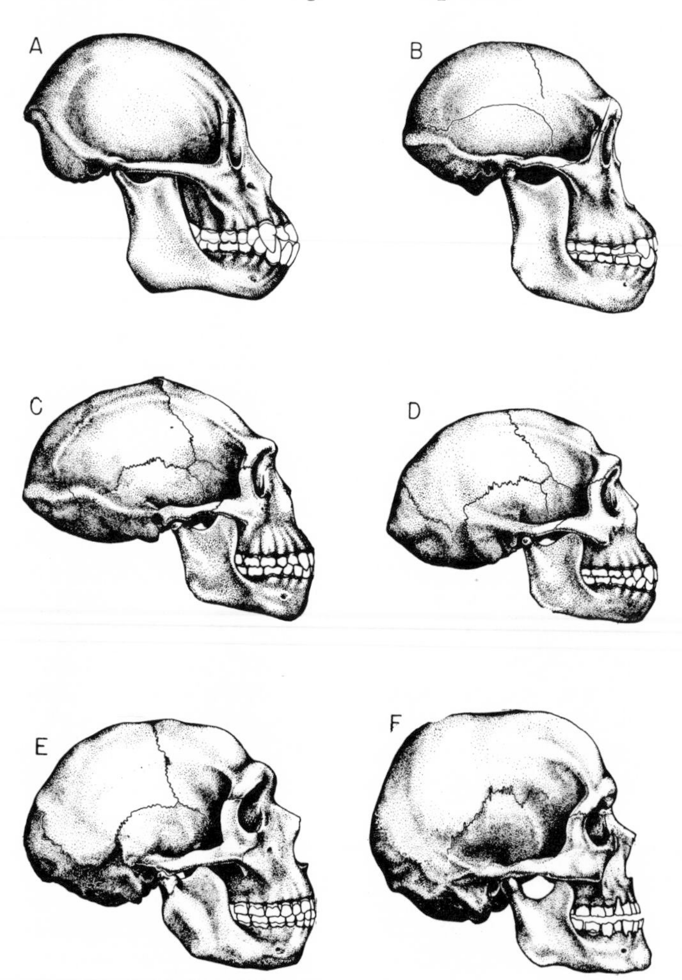

Fig. 332.—Skulls of higher primates and man. *A*, *"Proconsul"* (*Dryopithecus*); *B*, *Australopithecus*, adult; *C*, *"Pithecanthropus"*; *D*, *"Sinanthropus"*; *E*, Neanderthal man; *F*, Cro-Magnon man. (*A* about 1/3 natural size; *B* about 1/4 natural size; others about 1/6 natural size.) (*A* after Napier and Le Gros Clark; *B* after Robinson; *C, E, F* after McGregor; *D* after Weinert.)

Our knowledge of the fossil history of these higher apes and of presumed human ancestors on this level is tantalizingly poor—due, presumably, to the fact that, like the modern forms, the ancestral types inhabited tropical regions from which few fossils are known. In the Miocene and Pliocene of Eurasia and Africa there have been found, from time to time, specimens of advanced apes—most of which, however, consist of no more than a few teeth. Nearly a score of generic names have been given to these scraps, but nearly all of them appear to represent a single genus, *Dryopithecus* (the "oak ape," so called because oak-leaf impressions were present in the deposit with the first specimen found). *Dryopithecus* (Figs. 325*C*, 326*E*, 332*A*) appears to have ranged widely over much of Europe, southern Asia, and Africa in the Miocene and early Pliocene. Its structure, as far as can be determined, was of a sort which would make it a possible ancestor of the chimpanzee and gorilla—and of man.

The best *Dryopithecus* material is that found in recent years in East Africa. Several species, large and small, appear to have been present. The remains have been given the distinctive name of *"Proconsul,"* but the differences between the East African specimens and those here attributed to *Dryopithecus* from other regions are negligible. In the case of *"Proconsul"* we are fortunate in that we know not merely teeth, but much of the skull and parts of the postcranial skeleton. The skull, as expected, is that of a general anthropoid-ape type, with a low, rounded skull vault, a moderate development of supraorbital ridges, and a moderate forward projection of the face and jaws. The skeletal material, although fragmentary, is of interest in that it shows that the limbs were only modestly developed in the direction of arboreal brachiation and, further, the foot material indicates a trend toward bipedal posture.

East Africa today appears to have an ecology similar to that which the geological evidence suggests was present in the Miocene when *"Proconsul"* lived there. This is—and was—a savanna type of country, in which there are clusters of trees separated by open grasslands. Ground dwelling and bipedalism may have had their origin in the fact that, unless unduly restricted in range, an ape living under these conditions was forced to develop efficient terrestrial locomotion.

With the presumed development from *Dryopithecus* of the chimpanzee and gorilla, the story of the great apes, as such, is, as far as we know, concluded, except for the presence in southeastern Asia of a large gorilla-like form, *Gigantopithecus.* The remainder of primate history is the development of human types. A first stage toward man from the *Dryopithecus* level may be represented by *Ramapithecus,* a form which appeared in southern Asia and Africa at about the Miocene-Pliocene boundary. *Ramapithecus* is known only from the dentition; but as far as can be told from the evidence, this genus may be transitional to australopithecines and man himself.

Human characters.—Man and his close fossil relatives are usually placed in a separate family; but, as a matter of fact, the anatomical features which distinguish men from apes are comparatively few, although, from our own point of view, important. Of great significance has been the growth of the brain. The average European male has a brain capacity of about 1,500 cubic centimeters—about triple that of any of the great apes. This increase is mainly associated with an expansion of the neopallium, the "gray matter" covering the cerebral hemispheres. Here new centers are developed, such as those having to do with speech; and especially prominent are areas, absent in the apes, which seem to be the seat of higher human faculties. This enlargement of the brain results in a much expanded braincase and the consequent disappearance of crests and supraorbital ridges. In lower primates the cheek teeth form two parallel rows, with the prominent canines at each front corner; in man the total tooth row (Figs. 325*D;* 326*F*) is shorter and rounded, with the canines much reduced. In relation to this reduction of the tooth row, the face is less projecting, although there is a tendency in modern man for the retention of a long lower jaw in the development of a chin. With reduction of the face in general, the nasal region tends to develop into an isolated projecting structure.

Other important differences are associated with the change in locomotor habits. Primitive primates were arboreal, and the living great apes, we have seen, tend to specialize in this mode of life; man, in contrast, is a terrestrial biped. The modern apes during millions of years of tree-living have lengthened their arms and shortened their legs to varying degrees, while man has retained more of the original proportions. The human hand, except for somewhat greater breadth and flexibility, is of a type probably present in the ancestral apes; the living great apes have tended to elongate the fingers, except the thumb, for hooking over limbs. A greater difference is in the foot. The usual primate foot is similar to the hand. This, however, is an awkward structure for ground life. In man the toes have greatly shortened, the hallux has lost its primitive opposability, and the calcaneum expands to form a prop at the back (the last is true of the higher anthropoid apes as well). In further adaptation to the bipedal gait, the back of man is much more sinuously curved than in the apes, with the effect of swinging the center of gravity up above the hips and raising the head.

The features listed above are among those which distinguish modern human types from the apes. Fossil human types tend to bridge this structural gap and exhibit intermediate conditions; but our knowledge of human ancestors is still a very imperfect one, and we know almost nothing of the evolution of the postcranial skeleton.

Early human cultures.—Man is a maker and user of tools, and this trait may well have appeared early in his development. Flint is a material which appears to have found favor for toolmaking because it is easily fractured and worked and yet is extremely hard. Flints, thought by some to have been used by manlike creatures, have been reported from Tertiary rocks. While such finds are of dubious nature, flints were certainly being worked in early Pleistocene times, and a series of human cultural stages, based primarily on stone implements, has been established for the Pleistocene of Europe and, less surely, for other areas of the Old World. In the early Pleistocene many finds of implements are doubtful in nature, and even tools believed to have been used (if not actually made by human forerunners) are very crude in form and nature, and are often termed "eoliths." This rather dubious stage may be termed (1) the Pre-Paleolithic. The Middle Pleistocene, covering much of the ice age, is the time of (2) the Lower Paleolithic, the first accepted subdivision of the Old Stone Age. In much of the Old World this was the time of the use of the large hand ax or *coup de poing,* associated with Abbevillian (Chellean) and Acheulean cultures; in other parts of Europe and Asia the time was one of the use of simple chopping tools and flint flakes, such as may have been used by *Pithecanthropus* and the related Peking man. (3) The Middle Paleolithic is the time of the third interglacial stage and the early part of the final (Würm) glaciation of Europe. The dominant culture is that of the Mousterian, characterized by well-shaped flint flakes and often found associated with remains of Neanderthal man. Toward the close of the Würm glaciation there is in Europe a characteristic series of cultures—Aurignacian, Solutrean, and Magdalenian—which constitute (4) the Upper Paleolithic (related cultures are found in other areas). In addition to expertly made flakes, bone has become an important cultural material. These cultures are the work of our own species, *Homo sapiens.* To Recent times rather than to the Pleistocene (and hence beyond our province) are to be assigned later human cultures—the Mesolithic; the Neolithic, in which pottery, agriculture, and domestication of animals were added to domestic economy; and the Bronze and Iron ages of "historic" times.

Australopithecines.—Several decades ago the skull of an infant primate (Fig. 330*D*), to which the name

Australopithecus was given, was discovered in a South African limestone cave deposit of Pleistocene age. Although the young of apes and men differ less than do the adults, the specimen was suggestive of an evolutionary stage structurally antecedent to true men. Subsequent finds of adult specimens of these "man-apes" have fully confirmed these indications. The remains mainly consist of skulls (Fig. 332*B*). (As is the case also with the later true human skulls of the Middle Pleistocene of Asia, the base was invariably broken off, suggesting a cannibalistic trait, with a fondness for brains.) The braincase was small, little larger than that of the larger great apes, with capacities estimated at 600 to 700 cubic centimeters; the vault, however, is rather higher and rounded, and only exceptionally, it would seem, was a sagittal crest developed. The occipital condyles are turned far forward beneath the skull, a feature strongly indicative of upright posture. The face is somewhat projecting, but much less so than in great apes. The molar teeth are large compared with either great apes or modern man; this appears to be a primitive trait in the human line. The dentition is essentially human rather than apelike in almost every regard. The molar cusps are blunt rather than sharp as in apes, the premolars are unspecialized, the canines do not project, the incisors are small, the tooth row relatively short and curved in the human parabolic fashion. The postcranial skeleton is very inadequately known. The ilium is broad, closely comparable to that of man and in contrast with the slender iliac blade of apes, this again indicating erect posture. Skeletal materials suggest that the australopithecines were of short stature; a female (apparently there were marked sex differences) is estimated to have been no more than 4 feet high, with a weight of 50 pounds or so. Simple pebble tools and split pebbles have been found in caves containing australopithecine remains; possibly these man-apes had begun tool-using. The South African man-apes have been ascribed to several genera and species. Most are included in *Australopithecus; Paranthropus* has been used for certain larger and more massive skulls. All the South African forms are definitely of Pleistocene age, and it seems probable that many, at least, are attributable to the Lower Pleistocene—a time well before that of the appearance of true men in the Middle Pleistocene of Asia. While the best material of australopithecines comes from South Africa, this may mean little except that the region was fortunate in having good cave sites well developed in the early Pleistocene and that such sites have been carefully explored. An early Pleistocene skull of the same type (termed *Zinjanthropus*) has been recently found in East Africa; two lower jaws from Java, with large teeth, which have been termed *Meganthropus,* may be eastern representatives. Most students agree that the australopithecines are structurally antecedent to later men, and although some members of the group may have survived to overlap true man in time, these man-apes appear to be representatives of the ancestral human stock.

Early true men.—Most adequately known of early human types are the finds from Java and China customarily termed the "Java ape man" and "Peking man." *"Pithecanthropus" erectus* (Fig. 332*C*) was first discovered, associated with an abundant mammalian fauna of Middle Pleistocene date, more than half a century ago on the banks of a Javanese river near the little village of Trinil. The original find consisted merely of a skull cap, with which were more or less doubtfully associated a thigh bone, several teeth, and jaw material (from another locality). In later years several other partial skulls and jaws were found, so that the cranial anatomy is fairly well known. The braincase, including the forehead region, was extremely low, and the brow ridges over the orbits enormous. Further, the skull is rather narrow compared to its length, a feature common to most early human forms (skulls in which, as here, the breadth is 75 per cent or less of the length are termed dolichocephalic; short-headed types, with the figure over 80 per cent, are brachycephalic; intermediates are mesocephalic).

Endocranial casts suggest an essentially human type of brain, although a small one; the average figure for endocranial capacity of 890 cubic centimeters is far below the mean of modern human types, although well above the highest of apes. In contrast with modern men, the face was projecting and broad, with a large (and probably flattened) nose. The teeth were in most respects of human type, but there is evidence suggesting that the males had somewhat enlarged canines. There was no chin projection. If the femur is properly associated, *"P." erectus* (as his specific name would imply) had already assumed an upright posture.

During the 1930's a considerable number of skulls and jaws of a primitive human type were recovered from Middle Pleistocene deposits at Choukoutien, near Peking, China (Fig. 332*D*), and study revealed another early type of man, very similar to *"Pithecanthropus"* in many respects. The only conspicuous difference lies in the fact that the average of four endocranial cavities measured is about 1,050 cubic centimeters in the Peking man, compared with a much lower figure in the case of the Java form. This is, however, thought to be due to the fact that the Chinese specimens may be for the most part males, the East Indian skulls those of females; sex differences in brain size are marked in many human races. The Chinese form was given a separate name—*"Sinanthropus"*—but students of these forms agree that they are closely related, if not identical, races and that the general *"Pithecanthropus"* stock was a widespread early human type, in Asia at least.

It was once thought that this primitive *"Pithecanthropus-Sinanthropus"* type of man was characteristic of eastern Asia, and some different sort of man inhabited Europe where, except for a single jaw from a sandpit near Heidelberg, no remains were known before a relatively late Pleistocene stage. But this, it seems, was not the case. Jaws of "*Pithecanthropus*" type (named, redundantly, "*Atlanthropus*") have recently been found in Algeria, and equally primitive (although inadequately known) materials have been found in East Africa.

A word may be added here with regard to the nomenclature of human finds. We have freely used several *generic* terms for various early human finds. Such a usage implies that the forms differed widely from one another, had independent evolutionary histories, and did not interbreed—that the differences between them were not merely of species value but of such a magnitude as, for example, those between a cow and an antelope, a dog and a fox. This is absurd. Because they are so close to us, we tend to magnify differences. Actually, the differences between a modern man and "*Pithecanthropus*" are, viewed impersonally, rather minor ones (particularly if we keep in mind the considerable variations found even today), and quite surely all types on the human line above the *Australopithecus* level pertain to our own genus *Homo*. Further, while communications between the various Old World areas in which man was early present were obviously poor, and there presumably was little interbreeding and consequently (as today) a tendency for the differentiation of regional races, it seems fair to assume that throughout our long Pleistocene history, our human ancestors formed at all stages a single, if variable, group.

A note must be inserted here on "Piltdown man"—the greatest of all hoaxes ever palmed off on honest (and hence unsuspecting) scientists. In an old Pleistocene gravel deposit in southern England were said to have been discovered, half a century ago, pieces of a human skull, and in the gravel were found other fragments, including a lower jaw, crude implements, and a number of Pleistocene mammal specimens. The skull is essentially modern in type, but the jaw a typical ape jaw. This paradoxical association was flatly contradictory to all the rest of our accumulating knowledge of the trend of human evolution, and hence "Piltdown man" was long an unsolved and perplexing puzzle. It implied that the ascent of man must have been by two routes—one *via* Piltdown man, where the braincase rapidly assumed a modern aspect at a time when the jaw was still purely apelike; the other, an ascent through a "*Pithecanthropus*" stage, where the jaw (except for the absence of a projecting chin) was approaching human standards but the skull vault was low. Recently this puzzle was solved by proof that the whole affair was a gigantic hoax. The skull pieces may be of some modest (if uncertain) antiquity, but the other materials were "salted" in the gravel, and the jaw is that of a modern orang, broken, stained, and with neatly filed teeth, to give an impression of antiquity.

Neanderthaloid men.—The characteristic human type of the Mousterian cultural period of the last interglacial stage and the early portion of the final European glaciation is *Homo neanderthalensis* (Figs. 325*D*, 326*F*, 332*E*, 333), named from a German site but now known from a great number of localities in Europe and even

represented in western Asia and northern Africa. The braincase had as high a cranial capacity as most living men (averaging about 1,550 cubic centimeters for males). This was mostly accounted for by the expansion in the occipital region of the brain, for the forehead was low and the supraorbital ridges greatly developed. These features, together with a face more elongate than in modern men and the undeveloped chin, must have given the creature in life a very apelike appearance. Here, for the first time, complete skeletal remains are known. The body was essentially modern in build, although with a few archaic features.

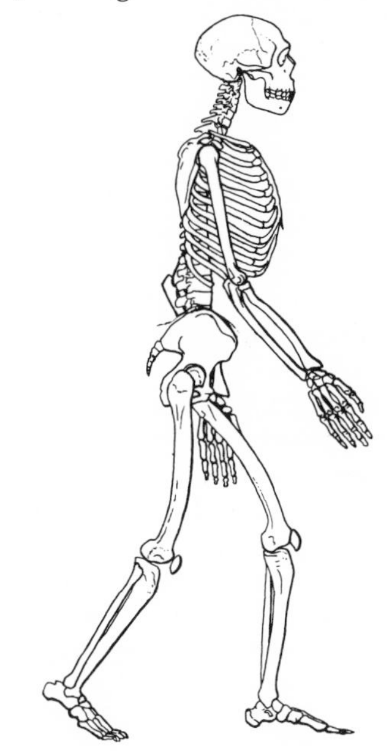

FIG. 333.—Skeleton of Neanderthal man. (After Boule, modified.)

A type of man definitely assignable to our own species, *Homo sapiens*, appeared in Europe well toward the end of the last glaciation, not more than 50,000 years or so ago. One would at first assume that he had arisen from his Neanderthal predecessor. But the contrasts are too great; there is (in Europe, at least) no evidence of transitional types; the appearance of modern man was, the evidence suggests, relatively sudden. There is every indication that the "modernized" invaders wiped out their predecessors (Tasmania is a modern parallel).

What, then, is the history of these two forms; where and how did the modern type originate? The answers are still none too clear, but they are to be sought in other regions and in earlier times.

Neanderthal man, as we see him in Europe during the last glaciation, is apparently a rather specialized race of a more generalized neandertha*loid* type of man which, during late Middle Pleistocene times, appears to have been advancing upward from the "*Pithecanthropus*" level. The story in Europe is very fragmentary, but in the Middle Pleistocene there have been found at Swanscombe in England and at Steinheim and Ehringsdorf in Germany incomplete skulls which appear to have belonged to men of a neanderthaloid type, but less specialized than the later European Neanderthal, and possibly close to the common ancestor of that race and his modernized successor. At Mount Carmel in Palestine there have been found a number of skulls and skeletons which (although difficult of interpretation) suggest a type of primitive, neanderthaloid nature but with advanced characters. It would seem that over much of the Old World there were regional advances in human structure, reaching the general Neanderthal evolutionary level, but with variations.

At Broken Hill, Rhodesia, a skull from a cave deposit of an unknown but possibly late Pleistocene date is that of a man with a massive face, heavy brow ridges, and low forehead, suggesting Neanderthal affinities. It is thought that Rhodesian man may be related, through other archaic types from South Africa, to the ancestry of the living Bushmen. Somewhat similar is *Homo solensis,* represented by numerous partial skulls from the late Pleistocene of the Solo River in Java. This man, it would seem, also represents a stage comparable to that of Neanderthal. It is also thought that Solo man may be related to the ancestry of the Australians; other skulls from Java and Australia are interpreted as transitional in type.

Modern man.—It is probably from some neanderthaloid stock of this general sort that there finally evolved, in late Pleistocene times, a man of modern type, such as we see appearing in Europe toward the end of the last glaciation, as creator of the Upper Paleolithic cultures. In such races, to which the term Cro-Magnon is often applied (Fig. 332*F*), the forehead is high, the cranial capacity about 1,500 cc, the supraorbital ridges gone, the face short, the nose well developed, the chin protruding. The area in which modern man first appeared—in Asia or Africa?—is uncertain, but Upper Paleolithic races of man which are apparently not too dissimilar to those in Europe are known to have appeared at about this time in areas as far removed from one another as Algeria, eastern Africa, and China. Several African finds which appear to be of some antiquity show primitive negroid features, and skeletons from the Upper Paleolithic of Grimaldi on the Riviera also show negroid characteristics. We have, however, little evidence as to the early history of the mongoloid type of man now dominant in eastern Asia.

The time of arrival of man in America has been a question of some interest. There is here no series of cultures comparable to those of the Old World, and no skeletons have been discovered which show marked differences from the modern Indian. Very probably man reached America only during the last retreat of the glaciers. But his arrival must have occurred at an early postglacial date, for finds in southwestern United States and in South America show that he was present at a time when ground sloths, horses, camels, and other types now extinct were still in existence in the Western Hemisphere.

CHAPTER 19

Carnivorous Mammals

Carnivore adaptations.—The development of carnivores is a logical consequence of the insectivorous habits of the ancestral mammals. Once herbivorous mammals had come into existence, it was only to be expected that certain of the early insectivores should, with increase in size, tend to prey upon them and take on various adaptations better fitting them for a flesh-eating existence. Modifications for this type of life markedly affect the dentition. A good set of incisors for biting off flesh and a pair of piercing canines are essentials which were already present in insectivores and are seldom modified in carnivores. The primitively sharp-cusped cheek teeth may undergo various changes. For flesh-eating, some sort of shearing apparatus to slice off the meat and cut tough sinews is a necessity. In many insectivores there is some shearing effect between the back edge of each upper cheek tooth and the front edge of the following lower one (cf. Fig. 298). Carnivores have generally tended to emphasize the shear between a pair of highly specialized cheek teeth called the carnassials. In the development of the opposed shearing surfaces, the originally diagonal line of meeting between the teeth tends to swing around to a fore-and-aft position, and the teeth tend to simplify in structure and become high, narrow, and elongated. In different carnivore groups the pair of teeth which developed into carnassials has varied, and some early forms never progressed far toward development of a specific carnassial pair. The teeth behind the carnassials tend to be reduced and lost in the more purely carnivorous types. In forms which have a mixed diet they are less reduced and in some cases are low-crowned crushing teeth capable of dealing with nuts, fruit, and the like.

A carnivore normally uses its claws for seizing its prey, and consequently there is but rarely any tendency to develop hoofs. The metacarpals and metatarsals are never greatly elongated, and there is never much reduction in the toes, except for the pollex and hallux.

A number of different mammalian types have become flesh eaters. We have already given an account of the extinct carnivorous marsupials which evolved in South America and the parallel development of Australian marsupial flesh eaters. Certain other early types which tended in this direction are here included in the condylarths, described in the next chapter. In this chapter we will consider two major orders of placental flesh eaters—the Creodonta, dominant in the early Tertiary, and the more "modernized" forms which have since dominated the scene as members of the order Carnivora.

Creodonts.—In early days the term "creodont" was generally applied to all placentals of the Paleocene and Eocene which appear to have been carnivorous in habits. Currently, however, more analytical consideration has suggested a rather different interpretation. Two of the families concerned are now believed to be related to the archaic ungulate order Condylarthra and are so assigned here; another family appears to be truly ancestral to the later members of the true Carnivora and will be described in connection with them. There remain, however, two families, the Oxyaenidae and Hyaenodontidae, which were the prominent carnivores of the Eocene, and the term Creodonta may still be reasonably applied to them.

The creodonts were rather archaic in their build, and their characteristics were mainly negative, primitive ones. The skull was generally low, the braincase small, with consequently well-developed sagittal and occipital crests. The molars were primitively tribosphenic; but there was considerable variation, and the carnassials developed farther back in the series than in the case with the later true carnivores. In contrast with most members of that group, the auditory bulla was unossified; and in the carpus the scaphoid, lunar, and centrale were distinct. There was usually no loss of toes. The terminal phalanges were fissured for insertion of the claw bases. In creodont habits there appears to have been considerable variation: some were probably still omnivorous, some small types perhaps still mainly insectivorous, others may have been carrion feeders, rather than typical carnivores.

The brain of the creodonts was generally of relatively small size and their intelligence presumably low. This may have been a main cause of the early extinction of almost all members of the group; for with replacement of the slow-footed and stupid herbivores of early Tertiary time by the swifter modernized ungulates, intelligent group pursuit (as in the wolf pack) or clever stalking (as in the case of the cats) became necessary for the capture of prey.

The two creodont families are closely related, and the taxonomic boundary between them has been variably drawn. The oxyaenids include a series of forms of which the earliest appear in the late Paleocene; most members of the family were of Eocene age. The skull was broad and short, the jaw deep and massive. M^1 over M_2 developed as shearing teeth, and the postcarnassial molars tended to disappear. Typical members of the family were rather long-bodied, short-legged types with spreading plantigrade feet.

Although tending to grow to large size in some cases, they seem to have been somewhat comparable to the modern mustelids in habits. *Oxyaena* (Figs. 334*B*, 335) of the Lower Eocene was a powerful wolverine-like animal which appears to be ancestral to the Middle Eocene *Patriofelis* of bearlike size and to *Sarkastodon* of the Upper Eocene of Asia, far larger still.

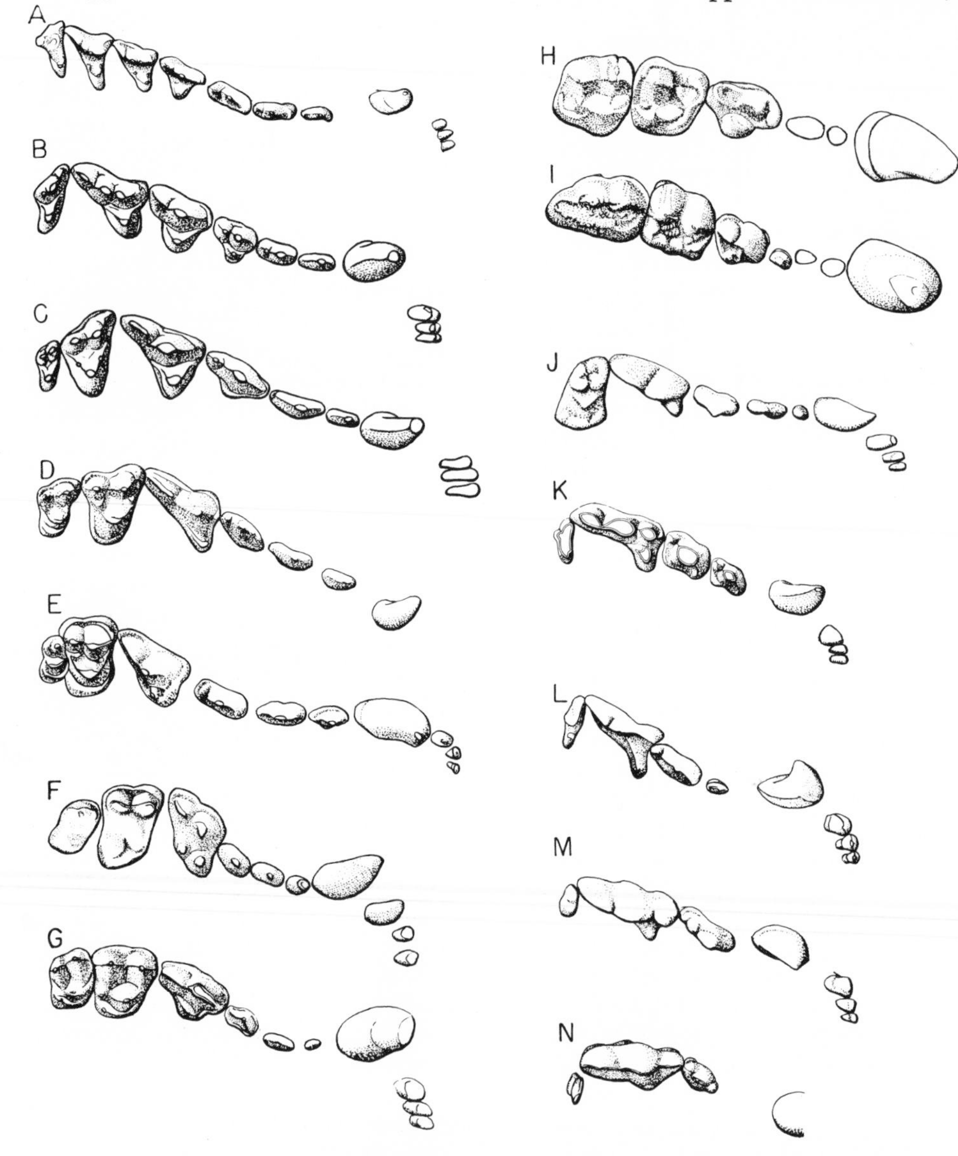

FIG. 334.—Right upper dentition of creodonts and carnivores. *A*, *Sinopa* (Hyaenodontidae), ×5/6; *B*, *Oxyaena*, ×1/2; *C*, *Vulpavus* (Miacidae), ×3/4; *D*, *Hesperocyon*, a primitive fissipede, ×1 approx.; *E*, *Temnocyon* (Canidae), ×1/2 approx.; *F*, *Phlaocyon*, a raccoon-like canid, ×5/4; *G*, *Hemicyon*, intermediate between dogs and bears, ×1/2; *H*, *Agriotherium*, a primitive bear, ×3/8; *I*, *Arctodus*, a Pleistocene bear, ×5/12; *J*, *Mustela palaeosinensis* (Pliocene, China), ×9/8; *K*, *Hyaena bosei*, from the late Tertiary of India, ×5/12; *L*, *Dinictis*, an Oligocene true cat, ×5/8; *M*, *Pseudaelurus*, a Pliocene true cat, ×5/9; *N*, *Smilodon*, a Pleistocene sabertooth, ×2/5. (*A–C* after Wortman; *D* after Scott; *E, F, K, L, N* after Matthew; *G* after Gaudry; *H* after Frick; *I* after Merriam and Stock; *J, M* after Zdansky.)

Palaeonictis, a rather catlike type from the Lower Eocene of both Europe and North America, is of interest in that the shearing function is shared by both M^1 and P^4 above and M_2 and M_1 below; it is thus structurally intermediate between the oxyaenids and the miacid ancestors of the true Carnivora, although not ancestral to the latter. No oxyaenids survived beyond the Eocene.

A much larger assemblage is that of the Hyaenodontidae, a group which appeared at the beginning of the Eocene and flourished during that period. These were forms in which the skull and tooth row tended to be more elongate than in oxyaenids, the jaw more slender, the body more slimly built. The limbs were rather longer, the feet more digitigrade in posture, and the nature of the terminal phalanges suggests the presence of typical claws in contrast to the flatter tips of the short toes of oxyaenids.

Some Eocene forms, such as *Limnocyon,* have a rather broad skull and M^1 over M_2 functioning as carnassials, as in the oxyaenids; they are included here, since in other features they are closer to the typical hyaenodonts. With this group are associated *Machaeroides* and *Apataelurus,* with jaws remarkably similar to those of the saber-toothed cats of later Tertiary epochs.

More typical hyaenodonts are *Sinopa* (Figs. 334*A*, 335) and *Hyaenodon* (Fig. 336). These animals have long, narrow skulls and well-developed carnassials formed by M^2 and M_3. *Sinopa* of the Lower and Middle Eocene is representative of a series of small, lightly built genera; its species range in size from weasel to fox. *Hyaenodon,* found in both North America and the Old World in the Oligocene, was a much larger and more heavily built type which may have been able to prey upon the larger ungulates of the time (titanotheres, for example).

In contrast to the oxyaenids, the typical hyaenodonts survived, in diminishing numbers, for most of the Tertiary. A number of forms were still present in the Oligocene, and relatives of both *Sinopa* and *Hyaenodon* were still present in the Old World tropics until Miocene and even Pliocene times.

Creodont ancestors.—It is highly probable that the creodonts have a pedigree distinct not only from that of the Carnivora proper, but, indeed, distinct from most other placentals, tracing back to the late Cretaceous. In the more primitive creodonts the upper molar is sharply V-shaped, with paracone and metacone close together and well in from the outer margin. Quite similar teeth are to be found in the members of a little group of small placentals of the earliest Tertiary and

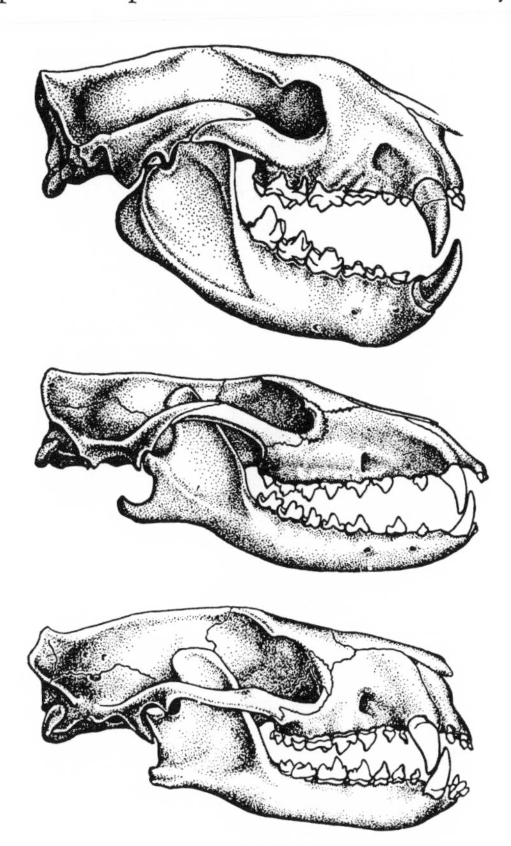

FIG. 335.—Skulls of early creodonts and carnivores. *Above, Oxyaena,* skull length 8.25 inches; *center, Sinopa,* skull length about 6 inches; *below,* the miacid *Vulpavus,* skull length about 3 inches. (After Wortman and Matthew.)

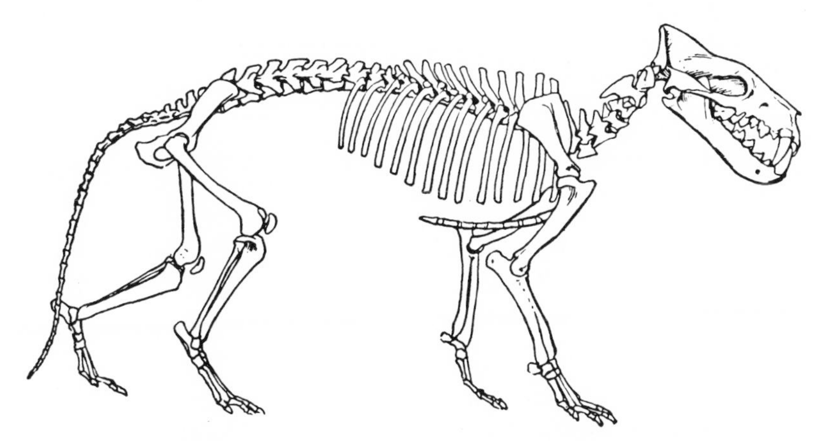

FIG. 336.—The creodont *Hyaenodon,* original about 4 feet long. (From Scott.)

late Cretaceous; *Deltatheridium* (317*A*, 318*A*, 319*A*) and *Didelphodus* (Fig. 298) are typical. These little mammals are customarily placed in the Insectivora. They are, indeed, primitive in many ways and (for example) show no indication of development of special carnassials, although their sharp, closely interlocking cheek teeth suggest a good shearing power. They appear to point the way toward the evolution of creodonts. There is no positive evidence that they are

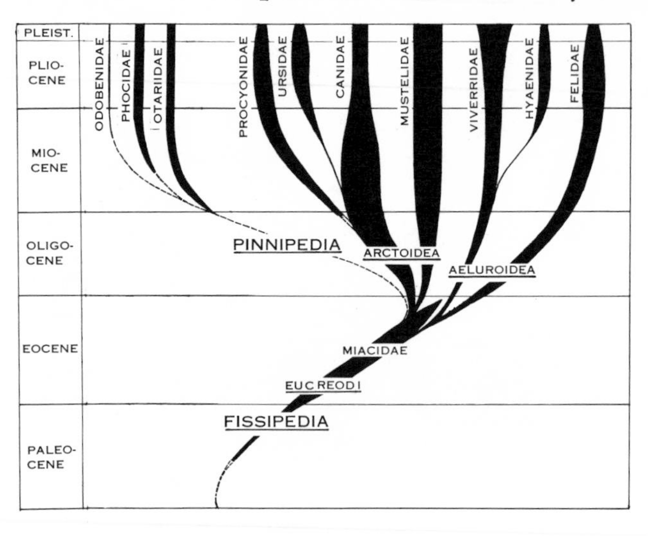

FIG. 337.—The phylogeny of the Carnivora

ancestral to any further major group, although some obscure Asian Oligocene forms, such as *Didymoconus*, may be relatively unchanged survivors of the ancient deltatheridians. Possibly the "zalambdodonts," which we have included in the Insectivora, are related, but more probably the development of narrow V-shaped upper molars is a parallelism.

Early Carnivora.—The more advanced flesh eaters, forming the order Carnivora, did not arise from the creodonts or, it would seem, from their deltatheridian forebears, but sprang from the ancestral insectivore stock by a distinct line, represented by the family Miacidae. *Didymictis* and several other miacid genera appeared in the Middle Paleocene, and the family, represented by such forms as *Miacis* and *Vulpavus* (Figs. 334*C*, 335, 338), continues through the Eocene. The miacids are, for the most part, poorly known, for almost all were of small size; and there is evidence suggesting that they were rather persistently arboreal forest dwellers—types which are infrequently fossilized. As in primitive carnivores generally, the body and tail were long, the limbs short but flexible, the pollex and hallux somewhat opposable (Fig. 338*A*). The tympanic bulla was unossified, and there was no fusion of carpal bones. In such features they resemble the creodonts. However, they differ in a number of respects. The terminal phalanges are not fissured, as they were in creodonts. More significant, presumably, is the fact that the miacids are thought to have had relatively larger brain capacities than their creodont contemporaries; the rapid evolution of modernized ungulates may have placed a premium on brains as an important factor in survival. Finally, a diagnostic feature lies in the carnassial teeth. These were well developed and are formed, as in later carnivores, by P^4 over M_1. The upper molars do not resemble those of the deltatheridians but have the broader triangular shape seen in such forms as the leptictids, with the paracone and metacone well separated and well toward the outer margin of the tooth.

In late Eocene and early Oligocene days there began the development, from the miacids, of the diversified families which have since become the dominant land carnivores. The skeletal alterations necessary to transform a miacid into a typical fissipede are few—fusion of scaphoid, lunar, and centrale into a single element in the carpus and ossification of the auditory bulla. These structures are seldom present in the fossil material (usually quite imperfect) of early fissipede genera. Deposits from the time of transition have

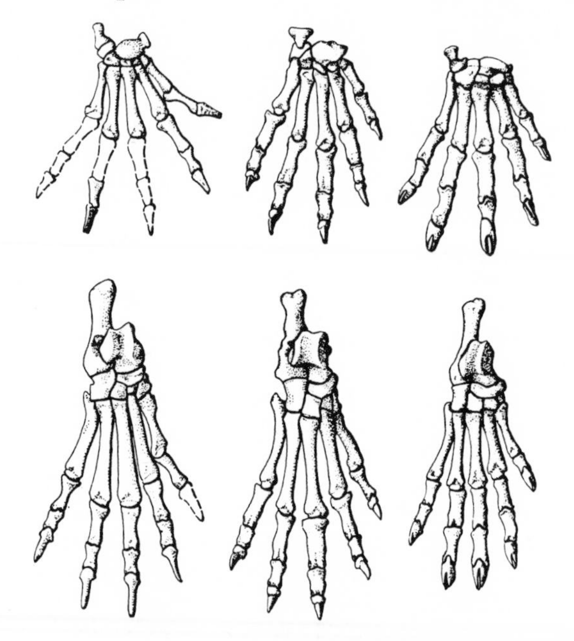

FIG. 338.—Feet of carnivores, manus above, pes below. *Left*, Miacids, manus of *Vulpavus*, pes of *Didymictis*; *center*, the Oligocene canid *Daphoenus*; *right*, *Hoplophoneus*, an Oligocene sabertooth cat. (*Left* after Matthew, *center* after Hatcher, *right* after Adams.)

yielded remains of a number of genera of small carnivores which are exceedingly difficult to place. They surely include the ancestors of later civets, weasels, and dogs; but these stocks were at the time almost impossible to separate, although there are slight but suggestive differences in the dentition. Were we living at the beginning of the Oligocene, we should probably consider all these small carnivores as members of a single family.

Cynodictis and *Hesperocyon* [*Pseudocynodictis*] (Figs. 334*D*, 341*A*, 342) of the early Oligocene are typical primitive forms. The skeleton was much like that of a civet or a weasel. The body was long and flexible, the limbs short. All five toes were present on the short and spreading feet, which appear to have been armed with retractile claws. As in most later carnivores, the last upper molar had already disappeared. These genera are customarily considered primitive dogs, but in most respects they seem not far from ideal ancestors of the whole series of later Carnivora.

The Carnivora are customarily divided into two suborders, the Fissipedia for all land forms, and the Pinnipedia for the seals and walruses. The fissipedes, in turn, may be divided into three infraorders, one of which, the Miacoidea, includes the ancestral miacids. In the late Eocene and Oligocene, miacid descendants began a ramification into the more advanced families characteristic of later Tertiary and Recent times. It has long been generally agreed that the later fissipedes may be conveniently divided into two infraorders which are often termed, as here, the Aeluroidea, including civets, hyenas and felids, and the Arctoidea, which includes such types as dogs, raccoons, bears, and weasels. The fossils tend to support this division, although the distinctions often are none too clear-cut. The aeluroids, which we shall consider first, may be defined by such technical features as the fact that the tympanic bulla is (in contrast to the arctoids) a compound structure; the tympanic bone forms only the outer margin of the bulla, the remaining portion being formed by a separate ossification, the entotympanic.

Civets.—The basal stock of the aeluroids lies in the family Viverridae, including civets, genets, the mongoose, and a number of other Old World types. The civet, *Viverra,* is a representative form. It is a small carnivore with a long skull, a slim body, long tail, and short legs. All the premolars are present in this genus, but the first is tiny and is lost in many relatives. The carnassial is a good shearing tooth. Only two molars are present in either jaw; the upper ones are triangular and never develop a hypocone; the second is considerably smaller than the first.

In dentition and many other respects the typical civets are very similar to the ancestral miacids from which they are descended and from which early viverrids are difficult to distinguish; so similar are they that it has been suggested that miacids and viverrids might even be considered as constituting early and later members of a single, long-lived family.

The viverrids are characteristically the small carnivores of the Old World tropics. Small size, forest-dwelling habits, and tropical distribution all have tended to make them relatively rare in the fossil record; a number of genera are, nevertheless, recorded in Tertiary and Pleistocene deposits; none ever reached America. Despite the relatively primitive position of the family as a whole, the viverrids nevertheless show great variability in both fossil and Recent genera. They are the only carnivores to have reached and flourished in Madagascar. Among the variant types are some which are strikingly suggestive of the felid family. The existing fossa (*Cryptoprocta*) of Madagascar is a large and rather catlike viverrid; *Stenoplesictis* of the Oligocene is an ancient type showing similar specializations. By some writers these forms are considered as ancestral felids and placed in the Felidae. They are here, however, retained conservatively in the Viverridae; for, although they are structurally antecedent to the cats, it is doubtful whether they represent the actual viverrids from which the felids took their origin.

Hyenas.—A side branch of the civets which has progressed so far as to rank as an independent family is that constituting the Hyaenidae (Figs. 334*K*, 341*F*). The living hyenas of the Old World tropics are large scavengers with heavy and blunt teeth for crushing bones. The carnassials are powerful, but of the post-carnassial teeth there remains only a tiny upper molar. The hyena is a digitigrade running type which has lost hallux and pollex, and has blunt nails rather than claws.

There are two living types—the striped hyena (*Hyaena*) of northern Africa and western Asia, and the spotted hyena of south and central Africa, often considered as a separate genus, *Crocuta;* both were present in Europe in the late Pliocene and the Pleistocene. The hyenas can be traced back to the viverrids, from which they are a Miocene offshoot. *Ictitherium* and similar genera of the later Miocene and Pliocene are intermediate in character, for although trending in the hyena direction, they have the complete viverrid dentition and the teeth are not so heavy as in the modern types. Although the hyenas are almost entirely inhabitants of the warmer regions of Eurasia and Africa, one rare form, *Chasmaporthetes,* reached America in the early Pleistocene.

Conventionally grouped with this family is the "aard wolf," *Proteles,* of South Africa, which superficially has the appearance of a small, slim hyena. It is, however, an insect eater, and in relation to this fact the jaws are (in marked contrast to the hyena type) slim in build and the teeth tiny and uniform in character. Like the hyenas the aard wolf is obviously a specialized offshoot of the civet group, but no connecting fossil forms are known.

Cats.—The most interesting of the aeluroids and in many ways the most specialized of all carnivores are the "cats," comprising the family Felidae. Although an almost innumerable series of names has been proposed for the various felines, they are so similar in structure that all modern felids, from the domestic form to the lion and tiger, may be included in the genus *Felis* in a broad sense (the hunting leopard, *Acinonyx,* is the one conspicuous exception). In contrast to the dogs, for example, most cats, although digitigrade, are not good running types but depend for success in catching prey on a sudden pounce; the skeleton is hence more flexible than that of the dog; the claws are sharper and are retractile. The hallux is lost in modern forms, although the pollex is retained. The skull is comparatively short and high. The canines are powerful. The premolars partake somewhat of the carnassial pattern, and one upper and two lower ones are functional (there are one or two additional small ones in the upper jaw). The carnassials are well developed as purely shearing structures of two blades. The cats are almost pure carnivores, and there is no grinding surface left in their dental battery; as in the hyenas, only a tiny first upper molar remains behind the carnassials.

In sharp contrast with the typical living felines are the extinct saber-toothed "cats," of which the Pleistocene *Smilodon* (Figs. 334*N*, 339*B*) and *Machairodus* of the Old World Pliocene are good examples. In the sabertooths the upper canines were exceedingly long stabbing and slicing structures; the lower canines were correspondingly reduced, and there was often a flange at the front end of the lower jaw serving as a guard for the upper canine tooth. Connected with this specialization were many related ones. The jaw was so constructed that it could be opened to about a right angle; the reduced ascending ramus of the jaw and the high-crested, V-shaped back part of the skull are related to the consequently changed position of the temporal muscle. The mastoid process was powerfully developed for the attachment of muscles which pull down the head. It is believed that the sabertooths preyed mainly upon thick-skinned animals such as mastodons and elephants; the sabers may have been effective in slicing wounds which would cause death by copious bleeding.

The felids had the most rapid development of all the fissipede families. In the late Eocene and early Oligocene, at a time when, as we have said, most of the modern families could hardly be told from one another

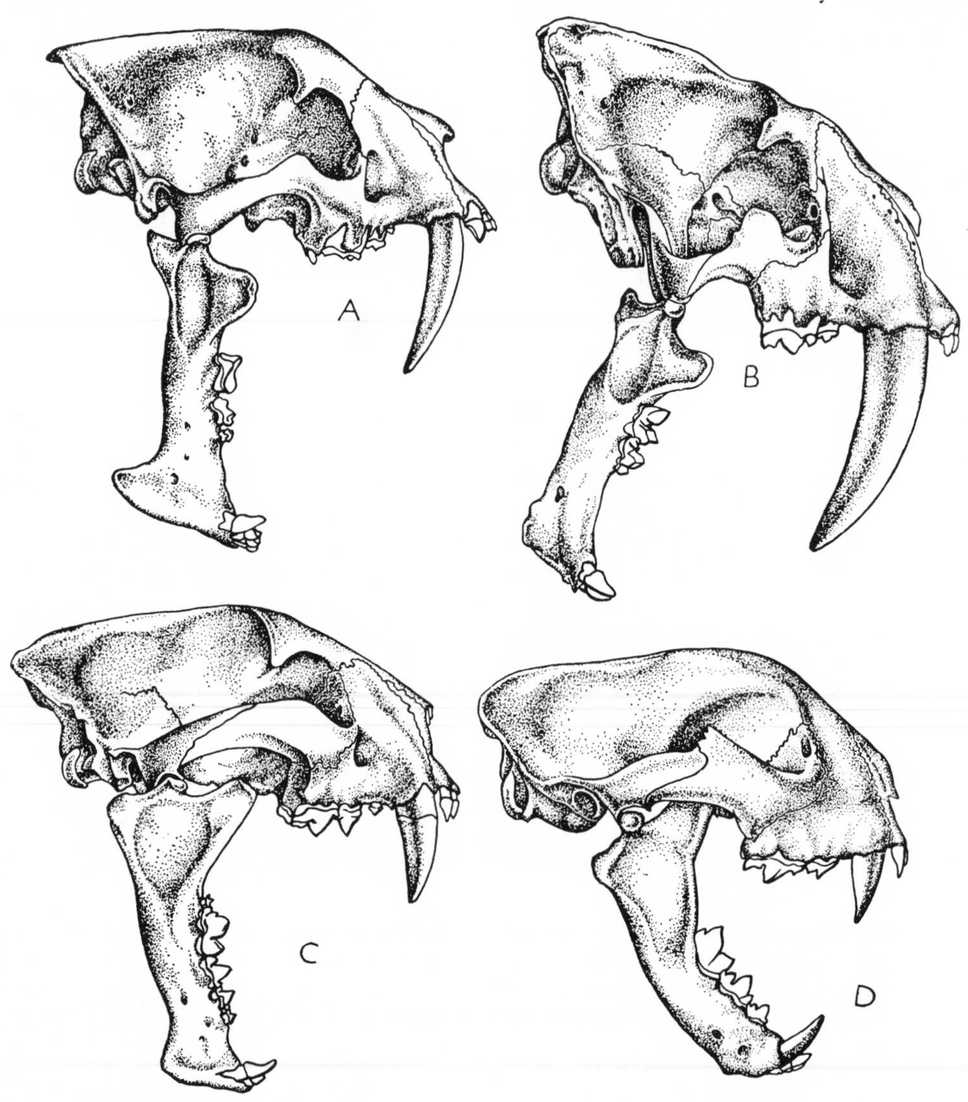

Fig. 339.—Skulls of sabertooths and true cats. *A*, *Hoplophoneus*, an Oligocene sabertooth, skull length about 6.25 inches; *B*, *Smilodon*, Pleistocene, skull length about 12 inches; *C*, *Dinictis*, an Oligocene form with small sabers possibly ancestral to the true cats, skull length about 6.12 inches; *D*, *Pseudaelurus minor*, a Pliocene true cat, skull length about 5.75 inches. (*A–C* after Matthew; *D* after Zdansky.)

or from their miacid ancestors, we find that not merely recognizable felids but even highly specialized sabertooths were already present. *Hoplophoneus* (Figs. 338, 339*A*), a characteristic American Oligocene form, shows all the essential characters of the later sabertooths, although they were here not so highly developed; the skull was not so high posteriorly, the saber not quite so long (although the jaw flange was even more prominent), and the premolars were less reduced than in later forms. Sabertooths were present at nearly every stage of the Tertiary in the northern continents; and in the Pleistocene *Smilodon* (accompanied by *Felis*) successfully "raided" South America, where the large native ungulates may have afforded an easy prey. The eventual extinction of sabertooths may perhaps be associated with the increasing rarity of the large "pachyderms" upon which they may have preyed. It is possible that the mastodons were a favorite victim of the later sabertooths. In North America these primitive proboscideans persisted until the end of Pleistocene times, and so did *Smilodon*. In Europe, however, mastodons disappeared in the early Pleistocene, and the sabertooths also disappeared early.

Modernized catlike types, *Felis* and related genera, with "normal" canines appeared in the early Pliocene. Their origin has been much debated, for all the better-known older felids have much larger canines and hence were sabertooths of one sort or another. Those who believe that evolutionary trends never reverse themselves are forced to believe that the later felines have descended from earlier Tertiary ancestors who remained in obscurity in Oligocene and Miocene times and are not readily identifiable, if present at all, in the fossil record.

An alternative solution, however, has been suggested. Paralleling the typical sabertooths from late Eocene to Pliocene, we find a series of forms which may be termed "false sabertooths," of which the Oligocene *Dinictis* (Figs. 334*L*, 339*C*, 340) is typical. The upper canine was considerably larger than in modern cats, the lower canines small, and there was a distinct (although small) flange on the lower jaw. These false sabertooths persist through the Miocene into the Pliocene. *Pseudaelurus* [*Metailurus*] (Figs. 334*M*, 339*D*) of the Miocene and Pliocene belongs to this series, but the jaw and dentition are less like the sabertooths and closer to the living felines. It is thus not improbable that the "normal" modern cat tribe has evolved from long-tusked ancestors.

Mustelids.—The arctoid group of modern carnivores, including weasels, dogs, raccoons, bears, and related types, may be distinguished from typical aeluroids by various technical characters. The claws, for example, are almost never retractile (as they are in the cats); the tympanic bulla is generally formed by the tympanic bone alone; there is a long canal, not developed in aeluroids, running fore and aft beneath the bulla for the carotid artery, which supplies blood to the brain and skull region (cf. Fig. 295).

A basal group of arctoids, comparable in position to that of the viverrids among the aeluroids, is that of the family Mustelidae. The mustelids include such familiar forms as skunk, weasel, badger, wolverine, and otter. These are the characteristic small carnivores of the north temperate zone, parallel to the viverrids of tropical habitat. In general form and habits typical mustelids are, like the viverrids, relatively primitive. They tend to remain comparatively small in size, with short, stocky limbs and a full complement of toes; a majority have remained forest dwellers, and many are still more or less arboreal; most mustelids have remained good carnivorous types. Continued emphasis on a flesh-eating diet has not only resulted in the retention of well-developed shearing teeth in the greater number of mustelids but also in the reduction of the molars; no living mustelid has more than a single postcarnassial

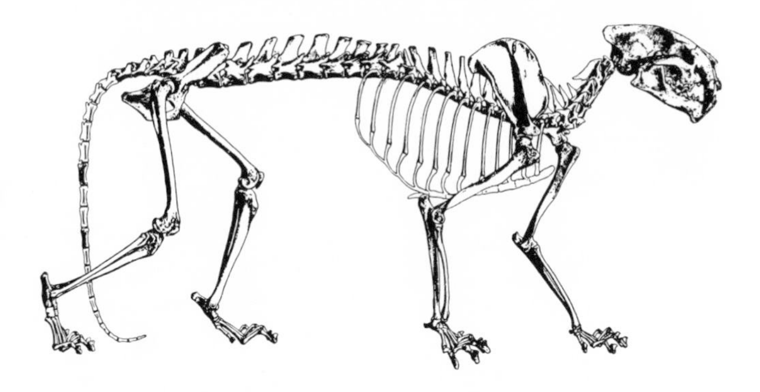

FIG. 340.—*Dinictis*, an Oligocene felid; original about 3.25 feet long. (From Matthew.)

molar in either jaw. In most forms the upper molar has a characteristic expansion of the inner portion with a "waist" between the two halves. The reduction in the molars (and often in the anterior premolars as well) has been accompanied by a shortening of the facial region of the skull; seldom is any diastema developed.

The taxonomy of the numerous modern forms is somewhat unsettled (some have proposed to divide them into as many as thirteen subfamilies); and the fossil history of mustelids is comparatively poorly known, for apparently most of them were forest-dwelling types, and although several scores of genera are known as fossils, the remains are usually fragmentary. The ancestral Oligocene forms, such as *Stenogale*, were small, sharp-toothed carnivores. Two upper molars were usually still present, while the peculiar shape of the first molar had not been developed. These early types seem to have been very close to the more primitive dogs and even to the contemporary civet ancestors.

We may distinguish among later mustelids several groups which appear to have been established before the close of the Miocene. One includes the weasels (*Mustela* [Figs. 334*J*, 341*E*]), martens, fishers, and so forth; these are the typical small, short-limbed, bloodthirsty carnivores of the family, and have retained a sharp, cutting type of dentition. The wolverine (*Gulo*) is a large representative of this group, and to it also belongs the largest of all known mustelids, an early Miocene American form, *Megalictis*, somewhat comparable to a wolverine and as large as a black bear.

The badgers, such as *Meles* of the Old World and *Taxidea* of the New, are members of a series traceable back to the Miocene; the second molars are of good size, and the carnassials have developed a hypocone, features indicating a diet which is less purely carnivorous. Farther away from pure flesh eating are the skunks (*Mephitis*) and their relatives, likewise traceable back to the Miocene, in which the carnassials have lost much of their shearing function, while the molars tend to take on a grinding character.

The otters (such as *Lutra*), with an Oligocene ancestry, have taken up an aquatic, fish-eating life; the Pleistocene and living *Enhydra* has become marine, and the Pliocene *Semantor* was a large aquatic form.

Dogs.—A second arctoid stock is that of the family Canidae. Represented today by a considerable number of dogs, wolves, and foxes, the group goes back to the very beginning of arctoid history, and fossil types are very numerous, presumably because the canids have been in great measure dwellers in open plains. They are essentially cursorial terrestrial types, with four digits well developed but with the pollex and hallux reduced, the toes close together, the claws blunt. The skull is long; the bulla typically large. The shearing function of the carnassials is well developed; but there is some grinding surface on the molars, two of which are normally retained above and two or even three below.

Cynodictis and *Hesperocyon* [*Pseudocynodictis*] (Fig. 342), mentioned earlier, are among a series of small carnivores which are very primitive in build but appear to lead to the dog group. Such forms appeared before the end of the Eocene. Early in the Oligocene there had evolved genera more definitely canid in

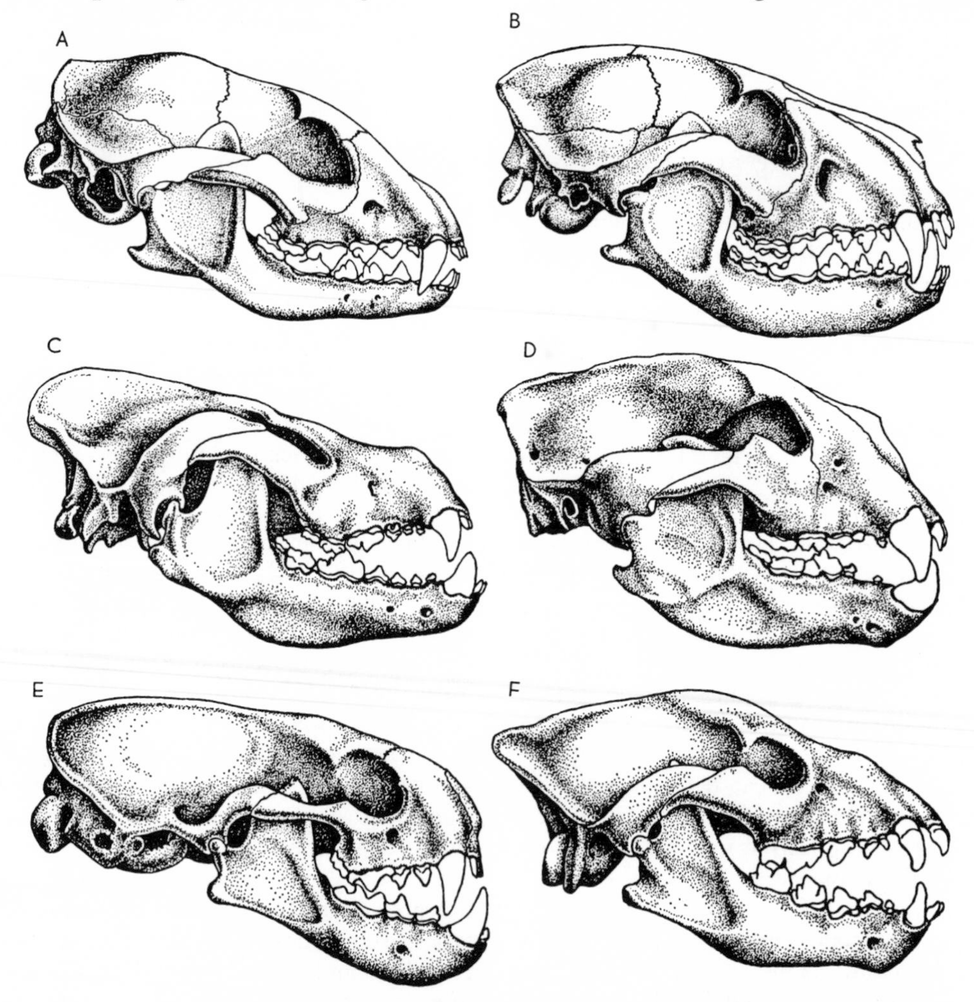

Fig. 341.—The skull in fissipede carnivores. *A*, *Hesperocyon*, a primitive Oligocene fissipede, length of original about 3.25 inches; *B*, *Cynodesmus*, a Miocene canid, length of original about 5.25 inches; *C*, *Hemicyon*, a Miocene form intermediate between dogs and bears, skull length about 13 inches; *D*, *Arctodus*, an American Pleistocene bear, skull length about 13 inches; *E*, *Mustela robusta*, an English Pleistocene weasel, skull length about 2.8 inches; *F*, *Hyaena variabilis*, a late Tertiary fossil hyena from China, skull length about 9 inches. (*A* after Scott; *B* after Matthew; *C* after Frick; *D* after Merriam and Stock; *E* after Reynolds; *F* after Zdansky.)

nature. Fifty genera of doglike animals are known from the Tertiary, representing a complex series of phyletic lines. A main sequence appears to lead through such forms as *Cynodesmus* (Fig. 341*B*) and *Temnocyon* (Fig. 334*E*) toward the modern dogs, wolves, and jackals (*Canis*) and the foxes (*Vulpes, Urocyon*). In this line there was little tendency toward increase in size; the limbs were elongating, pollex and hallux were in process of reduction, and the running gait was evolving. Although now present in the three southern continents, canids were unknown in South America before the Pleistocene, and the dingo of Australia is quite surely a type introduced by man. Dogs appear to have been the oldest animals domesticated by man, and remains of this camp follower are common in Neolithic sites; the original dog was presumably a wolf type, although the jackal has been advocated as an ancestor. *Otocyon*, the "large-eared fox" of South Africa, is of interest in that the number of molar teeth has secondarily increased so that there are often 4 molars above and 5 below—one of the few cases, apart from whales, in which the primitive placental number of teeth has been exceeded.

There were a number of lines of canid evolution which have failed to reach Recent times. One was a group of hyena-like dogs with peculiarly swollen foreheads, mainly American types of the Miocene and Pliocene (*Borophagus, Osteoborus*). In addition, there were several lines in which there was a tendency, in the Miocene particularly, for growth to considerable size and consequent ungainly form. Such canids are frequently termed "bear-dogs." In the case of many

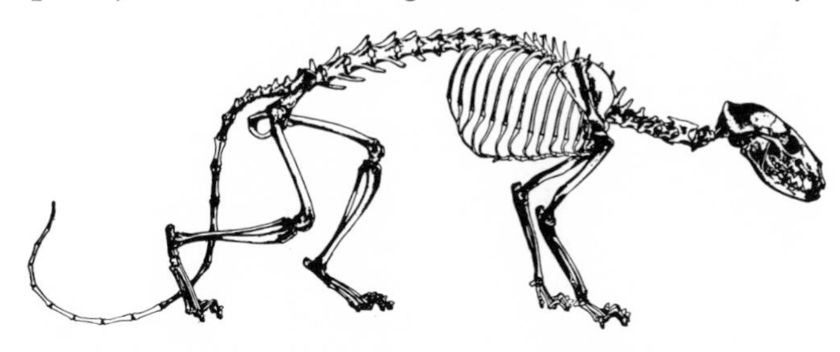

FIG. 342.—*Hesperocyon*, a small, primitive, Oligocene fissipede carnivore; original about 20 inches long. (From Matthew.)

genera, such as *Daphoenodon* (Fig. 343) and *Amphicyon* (which rivaled the bears in bulk), the term certainly is a misnomer, for, apart from size, there are few ursid features present; bears (and raccoons as well) are presumably descended from primitive canids, but not from this group (which appear to be, likewise, rather remote from the main line of canid descent). The build was heavy, the tail long and massive, the limbs short and powerful, the feet five-toed and rather spreading; unlike the bears, however, they were digitigrade types. The complete set of molars was retained and highly developed, while the shearing function of the carnassials was reduced. Obviously, these big forms were far from being pure carnivores.

Raccoons.—Fairly closely related to the canids are the Procyonidae (Fig. 344), of which the raccoon,

Procyon, is the type. This and most other members of the family are small, tree-living forms with semiplantigrade feet and a full complement of toes. The procyonids are mixed feeders; and, while there are only two molars, these have a good grinding surface. The carnassials have lost their shearing function, and the upper one has developed a hypocone. This obviously

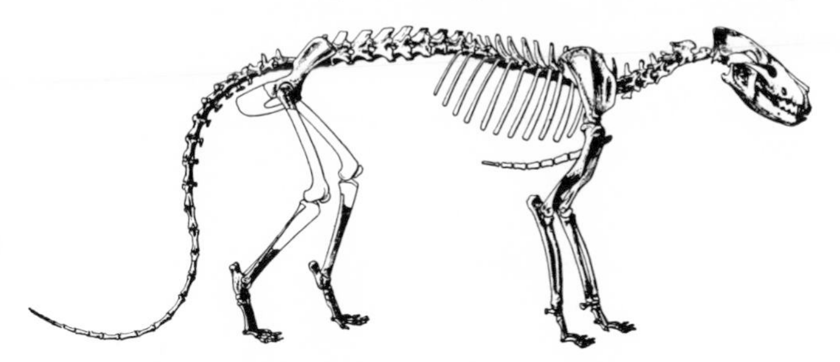

FIG. 343.—*Daphoenodon*, a Lower Miocene bear-dog; original about 5 feet long. (From Peterson.)

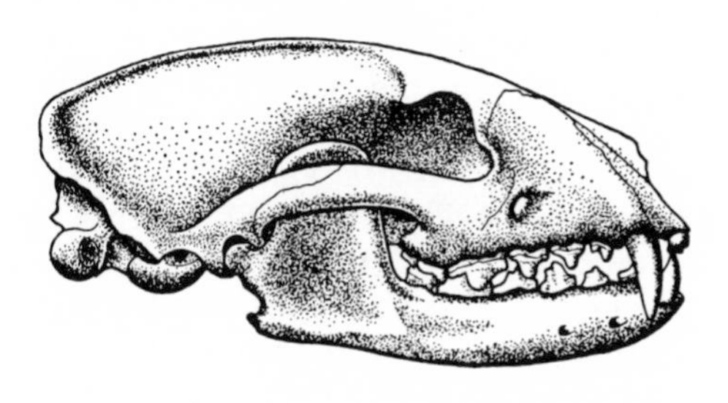

FIG. 344.—*Zodiolestes*, a Miocene procyonid, skull length about 4.5 inches. (After Riggs.)

seems to be a reversion from the primitive carnivorous canid adaptations back toward an omnivorous diet. In other respects, however, the procyonids are seemingly merely a series of persistently primitive relicts of the arboreal ancestors of the dogs. There are several other forms from Central and South America, such as the kinkajou and the coati, which are customarily included in the Procyonidae. A bit further from the typical dogs is the true Asiatic panda, *Ailurus*, a somewhat larger terrestrial type with a better development of the grinding teeth.

The procyonids appear to have been primarily a North American group. Fossils are, however, rare, a fact to be associated with their usual arboreal habits. Some of the mid-Tertiary canids, such as *Phlaocyon*, paralleled their raccoon cousins in dentition (and probable habits), and can only be distinguished by technical details of skull structure. Members of the family reached South America in the late Pliocene among the first invaders of that continent from the north after its long Tertiary isolation.

Bears.—The bears, constituting the family Ursidae, are a further development of the arctoid "dog" group away from a carnivorous mode of life to a mixed and mainly herbivorous diet. Their ancestors had lost the last molar before giving up flesh eating; but, in compensation, the remaining molars are much elongated, with wrinkled grinding surfaces. Here, as in the raccoons, the carnassial has lost much of its shearing character, but no hypocone has developed, and the protocone has migrated backward. All the digits are retained in the plantigrade feet. While in diet and certain structural features the bears show some similarity to the raccoons, they differ in other characters, such as, for example, their notably large size, their generally terrestrial habits, and (as a minor feature) the stub tail. Presumably their similarities are due merely to the descent of both groups from the primitive dog stock of the Oligocene.

Of modern bears, most—including the brown and grizzly bears, the black bears and the polar bears—are so similar structurally that they can be included in a common genus *Ursus,* although discrete generic names are sometimes used for each type. Somewhat more distinct are certain forms from southern Asia, such as the Malay sun bear and the honey bear of India, and a still wider departure from the pattern is the short-faced spectacled bear, *Tremarctos,* of South America. Relatives of the last, such as *Arctodus* (Figs. 334*I*, 341*D*), were present in the Pleistocene of North as well as South America, and bears of the *Ursus* type are common finds in the Pleistocene of Eurasia and North America. *Ursus spelaeus,* a giant race of brown bears, was a cave-dwelling contemporary of early man in Europe.

The bears appear to have arisen from the canid group during the Miocene. We have mentioned some large canids of that age which are sometimes termed "bear-dogs" but appear not to be related to bears. However, other large late Miocene and Pliocene dog-like forms such as *Hemicyon* (Figs. 334*G*, 341*C*) and *Dinocyon,* are suggestive of bear relationships in such technical features as the backward movement of the carnassial protocone, and *Ursavus* and *Agriotherium* [*Hyaenarctos*] (Fig. 334*H*) of the Pliocene show a definite beginning of the characteristic ursid enlargement of the molars. There is evidence suggesting that a line of ancestry of the bears may be traceable back among early canids to the earliest appearance of the dog family. *Ailuropoda,* the large "particolored bear" or "giant panda" of the Pleistocene and Recent faunas of southeast Asia, has often been considered to be, like the smaller true panda, a procyonid, but thorough study of the animal's anatomy indicates that it is a conservatively primitive offshoot of the early bear stock.

Pinnipeds.—All the Carnivora so far described are usually placed in a suborder Fissipedia; the seals and walruses, now to be considered, are classed as the Pinnipedia. In these water dwellers, in contrast with many other aquatic vertebrates, swimming is entirely accomplished by means of the limbs; in the seal ancestors, we may assume, the tail had been reduced, before aquatic habits evolved, to a slender, typically mammalian structure from which it was impossible to recreate a propulsive organ. The two hind legs in the seals have been turned backward to form a substitute for the tail, while the front legs are well-developed steering organs. In the limbs the upper segments are short and lie inside the surface of the trunk. The hand and foot are long, although there is no increase in the number of phalanges. All five toes are retained with webs of skin between them, the first digit often being quite strong and acting as a cutwater. Scaphoid, lunar, and centrale are fused in the hand, as in fissipedes (suggesting descent from that group). The dentition, for fish- or mollusk-eating, is regressive. The incisors tend to disappear, although the canines are usually well developed. The molars are usually reduced in number, but the premolars are preserved. There is no specialization of carnassials, and the cheek teeth form a uniform series of simple structure.

Most primitive in many ways are the eared seals (family Otariidae), in which group are included the fur seal and the sea lion. External ears are present, and the hind legs are still flexible enough to be brought around forward again from the swimming position for locomotion on land. The cheek teeth have but a single cusp and usually but a single root. The family goes back to Lower Miocene strata in which *Allodesmus* is a rather primitive type. Practically all known fossils are from the shores of the Pacific, which thus appears to have been the ancestral home of the family.

An offshoot of this group is *Odobenus,* the walrus of northern seas, placed in a distinct family. It appears to be structurally related to the eared seals but is larger and clumsier, earless, and with a dentition adapted to mollusk eating. There are huge, rootless upper canines, but most of the other teeth are small and tend to drop out early, leaving a few heavy pegs of premolars for shell crushing. The oldest forms, such as *Prorosmarus,* of the Upper Miocene, were somewhat more primitive and closer to the eared seals.

The Phocidae, earless seals, are the most abundant pinnipeds today. These forms are more completely adapted for water living; the hind legs cannot be turned forward at all. In contrast with the eared seals, however, the cheek teeth usually have two roots and accessory cusps.

As with the two other types, the earless seals may be traced back to the Miocene. Beyond this date, however, our record of pinniped ancestry cannot be traced. It has been suggested that they have descended directly from Eocene creodonts. But a number of features suggest that they have been derived from primitive fissipedes, probably of the arctoid group. It is further possible, considering the numerous differences between the two families, that otariids and phocids have acquired aquatic adaptations independently of one another.

CHAPTER 20

Archaic Ungulate Groups

Under the head of "ungulates," or hoofed mammals, may be ranged almost all the larger herbivorous members of the class. The name, however, is not entirely a distinctive one; for while typical forms, such as the horse and cow, have hoofs, there are included in the ungulate orders a number of animals with well-developed claws and even such types as the purely aquatic sea cows. Further, it is far from certain that the ungulates form a single natural group, for large, hoofed herbivores may well have developed along a number of independent lines.

Despite the possibly artificial nature of the assemblage, there are certain structural changes which have generally occurred during the transformation of a primitive placental of whatsoever group into a large herbivore—changes having chiefly to do with dentition and the locomotor apparatus.

Ungulate teeth.—The generalized dentition of primitive placentals was unfitted for a purely herbivorous diet; low-crowned, sharp-cusped molars are not suitable organs for undertaking the thorough mastication which leaves, grain, or grass must undergo before passing into the digestive tract; and the development of a large grinding area is an obvious necessity for a herbivore. The primitively triangular upper molars have tended to square up, usually by the development of a hypocone at the back inner corner, giving essentially a four-cusped tooth in which the two intermediate conules have played a varying role (Fig. 297*B*). In the lower molars the paraconid has disappeared, and the heel is built up to give a type of tooth with the talonid and trigonid of equal height, each contributing two of the four principal cusps—protoconid, metaconid, hypoconid, and entoconid.

The originally sharp tubercles generally softened down to low swellings, giving a bunodont condition found in many early ungulates (Fig. 350*A*, *B*) and still present in an exaggerated form in the swine. Further changes have taken place in most groups. There may be an elongation of the individual cusps into curved ridges, most characteristically seen in the selenodont (crescent moon) pattern of many even-toed ungulates, such as camels, deer, and cattle (cf. Fig. 400*A*, *B*, *C*). More common, however, has been the development of connections between adjacent cusps, forming varied patterns of ridges (lophs) on the tooth. In a lophodont upper molar (Fig. 299) three lophs are most common: (1) an ectoloph, forming the outer wall of the tooth, connecting paracone and metacone and often (if the styles external to the cusps develop) with a W-shaped contour; (2) a protoloph, including the protocone and protoconule; and (3) a metaloph, similarly formed from hypocone and metaconule. There are sometimes varied secondary connections of these crests, and additional spurs and ridges may develop. The cross crests may develop but not the ectoloph (primitive proboscideans, tapirs, etc.); conversely, only the ectoloph may develop, leaving isolated rounded cusps internally, a bunolophodont tooth (as in titanotheres, chalicotheres [Fig. 384*G*, *H*]). In the lower teeth a common development has been the retention and emphasis of the two V's of the original triangular pattern (Fig. 385). These ridges undergo various modifications; they may become crescentic in shape or develop into two hook-shaped ridges which may eventually become two cross-lophs very similar to those sometimes found in the upper teeth.

Food grinding, with the passage of one tooth across the surface of its opponent, implies considerable wear. This wear is greatly increased when an animal takes up highly abrasive siliceous food, such as grass; further, increased size of the individual only squares the grinding surface available while cubing the amount of food required. It is not surprising, therefore, to find that, although many early types had low-crowned (brachyodont) molars, many ungulates develop a high-crowned, prism-shaped hypsodont type of cheek tooth which will stand a very considerable amount of grinding before being worn down to the roots and exhausted.

In this process of development of hypsodont teeth, we find that the cement, originally confined to the roots of the teeth, plays a prominent part. If nature were to construct a high-crowned tooth merely by greatly elongating the original ridges and cusps, the product would be a series of thin parallel spires or columns, easily broken and full of crannies. One alternative "solution" would seem to be that of enormously thickening the block of dentine above the root, leaving the cusps as superficial structures. This, however, is seldom seen except in some archaic and comparatively unsuccessful mammals. Instead, in a number of groups, there has been a development of the cement so that it completely covers the tooth before eruption and fills all the interstices between the ridges.

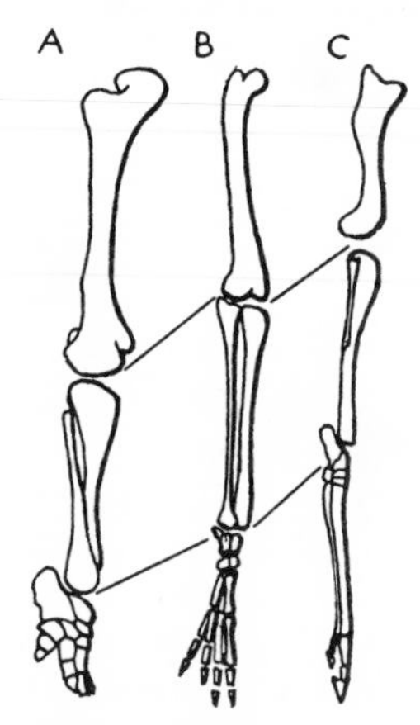

FIG. 345.—Hind limbs of elephant, opossum, and horse, to show changes in limb proportions from a primitive condition (*B*) to (*C*) that of a fast-running ungulate with short femur and long metapodials, and (*A*) a graviportal type with a long femur and short broad foot. (Mainly after Gregory.)

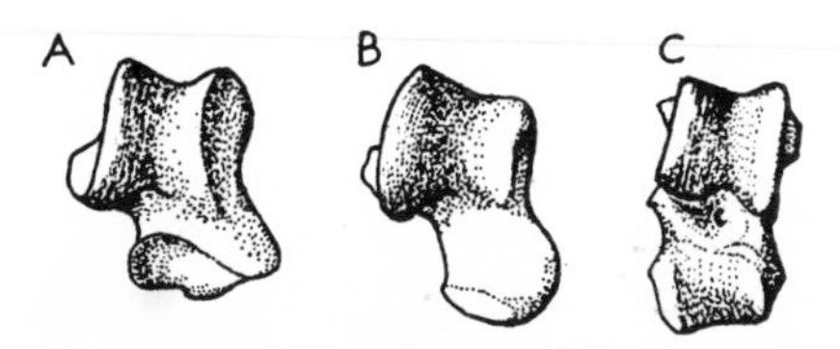

FIG. 346.—Astragali of Eocene ungulates: *A, Heptodon,* a perissodactyl; *B, Phenacodus,* a condylarth; *C, Homacodon,* an artiodactyl. (*A, B* after Osborn; *C* after Marsh.)

This results in a solid structure which, upon wear, exhibits all three tooth-forming substances on the crown—dentine in the center of the ridges, a hard enamel layer at their boundaries, and cement outside and filling all vacant spaces (Fig. 347).

Increased surface may also be attained by elongation of the row of grinding teeth. This may be accomplished by the increase in size of individual teeth, and in many cases (as in the horses) molars in later types are longer than in earlier ones. But a more common process by which this end has been attained is by expanding the premolars (usually of comparatively small size to begin with) and adding them functionally to the molar grinding series. Independently in a large number of groups may be traced a progressive increase in the surface area of the premolars, and in some cases (horses, for example) all except the first of these teeth have become almost identical in structure with the molars (cf. Fig. 384*A, B*).

The cropping of food is the function of the anterior portion of the dentition. For this duty the incisors are commonly retained, as in the horse. But the upper ones may be lost (as in the cow), the upper lip or a horny pad taking their place; and we find here and there various other specializations, such as the much enlarged, chisel-like incisors of South American ungulates. The canines are often functionless and reduced or absent; in most ungulate groups we find a gap, or diastema, developing between the front group of cropping teeth and the grinding series in the cheeks.

The ungulates have always been the main source of food for the larger contemporary flesh eaters. In connection with defense, horns or hornlike structures have developed in many groups. Sometimes (especially in early types) there may be large stabbing canines. In general, however, the best defense is flight; and rapid locomotion, both for escape from enemies and for migration from one feeding ground to another, is characteristic of many ungulates.

Ungulate limbs.—In the more typical ungulates the limbs are elongated and the limb bones usually slim. In fast-running forms the humerus and femur are relatively short, giving a speedy drive to the leg as a whole, while elongation of the second segment of the limb increases the scope of each stride. With this elongation there is often a tendency toward the reduction of the ulna and the fibula and the placing of the entire weight on radius or tibia. There is seldom, however, a complete loss of these bones, for they carry important muscle attachments and articulations, and the ends are usually present but fused with the remaining bone. There is practically no capability

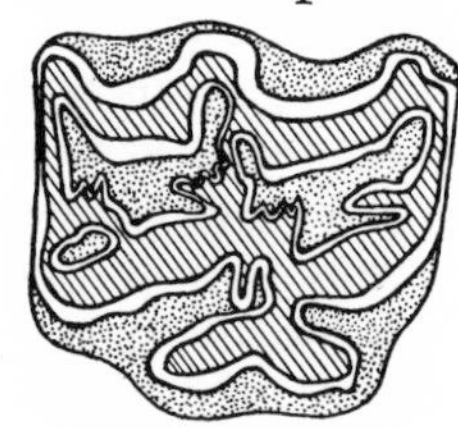

FIG. 347.—A worn upper molar of *Equus,* to show the arrangement of the three tooth materials—enamel (*unshaded*), dentine (*hatched*), and cement (*stippled*).

of rotating the limb segments; motion tends to be restricted to a highly efficient fore-and-aft drive. Most ungulates have an alternating type of carpus (and tarsus)—one, that is, in which the connections between the elements and with the metapodials are so arranged that each functional toe helps support at least two of the three proximal carpal elements. The astragalus (Fig. 346) usually develops a rounded and deeply grooved upper surface, over which the tibia glides without danger of turning or twisting.

The primitive placental was presumably more or less plantigrade. With faster locomotion, the metapodials (and thus the palm and sole) tend to be lifted off the ground, giving a digitigrade gait. With further development of speed the digits themselves are lifted, until the animal may touch the ground only with the tips of the toes—the unguligrade condition. With the result of attaining solid stance, hoofs are developed; the terminal phalanx of the toes broadens and is surrounded on the front and sides by a thick, nail-like modification of the original claw, while a horny layer padded by elastic tissues protects the flat lower surface.

With the lifting of the hand and foot into an unguligrade position, it is obvious that the shorter lateral toes would fail to reach the ground and would become nonfunctional. This was, in most groups, followed quite naturally by a reduction of these side toes and strengthening of the central ones. The hallux and pollex, which not only are short but originally diverged at an angle from the other toes, were the first to go. Beyond this stage reduction has taken place according to two schemes. In the majority of ungulates (as in the ancestors of the horses, for example) the third toe was the longest, and the axis of symmetry of the foot lies through this digit (mesaxonic). In such forms the fifth toe has usually been lost, giving a three-toed condition; and further emphasis of the central toe and reduction of lateral ones has led in the modern horses to the development of a one-toed form (Figs. 383, 392, 393). In other ungulates, such as the ancestors of the cow and pig, the third and fourth toes were primitively equal in length, and the axis of symmetry passed between them (the paraxonic type). As in mesaxonic ungulates, pollex and hallux usually disappear, and many forms are found to be in a four-toed state. Frequently, however, the two central toes have enlarged, and the lateral ones have dwindled or vanished, to leave a two-toed form; the cloven hoof of the cow or deer is, of course, really two appressed digits (Figs. 407, 413, 415).

Simultaneously with toe reduction, changes have usually occurred in the metapodials. In fast-running forms there has been a great lengthening of these elements, giving a third functional segment to the limb through which an additional upward and forward drive may be imparted to the body. With the reduction of the toes, metapodial reduction usually takes place as well. Splints of bone often represent the metapodials of lost digits, as in the horse; while, on the other hand, the metapodials may be incomplete, although small nonfunctional toes may remain as dew claws. In paraxonic forms the two main metapodials may fuse in advanced types into a single element—the cannon bone.

In typical ungulate running, the forward spring is given chiefly by the hind legs; the forelimbs bear the main impact on landing. The clavicle is usually absent, thus releasing the body from any solid connection with the front limbs and allowing the shock to be taken up by the elastic "give" of the muscles slinging the body between the shoulder blades.

The heavy ungulates, such as the elephants, have a different locomotor "problem" somewhat comparable with that of the dinosaurs. In an animal of this sort—the graviportal type—the limbs tend to be large straight columns, with the femur and humerus longer than the second segments (Fig. 345*A*), giving a powerful, rather than a speedy, muscular action. There is little elongation of the metapodials, and all the digits tend to be retained to form a semicircular clump presumably enclosed, in fossil forms as in the living elephant, in a thick pad. The pelvis tends to be so shaped that the head of the femur lies beneath it rather than at the side, and the ilia are typically expanded laterally into broad wings in relation to the changed position of the muscles.

Condylarths.—The term "ungulate" covers, as we have noted, a vast and varied assemblage. Most living forms are included in two orders—the mesaxonic (odd-toed) perissodactyls and the paraxonic (even-toed) artiodactyls—which have flourished in the northern continents since the beginning of the Eocene, while

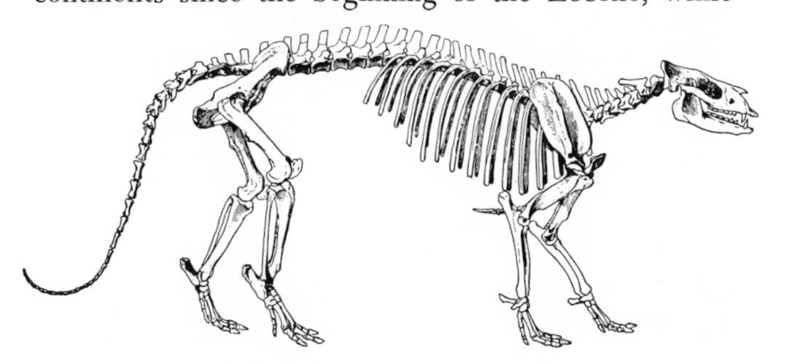

Fig. 348.—*Phenacodus,* a primitive ungulate (Condylarthra), original about 5.5 feet long. (From Osborn.)

the elephants, conies, and sea cows are surviving members of an ancient African group of subungulates. There are, in addition, a number of entirely extinct groups which will be treated in the present chapter and that which follows.

Certain of these extinct groups, as will be seen, were highly specialized in various directions; one ancient order, however, that of the Condylarthra, appears to represents a truly basic stock from which many, at least, of the further ungulate orders may well have been derived. The boundaries of this group are none too well defined. The condylarths appear to be an assemblage of forms transitional from ancestral insectivores to true ungulates, and hence it is not surprising that various genera and families which we will here include in this order have been assigned at one time or another not merely to other ungulate groups, but to the Insectivora, Carnivora, and even Primates. As might be expected from their primitive nature, the condylarths are confined to the early Tertiary. The great majority of known genera are from the Paleocene, where they constituted a large proportion of the known fauna. A relatively few members were present in the Eocene, and none is present beyond early Oligocene times.

We may begin our account of this important group by a description of the best-known form, the Lower Eocene *Phenacodus* (Figs. 346*B;* 348; 349; 350*A, B;*

351*C*, *D*), which reached a size rather considerable for the times, the largest species attaining the dimensions of a tapir. Its general appearance was that of a carnivore rather than an ungulate, for the tail was long and the limbs short and quite primitive in structure. The radius and tibia were about the same length as humerus and femur, and ulna and fibula were unreduced. There was little elongation of the metapodials; and all five toes were present, although (in mesaxonic fashion) the third toe was somewhat the longest and strongest. The carpus was serial in nature, each distal element lying directly beneath a proximal one, without the sharing of support characteristic of the alternating type of ungulate carpus. The clavicle, however, had already disappeared; and one diagnostic ungulate character was present in that there were obviously hoofs on the stumpy end phalanges.

The skull, again, was much like that of the creodonts—long and low, with orbits open behind and with a sagittal crest in correlation with the relatively small brain cavity. The dentition was complete, the canines were still large, and there was only a slight diastema. However, the cheek teeth were already partially adapted for a herbivorous, rather than a flesh-eating, mode of life. The molars were bunodont. Six cusps were present above and below, for a hypocone had already developed in two of the upper molars, while the paraconid was still present in the lower teeth. The premolars were still comparatively simple, although the last lower one had attained a molar-like appearance.

This interesting form was once believed by some to be the ancestor of the later ungulates, particularly the perissodactyls. But it is rather late in occurrence and much too large to occupy such an ancestral position, and some of its characters (as the serial carpus) are not those expected in an ancestor of the later groups. Even within the same family, however, Paleocene predecessors of *Phenacodus*, such as *Tetraclaenodon*, show more generalized and primitive characters; the dentition is more primitive in nature, and the feet appear to have terminated in broad claws rather than hoofs.

The phenacodonts are the "type" condylarths. With them, however, are to be associated several other early Tertiary families of a broadly similar nature, all of them, apparently, tending toward a herbivorous diet and ungulate characteristics.

Hyopsodus (Figs. 349; 350*C*, *D*), common in the American Eocene, is a late representative of a family almost entirely confined to the Paleocene of North America (although one far-flung genus appears to have reached South America). *Hyopsodus* was about the size and proportions of a hedgehog. In skull and dentition it was rather primitive and not unlike certain insectivores or lemurs. There was no diastema; but the canine was reduced, and the teeth were of uniform height, while the molars were of the bunodont type of many primitive ungulates. Primitive, again, were the clawed phalanges and the short and spreading feet; it is not improbable that *Hyopsodus* was a semiarboreal type. Despite these facts, it is to be regarded as a condylarth rather than as an insectivore or primate, with which groups many once placed it. Its Paleocene relatives, such as *Mioclaenus*, are mainly known from cheek teeth which are quite sharp cusped and similar to those of early creodonts.

Little *Meniscotherium* (Fig. 350*E*), of the late Paleocene and Lower Eocene of North America (with

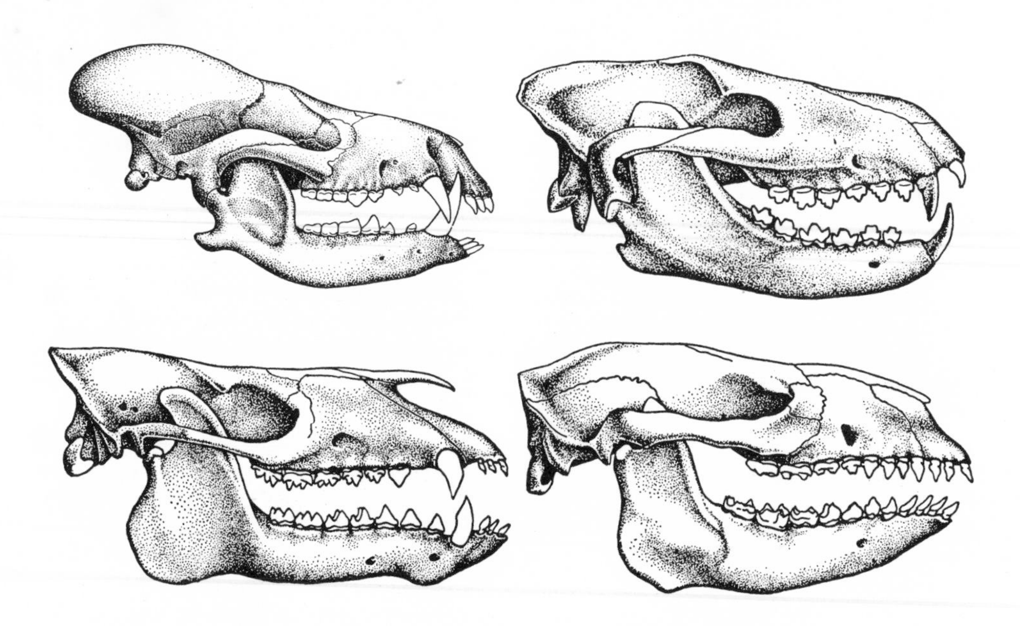

Fig. 349.—Condylarth skulls. *Upper left, Arctocyon*, length about 10.5 inches (after Russell); *upper right, Mesonyx*, length 11 inches (after Scott); *lower left, Phenacodus*, length about 9 inches (after Cope); *lower right, Hyopsodus*, length about 2.75 inches (after Matthew).

contemporary European relatives), represents still another type of condylarth, with a peculiar combination of characters. This form, locally common in the early Eocene deposits, was rather larger than members of the *Hyopsodus* group. The molars were advanced in structure, with a selenodont pattern, resembling that of some later artiodactyls in the lower molars and the outer cusps of the upper ones. The limbs were short and stout, as in *Phenacodus;* but the feet (as in *Hyopsodus*) appear to have had claw-like structures rather than hoofs.

Periptychus (Fig. 350*G, H*), *Ectoconus,* and a number of North American Paleocene allies form a still further family of primitive ungulates, generally assigned to the Condylarthra. *Periptychus* was nearly as large as *Phenacodus,* although occurring much earlier (in the Lower and Middle Paleocene), and, except for rather shorter limbs, would have had perhaps much the same general appearance in skull and skeleton. The dentition is distinctive. The cheek teeth are bunodont, with a trend toward the addition of supplementary cusps to the tooth pattern. In most ungulates the protocone tends to take up a position toward the anterior internal corner of the upper molars, with the hypocone behind it; here the prominent protocone lies centrally along the inner margin, flanked by the hypocone behind and an accessory cusp in front.

We here add to the roster of condylarths two further families of primitive placentals which show little in the way of ungulate specialization and have, in fact, been generally considered in the past to be allied in some fashion to the carnivores. But there are no positive evidences of relationship to the true carnivores; there is, for example, no development of the shearing ability in the cheek teeth characteristic of flesh eaters. Many of them (particularly of the first of the two families to be discussed) are difficult to distinguish from forms definitely included in the Condylarthra. They appear to be, essentially, insectivore descendants which tended in many cases to grow to large size, presumably had a mixed and varied diet, and were a stock out of which the true condylarths and, very probably, other ungulates may have developed.

The more generalized of these two families is that of the Arctocyonidae, a group almost entirely confined to the Paleocene but very abundant in that epoch; there are more than a score of recognized genera. The cheek teeth (Fig. 350*I*) are essentially of a primitive nature and moderately sharp cusped, but with a trend, in many forms, toward a squaring-up of the molars and development of a hypocone. The brain-case was small, the body slim, the limbs slender with clawed feet. Within the family there were several divergent lines. Some included forms of large size, such as the European *Arctocyon* (Fig. 349) and the American *Claenodon;* these grew as large as bears and were an early parallel to the bears in dentition and probably in omnivorous habits as well. Apart from advanced forms, however, the family included very primitive types; *Protungulatum* of the late Cretaceous is the oldest known mammal which appears to be definitely a member of the ungulate tribe.

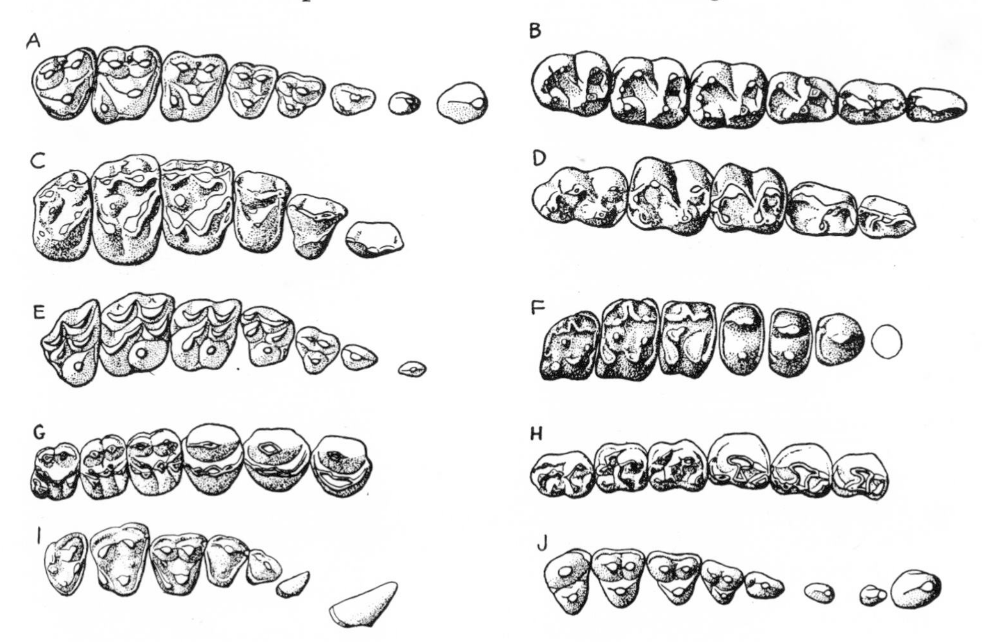

FIG. 350.—The cheek teeth of condylarths: *A, B,* right upper and left lower cheek teeth of the condylarth *Phenacodus,* ×5/6 approx.; *C, D,* same of *Hyopsodus,* ×3/2 approx.; *E,* right upper teeth of *Meniscotherium,* ×5/4 approx.; *F,* right upper teeth of *Didolodus,* a South American condylarth probably related to litoptern ancestry; *G, H,* right upper and lower left cheek teeth of *Periptychus,* ×2/3 approx.; *I,* right upper cheek teeth of the arctocyonid *Tricentes,* ×3/2 approx.; *J,* right upper cheek teeth of *Mesonyx,* ×1/2. (*A–E, G–J* after Matthew, *F* after Ameghino.)

The Mesonychidae, second of the two supposed carnivore families here appended to the Condylarthra, are of interest because of their specializations and large size. They appeared in the Paleocene, but were more common in the Eocene and lingered on into early Oligocene times. The upper cheek teeth were triangular, with three blunt cusps; the lower molars had curious shearing talonids. The feet (Fig. 351*A, B*) were wolflike in proportions but with fissured terminal phalanges which appear to have borne small hoofs rather than claws. *Mesonyx* (Figs. 349, 350*J*, 352) of the Middle Eocene had a skull a foot in length; later Eocene types grew to the size of a Kodiak bear, and the Mongolian *Andrewsarchus* had a skull a full yard long. The habits of these grotesque creatures are difficult to imagine; carrion feeding, mollusk-eating, a diet of some type of tough vegetable matter are among the guesses.

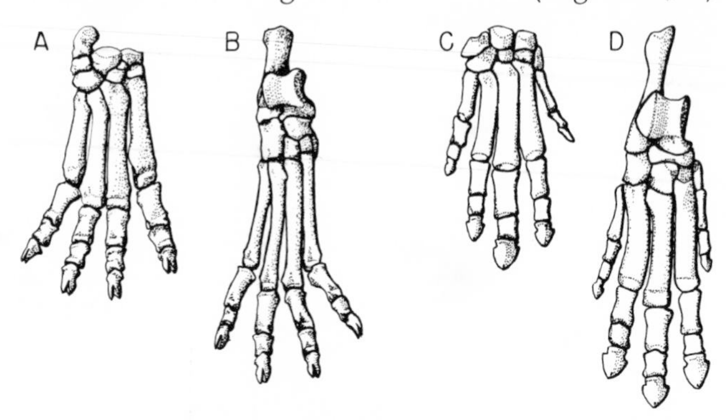

Fig. 351.—Condylarth feet. *A, B,* Manus and pes of the mesonychid *Synoplotherium* (after Wortman); *C, D,* manus and pes of *Phenacodus* (after Cope).

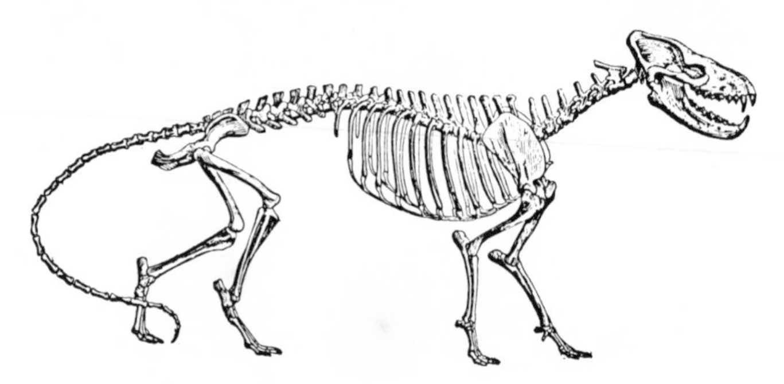

Fig. 352.—The skeleton of *Mesonyx,* original about 5 feet long. (From Scott.)

Discussed in a later chapter are the litopterns, an extinct group of South American ungulates. The older litopterns show many resemblances to the condylarths, and seem surely descended from them. This conclusion is strengthened by the presence in the earliest Tertiary faunas of South America (in addition to the hypsodont mentioned earlier) of the family Didolodontidae (Fig. 350*F*), condylarths similar to the phenacodonts of North America. Most of the didolodonts are confined to the Paleocene and Eocene, but a Miocene survivor of large size has been recently discovered.

Pantodonts.—Beyond the incipient ungulate stage represented by the condylarths, the more successful ungulate groups of the Tertiary, particularly the perissodactyls and artiodactyls, progressed steadily but essentially conservatively toward their varied goals. In contrast, however, we find in the early Tertiary, notably the Paleocene, a number of archaic ungulate groups which rapidly tended—prematurely, it would seem—to attain large size and ponderous build, only to face early extinction in competition with more progressive groups.

One series of forms of this sort, the elephants and other "subungulates," appears to have had its center in Africa, and will be dealt with in a succeeding chapter. We will here consider a number of other groups of archaic character which, while present to some extent in the Old World, were predominantly American in their history. These will include the Pantodonta, Dinocerata, Xenungulata, and Pyrotheria.

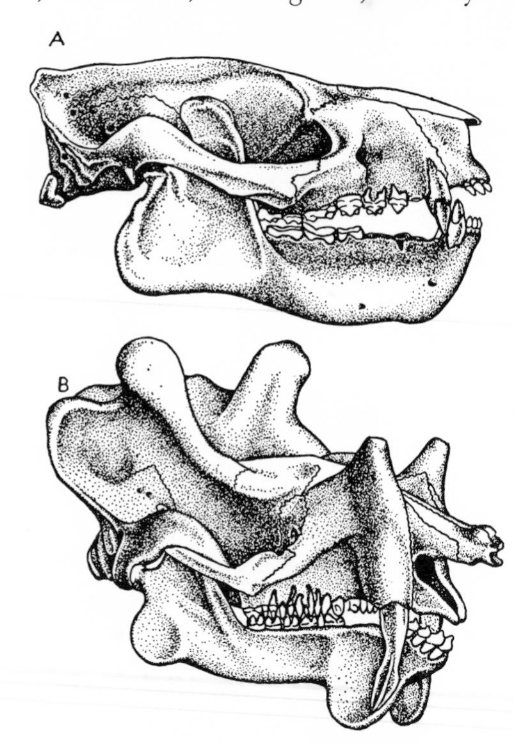

Fig. 353.—*A, Pantolambda bathmodon,* a Paleocene pantodont skull about 6 inches long; *B, Uintatherium,* an Upper Eocene member of the Dinocerata, skull about 2.5 feet in length. (*A* after Osborn; *B* after Marsh.)

Members of all four of these types tended to grow to large size and, further, to develop broad-surfaced low-crowned cheek teeth with prominent if varied cross-loph patterns. Many of the resemblances between the four may be attributed to parallelism, but it is not improbable that, as some have thought, they may be divergent members of a common stock, to which the appropriate ordinal name of Amblypoda has been applied.

Prominent in the Paleocene were the Pantodonta. *Pantolambda* (Figs. 353–55) of the Middle and Upper Paleocene was the first of the order to appear. This was a relatively small form, ranging from collie to sheep in comparative size, with short limbs and short, broad feet. The cheek teeth are of a simple pattern, the triangular upper molars with V-shaped cusps, the canines moderately large.

In the late Paleocene and Lower Eocene pantodonts had increased greatly in size and had diversified to become a dominant group of large and ungainly ungulates. In the late Paleocene, *Titanoides* and *Barylambda* (Fig. 356) are well-known American genera. The former had grown to dimensions about two and a half times those of *Pantolambda,* with greatly developed canines and clawed rather than hoofed feet. This quite un-ungulate feature we shall see developed in other later groups of "hoofed" mammals and here, as elsewhere, it is believed to have been associated with rooteating. *Barylambda* was about 8 feet in length, with a short but high body, heavy limbs and tail, and a small skull, in which the canines were not particularly prominent.

In the Lower Eocene, *Coryphodon* (Figs. 354, 357) ranged widely from North America to Europe. This form was about the size of *Barylambda* but quite different in many respects, such as the rather longer body and the long, slender tail. The skull was large, with a broad muzzle, broad arches, and a flaring occipital region. The canines were large and the upper molars peculiarly developed. Paracone and metacone form a diagonal posterior crest, and the protocone develops into a parallel ridge anterior to this.

This is the last appearance of pantodonts in North America, but in Asia genera fairly close to *Coryphodon* survived into the late Eocene and even the Oligocene.

Uintatheres.—Still more grotesque than the pantodonts were the uintatheres, or Dinocerata. These flourished especially in the later part of the Eocene of North America in such genera as *Uintatherium* (Figs. 353, 354, 355), which was as large as a modern African rhinoceros and was the giant of his day. In their ponderous build, short limbs and feet, and other weight-carrying adaptations, the uintatheres are similar to the pantodonts. As the name implies, the characteristic genera have developed pairs of hornlike bony swellings above the long, low skull on nasals, maxillae, and parietals; powerful upper canines were present in the males. The upper incisors were usually absent, the lower ones small. The upper molars were of peculiar construction, with two crests, converging internally in V-shaped fashion, occupying most of the surface of the tooth. It is believed that one ridge is formed by protoconule and paracone, the other by protocone and metacone. Outside of North America, uintatheres are unknown except for a primitive Asiatic late Paleocene genus and *Gobiatherium* of the Upper Eocene of Asia—the latter with a very long, low skull which lacks horns and, as well, the large upper canines of the American forms.

The oldest uintatheres are from the late Paleocene; forms of that age were hornless and somewhat smaller and more primitive in structure, but were, nevertheless, highly specialized forms, in which (for example) the peculiar upper molar pattern was already in evidence.

Xenungulata.—In the late Paleocene of South America there have been found fragmentary remains of an ungulate, *Carodnia,* paralleling in size and general build the contemporary pantodonts and uintatheres of North America. It has no known relatives or descendants, and has been made the type of an inde-

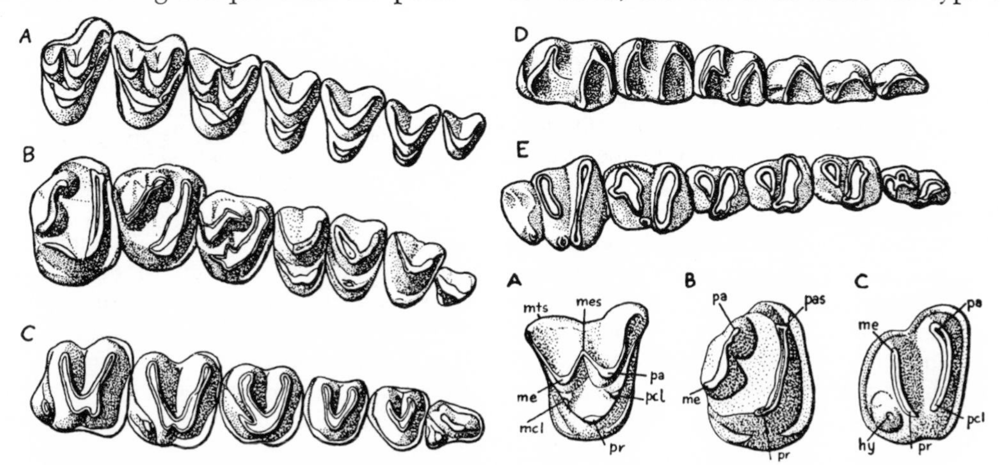

Fig. 354.—*Left,* Right upper cheek teeth of pantodonts and uintatheres. *A, Pantolambda,* about 3/4 natural size; *B, Coryphodon,* about 1/3 natural size; *C, Uintatherium,* about 1/2 natural size. *Upper right, D,* Left lower cheek teeth of *Coryphodon,* about 1/3 natural size; *E,* same of *Uintatherium,* about 1/2 natural size. (Mainly after Osborn.) *Lower right,* Right upper molars of amblypods and uintatheres. *A,* The Paleocene *Pantolambda; B,* the Lower Eocene *Coryphodon; C,* the late Eocene *Uintatherium.* In *Pantolambda* the tooth is of a simple tritubercular pattern. The homologies in *B* and *C* are disputed; the interpretation here adopted is that of Simpson. For abbreviations see Figures 297, 397.

pendent group, the Xenungulata. The large molar teeth develop, above and below, two cross crests in somewhat tapir-like fashion. Perhaps significant, however, is the fact that in the third molar the two crests converge inwardly as in the uintathere molars. This, as well as the general build (as far as known) suggests that *Carodnia* may be distantly related to the uintatheres and that the Xenungulata may be provisionally included—with that group and the pantodonts—in the Amblypoda.

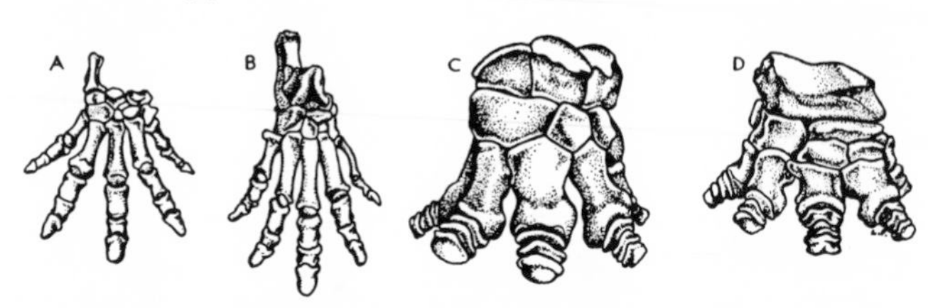

FIG. 355.—*A, B,* Manus and pes of the Paleocene *Pantolambda; C, D, Uintatherium* of the Upper Eocene. (*A, B* after Osborn; *C, D* after Marsh.)

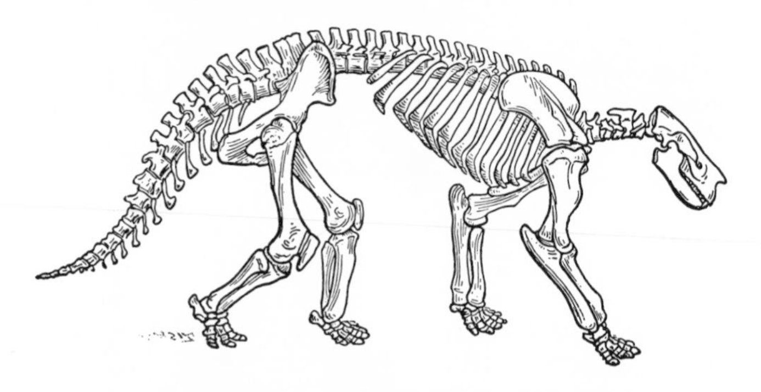

FIG. 356.—The late Paleocene pantodont *Barylambda,* length about 9 feet. (From Gregory.)

Pyrotheres.—*Pyrotherium* (Fig. 358), of the Oligocene, is a member of a small, short-lived, South American group, the Pyrotheria. This beast not only had grown at that early time in the Tertiary to the size of an elephant but anatomically had paralleled the proboscideans to a very remarkable degree. The dorsal nasal openings indicate the presence of a proboscis; the facial region was turned upward (rather than downward) on the cranial axis. Above the broad zygomatic arches lay small orbits, behind which no postorbital bar had developed. A heavy transverse crest lay along the top of the occiput for the attachment of strong neck muscles necessary to support the heavy head.

The dentition was strikingly like that of the early proboscideans. Chisel-like tusks were developing out of the incisors, of which but two were retained above (the lateral one large) and one below. Behind a long diastema lay the cheek teeth, six above and five below. As in the earliest proboscideans, these low-crowned teeth were each composed of two cross crests.

The many similarities in body, skull, and dentition all suggest that *Pyrotherium* was really related to the proboscideans, then developing to the east in Africa. But these resemblances are surely a case of exceedingly close parallelism, for the group apparently had

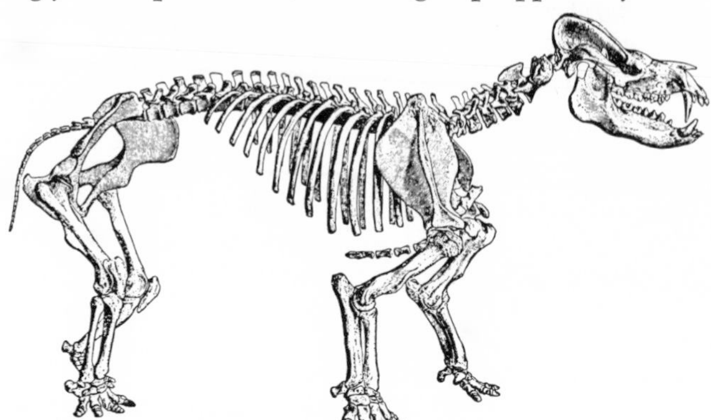

FIG. 357.—*Coryphodon,* a large Lower Eocene amblypod. Original about 8 feet long. (From Osborn.)

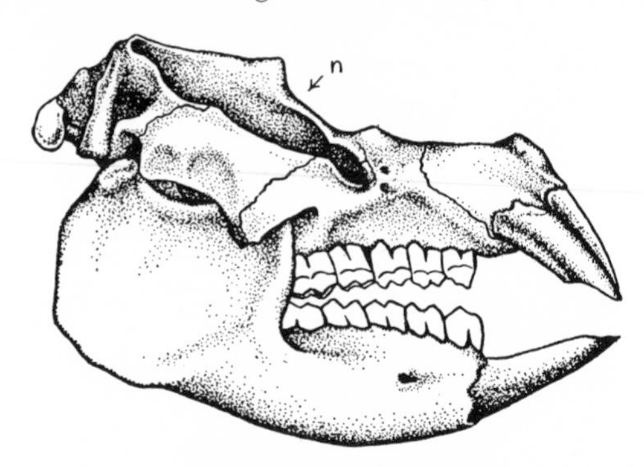

FIG. 358.—The skull of *Pyrotherium,* a South American ungulate paralleling the proboscideans. There is a tusklike development of the incisors, and the nostrils (*n*) are situated above the orbits, suggesting development of a proboscis. Skull length about 2 feet. (After Loomis and Gaudry.)

arisen in South America. Fragmentary remains of forms apparently related to *Pyrotherium* have been discovered in Eocene rocks of that continent. But beyond this fact we have no clue as to the origins of the pyrotheres, although obviously they come from some early ungulate stock, and it is not improbable that their ancestors were part of a general amblypod radiation.

CHAPTER 21

Subungulates

Often grouped as "subungulates" are several types of animals which at first sight appear to be unrelated. These include: (1) the hyraces or conies, small Old World hoofed mammals, rodent-like in appearance and habits; (2) *Arsinoitherium,* a huge horned fossil mammal from Egypt; (3) the proboscideans—the elephants and their extinct relatives; (4) the sirenians, or sea cows; (5) the peculiar amphibious fossil *Desmostylus* and its kin from the North Pacific region. It appears incongruous to place together a seeming jumble of land and sea forms, giant and small types, but early representatives of most of these varied groups show fundamental similarities which strongly suggest a common origin; and the fact that most of the earliest fossil subungulates are found in Africa suggests that that continent was their common ancestral home.

Because of the great diversity in adaptations which these forms have undergone, it is difficult to find many discernible features which hold true of all subungulates. There is usually a greatly enlarged pair of incisors or canines in either jaw, while other front teeth are often reduced; the grinding teeth tend to develop cross-lophs, the premolars become molarized. There is never a clavicle. Usually the land types retain most or all of the digits, with a mesaxonic symmetry, while the primitive claws have developed into structures more like nails than hoofs. Except for the hyracoids, the evidence suggests a swamp-dwelling existence for the ancestral subungulates.

Conies.—Most generalized subungulates in many respects are the Hyracoidea, represented today in Africa and Syria by *Hyrax* [*Procavia*] and related small forms, the conies of the Scriptures. They resemble the rabbits not only in size, general appearance, and habits, but also in many adaptive skeletal features. All are herbivores; some are dwellers in rocky country, while others are somewhat arboreal. The tail is short, the legs but moderately long. The gait is plantigrade, with the toes (four in front and three behind) bound together with a pad beneath. There is a centrale in the carpus, unusual in living ungulates.

The skull is of a rather normal construction. Alone among subungulates, the living hyraces have a complete postorbital bar. The lower jaw is very deep posteriorly. In living forms there is a considerable amount of reduction in the permanent teeth. In most there is a single, large, rootless upper incisor which meets two enlarged lower ones in rather rodent-like fashion. Behind this the lateral incisors and canines have been lost, leaving a diastema in front of the cheek teeth. The molars tend to be high crowned and rather resemble those of the earlier rhinoceroses in pattern, the upper ones having an ectoloph and two cross-lophs, and the lower teeth a double V. The posterior premolars are molarized; the anterior ones, simpler.

Although none too common today, the hyracoids appear to have played an important role in Africa in earlier times, for the group is represented by a considerable variety of forms, such as *Saghatherium* (Fig. 359*A, E*) and *Megalohyrax* (Fig. 359) in the Lower Oligocene beds of Egypt. These animals were already definitely hyracoid in structure and had even diverged considerably along various adaptive lines. They ranged in size from modest proportions to those of a lion; the teeth varied from brachyodont to high-crowned types, from bunodont to selenodont in pattern. However, they were all more primitive than the living forms, for the postorbital bar was incomplete, the brain small, and the dentition complete (although the teeth later lost were already reduced in size). The later Tertiary history of Africa is poorly known, but conies have been reported from various later deposits there. Except for one form, *Pliohyrax,* which invaded Europe in mid-Tertiary times, the conies in fossil form are known only from Africa, and they appear to have been characteristically African in origin and development.

Arsinoitherium.—A separate order, the Embrithopoda, is necessary for the reception of *Arsinoitherium* (Figs. 359*B, F;* 360), a peculiar form from the Lower Oligocene of Egypt. This great beast, perhaps a marsh dweller, was of rhinoceros size. The limbs were graviportal in structure with long and massive humerus and femur, short lower segments, and a broad, spreading, five-toed foot. The most striking feature of the animal was the presence of a huge pair of horns on the nasal bones, together with small ones on the frontals. The two great horns were fused at their bases, and, much as in some rhinoceroses, ossification of a partition descending between the nostrils aided in their support. The tooth row was complete, and there was not the enlargement of incisors seen in other subungulates. The molars were hypsodont (an unusual feature at that early Tertiary date), the upper ones having heavy protoloph and metaloph and a less developed ectoloph, while the lower ones had cross crests showing distinct traces of derivation from the double-V pattern.

This curious creature is quite isolated; we know nothing of its ancestors or of any possible descendants. Many points, especially the molar pattern, suggest a common origin with the hyracoids; but the relationship must be a distant one, for the development of the peculiar features of this form must have taken considerable time.

Proboscideans.—One of the most spectacular stories in mammalian evolution is that of the order of Proboscidea—the mastodons, elephants, and related types. Like the two preceding groups, they were of African origin but, in contrast with other subungulates, successfully invaded the other continents and by middle and later Cenozoic times were widespread in Eurasia and North America and even reached South America. Today, however, only two elephants survive. We may perhaps best treat this interesting group by describing the highly specialized structures found in the later elephants before taking up the earlier stages in their development.

The elephants and the related Pleistocene mammoths include the largest of late Cenozoic and living land mammals and, as such, have typical graviportal adaptations: an expanded ilium, columnar legs, a long humerus and femur, short lower limb segments retaining a well-developed ulna and fibula, and broad five-toed feet (cf. Fig. 364) with nail-like structures on the upper side of the digits and a pad beneath. The skull (Fig. 361*D*) is of huge size and roughly rounded shape, with a swollen top which is highly pneumatic. The brain, although reaching 11 pounds in weight in the Indian elephant, is, of course, small compared with the size of the skull. The comparatively small eyes have no bar behind them. The bony nasal orifice lies high up on the front of the skull between the eyes; from it projects the long, flexible trunk. Below this opening the premaxillae descend vertically, bearing the roots of the huge tusks. These are enormously enlarged second incisors. On emergence from the skull the tusks curve forward, upward, and outward. On the short palate the posterior development of the maxillae (in which the teeth develop) is notable. The lower jaw is correspondingly short; in front is a projecting chin superficially like that of man.

The dentition is remarkable. There are no anterior teeth other than the large upper tusks. In the cheek region the elephants have met the requirement of a large grinding surface in unique fashion. Six teeth develop in each jaw half; these consist of three milk premolars and the three molars (the permanent premolars never make their appearance). Each tooth (Fig. 362*D*) is elongate and exceedingly hypsodont and is formed of a large number of high, thin, crosswise ridges. The spaces between these "leaves" are filled by cement, so that with wear all three elements of the tooth are exposed in a regular pattern—dentine in the center of the ridges, an enamel band about this, and cement forming the outer portions. The number of

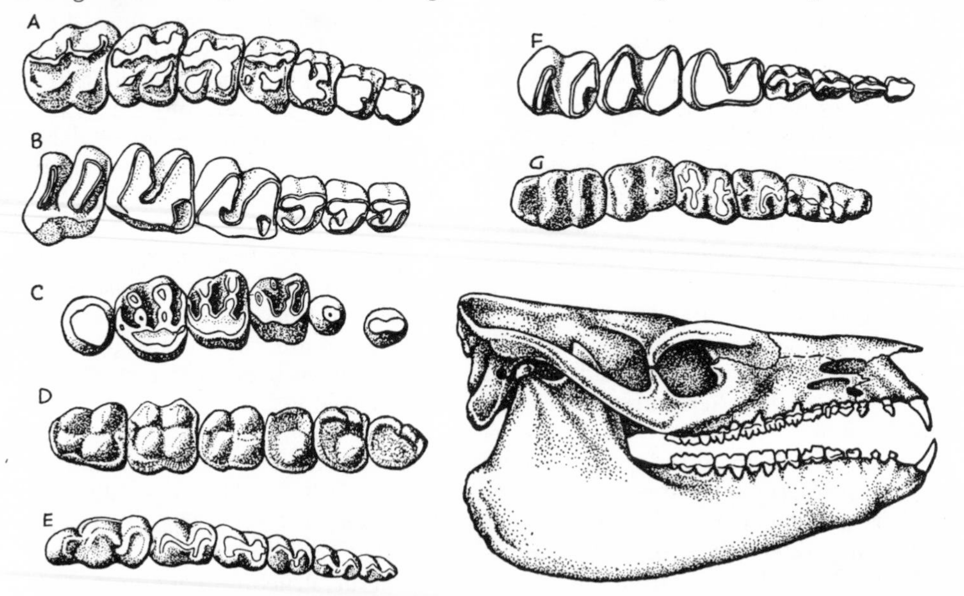

FIG. 359.—Subungulates. *A–D*, Right upper cheek teeth; *E–G*, left lower cheek teeth. *A, E*, *Saghatherium*, a hyracoid, ×5/8; *B, F*, *Arsinoitherium*, ×1/5 approx.; *C*, *Miosiren*, a Pliocene sirenian; *D, G*, *Moeritherium*, a primitive proboscidean, ×1/3 approx. *Lower right*, Skull of *Megalohyrax*, a Lower Oligocene hyracoid from Egypt, length about 12 inches. (*A, B, D–G* after Andrews; *C* after Abel; skull after Gregory and Schlosser.)

ridges increases considerably from front to back teeth, the milk premolars being much simpler in structure.

Instead of having all the teeth in place at once, as is usually the case in mammals, the elephants normally have exposed at any given time only four teeth in all, one in each half of each jaw. As these four teeth are worn down, they are pushed forward and the next group of teeth, which meantime has been forming in the maxilla or dentary, takes their place (Fig. 361*E*). This process is repeated until all members of a series are utilized.

Parenthetically, it may be noted that, while teeth are the most common remains of fossil elephants, their interpretation presents difficulties. The appearance of the ridges varies with the degree of wear. The number of plates not only differs from form to form but even more widely from tooth to tooth of the same individual; the premolars are relatively simple, the last molar the most complex.

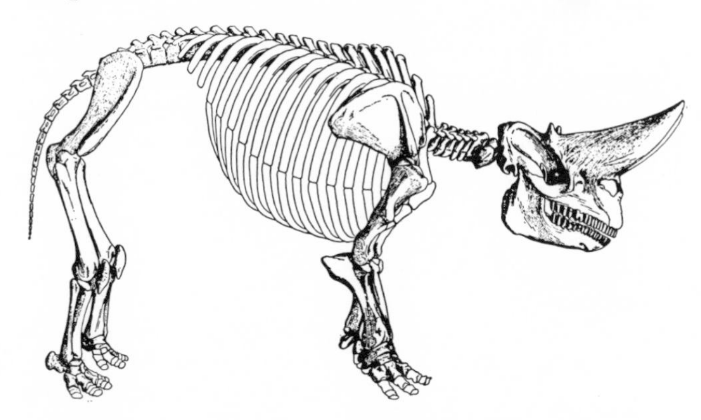

FIG. 360.—*Arsinoitherium*, a large horned subungulate from the Lower Oligocene of Egypt, about 11 feet in length. (From Andrews.)

Moeritherium.—Very different from the living elephants was *Moeritherium* (Figs. 359*D, G;* 361*A*) of the Upper Eocene and early Oligocene of Africa, which had already reached the size of a tapir. Quite in contrast to typical proboscideans, the trunk was long, the legs short. The skull was still fairly long, with the eyes far forward. The nasal opening was somewhat to the top; but the developing trunk was probably little more than a flexible, tapir-like snout. Not improbably, *Moeritherium* was a somewhat amphibious marsh dweller.

The dentition already shows the beginnings of the specializations found in later proboscideans. The formation of tusks had begun, for the second incisors were much enlarged, the upper one pointing down, the lower one projecting forward to meet it. A diastema was in process of development, and in connection with this the first premolars, the lower canine, and the lateral lower incisor had been lost. The molars were low crowned with two cross-lophs. The skull shows many interesting technical points of resemblance to that of the conies and sirenians.

There is considerable evidence that a number of types branched off early from the primitive probos-

cidean stock. For example, *Barytherium*, a contemporary of the last form, known only from a jaw and a few other fragments, possessed similar bilophodont molars and an enlarged lower incisor. But the jaw was peculiar in other respects, and it is possible that this genus was but distantly related to the proboscideans.

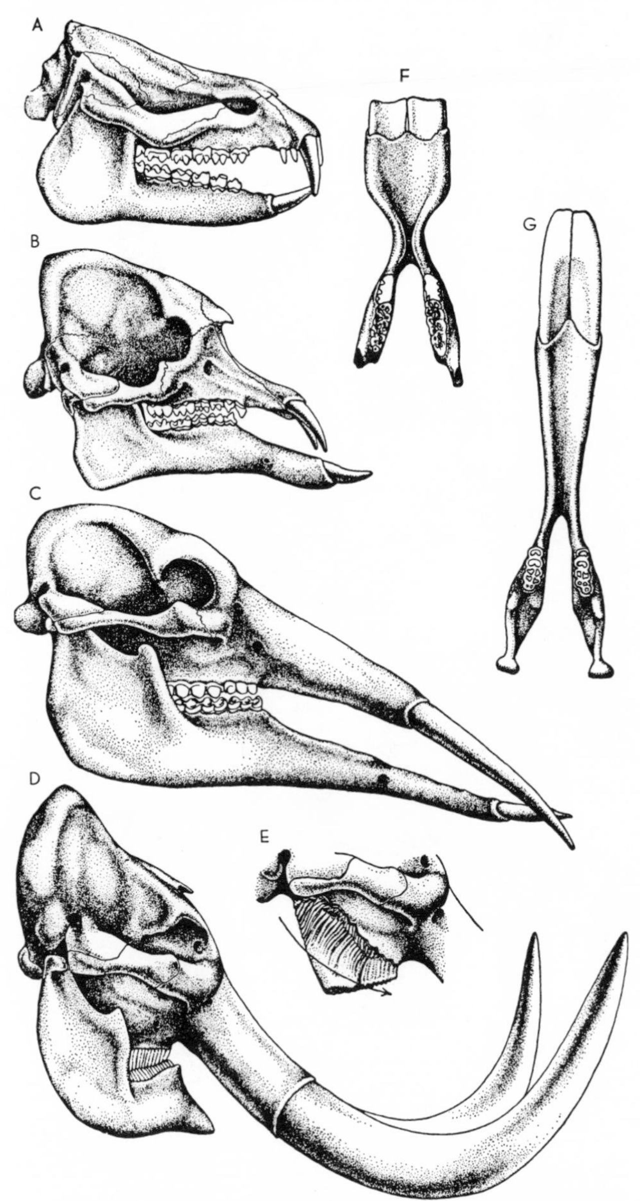

FIG. 361.—The skull and jaws of proboscideans. *A, Moeritherium*, from the late Eocene and early Oligocene of Egypt, ×1/8; *B, Phiomia wintoni*, a primitive bunomastodont from the Lower Oligocene of Egypt, ×1/16; *C, Gomphotherium* [*Trilophodon*] *augustidens*, from the Miocene of Europe, ×1/12; *D, Mammuthus primigenius*, the woolly mammoth, ×1/34; *E*, diagram of cheek region of the last, with the bone over the molars removed to show direction of tooth replacement; *F*, dorsal view of lower jaw of the late Tertiary mastodon *Platybelodon*, ×1/30; *G*, the lower jaws of *Amebelodon*, from the Pliocene of North America, ×1/33. (*A, B* after Andrews; *C, D, E* composite; *F* after Osborn; *G* after Barbour.)

Deinotherium.—Certainly a proboscidean, but obviously far off the main evolutionary path of the group as a whole, was *Deinotherium* (Fig. 363), a large form found fairly commonly in the Miocene and Pliocene of Eurasia. The general build was elephantine; and, although the earlier species were of modest size, later

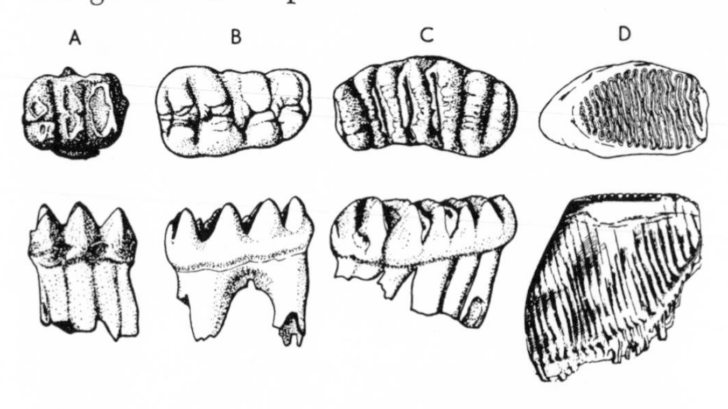

FIG. 362.—Crown views (*above*) and lateral views (*below*) of molar teeth of *A*, American *Mastodon; B, Tetralophodon longirostris; C, Stegodon; D, Mammuthus primigenius*. (*A*, about 1/10 natural size; *B* and *C*, about 1/6; *D*, about 1/12.) (*A* after Hay; *B* after Vacek; *C* after Matsumoto; *D* after Osborn.)

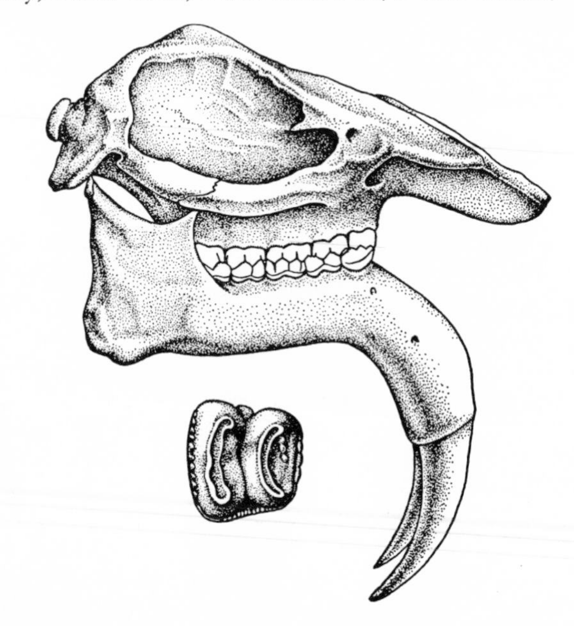

FIG. 363.—Skull of *Deinotherium*, a Miocene proboscidean, length about 47 inches. Molar tooth, ×1/5 approx. (After Gaudry and Andrews.)

types exceeded most of the true elephants in bulk. The cheek teeth (with but two premolars) were quite primitive, low crowned, and with but two cross ridges in the back two molars (the first had three). In persistently primitive fashion, too, all the cheek teeth were in place throughout life, and the permanent premolars succeeded their "milk" predecessors in a normal way. The nostrils were high up on the face, suggesting a long proboscis. While the premaxillae extended well forward, there was, in contrast with all other proboscideans, no trace of upper tusks. The lower tusks were, however, well developed and curved sharply downward and even backward. What purpose this curious structure could have served is a matter for speculation. *Deinotherium* disappeared from Eurasia during the Pliocene but, like many other primitive forms, survived much later in the tropics and was present in the Pleistocene in Africa.

Mastodons.—Leaving these divergent forms, we may now return to the consideration of the evolutionary main line represented by the mastodons. These proboscideans, although now extinct, were very numerous and diversified throughout the greater part of the Tertiary. The mastodons may generally be distinguished from elephants by the fact that the teeth were low crowned, with few ridges, and many or all of the cheek teeth were usually in place simultaneously. Almost always there was a lower, as well as an upper, pair of tusks.

In the Fayûm district of Egypt, where most of the remains of *Moeritherium* have been obtained, there appear, in the Lower Oligocene, the earliest and most primitive of known mastodons, *Palaeomastodon* (Fig. 364*A*) and *Phiomia* (Figs. 361*B*, 364*B*). These forms

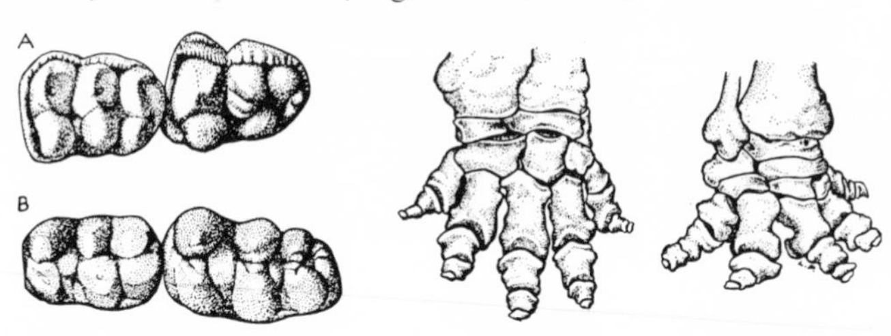

FIG. 364.—*A*, Right M^2–M^3 of *Palaeomastodon beadnelli; B*, left M_2–M_3 of *Phiomia wintoni;* ×1/4 approx. (After Andrews.) *Right*, Front and hind feet of the American *Mastodon*. (After Warren.)

were of larger size than *Moeritherium;* the largest had already reached the proportions of a modern elephant, although the average was much smaller. Structurally, too, they had progressed considerably beyond *Moeritherium*. The braincase was much shorter and higher; the nostrils were placed far up and back on the skull; and the snout must have been long. Of the anterior teeth there remained but a single pair of incisors. These were already well-developed tusks. The upper tusk was borne on the elongated premaxilla and curved downward and outward rather than directly forward. The short lower tusks formed a somewhat scoop-shaped affair at the end of the elongated jaw. Premolars (three above, but only two below) and molars were lophodont, low crowned, and all in place simultaneously.

In *Phiomia* the molars had advanced over those of *Moeritherium* in that three cross crests were usually present instead of two; and accessory cusps (in addition to the two which normally formed a crest) had begun to make their appearance, tending to make a somewhat bunodont and piglike tooth.

From *Phiomia* appear to have descended most of the later mastodons, characterized by teeth in which a multiplicity of accessory cusps tended to make their appearance. These forms with suid teeth, often called bunomastodonts, but technically known as the Gomphotheriidae, successfully invaded the northern continents. In these areas they were present in great and bewildering variety during the late Tertiary; only a few persisted into the Pleistocene.

Almost nothing is known of proboscideans in the middle and late Oligocene. *Gomphotherium* [*Trilophodon, Tetrabelodon*] was the characteristic Miocene genus (Fig. 361*C*), and was present, it would seem, throughout Eurasia and Africa in early Miocene times, and reached North America toward the end of that epoch. The lower jaw was enormously long, bearing short, broad tusks at the end, while the premaxilla was also elongated for the fairly long and slightly decurved upper tusks. The nostrils were high on the skull, but the free portion of the trunk (which presumably began at the end of the premaxilla) must have been comparatively short. Six cheek teeth were present above and five below, all in the jaw at one time, with permanent premolars still replacing the milk premolars. The pattern was becoming more piglike, with an increasing number of irregular cusps; and the number of ridges was increased slightly over *Phiomia*, for, although three were typical, four or five cross crests were present in the last molar.

Members of the bunomastodont group persisted in the Old World until fairly late in the Pliocene and are even found in the early Pleistocene in Africa. During this time, however, remarkable changes took place which were in many ways parallel to those which must have occurred in the ancestry of the true elephants. *Tetralophodon* was the characteristic Pliocene type in both the Old World and North America. The lower jaw had become greatly shortened, leaving only a very small lower tusk in the end of a chinlike projection of the dentary; the premaxilla, too, had shortened, while from it projected long, straight, upper tusks. With this shortening of the face the fleshy nose above was presumably freed to become a long proboscis such as that found in the modern elephant. The teeth had become somewhat high crowned and had gained some cement. Further elephant-like changes are seen in the loss of replacing teeth for the milk premolars and in the tendency for only a part of the dentition to come into play at once. There was, however, only a gain of a molar crest or so (Fig. 362*B*), four typically and five in the last molars; and the piglike multiplicity of cusps in the cheek teeth shows that in the later members of this group we are dealing with a line that was not ancestral but parallel to that of the true elephants. *Stegomastodon*, a genus which persisted into the early Pleistocene in North America (as did a related form in Asia) and even invaded South America, was still further advanced in jaw abbreviation and increase in cusps to a maximum of seven or eight cross crests.

The bunomastodonts discussed above appear to form the "main line" of evolution of this mastodon family; there were numerous side branches. *Cuvieronius*, representing another set of late Pliocene and Pleistocene survivors in the two Americas, had progressed in the loss of lower tusks and had developed peculiar, spirally twisted upper ones but had retained relatively primitive molars. *Rhynchotherium*, found from Africa to North America in the late Tertiary, retained a long-tusked lower jaw which was curved downward somewhat in the fashion of *Deinotherium*. In *Platybelodon* (Fig. 361*F*) of Asia and North America and *Amebelodon* (Fig. 361*G*) of the latter continent, the long, decurved lower jaws bore broad, flat, lower tusks

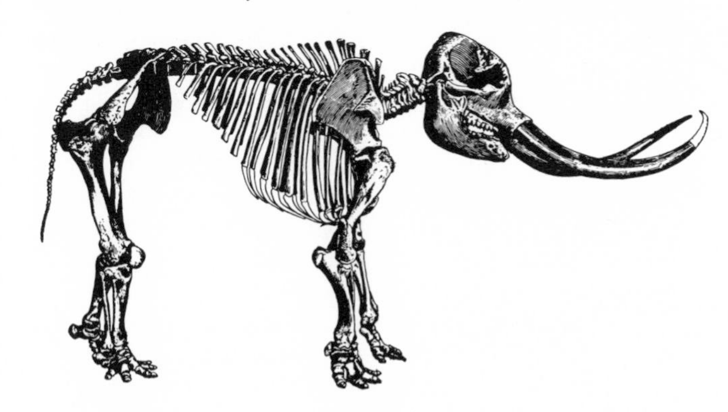

Fig. 365.—The American Pleistocene *Mastodon;* length about 10 feet. (From Hay.)

which formed an enormous shovel-like structure. In *Gnathabelodon* of the American Pliocene the lower tusks were absent, but an equally effective shovel was formed by the expanded rostrum of the lower jaw.

While *Phiomia* had three-lobed teeth which tended toward a complicated pattern, the contemporary *Palaeomastodon* retained rather simple, tapir-like, two-lobed molars. This genus may have given rise to a second, less prominent, but long-lived group of mastodons (family Mastodontidae) in which the teeth had a persistently simple pattern, always with low crowns and simple ridges and without cement. Such forms are found in the Miocene of Eurasia and Africa although less often than the gomphotheres. This group reached America before the end of the Miocene and had as an end form *Mastodon* proper (Figs. 362*A*, 365), which persisted in North America throughout the Pleistocene and (as numerous skeletons from postglacial swamps show) probably lived until not many thousands of years ago. The evolutionary history of this series is poorly known, but skull and tusk structure obviously must have changed in a fashion similar to that of the bunomastodonts, for the American mastodon never had more than a vestige of the lower incisor in the short jaw, there were huge upper tusks curving upward and outward, there were never more than two teeth at a time in each jaw half, and the skull, as in advanced bunomastodonts and elephants, was short and high.

Elephants.—The family Elephantidae, in which the two Recent proboscideans are included, is easily distinguished from a majority of the mastodons by the much higher, shorter head; the huge, curved upper tusks; the short, tuskless lower jaw; and especially by the cheek teeth with their high, cement-covered rows of numerous lamellae and peculiar mode of succession. The elephants are not descendants of any of the well-known late Tertiary mastodon types but appear to have had an independent history of some length. *Stegolophodon,* which appeared in Asia and Africa in the Miocene (and survived in the Indies to the Pleistocene), appears to be structurally transitional between bunomastodonts and elephants; in the cheek teeth the numerous cusps are aligned in transverse rows. *Stegodon* of the Pliocene and early Pleistocene of Asia is a primitive elephant. The skull was somewhat longer than in more advanced forms; but the teeth (Fig. 362*C*), while still low crowned, had sometimes as many as fourteen simple ridges with some traces of cement. While there were still traces of the permanent premolars, which have been lost in later elephants, the type of tooth succession was already advanced, for no more than two teeth were present at one time in a jaw half. Further, the lower-jaw symphysis had been reduced to a chin in which there were only tiny vestiges of the lower tusks, while the upper tusks were long and somewhat spirally curved.

In the early Pleistocene the first of the typical elephants, with short head, tuskless lower jaws, and high-crowned, cement-covered teeth, appeared in southern Asia and Europe; in later phases of the Pleistocene various elephant types usually referred to as "mammoths" were abundant in all the northern continents. Usage of generic terms for these forms is in a confused state. At one time all were included in the genus *Elephas,* using that term in a broad sense; recently there has been a tendency to split the elephants into a large number of genera. We shall here group the characteristic Pleistocene and Recent forms in three genera: *Loxodonta, Elephas* proper, and *Mammuthus.*

Typical of *Loxodonta* is the living African elephant; the molars are rather narrow and relatively low crowned, with a comparatively small number of ridges which, with wear, tend to show a rhomboidal pattern. The head is not as short or deep as in other late elephant types; the tusks are long but are little curved. *Loxodonta antiqua,* the straight-tusked "ancient elephant," was an inhabitant of southern Europe and northern Africa in the Pleistocene. Some specimens reached a height of 14 feet at the shoulder; on the other hand, dwarf races of this form, some no larger than a pig, have been found as fossils on Mediterranean islands. Other loxodonts occur in the Pleistocene of Asia and the East Indies.

Elephas, as used here, is typified by the living Indian elephant. In contrast with the loxodonts, the skull is very short and deep, with a rounded vault; the tusks are short and straight and, again contrasting with loxodonts, are directed more downward than forward from their sockets. The molars are broad and high and have numerous closely appressed ridge plates. Relatives of the living species were present in the Pleistocene of Asia.

Most interesting of fossil elephants is a great array of forms to which the term "mammoth" is usually applied and which are here gathered into the genus *Mammuthus.* In these mammoths the tusks are downwardly directed at their bases, as in *Elephas;* however, they are much elongated and curved and may even cross one another in old males. The skull tends to be even shorter and higher than in the Indian elephant and has a pointed roof. The teeth are constructed much as in *Elephas,* but the number of plates is variable. In some early and primitive Asiatic species even the last molars have but ten or a dozen plates; these teeth in some woolly mammoths have twenty-seven to thirty ridges. In this genus are included a number of prominent Pleistocene forms. *Mammuthus planifrons* of the early Pleistocene of Asia was a primitive type; *M. meridionalis,* the southern mammoth common in the Mediterranean region, was closely related; *M. imperator,* the imperial mammoth of southern North America, was more advanced in size and dental development. The form characteristic of temperate North America, usually termed the "Columbian mammoth," was not exceptionally large but had highly developed molars. *Mammuthus primigenius* (Figs. 361*D*, 362*D*), the woolly mammoth, was a form adapted to cold climates. It was common throughout the Pleistocene in the northern parts of both Eurasia and North America. It is known not only from skeletal remains but also through the many figures made by Paleolithic man on cave walls and through complete cadavers unearthed in the frozen tundras of Siberia.

Sirenians.—The Sirenia, manatees and dugongs, or sea cows, are not ungulates in any normal sense but are purely aquatic animals, found along the coasts and river mouths of various parts of the world. The skin of living sirenians is nearly naked and tough and leathery. The body has assumed the torpedo-like shape characteristic of many water vertebrates, with no distinct neck and with a laterally expanded tail. Here, as in other groups, the front legs have been transformed into flippers, while there is no surface indication of the vestigial hind limbs and pelvis. A characteristic feature of the skeleton of sirenians is that much of the bone has a solid, heavy structure—pachyostosis—also seen in certain other aquatic vertebrates (the mesosaurs, for example).

The brain (and consequently the cranial cavity) is small relative to the size of the skull in these large mammals. Only in the case of one living form is the postorbital bar developed. The premaxillae form a long, high, and narrow rostrum, usually curved downward, with the paired nostril openings placed above and well to the rear; the nasals are vestigial. The lower jaw is heavy and has a long symphysis in front. The modifications of the front limbs are not so great as in the whales, for there is still considerable freedom of movement at the elbow and wrist and in the fingers. There tends to be fusion of carpal elements, however. The thumb is reduced; but, while the other fingers may have but two joints, the number may increase to four.

Living forms possess a pelvis (Fig. 366) which is only a solid plate or rod without subdivisions and without bony connection with a sacral rib; nothing remains of the hind limb in modern genera.

Like their terrestrial subungulate relatives and in contrast with other types of aquatic mammals, the sea cows are purely herbivorous. The dentition of living forms is quite specialized and reduced anteriorly. The dugong has merely a pair of upper incisor tusks, while the manatee has no front teeth at all as an adult; in both types horny plates form a substitute cropping organ. The cheek teeth in primitive forms and in the living manatees have two cross crests, as in primitive proboscideans.

Primitive sirenians.—Remains of sea cows are not uncommon in the Eocene beds of Egypt. While this fact suggests an African origin, their migrations must have begun at an early date, for an early sirenian is also reported from Jamaica. Very primitive sirenians are *Protosiren* of Egypt and the Jamaican form, *Prorastomus; Eotheroides* ["*Eotherium*"], commonest of the Egyptian finds, is somewhat more advanced in the direction of the dugongs. The early genera show definite sirenian characteristics, such as the posterior retreat of the nostrils, the presence of two-ridged molars, and the beginning of pachyostosis. They were, however, more primitive in many ways than the modern forms. The snout is but little down-turned, and in *Prorastomus* the tooth row forms practically a straight line. Nasals and lacrimals, vestigial in later sirenians, were unreduced. The entire series of normal placental teeth is present; in specimens of *Prorastomus* and *Protosiren* one of the milk molars is retained in the adult, suggestive of a trend toward the increased number of teeth noted below for the manatees. In *Eotheroides* the pelvis was quite well developed (Fig. 366), while the hind legs appear to have been complete and still functioning, although of small size.

Dugongs.—The dugong, the "mermaid" of the Red Sea and Indian Ocean, is the sole survivor of a group very common as fossils throughout the Tertiary. There is a massive beak bent down above the heavy lower jaw. Of the front teeth, which are mainly replaced by a heavy, horny, rubbing pad, there remain only vestiges, except for upper incisor tusks in the male. The premolars are degenerate, and the permanent set never develops, while the molars are large and, with a wrinkled surface and many bunodont cusps, superficially resemble those of the pigs or the bunomastodonts.

A tendency in the dugong direction was already

apparent in some of the Eocene sirenians, such as *Eotheroides.* The beak was already slightly tilted; nearly the full complement of teeth was still present, but the first incisors were somewhat enlarged; the number of cusps in the molars was already tending to increase, and the intermediate teeth were much reduced in size. In later Tertiary deposits, especially in Europe, there have been found a considerable number of forms leading toward the living type. Some, such as *Halitherium* (Fig. 367) of the Oligocene and early Miocene, *Halianassa* of the late Miocene, and the Pliocene *Felsinotherium,* are known from nearly complete remains. In them a lower incisor tusk persisted for some time but

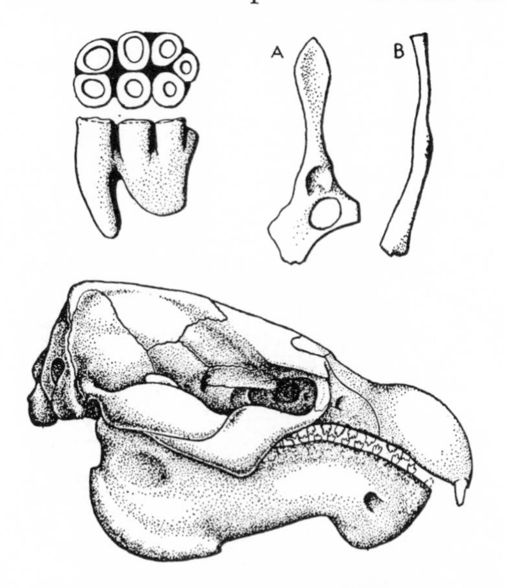

FIG. 366.—*Left,* Crown and side views of a molar tooth of the Miocene *Desmostylus,* ×1/3. (After VanderHoof.) *Right,* Pelvis of *A,* an early sirenian, *Eotheroides,* and *B,* the modern dugong. (After Abel.) *Lower,* Restored skull of the primitive Eocene sirenian *Protosiren;* length of original about 13 inches. (Mainly after Sickenberg.)

finally disappeared along with smaller remaining front teeth. The beak was bent downward in dugong fashion, the pelvis was reduced, and the hind legs dwindled and vanished.

Through most of the Tertiary various side lines of this family are known, of which perhaps the strangest was that which ended with Steller's sea cow, *Rytina,* found by the Russians in the Bering Sea and exterminated by them a century ago. Here both the tusks and the entire set of cheek teeth had disappeared, to be replaced by rubbing plates.

Manatees.—A second living group is that of the manatees, *Manatus,* inhabiting the Atlantic shores of Africa and America. A feature unique among mammals is that there are here but six cervical vertebrae. The beak, while long, is not markedly decurved. All the front teeth have disappeared. The bilobed cheek teeth have undergone a curious development. Their number in each jaw ramus has increased to twenty or more, of which five or six may function at one time. The method of replacement is quite peculiar; the teeth form at the back of the jaw, and as they function, push forward until they are worn down and disappear at the forward end of the tooth row. This type of tooth replacement is reminiscent of that of the later mastodons and elephants. Almost nothing is known of the history of the manatees.

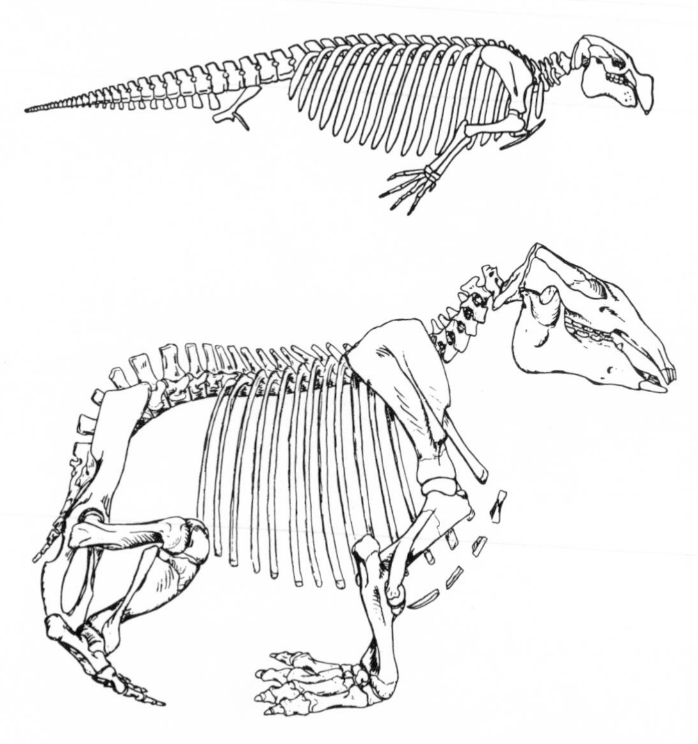

FIG. 367.—*Above, Halitherium,* an Oligocene sirenian, about 9 feet in length (after Strömer); *below, Paleoparadoxia,* a desmostylian, about 7.5 feet in length. (After Repenning.)

The earliest sirenians, as we have seen, showed very clearly their derivation from land (or at least amphibious) ancestors; and the fact that so many early remains are from Egypt suggests Africa as the continent of origin. The primitive complete dentition, with large incisors and molars with cross-lophs, is suggestive of that of the early mastodonts. Many details of skull structure are highly comparable with those of both proboscideans and conies. *Moeritherium* is universally included in the Proboscidea; but it is not a direct ancestor of later proboscideans, and may be as closely related to ancestral sirenias as to mastodons. The evolutionary tendencies shown in later members of the order are also similar to those of the proboscideans in such respects as the development of tusks, the tendency toward a polybunodont type of cheek tooth, and the tooth replacement of the manatee, basically similar to that of the modern elephant.

Desmostylus.—There have long been known from mid-Tertiary deposits on both shores of the North Pacific the teeth and (rarely) the skull of peculiar aquatic animals, *Desmostylus* (Figs. 366, 368), *Paleoparadoxia* (Fig. 367), and a few close relatives. The cheek teeth were unusual in structure, the large posterior ones consisting of a number of cusps developed as closely packed, heavily enameled cylinders. As in advanced proboscideans and manatees, the cheek teeth are replaced by means of a horizontal forward movement. The anterior teeth are present, in part, as tusks—an upper pair formed by the canines, two lower pairs formed by incisors and canines, in a fashion somewhat comparable to that of shovel-tusked mastodonts. The skull is low, and the snout is both long and broad. The external nares remain anteriorly placed; and, although the nasals remain well developed, they do not enter the border of the nostrils.

The dentition, particularly, suggested that *Desmostylus* was related to the subungulates, and the fact that the remains all came from marine deposits made it reasonable to believe that *Desmostylus* was a sirenian of sorts. Surprise, therefore, was great when skeletons of *Desmostylus* and *Paleoparadoxia* were discovered in recent years, and it was found that massively developed legs were present. The body had rather hippopotamus-like proportions, and we may reasonably assume that these animals spent their lives in and about shallow coastal waters. We have no clues, earlier than the early Miocene, as to the ancestry of *Desmostylus* and its relatives, but the evidence strongly suggests a common descent, with sirenians and proboscideans, from ancient amphibious—and probably African—ancestral subungulates.

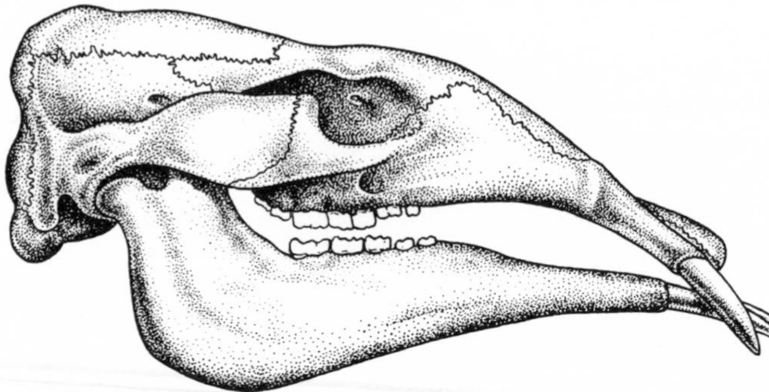

FIG. 368.—The skull of *Desmostylus,* restored; length (including tusks) about 32 inches. (After Gregory.)

All in all, our study of the various groups discussed in this chapter tends to confirm the seemingly odd assumption of the relationship of the diversified subungulate forms. But, although the early Egyptian fossils seem to be approaching the common type from which the subungulates have been derived, conies, proboscideans, and sirenians were already distinct groups at the time when they first appear in the fossil record. It is only through some future discovery of still earlier—Lower Eocene or Paleocene—fossil-bearing deposits in Africa that we may hope to find the common ancestor of the later subungulate orders.

CHAPTER 22

South American Ungulates

As mentioned previously, South America appears to have been completely separated from other continents during nearly the entire course of the Cenozoic, and few types of placental mammals appear to have been able, in early Tertiary times, to cross the water gap which presumably separated South America from North America. Of ungulates, we have already mentioned condylarths, xenungulates, and pyrotheres, none of which, however, played roles of importance in the history of this southern region. Until the close of Tertiary times no representative of either of the two great ungulate orders, perissodactyls and artiodactyls, which dominated in the Old World and in North America, ever crossed the water barrier. There developed, instead, native ungulate groups, notably the Notoungulata and Litopterna, both of which were present in the oldest South American fossil beds, of late Paleocene age.

In the isolation of South America, free from the depredations of the true carnivores and free from the competition of other hoofed mammals, these ungulates developed into an amazing variety of forms. In proportions they ranged from the size of a rat to that of an elephant. Some simulated the rodents in appearance and habits; others were analogous to the hippopotamus and the rhinoceros; still others were remarkably similar to horses in structure. These types were extremely interesting and often bizarre. But our consideration of them is hampered by the fact that, in the absence of existing representatives, we lack both popular names and that mental picture of the living form which in most cases furnishes us with a good starting point for the discussion of mammalian groups.

These ungulate groups reached the climax of their development in the Oligocene and Miocene; with the Pliocene and Pleistocene came the advent of placental carnivores, to which they probably fell an easy prey, and of higher ungulate types, with which they could not successfully compete for pasturage. By the end of the Pleistocene this once extremely numerous assemblage had vanished entirely.

It has proved difficult to establish the age of the beds in which these South American forms are found, since they differ in their mammalian fauna so greatly from those in other parts of the world. By some early workers these deposits were supposed to be considerably earlier in date than is now believed to be the case. In consequence of that belief, it seemed that forms which simulated the horses, elephants, and various other groups appeared in that continent considerably earlier than did these types themselves in other regions. Mistaking these parallelisms for real relationships, a patriotic Argentinian, to whom we owe much of our knowledge of these forms, laid claim to the origin of many mammalian groups for South America, a viewpoint which is frequently reflected in the names given to the various genera and groups.

Notoungulates.—By far the greater number of forms from South America are included in the order Notoungulata. The members of this group varied much in size and adaptive features, but many general structural similarities run through the various subdivisions of the order. The skull was typically rather short, flattened above, with a broad braincase and forehead, the nasals broad, the nostrils usually terminal. The zygomatic arches were broad and heavy, and there was never a postorbital bar. In all notoungulates there are peculiar features of the auditory region not found in any other order. Relatively few notoungulates show any reduction in the dentition or the development of a diastema; and while there were often strong chisel-like incisors, the rest of the teeth formed a series in which there was often a gradual transition from the small canine to the molars. Primitively brachyodont, the cheek teeth (Fig. 371) in many lines tended to become high prisms, often with persistently growing roots. Even in the earliest forms there was seldom any trace of the bunodont stage through which the ancestral forms presumably passed, and lophodont types of teeth were rapidly developed. The upper molars were lophodont, primitively triangular but coming to resemble superficially those of a rhinoceros, with a straight ectoloph, a long protoloph running back from this at an angle, and a metaloph perpendicular to the end of the ectoloph. A characteristic and even diagnostic notoungulate tendency is that toward the development of accessory cusps in the central valley of the upper molars. These may form secondary ridges or cause the tooth, on wear, to have a pattern of nearly continuous dentine with but a few notches or hollows lined with enamel. In the lower molars there was considerable resemblance to the perissodactyls in the development of two typically crescentic ridges; the front one was usually much the smaller. The most characteristic feature of the notoungulate dentition is the fact that the entoconid primitively lay isolated inside the curve of the posterior crescent (Fig. 372).

Although there is great variation, the feet were essentially mesaxonic in structure in a majority of cases, the number of toes tending to reduce from five to three. But there was never any great acquisition of true ungulate characters in the feet and never an attainment of the unguligrade position. A number of types appear to have had hoofs, but claws were present in many cases; the feet in general rather resembled those of rodents or the hyrax.

Ancestral notoungulates: the Notioprogonia.—With two exceptions, notoungulates are unknown beyond the boundaries of South America. These exceptions, however, are among the oldest records of the group. A single tiny jaw records the presence of a notoungulate, *Arctostylops* (Fig. 372), in the Lower Eocene of North America. Still more curious is the fact that in the late Paleocene of Asia the commonest animal so far discovered is a notoungulate, *Palaeostylops.* These discoveries suggest an Old World origin for the notoungulates and a long migration trail to South America. They do not prove this point, however; they merely indicate that notoungulates were widespread at an early stage of the Tertiary. It is further possible, although equally unproved, that the notoungulates developed in South America and that these records represent an early and unsuccessful invasion of other areas from a southern base.

Small archaic notoungulates, such as *Notostylops*

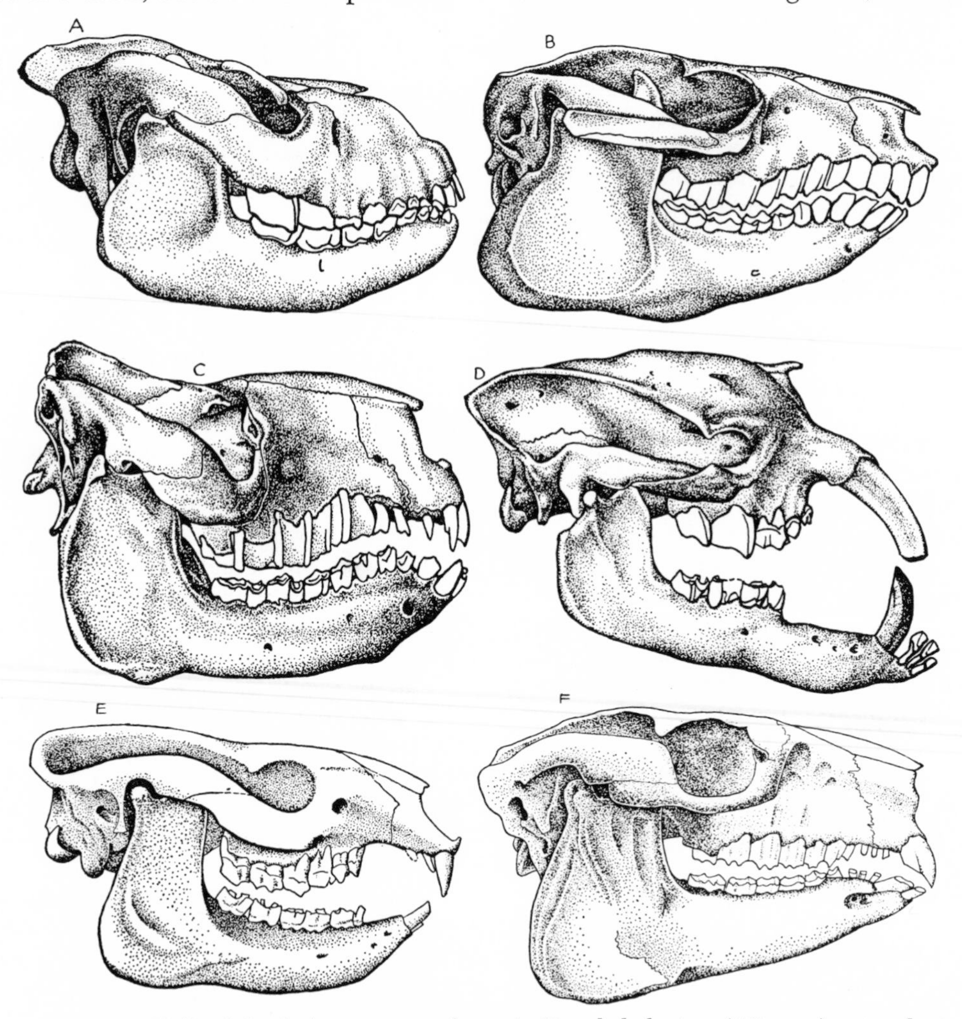

FIG. 369.—Skulls of South American ungulates. *A, Homalodotherium* (Miocene), a toxodont, skull length about 15 inches; *B, Protypotherium* (Miocene), a typothere, skull length about 4 inches; *C, Nesodon* (Miocene), a toxodont, skull length about 16 inches; *D, Astrapotherium* (Miocene), skull length about 27 inches; *E, Notostylops,* a primitive Eocene notoungulate (Notioprogonia), skull length about 6 inches; *F, Hegetotherium* (Miocene), skull length about 4.5 inches. (*A, C, D* after Scott; *B, F* after Sinclair; *E* after Simpson.)

(Fig. 369*E*), related to these forms, are the commonest of Eocene types. Apart from reduction of the anterior teeth in some instances, they show no marked specializations of any sort. The molars show a simple pattern. Except for a single diagonal accessory ridge in some forms (cf. Fig. 371*C*), the upper molars are still essentially triangular lophodont types, with protoloph and metaloph connected medially. These early forms are regarded as members of a basal suborder, the Notioprogonia, ancestral to the remaining notoungulate groups.

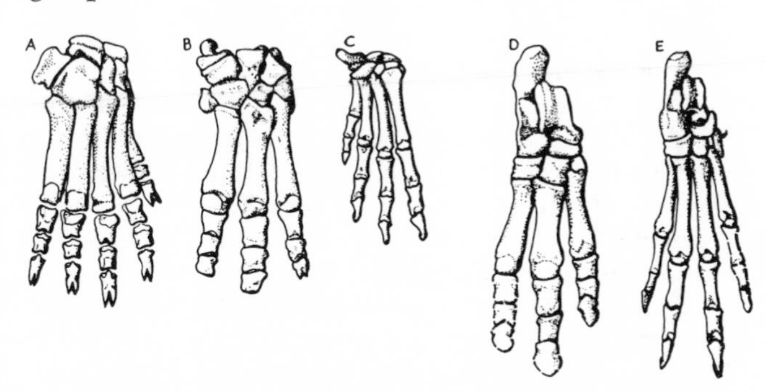

Fig. 370.—Feet of notoungulates. *A*, Manus of *Homalodotherium*, a clawed Miocene toxodont; *B*, manus of *Nesodon*, a Miocene toxodont; *C*, manus of *Protypotherium*, a Miocene typothere; *D*, pes of *Nesodon; E*, pes of *Protypotherium*. (*A, B, D* after Scott; *C, E* after Sinclair.)

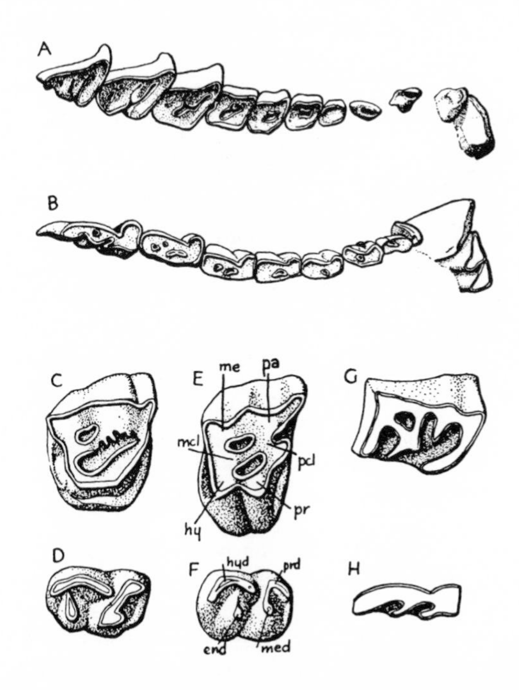

Fig. 371.—Teeth of notoungulates. *A*, Upper right, and *B*, lower left, tooth rows of the Miocene toxodont *Nesodon* (×1/4); *C*, upper right, and *D*, lower left, molars of *Pleurostylodon*, a primitive Eocene toxodont (×1); *E, F*, same of *Oldfieldthomasia*, a primitive Eocene toxodont (×3); *G, H*, same of *Proadinotherium*, an Oligocene toxodont (×1/3). Abbreviations, upper molar: *hy*, hypocone; *mcl*, metaconule; *me*, metacone; *pa*, paracone; *pcl*, protoconule; *pr*, protocone. Lower molar: *end*, entoconid; *hyd*, hypoconid; *med*, metaconid; *prd*, protoconid. (*A, B* after Scott; *C–F* after Schlosser; *G, H* after Loomis.)

The skeleton is imperfectly known in these primitive forms; however, *Thomashuxleya* (Fig. 373), while technically ranged in one of the more advanced groups of notoungulates, is an Eocene genus which shows a general structure apparently primitive for the order and retained in many of the less specialized later types. The build is that of a rather slow-moving, and essentially digitigrade type, analogous to that of the amblypods and Dinocerata of northern continents.

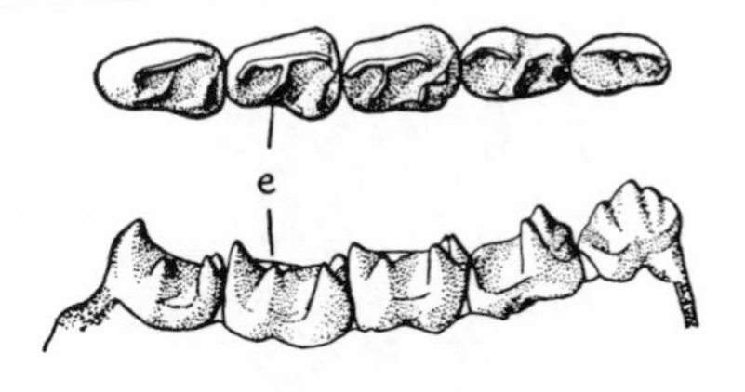

Fig. 372.—Fragment of lower jaw of *Arctostylops*, only known notoungulate from North America, left P_3–M_3, showing the characteristic development of the entoconid (*e*), crown and lateral views. (After Matthew.)

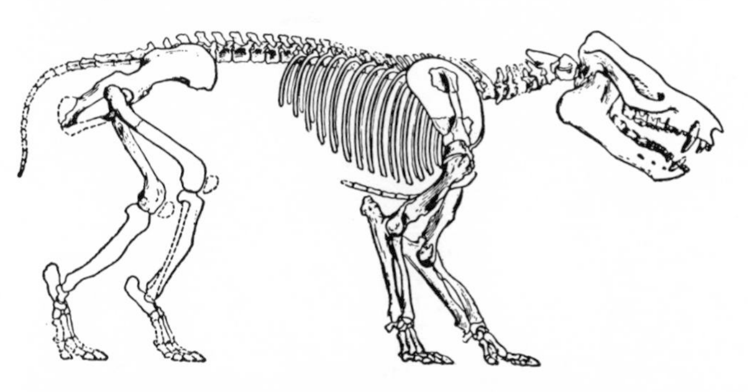

Fig. 373.—*Thomashuxleya*, a primitive Eocene notoungulate, about 5 feet long. (From Simpson.)

Toxodonts.—The largest and most conspicuous of later notoungulates are the toxodonts, which, with their relatives, form the suborder Toxodontia. Typical toxodonts were present in the Oligocene, flourished in the Miocene, and survived into Pleistocene times; the later genera were large, heavily built forms. Among the ungulate features of the skeleton may be mentioned the presence of but three hoofed toes on each foot and the absence of the clavicle (retained in some other notoungulates, however). But the ulna and fibula were always complete and heavy, although the latter sometimes fused with the tibia.

The teeth remained in a nearly closed row, seldom developing a diastema; and there was little loss except that the canines were weak or absent. The incisors were often expanded into chisel-like cropping structures, while the primitively low-crowned molars tended to become high crowned and rootless and sometimes were covered by a cement sheath. The upper molars curved in strongly toward each other, a feature to which the term "toxodont" (bow-tooth) refers; in surface view they are characteristically triangular, ectoloph and protoloph meeting at an acute angle (Fig. 371*A*).

Toxodon (Fig. 374) of the Pliocene and Pleistocene is the largest and one of the latest members of the group and is built like a short-legged rhinoceros. The dorsal position of the nasal opening suggests the presence of a large snout; the general appearance may have been that of a gigantic guinea pig. A diastema, absent in earlier types, had developed in the dentition.

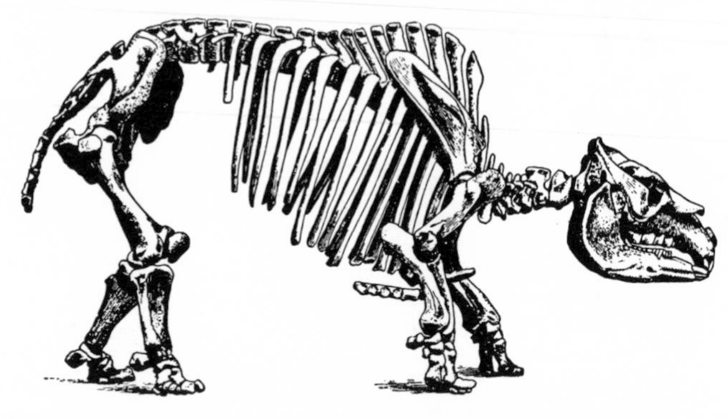

Fig. 374.—*Toxodon*, a large Pleistocene South American notoungulate, about 9 feet in length. (From Lydekker.)

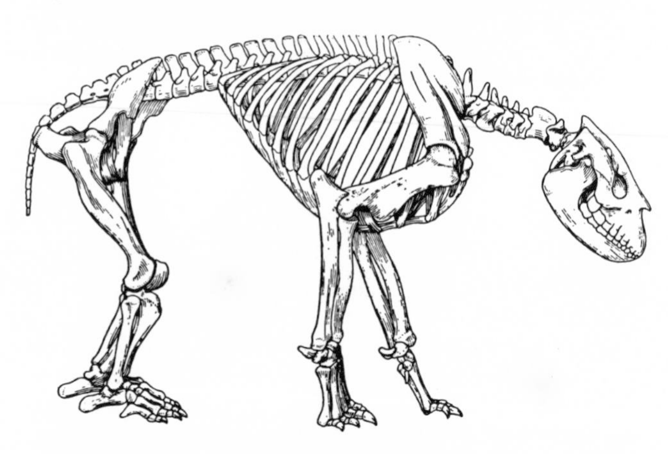

Fig. 375.—*Homalodotherium*, a clawed Miocene notoungulate, about 6 feet in length. (From Scott.)

Toxodon appears to have been the commonest large ungulate of the South American Pleistocene. The Miocene *Nesodon* (Fig. 369*C*; 370*B*, *D*; 371*A*) was a smaller and more slimly built toxodont, the size of a tapir. The dentition was comparatively primitive, for the tooth row was a closed one, although the intermediate teeth (canines and anterior premolars) were small, and the molars became rooted late in life. *Adinotherium* was another Miocene form of still smaller size, with a small horn on the frontal bones. Ancestral toxodonts, such as *Proadinotherium* (Fig. 371*G*, *H*), are present in the Oligocene and others, such as *Pleurostylodon* (Fig. 371*C*, *D*), small and primitive in character, in the Eocene and even the Paleocene (Fig. 371*E*, *F*).

A considerable amount of variation is present within the toxodont suborder, of which about three score genera, arranged in some eight families, are known. We may cite, as example, *Leontinia* of the Oligocene, in which the facial region was very short, and a high position of the nostril opening suggests the presence of an expanded muzzle or proboscis of some sort; *Notohippus* and its Eocene and Oligocene relatives, in which the complex teeth had some superficial resemblance to those of horses; the Miocene *Homalodotherium* (Figs. 369*A*, 370*A*, 375), a large animal of normal notoungulate build in most regards, but with powerful front legs armed with heavy claws, suggesting root-digging habits analogous to those of the chalicotheres of northern continents. The flowering of the toxodontoids was in the Eocene and Oligocene; in the later Tertiary none survived except forms closely related to the ancestry of *Toxodon* itself.

Typotheres.—The suborder Typotheria includes a series of forms which parallel the larger rodents to a considerable degree. *Typotherium* [*Mesotherium*] of the Pliocene and Pleistocene was as large as a black bear. Canines and premolars were reduced; the only teeth in the front of the mouth were incisors, one above and two below, the major pair being broad chisels which grew throughout life; the remaining cheek teeth were hypsodont and rootless. The postcranial skeleton was persistently primitive, the foot five-toed, with narrow terminal phalanges. *Protypotherium* (Figs. 369*B*; 370*C*, *E*; 376) of the Miocene was a more primitive

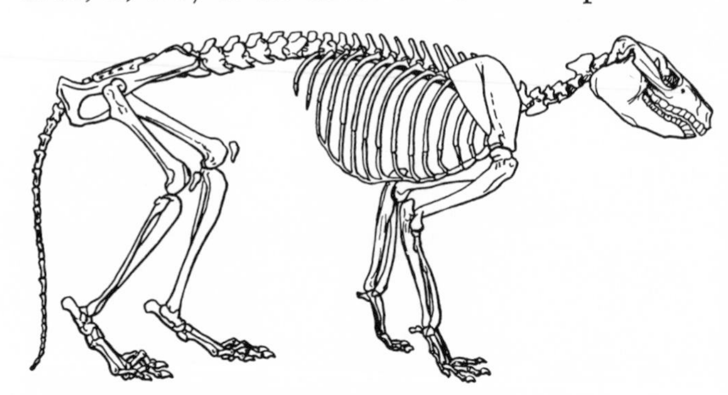

Fig. 376.—*Protypotherium*, a typotherian notoungulate from the Miocene of South America, about 20 inches in length. (After Sinclair.)

type, with a complete tooth row and normal incisors, and small and primitive typothere ancestors trace back to the early Eocene.

Hegetotheres.—Another rodent-like group is that of the hegetotheres (Fig. 369*F*), present from the Eocene to the Pliocene; these forms were often grouped with the typotheres but merit a separate suborder. Typical Miocene forms were *Pachyrukhos* and *Hegetotherium*. The former was remarkably small for a Miocene notoungulate. It was about the size of a rabbit and further resembled that animal in its stub tail and long hind legs. Its teeth show rodent-like adaptations similar to those seen in *Typotherium*, with loss of teeth and development of a diastema between the rodent-like incisors and the cheek-tooth battery. One cannot but believe that the mode of life of *Pachyrukhos* must have been similar to the rabbits and hares of northern continents. *Hegetotherium* was a larger and more heavily built form, more primitive in the retention (in a reduced stage) of the teeth lost in *Pachyrukhos.*

Notoungulate history.—The history of the notoungulates is indeed a remarkable one. Widespread distribution in the Paleocene—from Asia to Patagonia—suggests that they had evolved at an early stage of placental radiation; their almost complete absence from the Eocene deposits of the Northern Hemisphere suggests that they were unable to withstand the competition of more modernized ungulates, such as the perissodactyls and artiodactyls.

Fortunate in their inclusion among the few placental groups which reached South America, they flourished there in the comparative absence of herbivore competition or carnivore menace. The peak of their development was reached in the Oligocene; beyond the early Miocene dissolution set in. In the Pliocene many types of these ungulates disappeared, and the irruption into the continent in the late Pliocene and Pleistocene of the sabertooths and other carnivores and of competing ungulates of northern types may have caused havoc among the reduced notoungulate stocks. Today not a single form survives to give us any conception of the appearance, life, or habits of the notoungulates.

Astrapotheres.—*Astrapotherium* and its relatives of the Oligocene and Miocene of South America are members of a small group once thought to be a component of the notoungulates but now considered a separate order, the Astrapotheria. *Astrapotherium* (Figs. 369*D*, 377) was a large form, about 9 feet or so in length, remarkably specialized in every regard. The skull looks as if it had been amputated anteriorly, for the upper incisors had been lost and the premaxillae reduced to nubbins between the huge, curved, and persistently growing canines. The lower incisors were present, however, while the lower canines were also large. Following a marked diastema were vestigial posterior premolars and large molars, the last two of enormous size. The slitlike nasal opening was placed on the upper surface of the skull and was only partially roofed by the short nasals. The presence of large air sinuses in the frontal bones gave the forehead a swollen, domelike appearance.

The situation of the nostrils renders it probable that an elephant-like proboscis was present; the well-developed lower incisors indicate the presence of a tough upper lip to form a cowlike cropping mechanism. What peculiar habits were responsible for the presence of the great canines of this remarkable animal are not at all clear.

Equally puzzling is the postcranial skeleton. The anterior part of the trunk and front legs are rather stoutly built; the hind quarters and legs seem pitifully feeble, and the incompletely known feet defy comparison with those of any other group. All in all, the creature's mode of life is beyond reasonable conjecture, although amphibious habits have been suggested.

Albertogaudrya of the Eocene was a somewhat more primitive genus without the huge molars of later forms, but the development of the canines was already under way. This astrapothere was tending to gigantism; already of tapir size, *Albertogaudrya* was the largest member of the South American Eocene assemblage, and an Oligocene form was even larger than *Astrapotherium* itself. *Trigonostylops* of the Paleocene is a small form which is little specialized but appears to be a primitive member of the group. Possibly the astrapotheres had come from some primitive notoungulate type; but this is uncertain, and they may have arisen independently.

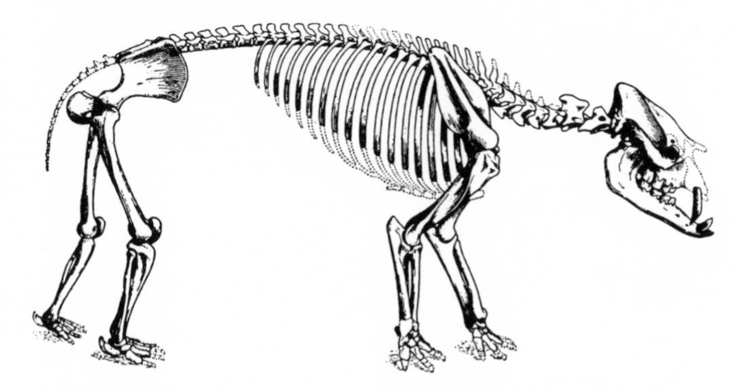

FIG. 377.—*Astrapotherium*, an Oligocene South American ungulate about 9 feet in length. (From Riggs.)

The litopterns.—Peculiar to South America and second only to the notoungulates in importance in the ungulate fauna of that continent were the members of the order Litopterna. This stock never developed into the multiplicity of types found in the notoungulates and was more orthodoxly ungulate in character. Hoofs were always present, and toe reduction took place on the mesaxonic plan with the development of three-toed types and even a one-toed horselike form. Except for occasional reductions in the incisors, a complete dentition and a closed tooth row were usual features in the order. The cheek teeth were usually low crowned but lophodont, with six-cusped upper molars in which the protocone was large and rather centrally placed. In contrast to the notoungulates the later types built up a postorbital bar.

It is highly probable that the litopterns are condylarth derivatives and in a sense to be regarded as a continuation of that group. We have here classed as condylarths *Didolodus* and its relatives of the Eocene of South America. These types were once regarded as litopterns and exemplify the lack of sharp demarcation between the two orders. Condylarth descent may explain, as parallelism, the resemblances between litop-

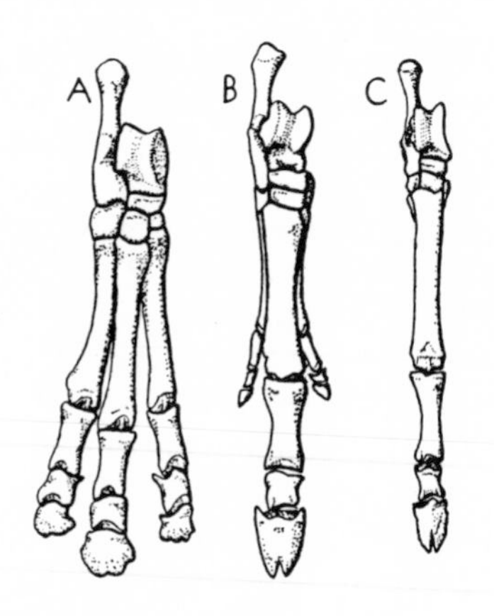

FIG. 378.—The pes of litopterns. *A, Macrauchenia; B, Diadiaphorus; C, Thoatherium.* (After Scott.)

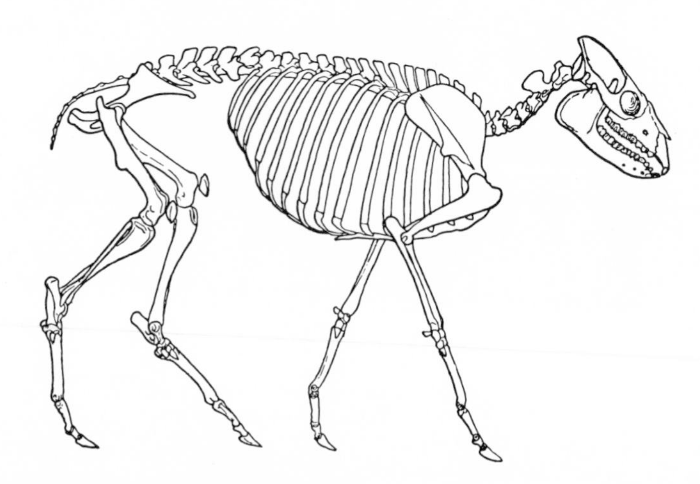

FIG. 379.—*Diadiaphorus,* a Miocene pony-like litoptern; original about 4 feet long. (From Scott.)

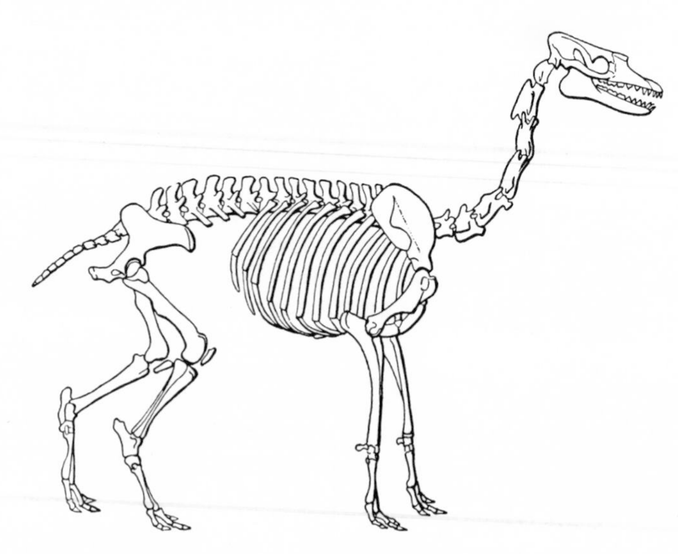

FIG. 380.—The skeleton of *Theosodon,* a Miocene litoptern; length of original about 6 feet. (After Scott.)

terns and perissodactyls, for the latter group is generally believed to have been derived from condylarth ancestors as well.

Two families were already distinguishable among the earliest litopterns. One, the proterotheres, developed into horselike types. *Diadiaphorus,* of the Miocene (Figs. 378*B*, 379, 381), was a three-toed form; but in the contemporary genus *Thoatherium* (Figs. 378*C*, 381*A*), we have an ungulate, about the size of *Mesohippus,* which seems more horselike than any true horse, for it was single toed with splints more reduced than those of modern equids. This pseudohorse was, however, comparatively unprogressive in other respects, for the cheek teeth were low crowned and the carpus was poorly adapted for monodactyl running.

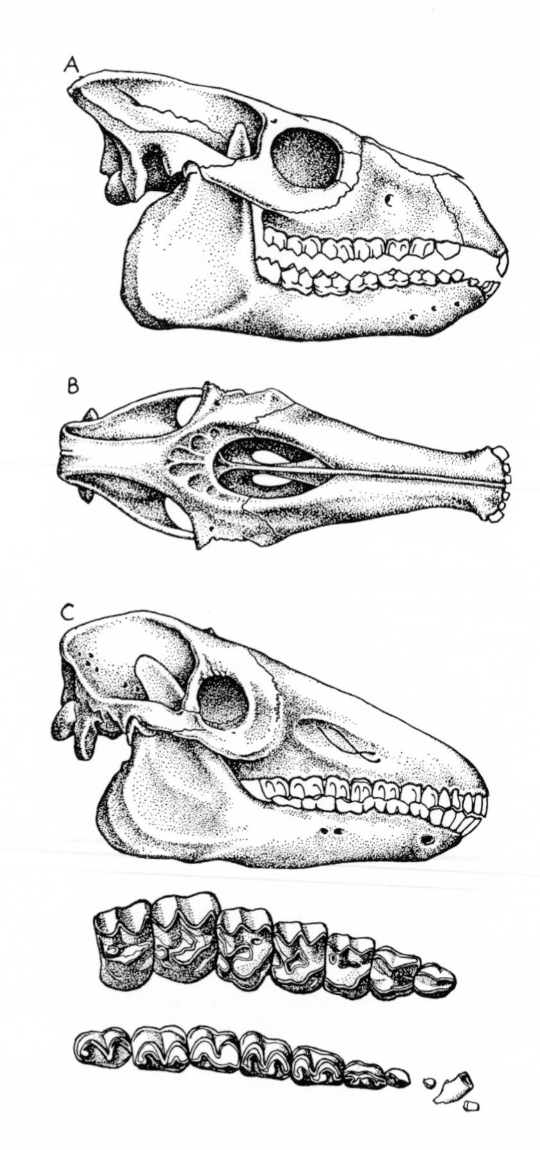

FIG. 381.—Litopterns. *A, Thoatherium,* skull length about 7 inches; *B, C, Macrauchenia,* dorsal and lateral views, skull length about 18 inches. *Lower,* right upper and left lower teeth of *Diadiaphorus* (×1 approx.). (*A* after Scott; *B, C* after Burmeister; teeth after Ameghino.)

Parallel to the typotheres was the development in *Thoatherium* of chisel-like incisors (one above and two below); these teeth, however, are more normally developed in other members of the family. This one-toed proterothere became extinct in the Miocene, and only some of the more conservative three-toed types lingered on to become extinct at the end of the Pliocene.

Macrauchenia (Figs. 378*A*; 381*B*, *C*) of the Pliocene and Pleistocene was the last surviving member of the order and the most highly developed of a second family of litopterns. The limbs and neck were elongated, and three functional toes persisted. The skull was peculiarly constructed. There was a very long snout which accommodated a well-developed battery of cheek teeth, high crowned, in contrast to other litopterns. The nasal opening was very far back on the top of the skull, much as in some sauropod dinosaurs; the nasal bones were vestigial. Some writers have suggested that a proboscis was present, but an alternative theory is that *Macrauchenia* was a swamp dweller with nostrils situated dorsally, whale fashion, for breathing in the water. This animal was about the size of a camel, and with camel-like proportions. Older and smaller members of this series, such as *Theosodon* (Fig. 380) of the Miocene, were more slenderly built, had lower-crowned teeth, and had the nasal opening farther forward.

23 CHAPTER

Perissodactyls

Among all vertebrates there is, perhaps, no group whose fossil history is better known than that of the order Perissodactyla, the "odd-toed" ungulates. Although represented at the present day by but a few species of tapirs, rhinoceroses, and horses, perissodactyls were numerous throughout the Tertiary and included not only the ancestors of these living groups but also such interesting extinct types as the huge, horned titanotheres and the grotesque, clawed chalicotheres.

Limbs.—As usual with ungulates, the locomotor apparatus has been of importance in the evolutionary history. In many perissodactyls the limbs have tended to become long and slim, with a lengthening of the distal segments as compared with the proximal ones, while the ulna and fibula may be reduced. There is, as usual in ungulates, no clavicle; but in contrast with the artiodactyls (or even-toed ungulates), the femur has a well-developed third trochanter. The carpus is of the alternating type. The upper end of the astragalus has a well-keeled surface for the tibia; but, unlike the artiodactyls, this is not duplicated at the lower end, which is saddle-shaped. The shape of the joints limits motion to a fore-and-aft drive (Fig. 346).

The feet (Figs. 383, 392, 393) are of the mesaxonic type, with reduction usually taking place to three toes or even to one in late horse types. In the hind foot, the earliest known forms had already reduced the number to three, with the first and fifth digits already lost. But the first stage in reduction in the front foot led to a four-toed condition common in the Eocene; for, while no trace of the thumb has been found in even the earliest of perissodactyls, typical early forms had retained all the other toes, although the outer one was somewhat weak. This stage in toe reduction, with four digits in front and three behind, is retained to this day by the tapirs. But beyond the Lower Oligocene no other surviving types had more than three toes on each foot.

Teeth.—Equally characteristic of the group is the tooth development (Figs. 384, 385). The full complement, except for a first lower premolar, is found in many early forms, but there may be various losses and specializations in the incisors and canines, and a diastema almost always develops, the first upper premolar being reduced or lost in the process.

The upper molars in the most primitive forms known had already gained a hypocone and (as seen in a very primitive horse [Fig. 297*B*]) showed six bunodont cusps, the intermediate conules being primitively well developed. Typically, a lophodont condition arises from this (Fig. 299). An ectoloph, sometimes W-shaped through the development of the external row of styles, occurs in most forms. Transverse ridges, protoloph and metaloph, are found in three of the five main groups; in titanotheres the inner cusps retain a bunodont condition, and in chalicotheres the protoloph fails to develop. In the lower molars a pair of V's usually develops from the original trigonid and talonid, and may become a W-shaped structure through union of their adjacent ends or have their main posterior limbs changed into transverse crests. The teeth may remain low crowned, as in the early forms, while in others (especially in the horses) the teeth may become markedly hypsodont.

Complete molarization of most of the premolars is general among later members of the group. In some cases, as in the horses, the process has been carried so far that there is almost complete identity in appearance. These highly developed teeth are in contrast with the small and undeveloped premolars found in artiodactyls.

A postorbital bar is developed only in the later horses. The nasals are broad posteriorly and usually extend freely well forward over the nasal opening. Hornlike structures have developed in the titanotheres and many rhinoceroses.

Horses.—The order may be divided into subordinal groups, of which we shall first consider the Hippomorpha, in which are placed the titanotheres and, most especially, the horses and their relatives, most progressive of all perissodactyls. In the horses there is almost no loss of teeth, the incisors being retained and forming chisel-shaped cropping organs. Cross-lophs slanting back from the ectoloph are developed on the upper cheek teeth; the intermediate conules tend to retain their individuality and thus make in many cases a very

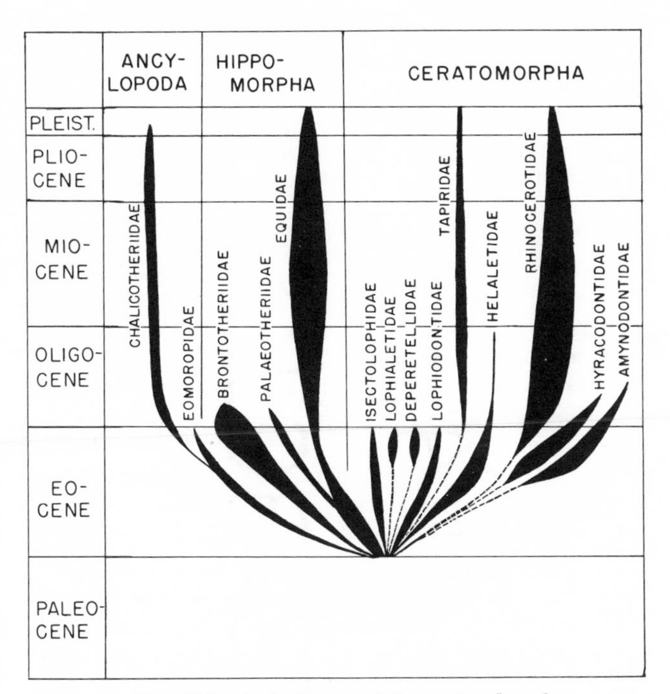

FIG. 382.—A phylogeny of the perissodactyls

complicated pattern. There was an early reduction of toes to three in the front as well as the hind foot, and in later types a progressive reduction toward and to a monodactyl condition (Fig. 383).

A discussion of the structure of the modern end forms constituting the genus *Equus* (horses, asses, zebras) gives a good introduction to horse history. The modern horse has the most highly specialized limb structure of any perissodactyl. In its mechanism it is almost perfectly adapted for rapid forward movement on hard ground and almost incapable of any other type of movement. The humerus and femur are short, pow-

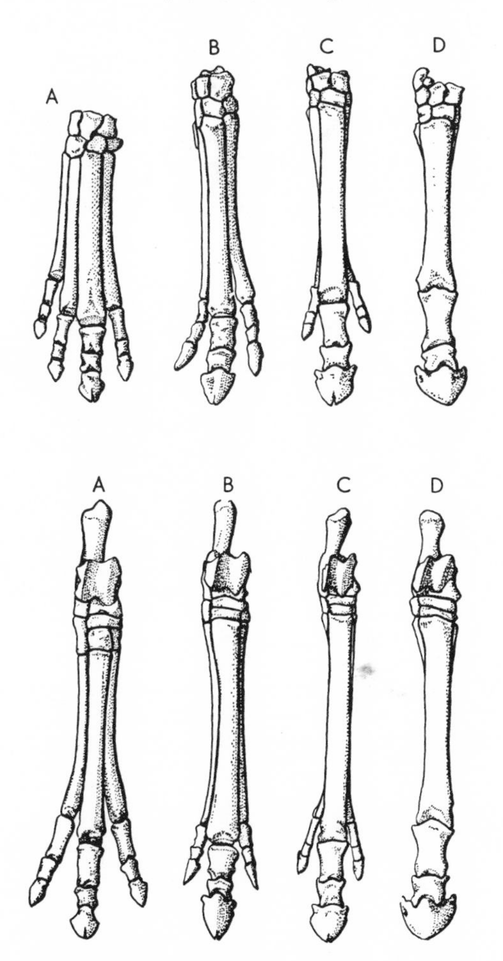

FIG. 383.—Feet of horses. Manus above, pes below. *A*, *"Eohippus,"* a primitive Lower Eocene perissodactyl with four toes in front, three behind; *B*, *Miohippus*, an Oligocene three-toed horse; *C*, *Merychippus*, a late Miocene form with reduced lateral toes; *D*, *Equus*. (*A* after Cope; *B*, *C* after Osborn.)

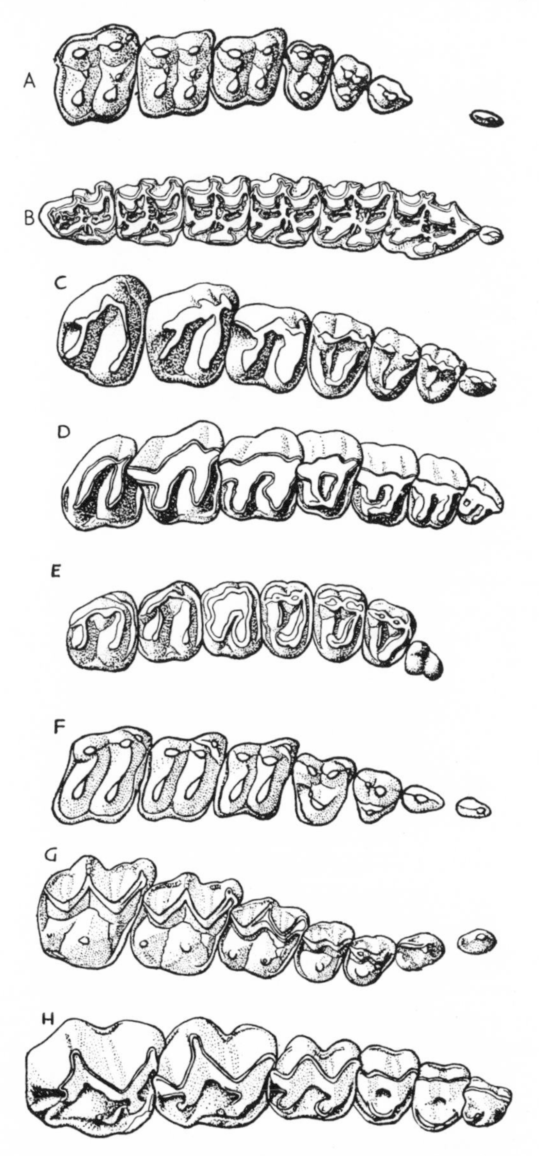

FIG. 384.—Right upper cheek teeth of perissodactyls. *A*, *"Eohippus,"* ×1 approx.; *B*, *Equus occidentalis* (Pleistocene), ×1/2 approx.; *C*, *Hyrachyus*, ×5/3 approx.; *D*, *Caenopus*, ×2/5 approx.; *E*, *Protapirus validus*, ×5/9 approx.; *F*, *Homogalax*, an Eocene tapiroid, ×1; *G*, *Palaeosyops*, ×5/12 approx.; *H*, *Moropus*, ×1/3. (*A*, *B*, *F* after Matthew; *C*, *D*, *G* after Osborn; *E* after Hatcher; *H* after Peterson.)

erful, driving segments from which depend the much longer radius and tibia, the main supporting elements of the front and hind feet, respectively. The shaft of the ulna is reduced and fused to the radius; while of the fibula there remains only a splint above and a vestige fused to the tibia below. Only one toe, the third, is well developed (Fig. 383*D*); this has a much elongated

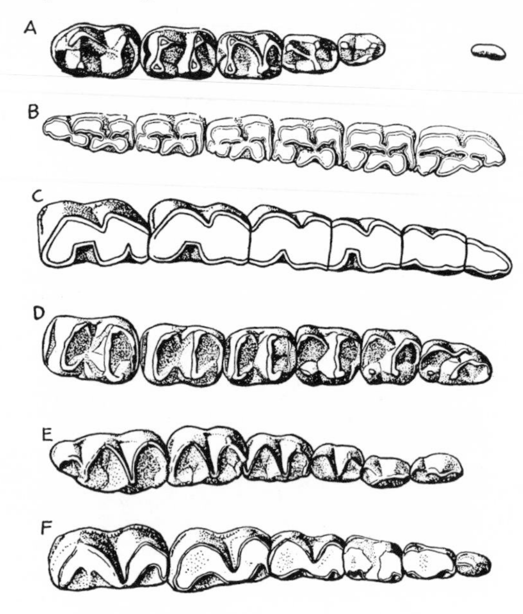

Fig. 385.—Left lower cheek teeth of perissodactyls. *A*, *"Eohippus,"* ×1 approx.; *B*, *Equus leidyi* (Pleistocene), ×1/2 approx.; *C*, *Diceratherium*, a Miocene rhinoceros, ×2/7 approx.; *D*, *Protapirus validus*, an Oligocene tapir, ×5/9 approx.; *E*, *Limnohyops*, an Eocene titanothere, ×5/12 approx.; *F*, *Moropus*, a Miocene chalicothere, ×1/3. (*A* after Wortman; *B* after Simpson; *C*, *F* after Peterson; *D* after Hatcher; *E* after Osborn.)

metapodial but short phalanges, the terminal one expanded into a semicircular structure supporting the hoof. Of the side toes, splints of the metapodials remain; nubbins on their ends may represent the lost digits. The carpals are arranged in two semicircular layers between the radius and the metacarpal, with the magnum (the normal support for the third finger) enlarged and the first carpal vestigial. Similarly in the tarsus the ectocuneiform plays a prominent part in carrying the weight from the well-keeled astragalus and the underlying navicular to the toe.

All the incisors are present in the modern horse; broad and in a closed row, they form an efficient cropping mechanism. Behind them is a long diastema, in which lies the small and variable canine. The small first upper premolar lies at the front end of the grinding battery (Figs. 384*B*, 385*B*). This is composed of six teeth in each jaw half, the molars and premolars almost identically constructed. The cheek teeth are very high crowned with a complicated prismatic pattern, the aspect of which changes somewhat with wear as the siliceous food grinds down across enamel and dentine and across the cement which fills the interstices and covers the tooth. The pattern, though complicated, can be identified with the primitive bunodont and lophodont conditions found in early equids (Fig. 386). In the upper molars the ectoloph is represented by two crescents for the paracone and metacone, with projections formed by the parastyle, mesostyle, and metastyle. Anteriorly a crescent swings inward as the protoconule. Internal to this is the nearly isolated protocone, while the posterior end of the protoconule connects with the metaloph, developed internally into a crescentic hypocone. From this last, in turn, a ridge representing the hypostyle runs outward toward the posterior end of the ectoloph. This much-folded pattern, presenting alternating layers of three different substances, forms a highly effective grinding organ. The lower teeth have had a similar history, but the pattern is obviously derivable from a primitive one in which a connecting link between the two V's is afforded by a metastylid adjacent to the metaconid.

The facial part of the skull (Fig. 387*B*) is elongated in relation to the long set of cheek teeth, and the diastema. In the modern horses there is a postorbital bar.

Eocene horses.—The modern horses and their relatives are almost perfectly adapted in both limbs and

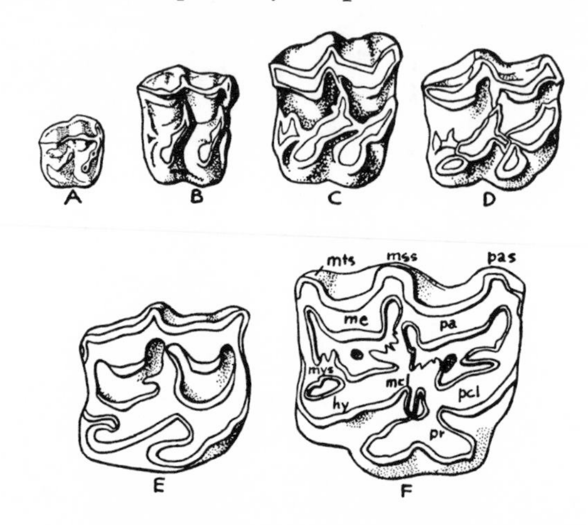

Fig. 386.—Right upper molar teeth of various horse types, all natural size. *A*, *"Eohippus"* (Lower Eocene); *B*, *Mesohippus* (Oligocene); *C*, *Parahippus* (Lower Miocene); *D*, *Merychippus* (Upper Miocene); *E*, *Pliohippus* (Pliocene); *F*, *Equus complicatus* (Pleistocene). The complicated cusp pattern of the modern horse tooth may be traced back to the comparatively primitive condition in *"Eohippus."* Abbreviations: *hy*, hypocene; *mcl*, metaconule; *me*, metacone; *mes*, *mss*, mesostyle; *mts*, metastyle; *pa*, paracone; *pas*, parastyle; *pcl*, protoconule; *pr*, protocone. (After Matthew.)

dentition for a grazing, plains-dwelling life. Far different in structure and probable habits were the earliest horses, found in the Lower Eocene. Best known are the American forms generally termed *"Eohippus."* Since, however, it has been discovered that these forms are generically identical with an obscure European form early described as *Hyracotherium*, the taxonomic purist applies the latter term to them. *"Eohippus"* (Figs. 383*A*, 384*A*, 385*A*, 386*A*, 387*A*, 388) was a small form, some species no larger than a fox terrier. In some specimens the face is primitively short, the orbits (lacking a postorbital bar) lying in the middle of its length. There was but a short diastema. The low-crowned molars were already squared up with six cusps present above and four below. There was here just the beginning of a tendency toward loph formation, but the tooth was still essentially bunodont in character. The premolars were still comparatively simple, none of the upper ones having progressed beyond a triangular shape. The limbs were moderately long and slim, with the metapodials rather elongated and with definite hoofs on the end phalanges (Fig. 383). The toes were already reduced in number. In the front foot all traces of the thumb had disappeared; but the other four toes were all functional, although the outer one was comparatively small. In the hind foot three toes were well developed. This small, forest-dwelling browser is of great interest. Unquestionably, it stands at the base of

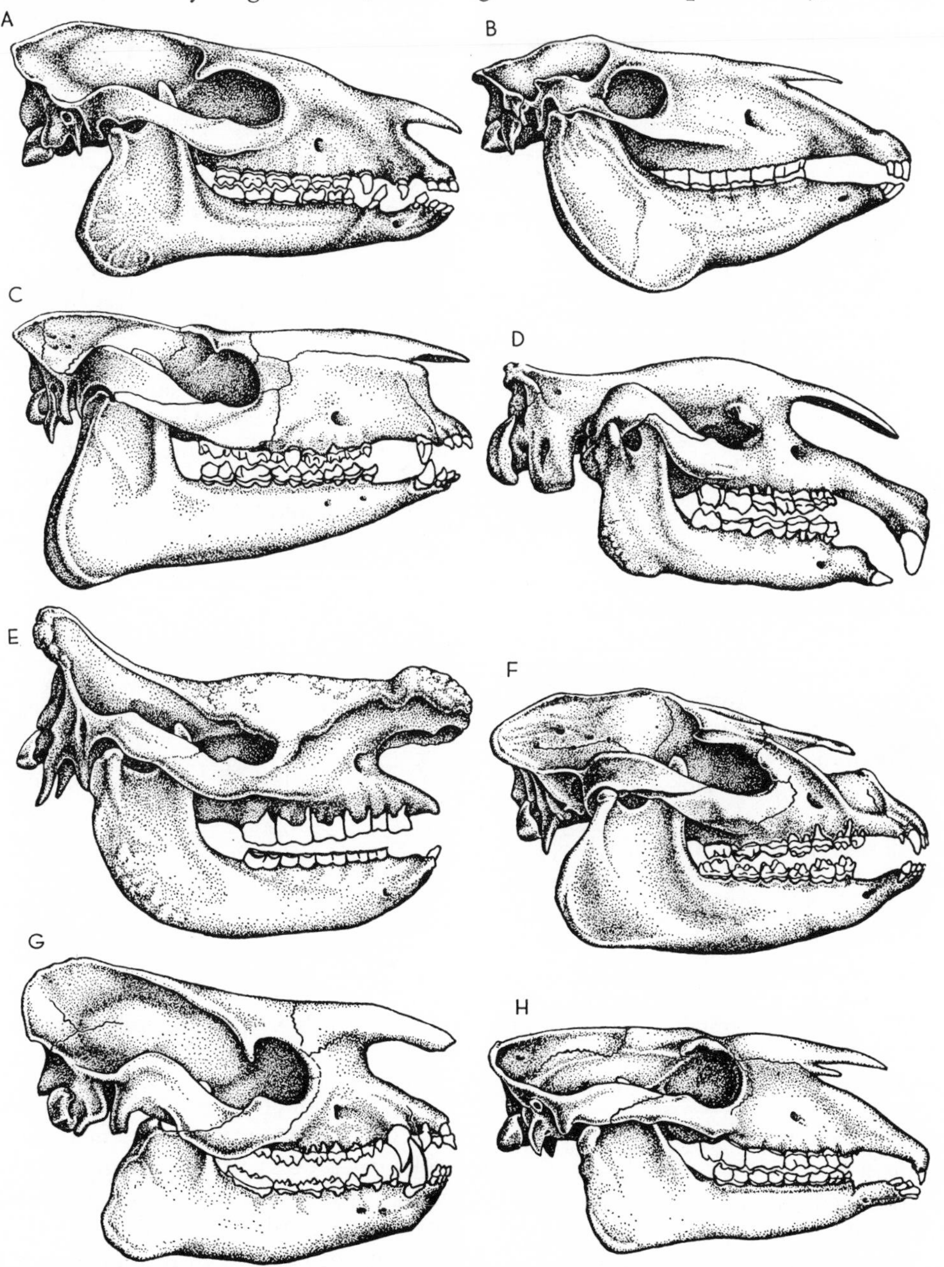

Fig. 387.—Skulls of perissodactyls. *A*, *"Eohippus,"* skull length about 5.25 inches; *B*, *Equus niobrarensis*, a Pleistocene American horse, skull length about 23 inches; *C*, *Hyrachyus*, an Eocene hyracodont, skull length about 11.5 inches; *D*, *Baluchitherium*, a giant hornless Oligocene rhinoceros, skull length about 4.5 feet; *E*, *Diceros pachygnathus*, a Pliocene horned rhinoceros, skull length about 2 feet; *F*, *Protapirus validus*, an Oligocene American tapir, skull length about 1 foot; *G*, *Limnohyops*, an Eocene hornless titanothere, skull length about 15 inches; *H*, *Moropus*, an American Lower Miocene chalicothere, skull length about 2 feet. (*A* after Cope; *B* after Hay; *C* after Scott; *D*, *G* after Osborn; *E* after Gaudry; *F* after Hatcher; *H* after Peterson.)

the horse series. But, even more than that, it exhibits many features to be expected in the ancestors of other perissodactyl families and, without doubt, was very close to the roots of the whole perissodactyl stock.

In many instances the derivation of advanced or specialized ungulate types from more primitive ancestors is none too clear. In the case of the perissodactyls, however, there is general agreement that they have been derived from the condylarths; some of the Paleocene genera of that order are, as far as known, quite similar to "*Eohippus*" and other early Eocene perissodactyls.

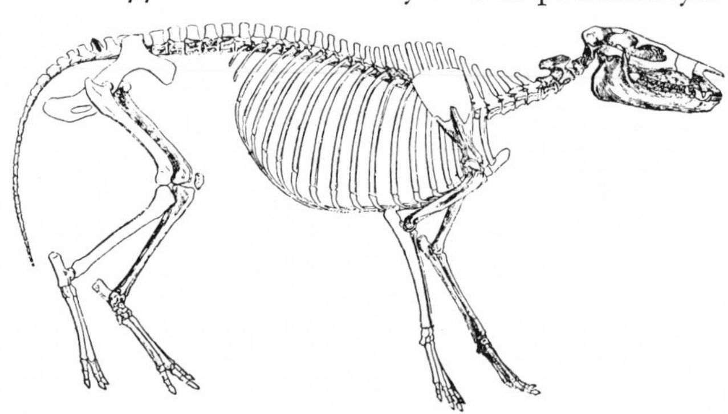

FIG. 388.—"*Eohippus*," the "dawn horse" of the Lower Eocene, probably close to the stem of the perissodactyls; length of this species about 18 inches. (After Cope, from Smith Woodward.)

In the Middle and Upper Eocene of North America we find, in *Orohippus* and *Epihippus*, descendants of this primitive equid in which the structure was essentially the same and in all of which four toes were retained in the front foot. A few advances, however, did take place; for example, the last two premolars took on more of the molar pattern.

Mesohippus and related types.—The next main stage in horse evolution begins in the Lower Oligocene, again in North America. There, in *Mesohippus* (Figs. 386*B*, 389), typically the size of a collie, and the somewhat larger *Miohippus* (Fig. 383*B*) we find the beginning of a series of functionally three-toed horses. In the front foot the outer toe has gone, although a vestige of its metatarsal persisted in many later equids. The metapodials of the three remaining toes were much more elongated than in Eocene types; but while the central toe was somewhat larger, all three undoubtedly reached the ground.

The diastema had continued its development; and, while the cheek teeth were still low crowned, all the premolars except the small first one had become completely converted to the molar pattern. This pattern was now definitely lophodont, with a typical W-shaped ectoloph above but with the cusps constituting the cross-lophs retaining their individuality. At the back edge of the upper molars the hypostyle was beginning to be prominent. In the lower jaw there were two crescents with a distinct double cusp (metaconid and metastylid) at their junction.

The teeth and feet of these typical Oligocene horses suggest that, like their ancestors (and the tapirs today), they were essentially browsing forms, living on soft vegetation in forests and glades where their spreading, three-toed feet would enable them to go over soft ground difficult for the later, single-toed horse. A tendency in some members of this stock to grow to larger size without much change in structure led, at the beginning of the Miocene, to the development of *Anchitherium*, about the size of a pony but with persistently primitive three-toed feet and low-crowned teeth. This type migrated to Europe (where true horses had been absent since the Eocene) and was common there in the early Miocene. Members of this group of relatively primitive forest horses persisted into Pliocene times (one of them reaching China), despite the fact that other and more progressive horses were developing in the meantime. The larger browsing forms of the late Miocene and Pliocene are often separated as a distinct genus—*Hypohippus*—and a giant form from the Pliocene, which may have been the size of an African rhinoceros, has been sometimes appropriately termed *Megahippus*. *Archaeohippus* represents another series of conservative Miocene horses, which, however, were of very small size.

Miocene horses.—The beginning of a third main structural stage is found in *Parahippus* of the Miocene of North America. In this horse the diastema was well developed, and the molar teeth were lengthening. The face was consequently somewhat elongated, the orbit

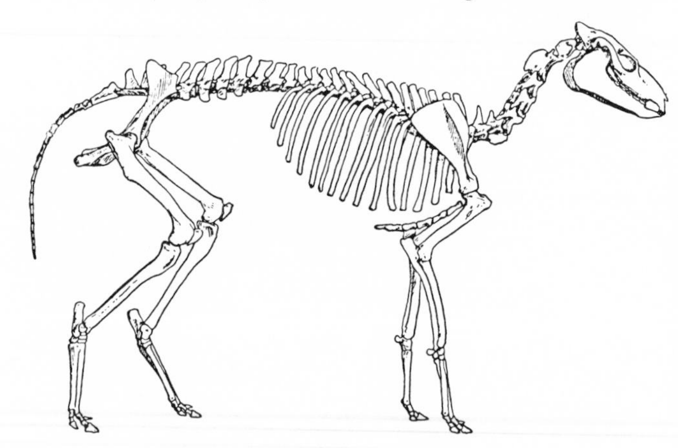

FIG. 389.—*Mesohippus*, a small, three-toed Oligocene horse, about 40 inches long. (From Scott.)

being considerably back of the middle of the skull length. The eye was partly enclosed behind by a developing process from the frontal; this formed a complete bar in all later types. While the molar teeth (Fig. 386*C*) were still fairly low crowned, cement was beginning to make its appearance on their surfaces. In connection with this feature, a posterior ridge develops on the upper molars connecting the hypocone with the ectoloph, and a cross connection (crochet) tends to connect protoloph and metaloph; these new structures result in the formation of "lakes" in the center of the tooth in which cement is deposited.

In the feet the side toes were somewhat reduced, so that they would probably not have touched the ground during rapid locomotion. Thus, it would seem that in *Parahippus* we have a form transitional from the old forest dwellers to modern plains-dwelling grazing horses.

Directly descended from this genus was *Merychippus* (Figs. 383*C*, 386*D*), characteristic of the later Miocene of North America and the undoubted ancestor of the later horse types. Here the teeth were definitely high crowned and prismatic, with a good cement covering and a molar pattern quite comparable to that of living horses. The development of hypsodont teeth is usually regarded as associated with grass feeding; this highly siliceous material would rapidly abrade low-crowned structures. This may well be the case. But we may note that Miocene horses were characteristically of larger size than their Oligocene predecessors. And simple mathematical considerations indicate that, even without change of diet, larger size would necessitate either larger grinding areas on the teeth or higher crowns.

Later horse evolution.—The species customarily included in *Merychippus* are highly varied, and the genus appears to include a number of lines leading to the varied advanced genera which comprise the horse fauna of the Pliocene. *Hipparion, Neohipparion,* and *Nannippus* are a group of closely related forms of rather light build. *Hipparion,* about the size of a pony, spread through Asia to Europe (in which area *Anchitherium* had already disappeared), was characteristic

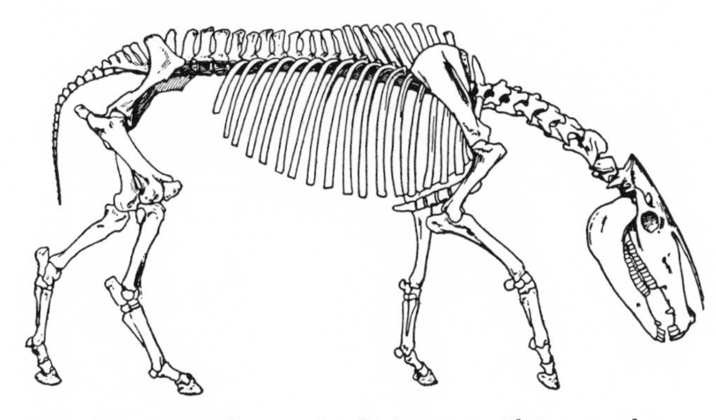

Fig. 390.—*Hippidion,* a South American Pleistocene horse. (From Scott, *Land Mammals of the Western Hemisphere,* by permission of Macmillan Co., publishers.)

of the Pliocene fauna there, and may have persisted in Africa with other relic types until the Pleistocene. *Neohipparion* and *Nannippus* (the latter a small form) were confined to America, as was *Calippus*—a small relative of *Pliohippus.*

This last genus (Fig. 386*E*) is the Pliocene American stock from which it appears that the modern horses and their relatives have come. There is a trend toward larger size, tooth patterns resembling those of living equids, and a rather stouter body build than in the *Hipparion* group. The feet are poorly known, but complete (if very slender) side toes were present in some forms. Some descendants of this group which reached South America became specialized types, such as *Hippidion* (Fig. 390) and *Onohippidium.* In these small but heavily built South American types, as in the true modern horses, the side toes were reduced to splints, but the limbs were short (suggesting a mountain habitat); and they differed in the excessively long and slim nasal bones. These forms did not survive the Pleistocene.

The major line of descent from *Pliohippus,* however, is that which led to the genus *Equus,* the living group

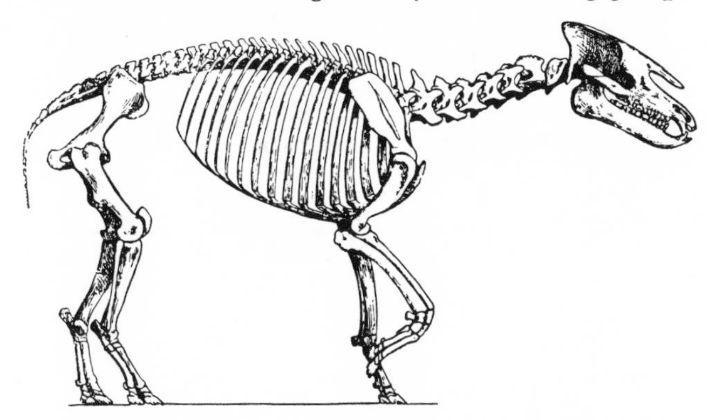

Fig. 391.—*Palaeotherium magnum,* a late Eocene giant horse-like mammal, about 1/20 natural size. (From Abel.)

of horselike animals. The rather slight remaining modifications, apart from the definite reduction of the side toes, appear to have taken place close to the Pliocene-Pleistocene time boundary. *Equus* is characteristically a Pleistocene genus, and its appearance is often considered as a time marker for that epoch. The genus appeared in North America but spread with remarkable rapidity to every continent (except, of course, Australia). Minor differentiations, hardly of generic worth, led to the development of the various modern horselike animals. The earliest American horses and some of those present in the Pleistocene of Europe show features suggestive of the zebras, now confined to Africa. Other horse types are the asses still found wild in some of the more barren tropical regions, and the true horses, of which a wild species still exists in central Asia. Strangely enough, while North America had been, all through the Tertiary, the center of horse evolution, with only a few occasional migrants reaching Eurasia, the group had become extinct in the New World by the end of the Pleistocene. The American plains and pampas, which proved to be perfectly suitable for horses when reintroduced by man, were barren of equid life when first seen by Europeans.

Palaeotheres.—Although the main line of horse evolution beyond the Lower Eocene was confined to North America, the European faunas of the later Eocene show the survival there of descendants of *"Eohippus"* which tended to become prematurely large and advanced in structure. *Palaeotherium* (Fig. 391), which survived to the beginning of the Oligocene, reached the size of a rhinoceros. The toes were reduced to three in both front and hind feet, but the legs were rather short and stout. Three premolars had become somewhat precociously molarized in the course of development of the genus; and the cheek teeth became markedly lophodont, although differing in details from those of the true horses. Other European horses of Middle to Upper Eocene age were generally smaller, with somewhat more primitive teeth and less premolar development. One form, however, had already acquired cement on the molar crowns and had very slim lateral metapodials—features which were not attained in the true horses until the beginning of the Miocene. Despite these progressive features, these early "native" European horses did not survive; the characteristic ungulates of the Oligocene and later Tertiary of Europe were artiodactyls, not perissodactyls.

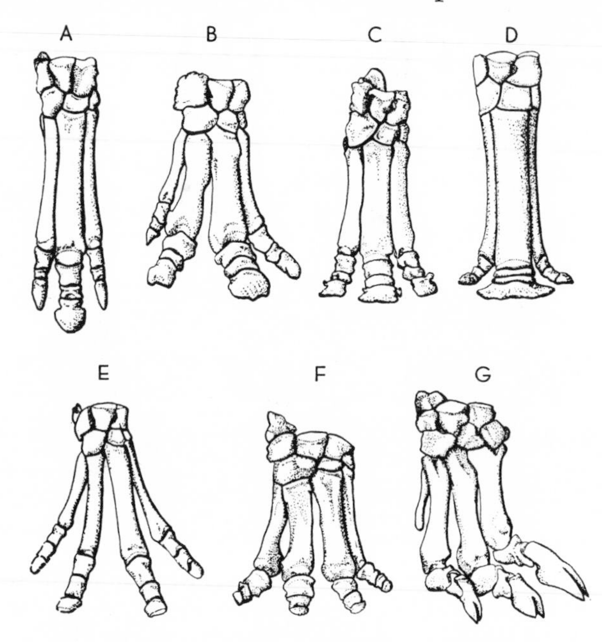

Fig. 392.—Manus of various perissodactyls. *A*, *Hyracodon*, an Oligocene running rhinoceros; *B*, *Trigonias*, an Oligocene four-toed true rhinoceros; *C*, *Diceratherium*, a three-toed primitive rhinoceros; *D*, *Baluchitherium*, a large Oligocene Asiatic rhinoceros with a pillar-like limb; *E*, *Protapirus validus*, an Oligocene tapir; *F*, *Brontotherium*; *G*, *Moropus*, a chalicothere. (*A* after Scott; *B* after Hatcher; *C* after Peterson; *D*, *F* after Osborn; *E* after Wortman and Earle; *G* after Holland and Peterson.)

Titanotheres.—Quite unlike the horses in their adaptations, but apparently closely related and derived from forms similar to "*Eohippus*," were the titanotheres, Brontotheriidae. Their evolutionary center lay in North America, where titanotheres were present throughout most of the Eocene and into the early Oligocene, but they were abundant in the late Eocene and early Oligocene of Asia as well, and late Eocene stragglers even reached eastern Europe. End forms of the titanotheres are found in such genera as *Brontops* (Fig. 394), *Brontotherium*, *Menodus* [*Titanotherium*], and *Megacerops*, of the early Oligocene. These creatures were of great size, the height at the shoulder reaching 8 feet in one type. The skull (as in almost all titanotheres) was long and low; the brain was relatively small; and the braincase had a characteristic sag on top rather than any expansion of the skull. A striking feature was the development at the front of a pair of large, rugose, hornlike processes. These horns were always of rather good size in end forms (although smaller in the females); but they varied enormously in the different genera, sometimes being long, diverging structures, sometimes (in *Embolotherium*, of the Oligocene of Mongolia) rising up sharply as a single broad, fused mass dividing at the top. The teeth (Figs. 384*G*, 385*E*), on the other hand, were strangely unprogressive. The incisors, as well as a premolar, might be lost entirely. The premolars were only partially molarized in pattern and small in size. The molars were large in surface area but crowned with a bunolophodont pattern. In the upper molars there was a W-shaped ectoloph like that of early equids, on the inner side of which lay the main grinding surface, and isolated bunodont internal cones representing protostyle, protocone, and hypocone. Below was a double V of somewhat the same pattern as that of the horses but without the reduplication of the cusp at the union of the V's which is found in most equids. The limbs, in correlation with the huge bulk, were of the graviportal type, short and massive with the conservative number of four and three toes, respectively, in the stubby front and hind feet (Figs. 392*F*, 393*D*).

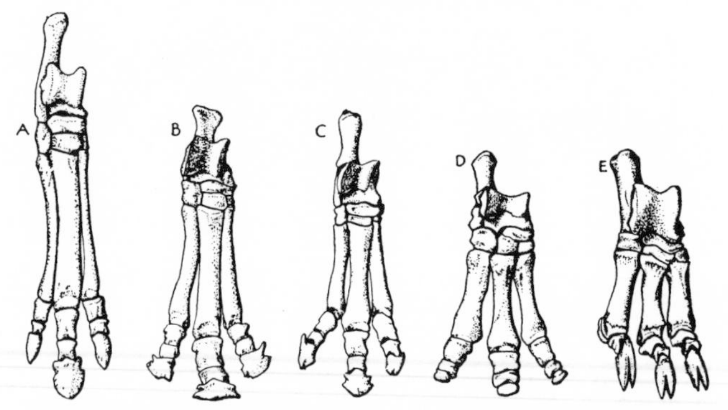

Fig. 393.—The pes of various perissodactyls. *A*, *Hyracodon*, an Oligocene running rhinoceros; *B*, *Diceratherium*, a Miocene rhinoceros; *C*, *Protapirus*; *D*, *Brontotherium*; *E*, the chalicothere *Moropus*. (*A* after Scott; *B* after Peterson; *D* after Osborn; *E* after Holland and Peterson.)

In the Upper Eocene of America and Asia were somewhat more primitive types, such as *Dolichorhinus*, in which there were never more than slight traces of horns on the long nasal bones. In size the Upper Eocene types averaged but about two-thirds that of the end forms; and the body was, in consequence, less massive in its proportions. All the teeth were present, the upper molars still possessed the two primitive intermediate cusps lost later, and the premolars were all simple and unmolarized. A still earlier stage is found in *Palaeosyops* and *Limnohyops* (Fig. 387*G*), of the Middle Eocene of North America, which had reached only the size of a tapir and were hornless. In the late Lower Eocene there was present in North America not only a primitive true titanothere (*Eotitanops*) but also a small form, *Lambdotherium*, intermediate between titanotheres and the primitive equoid types.

The main evolutionary trends seen in this series of forms seem to point toward the rapid attainment of large size. The development of horns may have been a somewhat ineffective "attempt" at protection against the larger contemporary carnivores, or, more probably, developed for male combat in the mating season (they seem to have been larger in supposed males); it is to be noted that in Upper Eocene times rudimentary horns were developing independently in a number of separate lines of titanotheres. Lack of good teeth seems to have been a main factor in their failure to succeed. Their diet must have consisted of extremely soft vegetation; for, despite the large amounts of food which must have been required for the sustenance of the huge body, the teeth could not undergo any great amount of wear. Any slight change in the vegetation and curtailment of soft food would readily have destroyed their hold on existence.

Chalicotheres.—The characteristics of those most curious perissodactyls, the chalicotheres, or Ancylopoda, are well illustrated in *Moropus* (Fig. 395), of the American Lower Miocene. In general appearance (as well as in size) *Moropus* was probably rather horselike, although the front legs were somewhat longer than the hind. The perissodactyl nature of this beast is indicated by many features of the skull and skeleton. In the slim skull the general proportions were those of the horses of that age; the long slender nasals were a perissodactyl character. The cropping teeth were weak or absent in chalicotheres. Unlike the typical perissodactyls, the premolars had remained comparatively small and simple; but the molars (which tended to remain low crowned) were rather similar to those of the titanotheres, with a double V in the lower molars and a W-shaped ectoloph above (Figs. 384*H*, 385*F*); but while protocone and protoconule remained bunodont, as in titanotheres, there was a well-developed metaloph.

In the feet (Figs. 392*G*, 393*E*) we have quite another story. There were in this genus three toes in each foot, and the manus was primitive in retaining a well-developed metacarpal for a small fifth digit (lost in almost every other perissodactyl group by that date). So far all is still orthodoxly perissodactyl. But the digits, instead of bearing hoofs, were terminated by large fissured ungual phalanges, undoubtedly bearing not hoofs but stout claws.

Such a feature is quite unlooked for in a form which otherwise agrees so well with the hoofed groups considered in this chapter; for half a century or so after the first fragmentary remains of chalicotheres were discovered it was not imagined that skull and feet could possibly have pertained to the same form. Because of the claws, some workers were inclined to place the chalicotheres in a separate order.

But the rest of the skeleton is so typically perissodactyl that we cannot refuse chalicotheres admittance to this group. We must regard them as a specialized side branch of the perissodactyls, a suborder Ancylopoda, in which the high development of claws is associated with some specialized habit of these forms. The long front legs of *Moropus* and other members of the chalicotheres have suggested to some a browsing habit in which the front feet may have been used in dragging down branches. But a more probable explanation is that the food consisted of roots and tubers and that the claws were used for digging.

More primitive chalicotheres were present in the Eocene. *Paleomoropus* of the Lower Eocene appears to have been similar to *"Eohippus,"* and can be distinguished from that form by little except the cusp pattern of the upper molars, and even *Eomoropus* of the late Eocene (the size of sheep) was persistently primitive, having a well-developed fifth toe in the hind foot, lacking the clawed feet characteristic of proper chalicotheres of the Oligocene and later epochs, and retaining a full complement of teeth. This latter genus appears to have been ancestral to the later members of the group. In the Eocene, chalicotheres were present in both Eurasia and North America; the later evolution of the group appears, however, to have taken place in the Old World. Chalicotheres are rare beyond the Eocene in North America, and none survived beyond Middle Miocene times, whereas they persisted into the Pleistocene in Asia and Africa.

Tapirs and related types.—It has become clear in recent years that if the perissodactyl families be grouped into larger categories, one major assemblage should include (despite their diversity in appearance) the horses and titanotheres, a second the chalicotheres, and the third as the suborder Ceratomorpha, the tapirs and rhinoceroses and their extinct kindred. One feature indicative of tapir-rhinoceros relationship is to be found in the construction of the upper molars (Fig. 384; cf. *A*, *B*, *G*, *H* with *C–F*). Horses, titanotheres, and chalicotheres early acquired a powerful W-shaped ectoloph, and transverse ridges are either slow to appear (as in equids) or fail to develop at all. In the tapir-rhinoceros group, in contrast, the ectoloph is a simple structure, and emphasis is laid on the development of transverse protoloph and metaloph.

Primitive living representatives of the tapir-rhinoceros group and forms which are still very close in many respects to the common ancestors of all perissodactyls are the tapirs (*Tapirus*) of tropical America and the oriental region. They have persistently retained the primitive limb structure of their Eocene ancestors: the legs and feet (Figs. 392*C*, 393*C*) are short; ulna and fibula are still complete and unfused. As in the Eocene perissodactyls, there are four toes in the front foot and three behind, and although hoofs are present, thick pads on palm and sole aid in support. The tapirs are still browsers, with low-crowned teeth which have no cement covering. The upper canines are reduced, but a lateral incisor develops in caniniform fashion to oppose a well-developed lower canine; there has been no loss of teeth, although a moderate diastema is developed. The modern forms have molarized three premolars, as have the horses. All the low-crowned cheek teeth (Figs. 384*E*, 385*D*) have a very simple pattern, the upper molars having a pair of simple cross-lophs and a short ectoloph, while the lower teeth have developed two cross crests. The one noteworthy specialization of the modern tapir is the presence of a short proboscis, with the accompanying backward migration of the bony opening of the nose and shortening of the nasal bones found in all forms which have developed a trunk (Fig. 387*F*).

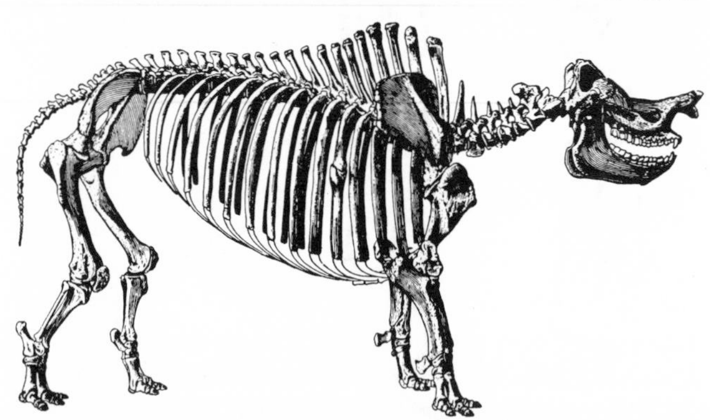

FIG. 394.—*Brontops*, a large Lower Oligocene titanothere. This specimen is a female, the short horns being a sex difference; original about 14 feet long. (From Osborn.)

True tapirs appeared, as the genus *Protapirus*, in the Oligocene of both Europe and North America. A number of Tertiary genera have been described, but all are very close to the modern type, and generic distinctions are slight. Specimens are few, presumably in relation to the forest-dwelling habits of the family, but they persisted in the present north temperate regions of both hemispheres until the Pleistocene. With the climatic vicissitudes of this last epoch they became restricted to the tropics. In the Old World their final refuge has been the Malayan region; in the New, they successfully invaded South America.

Although the tapirs in a narrow sense are not found before the Oligocene, there are numerous tapir-like forms in the Eocene, of which *Homogalax* (Fig. 384*F*), *Lophiodon*, and *Heptodon* are representative. These are sometimes grouped as "lophiodonts" but are currently thought better arrayed in several separate, but related, small families. The cheek teeth are of simple lophodont structure. In some there is little of the molarization of the premolars later acquired in true tapirs; others, however, show a precocious molarization. Some of these genera are close to the ancestry of true tapirs and to ancestral rhinoceroses; a majority, however, appear to represent short, sterile twigs of the perissodactyl family tree. Most were of rather modest size, but *Lophiodon* of the late Eocene of Europe was as large as a modern rhinoceros. This form was of ponderous build; others were more slender-legged, paralleling the early horses. *Hyrachyus* (Figs. 384*C*, 387*C*) and several of his relatives are frequently regarded as primitive members of the rhinoceros group, but their general build is so primitive that they are best regarded as tapiroids.

Rhinoceroses.—Among all the perissodactyl groups, the most complicated fossil history is that of the Rhinocerotoidea, the rhinoceroses and their relatives. At present there survive of the rhinoceroses only a few forms in the Old World tropics, but throughout most of the Tertiary they were exceedingly numerous in the northern continents. It seems certain that they were derived from early tapiroids, perhaps in several parallel lines; they have, however, tended to diverge considerably from that group. In many cases there have developed hornlike structures, which, however, are composed, not of bone or horn, but of a fused mass of modified hairlike material. As contrasted with most equids, rhinoceroses have tended to grow to large size, usually with comparatively short, stout limbs in which digital reduction has proceeded at a slow pace, for the fifth digit in the hand was present in some Oligocene and even Miocene forms, and a monodactyl stage has never been attained. The premolars have tended to molarize, but the row of cheek teeth does not greatly lengthen, and the face is usually comparatively short. The molars have a comparatively simple π-shaped pattern above, and in all later forms the last molar is reduced; the lower molars develop two asymmetrical crescents. The cheek teeth seldom tend to become very high crowned, and cement is almost never present. The

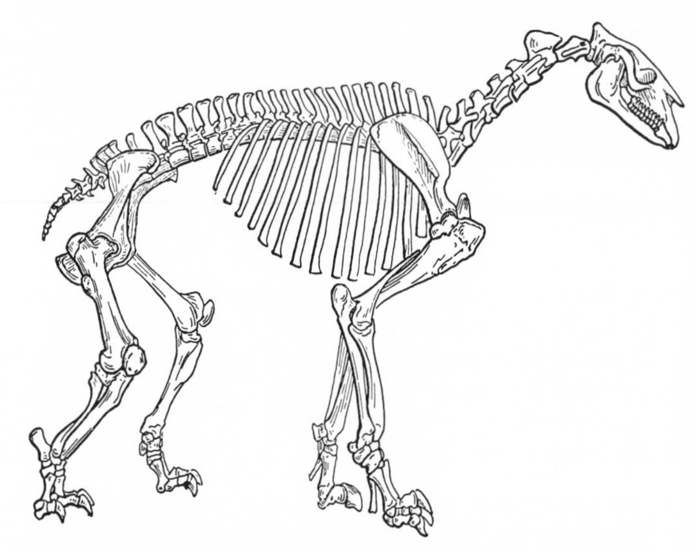

FIG. 395.—The chalicothere *Moropus*, length about 9 feet. (From Gregory.)

incisors and canines, however, are variable, and there are frequent losses and specializations.

Running rhinoceroses.—The rhinoceroses may be divided into three families, of which the most primitive and unspecialized are the members of the Hyracodontidae, the "running rhinoceroses," such as *Hyracodon* (Figs. 392*A*, 393*A*, 396) of the North American Oligocene and its forerunners in the late Eocene of both North America and Asia. This genus was rather large with the cheek teeth typically rhinocerotid and with the back premolars already molarized. *Hyracodon* was somewhat more specialized for a cursorial life than the tapiroids, with long slim legs and but three toes on front as well as hind feet. These evolutionary advances were very similar to those found in the contemporary horses but were not continued, for the group disappeared before the close of the Oligocene, perhaps because of unsuccessful competition with these rivals.

Amynodonts.—An early side branch, possibly derived from the primitive hyracodonts, was that of the family Amynodontidae. These forms are found in the late Eocene and Oligocene of both Eurasia and America and persisted in Asia until the Miocene. They had about the general size and proportions of a hippopotamus, and the conditions under which their remains are found suggest that they were river-living forms. *Metamynodon* (Fig. 397) of the American Oligocene had short, massive limbs, still retaining four short toes in front and three behind. In the heavy skull with a short "bulldog" muzzle, the premolars had failed to molarize, several had been lost, and the incisors were also reduced; on the other hand, the canines and molars were much enlarged.

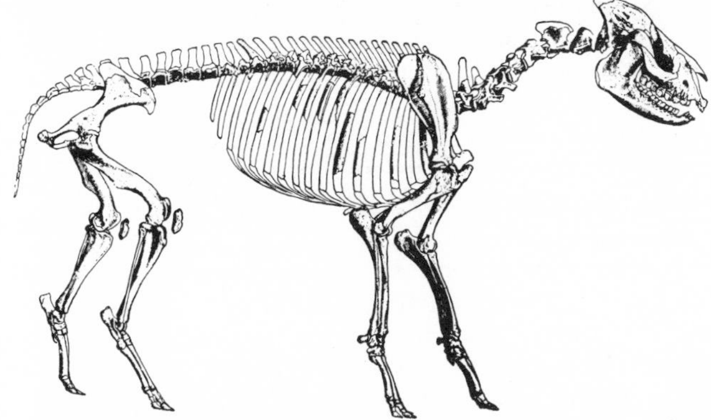

Fig. 396.—*Hyracodon*, an Oligocene running rhinoceros, about 5 feet in length. (From Osborn.)

Early true rhinoceroses.—All remaining forms are commonly placed in a third family as true rhinoceroses, the Rhinocerotidae. It seems certain that these types were derived from early running rhinoceroses, but Eocene forerunners are presented only by poorly known types. The true rhinoceroses first became prominent in the Oligocene and (in contrast with the contemporary running rhinoceros) tended to large size and stout limbs, while (in contrast with the amynodonts) the premolars became rapidly molarized. The cropping mechanism of true rhinoceroses is peculiar: a pair of incisors, the first upper and second lower, are

always enlarged cutting teeth (in some late forms they may be secondarily lost), and a narrow muzzle with a pointed lip develops. The last upper molar is always simpler than the others, with ectoloph and metaloph forming a single continuous crest (Fig. 384*D*).

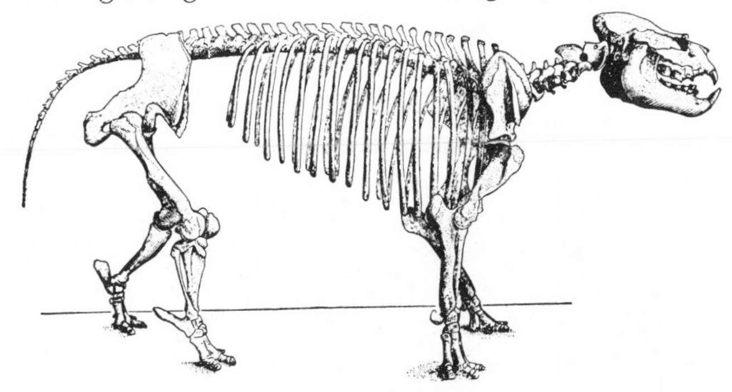

Fig. 397.—*Metamynodon*, an Oligocene amphibious rhinoceros (Amynodontidae), about 14 feet in length. (From Osborn.)

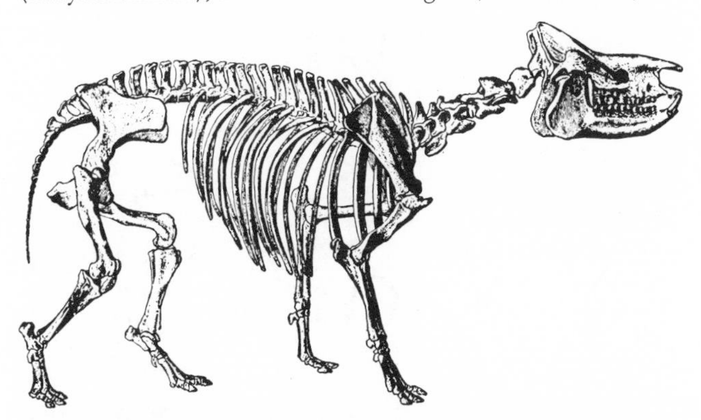

Fig. 398.—*Caenopus tridactylus*, an Oligocene hornless rhinoceros, about 8 feet in length. (From Osborn.)

Prohyracodon of the Eocene and *Trigonias* (Fig. 392*B*), *Subhyracodon*, *Caenopus* (Fig. 398), and related forms from the Oligocene of North America and Europe appear to represent the central stock of the true rhinoceroses. *Caenopus* had already attained fairly large size (the skull averaged more than a foot in length), and the limbs supporting the stocky body were stouter and shorter than in the hyracodonts and comparable to those of modern tapirs. Four toes were still present in front in *Trigonias*, but the outer toe was already small and had disappeared in *Caenopus*. All later types retained a three-toed foot. Incisor tusks were developed; and most of the other front teeth were retained, in contrast with later forms. The premolar teeth were at first simple but during the Oligocene tended to assume the molar pattern. The primitive Oligocene rhinoceroses were hornless (the matted hair of the "horn" does not fossilize, and our evidence for the presence of a horn usually consists of a roughened area on the skull for its attachment).

A number of later Miocene and Pliocene genera, such as *Aceratherium* of Eurasia and certain rarer American contemporaries, show relatively little change from the primitive pattern. These forms exhibit a tendency toward incisor reduction, but there is little evidence of horn development or other specializations.

The evolutionary history of the rhinoceroses offers a strong contrast to that of the equids. In the horses we have seen that there was a main stem on the evolutionary tree with but few short side branches; rhinoceros evolution, on the other hand, may be pictured as a branching bush. There is no main evolutionary stem, but a complex of sprouts, the components of which are difficult to disentangle.

An early branch from the *Caenopus* stock is that represented by *Diceratherium* of the American Miocene. This genus was quite similar to its Oligocene ancestors except for the fact that the males possessed small horns; but, unlike those of other later horned rhinoceroses, these horns were placed side by side at the tip of the nose.

Giant rhinoceroses.—Equally prompt to appear was a much more spectacular and aberrant group including *Baluchitherium* (Fig. 387*D*) and *Indricotherium,* huge hornless forms from the Oligocene and early Miocene of Asia. These were the largest of known land mammals. The head of *Baluchitherium* was about 4 feet in length but, even so, was small in proportion to the body size. This great beast must have stood about 18 feet high at the shoulders, with a long neck which, combined with long front legs, enabled him to browse on the higher branches of the trees. The grinding teeth were like those of *Caenopus,* while the single pair of blunt incisors, which were his only front teeth, are easily derivable from those of the more primitive Oligocene forms. The limbs, as would be expected, were massive but long, and pillar-like. There is even considerable elongation of the metapodials, which were stout and placed close together in a pillar above the three stubby toes (Fig. 392*D*); the lateral digits were more reduced than in any other rhinoceroses.

Later rhinoceros types.—Beyond the Oligocene, rhinoceroses became relatively rare in America and died out during the Pliocene; in Eurasia, on the other hand, there are numerous and varied genera in Miocene, Pliocene, and even Pleistocene deposits. The various phyletic lines are incompletely established; however, at least five end types can be distinguished and are listed below.

1. The short-legged rhinoceroses, such as *Teleoceras* and its relatives of the Miocene and Pliocene. These are round-bodied, stubby-limbed, and broad-footed beasts with a build like that of a hippopotamus and presumably similar amphibious habits. Technically they were horned, for there is a rugose area indicating the presence of at least a rudiment of this structure at the tip of the nasals. Alone of all the groups mentioned in this section, these rhinoceroses penetrated into North America, where abundant remains of *Teleoceras* have been found in early Pliocene deposits.

2. The term *Rhinoceros* in a narrow sense is confined to the large Indian "unicorn" and a smaller relative from Java, in which there is a single massive horn on the nasal bones and both incisor tusks have been retained. There are forms presumably ancestral in the later Tertiary of Eurasia.

3. Resembling the Indian rhinoceros in a single-horned condition but differing in the loss of incisors and other features, was *Elasmotherium* of the Pleistocene of the northern plains regions of Eurasia. This was a very large mammal; a great swelling, nearly a foot in diameter, on the frontal bone suggests the presence in life of a huge horn on the forehead. The name refers to the fact that the teeth, although possessing the basal rhinoceros pattern, have a wavy, strap-like outline of the enamel. A Pliocene Asiatic form of simpler structure and smaller size appears to be ancestral.

4. *Dicerorhinus* of Sumatra is a two-horned rhinoceros; but, in contrast to the early diceratheres, the horns are here placed one behind the other, tandem fashion, and in the living form are of small size. This form has retained incisor tusks. Recent research indicates that the Pleistocene *Coelodonta* is related to the Sumatran form. This was the woolly rhinoceros of northern Eurasia, adapted to a cold climate. Of it we know far more than the skeleton; it was a favorite subject for Old Stone Age artists, and specimens "embalmed" in a waxy material in an oil seep have been discovered in Galicia. Ancestors of these Eurasian tandem-horned rhinoceroses have been traced back through the Tertiary to Oligocene times.

5. A final group is that of the African tandem-horned genera, the living "white" and "black" rhinoceroses of Africa, *Diceros* or *Ceratotherium* (Fig. 387*E*). In contrast to *Dicerorhinus,* the horns are much elongated and incisors are absent. Their exact ancestry is uncertain; it is possible that they developed in Africa, and our knowledge of fossil rhinoceroses from that continent is extremely limited.

The history of the rhinoceroses has, in general, run a course parallel to that of the related horses but with emphasis on bulk rather than on speed. Eocene and early Oligocene members of both groups tended to vary widely, producing in the case of the rhinoceroses not only running types but the large amynodonts. In the Oligocene the rhinoceroses, in the main, settled down to become a group of forms characteristically much larger than the horses but more conservative in that most remained browsing forms. Like the horses, they were abundant in late Tertiary and Pleistocene times but are now limited to a few Old World types.

With the rhinoceroses we terminate our account of the perissodactyls. All five main groups had appeared early in the Eocene, and during the early part of the Tertiary they flourished greatly. By the Miocene, however, they were beginning to lose somewhat in relative importance. In Pleistocene times horses and rhinoceroses still flourished. Today the perissodactyls are but an insignificant part of the world's ungulate population.

CHAPTER 24

Artiodactyls

The order Artiodactyla—the even-toed ungulates—includes a great variety of living hoofed mammals, such as the pigs and peccaries, hippopotami, camels, deer, cows, sheep, goats, antelopes, and their relatives, together with many important extinct types. Exceeded in numbers in early Eocene formations by the perissodactyls, they have succeeded in far outdistancing their rivals to become the dominant hoofed mammals of later Tertiary and Recent times.

Structure.—The most obvious and characteristic feature of the group is the type of toe reduction; for, whereas almost all other ungulates tend to have a mesaxonic foot with the axis through the third toe, the artiodactyls are paraxonic, with the axis between the third and fourth toes. Pollex and hallux are reduced early, particularly in the hind foot (a five-toed manus survives in some Oligocene forms). Four-toed types are common in the earlier part of the Tertiary; the pigs and hippopotami are still four toed, and vestigial lateral "dew claws" are present in many other forms. The higher types tended early to reduce the toes to two on each foot, giving the typical "cloven hoof" in which the two principal metapodials fuse together into a cannon bone.

In contrast with perissodactyls, there is no third trochanter on the femur. The ulna is reduced and in higher forms fuses with the radius; and the fibula, too, is usually incomplete or fused with the tibia. In the carpus the three proximal elements are always separate, but in the distal row the magnum and trapezoid fuse in some advanced types to support the third metacarpal, while the unciform supports the fourth. The astragalus (Fig. 346*C*) is the most characteristic bone in the skeleton, for it has not only a rolling pulley surface above, but an equally developed lower pulley surface. This type of articulation gives very great freedom of motion to the ankle for flexion and extension of the limb and a potential springing motion, but limits movement to a straight fore-and-aft drive even more strictly than is the case in perissodactyls. The astragalus rests equally on the navicular and the cuboid (these two elements are fused in many types [cf. Fig. 415]). The cuboid lies above the fourth toe, the outer of the two principal digits, while the navicular transmits half the weight to the third toe through the ectocuneiform (the other cuneiform bones tend to be reduced).

The dentition was complete in many of the early types, as it is still in the pigs. The most primitive forms known had normal incisors and rather large, carnivore-like canines. The premolars were simple, and the molars in the simplest forms were of the primitive tritubercular pattern with rather bunodont cusps. From this primitive pattern, not so different from that of primitive carnivores, there have been great variations.

The incisors are often reduced; and the upper ones are absent in the more advanced types, their place as a cropping organ being taken functionally by the gums over the premaxillae. The upper canines often form defensive tusks; the lower ones sometimes take on the aspect of incisors. A diastema has usually developed, and the first premolar is frequently absent. The premolars, in contrast with those of perissodactyls, do not usually assume the full molar pattern but remain comparatively simple (cf. Fig. 401). The cheek teeth, primitively low crowned, have become hypsodont in many forms. As in other herbivores, the molars (Fig. 402) almost always take on a four-cusped pattern. The rather bunodont primitive condition may be emphasized, as in the pigs; but usually each cusp develops into a crescent, giving the typical selenodont pattern of the higher artiodactyls. In this process the outer cusps tended to become crescentic before the inner ones, giving a buno-selenodont condition in some fossil forms.

The nature of the four major cusps of the upper molars in advanced artiodactyls has been a matter of debate. In most advanced mammalian types the four major cusps (as in perissodactyls, cf. Fig. 384) are the protocone, paracone, metacone, and hypocone, the last being added to "square out" the original triangular tooth. In some artiodactyls there is definitely the addition of a hypocone (as in *Dichobune,* Fig. 402*B*), and it is probable that in piglike forms the four principal cusps include a hypocone. It is generally believed, however, that in the advanced selenodonts this is not the case, and that in the four-cusped upper molars of typical selenodonts (as in Fig. 402*L–P*) the postero-internal cusp is not a hypocone, but a highly developed metaconule. In some primitive selenodonts five cusps are present in the upper molars, with a single cusp representing either metaconule or hypocone.

There are typically nineteen thoracic and lumbar vertebrae. The clavicle is, as might be expected, rarely present.

Many forms have a postorbital bar, and there is always at least a postorbital process. The front part of the skull is sharply bent down on the braincase in some advanced forms; the frontals tend to be large, the parietals, on the contrary, reduced. There are interesting variations in the development of the mastoid region (Fig. 400). In primitive mammals this region of the ear ossification appears to have been well exposed on the surface of the skull at the posterolateral corner of the braincase, between the exoccipital and the squamosal. This is still the case in living ruminants; but in the "swine" the squamosal has grown back to meet the exoccipital over the area once occupied by the mastoid, and the mastoid no longer reaches the surface.

Classification (cf. Fig. 399).—The artiodactyls appear in the fossil record at the beginning of true Eocene times. These earliest types were quite primitive in many respects, with simple teeth implying a mixed diet and with features which suggest comparison with primitive carnivores. Nevertheless, such distinctive artiodactyl features as the double-pulleyed astragalus were already present, and it is obvious that the ancestral forms must have been undergoing development in some unknown area during the Paleocene. Possibly the ancestors were primitive condylarths. Rare at the beginning of the Eocene, the artiodactyls had developed into a great variety of types by the end of that period. Many of these have since become extinct, but others are indicative of the lines of descent leading to the existing forms.

The classification of living artiodactyls is simple. They comprise: (1) the suborder Suina, or swine in a broad sense—pigs, peccaries, and hippopotami; (2) the suborder Ruminantia (in a broad use of that term), cud chewers including (*a*) the Tylopoda, or camels, and (*b*) the Pecora, the deer-giraffe-antelope-cattle group; the little chevrotains (tragulids) of the Old World tropics are sometimes treated as a separate group, sometimes (as here) considered to be primitive pecorans. The "swine" have bunodont molar teeth, canine tusks, four-toed feet with separate metapodials in most cases, a simple stomach, and a concealed mastoid region. The living ruminants have selenodont molars; the upper incisors are small or absent; the upper canines are sometimes enlarged in males as weapons but are usually reduced or lost; the feet are usually two toed, with a cannon bone; there is a compound ruminating stomach; and the mastoid is exposed on the surface.

It is thus easy to classify the living types. But when the fossil forms are included, such an easy separation is impossible; we are confronted with a large array of families which are intermediate in many respects between existing types or belong to extinct side branches.

Numerous classifications have been made. These vary greatly. This reflects the fact that in the early artiodactyls the features of teeth, skull, and feet which are useful in distinguishing the major existing groups were present in variable and confusing combinations. However, when the Eocene forms are viewed broadly, it becomes evident that the forms present were, for the

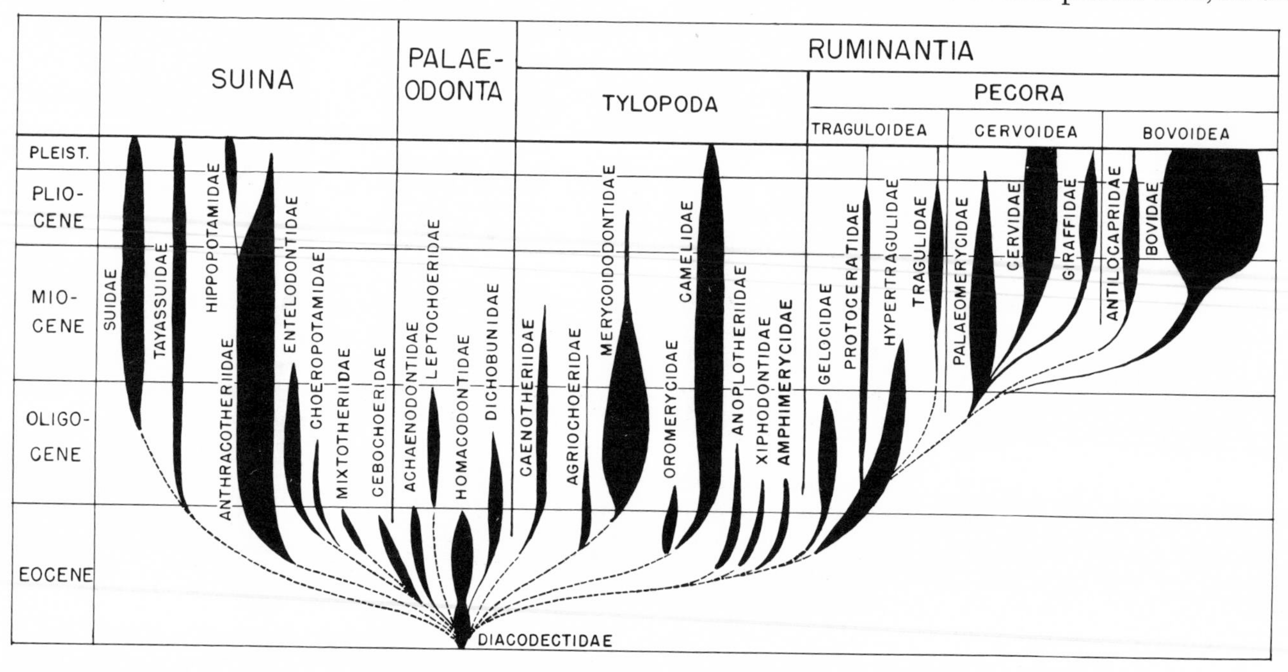

FIG. 399.—Provisional phylogeny of the artiodactyl groups

most part, tending in two major directions: on the one hand, toward types with bunodont dentitions, incisors, and canines of primitive type and an "amastoid" ear region and, on the other, toward animals with selenodont cheek teeth, modified front teeth, and a persistently exposed mastoid region. We are thus justified in adhering essentially to the type of classification used here for modern forms in making a major division into two suborders, Suina and Ruminantia, although adding a third category, Palaeodonta, for certain early and primitive types close to the point of origin of the order.

Archaic artiodactyls: the palaeodonts.—An assemblage of small artiodactyls from the Eocene and early Oligocene (mainly in North America) is that including such genera as *Diacodexis* and *Homacodon* of America and the European *Dichobune.* Many are inadequately known, and there is some variety in their structure, but this group includes the oldest and seemingly the most primitive of known artiodactyls. They are in many respects close to the primitive placental stock; and, were it not for the discovery in some instances of a typical artiodactyl astragalus associated with them, their position as artiodactyls (rather than insectivores or primates) might be questioned. There are few remains of the postcranial skeleton, but the limbs appear to have been generally short and primitive, with four functional toes and presumably a small pollex in some forms. The skull, as seen in *Homacodon* (Fig. 408*A*) was low and moderately elongate, and it is probable that the mastoid was generally exposed in primitive placental fashion. The molar teeth in the earlier Eocene genera were neither markedly bunodont nor markedly selenodont but showed a pattern of moderately blunt cusps; selenodont tendencies appear in most later genera. In *Diacodexis* (Fig. 402*A*) the upper molars are still in the primitive tritubercular stage seen in many Paleocene mammals; and this condition persists in a few rare Oligocene forms, such as *Leptochoerus.* Some later Eocene and early Oligocene forms developed a hypocone, giving a six-cusped tooth (Fig. 402*B*) in the fashion of "normal" placentals and in contrast with the "hypocone-less" pattern believed present in most artiodactyls. *Homacodon* and other American types retained primitively enlarged canines, as did the later swinelike artiodactyls; *Dichobune,* on the other hand, reduced its canines so that its teeth formed a rather uniform and continuous row, as in many early ruminants.

The late Eocene and Oligocene members of this

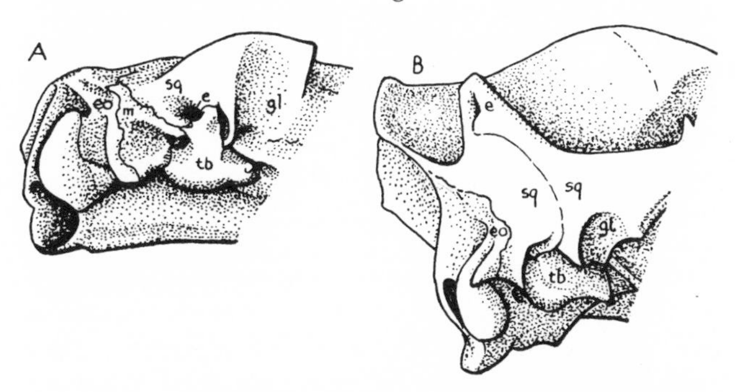

FIG. 400.—Lateral (and somewhat ventral) views of the posterior part of the skull in *A,* the anoplothere *Diplobune,* and *B,* the peccary *Tayassu,* to show the contrast between mastoid and amastoid artiodactyls. In *A* the mastoid is widely exposed on the surface of the skull between the squamosal and the exocccipital; in *B* this area is occupied by an extension of the squamosal. Abbreviations: *e,* external auditory opening; *eo,* exoccipital; *gl,* glenoid cavity; *m,* mastoid; *sq,* squamosal; *tb,* tympanic bulla. (After Pearson.)

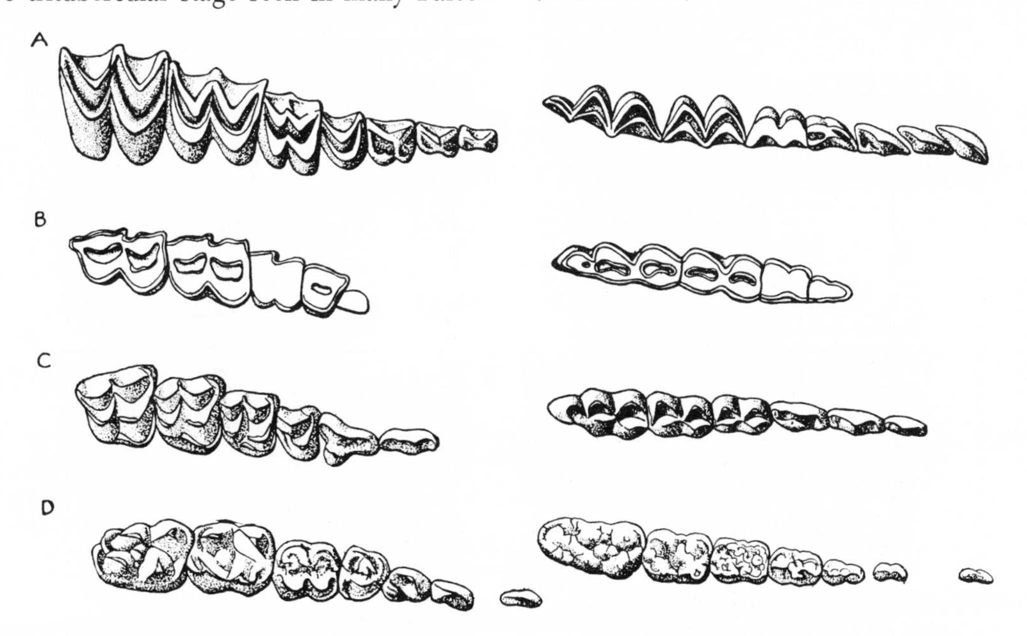

FIG. 401.—The cheek teeth of artiodactyls. *Left,* Teeth of upper right side; *right,* teeth of lower left side. *A, Merycochoerus,* a Miocene oreodon, ×1/3 approx.; *B, Camelops,* a Pleistocene camelid, ×2/7; *C, Archaeomeryx,* a primitive fossil pecoran, ×3/2 approx.; *D, Chleuastochoerus,* a Pliocene pig, ×5/8. (*A* after Loomis; *B* after Merriam; *C* after Matthew; *D* after Pearson.)

archaic assemblage appear to be end forms, without descendants; the earlier Eocene genera, however—specifically *Diacodexis*—may well be the true ancestors of later artiodactyls. Sometimes all (or nearly all) of these archaic artiodactyls are lumped in a single family, Dichobunidae. They show considerable variation, however, and are here arrayed in a series of related families. How to group them formally is a problem without any satisfactory solution; they are neither proper swine nor proper ruminants. One solution, adopted here, is to erect for them a separate basal artiodactyl group as the suborder Palaeodonta.

Primitive swine.—In the suborder Suina we shall here treat of the pigs, peccaries, hippopotami, and a number of fossil groups which may be associated with them. In almost none of these forms is there any development of a selenodont pattern in the teeth; on the contrary, there is usually a highly developed bunodont condition. The limbs are short and quite primitive in nature, generally with four well-developed toes. A cannon bone is almost never formed, and the lower leg bones (in contrast to higher ruminants) almost always remain separate. The stomach is always simple, and the diet of the swine is a mixed one. In typical members of the group the skull shows an amastoid condition. Applying these diagnostic characters to fossil types, we find that in the Suina, in a broad sense, we may include not only the surviving groups mentioned but also the entelodonts or "giant pigs" and the anthracotheres, both prominent in Middle Tertiary faunas.

The Suina are first seen in the Eocene in forms which presumably arose from members of the palaeodont stock. In the middle and later Eocene there are various genera which show evidence of a trend toward characteristic features of the suborder. The teeth are definitely bunodont and often show a pattern of five cusps (reduced in most cases to four at about the end of the period); the canines remain prominent; most show an amastoid condition. Some, like *Cebochoerus,* are short-faced forms, not far removed from such advanced palaeodonts as *Achaenodon* (Fig. 403*A*); others, such as *Choeropotamus*, are longer faced and may be close to the ancestors of the characteristic Suina of later periods.

Entelodonts.—One offshoot of the primitive "swine" stock which early rose to prominence and early disappeared was the family Entelodontidae, the "giant pigs" which flourished in Oligocene times. Common Oligocene genera were *Entelodon* in Europe and the similar *Archaeotherium* (Figs. 402*D,* 403*B*) in America. Last and largest of the group was *Dinohyus* (Figs. 404, 407*C*) of the early Miocene of America. The entelodonts were large forms, with huge skulls often reaching nearly a yard in length. The incisors were long and pointed; the canines, heavy and showing wear-grooves, suggest a root-eating diet. The premolars were simple; the molars were relatively small, with bunodont cusps. In the upper molars there is thought to be a true hypocone. The skull was much elongated, especially the facial region, while the braincase was small; a postorbital bar had already developed. A large flange was present on the zygomatic arch and two tuberosities on the lower jaw. Their functions are uncertain; perhaps they were for muscle attachments. The neck was short; there were high spines in the anterior thoracic region for the support of the heavy head, giving the back a humped appearance. Radius and ulna were fused; and, although no cannon bone was formed, the lateral toes were reduced to vestiges. These large beasts were not closely related to the true swine but resembled them in many features and may have been similar in habits.

Pigs and peccaries.—The most widespread of Suina in late Tertiary times and, except for the hippopotamus, the only survivors of the group are the pigs and peccaries. In these forms the long skull (Fig. 403*E, F*), low in front, ascends steeply toward the back of the head. The orbits are open behind. The canines have a persistent growth and are especially powerful in the males. The dentition (Fig. 401*D*) is often complete, but an incisor or the first premolar may drop out. The

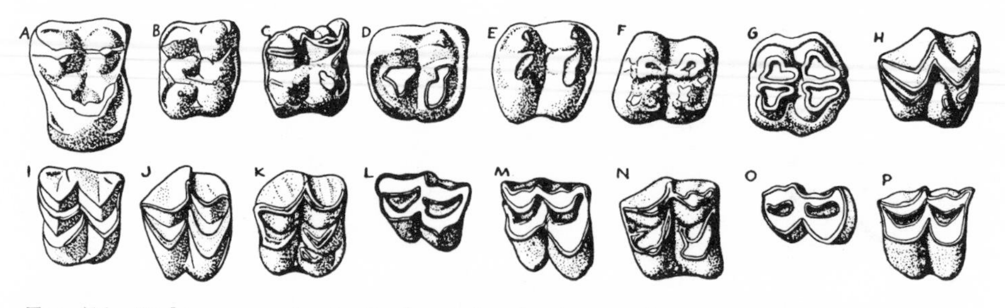

FIG. 402.—Right upper molar teeth of artiodactyls. *A, Diacodexis,* a Lower Eocene form with tritubercular molars; *B, Dichobune,* a hypocone present; *C, Anthracotherium,* a bunoselenodont, five-cusped molar; *D, Archaeotherium,* an entelodont, with a piglike tooth, hypocone developed; *E, Platygonus,* a peccary; *F, Sus erymanthus,* a Pliocene pig; *G,* a Pleistocene *Hippopotamus; H, Anoplotherium,* a bunoselenodont type with protoconule still present; *I, Cainotherium,* with protocone migrated to back half of tooth; *J, "Oreodon,"* a simple, four-cusped selenodont tooth; *K, Xiphodon; L, Alticamelus,* a Pliocene camel; *M, Samotherium,* a Pliocene giraffe; *N, Dicrocerus,* a primitive deer; *O, Tetrameryx,* a Pleistocene prongbuck; *P, Tragocerus,* a Pliocene bovid. (*A* after Sinclair; *B, I* after Stehlin; *C, F, H, P* after Gaudry; *D, J* after Scott; *E* after Osborn; *G* after Cuvier; *K, N* after Schlosser; *L* after Cook; *M* after Ringström; *O* after Lull.)

lateral toes, although usually complete, are small and function only on soft ground.

The typical pigs are Old World forms and have never penetrated America. A characteristic pig feature is that the upper canines curve outward and upward, a peculiarity which reaches its extreme in the babirussa of the East Indies, in which these teeth coil upward in hornlike fashion over the forehead. Besides the typical wild boar and the domesticated varieties (*Sus*), various specialized types are present in the Old World tropics. True pigs are unknown in the Eocene, but some of the rather poorly known Suina of that period may be reasonably regarded as swine ancestors. Appearing in the Oligocene, they became highly successful in later epochs. Canine tusks, little developed in the earliest forms, became prominent in the Miocene; in *Listriodon,* common in that epoch, it is, exceptionally, the lower rather than the upper tusks which are most highly developed. The molar teeth, primitively with a simple bunodont pattern, tended to a multiplication of cusps in the late Tertiary; in *Phacochoerus* and its relatives of Pleistocene and modern times in Africa, the last molars are elongated, with two dozen cusps or so. *Kubanochoerus* of the Miocene of the Caucasus was a giant form in which a median bony horn extended forward from the forehead.

The living peccaries of the New World, *Tayassu* (*Dicotyles*), are related to the pigs but have had a long separate history. A key character is found in the fact that the upper canines, although well developed, have remained in the normal vertical position. The molars (Fig. 402*E*) are comparatively short and simple in appearance in contrast with the wrinkled teeth of the Old World hogs (Fig. 402*F*). In some peccaries the side toes are much reduced (Fig. 407*B*). There is the beginning of the formation of a cannon bone, and radius and ulna become fused. In these features the peccaries are much more progressive than their Old World relatives.

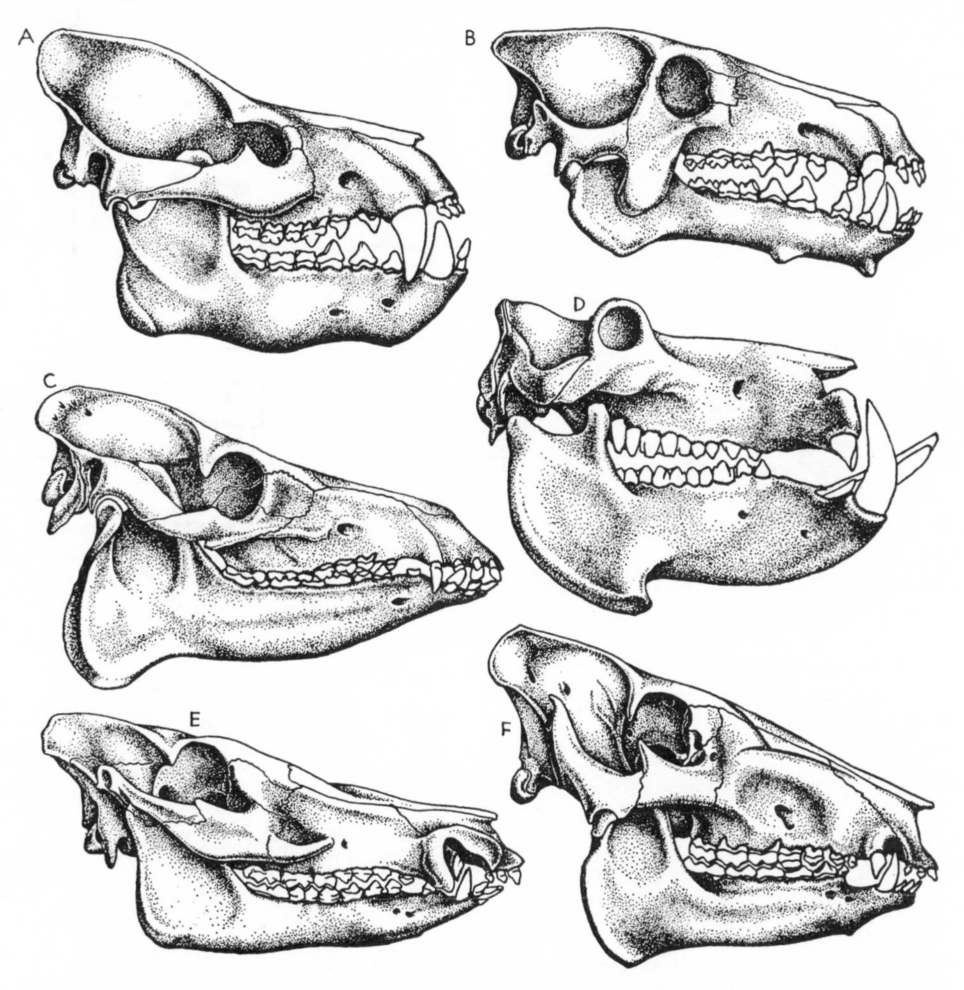

Fig. 403.—Skulls of piglike artiodactyls. *A, Achaenodon,* an upper Eocene "short-faced pig," skull length about 15 inches; *B, Archaeotherium,* an Oligocene entelodont, average skull length about 18 inches; *C, Bothriodon,* an Oligocene anthracothere, length of skull about 17 inches; *D,* a Pleistocene European *Hippopotamus,* skull length about 2 feet; *E, Chleuastochoerus,* a late Tertiary pig from China (boar), skull length about 12 inches; *F, Perchoerus,* an Oligocene and Miocene peccary, skull length about 10.5 inches. (*A, F* after Peterson; *B, C* after Scott; *D* after Reynolds; *E* after Pearson.)

The modern peccary ranges through South America, which it reached only at the end of the Pliocene, and as far north as Texas. In the Pleistocene, *Platygonus* (Fig. 406) and several other forms were still widespread in temperate North America. The oldest known North American peccary is *Perchoerus* [*Thinohyus*] (Fig. 403*F*), of the Oligocene. It seems certain that the peccaries are an offshoot of the primitive pigs, and certain Oligocene and Miocene Eurasian genera appear to represent persistent Old World peccaries.

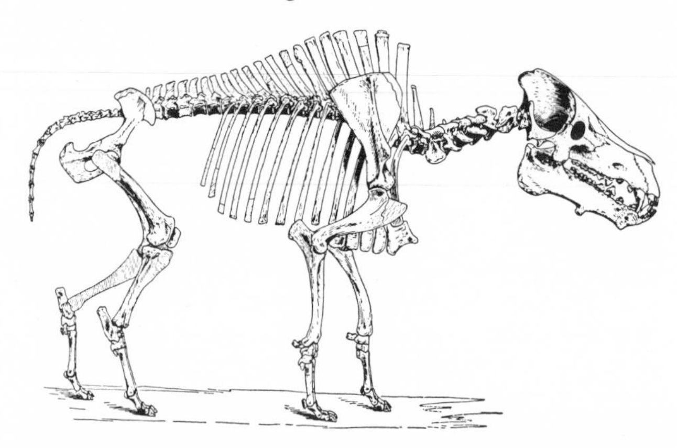

Fig. 404.—*Dinohyus*, a giant piglike artiodactyl (Entelodontidae) from the Lower Miocene of America; original about 10.5 feet long. (From Peterson.)

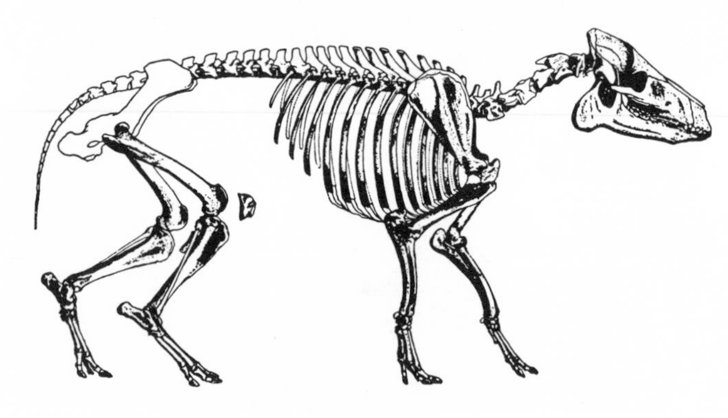

Fig. 405.—*Bothriodon*, an Oligocene anthracothere, about 5 feet in length. (From Scott.)

Anthracotheres.—*Anthracotherium* and *Bothriodon* (Figs. 403*C*, 405, 407*A*) were typical members of a group of rather large but primitive artiodactyls common in the Middle Tertiary of the Old World. The general impression is that of a rather piglike form. The limbs were short, with four functional toes and with a persisting pollex in early genera. The skull was low and generally had a long facial region. Except in some Eocene genera, the mastoid exposure had been reduced, as in the Suina generally. The dentition was complete, with incisors and canines normally developed. The molars were low crowned; some early anthracotheres were bunodont, but later forms (exceptional in the present suborder) tend to develop a somewhat selenodont pattern. In most anthracotheres a five-cusped upper-molar pattern was present, and the evidence from Eocene forms indicates that one of these cusps is a hypocone. The position of the anthracotheres has been much debated; but, except for the selenodont tendency in the cheek teeth, their affinities seem clearly to lie with the swine group.

The conditions under which anthracothere remains are found strongly suggest that they were amphibious, comparable in habits to the later hippopotami. Anthracotheres appear in the Middle Eocene; numerous remains have been found in the Upper Eocene of Asia, and that continent appears to have been the center of development of the group. The early forms appear to be still close to the ancestral palaeodonts in structure.

Anthracotherium was a long-lived Old World genus.

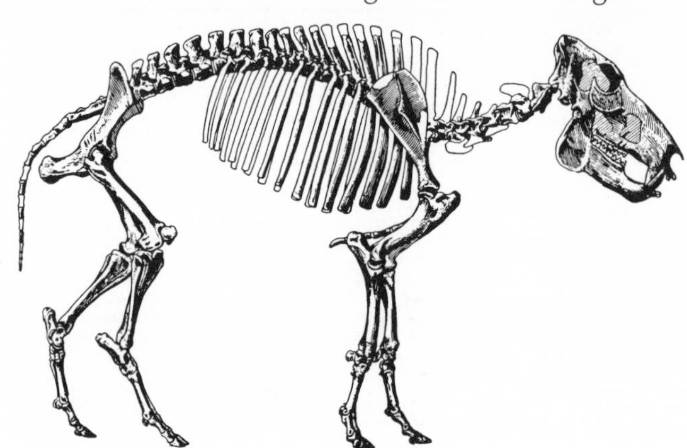

Fig. 406.—*Platygonus*, a Pleistocene peccary, about 3.5 feet long. (From Hay.)

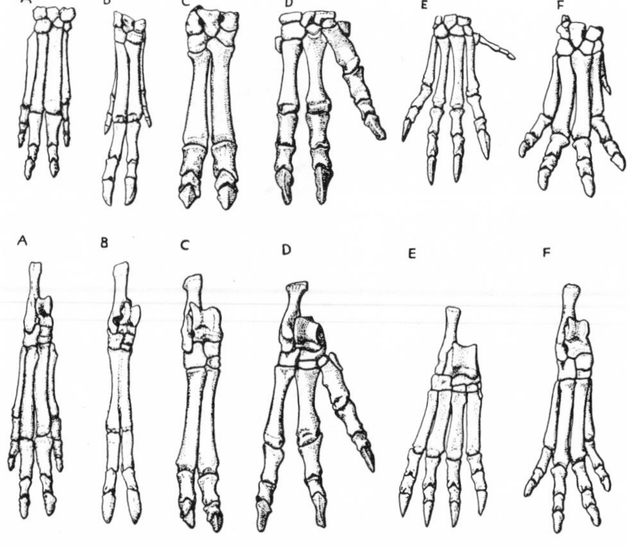

Fig. 407.—Feet of various artiodactyls. Manus above, pes below. *A*, *Bothriodon*, an anthracothere; *B*, *Mylohyus*, an advanced Pleistocene type of peccary, with side toes reduced in front, lost behind, and a cannon bone formed in the pes; *C*, *Dinohyus*, a Lower Miocene entelodont; *D*, *Diplobune*, an aberrant three-toed anoplothere; *E*, *Agriochoerus*, a clawed Oligocene American selenodont. In this and in *F*, *"Oreodon,"* the pollex is preserved. (*A* after Kowalevsky; *B* after Brown; *C* after Peterson; *D* after Schlosser; *E* after Wortman; *F* after Osborn.)

Bothriodon [*Ancodus*] was one of several genera which invaded North America in the Oligocene and Miocene; the group made little headway in the Western Hemisphere, however, possibly owing to competition with oreodonts of similar habits. In the Old World, *Merycopotamus* of the late Tertiary survived in the Pleistocene of the tropics.

Hippopotami.—The hippopotamus is the only living amphibious artiodactyl. All known members of the family may be reasonably included in the single genus *Hippopotamus* (Figs. 402*G*, 403*D*) and have been confined to the Old World. The hippopotamus lives on soft, water vegetation which is cropped with the heavy anterior teeth and lips. The incisors, flanked by the very stout canines, are set in a transverse row in the broad snout. All the teeth are present in the hippopotamus except the outer incisors, and even these were present in some fossil types. The molars are rather piglike, bunodont with four major cusps. The eyes are small and, in correlation with the amphibious habits, are set high up on the skull; there is the beginning of a postorbital bar. The plump body is supported by short, stout legs; four toes, all functional, are present in the broad foot. Today the hippopotami are confined to Africa, but in the Pleistocene they were widespread in the warmer regions of the Old World. A form similar to the larger African races was abundant in Europe, while pigmy types were present on some of the Mediterranean islands and in Madagascar (a pigmy is found today in Liberia). In the earliest Pleistocene, hippopotami were present in the Mediterranean area and India, those from the latter region being the more primitive in the possession of the complete set of six incisors rather than the four present today. There is considerable evidence that they are descendants of the anthracotheres, which they resemble in their mode of life and in some structural features. Most anthracotheres were long snouted, but *Merycopotamus* of the later Tertiary has a broad snout, flanked at either margin by stout canines, suggesting the initiation of hippopotamus specializations; further, while most anthracotheres had five-cusped upper molars, this genus had reduced to a four-cusped type, from which the hippopotamus tooth might have been derived.

Primitive ruminants: the tylopod assemblage.—Much more numerous than the Suina in both fossil and Recent states is a vast array of forms, of which the camels, deer, giraffes, and cattle are but a few conspicuous representatives. The cud-chewing habit, associated with the development of a complicated stomach, is a diagnostic feature. Numerous common characters or trends are likewise present in the skeleton. The cheek teeth are invariably selenodont in pattern; the anterior portion of the dental series is much modified, usually with reduction of the canines; the mastoid region remains exposed on the surface of the skull; the limbs are elongated in modern genera, with marked reduction of the lateral toes and with cannon bone formation. What names to apply to this group and its subdivisions is a problem. The term "Ruminantia" is sometimes restricted to advanced types; here we will instead utilize this term in a wide sense to embrace this entire series of cud-chewing selenodonts.

The living ruminants are divisible into two groups, or infraorders, Tylopoda and Pecora, the first including only the camels and llamas, the second all the remaining forms. Of the two, it is obvious that the Tylopoda are the more primitive, despite specializations seen in living genera. Carpal and tarsal bones, fused to a greater or lesser degree in Pecora, remain distinct. The complexity of the ruminating stomach is greater in pecorans than in camels. In pecorans the lower canine has become incisiform, so that there appear to be four spoon-shaped incisors in each jaw half, and in all living members of the group the upper incisors are lost; the modern camel dentition has been modified to a lesser extent, although there may be reduction and change. If we trace the ancestry of the Pecora back through the Tertiary, we find that they are relatively rare in the Oligocene and almost unknown in the Eocene. The camels, on the other hand, when followed back to the Eocene, are seen to take origin from a varied array of early and primitive ruminant-like artiodactyls which decrease in importance in later Tertiary stages. This complex of primitive ruminants is sometimes included, in whole or in part, under such terms as Bunoselenodontia or Ancodonta. We shall here include them in the Tylopoda in a broad sense, although recognizing that the modern camels are specialized remnants rather than generalized representatives of these archaic ruminant groups.

In Eocene groups included in the Suina we have seen the appearance of genera which tended to preserve a primitive short-limb structure and a primitive disposition of the incisors and canines, coupled with the development of cheek teeth with a bunodont pattern. Most of the early ruminants here included in the Tylopoda agree with them in a conservative limb structure. But in most respects even the early ruminants tended to differ sharply from the swinelike forms. The cheek teeth rapidly assumed the selenodont pattern of modern ruminants, and (perhaps with the adoption of a purely herbivorous diet) the anterior part of the tooth row changed markedly. Instead of the generalized condition of sharply accentuated canines separating incisors from premolars, the teeth tended to become evenly graded without break from incisors to the canines and to the adjacent premolars. The lower canines are always reduced and are difficult to distinguish from their neighbors. The upper canines are sometimes prominent; but when this condition is found in later camels, it is surely due to secondary growth of the canines, and the same history may be true in other instances. In many early members of this group the upper molars have five cusps, three in the front half of the tooth, two in the posterior row. The three anterior cusps are surely protocone, protoconule, and paracone.

In later forms the protoconule is reduced. In the posterior half of the tooth the outer cusp is, of course, the metacone; the question as to the nature of the inner cusp—metaconule or hypocone?—was discussed earlier.

In this tylopod group of primitive ruminants we include the Anoplotheriidae, Cainotheriidae, Xiphodontidae, Amphimerycidae, Agriochoeridae, Merycoidodontidae, Oromerycidae, and Camelidae. These families were prominent in the late Eocene and Oligocene. However, only the oreodonts and camels of America persisted into the later Tertiary, and of the entire assemblage only the highly specialized camel family has survived.

Anoplotheres.—Assemblages of primitive ruminants appear to have developed, independently of one another, in the Old and New Worlds, in late Eocene and Oligocene times. We shall first consider four families confined to the Old World before discussing North American types.

The anoplotheres, whose careers began in the Middle Eocene and ended before the close of the Oligocene, are in some ways the most archaic of all the forms we shall include in the ruminant assemblage. In the molar teeth (five-cusped above) the outer line of cusps were of a selenodont pattern, the inner were persistently bunodont. The body and limbs were stoutly built. The feet, seldom preserved, appear to have had a varied pattern; *Diplobune,* for example, developed a peculiar three-toed type (Fig. 407*D*). *Anoplotherium* (Figs. 402*H*, 408*B*), common in Europe on both sides of the Eocene-Oligocene boundary, grew to be about 3 feet high at the shoulders.

Cainotheres.—*Cainotherium* and related genera, likewise characteristic of the late Eocene and Oligocene of Europe, form a family of a very special type. *Cainotherium* (Figs. 402*I,* 408*D,* 409) was about the size of a hare and curiously suggestive of hares and rabbits in a number of structural features. This does not, of course, imply any relationship, but indicates the possibility that cainotheres (and perhaps small ancestral ruminants in general) were not unlike hares in habits and adaptations. The limbs were slender and, although five toes were present in the manus in some cases, there was a strong trend (comparable to that in the camels and Pecora) toward reduction of the side toes. The hind legs were markedly elongated, suggesting a bounding (although perhaps not a truly hopping) gait. In the skull the auditory bullae were enormous (as they tend to be in many small mammals), and the antorbital region was fenestrated much as in lagomorphs. The tooth row was an even and continuous one, with the canines indistinguishable from their neighbors. The one nontylopod specialization of the cainotheres is that, quite the opposite of other early selenodonts, there are three cusps in the back of the upper molars, two in front. One would tend to think that in this case, at least, the posterointernal cusp is a hypocone, but it is claimed that it is actually the protocone shifted posteriorly.

Xiphodonts.—Often regarded as close relatives of the camels are *Xiphodon* and a few related forms which were very common in Europe on the Eocene-Oligocene boundary. They closely resembled the early camels in both limbs and teeth. They, too, were precociously didactyl, with long, slim limbs retaining only small splints of the side metapodials. As in the ancestral camels, the dentition was complete and without a

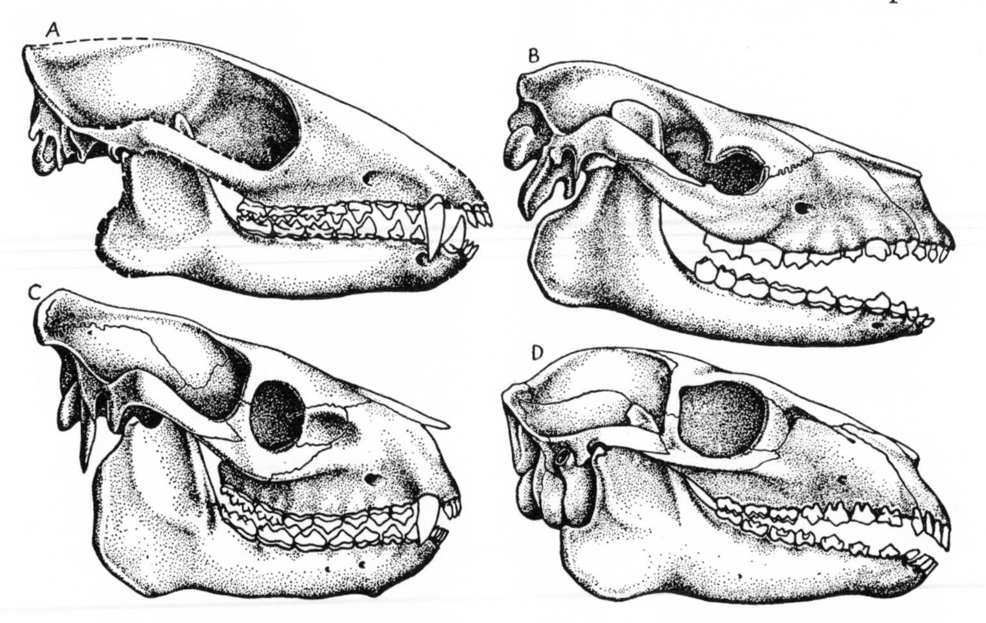

FIG. 408.—Skulls of some early artiodactyls. *A, Homacodon,* a primitive Eocene type (jaw restored from a related genus); *B, Anoplotherium,* a late Eocene anoplothere, skull length about 14 inches; *C, Oreodon culbertsoni,* an Oligocene oreodont, length of skull about 5 inches; *D, Cainotherium,* length of skull about 3 inches. (*A* after Sinclair; *B* after Blainville; *C* after Leidy; *D* after Schlosser and Hürzeler.)

diastema, and the teeth graded evenly from incisors into the cheek teeth. The upper molars (Fig. 402*K*) still had the primitive (not cainothere!) five-cusped pattern, the protoconule remaining distinct. It is probable that the xiphodonts are a European parallel to the camels, developed, like them, from an anoplothere-like stock.

Amphimerycidae.—A final Old World group of primitive ruminants is that including only *Amphimeryx* and *Pseudamphimeryx* of the later Eocene and early Oligocene. In these little animals the molars were fully crescentic, although still five-cusped above, and it is thought by some that they may be ancestral to the advanced ruminants of the traguloid and higher pecoran groups.

Oreodonts.—The four families described above are unknown in North America. In the late Eocene of this continent, however, there developed, in parallel with the Old World forms, a varied series of selenodont artiodactyls in which short limbs, five-cusped upper molars, and a tendency toward an evening-up of the tooth row are features frequently encountered. These include forerunners of the camels, and of the most successful of American ungulates of the Middle Tertiary—the oreodonts of the family Merycoidodontidae (Oreodontidae).

Typical oreodonts tended to be rather heavily built animals (Fig. 410), somewhat piglike in general appearance, with short limbs, always four toed and in some cases even with a small pollex (cf. Fig. 407*E*).

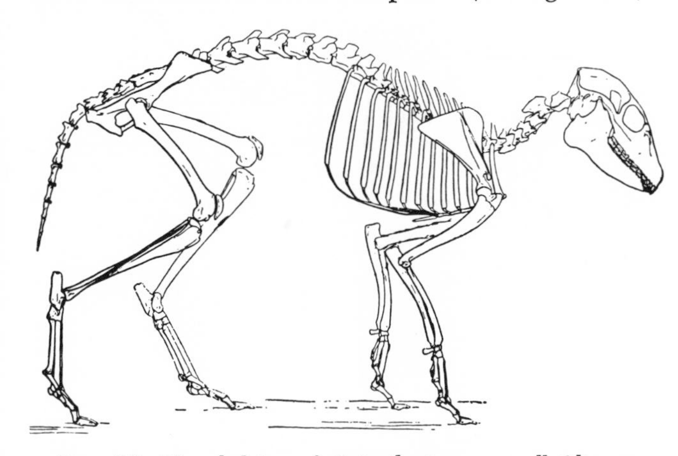

Fig. 409.—The skeleton of *Cainotherium*, a small Oligocene ruminant; length of original about 1 foot. (After Hürzeler.)

The teeth (Fig. 401*A*), on the other hand, indicate relationship to the ruminants. The tooth row was unbroken, without diastema or loss of teeth; the upper canine was a stout and moderately projecting chisel, but except for this the dentition was comparable to those of the other primitive tylopods just described. The lower canine had been reduced to the size of the incisors and had been functionally replaced by a somewhat enlarged first premolar—a feature which we shall see repeated in some more progressive ruminants. The cheek teeth were strongly selenodont in pattern, usually brachodont, but rather high crowned in some large

and late genera. The upper molars are four cusped, with the disappearance of the protoconule. The skull was large, the orbit usually closed behind; there was often an antorbital pit or opening presumably containing a facial gland as in higher ruminants.

This combination of a piglike body build with a more progressive dental structure has led to the popular name of "ruminating swine" for the oreodonts. They seem surely, however, to belong to the ruminant stock and in almost every way to fit into the general group of

Fig. 410.—*Promerycochoerus*, a Lower Miocene oreodon; original about 5.5 feet long. (From Peterson.)

archaic ruminants with such forms as the caenotheres and anoplotheres. But while these European families were rapidly eliminated, the oreodonts of North America flourished for most of the Tertiary, perhaps owing to less-intensive competition from higher artiodactyls. The first true oreodonts are found in the early Oligocene. They rapidly grew to prominence, and during the Oligocene and early Miocene oreodont fossils outnumber all other mammals combined in North American continental strata; the major collecting grounds in the Big Badlands of South Dakota, for example, are known as the "Oreodon beds" because of the abundance of the characteristic Oligocene genus *Merycoidodon* [*Oreodon*]. More than a score of genera were developed. For the most part they remained conservative in structure, but such Miocene genera as *Promerycochoerus* (Fig. 410), *Merychyus,* and *Brachycrus* [*Pronomotherium*], as well as Pliocene survivors, showed specialization in increase in size, development of hypsodont teeth, loss of incisors, and (in the last-named genus) an upward and backward retreat of the nasal opening, indicating proboscis development.

Agriochoeres.—Contemporary with the early oreodonts of the Oligocene was the genus *Agriochoerus.* In many regards the two were very similar. But *Agriochoerus* was more primitive in such features as the retention of five-cusped upper molars and a long tail, and —most exceptional for an artiodactyl—had redeveloped claws on its feet. We have little idea of the nature of the life of this aberrant form; it has been pictured both as a tree dweller and a digger for roots and tubers. *Protoreodon* of the Upper Eocene has often been included (as the name suggests) in the oreodont family.

It is currently placed in the Agriochoeridae but is quite surely close to the common stem of both groups.

Camels.—The sole surviving family of the tylopod assemblage is that of the Camelidae, represented today by the Old World camels and the llamas of South America. These are highly specialized forms which, however, can be traced back in the Tertiary of North America to primitive Upper Eocene ancestors, similar in structure to the tylopod families previously considered.

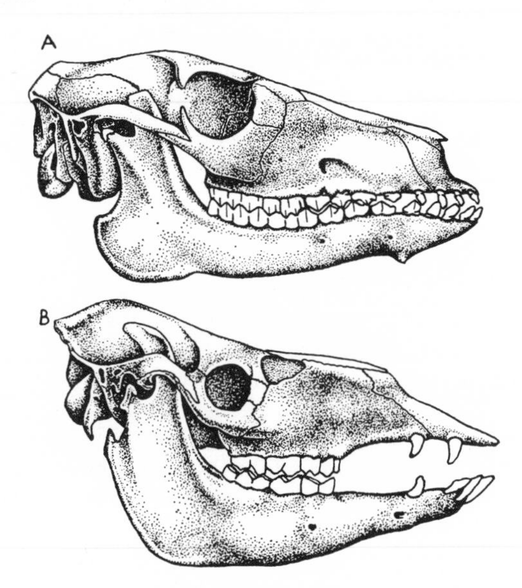

FIG. 411.—The skull of camelids. *A, Poëbrotherium,* a primitive Oligocene form, skull length about 6.25 inches; *B, Camelops,* a Pleistocene American type, skull length about 22.5 inches. (*A* after Wortman; *B* after Merriam.)

Despite their dissimilarity in superficial appearance, the camel and llama are essentially alike in structure, the hump of the camel and the heavier hairy covering of the South American llamas being probably recently acquired features associated with the habitats of these modern forms. Unlike the majority of the Pecora, members of the camel group are never horned. There is a postorbital bar in the living forms; the tympanic bulla is filled with spongy bone.

Modern camelids have a cropping mechanism similar to that of the Pecora, in which the lower incisors extend forward and the herbage is pressed between these and the gums over the premaxillae. The reduction of the upper incisors has taken place at a slower tempo than in the other living ruminants, however, and the lateral one is still retained. A small canine is present, and the premolars have been much reduced with the development of the diastema.

The neck is elongated, and the limbs are long. Ulna and fibula are, of course, much reduced, the latter to a tarsal-like nubbin, and there are no traces of lateral toes. As in the Pecora, the trapezium has gone from the carpus, and the mesocuneiform and ectocuneiform have fused in the tarsus; but the other elements are still separate. The feet (Fig. 413*B*) are exceedingly characteristic; for, unlike normal ungulates, the living camelids are digitigrade, with spreading toes set nearly flat on the ground, and, instead of a hoof, there is a small nail and a heavy pad beneath the toes.

The division of the stomach into chambers has not proceeded so far in camels as in the Pecora, but the development of stomach pockets for water storage is peculiar to this group.

An early camelid is *Poëbrotherium* (Fig. 411*A*) of the North American Oligocene, somewhat smaller than a sheep. The dentition was complete, with canines and incisors of similar build and grading over into the elongated premolars with almost no trace of a diastema, as in anoplotheres and cainotheres. The orbit was still open behind, although the two processes which were to bridge the gap were already long. The side toes had already been lost at this early date; the two remaining metapodials were still separate but already showed a tendency toward distal divergence of the two toes so prominent in later camels; the animal presumably had a hoof rather than the modern padded camel foot (Fig. 413*A*).

There were a large number of camelid genera in the later Tertiary, almost all of them until near the end of that period exclusively North American. Along the main evolutionary line there was a gradual increase in

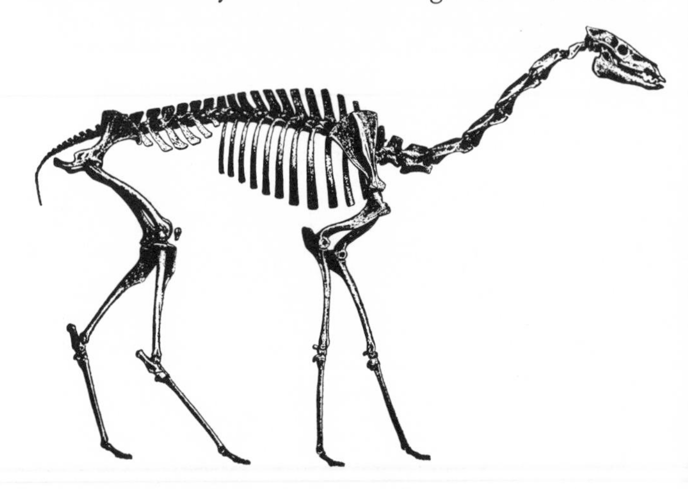

FIG. 412.—*Oxydactylus,* a long-limbed Miocene camel; original about 7.5 feet long. (From Peterson.)

size in Miocene and Pliocene forms such as *Procamelus* and *Pliauchenia,* the orbit became enclosed by bone, a diastema developed, the two inner upper incisors dropped out, and the metapodials fused to form a cannon bone. During the later half of the Tertiary, when camels were numerous in North America, there were many side branches, including such forms as the "gazelle camel," *Stenomylus,* a small and graceful type from the Lower Miocene, and the group of "giraffe camels," such as *Oxydactylus* (Fig. 412), with very long neck and legs.

In the Pliocene appeared forms much closer to the living camelids in which the true ungulate type of hoof was abandoned and the flat, spreading type of toe was developed. The differentiation of modern types was then well underway; at the end of the Tertiary, true camels, *Camelus,* reached the Old World to spread widely there during the Pleistocene, while relatives of the llama, *Lama* [*Auchenia*], reached their South American home at about the same time. In North America, *Camelops* (Fig. 411*B*) and other forms not only persisted into the Pleistocene but lived in the Southwest until comparatively recent times. As in the case of the horses, the cause of the extinction of this group in the region which had been so long their home is not obvious, for, when reintroduced during the last century, camels were well able to survive in a wild state in the western deserts.

Oromerycids.—Fragmentary remains of an ancestral camel similar to *Poëbrotherium* are known from the Upper Eocene. Much more common, however, at that time were forms such as *Eotylopus, Protylopus,* and *Oromeryx,* formerly included in the Camelidae, but now considered to form a distinct, although probably related, family. As would be expected, the oromerycids were more primitive than true camels in numerous regards: the limbs were short and the front feet four toed; and although there were but four upper molar cusps, the protocone had a peculiar forked shape, suggesting that it had recently fused with the "extra" protoconule.

The Pecora.—All remaining artiodactyls, including the majority of living members of the order, may be included in the Pecora in a broad use of that term. There are five generally recognized modern families and, in addition, several extinct groups. These are arranged here in three superfamilies, one primitive and two advanced, as follows: (1) superfamily Traguloidea, with families Hypertragulidae, Protoceratidae, Gelocidae, and Tragulidae (chevrotains); (2) superfamily Cervoidea, with families Palaeomerycidae, Giraffidae (giraffes), and Cervidae (deer); (3) superfamily Bovoidea, with families Antilocapridae (American prongbuck) and Bovidae (cattle, sheep, goats, antelopes).

In these specialized and very successful ungulates the stomach is complicated in structure and the ruminating habit highly developed. The build is, in general, a graceful one; and the limbs are long and slender. All are functionally two-toed types; but reduction of the side digits has gone on at a slower pace than in the tylopods, for small lateral toes are present in a number of living pecorans as well as in many fossil forms. The two principal metapodials have usually fused into an elongated cannon bone. There has been considerable fusion in carpal and tarsal elements, and the union of navicular and cuboid in the ankle is a feature diagnostic of the group. The ulna is reduced, and of the fibula there remains in typical living forms only a nubbin of bone in the ankle region.

Most living pecorans possess weapons in the shape of horns or similar bony outgrowths from the skull for defense or, in males, for breeding-season fights. In

many primitive types, both living and extinct, these are absent; but long, stabbing upper canines are frequently present in their stead. The upper incisors have given way to a horny pad as part of the cropping mechanism. There appear, at first glance, to be four lower incisors, for the canine has been taken over into the incisor group. One premolar disappeared early with the development of a diastema. The molars have, since early times, been selenodont in pattern with but four cusps in the upper molars, the protoconule having been lost, as in camels and oreodonts.

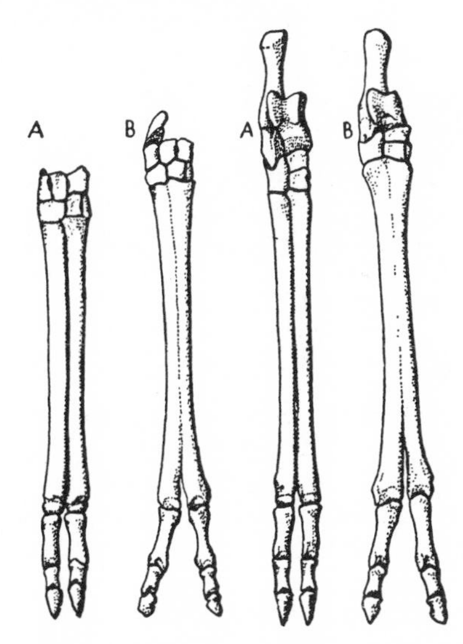

FIG. 413.—Manus (*left*) and pes (*right*) of camels. *A,* The Oligocene *Poëbrotherium; B,* the Pliocene *Procamelus.* (After Wortman.)

The chevrotains.—The older pecorans are here included in a superfamily Traguloidea, taking its name from the chevrotains, which alone survive. These forms from the Old World tropics give us some idea of the general nature of ancestral pecorans. Only two genera are present, *Tragulus* of the oriental region and *Hyemoschus* of tropical Africa. Both are small animals about a foot in height and weighing but half a dozen pounds. In their small size, retiring habits, and even, at first glance, their general appearance, one would tend to mistake them for rather large rodents, such as the agoutis of South America. But in every structural feature they are proper pecorans, comparable to deer or antelope, although with a number of more primitive characters.

The chevrotains are cud chewers, but the associated subdivisions of the stomach are less complex than in other living pecoran families. Unlike most progressive pecorans, they have no horns or antlers; instead there are (especially in the males) large upper canines as weapons. The upper incisors are absent, as in pecorans generally, and the typical pecoran cropping structure of lower incisors and canines is well developed. The limbs are long and functionally two-toed structures, much more progressive than in any forms considered earlier except the later camels and the xiphodonts; in the tarsus, the navicular, cuboid, and ectocuneiform bones are fused into a unit, a feature known only in pecorans. On the other hand, the limbs are less specialized than in advanced pecoran types. The fibula, reduced in higher families, is still a complete bone, although fused to the tibia. Each foot is four toed, and the lateral toes, although short and slender, are complete structures with respect to both phalanges and metacarpals, whereas the metacarpals are never complete in higher pecorans and the toes themselves are often reduced. In all higher pecorans the middle metapodials are fused in both front and hind legs into a cannon bone. In the living chevrotains such fusion has taken place in the hind limb in both genera; but in the African genus the two metacarpals are still separate, and they are but partially fused in the Asiatic form. In advanced pecorans the keels on the distal ends of the metapodials are present all the way from the dorsal to the ventral surface of the bone; in chevrotains and their fossil relatives the keels are present only on the ventral surfaces, suggesting a different and more limited movement of the toes (Fig. 415*A*). The front feet are noticeably shorter than the hind, whereas there is little difference in length in higher pecorans.

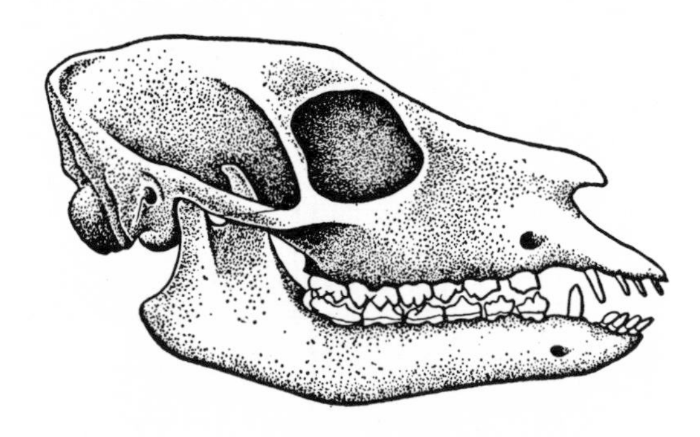

Fig. 414.—The skull of *Archaeomeryx*, a primitive pecoran from the Eocene of Mongolia. (After Colbert.)

In various skeletal features, then, the chevrotains are demonstrably on a lower plane of organization than typical pecorans. Absence of horns, large upper canines in the males, limbs relatively unprogressive, particularly the anterior ones, complete side toes, and cannon bones in process of formation—these are characters which we may expect in ancestral fossil pecorans.

Early traguloids.—Primitive pecorans are not uncommon in the Oligocene; in the Eocene, however, they are rare, and the origins of the group are still obscure. It is possible that the ancestral pecorans passed through a tylopod stage in their evolution; on the other hand, they may have arisen directly from palaeodont ancestors.

In the late Eocene there are fairly numerous remains of primitive traguloids which had already lost the extra upper molar cusp but which exhibit in other respects a condition as primitive as that seen in the living traguloids and appear to be very close to the actual ancestry of the existing Pecora. *Archaeomeryx* (Figs. 401*C*, 414) of Mongolia is such a genus; many of the features seen are still preserved in Oligocene types such as *Leptomeryx*. *Archaeomeryx* was about the size of the existing chevrotains and, except for a long tail—presumably primitive—had much of the proportions and probably much of the appearance of these modern genera. In the skull the upper canines were modestly developed. The front limbs were considerably shorter than the hind; the lateral toes and their metapodials were complete. In one item, *Archaeomeryx* has gone beyond the living traguloids, for its fibula had already become reduced by loss of its shaft. On the other hand, it was more primitive than the living genera in two respects: the metapodials were still separate in both front and hind limbs and—most unusually—a full set of small incisors was still present in the premaxilla.

Hypertragulids.—*Archaeomeryx*, as we have just seen, is in most respects a primitive pecoran which may lie close to the stem of the suborder. In one feature, however, it shows a specialization not expected in an ancestor of the higher Pecora. Here, as in the oreodons, an enlarged lower premolar has taken the functional place of the canine. This does not occur in the true tragulids or their descendants but is a diagnostic character of the family Hypertragulidae, a group mainly characteristic of the Oligocene and early Miocene of North America, although with a few representatives in the Eurasian late Eocene and Oligocene. Most of these forms have lost the upper incisors in typical pecoran fashion except for occasional vestiges; but in many genera, such as *Leptomeryx* (Figs. 415*A*, 416, 417*A*) and *Hypertragulas*, the general structure seen in *Archaeomeryx* has been retained. There are variations, however. The upper canine and the associated lower premolar vary in size—partly, perhaps, as a sexual feature. The lateral metatarsals may be reduced and the medial ones form a cannon, as in *Leptomeryx*. No hypertragulids were of any great size, and *Hypisodus* was a tiny creature no larger than a cottontail rabbit.

The Protoceras group.—A series of odd forms, developed in North America from the hypertragulids, is that best known from *Protoceras* of the Oligocene. In this genus the postcranial skeleton and many skull characters, including the characteristic canine-like lower premolar, are very similar to those of the hypertragulids. Advances, however, have taken place in other features, such as a considerable elongation of the face and a down-turning of the face on the braincase like that seen in some of the higher pecorans. Most characteristic, however, is the development in the males of hornlike structures—peaks of bone rising upward in the nasal region, above the orbits and above the braincase roof. Descendants of *Protoceras* persisted into the Pliocene, where *Synthetoceras* (Fig. 417*B*) shows a grotesque elaboration of two of these horn pairs, the nasal pair forming a curious compound structure.

Old World traguloids.—Presumably much more important for the evolutionary story than the hypertragulids and their protoceratid descendants were the Old World traguloids, a stock from which the higher pecorans appear to have descended. In this group, absence of the canine-like lower premolar is a diagnostic character; in other respects the early members of the group appear to be almost indistinguishable from the contemporary hypertragulids.

Generally considered as forming an independent family Gelocidae, almost entirely confined to the Oligocene, are such forms as *Gelocus* and *Prodremotherium,* in which limbs—with side toes sharply reduced in front as well as hind feet—and other features are suggestive of a transition toward the more advanced pecoran families. Presumably the living tragulids are a side branch of this progressive stock, but definite members of the modern family first appear in *Dorcatherium* of the Miocene.

Higher pecorans.—The remaining pecoran families show progressive features not present in the traguloids. The characteristically exaggerated upper canines of traguloids persist in some primitive forms, but they are

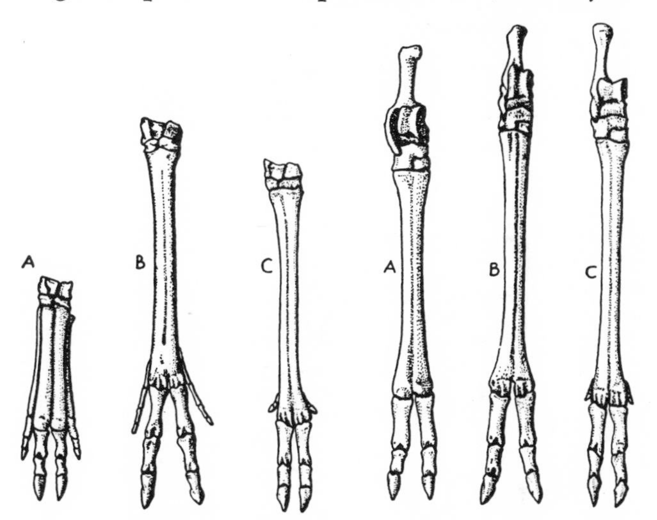

FIG. 415.—Manus (*left*) and pes (*right*) in various Pecora. *A,* The primitive Oligocene *Leptomeryx; B,* the Miocene palaeomerycid *Blastomeryx; C,* the Miocene antilocaprid *Merycodus.* (*B* after Scott; *C* after Matthew.)

typically reduced or absent, and in their stead horns or antlers of some type are usually present as weapons. The progressive development of elongate, two-toed limbs has continued. The ulna is much reduced, except for its head, and is fused to the radius; the fibula has gone, except for a distal nodule—a malleolar bone which looks like an accessory tarsal. Front and hind feet are subequal in length. A cannon bone is universally present and the lateral toes much reduced; never are there complete metapodials, and even the tiny phalanges are likely to be reduced. The keels on the distal ends of the cannon bones are found on the dorsal, as well as the ventral, surfaces. Among living forms the presence of

four chambers in the stomach offers a contrast to the three-chambered stomach of the chevrotains.

These high ruminants appear in the Miocene. Even at the beginning of their history, however, they show a distinct cleavage into two groups, which we shall consider as the superfamilies Cervoidea and Bovoidea. The former include the giraffe and deer families and their common ancestors; the latter include the American prongbucks and the bovid assemblage. The cervoids are browsing types, with low-crowned teeth, typically dwellers in forest or brush country; the bovoids are grassland grazers with high-crowned teeth suitable for

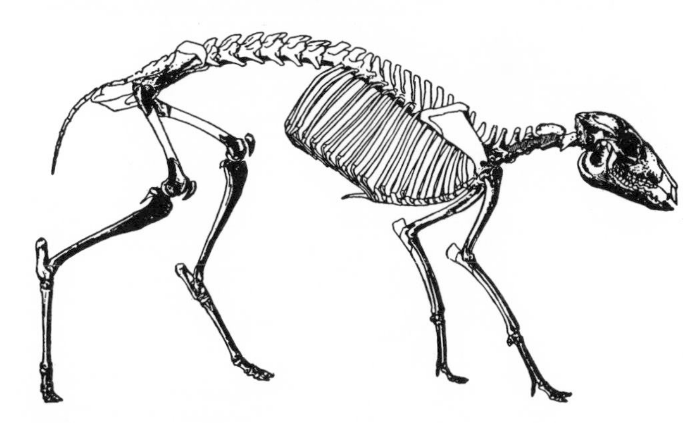

FIG. 416.—*Leptomeryx,* a primitive Oligocene pecoran about 2 feet in length. (From Scott.)

siliceous food. Most have "horns" of some type or other. In the bovoids true horns—bony cores covered by horn (keratin)—are present; in the cervoids these outgrowths lack a horny covering and may be skin covered, as in giraffes or in growing antlers, or may be bare, as in the mature antlers of the deer.

Palaeomerycids.—In the late Oligocene and Miocene of Eurasia and North America are found numerous pecorans which show many of the features to be expected in ancestors of the deer and giraffes but are difficult to assign to either family and appear to represent a common ancestral stock. Certain, at least, of these forms can reasonably be considered as constituting a basal cervoid family Palaeomerycidae. The limbs have the progressive features noted for both cervoids and bovoids; the teeth, as in later giraffes and deer, are low crowned and adapted for browsing. Such forms as the European *Dremotherium* and *Blastomeryx* (Figs. 415*B,* 417*D,* 419) of North America have no horns of any type and have instead (as have traguloids) stout upper canine tusks. (It must be noted, however, that in cervoids "horns" are frequently confined to the males, and hornless specimens sometimes prove to be females of horned types.) I have here included in this primitive family (albeit with some doubt) a series of American cervoids, mainly Miocene, of which *Dromomeryx* and *Cranioceras* (Fig. 418) were typical. Here hornlike structures were present as

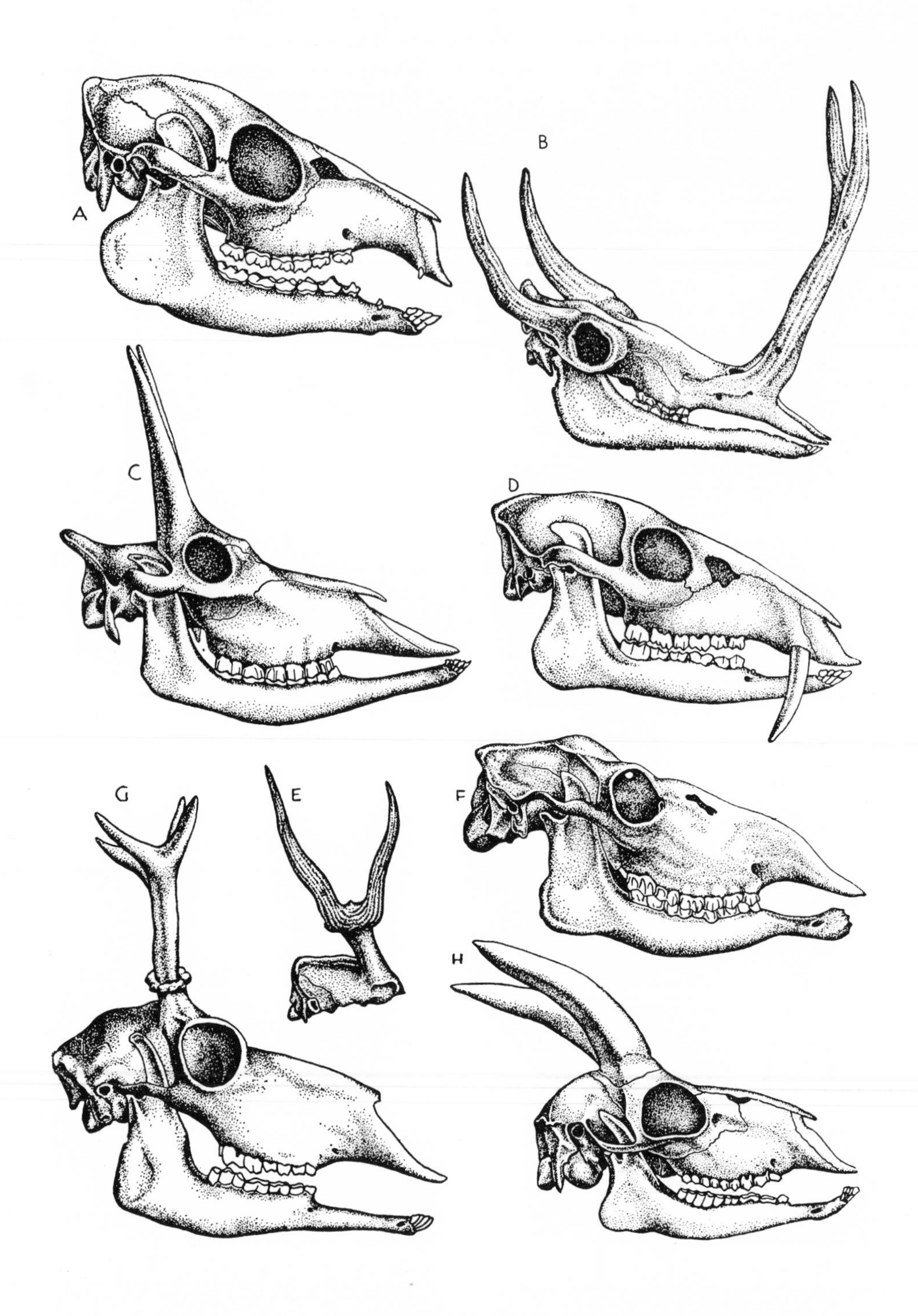

Fig. 417.—Skulls of pecorans. *A, Leptomeryx,* a primitive American Oligocene hypertragulid, skull length about 4.5 inches; *B, Synthetoceras,* a grotesque American Pliocene protoceratid, skull length about 18 inches; *C, Samotherium,* a Pliocene giraffe (a male; the "horns" shorter in the female), skull length about 2 feet; *D, Blastomeryx,* a Miocene hornless palaeomerycid, skull length about 14 inches; *E,* horns of *Dicrocerus,* a European late Miocene cervid; *F,* female of *Megaloceros,* the "Irish elk," skull length 20 inches; *G, Merycodus,* a Miocene antilocaprid, skull length about 7 inches; *H, Gazella brevicornis,* a Pliocene antelope, skull length about 6 inches. (*A, D, G* after Matthew; *B* after Stirton; *C* after Bohlin; *E, H* after Gaudry; *F* after Owen.)

straight bars of bone extending upward at various angles over the orbits. These structures were not, it seems, ever shed, nor is there any evidence of a horny covering. They appear to correspond to the skin-covered bony outgrowths on the skull of a modern giraffe or to the pedicels (usually short) at the base of a deer antler. There are, however, variations in structure; *Cranioceras*, for example, had developed a third median horn extending back from the occiput.

With still greater hesitation we shall here consider as members of this basic stock a small series of Old World Miocene cervoids such as *Lagomeryx*, sometimes considered as forming a separate family paralleling the contemporary *Dromomeryx* group of North America. *Lagomeryx* has progressed further than *Dromomeryx* in "horn" development, for in addition to paired vertical bony spikes extending upward from the skull, there are short side branches, as in the simpler deer antlers. But there is no indication that the "horn" was shed annually as in true deer; and branched "horns," never shed, are found among the extinct

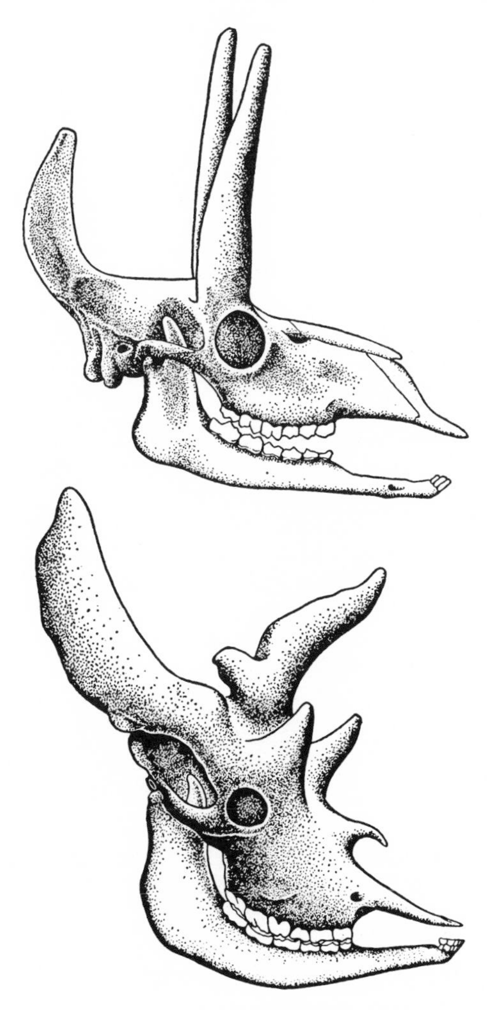

FIG. 418.—*Above, Cranioceras*, a Pliocene palaeomerycid with a median occipital "horn" as well as paired structures; skull length about 1 foot. (After Frick.) *Below, Sivatherium*, a Pleistocene giraffid with large branched horns; size 1/10 that of original. (After Colbert.)

giraffids. Possibly *Lagomeryx* and its relatives lie at the point of origin of the deer and giraffe families.

A few palaeomerycid genera persisted into the Pliocene. The group is generally believed to be extinct. A possible survivor, however, is *Moschus*, the "musk deer" of Pliocene to Recent times in central Asia. *Moschus* is usually included among the cervids. It lacks antlers and has large canines, as had palaeomerycids. This in itself does not prove the point, for at least one small true deer is hornless. Various other anatomical features, however, are in contrast with those found in true deer, and it is possible that the "musk deer" is really a palaeomerycid.

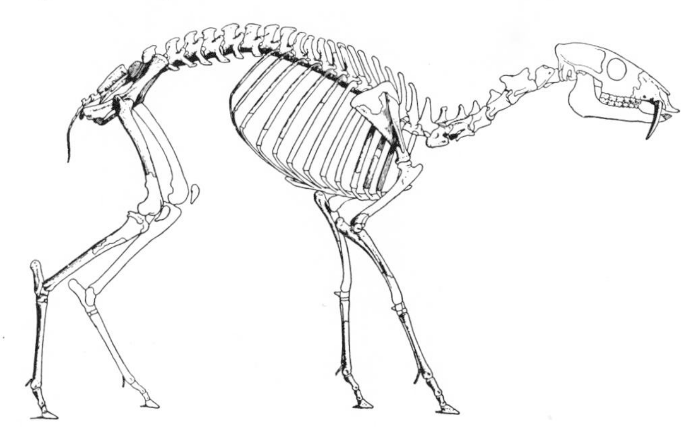

FIG. 419.—*Blastomeryx*, a Miocene palaeomerycid; original about 2.5 feet long. (From Matthew.)

Deer.—The Cervidae are a group of browsers primarily inhabiting the forests of the north temperate zone, in contrast to the more tropical environment favored by the related giraffe group. The most striking feature is the presence in the males (seldom the females), of almost all forms, of antlers—branching structures of solid bone which project from the skull posterior to the orbits. There is no horny covering; during growth they are surrounded by skin covered with downy hair, the velvet, which dries up and is rubbed off when the antler reaches full growth. Yearly the antler is shed (the place of resorption of bone at its base is marked by a roughened burr) and a new and usually more complex antler developed. This type of structure is in marked contrast to the hollow, horn-covered, unbranched, and permanent true horn of the bovids. The antler is reasonably considered to be a further development of the "horns" of the palaeomerycids. Their permanent spikes of bone are represented by the much-shortened unshed pedicels forming the base of the deer antler; the shed portion—the true antler—is a new cervid addition.

Early small cervids, such as *Stephanocemas* and *Dicrocerus* (Figs. 402*N*, 417*E*), appeared in the Miocene and early Pliocene of Eurasia, a region in which the deer seem to have had their origin; and by the end of the Pliocene there were present numerous types, many of which have survived to modern times. Relatively primitive and small deer, little modified from the *Dicrocerus* type, are the muntjacs, *Muntiacus* [*Cervulus*] of southeastern Asia (Pliocene to Recent) and the related *Elaphodus* of India, with very small and simple antlers and persistently large canines. Antlers tend to be more prominently developed in forms of large size, and deer with larger complex antlers are known from the Middle Pliocene onward.

FIG. 420.—*Megaloceros*, a giant deer ("Irish elk") from the Pleistocene of Europe. About 1/38 natural size. (From Reynolds.)

The late Cenozoic and living deer types were formerly all included in the single genus *Cervus*, broadly defined, but are now commonly regarded as constituting a score or more of related genera. The name *Cervus* is generally confined to the European red deer, the American wapiti ("elk"), and a few related types; *Alce* includes the European true elk and the American moose; *Rangifer* the reindeer and caribou (notable for the presence of antlers in the females). The fallow deer (*Dama*) and roe (*Capreolus*) are examples of a number of purely Eurasian types. Except for its northern fringes, Africa has not been invaded by the cervids, but in the Pleistocene the family sent numerous migrants to America. In addition to the forms common to both hemispheres, North America boasted a giant moose, *Cervalces*, in the Pleistocene; and *Odocoileus*, the "Virginia deer," is a purely American development. Pleistocene deer of this last type reached South America and still flourish there in the form of several derived genera. As in many other animal groups, the Pleistocene tended to be a time of production of giant forms; North America, we have noted, developed a giant moose, and in the Old World the "Irish elk," *Megaloceros* (Figs. 417*F*, 420), had the largest antlers of any known deer—a specialization useful to the male in the rutting season, but otherwise seemingly a hindrance rather than a help.

Giraffes.—The Giraffidae today include only the giraffe and okapi of Africa. These are browsing types, with low-crowned but heavy and rugose cheek teeth. Skin-covered "horns" of simple structure, comparable to those of the ancestral palaeomerycids, are present in the living genera and in a number of fossil types. The lateral digits have completely disappeared (whereas cervids persistently retain, as "dew claws," vestiges of the side toes in the manus). The long neck and legs are the most obvious specialization of the modern giraffe and are clearly associated with the tree-browsing habits of the animal. *Samotherium* (Figs. 402*M*, 417*C*), *Palaeotragus*, and other forms with a more normal build have long been known from the early Pliocene of southern Europe and Asia. It was hence of great interest when, early in the present century, there was discovered in the forests of the Congo the okapi, a comparatively short-legged, short-necked giraffid, almost indistinguishable from such ancestral Tertiary types.

As a side branch of the family may be included *Sivatherium* (Fig. 418) and other forms likewise from the Pleistocene of southern Eurasia and Africa. These were gigantic, heavily built ungulates in which there developed in the males a variety of large "horns," frequently in two pairs, from frontals and parietals. The hinder pair were sometimes massively branched in deerlike fashion but appear to have been nondeciduous in nature.

The giraffes are a tropical parallel to the cervid family. They have, however, failed to attain the success reached by their northern relatives.

Bovoids.—The cervoids are browsers; the Bovoidea, the superfamily now to be considered, are the characteristic grazers among the higher pecorans. In them we find the same advanced limb characters as in the cervoids; bovoids, however, tend to be even more progressive than deer in the reduction of lateral digits, for nothing remains of the toe skeleton except nodules of bone in some cases, and even the hoof vestiges may disappear. The upper canines are always reduced or absent, and in correlation with the grazing mode of life the cheek teeth are high crowned. Hornlike structures are universally present; but, in contrast to the cervoids, they are usually present in both sexes, although those of the males are often more highly developed. Further, the term "horn" is properly applicable to the structures found here, for a sheath of horn covers the bony core, which is never shed.

We have in the case of the bovoids no basal stock comparable to the palaeomerycids in the cervoids. Presumably the bovoids sprang from a traguloid stem. There are, however, no connecting forms; both surviving families appear fully developed in the Miocene—the Antilocapridae in North America, the Bovidae in the Old World.

Prongbucks.—*Antilocapra,* the prongbuck of the western plains, usually, but erroneously, called an "antelope," is today an isolated and interesting type. While in many features, such as the high-crowned cheek teeth and complete loss of side toes, this form is quite similar to the antelopes, the horns are radically different in character. These structures are, it is true, covered with horn, as in bovids; and the simple bony core is never shed. But the horn is somewhat forked, much as in primitive deer; and, curiously, the horny covering is shed yearly.

An early ancestor appears to be *Merycodus* (Figs. 415*C*, 417*G*) of the middle and late Miocene and Pliocene. This form at first sight would appear to be a deer rather than a prongbuck, for the bony outgrowths from the skull were long, forked (sometimes with several tines), and had a burrlike outgrowth at the base. But these structures were present and complete in all known specimens; it thus seems obvious that, in contrast with the deer, there were "horns" in both sexes and that the bony cores were never shed. The burr may be explained as the point at which an annual shedding of the horny covering occurred.

Merycodus is but one of a series of late Tertiary and Pleistocene types of antilocaprids which appear to have been abundantly present in the American plains in the latter part of the Cenozoic, playing there much the same role (although on a more modest scale) enacted by their antelope cousins in the Eastern Hemisphere. As far as known, all the group appears to have been quite uniform in skeletal and dental structures generally, but there is wide variation in horns. It seems certain that the rather simple horn of the modern prongbuck is a degenerate structure, for in all the fossil types the bony horn is more highly developed than in *Antilocapra* and typically strongly branched—so much that in the Pleistocene *Tetrameryx* we have an essentially four-horned type. The reduction of the group to a single living species may be related to invasion from Asia by the bison during the Pleistocene.

There are no apparent ancestors for the prongbucks in the Oligocene or early Miocene of North America. Perhaps they were Miocene invaders from Eurasia; but it is possible that they had an origin from American traguloids independent of the Old World bovids.

Bovids.—By far the largest group of living artiodactyls is that included in the family Bovidae, in which are placed such forms as the cattle, bison, musk ox, sheep, and goats, and that great and varied assemblage of forms termed "antelopes." In all bovids there are present (usually in both sexes) true horns consisting of a simple unbranched core of bone covered with a sheath of horn; no part of this structure is ever shed.

The bovids have been not only the most successful and numerous of artiodactyls but of ungulates in general in late Tertiary and Recent times; and, as this fact might suggest, their development has taken place at a comparatively late date. They have undoubtedly been derived from the traguloids of the Oligocene, but the first representatives of the family appear in the Old World only during the Miocene, in the shape of *Eotragus* and a few other rare forms. Eurasia appears to have been the major center of dispersal; literally dozens of bovids have been described from the Old World Pliocene, these including the ancestors of many living types, and about one hundred genera have been identified in the Pleistocene.

Before the close of the Pleistocene most bovids had disappeared from Europe; but in this case, as in many others, Africa and southern Asia have been havens of refuge for forms which increasing cold had forced from the north temperate region; Africa alone contains some twenty genera of living antelopes.

But few members of this group have reached the New World. Most bovids are plains dwellers in warm climates; the dispersal of the group seems to have taken place at such a late date that the cold climate heralding the approach of the Pleistocene glaciation may have rendered the passage through Siberia and Alaska difficult. Only the bison, the mountain sheep, the mountain "goat," and musk oxen, all of which seem to be able to withstand rigorous climatic conditions, have successfully invaded North America; and no bovid has ever reached South America, although the bison ranged southward to El Salvador.

A primitive group of antelopes is that which appeared in the Miocene and Pliocene in such forms as *Tragocerus,* with horns which were directed upward and backward and were straight, or at the most but slightly curved. Probably rather directly descended from them are a number of large types which have very long but nearly straight horns, also directed backward. First present in the Pliocene, there are still a number of African survivors, such as the oryx and addax. Side branches of the primitive antelope stock, of which the fossil history is almost unknown, include a number of tiny African forms, some no larger than rabbits, such as the duikerboks and klippspringers. In these forms the horns, still small and backwardly directed, are present only in the males.

The gazelles and related antelopes from northern Africa and southern Asia are steppe and desert forms of moderate size and graceful build, with horns slightly curved but still of a simple primitive type. *Gazella* (Fig. 417*H*) can be traced back to the late Miocene. Related to the gazelles is the saiga antelope, with a heavier build and peculiar swollen muzzle; now confined to the arid regions of Asia, it penetrated, in the Pleistocene, westward far into Europe, where its presence is indicative of steppe conditions, and eastward to Alaska.

A northern representative of the bovids is the musk ox, *Ovibos,* heavily built, with massive, laterally directed horns (much smaller in the females). The musk ox is now confined to the arctic zone; but in the Pleistocene, musk oxen penetrated as far south as France and Kentucky, and in North America there were also related forms, now extinct.

Two further lines of bovid evolution have led to the development of familiar types—to the sheep and goats, on the one hand, and to the cattle and bison, on the other. Pointing out the evolutionary road to the sheep and goats are the types often called "goat antelopes,"

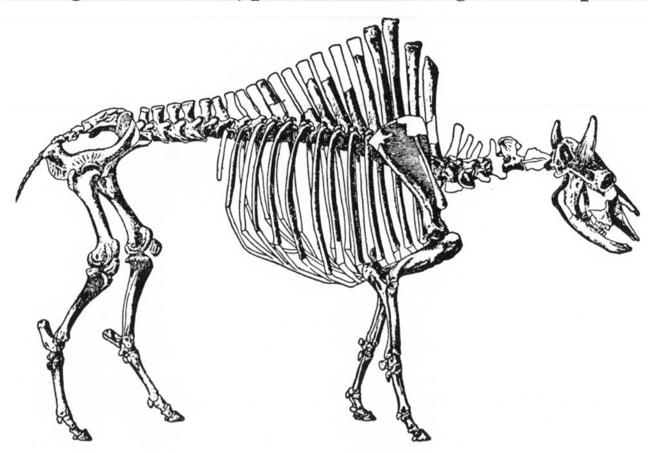

FIG. 421.—*Bison occidentalis*, a Western Pleistocene species, about 7 feet in length. (From Hay.)

such as chamois (*Rupicapra*) of Europe and the Rocky Mountain "goat" (*Oreamnos*), with short but sharply curved horns. In the sheep (*Ovis*) and goats (*Capra*) there has been a great development of the horns, which tend to become very large and often spirally coiled in the males (those of the females are much smaller). These forms are upland dwellers (which may account for the fact that their fossil history is relatively unknown), inhabitants of the mountain chains from the Mediterranean region eastward through Asia, and have, as we have noted, even penetrated to the Rocky Mountains.

The evolution of oxlike bovids appears to have begun well back in Pliocene times with the appearance of large antelopes with teeth rather resembling those of modern cattle and with horns which (although still directed backward) had developed a lyre-shaped form with a spiral twist. Living representatives are the eland of Africa and the nilghai of India. A type still more cowlike and also dating from the early Pliocene is that now represented by the large gnus and hartebeests of Africa, in which the lyrate horns spread sidewise and are, in the former, directed forward rather than backward.

From these cowlike antelopes the development of the cattle has been but a short step. A rather slim and antelope-like ox, *Leptobos*, is found in the early Pleistocene of Eurasia; cattle, most of which may be included in the genus *Bos*, were widespread in the Pleistocene. Common domesticated cattle are but one of several members of this group which have been utilized by man. A large wild ox survived in Europe into medieval times. Closely related to the cattle are the true buffaloes of southern Asia and Africa, now including domesticated types as well as wild forms.

Bison (Fig. 421) is a genus closely related to the true cattle, whose members were dwellers in the temperate regions of the Old World and which, alone of plains-dwelling bovids, has successfully invaded North America. The Old World bison has persisted in a wild state; a few still survive in eastern Europe. In America the bisons became exceedingly numerous in the Pleistocene and seemingly branched out into a number of types, to judge by the variation in fossil horn cores discovered (one specimen had a spread of horns of some 10 feet). Although essentially dwellers in the plains region, Pleistocene bisons were present from Atlantic to the Pacific.

CHAPTER 25

Edentates

Grouped as edentates in many of the older natural histories is a heterogeneous assemblage of forms from the tropics of both the New and the Old Worlds—the tree sloths, armadillos, and anteaters of South America; the aardvark of Africa; and the scaly pangolins of Africa and southern Asia. In great measure the bonds which were supposed to unite these varied types lay in the nature of the dentition. Some, such as the pangolins and the American anteaters, are toothless; in the others teeth are practically confined to the cheek region and lack the hard enamel covering found in other mammals.

But many of these edentates are ant eaters, and the reduction or loss of teeth in such types is obviously a functional adaptation and is no argument for their relationship to one another; we have seen that in the aard wolf, an ant-eating marsupial, and the monotreme *Echidna* similar habits have been associated with a similar reduction of the teeth. The aardvark is, it is now generally agreed, quite unrelated to the other edentates. The pangolins, as well, seem to occupy an isolated position; aside from specializations for a diet of ants, there is little to connect them with other forms.

True edentates.—As regards the other living "edentates," however, a different situation exists. Comprising, as they do, the South American anteaters, the armadillos, and the tree sloths, as well as the extinct armored glyptodons and giant ground sloths, these New World tropical types exhibit striking superficial differences. But in internal structure there are many features which tend to show that these varied forms really constitute a natural group to which the ordinal name Edentata may be restricted. Xenarthra is an essentially synonymous term but one which may be used for the typical South American families, to the exclusion of some archaic and ancestral forms.

A striking feature common to all living and most fossil members of the group and not found in any other mammals is the presence of extra (xenarthrous) articulations between the successive arches of the posterior trunk vertebrae in addition to the normal zygapophyses (Fig. 422*C*). Here, too, are found the only cases (apart from sirenians) in which there is variation from the normal mammalian number of cervical vertebrae; from six to nine may be present. A fusion of the cervical vertebrae may occur.

There are many peculiarities in the limb skeleton of South American edentates. In the shoulder the acromion and the coracoid process are generally much more developed than in other placentals and may join one another (Fig. 422*A*); the ischium articulates with the proximal caudal vertebrae to form a peculiarly elongated sacrum (Fig. 422*B*). The limb bones are usually short, stout, and massive, with strong muscular processes. Radius and ulna remain separate, but tibia and fibula have fused in armadillos and ground sloths. Claws are often excessively developed; in several groups there is a tendency to walk on the outer side of the front foot, and the hind feet are oddly constructed. Perhaps these peculiarities of the feet are due to descent from arboreal ancestors which suspended themselves from the branches, as the tree sloths still do today.

The cranial region is characteristic. The brain is small and on a low plane of organization; the braincase is a long, cylindrical tube. The premaxilla is usually reduced, although there is sometimes a supernumerary prenasal bone strengthening the snout. The palate is elongate and may even be bridged over between the pterygoids. Except in armadillos and some glyptodons, the zygomatic arch is incomplete, the jugal often ending posteriorly in a fan-shaped expansion but failing to reach the squamosal.

Only in anteaters have the teeth been altogether lost. In other edentates they are usually absent from the front of the mouth, while roots are never formed in the cheek and there is no enamel covering except in a few early genera.

A tendency toward the development of protective armor in the skin appears to have been inherent in the group. In the armadillos and the extinct glyptodons we find a bony carapace covered by horny scales. No bones are present in the skin of tree sloths and anteaters, but in some of the extinct ground sloths the skin was reinforced by bony nodules lodged in the thick hide.

The center of evolution of these types is unquestionably South America, the present home of practically all members of the group. Ground sloths, glyptodons, and armadillos are known as fossils from the late Cenozoic of the United States; but it is generally agreed that these forms were immigrants from the south. It seems highly probable that the ancestral edentates, like the peculiar ungulates of that region, inhabited South America at the very beginning of the Age of Mammals and there underwent an isolated development during most of the Tertiary.

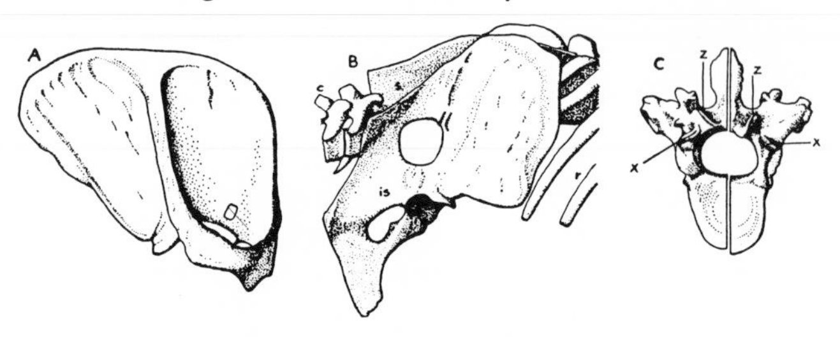

Fig. 422.—*Mylodon. A*, Scapula, to show bridge from acromion to coracoid; *B*, pelvic region, showing upward growth of ischium (*is*) to join with sacrum (*s*); *r*, ribs; *c*, caudal vertebrae. *C*, a posterior dorsal vertebra, to show "xenarthral" articulations (*x*) in addition to the normal zygapophyses (*z*). On the left side the anterior view, posterior on the right. (After Stock.)

Palaeanodonts.—If the edentates are assumed to have originated elsewhere and to have reached their southern home by way of North America, it might be expected that some trace of the group would be found in the early Tertiary of the north. This expectation has been fulfilled; three North American genera, ranging from late Paleocene to Oligocene, appear to represent a primitive edentate stock. All three are rare, but *Metacheiromys* (Fig. 423) of the Eocene is known from fairly complete material.

In many ways these forms were much more primitive and generalized than any of the later South American edentates. *Metacheiromys* resembles the armadillos in general proportions and even in many structural details, but there is no trace of dermal armor, and the accessory xenarthral articulations are absent in the backbone. Long, compressed claws were present in the front feet, somewhat shorter and broader ones behind, and even here there was already some tendency for a twist in the wrist.

The teeth show many of the characteristics to be expected in the ancestral edentates, for the incisors had disappeared except for one small lower tooth, the cheek teeth were, at the most, pegs, and nearly all of the enamel covering had been lost. There were, however, some dental specializations not in line with the direct ancestry of South American forms. The canines were large; in *Metacheiromys* the cheek teeth have nearly completely disappeared and may have been functionally replaced by horny pads.

The clawed feet and peculiar dental apparatus of the palaeanodonts suggest that they may have fed upon small terrestrial invertebrates, digging for grubs, insects, and worms much as do modern armadillos.

These North American genera, constituting the suborder Palaeanodonta, appear to represent in most of their features the ancestral edentate type from which the South American forms have sprung. It is obvious, however, that they are too late in time to have been the actual ancestors, for South America was presumably isolated at the time of their appearance. They were merely wayside stragglers, destined not to survive in competition with the numerous other orders already appearing in North America.

Xenarthrans.—The continuation of the story of edentate development is to be sought in South America. There edentates were already present in the oldest known Tertiary beds. Although most Eocene remains are fragmentary, it is reasonable to believe that the earliest South American representatives possessed many of the features absent in the palaeanodonts but characteristic of later groups, such as the additional vertebral articulations, union of ischium and backbone, and peculiar skull structure.

Among these South American forms, the Xenarthra, there have evolved five distinct types which may be placed in three infraorders. One group, the Loricata (or Cingulata, armored), comprises the "shelled" forms, the living armadillos and the extinct glyptodons; a second, termed the Pilosa (hairy), includes the tree sloths and the extinct ground sloths; a third, the Vermilingua, the South American anteaters.

Loricates.—The great development of dermal armor is the most obvious feature uniting the armadillos and the now extinct glyptodons. A large number of bony plates covered with horny scutes form a protective carapace over the trunk, while plates may develop on the head and tail as well. The teeth tend to be more numerous than in the sloths; there are always as many

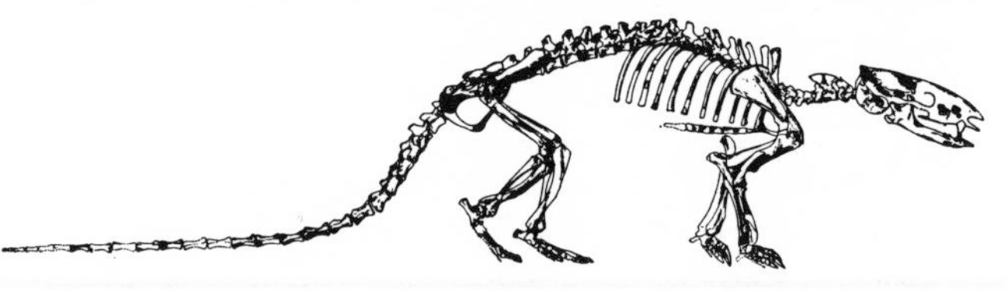

Fig. 423.—*Metacheiromys*, a small Eocene edentate, about 18 inches long. (From Simpson.)

as seven cheek teeth, often a higher number. The zygomatic arch is usually complete. There is a tendency for a fusion of vertebrae under the shield; and some of the cervicals, at least, are fused in all forms.

In the armadillos, the family Dasypodidae, the carapace never becomes a single solid shield, as in glyptodons. Instead, the bony scutes usually tend to form solid plates over the shoulder and pelvis, between which are transverse movable bands. Beneath the pelvic plates is a long, heavy sacrum, while fused neck vertebrae lie beneath the shoulder plate. The cheek teeth of these insect and carrion feeders are usually simple pegs, about eight to ten in number in each jaw half. There is a tendency to exceed the normal placental number, an extreme being reached in one living form with as many as twenty-five.

These armored types at first sight would seem to be very aberrant. But, curiously enough, they are the oldest known of South American edentates, for they are the only types represented in the Paleocene and early Eocene beds of that continent. In these beds, there are not merely numerous isolated armadillo scutes but also much skull and skeletal material of the genus *Utaetus*. Furthermore, the structural tendencies seen in palaeanodonts seem to lead more directly to the armadillos than to other edentates. It thus appears probable that the armadillos represent the main stem of the South American edentates, although one dislikes the implication that the "hairy" edentates have descended from armor-bearing ancestors.

By the Miocene, armadillos, such as *Proeutatus* (Figs. 424*A*, 425*A*) and *Stegotherium*, were abundant and varied. A peculiar side branch is that of *Peltephilus*

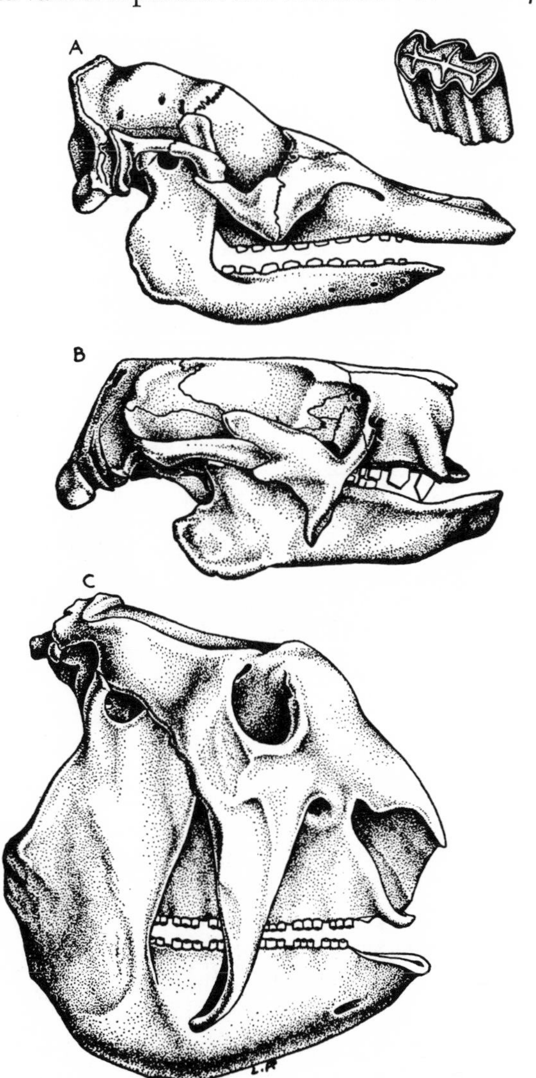

FIG. 424.—Skulls of edentates. *A*, *Proeutatus*, a Miocene armadillo, skull length about 5 inches; *B*, *Paramylodon*, a ground sloth, skull length about 20 inches; *C*, *Glyptodon*, skull length about 11 inches. *Above*, A glyptodont tooth. (*A* after Scott; *B* after Stock; *C* after Burmeister.)

and its relatives, which contrast with normal armadillos in their short, broad skulls and the development of hornlike structures in their head armor. With the re-establishment of continental connections, the armadillos pushed north, reaching the Gulf Coast in Pleistocene times; one form still flourishes in that region. In the Pleistocene of both continents there were present giant armadillos, now extinct—one as large as a rhinoceros.

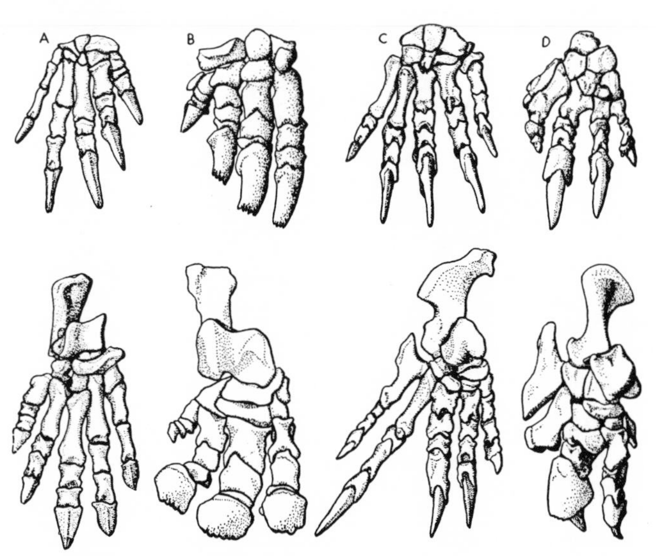

FIG. 425.—Feet of edentates, manus above, pes below. *A*, *Proeutatus*, a Miocene armadillo; *B*, *Panochthus*, a Pleistocene glyptodon; *C*, *Hapalops*, a primitive Miocene ground sloth; *D*, Pleistocene ground sloths, manus of *Paramylodon*, pes of *Nothrotherium*. (*A*, *C* after Scott; *B* after Burmeister; *D* after Stock.)

The extinct glyptodons (*Glyptodon* [Figs. 424*C*, 426, 427], *Panochthus* [Fig. 425*B*]) were a group related to the armadillos and, like them, had developed a protective armor, but in a different fashion. Originally, it would seem, the bony plates on the back may have been placed in cross rows, as in the primitive armadillos. But typical glyptodons tended to fuse the entire mass into a solid, turtle-like carapace composed of a mosaic of countless small polygonal plates. There was much fusion of the vertebrae, mainly in connection with support of the shield. Most of the cervicals, except the first, were fused; a second solid mass included most of the dorsals; while the last dorsals, lumbars, and sacrals were fused and connected with the posterior part of the carapace. The tail was sheathed in armor plates, sometimes with projecting spikes. Like the armadillos, these forms were generally five toed and clawed; but the toes were short and stubby, the claws broad and rather hooflike in the hind feet (the front feet were more variable).

The skull was covered with a bony casque; the face and jaws were excessively deep. In these forms, alone among edentates, there was a postorbital bar, and the zygomatic arch had a long downward-projecting process of the jugal. There was no clavicle, in contrast to most other edentates. The teeth numbered eight in each jaw ramus. They were typically very high crowned and had a peculiar three-lobed pattern analogous to that of some of the ratlike rodents.

It seems certain that the glyptodons were derived from the primitive armadillos. They appeared in the later Eocene and were well represented in the Miocene beds of South America. The earlier forms were small compared to the later members of the group but

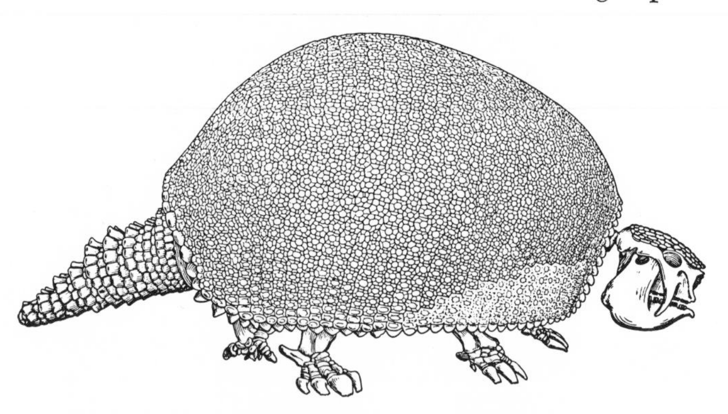

FIG. 426.—Carapace of *Glyptodon,* from the Pleistocene of South America; length about 9 feet. (After Burmeister and Hoffstetter.)

were already of larger size than most of the contemporary armadillos. There were still present many primitive armadillo-like characters, such as the comparatively long skull, low-crowned teeth, and the persistently primitive arrangement of some anterior plates in more or less transverse rows. By the Pliocene, large size and characteristic glyptodon structure had been attained. These huge armored forms migrated northward early, and one genus had reached Texas at the end of the Pliocene. In the Pleistocene the glyptodons were numerous from Florida to Argentina. Here again, however, as in the case of most large land mammals in the Western Hemisphere, the end of the Pleistocene witnessed the extinction of the group.

Sloths.—In the remaining South American edentates the zygomatic arch is always incomplete, although the jugal is usually much expanded posteriorly. There are never more than four or five simple cheek teeth. Among the peculiar features in the skeleton may be noted the fact that the expanded acromion arches forward to unite with the coracoid region (Fig. 422*A*). The greater part of such edentates are sloths—living tree sloths and fossil ground sloths, forming the infraorder Pilosa.

The living tree sloths of South America included in the family Bradypodidae are among the most curious of mammals. They are small, arboreal, nocturnal forms with a lichenous growth which often gives a green tinge to their gray hair. They are very slow and clumsy, spending much of their time hanging upside down from branches and holding on by their long curved claws, two or three in number. The front legs are longer than the hind; the body is elongated; the tail is vestigial. These leaf-eating types have but four or five cylindrical cheek teeth, one of which is anteriorly placed and may be a canine. The skull is very short in the facial region, and there is only a tiny premaxilla. The jugal has a flaring termination posteriorly.

The tree sloths are unknown as fossils. Remains of these forest dwellers are not to be expected in the deposits of the Argentinian plains from which most of the older South American fossils are obtained. They appear, on structural grounds, to be related to the ground sloths of the *Megatherium* and *Megalonyx* families.

The large and numerous forms of the ground-sloth group in the Pleistocene of both North and South America are among the most interesting of extinct mammals. The first definitely identifiable ground sloths appear in the Oligocene of South America, and small and relatively primitive members of the group are common in the early Miocene. The characteristic Miocene *Hapalops* was (for a ground sloth) small in size, with a length of but four feet or so, including the elongate tail. The body was moderately long and rather slender in its build. A full complement of clawed toes was present (Fig. 425*C, D*). The claws were particularly well developed on the front foot, which was twisted over until the weight rested on the outer knuckles, while the stocky hind foot was turned

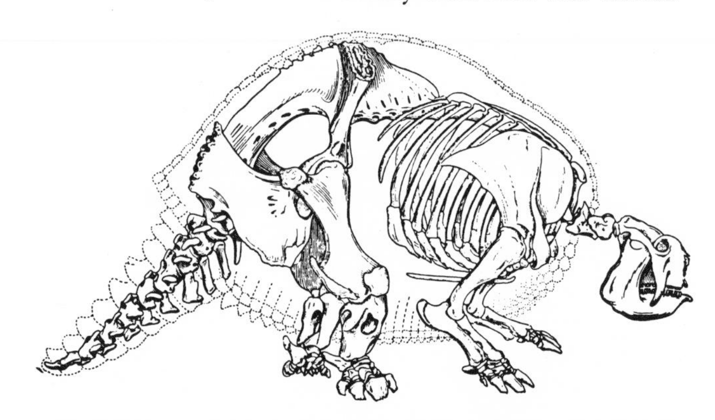

FIG. 427.—*Glyptodon,* carapace removed. (After Burmeister and Hoffstetter.)

downward laterally so that the weight rested on its outer margin. The skull was moderately elongated; slim premaxillae above and a spoutlike process from the jaws presumably supported horny cropping plates; in the cheek there were present, as in ground sloths generally, five simple teeth above and four below, the first resembling a canine. The zygomatic arch was incomplete; but the huge jugal, which splayed out into a number of projections posteriorly, came close to the anterior end of the squamosal.

Nothrotherium (Fig. 428) was a descendant of *Hapalops* found in the Pleistocene of both North and South America. Its size, about that of a tapir, was relatively small for a ground sloth of that age, and there were few further modifications except for the loss of the "canines." It persisted to a very late date in the southwestern United States, for not only has a skeleton been found with much of the skin and tendons still preserved, but it also appears that this form was a contemporary of early man.

Hapalops and *Nothrotherium* represent a line of sloths of persistently small size (relatively speaking) which are currently considered to be a branch of the Megatheriidae. However, *Megatherium*, which gives the name to the family, was strikingly different, particularly in its proportions. This animal, which ranged from South America to the southeastern United States in the Pleistocene, was the largest of ground sloths, exceeding an elephant in size, with a length of 20 feet and a very massive build. The typical megatheres also had roots in small Miocene forms, but differed from the *Hapalops* group (and from the *Megalonyx* group as well) in the lack of specialization of the "canines," reduction of the two inner toes, and very deep-rooted cheek teeth which simulate a bilophodont condition.

Related to the megatheres, derived from common Miocene ancestors, but differing in such features as absence of the spoutlike terminus of the lower jaws and presence of "canines," were the Megalonychidae. This family played almost no role in the edentate life of South America. They were, however, successful invaders of more northern regions. In some fashion (perhaps via Central America) members of this group reached the West Indies, where there were a number of Pleistocene representatives. These forms ranged in size from bear to cat—a dwarfed condition presumably related to insular life (there are parallel cases amongst elephants and hippopotami). On the other hand, *Megalonyx*, which reached North America and flourished there, was about the size of a large ox, with a skull a foot in length.

Quite distinct from the two other families of ground sloths were the Mylodontidae. Diagnostic features include partial development of upper "canines," a rather triangular shape in the cheek teeth, and a hind foot in which the first toe (but not the second) had been reduced. In the mylodonts, round ossicles deeply embedded in the skin seem to have afforded a defense against enemies. The mylodonts appear to have had a separate history from the other families since the Oligocene, but it was only in the Pliocene and Pleistocene that they increased in numbers and variety to become the dominant ground sloth group in South America. Here, as in the case of the megatheres and *Megalonyx*, there was a marked trend toward increased size and consequent stouter (and clumsier) build in late members of the group. *Mylodon* of the South American Pleistocene and the related *Paramylodon*, which reached and spread widely in North America, were half again as large as the contemporary *Megalonyx*. *Paramylodon* (Fig. 424*B*), like *Nothrotherium* (Fig. 428), appears to have been a late survivor in the southwestern United States, and a *Mylodon* specimen found in a Patagonian cave had been killed by man.

Ground sloths flourished in the Pleistocene. But at the close of that period they were wiped out. Why, we cannot say. Unlike the South American ungulates, which seem to have disappeared when faced by progressive competitors, the ground sloths had not only held their ground but had successfully invaded North America. The factors causing their extinction are as mysterious as those which destroyed most of the other larger mammals of the Western world.

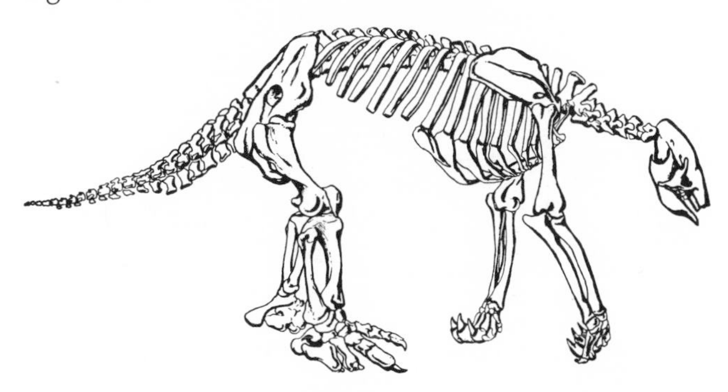

FIG. 428.—The skeleton of *Nothrotherium*, a small Pleistocene ground sloth; length about 7.5 feet. (From Stock.)

A priori, one would assume that ancestral sloths, as presumably with ancestral edentates as a whole, would have been ground dwellers, and hence that the tree sloths, with their purely arboreal habits, represent a side branch of the sloth tribe. But the story is probably not this simple. The structure of the ungainly sloth limbs and, particularly, of their feet, is not of a sort well adapted for ground locomotion, and strongly suggests that the limbs had been at one time adapted for tree life. It seems highly probable that the small "ground" sloths of the earlier Tertiary were not purely terrestrial, but were, rather, semiarboreal in habits. Along the "main line" of ground sloth evolution, arboreal habits were of necessity abandoned with increasing size. And, on the other hand, it seems probable that the living tree sloths represent a side branch which was persistently tree dwelling and became increasingly specialized in limb structure.

Anteaters.—The members of the infraorder Vermilingua—*Myrmecophaga* and related forms—exhibit specializations which seem but a logical outcome of the presumed insect-eating habits of the palaeanodonts. In the front feet are developed large claws (particularly on the middle toe), which are used for digging into termite nests; the lateral toes are reduced. With the presence of these large excavating tools it has become impossible for the foot to be placed flat upon the ground, and the weight rests on a pad on the outer side of the knuckles with the toes turned inward. In the hind foot the claws are smaller, and the foot is plantigrade in the more primitive types.

The postcranial skeleton of the anteaters seems to be of the type which would be expected in the ancestors of the remaining groups of edentates. But the highly developed termite-eating habit has been associated with great modification in the head region. There is a very long, tubular snout containing a protrusile, whiplike tongue; the jaw is weak; and these forms, alone among true edentates, have entirely lost their teeth.

Pangolins.—The Old World "edentates," the pangolins and aardvark, may be considered here for want of a better connection.

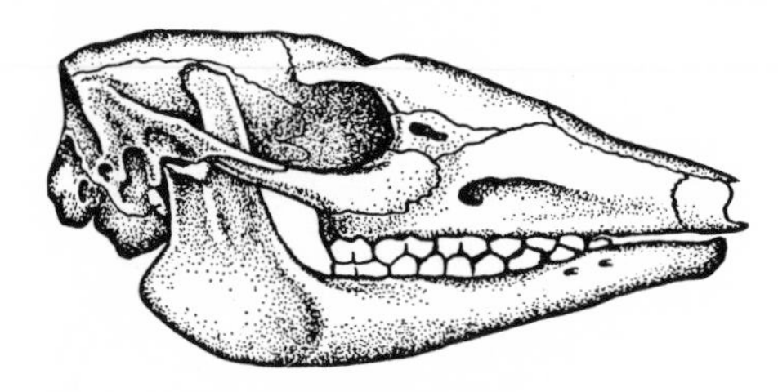

FIG. 429.—*Orycteropus gaudryi,* a fossil aardvark from the Lower Pliocene of Samos; skull length about 6.25 inches. (After Andrews.)

The scaly anteaters or pangolins of tropical Asia and Africa, constituting the genus *Manis,* are the sole representatives of the order Pholidota. The horny overlapping scales completely covering the body form the most notable peculiarity of these animals. This covering is surely a secondary protective device and not due to direct inheritance from reptilian ancestors. With the adoption of an ant-eating diet, there have developed many cranial adaptations similar to those of American termite eaters. Teeth are completely absent; there is a long snout, a slim jaw, and a long tongue. The temporal arch is incomplete and the jugal lost; the eyes are much reduced. The tail is usually long, sometimes prehensile. All toes are present; but the hand is functionally tridactyl, with the development of powerful digging claws.

Representatives of the living genus have been found in the Pleistocene of southern Asia. A few bones from the Miocene and Oligocene of Europe are thought to belong to this group but tell us nothing of their pedigree. Relationship with American edentates has been often suggested, but the similarities are such as are obviously related to digging and to eating ants; there are no positive resemblances to the xenarthrans. The pangolins may have evolved quite independently from some early primitive placental stock.

The aardvark.—Quite isolated, too, is the "earth pig," *Orycteropus,* sole living representative of its own order, the Tubulidentata. This is an uncouth-looking African mammal which lives on termites. There is a long snout, a small mouth, and a long tongue, as in other ant-eating mammalian types; but there is comparatively little reduction of the jaws, and the temporal arch is well developed in contrast to other anteaters. In the embryo, teeth are numerous, exceeding the usual placental number; but in the adult only four or five peglike cheek teeth remain. These lack enamel but have cement on the outside; while, instead of a pulp cavity, the dentine is traversed by a large number of small tubules, a feature to which the group owes its name.

There are a number of primitive skeletal features. Tibia and fibula, however, are fused proximally, and the pollex has disappeared. The feet are semiplantigrade; and there are stout "nails," intermediate between claws and hoofs in nature, inserted in fissured end phalanges, much as in some early carnivores.

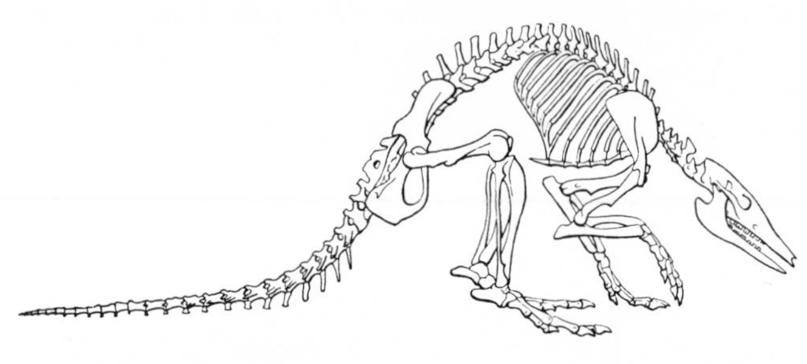

FIG. 430.—The skeleton of *Orycteropus gaudryi,* length about 3.5 feet. (After Colbert.)

Fossil aardvarks (Figs. 429, 430) are known from the African Miocene and the Pliocene of southern Europe and Asia. There are a few fragments of doubtful nature in the Eocene and Oligocene of Europe. In the Lower Eocene of North America have been found fragmentary remains of *Tubulodon,* a form which shows a tubular structure of the teeth similar to that of aardvarks. The aardvark skeleton is strikingly similar to that of some early condylarths, and it is possible that it is descended from early ungulates.

CHAPTER 26

Whales

Among mammals which have turned to an aquatic life, the whales—the order Cetacea—constitute the largest and most important group and that best adapted to an existence in the water. Both structurally and functionally they have become completely divorced from their former land life and are helpless if stranded. Only in their need for air breathing do they exhibit any functional reminiscence of their former terrestrial existence.

Modern whales have reassumed the torpedo-like, streamlined shape of primitive aquatic vertebrates, partly through a shortening of the cervical vertebrae and the consequent absence of a neck. However, the body is thick and rounded in section, and hence (unlike the typically slimmer fish) the main propulsive force is confined to the tail fin alone. As in other aquatic mammals, the tail has failed to resume its original fin structure, and (as in the sirenians) horizontal flukes supported by fibrous tissue supply the locomotive power. A dorsal fin has usually redeveloped. The hind limbs are lost completely, as far as any superficial indication goes, although vestiges may be present internally. The front legs have been transformed into short, broad, steering flippers. Although (in contrast to some ichthyosaurs) there are no extra digits, as many as a dozen extra phalanges may be present. Hair has been lost and may be absolutely lacking on the skin of the adult whale; a thick layer of blubber affords protection against cold.

Marine life has been accompanied by many internal modifications. The original whales appear to have been fish-eating carnivores. The majority of modern whales are still toothed, but, as in the seals, the teeth have been simplified, usually into simple pegs. The number has in many cases increased greatly over the primitive placental forty-four; in others teeth have been reduced in number or entirely abandoned for a straining apparatus of whalebone. The anterior portion of the skull has been elongated from the first. But in correlation with the breathing problem in diving types, the nostrils have moved backward in the skull and in typical living whales are placed, as the blowhole, on the top of the head (Fig. 431*B–D*). In this process the premaxillary and maxillary bones have been dragged back over the more posterior elements, often in an asymmetrical fashion. This results in a peculiar, telescoped effect, heightened by the fact that usually the occipital bones have pushed forward over the top of the skull. Osteologically there is no top to the skull; it is all front and back.

The braincase has been much modified in shape in relation to these odd specializations and is short but broad and high. The orbit is always open behind, although typically covered by a broad supraorbital process of the frontal; and the jugal is small. The premaxilla of modern whales is usually toothless (there are small teeth in a few porpoises); the bone, however, is well developed on the skull roof. The ears are much changed for use in such "submarine" forms; the external tube and drum tend to be modified and reduced; and hearing, apparently acute (in contrast to the lost sense of smell), appears to be accomplished through the instrumentality of vibrations set up within the heavy, shell-like bulla, which, fused with the periotic, is but loosely attached to the remainder of the skull.

Whales may be divided into three suborders: the Archaeoceti, of the early Tertiary; and two living groups—the Odontoceti, or toothed whales, and the Mysticeti, or whalebone whales (Fig. 434).

Primitive whales.—The archaeocetes are the oldest and most primitive of the cetacean groups, flourishing in the Eocene. A large fraction of the known Eocene specimens are from northern Africa, suggesting that (like the subungulates) they may have originated on that continent. Many features of their structure suggest their origin as a branch of a primitive carnivorous or pre-carnivorous stock which had taken up a fish-eating life, but a number of important modifications had already occurred in the primitive Middle Eocene *Protocetus* and *Prozeuglodon* (Fig. 431*A*) of the Upper Eocene. These were relatively small cetaceans, comparable in size to the smaller of modern porpoises. The snout was elongate (as in many fish-eating reptiles before them), and the nostrils had already accomplished half of their migration backward onto the top of the skull. In other respects, however, the skull was still much like that of a primitive carnivore; there was a long, low braincase, and there was no trace of the telescoping of elements which was to be the most marked peculiarity of later whale skulls. The dentition, too, was essentially primitive; for, while the front teeth were peglike, the cheek teeth were still much like those of early carnivores in appearance, and the primitive placental tooth count of forty-four was not exceeded. But while the skull was still quite primitive, the body skeleton seems already to have advanced far in aquatic adaptations. Skeletal remains are rare; but by Upper Eocene times, if not earlier, the hind legs had been reduced to vestiges which did not project from the body. Obviously, the earliest whales were already more highly adapted to marine life than are the living seals, although the story of whale specialization was far from finished.

The peak of archaeocete development was reached in *Basilosaurus* (or *Zeuglodon* [Fig. 432]) and its relatives, which were widespread and common in the Upper Eocene seas. These were the giants among primitive whales. The long low skull (which reached a maximum length of 5 feet) was still primitive in

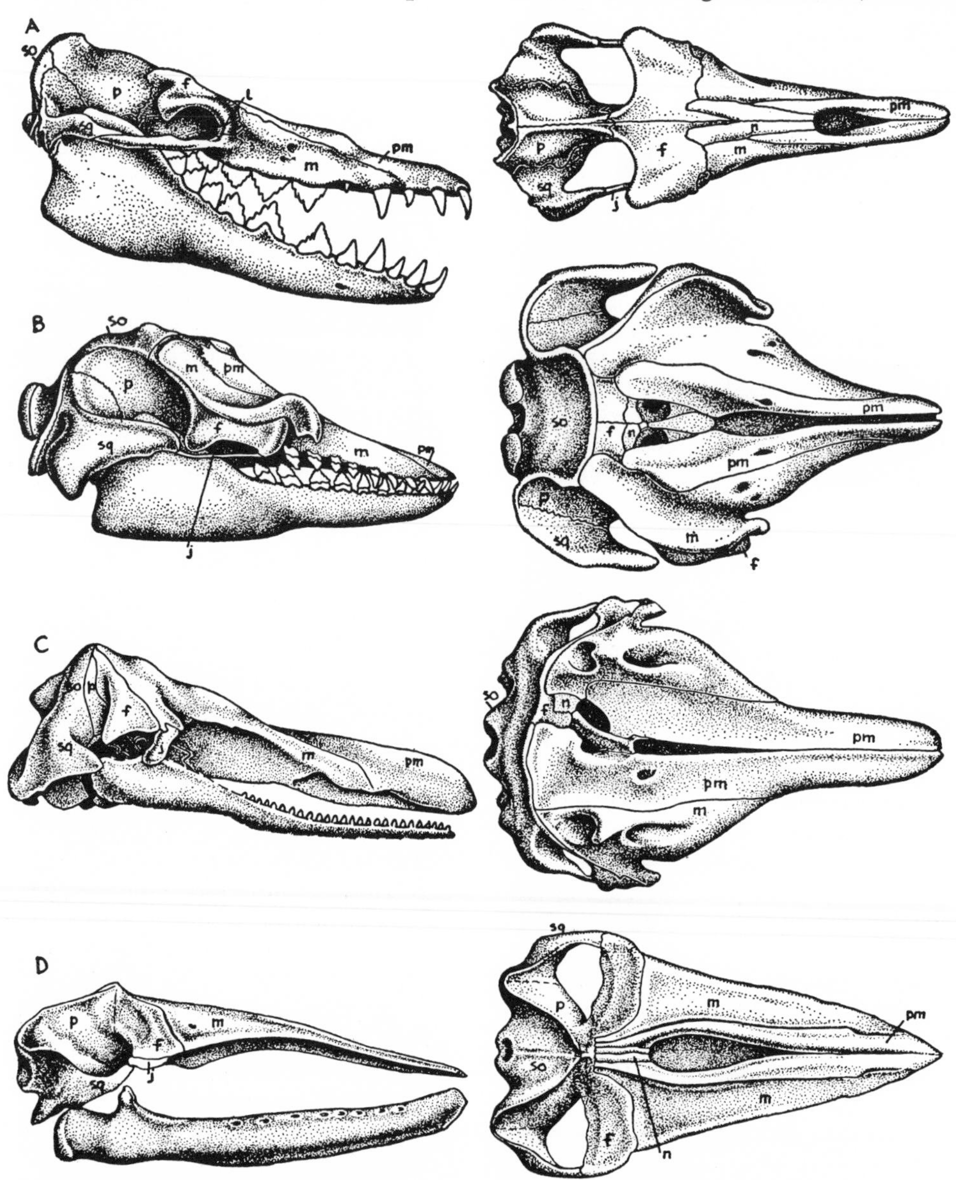

FIG. 431.—Skulls of cetaceans, lateral and dorsal views. *A, Prozeuglodon atrox,* an Eocene archaeocete, skull length about 2 feet; *B, Prosqualodon,* a Miocene squalodont porpoise, skull length about 18 inches; *C, Aulophyseter,* a Miocene sperm whale, skull length about 4 feet, jaw restored; *D, Cetotherium samarinense,* a Miocene whalebone whale, skull length about 22 inches. Abbreviations: *f,* frontal; *j,* jugal; *l,* lacrimal; *m,* maxilla; *n,* nasal; *p,* parietal; *pm,* premaxilla; *so,* supraoccipital; *sq,* squamosal. (*A* after Andrews; *B* after Abel; *C* after Kellogg; *D* after Capellini.)

many features, but the much-compressed and serrated cheek teeth were departing further from the primitive carnivore type. While modern cetaceans are stockily built, *Basilosaurus* was slim and elongate, reaching as much as 70 feet in length, with the proportions which modern imagination ascribes to sea serpents.

Basilosaurus and other primitive types did not, in general, survive the end of the Eocene. Of the archaeocetes, some comparatively small, short-bodied forms—the *Dorudon* group—persisted in lessening numbers through the Oligocene and into the beginning of Miocene times.

Primitive toothed whales.—Odontocetes, or toothed whales, comprise the great majority of living cetaceans, ranging from small porpoises and dolphins to the huge sperm whale. The body is short and stocky in contrast with the elongation in *Basilosaurus*. The teeth are usually simple pegs or wedges, are not differentiated into incisors, canines, and cheek teeth, and may far exceed the primitive placental number (as many as three hundred in one porpoise). Most marked feature of specialization, however, is the extreme telescoping of the skull roof, mentioned above. The nostrils have moved far back over the top of the skull, forming a single vertically placed blowhole, unroofed by the reduced nasals. A backward elongation of the premaxillae and maxillae has accompanied this process. The maxilla has spread out sidewise over the frontal in a great shelf over the orbit and temporal region and, pushing the parietal entirely out of the midline of the skull roof, often comes in contact with the supraoccipital.

No trace of this telescoping process is seen in typical archaeocetes, and although the odontocetes presumably derive from a stock related to the archaeocetes, the two groups must have diverged at an early stage. The development of the modern whale type proceeded rapidly. Two ancestral forms, of which *Agorophius* is the better known, appeared in Upper Eocene seas as contemporaries of *Basilosaurus*. The skeleton is unknown, but in their skulls they definitely show an early stage in the telescoping process; for, while there was little shifting of the occipital bones, and the temporal region was still open above, the nostrils were already in position above the orbits, and the maxillae were beginning to extend back and cover the frontals.

These primitive forms appear to have been ancestral to the squalodonts, which, first appearing in the late Oligocene, became the characteristic worldwide cetaceans of early Miocene times. Such forms as *Squalodon* and *Prosqualodon* (Fig. 431*B*) probably resembled greatly the modern porpoises in habits and appearance. These types owe their name to the triangular, shark-like teeth found in the posterior part of the beak (which was often much elongated). The squalodonts were much more modernized than their Eocene forebears, for the telescoping process in them had been completed; the parietals were eliminated from the top of the skull, the maxillae had formed a contact with the supraoccipital, and the blowhole had reached its most posterior position above and behind the eyes. The temporal opening, however, had not yet been roofed over, as in many later odontocete types. The squalodonts were not long destined to retain their importance, for even by middle Miocene times they had been largely replaced by other porpoise-like types, and only a single survivor lingered on into the Pliocene.

Perhaps most closely related of living whales to the squalodonts are the long-beaked river porpoises of the Ganges (*Platanista*), the Yang-Tse (*Lipotes*), and the Amazon (*Inia*). Here, as in most later odontocetes (and in contrast with squalodonts), all the teeth are simple peglike structures, and the premaxilla has tended to become toothless. But a primitive feature is found in the fact that in these porpoises, as in the squalodonts, the temporal opening is still unroofed. Their ancestors were undoubtedly sea dwellers; a number of presumed relatives were present in the Pliocene, and even in early Miocene times.

Beaked whales.—While the squalodonts were the first toothed whales to attain prominence, other porpoise-like odontocetes became increasingly important during the Miocene. Among these forms are found the ancestors of the modern beaked whales, the Ziphiidae. In modern beaked types, the teeth have been lost except for one or two tusks in the lower jaw and nonfunctional vestiges in the maxilla. There are deep pockets on the top of the skull at either side of the blowhole. The temporal opening is roofed over by the maxilla in this family and in the remaining odontocete types considered; and, in addition, there is some asymmetry of the skull, a twisting of the elements about the blowhole. In modern members of this family, too, we find the beginnings of a fusion of the cervical vertebrae characteristic of the larger whales.

The beginnings of the beaked whales are to be found as far back as the early Miocene in the small dolphin, *Notocetus* [*Diochotichus*], which showed ziphioid characters in many respects but still resembled the squalodonts in the retention of a good set of teeth and in other primitive characters. True but primitive beaked types (such as *Choneziphius*) with a much reduced dentition were developed during the Miocene, and, in the late Miocene, ziphiids were an exceedingly flourishing group. They declined greatly in impor-

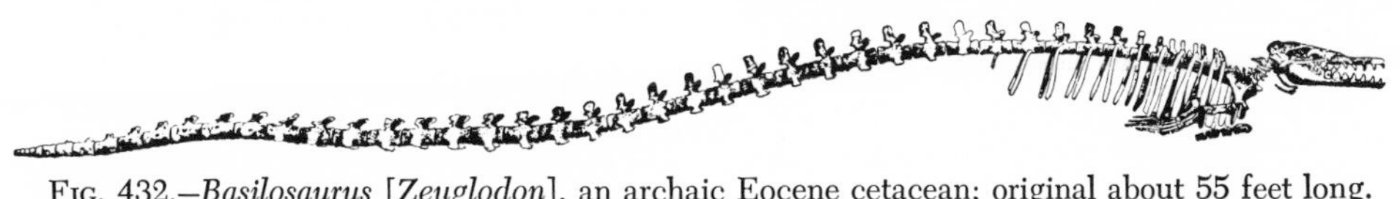

Fig. 432.—*Basilosaurus* [*Zeuglodon*], an archaic Eocene cetacean; original about 55 feet long. (From Gidley.)

tance, however, in the Pliocene, and only two types survive today.

Porpoises.—The most numerous of living small cetaceans are those included in the family Delphinidae, the modern oceanic porpoises and dolphins and their relatives. They are advanced in such structural features as the completely telescoped skull and the temporal region. But in contrast to most other living whales, most delphinids have retained an efficient battery of teeth in both upper and lower jaws. The dolphins must have developed rapidly out of the primitive toothed-whale stock, for forms essentially modern in build

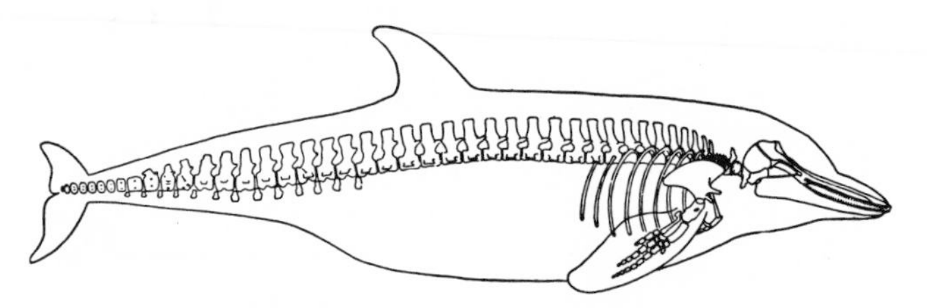

FIG. 433.—*Kentriodon,* a Miocene porpoise, about 5.5 feet long. (From Kellogg.)

were not uncommon in the early Miocene when the squalodonts still flourished, and the family had become the most abundant group of odontocetes by middle Miocene times, when nearly a score of delphinid genera appear to have flourished (Fig. 433). Apart from the Delphinidae, the general porpoise-dolphin group includes a variety of specialized types, recent and fossil, arrayed in half a dozen discrete families. These include, for example, *Eurhinodelphis* and other long-snouted Miocene forms, in which the rostrum makes up four-fifths or more of the skull length; the killer whales, which prey upon seals, penguins, and even larger cetaceans; the narwhal, with its "horn" formed from a sole remaining tooth.

Sperm whales.—Largest of all odontocetes and exceeded among all animals only by certain whalebone whales is the living sperm whale, *Physeter,* type of the family Physeteridae; *Kogia,* the pigmy sperm whale, is a much smaller form. The huge snout of these whales is mainly occupied by a great reservoir of sperm oil; the upper jaws are slender and depressed, but farther back the skull rises to a great cross crest cupping the hinder end of the sperm pocket. The asymmetrical twisting of the bone about the blowhole is especially noticeable in these forms, particularly in the greater backward expansion of the right premaxilla. In the upper jaw there are only a few functionless vestiges of teeth in the gums, but the lower jaw retains a complete and well-developed battery of teeth. In the postaxial skeleton a specialization lies in the fact that all the short cervical vertebrae are fused (with the exception of the atlas) into a solid mass. This feature is seemingly associated with the enormous head and repeated in some large-skulled whalebone whales. Although the living *Physeter* was already present when the family first appeared in the Miocene, there were also present a number of more primitive types—smaller, closer to the ancestral porpoise stock, still retaining teeth in the upper jaw but with distinct evidences of the development of the pocket for the sperm-oil reservoir. *Aulophyseter* (Fig. 431*C*) of the Middle Miocene is transitional in structure.

Whalebone whales.—The Mysticeti, the whalebone whales, today include only a small number of types, almost all of which, however, are of enormous size and include the largest of vertebrates, living or extinct. These greatest of cetaceans live on the smallest of prey, tiny floating invertebrates; it is impossible for such a whale to swallow any large object, for the gullet in the largest form does not exceed 9 inches in width. Whalebone consists of ridges of hardened skin which extend down from the roof of the mouth in parallel crosswise rows like the leaves of a book. They are fringed at the edge with "hairs" upon which food particles catch, to be promptly licked off by the huge tongue. The teeth have been lost with the development of this peculiar apparatus: the mouth is enormously enlarged, the jaws bending out and down (and failing to meet in a symphysis), and the rostrum sometimes arched up. Telescoping of the skull has taken place in this group also, but in a somewhat different fashion from that of the toothed whales; for, while even in modern types the blowholes (here double) are still situated in front of

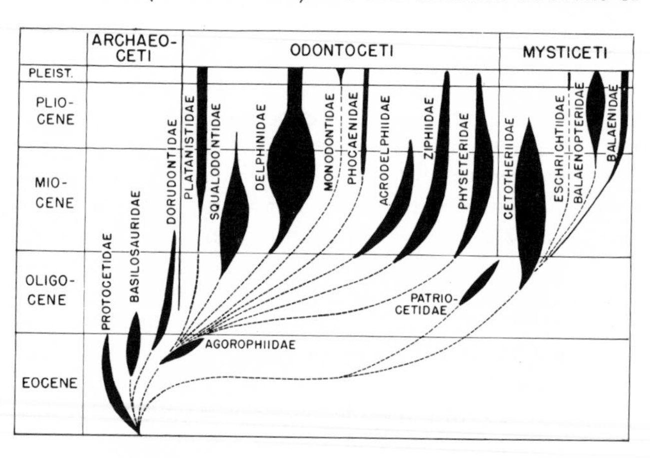

FIG. 434.—The phylogeny of the whales

the orbits, the normal roofing elements have been almost excluded from dorsal view by a forward migration of the supraoccipital region. The maxilla is braced against the frontal bone but does not override that element, although sending a finger-like process back over the skull roof. There is, in contrast with toothed whales, no trace of asymmetry.

The whalebone whales seem to be derivable from archaeocetes, although not from the more specialized members of that group. *Archaeodelphis* of the late Eocene and *Patriocetus* of the Oligocene appear to be transitional; in these forms there has begun a backward movement of the nostrils and a trend, little advanced, toward telescoping in symmetrical mysticete fashion. Certain mysticete history begins in the Oligocene with the appearance of the cetotheres, such as *Cetotherium* (Fig. 431*D*), primitive whalebone whales which were very common forms in the Miocene. The cetotheres, including such forms as *Aglaocetus* and *Mesocetus,* were already toothless types in which telescoping was well underway. This process, however, had not progressed so far as in living types, for dorsally a considerable gap still existed between the backward, finger-like process of the maxilla and the supraoccipital.

Only five genera of whalebone whales, representing families developed in the later Tertiary, survive today. Familiar forms are the rorquals, such as *Balenoptera,* and the right whales, *Balaena,* now confined to the Arctic and Antarctic regions. The rorquals have small skulls, free cervical vertebrae, and a low and broad rostrum; in contrast, the right whales have large skulls, fused cervicals, a curved rostrum, and a deep jaw. Of the rorquals, the blue or sulphur-bottom whale is estimated to reach a maximum weight of 150 tons or more, thus being far larger than any dinosaur.

27

CHAPTER

Rodents

Among the gnawing animals—the order Rodentia—are included the squirrels and beavers, rats and mice, porcupines and guinea pigs, and hosts of less familiar forms (Fig. 435). The rodents are, without question, the most successful of all living mammals. In number of genera and species they exceed all other orders combined; they are found in almost every habitable land area of the globe and seem to thrive under almost any conditions, flourishing even in towns and cities. The range of adaptations is a wide one. A majority of rodents are terrestrial and often burrowing types. No purely aquatic forms have developed, but such forms as the beaver and muskrat have progressed far in water life. Others, like the squirrels, are arboreal; and while there are no flying rodents, the "flying" squirrels have tended in this direction.

In size, however, the rodents are modest animals. Most forms are small, and the rat or squirrel may be taken as an average; the capybara of South America, the size of a pig, is the giant of the group.

Dentition.—The rodents are essentially herbivores, although certain of them, such as the rats, will accept a wide variety of food; the peculiar gnawing and grinding dentition and adaptations related to its use are highly characteristic of the group. A single pair of incisors is present in the lower jaw, and it is opposed by a principal pair above. These huge teeth never form roots and have bases which curve far back inside the bones of the upper and lower jaws; their continual growth counterbalances the wear to which they are subjected in gnawing. The other incisors, the canines, and a number of premolars have been lost in all known rodents, leaving a long diastema between the gnawing teeth and the grinding series. There are never more than two upper premolars and a single lower one, and in many of the modern ratlike forms only the molars remain in the cheek region. The cheek teeth often become high crowned, and in many cases they fail to close their roots and, like the incisors, continue to grow throughout life. The two upper rows of grinders are closer together than the lower ones and are tilted so that they meet the lower teeth in a plane that faces somewhat outward as well as downward.

The rodent molar pattern appears primitively to have followed the general mammalian herbivore tendency toward the development of rectangular, four-cusped cheek teeth, and in a few cases a simple pattern readily interpretable in terms of familiar placental cones has persisted (Fig. 439*A*); usually, however, such interpretation is difficult, and a wide degree of variation exists, even within the limits of single families. Much of this variation is associated with the development of high-crowned teeth which tend to be formed of a complicated series of ridges. These often become united as the tooth is worn; the original boundaries may be indicated by grooves and loops in the enamel pattern or may be completely obliterated in the formation of peglike structures.

A molar pattern found in some members of all major rodent groups and perhaps antecedent to many of the more complicated types is that illustrated in Figure 436. The shape is roughly that of the letter E. Three transverse ridges project laterally in the upper molar; the anterior and posterior ridges are formed primarily by paracone and metacone; the intermediate one is a newly developed mesoloph. Additional cross-lophs may be formed anteriorly and posteriorly from cingulum upgrowths or by elaboration and subdivision of those already present to give a total of five lophs. Toward the medial side are protocone and hypocone; a new mesocone appears between them on the longitudinal crest of the tooth. The lower molar develops in comparable fashion.

Skeleton.—With these adaptations for grinding and gnawing has come a series of peculiar adaptations in the skull and jaws. In most rodents there occur both fore-and-aft and transverse grinding movements of the lower jaws on the skull. In consequence the glenoid fossa of the skull (in which the lower-jaw condyle is received) is long and shallow, permitting considerable freedom of motion. In connection with powerful jaw musculature, the angular process of the lower jaw is always strongly developed; in many rodents (notably South American forms) it is sharply inflected outward. The rodent skull is usually long and low, the orbit always open behind.

There are comparatively few specializations in the postcranial skeleton. The clavicle is usually retained;

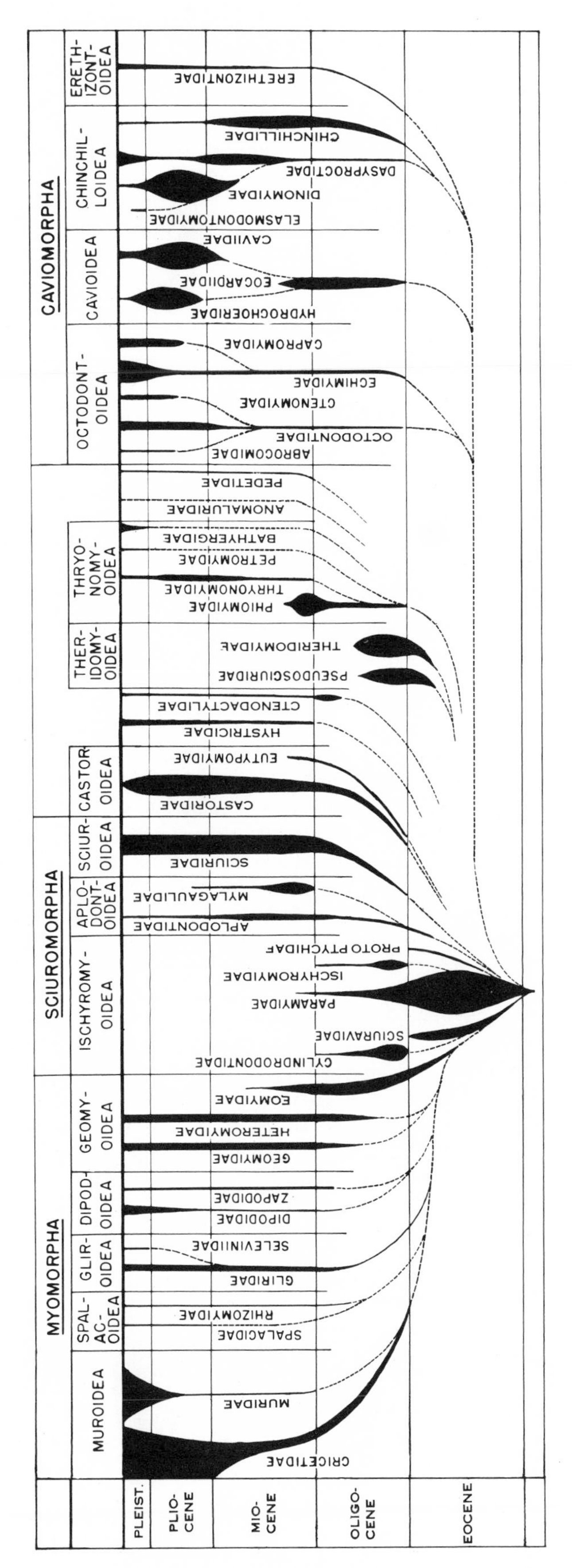

Fig. 435.—The phylogeny of the rodents

claws are always present. The front leg is usually flexible and often used as an aid in bringing food to the mouth, and there is little tendency toward loss of digits in the manus except for the occasional reduction of the pollex. The hind leg, however, is less flexible in its movements and confined to a fore-and-aft motion; tibia and fibula often fuse distally, and there are varying degrees of reduction of the toes with the development of a hopping or springing type of locomotion.

Major criteria in rodent classification lie in skull modifications related to the jaw musculature. In addition to the temporal muscles closing the jaw, all mammals possess another series termed the masseters. These primitively pass from the zygomatic arch downward to the outer surface of the mandible (Fig. 437). In rodents the masseters consist of three parts. A superficial portion is essentially uniform in all rodents, arising far forward on the side of the face (often from a distinct tubercle) and running backward and a bit downward to insert on the angle of the jaw. The two other parts show important variations (Fig. 437). The middle layer primitively arises from the under edge of the zygomatic arch and inserts along the lower margin of the jaw; the deep layer primitively arises from the inner surface of the arch and inserts higher up on the outer jaw surface.

In squirrels and beavers the deep masseteric layer retains its original position, but the origin of the middle layer has moved forward onto the side of the snout, and a channel develops in the skull in front of and below the orbit to carry the muscle back and down to its insertion on the jaw. In the great group of ratlike rodents this same channel develops, but the deep layer has shifted as well. In mammals in general an opening—the infraorbital canal—runs forward from the orbit carrying small nerves and blood vessels onto the side of the snout. In the rat group this canal is enlarged, and the origin of the deep layer pushes forward through it onto the snout. In South American rodents and a variety of Old World forms the canal is enormously enlarged and, while the middle masseter retains its primitive position, the greatly developed deep masseter has pushed forward and upward through this great opening to gain its origin from the lateral surface of the snout.

The origin of the rodents is obscure. When they first appear, in the late Paleocene, in the genus *Paramys*, we are already dealing with a typical, if rather primitive, true rodent, with the definitive ordinal characters well developed. Presumably, of course, they had arisen from some basal, insectivorous, placental stock; but no transitional forms are known. To perfect the dental and other features of the order, a considerable period of time—perhaps the whole extent of the Paleocene—seems necessary. But in what region or environment this occurred, we do not know.

Once they appeared, the rodents almost immediately took over an important part of terrestrial economy, and have continuously flourished, and flourished increasingly, from the Eocene to the present.

Classification.—So numerous and varied are the rodents, living and fossil, that they are exceedingly difficult to classify, and their systematic arrangement and phylogenetic history have long been—and continue to be—matters of dispute. It was once thought by many that all rodents could be included in three suborders. This is no longer held to be the case. From the ancestral rodent stock there appears to have been a bushlike evolutionary radiation, with a number of short sprouts (so to speak) developing, as well as a few main trunks. Most workers today believe that there are a number of minor groups of living and fossil rodents that show no clear relationship to any other members of the order. Probably, however, a large fraction of members of the order can still be assigned to three subordinal assemblages, as follows:

Suborder Sciuromorpha (or Protrogomorpha). Includes primitive rodents, with the squirrels as peripheral modern representatives.

Suborder Caviomorpha. A great group developed in South America, of which the guinea pig (*Cavia*) is representative.

Suborder Myomorpha. Rats and mice and their relatives.

Sciuromorphs.—We shall here use the term Sciuromorpha for a suborder including the most primitive rodent types; workers on these forms, however, prefer the term Protrogomorpha, since the squirrels (*Sciurus*) are a specialized offshoot rather than central to the group. The most primitive of known rodents is a family of which *Paramys* (Figs. 438*A*, 439*A*, 440, 441) is representative. This is the oldest known rodent genus, appearing in the late Paleocene and flourishing until the Middle Eocene; a score or so of other genera of Eocene rodents and a few Oligocene survivors are closely enough related to be classed in the same family. *Sciuravus, Ischyromys, Protoptychus,* and *Cylindrodon* are representatives of further Eocene and Oligocene families which are closely related. Of these, the sciuravids are of special interest as probable ancestors of the myomorphs. The characters possessed by *Paramys* and its relatives are primitive and, hence, essentially negative. With two premolars above and one below, plus the usual complement of molars, the sciuromorphs have the maximum number of teeth found in any rodent group; however, the anterior upper premolar is already reduced and, exceptionally, may be lost in a few members of the suborder. The molars of *Paramys* and its close kin are the most primitive of those of any known rodents, and even those of the somewhat more advanced *Ischyromys* (Fig. 439*B*) are still quite primitive. The paramyid molars, low crowned, have departed relatively little from the basal placental tribosphenic pattern. There has been some increase in the number of bunodont cusps, but there is little trend toward the complicated series of lophs characteristic of rodents generally. The masseters of most sciuromorphs appear to have been relatively undeveloped, and the skull (Fig. 437*A*) shows none of the enlargement of the infraorbital canal seen in other rodent groups; nor is there in most members of the suborder any development of the channel below the orbit for the middle masseteric layer. In this suborder the jaw angle is of normal construction. The paramyid skeleton is relatively unspecialized; *Paramys* and its kin were apparently, in life, small, rather squirrel-like scamperers.

The most direct descendant of the early sciuromorphs, and perhaps deserving of being considered the most primitive of living rodents, is *Aplodontia,* the sewellel or "mountain beaver" of the American Northwest. In this small ground-dwelling rodent the skeleton is essentially similar to those of the paramyids, but the teeth tend toward a hypsodont condition. Although reduced to a single form today, the aplodontids achieved modest success in the Miocene. An early form, presumably of paramyid descent, appeared in the late Eocene. From the aplodontids, it is believed, arose in the Miocene and Pliocene of North America one of the most specialized of rodent types, the Mylagaulidae. *Mylagaulus* was about the size of a marmot; the powerful front legs suggest digging habits. The dentition is highly aberrant. The cheek teeth are high crowned and lophodont; except in one primitive genus there are but two upper molars and but a single premolar in each jaw; the premolars are enormously enlarged, equalling or exceeding the molars behind them.

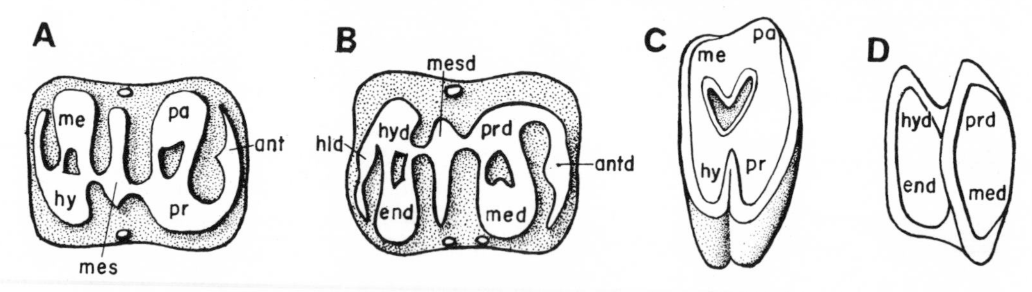

Fig. 436.—Diagrams of molar patterns in rodents and lagomorphs. *A, B,* Right upper and left ower molars of a type common in rodents (particularly myomorphs); *C, D,* right upper and left lower molars of lagomorphs. Abbreviations: *ant,* anterocone; *antd,* anteroconid; *mes,* mesocone; *mesd,* mesoconid. For other abbreviations see Figure 297, with which these patterns may be compared. (After Wood and Wilson.)

Most, at least, of the mylagaulids are unique among rodents in carrying a well-developed median bony horn above the forehead.

Although departing somewhat from the typical pattern of the suborder, we include here the squirrels and relatives, the Sciuridae. These amiable creatures are relatively unspecialized in most regards, with, for example, cheek teeth which are low crowned and of a simple, although lophodont, pattern. They are, however, divergent in one very distinctive feature. Parallel to the beavers and myomorphs, the middle masseteric layer extends forward on to the rostrum in a channel lying below and anterior to the orbit (Fig. 437*B*).

The main group of squirrels, including *Sciurus* proper, are arboreal forms, with the "flying" squirrels of the Northern Hemisphere as an offshoot. Others, as

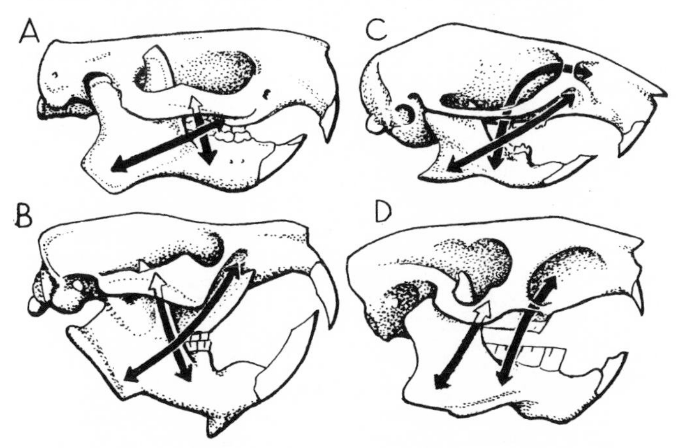

Fig. 437.—Diagrams to show the development of the middle and deep portions of the masseteric muscle in various rodent types (a superficial layer omitted). *A*, A primitive sciuromorph; the masseter originates mainly from the lower edge of the zygomatic arch. *B*, An advanced sciuromorph; the middle masseter originates from the outer side of the skull in front of the orbit. *C*, A myomorph; the middle masseter is similar, but the deeper portion has pushed up through the orbit and arises in a pocket on the face, formed from the infraorbital foramen. *D*, A hystricomorph; the middle masseter is unspecialized, but the foramen is enormously developed to accommodate the deeper portion.

ground squirrels, have become primarily terrestrial, including such forms as the chipmunks. A group of large, short-tailed burrowing forms includes the prairie dog, common on the western American plains, and *Marmota* (*Arctomys*), the woodchuck or marmot, common as a fossil in the colder phases of the European Pleistocene. *Sciurus* and other members of the family, presumably derived from the early paramyids, appeared in the Oligocene and spread widely in Eurasia and North America. They never reached Australia but successfully invaded South America in the Pleistocene.

Caviomorphs.—In the Caviomorpha are included forms which constitute nearly the whole rodent population of South America, but are confined to that continent except for the North American porcupine, *Erethizon.* The classification of the group is not too well determined, but there are perhaps a dozen families, nearly all of which are represented by both living and fossil forms, with a total of about 150 genera. From sciuromorphs they are distinguished readily by the presence of an enlarged infraorbital canal for the deep masseteric muscle; they differ from the myomorphs in the very large size of this canal and by the absence of an anterior channel for the middle masseteric layer (Fig. 437*D*, 438*E*), and are further distinguished from both these suborders by the flaring angle of the jaw. In most families the teeth (Fig. 439*E*) are hypsodont, with a strong trend toward the development of numbers of cross-lophs of simple appearance. There is but one upper premolar.

Few of the caviomorphs are familiar to non-natives of South America, and it is difficult to picture briefly the nature of the various family groups. The members of the Octodontidae and the related Echimyidae, Abrocomidae, Ctenomyidae, and Capromyidae have in general a rather ratlike build; however, *Abrocoma* is a furry form and the tuco-tuco, *Ctenomys,* is a burrower somewhat analogous to North American pocket gophers. Members of this group of families appear to have reached the West Indies in considerable numbers; about a dozen genera of this sort are known as Pleistocene or sub-Recent forms, and two survive. The Echimyidae are currently the most flourishing of all caviomorph families. A second group of families is that of the Chinchillidae, with a number of furry, bush-tailed and somewhat squirrel-like burrowers, and the Dinomyidae. *Dinomys,* the paca, the surviving member of the latter family, is a large form; there are a considerable number of related Pliocene types and a number of extinct West Indian representatives. Chinchillids and dinomyids are notable for multiple cross-lophed hypsodont molars.

A third family group has as its principal component the Caviidae, including the familiar guinea pig. Most, like *Cavia,* are stub-tailed, stockily built forms of modest size. A related family, well represented in the Pliocene and Pleistocene, has as its living member the great capybara, *Hydrochoerus,* giant among living rodents, with a body length of 4 or 5 feet.

A final group includes only the New World porcupines, Erethizontidae, spiny arboreal types with a muscular and sometimes prehensile tail.

The caviomorphs appear abruptly in the Oligocene, and from the first show considerable variation, for several families were already represented—Octodontidae, Echimyidae, Chinchillidae, Eocardiidae (related to the Caviidae), and Erethizontidae. For the most part, however, the early members of these various families were still fairly close to one another. They appear to be new arrivals, for had they been part of the original stock of Paleocene-Eocene placentals of South America, some trace of them probably would have been found earlier. It was formerly believed that a number of African rodents were caviomorph relatives, and it was suggested that the South American rodents had crossed the Atlantic from Africa. But such a crossing, while not impossible, is improbable; and, further, it is now thought that there is no real relationship between the rodents of the two continents. It is more reasonable to believe that the caviomorphs are of North American ancestry, derived from paramyid descendants who managed to cross a relatively short water gap between Central America and the South American continent, perhaps in the late Eocene. In the absence of competitors, the caviomorphs flourished in Miocene and, especially, Pliocene days. By the Pleistocene, at least, if not earlier, a considerable number of caviomorphs had reached the West Indies, where a few survive. Some of these insular forms were large; the giant of fossil rodents, however, was the mainland dinomyid *Eumegamys* of the Pliocene, with a skull close to 2 feet in length, and a bulk presumably comparable to that of a wild boar. The caviomorphs do not appear to have been seriously reduced by the Pleistocene invasion from the North. On the other hand, except for the porcupine, they made little progress toward an invasion of North America, although the capybaras reached the Gulf states in the Pleistocene.

Myomorphs.—Third of the three well-defined suborders of the rodents is that of the Myomorpha, of which the "domestic" rats and mice are typical. In them (Figs. 437*C*, 438*D*) the infraorbital canal is enlarged for the transmission of masseteric musculature, in contrast to sciuromorphs, but this canal is much less enlarged than in caviomorphs. They have the suborbital channel for the middle masseter found in squirrels and beavers but not found in more typical sciuromorphs or in the South American forms. A further difference from the caviomorphs is the absence of the out-turned angular process. In typical myomorphs the premolars have vanished, leaving, above and below, only the three molars (the first usually much larger than the others) to take over the entire burden of grinding. In the more primitive genera the teeth are low crowned and rooted; in others there is a development of high-crowned, rootless teeth.

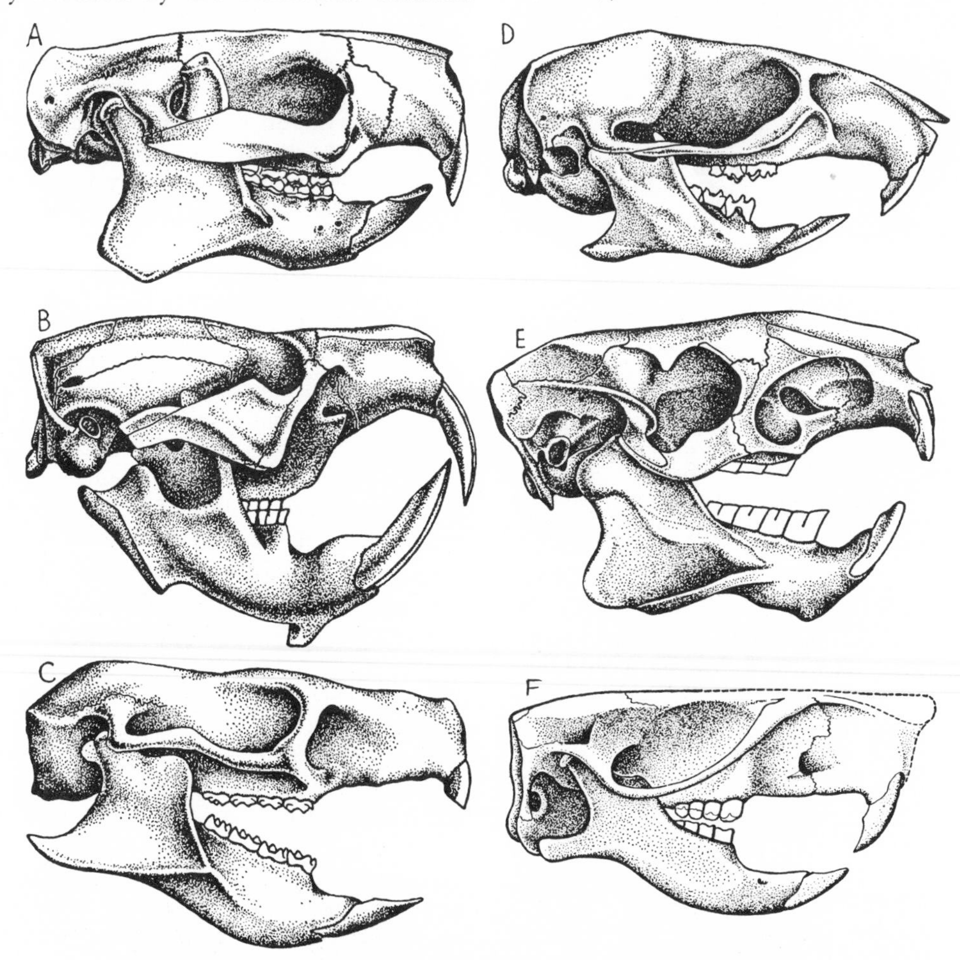

Fig. 438.—Skulls of rodents. *A, Paramys,* a primitive sciuromorph, skull length about 3.5 inches; *B, Palaeocastor,* a Lower Miocene castorid, skull length about 2.75 inches; *C, Pseudosciurus,* an Eocene theridomyoid, skull length about 2.5 inches; *D, Cricetops,* a Tertiary murid, skull length about 1 inch; *E, Neoreomys,* a Miocene South American caviomorph, skull length about 4 inches; *F, Florentiamys,* a Miocene geomyoid, about natural size. (*A* after Matthew; *B* after Peterson; *C* after Schlosser; *D* after Schaub; *E* after Scott; *F* after Wood.)

The central rat-mouse stock of the suborder constitutes the superfamily Muroidea. This is an old and highly successful group. The first definitely recognizable genera, apparently derived from the *Sciuravus* group of primitive sciuromorphs, date from the early Oligocene, when *Cricetodon* and *Eumys* (Fig. 439*F*) appear, but it is not until the Pliocene that they become common and appear to have undergone an explosive evolutionary radiation. In late Cenozoic and Recent times they have become the most widespread and abundant of rodents—indeed, of all mammals. Recent muroids outnumber all other rodents in the number of genera and species and, it seems certain, in the number of individuals as well, and their dominance in Pliocene and Pleistocene days was nearly as great as now. Their geographical range is greater than any other rodent group, for they not only invaded South America when that continent became readily accessible in the Pleistocene, but are also present in Australia, a region to which no other terrestrial placentals were able to make their way before the coming of man. Two major groups of muroids may be distinguished—the Cricetidae and Muridae. The center of murid evolution appears to have been the Eastern Hemisphere, but the cricetids have been present in North America since the Oligocene, and they constitute the entire native assemblage of rats and mice. The murid division is native to the Old World tropics, but in recent times the common mouse (*Mus*) and rat (*Rattus*) have accompanied man to all corners of the world.

Four further superfamilies may be included in the myomorph suborder. The Dipodoidea include the jerboas (*Dipus* and various other Old World genera) and the related jumping mouse, *Zapus*, of North America and Eurasia. The modern forms are small desert

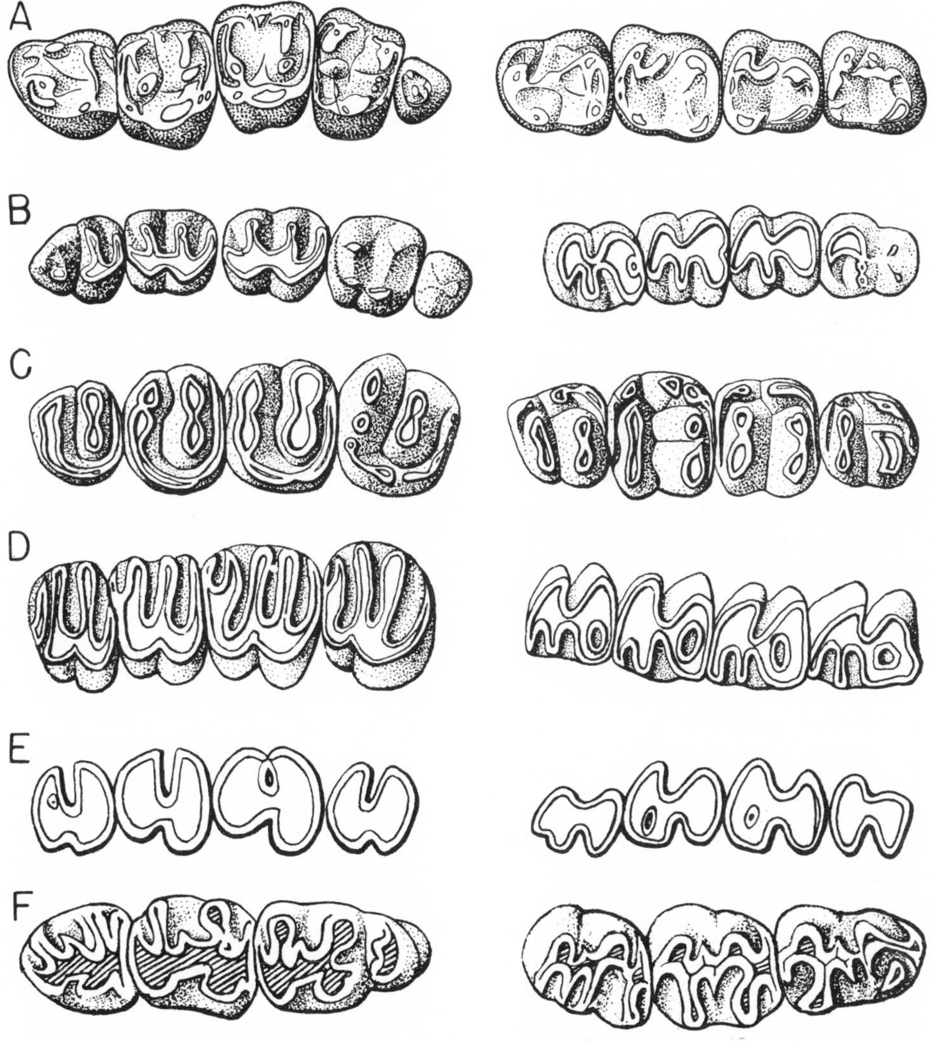

FIG. 439.—Teeth of rodents. *Left,* Upper cheek teeth of right side; *right,* lower cheek teeth of left side. *A, Paramys,* a primitive Eocene rodent (×4); *B, Ischyromys,* an Oligocene sciuromorph (×4); *C, Florentiamys,* a Miocene geomyoid (×7.5 approx.); *D, Theridomys,* an early Tertiary theridomyoid (×9 approx.); *E, Sciamys,* a Miocene South American caviomorph (×7); *F, Eumys,* an Oligocene murid (×7.5 approx.). (*A, C* after Wood; *B, F* after Matthew; *D* after Schlosser; *E* after Scott.)

dwellers, most of which have assumed a hopping gait, with reduced front limbs and long hind legs with curiously birdlike feet in which there are three symmetrically arranged toes with the metapodials fused into a single long element. Their basic structure is typically myomorph, apart from such minor features as the retention of a vestigial upper premolar and an enlargement of the infraorbital canal to a greater degree than in typical rats and mice. Fossil representatives are rare, but an ancestral dipodoid, *Simimys,* not too far removed, it would seem, from ancestral cricetids, is known from the late Eocene. Both modern families are present from the Oligocene on—the jerboas restricted to Eurasia.

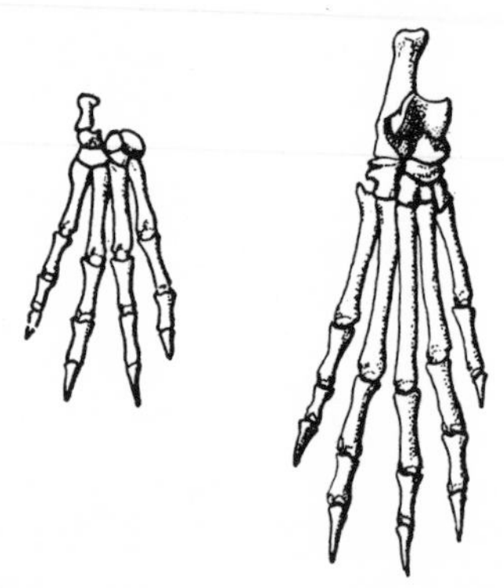

FIG. 440.—Right manus and pes of *Paramys,* a primitive Eocene rodent.

A second group not too far removed phylogenetically from the muroids is that of the superfamily Spalacoidea, Old World burrowing rats such as the modern *Spalax* and *Rhizomys.* The basic structure is typically that of a myomorph, but in their powerful digging limbs and other burrowing adaptations they have to a considerable degree paralleled the moles. Of the few known fossil forms, none is older than the late Miocene.

The Gliroidea are an Eastern Hemisphere group; the dormice, or sleepers, such as *Glis* [*Myoxus*], are typical. In these small, arboreal fruit eaters the teeth have, in contrast to most myomorphs, remained persistently low crowned, and a further primitive feature is the persistence of a small premolar in both upper and lower cheek batteries. Many Eurasian fossil dormouse genera enable us to follow this group back to the Upper Eocene. They appear to be of sciuravid descent.

We shall also include here, although with much doubt, the superfamily Geomyoidea, represented by the American pocket gophers, *Geomys,* and the kangaroo rats, such as *Heteromys.* The pocket gophers are burrowing types, rather paralleling the Old World spalacoids, which owe their name to their large cheek pouches; the mouselike kangaroo rats are small hopping desert rodents with elongated hind legs. Despite the disparity in adaptive features, there are enough common features to merit associating the two types in a common superfamily. The teeth (Fig. 439*C*), low crowned in early forms, are hypsodont in modern genera and, in the pocket gophers, rootless. As in myomorphs, but in no other rodents except squirrels and beavers (Fig. 438*F*), there is a well-developed channel below the orbit for the middle masseteric muscle layer; but in contrast with proper myomorphs, the deep masseter does not penetrate the infraorbital canal, and there is a well-developed premolar in both upper and lower jaws. Their relationships to the other myomorphs may, hence, be rather remote. Both modern families are represented as far back as the Oligocene by American fossil forms, and there is even a late Eocene genus. Related to the ancestry of modern geomyoids is the extinct family Eomyidae, represented in both hemispheres from the Eocene on.

Beavers.—Having now considered those groups which can with reason be assigned to one or another of the three well-defined suborders, let us proceed to a consideration of a variety of types which do not toe in well with these major groups. To begin with, the Castoroidea, of which the beavers, *Castor,* are the modern representatives. In the past these were generally assigned to the Sciuromorpha, but they have departed so far from the general structural pattern of that group that their relationships are obscure. In the high-crowned cheek tooth series, a single well-developed premolar is present above and below, which at once rules out close relationships to myomorphs; in contrast to the caviomorphs, there is no enlargement of the infraorbital canal. The dentition differs basically from that of sciuromorphs only in the loss of the small upper premolar; but there is a well-developed masseteric channel below the orbit, which in that group appears only in the "modernized" squirrel family. In short, the beavers are presumably derived from some primitive sciuromorph stock, but there are no annectant types between such forms and the oldest Oligocene castoroids to prove direct relationship.

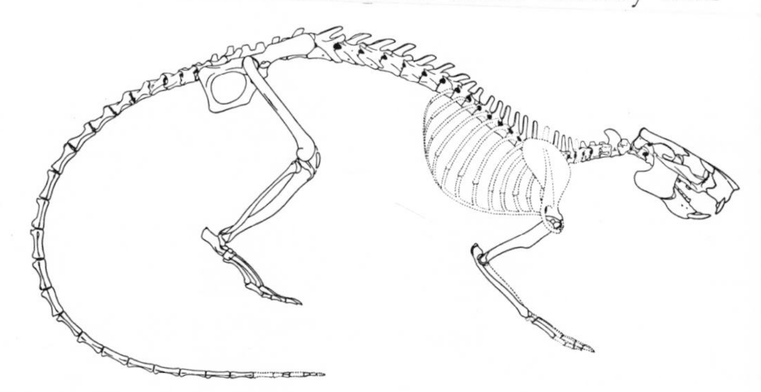

FIG. 441.—The primitive Eocene rodent *Paramys;* length about 2 feet. (From Wood.)

The living beavers are aquatic, and this may well have been the mode of life in some of the older castoroids. Some, however, were burrowers. *Palaeocastor* (Fig. 438*B*), *Steneofiber,* and their relatives are common late Oligocene and Miocene fossils in both Europe and North America; in western Nebraska, peculiar upright "devil's corkscrews" seen in the Lower Miocene beds are thought to be their mud-filled burrows. Giant beaver types lived in both northern continents in the Pleistocene; *Castoroides* (Fig. 442) of North America attained the size of a half-grown bear; *Trogontherium* was a large European parallel.

Old World "hystricomorphs."—It was long believed that a varied series of fossil and Recent Old World rodents which resembled the caviomorphs in the presence of a large infraorbital canal and, usually, of a cheek series of four well-developed teeth were actually related to the South American forms. To this vast assemblage the term Hystricomorpha (based on the Old World porcupines) was applied. Recent work, however, strongly indicates that the Eastern Hemisphere forms have merely paralleled the caviomorphs in the development of common features and that, further, there is no proof that these supposed "hystricomorphs" are in themselves a single unit; they seem to be, on the contrary, a series of half a dozen or so units with uncertain relationships to one another and to other major rodent groups. We shall review them briefly.

First there are the Hystricidae, the Old World porcupines. They resemble the New World porcupines in their spiny covering (but not in habits, for they are burrowers, rather than arboreal), and there is no evidence in the fossil record for connecting the two groups. There are a few fossil forms, back to the Miocene and possibly late Oligocene, but these give no indication of relationship of hystricids to other rodent types.

Thryonomys, the cane "rat," and *Petromus* [*Petromys*], the rock "rat," are two African rodents which show a degree of relationship to one another, and may thus be bracketed in a common superfamily Thryonomyoidea. Apart from the "hystricomorph" features of a large infraorbital canal and a cheek battery of four teeth, they show no indications of relationships to other major groups. Almost nothing is known of the ancestry of *Petromus; Thryonomys* is recorded in the African Miocene, and there appears to be a Pliocene Asian relative. Possibly ancestral to these forms is the extinct family Phiomyidae, members of which appear in the early Oligocene Fayûm beds and are common in the African Miocene. We appear to be dealing with a group which arose in Africa.

The superfamily Theridomyoidea includes a number of European fossil forms of the late Eocene and Oligocene, such as *Theridomys* (Fig. 439*D*) and *Pseudosciurus* (Fig. 438*C*), which show the same "hystricomorph" characters as the African thryonomyoids and appear to have been a northern parallel to that group. At present we know nothing of their ancestry or possible descendants, although the flying "squirrels" of Africa, the Anomaluridae (not known with certainty as fossils) have been thought related.

Still further African forms once grouped with the "hystricomorphs" include *Pedetes,* the Cape "jumping hare," and *Ctenodactylus* and relatives, the desert-dwelling "gundis." The *Pedetes* group can be traced to the African Miocene; the gundis have Asiatic relatives back to the Oligocene; for neither are there indications of relationships to other groups.

We conclude our long journey through the rodents with mention of the Bathyergidae—subterranean African dwellers unknown as fossils except in the Pleistocene, and showing few indications of relationship to any group whatever.

Lagomorphs.—The hares and rabbits were long con-

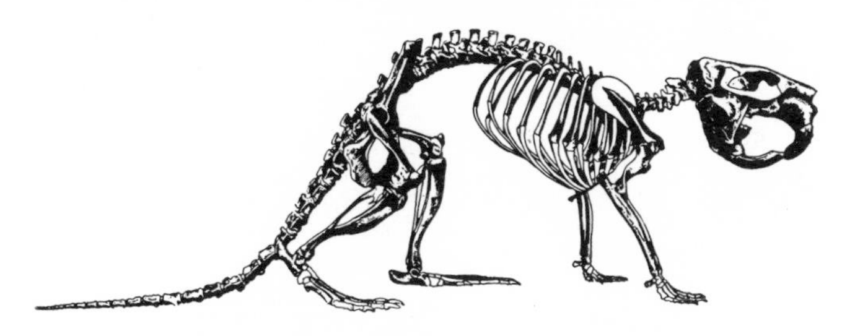

Fig. 442.—*Castoroides,* a giant beaver, about 7.5 feet in length, from the Pleistocene of North America.

sidered to be a suborder of the rodents (as Duplicidentata); but, apart from the fact that they have gnawing incisors and grinding cheek teeth, they have little in common with the rodents and are now regarded as constituting a separate order Lagomorpha.

As in true rodents, the functionally important gnawing teeth are a single pair of incisors above and below, but in the lagomorphs an accessory pair of small upper incisors is present behind the principal ones. There is, as in rodents, a long diastema; the outer incisors, canines, and some premolars have been lost in its development. The number of cheek teeth present is higher than in any rodent, for there are always three upper premolars and two lower ones, in addition to three molars. The cheek teeth (Figs. 436*C, D;* 443) are high crowned and rootless, with patterns presumably derived from those seen in early placentals. The upper molars seem basically comparable with the primitive tritubercular pattern; the lower ones have, as in many other groups, developed two cross-lophs. In contrast with rodents, the two upper tooth rows are farther apart than the lower ones, and the teeth have more of a chopping than a grinding motion. There is never any enlargement or muscular invasion of the infraorbital canal. A peculiar lattice-work fenestration of the side of the snout is a noticeable feature of the lagomorph skull. The tail is short or absent in all known forms.

Best known of the group are the leporids, including the living hares (*Lepus*) and rabbits (*Oryctolagus*) and their relatives, leaping forms with much-elongated hind legs. A less prominent type is that represented today by the pika, *Ochotona,* of the mountains of western America and Asia, with short legs and short ears.

The lagomorphs show no close approach to other placental groups, and the ordinal characters are well developed in even the oldest known forms. The appearance of chisel-like front teeth was, however, a common development in many early Tertiary mammal groups (as well as in the true rodents); and although obscured by hypsodonty, the cheek-tooth pattern is readily derivable from that of primitive placentals.

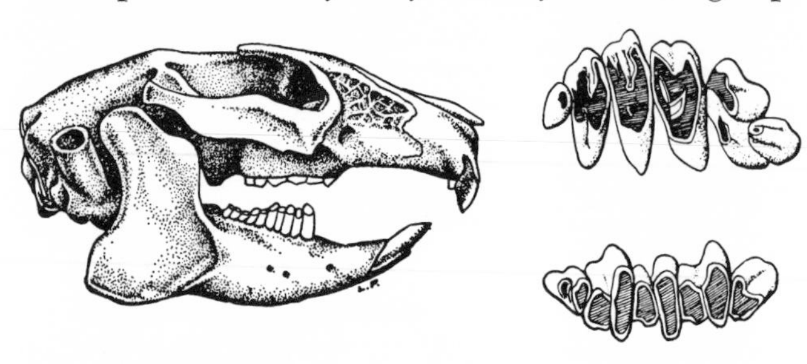

Fig. 443.—*Palaeolagus,* an Oligocene hare; skull, right upper and left lower cheek teeth; skull about 2 inches in length; teeth ×3 approx. (After Troxell and Wood.)

On present evidence, the lagomorphs were "late starters" in the race for success among placental orders. Before the late Eocene we know nothing of the group except a single, somewhat aberrant form, *Eurymylus,* from the late Paleocene of Mongolia. Following an almost complete gap in the Eocene, the lagomorphs became well established in both northern continents in Oligocene days, when, amongst other forms, *Palaeolagus* (Fig. 443) is a familiar leporid, and ochotonids were also to be found in both Eurasia and North America.

In North America there appeared, in *Archaeolagus* of the Lower Miocene, a group of hares which flourished in this country until the early Pliocene. However the modern types of hares and rabbits appear to have arisen in Eurasia in the late Tertiary, and in Pleistocene and Recent times *Lepus* and related hare and rabbit forms have infested every continent of the globe —even, with man's help, Australia.

CHAPTER 28

Vertebrate History: Introductory Paleozoic Vertebrates

In previous chapters we have considered the history of the various vertebrate groups with especial regard to structure and phylogeny. In the present chapter and in chapters 29 and 30 we shall review briefly the knowledge thus gained in a discussion of the faunal succession of the vertebrates, their distribution in time and space.

Geologic periods.—The conventional subdivisions of geologic time—eras and periods—have, in general, been established on the basis of events in the marine waters of Europe and North America; and boundaries between periods were set up at points where there appeared to be either major breaks in the record of deposition or marked changes in the invertebrate life. But it is obvious, a priori, that a break in the record of sedimentation in one region by no means proves that a similar cessation will appear in others, and it becomes increasingly apparent that sharp lines of demarcation do not in reality exist. Furthermore, a comparatively sudden change in marine invertebrates by no means implies a marked simultaneous change on the continents, and many of our most interesting vertebrate faunas lie on the borderline between periods or even eras. The earliest Devonian vertebrates, for example, are more like those of the late Silurian than like later Devonian fishes; the late Pennsylvanian tetrapods are almost identical with those of the Lower Permian. We cannot, however, readjust the period boundaries to suit the vertebrates alone, and we may retain the current nomenclature, while recognizing its artificiality (cf. Table 1).

Correlation of deposits.—In assigning an age to vertebrate-bearing beds, the evidence obtained from associated invertebrates is exceedingly useful. Many of the more important deposits, however, are of continental origin and contain almost no invertebrate remains. Here recourse is often to a direct comparison with vertebrates of other regions. If two series of deposits contain a considerable number of identical or closely related forms, it is highly probable that they are contemporaneous. But if the forms compared are few in number or but distantly related, the argument for contemporaneousness is greatly weakened.

Negative evidence.—Any contrast between the life of different periods or different geographical areas is based upon the presence at one time or place of animals apparently absent at an earlier or later date or from another region. But it must be pointed out that, while the presence of an animal may be definitely shown, absence is incapable of absolute proof; all that we can really say is that remains have not been discovered. Until recent decades no placental mammal had been identified in collections from the Cretaceous. But our knowledge of later mammalian evolution led us strongly to believe, despite this negative evidence, that there must have been Mesozoic placentals. Today, owing to work first in Mongolia and later in the American West, a number of placentals are known in the Cretaceous. Negative evidence of absence has been replaced by positive knowledge of the presence of these forms. Again, no trace of coelacanth crossopterygians has ever been discovered in Cenozoic strata; and it was long believed, with confidence, that the coelacanths had become extinct at the end of the Cretaceous. The recent discovery of a living specimen has rudely shattered this belief.

It is probable that the range in time or space of many forms with which we are familiar was greater than is now known. It is further probable that there may have existed many interesting vertebrate types of which no remains have as yet been discovered. But such probabilities are decreasing as our knowledge expands. During the present century a very great amount of paleontological work has been done, and many strange forms have been brought to light. These, however, have been almost always members of groups already known or forms tending to connect such groups.

Environmental differences.—In a study of past faunas the great variety of possible habitats must be taken into consideration. Often the structure of an animal gives much evidence concerning its environment. The teeth may afford a basis for distinguishing between forest-living, browsing types and plains-dwelling grazers; the limbs may show us whether we are dealing with an aquatic, terrestrial, arboreal, or aerial form. The nature of the sediments and the associated plants and invertebrates may give us important additional information as to its surroundings in life or death.

Among fishes the question of a fresh-water or salt-water habitat is often of great interest, as we have noted. If a fossil type is found in strata in which marine invertebrates do not occur, it is not unreasonable to assume that it inhabited fresh waters. If it is accompanied by such invertebrates, it may well be a marine form; but this is none too certain if the association is a rare one or if the vertebrate remains are badly broken, for individuals of a river-dwelling type may drift out and become buried in the sea.

TABLE 1 **The Geologic Periods**

ERA	PERIOD		APPROXIMATE TIME (In Millions of Years)	
			Duration	Since Beginning
CENOZOIC (age of mammals and man)	Quaternary		2+	2+
	Tertiary		60	62
MESOZOIC (age of reptiles)	Cretaceous		68	130
	Jurassic		50	180
	Triassic		50	230
PALEOZOIC (age of invertebrates and primitive vertebrates)	Permian		50	280
	Pennsylvanian	Carboniferous	30	310
	Mississippian		30	340
	Devonian		60	400
	Silurian		50	450
	Ordovician		50	500
	Cambrian		70	570

The older eras, in which vertebrates are unknown, are not included.

Among land animals, distinctions may be made between lowland and upland types, forest and plains dwellers. Here an unfortunate difference in the probability of preservation appears. Wooded areas are obviously regions in which little deposition of sediments takes place and in which skeletons are usually rapidly destroyed by insects, grass and tree roots, or acids in the soil. Hence our knowledge of the evolution of arboreal and forest-dwelling animals is relatively meager; in contrast, deposits are common and skeletons often preserved in plains areas. Of the evolution of primates and raccoons we know relatively little; horses and dogs, mainly dwellers in open country, are abundant as fossils. Deposition has been common in coastal lowlands, but highland regions are areas of erosion; in consequence, our knowledge of the evolution of upland forms is slight—particularly for the older geological periods, from which little of inland or upland sediments has been preserved.

Climates.—At the present time differences in temperature are a major factor in the distribution of vertebrate life. Many groups are confined to the tropics, others to arctic or temperate regions. This zonal distribution forms an effective check to migration. Penguins presumably would flourish in the arctic regions, polar bears in the south; but they cannot cross the tropics. The animal life of Old and New World tropics differs greatly, for intermigration in the more recent geological epochs necessitated a passage through the arctic zone. Similar situations may have strongly influenced the migration and distribution of vertebrates in the past.

Ancient climatic conditions, however, appear to have varied greatly. There is considerable evidence suggesting warmer climates at times in the past in regions now in the temperate or arctic zones, while ancient glacial deposits are known in areas that are now temperate or tropical. Presumably, there have been, throughout geologic history, climatic rhythms—times in which conditions were mild and equable have alternated with periods of sharply zoned climates. There are, however, other possible explanations. Some have suggested a wandering of the poles and consequent readjustments of climatic bands, others (as discussed later) an actual movement of land masses.

The vertebrates themselves offer evidence as to the climates in which they lived. Living "cold-blooded" reptiles and amphibians cannot exist in arctic or subarctic regions, and most of the extinct ones may have had similar limitations. But this type of reasoning must not be carried too far. Elephants and rhinoceroses today are tropical animals; but we know that, in the Pleistocene, hairy types closely related to living forms in both groups existed near the glacial front. The evidence of plant and invertebrate life is important in determining climatic conditions. Plant types now characteristic of the tropics once grew in Greenland. It is probable that this indicates a former warm climate there; but it is possible that plant habits have changed. The sediments themselves commonly suggest the nature of the climate under which they were formed. For example, striated rock surfaces and deposits similar to those made by recent glaciers are indicative of ancient ice ages. But the evidence of sediments is not always beyond doubt. Red deposits, for example, are now being laid down at the mouth of the Amazon and in other warm regions of abundant rainfall; but in many ancient red beds in which vertebrates are found, there are strong indications, from floras and from the nature of the sediments, that arid conditions were present. Quite probably there were alternating wet and dry seasons, as in various tropical regions today.

Migrations.—No matter how potentially successful a

newly evolved form may be, obstacles are often present which prevent its invasion of regions in which it might flourish. The greater part of the world has always been covered by seas, and migration is comparatively easy for marine forms. Even here, however, barriers may exist, due, for example, to differences in water temperature or salinity. For terrestrial forms there are numerous types of barriers. Forests, deserts, and mountains, as well as temperature differences, are often effective bars to migration. The greatest obstacle, however, is the lack of a land connection, the interposition of a broad body of water between continental areas. Many oceanic islands have no native land vertebrates of any sort; no terrestrial mammal (before man) ever reached New Zealand.

Viewed on the ordinary Mercator map, intercontinental migration appears a complicated process. On a north polar projection, however, the situation is much simplified. Europe and Asia form a unit, and North America is separated from them only by a narrow strait across which dry land is believed to have stretched at many times. From this northern land mass depend three great southern peninsulas, South America, Africa, and Australia. The first two are connected with the north by narrow isthmuses believed to have been under water in earlier times, while Australia is separated by an island-studded region which may once have been dry land. Many of the facts of vertebrate distribution, during Cenozoic and Recent times are readily explained by a make or break of these four connections.

Land bridges have often been built in the minds of scientists to account for certain facts of distribution. Some of these structures are quite fanciful, made in an attempt to explain the real or supposed resemblances between certain animal types in two disconnected regions but disregarding the divergences between the two regions in other animal groups. Others, however, deserve more serious consideration. This is especially true of bridges advocated between the three southern continents. The presence of carnivorous marsupials in Australia and in the South American Tertiary has suggested the possibility of a former land connection between these two areas. South America and Africa show resemblances not only in some mammalian types but also in some groups of freshwater teleostean fishes and in the presence of related lungfish genera, and a connection between them in the early Tertiary has been strongly advocated. Elsewhere it has been pointed out that many of these facts may be accounted for by the former presence of the common types or their ancestors in northern areas from which these continents might have been reached without any major change in land areas. One must not "erect" a bridge for the convenience of one group of animals without taking into account its effect upon other elements of the fauna. For example, hystricoid rodents with supposed relatives in Africa appear suddenly in South America in the Oligocene. A transatlantic land bridge would afford easy access for the rodents and an easy solution for the problem concerning them. But in both continents there were numerous other animals that one would think equally capable of crossing such a bridge. There is, however, no evidence that they did cross. Land bridges that transmit only one type of animal, and in but one direction, may have existed; but their existence is highly improbable.

Such bridges, however, may have been somewhat selective in their nature in many cases—"filter-bridges." The present "bridge" between the two Americas has allowed the passage of many animal types. Among the ungulates, deer, peccaries, tapir, and camelids have made successful southern migrations, but the plains-loving bison never penetrated beyond Honduras into this region of tropical jungles and mountains. The Bering Straits connection between Asia and Alaska was a well-traveled highway of migration in the Pleistocene. But heat-loving animals, such as most of the antelopes, the giraffes, and hippopotami, were unable to cross.

"Island hopping" is a partial solution for some problems of migration. Asia and Australia have not been in direct contact, it seems certain, at any time during the Cenozoic. Between them, however, stretch the East Indies, separated by short gaps, certain of which may have been, at times, dry land. Down the island chain is scattered a decreasing percentage of placental mammals of Asiatic origin. Certain rat types have made the whole series of "hops" to Australia; a few marsupials have worked part of the reverse journey.

It seems certain that land animals do at times cross considerable bodies of water where land connections are utterly lacking. The West Indies and Madagascar, for example, have sparse mammalian faunas which are composed of a few odds-and-ends of the numerous mammal groups found on the neighboring continental areas. Obviously, these forms have reached their island homes by some chance type of overseas transport. Floating masses of vegetation, such as are sometimes found off the mouths of the Amazon, may be one means of effecting this type of migration.

Continental drift?—In the past it was almost universally accepted by geologists that the continents have always been in the positions they occupy today, and the physicists viewed with horror any advocacy of theories that there had ever been any movement of them (barring, of course, tectonic vertical movement, minor crumpling, and so on). More than half a century ago, however, there was advocated a theory to the effect that such major movements had actually taken place. In its extreme form, it assumed that all the land areas of the world had been, far back in geologic times, a single mass; that Australia and Antarctica represent pieces broken off to the southward and—most especially—that at one time the Atlantic Ocean did not exist, so that North America and Europe once lay cheek by jowl, and South America and Africa formed a joint continent. The argument regarding the Atlantic is particularly tempting, for, if one traces not merely the present shores but, better, the edges of the continental shelves on the two sides of the Atlantic, their juxtaposition gives in most regions a remarkably close fit; and both in North America and in southern South America the geologic structures show a close correlation in many regards with those opposite them in Europe and Africa.

In recent years the geophysical objections seem less important, and a considerable number of geologists are willing to admit the possibility of continental movements. If, however, there has occurred such a drifting apart of once-united continental areas as this theory assumes, much of this movement must have occurred quite far back in geologic time. For the Cenozoic, the evidence of fossil mammals clearly shows that there can have been no juxtaposition of land areas other than those now existing, and for the Cretaceous and Jurassic, there is no vertebrate evidence, pro or con, of importance.

But for the Triassic and earlier periods, the situation is a different one. It is becoming increasingly apparent that in Triassic times South America and Africa had very comparable reptilian faunas. The faunas of North America and western Europe appear to be exceedingly similar in the amphibians and reptiles in Pennsylvanian and early Permian times. It is, of course, not impossible that the similarities (much greater than today) may have been due to migration via the routes possible today. But it seems, on the whole, more probable that in early times the Atlantic Ocean did not exist.

Another series of facts disturbing to believers in continental fixity relates to evidence that climates in the past differed greatly from those of today in various regions. As was said earlier, some problems encountered when warm-climate animals and plants are found in arctic regions can be explained by assuming that climates in the past have sometimes been more equable than now. But how does one explain the fact that there is unmistakable evidence of strong glaciation in the late Paleozoic in southern Africa, and comparable evidence in Argentina, peninsular India, and western Australia? Either these areas have shifted position or—less probably—the poles have shifted.

A further series of facts disturbing to the belief in fixed continents has recently come to light with paleomagnetism. It has been found that many sediments still retain, today, the magnetic orientation true of the time of their deposition. Studies of this sort, currently being intensively pursued, indicate that the magnetic poles (and, it is assumed, the true poles of rotation) in early geologic periods had a quite different position from those today with regard to various land masses. As in southern glaciations, we must assume either that the poles have shifted or, more probably, that the continental areas have changed position. The evidence is currently not conclusive, but it strongly reinforces theses of continental movements and fragmentation.

CAMBRIAN

During the Cambrian, earliest of the Paleozoic periods, almost every major type of invertebrate life made its appearance, and many groups had already become so highly organized and diversified that we may be sure that they had already been in existence for a long period of time. But vertebrates are absent from the record. This might perhaps be accounted for by the fact that ancestral vertebrates were soft-bodied forms which we would not expect to find fossilized under ordinary conditions. But even in the case of a remarkable Middle Cambrian deposit in British Columbia, in which the shales have preserved very delicate impressions of soft-bodied invertebrate types such as jellyfishes, no chordates of any sort have been recognized. Either vertebrate ancestors were extremely rare or were absent from the salt waters in which all well-known Cambrian deposits were laid.

ORDOVICIAN

Ordovician sediments are likewise predominantly marine in origin. Invertebrates are abundant and diversified, but vertebrates have been reported in only two regions. Near Leningrad, Russia, a near-shore deposit of glauconitic sands of early Ordovician age has yielded a few denticle-like structures which appear to be fragments of the surface ornament of dermal armor. More interesting and tantalizing are numerous fragments of vertebrate dermal armor to be found in the Harding sandstone and other formations, somewhat above the middle of the Ordovician, in a line running from central Colorado north through Wyoming to Montana. In no case have we enough to determine much of the shape or nature of the animal concerned, but the microscopic structure shows that we are dealing with members of the ostracoderm order Heterostraci, common in late Silurian and early Devonian faunas. The sediments appear to be near-shore deposits; it is believed that an Ordovician land mass lay to the east of the region where these remains are found. One may believe that the Ordovician heterostracans can be interpreted as marine forms which lived in coastal shallows or, alternatively, as freshwater forms whose remains have been washed into coastal sediments.

SILURIAN

Although richly fossiliferous sediments of Silurian age are abundant in both Europe and North America, and although the evidence from the Silurian shows that vertebrates capable of fossilization were in existence, finds of vertebrates are still exceedingly limited. This situation seems clearly correlated with the fact that, as in the Ordovician, the known sediments of this period were still predominantly marine. In the United States, the midwestern region was covered by an epicontinental sea during much of the Silurian; along the Atlantic shores, it is believed, were mountains draining west into this inland basin. Not a trace of a vertebrate has ever been found in the richly fossiliferous marine beds

TABLE 2 Paleozoic Vertebrate Localities

Period	European Stages	European Localities	N. American Stages	N. American Localities	Other Continents
PERMIAN	Thuringian or Tartarian	N. Dvina R. (Russia); Cutties Hillock (Scotland); Zechstein of Germany	Ochoan		*Cisticephalus* and *Endothiodon* zones of Beaufort series; Ruhuhu; Tanga, Chiweta beds (Af.); Bijori beds (India)
	Saxonian or Kazanian and Kungurian	Continental Russian beds of Zones I–III; Kupferschiefer (Ger.); Magnesian ls. and Marl slate of Durham (Eng.)	Guadalupian	Flower Pot and San Angelo fms. (Tex.-Okla.); Phosphoria fm. (Idaho) *M*	*Tapinocephalus* zone of Beaufort series (S.Af.)
	Autunian or Artinskian, Sakmarian	Rotliegende and equivalents, Germany (Niederhässlich, Lebach, etc), France (Autun), England (Warwickshire)	Leonardian, Wolfcampian	Clear Fork, Wichita groups of Texas, E. Greenland *M*, Cutler and Abo (N.M., Colo., Utah), Dunkard (Pa., W.Va., Ohio)	Ecca (barren); Dwyka (S.Af.) Itarare (Brazil)
PENNSYLVANIAN	Stephanian or Uralian	Kounová, Nýřany (Bohemia)	Monongahela and Conemaugh or Virgil and Missouri series	Danville (Ill.); Monongahela and Conemaugh localities (Pa.-Ohio)	
	Westphalian C and D	Middle and Upper Coal Measures of Newsham, Swannick, Airdrie, Pirnie, etc.; Jarrow, Kilkenny (Ireland)	Allegheny or Des Moines series	Mazon Cr. (Ill.), Linton (Ohio); Sydney, Morien, etc. coal group of Nova Scotia	
	Westphalian A and B and Middle and Upper Namurian (pt.) (Lanarkian)	Lower English Coal Measures	Pottsville (Kanawha and Lee) or Lampasas and Morrow series	Joggins (N.S.)	
MISSISSIPPIAN	L. Namurian		Chesterian	Mauch Chunk at Greer, Hinton (W.Va.); Pt. Edward fm. (N.S.)	
	Visean-Dinantian (pt.)	Edinburgh coal field (Gilmerton, Loanhead, etc.) and oil shale group (Burdiehouse, Wardie, etc.); Carboniferous lss. of Bristol (Eng.), Armagh (Ire.)	Meramecan and Osagean	St. Louis, Burlington and Keokuk lss. *M*	
	Tournaisian-Dinantian (pt.)		Kinderhookian	Albert Mines (N.B.)	
DEVONIAN	Famennian and Frasnian	U. "Old Red ss." of central Scotland and E. Baltic; Wildungen (Ger.) *M*	Bradfordian and Chatauquan	"Catskill" redbeds; "Chemung" *M;* Scaumenac Bay; Cleveland shale *M;* E. Greenland	Victoria and N.S. Wales (Aus.); E. Asia; Antarctica
	Givetian and Eifelian	Middle "Old Red" of N. Scotland (Caithness, Orkneys, Elgin, etc.); E. Baltic region; Wijhe Bay, Grey Hoek (Spits.)	Erian	Delaware ls. *M;* Ellesmere Land	N. S. Wales
	Emsian, Siegenian, Gedinnian of Rhineland; Breconian, Dittonian, Downtonian of Gt. Britain	Red Bay, Wood Bay (Spits.), Lower "Old Red" and Downtonian series of S. Scotland and Welsh border; Kosor (Bohemia); "Old Red" and Czartków (Podolia); Rhineland *M;* Ludlow bonebed	Ulsterian	Beartooth Butte (Wyo.); Blacksmith Fork (Utah); Knoydart fm. (N.S.); Campbellton (N.B.); Onondaga and Columbus lss. *M*	
SILURIAN	Ludlovian	Lanarkshire (Scot.); Oesel, Beyrichienkalk of Baltic region *M;* Ringerike (Norway)	Cayugan	Salina group (N.Y.); Bloomsburg redbeds (Pa.); Read Bay and Cape Phillips fms. (Cornwallis Is.)	
	Wenlockian		Niagaran	Shawangunk congl. (N.Y.); Allen Bay fm. (Cornwallis Is.)	
	Llandoverian or Valentian		Albian (or Medinan)		

M = marine

of the inland area. But in late Silurian days, a series of redbeds which appear to represent deltaic deposits was formed along a line from central New York State south through Pennsylvania. Here, accompanied by few other fossils except eurypterids, there have been made a number of finds of cyathaspid heterostracans. Apart from this, the only American find of vertebrates of any note is one of heterostracans from the Silurian on Cornwallis Island in the Arctic. The formations in which these ostracoderms were found are dominantly marine, but it would appear that few, if any, of the vertebrate finds were in direct association with marine invertebrate faunas.

In Europe we meet with a similar dearth of vertebrate finds in Silurian deposits, which are predominantly marine. Toward the end of the Silurian began the Caledonian revolution, a time of the building of mountain ranges, erosional materials from which formed numerous sedimentary formations. In but few areas, however, were such sediments formed before the Silurian came to a close. The situation in the English-Welsh border country is typical. Central Wales was a deep ocean basin; high and dry land was presumably present in central England; between there was, for most of the Silurian, a littoral region of shallow seas, in which are occasionally found fragments of eurypterids and small vertebrate scales or spines, but never a complete fish. At the Silurian-Devonian boundary is the famous Ludlow bone bed, readily identified over large areas of country. Here there are present numerous vertebrate remains, mainly heterostracans of the *Pteraspis* type. The deposition of deltaic and continental sediments now begins and continues into the Devonian.

In the southern uplands of Scotland and at Ringerike, Norway, the advent of continental sediments occurs before the close of the Silurian. In both these regions there are thick series of sandstones and shales, predominantly red in color, the lower part Silurian, the upper early Devonian. The basal portion of the series in both cases is predominantly marine, but as we proceed upward, marine fossils become increasingly rare, and there begin to appear, before the Devonian is reached, beds in which marine invertebrates are absent and there are found eurypterids and a variety of ostracoderms, mainly osteostracans and anaspids.

A different condition, best shown on the island of Oesel, at the mouth of the Gulf of Riga, appears to have been present in the Baltic region. Here a sequence of four late Silurian formations carry vertebrate remains. The lowest of the four includes limestones with almost no invertebrate remains except eurypterids, but with ostracoderms, mainly *Tremataspis* and other osteostracans. The fauna appears to be essentially a freshwater one, perhaps that of a coastal lagoon. The two following formations are definitely marine; they include numerous isolated vertebrate scales and spines but no articulated skeletons. A final formation includes a large number of successive layers, most of which are marine; but at a few horizons there are thin bone beds containing masses of ostracoderm and acanthodian fragments. The conditions suggest that these have been washed out from some nearby coast.

If the known remains of Silurian vertebrates be added up, it will be seen that a considerable variety of primitive jawless fishlike vertebrates were then already in existence. Heterostracan ostracoderms, we have seen, had already evolved in the Ordovician; and, in the Silurian, typical members of that group, on the *Cyathaspis* pattern, are the commonest of finds. Present, but not as common, are *Tremataspis* and other members of the Osteostraci, anaspids, and coelolepids. But of jawed vertebrates there are as yet no remains, except that scales and spines show that acanthodians were already in existence.

But where are the ancestors of the placoderms and higher bony fishes which presently, before the following Devonian period was far advanced, were to become the dominant elements in the vertebrate fauna? All—or nearly all—of the known ostracoderms were too highly specialized to be considered as ancestral to them, and the acanthodians, as far as present knowledge goes, are likewise too specialized to be direct ancestors of other jaw-bearing forms. The most reasonable hypothesis is that the ancestors of more progressive fishes were evolving in some inland area—the higher reaches of streams in the uplands formed in the process of Caledonian mountain building. But it is almost beyond hope that these ancestors will ever be found as fossils, since such upland areas became regions of erosion rather than deposition.

DEVONIAN

Although amphibians had evolved by the end of the Devonian, this period is often, and reasonably, called the Age of Fishes. The short list of Silurian vertebrate localities was confined essentially to a few areas of Europe and North America. In the Devonian our perspective widens. As will be seen, new continental areas are added to the list before the close of the period. Further, some fish groups became abundant in saltwater environments, and hence Devonian marine formations frequently include vertebrate remains.

Devonian localities.—The classic localities for Devonian freshwater fishes are those of the "Old Red Sandstone" of Great Britain. These continental deposits embrace nearly the whole extent of the Devonian, and are customarily divided into lower, middle, and upper portions. The lower "Old Red" continues without marked break from the Upper Silurian along the English-Welsh border and in southern Scotland (the lowest portion of the beds in these areas, termed the Downtonian, were formerly assigned to the Silurian). Similar early Devonian deposits are known in Poland and Podolia. Exceedingly rich Lower Devonian freshwater deposits are present in Spitsbergen. In Scotland, middle "Old Red Sandstone" fossil beds are located in the north, particularly Caithness, with limited upper "Old Red" in central Scotland. Freshwater beds of the middle and late Devonian are present in the Dvina River Basin of Latvia and adjacent regions, and there are other scattered continental beds. In North America the most prominent continental beds are those of the later part of the period and are found in eastern regions. As in the Silurian, the central United States was, for most of the period, occupied by seas, with mountains along the present Atlantic Coast. In the then lowlands west of the mountains, in the present Catskills and on southwest into central Pennsylvania, there were deposited thick masses of continental redbed sediments. A further American deposit of note is near Scaumenac Bay, on the Gaspé Peninsula of eastern Canada, where richly fossiliferous late Devonian freshwater shales are exposed for some miles along a sea cliff. Other continental Devonian deposits are present in the west—at Beartooth Butte, Wyoming, for example—and even in Ellesmere Land and East Greenland in the Arctic. In contrast with the Silurian, Devonian deposits are known to some extent in continents other than Europe and North America—Australia, for example, and even Antarctica.

Freshwater fishes.—Freshwater Devonian vertebrates are varied in nature and shift radically between the beginning and end of the period. In the early Devonian, ostracoderms are abundantly present, notably in Spitsbergen; in late phases of the Devonian they are sharply reduced in numbers, and none is known to survive beyond the Devonian. Acanthodians are represented in the Silurian by scales and spines; in the Lower Devonian we find complete specimens of these little fishes for the first time. This appears to have been the peak of abundance for this group, found only rarely in later Devonian phases.

New groups now make their appearance. No placoderms are known with certainty in the Silurian, but in Lower Devonian freshwater deposits there are found abundant remains of the grotesque arctolepid arthrodires, with long scimitar-like pectoral spines. Later arthrodires are mainly marine, but a few more advanced arthrodires, such as *Coccosteus* of the mid-Devonian, persisted in fresh waters in the later phases of the period. The antiarchs, with their flipper-like appendages of bone, were almost entirely a freshwater group, appearing in the mid-Devonian. They became exceedingly abundant; in such a deposit as that of Scaumenac Bay, half or more of all specimens found are those of the common antiarch genus *Bothriolepis.*

The Devonian is, further, the time of appearance of the higher bony fishes, the Osteichthyes. A single genus of crossopterygian, *Porolepis,* is known from fragmentary materials from the Lower Devonian, and there are a few traces of lungfishes. But in the middle "Old Red" of Scotland and equivalent formations, the bony fishes appear in startling numbers and variety. There are representatives of all three major groups—actinopterygians of the palaeoniscoid type, particularly the primitive genus *Cheirolepis;* rhipidistian Crossopterygii such as *Osteolepis;* and primitive Dipnoi such as *Dipterus.* All three types are fully developed and fully differentiated. Obviously they already had had a long evolutionary history. But as to where and how this occurred, we know nothing. In the Middle and Upper Devonian the bony fishes form a large fraction of all finds in freshwater formations. But while in Carboniferous and later times actinopterygians are by far the most common, they are rare in the Devonian, and it is crossopterygians and dipnoans which are the abundant forms.

Marine occurrences.—During the course of the Devonian there appears to have been a considerable expansion of fish life in salt waters. Such occurrences are relatively rare in the Lower Devonian, although we find, for example, puzzling rhenanid and petalichthyid placoderms, such as *Gemuendina,* in Rhineland early marine beds. Finds of marine fishes become more common in the Middle Devonian in such formations as the Columbus and Delaware limestones of Ohio and comparable deposits of western Germany. Placoderms of a number of types are the major elements in this developing saltwater fauna, and *Macropetalichthys,* for example, appears to have been worldwide in distribution in mid-Devonian days. There are famous late Devonian marine fish deposits in which the arthrodires reached the peak—and end—of their evolutionary history. The black Cleveland shales of northern Ohio have, for a century, yielded numerous remains of arthrodires, some of gigantic proportions; Wildungen, in western Germany, has produced an abundance of highly varied, advanced arthrodires.

Last of all major groups of fishes to appear were the elasmobranchs. A few shark teeth are present in mid-Devonian limestones in Ohio, and a number of Upper Devonian formations have yielded fragmentary shark remains and a few tooth plates of "bradyodonts." Our only good view of early sharks, however, is given by specimens from the Cleveland shales, where unusual conditions of some sort have preserved specimens of *Cladoselache* and *Ctenacanthus*-like forms in which there are present not only teeth and cartilages but soft parts, including fossil musculature. In the late Devonian, too, there appear to have been a few marine palaeoniscoids, and it appears that at this time the coelacanth crossopterygians were beginning a marine career.

Fish evolution in the Devonian.—The close of the Devonian witnessed the decline or disappearance of many important fish groups. The ostracoderms became reduced in numbers well before the end of the period and did not survive its close. Antiarchs and arthrodires, although numerous in the Upper Devonian, disappeared before the opening of the Carboniferous. In the space of a single period, almost every fish type present in the Lower Devonian was wiped out except for a few acanthodians.

The more important events in Devonian fish history

include (1) the decline and fall of the ostracoderms; (2) the rise and decline of the acanthodians; (3) the rise and disappearance of the arthrodires; (4) the appearance in the middle "Old Red" of the flourishing antiarchs and their extinction at the close of the period; (5) the rise of sharklike forms in the late Devonian; and (6) the blossoming in the middle "Old Red Sandstone" of the bony fishes.

For completeness, this body of knowledge leaves much to be desired. A chronological treatment tends only to throw into stronger relief the fact that we know almost nothing of the origin or real relationships of most of the early fish groups.

Amphibians.—It was noted in earlier chapters that the Devonian was probably a time of marked seasonal droughts; much of the success of the bony fishes (and perhaps the antiarchs) may have been due to their possession of lungs and other drought-resisting adaptations. (We find, for example, that at such a classic freshwater locality as that of Scaumenac Bay, perhaps 90 per cent or over of recovered specimens are of fishes which presumably possessed lungs.) The amphibians appear to have been, in their initial period, merely another fish group which had developed terrestrial locomotion as a further device helpful under Devonian conditions of life, and their early development from the crossopterygians occurred in that period. The first stages in their history are, however, poorly known; for most of the course of the Devonian we possess only a skull roof (*Elpistostege*) from the early Upper Devonian of Scaumenac Bay, which is that of a very fishlike amphibian or very amphibian-like fish. However, from a Greenland formation close to the Devonian-Carboniferous border, there have been recovered skulls and skeletal materials of *Ichthyostega* and its relatives, definitely amphibians, but amphibians of a very archaic nature.

CARBONIFEROUS

The time covered by the deposits frequently grouped as the Carboniferous Period is very great. In North America a division into two periods—Mississippian and Pennsylvanian—is now customary, for there is generally a decided contrast between a lower division in which marine limestones predominate and an upper one characterized by coal seams and shallow-water deposits. Since the fish faunas which still constitute the bulk of vertebrate remains are not sharply contrasted, we shall here consider the Carboniferous faunas as a unit.

Vertebrate localities.—The Lower Carboniferous, or Mississippian, is predominantly marine in both Europe and North America and is represented, for the most part, by thick series of limestones. These frequently contain rich invertebrate faunas; less frequently they contain the remains of marine fishes, usually in the form of isolated spines or teeth. Formations and localities of this sort include the St. Louis, Burlington, Keokuk, and other limestones of the Mississippi Valley; a rich horizon in the Carboniferous limestone series near Bristol, England; a similar formation at Armagh in northern Ireland; and numerous other limestone localities in western and central Europe and Russia.

Mississippian continental deposits are, on the other hand, rare. In North America there are few localities of this sort; exceptional are those at Albert Mines, New Brunswick, where large numbers of a few species of freshwater fishes are present in oil shales, and in West Virginia, where a few fishes and amphibians have been recovered from late Mississippian shales. In other continents there is likewise an almost complete lack of Mississippian fossils of continental origin. Scotland is the sole exception. In the southern part of that country the Lower Carboniferous is highly developed. Some limestones are present, but materials of a more continental type are prominent: calciferous sandstones, oil shales, and even coal seams. At various localities, often seemingly estuarine in nature, there are rich fish faunas and, in addition, near the Firth of Forth, occasional finds of amphibians in oil shales and in a coal field have been reported.

Pennsylvanian fossil deposits are, in contrast to those of the Mississippian, predominantly continental in type. This was the time at which the most important coal seams of Europe and North America were formed; and numerous inhabitants of the coal-swamp pools have been preserved in the shales, ironstones, and impure cannels associated with the coals. In Europe the typical Coal Measures are included in the Westphalian floral zone. Although coals of this age are abundant on the Continent, Westphalian fossil finds are almost entirely confined to the British Isles, where Newsham near Newcastle and Kilkenny in Ireland are important localities. In the late Pennsylvanian appears the Stephanian flora, transitional to that of the Permian. The coals of this horizon are well represented on the Continent, and Stephanian vertebrates are richly represented at Nýřany (Nürschan) and Kounová in Bohemia.

In America the earliest Pennsylvanian formations—those of the Pottsville—are, so far as is known, barren. In the somewhat later formations, equivalent to the Westphalian of Europe, there are several rich deposits, including a pocket of fossiliferous cannel coal at Linton, Ohio; nodules containing vertebrates, as well as plants, in the coal shales at Mazon Creek, Illinois; and material filling the hollow stumps of coal-swamp trees at the Joggins and North Sydney in Nova Scotia. Other scattered North American localities are of late Pennsylvanian age (Stephanian). In some areas and horizons typical Coal Measures conditions persist to the end of the period; in others there is a transition to red sediments, which continue without break into the Permian.

Despite the predominance of discoveries of continental forms of life, marine vertebrates are not uncommon finds in the Pennsylvanian. Some are present in typical marine beds. More important, however, is the fact that coal deposition was not a continuous process but a cyclical one, and layers of marine origin containing remains of sea dwellers are frequently present between successive coal horizons.

Freshwater fishes.—The character of Carboniferous fish life in inland waters was markedly different from that of the Devonian. Of lower fish groups, the ostracoderms were extinct; gone, too, were the antiarchs so numerous in the late Devonian. Acanthodians persisted in small numbers, and, of the sharklike fishes, the pleuracanths were common in many localities. All other freshwater forms were members of the Osteichthyes. But, although the groups represented were the same as in the later Devonian—crossopterygians, lungfish, and palaeoniscoids—their relative abundance had changed greatly in character. In the Devonian the two fleshy-finned groups were dominant, the ray fins rare. In the Carboniferous, perhaps in correlation with changed environmental conditions, the proportions were reversed. *Megalichthys, Rhizodus,* and a few other crossopterygians, including freshwater coelacanths, survived in reduced numbers; the dipnoans, as *Sagenodus* and *Ctenodus,* were a modest group. They were vastly outnumbered by a host of palaeoniscoids, which appear to have swarmed in every Carboniferous pool and river. About two-score genera have been described; some were still rather primitive in nature, others specialized or advanced in various degree.

Marine fishes.—In the sea as on land, the archaic fish groups were reduced in numbers. There are no ostracoderms; of the primitive jawed fishes, fin spines suggest the presence of large acanthodians, but the arthrodires are absent. Nor was there any notable tendency, as yet, for a major marine invasion by the higher bony fishes.

The Chondrichthyes, however, developed greatly in the seas of this age. Seldom are entire specimens preserved, but abundant teeth and spines testify to the number and variety of sharks and sharklike fishes in these oceans. Most abundant were the so-called pavement-toothed sharks, including the cochliodont family, perhaps ancestral to chimaeras, petalodonts, psammodonts, and others. These were mollusk-eating forms which appear to have been depressed bottom dwellers analogous in habits to the later skates and rays. These fishes were presumably a major prey of the predaceous shark types which were also common inhabitants of the Carboniferous seas. Of these sharks, some continue to represent the primitive *Cladoselache-Ctenacanthus* group, which had appeared in the later Devonian; others, however, appear to have become more "modernized" in fin structure and belong to the hybodont assemblage, destined to continue into the Mesozoic.

Amphibians.—The Carboniferous is the time of greatest development of the amphibians. Once evolved, with the possibilities of amphibious, if not terrestrial, life before them, they had spread into a host of types varying greatly both in structural features and in adaptations. Of the two major groups present, the labyrinthodonts are the better known, partly because of their tendency to grow to large size, with consequent better chances of preservation and recovery. There is a single surviving ichthyostegid, *Otocratia,* in the Mississippian. Abundant in the Pennsylvanian were various rhachitomous temnospondyls, preserved mainly in coal-swamp shales. Of anthracosaurians, remains of embolomeres, essentially water-dwelling fish predators, are common finds in the Coal Measures.

Less conspicuous because of smaller size but abundant and varied were the lepospondyls. They appear in early Mississippian horizons and had already reached and passed their time of major development by the close of the Carboniferous. There were snakelike aistopods, *Dolichosoma* and *Ophiderpeton;* numerous nectrideans, including "horned" forms such as *Keraterpeton,* and the elongate *Urocordylus;* and a host of varied microsaurs, such as *Microbrachis* and *Coccytinus.* Both among the labyrinthodonts and the lepospondyls the faunas are exceedingly similar on the two sides of the Atlantic.

Reptiles.—The development of reptiles, primarily characterized by their improved reproductive processes, took place during the Carboniferous. Unfortunately, our record of their early history is poor, owing, one may believe, to the fact that most Coal Measures localities are deposits of a nature which tends to include pool dwellers rather than inhabitants of the dry land. However, the Lower Pennsylvanian tree stumps of the Joggins contain the oldest known reptile, the captorhinomorph cotylosaur *Hylonomus,* and, less certainly, remains of a first pelycosaur. In the late Pennsylvanian there appeared varied pelycosaurs—even the long-spined *Edaphosaurus.*

PERMIAN

Sediments.—Toward the close of the Carboniferous a gradual change in sediments occurred in many portions of western Europe and North America. Coal seams and intervening marine beds gradually gave way to terrestrial deposits prevailingly red in color, their materials derived from the great mountain masses then being formed, such as the ancestral Alps in Europe and the ancestral Rockies and the earliest Appalachians in North America. This change in sediments was gradual, and only arbitrary lines of division can be made between the Upper Carboniferous and the Permian redbeds; correspondingly, we find that the vertebrate life of the late Pennsylvanian is almost indistinguishable from that of the early Permian. In North America an abundant Permo-Carboniferous land fauna appeared in late Pennsylvanian times and persisted well into the Lower Permian. Redbeds containing these forms are found in regions as far apart as Utah, New Mexico, and Prince Edward Island, but are best developed as the Wichita and Clear Fork groups in northern Texas. The many specimens collected from the latter region give us our best picture of the world's early Permian life.

In Europe the Lower Permian—the Rotliegende—is a series comparable in nature to the American redbeds; the known fauna, however, is more restricted, presumably because of fewer exposures. A few specimens have been recovered from beds of this age in England, France, and Czechoslovakia; the most famous locality is that at the Plauen'schen Gründe (Niederhässlich) near Dresden, where slabs containing remains of numerous small amphibians and reptiles have been obtained.

These fossiliferous redbed deposits cover but the earliest portion of the Permian. In America further redbed deposition continued through much of the period, but at about the beginning of middle Permian times the vertebrate record ceases. In western Europe the Rotliegende is followed by the Kupferschiefer and the marine phases of the Zechstein, with a sparse fish fauna and almost no tetrapods. For land vertebrates of the Middle and Upper Permian we must turn to other areas—South Africa and Russia.

In Africa pre-Permian vertebrates are unknown; there are no fossiliferous early Paleozoic sediments of any sort, and those of the Devonian and Carboniferous are almost exclusively marine. At the close of the Carboniferous, however, there began the formation of the continental deposits of the Karroo system, some thirty thousand feet thick, which cover much of the Union of South Africa and have outliers in central and eastern Africa and Madagascar. At the base is the Dwyka series, apparently earliest Permian in age, containing glacial beds and with a sparse vertebrate fauna. Following this are some 6,000 feet of the Ecca series, a barren equivalent of the Lower Permian redbeds of Europe and North America. The Beaufort beds, many thousands of feet thick, begin with the Middle Permian and continue without marked break, lithologically or faunally, into the Triassic. Their exposures cover a great area centering in the Karroo desert of northern Cape Colony and have a rich tetrapod fauna. Three zones, named after index reptiles (*Tapinocephalus*, *Endothiodon*, *Cisticephalus*), are distinguishable in the Permian portion of the series. Equivalents of the Upper Permian portion of the Beaufort beds are present in East Africa. In South America there are numerous continental Permian beds, but very few vertebrates have been found in them, and Asian and Australian beds of this period are likewise nearly barren.

More than a century ago a few fossils were found in Permian beds in Russia west of the Ural Mountains, and at about the turn of the century further remains were discovered in far northern Russia. In recent decades intensive work has revealed a series of beds with faunas of great interest, which parallel those of the Permian Beaufort beds of South Africa and extend somewhat below the base of the Beaufort, so that there now appears to be some overlap between them and the Lower Permian of the American redbeds.

Marine Permian formations are abundant, but most are barren of vertebrates. A modest number of fishes have been found in early Permian limestones in scattered areas of North America (Texas, Nebraska, Wyoming, East Greenland), Russia, southern Asia, and western Australia. In the later Permian the Kupferschiefer deposits of Germany appear to have been laid down in brackish lagoons.

Fishes.—As our discussion of localities suggests, marine fish life is poorly known in the Permian. In the earliest horizons there are a few representatives of common Carboniferous chondrichthyan groups—the last cladodont sharks, "bradyodonts," and hybodonts; there is little evidence of any saltwater bony fish in the early Permian. *Menaspis* and *Janassa* in the Kupferschiefer are the last of the once abundant pavement-toothed sharks, and there were a few surviving hybodonts. The old vertebrate fauna of the oceans appears to have been almost entirely wiped out at the close of the Paleozoic, although it is certain that some hybodonts and cochliodont descendants must have survived to give rise to the sharks and chimaeras of later times. The first marine radiation of the vertebrates was at an end.

In fresh waters, *Acanthodes* survived into the Lower Permian as the last of the acanthodians. *Pleuracanthus* was a common early Permian freshwater shark but is unknown in later times except in Australia, where the group appears to have persisted into the Triassic. The Lower Permian also sees the end of rhipidistian crossopterygians; *Ectosteorhachis* in America is the last survivor of this phylogenetically important group. Coelacanths, however, persisted. Lungfishes, particularly *Sagenodus*, were common at the beginning of the period, and, in North America, *Gnathorhiza* of the Lower Permian appears to be the first member of the *Protopterus-Lepidosiren* group.

The dominant types are palaeoniscoids. Relatively few Permian genera are known, but they appear to have been varied in nature and included such unusual and progressive types as *Dorypterus*. A single holostean in the late Permian marks the beginning of the change in actinopterygian faunas to be witnessed in the Triassic. The Kupferschiefer deposits contain actinopterygians which may have been able to withstand brackish, if not strongly saline, waters and may represent the beginning of the first major invasion of the seas by the ray-finned fishes.

Amphibians.—There is no marked break between the late Pennsylvanian and early Permian in amphibian faunas; the redbeds exhibit a final phase in Carboniferous amphibian history. There are surviving embolomeres and seymouriamorphans, such as *Archeria*, *Seymouria*, and *Diadectes*. The rhachitomes, already present in the Carboniferous, reached the peak of their development in the Lower Permian. *Eryops* was the common and characteristic large form in the American redbeds; *Actinodon* and related types were comparable European genera. In addition, there were a score or more of rhachitomes of varied character—the dissorophids, including armored types; long-snouted aquatic types such as *Archegosaurus* and *Chenoprosopus;* the flat-headed *Trimerorhachis.* Lepospondyls were reduced greatly in numbers, but there were a few early Permian survivors, such as *Diplocaulus, Lysorophus,* and the gymnarthrids, including *Pantylus.*

In the later Permian, amphibian life was much reduced. The degenerate genus *Kotlassia* represents the seymouriamorphs in the late Permian, but the embolomeres have disappeared. Gone, too, are the lepospondyls. The temnospondyls alone persisted in moderate numbers, in such forms as *Rhinesuchus,* which, with flatter heads and broadly opened palatal vacuities, were entering on a "neorhachitomous" condition foreshadowing the advanced stereospondyls of the Triassic.

Cotylosaurs.—Reptile origins took place in the Carboniferous, but it is only with the development of more terrestrial conditions in the redbeds that we can gain a comprehensive picture of early reptilian life. The "stem reptiles" were an important factor in the land fauna of the early Permian. Some types, such as *Limnoscelis* and *Romeria,* appear to be very primitive; most cotylosaurs, however, are advanced or specialized in nature, showing that their point of common divergence must have been far back in the Pennsylvanian. *Captorhinus* and *Labidosaurus* are examples of the advanced American captorhinomorphs of the early Permian. In the later Permian, captorhinomorphs are extinct, and the remaining cotylosaurs are herbivorous procolophonians. *Nyctiphruretus* of the Russian Middle Permian is a small, short-jawed form which might well be ancestral to both procolophonids and pareiasaurs. Abundant in the middle and late Permian were the pareiasaurs. These ungainly-looking creatures are the commonest of finds in the Middle Permian of South Africa. Their remains are frequently found as if in the position in which they had become bogged down in the swamps where they found their food and met their death. In the Upper Permian of South Africa, pareiasaurs are relatively rare; they were, however, widespread at this time, for there are Upper Permian pareiasaurs in Russia and even in Scotland.

Pelycosaurs.—It is of interest that the most prominent of early reptilian orders is not a primitive group or one which led to later typical reptiles, but that of the synapsids, destined, after many changes and vicissitudes, to give rise to the mammals. Pelycosaurs had arisen in the Carboniferous as perhaps the earliest offshoot from the stem reptiles and still retained many primitive features. They were highly diversified in the redbeds. Some, like *Ophiacodon,* were long-snouted, semiaquatic fish eaters. *Edaphosaurus,* with long spines and crossbars, *Casea,* and *Cotylorhynchus* were representative of a side line of herbivores. More important phylogenetically were *Dimetrodon, Haptodus,* and their relatives, the dominant carnivores of the early Permian, from which the later therapsids appear to have been derived.

Therapsids.—These more advanced, mammal-like reptiles replace their ancestors, the pelycosaurs, as the common reptiles of the middle and late Permian; they are abundantly represented in the Beaufort beds of South Africa and are present in Russia as well. Some of the earliest and most primitive Russian forms are little advanced over the pelycosaurs of the early Permian. The great dinocephalians, including the carnivorous titanosuchids and the herbivorous tapinocephalids, are confined to, and common in, the Middle Permian. Dicynodonts are present in the Middle Permian of South Africa but are relatively small and rare; in the Upper Permian a great variety of these nearly toothless herbivores are the commonest of vertebrates. There is, in addition, a variety of more progressive carnivores, gorgonopsians, and therocephalians, in the Middle and Upper Permian of South Africa and the Upper Permian of Russia.

Other reptiles.—It is curious that in the Permian there are few evidences of the ancestry of the reptilian groups which were to become prominent in the Mesozoic. *Eunotosaurus* of the Middle Permian of South Africa is questionably a forerunner of the turtles; there are no clues as to the ancestry of the ichthyosaurs. Of the euryapsids, we find no forms resembling the aquatic sauropterygians of the Mesozoic, but there are a number of relatively rare euryapsids, such as *Araeoscelis* of the American beds and perhaps *Protorosaurus* of the European Kupferschiefer. Diapsid reptiles are notably absent in the Lower and Middle Permian; in the Upper Permian of South Africa, however, there appear the millerettids, apparently antecedent to them, and small primitive eosuchians, such as *Youngina.*

The early Permian witnesses the only appearance of that problematical order of small aquatic reptiles, the Mesosauria. *Mesosaurus* is known from the Dwyka of South Africa and an equivalent formation across the South Atlantic in Brazil; no trace of this group has ever been found in northern continental deposits.

Gondwanaland.—In all three southern continents and in peninsular India as well, sediments of similar types are found in the late Paleozoic and early Mesozoic; in all three continents there is evidence of pronounced glaciation in late Carboniferous or early Permian time; in all these areas there is found a common flora characterized by the genera *Glossopteris* and *Gangamopteris.* This has led to a belief that these regions were then parts of a common land mass named Gondwanaland, separated from the Eurasian land areas by an east-west Tethys Sea. The distribution of the vertebrate life of the Permian, especially the Karroo therapsid fauna, has been held to support this theory. Although the Triassic tells a rather different story, the Permian vertebrate record offers little in confirmation since there are few Permian vertebrates in the southern continents apart from South Africa, and the presence in Russia of a Permian therapsid fauna comparable to that of South Africa argues against the isolation of the southern continents. The only positive evidence for this theory furnished by Permian vertebrates is the presence of *Mesosaurus* on both shores of the South Atlantic.

Permian: Summary.—The chief characteristics and events of Permian vertebrate history are (1) the disappearance during the period of many types of fishes, including cladoselachians, acanthodians, most pavement-toothed sharks, and primitive crossopterygians; (2) continued dominance of palaeoniscoids and first appearance of the Holostei; (3) disappearance of most of the amphibian types and survival (as far as known) only of advanced labyrinthodonts; (4) abundance of cotylosaurs; (5) dominance throughout of synapsid reptiles—pelycosaurs in early times, therapsids in the later Permian; and (6) presence of rare representatives of a few other reptilian groups.

The Paleozoic as a whole.—In the Paleozoic is included by far the greater portion of the vast period of time which has elapsed since life first appeared in abundance. From the point of view of the student of invertebrates, the Paleozoic may form a natural unit, since most of these various lower animal groups appeared full fledged at its beginning and at its close underwent marked change. But with the vertebrates it is otherwise. These forms appeared at a relatively late date and evolved rapidly through a series of progressive stages; and at the end of the era our story is broken, so to speak, in the middle of a sentence. Looked at in its broadest outlines, the history of vertebrates in the Paleozoic is not that of a single chapter in their development but of two full cycles and the beginning of a third.

A first major phase in vertebrate history seemingly occurred in the earlier half of the Paleozoic. The earliest vertebrates are little known except at the very end of the cycle—at the Silurian-Devonian boundary. They were then a rather obscure group of small, bottom-living river dwellers, playing but a minor part in the drama of life development. Their lack of biting mouth parts rendered impossible the trituration of food of any degree of size or toughness; this appears to have been a major factor militating against their development. Mud grubbers they were, and mud grubbers, seemingly, they were destined to remain.

But apparently toward the end of the Silurian there took place a great advance in vertebrate structure which initiated a new cycle in the history of the group. This advance was the development of jaws; from this resulted a true conquest of the waters by vertebrates. With the development of jaws and teeth came the possibility of a predaceous existence, and active swimming succeeded the sedentary bottom-dwelling habits of the ancestral types. Jawed fishes were exceedingly abundant in fresh waters before the Devonian was far advanced and were widespread in the oceans in the Carboniferous, although a decisive and permanent conquest of the salt waters was not to be until Mesozoic times. To this great cycle of aquatic conquest belongs the development of the amphibians as well, for lungs and tetrapod limbs seem originally to have been merely adaptations for a more successful life in inland waters.

In the Carboniferous there took place a second great advance among the more progressive vertebrates, resulting in the initiation of a third stage in their history. This advance was the release of tetrapods from the water through changes which eliminated the aquatic stage in individual development. The resulting new cycle of development was the conquest of the land by the reptiles.

Primitive reptiles, little known in the Carboniferous, were abundant by the beginning of the Permian. Most of them were members of two ancient groups—the cotylosaurs, stem stock of all true land vertebrates, and the synapsids, destined to give rise to mammals. The close of the Paleozoic still saw these two groups dominant. Only during the course of the Triassic did this archaic land radiation come to an end and these forms give way to newer reptilian types which held sway during the rest of the Mesozoic.

CHAPTER 29

Mesozoic Vertebrates

The Mesozoic era is usually known as the Age of Reptiles, for it was during this time that there flourished the great reptilian groups now extinct: the varied dinosaurs on land, the flying reptiles in the air, and ichthyosaurs, plesiosaurs, and other marine types in the seas. The Mesozoic witnessed, as well, the more modest beginning of existing reptile groups—turtles, lizards, snakes, crocodiles. Further, it was during this era that birds and mammals, both derived from reptilian ancestors, made their appearance. Among amphibians, the Mesozoic saw the change from old to new—extinction of the labyrinthodonts and appearance of typical frogs and urodeles. And although our attention naturally centers on higher groups, the fishes of the Mesozoic had an interesting development. During this time there developed modern groups of sharks and rays and the chimaeras; the ray-finned bony fishes progressed from the chondrostean to the holostean grade of evolution and finally to the modern teleost type.

TRIASSIC

The Triassic forms an introduction to the Mesozoic vertebrate story. At its beginning we find faunas both on land and in the water comparable to those of the late Paleozoic. During the period, however, there were marked changes among both fishes and reptiles; by the close of the Triassic there had appeared almost every one of the striking vertebrate groups which were to play major roles in the Mesozoic story.

Vertebrate beds.—The Triassic owes its name to the sequence of fossiliferous deposits characteristic of the period in central and western Europe. It opens with the Bunter—variegated continental beds with sparse vertebrate remains, essentially mud flats deposited on the northern shores of the seas covering the present Alpine region. These seas advanced northward; their sediments form the Middle Triassic Muschelkalk, a limestone group with a rich fauna. In the Upper Triassic, with retreating seas, there is a return to dominantly continental beds in the Keuper, a series mainly red in color, supposedly deposited under arid conditions and with a good assemblage of vertebrate remains. The Upper Triassic closes with the Rhaetic, forming a transition to the Jurassic and including a number of bone beds. The Triassic areas described are best represented in southern Germany and eastern France; Middle and early Upper Triassic aquatic and shore-dwelling vertebrates are abundant at Perledo and other localities in the southern foothills of the Alps, and in Great Britain the Upper Triassic is represented by redbeds of "Keuper" type, notably in the Bristol district, and in a limited but richly fossiliferous deposit near Elgin, Scotland.

A second important Triassic sequence is that of South Africa. The earlier Triassic, poor in fossils in most of Europe, is here represented by the fossiliferous upper zones of the Beaufort series—the *Lystrosaurus* zone (of which the *Procolophon* "zone" is a facies) and the *Cynognathus* zone. The succeeding Stormberg series covers the remainder of the Triassic; the Middle Triassic beds (Molteno) have, until recently, yielded no vertebrates, but the redbeds and cave sandstones of the Upper Triassic have yielded vertebrates similar to those of the Keuper and the Rhaetic. In Tanganyika the Manda beds, equivalent to the Molteno, have yielded a varied Middle Triassic fauna.

In southern Brazil and western Argentina there are thick Triassic deposits comparable to those of the Beaufort and Stormberg of Africa. Most were long considered barren, but exploration in recent decades is bringing to light a variety of fossiliferous zones and areas. The Santa Maria beds of southern Brazil and those of the Ischigualasto region of western Argentina have produced rich reptilian faunas of Middle Triassic type. The Cacheuta series of Argentine freshwater beds, with fishes and brachyopid amphibians, are apparently early Triassic, and recently a rich early Triassic reptile assemblage has been found in the Argentinian Chañares formation. There are notable resemblances to the African faunas, in corroboration of the general belief of South American and African geologists that the two continents were intimately connected in Triassic days.

In North America the early Triassic Moenkopi beds of Arizona are comparable to the Bunter of Europe. No continental beds of Middle Triassic age are present, but in the Upper Triassic there are widespread redbeds comparable to the European Keuper—such for-

TABLE 3 **Mesozoic Vertebrate Localities**

Period	European Stages	Europe	North America	South America	Asia	Africa	Australia
CRETACEOUS	Upper Cretaceous inc. Senonian, Turonian, Cenomanian (U. Greensand) (Planerkalk)	Transylvania dinosaur beds; Chalk of S. Eng., Belg., Maestricht (Neth.), N. France; Westphalia, Lesina (Dalmatia), Comen (Istria)	Lance, Edmonton, Belly R. (Alberta-Mont.-Wyo.); Fruitland, Kirtland, Ojo Alamo (N.M.); Niobrara (Kan.) *M; Pierre* (S.D.) *M; Chico* (Calif.) *M;* Austin chalk (Tex.) *M;* Monmouth, Matawan (N.J.) *M;* Selma (Ala., Miss.) *M;* Woodbine (Tex.) *M*	Redbeds of Patagonia; Neuquen fm. (Arg.); Ceará *M*, Bahia *M*, Baurú (Brazil) *M*	Djadochta (Mongolia); Lameta series, Trichinopoly and Ariyalur stages (Ind.); Mt. Lebanon (Syria) *M*	Baharije stage (Egypt); Madagascar; Morocco (*M*)	Opal beds (N. S.W.); New Zealand *M*
	Lower Cretaceous inc. Albian (Gault), Aptian (Lower Greensand), Neocomian (Barremian, Hauterivian, Valangenian, Wealden)	Wealden (Eng., Belg., Hanover); Neocomian of Voirons (Switz.) *M;* L. Greensand, Gault of Channel region (Eng.) *M*	Cloverly (Wyo.); Arundel (Md.); Trinity (Tex.)		Iren Dabasu, Oshih, Ondar Sair (Mongolia); Shantung (China)	Uitenhage (S.Af.)	Tambo ser., Rolling Downs (Queens.)
JURASSIC	Upper Jurassic Malm or U. Oölites inc. Portlandian (Purbeckian, Bononian), Kimmeridgian, Lusitanian, Oxfordian, Callovian	Purbeck beds (Eng.); Lithographic ls. of Solenhofen, Kelheim (Ger.), Cirin (France) *M;* Oxford clays and Kimmeridge beds of S. Eng. and N. France *M*	Morrison fm. of western states (Como, Lost Cabin, Vernal, etc.); Sundance (Wyo.) *M;* western Cuba *M*			Tendaguru (E.Af.)	
	Middle Jurassic Dogger or L. Oölites, incl. Bathonian, Bajocian	Stonesfield slate, Forest Marble (Eng.)	Navajo ss. (Ariz.)			Madagascar	
	Lower Jurassic Lias (Aalenian, Toarcian, Charmouthian, Sinemurian, Hettangian)	Lias of Yorkshire, Dorset (Eng.), Caen region (France), Holzmaden (Bavaria) *M*	Wingate (Ariz.)		Kota (India)		Talbraggar (N.S.W.); Lower Walloon, Durham Downs (Queens.)
TRIASSIC	Rhaetian, Norian, Carnian	Rhaetic (S.W. Eng., S. Ger., Scania); Keuper and Lettenkohle (Ger.)	Chinle (Ariz.); Chugwater (Wyo.); Dockum (Tex.); Newark ser. (Nova Scotia to N.C.); Hosselkus ls. (Calif.-Nev.) *M*	Colorados fm. (Arg.)	Maleri (Ind.); Lufeng (Yunnan, China)	Cave ss. and redbeds of Stormberg series (S.Af.)	Wianamatta incl. St. Peter's beds (N.S.W.)
	Ladinian, Anisian	Muschelkalk of Germany and equivalents in Alpine region (Besano, Perledo, Monte San Giorgio, etc.) *M;* Spitsbergen *M*	Shinarump (Ariz.); W. Humboldt Range (Nev.) *M*	Santa Maria (Brazil); Ischigualasto fm. (Arg.)		Upper Ruhuhu bonebed, Manda beds (E.Af) Molteno beds of Stormberg (S.Af.)	Hawkesbury beds of Brookvale, etc. (N.S.W.)
	Scythian ("Eotriassic")	Bunter of Central Europe; Lower Triassic (Vetlugian) of E. Russia; Spitsbergen	Moenkopi (Ariz.); E. Greenland	Chañares fm., Cacheuta group (Arg.)	Sinkiang, Shansi (China); Luang Probang (Indo-China); Yerrapalli, Panchet (Ind.)	*Lystrosaurus* (+*Procolophon*) and *Cynognathus* zones of U. Beaufort (S.Af.); Madagascar	Narrabeen, incl. Gosford (N.S.W.)

M = marine

mations including the Newark series of the Atlantic Coast (famous for footprints), the Dockum of Texas, the Chinle of Arizona, and the Chugwater of Wyoming. In recent years productive continental beds of both early and late Triassic age have been discovered in western China, and similar faunas are present in the Panchet and Maleri beds of India.

In areas as far apart as Madagascar, Spitsbergen, and East Greenland there are highly productive freshwater early Triassic (Scythian) beds with numerous fishes (mainly actinopterygians) and labyrinthodonts. Other notable fish faunas are found in varied Australian beds—Narrabeen, Hawkesbury, and Wianammata. Marine faunas of middle and late Triassic age, with ichthyosaurs as a prominent element, are found in Spitsbergen and in Nevada and California, as well as in the Muschelkalk of Central Europe and the southern Alps.

Fishes.—We have noted a marked decrease in the numbers and variety of fishes, notably marine fishes, present in the last phases of the Paleozoic: acanthodians had become extinct, elasmobranchs and sarcopterygians reduced to a few types, and only the actinopterygians had remained flourishing. This situation still prevailed in the Triassic. Of the cartilaginous fishes, primitive chimaeras were presumably present but no specimens are known with certainty; the last lingering pleuracanths are found in the freshwater Triassic of Australia; *Hybodus* and a few related shark types are the only known chondrichthyan inhabitants of the Triassic seas. Of the sarcopterygians, the typical crossopterygians were extinct, but the coelacanths had survived in modest numbers; *Ceratodus* represents the lungfishes and is common in some formations presumably deposited under arid conditions.

The history of actinopterygian dominance and progress seen in the later Paleozoic is continued in the Triassic. The primitive palaeoniscoids are still prominent in the early Triassic. The characteristic ray fins of the period, however, are those to which the term "subholosteans" is frequently applied—forms progressive in character, structurally intermediate between palaeoniscoids and holosteans, and giving rise, very likely in polyphyletic fashion, to fishes on the holostean level of evolution. There are perhaps a half-a-hundred genera and half-a-dozen families of "subholosteans" in the Triassic. *Dictyopyge, Redfieldia* [*Catopterus*], *Parasemionotus, Perleidus, Saurichthys,* and *Cleithrolepis* are representative. By the end of the period, however, the subholosteans were, in turn, becoming much reduced in numbers and were giving way to typical holosteans—such as *Semionotus* and *Furo* [*Eugnathus*]—which were to dominate in the Jurassic.

For the most part, the older actinopterygians appear to have been freshwater forms. However, in the Triassic a strong trend is apparent among ray fins toward a marine existence. From this time onward the sea is a major center of actinopterygian life and evolution.

Amphibians.—The Rhachitomi had been the dominant labyrinthodonts of the Permian. In the late Permian the main line of temnospondyl evolution had progressed to the "neorhachitomous" stage, leading to the typical Stereospondyli and perhaps worthy of inclusion in that group as primitive members. Neorhachitomes are abundant in the early Triassic, but there soon appear typical stereospondyls, such as *Capitosaurus* and *Cyclotosaurus,* and in northern continents. *Metoposaurus* [*Eupelor, Buettneria*], long-skullec forms with the eyes far forward. In the earlier Triassic of the southern continents and of India are found the short-skulled brachyopids; in the late Triassic are present the very different short-skulled plagiosaurids. The early Triassic was the time when there flourished *Trematosaurus* and related long-snouted forms which appear, alone among amphibians, to have become marine fish eaters. These varied temnospondyl types flourished during the Triassic, but were doomed to extinction at the end of the period. Presumably the evolution of the modern amphibian orders was under way at this time, but we know nothing of them except for the presence in Madagascar of the primitive "frog" *Triadobatrachus* [*Protobatrachus*].

Primitive reptile groups.—In reptilian evolution, the Paleozoic-Mesozoic boundary is almost without meaning. By the end of the Triassic there had appeared the typical reptilian groups that were to dominate the later Mesozoic; at the beginning of the period, however, the fauna was still essentially a Paleozoic one.

Dominant in this older fauna were the therapsids. In the Lower Triassic the best representation is in the Upper Beaufort beds of South Africa and the Chañares formation of Argentina. Gorgonopsians, therocephalians, and dinocephalians were extinct, but major carnivorous lines were well represented by cynodonts such as *Cynognathus* and a parallel group of therocephalian ancestry, of which *Bauria* is best known. In later phases of the Triassic the typical carnivores dwindle and disappear. During the Triassic, forms which had departed from the primitive carnivorous mode of life arose from the theriodonts. In the Lower Triassic, *Diademodon* and other members of the cynodont group had developed a "molar" series with masticatory potentiality. In recent years, forms of this sort, the "gomphodonts," have proved to be abundant in the Middle Triassic of Africa and South America. In the late Triassic a further theriodont offshoot is that of the tritylodonts, with a highly specialized dentition, represented by such forms as *Tritylodon* of South Africa, *Oligokyphus* of Europe, *Bienotherium* of China, and an additional further North American type. Dicynodonts became much reduced in variety in the Triassic, but survived throughout the period, mainly as large forms which appear to have been worldwide in distribution.

Apart from sterile side branches, such as the "gomphodonts" and tritylodonts, carnivorous therapsids advanced upward toward the mammalian condition as the Triassic progressed. Such forms as *Diarthrognathus* of South Africa lie very close to the therapsid-mammal boundary, and in the closing phases of the Triassic are found a few forms which can be identified as being surely on the mammalian side of the dividing line.

Cotylosaurs are present in the form of *Procolophon* and its relatives, the last survivors of the order. Euryapsids, with a single upper temporal opening, had appeared in the Permian; in addition to the marine forms mentioned below, there are such odd Triassic members of the class as *Trilophosaurus;* possibly the peculiar, long-necked *Tanystropheus* pertains to this group.

Newer reptile groups.—Other reptile types, however, were making their appearance even in the Lower Triassic, to mark the beginning of the new Mesozoic radiation of reptiles. The little eosuchians, which had been present in the late Permian, continue into the Triassic and are presumably ancestral to the rhynchocephalians and definitely ancestral to the lizards. In the middle and late Triassic are a number of genera—some highly specialized, and even including "flying" forms—which lie close to, and a bit over, the evolutionary boundary between the ancestral eosuchians and their lacertilian offspring. Rhynchocephalians are represented in the early Trias by a few rare primitive types; later in the period there developed, in *Rhynchosaurus* and its allies, a peculiar short-lived branch of rather large, strong-beaked reptiles which flourished particularly in the mid-Triassic of Africa and South America. Turtles such as *Proganochelys* [*Triassochelys*], which is a characteristic although somewhat primitive member of their order, appear—seemingly suddenly—in the late Triassic.

Marine reptiles.—In the marine Trias (Muschelkalk) of Germany and in other formations of middle and late Triassic age in Spitsbergen and western North America, we see an early stage in the development of marine reptiles. A number of groups are already represented, and in no case have we any adequate knowledge of their ancestry. Most striking are the early ichthyosaurs, such as *Mixosaurus* and *Shastasaurus,* already well adapted for a marine life but somewhat less advanced than their Jurassic descendants in development of paddles and tail fin; some were of gigantic size. Among euryapsids, the sauropterygians were present in the form of nothosaurs, forms already aquatic but much less specialized than the plesiosaurs which were to replace them in the Jurassic. The Triassic was the period during which there developed the placodonts—amphibious mollusk-eating euryapsids with curious resemblances to the turtles. *Thalattosaurus* of the American Triassic and *Askeptosaurus* of Europe are long-bodied aquatic offshoots of the Eosuchia.

Archosaurs.—The archosaurs, or ruling reptiles, were destined to be the dominant land vertebrates of the later Mesozoic; the stem order of this subclass, the Thecodontia, made its appearance with a few genera in the early Triassic and became prominent in the later part of the period but was destined to disappear at its close. Most representative of thecodonts were the pseudosuchians, a central stock consisting of small reptiles with bipedal tendencies, such as *Euparkeria.* Heavily armored quadrupeds include, among others, *Aetosaurus* and *Stagonolepis* of Europe (forms of modest size) and the giant *Desmatosuchus* of North America. Common in northern continents in the late Triassic but unknown in the Southern Hemisphere were equally well-armored quadrupeds, the long-snouted amphibious phytosaurs, ecological predecessors of the crocodilians. Before the end of the period, however, there are also present ancestral crocodilian types.

Primitive dinosaurs.—The dividing line between progressive bipedal thecodonts and primitive dinosaurs (particularly saurischians) is difficult to draw. But it is clear that true dinosaurs had begun their evolution toward the end of middle Triassic times. Triassic ornithischian remains are rare, but have been reported from Upper Triassic deposits in regions as far removed as China and South Africa. Much more common are saurischians. By Upper Triassic times there were already typical lightly-built coelurosaurs, such as *Coelophysis,* and ancestral carnosaurs, such as *Ornithosuchus.* More common were large, heavily-built prosauropods, only partially bipedal, such as *Plateosaurus.*

Triassic: Summary.—The vertebrate life of the Triassic is characterized by (1) rarity of cartilaginous fishes; (2) replacement of chondrosteans by "subholosteans" and holosteans as the dominant fish group; (3) a trend toward marine life among the ray-finned fishes; (4) last appearance of the labyrinthodonts; (5) appearance of rhynchocephalians, including abundant rhynchosaurs, and of ancestral lizards and turtles; (6) initial development of reptilian marine life, including ichthyosaurs and varied sauropterygians; (7) the abundance and radiation of the thecodonts, followed by their extinction; (8) appearance of dinosaurs and rapid development of saurischians; (9) extinction of typical mammal-like reptiles, after a mid-Triassic "gomphodont" phase, but (10) persistence of forms approaching and finally reaching the mammalian level.

JURASSIC

The Jurassic was a time of widespread seas and reduced land areas. For most of the duration of the period the principal deposits are of marine origin, and it is only toward the close of the Jurassic that continental sediments become prominent. In consequence, our knowledge of aquatic vertebrates is much more complete than is that of terrestrial forms.

Vertebrate localities.—The typical European Jurassic section is almost exclusively marine in nature. The Lower Jurassic—the Lias—contains numerous localities with marine vertebrate remains, with a few land forms that drifted in. Notable areas are the Yorkshire coast; the region of Lyme Regis on the English Channel; the opposite coastal region of Normandy; southern Germany, particularly the famous Holzmaden quarries in Bavaria. The later European Jurassic—frequently termed the Dogger and Malm or (combined) the Oölites—in general contains far fewer remains of vertebrates. There are, however, important exceptions. The Oxford and Kimmeridge clays are rich in fish and marine reptilian remains. Prominent late Jurassic localities are the lithographic limestone deposits of Solenhofen and Kelheim in Germany and Cirin in France. Here fine-grained sediments, presumably laid down in quiet, coral-reef lagoons, have preserved numerous specimens of fishes and occasional tetrapod intruders. Jurassic marine deposits in other parts of the world occasionally produce vertebrates.

Land vertebrates are, in general, notable for their absence. In the lower Jurassic, the only known truly continental fauna is that recently discovered in the Kota beds of India. A few vertebrates are present in the marine European records, as at Solenhofen, and in Middle and Upper Jurassic estuarine localities in England, at Stonesfield, and in the Purbeck beds. An abundant record of terrestrial forms is found at the very end of the period in two regions. The Morrison beds, exposed over large areas of western North America from Montana through Wyoming to Utah, Colorado, and New Mexico, appear to have been laid down under conditions similar to those of the modern Mississippi delta, and contain a particularly rich dinosaur fauna. Very similar in nature and in fauna are the Tendaguru beds in Tanganyika.

Fishes.—The Jurassic, with seas full of invertebrate food, saw the beginning of a great recrudescence of cartilaginous fish life. There appear in the Lias *Squaloraja*, *Myriacanthus*, and other primitive members of the mollusk-eating chimaeras, destined to play a modest part in marine life from this time forth. Much more important were developments among the selachians. From the surviving hybodonts there began to branch off various types of modern sharks. *Heterodontus* and *Hexanchus*, relatively primitive living forms, had appeared before the end of the Jurassic, as had half-a-dozen genera representing more modern shark families. Somewhat slower to develop were the skates and rays, but relatives of the guitarfish, *Rhinobatis*, a primitive skate type, had already evolved by the late Jurassic.

Of freshwater bony fishes we know almost nothing. The sea, however, was swarming with members of this class. Apart from *Holophagus* [*Undina*] and a few other coelacanths, all these forms were actinopterygians. There were surviving palaeoniscoids; the earliest sturgeon, *Chondrosteus*, and a few "subholosteans." Almost all, however, were holosteans. The semionotids *Dapedius* and *Lepidotus;* pycnodonts such as *Gyrodus* and *Microdon; Aspidorhynchus;* amioids such as *Caturus*, *Furo* [*Eugnathus*], *Pachycormus*, and *Macrosemius;* advanced types such as *Pholidophorus*—these are but a few of the many common forms. And, in addition, we find, particularly toward the end of the period, swarms of small primitive teleosts, such as *Leptolepis*, which herald the approaching rise of this final major division of the ray-finned fishes.

Amphibians.—The Jurassic marks the low point in our knowledge of amphibian life. We have only a few finds of true (if primitive) anurans, notably in Patagonia, to represent the entire class.

Marine reptiles.—The nothosaurs and placodonts and the obscure thalattosaurs of the Triassic were absent in Jurassic waters, but the seas contained an abundance of marine reptiles. The ichthyosaurs of the period were far advanced in their readoption of a fusiform fish body and very fishlike fins. In the Lias they appear already to have reached the peak of their development. Most of the ichthyosaurs of the period were long "lumped" in the single genus *Ichthyosaurus*, but they appear to have actually included a wide range of genera varying in paddle and jaw development; *Eurhinosaurus*, for example, is a form closely analogous to the modern swordfish. The sauropterygians are represented by a variety of plesiosaurs, including short-necked types, such as *Pliosaurus* and *Peloneustes*, and more abundant longer-necked genera, such as *Plesiosaurus* and *Cryptocleidus*. The only instance of invasion of the seas by the archosaurs is the development in the later Jurassic of the short-lived group of marine crocodiles, such as *Metriorhynchus* and *Geosaurus*.

Dinosaurs.—Few remains of land vertebrates are known from the earlier deposits of the period; most of our knowledge of dinosaurs is gained from the faunas of the Morrison of North America and the Tendaguru beds, both of late Jurassic age. By this time the dinosaurian groups had developed in spectacular fashion. Among the saurischians, the coelurosaurs were present in forms of modest size, as *Compsognathus* and *Ornitholestes*. Other bipeds had, however, increased in size to produce such large types as "*Allosaurus*," *Ceratosaurus*, and *Megalosaurus*. Parallel to this development had been the evolution of their gigantic cousins, the amphibious sauropods. These appear to have been at the peak of their development in the late Jurassic—*Diplodocus*, *Apatosaurus*, *Brontosaurus*, *Camarasaurus*, *Cetiosaurus*, and *Brachiosaurus* are among the familiar names.

The development of the ornithischians appears to have been a slower process than that of the saurischians but is poorly documented. By the time of deposition of the Morrison and its equivalents a variety of these forms was in existence. There were bipedal ornithopods of modest size, of which *Camptosaurus* is representative. A striking Morrison form is the quadruped *Stegosaurus*, with a double row of plates and spines on its back; *Kentrurosaurus* is a closely related East African genus.

Other reptiles.—Our knowledge of the smaller reptiles of the Jurassic is very limited. We find a variety of turtles, mainly of the primitive amphichelydian group but including, in *Thalassemys* and its allies, primitive forerunners of the cryptodires. Most were presumably amphibious in nature; it is possible that some were well along in adaptation to marine existence.

Homoeosaurus is a small Jurassic reptile exceedingly close to the surviving rhynchocephalian *Sphenodon* of New Zealand; *Pleurosaurus* and *Sapheosaurus* are small types of doubtful position but often included in the Rhynchocephalia. Lizard remains are rare.

Crocodiles are abundant as fossils. We have already noted the development of a marine group; there were four other families showing more normal amphibious structures. All were in the mesosuchian stage of evolution, with the palate less developed than in modern forms.

The Jurassic was the time of the appearance and greatest development of the pterosaurs. In the Lias several genera of relatively primitive long-tailed forms, such as *Dimorphodon,* are present and are already completely adapted for flight. In the late Jurassic, long-tailed pterosaurs, such as *Rhamphorhynchus,* are still abundant, but we find, in addition, numerous specimens of *Pterodactylus,* with a short tail and more elongated wings.

Birds.—Our first glimpse of bird life comes from three specimens of *Archaeopteryx* from the Solenhofen Upper Jurassic lithographic stone. It is certain that birds are of thecodont derivation, but earlier stages in their development are unknown and would be difficult to recognize if found in the absence of feathers. These typical avian structures are here present to prove the nature of the specimens despite their retention of teeth, long bony tails, and typically dinosaurian "arms."

Mammals.—In the Jurassic, there are mammals of primitive but varied types; they are, however, small, rare, and fragmentary. Almost all known specimens come from three sites—Stonesfield in the Middle Jurassic of England, the late Jurassic Purbeck beds of that country, and a single small pocket in the Morrison at Como Bluff, Wyoming. In the Stonesfield locality there is a last therapsid survivor, a tritylodont; its definitely mammalian fauna is extremely limited and includes but two triconodonts and a single pantothere, *Amphitherium.* The two Upper Jurassic sites have a much more extensive fauna; those of the two sides of the Atlantic show relatively small differences from one another. There are further triconodonts, docodonts, and numerous pantotheres; the symmetrodonts make their only known appearance at this time. *Plagiaulax* and its relatives are the earliest representatives of the curious, aberrant multituberculates, destined to persist to Eocene times.

Jurassic: Summary.—Prominent features of Jurassic vertebrates were (1) among fishes, the height of the holostean supremacy, the appearance of teleosts, and the beginning of the radiation of modern elasmobranch types; (2) presence of true frogs, first of the modern amphibians; (3) among dinosaurs, the continued expansion of bipedal types in both orders and the evolution of armored forms and of sauropods; (4) a major radiation of marine reptiles, especially ichthyosaurs and plesiosaurs; (5) appearance of typical crocodiles and pterosaurs; (6) the first birds; (7) development of small primitive mammals of several types.

CRETACEOUS

The Cretaceous is among the longer periods of the earth's history, and there is considerable contrast between the vertebrates of the earlier and the later parts of the period.

Vertebrate localities.—Lower Cretaceous marine life is relatively poorly known, although represented in the Neocomian beds of some regions of Europe and equivalent deposits containing fish faunas and occasional aquatic reptiles in North America, Bahia (Brazil), and other regions. Much more prolific are Upper Cretaceous marine beds: the Chalk and related deposits of southern England and Westphalia; the similar Niobrara Chalk of North America; Australian marine deposits; Ceará (Brazil); and so on. Notable are the limestones of Mount Lebanon, Syria, with a wealth of varied fishes, and comparable beds in northern Africa.

In the Lower Cretaceous, continental deposits are none too abundant. The Wealden of southern England, northern France, and Belgium has a terrestrial fauna of very early Cretaceous age, and there are other, if less important, areas of this age in North America, China, and Mongolia. In the Upper Cretaceous the most famous of continental formations are the great dinosaur beds of western North America—Belly River, Edmonton, Lance, and equivalents—which extend over great areas from Alberta down the flanks of the Rockies to New Mexico. In Europe, Upper Cretaceous continental sediments are rare, except for an area in Transylvania. Mongolia and China have produced numerous dinosaurs of this age, and lesser faunas are found in India, Patagonia, and eastern Australia.

Fishes.—The selachians have expanded greatly in variety. The ancient hybodonts make their final appearances in the Cretaceous. Their advanced shark descendants are numerous. Of some sixteen or so families of living sharks, all but four had representatives in the Cretaceous seas. Many of the genera living today, such as *Carcharias, Isurus, Carcharodon, Galeocerdo,* were already identifiable; since the opening of the Jurassic the major part of modern shark evolution had been accomplished. The skates and rays had also reached essentially modern conditions before the end of the Cretaceous, and there were already present such familiar living types as *Pristis,* the sawfish; *Raja,* the common skate; *Trygon,* the sting ray; and *Myliobatis,* the great eagle ray. Chimaera teeth are not uncommon.

In the Cretaceous, *Mawsonia* and a few other rare types are the last of fossil coelacanths and were, indeed, thought to represent the termination of crossopterygian existence before the discovery of a living coelacanth. This period sees the replacement of the holosteans by the teleosts as the dominant group of fishes. Of lower actinopterygians, the seas contained the last rare palaeoniscoid and a few holosteans. Most of these holosteans disappeared before Upper Cretaceous times, and only the pycnodonts remained in a flourishing condition. The teleosts were diversifying rapidly and replacing them. Most were relatively primitive types, ancestral to the tarpons and to the relatively primitive herring-like and salmon-like fishes. Early Cretaceous genera were few, but in the late Cretaceous there were swarms of such fishes, some, such as *Xiphactinus* [*Portheus*], of very considerable size. In the Upper Cretaceous more specialized teleosts of various types were present. A dearth of freshwater deposits is perhaps responsible for our lack of knowledge of the important Ostariophysi—the carp-catfish group. There are, in the Upper Cretaceous, primitive eels, numerous stomatioids and myctophoids, and even representatives of the great group of the Acanthopterygii. These spiny-finned fishes are, however, rare and are represented almost entirely by primitive Beryciformes.

Amphibians.—To the meager record of the history of modern amphibian groups, we may add in the Lower Cretaceous the first appearance of a urodele.

Marine reptiles.—Aquatic reptiles continue in abundance in the Cretaceous, although with some change in the membership of the groups concerned. The marine crocodiles disappeared early in the period. Ichthyosaurs, although seemingly highly adapted to a marine life, were on the downgrade. They appear to have become quite rare by Upper Cretaceous times; not one, for example, has ever been found in the richly fossiliferous Niobrara Chalk of North America. Plesiosaurs, on the other hand, flourished vigorously, and the Upper Cretaceous chalk deposits contain abundant remains of a score of genera, among which *Polycotylus, Trinacromerum,* and *Elasmosaurus* are familiar forms; *Kronosaurus* of Australia, with a skull about 10 feet long, was a giant of the order.

The turtles, amphibious by nature, took strongly to the sea in Upper Cretaceous times. Several families of cryptodires, including such genera as *Archelon, Protostega,* and *Toxochelys,* adopted this habitat and tended to modify themselves to it by paddle-like limb developments and lightening of armor.

Another interesting Cretaceous development was the appearance of marine lizards. *Dolichosaurus* and *Aigialosaurus* are representatives of monitor-like lizard families which were becoming adapted to an aquatic habitat. From such forms sprang the mosasaurs, such as *Mosasaurus* of Europe and *Tylosaurus* and *Clidastes* of North America, large and common predaceous reptiles of the Upper Cretaceous.

Dinosaurs.—Of Lower Cretaceous dinosaurs we know but little, apart from those of the Wealden of western Europe, where *Iguanodon,* a large ornithopod, was common, and of eastern Asia, where carnivores, sauropods, and ornithopods were present. Apparently, the fauna was, in general, similar in content to that of the Morrison.

Late Cretaceous dinosaur beds are highly developed in western North America, and more scattered beds and finds are present in every other continent. In many respects this Upper Cretaceous fauna is a very different assemblage from that seen a period earlier. Coelurosaurs still survive, but even these slenderly built saurischians have tended to grow in size and specialize into such forms as the toothless ostrich dinosaur, *Ornithomimus.* The larger carnivores are now represented by ponderous giants, such as *Tyrannosaurus.* The sauropods had regressed in numbers and variety during the Cretaceous. Among all the numerous dinosaur finds from North America, there are but two fragmentary specimens of these amphibious forms. It appears, however, that this paucity of sauropods is less pronounced in other regions, for there are a number of records of *Titanosaurus* and related sauropods from the Upper Cretaceous of the Southern Hemisphere and even in India and China.

Ornithischians abound in the Upper Cretaceous. Little *Thescelosaurus* is a survivor of a very primitive ornithopod group. Of these bipeds, however, the dominant forms of the times were the amphibious duckbills, the hadrosaurs or trachodonts, abundantly represented by a score of genera, mainly in North America and eastern Asia, which show numerous variations in the curious crests which frequently crown their heads. The stegosaurs had disappeared but have been replaced by another group of armored dinosaurs—*Ankylosaurus* and related types—low-bodied, large-skulled genera covered with a solid dorsal armor of bony plates. A final group of ornithischians is that of the horned dinosaurs, the ceratopsians, with a bony neck frill in the primitive genus *Protoceratops,* and with large horns developed in addition in more progressive genera, such as *Triceratops* and *Centrosaurus.*

Other reptiles.—Turtles are numerous in Upper Cretaceous deposits. Many are still amphichelydians of Mesozoic type, but representatives of the two existing suborders are present as well. There are a few pleurodires, including even representatives of the living South American genus *Podocnemis.* More numerous are cryptodires. In addition to the marine forms already mentioned and survivors of the Jurassic *Thalassemys* group, there are relatives of the living marsh turtles and the soft-shelled trionychids. *Champsosaurus,* a small, long-snouted, amphibious diapsid of dubious ancestry, appears in the late Cretaceous and persists into the early Tertiary (paying no attention to the era boundary!). In the Upper Cretaceous, lizards are, for the first time, abundant as fossils. Apart from the mosasaurs, half-a-dozen families are represented, including forms typical of most of the major subdivisions of the group. In this period, too, there appear, in *Pachyophis* and *Simoliophis,* forms transitional to the snakes, last of reptilian groups to make their appearance.

As in the Jurassic, the crocodiles are well represented. Most are still in the mesosuchian stage of development, but there are a number of more progressive genera leading to the modern crocodiles and alligators.

Pterosaur remains are rare except in the Niobrara Chalk of the Upper Cretaceous. The rhamphorhynchoid type is extinct, and by the Upper Cretaceous we find surviving only short-tailed pterodactyloids, such as *Nyctosaurus* and the giant, toothless *Pteranodon.*

Reptilian extinction.—By the end of the Cretaceous the greater part of the reptilian life of the Mesozoic had become extinct—all ichthyosaurs, all sauropterygians, all the dinosaurs and pterosaurs. We find only the turtles, lizards, snakes, crocodilians, and a lone rhynchocephalian surviving into the typical Cenozoic. We must not overemphasize the rapidity of this extinction. Some of the groups—such as the ichthyosaurs and perhaps the pterosaurs—had become reduced in numbers well before the end of the period. Many dinosaur groups, however, appear to have flourished in very late Cretaceous deposits, and their extinction at the Mesozoic-Tertiary boundary is one of the most dramatic events in vertebrate history.

The reasons for this reptilian catastrophe have been much debated, and the causes may have been complex in nature. It has, however, been reasonably argued that geological processes may be, in great measure, fundamentally responsible. In the late Cretaceous there came about the Laramide revolution, a time of mountain building during which there began the elevation of the Rockies and other mountain chains. This condition of rising lands might well have affected markedly the more amphibious dinosaur groups, such as the sauropods and duckbills, by limiting the areas of the swamps and lagoons in which they made their livelihood. Food, too, would be affected, for these geological disturbances brought about climatic shifts, which are seen reflected in marked changes in vegetation in the late Cretaceous. Many herbivores are narrowly restricted in their diets, and floral changes may be responsible for much extinction. Finally, the disappearance of carnivorous dinosaurs would inevitably follow the disappearance of their herbivorous cousins, upon which they fed.

The reasons for the elimination of marine forms are more difficult to deduce. It may be, however, that the replacement of holostean fishes by progressively higher and presumably more efficient teleost types was a factor, in addition to changes in invertebrate life.

Birds.—There is a gap in our history of birds between the late Jurassic and the late Cretaceous. Even in deposits of the latter age our knowledge of birds is almost entirely confined to oceanic or water-dwelling forms, such as the ternlike *Ichthyornis* and the swimmer *Hesperornis* of the Niobrara Chalk. Fragmentary remains, however, appear to belong to several of the existing orders, and it is probable that the radiation of birds of modern types was already under way.

Mammals.—Mammalian remains are rare in the Lower Cretaceous. A few multituberculate teeth are known from the Wealden of England; in the Trinity sands of Texas there have been found a quantity of materials, but materials of a very fragmentary nature. Multituberculates are common; symmetrodonts are reported from both localities, but there is no definite evidence of survival of other Mesozoic types. Of great interest, however, is the discovery of teeth indicating the presence of true therians—forms presumably descended from Jurassic pantotheres which, as far as the scant evidence goes, may be the common ancestors of both marsupials and placentals.

In the late Cretaceous, mammalian remains, although almost invariably fragmentary, are much more abundant. Materials including skulls, as well as teeth, are present in Mongolia. Mammal remains have long been known from the Lance Formation of Wyoming, and there have been further finds in the American West, notably a rich deposit at "Bug Creek," Montana. Multituberculates still flourish, but the other ancient orders are extinct. Instead, we find early representatives of the two great modern divisions of therian mammals—marsupials, represented by forms related to the modern opossums, and a modest array of primitive placentals, including archaic insectivores and types suggestive of the ancestry of several of the orders which were to gain prominence in the Cenozoic.

Cretaceous: Summary.—Among the principal features of Cretaceous vertebrate history may be cited the following: (1) among fishes, further deployment of the selachians and a sharp dwindling in importance of the holosteans, which were being replaced by primitive teleosts and, at the end of the period, by the first spiny-finned teleosts; (2) appearance of the first urodeles; (3) among marine reptiles, continued importance of plesiosaurs, appearance of marine lizards and turtles, reduction of ichthyosaurs; (4) progressive modernization of the crocodiles and turtles and development of snake ancestors; (5) climax in size of pterosaurs; (6) among dinosaurs, development of the largest carnivores, dwindling importance of sauropods, appearance of duckbilled ornithopods, heavily armored ankylosaurs, and horned dinosaurs; (7) birds reaching modern development except for the retention of teeth in one or more types; (8) replacement of Jurassic mammal groups, except multituberculates, by small primitive marsupials and placentals; (9) and, finally, the complete extinction by the end of the period of most of the characteristic reptilian types of the Mesozoic, including all the dinosaurs, the ichthyosaurs, plesiosaurs, marine lizards, and pterosaurs.

Mesozoic history.—The reptiles are the most interesting and most important vertebrates of the era. The

Mesozoic includes the full cycle of existence of many major reptilian types—the dinosaurs, pterosaurs, and the varied marine reptiles—as well as the age of origin and expansion of still surviving groups, such as the turtles, lizards, and crocodiles. But it includes as well, in the early Triassic, the close of an earlier cycle in reptilian history which had begun before the end of the Carboniferous, with cotylosaurs and mammal-like reptiles as its chief characters. The earliest Mesozoic is an age of transition from the old to the new. On the other hand, the end of the era is a seemingly natural point of cleavage as regards vertebrate life.

With other groups, too, it seems that the Mesozoic includes not merely one typical cycle but, at its beginning, the final chapter in the old order. Among the bony fish the palaeoniscoids were still dominant at the beginning of the Triassic, just as they had been since the beginning of the Carboniferous. Before the close of the Mesozoic not only had the holosteans replaced them, but they in turn had been driven into insignificance by the progress of the teleosts. Again, the only known Triassic amphibians (apart from a pre-frog) were the last survivors of the old labyrinthodonts of the Paleozoic, and only in the Jurassic and Cretaceous do we find the first characteristic members of the modern groups. Among cartilaginous fishes, also, the Triassic was a transitional period, while the Jurassic and Cretaceous witnessed a great revival in shark development. Primitive birds and primitive mammals are present in the Mesozoic; but mammals appear only at the end of the Triassic, and the first bird not until nearly a full period later.

The Triassic was thus a boundary period—a time of transition. In the Jurassic and Cretaceous we witness a full and characteristic Mesozoic life. And with the coming of the Cenozoic we enter a new age, the Age of Mammals.

30 CHAPTER

Cenozoic Vertebrates

With the extinction of the great reptiles there began a new chapter in the history of land vertebrates. The mammals had until then remained small, rare, and inconspicuous, but had seemingly progressed far in structural organization. Once the way was cleared, they emerged from obscurity and began a spectacular radiation into a host of types, many of which survive to the present day. The account of the Cenozoic given in this chapter will be based mainly on the history of the Mammalia; other groups are treated but briefly.

This last phase of the earth's history, comprising the Cenozoic era, is usually divided into a Tertiary period of considerable magnitude and a second period, the Quaternary, barely begun. The Tertiary may be divided into some five epochs—Paleocene, Eocene, Oligocene, Miocene, and Pliocene; the Quaternary includes only the short Pleistocene epoch and Recent times. These divisions are based in most cases upon the degree of modernity reached by invertebrates in European marine deposits. In Europe these epochs have been subdivided into a series of ages listed in Table 4. In many instances the terrestrial deposits of Europe can be correlated with marine beds and hence placed in the appropriate stage. But in North America this is not so readily done, for most American mammalian deposits are in the interior western states, far from marine fossil-bearing sediments. A series of ages based on the mammals contained in the various North American beds has, however, been established, and approximate correlations can be made between these ages and the European ones. Similar series have been erected for Asiatic and South American deposits. These series of ages are given in Table 1, together with data on the best-known localities, areas, or formations in the various continental areas.

Cenozoic deposits.—The best and most complete series of mammal-bearing beds is that found in western North America, where sediments, spreading out mainly from the Rocky Mountains, have preserved a good record of numerous faunas ranging from the beginning of the Cenozoic to the Pliocene. Europe in Tertiary times was a peninsular area frequently flooded by shallow marine waters; the deposits along the shores of these seas or derived from the rising Alps include numerous fossiliferous formations which have been studied intensively for more than a century. In Asia, early Tertiary deposits are well represented in Mongolia; later epochs are best represented in the Siwalik Hills and other areas of sediments from the rising Himalayas; Burma supplies further data for the early Tertiary and Pleistocene, China for the later Cenozoic, Java for the Pleistocene. In Australia, Pleistocene fossils are rather plentiful, but few Tertiary finds have as yet been made. Africa appears to have been a stable land mass during most of the Cenozoic; apart from later Eocene and early Oligocene faunas from Egypt, and areas of Miocene deposition in Kenya, Tertiary continental deposits are extremely restricted in area and faunas. In South America there are extensive sediments in Patagonia and the Andean foothills which have preserved much of the curious fauna of mammals which evolved here during Tertiary times.

Cenozoic conditions.—We have little reason to believe that there has been much alteration in the extent or position of the continental areas during the Cenozoic, and almost all known facts in the distribution of mammals may be accounted for by small changes in the present continental relations. It is probable that Australia has been in its present condition of isolation since late in the Mesozoic, for almost none of the Cenozoic types of terrestrial mammals have reached that region. The independent evolution seen in South American mammals indicates that that continent was difficult to reach from North America in the early Cenozoic (and probably in the Cretaceous as well), and that an easily traversed connection was not rebuilt until about the end of Tertiary times. Although, as noted above, the history of Tertiary Africa is poorly known, it is very probable that this region was isolated from Eurasia during the earliest Cenozoic and that opportunity was given there for the initiation of peculiar and characteristic placental groups. Comparison of successive North American and Eurasian faunas indicates that the Bering Straits connection between the two areas was made and broken a number of times.

Various Tertiary land bridges, we have noted, have

been proposed between continental areas to account for real or fancied similarities between the regions so connected. Most of them rest on flimsy evidence and need not be considered here. Stronger support was at one time given to a proposed connection between South America and Australia, either directly or via Antarctica. This was based upon the similarity in the marsupials of the two continents; but this similarity can be as readily explained as being due to parallel evolution from common ancestors, which were presumably widespread in northern continents in late Mesozoic days. Arguments have been made for a Tertiary bridge between South America and Africa, but this, when considered from a broad point of view, has little in its favor. Such a bridge would "explain" the appearance of rodents and monkeys in the Oligocene of South America as African immigrants and would account for the presence in the two continents of certain similar groups of freshwater fishes. But in almost every other regard the faunas of the two continents are radically different, as would not have been the case had there been a land connection.

Early Tertiary times seem to have had an equable, warm climate, which gradually gave place to the extreme temperature differences found in the Ice Age, with a cold arctic zone grading rapidly southward into warm tropics, a situation still existing at the present time. This climatic change seems to have resulted in a slow southward retreat of many forms from the better-known regions of Eurasia and North America and the survival in the tropics of many groups long after their extinction in the present north temperate zone. The peccaries and tapirs of South America are relatively recent migrants from the north; the antelopes and associated mammals of tropical Africa form a fauna strikingly similar to that of Eurasia in the early Pliocene.

The geographic regions customarily used by zoogeographers in interpreting the distribution of life today reflect accurately the geological history. Some six regions are generally accepted: the Palaearctic, for Europe and the Mediterranean region and northern and central Asia; Nearctic, for North America; Oriental, for southern Asia and the East Indies; Ethiopian, including most of Africa; Neotropical, for South and Central America; and Australian, including Australia, New Guinea, and certain adjacent islands.

These six regions are not all equally distinct from one another. The Australian region stands in sharp contrast with all five other regions in its faunal content, a feature associated with the separate Cenozoic history of that continent. The Neotropical region also shows very individual characteristics, due to the development of a peculiar fauna during the Tertiary isolation of South America. The Palaearctic and Nearctic regions are identical in climate and separated only by the variable make-and-break between Alaska and Siberia. Their faunas are so similar that they are frequently considered as a single Holarctic area. The distinction of Ethiopian and Oriental regions from the Palaearctic is due primarily to the relatively recent establishment of sharp temperature gradients; during much of the Tertiary the three were essentially a single faunal unit.

The major interest in the Cenozoic lies in mammalian history; in most of the other vertebrate groups relatively little evolution took place during this era, and a treatment of them by successive epochs is unnecessary.

Fishes.—Shark and ray teeth are common in many Tertiary marine beds, and chimaeroid tooth plates are occasionally present; only in exceptionally favorable circumstances are skeletons of cartilaginous fishes discovered. All, or nearly all, existing groups of sharks, skates, and rays appear to have been established in the early Tertiary, and but little further evolution appears to have taken place. Actinopterygians below the teleost level are rare in the Cenozoic. The pycnodonts persisted into the Eocene; in later Tertiary epochs and today we find, of lower ray-finned fishes, only the African polypterids, the sturgeons, paddlefishes, garpike (*Lepisosteus*), and *Amia,* the bowfin. The last three are represented by characteristic specimens from the Eocene, the first two by fragmentary remains of equally early date. There is a single living coelacanth crossopterygian; but no Tertiary fossils are known, presumably because this group had taken up life in deep waters, an environment of which there are few known geological records. Of the three surviving lungfishes, there are Tertiary records of the Australian and African genera but not of the South American form.

The teleosts are the dominant Tertiary fishes and are found in a variety of deposits both fresh and salt in origin. The shales of the Green River region of North America and of Monte Bolca in Italy are famous Eocene fish localities; diatom beds in California and marine deposits in Algeria and at Licata in Sicily are among prominent late Tertiary deposits. In fresh waters the dominance of the Ostariophysi, the carp-catfish group, is a striking feature of the Cenozoic; in marine deposits there is an increasing variety of higher, spiny-finned teleosts, although the more old-fashioned clupeoids are still abundant.

Amphibians and reptiles.—Amphibian remains are relatively rare in the Tertiary and shed little light on the history of the frog and salamander groups represented. Turtles are abundant in many formations, and most of the living types are already represented by close relatives at the very beginning of the era. In the Eocene the primitive amphichelydians became extinct. The pleurodires, subsequently rare as fossils, have been confined to the southern continents since the earlier Cenozoic. Lizards are not infrequently discovered. The snakes are the one progressive group of reptiles during these latter times; poisonous snakes are apparently a Tertiary reptilian innovation. Crocodiles, mostly related to living forms, are fairly numerous throughout the period. *Champsosaurus* of the late Cretaceous, contemptuous, it would seem, of era boundaries, persisted into the

TABLE 4 Cenozoic Vertebrate Localities

Epoch	European Ages	European Localities	Asiatic Ages	Asiatic Localities	African Localities	North American Ages	North American Localities	South American Ages	South American Localities
PLEISTOCENE	Glacial period	Caves, river terraces, loess, etc.	Pinjorian	Pinjor of Siwaliks; loess of China; Irrawady beds (Burma); volcanic-ash deposits of Java	North African caves; S. African river deposits; Olduvai (pt.) (Tang.); Ternifine, Lac Karar (Alg.)	Glacial period	Caves, postglacial bogs; La Brea, McKittrick, Irvington (Cal.); Sheridan beds (incl. Hay Springs), Rock Cr. (Neb.); Fossil Lake (Ore.)	Pampean	Pampas of Argentina, Uruguay; Tarija (Bolivia); Brazilian caves
	Villafranchian, Calabrian	Arno Valley (Italy); Cromer (Eng.); Tegelen (Holl.); Süssenborn (Ger.); Perrier, St. Vallier, Senèze (Fr.); Villaroya (Sp.)	Tatrotian	Tantrot of Siwaliks; Nihowan (China)	Omo (Ethiopia); Kanam (Kenya); Kaiso, Kazinga (Uganda); Olduvai (pt.) (Tang.); S. African caves and Vaal R. terraces	Blancan	Blanco (Tex.); Hagerman (Idaho); Rexroad (Kans.); San Joaquin, Tehama (Cal.); Broadwater (Neb.); San Pedro (Ariz.)	Chapadmalalian	Chapadmalal (Arg.)
PLIOCENE	Plaisancian, Astian	Montpellier, Perpignan, Trevoux (Fr.); Red Crag (Eng.); Piemonte region *M* (Italy)	Dhokpathanian	Dhok Pathan (India)	Bon Hanifa, Qasir-es-Saha; Vaal R. (S.Af.)	Hemphillian	Hemphill (Tex.); Long Island (Kans.); Ash Hollow, U. Ogallala (Neb.); Wray Co. (Colo.); Rattlesnake (Ore.); Thousand Cr. (Nev.); Kern R. (Calif.)	Montehermosan, Tunuyanian	Monte Hermoso, Tunuyan (Arg.)
	Pontian, Pannonian	Teruel, Concud (Sp.); Curcuron, Mt. Léberon (Fr.); Antwerp (Belg.); Eppelsheim (Ger.); Vienna basin, *M* in part (Aus.); Pikermi, Samos (Gr.)	Nagrian	Nagri (Siwaliks); Maragha (Persia); Honan, Shansi, redbeds (China)	Wadi Natrun, Maghreb (Egypt); Beni Mallal (Morocco); Ben Hanifa Oued-el-Hammam (Alg.)	Clarendonian	Clarendon (Tex.); Burge, M. Ogallala (Neb.); Alachua (Fla.)	Mesopotamian, Huayquerian	Entre Ríos; Catamarca, Huayqueria beds (Arg.)
MIOCENE	Sarmatian, Tortonian, U. Vindobonian	Grive St. Alban (Fr.); Antwerp (Belg.); Sevastopol (USSR)	Chinjian	Chinji (Siwaliks); Tung Gur (Mongolia)	Ft. Ternan (Kenya)	Barstovian	U. Barstow (Cal.); L. Ogallala, Valentine (Neb.); Madison Valley (Mont.); Santa Fe (N.M.)	Chasicoan, Friasian	Chasico; Rio Frias (Arg.); La Ventana (Colomb.)
	L. Vindobonian, Helvetian, U. Burdigalian	Sansan, Simorre, St. Gaudens (Fr.); Eibiswald, Eggenberg, etc. (Aus.); Molasse in part, Steinheim, Baltringen, Oeningen (Ger.); Mte. Bamboli (It.)	Kamlialial	Kamlial (Siwaliks)		Hemingfordian	Hemingford, Marsland, Sheep Cr., Snake Cr. (Neb.); Mascall (Ore.); Pawnee Cr. (Colo.); Calvert (Md.) *M;* Temblor, L. Barstow (Cal.)		
	L. Burdigalian, Aquitanian	Faluns of Touraine, St. Gérand-le-Puy, Agen region, Leognan, Orleans sands (Fr.); Molasse in part, Mainz basin, Ulm (Ger.); Vallés penedés (Sp.)	Bugtian	Hsanda Gol (pt.), Shara Gol (pt.), Loh (Mongolia); Bugti hills (W. Pakis.); Turgai (W. Turkestan)	Moghara (Egypt); Namib (SW.Af.); Rusinga, Songhor, L. Rudolf (Kenya), Napak (Uganda)	Arikareean	Harrison, Gering, Monroe Cr. (Neb.); M.–U. John Day (Ore.); Rosebud (S.D.); Thomas farm (Fla.)	Santacrucian, Patagonian fm. *M*	Santa Cruz, Patagonia (Arg.) *M*

OLIGOCENE	U. Stampian (Chattian)	Sandstones of Calaire de Beauce *M*, La Rochette, La Milloque, Agen region (Fr.); Mainz basin (Ger.); Cadibona (It.)	Hsandagolian	Shara Gol (pt.), Hsandsa Gol (pt.) (Mongolia); Kazakstan (W.As.)		Whitneyan	Upper Brulé (*Protoceras* beds) of White River series (S.D., N.D., Neb., Wyo., Colo.)	Colhuehuapian	Colhué Huapí beds (*Colpodon* fauna) of Patagonia
	L. and M. Stampian (Rupelian)	La Ferte–Aleps, Bournoncle–St. Pierre, Aveyron, Quercy phosphorites in part (France); Flonheim, Weinheim (Ger.)	Houljinian	Hsanda Goł (pt.), Houljin gravels (Mongolia)		Orellan	Lower Brulé (*Oreodon* beds) of White River series (S.D., N.D., Neb., Wyo., Colo.); Cedar Cr. (Colo.)		
	Sannoisian (Lattorfian)	Quercy phosphorites in part, Ronzon, Velay, Brie limestone, Lobsann (Fr.); Hampstead beds (Eng.)	Ulangochuian	Ulan Gochu, Ardyn Obo (pt.) (Mongolia)	Gebel Qatrani (Fluvio-Marine beds) of Fayûm (Egypt)	Chadronian	Chadron (Titanothere beds) of White River series (S.D., N.D., Neb., Wyo., Colo.); Pipestone Spr. (Mont.); Cypress Hills (Sask.)	Deseadan	Deseado beds (*Pyrotherium* fauna) of Patagonia
EOCENE	Ludian, Bartonian, Anversian	Gypsum beds of Montmartre, Quercy phosphorites in part, Débruge, Gargas, St. Hippolyte de Caton, Robiac, Castres, Calcaire de St. Ouen (France); Mormont (Switz.); Isle of Wight (Eng.)	Sharamarunian, Irdinmanhan	Ardyn Obo (pt.), Sharu Marun, Irdin Manha, Ulan Shireh (Mongolia); Pondaung (Burma)	Qasr-el-Sagha beds of Fayûm, Upper Mokattam *M*, Birket-el-Qurûn *M* (Egypt)	Duchesnean, Uintan	Duchesne River, Uinta (Utah); Beaver Divide, Washakie in part (Wyo.); Sespe in part (Calif.); Jackson of southeastern states *M*; Clarno (Calif.)		Divisadero Largo (Arg.)
	Lutetian	Calcaire grossier of Paris region, Argenton Issel, Buchsweiler (Fr.); Bracklesham (Eng.); Egerkingen (Switz.); Geiselthal, Messel (Ger.); Mte. Bolca *M* (It.)	Arshantan	?Arshanto (Mongolia)	Lower Mokattam *M* (Egypt; Nigeria *M*)	Bridgerian	Bridger, Washakie in part (Wyo.); Green River in part (Wyo., Colo., Utah); Huerfano in part (Colo.)	Mustersan	Musters beds with *Astraponotus* fauna (Patagonia)
	Ypresian, Sparnacian, Cuisian	Soissons, Meudon, Epernay (Fr.); London Clay (Eng.); Erquelines, Orsmael (Belg.)	Ulanbulakian			Wasatchian	San Jose, Largo, Almagre (N.M., Colo.); Huerfano in part (Colo.); Gray Bull of Bighorn basin, Lysite and Lost Cabin of Wind River basin (Wyo.)	Casamayoran	Casa Mayor beds with *Notostylops* fauna (Patagonia)
PALEOCENE	Thanetian	Cernay (Fr.); Walbeck (Ger.); Thanet Sands (Eng.) *M*	Gashatan	Gashato (Mongolia)		Tiffanian	Clark Fork (pt.), Silver Coulee fm. of Polecat Bench, Bison Basin (Wyo.); Sentinel Butte fm. of Ft. Union (Wyo., Mont.); Plateau Valley (Colo.); Tiffany (N.M.); Paskapoo (Alberta)	Riochican	Rio Chico (Patagonia); Itaborai, Pernambuco (Brazil) *M*
	Montian					Torrejonian, Dragonian	Torrejon (N.M.); Polecat Bench in pt. (Wyo.); Lebo fm. of Ft. Union, Tongue R. (Mont.); Dragon (Utah)		
	Danian					Puercan	Puerco (N.M.); Mantua lentil of Polecat Bench (Wyo.)		

M = Marine

early Eocene. Fossil representatives of the typical Rhynchocephalia are unknown.

Birds.—It has been said that the Cenozoic is as much the Age of Birds as it is the Age of Mammals; the birds are the dominant group in the air during the entire era. The fragile nature of their bones and their usually small size have rendered the unraveling of their history difficult. The principal points in their Tertiary history, including the sporadic development of large flightless forms, have been noted in an early chapter. Except for some tropical types, almost every large group of modern birds, and even most of the major families, were already present in the early Tertiary.

PALEOCENE

This oldest of Tertiary epochs is relatively poorly represented in the marine sequence, but the existence of a pre-Eocene chapter is abundantly proved by numerous older continental deposits, particularly well developed in North America. In the San Juan basin of New Mexico and southern Colorado, Lower, Middle, and Upper Paleocene deposits are represented by the widespread Puerco and Torrejon beds and the small Tiffany pocket; the Polecat Bench formation of northwestern Wyoming appears to cover the entire time interval between the end of the Cretaceous and the true Eocene; "Fort Union" beds in eastern Montana and adjacent states have produced Paleocene mammals in various areas, particularly in Middle Paleocene deposits near the Crazy Mountains; the Dragon fauna of Utah is an early Middle Paleocene assemblage.

In other continents Paleocene mammals are unknown, rare, or of late date. In Australia the Paleocene fauna is unknown, as is that of much of the Tertiary, and there are no African Paleocene mammals. In Europe there is a modest late Paleocene fauna from Cernay near Rheims, and a German locality of somewhat earlier date; in Asia a small assemblage, apparently equally late in time, is found in the Gashato of Mongolia. In South America a handful of late Paleocene specimens are known from the Rio Chico of Patagonia; recently abundant materials of similar age have been found in limestone fissures and caves at Itaborai, near Rio de Janiero. Our discussion below of the Paleocene is based mainly on the American record as seen in the Puerco and Torrejon and their equivalents; the late American Paleocene and that of other continents is considered separately.

The terrestrial fauna of the Paleocene, with the dinosaurs absent, is mainly a mammalian one. But if we were able to visit the Paleocene, the aspect of the fauna would be of a strange and unfamiliar sort. Many of the mammals then present are assigned to orders still existing, such as the Insectivora and Primates, but these Paleocene representatives of the modern orders were for the most part archaic or aberrant types. Not a single living family is present in the characteristic Paleocene, and much of the fauna belonged to orders now entirely extinct.

The few mammalian types known in the late Cretaceous are still present. Opossum-like marsupials persist in insignificant numbers. The archaic multituberculates, on the other hand, are present in abundance; most were small members of the *Ptilodus* group; but, in addition, we find *Taeniolabis* and its allies, forms of considerable size and still higher specialization.

Insectivores are prominent. Some are persistently generalized placental types; others, however, are diversifying and tending toward the structural patterns of later carnivores, primates, and ungulates. Some side branches, such as *Mixodectes* and the pantolestids, had departed far from the typical insectivore pattern, but were hardly important enough to merit ordinal separation.

The typical creodonts characteristic of the Eocene are not present until late in the Paleocene, although primitive related forms are to be found; progressive miacids, destined to give rise to modern carnivores, were represented from the Middle Paleocene onward.

Primates appear in the Middle Paleocene. The forms are more or less transitional in their cheek teeth between the insectivores, on the one hand, and lemurs and tarsioids, on the other, and ordinal assignment is sometimes difficult. A puzzling feature is that nearly all the Paleocene genera (and certain of the Eocene forms as well) have developed large and often rodent-like incisors, perhaps in relation to some type of feeding habit now difficult to interpret.

Ungulates were abundant, but ungulates of unfamiliar types, of primitive structure, and, for the most part, of small size. Most may be included in the Condylarthra as a basal order of hoofed mammals; the condylarths (in a broad sense) make up a large fraction of the Paleocene fauna. The low-crowned and often squared-up molars indicate a shift to or toward a herbivorous diet; hoofs are present in some cases, but not all; in many respects, however, the early condylarths were still close to the insectivore-carnivore stock. Half a dozen families are present, of which *Oxyclaenus, Dissacus, Mioclaenus, Periptychus, Meniscotherium,* and *Tetraclaenodon* are representative. The first two are members of families (Arctocyonidae, Mesonychidae) which are so little advanced that they have in the past been considered as perhaps leading toward the carnivores rather than ungulates.

The orders so far mentioned are still so close to the ancestral placentals that the group boundaries are difficult to determine; were we ignorant of their later histories, we should be justified in including almost the entire assemblage in a single ordinal group.

More divergent forms were, however, already making their appearance. The distinction is often made between "archaic" and "progressive" mammal groups, the former being types which tended to a high degree of specialization at an early stage but were destined for early extinction; the latter, forms which tended to evolve more slowly but with greater adaptability and greater success. Among the Paleocene mammals, archaic types are already distinguishable in the taeniodonts, such as *Conoryctes,* giants of their times, and *Pantolambda,* an ungulate which had already grown to the size of a sheep and which in its relatively heavy build foreshadows the larger pantodonts of the late Paleocene and Eocene. A tendency toward prematurely large size is also seen in some of the condylarths.

Following the typical faunas of the epoch, we find in the late Paleocene of North America a fauna in which most of the groups previously present still persist but are often represented by larger forms, and in which the mammalian assemblage is more varied and foreshadows that of the true Eocene. The more primitive condylarth families are already reduced in numbers. The presence of *Oxyaena* marks the appearance of the typical creodonts. The early ungulates were increasing in size and variety; particularly prominent were great pantodonts, such as *Barylambda* and, at the very end of the Paleocene, *Coryphodon.* Several new mammalian groups, whose earlier history is unknown, appear. These include the first of the uintatheres among archaic ungulates; the first of the peculiar tillodonts; *Palaeanodon,* an ancient relative of the South American edentates; the first tarsioid primates; and *Paramys,* an ancestral rodent.

The few Paleocene mammals of Europe, late in date, add little to our knowledge of mammalian evolution. Such forms as have been discovered appear to be close to American mammals of similar age, although nearly all are generically distinct, and indicate that there was a common fauna widespread in northern regions. The known Paleocene of Asia, equally late in date, is puzzling and disappointing. It was once believed that Asiatic beds of this age, when discovered, would reveal the ancestors of some of the groups, particularly of "modern" orders of ungulates, absent in the American Paleocene. The single fauna so far found, the Gashato, has not realized these hopes. It reveals, much as in North American beds of the same horizon, multituberculates, insectivores, creodonts, condylarths, and a uintathere. A welcome find is the oldest fossil "hare," *Eurymylus.* The commonest fossil is an ungulate, *Palaeostylops;* but, startlingly, this belongs not to any normal northern group but to the characteristic South American order of Notoungulata.

In South America itself the Rio Chico and Itaborai finds appear to be late in date. The material shows that most of the groups which were to characterize the later Tertiary of that continent were already present. These forms include marsupials, edentates (represented by armadillo scales), condylarths and litopterns evolving from them, notoungulates, astrapotheres, and the archaic xenungulate *Carodnia.* This is a peculiar "sample" of the variety of mammals which we have seen evolving during the Paleocene in northern continents. Condylarths were, of course, common in the northern Paleocene; but marsupials were not common, edentates and notoungulates very rare, and astrapotheres and *Carodnia* unknown. There are lacking, for instance, rodents and primates (which only reached South America much later) and placental flesh-eaters. Some of the groups present may have evolved in South America; but militating against that is the fact that there is present no basic stock of insectivores from which advanced forms might have developed. The situation suggests that even in the Paleocene (and perhaps earlier) South America was isolated from the north by a water gap; the groups which were to form the basis of the spectacular Tertiary faunas of South America are merely the few odds-and-ends of northern groups which chanced to succeed in crossing that gap.

It is obvious that our known Paleocene faunas, while giving us interesting chapters in the early deployment of the placentals, do not give us a complete story. No fossils are known from any early Tertiary horizon in Australia, and we are in the dark as to the early history of the interesting fauna of that continent. There are no Paleocene deposits in Africa, where one may expect that the early development of subungulates and possibly other types, first known from the late Eocene and early Oligocene, was under way.

We have, further, no knowledge of stages in the evolution of rodents and lagomorphs, which appear fully developed at the close of the Paleocene, or of the perissodactyls and artiodactyls, which were to appear abruptly at the beginning of the Eocene. Possibly these orders may have been evolved in some area of Eurasia or North America unrepresented in our records; possibly in familiar regions but in environments not represented by fossiliferous deposits.

EOCENE

In the Eocene, North America is, again, the continent in which the faunal sequence is most abundantly represented. Lower Eocene faunas, usually termed "Wasatch," are present in abundance in the Bighorn and Wind River basins in Wyoming; the Bridger basin of southwestern Wyoming is the classical Middle Eocene collecting ground; that for the Upper Eocene is to the south in the Uinta basin of northeastern Utah.

In Europe the Lower Eocene is known from relatively poor beds in the London basin and in related deposits in northern France and Belgium; later Eocene horizons are increasingly fossiliferous, including such well-known localities as Egerkingen in the Swiss Middle Eocene and the Upper Eocene gypsum beds of Montmartre (Paris). The Quercy phosphorites of south-central France are cave fillings in older limestones containing numerous small mammals. Part are late Eocene; the later deposits extend into the Oligocene. In Asia there are few fossils before the latter part of the epoch, when several faunas are present in Mongolia and in Burma. In South America the Casa Mayor beds are of early Eocene age; the Musters, also in Patagonia, rather later. Africa now enters the Tertiary terrestrial record. The Mokattam beds of Egypt contain a marine mammalian fauna of middle to late Eocene age, and continental deposits begin in the Fayûm of western Egypt in late Eocene times.

In the Lower Eocene, it appears, the northern continental areas formed, much as today, a single zoogeographic region, for Europe and North America had many genera in common. There was no sharp break between the life of the Paleocene and the new epoch; for, with the exception of the *Periptychus* type of condylarths, every major group present in the Paleocene was still present. Most of the older groups were destined to become extinct; but their disappearance was a gradual one. The Eocene contains the last survivors of the long-lived multituberculates; opossums (*Peratherium*) persist into the Eocene and onward into the Miocene of both Europe and North America. There are insectivores—primitive, typical, and aberrant. Creodonts of the oxyaenid and hyaenodont families are the dominant carnivores of the Eocene, but miacids persist, and had progressed to give rise to the first representatives of modern carnivore families at the close of the epoch. The aberrant primates of the Paleocene and, in addition, more typical lemurs, such as *Notharctus,* and numerous tarsioids are still present. Condylarths, abundant in the Paleocene, are now sharply reduced in numbers. However, *Phenacodus* is a familiar Lower Eocene ungulate; *Hyopsodus,* a late survivor, represented the group in America until the end of the epoch. The blunt-toothed mesonychids persisted until the end of the epoch, *Andrewsarchus* of Mongolia being a giant form; and there are still later forms which may be relict survivors of the mesonychids. *Paramys* and other rodents are present in both continents, but the order is relatively rare in the earlier European Eocene. The pantodont *Coryphodon* is found in the Lower Eocene in Europe and America alike. A second curious appearance of the notoungulates (in addition to *Palaeostylops* of the late Paleocene of Asia) is that of a single fragment from the Lower Eocene of Wyoming; except for these occurrences, notoungulates have never been reported beyond the confines of South America. Uintatheres were never present in Europe but developed further in North America and eastern Asia. The tillodont *Esthonyx* was present in both hemispheres; taeniodonts were present in the American Lower Eocene but not in Europe.

A diagnostic feature marking the beginning of the Eocene is the presence of the two great modern groups of ungulates—the perissodactyls and the artiodactyls. Both appear simultaneously and seemingly abruptly in Europe and North America in Lower Eocene times. In both groups the ordinal characters are fully established, but their areas of development are unknown, although the perissodactyls surely, and the artiodactyls probably, are of condylarth ancestry. In the early Eocene the artiodactyls are rare. The perissodactyls, on the other hand, became at once abundant; "*Eohippus*" and primitive tapiroids are common finds in the Wasatch, and close relatives were present in Europe; in North America the first titanotheres appeared in the early Eocene, and chalicotheres appeared at that time in both Europe and North America.

For the later history of the Eocene in the Holarctic area, Asiatic evidence is available, as well as that from the European and American extremities of the region. In many groups—notably the ungulates—there are marked differences between the two hemispheres which suggest regional isolation in the Middle Eocene, partially corrected by intermigration toward the end of the epoch.

Bats appear full fledged in both hemispheres in the Middle Eocene; various lemuroid, tarsioid, and aberrant primates persisted throughout the epoch, but were much reduced toward its close. Creodonts are the characteristic carnivores; some remained small but others, such as *Pterodon* and *Hyaenodon* of both hemispheres and the oxyaenid *Sarkastodon* of Mongolia, tended to grow to great size. Miacids are common in both land masses. In the late Eocene this family comes to an end, seemingly by the evolution from it of the first of the more typical fissipedes, with such forms as *Palaeoprionodon, Amphictus, Cynodictis,* and *Aelurogale* representing the forerunners of civet, weasel, dog, and cat families. Primitive sciuromorph rodents, notably the Paramyidae, are present in both hemispheres but are more common in North America; however, somewhat more advanced types appeared in both areas before the end of the Eocene. We have noted the occurrence of a rare lagomorph in the Paleocene. The next record of hares is in the Upper Eocene of eastern Asia and North America; the former continent may well have been the place of their development.

Many features in the history of nonungulate mammals in the later Eocene are thus similar in Eurasia and North America. But there were notable differences; the former region shows, on the whole, the more progressive situation. Shrews and moles are present in Europe by the end of the Eocene but do not appear in North America until later; on the other hand, the aberrant mixodectid insectivores persisted in North America in the later Eocene, as did the taeniodont *Stylinodon,* tillodonts, and palaeanodont edentates.

Although free interchange between Eurasia and North America appears to have been possible in the late Eocene, lack of connection in Middle Eocene times resulted in rather different developments in the ungulates of the two areas. Perissodactyls flourished in both regions but followed different lines of development in many respects. In North America the horses, in the form of the relatively rare genera, *Orohippus* and *Epihippus,* remained conservative in size and structure; in Europe, on the other hand, there was a rapid trend toward large size and specialization, some specimens of the three-toed *Palaeotherium* of the late Eocene being even larger than modern horses. Primitive tapiroids were present in both continents. In Europe *Lophiodon* was a typical large form; *Hyrachyus* and its relatives, the so-called running rhinoceroses, were prominent in North America. The hyracodonts, slender-limbed rhinocerotoids, and such amphibious forms as *Amynodon* inhabited both North America and Asia. In the late Eocene of all three northern continents are found the first rare specimens of the true Rhinocerotidae which were to become abundant in the Oligocene. In the late Eocene the chalicotheres and titanotheres—earlier confined to North America—flourished in Asia, and titanothere stragglers have been discovered in the late Eocene of eastern Europe.

The artiodactyls, rare in the early Eocene, show a remarkable expansion in both Eurasia and North America. Among the forms developed in the former area were anthracotheres and bunodonts suggestive of later pigs and peccaries, although apparently not the actual ancestors of these forms. Most, however, are primitive selenodonts. Cainotheres, anoplotheres, xiphodonts, and amphimerycids are the dominant European types. In the Upper Eocene there appears in North America a parallel series of artiodactyls, which belong to different families. They include some rather piglike types, primitive selenodonts ancestral to the camels, the agriochoeres, and a considerable number of pecorans of the hypertragulid group.

We have noted above some features in which the later Eocene of Asia is comparable faunally with Europe or North America or both. Imperfect, however, as our knowledge of Eocene history there is, we can see that in certain respects Asia had developed differently from the regions to west or east. Among the primitive ungulates, large condylarths and uintatheres persisted there through the Eocene. Horses were absent, but tapiroids were apparently common. Anthracotheres were very abundant in the late Eocene and may have developed in that area. *Archaeomeryx* of Asia is an exceedingly primitive ruminant.

In the Eocene appear the first representatives of the two purely marine orders of mammals, the Cetacea and the Sirenia. We are ignorant of their terrestrial forebears and cannot be sure of their place of origin. But although archaeocete whales (such as *Basilosaurus*) and a sirenian are found in the North American Eocene, most of the earliest and most primitive types are from the Mediterranean region and especially from the Mokattam beds of Egypt. This suggests that these groups originated in Africa during the earliest Tertiary. The only known terrestrial African mammals of the Eocene are the early proboscideans *Moeritherium* and *Barytherium* from the Fayûm. Not improbably, proboscideans, sirenians, and hyracoids (which appear slightly later in that area) are products of early Tertiary African evolution from a primitive subungulate stock.

In South America the Casa Mayor beds of the early Eocene give us our first adequate picture of the specialized Tertiary faunas of that continent; the Muster beds of the Middle Eocene show a slightly more advanced stage. Except for rodents and primates, all members of the typical South American fauna of later Tertiary epochs are represented. Marsupials were abundant and include not only opossums but larger and more strictly carnivorous borhyaenids (one already as large as a wolf), caenolestids, and *Polydolops* and its relatives, which paralleled the multituberculates in dentition. Edentates are represented by armadillos of a primitive nature, such as *Utaetus;* the first glyptodonts appear later in the Eocene, and sloths not until the Oligocene. *Didolodus* and similar types are condylarths related to the ancestry of the litopterns; of the latter there are genera ancestral to the two families prominent in the later Tertiary. The Notoungulata are, as later, the commonest of ungulates; the Eocene genera appear to include forms antecedent to the various subgroups of later epochs as well as numerous primitive notioprogonians. All are small, primitive, little differentiated, and with low-crowned teeth. The astrapotheres and pyrotheres, as archaic ungulates, appear to have played in South America the role of pantodonts and uintatheres of the Northern Hemisphere; large representatives of these two groups were the giants of the fauna.

OLIGOCENE

European Oligocene deposits are numerous. A portion of the Quercy phosphorites are of early Oligocene date. Ronzon in France and the Mainz basin in Germany are among the better-known localities, a large proportion of which are located in France. In Asia, Oligocene continental beds are almost entirely confined to Mongolia, with Ulan Gochu and Hsanda Gol among the main sites. The earliest of important Tertiary terrestrial African deposits is that of the Lower Oligocene of the Egyptian Fayûm. In North America the main fossil localities are those in the White River series in South Dakota and adjacent states, with successive titanothere (Chadron) and *Oreodon* and *Protoceras* beds (Brulé). In South America, Patagonia continues to be the area of interest. The Deseado beds, with the *Pyrotherium* fauna, are Lower Oligocene; the late Oligocene Colhué Huapi beds contain essentially a Miocene assemblage.

In the northern continents the Oligocene fauna has a much more modern appearance than that of the Eocene. This is not due in any measure to the appearance of new types, for hardly a third of the living families of terrestrial mammals were then present and, except in the case of the carnivores, there were few new groups introduced at this time. This more modern aspect is rather due to the extinction, during or at the close of the Eocene, of a great number of families and even orders which were characteristic of Paleocene and early Eocene days. To cite this list is to cite the greater part of the roster of the typical Paleocene mammals, including multituberculates, primitive insectivore groups, condylarths, and plesiadapid primates and, in addition, a number of the late Paleocene and Eocene families, such as oxyaenid creodonts, adapid lemurs, uintatheres, and a few of the more primitive forms among the modern ungulate orders, such as dichobunids and hyrachyids.

In the Holarctic land areas, the early Oligocene shows a considerable degree of similarity between continents, suggesting a rather free interchange, as in the late Eocene. Europe and North America offer two extremes; Asia has characteristics of both, although showing greater affinities with Europe. There are few features suggesting any great intermingling of faunal elements between the two hemispheres during the later phases of the Oligocene; most of the common characteristics appear to have resulted from late Eocene and early Oligocene interchange.

Common types include such forms as opossums, the major families of modern insectivores, and primitive beavers, which appear in the late Oligocene. In both hemispheres there appear in the Oligocene the first representatives of the great muroid group, destined to play an increasingly important role in the later epochs. Among carnivores, both hemispheres show the persistence in the early Oligocene of the last surviving group of the creodonts, in *Hyaenodon* and its relatives, later confined to tropical regions. A new development of importance is the rapid deployment of progressive fissipede families developed from the miacids. Canids of various types, mustelids, and felids, both sabertooths and false sabertooths of the *Dinictis* group, were abundant in both hemispheres.

A limited number of ungulates are common to both hemispheres. Among the perissodactyls, chalicotheres were present in Eurasia and doubtfully in North America; *Protapirus,* first of true tapirs, also appears in both areas; amynodont rhinoceroses are widespread, with *Cadurcotherium* as the European representative, *Metamynodon* the American genus. A notable perissodactyl development is the expansion of the true rhinoceroses during the Oligocene in such primitive genera as *Caenopus* in North America and *Aceratherium* in Eurasia. Among the piglike artiodactyls there are common bunodont families. *Anthracotherium* and other members of its family are abundant in Eurasia, and *Bothriodon* penetrated to America, where the group failed to survive. Pigs and peccaries make their appearance; both are found in Europe, and the peccaries reached their later American center in the Oligocene.

Among the nonungulates the Eurasian Oligocene reveals various groups absent from America: the first of viverrid carnivores, a group which never reached the Americas; ochotonids, which arrived there much later; the first dormice and dipodoids. Among the perissodactyls the only distinctive European forms are *Palaeotherium* and its relatives, which persisted until the Middle Oligocene. There were, however, numerous selenodont artiodactyls peculiar to Eurasia. *Dichobune, Cainotherium, Anoplotherium,* and *Xiphodon* are examples of archaic types, for the most part exterminated before the close of the epoch. More important, however, were forms such as *Amphitragulus* which foreshadow the later development of the higher ruminant groups of the later Tertiary.

North America shows equally distinctive features. There are various rare holdovers here of the archaic Paleocene and Eocene groups—omomyid tarsioids, apatemyid and leptictid insectivores, palaeanodont edentates, surviving paramyid rodents and their derivatives. There are, as well, ancestral pocket gophers and kangaroo rats.

Among ungulates, the perissodactyls are more prominent in North America than in Oligocene Europe. Small three-toed horses, *Mesohippus* and *Miohippus,* flourished; *Hyracodon* is a common rhinocerotoid; primitive tapiroids survived; the lowest Oligocene beds are characterized by the presence of numerous giant titanotheres, which disappeared abruptly before the Middle Oligocene. Artiodactyls are increasingly important, however. Camels, such as *Poëbrotherium,* are common; *Protoceras* represents a peculiar American family of primitive pecorans; *Leptomeryx* and other hypertragulids are common Oligocene finds; *Agriochoerus* is present. The oreodonts appear in the Oligocene and become the commonest of American ungulates; *Merycoidodon* [*Oreodon*] and other members of the family are the most frequent fossil finds in middle and late Oligocene beds.

Known Asiatic deposits show that many of the European Oligocene groups were represented in that continent, and accidents of preservation and collection may be responsible for the seeming absence of other European families. Distinctive Asiatic features include the absence of equoids of any sort, with a variety of surviving tapiroids in their place; the presence of a last lingering pantodont; the development of giant rhinoceroses, such as *Baluchitherium,* which persisted until the Miocene. Other features, implying a connection with North America in the late Eocene or early Oligocene, include the common presence of hares and of abundant giant titanotheres.

We have noted the almost complete lack of Paleocene or Eocene African terrestrial records. In the Lower Oligocene we at last find an African fauna; this, unfortunately, lies at the northern margin of the continent, in the Egyptian Fayûm, and shows evidences of invasion from Eurasia. In the known fauna there are no fissipedes, but hyaenodont creodonts such as *Pterodon* and *Hyaenodon* and a considerable variety of anthracotheres are present. These forms may be recent immigrants, since they are members of contemporary

Eurasian families. More doubtful as to origin are rodents of distinctive type, such as *Phiomys*, and catarrhine primates, unknown earlier (except for possible Burmese forms). These include primitive catarrhines such as *Parapithecus*, an early, manlike ape. Definitely autochthonous forms are represented by *Arsinoitherium*; a considerable variety of hyracoids; and proboscideans, which now include not only the archaic *Moeritherium* but the mastodonts, *Palaeomastodon* and *Phiomia*.

The Deseado beds of Patagonia show an important phase in the development of the isolated South American fauna. Except for marsupials of the *Polydolops* type and a few families of ungulates, all Eocene groups are still present. Marsupial carnivores are numerous and varied (one was larger than a cave bear). Edentates were still rare but were increasing in variety, for glyptodons, in addition to armadillos and ground sloths of modest size, now made their appearance. The native ungulates had already reached a climax in size and variety; more than a dozen families were represented. Both major families of litopterns were advanced in their development. Notoungulates were numerous and exceedingly varied; high-crowned teeth had developed in many lines. Most characteristic of the notoungulates of the age were *Leontinia* and related types. Huge astrapotheres were present, and the giant of the fauna was the elephant-like *Pyrotherium*.

A startling phenomenon is the abrupt appearance of caviomorph rodents. There is no trace of rodents in the Eocene of the continent, and it is highly improbable that these forms, so successful from this time on, could have been present earlier and escaped discovery. Presumably they had successfully crossed a water gap from North America.

Oligocene marine mammals are rare, perhaps because of the rarity of appropriate types of sediments. Sirenians are known from the European Oligocene. There is a surviving archaeocete. We are sure that the differentiation of more modern whale groups was well advanced, for a lone squalodont represents a primitive toothed-whale group, and a cetothere announces the presence of early whalebone whales.

We may note here what little is known of the Tertiary of Australia. Until recently nothing was known of Australian mammals earlier than the Pleistocene, except for the skull of a phalanger, *Wynyardia*, from Tasmanian beds possibly of Oligocene age. In recent years, however, Tertiary beds have been explored, and a modest number of specimens found of Pliocene, Miocene, and possibly Oligocene age. All, however, appear to belong to families still present in Australia and tell us little of the early radiation of the marsupials which compose most of the Pleistocene and Recent fauna, and nothing of the history of the curious monotremes.

MIOCENE

The Miocene is an epoch of considerable length and one represented by numerous and varied faunas in a number of regions. European localities are most abundant in France and Germany. St. Gérand, Sansan, and Simorre are among famous French localities of varied ages, and the various Molasse deposits and those of the Mainz basin and Steinheim are among the familiar fossiliferous beds of the German region. In recent years a considerable Miocene fauna has been found on Rusinga Island and other regions of Kenya, and there are a few other African localities. In Asia the Miocene is present in but few places. The Loh and Tung Gur beds represent the last of the sequence of Mongolian deposits; the Bugti beds of Baluchistan have yielded a few mammals. In the late Miocene, however, we see the beginning of the deposits of the Siwalik Hills in India, which continue from this date to the Pleistocene.

In North America a major portion of the Miocene beds lies in the Great Plains area; they include such well-known deposits as the Harrison, Rosebud, Sheep Creek, Marsland, and Pawnee Creek. In Oregon the fossiliferous portion of the John Day beds is early Miocene. Florida in the Miocene has, at Thomas Farm, yielded the only good fauna found in eastern America in any stage of the Tertiary.

In South America the Miocene opened with a marine invasion represented by the Patagonian formation; following this were deposited the Santa Cruz of Patagonia, with a rich and varied fauna, and, after a gap, late Miocene beds in both Argentina and Colombia.

The Miocene faunas are greatly "modernized" as compared with conditions at the beginning of the preceding epoch. During or at the end of the Oligocene there had become extinct a score of families of more or less archaic types, particularly among the ungulates. Among the victims were the last amblypods in Asia; the pyrotheres in South America; older proboscidean types; titanotheres, hyracodonts, and lophiodonts; primitive artiodactyls and ruminants such as anoplotheres, xiphodonts, and amphimerycids; several primitive rodent families; the last of the archaic primate groups which characterized the Paleocene and Eocene. As a result of these extinctions and of new evolutionary developments we find that, in the known Miocene faunas as a whole, half of the living families of terrestrial mammals were then in existence and that these likewise constituted a good half of the faunal assemblage. If the peculiar South American types be excluded, the modern note is more marked; in the Northern Hemisphere only a quarter of the fauna is made up of families which have failed to survive.

We have noted the nearly complete isolation of the two hemispheres during much of the Oligocene; as a result, the Palaearctic and Nearctic faunas at the opening of the epoch differed widely in many respects. There are, however, indications of a modest degree of intermigration during the early Miocene and a very considerable interchange toward its close, which tended to reduce to some degree the faunal differences. The evidence suggests free communication between Africa and Eurasia during this epoch, but South America and Australia were still separate evolutionary areas.

As before, the differences between North America and Eurasia were relatively small among the nonungulate orders, although even in cases where families were common to both areas the genera present were usually distinct. The creodont carnivores had disappeared except for lingering hyaenodonts in the Old World tropics. Varied dogs were present, including large "bear dogs" and, in the late Miocene, such forms as *Hemicyon*, which appear close to the origin of the Ursidae. Mustelids and felids, including true sabertooths and pseudo sabertooths, such as *Nimravus* of North America and *Pseudaelurus* in Europe, flourished. Procyonids appear in both areas. In the Old World the civets persisted, and from them had evolved, by the end of the Miocene, the first hyenas. Lagomorphs, rats, and primitive beaver-like forms (*Steneofiber, Palaeocastor*) are among the common rodents.

A single tarsioid in the early Miocene is the last native primate in North America. In the Eastern Hemisphere, however, the evidence, scanty though it is, tells of a considerable amount of evolution among the higher primates during the Miocene. *Pliopithecus*, a primitive gibbon, and *Dryopithecus*, an advanced manlike ape, appear in the European Miocene, and the discovery of genera of apes in the early Miocene of Africa suggests that this continent may have been an important center of primate evolution.

As in the Oligocene, there is a marked difference between the two northern continental areas in ungulate evolution, and there are marked differences between early and late faunas. Among the perissodactyls, rhinoceroses are present in both hemispheres. *Diceratherium* is the common American genus in the early Miocene; *Baluchitherium* persisted into the Miocene of Asia; more normal genera, *Aceratherium* and similar forms, are found in both areas but are more common in Eurasia. *Macrotherium* and *Moropus* represent the chalicotheres in Eurasia and North America, respectively; tapirs are present in both continents.

Miocene horse history is mainly an American story. *Miohippus* persisted into the early Miocene and gave rise to *Anchitherium* and *Hypohippus* as larger forms of the same primitive type. *Parahippus*, first of horses with high-crowned teeth and reduced side toes, appeared at the beginning of the Miocene, and still more advanced genera, *Merychippus* and *Pliohippus*, are characteristic of the American late Miocene. With the extinction of the palaeotheres, Eurasia had become barren of equoids, but *Anchitherium* successfully migrated to the Eastern Hemisphere and is present there in many Miocene localities.

The two areas have little in common as regards artiodactyls. *Dinohyus* of the North American early Miocene is the last entelodont. The common bunodonts were swine in Eurasia, peccaries in North America; *Listriodon* and *Palaeochoerus* were representative piglike forms. Anthracotheres are absent in North America and in Europe, as well, in the later Miocene, but they survived in southern Asia and in Africa. In Europe, *Cainotherium* was present as the last of nonruminant selenodonts in that area; in America, in sharp contrast, there were abundant and varied camels, such as *Stenomylus* and *Oxydactylus*, and numerous oreodonts, including *Promerycochoerus*, and *Merychyus*. In both hemispheres, primitive ruminants of the tragulid or hypertragulid groups survive but are uncommon.

In the Miocene are found the first stages in the deployment of higher ruminants. The earliest Miocene of both hemispheres witnesses the appearance of primitive cervoids: in Europe, *Dremotherium* and *Amphitragulus* and, somewhat later, *Palaeomeryx;* in America, *Blastomeryx*, followed by other deerlike forms. Later in the Miocene appear the first (and relatively rare) representatives of the other living pecoran families. In Eurasia are primitive antelopes such as *Protragocerus;* primitive deer, as *Dicrocerus;* and *Palaeotragus*, the first giraffid. These forms are absent from America; instead, we find *Merycodus* as the first of the prongbuck family.

A new element in the ungulate fauna of the northern continents is the advent of the proboscideans, previously known only from Africa. *Dinotherium*, with its peculiar lower tusks; bunodont mastodonts, such as *Gomphotherium* and *Serridentinus;* and the lophodont *Mastodon* all appeared in Europe early in the Miocene and persisted throughout that epoch. In America their advent was retarded until the late stages of the Miocene; *Dinotherium* failed to reach America.

The Miocene Santa Cruz fauna is the best known of South American mammal assemblages. Marsupial carnivores and caviomorph rodents were numerous and varied. Edentates were now abundant and were a large proportion of the fauna; forms, mostly of relatively small size, now represented not only armadillos, but anteaters, glyptodonts, and the major families of ground sloths. Litopterns were at the peak of their development, including, for example, *Diadiaphorus* and the pony-like *Thoatherium* among the proterotheres and *Theosodon* of the more heavily built macraucheniid family. Pyrotheres were extinct, but *Astrapotherium* was a giant end form of another archaic line. Notoungulates were now much less varied in the number of families represented and were, in general, still small in size; in number of species and individuals present, however, they bulk large in the faunal picture. *Homalodotherium*, the toxodonts *Adinotherium* and *Nesodon, Pachyrukhos*, and *Protypotherium* are representative genera.

In *Homunculus* and its contemporaries we see representatives of the platyrrhine monkeys of South America. The origin of this group is as puzzling as that of the caviomorph rodents. Presumably the advent of the platyrrhines took place by island hopping from the north in the late Oligocene, as in the case of the rodents earlier. But primate remains are rare also in later South American sediments, and it is possible that ancestral primates were present earlier in the Tertiary in more wooded tropical areas.

Marine mammal-bearing deposits are present in many regions of the world; the Calvert Cliffs of Maryland, the Patagonian beds of South America, the Aquitanian basin of France, and the Antwerp basin of Belgium are among the prominent sources of fossil marine mammals. Lingering archaeocetes were present in the early Miocene, but the epoch was one of rapid expansion of higher cetacean groups; the order was seemingly never more abundant or varied than at that time. Nearly every family of toothed and whalebone whales had appeared by the end of the Miocene. The squalodonts, primitive toothed whales, are common even in the earliest Miocene; and dolphins, ziphiids, and sperm whales are found in increasing numbers in later formations. The cetotheres, primitive whalebone whales, were abundant in the late Miocene, and there were even rare representatives of the living families of this suborder. *Halianassa* and other sirenians were present. The Miocene is further notable for the appearance of the first pinnipeds, including both eared and earless seals, and the first walrus, *Prorosmarus*.

PLIOCENE

The Pliocene is a relatively short epoch. The most characteristic Pliocene fauna is that of the early part of the epoch, known in Eurasia as the Pontian age (and considered by French workers as late Miocene). Deposits containing this fauna are found widely distributed throughout the Mediterranean region and in eastern and central Europe; Eppelsheim, various localities in the Vienna basin, and Pikermi and Samos in the Aegean are among the best-known localities. To the eastward this same fauna extends through Persia (Maragha) and the Siwaliks to rich early Pliocene deposits in Honan and Shansi in China. Somewhat later European levels are best represented in southern France, as at Perpignan and Montpellier; in India the Siwalik series continues throughout the Pliocene. The Pliocene of Africa is poorly known.

In North America there are numerous Pliocene deposits, the greater part of them in the Great Plains region. The earlier part of the epoch is best represented in the Clarendon beds of Texas; a later stage is represented by such formations as the Hemphill of the Texas Panhandle and equivalent beds in Kansas and Nebraska.

In South America the long series of Patagonian sediments ceases at about the end of the Miocene, and the principal Pliocene areas are found farther to the north in Argentina. In the Andes foothills are beds which cover nearly the whole length of the Pliocene; Monte Hermoso in the southern Pampas and the Entrerian region of northeastern Argentina also have characteristic deposits of this epoch.

The Pliocene fauna of the Northern Hemisphere is in great measure one of modern type. There had been a considerable elimination of old families, particularly of ungulates, during or at the close of the Miocene. The last amynodonts, entelodonts, caenotheres, and hypertragulids had been eliminated in the northern continental areas; the last archaic archaeocetes had disappeared from the oceans; and in South America astrapotheres, homalodotheres, notohippids, and interatheres are absent from the ungulate roster. The fauna is in the main one of the modern families. Excluding South America and Australia, 80 per cent of the families of terrestrial mammals then present are types still extant; and, while a number of living families are not represented by Pliocene fossils, these are mainly rare and obscure groups. The Pliocene faunas further show much similarity generically to modern times; many familiar living genera, particularly among the carnivores and rodents, had arisen before the period was over, and in other cases generic differences between modern and Pliocene types are small.

There is little evidence of faunal interchange between Eurasia and North America at the beginning of the Pliocene. Old and New Worlds show many faunal similarities, but these can be attributed in the main to persistence of types interchanged during the late Miocene, and the genera present are usually distinct in the two regions.

The carnivores have run a parallel course in the two continents. There are numerous canids and mustelids. Felids close to the living genera are present, in addition to sabertooths. Primitive bears have appeared. Hyenas are prominently represented in Eurasia. There is a variety of bunodont mastodonts in both continents. *Hypohippus*, descended from *Anchitherium* of the Miocene, is present in Asia as in North America; *Hipparion*, a slightly built, three-toed horse, reached Eurasia from North America and became widespread and common there. Rhinoceroses are present in both regions but are rare in North America; a few palaeomerycids survive in both hemispheres.

In Pontian times a nearly homogeneous fauna is found all the way from Spain to China and may well have been characteristic of Africa as well, for it is one quite similar in many respects to that found in Africa today. Elements in common with North America were noted above, but the presence of *Hipparion* as a characteristic Pontian form deserves emphasis. A striking feature is the abundance of bovids; several score of genera of antelopes have been described from the Pontian. Primitive genera of deer are common, as are the pigs and, in Asia, varied giraffids, including giant-horned forms. In Eurasia, chalicotheres and the peculiar proboscidean *Dinotherium* were still present, and in the eastern area primitive elephants, such as *Stegodon*, make their appearance.

In North America we find an ungulate assemblage still differing markedly from that of Europe. Not a bovid is present, or a true cervid. Instead, there is a great radiation of antilocaprids with varied types of horns, a last protoceratid, and a few final representatives of the oreodonts. Camels still flourished in a variety of forms, and peccaries were not uncommon. Rhinoceroses persisted, and horses of varied types were especially abundant, including holdovers from the Miocene, as well as the *Hipparion* group and types progressing toward the modern one-toed forms.

There is evidence that the South American land bridge was now in process of re-establishment after the lapse of most of the Tertiary, for a first ground sloth had appeared in North America by mid-Pliocene times and a glyptodont at the end of the epoch. In South America the Pliocene faunas show evidence of interchange in the presence of procyonids, which may have used a Central American island chain as stepping stones. The fauna was otherwise purely native in character. A number of Miocene families of ungulates had disappeared, but other types of hoofed mammals still flourished, and the edentates and the caviomorph rodents were highly varied, constituting about a third of the entire fauna; in many instances there was a strong tendency toward large size. The borhyaenid marsupials still functioned as the carnivores of the faunas, and the evolution of a marsupial sabertooth is an interesting development.

In the sea the archaeocetes were extinct, and the primitive squalodont whales were reduced and had vanished by the end of the Pliocene; ziphiids were fewer, and the various dolphin types were less numerous. Sperm whales, however, continued in fair numbers. Among the whalebone whales, a few final members of the cetotheres survived; they were being replaced by more abundant balaenids and balaenopterids.

PLEISTOCENE

In the past, the term Pleistocene was most commonly restricted to all or a part of the Ice Age, with its succession of glacial and interglacial stages. Recently, however, there has been rather general agreement that the term be extended to include earlier, pre-glacial beds more or less transitional from the Pliocene. Even so, the Pleistocene remains the shortest of all the Cenozoic epochs, with an extent of not more than two million years. We may consider this early stage briefly before turning to the more typical Pleistocene.

The early Pleistocene of Europe is the Villafranchian stage, best represented in the Arno Valley of Italy and other localities in the southern part of that continent; beds of similar age are present in the Siwalik Hills of India and in China. In America this is the Blancan stage, represented by such deposits as the Blanco of Texas and those of Hagerman, Idaho, and Rexroad, Kansas. The fauna includes numerous surviving Pliocene forms. The period was apparently one of rising land levels, with the re-establishment of the Bering Straits bridge, and considerable faunal exchange of temperate-zone faunal elements. The "index fossil" of the Villafranchian-Blancan stage was the modern horse genus, *Equus*, which, arising in North America, spread rapidly through the Old World.

In South America the Chapadmalal beds of the southern pampas likewise represent a time of transition (although workers there prefer to consider them as a final phase of the Pliocene, rather than early Pleistocene). The earlier Pliocene "natives" are still present, but in these deposits are also found numerous invaders from the north. Felids, normal and sabertoothed, and bears, as well as procyonids, are among the invading carnivores; horses, peccaries, and deer have entered to compete with native ungulates, and cricetid rodents to rival the caviomorphs. The end is in sight for much of the native population.

Following the Villafranchian-Blancan stage, we enter the period of the Ice Age of northern regions. In the Pleistocene we reach a climax in the development of steep climatic gradients—a development which appears to have been in process of establishment during the later Tertiary. Arctic, temperate, and tropical zones were well differentiated. Their boundaries fluctuated considerably during the course of the Pleistocene. The evidence indicates that there were four successive southward advances of ice caps which covered much of Europe and North America. Contemporaneous with these advances were the development of great areas of tundra, local glaciations in more southerly mountain masses, and a shifting of climatic zones so that areas such as the Sahara and the deserts of the southwestern United States were, for the time, well watered. Between the periods of glaciation were three interglacial stages, during which zonal boundaries shifted far to the north, so that essentially tropical conditions were to be found over great parts of Europe and North America.

As a consequence of these climatic conditions, our treatment of faunal assemblages in the typical Pleistocene must be on a somewhat different basis from that used in the discussion of earlier Tertiary epochs. There we were able to consider the fauna of a continental area as essentially a unit; in the Pleistocene, however, divisions by climatic zones are prominent in the faunal picture. In both Eurasia and America the Arctic fauna was (and is) a very meager one; the temperate-zone fauna was much restricted in many ways, particularly as regards ungulates; various groups once widespread over much of the northern continental areas had retreated to the tropics, to reappear in the north, if at all, only during the warm interglacial stages.

Complete extinction of mammalian groups—families or higher categories—had become relatively rare during the Pliocene or Villafranchian. Definite extinction may be noted only in the case of borhyaenid marsupials in South America, which were replaced by incoming fissipede placentals; of the last tropical survivor of the old creodonts; of the proterotheriid litopterns in South

America; and, in North America, of two native ruminant families—the oreodonts and the protoceratids. There is, however, a very significant reduction in many groups—a reduction in numbers of genera or of areas of distribution. Rhinoceroses had made their last appearance in North America; *Dinotherium* had gone from Eurasia; in South America the macraucheniid litopterns were reduced in numbers; and still further reduction took place among the notoungulates. The variety of mastodonts had decreased in the early Pleistocene. In the Old World, anthracotheres were close to extinction, rhinoceroses, tragulids, and giraffids reduced. The typical Pliocene horses had become extinct, except for sparse records of *Hipparion* in the Villafranchian.

In any region where the fauna is adequately known, we find close relatives of mammals now living there, although the Pleistocene forms are often larger than their modern descendants. But, in addition, there are invariably numerous animals now extinct; and most of these are of large size.

The Pleistocene giants include representatives of almost every order of mammals: giant deer and giant beaver in both Eurasia and North America; the largest of proboscideans and of edentates. Large types were present among the South American rodents. In Australia there were giant kangaroos and wombat-like forms and even an "outsized" *Ornithorhynchus.* Nor were the primates exempt, for Madagascar lemurs included forms as large as the great apes.

Apart from these large forms, since extinct, the Pleistocene assemblage—particularly that of the later Pleistocene—was one with an air of familiarity about it. The time since the close of that epoch has been too short for any major evolutionary advance to have taken place, and hence it is not surprising that not merely most of the genera but also many of the species living today were common members of the Pleistocene faunas.

In the Pleistocene the northern continents were more united faunally than they had been since the earliest Tertiary. The Bering Strait bridge appears to have been in existence during much of the Pleistocene, permitting considerable intermigration. Climatic conditions, however, were a strong limiting factor here. Arctic types and temperate-zone forms accustomed to cold climates and rugged terrain could cross this region readily; plains animals would find passage difficult; tropical forms were barred.

The Palaearctic and Nearctic regions had, in the Pleistocene as today, many groups in common and with generic identity in many cases. A few, such as the woolly mammoth and the musk oxen, found from England to New Mexico, were arctic types; others were groups familiar in temperate zones or climatically ubiquitous. Among essentially temperate-zone types of Eurasian origin we may note the deer and more northerly types of bovids, such as sheep, bison, and goats; the first reached North America in early Pleistocene times, and the bison, a mid-Pleistocene immigrant, became the commonest of American ungulates; and, while true goats did not make the passage, the related *Oreamnos,* the mountain "goat," did.

Groups already present in both areas and, in general, more tolerant of changing climatic surroundings are numerous. Such include varied sciuromorph and myomorph rodents; lagomorphs; canids; true cats, large and small; persistent sabertooths; many mustelids. Pliocene mastodont types persisted into the early Pleistocene in both continents, and the true *Mastodon* was a late survivor in North America. Elephants, presumably of Asiatic origin, made the crossing to North America, and several types of mammoths, in addition to the woolly variety, were present in Old and New Worlds. Tapirs were present in both Eurasia and the Americas; true horses of the genus *Equus* flourished in both areas; camelids, still present in their American homeland throughout the Pleistocene, made the passage to Eurasia, as well as to South America.

Despite the many resemblances due to earlier as well as to Pleistocene interchanges, Old and New World faunas differed in a number of respects. Some of the contrasting types, such as the woolly rhinoceros and the giant tundra-dwelling *Elasmotherium* of Eurasia, were cold-climate forms. Most, however, were heat-loving animals, many of them relatively recent in their development; even in the Pliocene, it would seem, the Siberian-Alaskan bridge was a difficult route for such animals to travel and in the Pleistocene was impossible.

With the coming of the Ice Age, many elements of the Pliocene Eurasian fauna were driven southward. A few of these forms, such as the hippopotami, hyenas, and southern genera of rhinoceroses ventured north again during the warm interglacial stages. Others never returned. They were, for the most part, confined to the southern regions of Asia and of Africa. Many of the types found in the Pleistocene in these areas have survived to modern times. Among these may be mentioned a varied series of primates, including great apes, cercopithecid monkeys, and lemurs; abundant antelopes; viverrids; pangolins; aardvarks. As has been said, this tropical fauna is markedly similar to the older Pontian fauna once widespread in more northerly regions. This similarity is enhanced by the persistence in the tropics during the Pleistocene of a number of archaic mammalian types which have since become extinct, such as *Dinotherium* and the last anthracotheres and chalicotheres.

In North America, as in Eurasia, indigenous mammalian types persisted in the Pleistocene. Such forms were, however, fewer in number than in the Old World; the more prominent forms are the peccaries, native geomyoid rodents, and antilocaprids, of which several genera were present. More notable was a considerable influx of South American mammals which had traveled northward over the Central American land bridge. A variety of ground sloths—*Megatherium, Paramylodon, Megalonyx, Nothrotherium*—were common mammals; and glyptodonts and armadillos, large and small, were present in the Gulf region. Of caviomorph rodents, the capybara gained a momentary foothold in the north, but only the porcupines succeeded in establishing themselves. It is not improbable that the Pleistocene and Recent opossums of North America are reimmigrants from the south.

A great host of northern invaders entered South America in the typical Pleistocene and are abundant fossils in such superficial deposits as those of the Argentine and Uruguayan pampas and the Tarija beds of Bolivia. Of carnivores, the mustelids and dogs had joined the procyonids, cats, and bears already present and presumably made havoc among the native animals. Competition of newcomers with the native herbivores was increased with the incoming, during the typical Pleistocene, of tapirs, peccaries, and llamas to join the horses and deer, which had entered somewhat earlier. Mammoths failed to gain entry, but mastodonts were present, and among the smaller mammals there were numerous invading cricetids, squirrels, and lagomorphs.

Of the old natives of South America, the caviomorph rodents, the monkeys, and the opossums appear to have been little affected. The edentates were seemingly in vigorous condition; for, in addition to the smaller types—armadillos, anteaters, tree sloths—the Pleistocene fauna included a variety of ground sloths and glyptodons, unaware (so to speak) of their approaching extinction. On the other hand, the native ungulates were already close to the vanishing point. Of the great array of families and orders present in the earlier epochs, only a few genera were still present—these included the litoptern *Macrauchenia* and the notoungulates *Toxodon* and *Typotherium.* Pleistocene and Subrecent cave deposits in the West Indies indicate that by some route a variety of small sloths and caviomorph rodents had reached those islands, probably during the late Tertiary, and flourished during the Pleistocene; except for a few rodents, all are now extinct.

As noted earlier, the Tertiary fauna of Australia is poorly known. The fauna as seen in the Pleistocene was in great measure that found today: monotremes; a great variety of marsupials, both carnivorous and herbivorous; of terrestrial placentals only a series of rats which had filtered down the East Indian island chain during the late Tertiary; and—presumably fellow entrants in the late Pleistocene—man and dog. A feature of the Pleistocene here, as elsewhere, is the development of giant types, such as the marsupials *Diprotodon* and *Thylacoleo.*

Man appears to have been essentially a Pleistocene development. His homeland, in both australopithecine and true human stages, was the Eurasian-African land mass, and this area marks the limit of his spread for most of the Pleistocene. The date of human entry into Australia is uncertain; America, as far as known, was not reached until late Pleistocene times.

RECENT

The Recent fauna of the world is marked, apart from the spread of mankind, by a relative paucity in its mammalian population. There are presumably no living types which were not also present in the Pleistocene, but many forms then present are now extinct. These were mainly mammals of large or more than average size. We may list the more conspicuous of the absent groups: the giant marsupials and monotremes of Australia; the sabertooths; the last of the South American ungulates; the ground sloths and glyptodons; all the proboscideans except the two surviving elephants; the chalicotheres; the sivatheres; the anthracotheres; giant cervids; giant beavers. Regionally, too, there has been much extinction among groups formerly widespread. The range of the surviving proboscideans is a greatly restricted one. Camels and tapirs disappeared from North America, peccaries from much of this region, horses from both Americas, rhinoceroses and hippopotami from much of Eurasia.

The reasons for this extinction of large mammals is difficult to discern. It was not altogether a matter of climatic vicissitude due to the Ice Age, for many of the extinct proboscideans survived to the end of the Pleistocene; and horses, camels, and ground sloths persisted in the American Southwest until very late times. Nor can we believe that disease can have been a major factor; for a disease which would attack such a variety of large animals and leave small ones untouched would be peculiar indeed. The spreading of mankind, *Homo sapiens,* is the only new feature seen in the late Pleistocene. Man may perhaps be responsible in some degree for this catastrophe; but his influence at the most must have been an indirect one.

CLASSIFICATION OF VERTEBRATES

Below is given a classification of vertebrates which includes a comprehensive list of genera known as fossils, together with a brief record of the ages and areas of their occurrences. It will be understood that such a compilation cannot be completely critical or exhaustive and hence undoubtedly contains inaccuracies and omissions. To place the fossil forms in proper perspective, all families of vertebrates are listed (with their present distribution) whether they include fossil forms or not. Fossil genera are listed alphabetically in each family.

ABBREVIATIONS

Geological

Carb., Carboniferous; Cret., Cretaceous; Dev., Devonian; Eoc., Eocene; Jur., Jurassic; L., Lower; M., Middle; Mioc., Miocene; Miss., Mississippian; Olig., Oligocene; Ord., Ordovician; Paleoc., Paleocene; Penn., Pennsylvanian; Perm., Permian; Pleist., Pleistocene; Plioc., Pliocene; R., Recent; Sil., Silurian; SubR., Subrecent; Tert., Tertiary; Trias., Triassic; U., Upper.

Geographical

Af., Africa; Ant., Antarctica; Arc., Arctic; As., Asia; Atl., Atlantic; Aus., Australia; CA., Central America; CAs., Central Asia; Cos., Cosmopolitan; EAf., East Africa; EAs., Eastern Asia; EInd., East Indies; EEu., Eastern Europe; Eu., Europe; Gr., Greenland; Ind., India; Indo-Pac. Oc., Indo-Pacific Ocean; Mad., Madagascar; Maurit., Mauritius; Med., Mediterranean; NA., North America; NAf., North Africa; NAs., Northern Asia; NAtl., North Atlantic; NNA., Northern North America; NOc., Northern Oceans; NPac., North Pacific; NZ., New Zealand; Oc., Ocean(s); Pac., Pacific; SA., South America; SAf., South Africa; SAs., Southern Asia; SAtl., South Atlantic; SOc., Southern Oceans; SPac., South Pacific; Spits., Spitsbergen; SWAs., Southwest Asia; WAf., West Africa; WAs., Western Asia; WInd., West Indies; WNA., Western North America.

CLASS AGNATHA

SUBCLASS MONORHINA

ORDER OSTEOSTRACI. **Tremataspidae,** *Timanaspis* L. Dev. Eu. *Tremataspis* [*Odontotodus Stigmolepis*] U. Sil. Eu. **Dartmuthiidae,** *Dartmuthia* [*? Lophosteus*] U. Sil. Eu. *Didymaspis* L. Dev. Eu. *Oeselaspis* [*? Trachylepis*] U. Sil. Eu. *Saaremaaspis* [*Dasylepis ?Dictyolepis Rotsikuellaspis*] U. Sil.-L. Dev. Eu. **Hemicyclaspididae (Ateleaspidae),** *Aceraspis Hemicyclaspis* [*Hemiteleaspis*] *Hirella* [*Micraspis*] L. Dev. Eu. *Tuvaspis* L. Dev. NAs. *?Witaaspis* U. Sil. Eu. **Sclerodontidae,** *Sclerodus* [*Eukeraspis*] L. Dev. Eu. **Cephalaspidae,** *Benneviaspis* L. Dev. Eu. Spits. *Boreaspis* L. Dev. Spits. *Cephalaspis* [*Alaspis Camptaspis Escuminaspis Eucephalaspis Mimetaspis Pattenaspis Scolenaspis Zenaspis*] U. Sil.-L. Dev. Eu., L., ?M. Dev. Spits., L. Dev. EAs., L.-U. Dev. NA. *Ectinaspis Hoelaspis* L. Dev. Spits. *Nanpanaspis* L. Dev. EAs. *Procephalaspis* U. Sil. Eu. *Securiaspis* L. Dev. Eu. Spits. *Stensiöopelta* L. Dev. Eu. *? Tannuaspis* L. Dev. NAs. *Tegaspis* L. Dev. Spits. *Thyestes* [*Auchenaspis*] U. Sil.-L. Dev. Eu. **Kiaeraspididae,** *Acrotomaspis Axinaspis* L. Dev. Spits. *?Ilemoraspis* L. Dev. EEu. *Kieraspis* L. Dev. Spits. *? Nectaspis* L.-M. Dev. Spits. **Galeaspidae,** *Galeaspis* L. Dev. EAs.

OSTEOSTRACI INCERTAE SEDIS: *Turinia* [*Cephalopterus*] L. Dev. Eu.

ORDER ANASPIDA. **Jamoytiidae,** *Jamoytius* M. Sil. Eu. **Birkeniidae,** *Birkenia* M.-U. Sil., L. Dev. Eu. *Ctenopleuron* U. Sil. NA. *Pharyngolepis Pterygolepis* [*Pterolepidops Pterolepis*] *Rhyncholepis* L. Dev. Eu. *Saarolepis* [*Anaspis*] U. Sil. Eu. **Lasaniidae,** *Lasanius* U. Sil.-L. Dev. Eu. **Euphaneropsidae,** *Euphanerops* U. Dev. NA. **Endeiolepidae,** *Endeiolepis* U. Dev. NA.

ORDER CYCLOSTOMATA

SUBORDER PETROMYZONTOIDEI. **Petromyzontidae** R. Oc. Eu. Aus. NZ. NA. SA.

SUBORDER MYXINOIDEI. **Myxinidae (Bdellostomatidae Eptatretidae)** R. Oc.

SUBCLASS DIPLORHINA

ORDER HETEROSTRACI (PTERASPIDOMORPHA). **Astraspidae,** *Astraspis* M., ?U. Ord. NA. *?Palaeodus* L. Ord. Eu. *Pycnaspis* M.-U. Ord. NA. **Eriptychiidae,** *?Archodus* L. Ord. Eu. *Eriptychius* M. Ord. NA. **Cyathaspidae,** *Allocryptaspis* [*Cryptaspidisca Cryptaspis*] L. Dev. NA. *Americaspis* [*Palaeaspis*] U. Sil. NA. *Anglaspis* [*Fraenkelaspis*] L. Dev. Eu. Spits. *?Aphataspis* L. Dev. NAs. *Archegonaspis* [*Aequiarchegonaspis Lanaspis*] U. Sil., ?L. Dev. Eu., U. Sil. NA. *Ariaspis* L. Dev. NA. *Bothriaspis* L. Dev. Eu. *Ctenaspis* L. Dev. Eu. Spits. *Cyathaspis* [*Diplaspis*] U. Sil. NA., U. Sil.-L. Dev. Eu., L. Dev. NAs. *Dikenaspis* L. Dev. NA. *Dinaspidella* [*Dinaspis*] *Homaspidella* [*Homalaspis Homaspis*] L. Dev. NA. Spits. *Irregulareaspis* [*Dictyaspidella Dictyaspis*] L. Dev. Eu. Spits. *?Kallostracon* L. Dev. Eu. *Kiangsuaspis* ?U. Sil. EAs. *Lauaspis* U. Sil. Eu. *Listraspis Pionaspis* L. Dev. NA. *Poraspis* [*Holaspis*] L. Dev. Eu. Spits. *Ptomaspis* L. Dev. NA. *Putoranaspis* L. Dev. NAs. *Seretaspis Steinaspis* L. Dev. Eu. *Tolypelepis* [*Tolypaspis*] U. Sil.-L. Dev. Eu. *Vernonaspis* [*? Anatiftopsis Eoarchegonaspis*] U. Sil.-L Dev. NA. **Amphiaspidae,** *Amphiaspis* L. Dev. NAs. *Eglonaspis* L.-M. Dev. NAs. *Hibernaspis Pelurgaspis* L. Dev. NAs. **Corvaspidae,** *Corvaspis* U. Sil.-L. Dev. Eu., L. Dev. Spits. **Pteraspidae,** *Doryaspis Dyreaspis Ennosviaspis Giganthaspis* L. Dev. Spits. *Glossoidaspis* L. Dev. NA. *Grumantaspis* L. Dev. Spits. *Protaspis* [*Cyrtaspidichthys Cyrtaspis Eucyrtaspis Europrotaspis*] L. Dev. NA. Eu. Spits. *Pteraspis* [*Archaeoteuthis Althaspis Belgicaspis Brachipteraspis Brotzenaspis Cymripteraspis Lerichaspis Loricopteraspis Mylopteraspis Palaeoteuthis Parapteraspis Penygaspis Plesiopteraspis Podolaspis Protopteraspis Pseudopteraspis Rhinopteraspis Scaphaspis Simopteraspis Steganodictyum Zascinaspis*] L. Dev. Eu. Spits. NAs. NA. *Yoglinia* M. Dev. Eu. **Traquairaspidae,** *Traquairaspis* [*Lophapiscis Lophaspis Orthaspis Phialaspis*] U. Sil. Eu., U. Sil.-L. Dev. NA., L. Dev. Spits. *? Yukonaspis* L. Dev. NA. **Cardipeltidae,** *Cardipeltis* L. Dev. Spits. NA. **Drepanaspidae (Psammosteidae),** *Aspidosteus* [*Aspidophorus Obruchevia*] U. Dev. EEu. *Drepanaspis* L. Dev. Eu. *Ganosteus* M. Dev. Eu. *Karelosteus* U. Dev. Eu. *Psammolepis* M. Dev. Eu. Spits. *Psammosteus* [*Dyptychosteus Megalopteryx Placosteus*] M.-U. Dev. Eu., ?M. Dev. Spits., U. Dev. NA. *Psephaspis* L. Dev. NA. *Pycnosteus Schizosteus* [*Cheirolepis Microlepis*] M. Dev. Eu. *Strophispherus* [*Oniscolepis*] U. Sil. Eu. *Tartuosteus* M.-U. Dev. Eu. *Tesseraspis Weigeltaspis* L. Dev. Eu. **Obliaspidae** L. Dev. NAs.,.*Gunaspis* L. Dev. Eu. *Obliaspis* [*Menneraspis*] L. Dev. NAs. *Sanidaspis* L. Dev. NAs. *Siberiaspis* L. Dev. NAs. **Polybranchiaspidae,** *Polybranchiaspis* L. Dev. EAs.

?ORDER COELOLEPIDA (THELODONTI). **Thelodontidae,** *Bystrowia* U. Sil. Eu. *Coelolepis* U. Sil.-L. Dev. Eu. *Lanarkia Phlebolepis* U. Sil. Eu. *Thelodus* [*Thelolepis Thelolepoides Thelyodus*] M. Sil.-L. Dev. Eu., U. Sil.-L. Dev. NA., L. Dev. Spits.

?AGNATHA INCERTAE SEDIS

Coscinodus Dipnoites Gompholepis Gyropeltus Palaeosteus Rabdacanthus Rhabdiodus Tylodus U. Sil. Eu.

CLASS PLACODERMI

ORDER PETALICHTHYIDA. **Macropetalichthyidae,** *Ellopetalichthys* U. Dev. NA. *Epipetalichthys* M.-U. Dev. Eu. *Lunaspis* L. Dev. Eu. *Macropetalichthys* [*Acanthaspis Agassizichthys Heintzaspis Macropetalicthys Ohiodurulites Parapetalichthys Physichthys*] M. Dev. Eu. NA. *Notopetalichthys* M. Dev. Aus. *Wijdeaspis* M. Dev. Spits. ?**Stensioellidae,** *Nessariostoma Paraplesiobatis Pseudopetalichthys Stensioella* L. Dev. Eu. ?**Cratoselachidae,** *Cratoselache* Miss. Eu.

ORDER RHENANIDA. **Palaeacanthaspidae,** *Dobrowlania Kosoraspis Palaeacanthaspis* L. Dev. Eu. **Gemuendinidae,** *Asterosteus* [*?Ohioaspis*] M. Dev. NA. *?Farnellia Gemuendina Hoplopetalichthys* L. Dev. Eu. *Jagorina* U. Dev. Eu. *Kolymaspis* L. Dev. NAs. *Radotina* L. Dev. Eu.

ORDER ARTHRODIRA

SUBORDER ARCTOLEPIDA (DOLICHOTHORACI). **Phlyctaenaspidae (Arctolepidae),** *Actinolepis* [*Lataspis Plataspis*] L.-M. Dev. Eu., L. Dev. Spits. *Aethaspis* L. Dev. NA. *Aggeraspis* L. Dev. Eu. *Anarthraspis* L. Dev. NA. *Arctaspis* L. Dev. Spits. *Arctolepis* [*"Acanthaspis" Jaekelaspis*] L. Dev. Eu. Spits. NA. *Bryantolepis* [*Bryantaspis Euryaspidichthys Euryaspis*] L. Dev. NA. *Diadsomaspis* L. Dev. Eu. *Elegantaspis* L. Dev. Spits. *Heterogaspis* [*Monaspis*] L.-M. Dev. Spits. *Huginaspis* M. Dev. Spits. *Kujdanowiaspis* L. Dev. Eu. *? Lophostracon Mediaspis* L. Dev. Spits. *?Murmur* [*Euptychaspis Ptychaspis*] L. Dev. NA. *Overtonaspis* L. Dev. Eu. *Phlyctaenaspis* [*Phlyctaenius*] L.-M. Dev. NA., ?L. Dev. Eu., ?NAf. *Polyaspis* M. Dev. Spits. *Prescottaspis Prosphymaspis* L. Dev. Eu. *Rotundaspis ?Sedowichthys* M. Dev. Eu. *Simblaspis* L. Dev. NA. *Stuertzaspis* L. Dev. Eu. *Svalbardaspis* L. Dev. Eu. Spits. *? Taunaspis Tiaraspis Wheathillaspis* L. Dev. Eu. **Williamsaspidae,** *Williamsaspis* M. Dev. Aus. **Holonemidae,** *Anomalichthys* U. Dev. Eu. *Aspidichthys* [*Aspidophorus*] M.-U. Dev. NA., U. Dev. Eu. *Deitosteus* U. Dev. Eu. NA. *Deveonema* U. Dev. Eu. *Glyptaspis* M., ?U. Dev. Eu., U. Dev. NA. *Gyroplacosteus* [*Operchallosteus*] U. Dev. Eu. *Holonema* M.-U. Dev. Eu. NA., M. Dev. As. Spits., U. Dev. Aus. *Megaloplax* U. Dev. Eu. *Rhenonema* M. Dev. Eu. **Groenlandaspidae,** *Grazosteus* M. Dev. Eu. *Groenlandaspis* U. Dev. Gr. Aus. *? Tropidosteus* M. Dev. Eu.

SUBORDER BRACHYTHORACI. **Gemuendenaspidae,** *? Euleptaspis* [*Leptaspis*] U. Dev. Eu. *Gemuendenaspis* L. Dev. Eu. **Coccosteidae,** *Buchanosteus* M. Dev. Aus. *Clarkosteus* M. Dev. NA. *Coccosteus* M.-U. Dev. Eu., M. Dev. NA., Dev. Nova Zembla, ?U. Dev. NAs. *Dickosteus ?Goniosteus* M. Dev. Eu. *? Hussakofia* [*Brachygnathus Copanognathus*] U. Dev. NA. *Livosteus* M. Dev. Eu. *?Machaerognathus* U. Dev. NA. *Millerosteus* M. Dev. Eu. *Ostophorus* [*Sphenophorus*] U. Dev. NA. *Plourdosteus* [*? Pelecyphorus ? Tomaiosteus*] M.-U. Dev. Eu., U. Dev. NA. *Protitanichthys* [*Liognathus Lispognathus Woodwardosteus*] M. Dev. NA., ?M., U. Dev. Eu. *Rhachiosteus Taemasosteus* M. Dev. Eu. *? Trachosteus* U. Dev. NA. *Watsonosteus* M. Dev. Eu. **Pholidosteidae,** *Malerosteus* U. Dev. Eu. *Pholidosteus* U. Dev. Eu. NA. *Tapinosteus* U. Dev. Eu. **Heterosteidae,** *Heterosteus* [*Asterolepis Chelonichthys Ichthyosauroides*] L.-M. Dev. Gr. Spits. Dev. Eu. **Homostiidae,** *Angarichthys* L. Dev. Spits., M. Dev. NAs. *Homostius* [*Homosteus*] M.-U. Dev. Eu., M. Dev. Gr. *? Luetkeichthys ? Sedowichthys* M. Dev. EEu. *Tityosteus* L. Dev. Eu. *Tollichthys* M. Dev. EEu. **Brachydeiridae,** *Brachydeirus* [*Brachydirus*] U. Dev. Eu. *? Kiangyousteus* M. Dev. EAs. *Oxyosteus* [*Platyosteus*] *Synauchenia* [*Synosteus*] U. Dev. Eu. **Trematosteidae,** *Belosteus Brachyosteus Braunosteus Cyrtosteus Helmerosteus Tre-*

matosteus U. Dev. Eu. **Pachyosteidae,** *Enseosteus Erromenosteus Leiosteus Menosteus Microsteus Ottonosteus Pachyosteus Paraleiosteus Parawalterosteus Rhinosteus* [*Platyrhinosteus*] U. Dev. Eu. **Titanichthyidae,** *Titanichthys* [*Brontichthys*] U. Dev. NA. NAf. **Selenosteidae,** *Callognathus Gymnotrachelus Paramylostoma Selenosteus Stenosteus* U. Dev. NA. **Hadrosteidae,** *?Diplognathus* U. Dev. NA. *Hadrosteus* U. Dev. Eu. **Leptosteidae,** *Leptosteus* U. Dev. Eu. NA. **Mylostomidae,** *Dinognathus Dinomylostoma Mylostoma* U. Dev. NA. **Dinichthyidae,** *?Bungartius* U. Dev. NA. *Dinichthys* [*Perissognathus Ponerichthys*] U. Dev. ?Eu. NAf. NA. *Eastmanosteus* U. Dev. NA. *Dunkleosteus* U. Dev. ?Eu. NAf. NA. *Gorgonichthys* U. Dev. NA. ?NAf. *Heintzichthys* [*Stenognathus*] *?Holdenius* U. Dev. NA. *?Tafilalichthys* U. Dev. NAf. *?Timanosteus* U. Dev. NEu.

ORDER PHYLLOLEPIDA. **Phyllolepidae,** *Phyllolepis* [*Pentagonolepis*] M.-U. Dev. Eu., U. Dev. NA. Aus. Gr.

ORDER PTYCTODONTIDA. **Ptyctodontidae,** *Chelyophorus* U. Dev. Eu. *Ctenurella Deinodus* M. Dev. Eu. *Eczematolepis* [*Acantholepis Palaeomylus Phlyctaenacanthus*] M.-U. Dev. NA. *?Gamphacanthus* [*Heteracanthus*] M.-U. Dev. Eu. NA. *Goniosteus Paraptyctodus ?Pseudodontichthys* M. Dev. NA. *Ptyctodus* [*Aulacosteus Rinodus*] M.-U. Dev. Eu., M. Dev.-L. Miss. NA., U. Dev. NAs. *Rhamphodopsis* M. Dev. Eu. *Rhynchodus* [*Ramphodus Rhamphodontus Rhamphodus Rhynchodontus Rhynchognathus Ringia Rynchodus*] M.-U. Dev. Eu. NA.

ORDER ANTIARCHI. **Astrolepidae,** *?Allolepis* L. Dev. Eu. *Astrolepis* [*?Acintolepis Asserolepis Asterolepis Asteroplax Chelonichthys ?Narcodes ?Odontacanthus*] ?L., M. Dev. Spits., M. Dev. As. Gr. Aus., M.-U. Dev. Eu. *Belemnacanthus* [*Ceraspis Cornaspis Grossaspis*] M. Dev. Eu. *Bothryolepis* [*Bothriolepis Glyptosteus Homothorax Macrobrachius Pamphractus Phoebammon Placothorax Shurcabroma Stenacanthus*] M.-U. Dev. Eu. EAs. Nova Zembla Aus. Af. NA. Ant. Gr. *?Byssacanthoides* U. Dev. Ant. *Byssacanthus* M.-U. Dev. Eu. *Ceratolepis* [*Lepadolepis*] U. Dev. Eu. *Cypholepis* M.-U. Dev. Eu. *Dianolepis* M. Dev. EAs. *Gerdalepis* M. Dev. Eu. *Grossilepis* U. Dev. Eu. *Microbrachius* [*Microbrachium*] M. Dev. Eu. *?Orelodus* U. Dev. Eu. *Pterichthyodes* [*Millerichthys Pterichthys*] M. Dev. Eu. *Sinolepis* U. Dev. EAs. *Taeniolepis* U. Dev. Eu. **Remigolepidae,** *Remigolepis* U. Dev. Gr. ?Aus. ?**Wudinolepidae,** *Wudinolepis* M. Dev. EAs.

SYSTEMATIC POSITION UNCERTAIN: *Palaeospondylus* M. Dev. Eu. *Palaeomyzon* U. Perm. Eu.

CLASS CHONDRICHTHYES

SUBCLASS ELASMOBRANCHII

ORDER CLADOSELACHII (PLEUROPTERYGII). **Cladoselachidae,** "*Cladodus*" (*partim*) M. Dev.-L. Perm. NA., U. Dev.-U. Perm. Eu., Carb. EAs., Miss. Aus., U. Dev., ?U. Perm. Gr. *?Cladolepis* M. Dev. NA. *Cladoselache* U. Dev. Eu. NA. *?Deirolepis* M. Dev. NA. *?Dicentrodus* Miss.-Penn. Eu. *?Ohiolepis* M. Dev. NA. *?Tamiobatis* Miss. NA. **Ctenacanthidae,** *Ctenacanthus* [*?Anaclitacanthus "Cladodus" (partim) ?Eunemacanthus*] M. Dev.-L. Perm. Eu., U. Dev.-L. Perm. NA., U. Dev. As., Penn.-L. Perm. SA. *Goodrichthys* [*Goodrichia Moythomasina*] Miss. Eu. *?Phoebodus* [*Bathycheilodus*] M. Dev.-Miss. NA.

ORDER PLEURACANTHODII. **Xenacanthidae (Pleuracanthidae),** *Anodontacanthus* Penn. Eu., ?L. Perm. NA. *Compsacanthus* Penn. NA. *?Hypospondylus* L. Perm. Eu. *Thrinacodus* Miss. NA. *Xenacanthus* [*Aganodus ?Brachyacanthus Diacranodus Didymodus Diplodus Dissodus Dittodus Ochlodus Orthacanthus ?Platyacanthus Pleuracanthus Pternodus ?Stemmatias ?Stemmatodus Triodus*] U. Dev.-M. Perm. NA., Miss.-L. Perm., ?U. Trias. Eu., L. Perm. SA., U. Trias. Aus.

ORDER SELACHII

SUBORDER HYBODONTOIDEA. **Coronodontidae,** *Coronodus* U. Dev. NA. *?Denaea* Miss. Eu. *Diademodus* [*Tiarodontus*] U. Dev. NA. **Tristychiidae,** *Tristychius* [*Ptychacanthus*] Miss. Eu. **Hybodontidae,** *Acrodonchus* M.-U. Trias. Eu. *Acrodus* [*Adiapneustes ?Psilacanthus Sphenonchus Thectodus*] L. Trias.-U. Cret. Eu., Trias. SAs., M. Trias.-U. Cret. NA., U. Trias. Spits., U. Cret. SA. *Arctacanthus* [*Dolophonodus Hamatus*] M. Perm. NA. Gr. *Asteracanthus* [*Curtodus Strophodus*] U. Trias.-L. Cret. Eu., U. Trias. NA., Jur. EAs. Mad., Jur.-Paleoc. NAf. *Bdellodus* L. Jur. Eu. *Carinacanthus* U. Trias. NA. *Coelosteus* Miss. NA. *Dicrenodus* [*Carchariopsis Pristicladodus*] Miss.-Penn. Eu., Miss. NA. *Doratodus* M.-U. Trias. Eu. *Echinodus* Penn. Eu. *Eoörodus* U. Dev. NA. *Hybocladodus* Miss. NA. *Hybodonchus* M.-U. Trias. Eu. *Hybodus* [*Leiacanthus Meristodon Orthybodus Parhybodus ?Selachidea*] ?U. Perm., M. Trias.-U. Cret. Eu., L. Trias. Spits., L.-U. Trias. Gr., L. Trias.-U. Cret. NA., U. Trias. EAs., U. Trias.-Paleoc. Af., Jur.-Cret. Aus. *?Lambdodus* Miss. Eu. NA. *Lissodus* L. Trias. SAf. *Lonchidion* U. Cret. NA. *Mesodmodus* Miss. NA. *?Monocladodus* Penn. NA. *Nemacanthus* [*Desmacanthus Nematacanthus*] ?L. Trias. Gr., M. Trias.-U. Jur. Eu., U. Trias. Spits. NA. *Orthacodus* U. Jur.-L. Cret. Eu. *Palaeobates* M.-U. Trias. Eu. *?Petrodus* [*Octinaspis Ostinaspis*] Miss.-Penn. Eu., Penn. NA. *Polyacrodus* L. Trias. Gr., M.-U. Trias. Eu. *Priorybodus* U. Jur.-L. Cret. Af. *Pristacanthus* U. Jur. Eu. *Prohybodus* L. Cret. NAf. *Protacrodus* U. Dev. Eu. *Scoliorhiza* U. Trias. NA. *Sphenacanthus* Miss.-Perm. Eu. *Styracodus* [*Centrodus*] Penn. Eu. *Symmorium* Penn. NA. *Wodnika Xystrodus* U. Perm. Eu. **Edestidae,** *Campodus* [*Agassizodus Arpagodus Lophodus*] Miss.-L. Perm. Eu. NA., M. Perm. Gr. *Campyloprion* Penn. Eu. NA. *Edestodus* Penn. NA. *Edestus* [*Edestes Protopirata*] Miss.-Penn. Eu., Penn.-L. Perm. NA. *Erikodus* M. Perm. Gr. *Fadenia* Penn. NA., M. Perm. Gr. *Helicampodus* U. Perm.-L. Trias. SAs. *Helicoprion* [*Lissoprion*] Penn.-M. Perm. NA., L.-M. Perm. As., L. Perm. Eu. EInd. Aus. *Lestrodus* Miss. Eu. *Metaxyacanthus* Penn. Eu. *Ornithoprion* Penn. NA. *Orodus* [*?Chiastodus ?Desmiodus ?Hybodopsis ?Leiodus*] Miss.-Penn. Eu., Miss.-L. Perm. NA. *Parahelicampodus* U. Trias. Gr. *Parahelicoprion* L. Perm. Eu. Gr. *Physonemus* [*Batacanthus Drepanacanthus Xystracanthus*] Miss.-L. Perm. Eu., ?Miss. Aus., Miss.-Penn. NA., M.-U. Perm. SAs. *Prospiraxis* U. Dev. NA. *?Pseudodontichthys* M. Dev. NA. *Sarcoprion* M. Perm. Gr. *Syntomodus* U. Perm. Eu. *Toxoprion* Penn. NA. **Ptychodontidae,** *Hemiptychodus* U. Cret. Eu. NA. *Heteroptychodus* U. Cret. EAs. *Hylaeobatis* L. Cret. Eu. WAf. *Ptychodus* [*Aulodus ?Platychodus Sporetodus*] L.-U. Cret. Eu., U. Cret. As. NAf. NA.

SUBORDER HETERODONTOIDEI. **Palaeospinacidae,** *Palaeospinax* L. Jur. Eu. **Heterodontidae (Cestraciontidae)** R. Indo-Pac. Oc., *Heterodontus* [*Cestracion Drepanephorus Gyropleurodus Platyacrodus Pseudacrodus Tropidotus*] U. Jur.-Mioc. Eu., U. Cret.-Mioc. SA., U. Cret.-Eoc. Af.,

Mioc. Aus. NZ., R. Indo-Pac. Oc. *Strongyliscus* Mioc. WNA. *Synechodus* L. Cret.-Eoc. Eu., U. Cret. NAf., U. Cret.-Eoc. NA. SA., Eoc. As. NZ.

SUBORDER HEXANCHOIDEA (NOTIDANOIDEA). **Hexanchidae (Notidanidae)** R. Oc., *Hexanchus* [*Holodus Notidanion Notidanus Xiphidolamia*] L. Jur.-Plioc. Eu., U. Cret. SWAs. Mad. NZ., U. Cret.-Olig. SA., U. Cret.-Mioc. Af., U. Cret.-Pleist. NA., R. Oc. **Chlamydoselachidae** R. NAtl. NPac., *Chlamydoselache* Mioc. WInd., Plioc. Eu., R. NAtl. NPac.

SUBORDER GALEOIDEA. **Carchariidae (Odontaspidae)** R. Oc., *Anomotodon* U. Cret. Eu., U. Cret.-Paleoc. Af. *? Anotodus* Olig.-Plioc. Eu. Af. NA. *Carcharias* [*Hypotodus ?Iekelotodus Odontaspis ? Palaeohypotodus Parodontaspis Priodontaspis Striatolamia Synodontaspis Triglochis*] L. Cret.-Plioc. Eu., U. Cret.-Mioc. As., U. Cret.-Plioc. SA. NZ. Af., U. Cret.-Pleist. NA., Pleist. EInd., R. Oc. *Palaeocarcharias* U. Jur. Eu. *Scapanorhynchus* [*Rhinognathus*] L. Cret.-Paleoc. Eu., U. Cret. As. Aus. NZ. NA. SA., U. Cret.-Paleoc. Af., R. NAtl. NPac. **Orthacodontidae,** *Orthacodus* [*Parorthacodus Sphenodus*] L. Jur.-Eoc. Eu., U. Jur. Af., U. Cret. NA. **Isuridae (Carcharodontidae Lamnidae)** R. Oc., *Carcharoides* Mioc. Aus. SA. *Carcharodon* [*Agassizodon Carchariolamna Eocarcharodon Macrorhizodus Megaselachus Palaeocarcharodon Procarodon*] ?L. Cret., Paleoc.-Pleist. Eu., Paleoc.-Plioc. Af., Eoc.-Pleist. NA., Olig.-Pleist. Aus., Mioc. SAs. SA. EInd., Mioc.-Plioc. NZ. WInd., Mioc.-Pleist. Aus., R. Oc. *Isurus* [*Carcharocles Cosmopolitodus Cretoxyrhina Isuropsis Oxyrhina Paraisurus*] I. Cret.-Pleist. Eu., U. Cret.-Mioc. WInd. As. Aus. SA., U. Cret.-Pleist. NA., Mioc. Mad.,Mioc.-Plioc. NZ., Pleist. EInd., R. Oc. *Lamiostoma* Mioc. Eu., R. Oc. *Lamna* [*Cretolamna ? Euchlaodus Jekelotodus Leptostyrax Otodus*] L. Cret.-Plioc. Eu., U. Cret. Aus. As., U. Cret.-Eoc. Af., U. Cret.-Pleist. NA., Tert. NZ. *Palaeocorax* U. Cret. Eu. WAs. *Parisurus Pseudocorax* U. Cret. Eu. NA., U. Cret.-Paleoc. Af. *Pseudoisurus* U. Cret. Eu. NAs. *Squalicorax* [*Anacorax Corax ?Xenolamia*] U. Cret. Eu., U. Cret.-Paleoc. NA. **Cetorhinidae** R. Oc., *Cetorhinus* [*Hannoveria*] Olig.-Plioc. Eu., Mioc.-Plioc. NA., R. Oc. **Alopiidae** R. Oc., *Alopias* [*Alopecias Vulpecula*] Eoc. NA. Af., Olig.-Mioc. Eu., Mioc. WInd., Plioc. SA., R. Oc. **Orectolobidae** R. Oc., *Cantioscyllium* U. Cret. Eu. *Chiloscyllium* U. Cret. SWAs. NAf. WInd., Mioc. Eu., R. Oc. *Corysodon Crossorhinops* U. Jur. Eu. *Ginglymostoma* [*Plicodus*] U. Cret.-Mioc. Eu. Af., U. Cret. WInd., Eoc.-Mioc. NA., Eoc. As., R. Oc. *Orectolobus* [*Crossorhinus Palaeocrossorhinus*] U. Jur. Eu., R. Oc. *Phorcynus* U. Jur. Eu. *Squatirhina* U. Cret. NA., U. Cret.-Eoc. Eu., Paleoc.-Eoc. Af. *Squatirhynchus* U. Cret. SWAs. **Rhincodontidae** R. Oc. **Scyliorhinidae (Scylliidae)** R. Oc., *Galeus* U. Jur.-Plioc. Eu., Eoc. NA. Af., R. Oc. *Mesiteia* U. Cret. SWAs., Eoc. Eu. *Palaeoscyllium* U. Jur.-U. Cret. Eu. *Pristiurus* U. Jur. Eu., R. Oc. *Protogaleus* Paleoc.-Eoc. NAf. *Scyliorhinus* [*Scyllium Thyellina*] U. Cret. SWAs., U. Cret.-Mioc. NA., U. Cret.-Plioc. Eu., Paleoc.-Eoc. Af., R. Oc. *?Scylliodus* U. Cret. Eu. *? Trigonodus* Eoc. Eu. **Pseudotriakidae** R. Oc. **Mustelidae (Triakidae)** R. Oc., *Mustelus* [*Galeus*] Olig.-Plioc. Eu., R. Oc. *Triakis* U. Cret. SWAs., Paleoc. WInd. **Carcharhinidae** R. Oc., *Alopiopsis* [*Pseudogaleus*] Eoc. NAf., Eoc.-Olig. Eu., R. Oc. *Aprionodon* [*Aprion*] Eoc. Af., Eoc.-Plioc. Eu., Mioc. As., R. Oc. *Carcharhinus* Mioc.-Plioc. Eu. Af., Mioc.-Pleist. NA., ?Olig., Mioc. SA., Mioc. Aus., Plioc. EInd., R. Oc. *Galeocerdo* Eoc.-Mioc. NAf., Eoc.-Plioc. NA. Eu., Mioc. WInd. SA., Mioc.-Plioc. As. Aus., Pleist. EInd., R. Oc. *Galeorhinus* Paleoc.-Eoc. NAf., Eoc.-Pleist. Eu. NA., R. Oc. *Hemipristis* Eoc.-Mioc. NA. Af., Eoc.-Plioc. Eu., Mioc. As. Aus. SA., Mioc.-Pleist. EInd., R. Red Sea Indian Oc. *Hypoprion* Eoc. NA., Mioc. Eu. Af., R. Oc. *Negaprion* Eoc.-Mioc. NA., R. Oc. *Physodon* Eoc. Af., Eoc.-Mioc. Eu., R. Oc. *Prionace* Pleist. WNA., R. Oc. *Prionodon* [*Glyphis*] Eoc. Af., Eoc.-Plioc. Eu. NA., Olig.-Mioc. SA., Mioc. As. WInd., R. Oc. *Scoliodon* [*Loxodon Rhizoprinodon*] Eoc. Af., Eoc.–Mioc. Eu., Eoc.-Pleist. NA., R. Oc. **Sphyrnidae,** *Sphyrna* [*Zygaena*] Eoc.-Plioc. Eu., Eoc.-Pleist. NA., Mioc. Af. Aus., Plioc. EInd. SA., R. Oc.

SUBORDER SQUALOIDEA. **Squalidae** R. Oc., *Centrophorus* U. Cret. SWAs., ?U. Cret., Mioc. Eu., R. Oc. *Centropterus* U. Cret. Eu. *Centroscymnus* Mioc. WInd., R. Oc. *Centrosqualus* U. Cret. SWAs. *Cheirostephanus* Mioc. WInd. *Etmopterus* [*Spinax*] ?U. Cret., Mioc. Eu., R. Oc. *Oxynotus* [*Centrina*] Mioc.-Plioc. Eu., R. Oc. *Protospinax* U. Jur. Eu. *Squalus* [*Acanthias Centrophoroides*] U. Cret. SWAs., U. Cret.-Plioc. Eu., Paleoc.-Eoc. NAf., Olig. SA., Mioc. Aus., Mioc.-Pleist. NA., R. Oc. **Dalatiidae (Scymnorhinidae)** R. Oc., *Dalatias* [*Scymnus Scymnorhinus*] U. Cret. SWAs. NA., Eoc.-Plioc. Eu., R. Oc. *Isistius* U. Cret.-Eoc. Af., Eoc.-Mioc. Eu., R. Oc. *Somniosus* [*Laemargus*] Eoc. NAf., R. Oc. **Echinorhinidae** R. Oc., *Echinorhinus* [*?Goniodus*] Eoc.-Mioc. NA., Plioc. Eu., R. Oc. **Pristiophoridae** R. Oc., *Pliotrema* Tert. NZ., R. Oc. *Pristiophorus* U. Cret. SWAs., Mioc. Eu. NA., Mioc.-Plioc. Aus., R. Oc. *Propristiophorus* U. Cret. SWAs. **Squatinidae (Rhinidae)** R. Oc., *Squatina* [*Rhina Thaumas Trigenodus*] U. Jur.-Plioc. Eu., U. Cret. SWAs., U. Cret.-Pleist. NA., Eoc. Af., Mioc. Aus., R. Oc.

ORDER BATOIDEA

SUBORDER PRISTOIDEA. **Pristidae** R. Oc., *Anoxypristis* [*Oxypristis*] Eoc. Eu. *Ctenopristis* U. Cret. SWAs., Paleoc. Af. *Marckgrafia* L. Cret.-Paleoc. NAf. *Onchopristis* L.-U. Cret. NAf., U. Cret. NA. *Onchosaurus* [*Dalpiazia Gigantichthys Ischyrhiza*] U. Cret. Eu. NA. SA., U. Cret.-Paleoc. NAf. *Peyeria* U. Cret. NAf. *Pristis* [*Myriopristis Pristibatus*] ?U. Cret., Eoc. NA., Eoc.-Mioc. Eu. As., Eoc. Af., Mioc.-Plioc. Aus. *Propristis* [*Amblypristis Eopristis*] Eoc. NAf. NA. *Pucapristis* U. Cret. SA. *Schizorhiza* U. Cret. SWAs. NA. SA., U. Cret.-Paleoc. Af. *Sclerorhynchus* [*Ganopristis*] U. Cret.-Paleoc. Eu. Af., U. Cret.-Eoc. As.

SUBORDER RHINOBATOIDEA. **Rhynchobatidae** R. Oc., *Rhynchobatus* Cret. Af., Eoc.-Mioc. Eu., R. Oc. **Rhinobatidae** R. Oc., *Aellopos* [*Euryarthra Spathobatis*] *Asterodermus Belemnobatis* U. Jur. Eu. *Cyclarthrus* L. Jur. Eu. *Platyrhina* Eoc. Eu., R. NPac. *Rhinobatos* [*Rhinobatus*] L. Cret.-Mioc. Eu., U. Cret. SWAs., U. Cret.-Eoc. NAf., R. Oc. *Trygonorrhina* Eoc.-Mioc. Eu., R. Oc.

SUBORDER TORPEDINOIDEA. **Torpedinidae** R. Oc., *? Eotorpedo* Paleoc.-Eoc. Af., Eoc. Eu. *Narcine* Eoc. Eu., R. Oc. *Narcopterus* Eoc. Eu. *Torpedo* [*Narcobatus*] ?Eoc. Eu., R. Oc. **Narkidae Temeridae** R. Oc.

SUBORDER RAJOIDEA. **Rajidae** R. Oc., *Acanthobatis* Mioc. Eu. *Cyclobatis* U. Cret. SWAs. *?Dynatobatis* ?Plioc. SA. *Oncobatis* Plioc. NA. *Platyspondylus* L.-U. Cret. NAf. *Raja* [*Actinobatis Raia*] U. Cret. SWAs. WInd., U. Cret.-

Plioc. Eu., U. Cret.-Eoc. NAf., Eoc.-Mioc. NA., R. Oc. **Pseudorajidae Anacanthobatidae** R. Oc.

SUBORDER MYLIOBATOIDEA. **Urolophidae** R. Oc., *Urolophus* [*Leiobatis*] Eoc. Eu., Pleist. NA., R. Oc. **Dasyatidae (Trygonidae)** R. Oc., *Dasyatis* [*Dasibatus Dasybatus Heliobatis Palaeodasybatis Pastinachus Trygon Xiphotrygus*] ?L. Cret., Eoc.-Plioc. Eu., U. Cret. NAf., Eoc.-Pleist. NA., Mioc. As. Aus., Pleist. EInd., R. Oc. *Gryphodobatis* Plioc. NA. *Hypolophites* U. Cret.-Eoc. Af. *Parapalaeobates* U. Cret. SWAs. NAf. *Ptychotrygon* Eoc. Eu. *Rhombodus* U. Cret. Eu. SWAs. NAf. SA. *Taeniura* Eoc.-Plioc. Eu., R. Oc. **Potamotrygonidae** R. Af. SA., *Potamotrygon* Pleist.-R. Af., R. SA. **Gymnuridae** R. Oc., *Gymnura* [*Pteroplatea*] Mioc. NA., R. Oc. **Mobulidae** R. Oc., *Manta* Plioc. NA., R. Oc. **Rhinopteridae** R. Oc., *Rhinoptera* [*Zygobatis*] U. Cret.-Plioc. Eu., U. Cret.-Mioc. Af., U. Cret., Mioc.-Plioc. SA., Eoc.-Mioc. NA., Mioc. As. **Myliobatidae** R. Oc., *Aetobatus* [*Plinthicus*] Paleoc.-Plioc. Eu., Eoc.-Mioc. Af., Eoc.-Pleist. NA. Mioc. SAs. SA., R. Oc. *Apocopodon* U. Cret. NA. SA. *Mesibatis* Plioc. NA. *Myliobatis* [*Ichthyaetus Ptychopleurus*] U. Cret.-Plioc. Eu. NA., Paleoc.-Eoc. Af., Eoc.-Mioc. As., Mioc.-Pleist. Aus., Mioc.-Plioc. SA., Tert. NZ., R. Oc. *Promyliobatis* Eoc. Eu.

ELASMOBRANCHII INCERTAE SEDIS

Crookalia Jur. Eu. *Raineria* U. Trias. Eu. *Rhaibodus* Eoc. Af. *Styptobasis* Penn.-L. Perm. NA.

INCERTAE SEDIS, ?ORDER BRADYODONTI. **Petalodontidae,** *Ageleodus* [*?Callopristodus*] Miss.-Penn. Eu., Penn. NA. *Brachyrhizodus* L. Perm. NA. *Chomatodus* [*? Antliodus Tanaeodus*] Miss. Eu. NA. Aus. *Ctenoptychius* [*Ctenopetalus Cymatodus Harpacodus Paracymatodus Peripristis Petalodopsis Serratodus*] Miss.-Penn. NA., Miss.-U. Perm. Eu. *Cynopodius Euglossodus* [*Glossodus*] Miss. Eu. *Fissodus* [*Cholodus Peltodus*] Miss. Eu., Miss.-Penn. NA. *Hoplodus* [*?Diodontopsodus ?Pristodus*] Miss. Eu. *Janassa* [*Byzenos Climaxodus Dictea Strigilina Thoracodus Trilobites*] Miss.-U. Perm. Eu., Penn.-M. Perm. NA., M. Perm. Gr. *?Megactenopetalus* L. Perm. NA. *Mesolophodus* Miss. Eu. *Petalodus* [*Antliodus ?Calopodus Getalodus Glyphanodus ?Lisgodus Sicarius*] Miss.-L. Perm. Eu. NA., U. Penn. EAs. *Petalorhynchus Polyrhizodus* [*Dactylodus ?Rhomboderma*] Miss. Eu. NA. **Psammodontidae,** *Lagarodus* Miss.-Penn. Eu. *Mazodus* Miss. NA. *Psammodus* [*Archaeobatis Astrobodus Homalodus*] ?U. Dev., Miss. NA., Miss. Eu. **Copodontidae,** *Acmoniodus* U. Dev. NA. *Copodus* [*Characodus ?Dimyleus Labodus Mesogomphus Mylacodus Mylax Pinacodus Pleurogomphus Rhymodus*] Miss. Eu. NA. *?Solenodus* Penn. Eu. **Chondrenchelyidae,** *Chondrenchelys ?Eucentrurus* Miss. Eu. **Helodontidae,** *Helodus* [*Diclitodus Pleurodus Pleuroplax*] U. Dev.-L. Perm. NA., Miss.-Penn. Eu., Perm. Aus. *Venustodus* [*Lophodus Oxytomodus ?Rhamphodus Tomodus*] Miss. Eu. NA.

SUBCLASS HOLOCEPHALI

ORDER CHIMAERIFORMES

SUBORDER COCHLIODONTEI. **Cochliodontidae,** *Cochliodus* [*Chitinodus Cyrtonodus*] Miss.-Penn. Eu., Miss. NA. *?Cranodus* Miss. Eu. *Crassidonta* L. Perm. Aus. *?Cymatodus* Penn. NA. *Deltodus* [*Deltodopsis Stenopterodus Taeniodus*] Miss. Eu., Miss.-Penn. NA., Perm. Aus. *Deltoptychius* [*? Antacanthus Listracanthus ?Lophacanthus Phigeacanthus ?Phricacanthus Platacanthus Platycanthus Streblodus*] Miss.-Penn. Eu. NA. *Dichelodus Diplacodus* Miss. Eu. *?Erismacanthus* [*Cladacanthus Dipriacanthus Gampsacanthus Lecracanthus*] Miss. Eu. NA. ?Aus. *Helodopsis* Perm. SAs. *Icanodus* [*Enniskillen Eutomodus*] Miss. Eu. NA., Perm. Aus. *?Macrodontacanthus* L. Perm. NA. *?Menaspacanthus* Miss. Eu. *Menaspis* [*? Asima ?Radamus*] U. Perm. Eu. *Platyodus* Miss. NA. *Platyxystrodus* [*Xystrodus*] Miss.-Penn. NA. *Poecilodus* Miss. Eu., Miss.-Penn. NA., ?Perm. Aus. *Psephodus* [*Aspidodus*] Miss.-Penn. Eu., Miss.-L. Perm. NA., Penn. SAs. *Sandalodus* [*?Orthopleurodus Trigonodus Vaticinodus*] U. Dev.-Penn. NA., Miss. Eu. *Synthetodus* U. Dev. NA. *Thoralodus* U. Dev. Eu. *Xenodus* [*Goniodus*] U. Dev. NA.

SUBORDER MYRIACANTHOIDEI. **Acanthorhinidae,** *Acanthorhina* L. Jur. Eu. **Chimaeropsidae,** *Chimaeropsis* L.-U. Jur. Eu. **Myriacanthidae,** *Myriacanthus* [*Metopacanthus Prognathodus*] L.-M. Jur. Eu.

SUBORDER SQUALORAIIFORMES. **Squaloraiidae,** *Squaloraia* [*Spinocorhinus*] L. Jur. Eu.

SUBORDER CHIMAEROIDEI. **Chimaeridae** R. Oc., *Brachymylus* [*Aletodus*] L.-U. Jur. Eu. *Chimaera* [*Plethodus*] U. Cret. Aus., Eoc.-Plioc. Eu., Tert. EInd. NZ., R. Oc. *Edaphodon* [*Bryactinus Diphrissa Dipristis Eumylodus ?Isotaenia ?Leptomylus Loxomylus Mylognathus Passalodon Psittacodon ?Sphagepoea*] L. Cret.-Plioc. Eu., U. Cret.-Eoc. NA., Mioc.-Plioc. Aus. *Ganodus* [*Leptacanthus*] U. Jur. Eu. *?Ichthypriapus* U. Cret. NA. *Ischyodon* ?Mioc. Aus. *Ischyodus* [*Auluxacanthus Chimaeracanthus*] M. Jur.-Paleoc. Eu., Cret. NZ. *Myledaphus* U. Cret. NA. *Pachymylus* M. Jur. Eu. *Psaliodus* Eoc. Eu. **Rhinochimaeridae** R. Oc., *Amylodon* Olig. Eu. *Elasmodectes* [*Elasmognathus*] U. Jur.-U. Cret. Eu. *Elasmodus* U. Cret.-Eoc. Eu. **Callorhinchidae** R. Oc., *Callorhinchus* [*Callorhynchus*] U. Cret. NZ., Mioc. SA., R. Oc.

CHONDRICHTHYES OR PLACODERMI

INCERTAE SEDIS—ICHTHYODORULITES

Acondylacanthus Miss.-Penn. Eu. NA. *Aganacanthus* Miss. Eu. *Alienacanthus* U. Dev. Eu. *Ancistriodus* ?U. Penn., L. Perm. NA., L. Perm. EInd. *Apateacanthus* U. Dev. NA. *Asteroptychius* Miss. Eu., Miss.-Penn. NA. *Bulbocanthus* L. Dev. NA. *Bythiacanthus* Miss. NA. *Chalazacanthus* Miss. Eu. *Cyrthacanthus* M. Dev. NA. *Euctenius* Miss.-Penn. Eu., Penn. NA. *Euctenodopsis* Penn. Eu. *Euphyacanthus* Miss. Eu. *Glymmatacanthus* Miss. NA. *Gnathacanthus* Miss. Eu. *Harpacanthus* Miss. Eu. NA. *Lepracanthus* Penn. Eu. *Lispacanthus Margaritacanthus* [*Euacanthus*] Miss. Eu. *Marracanthus* Miss. NA. *Ostracanthus* Penn. Eu. *Rapidentichthys* M. Perm. NA. *Sentacanthus* U. Dev. Eu. *Stethacanthus* U. Dev.-Miss. NA. *Thaumatacanthus* Perm. SAs. *Tubulacanthus* L. Perm. Eu.

CLASS OSTEICHTHYES

?SUBCLASS ACANTHODII

ORDER CLIMATIFORMES. **Climatiidae,** *Arhaeacanthus* [*Devononchus*] U. Sil.-U. Dev. Eu. *Asiacanthus* L. Dev. EAs.

Brachyacanthus L. Dev. Eu. *Climatius* U. Sil.-L. Dev. Eu. NA., L. Dev. Spits. *Euthacanthus* L. Dev. Eu. *Nodocosta* L.-U. Dev. Eu. *? Nodonchus* L. Dev. Eu. *Nostolepis* [*Diplacanthoides*] U. Sil. NAf. NA., U. Sil.-L. Dev. Eu., L. Dev. Spits. *Parexus* L. Dev. Eu. *Sinacanthus* L. Dev. EAs. **Diplacanthidae,** *?Dendracanthus* L. Dev. Eu. *Diplacanthus* [*Rhadinacanthus*] M. Dev. Eu., U. Dev. NA. *?Dontacanthus* L. Dev. Eu. *Homacanthus* [*? Amacanthus Homocanthus Hoplonchus*] L. Dev.-Miss. NA., M. Dev.-Penn. Eu. *? Nodacanthus* L.-U. Dev. Eu., M.-U. Dev. NA. **Gyracanthidae,** *? Gyracanthides* [*Chiraropalus Rhytidaspis*] Miss. Aus. *Gyracanthus* Miss.-Penn. Eu., L. Dev.-L. Perm. NA. *Periplectrodus* Miss. NA.

ORDER ISCHNACANTHIFORMES. **Ischnacanthidae,** *Acanthodopsis* Penn. Eu. *Atopacanthus* M. Dev. Eu. Spits. M.-U. Dev. NA., *? Byssacanthoides* U. Dev. Ant. *Doliodus* L. Dev. NA. *Ischnacanthus* [*Ictinocephalus*] L.-M. Dev. Eu. *Marsdenius* Miss. Eu. *?Monopleurodus* [*Ancistrodus Campylodus*] U. Sil. Eu. *Onchus* [*Archaeacanthus Leptocheles Oncus*] U. Sil.-L. Dev. Eu., U. Sil.-U. Dev. NA., L. Dev. Spits. *Plectrodus* U. Sil.-L. Dev. Eu., L. Dev. Spits., M. Dev. NA. *Protodus* [*Gomphodus Poracanthodes*] ?L. Sil., U. Sil.-L. Dev. Eu., L. Dev. NA. Spits.

ORDER ACANTHODIFORMES. **Mesacanthidae,** *Mesacanthus* L.-M. Dev. Eu., U. Dev. NA. **Acanthodidae,** *Acanthodes* [*Acanthodis Acanthoessus Holacanthodes ? Holmesella Pelonectes*] U. Dev.-L. Perm. NA., Miss. Aus., Miss.-L. Perm. Eu. As., U. Perm. Gr. *? Acanthoides* L. Dev. Eu., M. Dev. NA. *Cheiracanthoides* M. Dev. NA. *Cheiracanthus* [*Chiracanthus*] L. Dev. NA., L.-U. Dev. Eu., U. Dev. Ant., Miss. Aus., Penn. NAs., *Eupleurogmus* Miss. Aus. *Haplacanthus* L.-U. Dev. Eu. *? Helolepis* M. Dev. NA. *Homalacanthus* [*Paracanthodes*] U. Dev. NA. *Pseudacanthodes* [*Protacanthodes*] *Traquairichthys* [*Traquairia*] Penn. Eu.

ACANTHODII INCERTAE SEDIS

Ancistrodon [*Grypodon*] U. Sil. Eu. *Balacanthus* L. Dev. Eu. *Chelomodus* U. Sil. Eu. *Conolepis* L. Dev. Eu. *Cosmacanthus* U. Dev. Eu. *Geisacanthus* Miss. Eu. NA. *Gomphacanthus* Miss. Eu. *Helenacanthus* L. Dev. NA. *Lophosteus* L. Dev. Eu. *Machaeracanthus* [*Machaerius*] L.-M. Dev. Eu., ?L. Dev. NAf., L.-U. Dev. NA. *Naulas* M. Dev. Eu. *Oracanthus* [*Phoderacanthus Stichacanthus*] Miss. Eu. NA. *? Pinnacanthus* L. Dev. NA. *Stigmodus* L. Dev. Eu. *?Striacanthus* U. Dev. Aus.

SUBCLASS ACTINOPTERYGII

INFRACLASS CHONDROSTEI

ORDER PALAEONISCIFORMES

SUBORDER PALAEONISCOIDEI. **Cheirolepidae,** *Cheirolepis* M. Dev. Eu., U. Dev. NA. **Stegotrachelidae,** *? Borichthys* Miss. Eu. *Kentuckia* Miss. NA. *Moythomasia* M.-U. Dev. Eu., U. Dev. NA., ?NAs. *Orvikuina* M. Dev. Eu. *Stegotrachelus* Dev. Eu., ?U. Dev. NA. Ant. **Tegeolepidae,** *Tegeolepis* [*Actinophorus*] U. Dev. NA. **Rhabdolepidae,** *Osorioichthys* [*Stereolepidella Stereolepis*] U. Dev. Eu. *Rhabdolepis* L. Perm. Eu. **Rhadinichthyidae,** *Aetheretomon* Miss. Eu. *Cycloptychius* Miss. NAs., Miss.-Penn. Eu. *? Eurylepidoides* L. Perm. NA. *Rhadinichthys* Miss. Af., Miss.-Penn. Eu., Penn., ?M. Trias. SA. *Rhadinoniscus Strepheoschema* Miss. Eu. **Carbovelidae,** *Carboveles* Miss. Eu. *Phanerosteon* [*Gymnoniscus Sceletophorus*] Miss.-L. Perm. Eu. *?Sphaerolepis* [*Trissolepis*] L. Perm. Eu. **Canobiidae,** *Canobius Mesopoma* Miss. Eu. *Whiteichthys* Penn. Gr. **Cornuboniscidae,** *Cornuboniscus* Miss. Eu. **Styracopteridae,** *Benedenius* [*Benedenichthys*] *Styracopterus* [*? Fouldenia*] Miss. Eu. **Cryphiolepidae,** *Cryphiolepis* Miss. Eu. **Holuriidae,** *Holuropsis* U. Perm. Eu. *Holurus* Miss. Eu. **Cosmoptychiidae,** *Cosmoptychius* Miss. Eu. *Watsonichthys* Miss. Eu., L. Perm. SAf. **Pygopteridae,** *Nematoptychius* Miss. Eu. *Pygopterus* M. Perm. Gr., U. Perm.-L. Trias. Eu., L. Trias. SAf. Spits. **Elonichthyidae,** *Drydenius* Miss.-Perm. Eu. *Elonichthys* [*Ganacrodus ? Pariostegus ? Propalaeoniscus*] Miss.-U. Perm. Eu., Miss.-Penn., ?U. Trias. NA., Miss., U. Perm., ?U. Trias. Aus., L. Perm. SAf., M. Perm. Gr., U. Perm. SA., ?Trias. WAf. *Ganolepis* Perm. NAs. *Gonadotus* Miss. NA., Miss.-Penn. Eu., **Acrolepidae,** *Acrolepis* Miss.-L. Perm. NA., Miss.-U. Perm. Eu., Penn. SA., Penn.-L. Perm. NAs., ?L., U. Perm. SAf., ?Trias. Aus. *Acropholis* M. Perm. Gr. *Acrorhabdus* L. Trias. Spits. Gr. *Boreosomus* [*Diaphorognathus*] L. Trias. Gr. Spits. Mad., ?M. Trias. NA. *Hyllingea* U. Trias. Eu. *Leptogenichthys* M. Trias. Aus. *Mesonichthys* Miss. Eu. *Namaichthys* L. Perm. SAf. *Plegmolepis* M. Perm. Gr. *Reticulolepis* ?Miss., U. Perm. Eu. *Tholonotus* L. Perm. SA. **Coccocephalichthyidae,** *Coccocephalichthys* [*Coccocephalus Cocconiscus*] Penn. Eu. **Haplolepidae,** *Haplolepis* [*Eurylepis Mecolepis Mekolepis Parahaplolepis*] *Pyritocephalus* [*Teleopterina*] Penn. Eu. NA. **Amblypteridae,** *Amblypterina* U. Perm. Eu. *Amblypterus* L.-U. Perm. Eu., ?L. Perm. NA. As., ?M. Trias. SA. *? Lawnia* L. Perm. NA. **Commentryidae,** *Commentrya* [*Elaveria*] *Paramblypterus* [*Amblypterops Cosmopoma Dipteroma*] L. Perm. Eu. **Palaeoniscidae,** *? Aegicephalichthys* M. Trias. Aus. *Cosmolepis* [*Oxygnathus Thrissonotus*] L. Jur. Eu. *? Gyrolepidoides* M. Trias. SA. *Gyrolepis* L. Trias. EAs., M.-U. Trias. Eu., U. Trias. NA. *Palaeoniscum* [*Eupalaeoniscus Geomichthys Palaeoniscus Palaeothrissum*] ?L. Perm. NA., ?L. Perm., U. Perm., ?L. Trias. Eu., M. Perm. Gr., L. Trias. Spits. ?EAs., ?U. Trias. Aus. *Progyrolepis* L. Perm. Eu. NA. *Pteronisculus* [*Glaucolepis*] L. Trias. Spits. Gr. Mad. Aus. *? Trachelacanthus* L. Perm. Eu. *Turseodus* [*Eurecana Gwyneddictis*] U. Trias. NA. **Aeduellidae,** *Aeduella* L. Perm. Eu. *Westollia* [*Lepidopterus*] L. Perm. Eu. **Dicellopygidae,** *? Aneurolepis* [*Urolepis*] M.-U. Trias. Eu. *? Brachydegma* L. Perm. NA. *Dicellopyge* L. Trias. SAf. **Boreolepidae,** *Boreolepis* M. Perm. Gr. **Birgeriidae,** *Birgeria* [*Xenestes*] L. Trias. Spits. Gr. NA. Mad., L.-U. Trias. Eu. *Ohmdenia* L. Jur. Eu. *? Psilichthys* Trias. or Jur. Aus. **Scanilepidae,** *Scanilepis*. L. Trias. Spits., U. Trias. Eu. **Centrolepidae,** *Centrolepis* L. Jur. Eu. **Coccolepidae,** *Browneichthys* L. Jur. Eu. *Coccolepis* [*? Palaeoniscionotus*] L. Jur.-L. Cret. Eu., M. Jur. As., Jur. Aus.

SUBORDER PLATYSOMOIDEI. **Platysomidae,** *Mesolepis* [*Pododus*] Miss.-Penn. Eu. *Paramesolepis* Miss. Eu. *Platysomus* [*? Tonipoichthys Uropteryx*] Miss.-U. Perm. Eu., Miss.-L. Perm., ?L. Trias. NA., M. Perm., M. Trias. Gr., ?L. Trias. Spits., U. Trias. Aus. *Wardichthys* Miss. Eu. **Chirodontidae (Amphicentridae),** *Cheirodopsis* Miss. Eu. *Chirodus* [*Amphicentrum Cheirodus Hemicladodus*] Miss.-Penn. Eu. NA. *Eurynothus* [*Eurynotus Notacmon Plectrolepis*] Miss.-Penn. Eu. NA., Penn. NAs., ?M. Trias. SA. *Globulodus* [*Eurysomus ? Leukanichthys*] U. Perm. Eu. *Paraeurynotus* L. Perm. Eu. *Proteurynotus* Miss. Eu. **Bobasatraniidae,** *Bobasatrania* [*? Haywardia*] L. Trias. Spits. Gr. Mad., L., ?U. Trias. NA. *? Caruichthys* L. Trias. SAf. *Ecrinesomus* L. Trias. Mad. *Lambeichthys*

L. Trias. NA. **Dorypteridae,** *Dorypterus* L. Perm. EAs. U. Perm. Eu.

SUBORDER TARRASIOIDEI. **Tarrasiidae,** *Tarrasius* Miss. Eu.

SUBORDER PTYCHOLEPOIDEI. **Ptycholepidae,** *Ptycholepis* M. Trias.-L. Jur. Eu., U. Trias. NA.

SUBORDER PHOLIDOPLEUROIDEI. **Pholidopleuridae,** *Arctosomus* [*Neavichthys*] L. Trias. NAs. *Australosomus* L. Trias. Af. Mad. Gr., M. Trias. Spits. *Macroaethes* M. Trias. Aus. *Pholidopleurus* M.-U. Trias. Eu.

SUBORDER LUGANOIOIDEI. **Luganoiidae,** *Besania Luganoia* M.-U. Trias. Eu.

SUBORDER REDFIELDOIDEI. **Redfieldiidae (Catopteridae Dictyopygidae),** *Atopocephala* L. Trias. SAf. *Beaconia Brookvalia* M. Trias. Aus. *Daedalichthys* L. Trias. SAf. *Dictyopleurichthys* M. Trias. Aus. *Dictyopyge* L.-U. Trias. Eu., U. Trias. NA., ?SAf. Aus. *Geitonichthys* M. Trias. Aus. *Helichthys* L. Trias. SAf. *?Ischnolepis* ?L. Trias. SAf. *Molybdichthys Phlyctaenichthys* M. Trias. Aus. *Pseudobeaconia* M. Trias. SA. *Redfieldia* [*Catopterus*] U. Trias. NA. *Sakamenichthys* L. Trias.Mad. *Schizurichthys* M. Trias. Aus. *Sinkiangichthys* Trias. EAs.

SUBORDER PERLEIDOIDEI. **Perleididae (Colobodontidae),** *Chrotichthys* L. Trias. Aus. *Colobodus* [*Asterodon ?Cenchrodus ?Charitodon ?Charitosaurus Dacylolepis Eupleurodus ?Hemilopus ?Nephrotus ?Omphalodus*] L. Trias. SAf. NAs. EAf., L.-U. Trias. Aus., M.-U. Trias. Eu. *Crenolepis* [*Crenilepis Crenilepoides*] M. Trias. Eu. *Dimorpholepis* Trias. WAf. *Dollopterus* L. Trias. NA., M. Trias. Eu. *Engycolobodus* M. Trias. Eu. *?Gigantopterus* U. Trias. Eu. *?Helmolepis* L. Trias. Gr. *Manlietta* M. Trias. Aus. *Meidiichthys* L. Trias. SAf. *Mendocinichthys* [*Mendocinia*] M. Trias. SA. *Meridensia* M.-U. Trias. Eu. *Perleidus* L. Trias. EAs. Mad. Gr. Spits., M.-U. Trias. Eu., Trias. WAf. *Pristisomus* L. Trias. Aus. Mad. *Procheirichthys* M. Trias. Aus. *?Thoracopterus* [*?Pterygopterus ?Urocomus*] ?L. Trias. NA., M. Trias. Aus., U. Trias. Eu. *Tripelta* L. Trias. Aus. *Zeuchthiscus* L. Trias. Aus. **Cleithrolepidae,** L. Trias. SAf. *Cleithrolepis* L.-U. Trias. Aus., M. Trias. SA., U. Trias. Eu. *Dipteronotus* U. Trias. Eu. *Hydropessum* L. Trias. SAf. **Peltopleuridae,** *?Habroichthys* U. Trias. Eu. *Peltopleurus* M. Trias. EAs., M.-U. Trias. Eu. *Placopleurus* M.-U. Trias. Eu. **Platysiagidae,** *Platysiagum* M. Trias.-L. Jur. Eu. **Cephaloxenidae,** *Cephaloxenus* M.-U. Trias. Eu. **Aethodontidae,** *Aethodontus* M.-U. Trias. Eu.

PALAEONISCIFORMES NOT ASSIGNED TO FAMILIES. *Aldingeria* Penn. Gr. *?Anaglyphys* L. Perm. Eu. *Anatoia* L. Trias. SA. *Apateolepis* M. Trias. Aus. *? Atherstonia* [*Hypterus*] U. Perm. Eu., L. Trias. SAf. Mad., U. Trias. NA. *Belichthys* M. Trias. Aus. *Broometta* L. Trias. SAf. *Caminchaia Cenechoia* L. Trias. SA. *Challaia* L. Trias. SA. *Diphyodus* Miss. NA. *Disichthys* U. Perm. SAf. *Echentaia* L. Trias. SA. *Elpisopholis* U. Trias. Aus. *Eurynotoides* U. Perm. Eu. *Evenkia* L. Trias. NAs. *Guaymayenia* L. Trias. SA. *Gyrolepidotus* Miss. NAs. *Isodus* Miss. Eu. *Leighiscus Megapteriscus Mesembroniscus* M. Trias. Aus. *Myriolepis* ?Penn. Eu., L.-U. Trias. Aus., ?L. Trias. SA. *Neochallaia* L. Trias. SA. *Oxypteriscus* Miss. NAs. *Palaeobergeria* Miss. NAs. *Pasamhaya* L. Trias. SA. *Peleichthys* U. Perm. SAf. *Pteroniscus* Jur. WAs. *?Schizospondylus* U. Cret. Eu. *Urosthenes* Penn. Aus.

ORDER POLYPTERIFORMES. **Polypteridae** R. Af., *Polypterus* Eoc.-R. Af.

ORDER ACIPENSERIFORMES. **?Phanerorhynchidae,** *Phanerorhynchus* Penn. Eu. **?Errolichthyidae,** *Errolichthys* L. Trias. Mad. **Chondrosteidae,** *Chondrosteus Gyrosteus* [*?Strongylosteus*] L. Jur. Eu. *?Stichopterus* L. Cret. As. **?Saurichthyidae,** *Saurichthys* [*Acidorhynchus Belonorhynchus Giffonus Gymnosaurichthys Ichthyorhynchus Saurorhynchus Stylorhynchus*] L. Trias. EAs. Spits. Gr. Mad., L.-M. Trias. Aus., L. Trias.-L. Jur. Eu., U. Trias. NA. **Acipenseridae** R. Eu. As. NA. *Acipenser* U. Cret.-R. NA. Eu., R. As. *Huso* Plioc.-R. Eu., R. NAs. *Protoscaphirhynchus* U. Cret. NA. **Polyodontidae** R. NA. EAs., *Crossopholis* Eoc. NA. *Paleopsephurus* U. Cret. NA. *? Pholidurus* U. Cret. Eu.

INFRACLASS HOLOSTEI

ORDER SEMIONOTIFORMES

SUBORDER SEMIONOTOIDEI. **Semionotidae (Lepidotidae),** *Acentrophorus* U. Perm. Eu., U. Trias. NA. *Aetheolepis* Jur. Aus. *Alleiolepis* [*Lei, Leiolepis*] M. Trias. Eu. *Angolaichthys* Trias. SWAf. *Aphelolepis* Trias. Eu. *Aphnelepis* Jur. Aus. *Archaeolepidotus* L. Trias. Eu. *Asialepidotus* M. Trias. EAs. *Corunegenys* U. Trias. Aus. *Dapedium* [*Aechmodus Amblyurus Dapedius Omalopleurus Pholidotus*] U. Trias.-L. Jur. Eu. SAs. *Enigmatichthys* M. Trias. Aus. *Eosemionotus* L.-M. Trias. Eu. *Heterostrophus* [*Heterostichus*] U. Jur. Eu. *Lepidotes* [*Lepidosaurus Lepidotus Plesiodus Prolepidotus Scrobodus Sphaerodus*] U. Trias.-L. Cret., ?U. Cret. Eu., U. Trias.-L. Cret. Af. NA., U. Trias.-U. Cret. As., ?Trias., L. Jur.-U. Cret. SA., L. Jur. Mad. *?Orthurus* U. Trias. Eu. *Paracentrophorus* L. Trias. Mad. *Paralepidotus* U. Perm.-U. Trias. Eu. *Prionopleurus* [*Pantelion*] Jur. Eu. *Pristiosomus* U. Trias. Aus. *Sargodon* U. Trias. Eu. *Semionotus* [*? Archaeosemionotus Ischypterus*] L.-U. Trias. Eu., ?L. Trias. SA., U. Trias. Aus. SAf. NA. *Serrolepis* M.-U. Trias. Eu. *Sinosemionotus* M. Trias. EAs. *Tetragonolepis* [*Homoeolepis Pleurolepis*] L.-U. Jur. Eu., L. Jur. As. *Woodthorpea* U. Trias. Eu.

SUBORDER LEPISOSTOIDEI. **Lepisosteidae** R. NA., *Lepisosteus* [*Atractosteus Clastes Clastichthys Clestes Cylindrosteus Lepidosteus Litholepis Pneumatosteus Psalliostomus Sarchirus Sarcochirus*] U. Cret.-Mioc. Eu., U. Cret.-R. NA., Eoc. SAs. WAf., R. WInd. CA. *?Paralepidosteus* L.-U. Cret. WAf.

ORDER PYCNODONTIFORMES. **Pycnodontidae,** *Acrotemnus* U. Cret. Eu. WAf. *Anomaeodus* L.-U. Cret. Eu. NA., U. Cret. Af. *Athrodon* U. Jur.-U. Cret. Eu. *Coccodus* U. Cret. SWAs. *Coelodus* [*Anomiophthalmus Cosmodus Glossodus*] U. Jur.-U. Cret. Eu., L.-U. Cret. NA. Af., U. Cret. Mad., U. Cret.-Eoc. As. *Ellipsodus* L. Cret. Eu. *Eomesodon* U. Trias.-U. Jur. Eu. *Grypodon* [*Ancistrodon Ankistrodus*] U. Cret. Eu. NAf. NA. *Gyrodus* [*Stromateus*] M. Jur.-U. Cret. Eu., U. Jur. WInd., U. Jur., ?U. Cret. NA., U. Cret. Af. *Gyronchus* [*Gynoncus ?Gyroconchus Macromesodon Mesodon Scaphodus Typodus*] M. Jur.-L. Cret. Eu. *Mesturus* U. Jur. Eu. *Micropycnodon* [*Pycnomicrodon*] U. Cret. NA. *Palaeobalistum* [*Palaeobalistes*] U. Cret. As. SA., U. Cret.-Eoc. Eu. Af., *Polygyrodus* U. Cret. Eu. *Proscinetes* [*Microdon Polysephis*] M. Jur.-L. Cret. Eu., L. Cret. NA. *Pycnodus* [*Periodus Pychnodus*] ?M. Jur., L. Cret.-Eoc. Eu., U. Jur. WInd.,

L. Cret. Aus., L.-U. Cret. NA., Eoc. Af. As. *Stemmatodus* ?U. Jur., L. Cret. Eu. *Tibetodus* U. Jur. CAs. *Trewavasia* [*Xenopholis*] U. Cret. WAs. *Uranoplosus* L. Cret. Eu. NA.

ORDER AMIIFORMES

SUBORDER PARASEMIONOTOIDEI. **Parasemionotidae (Ospiidae),** *Broughia ?Helmolepis* L. Trias. Gr. *Jacobulus* L. Trias. Mad. *Ospia* L. Trias. Gr. *Parasemionotus* L. Trias. Mad. Gr. *Promecosomina* M.-U. Trias. Aus. *Stensiöonotus Thomasinotus* L. Trias. Mad. *Tungusichthys* L. Trias. NAs. *Watsonulus* [*Watsonia*] L. Trias. Mad. Gr. **Catervariolidae,** *Catervariolus Lombardina ?Signeuxella* U. Jur. WAf.

SUBORDER AMIOIDEI. **Caturidae (Eugnathidae, Furidae),** *Allolepidotus* [*Plesiolepidotus*] M.-U. Trias. Eu. *Callopterus* U. Jur., ?L. Cret. Eu. *Caturus* [*? Amblysemius Conodus Ditaxiodus Endactis Strobilodus Thlattodus Uraeus*] U. Trias.-L. Cret. Eu., U. Jur. WAf., ?WInd. *Dandya* [*Spaniolepis*] U. Trias. Eu. *Eoeugnathus* M.-U. Trias. Eu. *Eurycormus* U. Jur. Eu. *Furo* [*Eugnathus Isopholis Lissolepis*] M. Trias.-U. Jur. Eu., Jur. As. *Heterolepidotus* [*Brachyichthyes Eulepidotus*] M. Trias.-U. Jur. Eu., Jur. CAs. *Ionoscopus* [*Attakeopsis Macrorhipis Oenoscopus Oeonoscopus Opsigonias*] U. Jur.-L. Cret. Eu. *Lophiostomus* U. Cret. Eu. *Macrepistius* L. Cret. NA. *Neorhombolepis* L.-U. Cret. Eu., L. Cret. SA. *Osteorachis* [*Harpactes Harpactira Isocolum*] L.-U. Jur. Eu. *Otomitla* L. Cret. NA. *Sinoeugnathus* M. Trias. EAs. **Amiidae,** *Amia* [*Amiatus Cyclurus Hypamia ? Kindleia Notaeus Pappichthys Paramiatus Protamia Stylomyleodon*] U. Cret.-R. NA., Paleoc.-Mioc. Eu., Eoc.As. Spits. *Amiopsis* U. Jur.-L. Cret. Eu., ?Jur. NA. *Enneles* U. Cret. SA. *Ikechaoamia* L. Cret. EAs. *Liodesmus* [*Lophiurus*] U. Jur. Eu. *Platacodon* U. Cret. NA. *Pseudamiatus* [*Pseudamia*] Eoc. Spits. *Sinamia* U. Jur. EAs. *Urocles* [*Megalurus Synergus*] U. Jur. Eu., L. Cret. SA. **Macrosemiidae,** *Enceliolepis* U. Jur. Eu. *Eusemius* U. Jur. Eu. NA. *Histionotus* U. Jur.-L. Cret. Eu. *Legnonotus* U. Trias.-L. Cret. Eu. *Macrosemius* [*Disticholepis*] U. Jur. Eu. WAf. *Notagogus* [*Blenniomoeus Blenniomogeus Calignathus*] U. Jur.-L. Cret. Eu. *Ophiopsis* M. Trias.-L. Cret. Eu., U. Jur. WAf. *?Orthurus* U. Trias. Eu. *Petalopteryx* [*Aphanepygus*] L.-U. Cret. Eu., U. Cret. SWAs. *Propterus* [*Rhynchoncodes*] U. Jur.-L. Cret. Eu. *Songanella* U. Jur. WAf. *?Stromerichthys* U. Cret. NAf. *Uarbyichthys* Jur. Aus. **Pachycormidae,** *Asthenocormus* [*Agassizia*] U. Jur. Eu. *? Eugnathides* U. Jur. NA. *Euthynotoides Euthynotus* [*Cyclospondylus Heterothrissops Parathrissops Pseudothrissops*] L. Jur. Eu. *Hypsocormus* U. Trias.-U. Jur. Eu. *? Leedsichthys* [*Leedsia*] *Orthocormus* U. Jur. Eu. *Pachycormus* [*Cephenoplosus ? Pachylepis ? Lycodus*] *Prosauropsis* [*Protosauropsis Saurostomus*] L. Jur. Eu. *Protosphyraena* [*Erisichthe Pelecopterus*] U. Cret. Eu. As. NA. SA. *Sauropsis* [*Diplolepis*] L.-U. Jur. Eu., Jur. NA.

ORDER ASPIDORHYNCHIFORMES. **Aspidorhynchidae,** *Aspidorhynchus* M.-U. Jur. Eu., Cret. SA. Aus. *Belonostomus* [*Belonostmus Dichelospondylus Hemirhynchus Ophirachis Vinctifer*] U. Jur.-U. Cret. Eu., L.-U. Cret. NA. As. SA. Aus., U. Cret. NAf.

ORDER PHOLIDOPHORIFORMES (HALECOSTOMI). **Pholidophoridae,** *? Ceramurus* U. Jur. Eu. *? Flugopterus* [*Megalopterus*] U. Trias. Eu. *Hungkiichthys* U. Jur. EAs. *Pholidophorides* L. Jur. Eu. *Pholidophoristion* U. Jur. Eu. *Pholidophorus* [*Baleiichthys ? Microps ? Nothosomus Phelidophorus Poriergus*] ?L. Trias., U. Jur. SA. M. Trias.-U. Jur. Eu., U. Trias., U. Jur. WAf., L. Jur. NA., U. Jur. As., *? Prohalecites* M.-U. Trias. Eu. **Ichthyokentemidae,** *Ichthyokentema* U. Jur. Eu. NAs. **Majokiidae,** *Majokia* U. Jur. WAf. **?Ligulellidae,** *Ligulella* U. Jur. WAf. **Pleuropholidae,** *Austropleuropholis Parapleuropholis* U. Jur. WAf. *Pleuropholis* U. Jur.-L. Cret. Eu., U. Jur. WAf. **Archaeomaenidae,** *Archaeomaene* M. Trias.-Jur. Aus. *Madariscus* Jur. Aus. **Oligopleuridae,** *Calamopleurus* U. Jur.-L. Cret. Eu., L. Cret. SA. *Galkinia* Jur. WAs. *Oligopleurus* U. Jur.-L. Cret. Eu., U. Cret. SA. *Spathiurus* [*Amphilaphurus*] U. Cret. SWAs.

INFRACLASS TELEOSTEI

SUPERORDER LEPTOLEPIMORPHA

ORDER LEPTOLEPIFORMES. **Leptolepidae,** *Anaethalion* [*Aethalion*] U. Jur.-U. Cret. Eu. *Carsothrissops* U. Cret. Eu. *Cearana* L. Cret. SA. *? Clupavus* U. Jur.-L. Cret. Eu., Cret. Af. NA. *Cteniolepis* U. Jur. Eu. *Eurystichthys* [*Eurystethus*] U. Jur. Eu. *Haplospondylus* L. Cret. SA. *Leptolepis* [*Ascalabos Liassolepis Megastoma Sarginites Tharsis*] U. Trias.-U. Cret. Eu. NA., ?Trias. Mad., Jur. Spits. Aus., Jur.-Cret. As., ?Jur., L.-U. Cret. SA., L.-U. Cret. Af. *Luisichthys* U. Jur. WInd. *Tharrias* U. Cret. SA. *Vidalamia* [*Vidalia*] U. Jur.-L. Cret. Eu. **Lycopteridae,** *Lycoptera* [*Asiatolepis*] U. Jur. EAs., ?Eu. *? Neolycoptera* ?U. Jur. SA. *Manchurichthys* L.-M. Cret. EAs. *Mesoclupea Sungarichthys* L. Cret. EAs.

SUPERORDER ELOPOMORPHA

ORDER ELOPIFORMES

SUBORDER ELOPOIDEI. **Elopidae + Megalopidae** R. Oc., *? Acrogrammatolepis* U. Cret. Eu. *Brannerion* U. Cret. SA. *Broweria* Mioc. EInd. *Caeus* (*Cacus*) U. Cret. Eu. *Camalopleurus* L. Cret. SA. *Coryphaenopsis Dinelops* U. Cret. Eu. *Ectasis* Plioc. NA. *Elopopsis* L.-U. Cret. Eu., U. Cret. Af. *Elops* Eoc. Eu., R. Oc. *Eoprotelops* U. Jur. Eu. *Esocelops* [*Eurygnathus*] Eoc. Eu. *Flindersichthys* U. Cret. Aus. *Helmintholepis* U. Cret. NA. *Histialosa* [*? Hemielopopsis*] L. Cret. Eu. *Hypsospondylus* U. Cret. Eu. *Laimingia Laminospondylus* U. Cret. NA. *? Lastroichthys* L. Cret. SA. *Lyrocephalus* Eoc. Eu. WAs. *Megalops* Eoc. Eu., L. Tert. Aus., ?Mioc. NA., R. Oc. *Notelops* U. Cret. SA. *Osmeroides* [*Dermatoptychus Dypterolepis Holcolepis ? Kymatopetalolepis Leptogrammatolepis ? Micropetalopsis Petalolepis Rhabdolepis Sardinoides*] U. Cret. As. Af. NA. SA., U. Cret.-Paleoc. Eu. *Parelops Protelops* U. Cret. Eu. *Rhacolepis* U. Cret. SA. *Sauropsidium* [*Hyptius*] U. Cret. Eu. *Spaniodon* [*Lewisia*] U. Cret. SWAs. ?NA. *? Thrissops* [*Eubiodectes*] ?M. Jur., U. Jur.-L. Cret. Eu. *Thrissopteroides* U. Cret. Eu. SWAs. **Apsopelicidae,** *Apsopelix* [*Helmintholepis Leptichthys Palaeoclupea Pelycorapis Syllaemus*] U. Cret. Eu. NA. **Pachyrhizodontidae (Thrissopateridae),** *? Conosaurus* [*Conosaurops*] Eoc. NA. *Cyclotomodon* U. Cret. NA. *Pachyrhizodus* [*Acrodontosaurus ? Kansanus ? Megalodon Oricardinus Raphiosaurus*] U. Cret. Eu. NAf. Aus. NA. *Thrissopater* U. Cret. Eu. NAf. NA.

SUBORDER ALBULOIDEI. **Albulidae** R. Oc., *Albula* [*Glossodus Pisodus*] Paleoc-Eoc. Eu. NAf., Eoc. NA., R. Oc. *? Ancylosteus* U. Cret. Eu. *Chanoides* Eoc. Eu. *Chicolepis*

U. Cret. NA. *Eucoelogaster* [*Coelogaster*] Eoc. Eu. *Haljulia* U. Cret. SWAs. *Istieus* [*Histieus*] U. Cret. Eu. *Kleinpellia* U. Cret. NA. *Monopterus* Eoc. Eu. *Paralbula* Mioc. NA. *Pterothrissus* ?M. Jur., U. Jur.-U. Cret. Eu., Eoc. Eu., Olig. Aus.

ORDER ANGUILLIFORMES (APODES)

SUBORDER ANGUILLOIDEI. **Anguillavidae,** *Anguillavus* U. Cret. SWAs., ?NA. **Anguillidae** R. Cos., *Anguilla* Eoc.-R. Eu., R. EAs. NAf. NA. Aus. Oc. *Mastygocerus* Mioc.-R. EInd. **Moringuidae** R. Oc. **Myrocongridae** R. SAtl. Oc. **Xenocongridae** R. Oc., *Echelus Eomyrophis* [*Eomyrus*] Eoc. Eu., R. Oc. *Mylomyrus* Eoc. NAf. *Paranguilla* [*Enchelyopus*] *Rhynchorinus* Eoc. Eu. **Muraenidae** R. Oc., *?Deprandus* Mioc. NA. *Muraena* Plioc. NAf., R. Oc. **Heterenchelyidae** R. Oc., *Heterenchelys* Mioc. Aus., R. Oc. **Dyssomminidae** R. Oc. **Muraenesocidae** R. Oc., *Muraenesox* Eoc. Eu., Mioc. Aus., R. Oc. **Neenchelyidae** R. Ind. Oc. **Nettastomatidae** R. Oc., *Nettastoma* Eoc. Eu., R. Oc. **Nessorhamphidae** R. Oc. **Congridae** R. Oc., *Ariosoma* [*Congermuraena*] Olig.-Plioc. Eu., R. Oc. *Astroconger* Olig. Aus. *Conger* Eoc.-Olig. NA., Eoc.-Mioc. Eu., Mioc. NZ., R. Oc. *Enchelion* U. Cret. SWAs. *Parbatmya* Eoc. NA. *Uroconger* Eoc.-Mioc. Eu., Mioc.-Plioc. Aus., R. Oc. **Ophichthidae** R. Oc., *Caecula* Mioc. Eu., R. Oc. *Mystriophis* ?Mioc. NZ., R. Oc. *Ophichthus* Eoc. Eu., R. Oc. **Todaridae Synaphobranchidae** R. Oc. **Simenchelyidae** R. NAtl. NPac. Oc. **Dysommidae Derichthyidae Macrocephenchelyidae Serrivomeridae** R. Oc. **Nemichthyidae** R. Oc., *Nemichthys* Eoc.-Olig. WAs., R. Oc. **Cyemidae Aoteidae** R. Oc. **Urenchelyidae,** *Urenchelys* U. Cret. Eu. SWAs.

SUBORDER SACCOPHARYNGOIDEI (LYOMERI). **Saccopharyngidae Eurypharyngidae Monognathidae** R. Oc.

ORDER NOTACANTHIFORMES (LYOPOMI). **Halosauridae** R. Oc., *Echidnocephalus* U. Cret. Eu., ?NA. *Enchelurus* U. Crèt. Eu. SWAs. *Halosaurus* Mioc. NA., R. Oc. *Laytonia* ?Cret., Olig.-Plioc. NA. **Lipogenyidae** R. Oc. **Notacanthidae** R. Oc., *Pronotacanthus* U. Cret. SWAs., Olig. Eu.

SUPERORDER CLUPEOMORPHA

ORDER CLUPEIFORMES

SUBORDER DENTICIPITOIDEI. **Denticipitidae** R. Af., *Paleodenticeps* Tert. EAf.

SUBORDER CLUPEOIDEI. **Clupeidae + Engraulidae** R. Cos., *Alisea* Mioc. NA. *Alosa* [*Alausa Caspialos Clupeonella*] ?Eoc. NA., Olig.-Plioc. Eu., Plioc. NAf., Plioc.-R. WAs., R. Oc. *Austroclupea* Plioc. SA. *Bramlettia* U. Cret. NA. *Brevoortia* Plioc. NAf., R. Oc. *?Chanopsis* L. Cret. WAf. *Clupea* [*Alosina Clupeops Sahelinia*] Eoc.-Pleist. Eu., Olig. SA., Mioc. WAs. EInd., Plioc. NAf., R. Oc. *Clupeopsis* Eoc. Eu. *Crossognathus* L. Cret. Eu. *Diplomystus* [*Copeichthys Histiurus Hyperlophus ?Oncochetos*] U. Cret. SWAs., U. Cret.-Eoc. Af. NA. SA., U. Cret.-Mioc. Eu. *Driverius* U. Cret. NA. *Engraulis* Eoc.-Mioc. Eu., R. Oc. *Engraulites Epelichthys Etringus* Mioc. NA. *Etrumeus* [*Halecula Parahalecula*] Olig. WAs., Mioc.-Plioc. Eu., Plioc. NAf., R. Oc. *Ganoessus Ganolytes* [*Diradias*] Mioc. NA. *Gasteroclupea* U. Cret. SA. *Hacquetia Halecopsis* Eoc. Eu. *Hayina* [*Jobertina Smithites*] Mioc. NA. *Histiothrissa* U. Cret. Eu. SWAs. Af. *Iquius* U. Tert. EAs. *Knightia* [*Ellimma Ellipos*] Eoc. SA. *Lembacus* Mioc. NA. *Lygisoma* Mioc. NA. *Melettina Neohalecopsis* Olig. Eu. *Opisthonema* Mioc. NA., R. Oc. *?Ostariostoma* U. Cret. NA. *Paraclupavus* U. Jur.-U. Cret. Af. *Paraclupea* L. Cret. EAs. *Pateroperca* U. Cret. SWAs. *Pomolobus* Mioc. Eu., R. Oc. *Pseudoberyx* U. Cret. SWAs. *Pseudochilsa* Mioc. Eu., R. Oc. *Pseudoetringus* Eoc. NA. *Quisque* Mioc. NA. *Rhomarus* Mioc. NA. *Sardinella* Mioc. Eu., R. Oc. *Sarmatella* Mioc.-R. Eu. *Scombroclupea* L.-U. Cret. Eu., U. Cret. SWAs. SA. *Sprattus* [*Meletta*] Olig. Eu., R. Oc. *Steinbergia* Mioc. NA. *Stolephorus* [*Sprattelodes*] Mioc. NA., Plioc. NAf., R. Oc. *Syllaemus* U. Cret. Eu. NA. *Wisslerius* U. Cret. NA. *Xenothrissa Xyne Xyrinus* Mioc. NA. **Chirocentridae** R. Oc., *Allothrissops* U. Jur.-L. Cret. Eu. *Chirocentrum* ?U. Tert. EInd., R. Oc. *Pachythrissops* [*Parathrissops*] U. Jur., L.-U. Cret. Eu. *Platinx* [*Thrissopterus*] Eoc. Eu.

SUPERORDER OSTEOGLOSSOMORPHA

ORDER OSTEOGLOSSIFORMES

SUBORDER ICHTHYODECTOIDEI. **Ichthyodectidae,** *Chirocentrites* [*Andreiopleura*] L.-U. Cret. Eu., L. Cret. WAf. *Cladocyclus* [*Anaedopogon Chiromystus Eunelichthys*] L.-U. Cret. SA. *Ichthyodectes* [*Gillicus*] L.-U. Cret. Eu. NA., ?U. Cret. Af. *Notodectes* L. Cret. SA. *Proportheus* U. Cret. Af. SA. *Prymnetes* U. Cret. NA. *Spathodactylus* L. Cret. Eu. *Xiphactinus* [*? Hypsodon Portheus*] L.-U. Cret. Eu., U. Cret. SWAs. NAf. Aus. NZ. NA. **Saurocephalidae,** *Saurocephalus* L.-U. Cret. Eu., U. Cret. NA., ?NAf. *Saurodon* [*Daptinus*] U. Cret. Eu. ?NAf. NA. **Thryptodontidae (Plethodontidae),** *Bananogmius* [*Ananogmius Anogmius*] U. Cret. Eu. NA. *Martinichthys Niobrara* U. Cret. NA. *Paranogmius* U. Cret. NAf. *Plethodus* U. Cret. Eu. NAf. *Syntegmodus Thryptodus* [*? Pseudothryptodus*] *Zanclites* U. Cret. NA.

SUBORDER OSTEOGLOSSOIDEI. **Osteoglossidae** R. Af. SAs. EInd. Aus. SA. *? Brychetus* [*Platops Pomaphractus*] Eoc. Eu. NAf. *? Eurychir* U. Cret. NA., Paleoc.-Eoc. NAf. *? Genartina* Eoc. NA. *Musperia* L. Tert. EInd. *Phareodus* [*Dapedoglossus*] Eoc. NA., Olig. Aus. *Scleropages* Olig-R. EInd., Mioc.-R. Aus. **Pantodontidae** R. Af.

SUBORDER MORMYRIFORMES (SCYPHOPHORI). **Mormyridae** R. Af. **Gymnarchidae** R. Af.

SUBORDER NOTOPTEROIDEI. **Hiodontidae** R. NA. **Notopteridae** R. Af. SAs. EInd., *Notopterus* L. Tert.-R. EInd. SAs., R. Af.

SUPERORDER PROTACANTHOPTERYGII

ORDER SALMONIFORMES

SUBORDER SALMONOIDEI. **Salmonidae** R. Oc. Eu. As. NA., *Beckius* Eoc. NA. *Coregonus* Pleist. Eu., R. As. NA. *Cyclolepis* U. Cret. NA. *Cyclolepoides* Eoc. NA. *Goudkoffia Leucriteryops Natlandia* U. Cret. NA. *Parastenodus* Eoc. NA. *Procharacinus Prohydrocyon* Eoc. Eu. *Prosopium* Pleist.-R. NA. *Protothymallus* [*Prothymallus*] Mioc. Eu. *Salmo* [*Rhabdofario*] Mioc.-R. Eu. NA., R. Oc. *?Salmodium* Eoc. Eu. *Salvelinus* Pleist.-R. Eu., ?Pleist. NA. *Thaumaturus* Eoc.-Mioc. Eu. *Thymallus* Eoc.-Mioc. Eu., R. NOc. **Plecoglossidae** R. NOc. **Osmeridae** R. NOc., *Mallotus* Pleist. Eu. NA. Gr., R. NAtl. *Osmerus* Mioc. Eu., R. Oc.

SUBORDER ARGENTINOIDEI. **Argentinidae + Bathylagidae** R. NOc., *Argentina* Eoc.-Mioc. Eu., R. NOc. *Azalois* Mioc. NA. *Bathylagus* [*Quaesita*] Mioc. NA., R. NOc.

Lygisma Mioc. NA. *Proargentina* Mioc. Eu. *Sternbergia* Mioc. NA. **Opisthoproctidae** R. NOc.

SUBORDER GALAXIOIDEI. **Salangidae** R. EAs. **Retropinnidae** R. Aus. NZ. **Galaxiidae** R. SAf. Aus. NZ. SA., *Galaxias* Olig., Plioc.-R. NZ. **Aplochitonidae** R. Aus. SA.

SUBORDER ESOCOIDEI (HAPLOMI). **Palaeoesocidae,** *Palaeosox* Eoc. Eu. **Esocidae,** *Esox* [*Trematina*] Olig.-R. Eu., Tert.-R. As., R. NA. **Umbridae** R. Eu. NAs. NA., *Umbra* Olig.-R. Eu.

SUBORDER STOMIATOIDEI. **Gonostomidae** R. Oc., *Cyclothone* [*Rogenius*] Mioc. NA., R. Oc. *?Gonorhynchops* U. Cret. Eu. NA. *Gonostoma* Mioc. Eu., Plioc. NAf. *Indrissa* U. Cret. NAf. *Ohuus* Mioc. EAs. *Paravinciguerria* U. Cret. NAf. *Photichthys* Mioc. Eu., Plioc. NAf., R. Oc. *Scopeloides* [*Mrazecia*] *Vinciguerria* [*Zalarges*] Olig.-Mioc. Eu. **Sternoptychidae** R. Oc., *Argyropelecus* Olig.-Mioc. Eu., Mioc. NAf. NA., R. Oc. *Polyipnoides* Eoc. Eu. *Polyipnus Sternoptyx* Olig. Eu., R. Oc. **Astronesthidae Melanostomiidae Malacosteidae** R. Oc. **Chauliodontidae** R. Oc., *Chauliodus* Mioc. NA., R. Oc. **Protostomiatidae,** *Pronotacanthus* U. Cret. SWAs. *Protostomia* U. Cret. NAf. **Stomiatidae** R. Oc., *Eostomias* Mioc. NA. **Idiacanthidae** R. Oc.

SUBORDER ALEPOCEPHALOIDEI. **Alepocephalidae** R. Oc., *Palaeotroctes* Mioc. Eu. *Xenodermichthys* Mioc. Eu., R. Oc.

SUBORDER BATHYLACONOIDEI. **Bathylaconidae** R. Oc.

SUBORDER MYCTOPHOIDEI. **Enchodontidae,** *Apateodus* U. Cret. Eu., U. Cret.-Paleoc. NAf. *Cimolichthys* [*Empo Plinthophorus*] U. Cret. Eu. NA. SA., U. Cret.-Paleoc. Af. *Diplolepis* U. Cret. Eu. *Enchodus* [*Eurygnathus Holcodon Ischyrocephalus Isodon Phasganodus Solenodon Tetheodus*] U. Cret. SWAs. NA., U. Cret.-Paleoc. Eu. Af., U. Cret.-Eoc. SA. *Eurypholis* [*Saurorhamphus*] U. Cret. Eu. NAf. SWAs. *Halec* [*Archaeogadus Phylactocephalus Pomognathus*] U. Cret. Eu. SWAs. *Halecodon Leptecodon ?Luxilitos* U. Cret. NA. *Palaeolycus* U. Cret. Eu. *Pantophilus* U. Cret. SWAs. *Prionolepis* U. Cret. Eu. *Rharbichthys* U. Cret. NAf. *Volcichthys* U. Cret. Eu. **?Tomognathidae,** *Tomognathus* U. Cret. Eu. **Ipnopidae** R. Oc. **Paralepidae (Sudidae),** *Holosteus* [*Pavlovichthys*] Eoc.-Olig. Eu. *Iniomus* Eoc.-Olig. NA. *Lestichthys* Mioc. NA. *Paralepis* [*Anapterus Tydeus*] Mioc. Eu. NAf., R. Oc. *Parascopelus* Mioc. Eu. *Sudis* Mioc.-Plioc. Eu. NAf., R. Oc. *Trossulus* Mioc. NA. **Omosudidae Aleposauridae Anotopteridae Evermannellidae Scopelarchidae Scopelosauridae** R. Oc. **Myctophidae (Scopelidae)** R. Oc., *Acrognathus* U. Cret. Eu. SWAs. *Cassandra* [*Leptosomus*] U. Cret. Eu. As. *Ceratoscopus* ?Mioc. NA. *?Dactylopogon* U. Cret. Eu. *Diaphus* Olig.-Mioc. Eu., R. Oc. *Eomyctophum* Olig.-Mioc. Eu., Olig. WAs. *Hakelia* U. Cret. SWAs. *Hemisaurida* U. Cret. Eu. *Ichthyotringa* [*Rhinellus*] U. Cret. Eu. SWAs. NAf. NA. *Lampanyctus* Mioc. NZ. NA., Mioc.-Plioc. Eu. *Myctophum* [*Hygophum Scopelus*] Eoc.-Plioc. Eu., Mioc. NZ. NA., Plioc. NAf., R. Oc. *?Nematonotus* U. Cret. SWAs. *Nyctophus* Mioc.-Plioc. Eu. NA. *Omiodon* Eoc. Eu. *Opistopteryx* U. Cret. SWAs. *?Palimphemus* Mioc. Eu. *?Rhamphornimia* U. Cret. SWAs. *Sardinius* U. Cret. Eu. SWAs. NA. *Sardinoides* U. Cret. Eu. SWAs. *Sedenhorstia* [*Micrococelia*] U. Cret. Eu. SWAs. *Tachynectes* U. Cret. Eu. **Neoscopelidae** R. Oc. **Cheirothricidae,** *Cheirothrix* [*Megapus Megistopus*-U. Cret. Eu. SWAs. *Exocoetoides* U. Cret. SWAs. *Tele*] *pholis* U. Cret. Eu. SWAs. **Dercetidae,** *Benthesikyme* [*Leptotrachelus*] U. Cret. Eu. SWAs. NA. *Dercetis* U. Cret. Eu. As. *Pelargorhynchus* U. Cret. Eu. *Prionolepis* [*Apateopholis Aspidopleurus*] U. Cret. Eu. SWAs. *Rhynchodercetis* U. Cret. Eu. WAf. *Stratodus* U. Cret. SWAs. NA., U. Cret.-Paleoc. NAf. *Trianaspis* U. Cret. NA.

ORDER CETOMIMIFORMES

SUBORDER CETOMIMOIDEI. **Cetomimidae Barbourisiidae Rondeletiidae** R. Oc.

SUBORDER ATELEOPODOIDEI. **Ateleopodidae (Podatelidae)** R. Oc.

SUBORDER MIRAPINNOIDEI. **Mirapinnidae Eutaeniophoridae (Taeniophoridae)** R. Oc.

SUBORDER GIGANTUROIDEI. **Giganturidae Rosauridae** R. Oc.

ORDER CTENOTHRISSIFORMES. **Ctenothrissidae,** *Ctenothrissa* [*Aenothrissa*] U. Cret. Eu. SWAs. **Aulolepidae,** *Aulolepis* [*?Codonolepis ?Perigrammatolepis*] U. Cret. Eu. **Macristiidae** R. NAtl.

ORDER GONORHYNCHIFORMES

SUBORDER GONORHYNCHOIDEI. **Gonorhynchidae** R. Indo-Pac. Oc., *Charitosomus* [*Solenognathus*] U. Cret. Eu. SWAs. NA. *Gonorhynchops* U. Cret. Eu. *?Notogoneus* [*Colpopholis Phalacropholis Protocatostomus Sphenolepis*] Eoc.-Olig. Eu., Eoc. NA., Olig. Aus.

SUBORDER CHANOIDEI. **Chanidae** R. Indo-Pac. Oc., *?Ancylostylus* U. Cret. Eu. *?Chanopsis* L. Cret. WAf. *Chanos* Eoc.-Mioc. Eu., R. Indo-Pac. Oc. *Dartbile* U. Cret., ?Eoc. SA. *Parachanos* L. Cret. Af. SA., U. Cret. Eu. *Prochanos* L. Cret. Eu. **Kneriidae Phractolaemidae** R. Af.

SUPERORDER OSTARIOPHYSI

ORDER CYPRINIFORMES

SUBORDER CHARACOIDEI. **Characidae** R. Af. SA. CA., *Alestes* Tert.-R. Af. *Astyanax Brycon* Pleist.-R. SA. *Characilepis* Mioc. SA. *Eobrycon* L. Tert.-Plioc. SA. *?Erythrinolepis* U. Cret. NA. *Lignobrycon* Eoc. SA. *Pareobasis* Pleist.-R. SA. *Procharax* Plioc. SA. *Triportheus* Pleist.-R. SA. **Erythrinidae Ctenoluciidae** R. SA. **Hepsetidae** R. Af. **Cynodontidae** R. SA. **Lebiasinidae** R. SA. CA. **Parodontidae Gasteropelecidae Prochilodontidae** R. SA. **Curimatidae** R. SA., *Curimatus* Pleist.-R. SA. **Anostomidae Hemidontidae Chilodontidae** R. SA. **Distichodontidae Citharinidae Ichthyboridae** R. Af.

SUBORDER GYMNOTOIDEI. **Gymnotidae** R. SA.

SUBORDER CYPRINOIDEI. **Cyprinidae** R. Eu. As. Af. NA., *Abramis* [*Ballerus*] Mioc. NAs., Mioc.-R. Eu. *Acrocheilus* Plioc.-R. NA. *Alburnoides* Mioc.-R. As., R. Eu. *Alburnus* Mioc.-R. Eu., Plioc.-R. As. *Alisodon* Pleist. NA. *Amblypharyngodon* U. Tert.-R. EInd. *Aphelichthys* Pleist.-R. NA. *Aspiurnus* U. Tert. NAs. *Aspius* Mioc.-R. Eu. *Barbus* Olig.-R. Eu., Mioc.-R. Af. As., U. Tert.-R. EInd. *Blicca* Paleoc.-R. Eu., Mioc.-R. NAs. *Campostoma* Pleist.-R. NA. *Capitodus* Mioc. Eu. *Carassius* Plioc.-R. As., R. Eu. *Chela* Eoc. Eu., R. As. EInd. *Chondrostoma* Olig.-R. Eu., ?Mioc. As. *Chrosomus* Pleist.-R. NA. *Ctenopharyngodon* Plioc.-R. As. *Cyprinus* Mioc.-R. Eu., Plioc.-R. As. *Daunichthys* Tert.-R. EInd. *Diastichus* Pleist. NA. *Dionda* Pleist.-R. NA. *Enoplophthalmus*

Mioc. Eu. *Eocyprinus* L. Tert. EInd. *Evomus* Plioc. NA. *Gila* [*? Anchypopsis Siphateles*] Mioc.-R. NA. *Gobio* Mioc.-R. Eu. *Hemiculturella* Plioc.-R. As. *Hemitrichas* Olig. Eu. *Hexapsephus* Tert.-R. EInd. *Hybognathus* Pleist.-R. NA. *Hybophthalmichthys* Plioc.-R. EAs. *Hybopsis Ictiobus* Pleist.-R. NA. *Leuciscus* Olig.-R. Eu., Mioc.-R. As., R. Af. NA. *Leucus* Pleist. NA. *Mylocyprinus* Plioc.-Pleist. NA. *Mylopharodon* Plioc.-R. NA. *Mylopharyngodon* Plioc-R. EAs. *Notemigonus* Pleist.-R. NA. *Notropis* Plioc.-R. NA. *Oligobelus* Pleist.-R. NA. *Osteochilus* Tert.-R. EInd. *Paraleuciscus* Mioc. Eu. *Pelecus* [*Culter*] Plioc.-R. *Pimephales ? Proballostomus* Pleist.-R. NA. *Pseudorasbora* Mioc.-R. As. *Ptychocheilus* [*Squalius*] Plioc.-R. NA. *Puntius Rasbora* Tert.-R. EInd. *Rodeus* Mioc.-R. Eu. *Rutilus* Eoc.-R. Eu. *Scardinius* Olig.-R. Eu., Mioc. NAs. *Semotilus* Pleist.-R. NA. *Sigmopharyngodon* Pleist. NA. *Soricidens* Mioc. Eu. *Thynnichthys* Tert.-R. EInd. *Tinca* [*Tarsichthys*] Olig.-R. Eu. *Varicorhinus* Plioc.-R. Eu. As. *Xenocypris* Plioc.-R. As. **Gyrinocheilidae** R. EAs. **Psilorhynchidae** R. SAs. **Catostomidae** R. EAs. NA., *Amyzon* Eoc.-Mioc. NA. *Carpiodes* Pleist.-R. NA. *Catostomus* Eoc.-R. As., Pleist.-R. NA. *Chasmistes* ?Plioc., Pleist.-R. NA. *Deltistes* Pleist.-R. NA. *Moxostoma* Pleist.-R. NA. *Pantosteus* Plioc.-R. NA. **Homalopteridae** R. As. **Cobitidae** R. Eu. As. EInd. NAf. NA., *Cobitis* [*Acanthopsis*] Mioc.-R. Eu., R. As. *Nemacheilus* Olig.-R. Eu., R. As. NAf.

ORDER SILURIFORMES (NEMATOGNATHI). **Diplomystidae** R. SA. **Ictaluridae (Ameiuridae)** R. As. NA. CA., *Ictalurus* [*Ameiurus*] ?Eoc., Olig. As., Mioc.-R. NA. *Pylodictis* Pleist.-R. NA. **Bagridae (Porcidae Mystidae)** R. As. EInd. Af., *Aoria* Tert.-R. As. EInd. *Bagre* [*Bagrus Felichthys Porcus*] Pleist.-R. Af. *Bucklandium* [*Glyptocephalus*] Eoc. Eu. *Chrysichthys* Pleist.-R. Af. *? Claibornichthys* Eoc. NA. *Eaglesomia* Eoc. Af. *Eomacronas* [*Macronoides*] Eoc. WAf. *Fajumia* Eoc. NAf. *Heterobagrus* Plioc.-R. As. *Mystus* [*Macrones Macronichthys*] Pleist.-R. EInd. *Nigerium* Eoc. WAf. *Rita* Plioc.-R. As. *Socnopaea* Eoc. NAf. **Cranoglanidae** R. EAs. **Siluridae** R. Eu. As. Af., *? Bachmannia* Tert. SA. *Parasilurus* Plioc.-R. EAs. *Pliosilurus* Plioc.-R. As. *Silurus* Eoc.-R. Eu., Plioc.-R. As. **Schilbeidae + Pangasiidae** R. SAs. EInd. Af., *Pangasius* L. Tert.-R. As. *Pseudeutropius* [*? Brachyspondylus*] Tert.-R. EInd. **Amblycipitidae Amphiliidae** R. Af. **Akysidae** R. SAs. **Sisoridae (Bagariidae)** R. SAs. EInd., *Bagarius* Plioc.-R. SAs., Tert.-R. EInd. **Clariidae** R. SAs. EInd. Af., *Clarias* Plioc.-R. SAs. Af., Pleist.-R. EInd. *Clarotes* Pleist.-R. Af. *Heterobranchus* Plioc. Eu., Pleist.-R. Af. **Heteropneustidae** R. SAs. **Chacidae** R. SAs. EInd. **Olyridae** R. SAs. **Malapteruridae** R. SA. **Mochokidae (Synodontidae)** R. Af., *Synodontis* Mioc. Eu., Plioc.-R. Af. **Ariidae (Tachysuridae)** R. tropical coasts, *Arius* [*Tachysurus*] Eoc.-R. Af. SAs. EInd. WInd. SA. *Auchenoglanis* Pleist.-R. Af. *Eopeyeria* [*Ariopsis Peyeria*] Eoc. NAf. *Osteogeneiosus* Eoc. Af., R. EInd. *Rhineastes* [*Astephas*] Eoc.-Olig. NA., Eoc.-R. As. **Doradidae Auchenipteridae Aspredinidae (Bunocephalidae)** R. SA. **Plotosidae** R. coasts As. Aus. EAf., *Tandanus* Olig.-R. Aus. **Pimelodontidae** R. CA. SA., *Pimelodus* Tert.-R. SA., ?Eoc. Af. **Ageneiosidae Hypopthalmidae Helogeneidae Cetopsidae** R. SA. **Trichomycteridae (Stegophilidae Pygidiidae Eretmophilidae)** R. SA., *Propygidium* Eoc. SA. **Callichthyidae** R. SA., *Corydoras* Mioc.-R. SA. **Loricariidae** R. SA.

SUPERORDER PARACANTHOPTERYGII

ORDER AMBLYOPSIFORMES (SALMOPERCAE)

SUBORDER AMBLYOPSOIDEI. **Amblyopsidae** R. NA.

SUBORDER APHREDODEROIDEI. **Aphredoderidae** R. NA., *Amphiplaga Asineops Erismatopterus* Eoc. NA. *Trichophanes* Mioc. NA.

SUBORDER PERCOPSOIDEI. **Percopsidae** R. NA.

ORDER BATRACHOIDIFORMES (HAPLODOCI). **Batrachoididae** R. Oc. CA. SA., *Batrachoides* Mioc.-Plioc. NAf., R. Oc.

ORDER GOBIESOCIFORMES (XENOPTERI). **Gobiesocidae** R. Oc., *? Bulbiceps* Mioc. NA.

ORDER LOPHIIFORMES (PEDICULATI)

SUBORDER LOPHIOIDEI. **Lophiidae** R. Oc., *Lophius* Eoc. Eu., Plioc. NAf., R. Oc.

SUBORDER ANTENNARIOIDEI. **Brachionichthyidae** R. Oc. **Antennariidae** R. Oc., *Histionotophorus* [*Histiocephalus*] Eoc. Eu. **Chaunacidae Ogcocephalidae** R. Oc.

SUBORDER CERATIOIDEI. **Melanocetidae Diceratiidae Himantolophidae Oneirodidae Gigantactinidae Neoceratiidae Centrophrynidae Ceratiidae Caulophrynidae Linophrynidae** R. Oc.

ORDER GADIFORMES (ANACANTHINI)

SUBORDER MURAENOLEPIDOIDEI. **Muraenolepididae** R. SOc.

SUBORDER GADOIDEI. **Moridae + Bregmacerotidae + Gadidae + Merulcciidae** R. Oc. Eu. As. NA., *Arnoldites* [*Arnoldina*] Plioc. NA. *Bregmacernia* Mioc. Eu. *Bregmaceros* [*Podopteryx*] ?Eoc. NZ., Eoc.-Mioc. NA., Eoc.-Plioc. Eu., Olig. WAs., Mioc. Aus., Plioc. NAf., R. Oc. *Brosmius* [*Brosme*] Olig.-Plioc. Eu., Plioc. NAf., R. Oc. *Eclipes* [*Merriamina*] Mioc. NA. *Gadus* [*Morhua*] Paleoc.-Pleist. Eu., Olig. Aus., Pleist. Gr., R. Oc. *Lepidion* Mioc. EAs., R. Oc. *Lota* Plioc.-R. Eu., R. As. NA. *Lotella* Mioc.-R. Eu. *Melanogrammus Melanotus* Olig. Eu., R. Oc. *Merlangus* Mioc. Eu., R. Oc. *Merluccius* [*Spinogadus*] Eoc.-Mioc. Eu., Mioc. Aus. NZ., R. Oc. *Molva* Plioc. Eu., R. Oc. *Neobythites* Eoc. Eu., R. Oc. *Odontogadus* Mioc. Eu. *Onobrosmius* Mioc. WAs., Mioc.-Plioc. Eu. *Palaeogadus* [*Lotimorpha Megalolepis Nemopteryx Palaeobrosmius Pseudolota Ruppelianus*] Eoc. WAs., Olig.-Mioc. Eu. *Palaeomolva* Mioc. Eu. *Paratichthys* U. Cret. NA. *Petalolepis* U. Cret. Eu. *Physiculus* Mioc. Eu. NZ., R. Oc. *Progadius* Plioc. NA. *Promerluccius* Olig. WAs. *Raniceps* Eoc.-Mioc. Eu., R. Oc. *Strinsia* Mioc. Eu., R. Oc. *Urophycis* [*Phycis*] Eoc.-Mioc. Eu., R. Oc.

SUBORDER OPHIDIOIDEI. **Ophidiidae (Brotulidae)** R. Oc. WInd., *Bauzaia* Eoc. NA. *Brotula* Mioc.-Plioc. Eu., R. Oc. *Glyptophidium* Mioc. EAs. *Neobythites* Eoc. Eu., R. Oc. *Ophidion* [*Ophidium*] ?Paleoc., Eoc.-Mioc. Eu., Eoc. WInd., Mioc. Aus., R. Oc. *Preophidion* Eoc. NA. *Protobrotula* Olig.-Mioc. Eu. *Signata* Eoc. NA. **Carapidae (Fierasferidae)** R. Oc., *Carapus* [*Fierasfer Jordanicus*] Olig.-Mioc. NZ., Mioc. Aus., R. Pac. Oc. **Pyramodontidae** R. Oc.

SUBORDER ZOARCOIDEI. **Zoarcidae (Lycodontidae Lycididae)** R. Oc.

SUBORDER MACROUROIDEI. **Macrouridae (Coryphaenoididae)** R. Oc., *Amblygoniolepidus* Olig. NA. *Bolbocara* Mioc. NA. *Calilepidus* Olig.-Mioc. NA. *Coelorincus* Mioc. Aus.,

R. Oc. *Homeocoryphaenoides* Mioc. NA. *Homeomacrurus* Plioc. NA. *Homeonezumia* Olig. NA. *Hymenocephalus* Olig.-Mioc. Eu., R. Oc. *Leptacantholepidus* Mioc. NA. *Macrourus* [*Coryphaenoides*] Eoc.-Mioc. Eu., Mioc. NA. NZ., R. Oc. *Oxygoniolepidus* Olig. NA. *Palaeobathygadus* Olig.-Mioc. NA. *Probathygadus* Olig. NA. *Promacrurus* Olig.-Mioc. NA. *Pyknolepidus* Olig. NA. *Rankinian* U. Cret. NA. *Trachyrincus* Mioc. NZ., R. Oc. *Trichiurichthys* Mioc.-Plioc. Eu.

SUPERORDER ATHERINOMORPHA

ORDER ATHERINIFORMES

SUBORDER EXOCOETOIDEI (SYNNENTOGNATHI). **Exocoetidae (Hemiramphidae)** R. Oc., *Chirodus* Mioc. Eu. *Cobitopsis* U. Cret. SWAs., Olig. Eu. *Derrhias* Mioc. NA., R. Oc. *Exocoetus* [*Hemiexocoetus*] Eoc. NA. *Euleptorhamphus* [*Beltion*] Mioc. NA., R. Oc. ? *Hemilampronites* U. Cret. Eu. NA. *Hemiramphus* Eoc. Eu., R. Oc. *Rogenites* [*Rogenia*] *Zelosis* Mioc. NA. **Belonidae** R. Oc., *Belone* Eoc.-Plioc. Eu., Mioc. NAf., R. Oc. **Scomberesocidae** R. Oc., *Praescomberesox* Eoc.-Olig. NA. *Scomberersus* Mioc. NA. *Scomberesox* Mioc. Eu. NA., Plioc. NAf., R. Oc. **Forficidae,** *Forfex* Mioc. NA. *Zelotichthys* [*Selota Zelotes*] Mioc. NA. ?**Tselfatiidae,** *Protobrama* U. Cret. SWAs. *Tselfatia* U. Cret. NAf.

SUBORDER CYPRINODONTOIDEI. **Oryziatidae Adrianichthyidae** R. EInd. **Horaichthyidae** R. As. **Cyprinodontidae** R. Eu. Af. NA. SA., *Aphanius* Mioc.-R. As. *Brachylebias* Mioc. WAs. *Carrionellus* L. Tert. SA. *Cyprinodon* Olig.-R. Eu., L. Tert.-R. NA. *Empetrichthys* Plioc.-R. NA. *Fundulus* [*Gephyrura Parafundulus*] ?Mioc., Plioc.-R. NA., Mioc.-R. Eu. *Haplochilus* Olig.-R. Eu. *Lithofundulus* Tert. SAs. *Lithopoecilus* Tert. EInd. *Pachylebias* [*Anelia ? Physocephalus*] Olig.-Mioc. Eu. *Prolebias* [*Ismene Pachystetus*] Olig.-Mioc. Eu. **Goodeidae** R. NA. **Anablepidae Jenynsiidae** R. CA. SA. **Poeciliidae** R. NA. SA.

SUBORDER ATHERINOIDEI. **Melanotaeniidae** R. EInd. **Atherinidae** R. As. Aus. NA. Oc., *Atherina* Eoc.-R. Eu., Mioc.-R. As., R. NA. *Menidia* Plioc.-R. NA. *Prosphyraena* Mioc. Eu. *Rhamphognathus* [*Mesogaster*] Eoc. Eu. *Zanteclites* Mioc. NA. **Isonidae** R. Oc. **Neostethidae Phallostethidae** R. NPac. EAs.

SUPERORDER ACANTHOPTERYGII

ORDER BERYCIFORMES

SUBORDER STEPHANOBERYCOIDEI. **Stephanoberycidae** R. Oc. **Melamphaeidae** R. Oc., *Scopelogadus* Mioc. NA. **Gibberichthyidae** R. Oc.

SUBORDER POLYMIXIODEI. **Polymixiidae** R. Oc., *Berycopsis* [*Platycormus*] *Homonotichthys* [*Homonotus Stenostoma*] U. Cret. Eu. *Omosoma* U. Cret. Eu. SWAs. NAf. *Parapolymyxia* Eoc.-Olig. NA. *Polymixia* Eoc. Eu., R. Oc. *Pycnosterinx* [*Imogaster*] U. Cret. SWAs. **Sphenocephalidae,** *Sphenocephalus* U. Cret. Eu.

SUBORDER DINOPTERYGOIDEI. **Dinopterygidae,** *Dinopteryx* U. Cret. SWAs. **Aipichthyidae,** *Aipichthys* U. Cret. SWAs. **Pycnosteroididae,** *Pycnosteroides* U. Cret. SWAs. **Pharmacichthyidae,** *Pharmacichthys* U. Cret. SWAs.

SUBORDER BERYCOIDEI. **Diretmidae** R. Oc., *Absalomichthys* [*Abantis*] Mioc. NA. *Chalcidichthys* Mioc. NA. **Trachichthyidae,** *Acrogaster* [*Acanthophoria*] U. Cret. Eu. SWAs. NAf. *Gephyroberyx* Olig.-Mioc. Eu., R. Oc. *Hoplopteryx* [? *Goniolepis Hemicyclolepis Hemigonolepis ? Priconolepis*] U. Cret. Eu. SWAs. NAf. NA. *Hoplostethus* Plioc. Eu., R. Oc. *Rothwellia* U. Cret. NA. *Trachichthodes* Tert. Aus. *Tubantia* U. Cret. Eu. **Korsogasteridae Anoplogasteridae** R. Oc. **Berycidae** R. Oc., *Beryx* ?Cret., Eoc. Eu., Eoc.-Olig. NA., Olig. WAs., R. Oc. ? *Costaichthys* [*Heterolepis*] U. Cret. Eu. *Echinocephalus* Eoc. Eu. ? *Electrolepis* U. Cret. Eu. ? *Kemptichthys* Tert. NAf. ? *Lobopterus* [*Dictynopterus*] U. Cret. Eu. ? *Platylepis* Eoc. Eu. ? *Spinacites* U. Cret. Eu. **Monocentridae** R. Oc., *Brazosiella* Eoc. NA. *Cleidopus* Olig.-R. Aus. *Monocentris* [*Lepisacanthus*] Paleoc.-Eoc. Eu., Eoc. WInd., Eoc.-Plioc. Aus., R. Oc. **Anomalopidae** R. Oc. **Holocentridae** R. Oc., *Africentrum* [*Microcentrum*] Mioc. Eu., Plioc. NAf. *Caproberyx* U. Cret. Eu. NAf. *Holocentrites* Eoc.-Olig. NA. *Holocentroides* Olig. Eu. *Holocentrus* Eoc.-Mioc. Eu., R. Oc. *Kansius* U. Cret. NA. *Myripristis* Eoc.-Mioc. Eu., R. Oc. *Paraberyx* U. Cret.-Olig. NA. *Stintonia* Eoc.-Olig. NA., Eoc.-Mioc. Eu. *Trachichthyoides* U. Cret. Eu., ?Olig. Aus. *Weileria* Eoc.-Olig. NA., Eoc.-Mioc. Eu.

ORDER ZEIFORMES. **Parazenidae Macrurocyttidae** R. Oc. **Zeidae** R. Oc., *Cyttoides* Olig. Eu. *Palaeocentrotus* Paleoc. Eu. *Zenopsis* Olig.-Mioc. Eu. *Zeus* Olig.-Plioc. Eu., ?Plioc. NAf., R. Oc. **Grammicolepidae** R. Oc. WInd. SAf. EInd. **Oreosomidae** R. Oc. **Caproidae** R. Oc., *Capros* [*Glyphisoma Metapomichthys Proantigonia*] Olig.-Mioc. Eu., Plioc. NAf., R. Oc. *Caprovesposus* Olig.-Mioc. Eu.

ORDER LAMPRIDIFORMES

SUBORDER LAMPRIDOIDEI. **Lampridae** R. Oc., *Lampris* [*Diatomoeca*] Mioc. NA., R. Oc.

SUBORDER VELIFEROIDEI. **Veliferidae** R. Oc.

SUBORDER TRACHIPTEROIDEI. **Lophotidae** R. Oc., *Lophotus* Olig. SWAs., R. Oc. *Protolophotus* Olig. SWAs., R. Oc. **Trachipteridae Regalecidae** R. Oc.

SUBORDER STYLEPHOROIDEI. **Stylephoridae** R. Oc.

ORDER GASTEROSTEIFORMES

SUBORDER GASTEROSTEOIDEI (THORACOSTEI). **Gasterosteidae** R. Eu. As. NAf. NA., *Gasterosteus* [*Merriamella*] Plioc.-R. NA., R. Eu. As. NAf. *Pungitius* [*Gastrosteops*] Plioc.-R. NA., R. As. NAf. **Aulorhynchidae** R. Oc., *Aulorhynchus* [*Protosyngnathus*] Tert. SAs. EInd., R. Oc. *Protaulopsis* Eoc. Eu. **Indostomidae** R. SAs.

SUBORDER AULOSTOMOIDEI. **Aulostomidae** R. Oc., *Aulostomus* Eoc.-Mioc. Eu., R. Oc. *Urosphen* Eoc.-Olig. Eu. **Fistulariidae** R. Oc., *Fistularia* Olig.-Plioc. Eu., R. Oc. **Macrorhamphosidae (Centriscidae)** R. Oc., *Aeoliscus* Olig.-Mioc. Eu. *Centriscus* [*Amphisile*] Olig. WAs., Olig., ?Plioc. Eu., R. Oc.

SUBORDER SYNGNATHOIDEI. **Solenostomidae** R. Oc., *Solenorhynchus* Eoc. Eu. *Solenostomus* Eoc. Eu., R. Oc. **Syngnathidae** R. Oc., *Calamostoma* Eoc. Eu. *Dunckerocampus* [*Acanthognathus*] Olig.-Mioc. Eu., R. Oc. *Hipposyngnathus* Mioc. Eu. *Pseudosyngnathus* Eoc. Eu. *Syngnathus* [*Siphonostoma*] Eoc.-Plioc. Eu., Mioc. NA., Plioc. NAf., R. Oc.

ORDER CHANNIFORMES. **Channidae (Ophiocephalidae)** R. As. Af., *Channa* [*Ophiocephalus*] Plioc.-R. As., R. Af.

ORDER SYNBRANCHIFORMES. **Synbranchidae** R. SAs. Af. Aus. SNA. NSA.

ORDER SCORPAENIFORMES (SCLEROPAREI)

SUBORDER SCORPAENOIDEI. **Scorpaenidae** R. Oc., *Amphe-ristius* [*Goniognathus*] Eoc. Eu. *Ctenopomichthys* [*Cteno-poma Jemelka*] Mioc. Eu. *Rhomarchus Rixator* Mioc. NA. *Scorpaena* Olig.-Mioc. Eu., Mioc. NA., Plioc. NAf., R. Oc. *Scorpaenodes* [*Sebastodes*] Mioc.-Plioc. NA., Plioc. EAs., R. Oc. *Scorpaenoides* Eoc.-Olig. Eu. *Scorpaenop-terus* Mioc. Eu. *Sebastavus Sebastinus Sebastoëssus* Mioc. NA. **Triglidae** R. Oc., *Peristedion* Plioc.-Eu., R. Oc. *Trigla* [*?Trigloides*] Eoc.-Mioc. Eu., Mioc. WAs., Plioc. NAf., Pleist. NA., R. Oc. **Caracanthidae Aploactinidae** R. Oc. **Synancejidae** R. Oc., *Eocynanceja* Eoc. Eu. **Patae-cidae** R. Oc.

SUBORDER HEXAGRAMMOIDEI. **Hexagrammidae** R. Oc., *Achrestogrammus* Mioc. NA. *Zemigrammatus* Mioc. NA. **Anoplopomatidae** R. Oc., *Aenoscorpius* [*Eoscorpius*] Mioc. NA.

SUBORDER PLATYCEPHALOIDEI. **Platycephalidae** R. Oc.

SUBORDER HOPLICHTHYOIDEI. **Hoplichthyidae (Oplich-thyidae)** R. Af. Aus. SA.

SUBORDER CONGIOPODOIDEI. **Congiopodidae** R. Af. Aus. SA.

SUBORDER COTTOIDEI. **Cottidae (Icelidae)** R. Eu. As. NA. Oc., *Cottopsis* Eoc. WAs. *Cottus* Eoc.-R. As., Mioc. NZ., Mioc.-R. Eu., Plioc.-R. NA. *Eocottus* Eoc. Eu. *Hayia* Mioc. NA. *Lepidocottus* Olig.-Mioc. Eu. *Lirosceles* Mioc. NA. *?Paraperca* Olig. Eu. **Cottocomephoridae**, R. L. Baikal. **Normanichthyidae** R. SPac. **Agonidae** R. Oc., *Agonus* Eoc. Eu., R. Oc. **Cyclopteridae** R. Oc., *Cyclop-terus* Pleist. NA., R. Oc.

ORDER DACTYLOPTERIFORMES. **Dactylopteridae (Cephala-canthidae)** R. Oc.

ORDER PEGASIFORMES (HYPOSTOMIDES). **Pegasidae** R. Indo-Pac. Oc.

ORDER PERCIFORMES

SUBORDER PERCOIDEI. **Centropomidae** R. Eu. As. NAf. Aus. NA., *Centropomus* Eoc.-R. Eu., R. NA. *Lates* Eoc.-Mioc. Eu., Eoc.-R. Af., R. As. Aus. *Paralates Platylates* Olig. Eu. *Psammoperca* Eoc. Eu., R. Oc. **Serranidae** R. Oc., *Acanthroperca* Eoc. Eu. *Acanus* Olig. Eu. *Allomo-rone* Eoc. NA. *Amphiperca* Eoc.-Mioc. Eu. *Anthias* Mioc. Eu., R. Oc. *Arambourgia* [*Apogonoides*] Plioc. NAf. *Aritolabrax* Mioc. EAs. *Blabe* Eoc. NAf. *Centro-pristis* Olig.-Mioc. Eu., R. Oc. *Cyclopoma* Eoc. Eu. *Da-palis* [*Smerdis*] ?Eoc. NAf., Paleoc.-Mioc. Eu. *Dicentrar-chus* [*Labrax*] Eoc.-Plioc. Eu., R. Oc. *Eoserranus* Tert. SAs. *Epinephelus* [*Emmachaere*] Mioc. NA., Plioc. NAf., R. Oc. *Maccullochella* Tert.-R. Aus. *Morone* Eoc.-Mioc. Eu., R. Oc. *Niphon* Olig. Eu., R. Oc. *Paracentropristis* Tert. Eu. *Paramorone* L. Tert. NA. *Percalates* Olig.-R. Aus. *Percichthys* Eoc.-R. SA. *Percilia* Eoc. Eu., R. SA. *Phosphichthys* Eoc. NAf., NA. *Prolates* [*Pseudolates*] Paleoc. Eu. *Properca* Eoc.-Mioc. Eu. *Proserranus* Paleoc. Eu. *Protanthias* Mioc. NA. *Serranus* Eoc.-Mioc. Eu., Plioc. NAf., R. Oc. *Siniperca* Plioc.-R. As. **Plesiopidae** R. Indo-Pac. Oc. **Pseudoplesiopidae** R. Indo-Pac. Oc. **Aniso-chromidae** R. WInd. Oc. **Acanthoclinidae** R. Indo-Pac. Oc. **Glaucosomidae** R. Pac. Oc. **Therapomidae** R. Indo-Pac. Oc. **Banjosidae** R. NPac. Oc. **Kuhliidae** R. Oc. As. EInd. Polynesia **Gregoryinidae** R. Oc. **Centrarchidae** R. NA., *Ambloplites Archoplites* Pleist.-R. NA. *Borescen-tranchus* Mioc. NA. *Centrarchites* Eoc. NA. *Chaenobryt-tus* ?Mioc., Plioc.-R. NA. *Lepomis* Plioc.-R. NA. *Microp-terus Mioplarchus* Mioc. NA. *Oligoplarchus* Olig. NA. *Pomoxis* ?Mioc., Plioc.-R. NA. **Priacanthidae** R. Oc., *Priacanthus* [*Apostasis*] Olig.-Mioc. Eu., R., Oc. *Pristi-genys* [*Pseudopnacanthus*] Eoc.-Mioc. Eu., R. Oc. **Apogon-idae (Cheilodipteridae)** R. Oc., *Apogon* [*?Eretima*] Eoc.-Mioc. Eu., Mioc. NA., R. Oc. *Apogonoides* Plioc. NAf. *Praegaleagra* Olig. NA. **Percidae** R. Eu. As. NA., *Acerina* U. Tert.-R. As., R. Eu. *Anthracoperca Cristigerina* Eoc. Eu. *?Dasceles ?Erisceles* Mioc. NA. *?Guoyquichthys* Tert. SA. *Leobergia* Plioc. NAs. *Lucioperca* Pleist.-R. Eu. *Mio-plosus* Eoc. NA. *?Pachygaster ?Paralates* Olig. Eu. *Perca* [*?Coeloperca Eoperca ?Percostoma ?Plioplarchus ?San-droserrus*] Eoc.-R. Eu., Olig., Pleist.-R. NA., Mioc. WAs. *Percalates* Mioc. NZ. *Percarina* Olig.-R. Eu. *Podocys Pro-percarina* Olig. Eu. *Sandar* Pleist.-R. Eu. **Sillaginidae** R. Indo-Pac. Oc., *Sillago* Olig.-Plioc. Aus., R. Oc. **Branchio-stegidae (Latilidae)** R. Oc., *Branchiostegus* [*Latilus*] Plioc. NAf., R. Oc. **Labracoglossidae** R. Pac. Oc. **Lactari-idae** R. Indo-Pac. Oc. SAs. EInd., *Lactarius* Mioc. Au., R. SAs. EInd. **Pomatomidae (Scombropsidae)** R. Oc., *Lophar* Mioc. NA. *Scombrops* Plioc. EAs., R. Oc. **Rachy-centridae** R. Indo-Pac. Oc. **Echeneidae** R. Oc., *Echeneis* Olig.-Mioc. Eu., R. Oc. *Opisthomyzon* Olig. Eu. **Carangi-dae (Seriolidae)** R. Oc., *Acanthonemopsis* Mioc. Eu. *Acanthonemus* Eoc.-Olig. Eu. *Aliciola* Mioc. NA. *Archae-us* [*Archaeoides*] Olig.-Mioc. Eu. *Carangopsis* Eoc. Eu. *Caranx* [*Citula Parequula*] Eoc.-Plioc. Eu., Eoc.-Pleist. NA., Plioc. NAf., R. Oc., *Decapterus* [*Lompochites*] Mioc. NA., Plioc. NAf., R. Oc. *?Desmichthys* Mioc. Eu. *Ductor* Eoc. Eu. *Hypacanthus* [*Lichia*] Olig.-Pleist. Eu., R. Oc. *Irifera* Mioc. NA. *Oligoplites* [*Palaeoscomber*] Mioc. Eu., R. Oc. *Pseudoseriola* Mioc. NA. *?Pseudovomer* Mioc. Eu. *Seriola* [*Micropteryx*] Eoc.-Olig. Eu., Mioc. NA., Plioc. NAf., R. Oc. *Teratichthys Trachinotus* Eoc. Eu., R. Oc. *Trachurus* Eoc. Eu., Mioc.-Plioc. NAf., R. Oc. *Vomer* Tert. Eu., R. Oc. *Vomeropsis* Eoc. Eu. **Coryphaenidae** R. Oc. **Formionidae (Apolectidae)** R. Indo-Pac. Oc. **Menidae** R. Indo-Pac. Oc., *?Bathysoma* Paleoc. Eu. *Mene* [*?Gaste-racanthus*] Paleoc.-Plioc. NAf., Eoc. Eu., R. Oc., **Leiog-nathidae (Equulidae)** R. Oc., *Leiognathus* [*Equula*] Olig.-Mioc. Eu., R. Oc. **Bramidae Caristiidae** R. Oc. **Arripidi-dae** R. SPac. Oc. **Emmelichthyidae** R. Indo-Pac. Oc. **Lutjanidae** R. Oc., *Caesio* Eoc. Eu. *Lednevia* Mioc. Eu. *Lutjanus* [*Lutianus*] ?Eoc., Mioc. Eu., Olig. NA., Mioc. Aus., R. Oc. **Lobotidae** R. Oc., *Protolobotus* Olig. SWAs. **Gerridae** R. Oc. **Pomadasyidae (Pristipomidae)** R. Oc., *Orthopristis* Plioc. NAf., R. Oc. *Parapristopoma* Plioc. NAf. *Pomadasys* [*Pristipoma*] Eoc.-Mioc. Eu. NAf., R. Oc. **Sparidae (Nemipteridae Leithrinidae)** R. Oc., *Atkin-sonella* Mioc. NA. *Boops* [*Box*] Mioc. Eu., Plioc. NAf., R. Oc. *Crenidentex* Plioc. NAf., R. Oc. *Crommyodus* [*Pliaco-dus*] Mioc. NA. *Ctenodentex* Eoc. Eu. NAf. *Dentex* Eoc.-Plioc. Eu., Mioc. NZ., Plioc. NAf. R. Oc. *Kreyenhagenius* Eoc. NA. *Pagellus* Paleoc.-Mioc. Eu., Olig.-Mioc. NZ., NZ., R. Oc. *Pagrosomus* Plioc. Aus., R. Oc. *Pagrus* Pa-leoc.-Mioc. Eu., Plioc. NAf., R. Oc. *Paracalamus* Plioc. WAf. *Plectrites Rhythmias* Mioc. NA. *Sargus* [*Diplodus*] Eoc.-Plioc. Eu., ?Eoc., Mioc.-Plioc. NAf., Mioc. Aus., Plioc. NA. R. Oc. *Sparnodus* Eoc.-Mioc. Eu. *Sparosoma* [*Rhamnubia*] Olig. Eu. *Sparus* [*Aurata Chrysophrys*] Eoc. WInd., Olig.-Plioc. Eu., Plioc. NAf., R. Oc. *Spon-dyliosoma* [*Cantharus*] Eoc. Eu., R. Oc. **Sciaenidae**

(**Otolithidae**) R. Oc. Eu. NA. SA., *Aplodinotus* Pleist. R. NA. *Corvina* Mioc. Eu., Eoc.-Olig. NA., R. Oc. *Cynoscion* [*Aristocion*] Mioc. NA., R. Oc. *Eocilophyodus* U. Cret.-Eoc. WAf. *Eokokemia* Eoc. NA. *Ioscion* Mioc. Na. *Jefitchia* Eoc. NA. *Larimus* Mioc. Eu., R. Oc. *Lompoquia* Mioc. NA. *Otolithes* [*Otolithus*] Eoc. Eu., R. Oc. *Pogonias* Mioc.-Pleist. NA., R. Oc. *Sciaena* [*Sciaenops*] Olig.-Mioc. Eu. NA., R. Oc. *Pseudoumbrina* Plioc. Eu. *Umbrina* Plioc. Eu., R. Oc. **Mullidae** R. Oc., *Mullus* Mioc. Eu., R. Oc. **Monodactylidae** (**Psettidae**) R. Indo-Pac. Oc. coasts. **Pempheridae** R. Oc. **Bathyclupeidae** R. Indo-Pac. Oc. Caribbean. **Toxotidae** R. EInd., *Toxotes* L. Tert.-R. EInd. **Coracinidae** (**Dichistiidae**) R. SOc. **Kyphosidae** (**Scorpididae Girellidae**) R. Oc. **Ephippidae** (**Chaetodipteridae Platacidae**) R. Oc., *Amphistium* [*?Macrostoma ?Woodwardichthys*] Eoc.-Mioc. Eu. *Ephippites* Olig. Eu. *Ephippus* Eoc. Eu., R. Oc. *Exellia* [*Semiophorus*] Eoc. Eu. *Paraplatox* Olig. EEu. *Platax* Eoc.-Plioc. Eu., ?Plioc. NA., R. Oc. **Scatophagidae** R. Indo-Pac. Oc. coasts, *Scatophagus* Eoc.-Mioc. Eu., R. Oc. **Chaetodontidae** (**Pomacanthidae**) R. Oc., *Chaetodon* Olig.-Mioc. Eu., Plioc. NAf., R. Oc. *Chelmo* Mioc. EInd., R. Oc. *Holacanthus* Eoc.-Mioc. Eu., R. Oc. *Pomacanthus* Eoc. Eu., R. Oc. **Enoplosidae** R. SPac. *Enoplosus* Eoc. Eu., R. Oc. **Histiopteridae** R. Indo-Pac. Oc. **Nandidae** (**Polycentridae Pristolepidae**) R. SAs. Af. EInd. SA. **Oplegnathidae** R. EAs. Aus. SAf. SA., *Oplegnathus* Plioc.-R. Aus. **Embiotocidae** (**Ditremidae**) R. NPac., *Ditrema* Eoc. Eu., R. NPac. *Eriquius* Mioc. NA. **Cichlidae** R. Af. SAs. CA. SA. WInd., *Acara* Tert.-R. SA. *Acaronia* Pleist.-R. SA. *Aequideus* Plioc.-R. SA. *Cichlaurus* [*Cichlasoma*] Mioc. WInd., R. SA. *Macracara* Eoc. SA. *?Palaeochromis* Tert. NAf. *Tilapia* Tert.-R. Af., R. WAs. **Pomacentridae** (**Abudefdufidae**) R. Oc. Af., *Chromis* Mioc.-R. Af. *Cockerellites* Eoc. NA. *Izuus* Mioc. EAs. *Odonteus* Eoc. Eu. *?Palaeochromis* Tert. NAf. *Priscacara* Eoc. NA. **Gadopsidae** R. Aus. **Cirrhitidae** R. Indo-Pac. Oc. **Chironemidae** R. SOc. **Aplodactylidae** R. Oc. **Cheilodactylidae Latridae** R. Indo-Pac. Oc. **Owstoniidae** R. Oc., *Owstonia* Plioc. Eu. EAs., R. Oc. **Cepolidae** R. Oc., *Cepola* Eoc.-Plioc. Eu., Plioc. NAf., R. Oc.

SUBORDER MUGILOIDEI. **Mugilidae** R. Oc., *Mugil* Eoc.-Plioc. Eu., Plioc. NAf., R. Oc. **Sphyraenidae** R. Oc., *Sphyraena* Eoc. Af., Eoc.-Mioc. Eu., Mioc. EInd., Pleist. As. NA., R. Oc. **Polynemidae** R. Oc., *Polydactylus* Tert. EEu.

SUBORDER LABROIDEI. **Labridae** R. Oc., *Coris* Mioc. Eu., R. Oc. *Egertonia* Paleoc.-Eoc. Eu., Eoc. NAf. *Eodiaphyodus* Paleoc.-Eoc. Af. *Eolabroides Gillidia* Eoc. Eu. *Julis* Mioc. Eu., R. Oc. *Labrodon* [*Diaphyodus Nummopalatus Pharyngophilus*] Paleoc.-Plioc. Eu., Eoc. NAf. NZ., Mioc. NA., Mioc.-Pleist. As., R. Oc. *Labrus* Eoc.-Plioc. Eu., R. Oc. *?Phyllodus* [*Paraphyllodus*] Paleoc.-Mioc. Eu., Eoc. NAf., NA. *Platylaemus* Eoc. Eu. NAf. *Pseudoegertonia* Paleoc. SWAs., Paleoc.-Eoc. Af. *Pseudostylodon* Eoc. Eu. *?Pseudovomer Stylodus* Mioc. Eu. *Symphodus* [*Bodianus Crenilabrus*] Mioc. Eu., Plioc. NAf. *Taurinichthys* Mioc.-Plioc. Eu. **Odacidae** R. coasts Aus. NZ. **Scaridae** R. Oc. As. NA. SA., *Pseudoscaris* Eoc. Eu., R. As. NA. SA. *Scaroides* Tert. CA., R. Oc. *Scarus* [*Callyodon*] Eoc. Eu., Mioc. SAs., R. Oc.

SUBORDER TRACHINOIDEI. **Trichodontidae** R. Indo-Pac. Oc. **Opisthognathidae** R. Oc. **Bathymasteridae** R. NPac. **Mugiloididae** R. Indo-Pac. Oc. **Cheimarrhichthyidae** R. NZ. **Trachinidae** R. NEAtl. Med., *Callipteryx* Eoc. Eu. *Trachinopsis* Plioc. Eu. *Trachinus* Eoc.-Mioc. Eu., R. Oc. **Percophididae** R. SAtl. **Trichonotidae** R. NPac. **Creediidae** R. Aus. coasts. **Limnichthyidae Oxudercidae** R. Oc. **Leptoscopidae** R. SPac. Aus. NZ., *Neopercis* Mioc. NAf., R. SPac. **Dactyloscopidae** R. SAtl. SPac. **Uranoscopidae** R. Oc., *Uranoscopus* Eoc. Eu., R. Oc. **Champsodontidae** R. Indo-Pac. Oc., *Myersichthys* Olig. Eu. *Pseudoscopelus* Mioc. Eu., R. Oc.

SUBORDER NOTOTHENIOIDEI. **Bovichthyidae** R. SOc. **Nototheniidae** R. SOc., *Notothenia* Mioc. NZ., R. Oc. **Bathydraconidae Channichthyidae** R. SOc.

SUBORDER BLENNIOIDEI. **Blenniidae** R. Cos., *Blennius* Mioc. Eu., R. Oc. *Oncolepis* Eoc. Eu. *Problennius* Eoc. WAs. **Anarhichadidae** R. NAtl. NPac., *Anarhichus* Plioc. Eu., R. Oc. *Xenocephalidae* R. EInd. **Congrogadidae** R. Indo-Pac. Oc. **Notograptidae Peronedyidae** R. Aus. **Ophioclinidae** R. Ind. SPac. Oc. **Tripterygiidae** R. Oc., *Tripterygion* Plioc. NAf., R. Oc. **Clinidae** R. Oc., *Clinus* Mioc. Eu., Plioc. NAf., R. Oc. *Pterygocephalus* Eoc. Eu., R. Oc. **Stichaeidae** R. NOc., *Stichaeus* Tert. EAs., R. Oc. **Ptilichthyidae** R. NPac. **Pholidae** R. NOc. **Scytalinidae Zaproridae** R. NPac.

SUBORDER ICOSTEOIDEI. **Icosteidae** R. Oc.

SUBORDER SCHINDLERIOIDEI. **Schindleriidae** R. Oc.

SUBORDER AMMODYTOIDEI. **Ammodytidae** R. Oc., *Ammodytes* Eoc.-Mioc. Eu., R. Oc. *Rhamphosus* Eoc. Eu. **Hypoptychidae** R. Oc.

SUBORDER CALLIONYMOIDEI. **Callionymidae** R. Oc., *Callionymus* Eoc.-Plioc. Eu., R. Oc.

SUBORDER GOBIOIDEI. **Gobiidae** R. Cos., *Gobiopsis* Eoc. WAs., R. Oc. *Gobius* ?Eoc., Olig.-Plioc. Eu., Mioc. NA. WAs., Plioc. NAf., R. Oc. *Lepidogobius* Mioc.-R. NA. *Priskenius* Olig.-Mioc. Eu. **Rhyacichthyidae Gobioididae Trypauchenidae Microdesmidae Eleotridae** R. Oc.

SUBORDER KURTOIDEI. **Kurtidae** R. EInd.

SUBORDER ACANTHUROIDEI. **Acanthuridae** R. Oc., *Acanthurus* Eoc.-Mioc. Eu., R. Oc. *Apostasella* [*Apostasis*] Olig.-Mioc. Eu. *Aulorhamphus* Eoc. Eu. *Naso* [*Naseus*] Eoc. Eu., R. Oc. *Parapygaeus* Eoc. Eu. *Protautoga* Mioc. NA., U. Tert. SA. *Pseudosphaerodon Pygaeus* Eoc. Eu. *Zauchus* Eoc. Eu., R. Oc. *Zebrasoma* Tert. WInd., R. Oc. **Teuthidae** (**Siganidae**) R. Oc., *Archaeoteuthis* [*Protosigana*] Olig. Eu.

SUBORDER SCOMBROIDEI. **Gempylidae** R. Oc., *Acanthonotos* [*Hemithyrsites*] Mioc.-Plioc. Eu., Plioc. NAf., R. Oc. *?Bathysoma* U. Cret. Eu. *Eothyrsites* Olig. NZ. *Eutrichiurides* Eoc. Eu. *Euzaphleges* [*Zaphleges*] Mioc. NA. *Gempylus* Olig. Eu. SWAs., R. Oc. *Thyrsites* Mioc. NA., R. Oc. *Thyrsitocephalus* Olig. Eu. *Thyrsocles Zaphlegus* Mioc. NA. **Trichiuridae** R. Oc., *Eutrichiurides* Paleoc. Eu., Paleoc.-Eoc. Af. *Lepidopus* [*Acanthonotus Anenchelum Lepidopides*] Eoc. WAf. WInd., Olig. SA., Olig.-Plioc. Eu., Plioc. NAf. *Trichiurides* Paleoc.-Mioc. Eu. *Trichiurus* Eoc. Af. Eoc.-Plioc. Eu., Pleist. NA., R. Oc. **Scombridae** R. Oc., *Amphodon* [*Scombramphodon*] Eoc.-Olig. Eu. *Aramichthys* Eoc. WAs. *Ardiodus* Eoc. Eu. *Auxides* Mioc. NA. *Auxis* Eoc.-Mioc. Eu., R. Oc. *Eocoelopoma* Eoc. Eu. *Eoscombrus* Eoc.-Olig. NA. *Eothynnus* [*?Cariniceps ?Coelocephalus ?Phalacrus ?Rhonchus*] Eoc.

Eu., Plioc. NAf. *Eucoelopoma* [*Coelopoma*] Eoc. Eu. *Euthynnus* [*Katsuwonus*] Pleist. WInd., R. Oc. *Gymnosarda* [*Cybiosarda*] Eoc.-Olig. Eu., R. Oc. *Isurichthys* Olig. Eu. WAs. *Landanichthys* Eoc. WAf. *Matarchia* Mioc. Eu., R. Oc. *Miothunnus* Mioc. Eu. *Megalolepis* Olig. Eu. *Neocybium* Olig.-Mioc. Eu. *Ocystias Ozymandias* Mioc. NA. *Palaeoscomber* Olig. Eu. *Palimphyes* [*Krambergeria*] Olig. Eu. *Pelacybium Sarda* [*Pelamys*] Paleoc.-Olig. Eu., Mioc. NA., Plioc. NAf., R. Oc. *Sarmata* Mioc. Eu. *Scomber* [*Pneumatophorus*] Eoc.-Plioc. Eu. Af., Mioc. NA., Tert. EAs., R. Oc. *Scomberomorus* [*Cybium*] Paleoc. Af., Eoc.-Mioc. Eu., Olig. SWAs., R. Oc. *Scombraphodon* Eoc.-Olig. Eu. *Scombrinus* Eoc. Eu. *Scombrosarda* Eoc. EEu. *Sphyraenodus* [*Dictyodus*] Eoc. NAf., Eoc.-Mioc. Eu. *Starrias* Mioc. NA. *Stereodus* Mioc. Eu. *Thunnus* [*Orcynus Thynnus*] Eoc.-Pleist. Eu., Mioc. NA. Af., R. Oc. *Tunita* Mioc. NA. *Turio* Mioc. NA. *Xestias* Mioc. NA. *Xiphopterus* Eoc. Eu. **Xiphiidae** R. Oc., *Acestrus Blochius* Eoc. Eu. *Brachyrhynchus* Mioc. Eu. *Coelorhynchus* [*Cylindracanthus Glyptorhynchus*] U. Cret.-Eoc. As. NAf., U. Cret.-Olig. Eu., Eoc. NA. *Congorhynchus* U. Cret. Eoc. WAf. *Hemirhabdorhynchus* Eoc. Eu. WAf. *Xiphias* Eoc. WAf., Eoc.-Olig. Eu., Pleist. NA., R. Oc. *Xiphiorhynchus* [*?Ommatolampes*] Paleoc.-Plioc. NAf., Eoc. WAf., Eoc.-Mioc. Eu. **Luvaridae** R. Oc. **Istiophoridae** R. Oc., *Istiophorus* [*Histiophorus*] U. Cret.-Mioc. NA., Eoc.-Plioc. Eu., R. Oc. *Tetrapterus* Eoc. Eu., R. Oc. **Palaeorhynchidae,** *Homorhynchus* [*Hemirhynchus*] Eoc.-Mioc. Eu. *Palaeorhynchus* Eoc.-Mioc. Eu., Olig. WAs. *Pseudotetrapterus* Olig.-Mioc. Eu.

SUBORDER STROMATEOIDEI. **Centrolophidae** R. Oc. **Nomeidae** R. Oc., *Carangodes* Eoc. Eu. **Stromateidae** R. Oc., *Seserinus* [*Aspidolepis*] ?U. Cret. Eu., R. Oc. **Tetragonuridae** R. Oc.

SUBORDER ANABANTOIDEI. **Anabantidae** R. As. EInd. Af., *Anabas* Pleist.-R. EInd. **Belontiidae Helostomidae** R. EInd. **Osphronemidae** R. As. EInd., *Osphronemus* L. Tert.-R. EInd., R. As.

SUBORDER LUCIOCEPHALOIDEI. **Luciocephalidae** R. EInd.

SUBORDER MASTACEMBELOIDEI. **Mastacembelidae** R. SAs. EAf. **Chaudhuriidae** R. SEAs.

ORDER PLEURONECTIFORMES (HETEROSOMATA)

SUBORDER PSETTODOIDEI. **Psettodidae** R. Oc., *Joleaudichthys* Eoc. NAf.

SUBORDER PLEURONECTOIDEI. **Eucitharidae** R. Oc., *Eucitharus* [*Citharus*] Mioc.-Plioc. Eu., R. Cos. *Citharichthys* Olig. Eu., Plioc. NAf., R. Oc. **Scophthalmidae** R. Oc., *Scophthalmus* [*Rhombus*] Eoc.-Plioc. Eu., Mioc. WAs., R. Oc. **Bothidae** R. Oc., *Bothus* Eoc.-Olig. Eu., R. Oc. *Eobothus* Eoc. As., Eoc., ?Olig. Eu. *Evesthes* Mioc. NA. *Imhoffius* Eoc. Eu. *Paralichthys* [*Vorator*] Mioc. NA., R. Oc. **Pleuronectidae** R. Oc., *Arnoglossus* Mioc. NA., R. Oc. *Cleisthenes* [*Protopsetta*] Tert. EAs., R. Oc. *Hippoglossoides* Mioc. NA., R. Oc. *Limanda* Plioc. Eu., R. Oc. *Pleuronectes* [*Platessa*] Mioc. Aus., R. Oc. *Pleuronichthys* [*Zoropsetta Zororhombus*] Mioc. NA., Plioc. Eu., R. Oc. *?Propsetta* Mioc. Eu. **Soleidae** R. Oc., *Achirus* Plioc. NAf., R. Oc. *Anoterisma* Mioc. NAf. *Arambourgichthys* Mioc. NAf. *Eosolea* Eoc. NA. *Eubuglossus* Eoc. NAf. *Microchirus* [*Monochir*] Mioc. Eu., Plioc. NAf., R. Oc. *Solea* Paleoc.-Mioc. Eu., Eoc.-Plioc. NAf., Mioc. WAs., R. Oc. *Turahbuglossus* Eoc. NA. **Cynoglossidae** R. Oc., *Cynoglossus* Mioc. Eu., R. Oc.

ORDER TETRAODONTIFORMES (PLECTOGNATHI)

SUBORDER BALISTOIDEI. **Trigonodontidae,** *Eotrigonodon* L. Cret.-Eoc. Af., Eoc. Eu. *?Stephanodus* U. Cret.-Eoc. Eu. Af., U. Cret. SWAs. NA. *?Spinacanthus* [*Probalistum*] Eoc. Eu. *Trigonodon* Eoc.-Plioc. Eu., Mioc. NZ., Plioc. NAf. **Triacanthidae** R. Oc., *Acanthopleurus* Eoc.-Olig. Eu. *Pristigenys* [*Pseudotriacanthus*] Eoc. NAf., R. Oc. *? Protriacanthus* U. Cret. NAf. *Triacanthus* Eoc. NAf., R. Oc. **Balistidae** R. Oc., *? Agoreion* [*Acanthoderma*] Olig. Eu. *Balistes* Olig.-Plioc. Eu., Mioc. WInd. NAf., Plioc. EInd., R. Oc. *Marosichthys* [*Marosia*] Mioc.-R. EInd. *Oligobalistes* Olig.-Mioc. Eu. *Probalistium* Eoc. Eu. **Ostraciontidae** R. Oc., *Ostracion* Eoc.-Olig. Eu., Tert. WInd., R. Oc.

SUBORDER TETRAODONTOIDEI. **Tetraodontidae** R. Oc., *Tetraodon* [*Ovoides*] Eoc.-Plioc. Eu., R. Oc. **Triodontidae** R. Indo-Pac. Oc., *Triodon* Eoc. Eu. Af., R. Oc. **Diodontidae** R. Oc., *Chilomycterus* Mioc. Eu., R. Oc. *Diodon* [*Enneodon Gymnodus Heptadiodon Megalurites Progymodon*] Eoc.-Plioc. Eu. As. Af., Eoc.-Pleist. NA., Mioc. EInd. WInd., Mioc.-Plioc. Aus., R. Oc. *Eodiodon* Eoc. Eu. *Kyrtogymnodon* Plioc. Eu. *Oligodiodon* Mioc. Eu. *Prodiodon* ?Eoc., Mioc. Eu. **Molidae** R. Oc., *Mola* [*Orthagoriscus*] Mioc.-Plioc. Eu., Tert. SA., R. Oc.

SUBCLASS SARCOPTERYGII

ORDER CROSSOPTERYGII

SUBORDER RHIPIDISTIA

SUPERFAMILY OSTEOLEPIDOIDEA (OSTEOLEPIDIFORMES).
Osteolepidae, *Bogdanovia* U. Dev. Eu. *Canningius* M. Dev. Gr. *Ectosteorhachis* L. Perm. NA. *Glyptopomus* [*Glyptognathus Glyptolaemus Pennagnathus Platygnathus*] M.-U. Dev. Eu. NA. *Gyroptychius* [*Diplopterax Diplopterus Diptopterus*] M., ?U. Dev. Eu., M. Dev. Gr. *Latvius* U. Dev. Eu. *Megalichthys* [*Carlukeus Centrodus Parabatrachus ? Plintholepis Rhomboptychius ?Sporelepis*] ?Miss., Penn. Eu., Penn. NA. *Megistolepis* U. Dev. Eu. *Osteolepis* [*Pleiopterus Pliopterus Triplopterus Tripterus*] M. Dev. NAs. Ant., M.-U. Dev. Eu. *Panderichthys* [*? Cricodus ? Polyplocodus*] M.-U. Dev. Eu. *Thaumatolepis* U. Dev. Eu. NAs. *Thursius* M. Dev. Eu., ?U. Dev. NA. **Rhizodontidae,** *?Devonosteus* M.-U. Dev. Eu. *Eusthenodon* U. Dev. Gr. NA. Eu. *Eusthenopteron* U. Dev. Eu. NA. *Litoptychius* U. Dev. NA. *Platycephalichthys* U. Dev. Eu. *Rhizodopsis* [*Characodus Ganolodus Gastrodus Orthognathus*] U. Dev. As., U. Dev.-Carb. NA., Miss. NAs., Perm. Eu. *Rhizodus* [*Archichthys ? Coelosteus ? Colonodus Dendroptychius Labyrinthodontosaurus Polyporites ?Sigmodus Strepsodus*] Miss. Aus., Miss.-Penn. Eu. NA. *Sauripteris* [*Sauripterus*] U. Dev. Eu. NA. *Tristichopterus* M. Dev. Eu.

SUPERFAMILY HOLOPTYCHOIDEA (POROLEPIDIFORMES).
Holoptychidae, *Glyptolepis* [*?Hamodus Plyphlepis ?Sclerolepis*] M. Dev. Spits. Gr., M.-U. Dev. Eu., U. Dev. NA. NAs. *Holoptychus* [*? Apedodus ? Apendulus ?Dendrodus Holoptychius Lamnodus*] M. Dev.-Miss. Eu., U. Dev. NA. Gr. Aus. Ant. NAs. *Laccognathus* U. Dev. Eu. *Pseudosauripterus* U. Dev. Eu. **Porolepidae,** *Porolepis* [*Gyrolepis*] L.-M. Dev. Eu. Spits. NAs. **Onychodontidae,** *Onychodus* [*Protodus*] L. Dev. NAs.,

L.-M. Dev. Spits., M.-U. Dev. Eu. NA., U. Dev. Gr. *Strunius* U. Dev. Eu.

SUBORDER COELACANTHINI (ACTINISTIA). **Euporosteidae,** *Euporosteus* U. Dev. Eu. **Diplocercidae,** *Chagrinia* U. Dev. NA. *Dictyonosteus* U. Dev. Spits. *Diplocercides Nesides* U. Dev. Eu. *Rhabdoderma* [*Conchiopsis Holopygus*] Miss. NAf. NAs., Miss.-Penn. Eu. NA. *Synaptotylus* Penn. NA. **Coelacanthidae,** *Axelia* L. Trias. Spits. *Bunoderma* Jur. SA. *Coccoderma* [*Kokkoderma*] U. Jur. Eu. ?*Coelacanthopsis* Miss. Eu. *Coelacanthus* [*Hoplopygus*] Penn. U. Perm. Eu., L. Trias. Mad. *Cualabaea* U. Jur WAf. *Diplurus* [?*Holophagoides Osteopleurus Pariestegus Rhabdiolepis*] U. Trias. NA. *Graphiurichthys* [*Graphiurus*] U. Trias. Eu. *Heptanema* M.-U. Trias. Eu. *Libys* U. Jur. Eu. *Macropoma* [*Eurycormus Eurypoma Lophoprionolepis*] U. Cret. Eu. SWAs. *Macropomoides* U. Cret. SWAs. *Mawsonia* L. Cret. SA. WAs., L.-U. Cret. Af. *Moenkopia* L. Trias. NA. *Mylacanthus* L. Trias. Spits. *Piveteauia* L. Trias. Mad. *Rhipis* Jur. Af. *Sassenia* L. Trias. Spits. Gr. *Scleracanthus* L. Trias. Spits. *Sinocoelacanthus* L. Trias. EAs. *Spermatodus* L. Perm. NA. *Whiteia* L. Trias. Mad. Gr., ?U. Trias. NA. *Wimania* [*Leioderma*] L. Trias. Spits., ?Gr. **Laugiidae** ?*Holophagus* [*Trachymetopon Undina*] U. Trias.-U. Jur. Eu., ?Jur. Aus. *Laugia* L. Trias. Gr. **Latimeriidae** R. Ind. Oc.

ORDER DIPNOI. **Dipnorhynchidae,** *Dipnorhynchus* L. Dev. Eu., M. Dev. Aus., ?L. Dev. NA. *Ganorhynchus* M.-U. Dev. Eu. NA. *Griphognathus Holodipterus* [*Archaeotylus Holodus*] U. Dev. Eu. **Dipteridae,** *Chirodipterus* U. Dev. Eu. *Conchodus* [*Cheirodus*] M.-U. Dev. Eu., U. Dev. NA. *Dipteroides* U. Dev. Eu. *Dipterus* [*Catopterus Eoctenodus Paradipterus Polyphractus*] L.-U. Dev. Eu., M.-U. Dev. NA. Nova Zembla, U. Dev. NAs. Aus. *Grossipterus* U. Dev. Eu. ?*Palaedaphus* [? *Archaeonectes* ? *Heliodus* ? *Paloedaphus*] U. Dev. Eu. NA. *Pentlandia* M. Dev. Eu. *Rhinodipterus* M.-U. Dev. Eu. ?*Rhynchodipterus* U. Dev. Eu. Gr. **Phaneropleuridae,** *Fleurantia* U. Dev. NA. *Jarvikia Nielsenia Oervigia* U. Dev. Gr. *Phaneropleuron* U. Dev. Eu., ?Gr. *Scaumenacia* [*Canadiptarus Canadipterus*] U. Dev. NA., ?Eu. *Soederberghia* U. Dev. Gr. **Ctenodontidae,** *Ctenodus* [?*Campylopleuron Proctenodus Rhadamista*] Miss.-Penn. Eu. NA., Miss. Aus. *Tranodis* Miss. NA. **Sagenodontidae,** ?*Proceratodus* Penn.-L. Perm. NA. *Sagenodus* [*Megapleuron Petalodopsis Ptyonodus*] Miss.-Penn. Eu., Miss.-L. Perm. NA. *Straitonia* Miss. Eu. **Uronemidae,** *Uronemus* [*Ganopristodus*] Miss. Eu.? NA. **Conchopomidae,** *Conchopoma* [*Conchiopsis Peplorhina*] Penn.-L. Perm. NA., L. Perm. Eu. **Ceratodontidae,** *Ceratodus* [*Hemictenodus Metaceratodus Scropha*] L. Trias.-U. Jur. Eu., L. Trias.-L. Cret. As., L. Trias.-U. Cret. Mad. NA. Aus., L. Trias.-Paleoc. Af., Trias. Spits., U. Cret. SA. *Gosfordia* L. Trias. Aus. *Microceratodus* Trias. NAf., L. Trias. Mad. *Paraceratodus* L. Trias. Mad. *Ptychoceratodus* U. Trias. Eu. *Neoceratodus* [*Epiceratodus* ?*Ompax*] U. Cret.-R. Aus. **Lepidosirenidae,** *Gnathorhiza* Penn.-L. Perm. NA., U. Perm. Eu. *Lepidosiren* [*Amphibichthys*] Mioc.-R. SA. *Protopterus* [*Protomalus Rhinocryptis*] Eoc.-R. Af.

DIPNOI INCERTAE SEDIS: ?*Osteoplax* Miss. Eu. ?*Palaeophichthys* Penn. NA.

CLASS AMPHIBIA

SUBCLASS LABYRINTHODONTIA

ORDER ICHTHYOSTEGALIA. **Elpistostegidae,** *Elpistostege* U. Dev. NA. **Ichthyostegidae,** *Ichthyostega Ichthyostegopsis* U. Dev. Gr. **Otocratiidae,** *Acanthostega* U. Dev. Gr. *Otocratia* L. Miss. Eu.

ORDER TEMNOSPONDYLI

SUBORDER RHACHITOMI

SUPERFAMILY LOXOMMATOIDEA. **Loxommatidae,** *Baphetes* ?L. Penn. Eu., M. Penn. NA. *Loxomma* U. Miss.-L. Penn. Eu. *Megalocephalus* [*Megacephalus Orthosauriscus Orthosaurus*] L.-M. Penn. Eu., M. Penn. NA. *Spathicephalus* U. Miss. Eu. NA.

SUPERFAMILY EDOPOIDEA. **Edopidae,** *Edops* L. Perm. NA. ?*Lusor* L. Perm. Eu. **Dendrerpetontidae,** *Dendrerpeton* [*Dendryazousa Dendrysekos Platystegos Smilerpeton*] L. Penn. NA. *Erpetocephalus* L. Penn. Eu. *Eugyrinus* L. Penn. Eu. ?NA. **Cochleosauridae,** *Chenoprosopus* L. Perm. NA. EEu. *Cochleosaurus* [*Nyrania*] *Gaudrya Macrerpeton* [*Capetus Mytaras*] M. Penn. Eu. NA. ?**Colosteidae,** *Colosteus Erpetosaurus* M. Penn. NA.

SUPERFAMILY TRIMERORHACHOIDEA. **Trimerorhachidae,** *Acroplous* L. Perm. NA. *Chalcosaurus* M. Perm. Eu. ?*Dawsonia* U. Penn. Eu. ?*Enosuchus* M. Perm. EEu. *Eobrachyops* L. Perm. NA. ?*Lysipterygium* L. Perm. SAs. *Neldasaurus* L. Perm. NA. *Saurerpeton* [*Branchiosauravus* "*Pelion*"] M. Penn. NA. *Slaughenhopia Trimerorhachis* L. Perm. NA. **Dvinosauridae,** *Dvinosaurus* ?M., U. Perm. EEu.

SUPERFAMILY ERYOPOIDEA. **Eryopidae,** *Actinodon* [*Euchirosaurus* ?*Pleuroneura* ?*Protriton* ?*Salamandrella*] *Chelyderpeton* [*Chelydosaurus*] L. Perm. Eu. *Eryops* [? *Anisodexis Epicordylus Eryopsoides Rhachitomus*] L. Perm. NA. *Glaukerpeton* U. Penn. NA. ?*Haplosaurus* L. Perm. Eu. *Onchiodon* ["*Branchiosaurus*" *Pelosaurus Porierpeton*] U. Penn.-L. Perm. Eu. *Osteophorus Sclerocephalus* [?*Leptorophus* ?*Micromelerpeton* ?*Promelerpeton Weissia*] L. Perm. Eu. **Dissorophidae,** *Alegeinosaurus* L. Perm. NA. *Amphibamus* [?*Eumicrerpeton Mazonerpeton* ?*Micrerpeton Miobatrachus Mordex Pelion Peliontonias Platyrhinops Potamochoston Raniceps*] M. Penn. Eu. NA. *Arkanserpeton Aspidosaurus* L. Perm. NA. *Brevidorsum Cacops Conjunctio Dissorophus* [*Otocoelus*] L. Perm. NA. ?*Limnerpeton* M. Penn. Eu. ?*Micropholis* [*Petrophryne*] L. Trias. SAf. *Platyhystrix Tersomius* ? *Vaughniella* L. Perm. NA. *Zygosaurus* M. Perm. EEu. **Trematopsidae,** *Acheloma Trematops Trematopsoides* L. Perm. NA. **Parioxyidae,** *Parioxys* L. Perm. NA. **Zatracheidae,** *Acanthostoma Dasyceps* L. Perm. Eu. *Stegops* M. Penn. NA. *Zatrachys* L. Perm. NA. **Archegosauridae,** *Archegosaurus* L. Perm. Eu. SAs. *Platyops* [*Platyposaurus*] M. Perm. EEu. *Prionosuchus* L. Perm. SA. **Melanosauridae,** *Jugosuchus* M. Perm. Eu. *Melanosaurus* ? *Tryphosuchus* M. Perm. EEu. ?**Intasuchidae,** *Intasuchus Syndiodosuchus* M. Perm. EEu.

SUPERFAMILY TREMATOSAUROIDEA. **Trematosauridae,** *Aphaneramma* [*Lonchorhynchus*] L. Trias. Spits. ?EEu., ?Aus., ?NA. *Gonioglyptus* [*Glyptognathus*] L. Trias. SAs. *Inflectosaurus* L. Trias. EEu. ? *Laidleria* L. Trias. SAf. *Lyrocephaliscus* L. Trias. Spits. *Microposaurus* L. Trias. SAf. *Platystega* L. Trias. Spits. *Stoschiosaurus* L. Trias.

Gr. *Tertrema* L. Trias. Spits., ?Aus. *Trematosaurus* L. Trias. Eu. NAf. *Trematosuchus* L. Trias. SAf. *Wantzosaurus* L. Trias. Mad. **Rhytidosteidae,** *Deltasaurus* L. Trias. Aus. *Peltostega* L. Trias. Spits. *Rhytidosteus* L. Trias. SAf.

SUBORDER STEREOSPONDYLI

SUPERFAMILY RHINESUCHOIDEA. **Rhinesuchidae,** *Laccosaurus Muchocephalus* U. Perm. SAf. *Rhineceps* U. Perm. EAf. *Rhinesuchoides* M. Perm. SAf. *Rhinesuchus* M. Perm. SAf., Perm. SAs. *?Tryphosuchus* M. Perm. EEu. **Lydekkerinidae,** *Broomulus* [*?Putterillia*] *Limnoiketes Lydekkerina ?Ptychocynodon* L. Trias. SAf. **Uranocentrodontidae,** *Laccocephalus* L. Trias. SAf. *?Sclerothorax* L. Trias. Eu. *Uranocentrodon* [*Myriodon*] L. Trias. SAf.

SUPERFAMILY CAPITOSAUROIDEA. **Benthosuchidae,** *Benthosuchus* [*Benthosaurus*] L. Trias. EEu. Mad. *Deltacephalus* L. Trias. Mad. *Gondwanosaurus Pachygonia* L. Trias. SAs. *Parabenthosuchus* L. Trias. EEu. *Sassenisaurus* L. Trias. Spits. *Thoosuchus* L. Trias. Eu. **Capitosauridae,** *Capitosaurus* U. Trias. Eu. *Cyclotosaurus* [*Hemprichisaurus ?Hercynosaurus*] ?L. Trias. Spits, ?M. Trias. NA., M.-U. Trias. Eu. *Kestrosaurus* L. Trias. SAf. *Latiscopus* U. Trias. NA. *Mastodonsaurus* [*?Diadetognathus Heptasaurus Labyrinthodon Rhombopholis Salamandroides Xestirrhytias*] L.-U. Trias. Eu., ?U. Trias. NAf. *?Moenkopisaurus* M. Trias. NA. *Paracyclotosaurus* [*? Austropelor*] U. Trias. Aus. *Parotosaurus* [*?Mentosaurus ?Odontosaurus ?Ptychosphenodon ?Syphonodon*] L. Trias. Eu. Spits. SAf. NAf., M. Trias. EAf. *Procyclotosaurus* U. Trias. Eu. *Promastodonsaurus* M. Trias. SA. *Rhadalognathus Stanocephalosaurus* L. Trias. NA. *Stenotosaurus* L. Trias. Eu. *Subcyclotosaurus* L. Trias. Aus. *? Yarengia* L. Trias. EEu. *Wetlugasaurus* [*Volgasaurus Volgasuchus*] L. Trias. EEu. Gr. SAf. Mad.

SUPERFAMILY BRACHYOPOIDEA. **Brachyopidae,** *Batrachosuchus* L. Trias. SAf. *Blinasaurus* L. Trias. Aus. *Boreosaurus* L. Trias. Spits. *Bothriceps* [*Platyceps*] ?U. Perm. Aus., ?L. Trias. Mad. *Brachyops* L. Trias. SAs. *Enosuchus* M. Perm. EEu. *?Hadrokkosaurus* [*Taphrognathus*] L. Trias. NA. *Indobrachyops* L. Trias. SAs. *Pelorocephalus* [*Chigutisaurus Icanosaurus ?Otaminisaurus*] L. Trias. SA. *Plagiorophus* L. Trias. EEu. *Phrynosuchus* U. Perm. SAf. *?Tungussogyrinus* L. Trias. NAs. *?Tupilakosaurus* L. Trias. Gr. *Trucheosaurus* U. Perm. Aus.

SUPERFAMILY METOPOSAUROIDEA. **Metoposauridae,** *Metoposaurus* [*Anachisma Borborophagus Buettneria ?Calamops ?Dictyocephalus Eupelor ?Hyperokynodon Kalamoiketer Koskinonodon Metopias Trigonosternum*] U. Trias. Eu. SAs. NA.

SUBORDER PLAGIOSAURIA. **Peltobatrachidae,** *Peltobatrachus* U. Perm. EAf. **Plagiosauridae,** *Gerrothorax* U. *Plagiosaurus* U. Trias. Eu. *Plagiosternum* ?M. Trias. Spits., M.-U. Trias. Eu. *Plagiosuchus* M.-U. Trias. Eu.

ORDER ANTHRACOSAURIA

SUBORDER SCHIZOMERI. **Pholidogasteridae,** *? Papposaurus Pholidogaster* U. Miss. Eu.

SUBORDER DIPLOMERI. **Diplovertebrontidae,** *Diplovertebron* [*Eusauropleura Gephyrostegus*] M. Penn. Eu. NA.

SUBORDER EMBOLOMERI. **Anthracosauridae,** *Anthracosaurus* [*Eobaphetes Erpetosuchus Leptophractus*] L. Penn. Eu., L.-M. Penn. NA. *?Crassigyrinus* U. Miss. Eu. **Cricotidae,** *Archeria* L. Perm. NA. *Calligenethlon* [*Atopotera*] L. Penn. NA. *?Cricotillus* L. Perm. NA. *Cricotus* U. Penn. NA. *Ichthyerpeton* L. Penn. Eu. *Memonomenos* ?U. Penn.-L. Perm. Eu. *Neopteroplax* U. Penn., ?L. Perm. NA. *Nummulosaurus* M. Penn. Eu. *Palaeogyrinus* L. Penn. Eu. *Pholiderpeton* ?U. Miss. NA., L. Penn. Eu. *Pteroplax* [*Eogyrinus Leptognathus Macrosaurus Streptodontosaurus*] L.-M. Penn. Eu. *Spondylerpeton* M. Penn. NA.

SUBORDER SEYMOURIAMORPHA. **Seymouriidae,** *?Ghorhimosuchus* L. Perm. WAs. *Rhinosaurus* M. Perm. EEu. *Seymouria* [*Conodectes Desmospondylus*] L. Perm. NA. **Kotlassiidae,** *Buzulukia Bystrowiana Kotlassia* [*Karpinskiosaurus*] U. Perm. EEu. *?Nyctiboetus ?Rhipaeosaurus* M. Perm. EEu. **Discosauriscidae,** *Discosauriscus* [*? Apateon Discosaurus Phaierpeton ?Sparagmites*] L.-M. Perm. Eu. *Letovertebron ?Lusor* L. Perm. Eu. *Melanerpeton* [*?Promelanerpeton*] U. Penn.-L. Perm. Eu. **Chronisuchidae,** *Chronisuchus* U. Perm. EEu. **?Nycteroleteridae,** *Nycteroleter* M. Perm. EEu. **?Waggoneriidae,** *?Helodectes Waggoneria* L. Perm. NA. **?Lanthanosuchidae,** *Lanthanosuchus* M. Perm. EEu. **Tseajaiidae,** *Tseajaia* L. Perm. NA. **Diadectidae,** *Desmatodon* U. Penn. NA. *Diadectes* [*? Animasaurus Bolbodon Chilonyx Diadectoides Empedias Empedocles Metarmosaurus Nothodon*] *Diasparactus* L. Perm. NA. *Phanerosaurus* [*Stephanospondylus*] L. Perm. Eu.

?SEYMOURIAMORPHA, INCERTAE SEDIS: *Adenoderma* U. Penn. Eu. *Hesperoherpeton* U. Penn. NA.

SUBCLASS LEPOSPONDYLI

ORDER NECTRIDEA. **Urocordylidae,** *Crossotelos* L. Perm. NA. *Ctenerpeton* M. Penn. NA. *Ptyonius* [*?Hyphasma*] *Sauropleura* [*Oestocephalus*] M. Penn. NA. Eu. *Urocordylus* L. Penn. Eu. **Lepterpetontidae,** *Lepterpeton* L. Penn. Eu. **Keraterpetontidae,** *Batrachiderpeton* L. Penn. Eu. *Diceratosaurus* [*Eoserpeton*] M. Penn. NA. *Diplocaulus* [*?Platyops ?Permoplatyops*] U. Penn.-L. Perm. NA. *Diploceraspis* L. Perm. NA. *Keraterpeton* [*Ceraterpeton*] L.-M. Penn. Eu., M. Penn. NA. *?Sauravus* U. Penn. Eu. *Scincosaurus* M. Penn. Eu.

ORDER AISTOPODA. **Phlegethontiidae,** *Dolichosoma* L. Penn. Eu. *Phlegethontia* M. Penn.-L. Perm. NA. **Ophiderpetontidae,** *Ophiderpeton* [*? Anthracerpeton ? Anthrakerpeton Steenisaurus Thyrsidium*] L.-M. Penn. Eu., M. Penn. NA. **F. indet.** "*Ophiderpeton*" L. Miss. Eu.

ORDER MICROSAURIA. **Adelogyrinidae,** *Adelogyrinus Dolichopareias* L. Miss. Eu. **Molgophidae,** *Palaeomolgophis* L. Miss. Eu. *Molgophis* [*Brachydectes Pleuroptyx*] M. Penn. NA. *Megamolgophis* L. Perm. NA. **Lysorophidae,** *Cocytinus* M. Penn. NA. *Lysorophus* U. Penn.-L. Perm. NA. **Microbrachidae,** *?Hyloplesion* [*Stelliosaurus*] ?L. Penn. NA., M. Penn. Eu. *Microbrachis* [*?Odonterpeton Orthocosta Seeleya*] M. Penn. Eu., ?NA. *?Paramicrobrachis* L. Perm. Eu. **Gymnarthridae** (**Pariotichidae Pantylidae**), *Ambylodon* [*Leiocephalikon*] L. Penn. NA. *Cardiocephalus Euryodus Goniocara* [*Goniocephalus*] *Gymnarthrus* L. Perm. NA. *Hylerpeton* L. Penn. NA. *Isodectes Pantylus Pariotichus* L. Perm. NA. *Sparodus* L. Penn. NA., M. ?U. Penn. Eu.

Tuditanidae, *?Ostodolepis* L. Perm. NA. *Tuditanus* [*Eosauravus*] M. Penn. NA. *Asaphestera* L. Penn. NA. *?Ricnodon* ?L. Penn. NA., M. Penn. Eu.

Older Amphibia, Incertae Sedis: *Broilisaurus* L. Perm. Eu. *Erierpeton ?Erpetobrachium* M. Penn. NA. *?Palaeosiren* L. Perm. Eu.

SUBCLASS LISSAMPHIBIA

Superorder Salientia

ORDER PROTANURA. **Triadobatrachidae (Protobatrachidae),** *Triadobatrachus* [*Protobatrachus*] L. Trias. Mad.

ORDER ANURA

SUBORDER ARCHAEOBATRACHIA. ?**Notobatrachidae,** *Notobatrachus* M. Jur. SA. **Leiopelmatidae (Ascaphidae)** R. WNA. NZ., *?Eodiscoglossus* U. Jur. Eu. **Discoglossidae (Bombinidae)** R. Eu. As. NAf., *Alytes* L. Mioc.-R. Eu. *?Barbourula* U. Cret. NA. *Bombina* [*Bombinator*] ?Mioc., L. Pleist.-R. Eu., R. As. *Discoglossus* L. Mioc.-R. Eu., ?Mioc., R. NAf. *Latonix* [*Latonia*] *Pelophilus* M. Mioc. Eu. *Prodiscoglossus* U. Olig. Eu. *Zaphrissa* Mioc. Eu.

SUBORDER NEOBATRACHIA. **Pelobatidae** R. Eu. NAf. SEAs. NA., *Amphignathodontoides ?Archaeopelobates* Eoc. Eu. *?Eopelobates* Olig.-Mioc. NA., ?Olig., Mioc. Eu. *Macropelobates* Olig. EAs. *Miopelobates* Mioc. Eu. *?Palaeopelobates* Eoc. Eu. *Pelobates* Olig.-R. Eu., Olig.-Plioc. NA. *?Pelobatinopsis* Eoc. Eu. *Protopelobates* Mioc. Eu. *Scaphiophus* Plioc.-R. NA. *Spea* [*Neoscaphiophus*] L. Plioc.-R. NA. **Pelodytidae** R. Eu. NA. *Miopelodytes* Mioc. NA. *?Propelodytes* Eoc. Eu. **Leptodactylidae** R. Af. Aus. SNA. WInd. SA., *Calyptocephalella* [*Calyptocephala Caudiverbera Eophractus Gigantobatrachus*] Eoc.-R. SA. *Ceratophrys* U. Plioc.-R. SA. *Eorubeta* Eoc. NA. *Eupsophus* Olig.-R. SA. *Leptodactylus* Pleist.-R. SA. *Syrrhophus* Pleist.-R. NA. *?Teracophrys* U. Olig. SA. *Wawelia* Mioc. SA. **Bufonidae** R. Eu. As. Af. NA. SA. EInd., *?Bufavus* Mioc. Eu. *Bufo* [*Palaeophrys ?Pliobatrachus Neoprocoela*] Olig.-R. SA., ?Olig., Mioc.-R. Eu., Mioc.-R. Af. NA., Pleist.-R. As. *?Bufonopsis* Eoc. Eu. *Diplopelturus* Plioc. Eu. *?Eobufella* Eoc. Eu. *?Indobatrachus* Eoc. SAs. *?Parabufella* Mioc. Eu. *Platosphus* Mioc. or Pleist. Eu. *Protophrynus* Mioc. Eu. **Pseudidae** R. SA. **Hylidae** R. Eu. As. NAf. EInd. Aus. NA. SA., *Acris* Mioc.-R. NA. *?Amphignathodon* Mioc. Eu., R. SA. *Hyla* [*?Lithobatrachus*] Mioc.-R. Eu. NA., Pleist. WInd. WAs., R. Cos. *Proacris Pseudacris* Mioc.-R. NA. *Triprion* Pleist.-R. NA. **Centrolenidae Atelopodidae** R. SA. **Ranidae** R. Cos., *Amphirana* Olig. Eu. *Aspherion Batrachulina* [*Batrachus*] Mioc. Eu. *Ptychadena* Mioc.-R. Af. *Rana* [*Anchylorana*] Eoc.-R. Eu., Mioc.-R. As. Af. NA., R. Co. *?Ranavus* Mioc. Eu. **Rhacophoridae (Polypedatidae)** R. SEAs. EInd. Af., *Rhacophorus* [*Polypedates*] Pleist.-R. As. **Microhylidae (Brevicipitidae)** R. SEAs. EInd. Af. NA. SA., *Microhyla* [*Gastrophryne*] Mioc. Eu. Af., Mioc., Pleist.R. NA. R. EAs. **Phrynomeridae** R. Af.

INCERTAE SEDIS: *Comobatrachus* U. Jur. NA. *Germanobatrachus Halleobatrachus* Eoc. Eu. *Montsechobatrachus* U. Jur. Eu. *Opisthocoelellus Quinquevertebron* Eoc. Eu. *Vieraella* L. Jur. SA.

Superorder Caudata

ORDER URODELA. **Cryptobranchidae** R. NA. EAs., *Andrias* [*Megalobatrachus Plicagnathus Proteocordylus Tritogenius Tritomegas*] Olig.-Plioc. Eu., Mioc., ?Plioc. NA., Pleist.-R. EAs. **Hynobiidae** R. As. **Scapherpetontidae,** *Lisserpeton* U. Cret.-Paleoc. NA. *Scapherpeton* [*Hemitrypus Hedronchus*] ?L. Cret., U. Cret.-Paleoc. NA. **Ambystomatidae** R. NA., *Ambystoma* [*Amblystoma ?Ambystomichnus Lanebatrachus Ogalallabatrachus Plioambystoma*] ?Paleoc., Mioc.-R. NA. *Bargmannia* Mioc. Eu. *?Geyeriella* Paleoc. Eu. *?Wolterstorffiella* Paleoc.- Mioc. Eu. **Salamandridae,** R. Eu. As. NA., *Archaeotriton* Mioc., or Eoc. Eu. *Brachycormus Chelotriton Heliarchon* Mioc. Eu. *Heteroclitotriton* Eoc. or Olig. Eu. *Koalliella* Paleoc. Eu. *Megalotriton* Eoc. or Olig. Eu. *Notophthalmus* Mioc.-R. NA. *Oligosemia* Olig., Mioc. Eu. *Palaeopleurodeles Palaeosalamandra* Mioc. Eu. *Polysemia* ?U. Cret., Mioc. Eu. *Salamandra* Olig.-R. Eu. *Taricha* [*Palaeotaricha*] Olig. NA. *Tischleriella* Mioc. Eu. *Triturus* [*Molge Triton*] ?Eoc., ?Olig., Mioc.-R. Eu., R. NA. As. *Tylototriton* Eoc.-Mioc. Eu., R. EAs. *Voigtiella* Mioc. Eu. **Amphiumidae,** R. NA., *Amphiuma* ?Eoc. Eu., Paleoc., Pleist.-R. NA. **Plethodontidae** R. NA. NSA., *Batrachoseps* Plioc.-R. NA. *?Dehmiella* Mioc. Eu. *Desmognathus* Pleist.-R. NA. *Opisthotriton* U. Cret.-Paleoc. NA. *Plethodon* Pleist.-R. NA. *Prodesmodon* L.-U. Cret. NA., Mioc. Eu. **Batrachosauroididae,** *Batrachosauroides* Eoc., Mioc. NA. **Proteidae** R. Eu. NA., *?Comonecturoides* U. Jur. NA. *?Hylaeobatrachus* L. Cret. Eu. *Necturus* ?Olig., Pleist.-R. NA. *Orthophyia* Mioc. Eu. *Palaeoproteus* Eoc. Eu. **Sirenidae** R. NA., *Habrosaurus* [*Adelphesiren*] U. Cret.-Paleoc. NA. *Prosiren* L. Cret. NA. *Pseudobranchus* Plioc.-R. NA. *Siren* Eoc.-R. NA.

Superorder Gymnophiona

ORDER APODA (CAECILIA). **Caeciliidae** R. As. Af. CA. SA.

CLASS REPTILIA

SUBCLASS ANAPSIDA

ORDER COTYLOSAURIA

SUBORDER CAPTORHINOMORPHA. **Romeriidae,** *Archerpeton* L. Penn. NA. *Cephalerpeton* M. Penn. NA. *Hylonomus* [*Fritschia Hylerpeton*] L. Penn. NA. *?Ichthyacanthus* M. Penn. NA. *Melanothyris Paracaptorhinus* L. Perm. NA. *?Petrobates* L. Perm. Eu. *Protorothyris Romeria* L. Perm. NA. *Solenodonsaurus* ?M. Penn. NA., U. Penn. Eu. **Limnoscelidae,** *Limnoscelis ?Limnosceloides* L. Perm. NA. **Captorhinidae,** *Captorhinikos Captorhinoides Captorhinus* [*Ectocynodon Hypopnous "Pariotichus"*] *?Chamasaurus* L. Perm. NA. *Geocatogomphius* M. Perm. EEu. *Hecatogomphius* M. Perm. Eu. *Kahneria* M. Perm. NA. *Labidosaurikos Labidosaurus Pleuristion ?Puercosaurus* L. Perm. NA. *Rothioniscus* [*Rothia*] M. Perm. NA. *?Sphenosaurus* [*Palaeosaurus*] L. Perm. Eu. ?**Bolosauridae,** *Bolosaurus* L. Perm. NA.

SUBORDER PROCOLOPHONIA

SUPERFAMILY PROCOLOPHONOIDEA. **Nyctiphruretidae,** *Barasaurus* U. Perm. Mad. *?Broomia* M. Perm. SAf. *Nyctiphruretus* M. Perm. EEu. *Owenetta* U. Perm. SAf. **Procolophonidae,** *Anomoiodon* U. Perm. Eu. *Candelaria* M. Trias. SA. *?Estheriophagus* U. Perm. NAs. *Hypsognathus* U. Trias. NA. *Koiloskiosaurus* L. Trias. Eu. *Leptopleuron* [*Telerpeton*] U. Trias. Eu. *?Microcnemus* [*?Tichvinskia*] U. Perm. EEu. *Myocephalus Myognathus* L. Trias. SAf. *Neoprocolophon* L. Trias. EAs. *Paoteodon* U. Trias. EAs. *Phaanthosaurus* L.

Trias. EEu. *Procolophon* [*?Microthelodon ?Thelegnathus*] L. Trias. SAf. *Santaisaurus* L. Trias. EAs. *Sclerosaurus* [*Aristodesmus ?Saurosternon*] L. Trias. Eu. *Sphodrosaurus* U. Trias. NA. *Spondylolestes* L. Trias. SAf.

SUPERFAMILY PAREIASAUROIDEA. **Rhipaeosauridae,** *Leptoropha Parabradysaurus Rhipaeosaurus* M. Perm. EEu. **Pareiasauridae,** *Anthodon* U. Perm. SAf. EAf. EEu. *Bradysaurus* [*Brachypareia Bradysuchus Koalemasaurus Platyoropha*] M. Perm. SAf. *Elginia* U. Perm. Eu. *Embrithosaurus* [*Dolichopareia Nochelesaurus*] M. Perm. SAf. *Parasaurus* U. Perm. Eu. *Pareiasaurus* [*Nanopareia Pareiasuchus Propappus*] U. Perm. SAf. EAf. EEu. *Scutosaurus* [*Amalitzkia Proelginia*] U. Perm. Eu. *Shihtienfenia* U. Perm. EAs.

SUPERFAMILY MILLEROSAUROIDEA. **Millerettidae,** *?Elliotsmithia ?Heleosaurus* M. Perm. SAf. *Millerettoides Milleretta* [*Millerina*] *Millerettops Millerosaurus Nanomilleretta* U. Perm. SAf.

ORDER MESOSAURIA. **Mesosauridae,** *Mesosaurus* [*Ditrichosaurus Noteosaurus Notosaurus Stereosternum*] L. Perm. SAf. SA.

ORDER CHELONIA

SUBORDER PROGANOCHELYDIA

SUPERFAMILY PROGANOCHELYOIDEA. **Proganochelyidae,** *Proganochelys* [*Psammochelys Chelytherium ?Chelyzoon Stegochelys Triassochelys*] ?M., U. Trias. Eu. *Proterochersis ?Saurischiocomes* U. Trias. Eu.

SUBORDER AMPHICHELYDIA

SUPERFAMILY PLEUROSTERNOIDEA. **Pleurosternidae,** *? Archaeochelys Desmemys* L. Cret. Eu. *Glyptops* U. Jur., ?U. Cret. Eu. EAs. ?NA. *Helochelys Kallokibotion* U. Cret. Eu. *Naomichelys* U. Jur. NA. *Platychelys* [*Helemys*] U. Trias.-U. Jur. Eu. *Pleurosternon* [*Digerrhum Megasternon Megasternum*] U. Jur.-L. Cret. Eu. *?Probaena* U. Jur. NA. *?Protochelys* M. Jur. Eu. *Stegochelys* U. Jur. Eu. *?Trachydermochelys* [*Plastremys*] L. Cret. Eu. **Plesiochelyidae,** *Craspedochelys* U. Jur. Eu. *Plesiochelys* [*Brodiechelys Hylaeochelys Parachelys Tholemys Wincania*] U. Jur. EAs., U. Jur.-L. Cret. Eu. **Thalassemyidae,** *? Anaphotidemys* [*Chelonides*] U. Jur. Eu. *?Cimochelys* [*Cimoliochelys*] L. Cret. Eu. *Changisaurus* U. Jur. EAs. *Eurysternum* [*Achelonia Acichelys Aplax Euryaspis Hydropelta Palaeomedusa*] *Idiochelys* [*?Chelonemys*] U. Jur. Eu. *Pelobatochelys* U. Jur.-L. Cret. Eu. *?Proeretmochelys* L. Cret. Eu. *Pygmaeochelys Sontiochelys* U. Cret. Eu. *Thalassemys* [*Enaliochelys*] U. Jur. Eu. *Tropidemys* U. Jur.-L. Cret. Eu. *Yaxartemys* U. Jur. As. **Sinemydidae,** *Manchurochelys Sinemys* [*Cinemys*] U. Jur. EAs. **Apertotemporalidae,** *Apertotemporalis* U. Cret. NAf. *Chitracephalus* L. Cret. Eu.

SUPERFAMILY BAENOIDEA. **Neurankylidae,** *Boremys ?Charitemys Neurankylus Thescelus* U. Cret. NA. **Baenidae,** *Baena* U. Cret. EAs., U. Cret.-Eoc. NA. *Chengyuchelys* U. Jur. EAs. *Chisternon* Eoc. NA. *Macrobaena* Eoc. EAs. *?Polythorax* U. Cret. NA. *?Tienfuchelys* ?U. Jur. EAs. **Meiolaniidae,** *Crossochelys* Eoc. SA. *Meiolania* [*Ceratochelys Miolania*] Pleist. Aus. *Niolamia* U. Cret. SA. **Eubaenidae,** *Eubaena* U. Cret. NA.

SUBORDER CRYPTODIRA

SUPERFAMILY TESTUDINOIDEA. **Dermatemydidae** R. CA., *Adocus* U. Cret.-Paleoc. NA., Eoc. EAs. *Agomphus* [*Amphiemys*] U. Cret. NA. *? Alamosemys Baptemys* Eoc. NA. *Basilemys* U. Cret.-Eoc. NA. *Compsemys* U. Cret.-Paleoc. NA. *?Heishanemys* ?U. Cret. EAs. *?Homorophus* U. Cret. NA. *Hoplochelys* Paleoc.-Eoc. NA. *?Kallistira* U. Cret. NA. *Lindholmemys* U. Cret. NAs. *?Notomorpha* Eoc. NA. *Patanemys* Eoc. Eu. *Peishanemys* L.-U. Cret. EAs. *Sinochelys* U. Jur. EAs. *?Trachyaspis* Eoc.-Mioc. Eu. Af. *Tretosternon* [*Heolochelydra ?Peltochelys*] U. Jur.-L. Cret. Eu. *Tsaotanemys* U. Cret. EAs. *? Yumenemys* L. Cret. EAs. *?Zygoramma* U. Cret. NA. **Chelydridae,** R. NA. SA. EInd., *Acherontemys* Mioc. NA. *Chelydra* Plioc.-R. NA., R. SA. *Chelydrops* Mioc. NA. *?Chelydropsis* Olig.-Mioc. Eu. *?Gafsachelys* Eoc. NAf. *Kinosternon* [*Cinosternum, Sternotherus*] Plioc.-R. NA., R. SA. *Macroclemys* [*Macrochelys*] Mioc.-R. NA. *Xenochelys* Olig. NA. **Testudinidae (Emydidae)** R. Eu. As. Af. NA. SA., *Achilemys* Eoc. NA. *?Broilia* Olig. Eu. *Chrysemys* ?Eoc.-Mioc. Eu., Plioc.-R. NA. *Clemmydopsis* Tert. Eu. *Clemmys* [*Geoliemys ?Paralichelys Sharemys*] Paleoc.-R. NA., Eoc.-R. Eu. As., Pleist.-R. EInd., R. NAf. *Cyclemys* Pleist.-R. SAs., R. EInd. *Cymatholcus* Eoc. NA. *Deirochelys* Pleist.-R. NA. *?Dithyrosternon* Olig. Eu. *Echmatemys* Eoc. NA. *Emyderidea* Pleist.-R. NA. *Emys* [*Platemys*] Eoc.-R. Eu. As., Plioc.-R. NAf. *Epiemys* Plioc. EAs. *Floridemys* [*Bystra*] Plioc. NA. *Geoclemys* [*Chinemys Polyechmatemys*] Eoc.-R. As., Mioc. EAs., R. EInd. *Geoemyda* [*Nicoria*] Eoc.-Pleist. Eu., Eoc.-R. As., R. CA. SA. EInd. *Gopherus* Olig.-R. NA. *Graptemys* ?Olig., R. NA. *?Gyremys* U. Cret. NA. *Homopus* ?Eoc., Mioc. Eu., R. Af. *Isometemys* Eoc. EAs. *Kachuga* Plioc.-R. SAs. *Kansuchelys* ?Eoc., ?Olig. EAs. *Kinixys* [*Cheirogaster Cinixys Cinothorax*] ?Eoc., Olig. Eu., R. Af. *Macrocephalochelys* Plioc. Eu. *Malayemys* Plioc.-R. SAs. R. EInd. *Ocadia* Eoc.-Mioc. Eu., Mioc.-R. EAs., Plioc. NAf. *Palaeochelys* ?Olig. As., Olig.-Plioc. Eu. *?Palaeotheca* Eoc. NA. *?Promalacoclemmys* Mioc. Eu. *Pseudemys* [*Trachemys*] Olig.-R. NA., Pleist.-R. WInd., R. CA. SA. *Ptychogaster* Olig.-Mioc. Eu. *?Sakya* Plioc. Eu. *?Scutemys* U. Jur. EAs. *Senryuemys* Olig.-Mioc. EAs. *Shansiemys* Plioc. EAs. *?Sinohadrianus* Eoc. EAs. *Stylemys* Eoc.-Mioc. NA., Olig. CAs., Mioc. Eu. *Temnoclemmys* U. Tert. Eu. *Terrapene* [*Cistudo*] Mioc.-R. NA., ?SubR. EAs., R. CA. *Testudo* [*Caudochelys Colossochelys Eupachemys Geochelone Hadrianus Hesperotestudo Megalochelys*] Eoc.-Pleist. NA., Eoc.-R. Eu. As. Af., Mioc.-R. SA., Pleist. EInd.

SUPERFAMILY CHELONIOIDEA. **Toxochelyidae,** *Ctenochelys ?Cynocercus Lophochelys* U. Cret. NA. *Osteopygis* [*Euclastes ?Lytoloma ?Propleura Rhetechelys*] U. Cret. EAs., U. Cret.-Eoc. NA. *Peritresius Porthochelys Prionochelys* U. Cret. NA. *Protemys* Eoc. Eu. *Thinochelys* U. Cret. NA. *Toxochelys* [*Phyllemys*] M. Cret.-Paleoc. NA. **Protostegidae,** *Archelon Calcarichelys Chelosphargis Protostega* U. Cret. NA. *Pseudosphargis* Olig. Eu. *?Therezinosaurus* U. Cret. EAs. **Cheloniidae** R. Oc., *Allopleuron* U. Cret. Eu. *Argillochelys* Eoc. Eu. *? Atlantochelys* U. Cret. NA. *?Bryochelys* Olig. Eu. *Caretta* [*?Pliochelys ?Proganosaurus Thalassochelys*] U. Cret.-Eoc., Plioc. Eu., Eoc. NAf., Pleist. NA., R. Oc. *Carolinochelys* Olig. NA. *Catapleura* U. Cret. NA. *Chelonia* [*Chelone*] U. Cret., Olig.-Plioc.

Eu., ?Eoc., Mioc.-Plioc. NA., R. Oc. *Chelyopsis* Olig. Eu. *Corsochelys* U. Cret. NA. ?*Cratochelone* L. Cret. Eu. *Desmatochelys* U. Cret. NA. *Eochelone* Eoc. Eu. *Glarichelys* Eoc.-Olig. Eu. *Glaucochelone* U. Cret. Eu. *Glyptochelone* U. Cret. Eu. *Kurobechelys* Mioc. EAs. *Lembonax* Eoc. NA. ?*Neptunochelys* U. Cret. NA. ?*Notochelone* [*Notochelys*] U. Cret. Aus. *Oligochelone* Olig. Eu. ?*Pachychelys* Plioc. Eu. ?*Platychelone* U. Cret. Eu. *Procolpochelys* ?Olig., Mioc. NA. *Puppigerus* [*Erquelinnesia* ?*Glossochelys* ?*Pachyrhynchus*] U. Cret.-Eoc. Eu. *Rhinochelys* L.-U. Cret. Eu. *Syllomus* Mioc. NA. *Tomochelone* U. Cret. Eu.

SUPERFAMILY DERMOCHELOIDEA. **Dermochelyidae** R. Oc., *Cosmochelys* Eoc. WAf. *Dermochelys* [*Sphargis*] Mioc. Eu., R. Oc. *Eosphargis* Eoc. Eu. *Psephophorus* [*Macrochelys*] Eoc. NAf., Eoc.-Mioc. NA., Eoc.-Plioc. Eu.

SUPERFAMILY CARETTOCHELYOIDEA. **Carettochelyidae** R. New Guinea, *Akrochelys Allaeochelys* Eoc. Eu. *Anosteira* [? *Apholidemys Castresia Pseudotrionyx*] Paleoc.-Olig. Eu., Eoc. NA. NAf., Eoc.-Olig. EAs. *Carettochelys* Mioc.-R. New Guinea ?*Hemichelys* ?Eoc. SAs. *Palaeochelys* U. Cret. Eu. *Pseudoanosteira* Eoc. Eu. NA.

SUPERFAMILY TRIONYCHOIDEA. **Trionychidae** R. As. EInd. Af. NA., *Chitra* Pleist. EInd., Pleist.-R. SAs. *Cycloderma* Mioc.-R. Af. *Lissemys* Plioc.-R. SAs. *Trionyx* [*Amyda Asperidites Axestemys Conchochelys Paleotrionyx Plastomenus Platypeltis Temnotrionyx*] U. Jur.-Plioc. Eu. L. Cret.-R. As., U. Cret.-R. NA., Mioc.-R. Af., Pleist.-R. EInd. *Sinaspiderites* U. Jur. EAs.

SUBORDER PLEURODIRA. **Pelomedusidae** R. Af. SA., *Amblypeza* U. Cret.-Eoc. NA. *Anthracochelys* Olig. Eu. *Apodichelys* U. Cret. SA. *Bantuchelys* Paleoc. Af. *Bothremys* U. Cret. NA. *Carteremys* Eoc. EAs. *Cyclochelys* Olig. Eu. *Dacochelys* Eoc. Eu. *Dacquemys* Olig. NAf. *Elochelys* U. Cret.-Olig. Eu. *Euclastochelys* Olig. Eu. *Eusarkia* Eoc. Af. *Eustatochelys* Paleoc. NAf. *Naiadochelys* U. Cret. SA., ?NA. *Neochelys Palaeaspis* [*Palaeochelys Palemys*] Eoc. Eu. *Pelomedusa* Olig.-R. Af. *Pelusios* [*Sternothaerus*] Mioc.-R. Af. *Platyarkia* Eoc. Eu. *Platycheloides* L. Cret. EAf. *Podocnemis* [*Erymnochelys*] U. Cret. NA., U. Cret.-Mioc. Eu., U. Cret.-R. SA., Paleoc.-Pleist. Af., Eoc. SAs., R. Mad. ?*Polysternon* U. Cret.-Olig. Eu. ?*Rosasia* U. Cret. Eu. ?*Roxochelys* U. Cret. SA. *Shweboemys* Plioc. SAs. *Stereogenys* Eoc.-Olig. Af. *Taprosphys* [*Prochonias*] Paleoc. NA., Eoc. SA. **Chelidae** R. SA. Aus., *Chelodina* Olig.-R. Aus., ?Tert. EAs. *Chelus* [*Chelydra Chelys*] ?Mioc., Plioc.-R. SA. *Emydura* Pleist.-R. Aus. *Parahydraspis* Plioc. SA. *Pelocomastes* Pleist. Aus. *Phrynops* [*Acrohydraspis Rhinemys*] Plioc.-R. SA.

?SUBORDER EUNOTOSAURIA. **Eunotosauridae**, *Eunotosaurus* M. Perm. SAf.

CHELONIA, INCERTAE SEDIS: *Anthracochelys* Plioc. Eu. *Cautleya* Plioc. EAs. ?*Cyrtura* U. Jur. Eu.

SUBCLASS LEPIDOSAURIA

ORDER EOSUCHIA

SUBORDER YOUNGINIFORMES. **Younginiidae**, ? *Adelosaurus* ? *Aphelosaurus* U. Perm. Eu. ?*Galesphyrus Heleophilus* ? *Heleosaurus* ? *Heleosuchus* U. Perm. SAf. ?*Mesenosaurus* M. Perm. EEu. ? *Noteosuchus* [?"*Eosuchus*"] ? *Palaeagama* U. Perm. SAf. ?*Paliguana* L. Trias. SAf. ?*Saurosternon* [*Batrachosaurus*] *Youngina Youngoides Youngopsis* U. Perm. SAf. ?**Tangasauridae**, *Hovasaurus* U. Perm. Mad. *Tangasaurus* U. Perm. EAf.

SUBORDER CHORISTODERA. **Champsosauridae**, *Champsosaurus* [*Simoedosaurus*] U. Cret.-Eoc. Eu. NA. ?*Pachystropheus* U. Trias. Eu.

SUBORDER THALATTOSAURIA. **Thalattosauridae**, *Askeptosaurus* M. Trias. Eu. *Nectosaurus Thalattosaurus* [*Scenodon*] M. Trias. NA.

SUBORDER PROLACERTIFORMES. **Prolacertidae**, ?*Gwyneddosaurus* U. Trias. NA. *Macrocnemus* M. Trias. Eu. *Megacnemus Pricea Prolacerta* L. Trias. SAf.

ORDER SQUAMATA

SUBORDER LACERTILIA

INFRAORDER EOLACERTILIA. **Kuehneosauridae**, *Icarosaurus* U. Trias. NA. *Kuehneosaurus* U. Trias. Eu., *Perparvus* U. Trias. Eu. ?*Rabdopelix* U. Trias. NA.

INFRAORDER IGUANIA. ?**Bavarisauridae**, *Bavarisaurus* U. Jur. Eu. ?**Euposauridae**, *Euposaurus* U. Jur. Eu. **Iguanidae**, *Aciprion* Olig. NA. *Anolis* Olig. or Mioc., Pleist.-R. NA., R. CA. WInd. *Arretosaurus* Eoc. EAs. ? *Bavarisaurus* U. Jur. Eu. *Corytophanes* Pleist. NA., R. CA. NSA. *Crotaphytus* [*Gambelia*] ?Plioc., Pleist.-R. NA. *Cyclura* Pleist.-R. WInd. ?*Geiseltaliellus* Eoc. Eu. *Holbrookia* Pleist.-R. NA. *Iguana* Pleist.-R. WInd., R. CA. NSA. ?*Iguanosauriscus* [*Iguanosaurus*] Paleoc.-Eoc. Eu. *Leiocephalus* Mioc. NA., Pleist.-R. WInd. *Leiosaurus* U. Plioc.-R. SA., R. CA. *Parasauromalus* Eoc. NA. *Phrynosoma* [*Eumecoides*] Mioc.-R. NA., R. CA. *Sauromalus* Pleist.-R. NA. *Sceloporus* Plioc.-R. NA., R. CA. SA. *Tetralophosaurus* Mioc. NA. **Agamidae** R. SEEu. As. Af. EInd. Aus., *Agama* Eoc.-Olig. Eu., R. SEEu. SWAs. Af. *Chlamydosaurus* Pleist.-R. Aus. *Conicodontosaurus Macrocephalosaurus* U. Cret. EAs. *Tinosaurus* [*Paleochamaeleo*] Eoc. EAs. NA. Eu. **Chameleontidae** R. SEu. Af. Mad., *Chamaeleo* Pleist.-R. SWAs., R. SEu. Af. *Mimeosaurus* U. Cret. EAs.

INFRAORDER NYCTISAURIA (GEKKOTA). **Ardeosauridae**, *Ardeosaurus* [?*Eichstattosaurus*] U. Jur. Eu. ?*Teilhardosaurus Yabeinosaurus* U. Jur. EAs. **Broilisauridae**, *Broilisaurus* U. Jur. Eu. **Gekkonidae** R. As. Af. EInd. Aus. CA. WInd. SA. Oceania, *Aristelliger* Pleist.-R. WInd., R. CA. *Cadurcogekko* Eoc.-Olig. Eu. *Gerandogekko* Mioc. Eu. *Macrophelsuma* Pleist. Mascarenes *Rhodanogekko* Eoc. Eu. *Tarentola* Pleist.-R. WInd., R. SEu. NAf. SWAs. **Pygopodidae** R. Aus.

INFRAORDER LEPTOGLOSSA (SCINCOMORPHA). **Xantusiidae** R. SNA. CA. WInd., *Lepidophyma* [*Impensodens*] Pleist.-R. NA., R. CA. *Palaeoxantusia* Paleoc.-Eoc. NA. *Xantusia* Pleist.-R. NA. **Teiidae** R. SNA. CA. WInd. SA., *Chamops* [*Alethesaurus Lanceosaurus*] U. Cret. NA. *Cnemidophorus* Mioc.-R. NA., R. CA. SA. ?*Diasemosaurus* Mioc. SA. ?*Dicarlesia* [*Carlesia*] Cret. SA. *Dracaena* Mioc.-R. SA. *Haptosphenus Leptochamops Meniscognathus* U. Cret. NA. *Paradipsosaurus* ?Olig. NA. *Paraglyphanodon Polyglyphanodon* U. Cret. NA. *Teius* [*Tejus*] Olig., R. SA. *Tupinambis* Olig., Mioc.-R. SA., R. WInd. **Scincidae** R. SEu. As. Af. EInd. Aus. Oceania SNA. CA. WInd. SA., ?*Capitolacerta* Eoc. Eu. *Didosaurus* Pleist.-SubR. Mascarenes *Eumeces* [*Plestiodon*] Olig.-R. NA., R. CA. As. NAf. *Sauriscus* U. Cret. NA. *Thyrus* Pleist.-R. Mauritius. **Lacertidae** R. Eu. As. Af., *Dracaenosaurus* [*Cadurco-

saurus] Eoc.-Olig. Eu. *Eolacerta* Eoc. Eu. *? Palaeolacerta* U. Jur. Eu. *Pseudeumeces* Eoc. or Olig. Eu. *Plesiolacerta* Eoc. Eu. *Lacerta* Mioc.-R. Eu. **Cordylidae** (**Gerrhosauridae Zonuridae**) R. Af., *Gerrhosaurus* Mioc.-R. Af. *?Macellodus* [*Macellodon Saurillus*] U. Jur. Eu. *Pseudolacerta* Eoc. Eu. ?**Dibamidae** R. EAs. EInd. SNA.

INFRAORDER ANNULATA (AMPHISBAENIA). **Amphisbaenidae** R. Af. SNA. CA. SA. WInd. SWAs., *Changlosaurus* Olig. EAs. *Cremastosaurus* Olig. NA. *Gilmoreia Hyporhina* Olig. NA. *Jepsibaena* Eoc. NA. *Leposternon* Olig. NA., R. SA. *Lestophis* [*Limnophis Ototriton*] Eoc. NA. *Oligodontosaurus* Paleoc. NA. *Omoiotyphlops* Eoc.-Plioc. Eu. *Pseudorhineura* Olig. NA. *Rhineura* [*Platyrhachis*] Eoc.-R. NA. *Trogonophis* Pleist.-R. Af.

INFRAORDER DIPLOGLOSSA

SUPERFAMILY ANGUOIDEA. **Anguidae** R. Eu. As. NAf. NA. SA. WInd., *Anguis* Olig.-R. Eu., R. WAs. NAf. *Dimetopisaurus* Eoc. NA. *Gerrhonotus* [*Elgaria*] U. Cret.-R. NA., R. CA. *Haplodontosaurus* Paleoc.-Eoc. NA. *?Isodontosaurus* U. Cret. EAs. *Machaerosaurus* Paleoc. NA. *Melanosaurus* Eoc. NA., ?Eu. *Ophisauriscus* Eoc. Eu. *Ophisaurus* [*Propseudopus Pseudopus Sauromorus*] Eoc.-R. Eu., Plioc.-R. NA., R. As. NAf. *Paragerrhonotus* Plioc. NA. *Parapseudopus* [*Ophipseudopus*] *Paraxestops* Eoc. Eu *Peltosaurus* [*Odaxosaurus*] U. Cret., ?Mioc. NA. *Placosauriops Placosauroides* Eoc. Eu. *Placosaurus* [*Diacium Glyptosaurus Helodermoides Loricotherium Necrodasypus Placotherium Proiguana Protrachysaurus*] ?Paleoc.-Eoc. EAs., Eoc. Eu., Eoc.-Olig. NA. *Xestops* [*Oreosaurus*] Eoc., ?Olig. NA. **Anniellidae** R. SWNA. **Xenosauridae** R. EAS. CA., *Exostinus* [*Harpagosaurus Prionosaurus*] U. Cret.-Olig. NA.

SUPERFAMILY VARANOIDEA (PLATYNOTA). ?**Necrosauridae,** *Necrosaurus* [*Melanosauroides Neovaranus Odontomophis Palaeosaurus Palaeovaranus*] Paleoc.-Olig. Eu. **Parasaniwidae,** *Paraderma* U. Cret. NA. *Parasaniwa* U. Cret.-Eoc. NA. *Provaranosaurus* Paleoc. NA. **Helodermatidae** R. WNA., *Eurheloderma* Eoc. Eu. *Heloderma* Olig., R. WNA. **Varanidae** R. SAs. Af. EInd. Aus., *Chilingosaurus* U. Cret. EAs. *Megalania* [*Notiosaurus*] Pleist. SAs. Aus. *? Pachyvaranus* U. Cret. NAf. *Palaeosaniwa* [*Megasaurus*] U. Cret. NA. *Saniwa* [*Thinosaurus*] Eoc.-Olig. NA. Eu. *Telmasaurus* U. Cret. EAs. *Varanus* Mioc.-Pleist. Eu., Plioc.-R. SAs., Pleist.-R. Aus. Af., R. EInd. **Lanthanotidae** R. EInd. **Aigialosauridae,** *Aigialosaurus Carsosaurus* [*Opetiosaurus Mesoleptos*] *? Coniasaurus* M. Cret. Eu. *Proaigialosaurus* U. Jur. Eu. **Dolichosauridae,** *Acteosaurus* [*Adriosaurus*] M. Cret. Eu. *Dolichosaurus* U. Cret. Eu. *Eidolosaurus Pontosaurus* [*Hydrosaurus*] L. Cret. Eu. **Mosasauridae,** *Amphekepubis ? Amphorosteus* U. Cret. NA. *Angolosaurus* U. Cret. WAf. *Clidastes* [*Edestosaurus*] U. Cret. NA., ?Eu. *Compressidens Dollosaurus* U. Cret. Eu. *Ectenosaurus* U. Cret. NA. *Globidens* U. Cret. Eu. NA. SA. As. Af. *Hainosaurus* U. Cret. Eu. *Halisaurus* [*Baptosaurus*] *? Holcodus* U. Cret. NA. *Liodon* [*Leiodon*] U. Cret. Eu. NA. *Mosasaurus* [*Baseodon Batrachiosaurus Batrachotherium Drepanodon Lesticodus Macrosaurus Nectoportheus Pterycollosaurus*] U. Cret. Eu. WAf. NA. *Platecarpus* [*Holosaurus Lestosaurus Sironectes*] U. Cret. Eu. NA., ?Aus. *Plesiotylosaurus* U. Cret. NA. *Plioplatecarpus* [*Oterognathus Phosphorosaurus*] U. Cret. Eu. NA. *Plotosaurus* [*Kolposaurus*] U. Cret. NA. *Prognathodon* [*Ancylocentrum Brachysaurana Brachysaurus ? Elliptodon Prognathosaurus*] U. Cret. Eu. NA. *Taniwhasaurus* U. Cret. NZ. *Tylosaurus* [*Rhamposaurus Rhinosaurus*] U. Cret. NA. NZ. **Palaeophidae** (**Cholophidae Cholophidia**), *Anomalopsis* Eoc. Eu. *Palaeophis* [*Dinophis Titanophis*] Eoc. Eu. NA. Af. *Pterosphenus* [*Moeriophis*] Eoc. Af. NA. SA. **Simoliophidae,** *Lapparentophis* L. Cret. NAf. *Mesophis Pachyophis* L. Cret. Eu. *Simoliophis* [*Symoliophis*] U. Cret. Eu. NAf. Mad.

LACERTILIA INCERTAE SEDIS: *Araeosaurus* U. Cret. Eu. *Cteniogenys* U. Jur. NA. *Cuttysarkus* U. Cret. NA. *Goniosaurus* Cret. Eu. *Litakis* U. Cret. NA. *Naocephalus* Eoc. NA. *Patricosaurus* Cret. Eu. *Saurospondylus* U. Cret. Eu. *Progonosaurus* Plioc. Eu.

SUBORDER OPHIDIA (SERPENTES)

SUPERFAMILY TYPHLOPOIDEA (SCOLECOPHIDIA). **Typhlopidae** R. SEu. As. EInd. Aus. Af. CA. SA., *Typhlops* Eoc.-R. Eu., R. As. EInd. Aus. Af. CA. SA. **Leptotyphlopidae** R. SWAs. Af. NA. CA. WInd.

SUPERFAMILY BOOIDEA (HENOPHIDIA). **Dinilysiidae,** *Dinilysia* U. Cret. SA. **Aniliidae** (**Ilysiidae**) R. SAs. EInd. CA. SA. *Anilioides* Mioc. NA. *Coniophis* U. Cret.-Eoc. NA. **Boidae** R. As. Af. EInd. Aus. NA. SA., *Boavus* [*Protagras*] Eoc. NA. *Botrophis* [*Bothrophis*] Mioc. Eu. *Calamagras* [*Aphelophis*] Eoc.-Mioc. NA. *Charina* Mioc.-R. NA. *? Cheilophis* Eoc. NA. *Crythiosaurus* Olig. EAs. *Daunophis* Plioc. SAs. *Epicrates* Pleist.-R. WInd., R. CA. *Eryx* Olig.-Mioc. Eu., R. As. Af. *Gigantophis* Eoc. NAf. *Helagras* Paleoc. NA. *Heteropython* Mioc. Eu. *? Lithophis* U. Cret. NA. *Madtsoia* U. Cret.-Eoc. Mad., Paleoc.-Eoc. SA. *Ogmophis* Olig.-Mioc. NA. *Palaelaphis* Eoc. or Olig. Eu. *Paleryx* [*? Palaeopython*] Eoc., ?Mioc. Eu. *Paraepicrates* Eoc. NA. *? Peltosaurus* Mioc. NA. *Plesiotortrix* Eoc. or Olig. Eu. *Pseudoepicrates* Mioc. NA. *Python* Pleist.-R. As. Aus., R. Af. EInd. *Scytalophis Tachyophis* Eoc. or Olig. Eu.

SUPERFAMILY COLUBROIDEA (CAENOPHIDIA). ?**Archaeophidae,** *Archaeophis Anomalophis* Eoc. Eu. **Colubridae,** *Abastor* Pleist.-R. NA. *Alsophis* Pleist.-R. WInd. *Arizona Carphophis* Pleist.-R. NA. *Coluber* [*Masticophis*] ?Eoc., Mioc.-R. Eu., Plioc.-R. NA., R. As. NAf. EInd. CA. *Coronella* ?Pleist., R. Eu. *Diadophis* Mioc.-R. NA. *Dryinoides* Mioc. NA. *Drymarchon* Pleist.-R. NA., R. CA. *Elaphe* [*Paleoelaphe*] Mioc.-R. Eu., Plioc.-R. NA., Pleist.-R. As., R. EInd. CA. *Farancia* Pleist.-R. NA. *Haldea* Pleist. NA. *Heterodon* Plioc.-R. NA. *Lampropeltis* Plioc.-R. NA., R. CA. NSA. *Liodytes* Pleist.-R. NA. *Malpolon* Mioc.-R. Eu., R. SWAs. NAf. *Mionatrix* Mioc. EAs. *Natrix* Plioc.-R. NA., Pleist.-R. Eu. As., R. Af. EInd. Aus. *Opheodrys* Plioc.-R. NA., R. EAs. EInd. *Paleofarancia Paleoheterodon* Plioc. NA. *Paraoxybelis* Mioc. NA. *Pituophis* Pleist.-R. NA., R. CA. *Protropidonotus* Mioc. Eu. *Pseudocemophora* Mioc., ?Pleist., R. NA. *Ptyas* Pleist.-R. As. *Pylmophis* Mioc. Eu. *Rhadinaea* Pleist.-R. NA. *Sansanosaurus* [*Tamnophis*] *Scaptophis* Mioc. Eu. *Stilosoma* Plioc.-R. NA. *Storeria* Pleist.-R. NA., R. CA. *Tantilla* Pleist.-R. NA. *Thamnophis* Plioc.-R. NA., R. CA. NSA. *Tropidoclonion* Pleist.-R. NA. **Elapidae** R. As. Af. EInd. Aus. NA. SA., *Micrurus* Plioc.-R. NA., R. SA. *Naja* Plioc.

Eu., Pleist.-R. SAs., R. EInd. Af. *Palaeonaja* Mioc.-Plioc. Eu. **Hydrophiidae** R. Pac. and Ind. Oc. **Viperidae,** *Agkistrodon* Plioc.-R. NA., R. EEu. As. CA. *Bitis* Mioc.-Plioc. Eu., R. Af. *Crotalus* Plioc.-R. NA., R. SA. *Laophis Provipera* Mioc. Eu. *Sistrurus* Pleist.-R. NA. *Vipera* Plioc., R. Eu., R. As. EInd.

OPHIDIA INCERTAE SEDIS: *Dunnophis* Eoc. NA. *Ophidion.* Mioc. Eu.

ORDER RHYNCHOCEPHALIA. **Sphenodontidae** R. NZ., *Brachyrhinodon* U. Trias. Eu. *?Chometokadmon* U. Cret. Eu. *Clevosaurus* U. Trias. Eu. *Homoeosaurus* [*Leptosaurus Stelliosaurus*] *Meyasaurus* U. Jur. Eu. *Monjurosuchus* U. Jur. EAs. *Opisthias* [*Theretairus*] U. Jur. NA. *Palacrodon* L. Trias. SAf. *?Pachystropheus Polysphenodon* U. Trias. Eu. *Schargengia* L. Trias. EEu. **Rhynchosauridae,** *?Eifelosaurus* L. Trias. Eu. *Howesia* L. Trias. SAf. *Hyperodapedon* [*Stenometopon*] U. Trias. Eu. *Mesosuchus* L. Trias. SAf. *Parasuchus* [*Paradapedon*] U. Trias. SAs. *Rhynchosaurus* M.-U. Trias. Eu. *Scaphonyx* [*Cephalonia Cephalastron Cephalastronius Scaphonychimus*] M. Trias. SA. *Stenaulorhynchus* M. Trias. EAf. **Sapheosauridae (Saurodontidae),** *Sapheosaurus* [*Piocormus Sauranodon*] U. Jur. Eu. **?Clarazidae,** *Clarazia Hescheleria* M. Trias. Eu. **?Pleurosauridae,** *Pleurosaurus* [*Acrosaurus Anguillosaurus Anguisaurus Saurophidium*] U. Jur. Eu. **Of uncertain position:** *Colognathus* [*Xenognathus*] U. Trias. NA.

SUBCLASS ARCHOSAURIA

ORDER THECODONTIA

SUBORDER PROTEROSUCHIA. **Chasmatosauridae,** *Archosaurus* U. Perm. Eu. *Chasmatosaurus* [*? Ankistrodon ? Epicampodon ? Proterosuchus*] L. Trias. SAf. EAs. SAs. *Chasmatosuchus* L. Trias. EEu. *? Chigutisaurus* L. Trias. SA. *Elaphrosuchus* L. Trias. SAf. **Erythrosuchidae,** *Cuyosuchus* [*?Icanosaurus*] L. Trias. SA. *Erythrosuchus* [*Dongusia Garjainia Vjushkovia*] L. Trias. SAf. EEu. *? Hoplitosuchus* [*Hoplitosaurus*] *? Rauisuchus ?Saurosuchus* M. Trias. SA. *?Seemannia* L. Trias. Eu. *Shansisuchus* L. Trias. EAs. *? Arizonasaurus* M. Trias. NA.

SUBORDER PSEUDOSUCHIA. **Euparkeriidae,** *Euparkeria* [*Browniella*] L. Trias. SAf. **Erpetosuchidae,** *Dibothrosuchus* U. Trias. EAs. *?Dyoplax Erpetosuchus* [*Herpetosuchus*] U. Trias. Eu. *Parringtonia* M. Trias. EAf. *Hesperosuchus* U. Trias. NA. *Saltoposuchus* U. Trias. Eu. *Stegomosuchus* U. Trias. NA. *Strigosuchus* U. Trias. EAs. *Cerritosaurus ?Rhadinosuchus* M. Trias. SA. **Teleocrateridae,** *Teleocrater* M. Trias. EAf. **?Elachistosuchidae,** *Elachistosuchus* U. Trias. Eu. **Prestosuchidae,** *?Mandasuchus* M. Trias. EAf. *Prestosuchus ? Procerosuchus* M. Trias. SA. *?Stagonosuchus* M. Trias. EAf.

SUBORDER AETOSAURIA. **Aetosauridae (Stagonolepidae),** *? Acompsosaurus* U. Trias. NA. *Aetosauroides* M. Trias. SA. *Aetosaurus* U. Trias. Eu. *Argentinosuchus* M. Trias. SA. *Desmatosuchus* U. Trias. NA. *Ebrachosaurus* U. Trias. Eu., ?NA. *Stegomus* U. Trias. NA. *Stagonolepis* [*? Palaeosaurus Rileya*] U. Trias. Eu. *Typothorax* [*Episcoposaurus*] U. Trias. NA.

SUBORDER PHYTOSAURIA. **Phytosauridae,** *Angistorhinus* [*Brachysuchus*] U. Trias. NA. *Mesorhinus* ?L. or U. Trias. Eu. *Mystriosuchus* [*? Belodon ? Termatosaurus*] U. Trias. Eu. *Palaeorhinus* [*Ebrachosuchus Francosuchus Promystriosuchus*] U. Trias. Eu. NA. *Phytosaurus* [*Centemodon Clepsysaurus ? Coburgosuchus Compsosaurus ? Cubicodon ? Cylindrodon ? Heterodontosuchus Lophoprosopus Lophorhinus Nicrosaurus Omosaurus ? Pachysuchus Palaeoctonus Suchoprion*] U. Trias. Eu. NA. SAs. *Rutiodon* [*Angistorhinops ?Dolichobrachium Eurydorus Leptosuchus Machaeroprosopus Metarhinus Pseudopalatus Rhytidodon*] U. Trias. Eu. NA.

THECODONTIA INCERTAE SEDIS: *Anisodonsaurus* U. Trias. NA. *?Ocoyuntaia* L. Trias. SA.

ORDER CROCODILIA

SUBORDER PROTOSUCHIA. **?Sphenosuchidae** U. Trias. SAf., *? Pedeticosaurus ? Platyognathus* U. Trias. EAs. *Sphenosuchus* U. Trias. SAf. **Protosuchidae,** *Protosuchus* [*Archaeosuchus*] U. Trias. or L. Jur. NA., ?U. Trias. Eu.

SUBORDER ARCHAEOSUCHIA. **Notochampsidae,** *Erythrochampsa* U. Trias. SAf. *Microchampsa* U. Trias. EAs. *Notochampsa* U. Trias. SAf. **Proterochampsidae,** *Proterochampsa* M. Trias. SA.

SUBORDER MESOSUCHIA. **Teleosauridae,** *Aeolodon* [*Aëlodon Engyonimasaurus Engyomasaurus Glaphyrorhynchus Palaeosaurus*] L. Jur. Eu. *Gavialinum ? Gnathosaurus ? Haematosaurus* U. Jur. Eu. *Heterosaurus* L. Cret. Eu. *Mycterosuchus* U. Jur. Eu. *Pelagosaurus* [*Mosellaesaurus*] *Platysuchus* L. Jur. Eu. *Steneosaurus* [*Leptocranius Macrospondylus Mystriosaurus Sericodon Sericosaurus Streptospondylus*] L. Jur. SA., L.-U. Jur. Eu., M. Jur. Mad., Jur. NAf. *Teleidosaurus* M. Jur. Eu. *Teleosaurus* M. Jur. EAs., M.-U. Jur. Eu. **Pholidosauridae,** *? Anglosuchus Crocodilaemus* [*Chaelosaurus Crocodileimus*] U. Jur. Eu. *Dyrosaurus* [*Congosaurus*] U. Cret.-Eoc. Af. *Peipehsuchus* U. Jur. EAs. *Petrosuchus* U. Jur. Eu. *Pholidosaurus* [*Macrorhynchus*] U. Jur.-L. Cret. Eu. *Phosphatosaurus* Eoc. NAf. *Rhabdognathus* Eoc. WAf. *Rhabdosaurus* Eoc. NAf. *Sokotosaurus* Eoc. WAf. *Suchosaurus* L. Cret. Eu. *Sunosuchus* U. Jur. EAs. *Teleorhinus* [*Terminonaris*] U. Cret. NA. *?Wurnosaurus* Eoc. WAf. **Atoposauridae,** *Alligatorellus Alligatorium Atoposaurus* U. Jur. Eu. *Hoplosuchus* U. Jur. NA. *Shantungosuchus* U. Jur. EAs. **Goniopholidae,** *Baharijodon* U. Cret. NAf. *? Coelosuchus ?Dakotasuchus* U. Cret. NA. *Doratodon* [*Rhadinosaurus*] U. Cret. Eu. *Goniopholis* [*Amphicotylus Diplosaurus Hyposaurus*] U. Jur.-L. Cret. Eu., U. Jur.-U. Cret. NA. SA. *Hsisosuchus* U. Jur. EAs. *Itasuchus* U. Cret. SA. *Machimosaurus* [*Madrimosaurus*] U. Jur.-L. Cret. Eu., ?U. Cret. SA. *?Microsuchus* U. Cret. SA. *? Nannosuchus Oweniasuchus* [*Brachydectes*] U. Jur. Eu. *Paralligator* U. Cret. EAs. *Pinacosuchus Pliogonodon Polydectes* U. Cret. NA. *?Shamosuchus* L. Cret., ?U. Cret. EAs. *Symptosuchus* U. Cret. SA. *? Theriosuchus* U. Jur. Eu. **Notosuchidae,** *? Libycosuchus* U. Cret. NAf. *Notosuchus ?Sphagesaurus Uruguaysuchus* [*? Brasileosaurus*] U. Cret. SA. **Metriorhynchidae,** *Capellineosuchus* L. Cret. Eu. *Dakosaurus* [*Cricosaurus Plesiosuchus*] U. Jur.-L. Cret. Eu. *Enaliosuchus* L. Cret. Eu. *Geosaurus* [*Brachytaenius Itlilimnosaurus Neustosaurus*] U. Jur.-L. Cret. Eu. *Metriorhynchus* [*? Purranisaurus Rhachaeosaurus Suchodus*] M. Jur. SA., M.-U. Jur. Eu.

SUBORDER SEBECOSUCHIA. **Baurusuchidae,** *Baurusuchus Cynodontosuchus* U. Cret. SA. **Sebecidae,** *?Ilchunaia* Olig. SA. *Peirosaurus* U. Cret. SA. *Sebecus* Paleoc.-Mioc. SA.

SUBORDER EUSUCHIA. **Hylaeochampsidae,** *? Bernissartia* L. Cret. Eu. *Hylaeochampsa* [*Heterosuchus*] U. Jur.-L. Cret. Eu. **Stomatosuchidae,** *? Aegyptosuchus* [*Stromero-*

suchus] U. Cret. NAf. *Chiayusuchus* U. Cret. EAs. *Stomatosuchus* U. Cret. NAf. **Gavialidae,** *Gavialis* [*Leptorhynchus Rhamphostoma*] ?Eoc. Eu., ?Olig. SA., ?Mioc. Af., Mioc.-R. EInd. SAs., Pleist. EAs. *Rhamphostomopsis* ?Mioc., Plioc. SA. *?Rhamphosuchus* Plioc. SAs. **Crocodylidae,** *Alligator* [*Caimanoidea*] Olig.-R. NA., ?Olig., R. EAs. *Allodaposuchus* U. Cret. Eu. *Allognathosuchus* [*Hassiacosuchus*] Paleoc.-L. Olig. NA. *Arambourgia* Eoc.-Olig. Eu. *Asiatosuchus* ?Paleoc.-Eoc. EAs. *Balanerodus* Olig. SA. *Bottosaurus* U. Cret. NA. *Boverisuchus* Eoc. Eu. *Brachychampsa* U. Cret. NA. *Brachyuranochampsa* Eoc. NA. *Caiman* [*Brachygnathosuchus Dinosuchus Eocaiman Notocaiman Purrusaurus Xenosuchus*] Paleoc.-R. SA., ?Eoc.-Olig. Eu., R. CA. *Caimanosuchus* Eoc. Eu. *Ceratosuchus* Paleoc. NA. *Charactosuchus* Mioc. SA. *Colossoemys* [*Emysuchus, Enneodon*] Tert. or Pleist. SA. *Crocodylus* [*Crocodilus Thecachampsa*] U. Cret.-Mioc. Eu., U. Cret.-R. Af. NA. As., Mioc.-R. SA., Pleist.-R. Aus. EInd. CA. WInd. Mad. *Deinosuchus* [*Phobosuchus*] U. Cret. NA. Eu. *Dollosuchus* Eoc. Eu. *Eoalligator* Paleoc. EAs. *Eocenosuchus Eosuchus* Eoc. Eu. *Eotomistoma* U. Cret. EAs. *Euthecodon* Mioc.-L. Pleist. NAf. *Gavialosuchus* Mioc. Eu., Plioc. NA. *Gryposuchus* ?Pleist. SA. *Hispanochampsa* Olig. Eu. *Holops* U. Cret. NA. SA. *Kentisuchus* Eoc. Eu. *Leidyosuchus* U. Cret. SA., U. Cret.-Eoc. NA. *Leptorrhamphus* Plioc. SA. *Lianghusuchus* Eoc. EAs. *Limnosaurus* Eoc. NA. *Megadontosuchus ?Menatalligator* Eoc. Eu. *Mourasuchus* [*Nettosuchus*] Mioc., ?Plioc. SA. *?Navajosuchus* Paleoc. NA. *Necrosuchus* Paleoc. SA. *Orthogenysuchus* Eoc. NA. *Orthosaurus* [*Diplocynodon Plerodon Saurocainus*] Eoc. NA., Eoc.-Mioc. Eu., U.Tert. NAf. *?Oxyodonosaurus* [*Oxydontosaurus*] Olig. SA. *Paleosuchus* Plioc.-R. SA., R. CA. *Pallimnarchus* ?Olig., Pleist. Aus. *Pristichampsus* Eoc. Eu. *Proalligator* Plioc. SA. *Procaimanoidea* Eoc. NA. *Prodiplocynodon* U. Cret. NA. *Thoracosaurus* [*Sphenosaurus*] U. Cret. NA. Eu. ?SA. NAf. *Tienosuchus* Eoc. EAs. *Tomistoma* [*Rhynchosuchus*] Eoc.-Plioc. NAf., ?Eoc., Mioc. Eu., Eoc.-R. As., R. EInd. *Weigeltisuchus* Eoc. Eu.

ORDER PTEROSAURIA

SUBORDER RHAMPHORHYNCHOIDEA. **Dimorphodontidae,** *Campylognathoides* [*Camplyognathus*] *Dimorphodon Parapsicethalus* L. Jur. Eu. *Rhampocephalus* [*Dolichorhampus*] M. Jur. Eu. *Scaphognathus* [*Brachytrachelus Pachyrhamphus*] U. Jur. Eu. **Rhamphorhynchidae,** *Dermodactylus* U. Jur. NA. *?Doratorhynchus* M. Jur.-L. Cret. Eu. *Dorygnathus* L. Jur. Eu. *Odontorhynchus* U. Jur. Eu. *Rhamphorhynchus* U. Jur. Eu. EAf. **Anurognathidae,** *Anurognathus* U. Jur. Eu. *Batrachognathus* U. Jur. As.

SUBORDER PTERODACTYLOIDEA. **Pterodactylidae,** *Belonochasma* U. Jur. Eu. *Ctenochasma* U. Jur.-L. Cret. Eu. *?Doratorhynchus* M. Jur.-L. Cret. Eu. *Gnathosaurus* U. Jur. Eu. *Pterodactylus* [*Cycnorhamphus Cynorhamphus Diopecephalus Germanodactylus Macrotrachelus Ornithocephalus Ornithopterus Ptenodracon Pterodracon Pterotherium*], U. Jur. EAf., U. Jur.-L. Cret. Eu. **Ornithocheiridae (Nyctosauridae, Pteranodontidae),** *? Apatomerus* L. Cret. NA. *Criorhynchus* [*Coloborhynchus*] L. Cret. Eu. *Dsungaripterus* L. Cret. EAs. *Nyctosaurus* [*Nyctodactylus*] U. Cret. NA. SA. *Ornithocheirus* [*? Amblydectes Cimoliornis ? Cretornis ? Lonchodectes ?Osteornis ? Palaeornis Ptenodactylus*] ?U. Jur., L.-U. Cret. Eu., ?Cret. Af. *Ornithodesmus* L. Cret. Eu. *Pteranodon* [*Ornithochirus Ornithostoma*] U. Cret. Eu. As. NA. *Titanopteryx* U. Cret. SWAs.

ORDER SAURISCHIA

SUBORDER THEROPODA

INFRAORDER COELUROSAURIA. **Procompsognathidae,** *Avipes* U. Trias. Eu. *Coelophysis* U. Trias. NA. *Dolichosuchus Halticosaurus* U. Trias. Eu. *? Loukousaurus* U. Trias. EAs. *Podokesaurus* U. Trias. NA. *Procompsognathus* [*Pterospondylus*] *Saltopus Scleromochlus* U. Trias. Eu. *?Spinosuchus* U. Trias. NA. *Triassolestes* M. Trias. SA. *Velocipes* U. Trias. Eu. **?Segisauridae,** *Segisaurus* U. Trias. or L. Jur. NA. **Coeluridae (Coelurosauridae Compsognathidae),** *Agrosaurus* L. Jur. Aus. *Aristosuchus* L. Cret. Eu. *Brasileosaurus* L. Cret. SA. *Calamospondylus* [*Calamosaurus*] L. Cret. Eu. *Caudocoelus* U. Jur. Eu. *Coelosaurus* U. Cret. NA. *Coeluroides* U. Cret. SAs. *Coelurus* [*Ornitholestes*] U. Jur.-L. Cret. NA. *Compsognathus* U. Jur. Eu. *Compsosuchus* U. Cret. SAs. *?Dromaeosaurus* U. Cret. NA. *Elaphrosaurus* U. Jur. EAf. ?Eu., U. Cret. NAf. *? Hallopus* U. Jur. NA. *Jubbulpuria Laevisuchus* U. Cret. SAs. *Paronychodon* U. Cret. NA. *Sarcosaurus* L. Cret. Aus. *Saurornithoides* U. Cret. EAs. NA. *Sinocoelurus* U. Jur. EAs. *Thecocoelurus* [*Thecospondylus*] L. Cret. Eu. *Troödon* [*Polyodontosaurus*] U. Cret. NA. *Velociraptor* U. Cret. EAs. ?NA. *?Walgettosuchus* [*?Fulgurotherium ?Rapator*] L. Cret. Aus. **Ornithomimidae,** *? Betasuchus* U. Cret. Eu. *Cheirostenotes Macrophalangia* U. Cret. NA. *Ornithomimus* [*Struthiomimus*] ?L. Cret., U. Cret. NA., U. Cret. EAs. *Ornithomimoides* U. Cret. SAs. *Oviraptor* U. Cret. EAs. **?Caenagnathidae,** *Caenagnathus* U. Cret. NA.

INFRAORDER CARNOSAURIA. **Ornithosuchidae,** *? Clarenceia* U. Trias. SAf. *Ornithosuchus* [*Dasygnathoides Dasygnathus*] U. Trias. Eu. **?Poposauridae,** *Poposaurus* U. Trias. NA. **Megalosauridae,** *Allosaurus* [*Antrodemus Creosaurus ? Labrosaurus*] U. Jur.-L. Cret. NA., ?U. Jur. EAf., U. Jur., ?L. Cret. EAs. *Bahariasaurus* U. Cret. NAf. *Carcharodonsaurus* ?L. Cret., U. Cret. NAf. *Ceratosaurus* U. Jur. NA. *Chienkosaurus* U. Jur. EAs. *Chilantaiosaurus* L.-U. Cret. EAs. *?Dryptosauroides* U. Cret. SAs. *Dryptosaurus* [*Laelaps*] U. Cret. NA. *Embasaurus* L. Cret. WAs. *? Erectopus* L. Cret. Eu. *Eustreptospondylus* ["*Streptospondylus*"] M. Jur. Mad., U. Jur. Eu., Jur. SA. *Inosaurus* Cret. NAf. *Macrodontophion* Jur. Eu. *Megalosaurus* [*Aggiosaurus Iliosuchus Magnosaurus Nuthetes Poecilopleuron*] L. Jur. NA. L. Jur.-L. Cret., ?U. Cret. Eu., M. Jur.-L. Cret. NAf., U. Jur. EAf., *Metriacanthosaurus Proceratosaurus* U. Jur. Eu. *Sarcosaurus* L. Jur. Eu. **Spinosauridae,** *Acrocanthosaurus* L. Cret. NA. *Altispinax* L. Cret. Eu. *Spinosaurus* U. Cret. NAf. **Tyrannosauridae (Deinodontidae),** *Albertosaurus* U. Cret. NA. *Alectrosaurus ? Chingkankonsaurus* U. Cret. EAs. *Genyodectes* [*? Clasmosaurus Loncosaurus*] U. Cret. SA. *Gorgosaurus* [*? Aublysodon Deinodon Dinodon Teinurosaurus*] U. Cret. NA. EAs. *Majungasaurus* ?L. Cret. Mad., U. Cret. EAs. *Orthogoniosaurus* [*Indosaurus Indosuchus*] U. Cret. SAs. *Prodeinodon* U. Cret. EAs. *?Szechusanosaurus Tarbosaurus* U. Cret. EAs. *Tyrannosaurus* [*Dynamosaurus ?Manospondylus*] U. Cret. NA. EAs.

THEROPODA INCERTAE SEDIS: *Aeposaurus* U. Trias. or L. Jur. As. *Ammosaurus* U. Trias. NA. *Cladeiodon* [*Abalonia Cladyodon Kladeisteriodon Smilodon* ?U.

Trias. Eu. *Gresslyosaurus* [*Dinosaurus Pachysauriscus Pachysaurops Pachysaurus Picrodon*] U. Trias.-L. Jur. Eu. *Palaeosauriscus* [*Palaeosaurus*] M.-U. Trias. Eu. *Pneumatoarthrus Stenonychosaurus* U. Cret. NA. *Tichosteus* U. Jur. NA. *Zapsalis* U. Cret. NA. ?*Zatomus* U. Trias. NA, ?*Teratosaurus* U. Trias. Eu.

SUBORDER SAUROPODOMORPHA

INFRAORDER PROSAUROPODA. **Thecodontosauridae** (**Gryponychidae**), *Aetonyx* U. Trias. SAf. ? *Arctosaurus* U. Trias. NA. *Dromicosaurus Gryponyx* U. Trias. SAf. *Gryposaurus* [*Aristosaurus* ?*Hortalotarsus*] U. Trias. SAf. EAs. *Massospondylus* [*Leptospondylus Pachyspondylus*] U. Trias. SAf. EAf. EAs. SAs. *Spondylosoma* M. Trias. SA. *Thecodontosaurus* [*Anchisaurus*] M.-U. Trias. Eu., U. Trias. NA. SAf. Aus. ?EAs. *Yaleosaurus* [*Amphisaurus* ?*Megadactylus*] U. Trias. NA. **Plateosauridae,** ?*Herrerasaurus* [?*Ischisaurus*] M. Trias. SA. *Lufengosaurus* U. Trias. EAs. *Plateosaurus* [*Dimodosaurus Platysaurus Sellosaurus*] U. Trias. Eu. *Yunnanosaurus* U. Trias. EAs. **Melanorosauridae,** *Eucnemesaurus* ?*Euskelosaurus* [?*Basutodon Gigantoscelis Orinosaurus Orosaurus*] U. Trias. SAf. *Melanorosaurus* M.-U. Trias. SAf. *Plateosauravus* U. Trias. SAf. *Sinosaurus* U. Trias. EAs.

INFRAORDER SAUROPODA (OPISTHOCOELIA, CETIOSAURIA). **Brachiosauridae,** *Amygdalodon* M. Jur. SA. *Apatosaurus* U. Jur. NA. *Astrodon* [*Pleurocoelus*] U. Jur.-L. Cret. NA. Eu., L. Cret. NAf. ? *Austrosaurus* L. Cret. Aus. *Bothriospondylus* M. Jur. Mad., M.-U. Jur. Eu. *Brachiosaurus* U. Jur. NA. Eu. ?EAs., U. Jur.-L. Cret. Af. *Camarasaurus* [*Morosaurus Uintasaurus*] U. Jur. NA., L. Cret. Eu. *Cetiosaurus* [*Cardiodon Cetiosauriscus*] M. Jur. Af., M. Jur.-L. Cret. Eu. *Dystrophaeus Elosaurus* U. Jur. NA. *Euhelopus* [*Helopus*] L. Cret. EAs. *Haplocanthosaurus* [*Haplocanthus*] U. Jur. NA. *Omeisaurus* U. Jur.-L. Cret. EAs. ?*Parrosaurus* [*Neosaurus*] U. Cret. NA. *Pelorosaurus* [*Caulodon Chondrosteosaurus Dinodocus Eucamerotus Gigantosaurus Hoplosaurus Ischyrosaurus Morinosaurus Neosodon Oplosaurus Ornithopsis*] U. Jur.-L. Cret. Eu. *Rhoetosaurus* L. Jur. Aus. *Tienshanosaurus* U. Jur.-L. Cret. EAs. **Titanosauridae,** *Aegyptosaurus* U. Cret. NAf. *Aepisaurus* L. Cret. Eu. *Alamosaurus* U. Cret. NA. *Algoasaurus* L. Cret. SAf. *Amphicoelias* U. Jur. NA. *Antarctosaurus* U. Cret. SA. Af. SAs. *Argyrosaurus* [?*Campylodon*] U. Cret. SA. *Asiatosaurus* L. Cret. EAs. *Barosaurus* U. Jur. NA. EAf. *Brontosaurus* [? *Atlantosaurus*] U. Jur. NA. Eu. *Chiayusaurus* U. Cret. EAs. *Dicraeosaurus* U. Jur.-U. Cret. Af. *Diplodocus* U. Jur. NA. *Hypselosaurus* [*Magyarosaurus*] U. Cret. Eu. *Laplatasaurus* U. Cret. SA. SAs. Mad. *Macrurosaurus* U. Cret. Eu. *Mamenchisaurus* U. Jur. EAs. *Mongolosaurus* L. Cret. EAs. *Rebbachisaurus* L. Cret. NAf. *Succinodon* U. Cret. Eu. *Titanosaurus* L.-U. Cret. Eu., U. Cret. Af. SA. As. *Tornieria* [*Gigantosaurus*] U. Jur. EAf.

SAUROPODA INCERTAE SEDIS: *Epantherias* U. Jur. NA. ?*Clasmodosaurus* U. Cret. SA. *Kuangyuanpus* U. Jur. EAs. ?*Microcoelus Microsaurus* U. Cret. SA. ?*Sanpasaurus* U. Jur. EAs. *Symphyrophus* U. Jur. NA.

ORDER ORNITHISCHIA

SUBORDER ORNITHOPODA. **Heterodontosauridae,** *Heterodontosaurus* U. Trias. SAf. **Hypsilophodontidae,** ?*Geranosaurus* U. Trias. SAf. *Hypsilophodon* L. Cret. Eu. ?*Laosaurus* [*Dryosaurus*] U. Jur.-L. Cret. NA. ?*Lycorhinus* U. Trias. SAf. ?*Nannosaurus* U. Jur. NA. *Parksosaurus* U. Cret. NA. ?*Marcellognathus* U. Jur. NA. ?*Stenopelix* L. Cret. Eu. *Tatisaurus* U. Trias. EAs. *Thescelosaurus* U. Cret. NA. **Iguanodontidae,** *Anoplosaurus* [?*Eucerosaurus* ?*Syngonosaurus*] U. Cret. Eu. *Camptosaurus* [*Campotonodus Camptonotus Cumnoria*] U. Jur.-L. Cret. NA., U. Jur., ?L. Cret. Eu. *Craspedodon* U. Cret. Eu. *Cryptodraco* [*Cryptosaurus*] M. Jur. Eu. *Dysalotosaurus* U. Jur. EAf. *Iguanodon* [*Iguanosaurus Sphenospondylus Therosaurus*] U. Jur.-L. Cret. Eu., L. Cret. NAf. EAs. *Kangnasaurus* U.(?) Cret. SAf. *Rhabdodon* [*Mochlodon Ornithomerus*] U. Cret. Eu. *Vectisaurus* L. Cret. Eu. **Hadrosauridae** (**Trachodontidae**), *Anatosaurus* U. Cret. NA. *Bactrosaurus* U. Cret. As. *Brachylophosaurus* U. Cret. NA. *Cheneosaurus Claorhynchus Claosaurus Corythosaurus Dysganus Edmontosaurus* U. Cret. NA. *Hadrosaurus* U. Cret. NA., ?Paleoc. SA. *Hypacrosaurus Hypsibema Kritosaurus* [*Gryposaurus*] *Lambeosaurus* [*Stephanosaurus*] *Lophorhothon* U. Cret. NA. *Mandschurosaurus* [*Nipponosaurus*] U. Cret. EAs. *Ornithotarsus* U. Cret. NA. *Orthomerus* [*Hecatasaurus Limnosaurus Telmatosaurus*] U. Cret. Eu. *Parasaurolophus Procheneosaurus* [*Tetragonosaurus*] *Prosaurolophus* U. Cret. NA. *Saurolophus* U. Cret. NA. EAs. *Tanius* [*Tsintaosaurus*] U. Cret. EAs. *Thespesius* [*Cionodon Diclonius Didanodon Pterypelyx*] *Trachodon* U. Cret. NA. *Tsintaosaurus* U. Cret. EAs. *Yaxartosaurus* U. Cret. As. **Psittacosauridae,** *Protiguanodon Psittacosaurus* U. Cret. EAs. **Pachycephalosauridae** (**"Troödontidae"**), *Pachycephalosaurus* ?*Polyodontosaurus Stegoceras* ["*Troödon*"] U. Cret. NA.

SUBORDER STEGOSAURIA. **Scelidosauridae,** ?*Lusitanosaurus* L. Cret. Eu. *Scelidosaurus* [*Sarcolestes*] L. Jur. Eu. **Stegosauridae,** *Craterosaurus* L. Cret. Eu. ?*Chialingosaurus* U. Jur. EAs. *Kentrosaurus* [*Doryphorosaurus Kentrurosaurus*] U. Jur. EAf. ?EAs. *Lexovisaurus* M. or U. Jur. EAs. *Priconodon* L. Cret. NA. *Priodontosaurus* [*Dacentrurus Omosaurus Priodontognathus*] M.-U. Jur. Eu. *Saurechinodon* [*Echinodon*] U. Jur. Eu. *Stegosaurus* [*Diracodon Hypsirhophus Hysirophus*] U. Jur. NA., ?Eu.

SUBORDER ANKYLOSAURIA. **Acanthopholidae,** *Acanthopholis* U. Cret. Eu. *Hylaeosaurus* [*Hylosaurus Regnosaurus*] L. Cret. Eu. ?*Loricosaurus* U. Cret. SA. *Onychosaurus Rhodanosaurus Struthiosaurus* [*Crataeomus Danubriosaurus Leipsanosaurus Pleuropeltus*] U. Cret. Eu. **Nodosauridae** (**Ankylosauridae**), *Anodontosaurus* U. Cret. SA. ?*Brachypodosaurus* U. Cret. SAs. *Dyoplosaurus Edmontonia Euoplocephalus* [*Ankylosaurus Stereocephalus*] U. Cret. NA. *Heishansaurus* U. Cret. EAs. *Hierosaurus* U. Cret. NA. *Hoplitosaurus* L. Cret. NA. *Lametasaurus* U. Cret. As. *Nodosaurus Palaeoscincus Panoplosaurus* U. Cret. NA. ?*Paracanthodon* L. Cret. SAf. *Peishansaurus Pinacosaurus* U. Cret. EAs. *Polacanthoides Polacanthus* L. Cret. Eu. *Sauroplites* L. Cret. EAs. *Scolosaurus Silvisaurus Stegopelta* U. Cret. NA. ?*Stegosaurides Syrmosaurus Talarurus Viminicaudus* U. Cret. EAs.

SUBORDER CERATOPSIA. **Protoceratopsidae,** *Leptoceratops Montanoceratops* U. Cret. NA. ?*Notoceratops* U. Cret. SA.

Protoceratops U. Cret. EAs. **Ceratopsidae,** *Anchiceratops Arrhinoceratops Centrosaurus Ceratops* [*Proceratops*] *Chasmosaurus* [*Protorosaurus*] *Diceratops Eoceratops* U. Cret. NA. *Microceratops* U. Cret. EAs. *Monoclonius* [*Brachyceratops*] U. Cret. NA. *Pentaceratops* U. Cret. NA. ?EAs. *Styracosaurus Torosaurus Triceratops* [*Agathaumas Polyonax Sterrholophus*] U. Cret. NA. **Pachyrhinosauridae,** *Pachyrhinosaurus* U. Cret. NA.

SUBCLASS EURYAPSIDA (SYNAPTOSAURIA)

ORDER ARAEOSCELIDIA (PROTOROSAURIA). **Araeoscelidae,** ?*Aenigmasaurus* L. Trias. SAf. *Araeoscelis* [*Ophiodeirus Tomicosaurus*] L. Perm. NA. *Kadaliosaurus* [*Cadaliosaurus*] L. Perm. Eu. ?**Protorosauridae,** ?*Adelosaurus Protorosaurus* [?*Gracilosaurus Protosaurus*] U. Perm. Eu. *Trachelosaurus* L. Trias. Eu. ?*Trentinosaurus* U. Perm. Eu. ?**Tanystropheidae,** *Tanystropheus* [*Macroscelosaurus Pectenosaurus Procerosaurus Tribelesodon* ?*Zanclodon*] M. Trias. Eu. SWAs. ?**Weigeltisauridae,** *Coelurosauravus* U. Perm. Mad. *Weigeltisaurus* [*Palaeochamaeleo*] U. Perm. Eu. **Trilophosauridae,** ?*Anisodontosaurus* L. Trias. NA. *Gomphiosaurus* U. Trias. NA. ?*Toxolophosaurus* U. Cret. NA. *Trilophosaurus* U. Trias. NA. *Tricuspisaurus Variodens* U. Trias. Eu.

ORDER SAUROPTERYGIA

SUBORDER NOTHOSAURIA. **Nothosauridae,** *Ceresiosaurus Deirosaurus* M. Trias. Eu. *Keichousaurus* M. Trias. EAs. ?*Kwangsisaurus* L. Trias. EAs. *Lariosaurus* [*Macromerosaurus Macromirosaurus*] M. Trias. Eu. *Metanothosaurus* L. Trias. EAs. *Micronothosaurus* M. Trias. SWAs. ?*Nanchangosaurus* M. Trias. EAs. ?*Nothosauravus* U. Perm. Eu. *Nothosaurus* [*Chondriosaurus Conchiosaurus Dracontosaurus Dracosaururus Dracosaurus Kolposaurus Menodon Oligolycus*] L.-U. Trias. Eu., M. Trias. SWAs. NAf., ?U. Trias. EAs. *Paranothosaurus* ?*Parthanosaurus* [?*Microcleptosaurus*] M. Trias. Eu. *Pontopus* U. Trias. Eu. *Proneusticosaurus* [*Dolichovertebra* ?*Lamprosaurus Lamprosciuroides*] M. Trias. Eu. **Pachypleurosauridae,** *Neusticosaurus* [*Anarosaurus* ?*Anomosaurus Dactylosaurus* ?*Philotrachelosaurus* ?*Phygosaurus* ?*Psilotrachelosaurus*] M.-U. Trias. Eu. *Pachypleurosaurus* [*Pachypleura*] M. Trias. Eu. **Simosauridae,** ?*Corosaurus* U. Trias. NA. ?*Elmosaurus* M. Trias. Eu. *Simosaurus* [?*Opeosaurus*] M.-U. Trias. Eu.

SUBORDER PLESIOSAURIA

?SUPERFAMILY PISTOSAUROIDEA. **Pistosauridae,** *Pistosaurus* M. Trias. Eu. ?NAf. **Cymatosauridae,** *Cymatosaurus* [?*Eurysaurus Germanosaurus*] ?*Rhaeticonia* M. Trias. Eu. ?*Sulmosaurus* U. Trias. Eu.

SUPERFAMILY PLESIOSAUROIDEA (DOLICHODEIRA). **Plesiosauridae,** *Archaeonectrus* L. Jur. Eu. ?*Colymbosaurus* U. Jur. Eu. ?Gr. *Cryptocleidus* [*Apractocleidus*] U. Jur. Eu. ?*Eretmosaurus Microcleidus* L. Jur. Eu. *Muraenosaurus* U. Jur. Eu. WInd. *Picrocleidus* U. Jur. Eu. *Plesiosaurus* [*Halidragon Pentatarsostinus*] U. Trias.-M. Jur. Eu., L. Jur. NAf. ?SWAs. ?SA. *Tremamesacleis* U. Jur. Eu. **Thaumatosauridae,** *Eurycleidus Seeleyosaurus* L. Jur. Eu. *Simolestes* U. Jur. Eu. NAf. ?*Sthenarosaurus* L. Jur. Eu. *Thaumatosaurus* [*Enigmatosaurus Rhomleosaurus*] L.-M. Jur. Eu. **Elasmosauridae,** *Alzadasaurus* L. Cret. NA. SA. *Aphrosaurus* U. Cret. NA. *Brancasaurus* L. Cret. Eu., ?U. Cret. NAf. *Elasmosaurus* [?*Thalassonomosaurus*] *Fresnosaurus Hydralmosaurus* [?*Thalassiosaurus*] *Hydrotherosaurus Leuspondylus* U. Cret. NA. *Mauisaurus* U. Cret. NZ. ?Eu. *Morenosaurus* ?*Ogmodirus* U. Cret. NA. *Scanisaurus* U. Cret. Eu. *Styxosaurus Thalassomedon* U. Cret. NA. *Woolungasaurus* L. Cret. Aus.

SUPERFAMILY PLIOSAUROIDEA (BRACHYDEIRA). **Pliosauridae,** *Brachauchenius* U. Cret. NA. *Gymocetus* U. Cret. Eu. *Kronosaurus* L. Cret. Aus. *Liopleurodon* [*Ischyrodon*] U. Jur. Eu. ?*Macroplata* L. Jur. Eu. *Peloneustes* L. Jur. Eu. *Pliosaurus* [*Chelonosaurus* ?*Ischyrodon Sinopliosaurus Spondylosaurus*] L.-U. Jur. Eu., ?Cret. SA., U. Jur., ?L. Cret. As. *Stretosaurus* U. Jur. Eu. **Polycotylidae,** ?*Brimosaurus* U. Cret. NA. *Cimoliasaurus* [*Symoliosaurus*] L. Cret. Aus., U. Cret. NA., ?SA. *Discosaurus Embaphias Piptomerus Piratosaurus Polycotylus* U. Cret. NA. *Polyptychodon* U. Cret. Eu. Aus. NA. *Scarrisaurus* L. Cret. Aus. *Taphrosaurus Trinacromerum* [*Dolichorhynchus*] U. Cret. NA. **Leptocleididae,** ?*Aristonectes* U. Cret. SA. *Dolichorhynchops* L. Cret. Aus. *Leptocleidus* L. Cret. Eu., U. Cret. NAf. *Peyerus* L. Cret. SAf. ?*Tricleidus* U. Jur. Eu.

PLESIOSAURIA INCERTAE SEDIS: *Aptychodon* U. Cret. Eu- *Eurysaurus* M. Jur. Eu. *Hexatarsostinus* L. Jur. Eu. ?*Hunosaurus* ?*Iserosaurus* U. Cret. Eu. *Megalneusaurus* U. Jur. NA. ?*Nothosaurops Oligosimus Orophosaurus* U. Cret. NA. *Pantosaurus* [*Parasaurus*] U. Jur. NA., ?Eu., ?NAf. *Uronautes* U. Cret. NA.

ORDER PLACODONTIA. **Helveticosauridae,** *Helveticosaurus* M. Trias. Eu. **Placodontidae,** *Paraplacodus* M. Trias. Eu. *Placodus* [?*Anomosaurus Crurosaurus Pleurodus*] L.-M. Trias. Eu. **Placochelyidae (Cyamodontidae),** *Cyamodus Placochelys* M.-U. Trias. Eu. *Psephoderma* U. Trias. Eu. *Psephosaurus* ?M. Trias. NAf., U. Trias. Eu. SWAs. *Saurosphargis* M. Trias. Eu. **Henodontidae,** *Henodus* [?*Chelyposuchus*] U. Trias. Eu.

SUBCLASS ICHTHYOPTERYGIA

ORDER ICHTHYOSAURIA. **Mixosauridae,** *Mixosaurus* [*Phalarodon*] M. Trias. Eu. Spits. EInd. NA. SWAs. **Omphalosauridae,** *Grippia* M. Trias. Spits. *Omphalosaurus* [*Pessopteryx Tholodus*] M. Trias. NA. Spits. Eu. **Shastasauridae,** ?*Chonespondylus* M. Trias. NA. *Cymbospondylus* [?*Blezingenia*] M. Trias. NA. Eu. ?Spits. EInd. *Merriamia* [*Leptocheirus Rhachitrema*] U. Trias. NA. ?Eu. *Pessosaurus* [?*Ekbainacanthus*] M. Trias. Spits. Eu. ?NA. *Shastasaurus* [?*Pachygonosaurus*] M. Trias. Spits., M.-U. Trias. Eu., U. Trias. NA. *Toretocnemus* [*Californosaurus Delphinosaurus Perrinosaurus*] M. Trias. Eu., U. Trias. NA. **Ichthyosauridae,** *Brachypterygius* U. Jur. Eu. *Ichthyosaurus* [*Eurypterygius Gryphius Ichthyoterus* ?*Proteosaurus*] L. Jur. Gr., L.-M. Jur. Eu., L., ?M. Jur. SA., ?U. Jur. NA., ?L. Cret. NZ. *Macropterygius* M. Jur. SA., U. Jur.-L. Cret. Eu. *Myobradypterygius* U. Jur.-L. Cret. SA. *Myopterygius* [?*Ancanamunia Cetarthrosaurus* ?*Delphinosaurus*] U. Jur.-L. Cret. SA., L. Cret. SAs. EAs. EInd. Aus. NZ., L.-U. Cret. Eu., U. Cret. NA. *Ophthalmosaurus* [*Apatodontosaurus Baptanodon Microdontosaurus Sauranodon*] M.-?U. Jur., ?U. Cret. NA., M. Jur.-U. Cret. Eu. Gr. **Stenopterygiidae (Longipinnati),** *Eurhinosaurus* L. Jur. Eu. *Leptopterygius* [*Temnodontosaurus*] L. Jur. Eu., ?U. Jur. SA. *Nannopterygius* U. Jur. Eu. *Platypterygius* L. Cret. Eu. *Stenopterygius* [?*Streptospondylus*] L.-M. Jur. Eu., ?U. Jur. SA.

SUBCLASS SYNAPSIDA

ORDER PELYCOSAURIA (THEROMORPHA)

SUBORDER OPHIACODONTIA. **Ophiacodontidae,** *Basicranodon* L. Perm. NA. *Clepsydrops* [*Archaeobelus*] U. Penn. NA. *Ophiacodon* [*Arribasaurus Diopeus Poliosaurus Theropleura Therosaurus Winfieldia*] L. Perm. NA. *?Protoclepsydrops* L. Penn. NA. *Varanosaurus* [*Poecilospondylus*] L. Perm. NA. **Eothyrididae,** *Baldwinonus Bayloria Eothyris Stereophallodon* L. Perm. NA. *Stereorhachis* U. Penn. Eu. *?Tetraceratops* L. Perm. NA.

SUBORDER SPHENACODONTOIDEA. **Varanopsidae,** *Aerosaurus* L. Perm. NA. *?Anningia* [*?Galesphyrus*] M. Perm. SAf. *?Homodontosaurus* U. Perm. SAf. *Scoliomus Varanops* [*Varanoops*] L. Perm. NA. **Sphenacodontidae,** *Bathygnathus* L. Perm. NA. *?Ctenosaurus* L. Trias. Eu. *Ctenospondylus* L. Perm. NA. *Dimetrodon* [*Bathyglyptus Embolophorus*] L.-M. Perm. NA. *Haptodus* [*Callibrachion Datheosaurus Palaeohatteria Palaeosphenodon Pantelosaurus*] L. Perm. Eu. *Macromerion* U. Penn. Eu. *Neosaurus Oxyodon* L. Perm. Eu. *Secodontosaurus Sphenacodon* [*Elcabrosaurus*] L. Perm. NA. *?Steppesaurus* M. Perm. NA. *?Thrausmosaurus* L. Perm. NA.

SUBORDER EDAPHOSAURIA, **Nitosauridae,** *?Colobomycter ?Delorhynchus ?Glaucosaurus Mycterosaurus* [*Eumatthevia*] *Nitosaurus ?Oedaleops* L. Perm. NA. *?Petrolacosaurus* [*Podargosaurus*] U. Penn. NA. **Lupeosauridae,** *Lupeosaurus* L. Perm. NA. **Edaphosauridae,** *Edaphosaurus* [*Brachycnemius Naosaurus*] U. Penn.-L. Perm. Eu. NA. **Caseidae,** *Angelosaurus* M. Perm. NA. *Casea* L. Perm. NA. *Caseoides* M. Perm. NA. *Caseopsis* M. Penn. NA. *Cotylorhynchus* L.-M. Perm. NA. *Ennatosaurus ?Phreatophasma* M. Perm. EEu. *?Trichasaurus* [*Trispondylus*] L. Perm. NA.

ORDER THERAPSIDA

SUBORDER PHTHINOSUCHIA (EOTITANOSUCHIA). **Phthinosuchidae,** *Biarmosaurus Biarmosuchus* M. Perm. EEu. *Eotitanosuchus* M. Perm. EEu. *Gorgodon Knoxosaurus* M. Perm. NA. *?Phreatosaurus* M. Perm. Eu. *?Phreatosuchus Phthinosaurus Phthinosuchus* M. Perm. EEu.

SUBORDER THERIODONTIA

INFRAORDER GORGONOPSIA. **Galesuchidae,** *?Aelurosauroides Cerdodon Eoarctops Galesuchus Pachyrhinos* M. Perm. SAf. *?Scylacognathus* U. Perm. SAf. **Hipposauridae (Ictidorhinidae),** *Hipposauroides* U. Perm. SAf. *Hipposaurus* M. Perm. SAf. *Ictidorhinus Lemurosaurus Pseudohipposaurus* U. Perm. SAf. **Cynariopsidae,** *Cynarioides Cynariops* U. Perm. SAf. **Rubidgeidae,** *Clelandina* U. Perm. SAf. *Dinogorgon* [*Broomicephalus Dracocephalus*] U. Perm. SAf. EAf. *Pardocephalus Prorubidgea Rubidgea Smilesaurus* U. Perm. SAf. *?Tangagorgon* U. Perm. EAf. *Tigrisaurus* U. Perm. SAf. **Gorgonopsidae,** *Gorgonops Leptotrachelus* U. Perm. SAf. **Scymnognathidae,** *Arctops* U. Perm. SAf. *Chiwetasaurus Dixeya* U. Perm. EAf. *Lycaenoides* U. Perm. SAf. *Scymnognathus* [*Scymnosuchus*] U. Perm. SAf. EAf. **Aelurosauridae,** *Aelurosaurus* U. Perm. SAf. ?EAf. **Galerhinidae,** *Galerhinus* U. Perm. SAf. EAf. **Gorgonognathidae,** *Gorgonognathus* U. Perm. SAf. EAf. *Gorgonorhinus Tigricephalus* U. Perm. SAf. **Arctognathoididae,** *Arctognathoides Leontocephalus Leontosaurus* U. Perm. SAf. **Scylacopsidae,** *Cyniscopoides* U. Perm. SAf. *Cyniscops* U. Perm. SAf. ?EAf. *Cyonosaurus Galerhynchus Scylacops Sycocephalus* U. Perm. SAf. **Sycosauridae,** *Sycosaurus* U. Perm. SAf. **Arctognathidae,** *Arctognathus* U. Perm. SAf. Eu. EAf. *Lycaenodontoides* U. Perm. SAf. **Aelurosauropsidae,** *Aelurosauropsis* U. Perm. SAf. **Scylacocephalidae,** *Scylacocephalus* U. Perm. SAf. **Broomisauridae,** *Broomisaurus* [*Scymnorhinus*] M.-U. Perm. SAf. **Inostranceviidae,** *Inostrancevia* [*Amalitzkia*] *Pravoslavleria Sauroctonus* U. Perm. EEu. **?Burnetiidae (Burnetiamorpha),** *Burnetia* U. Perm. SAf. *Styracocephalus* M. Perm. SAf.

GORGONOPSIA INCERTAE SEDIS: *Aelurognathus* U. Perm. SAf. EAf. *Alopecorhynchus Aloposauroides* U. Perm. SAf. *Aloposaurus* M. Perm. SAf. *Arctosuchus* U. Perm. SAf. *Cerdognathus* U. Perm. SAf. *Cerdorhinus* M. Perm. SAf. *Cyniscodon* U. Perm. SAf. *Cynodraco* [*Cynodrakon*] *Delphaciognathus* [*Asthenognathus*] U. Perm. SAf. *Eriphostoma* M. Perm. SAf. *Genovum* U. Perm. EAf. *Lycaenodon Lycaenops Lycosaurus Nanogorgon* U. Perm. SAf. *Tetraodon* U. Perm. EAf. *Tigrisuchus* U. Perm. SAf.

INFRAORDER CYNODONTIA. **Procynosuchidae,** *Dvinia* [*Dwinia*] U. Perm. EEu. *Galeophrys* [*Galecranium*] *Leavachia* [*Aelurodraco*] *Nanictosuchus Paracynosuchus* U. Perm. SAf. *Parathrinaxodon* U. Perm. EAf. *Permocynodon* U. Perm. EEu. *Procynosuchus* U. Perm. SAf. EAf. *?Scalopocynodon* U. Perm. SAf. **Thrinaxodontidae (Galesauridae),** *Cynosaurus* [*Cynosuchoides Cynosuchus*] U. Perm. SAf. *Galesaurus* [*Glochinodon Glochinodontoides*] *?Micrictodon* L. Trias. SAf. *Nanictosaurus* U. Perm. SAf. *Notictosaurus Nythosaurus Platycraniellus* [*Platycranion*] L. Trias. SAf. *Sinognathus* L. Trias. EAs. *Sysphinctostoma Thrinaxodon* [*Ictidopsis*] L. Trias. SAf. **Cynognathidae,** *Chinquodon ?Colbertosaurus* M. Trias. SA. *Cynognathus* [*Cynidiognathus Cynogomphius Lycaenognathus Lycochampsa Lycognathus*] L.-M. Trias. SAf. *?Karoomys* L. Trias. SAf. *?Pachygenelus* U. Trias. SAf. *Tribolodon* L. Trias. SAf. **Diademodontidae,** *?Aleodon* M. Trias. EAf. *Diademodon* [*Cyclogomphodon ?Cynochampsa Diastemodon Gomphognathus Microhelodon Octagomphus*] *?Gomphodontoides* L. Trias. SAf. *Pascualgnathus* M. Trias. SA. *Protacmon* L. Trias. SAf. *Theropsodon* M. Trias. EAf. **Traversodontidae,** *?Belesodon* M. Trias. SA. *?Cricodon* M. Trias. EAf. *Exaeretodon Gomphodontosuchus Ischignathus* M. Trias. SA. *Luangwa* M. Trias. SAf. *Ordosiodon* L. or M. Trias. EAs. *Proexaeretodon* M. Trias. SA. *Scalenodon* M. Trias. EAf. *Scalenodontoides* M. Trias. SAf., ?U. Trias. NAf. *Theropsis Traversodon* M. Trias. SA. *Trirachodon* [*?Inusitatodon*] L. Trias. SAf. EAf. *Trirachodontoides* L. Trias. SAf.

?CYNODONTIA INCERTAE SEDIS: *Dromatherium* U. Trias. NA. *?Kunminia* U. Trias. EAs. *Microconodon* [*Tytthoconus*] U. Trias. NA. *Tricuspes* U. Trias. Eu. *Trithelodon* U. Trias. SAf.

INFRAORDER TRITYLODONTOIDEA. **Tritylodontidae,** *Bienotherium* U. Trias. EAs. *Chalepotherium* U. Trias.-L. Jur. Eu. *Likhoelia* U. Trias. SAf. *Lufengia* U. Trias. EAs. *Oligokyphus* [*Mucrotherium Uniserium*] U. Trias.-L. Jur. Eu. *Stereognathus* M. Jur. Eu. *Tritylodon* [*?Triglyphus*] U. Trias. SAf. ?Eu. *Tritylodontoides* U. Trias. SAf.

INFRAORDER THEROCEPHALIA. **Pristerognathidae,** *?Akidnognathus* U. Perm. SAf. *Alopecognathus Alopecorhinus* M. Perm. SAf. *Anna* U. Perm. EEu. *Chtho-*

nosaurus U. Perm. SAf. *Cynariognathus Glanosuchus ?Hyorhynchus Ictidoparia Ictidosaurus Karroowalteria* [*Walteria*] *Lycedops Maraisaurus* M. Perm. SAf. *Mirotenthis ?Notaelurodon* U. Perm. SAf. *Porostagnathus* M. Perm. EEu. *Pristerognathoides Pristerognathus Pristerosaurus Ptomalestes Scylacoides Scylacorhinus Scylacosaurus Scymnosaurus ?Tamboeria Therioides* M. Perm. SAf. **Alopecodontidae,** *Alopecodeops Alopecodon Pardosuchus Trochosuchus* M. Perm. SAf. *? Urumchia* U. Perm. EAs. **Trochosauridae (Lycosuchidae),** *Hyaenasuchus Lycosuchus Trochorhinus Trochosaurus* M. Perm. SAf. **Whaitsiidae,** *Alopecopsis ?Cerdosuchoides ?Cerdops ?Cerdosuchus Hewittia Hofmeyria Hyenosaurus Moschorhinus Moschorhynchus* U. Perm. SAf. *Moschowhaitsia* U. Perm. EEu. *Notaelurops Notosollasia* U. Perm. SAf. EAf. *Proalopecopsis Promoschorhynchus* U. Perm. SAf. *Theriognathus* U. Perm. SAf. EAf. *Whaitsia* [*? Aneugomphius*] U. Perm. SAf. **Euchambersiidae,** *Euchambersia* U. Perm. SAf.

INFRAORDER BAURIAMORPHA. **Lycideopsidae,** *? Arnognathus Lycideops* U. Perm. SAf. **Ictidosuchidae,** *Ictidosuchus* U. Perm. SAf. EAf. **Nanictidopsidae,** *Ictidosuchoides* U. Perm. SAf. *Ictidosuchops Nanictidops Pelictosuchus* U. Perm. SAf. **Silpholestidae,** *Ictidodraco Scaloporhinus* U. Perm. SAf. *Silphoictidoides* U. Perm. EAf. *Silpholestes Tetracynodon* U. Perm. SAf. **Scaloposauridae,** *Blattoidealestes* M. Perm. SAf. *Choerosaurus ?Haughtoniscus* [*Macroscelesaurus*] *?Homodontosaurus* U. Perm. SAf. *Icticephalus* M. Perm. SAf. *Ictidodon ?Ictidognathus* [*Ictidostoma*] *?Nanictocephalus Nanictosuchus ?Polycynodon* [*Octocynodon*] *Protocynodon Scalopocephalus Scaloposaurus Scaloposuchus Silphedestes Silphedocynodon* U. Perm. SAf. **Ericiolacertidae,** *Cistecynodon Cyrbasiodon* L. Trias. SAf. *Dongusaurus* L. Trias. EEu. *Ericiolacerta* L. Trias. SAf. **Bauriidae,** *Aelurosuchus Bauria* [*Baurioides*] *Melinodon Microgomphodon Sesamodon* [*Sesamodontoides*] *Watsoniella* L. Trias. SAf. ?**Rubidginidae,** *Mygalesaurus Mygalesuchus Rubidgina* U. Perm. SAf.
BAURIAMORPHA INCERTAE SEDIS: *Baurocynodon Herpetochirus Ictidochampsa.*

INFRAORDER ICTIDOSAURIA. **Diarthrognathidae,** *Diarthrognathus* U. Trias. SAf. ?**Haramiyidae (Microcleptidae),** *Haramiya* [*?Hypsiprymnopsis Microcleptes*] *Thomasia* [*Microlestes Plieningeria*] U. Trial.-L. Jur. Eu.

SUBORDER ANOMODONTIA

INFRAORDER DINOCEPHALIA

SUPERFAMILY TITANOSUCHOIDEA. **Brithopodidae (Titanophoneidae)** *Admetophoneus Archaeosyodon Brithopus* [*Dinosaurus Eurosaurus Orthopus*] *Chthomaloporus Doliosauriscus* [*Doliosaurus*] M. Perm. EEu. *Eosyodon* M. Perm. NA. *Syodon* [*Cliorhizodon*] *Titanophoneus* M. Perm. EEu. **Estemmenosuchidae,** *Estemmenosuchus Molybdopugus* M. Perm. EEu. **Anteosauridae,** *Anteosaurus* [*Broomosuchus Dinosuchus Titanognathus*] *Micranteosaurus Paranteosaurus Pseudanteosaurus* M. Perm. SAf. **Titanosuchidae (Jonkeriidae),** *Dinartamus Jonkeria* [*?Dinophoneus Dinopolus Dinosphageus ?Glaridodon*] *Titanosuchus* [*? Archaeosuchus Dinocynodon Enobius ?Lamiasaurus ?Parascapanodon Scapanodon Scullya*] M. Perm. SAf.

SUPERFAMILY TAPINOCEPHALOIDEA. **Deuterosauridae,** *Deuterosaurus* [*Mnemeiosaurus Uraniscosaurus*] M. Perm. Eu. ?**Driveriidae,** *Driveria* M. Perm. NA. ?**Mastersoniidae,** *Mastersonia* M. Perm. NA. ?**Tappenosauridae,** *Tappenosaurus* M. Perm. NA. **Tapinocephalidae,** *Agnosaurus Avenantia Criocephalus Delphinognathus Eccasaurus Keratocephalus Mormosaurus Moschognathus Moschoides Moschops Moschosaurus Pelosuchus Phocosaurus Pnigalion Riebeeckosaurus Struthiocephalellus Struthiocephaloides Struthiocephalus Struthionops Tapinocephalus Taurocephalus Taurops* M. Perm. SAf. *Ulemosaurus* M. Perm. EEu.

INFRAORDER VENYUKOVIAMORPHA. **Venyukoviidae,** *?Dimacrodon* M. Perm. NA. *Otsheria* M. Perm. EEu. *?Rhopalodon* M. Perm. Eu. *Venyukovia* [*Myctosuchus Venjukovia*] M. Perm. EEu.

INFRAORDER DROMASAURIA. **Galeopsidae,** *Galechirus* U. Perm. SAf. *Galeops Galepus ?Simorhinella* M. Perm. SAf.

INFRAORDER DICYNODONTIA. **Endothiodontidae,** *Brachyuraniscus* [*Brachyprosopus*] *Broilius* M. Perm. SAf. *Cerataelurus Chelyposaurus Compsodon* U. Perm. SAf. *Cryptocynodon* U. Perm. SAf. EAf. *Cteniosaurus Diictodontoides Emydops* [*Emydopsis Emydopsoides Emydorhynchus Emyduranus* U. Perm. SAf. *Endothiodon* [*Emydochampsa Endogomphodon Esoterodon*] U. Perm. SAf. EAf. *Eumantellia* [*Eumantella*] *Eurychororhinus Hueneus* U. Perm. SAf. *Koupia* M. Perm. SAf. *Myosauroides* U. Perm. SAf. *Myosaurus* L. Trias. SAf. *Newtonella* U. Perm. SAf. *Pachytegos* U. Perm. EAf. *Palemydops Parringtoniella* U. Perm. SAf. *Pristerodon* [*Diaelurodon Opisthoctenodon*] ?M., U. Perm. SAf. *Prodicynodon* U. Perm. SAf. *Robertia* M. Perm. SAf. *Storthyggognathus Synostocephalus Taognathus* U. Perm. SAf. *Tropidostoma* ?M., U. Perm. SAf. **Dicynodontidae,** *Aulacephalodon* [*Aulacocephalodon ? Aulacocephalus Bainia*] U. Perm. SAf. *Chelyrhynchus* U. Perm. SAf. *Dicynodon* [*Baiopsis Cirognathus Daptocephalus Dicranozygoma Eurycarpus Keirognathus Mastocephalus Orophicephalus Pylaecephalus Rhachiocephalodon Sintocephalus ?Theromus*] M.-U. Perm. SAf., U. Perm. Eu. SAs. *Dicynodontoides Digalodon Diictodon Dinanomodon Emydorhinus Eosimops* U. Perm. SAf. *Geikia Gordonia* U. Perm. Eu. *Haughtoniana* U. Perm. Eu. EAf. *Jimusaria* L. Trias. EAf. *Kingoria Kistecephalus* [*Cistecephalus*] U. Perm. SAf. EAf. *Oudenodon* [*Udenodon*] U. Perm. SAf., ?EEu. *Neomegacyclops* [*Kichingia Megacyclops*] *Pelanomodon* U. Perm. SAf. EAf. *Proaulacocephalodon Propelanomodon* U. Perm. SAf. *Rhachiocephalus* [*Eocyclops Pelorocyclops Platycyclops ? Platypodosaurus*] U. Perm. SAf. EAf. *Rhadiodromus* L. Trias. EEu. **Lystrosauridae,** *Lystrosaurus* [*Mochlorhinus Prolystrosaurus Ptychognathus Ptychosiagum Rhabdotocephalus*] L. Trias. SAf. EAf. SAs. EAs. **Kannemeyeriidae,** *Barysoma Ischigualastia* M. Trias. SA. *Kannemeyeria* [*Ptychocynodon Sagecephalus Uronautes*] L. Trias. SAf., L.-M. Trias. EAf., ?L. Trias. EEu., L. Trias. SA. *Parakannemeyeria* L. Trias. EAs. *Placerias* [*Brachybrachium Eubrachiosaurus*] U. Trias. NA. *Sinokannemeyeria* L. Trias. EAs. **Stahleckeriidae,** *Dinodontosaurus Stahleckeria* M. Trias. SA. **Shansiodontidae,** *Shansiodon* L. Trias. EAs. *Tetragonias* M. Trias. EAf.

CLASS AVES

SUBCLASS ARCHAEORNITHES (SAURIURAE)

ORDER ARCHAEOPTERYGIFORMES. **Archaeopterygidae,** *Archaeopteryx* [*Archaeornis Griphosaurus Gryphornis*] U. Jur. Eu.

SUBCLASS NEORNITHES

SUPERORDER ODONTOGNATHAE (ODONTOHOLCAE)

ORDER HESPERORNITHIFORMES. **Hesperornithidae,** *Coniornis Hesperornis* [*Hargeria Lestornis*] U. Cret. NA.

SUPERORDER PALAEOGNATHAE

ORDER TINAMIFORMES. **Tinamidae** R. SA., *Cayetanornis* Plioc. SA. *Crypturellus Nothoprocta Nothura* Pleist.-R. SA. *Querandiornis* U. Plioc. SA. *Rhynchotus Taoniscus* Pleist.-R. SA. *Tinamisornis* [*Roveretornis*] Plioc. SA. *Tinamus* Pleist.-R. SA.

ORDER STRUTHIONIFORMES. **Eleutherornithidae,** *Eleutherornis* Eoc. Eu. **Struthionidae** R. Af. WAs., *Struthio* [*Megaloscelornis Pachystruthio Palaeostruthio Struthiolithus*] Plioc.-Pleist. Eu., Plioc.-R. As., Pleist.-R. Af.

ORDER RHEIFORMES. **Opisthodactylidae,** *Opisthodactylus* Mioc. SA. **Rheidae** R. SA., *Heterorhea* Plioc. SA. *Pterocnemia Rhea* Pleist.-R. SA.

ORDER CASUARIFORMES. **Casuariidae** R. Aus., *Casuarius* Pleist.-R. Aus. **Dromaiidae** R. Aus., *Dromaius* [*Dromiceius*] Plioc.-R. Aus. **Dromornithidae,** *Dromornis* Pleist. Aus. *Genyornis* ?Mioc. NAf., ?Plioc., Pleist. Aus.

ORDER AEPYORNITHIFORMES. **Aepyornithidae,** *Aepyornis* [*Epyornis*] Pleist. Mad. *Eremopezus* [*?Psammornis*] Eoc. NAf. SWAs. *Mullerornis* [*Flacourtia*] Pleist. Mad. *Stromeria* Olig. NAf.

ORDER DINORNITHIFORMES. **Emeidae,** *Anomalopteryx* Mioc. or Plioc.-SubR. NZ. *Emeus* [*Meionornis Mesopteryx Syornis*] *Euryapteryx* [*Cela Celeus*] *Megalapteryx* [*Palaeocasuarius*] *Pachyornis Zelornis* Pleist.-SubR. NZ. **Dinornithidae,** *Dinornis* [*Megalornis Moa Mowia Owenia Palapteryx Tylopteryx*] Pleist.-SubR. NZ.

ORDER APTERYGIFORMES. **Apterygidae,** *Apteryx* Pleist.-R. NZ. *Pseudapteryx* Pleist. NZ.

SUPERORDER NEOGNATHAE

ORDER GAVIIFORMES. **Enaliornithidae,** *Enaliornis* [*Palaeocolymbus Pelagornis*] L. Cret. Eu. **Lonchodytidae,** *Lonchodytes* U. Cret. NA. **Gaviidae (Colymbidae)** R. Eu. As. NA., *Colymboides* [*Dyspetornis Hydrornis*] Eoc.-Mioc. Eu. *Eupterornis* Paleoc. Eu. *Gavia* [*Colymbus*] Plioc.-R. Eu. NA., R. As. *Gaviella* Olig. NA.

ORDER PODICIPEDIFORMES (COLYMBIFORMES). **Baptornithidae,** *Baptornis* U. Cret. NA. *Neogaeornis* U. Cret. SA. **Podicipedidae** R. Cos., *Aechmorphus* Pleist.-R. NA., R. SA. *Pliodytes* Plioc. NA. *Podiceps* [*Colymbus*] Mioc.-R. NA., Plioc.-R. Eu., Pleist.-R. As. WInd. SA. NZ. *Podilymbus* Pleist.-R. NA. WInd. SA.

ORDER PROCELLARIIFORMES (TUBINARES). **Diomedeidae** R. Oc., *Diomedea* Mioc.-Pleist. NA., Plioc.-Pleist. Eu., Pleist. EAs. NZ., R. Oc. *Gigantornis* Eoc. WAf. *?Manu* Olig. NZ. **Procellariidae** R. Oc., *Argyrodyptes* Mioc. SA. *Fulmarus* Mioc., Pleist. NA. Eu. EAs., R. Oc. *Macronectes* Pleist. NZ., R. Oc. *Plotornis* Mioc. Eu. *Pterodroma* Pleist. Bermuda WInd., R. Oc. *Puffinus* Olig.-Pleist. Eu., Mioc.-Pleist. NA., Pleist. EAs. WInd. Bermuda, R. Oc. **Oceanitidae** R. Oc., *Oceanodroma* Mioc. NA., Pleist. SA. **Pelecanoididae** R. Oc., *Pelecanoides* Pleist. SA., R. Oc.

ORDER SPHENISCIFORMES R. SOc. **Spheniscidae** R. SOc., *Anthropodyptes* Mioc. Aus. *Anthropornis* Mioc. Seymour Is. *Archaeospheniscus* Olig. NZ. *Arthrodytes* Mioc. SA. *Delphinornis* Mioc. Seymour Is. *Duntroonornis* Olig. NZ. *Eospheniscus* Mioc. Seymour Is. *Eudyptes Eudyptula* Pleist. NZ., R. Ant. *Ichtyopteryx* Mioc. Seymour Is. *Isotremornis* Mioc. SA. *Korora* Olig. NZ. *Megadyptes* Pleist.-R. NZ., R. Ant. Oc. *Neculus* Mioc. SA. *Notodyptes Orthopteryx* Mioc. Seymour Is. *Pachydyptes* Eoc. NZ. *Palaeeudyptes* Eoc.-Olig. NZ. *Palaeospheniscus* Mioc. SA., Plioc. NZ. *Paraptenodytes Paraspheniscus Perispheniscus* Mioc. SA. *Platydyptes* Olig. NZ. **Pseudospheniscus** Mioc. SA.

ORDER PELECANIFORMES (STEGANOPODES)

SUPERFAMILY SULOIDEA. **Elopterygidae,** *Argillornis* [*Megalornis*] Eoc. Eu. *Elopteryx* U. Cret. Eu. *Eostega* Eoc. Eu. **Phalacrocoracidae** R. Oc., *Actiornis* Eoc. Eu. *Graculavus* [*Limnosavis*] Paleoc. NA. *Phalacrocorax* [*Australocorax Miocorax Oligocorax Paracorax*] Olig.-R. NA., Mioc.-R. Eu., Pleist.-R. SA. Aus. EAs., R. Af. *Pliocarbo* Plioc. Eu. **Anhingidae** R. As. Af. Aus. NA. SA., *Anhinga* Plioc. Eu., Pleist. Maurit., Pleist.-R. Aus. Mad. NA., R. Af. As. SA. *Protoplotus* Eoc. EInd. **Sulidae** R. Oc., *Microsula* Mioc. Eu. NA. *Miosula* Mioc.-Plioc. NA. *Morus* Mioc.-R. NA., Pleist.-R. Eu., R. Oc. *Palaeosula* Mioc. NA. *Sula* Olig.-Mioc. Eu., Mioc.-Plioc. NA., Pleist. WInd., R. Oc.

SUBORDER PITAETHONTES. **Phaethontidae** R. Oc., *Phaethon* SubR. Rodriguez Bermuda *Prophaeton* Eoc. Eu.

SUBORDER ODONTOPTERYGIA. **Odontopterygidae,** *Odontopteryx* [*Odontornis*] Eoc. Eu. **Pseudodontornithidae,** *Osteodontornis* Mioc. NA. *Pelagornis* Mioc. Eu. *Pseudodontornis* Mioc. NA., ?SA. **Cyphornithidae,** *Cyphornis Palaeochenoides Tympanonesiotes* Mioc. NA.

SUBORDER PELECANI

SUPERFAMILY PELECANOIDEA. **Pelecanidae** R. Cos., *Liptornis* Mioc. SA. *Pelecanus* Mioc.-R. Eu., Plioc.-R. As., Pleist.-R. NA. Aus. WInd., R. SA.

SUBORDER FREGATAE. **Fregatidae** R. Oc., *Fregata* SubR.-R. WInd., R. Oc.

?SUBORDER CLADORNITHES. **Cladornithidae,** *Cladornis* Olig. NA.

ORDER CICONIIFORMES (ARDEIFORMES GRESSORES)

SUBORDER PHOENICOPTERI. **Torotigidae,** *?Parascaniornis* U. Cret. Eu. *Torotix* U. Cret. NA. **Scaniornithidae,** *?Gallornis* L. Cret. Eu. *Scaniornis* Paleoc. Eu. **Telmabatidae,** *Telmabates* Eoc. SA. **Agnopteridae,** *Agnopterus* [*Ptenornis*] Eoc.-Olig. Eu., Olig. WAs. **Phoenicopteridae** R. Eu. As. Af. NA. SA., *Elornis* [*Helornis*] Eoc.-Olig. Eu. *Pheniconaias* Pleist. Aus. *Phoenicopterus* Mioc.-R. Eu., ?Mioc. Aus., Plioc.-R. NA., Pleist.-R. WInd. SA., R. As. Af. **Palaelodidae,** *Megapalaelodus* Mioc.-Plioc. NA. *Palaelodus* Mioc. Eu.

SUBORDER PLATALEAE. **Plegadornithidae,** *Plegadornis* U. Cret. NA. **Plataleidae (Threskiornithidae)** R. Cos., *Ajaia* Pleist.-R. NA., R. SA. *Carphibis* Pleist.-R. Aus. *Eudocimus* [*Guara*] Mioc. Eu., Pleist.-R. NA. WInd. SA.

Ibidopodia Mioc. Eu. *Ibidopsis* Eoc. Eu. *Nipponia* Pleist.-R. EAs. *Platalea* Pleist.-R. Aus. Mad., R. Eu. As. Af. *Plegadis* Pleist.-R. NA. WInd., R. Cos. *Protibis* Mioc. SA. *Theristicus* Pleist.-R. SA.

SUBORDER ARDEAE. **Ardeidae** R. Cos., *Ardea* Mioc.-R. Eu., Plioc.-R. NA., Pleist.-R. WInd. SA., R. As. Af. Aus. *Ardeacites* Mioc. Eu. *Ardeola* Pleist.-R. Eu., R. As. Af. EInd. *Botaurites* Mioc. Eu. *Botauroides* Eoc. NA. *Botaurus* Pleist.-R. Eu. NA., R. Cos. *Butorides* Pleist.-R. NA. Maurit., R. As. Af. Aus. SA. *Casmerodius* Pleist.-R. NA. WInd. SA., R. Cos. *Egretta* Pleist.-R. Eu., R. As. Af. Aus. *Eoceornis* Eoc. NA. *Florida* Pleist.-R. NA., R. SA. *Goliathia* Eoc. or Olig. NAf. *Hydranassa* Pleist.-R. NA., R. SA. *Ixobrychus* Pleist.-R. Eu. NA. SA., R. Cos. *Mesophoyx* Pleist. Mad., R. As. Af. Aus. *Nyctanassa* Pleist.-R. NA. WInd., R. SA. *Nycticorax* Plioc.-R. NA., Pleist. Rodriguez, R. Eu. As. Af. EInd. *Palaeophyx* Pleist. NA. *Proardea* Eoc. or Olig. Eu. *Proherodius* Eoc. Eu. **Cochlearidae** R. CA. SA. **Balaenicipitidae** R. Af.

SUBORDER CICONIAE. **Scopidae** R. Af. **Ciconiidae** R. Cos., *Amphipelargus* Plioc. SWAs. *Ciconia* Plioc.-R. Eu., Pleist. NA. WInd., R. As. Af. *Ciconiopsis* Olig. SA. *Euxenura* Pleist.-R. SA. *Ibis* Mioc., ?Pleist. Eu., R. As. Af. EInd. *Leptoptilos* Plioc.-Pleist. Eu., Plioc.-R. As. EInd., R. Af. *Mycteria* Pleist.-R. NA. SA. *Palaeoephippiorhynchus* Olig. NAf. *Palaeopelargus* Pleist. Aus. *Pelargopappus* [*Pelargocrex Pelargoides Pelargopsis*] Eoc. or Olig., Mioc. Eu. *Pelargosteon* Pleist. Eu. *Prociconia* Pleist. SA. *Propelargus* Eoc. or Olig., Mioc. Eu., Mioc. NA. *?Xenorhynchopsis* Pleist. Aus. *Xenorhynchus* Pleist.-R. Aus., R. SAs. EInd.

ORDER ANSERIFORMES

SUBORDER ANSERES. **Parancyrocidae,** *Parancyroca* Mioc. NA. **Anatidae** R. Cos., *Aix* Pleist.-R. NA. *Alopochen* Pleist-R. Mad. Maurit., R. Af. *Amazonetta* Pleist.-R. SA. *Anabernicula* Plioc.-R. NA. *Anas* Olig.-R. Eu. As., Mioc.-R. Af., Pleist.-R. NZ. NA. WInd., R. Aus. *Anser* Mioc.-R. Eu., Pleist.-R. NA. As. *Aytha* Mioc.-R. Eu., ?Plioc., Pleist.-R. NA., Pleist.-R. As. Aus. NZ. *Biziura* Pleist.-R. Aus. NZ. *Branta* Mioc.-R. NA., Pleist.-R. Eu., ?Pleist. As. *Brantadorna* Pleist. NA. *Bucephala* Plioc.-R. NA., Pleist.-R. Eu., R. As. *Cairinia* Pleist.-R. SA., R. CA. *Centrornis* Pleist.-R. Mad. *Chendytes* Pleist. NA. *Chenonetta* Pleist.-R. NZ., R. Aus. *Chenornis* Mioc. Eu. *Clangula* Pleist.-R. Eu. NA., R. As. *Cnemiornis* Pleist. NZ. *Coscoroba* Pleist.-R. SA. *Cygnavus* Mioc. Eu. *Cygnanser* Plioc. Eu. *Cygnopsis* Pleist.-R. As. *Cygnopterus* Olig. Eu. *Cygnus* Mioc.-R. Eu., Pleist.-R. As. Aus. NZ. NA. *Dendrochen* Mioc. NA. *Dendrocygna* Pleist.-R. Aus. NA. SA. WInd. *Eoneornis* Mioc. SA. *Eonessa* Eoc. NA. *Eremochen* Plioc. NA. *Euryanas* Pleist. NZ. *Eutelornis* Mioc. SA. *Geochen* Pleist. Hawaii *Histrionicus* Pleist. Eu., Pleist.-R. NA., R. As. *Lophodytes* Pleist.-R. NA. *Loxornis* Olig. SA. *Malacorhynchus* Pleist.-R. NZ., R. Aus. *Melanitta* Pleist.-R. Eu. NA., R. As. *Mergellus* Pleist.-R. Eu., R. As. *Mergus* Pleist.-R. Eu. NZ. NA. SA., R. As. *Neochen* [*Chenalopex*] Pleist.-R. SA. WInd. *Nettapus* ?Plioc. Eu., Pleist. Aus., R. Af. *Nettion* Mioc.-R. Eu., Plioc.-R. Aus. NA., Pleist.-R. As. NZ. SA. *Nomonyx* Pleist.-R. SA., R. CA. WInd. *Ocyplonessa* Plioc. NA. *Olor* Pleist.-R. NA., R. Eu. As. *Oxyura* Pleist.-R. NA. SA., R. Cos. *Pachyanas* Pleist. Chatham Is. *Philacte* Pleist.-R. NNA., R. NAs. *Polysticta* Pleist.-R. NA., R. As. *Presbychen* Mioc. NA. *Querquedula* [*Archeoquerquedula*] Mioc. Eu. NA., Pleist. SA. *Romainvillia* Eoc. Eu. *Sarkidiornis* Pleist.-R. SA., R. As. Af. *Somateria* Pleist.-R. Eu., R. NAs. NNA. *Tadorna* Pleist.-R. Eu. As. Aus. NZ., R. Af. *Teleornis* Olig. SA.

SUBORDER ANHIMIA. **Anhimiidae** R. SA.

ORDER FALCONIFORMES (ACCIPITRIFORMES)

SUBORDER SARCORAMPHI (CATHARTIDES). **Vulturidae** R. NA. SA., *Breagyps* Pleist. NA. *Cathartes* Pleist.-R. NA. SA. *Cathartornis* Pleist. NA. *Coragyps* Pleist.-R. NA., R. SA. *Diatropornis* [*Tapinopus*] ?U. Eoc. or L. Olig. Eu. *Eocathartes* Eoc. Eu. *Gymnogyps* Plioc.-R. NA. *Lithornis* Paleoc. Eu. *Palaeogyps Phasmagyps* Olig. NA. *Plesiocathartes* ?U. Eoc. or L. Olig. Eu. *Pliogyps* Pleist. NA. *Sarcorhamphus* Mioc.-R. NA., Pleist.-R. SA. *Teracus* Olig. Eu. *Teratornis* [*Pleistogyps*] Pleist. NA. *Vultur* Plioc.-R. SA., R. NA. **Neocathartidae,** *Neocathartes* Eoc. NA.

SUBORDER ACCIPITRES (FALCONES). **Sagittariidae** R. Af., *Amphiserpentarius* U. Eoc. or L. Olig.-Mioc. Eu., R. Af. **Pandionidae** R. Cos., *Pandion* Pleist.-R. Eu. NA., R. Cos. **Accipitridae** R. Cos., *Accipiter* Pleist.-R. Eu. As. Maurit. NA., R. Cos. *Aegypius* Pleist.-R. Eu. As., R. NAf. *Aquila* Mioc.-R. Eu., Pleist.-R. As. Af., R. NA. *Aquilavus* U. Eoc. or L. Olig.-Mioc. Eu. *Aviceda* Pleist.-R. Aus., R. As. Af. *Buteo* Olig.-R. NA., Pleist.-R. Eu. As. SA. *Buteogallus* Pleist.-R. NA., R. CA. SA. WInd. *Calohierax* Pleist. Bahamas *Chondrohierax* Pleist.-R. SA., R. CA. WInd. *Circaetus* Pleist.-R. Eu., R. As. Af. *Circus* Pleist.-R. Eu. NZ. NA., R. Cos. *Climacarthrus Cruschedula* Olig. SA. *Elanus* Pleist.-R. NA., R. As. Af. Aus. SA. *Foetopterus* Pleist. SA. *Gypaetus* Pleist.-R. Eu. As., R. Af. *Gyps* Plioc.-R. Eu., Pleist.-R. As. *Haliaeetus* Mioc. Eu., Pleist.-R. Af. NA., R. As. *Harpagornis* Pleist. NZ. *Harpia* Pleist.-R. SA., R. CA. *Hieraaetus* Mioc.-R. Eu., R. As. Af. EInd. Aus. *Hypomorphnus* Mioc.-R. NA., R. SA. *Lagopterus* [*Asthenopterus*] Pleist. SA. *Milvus* Mioc.-R. Eu., R. As. Af. Aus. *Miohierax* Mioc. NA. *Morphnus* Pleist.-R. NA., R. SA. *Necraster* Pleist. Aus. *Neogyps* Pleist. NA. *Neophrontops* Mioc.-Pleist. NA. *Palaeastur* Mioc. NA. *Palaeoborus* Mioc.-Plioc. NA. *Palaeocircus Palaeohierax* Mioc. Eu. *Palaeoplancus* Olig. NA. *Parabuteo* Pleist.-R. NA., R. SA. *Pernis* Pleist.-R. Eu. As., R. Af. EInd. *Proictinia* Plioc. NA. *Promilio* Mioc. Eu. NA. *Spizaetus* Pleist.-R. NA., R. As. SA. *Thegornis* Mioc. SA. *Titanohierax* Pleist. Bahamas *Torgos* Pleist. Eu., R. Af. *Uroaetus* [*Taphaetus*] Pleist.-R. Aus. *Wetmoregyps* Pleist. NA. **Falconidae** R. Cos., *Badiostes* Mioc. SA. *Caracara* Pleist.-R. NA. WInd. SA. *Falco* Mioc.-R. NA. *Micrastur Milvago* Pleist.-R. SA., R. CA. *Plioaetus* Pleist. Aus. *Sushkinia* Plioc. As.

ORDER GALLIFORMES (GALLI). **Cracidae** R. NA. SA., *Anisolornis* [*Anissolornis*] Mioc. SA. *Boreortalis* Mioc.-Plioc. NA. *Crax* Pleist.-R. SA., R. CA. *Filholornis* U. Eoc. or L. Olig. Eu. *Gallinuloides* [*Palaeobonasa*] Eoc. NA. *Lucliortyx* Eoc., ?L. Olig. Eu. *Ortalis* Pleist.-R. SA., R. NA. CA. *Palaeonossa* Olig. NA. *Palaeophasianus Palaeortyx* Eoc. Eu. *Paracrax* Olig. NA. *Paraortyx* U. Eoc. or L. Olig. Eu. *Penelope* Pleist.-R. SA., R. CA. *Prirortyx* U. Eoc. or L. Olig. Eu. *Procrax* Olig. NA. *Taoperdix* Olig.-Mioc. Eu. **Opisthocomidae** R. CA. SA., *Hoazinoides* Mioc. NA. SA. **Megapodidae** R. EInd. Aus., *Alectura* Pleist.-R. Aus. *Chosornis* Pleist. Aus. **Numididae** R. Af., *Numida* Pleist. Eu., R. Af. **Phasi-**

anidae R. Cos., *Agriocharis* Pleist. NA., R. CA. *Alectornis* Plioc.-R. Eu., R. As. NAf. *Ammoperdix* Plioc. Eu., Pleist.-R. As., R. NAf. *Archaeophasianus* Mioc. NA. *Bonasa Callipepla Canachites Centrocercus* Pleist.-R. NA. *Chrysolophus* Pleist.-R. As. *Colinus* Plioc.-R. NA., R. CA. WInd. *Coturnix* Pleist.-R. Eu. As. NZ., R. Aus. *Crossoptilon* Pleist.-R. As. *Cyrtonyx* Mioc.-R. NA. *Dendragapus* Pleist.-R. NA. *Francolinus* Pleist.-R. Eu. As. Af. *Gallus* Plioc.-Pleist. Eu., R. As. EInd. ?Af. NZ. ?NA. ?WInd. *Lagopus* Pleist.-R. Eu., R. As. NA. *Lophortyx* Plioc.-R. NA. *Meleagris* Pleist.-R. NA. *Miogallus Miophasianus* Mioc. Eu. *Miortyx* Mioc. NA. *Nanortyx* Olig. NA. *Neortyx* Pleist. NA. *Odontophorus* Pleist.-R. SA., R. CA. *Oreortyx* Pleist.-R. NA. *Palaealectornis* Mioc. NA. *Palaeocryptonyx* Mioc.-Plioc. Eu. *Palaeoperdix* Mioc. Eu. *Palaeotetrix Parapavo* Pleist. NA. *Pedioecetes* Pleist.-R. NA. *Perdix* Pleist.-R. Eu. As. *Phasianus* Plioc.-R. Eu., Pleist.-R. As. *Pliogallus* Pleist. Eu. *Plioperdix* Mioc.-Plioc. Eu. *Proalector* Mioc. Eu. *Pucrasia* Pleist.-R. As. *Schaubortyx* Olig. Eu. *Syrmaticus* Pleist.-R. As. *Tetrao Tetrastes* Pleist.-R. Eu., R. As. *Tragopan* Pleist.-R. As. *Tympanuchus* Mioc.-R. NA.

ORDER RALLIFORMES (GRUIFORMES)

SUBORDER RALLI. **Rallidae** R. Cos., *Aphanapteryx* Pleist.-R. Mascarenes *Aphanocrex* Pleist. St. Helena *Aptornis* Pleist.-R. NZ. *Aramides* Pleist.-R. SA., R. NA. *Capellirallus* Pleist. NZ. *Coturnicops* Pleist.-R. NA., R. As. Af. SA. *Creccoides* L. Pleist. NA. *Crex* Pleist.-R. Eu. *Diaphorapteryx* Pleist. Chatham Is. *Eocrex* L. Eoc. NA. *Epirallus* Pleist. NA. *Euryonotus* Pleist. SA. *Fulica* L. Plioc.-R. NA., R. Cos. *Fulicaletornis* M. Eoc. NA. *Gallinula* Pleist.-R. Eu. Aus. NA., R. Cos. *Gallirallus* Pleist.-R. NZ. *Gypsornis* U. Eoc. Eu. *Hovacrex* Pleist. Mad. *Laterallus* Pleist.-R. NA., R. SA. *Miofulica* M. Mioc. Eu. *Mioralllus* M., ?U. Mioc. Eu. *Nesophalaris* Pleist. Chatham Is. NZ. *Nesotrochis* Pleist. WInd. *Notornis* Pleist.-R. NZ. *Ortygonax* Pleist.-R. SA. *Palaeoaramides* L. Mioc. Eu. *Palaeocrex* L. Olig. NA. *Palaeolimnas* Pleist. Maurit. *Palaeorallus* L. Eoc. NA. *Paraortygometra* L. Mioc. Eu. *Pararallus* U. Mioc. Eu. *Porphyrio* Pleist.-R. Aus., R. Eu. Af. As. *Porphyrula* Pleist.-R. NA. SA., R. Af. *Porzana* Pleist.-R. NA., R. Cos. *Pyramida* Pleist. NZ. *Quercyrallus* U. Eoc. Eu. *Rallicrex* U. Olig. Eu. *Rallus* L. Plioc.-R. Eu. *Telecrex* U. Eoc. EAs. *Telmatornis* U. Paleoc. NA. *Thiornis* L. Plioc. Eu. *Tribonyx* Pleist.-R. Aus. **Idiornithidae,** *Elaphrocnemus* ?U. Eoc. Eu., ?NA. *Idiornis* [*Orthocnemus*] U. Eoc. or L. Olig. Eu. **Heliornithidae** R. SAs. Af. SA.

SUBORDER MESITORNITHIDES. **Mesitornithidae** R. Mad.

SUBORDER TURNICES (HEMIPODES). **Turnicidae** R. Af. As. Aus., *Turnix* Pleist.-R. As., R. Af. Aus. **Pedionomidae** R. Aus.

SUBORDER GRUES. **Gruidae** R. Eu. As. Af. Aus. NA., *Aletornis* [*Protogrus*] M. Eoc. NA. *Anthropoides* Pleist. Eu., R. Af. *?Baeopteryx* Pleist. Bermuda *Bugeranus* Pleist.-R. Af. *Eobalearica* Eoc. WAs. *Eogrus* U. Eoc.-U. Mioc. EAs. *Ergilornis* Olig. EAs. *Geranoides* L. Eoc. NA. *Geranopsis* U. Eoc., ?L. Olig. Eu. *Grus* L. Plioc.-R. NA. R. Eu. As. Af. *Ornithocnemus* [*Palaeogrus*] M. Eoc.-L. Mioc. Eu. *Pliogrus* L. Plioc. Eu. *Probalearica* L. Mioc. Eu. NA. *Proergilornis* Olig. EAs. *Urmiornis* L. Plioc. Eu. WAs. **Aramidae** R. NA. SA., *Aminornis* L. Olig. SA. *Aramornis* M. Mioc. NA. *Aramus* Pleist.-R. NA. SA. *Badistornis* M. Olig. NA. *Gnotornis* U. Olig. NA. *Loncornis* L. Olig. SA. **Psophidae** R. SA.

SUBORDER RHYNOCHETI. **Rhynochetidae** R. Oc.

SUBORDER EURYPYGAE. **Eurypygidae** R. CA. SA.

SUBORDER CARIAMAE. **Phororhachidae,** *Andalgalornis* L.-M. Plioc. SA. *Andrewsornis* L. Olig. SA. *Devincenzia* ?U. Mioc. SA. *Onactornis* L., ?M. Plioc. SA. *Phororhacos* L.-M. Mioc. SA. *Titanornis* Pleist. NA. *Tomodus* [*? Palaeocicomia*] M. Mioc. SA. **Brontornithidae,** *Aucornis* L. Olig. SA. *Brontornis* L.-M. Mioc. SA. *Lophiornis* M. Mioc. SA. *Physornis* L. Olig. SA. *Pseudolarus* L. Olig.-M. Mioc. SA. *Rostrornis* L.-M. Mioc. SA. *Staphylornis* M. Mioc. SA. *Stephanornis* Mioc. SA. **Dryornithidae,** *Dryornis* U. Plioc. SA. *Hermosiornis* L. Plioc.-Pleist. SA. *Procariama* M. Plioc. SA. *Psilopterus* L.-M. Mioc. SA. *Smiliornis* L. Olig. SA. **Cunampaiidae,** *Cunampaia* U. Eoc. SA. **Bathornithidae,** *Bathornis* L.-U. Olig. NA. **Cariamidae** R. SA., *Cariama* Pleist.-R. SA. *Riacama* L. Olig. SA.

SUBORDER OTIDES. **Otididae** R. EAs. Af., *Chlamydotis* L. Mioc.-L. Plioc. Eu., R. As. Af. *Otis* L. Pleist.-R. Eu. *Palaeotis* M. Eoc. Eu. *Tetrax* Pleist.-R. Eu., R. As. NAf.

ORDER DIATRYMIFORMES. **Diatrymatidae,** *Diatryma* U. Paleoc.-M. Eoc. NA. Eu. *Omorhamphus* U. Paleoc. Eu. **Gastornithidae** *Dasornis Gastornis Remiornis* U. Paleoc. Eu.

ORDER ICHTHYORNITHIFORMES. **Ichthyornithidae,** *Ichthyornis* U. Cret. NA. **Apatornithidae,** *Apatornis* U. Cret. NA.

ORDER CHARADRIIFORMES

SUBORDER CHARADRII. **Cimolopterygidae,** *Cimolopteryx Ceramornis* U. Cret. NA. **Scolopacidae** R. Cos., *Actitis* Pleist.-R. Eu. NA., R. Cos. *Arenaria* Pleist. Eu. *Bartramia* Plioc.-R. SA. *Belonopterus* Pleist.-R. SA. *Calidris* Plioc. NA., Pleist. Eu., R. Cos. *Capella* Pleist.-R. Eu. NA. WInd. SA., R. Cos. *Catoptrophorus* Pleist.-R. NA., R. WInd. SA. *Charadrius* Olig.-R. NA., Pleist.-R. Eu., R. Cos. *Coenocorypha* Pleist.-R. NZ., Pleist. Chatham Is. *Crocethia* Pleist.-R. NA., R. Cos. *Dolichopterus* Olig. Eu. *Dorypaltus* Pleist. NA. *Elorius* Mioc. Eu. *Erolia* Mioc.-R. Eu., Plioc.-R. NA., Pleist.-R. SA., R. As. Af. Aus. *Ereunetes* Plioc.-R. NA., Pleist.-R. SA. *Eudromias* Pleist.-R. Eu., R. As. NAf. *Eupoda* Pleist.-R. NA., R. As. Af. Aus. *Haematopus* Pleist.-R. Eu., Pleist. Chatham Is., R. Cos. *Limicolavis* Mioc.-R. NA., R. Cos. *Limnodromus* Pleist.-R. NA., R. EAs. *Limosa* ?Eoc., Pleist.-R. Eu., Mioc.-R. NA., R. Cos. *Lobibyx* ?Pleist., R. Aus., R. SWPac. *Lobipes* Pleist.-R. NA., R. EAs. NPac. *Lymnocryptes* Pleist.-R. Eu., R. As. *Micropalama* Plioc.-R. NA., R. SA. *Numenius* Mioc.-R. Eu., Pleist.-R. NA., R. Cos. *Palaeotringa* U. Paleoc., ?Eoc. NA. *Palnumenius* Pleist. NA. *Palnumenius* Pleist. NA. *Palostralegus* Plioc. NA. *Paractictis* Olig. NA. *Paractiornis* Mioc. NA. *Phalaropus* Pleist.-R. NA., R. Oc. *Philohela* Pleist.-R. NA. *Philomachus* Pleist.-R. Eu. As., R. Af. *Pluvialis* Pleist. NA., R. Eu. As. Aus. *Rhynchaeites* M. Eoc. Eu. *Scolopax* Pleist.-R. Eu., R. As. EInd. *Squatarola* Pleist.-R. Eu. NA., R. Cos. *Totanus* Eoc. or Olig.-R. Eu., Pleist.-R. NA., R. Cos. *Tringa* Pleist.-R. Eu. NA. SA., R. Cos. *Vanellus* Olig.-R. Eu., Pleist.-R. As. **Recurvirostridae** R. Cos., *Coltonia* Eoc. NA. *Himantopus* Pleist.-R. Eu. NA. Aus., R. Cos. *Presbyornis* Eoc. NA. *Recurvirostra* Mioc.-R. NA., R. Cos. **Jacanidae** R. Af. SAs. Aus. CA. SA., *Jacana* Pleist.-R. SA. *Rhegminornis* Mioc. NA. **Burhinidae** R. Eu. As.

Af. Aus. CA. SA. WInd., *Burhinus* Pleist.-R. Eu. WInd., R. As. Af. Aus. CA. SA. *Milnea* Mioc. Eu. **Dromadidae** R. Ind. Oc. **Glareolidae** R. Eu. As. Af. Aus. **Thinocoridae** R. SA. **Chionididae** R. Ant.

SUBORDER LARI. **Laridae** R. Oc., *Anous* SubR. WInd., R. Cos. *Gaviota* Mioc. NA. *Halcyornis* Paleoc. Eu. *Hydroprogne* SubR.-R. NA., R. Cos. *Larus* Mioc.-R. Eu. NA., Pleist.-R. SA., R. Cos. *Ocyplanus* Pleist. Aus. *Pseudosterna* Pleist. SA. *Rissa* Pleist.-R. Eu., ?Pleist., R. NA., R. Cos. *Rupelornis* Olig. Eu. *Sterna* Mioc.-R. As., Pleist.-R. Eu. NA. WInd., R. Af. Aus. SA. *Thalasseus* Pleist.-R. Eu. WInd., R. Cos. *Xema* Pleist.-R. NA., R. Cos. **Rynchopidae** R. NA. SA. Af. SAs. **Stercorariidae** R. Oc., *Stercorarius* Pleist. NA., R. Oc.

SUBORDER ALCAE. **Alcidae** R. Eu. As. NA. NAtl. NPac. Arc., *Aethia* Pleist.-R. NA., R. NEAs. *Alca* Pleist.-R. Eu. NA., R. NAtl. *Australca* Plioc. NA. *Brachyramphus* Plioc.-R. NA., R. As. *Cepphus* Pleist.-R. Eu. NA., R. As. *Cerorhinca* Mioc.-R. NA., R. NPac. *Cyclorrhynchus* Pleist. NA., R. NPac. *Fratercula* Pleist.-R. Eu. NA., R. NAtl. NPac. *Hydrotherikornis* Eoc. NA. *Lunda* Pleist.-R. NA., R. EAs. *Mancalla* Plioc. NA. *Miocepphus* Mioc. NA. *Nautilornis* Eoc. NA. *Pinguinus* Pleist. Eu. NA. *Plautus* Pleist. Eu. *Ptychoramphus* Plioc.-R. NA. *Synthliboramphus* Pleist. NA., R. NAs. NPac. *Uria* Mioc.-R. NA., Plioc.-R. Eu., R. NAtl. NPac.

ORDER COLUMBIFORMES

SUBORDER PTEROCLETES. **Pteroclidae** R. Eu. As. Af., *Pterocles* U. Eoc. or L. Olig.-R. Eu., R. As. Af. *Syrrhaptes* Pleist. Eu., Pleist.-R. As.

SUBORDER COLUMBAE. **Columbidae** R. Cos., *Claravis* Pleist.-R. SA., R. CA. *Columba* Pleist.-R. Eu. As. NA. WInd. SA., R. Af. *Columbigallina* Pleist.-R. WInd. SA., R. NA. *Ectopistes* Pleist. NA. *Geotrygon* Pleist.-R. WInd. SA., R. CA. *Gerandia* Mioc. Eu. *Hemiphaga* Pleist.-R. NZ. *Leptoptila* Pleist.-R. SA., R. CA. *Leucosarcia* Pleist.-R. Aus. *Lithophaps Progoura* Pleist. Aus. *Nesoenas* Pleist.-R. Maurit. *Scardafella* Pleist.-R. SA., R. SNA. *Streptopelia* Pleist. Eu. WAs. Rodriguez *Uropelia* Pleist.-R. SA. *Zenaida* Pleist.-R. WInd., R. SNA. SA. *Zenaidura* Pleist.-R. NA. SA. **Raphidae,** *Pezophaps* Pleist. Rodriguez *Raphus* [*Didus*] R. Maurit.

ORDER PSITTACIFORMES. **Psittacidae** R. SAs. Af. Aus. CA. SA., *Amazona* Pleist.-R. WInd. SA. *Ara* Pleist.-R. WInd. SA., R. CA. *Aratinga* [*Protoconurus*] Pleist.-R. SA., R. CA. WInd. *Archaeopsittacus* Mioc. Eu. *Conuropsis* Mioc.-Pleist. NA. *Coracopsis* Pleist.-R. Mad. *Cyanoramphus* Pleist.-R. NZ. *Forpus* Pleist.-R. SA., R. CA. *Lophopsittacus* SubR. Maurit. *Necropsittacus* SubR. Rodriguez *Nestor* Pleist.-R. NZ., Pleist. Chatham Is. *Rhynchopsitta* Pleist.-R. CA. *Strigops* Pleist.-R. NZ., Pleist. Chatham Is.

ORDER CUCULIFORMES. **Musophagidae** R. Af. **Cuculidae** R. Cos., *Coua* Pleist.-R. Mad. *Crotophaga* Pleist.-R. WInd. SA., R. NA. *Cuculus* Pleist.-R. Eu., R. As. Af. Aus. *Dynamopterus* Eoc. or Olig. Eu. *Geococcyx* Pleist.-R. NA. *Necrornis* Mioc. Eu. *Neococcyx* Olig. NA. *Neomorphus* ?Pleist. NA., R. CA. SA. *Piaya* Pleist.-R. SA. *Saurothera* Pleist.-R. WInd. SA. *Tapera* Pleist.-R. SA., R. CA.

ORDER STRIGIFORMES. **Protostrigidae,** *Protostrix* Eoc. NA. **Strigidae** R. Cos., *Asio* U. Eoc. or L. Olig.-R. Eu., Pleist.-R. NA. *Athene* Pleist.-R. Eu. As. Rodriguez, R. Af. *Bubo* Eoc. or Olig.-R. Eu., Pleist.-R. As. NA. Rodriguez, R. Af. SA. *Ciccaba* Pleist.-R. CA., R. SA. Af. *Glaucidium* Pleist. WInd., Pleist.-R. Eu. NA. SA., R. As. Af. *Ketupa* Pleist.-R. As. *Necrobyas* U. Eoc. or L. Olig. Eu. *Ninox* Pleist.-R. NZ., R. As. Aus. Mad. *Nyctea* Pleist.-R. Eu., R. As. NA. *Ornimegalonyx* Pleist. WInd. *Otus* Pleist.-R. Eu. As. NA., Pleist. WInd., R. Af. SA. *Rhinoptynx* SubR.-R. SA., R. CA. *Sceloglaux* Pleist.-R. NZ. *Speotyto* Pleist.-R. NA. WInd. SA. *Strigogyps* U. Eoc. or L. Olig. Eu. *Strix* Mioc.-R. NA., Pleist.-R. Eu., R. As. SA. *Surnia* Pleist.-R. Eu., R. As. NA. **Tytonidae** R. Cos., *Tyto* Mioc.-R. Eu., Pleist.-R. NA. WInd. As. SA., R. Af.

ORDER CAPRIMULGIFORMES

SUBORDER STEATORNITHES. **Steatornithidae** R. SA.

SUBORDER CAPRIMULGI. **Podargidae Aegothelidae** R. Aus. SPac. **Nyctibiidae** R. CA. SA. WInd., *Nyctibius* Pleist.-R. SA., R. CA. WInd. **Caprimulgidae** R. Cos., *Caprimulgus* Pleist.-R. Eu. WInd., R. Cos. *Chordeiles* Pleist.-R. NA., R. WInd. SA. *Eleothreptus Hydropsalis* Pleist.-R. SA. *Nyctidronus* Pleist.-R. SA., R. NA. *Phalaenoptilus* Pleist.-R. NA., R. CA.

ORDER APODIFORMES

SUBORDER APODI. **Aegialornithidae,** *Aegialornis* Eoc. or Olig. Eu. **Apodidae** R. Cos., *Aeronautes* Pleist.-R. NA., R. SA. *Apus* Mioc.-R. Eu., Pleist.-R. As., R. Aus. SPac. Ind. Oc. *Collocalia* Mioc. Eu., R. As. Aus. SPac. Ind. Oc. *Clypselavus* Eoc. or Olig.-Mioc. Eu. *Streptoprocne* Pleist.-R. SA., R. CA. WInd. **Hemiprocnidae** R. As. SPac.

SUBORDER TROCHILI. **Trochilidae** R. NA. WInd. SA., *Clytolaema* Pleist.-R. SA.

ORDER COLIIFORMES. **Coliidae** R. Af.

ORDER TROGONIFORMES. **Trogonidae** R. As. Af. NA. WInd. SA., *Archaeotrogon* Eoc. or Olig. Eu. *Paratrogon* Mioc. Eu. *Trogon* Pleist.-R. SA., R. NA.

ORDER CORACIIFORMES

SUBORDER HALCYONES. **Momotidae** R. CA. SA., *Baryphthengus* Pleist.-R. SA., R. CA. *Eumomota* SubR.-R. CA. *Momotus* SubR.-R. CA., R. SA. *Uintornis* Eoc. NA. **Halcyonidae** R. Cos., *Alcedo* Pleist.-R. Eu., R. As. Af. Mad. SPac. *Cercyle* Pleist.-R. As. NA., R. Af. SA. *Chlorocercyle* Pleist.-R. SA., R. CA. *Protornis* Olig. Eu. **Todidae** R. WInd.

SUBORDER BUCEROTES. **Bucerotidae** R. As. Af. SPac., *Cryptornis Geiseloceros* Eoc. Eu. *Homalopas* Mioc. Eu.

SUBORDER CORACIAE. **Coraciidae** R. Eu. As. Af. Mad. Aus. SPac., *Coracias* Pleist.-R. Eu. As., R. Af. *Geranopterus* Eoc. or Olig. Eu. **Leptosomatidae** R. Mad. **Phoeniculidae** R. Af., *Limatornis* Mioc. Eu. **Upupidae** R. Eu. As. Af. Mad., *Upupa* Pleist.-R. Eu. As., R. Af. Mad.

SUBORDER MEROPES. **Meropidae** R. Eu. As. Af. Mad. Aus. SPac., *Merops* Pleist.-R. Eu., R. As. Af. Mad. Aus. SPac.

ORDER PICIFORMES

SUBORDER PICI. **Picidae** R. Eu. As. Af. NA. SA., *Asyndesmus* Pleist.-R. NA. *Bathoceleus* Pleist. WInd. *Campephilus* SubR.-R. NA., R. SA. *Chrysoptilus* Pleist.-R. SA. WInd. *Colaptes* Pleist.-R. NA. WInd. SA. *Dendrocopos* Pleist.-R. Eu. NA., R. As. Af. *Dryocopus* Pleist. Eu. NA., R. As. SA. *Jynx* Pleist.-R. Eu., R. As. Af. SA. *Leuconer-*

pes Pleist.-R. SA. *Melanerpes* Pleist.-R. NA. WInd. SA. *Palaeopicus* Mioc. Eu. *Picus* Mioc.-R. Eu., R. As. *Sphyrapicus* Pleist.-R. NA., R. SA. WInd. *Veniliornis* Pleist.-R. SA., R. CA.

SUBORDER GALBULAE. **Galbulidae** R. CA. SA. **Bucconidae** R. CA. SA., *Malacoptila* Pleist.-R. SA., R. CA. *Nystalus* Pleist.-R. SA.

SUBORDER CAPITONES. **Capitonidae** R. As. Af. CA. SA. **Indicatoridae** R. As. Af.

SUBORDER RAMPHASTIDES. **Ramphastidae** R. CA. SA., *Rhamphastes* Pleist.-R. SA., R. CA.

ORDER PASSERIFORMES

SUBORDER EURYLAIMI. **Eurylaimidae** R. As. Af.

SUBORDER TYRANNI. **Scytalopidae (Rhinocryptidae Pteroptochidae)** R. CA. SA., *Neanis* [*Hebe*] Eoc. NA. **Furnariidae** R. CA. SA., *Lepidocolaptes Xiphocolaptes* Pleist.-R. SA., R. CA. **Formicariidae** R. CA. SA., *Chamaeza* Pleist.-R. SA. **Tyrannidae** R. NA. WInd. SA., *Blacicus* Pleist.-R. WInd. *Contopus* Pleist.-R. NA., R. WInd. SA. *Empidonax* Pleist.-R. NA., R. WInd. CA. *Myiarchus* Pleist.-R. WInd., R. NA. SA. *Sayornis* Pleist.-R. NA., R. WInd. SA. *Tolmarchus* Pleist.-R. WInd. *Tyrannus* Pleist.-R. NA. WInd., R. SA. **Oxyruncidae** R. CA. SA. **Phytotomidae** R. SA. **Cotingidae** R. WInd. CA. SA. **Pipridae** R. CA. SA. **Pittidae** R. As. Af. Aus. SPac. **Xenicidae (Acanthisittidae)** R. NZ. **Philepittidae** R. Mad.

SUBORDER MENURAE. **Menuridae Atrichiidae (Atrichornithidae)** R. Aus.

SUBORDER PASSERES. **Alaudidae** R. Cos., *Alauda* Plioc.-R. Eu., Pleist.-R. As., R. Af. *Eremophila* Pleist.-R. Eu. NA., R. As. Af. SA. *Galerida* Pleist.-R. Eu. As., R. Af. *Lullula* Pleist.-R. Eu., R. As. Af. *Melancorypha* Pleist.-R. As., R. Eu. Af. **Palaeospizidae,** *Palaeospiza* Olig. NA. **Hiruninidae** R. Cos., *Delichon* Pleist.-R. Eu., R. As. Af. *Hirundo* Pleist.-R. Eu. As. SA., R. Cos. *Petrochelidon* Pleist.-R. NA. ?WInd., R. Af. Aus. SPac. NA. SA. *Progne* Pleist. NA. SA., R. WInd. *Riparia* Pleist.-R. Eu., R. As. Af. Mad. NA. SA. *Tachycineta* Pleist.-R. NA., R. WInd. SA. **Dicruridae** R. As. Af. Aus. SPac. Mad., *Dicrurus* Pleist.-R. As., R. Af. Mad. Aus. SPac. **Oriolidae** R. Eu. As. Af. Aus. SPac., *Oriolus* Pleist.-R. Eu. As., R. Af. Aus. SPac. **Corvidae** R. Cos., *Aphelocoma* Pleist.-R. NA., R. CA. *Cissilopha* SubR.-R. CA. *Corvus* Plioc.-R. Eu., Pleist.-R. As. NA. WInd., R. Cos. *Cyanocitta* Pleist.-R. NA., R. CA. *Cyanocorax* SubR.-R. CA., R. SA. *Garrulus* Pleist.-R. Eu. As., R. Af. *Gymnorhinus* Pleist.-R. NA. *Henocitta* Pleist. NA. *Miocitta* Mioc. NA. *Miocorvus* Mioc. Eu. *Nucitraga* Pleist.-R. Eu. NA., R. As. *Palaeocorax* Pleist. NZ. *Perisoreus* Pleist.-R. NA., R. Eu. As. *Pica* Pleist.-R. Eu. As. NA., R. Af. *Protocitta* Pleist. NA. *Pyrrhocorax* Pleist.-R. Eu. As., R. Af. *Uroleuca* Pleist.-R. SA. *Xanthoura* SubR.-R. CA., R. NA. SA. **Gymnorhinidae (Cracticidae) Grallinidae** R. Aus. **Callaeatidae** R. NZ., *Callaeas Heterolocha Philesturnus* Pleist.-R. NZ. **Ptilonorhynchidae** R. Aus. **Paradiseidae** R. SPac. Aus. **Paridae** R. Eu. As. Af. NA. CA., *Aegithalus* Pleist.-R. Eu., R. As. *Palaegithalus* Eoc. Eu. *Parus* Pleist. Eu. NA., R. As. Af. CA. **Sittidae** R. Eu. As. NA., *Sitta* Plioc.-R. Eu., Pleist.-R. NA., R. As. **Hyposittidae** R. Mad. **Certhiidae** R. Eu. As. Af. Aus. SPac. NA. CA., *Certhia* Pleist.-R. Eu., R. As. NA. *Tichodroma* Pleist.-R. Eu., R. As. **Chamaeidae** R. NA., *Chamaea* Pleist.-R. NA. **Timaliidae** R. Eu. As. Af. Mad. Aus. SPac., *Pterorhinus* Pleist.-R. As. *Turdoides* Pleist. WAs., R. Af. **Campephagidae** R. As. Af. Aus. SPac. **Pycnonotidae** R. As. Af. Mad. SPac., *Pycnonotus* Pleist.-R. As., R. Af. **Chloropseidae** R. As. **Palaeoscinidae,** *Palaeoscinus* Mioc. NA. **Cinclidae** R. Eu. As. NA. SA., *Cinclus* Pleist.-R. Eu., R. As. NA. SA. **Troglodytidae** R. Eu. As. NA. WInd. SA., *Catherpes* Pleist.-R. NA., R. CA. *Cistothorus* Pleist.-R. NA., R. SA. *Salpinctes* Pleist.-R. NA., R. CA. *Troglodytes* Pleist.-R. Eu. NA. SA., R. As. WInd. **Mimidae** R. NA. WInd. SA., *Margarops* Pleist.-R. WInd. *Mimus* Pleist.-R. WInd., R. NA. SA. *Oreoscoptes Toxostoma* Pleist.-R. NA. R. CA. **Turdidae** R. Cos., *Erithacus* Pleist.-R. Eu. As. *Hylocichla* Pleist.-R. NA., R. CA. *Luscina* Pleist.-R. Eu. As., R. Af. *Mimocichla* Pleist.-R. WInd. *Monticola* Pleist.-R. Eu., R. As. Af. *Oenanthe Phoenicurus* Pleist.-R. Eu. As., R. Af. *Saxicola* Pleist.-R. Eu., R. As. Af. Mad. *Sialia* Pleist.-R. NA., R. CA. *Turdicus* Pleist. Eu. *Turdus* Pleist.-R. Eu. As. NA., R. Af. SA. **Muscicapidae** R. Eu. As. Af. Mad. Aus. NZ. SPac., *Miro* Pleist.-R. NZ. *Muscicapa* Pleist.-R. As., R. Eu. Af. *Turnagra* Pleist.-R. NZ. **Sylviidae** R. Cos., *Acrocephalus Agrobates Phylloscopus Sylvia* Pleist.-R. Eu., R. As. Af. **Prunellidae** R. Eu. As. Af., *Prunella* Pleist.-R. Eu., R. As. Af. **Motacillidae** R. Cos., *Anthus* Plioc.-R. Eu., Pleist.-R. As. R. Cos. *Motacilla* Mioc.-R. Eu., Pleist.-R. As., R. Af. Mad. NA. **Bombycillidae** R. Eu. As. NA. CA., *Bombycilla* Pleist.-R. Eu. NA., R. As. CA. **Ptilogonatidae** R. NA. CA. **Dulidae** R. WInd. **Artamidae** R. As. Aus. SPac. **Vangidae** R. Mad. **Laniidae** R. Eu. As. Af. NA. CA., *Lanius* Mioc.-R. Eu., Pleist.-R. As. NA., R. Af. **Sturnidae** R. Eu. As. Af. Aus. SPac., *Laurillardia* Eoc. Eu. *Necropsar* Pleist.-R. Rodriguez *Onychognathus* Pleist. WAs., R. Af. *Pastor* Pleist.-R. Eu., R. As. *Sturnus* Pleist.-R. Eu. As., R. Af. SPac. **Meliphagidae** R. Af. Aus. SPac. NZ. Hawaii, *Prosthemadera* Pleist.-R. NZ. **Nectarinidae** R. As. Af. Aus. SPac. **Dicalidae** R. As. Aus. SPac. **Zosteropidae** R. As. Af. Mad. Aus. NZ. SPac. **Vireonidae** R. NA. WInd. SA., *Cyclarhis* Pleist.-R. SA., R. CA. *Vireo* Pleist.-R. WInd., R. NA. SA. **Coerebidae** R. WInd., CA. SA., *Coereba* Pleist.-R. WInd., R. CA. SA. **Drepanidae** R. Hawaii **Parulidae** R. NA. WInd. SA., *Dendroica* Pleist.-R. WInd., R. NA. SA. *Geothlypis* Pleist.-R. NA., R. WInd. SA. *Mniotilta* Pleist.-R. WInd., R. NA. SA. **Icteridae** R. NA. WInd. SA., *Agelaius* Pleist.-R. NA. WInd., R. SA. *Cassidix* Pleist.-R. NA., R. SA. *Cremaster* Pleist. NA. *Euphagus* Pleist.-R. NA., R. CA. *Gnorimopsar* Pleist.-R. SA. *Holoquiscalus* Pleist.-R. WInd. *Icterus* Pleist.-R. NA. WInd., R. SA. *Molothrus* Pleist.-R. NA., R. SA. *Ostinops* Pleist.-R. SA., R. CA. *Pandanaris* Pleist. NA. *Pseudoleistes* Pleist.-R. SA. *Pyelorhamphus* Pleist. NA. *Quiscalus* Pleist.-R. NA., R. CA. *Sturnella* Pleist.-R. NA., R. WInd. SA. *Xanthocephalus* Pleist.-R. NA. **Ploceidae** R. As., *Acanthis* Pleist.-R. Eu., R. As. *Carduelis* Pleist.-R. Eu. As., R. Af. *Carpodacus* Pleist.-R. NA., R. Eu. As. CA. *Chloris* Pleist.-R. Eu. As. *Coccothraustes* [*Hesperiphona*] Pleist.-R. Eu. NA., R. As. CA. *Loxia* Pleist.-R. Eu. NA., R. As. CA. WInd. *Montifringilla* Pleist.-R. Eu., R. As. *Passer* Pleist.-R. Eu. As., R. Af. *Petronia* Pleist.-R. As., R. Eu. Af. *Pinicola* Pleist.-R. Eu., R. As. NA. *Pyrrhula* Pleist.-R. Eu., R. As. *Serinus* Pleist.-R. Eu. As., R. Af. *Spinus* Pleist.-R. Eu. As. NA., R. Af. SA. **Tangridae** R. NA. WInd. SA., *Nesospingus* Pleist.-R. WInd. *Pheucticus* Pleist.-R. NA., R. WInd. SA. *Richmondena* Pleist.-R. NA., R. SA. *Saltator* Pleist.-R. SA., R. CA. WInd. *Spindalis* Pleist.-R. WInd.,

R. CA. **Fringillidae** R. Eu. As. Af. NA. WInd. SA., *Ammodramus* Pleist. NA. WInd., R. SA. *Amphispiza Calamospiza Chondestes* Pleist.-R. NA. *Emberiza Fringilla* Pleist.-R. Eu. As., R. Af. *Junco Melospiza* Pleist.-R. NA., R. CA. *Palaeostruthus* Plioc. NA. *Paserculus Passerella* Pleist.-R. NA., R. CA. *Passerherbulus* Pleist.-R. NA. *Pipilio* Pleist.-R. NA., R. CA. *Plectrophenax* Pleist. Eu., R. As. NA. *Pooecetes* Pleist.-R. NA. *Spizella* Pleist.-R. NA., R. CA. *Tiaris* Pleist.-R. WInd., R. CA. SA. *Zonotrichia* Pleist.-R. NA., R. WInd. SA.

CLASS MAMMALIA

SUBCLASS PROTOTHERIA

ORDER MONOTREMATA. **Ornithorhynchidae** R. Aus., *Ornithorhynchus* [*Platypus*] Pleist.-R. Aus. **Tachyglossidae (Echidnidae)** R. Aus. New Guinea, *Tachyglossus* [*Echidna*] *Zaglossus* Pleist.-R. Aus.

SUBCLASS UNCERTAIN

ORDER DOCODONTA. **Docodontidae,** *Docodon* [*Dicrocynodon Diplocynodon Ennacodon Enneodon Peraiocynodon*] U. Jur. Eu. ?**Morganucodontidae,** *Eozostrodon* U. Trias. Eu. *Erythrotherium* U. Trias. SAf. *Morganucodon* U. Trias. Eu. EAs.

ORDER TRICONODONTA. **Amphilestidae,** *Amphilestes* M. Jur. Eu. *? Aploconodon* U. Jur. Eu. *? Phascolodon* U. Jur. NA. *Phascolotherium* M. Jur. Eu. **Triconodontidae,** *Astroconodon* L. Cret. NA. *Priacodon* U. Jur. NA. *Sinoconodon* U. Trias. EAs. *Triconodon* U. Jur. Eu. *Trioracodon* U. Jur. Eu. NA.

SUBCLASS ALLOTHERIA

ORDER MULTITUBERCULATA

SUBORDER PLAGIAULACOIDEA. **Plagiaulacidae,** *Ctenacodon* [*Allodon*] U. Jur. Eu. NA. *Loxaulax* L. Cret. Eu., ?NA. *? Paulchoffatia Plagiaulax* [*? Bolodon*] U. Jur. Eu. *Psalodon* U. Jur. NA.

SUBORDER PTILODONTOIDEA. **Ectypodidae,** *Cimexomys* U. Cret. NA. *Ectypodus* ?M., U. Paleoc.-U. Eoc. NA., U. Paleoc. Eu. *Mesodma* [*Parectypodus*] U. Cret.-L. Paleoc. NA. *Neoplagiaulax* U. Paleoc. Eu. *? Mimetodon* M.-U. Paleoc. NA. **Cimolodontidae,** *Anconodon* M.-U. Paleoc. NA. *Cimolodon ? Essonodon* U. Cret. NA. *Liotomus* [*Neoctenacodon*] U. Paleoc. Eu. **Ptilodontidae,** *Kimbetohia* L. Paleoc. NA. *Prochetodon* U. Paleoc.-L. Eoc. NA. *Ptilodus* M.-U. Paleoc. NA.

SUBORDER TAENIOLABOIDEA. **Cimolomyidae,** *Cimolomys Meniscoessus* U. Cret. NA. *Sphenopsalis* U. Paleoc. EAs. **Eucosmodontidae,** *Djadochtatherium* U. Cret. EAs. *Eucosmodon* U. Cret., ?U. Paleoc. NA. *Microcosmodon* U. Paleoc. NA. *Neoliotomus* ?M. Paleoc., U. Paleoc.-L. Eoc. NA. *Pentacosmodon* U. Paleoc. NA. *Proliotomus Stygimys* U. Cret. NA. **Taeniolabidae,** *Catopsalis* U. Cret.-U. Paleoc. NA. *Prionessus* U. Paleoc. As. *Taeniolabis* [*Polymastodon*] L. Paleoc. NA.

MULTITUBERCULATA INCERTAE SEDIS: *Neoctenodon* U. Paleoc. Eu. *Paronychodon* U. Cret. NA.

ORDER SYMMETRODONTA. **Spalacotheriidae,** *Eurylambda* U. Jur. NA. *Spalacotherium* [*? Peralestes*] U. Jur., ?L. Cret. Eu. *Spalacotheroides* L. Cret. NA. *Tinodon* U. Jur. NA. **Amphodontidae,** *Amphodon* U. Jur. NA. *Manchurodon* L. Cret. EAs.

ORDER PANTOTHERIA. **Amphitheriidae,** *Amphitherium* M. Jur. Eu. *? Peramus* U. Jur. Eu. **Paurodontidae,** *Araeodon Archaeotrigon* U. Jur. NA. *Brancatherulum* U. Jur. EAf. *Paurodon Tathiodon* U. Jur. NA. **Dryolestidae,** *? Aegialodon* L. Cret. Eu. *Amblotherium* [*Stylacodon Stylodon*] U. Jur. Eu. NA. *Dryolestes Duthlastus Herpetairus Kepolestes* U. Jur. NA. *Kurtodon* U. Jur. Eu. *Laolestes Malthacolestes* U. Jur. NA. *Melanodon* U. Jur. NA., ?L. Cret. Eu. *Miccylotyrans* U. Jur. NA. *Peraspalax Phascolestes* U. Jur. Eu.

INFRACLASS METATHERIA

ORDER MARSUPIALIA

SUBORDER POLYPROTODONTA. **Didelphidae** R. NA. SA., *Alphodon Boreodon* U. Cret. NA. *Caluromys* Pleist.-R. SA. *Campodus* U. Cret. NA. *Chironectes* Pleist.-R. SA. *Cladodidelphys* Plioc. SA. *Coöna* L. Eoc. SA. *Delphodon* U. Cret. NA. *Derorhynchus* U. Paleoc. SA. *Diaphorodon* U. Cret. NA. *Didectodelphis Didelphidectes* M. Olig. NA. *Didelphis* Pleist.-R. NA., L. Pleist.-R. SA. *Didelphodon* U. Cret. NA. *Didelphopsis* U. Paleoc. SA. *Ectonodon* U. Cret. NA. *Eodelphis* U. Cret. NA. *? Gashternia Gaylordia Guggenheimia Ischyrodelphis* U. Paleoc. SA. *Lutreolina Marmosa* [*Marmosops*] U. Plioc.-R. SA., Pleist.-R. CA. *Metachirus* Pleist.-R. SA. *Microbiotherium* [*Hadrorhynchus Pachybiotherium Prodidelphys Proteodidelphys Notictis*] Olig.-L. Mioc. SA. *Mirandatherium* [*Mirandaia*] *Microbiotheridium Monodelphopsis* U. Paleoc. SA. *Nanodelphys* U. Eoc.-M. Olig. NA. *Paradidelphys* U. Plioc. SA. *Pediomys* U. Cret. NA. *Peradectes* U. Paleoc.-L. Eoc. NA. *Peramys* Pleist.-R. SA. *Peratherium* [*Herpetotherium Oxygomphius*] L. Eoc.-Mioc. NA., M. Eoc.-L. Mioc. Eu. *Perazoyphium* U. Plioc. SA. *Philander* ?M. Plioc., Pleist.-R. SA., R. CA. *Protodidelphis Schaefferia* U. Paleoc. SA. *Sparassocynus* U. Plioc. SA. *Thlaeodon* U. Cret. NA. *Thylacodon* L. Paleoc. NA. *Thylatheridium Thylophorops* U. Plioc. SA. *Zygolestes* ?M. Plioc. SA. **Caroloameghiniidae,** *Caroloameghinia* L. Eoc. SA. **Borhyaenidae,** *Achlysictis* Plioc. SA. *Acrocyon* L. Mioc. SA. *Acrohyaenodon* Plioc. SA. *Agustylus* L. Mioc. SA. *Apera* Mioc. SA. *Aminiheringia* [*Dilestes*] ?U. Paleoc., L. Eoc. SA. *Argyrolestes* L. Eoc. SA. *Borhyaena* L. Mioc. SA. *Borhyaenidium* M. Plioc. SA. *Chasicostylus* L. Plioc. SA. *Cladosictis* [*Hathlyacynus*] U. Olig.-L. Mioc. SA. *Conodonictis* L. Mioc. SA. *Eobrasilia* U. Paleoc. SA. *Hyaenodonops* U. Plioc. SA. *Ictioborus Lycopsis Napodonictis* L. Mioc. SA. *Nemolestes* ?U. Paleoc., L. Eoc. SA. *Notocynus Notosmilus* U. Plioc. SA. *Notogale* L. Olig. SA. *Palaeocladosictis* U. Paleoc. SA. *Parahyaenodon* M. Plioc. SA. *Patene* U. Paleoc.-L. Eoc. SA. *Perathereutes* L. Mioc. SA. *Pharsophorus* [*Plesiofelis*] ?U. Eoc., L. Olig. SA. *Proborhyaena* L.-U. Olig. SA. *Prothylacynus* L. Mioc. SA. *Pseudoborhyaena* U. Olig. SA. *Sipalocyon* L. Mioc. SA. *Stylocinus* Plioc. SA. *Thylacodictis* [*Amphiproviverra*] L. Mioc. SA. *Thylacosmilus* Plioc. SA. **Necrolestidae,** *Necrolestes* L. Mioc. SA. ?**Microtragulidae,** *Microtragulus* [*Argyrolagus*] U. Plioc.-L. Pleist. SA. **Dasyuridae** R. Aus., *Dasyurus* Pleist.-R. Aus., R. New Guinea *Glaucodon* Plioc. or Pleist. Aus. *Phascogale* [*Antechinus*] Pleist.-R. Aus., R. New Guinea *Sarcophilus* Pleist. Aus., R. Tasmania *Thylacinus* Pleist. Aus. New Guinea, R. Tasmania. **Notoryctidae** R. Aus.

SUBORDER PERAMELIDA. **Peramelidae** R. Aus., *Ischnodon* Plioc. Aus. *Perameles* Pleist.-R. Aus., R. New Guinea *Thylacis* [*Isoödon*] *Thylacomys* [*Paragalia*] Pleist.-R. Aus.

SUBORDER CAENOLESTOIDIA. **Caenolestidae** R. SA., *Abderites* [*Homunculites*] U. Olig.-L. Mioc. SA. *Acdestis* [*Callomenus*] L. Olig.-L. Mioc. SA. *Cladoclinus Dipilus* [*Decastis*] *Halmadromus* L. Mioc. SA. *Micrabderites* U. Olig. SA. *Palaeothentes* [*Epanorthus Metaepanorthus Palaepanorthus Paraepanorthus Prepanorthus*] *Parabderites* L. Olig.-L. Mioc. SA. *Pilchenia* L. Olig. SA. *Pitheculites ? Pitheculus* U. Olig. SA. *Pliolestes* U. Plioc. SA. *Progarzonia* L. Eoc. SA. *Pseudhalmarhiphus* [*? Clenialites*] Olig. SA. *Stilotherium* [*Garzonia ? Parhalmarhiphus ? Phonocdromus*] U. Olig.-L. Mioc. SA. **Groeberiidae,** *Groeberia* L. Olig. SA. **Polydolopidae,** *Amphidolops* [*Anadolops*] L. Eoc. SA. *Epidolops* U. Paleoc. SA. *Eudolops* [*Promysops Propolymastodon*] *?Odontomysops* L. Eoc. SA. *Polydolops* [*Anissodolops Archaeodolops Orthodolops Pliodolops Pseudolops*] U. Paleoc.-L. Eoc. SA. *Seumadia* Paleoc. SA.

SUBORDER DIPROTODONTA. **Phalangeridae** R. Aus., *Archizonurus Burramys* Pleist. Aus. *Cercaërtus* [*Eudromicia*] Pleist.-R. Aus., R. New Guinea *Dromicia* Pleist.-R. Aus. *Endromicia* Pleist. Aus., R. Tasmania *Gymnobelidius* [*? Palaeopetaurus*] ?Pleist., R. Aus. *Koalemus* Pleist. Aus. *Perikoala* Plioc. Aus. *Petaurus* Pleist.-R. Aus., R. New Guinea *Phalanger* [*Cuscus*] Plioc.-R. Aus., Pleist.-R. EInd. *Phascolarctos* Pleist.-R. Aus. *Pseudocheirus* Pleist.-R. Aus., R. New Guinea *Trichosurus* Pleist.-R. Aus. *Wynyardia* Plioc. Tasmania. **Thylacoleonidae,** *Thylacoleo* ?Plioc., Pleist. Aus. **Phascolomidae** (**Phascolomyidae**) R. Aus., *Phascolomis* [*Phascolomys*] Pleist.-R. Aus. *Phascolonus Ramsayia* Pleist. Aus. **Macropodidae** R. Aus., *Aepyprymnus* Pleist.-R. Aus. *Bettongia* ?Mioc., Pleist.-R. Aus. *Brachalletes* Pleist. Aus. *Macropus* [*Halmaturus Protemnodon*] Plioc.-R. New Guinea, Pleist.-R. Aus. *Palorchestes* ?Plioc., Pleist. Aus. *Petrogale Potorous* [*Hypsiprymnodon*] Pleist.-R. Aus. *Prionotemnus* Plioc. Aus. *Procoptodon Propleopus* Pleist. Aus. *Setonyx* ?Pleist., R. Aus. *Sthenurus* ?Plioc., Pleist. Aus. *Synaptodon* Pleist. Aus. *Thylogale* Pleist.-R. Aus., R. New Guinea. **Diprotodontidae,** *Diprotodon, Euowenia* [*Owenia*] *Euryzygoma* Pleist. Aus. *Meniscolophus* Plioc. Aus. *Nototherium* ?Plioc., Pleist. Aus. *Sceparnodon Sthenomerus Zygomaturus* Pleist. Aus.

MARSUPIALIA INCERTAE SEDIS: *Florentinoameghinia* L. Eoc. SA.

INFRACLASS EUTHERIA

ORDER INSECTIVORA

SUBORDER PROTEUTHERIA

?INCERTAE SEDIS: ?**Endotheriidae,** *Endotherium* M. Cret. EAs. ?**Pappotheriidae,** *Pappotherium* L. Cret. NA.

SUPERFAMILY TUPAIOIDEA. **Leptictidae,** *Adunator* M.-U. Paleoc. Eu. *Diacodon* U. Paleoc.-M. Eoc. NA. *Diaphyodectes* U. Paleoc. Eu. *Hypictops* M. Eoc. NA. *Ictops* [*Mesodectes*] L.-M. Olig. NA. *Leptictidium* M. Eoc. NA. *Leptictis* M. Olig. NA. *Myrmecoboides* M. Paleoc.-L. Eoc. NA. *Palaeictops* [*Parictops*] ?U. Paleoc., L.-M. Eoc. NA. *Procerberus* U. Cret.-L. Paleoc. NA. *Prodiacodon* [*Palaeolestes*] M. Paleoc. NA. **Zalambdalestidae,** *Zalambdalestes* U. Cret. EAs. **Anagalidae,** *Anagale* L. Olig. EAs. *Anagalopsis* ?U. Olig. EAs. *Pseudictops* L. Eoc. EAs. **Paroxyclaenidae,** *Dulcidon* M. Eoc. SAs. *Kochictis* M. Olig. Eu. *Kopidodon* M. Eoc. Eu. *Paroxyclaenus* U. Eoc. Eu. *Pugiodens* [*Vulpavoides*] *Russellites* M. Eoc. Eu. **Tupaiidae** R. SAs. EInd., *? Adapisoriculus* U. Paleoc. Eu. **Pantolestidae,** *Apheliscus* U. Paleoc.-L. Eoc. NA. *Chadronia* L. Olig. NA. *Dyspterna* L. Olig. Eu. *Galethylas Opsiclaenodon* L. Eoc. Eu. *Pagonomus* U. Paleoc. Eu. *Palaeosinopa* L. Eoc. Eu. NA. *Pantolestes* M. Eoc. NA. *Phenacodaptes Propalaeosinopa* [*Bessoecetor*] M.-U. Paleoc. NA., ?Paleoc. Eu. **Ptolemaiidae,** *Ptolemaia* L. Olig. NAf. **Pentacodontidae,** *Amaramnis* L. Eoc. NA. *Aphronorus* M. Paleoc. NA. *Bisonalveus* U. Paleoc. NA. *Coriphagus* [*Mixoclaenus*] *Pentacodon* M. Paleoc. NA. *Protentomodon* U. Paleoc. NA.

SUPERFAMILY APATEMYOIDEA. **Apatemyidae,** *Apatemys* [*Teilhardella*] U. Paleoc.-U. Eoc. NA. *Eochiromys* L. Eoc. Eu., ?NA. *Heterohyus* [*Amphichiromys Heterochiromys ? Necrosorex*] M. Eoc.-L. Olig. Eu. *Jepsenella* M. Paleoc. NA. *Labidolemur* U. Paleoc. NA. *Sinclairella* L.-M. Olig. NA. *Stehlinella* [*Stehlinius*] U. Eoc. NA. *Unuchinia* [*Apator*] U. Paleoc. NA.

SUBORDER MACROSCELIDEA. **Macroscelididae** R. Af., *Elephantulus* [*Elephantomys*] Pleist.-R. Af. *Metoldobotes* L. Olig. NAf. *Mylomygale* Pleist. SAf. *Myohyrax* L. Mioc. SAf. EAf. *Myomygale* L. Pleist. SAf. *Palaeothentoides* ?L. Pleist. SAf. *Protypotheroides* L. Mioc. SAf. *Rhynchocyon* L. Mioc.-R. Af.

SUBORDER DERMOPTERA

SUPERFAMILY MIXODECTOIDEA. **Mixodectidae,** *Dracontolestes* M. Paleoc. NA. *Elpidophorus* M.-U. Paleoc. NA. *Eudaemonema* M. Paleoc. NA. *Mixodectes* [*Indrodon Oldobotes*] M.-U. Paleoc. NA. *Remiculus* U. Paleoc. Eu.

SUPERFAMILY GALEOPITHECOIDEA. **Plagiomenidae,** *Plagiomene* L. Eoc. NA. *Planetetherium* U. Paleoc. NA. *? Thylacaelurus* U. Eoc. NA. **Galeopithecidae** (**Cynocephalidae**) R. SEAs.

INCERTAE SEDIS: **Picrodontidae,** *Picrodus* [*Megopterna*] M. Paleoc. NA. *Zanycteris* U. Paleoc. NA.

SUBORDER LIPOTYPHLA

SUPERFAMILY ERINACEOIDEA. **Adapisoricidae,** *Adapisorex* M.-U. Paleoc. Eu. *Amphidozotherium* U. Eoc. Eu. *Amphilemur* M. Eoc. Eu. *Ankylodon* U. Eoc.-M. Olig. NA. *Centetodon* [*Hypacodon*] ?L., M. Eoc. NA. *Clinopternodus* L. Olig. NA. *Creotarsus* L. Eoc. NA. *Dormaalius* U. Paleoc. Eu. *Entomolestes* [*Leipsanolestes*] U. Paleoc.-M., ?U. Eoc. NA. *Geolabis* [*Metacodon Protictops*] M. Eoc.-M. Olig. NA. *Gypsonictops* [*Euangelistes*] U. Cret. NA. *Hyracolestes* L. Eoc. EAs. *Ictopidium* L. Olig. EAs. *Leptacodon* M. Paleoc.-L. Eoc. NA. *Litolestes* U. Paleoc. NA. *Macrocranion* [*Aculeodens*] *Messelina* M. Eoc. Eu. *Myolestes* M. Eoc. NA. *Nyctitherium* L.-U. Eoc. NA. *Ocajila* L. Mioc. NA. *Opisthopsalis* L. Eoc. EAs. *Paschatherium* U. Paleoc. Eu. *Praolestes* L. Eoc. NA. *Proterixoides* U. Eoc. NA. *Scenopagus* M., ?U. Eoc. NA. *Sespedectes* U. Eoc. NA. *Stilpnodon* M. Paleoc. NA. *Talpavus* L.-U. Eoc. NA. *Tupaiodon* U. Olig. EAs. *Xenacodon* U. Paleoc. NA. **Erinaceidae** R. Eu. As. Af., *Amphechinus* [*Palaeoeri-*

naceus Palaeoscaptor] L. Mioc. Af., L. Olig.-U. Mioc. Eu., U. Olig.-L. Mioc. As. *Brachyerix* U. Mioc. NA. *Dimylechinus* L. Mioc. Eu. *Embassis* M. Olig. NA. *Erinaceus* [*Aethechinus Atelerix*] U. Mioc.-R. Eu., Pleist.-R. As., ?Pleist., R. NAf. *Galerix* L. Mioc. EAf., U. Mioc.-L. Plioc. Eu. *Gymnurechinus* L. Mioc. EAf. *Hemiechinus* Pleist.-R. WAs., R. NAf. *Lanthanotherium* [*Rubitherium*] M.-U. Mioc. Eu., U. Mioc.-L. Plioc. NA. *Metechinus* L. Mioc.-L. Plioc. NA. *Mioechinus* L.-U. Mioc. Eu. *Neurogymnurus* [*Necrogymnurus*] U. Eoc.-M. Olig. Eu. *Parvericius* U. Mioc. NA. *Postpalerinaceus* L. Plioc. Eu. *Proterix* M. Olig. NA. *Pseudogalerix* U. Mioc. Eu. *Tetracus* L. Olig. Eu. **Dimylidae,** *Cordylodon* U. Olig.-L. Mioc. Eu. *Dimyloides* U. Olig. Eu. *Dimylus* L. Mioc. Eu. *Exodaenodus* M. Olig. Eu. *Metacordylodon* U. Mioc. Eu. *Plesiodimylus* M. Mioc.-L. Plioc. Eu. *Pseudocordylodon* L. Mioc. Eu. **Talpidae** R. Eu. As. NA., *Condylura* Pleist.-R. NA. *Cryptoryctes* L. Olig. NA. *Desmana* [*Desmagale Mygale Myogale*] L. Plioc.-R. Eu., R. As. *Domninoides* L. Mioc.-L. Plioc. NA. *Galemys* ?L. Pleist., R. Eu. *Geomana* Pleist. Eu. *Geotrypus* M. Eoc.-M. Olig. Eu. *Hesperoscalops* U. Plioc. NA. *Hydroscapheus* L. Plioc. NA. *Mesoscalops* M. Mioc. NA. *Mydecodon* L. Mioc. NA., ?M. Mioc. Eu. *Mygalea* U. Mioc. Eu. *Mygalinia* L. Plioc. Eu. *Mygatalpa* M.-U. Olig. Eu. *Oligoscalops* M. Olig. NA. *Parascalops* Pleist.-R. NA. *Paratalpa* M. Olig. Eu. *Proscalops* [*Arctoryctes*] L., M. Olig.-L. Mioc. NA. *Proscapanus* M.-U. Mioc. Eu. *Scalopoides* L. Mioc. NA., ?M. Mioc. Eu. *Scalopus* [*Scalops*] L. Plioc.-R. NA. *Scapanulus* Plioc.-R. As. *Scapanus* L. Plioc.-R. NA. *Scaptochirus* Pleist. Eu., Pleist.-R. As. *Scaptogale* [*Echinogale*] L. Mioc. Eu. *Scaptonyx* M.-U. Mioc. Eu., R. EAs. *Talpa* [*Mogera*] L. Mioc.-R. Eu., Pleist.-R. As. *Urotrichus* Pleist.-R. EAs.

SUPERFAMILY SORICOIDEA. **Plesiosoricidae,** *Entomacodon* M. Eoc. NA. *Meterix* L. Plioc. NA. *Plesiosorex* M. Olig.-U. Mioc. Eu., L. Mioc. NA., ?As. *Saturninia* U. Eoc. Eu. **Soricidae,** *Amblycoptus* L. Plioc. Eu. *Anourosorex* [*Shikamainosorex*] L. Plioc.-R. As. *Beremendia* U. Plioc.-L. Pleist. Eu. *Blairinoides* U. Plioc.-L. Pleist. Eu. *Blarina* U. Plioc.-R. NA. *Blarinella* L. Plioc.-R. As., Pleist. Eu. *Crocidosorex* M. Olig. Eu. *Crocidura* ?U. Mioc., L. Plioc.-R. Eu., L. Plioc.-R. As., ?L. Mioc., L. Pleist.-R. Af., L. Pleist.-R. EInd: *Cryptotis* U. Plioc.-R. NA., Pleist.-R. NSA. *Domnina* [*Protosorex*] ?U. Eoc., L. Olig.-L. Mioc. NA. *Hesperosorex* ?L., U. Plioc. NA. *Limnoecus* L.-U. Mioc. Eu., L. Mioc.-L. Plioc. NA. *Macrosorex* Olig. Eu. *Microsorex* Pleist.-R. NA. *Myosorex* L. Pleist.-R. SAf. *Mystipterus* L. Plioc. NA. *?Necrosorex* L. Olig. Eu. *Neomys* [*Crossopus*] U. Plioc.-R. Eu., Pleist.-R. As. *Nesiotites* Pleist. Eu. *Notiosorex* U. Plioc.-R. NA. *Oligosorex* L. Mioc. Eu. *Paracryptotis* U. Plioc. NA. *Petenyia* U. Plioc.-L. Pleist. Eu. *Petenyiella* L. Pleist. Eu. *Pleisorex* Pleist. EAs. *Sorex* [*Drepanosorex*] M. Olig.-R. Eu., L. Plioc.-R. NA. *Soricella* L. Mioc. Eu. *Soriculus* U. Plioc.-Pleist. Eu., R. As. NAf. *Suncus* [*Pachyura*] L. Plioc.-R. Eu., Pleist.-R. Af., R. As. *Trimylus* [*Heterosorex*] M. Olig.-M., ?U. Mioc. NA., U. Olig.-L. Plioc. Eu. *Zelceina* L. Pleist. Eu. ?**Apternodontidae,** *Apternodus* ?U. Eoc., L. Olig. NA. *Eoryctes* M. Eoc. NA. *Oligoryctes* U. Eoc.-L. Olig. NA. ?**Nesophontidae,** *Nesophontes* Pleist.-SubR. WInd. ?**Solenodontidae** R. WInd., *Antillogale* Pleist. or SubR. WInd. *Solenodon* [*Atopogale*] Pleist.-R. WInd.

?SUBORDER ZALAMBDODONTA

SUPERFAMILY TENRECOIDEA. **Tenrecidae (Centetidae)** R. Mad. WAf., *Erythrozootes* L. Mioc. EAf. *Geogale* L. Mioc. EAf., R. Mad. *Oryzorictes* ?Pleist., R. Mad. *Protenrec* L. Mioc. EAf. *Tenrec* [*Centetes*] Pleist.-R. Mad.

SUPERFAMILY CHRYSOCHLOROIDEA. **Chrysochloridae** R. SAf., *Chlorotalpa Chrysochloris* [*?Chrystotricha*] Pleist.-R. SAf. *Proamblysomus* Pleist. SAf. *Prochrysochloris* L. Mioc. EAf.

INSECTIVORA, GENERA INCERTAE SEDIS: *Glasbius* U. Cret. NA. *Microtarsioides* M. Eoc. Eu. *Phenacolophus* U. Paleoc. EAs. *Pseudorhynchocyon* Eoc. or Olig. Eu.

ORDER TILLODONTIA. **Esthonychidae,** *Anchippodus* M. Eoc. NA. *Adadidium* U. Eoc. EAs. *Esthonyx* [*Plesesthonyx*] U. Paleoc.-L. Eoc. NA., L. Eoc. Eu. *Tillodon* M. Eoc. NA. *Trogosus* [*Tillotherium*] L.-M. Eoc. NA. *Kuanchuanius* M. Eoc. EAs.

ORDER TAENIODONTIA. **Stylinodontidae,** *Basalina* M. Eoc. SAs. *Conoryctella Conoryctes* M. Paleoc. NA. *Ectoganus* [*Calamodon*] L. Eoc. NA. *Lampadophorus* U. Paleoc. NA. *Onychodectes* L. Paleoc. NA. *Psittacotherium* M. Paleoc. NA. *Stylinodon* L.-U. Eoc. NA. *Wortmania* L. Paleoc. NA.

ORDER CHIROPTERA

SUBORDER MEGACHIROPTERA. **Pteropodidae** R. As. Af. Aus., *Archaeopteropus* M. Olig. Eu. *Pteropus* Pleist-R. Mad. EInd., R. SAs. Aus. *Rousettus* Mioc. Eu., Pleist.-R. EInd., R. Af. SAs.

SUBORDER MICROCHIROPTERA

SUPERFAMILY EMBALLONUROIDEA. **Rhinopomatidae** R. NAf. SAs. **Emballonuridae** R. SAs. Af. Aus. SA., *Saccopteryx* Pleist.-R. SA. *Taphozous* [*Saccolaimus*] ?Eoc., Olig. Eu., L. Mioc.-R. Af., R. SAs. Aus. *Vespertiliavus* U. Eoc. or L. Olig., Plioc. Eu. **Noctilionidae** R. CA. SA.

SUPERFAMILY RHINOLOPHOIDEA. **Nycteridae** R. Af. EInd. **Megadermatidae** R. SAs. Af. Aus., *Megaderma* Mioc. Eu., Pleist.-R. As., R. Aus. *Miomegaderma* M. Mioc. Eu. *Necromantis* U. Eoc. or L. Olig. Eu. **Rhinolophidae** R. As. Aus., *?Palaeonycteris* U. Olig. Eu. *Rhinolophus* U. Eoc.-R. Eu., Pleist.-R. As., R. Af. Aus. **Hipposideridae** R. SAs. Af. Aus., *Asellia* M. Mioc. Eu. R. NAf. SAs. *Hipposideros* Mioc. Eu., Pleist.-R. EInd., R. As. Af. Aus. *Palaeophyllophora* U. Eoc.-Mioc. Eu. *Paraphyllophora* U. Eoc. or L. Olig.-Mioc. Eu. *Pseudcrhinolophus* M. Eoc.-L. Olig., ?Mioc. Eu.

SUPERFAMILY PHYLLOSTOMATOIDEA. **Phyllostomatidae** R. NA. SA., *Artibeus* Pleist.-R. WInd. SA. *Brachyphylla* Pleist.-R. WInd. *Carollia* Pleist.-R. SA. *Chilonycterus* Pleist.-R. WInd., R. SA. *Chiroderma Chrotopterus* Pleist.-R. SA. *Erophylla* Pleist.-R. WInd. *Glossophaga* Pleist.-R. SA. *Leptonycteris* Pleist.-R. CA. *Lonchoglossa* Pleist.-R. SA. *Macrotus* Pleist.-R. WInd., R. NA. *Micronycteris* [*Schizostoma*] *Mimon* [*Anthorhina*] Pleist.-R. SA. *Monophyllus* Pleist.-R. WInd. *Mormoöps* Pleist.-R. WInd., Pleist. NA., R. SA. *Notonycteris* U. Mioc. SA. *?Paleunycteris* M. Eoc.-L. Olig. Eu. *Phyllonycteris* [*Reithronycteris*] Pleist.-R. WInd. *Phyllostomus* Pleist.-R. SA. *Provampyrus* Olig. Af. *Pygoderma* Pleist.-R. SA. *Stenoderma* [*Ariteus Phyl-*

lops] Pleist.-R. NA. WInd., R. SA. *Sturnira* Pleist.-R. SA. *Tonatia* [*Lophostoma*] Pleist.-R. WInd. SA., R. CA. ? *Vampyravus* L. Olig. NAf. *Vampyrops* Pleist.-R. SA. *Vampyrum* [*Vampyrus*] Pleist.-R. SA., R. CA. **Desmodontidae** R. CA. SA., *Desmodus* Pleist. SNA., R. SA.

SUPERFAMILY VESPERTILIONOIDEA. **Natalidae** R. CA. SA. WInd., *Natalus* Pleist.-R. WInd., R. CA. SA. **Furipteridae** R. SA. **Thyropteridae** R. SA. CA. **Myzopodidae** R. Mad. **Vespertilionidae** R. Cos., *Antrozous* Pleist.-R. NA. *Barbastella* Pleist.-R. Eu. NAf. WAs. *Eptesicus* [*Hesperoptenus Histiotus*] L. Mioc., Pleist.-R. NA., Pleist.-R. SA. Eu. WInd. As., R. SA. Af. Aus. *Io* Pleist.-R. EAs. *Lasiurus* [*Atalapha Dasypterus*] U. Plioc.-R. NA., Pleist.-R. SA., R. WInd. Hawaii *Miniopterus* Mioc.-R. Eu., Pleist.-R. As. *Miomyotis* L. Mioc. NA. *Myotis* M. Olig.-R. Eu., Pleist.-R. As. Af. NA., R. SA. Aus. *Oligomyotis* M. Olig. NA. *Paraptesicus* Mioc. Eu. *Pipistrellus* [*Nyctalus Vesperugo*] ?Plioc., Pleist.-R. Eu., Pleist.-R. As. NA., R. Af. EInd. *Plecotus* [*Corynorhinus*] Pleist.-R. Eu. As. NA., R. NAf. *Samonycteris* L. Plioc. WAs. *Simonycteris* L. Plioc. NA. *Stehlinia* [*Nycterobius Revilliodа*] U. Eoc. or L. Olig. Eu. *Suaptenos* L. Mioc. NA. *Vespertilio* Mioc.-R. Eu., ?Pleist. NA., Pleist.-R. Af., R. As. **Mystacinidae** R. NZ. **Molossidae** R. SAs. Af. Aus. SNA. SA., *Cheiromeles* Pleist.-R. EInd., R. SAs. *Molossus* [*Molossides*] Pleist.-R. NA. SA. *Tadarida* [*Chaerophon Nyctinomus*] U. Olig.-R. Eu., Pleist.-R. NA. WInd. EInd., R. SAs. Af. Aus. SA.

SUPERFAMILY UNCERTAIN: **Archaeonycteridae**, *Archaeonycteris* M. Eoc. Eu. **Palaeochiropterygidae**, ? *Cecilionycteris Palaeochiropteryx* M. Eoc. Eu.

CHIROPTERA INCERTAE SEDIS: *Alastor* U. Eoc. or L. Olig. Eu. *Leptomycteris* Pleist. NA. *Paradoxonycteris* U. Eoc. Eu.

ORDER PRIMATES

?SUBORDER PLESIADAPOIDEA. **Phenacolemuridae** (**Paromomyidae**), *Palaechthon Palenochtha* M. Paleoc. NA. *Paromomys* M., ?U. Paleoc. NA. *Plesiolestes* M. Paleoc. NA. *Phenacolemur* U. Paleoc.-L. Eoc. NA. **Carpolestidae**, *Carpodaptes* U. Paleoc. NA. *Carpolestes* [*Litotherium*] M. Paleoc.-L. Eoc. NA. *Elphidotarsius* M. Paleoc. NA. *Saxonella* U. Paleoc. Eu. **Plesiadapidae**, *Chiromyoides* U. Paleoc. Eu. *Platychoerops* L. Eoc. Eu. *Plesiadapis* [*Ancepsoides Nothodectes*] U. Paleoc. Eu., U. Paleoc., ?L. Eoc. NA. *Pronothodectes* M.-U. Paleoc. NA.

SUBORDER LEMUROIDEA. **Adapidae** (**Notharctidae**), *Adapis* [*Aphelotherium Leptadapis Palaeolemur*] *Anchomomys* M.-U. Eoc. Eu. *Caenopithecus* M. Eoc. Eu. *Gesneropithex* U. Eoc. Eu. *Lantianius* U. Eoc. EAs. *Notharctus* [*Hipposyus Limnotherium Prosinopa Telmalestes Telmatolestes Thinolestes Tomitherium*] L.-M. Eoc. NA. *Pelycodus* L. Eoc. NA. *Pronycticebus* U. Eoc. Eu. *Protoadapis* [*Europolemur Megatarsius*] ?L., M.-U. Eoc. Eu. *Smilodectes* [*Aphanolemur*] M. Eoc. NA. **Lemuridae** R. Mad., *Archaeoindris Archaeolemur* [*Bradylemur Globilemur Lophiolemur Nesopithecus Protindris*] Pleist. Mad. *Cheirogaleus* Pleist.-R. Mad. *Hadropithecus* Pleist. Mad. *Indri* [*Indris*] *Lemur Lepilemur Lichanotus* [*Avahi*] Pleist.-R. Mad. *Megaladapis Megalindris Mesopropithecus Neopropithecus Palaeopropithecus* [*Bradytherium*] *Peloriadapis Prohapalemur* Pleist. Mad. *Propithecus* Pleist.-R. Mad. **Daubentoniidae** (**Cheiromyidae**) R. Mad., *Daubentonia* [*Cheiromys*] Pleist.-R. Mad. **Lorisidae** R. As. Af., *Galago* [*Galagoides*] L. Mioc.-R. Af. *Indraloris* Plioc. SAs. *Progalago* L. Mioc. EAf.

SUBORDER TARSIOIDEI. **Anaptomorphidae**, *Absarokius* L. Eoc. NA. *Anaptomorphus* ?L. Eoc., M. Eoc. NA. *Anemorhysis* L.-M. Eoc. NA. *Berruvius* U. Paleoc. Eu. *Tetonius* [*Paratetonius*] *Tetonoides* L. Eoc. NA. *Trogolemur* M. Eoc. NA. *Uintalacus* L. Eoc. NA. *Uintanius* M. Eoc. NA. *Uintasorex* L.-M. Eoc., ?U. Eoc. NA. **Omomyidae**, *Cantius* L. Eoc. Eu. *Chlororhysis* L. Eoc. NA. *Chumashius Dyseolemur* U. Eoc. NA. *Ekgmowechashala* L. Mioc. NA. *Hemiacodon* M. Eoc. NA. *Hoanghonius* ?U. Eoc. EAs. *Loveina* L. Eoc. NA. *Lushius* M. Eoc. EAs. *Macrotarsius* U. Eoc.-L. Olig. NA. *Navajovius* U. Paleoc. NA. *Niptomomys* L. Eoc. NA. *Omomys* [*Euryacodon Palaeacodon*] ?L., M. Eoc. NA. *Ourayia* U. Eoc. NA. *Periconodon* M. Eoc. Eu. *Rooneyia* L. Olig. NA. *Shoshonius* L. Eoc. NA. *Stockia* U. Eoc. NA. *Teilhardina* L. Eoc. Eu. *Utahia* M. Eoc. NA. *Washakius* [*Yumanius*] M.-U. Eoc. NA. **Tarsiidae** R. EInd., *Microchoerus* U. Eoc. Eu. *Nannopithex* M. Eoc. Eu. *Necrolemur* M.-U. Eoc. Eu. *Pseudoloris* U. Eoc. Eu. ?**Microsyopsidae**, *Alsaticopithecus* M. Eoc. Eu. *Craseops* U. Eoc. NA. *Cynodontomys* L. Eoc. NA. *Microsyops* M. Eoc. NA.

SUBORDER PLATYRRHINI. **Callithricidae** (**Hapalidae**) R. SA., *Callithrix* [*Hapale*] Pleist.-R. SA. *Dolichocebus* U. Olig. SA. **Cebidae**, R. SA. CA., *Alouatta* [*Mycetes*] Pleist.-R. SA., R. CA. *Brachyteles* [*Eriodes*] Pleist.-R. SA. *Callicebus* Pleist.-R. SA. *Cebupithecia* U. Mioc. SA. *Cebus* Pleist.-R. SA., R. CA. *Homunculus* ?U. Olig., L. Mioc. SA. *Neosaimiri* ? *Pitheculus* U. Mioc. SA. ? *Xenothrix* Pleist. WInd.

SUBORDER CATARRHINI

SUPERFAMILY PARAPITHECOIDEA. **Parapithecidae**, *Apidium Parapithecus* L. Olig. NAf.

SUPERFAMILY CERCOPITHECOIDEA. **Cercopithecidae** R. As. Af., *Cercocebus* Pleist.-R. Af. *Cercopithecoides* Pleist. SAf. *Cercopithecus* U. Plioc. SAs., Pleist.-R. Af. *Dolichopithecus* Plioc.-L. Pleist. Eu. *Libypithecus* U. Plioc. NA. *Macaca* [*Aulacinus Macacus Rhesus*] U. Mioc.-R. NAf., U. Plioc.-L. Pleist. Eu., U. Plioc.-R. As., Pleist.-R. EInd. *Mesopithecus* ?L. Mioc. EAf., L. Plioc. Eu. WAs. *Papio* [*Brachygnathopithecus Choeropithecus Cynocephalus Dinopithecus Gorgopithecus Parapapio Simopithecus*] Plioc.-Pleist. SAs. Af. ? *Prohylobates* [*Victoriapithecus*] L. Mioc. EAf. *Presbytis* [*Paradolichopithecus Semnopithecus*] Pleist.-R. SAs. EInd., Pleist. Eu. Af. *Procynocephalus* Pleist. As. *Pygathrix* U. Plioc.-R. As. *Rhinopithecus* Pleist.-R. As. *Szechuanopithecus* Pleist. EAs. *Trachypithecus* Pleist.-R. EInd.

SUPERFAMILY HOMINOIDEA. **Oreopithecidae**, *Oreopithecus* U. Mioc. EAf., L. Plioc. Eu. **Pongidae** (**Simiidae**), *Aegyptopithecus Aeolopithecus* L. Olig. NAf. ? *Amphipithecus* U. Eoc. SAs. *Dryopithecus* [*Adaetontherium Ankarapithecus Anthropodus Griphopithecus* ? *Hylopithecus Indopithecus Neopithecus Paidopithex Paleopithecus Paleosimia Proconsul Rhenopithecus Sivapithecus Sugrivapithecus Udabnopithecus Xenopithecus*] L.

Mioc. Af., M. Mioc.-L. Plioc. Eu., U. Mioc.-Plioc. As. *Gigantopithecus* [*Gigantanthropus*] L. Pleist. EAs. *Hylobates* Pleist.-R. EInd., R. SEAs. *Moeripithecus Oligopithecus* L. Olig. NAf. *Pan* [*Gorilla Proconsuloides*] Pleist.-R. Af. *Piopithecus* [*Epipliopithecus Limnopithecus Plesiopliopithecus*] L. Mioc. Af., M. Mioc.-L. Plioc. Eu., ?Pleist. EAs. *?Pondaungia* U. Eoc. SAs. *Pongo* [*Simia*] L. Pleist. SAs., Pleist.-R. EInd. *Propliopithecus* L. Olig. NAf. **Hominidae,** *Australopithecus* [*Hemanthropus ?Meganthropus Paranthropus Plesianthropus Zinjanthropus*] L.-M. Pleist. SAf., ?L. Pleist. EAs., ?M. Pleist. EInd. *Homo* [*Atlanthropus Cyphanthropus Eoanthropus* (*partim?*) *Nipponanthropus Palaeanthropus Pithecanthropus Protanthropus Sinanthropus Telanthropus*] Pleist.-R. Eu. As. Af. Aus. NA., R. SA. SAf. *?Ramapithecus* [*Bramapithecus ?Kenyapithecus*] U. Mioc.-L. Plioc. SAs., L. Plioc. EAf. Eu.

?PRIMATES, INCERTAE SEDIS: *Ceciliolemur* M. Eoc. Eu. *?Menatotherium* Eoc. Eu.

ORDER CREODONTA

SUBORDER DELTATHERIDIA. **Deltatherididae (Palaeoryctidae),** *Aboletylestes* U. Paleoc. Eu. *Acmeodon Avunculus* M. Paleoc. NA. *?Cimolestes* [*Nyssodon Puercolestes*] U. Cret.-L. Paleoc. NA. *Deltatheridium Deltatheridoides* U. Cret. EAs. *Didelphodus* [*Didelphyodus Phenacops*] L.-M. Eoc. NA. *Gelastops* [*Emperodon*] M. Paleoc. NA. *Hyotheridium* U. Cret. EAs. *Palaeoryctes* M.-U. Paleoc. NA. *Pararyctes* U. Paleoc. NA. *Sarcodon* U. Cret. EAs. ?Didymoconidae (Tshelkariidae), *Ardynictis* L. Olig. EAs. *Didymoconus* [*Tshelkaria*] U. Olig. EAs. *Mongoloryctes* U. Eoc. EAs. **Micropternodontidae,** *Micropternodus* [*Kentrogomphius*] M. Eoc.-L. Olig. NA.

SUBORDER HYAENODONTIA. **Hyaenodontidae,** *Apataelurus* U. Eoc. NA. *Arfia* L. Eoc. NA. *Cynohyaenodon* [*?Pseudosinopa*] M.-U. Eoc. Eu. *Dissopsalis* L. Plioc. SAs. *Galethylax* U. Eoc. Eu. *Hemipsalodon* L. Olig. NA. *Hyaenodon* [*Neohyaenodon Pseudopterodon Taxotherium*] U. Eoc.-M. Olig. Eu. NA., L. Olig. NAf., L.-U. Olig. As. *Imperatoria* [*Prodissopsalis*] M. Eoc. Eu. *Ischnognathus* L. Olig. NA. *Limnocyon* [*Telmatocyon*] M.-U. Eoc. NA. *Machaeroides* M. Eoc. NA. *Metapterodon* L. Mioc. Af. *Metasinopa* U. Eoc. Eu., L. Olig. NAf. *Oxyaenodon* U. Eoc. NA. *Paracynohyaenodon* U. Eoc. Eu. EAs. *Propterodon* M. Eoc. Eu., U. Eoc. As., ?L. Olig. NA. *Protoproviverra* L. Eoc. Eu. *Prototomus* [*Prolimnocyon*] L. Eoc. NA., L.-M. Eoc. Eu. *Proviverra* [*Geiselotherium Leonhardtina Prorhyzaena*] ?L., M. Eoc. Eu. *Pterodon* U. Eoc. As. NA., U. Eoc.-L. Olig. Eu., L. Olig. NAf. *Quercytherium* U. Eoc. Eu. *Sinopa* [*Stypolophus ?Triacodon*] L.-M. Eoc. Eu. NA. *Thereutherium* U. Eoc. Eu. *Thinocyon* [*Entomodon*] M. Eoc. NA. *Tritemnodon* L.-M. Eoc. NA. *?Tylodon* Olig. Eu. **Oxyaenidae,** *Ambloctonus* [*Amblyctonus*] L. Eoc. NA. *Argillotherium* L. Eoc. Eu. *Dipsalidictides* L. Eoc. NA. *Dipsalodon* U. Paleoc. NA. *Oxyaena* [*Dipsalidictis*] U. Paleoc.-L. Eoc. NA., L. Eoc. EAs. *Palaeonictis* L. Eoc. Eu. NA. *Paroxyaena* U. Eoc. Eu. *Patriofelis* [*Aelurotherium Limnofelis Oreocyon*] M. Eoc. NA. *Protopsalis* L. Eoc. NA. *Sarkastodon* U. Eoc. EAs.

ORDER CARNIVORA

SUBORDER FISSIPEDIA

INFRAORDER MIACOIDEA. **Miacidae,** *Didymictis* M. Paleoc.-L. Eoc. NA. *Ictidopappus* M. Paleoc. NA. *Miacis* [*Mimocyon*] L.-U. Eoc. NA., U. Eoc. Eu. As. *Oödectes* M. Eoc., ?U. Eoc. NA. *Palaearctonyx* M. Eoc. NA. *Petersonella* [*Pleurocyon*] *Plesiomiacis* U. Eoc. NA. *Simpsonictis* M., ?U. Paleoc. NA. *Tapocyon* U. Eoc. NA. *Uintacyon* L.-U. Eoc. NA. *Vassocyon* L. Eoc. NA. *Viverravus* U. Paleoc.-U. Eoc. NA., U. Eoc. Eu., ?L. Olig. As. *Vulpavus* L.-M. Eoc. NA.

INFRAORDER AELUROIDEA. **Viverridae** R. Eu. As. Af., *?Amphicticeps* L.-U. Olig. EAs. *Atilax Crossarchus* Pleist.-R. Af. *Cryptoprocta* Pleist.-R. Mad. *Cynictis* Pleist.-R. Af. *Fossa* Pleist.-R. Mad. *Genetta* Pleist.-R. Af., R. WAs. SEu. *Haplogale* Eoc. or Olig. Eu. *Herpestes* [*Calogale*] M. Olig.-R. Eu., L. Mioc.-R. As. Af. *Jourdanictis* M. Mioc. Eu. *Leecyaena* L. Pleist. EAs., Pleist. SAf. *Leptoplesictis* M. Mioc. Eu. *Macrogalidia* Pleist.-R. EInd. *Mungos* Pleist.-R. Af. *Paguma* Pleist.-R. As. *Palaeoprionodon* U. Eoc. Eu., L. Olig. As. *Paradoxurus* Pleist.-R. EInd., R. As. *Semigenetta* L.-M. Mioc. Eu. *Stenoplesictis* M. Olig. Eu. *Suricata* Pleist.-R. Af. *Vishnuictis* M. Plioc.-L. Pleist. As. *Viverra* [*Anictis*] M. Mioc.-R. Eu., L. Plioc.-R. As., Pleist.-R. EInd. Af. **Hyaenidae,** *Chasmaporthetes* [*Ailurena*] L. Pleist. NA. *Crocuta* [*Crocotta Eucrocuta Percrocuta*] L. Plioc.-Pleist. Eu., U. Mioc.-Pleist. As., Pleist. EInd., Pleist.-R. Af. *Hyaena* [*? Adcrocuta Pachycrocuta Plesiocrocuta*] L. Plioc.-Pleist. Eu., L. Plioc.-R. As., ?Plioc., Pleist.-R. Af. *Hyaenictis* M. Mioc.-L. Plioc. Eu., ?U. Plioc. As., Pleist. As. *Ictitherium* [*Galeotherium Hyaenalopex Palhyaena Thalassictis*] M. Mioc.-L. Plioc. Eu. As., Plioc. NAf. *Lycyaena* U. Mioc.-U. Plioc. Eu., L.-U. Plioc. As. *Lycyaenops* [*Euryboas*] L. Pleist. Eu. Af. *Progenetta* [*Miohyaena*] Mioc.-Plioc. Eu. *Tunguricitis* U. Mioc. EAs. **Felidae,** *Acinonyx* [*Brachyprosopus Cynaelurus Schaubia*] L.-U. Pleist. Eu. As., Pleist.-R. Af. *? Aeluropsis* U. Plioc. As. *Aelurogale* [*Ailurictis Ictidailurus Nimraviscus*] U. Eoc. EAs., U. Eoc.-M. Olig. Eu., ?L. Olig. Af. *Archaelurus Dinaelurus* L. Mioc. NA. *Dinailurictis* M. Olig. Eu. *Dinictis* L. Olig.-L. Mioc. NA. *Dinobastis* Pleist. Eu. As. NA. *Dinofelis* L. Plioc. As. *Ekgmoiteptecela* L. Mioc. NA. *Eofelis* U. Eoc.-M. Olig. Eu. *?Eosictis* U. Eoc. NA. *Eusmilus* [*Paraeusmilus*] U. Eoc. EAs., U. Eoc.-U. Olig. Eu., M.-U. Olig. NA. *Felis* [*Leptailurus Lynx Neofelis Panthera Pristinofelis* etc., etc.] L. Plioc.-R. Eu. As., Plioc., Pleist.-R. NA., Pleist.-R. EInd. Af. SA. *Homotherium* [*Epimachairodus*] Plioc.-Pleist. As., U. Plioc. EInd., L. Pleist. Eu., Pleist. Af. *Hoplophoneus* L. Olig.-L. Mioc. NA. *?Hyainailouros* [*Hyaenaelurus*] L.-M. Mioc. Eu., U. Mioc. As. Af. *Ischyrosmilus* L. Plioc.-L. Pleist. NA. *Machairodus* [*Heterofelis Therailurus*] U. Mioc.-L. Plioc. Eu., L. Plioc. NA. As. Af. *Meganterion* [*Toscanius*] U. Plioc.-U. Pleist. As., L. Pleist. Eu., U. Pleist. Af. EInd. *Mellivorodon* U. Plioc. As. *Nimravides* M. Plioc. NA. *Nimravus* [*Nimravinus*] U. Olig.-L. Mioc. NA. *Paramachaerodus* [*Propontosmilus Sivasmilus*] L. Plioc. Eu., L.-U. Plioc. As. *Pogonodon* L. Mioc. NA. *Proailurus* [*Brachictis Stenogale*] L. Olig.-L. Mioc. Eu. *Pseudaelurus* [*Ailuromachairodus Metailurus Sansanailurus Schizailurus*] L. Mioc. Af., L. Mioc.-

L. Plioc. Eu., M. Mioc.-U. Plioc. NA., U. Mioc.-L. Plioc. As. *Sansanosmilus* [*Albanosmilus Grivasmilus*] M. Mioc. Eu., L. Plioc. As. *Sivaelurus* L. Plioc. As. *Sivapanthera* [*Sivafelis*] Pleist. As. *Smilodon* [*Smilodontidion Smilodontopsis Trucifelis*] U. Plioc.-Pleist. SA., U. Pleist. NA. *Vinayakia* L.-U. Plioc. As. *Vishnufelis* L. Plioc. As.

INFRAORDER ARCTOIDEA. **Mustelidae** R. Eu. As. Af. NA. SA., *Aelurocyon* L. Mioc. NA. *Aonyx* Pleist.-R. Af. *Amphictis* Eoc.-Olig. Eu. *Arctomeles* U. Plioc. Eu. *Arctonyx* Pleist.-R. As. *Baranogale* U. Plioc.-L. Pleist. Eu. *Brachyopsigale* U. Plioc. NA. *Brachyprotoma* L.-U. Pleist. NA. *Brachypsalis* U. Mioc.-U. Plioc. NA. *Broiliana* L. Mioc. Eu. *Buisnictis* U. Plioc. NA. *Canimartes* L. Pleist. NA. *Cernictis* U. Plioc. NA. *Charronia* Pleist.-R. As. *Conepatus* U. Plioc.-R. SA., Pleist.-R. NA. *Craterogale* U. Mioc. NA. *Cyrnaonyx* Pleist. Eu. *Dinogale* U. Mioc.-L. Plioc. NA. *Drassonax* M. Olig. NA. *Enhydra* L. Pleist. Eu., Pleist. WNA., R. NPac. *Enhydrictis* [*Pannonictis*] ?U. Mioc., Pleist. EAs., L. Plioc.-Pleist. Eu. *Enhydriodon* L. Plioc. Eu., M. Plioc. Af., L. Pleist. As. *Eomellivora* [*Sivamellivora*] L. Plioc. Eu. As. NA. *Galera* Pleist.-R. SA., R. CA. *Grison* [*Galictis*] Pleist. NA., Pleist.-R. SA., R. CA. *Grisonella* Pleist.-R. SA. *Gulo* Pleist.-R. Eu. NA. As. *Hadrictis* L. Plioc. Eu. *Hypsoparia* U. Mioc. NA. *Laphyctis* [*Ischyrictis*] L.-U. Mioc. Eu. *Leptarctus* [*Mephititaxus*] L. Mioc.-L. Plioc. NA., U. Mioc. EAs. *Limnonyx* M. Mioc. Eu. *Lutra* [*Basarabictis Plesiolatax*] L. Plioc.-R. Eu. NA., Pleist.-R. As. SA. NAf. *Lutravus* L. Plioc.-L. Pleist. NA. *Lyncodon* Pleist.-R. SA. *Martes* ["*Hydrocyon*" *?Paramartes Sansanictis*] M. Mioc.-R. Eu., L. Plioc.-R. As., L. Pleist.-R. NA. *Martinogale* L. Plioc. NA. *Megalictis* L. Mioc. NA. *Meles* [*Heterictis Iranictis*] L. Plioc.-R. As., L. Pleist.-R. Eu. *Melidellavus* M. Mioc. Eu. *Mellivora* [*Ursitaxus*] U. Plioc.-Pleist. As., Pleist.-R. Af. *Melodon* U. Mioc.-L. Plioc. As. *Mephitis* Pleist.-R. NA. *Miomephitis* L. Mioc. Eu. *Miomustela* U. Mioc. NA. *Mionictis* M. Mioc. Eu., M.-U. Mioc. NA. *Mustela* [*Paratanuki Putorius*] U. Mioc.-R. Eu. NA., L. Plioc.-R. As., Pleist.-R. SA., R. NAf. *Nesolutra* Pleist. Eu. *Oligobunis* L. Mioc. NA. *Osmotherium* [*Pelycictis*] Pleist. NA. *Palaeogale* [*Bunaelurus*] L. Olig.-L. Mioc. Eu. NA., M. Olig. As. *Palaeomeles Paralutra* M. Mioc. Eu. *Parameles* Pleist. EAs. *Parataxidea* L. Plioc. Eu. As. *Parenhydriodon* ?Plioc. Eu. *Paroligobunis* L. Mioc. NA. *Plesiogale* U. Olig.-L. Mioc. Eu. *Plesiogulo* [*Perunium*] L. Mioc.-U. Plioc. Eu., L. Plioc.-Pleist. As., Plioc. NA. *Plesiomeles* M. Mioc. Eu. *Pliogale* L.-U. Plioc. NA. *Plionictis* M.-U. Mioc. NA. *Pliotaxidea* L.-U. Plioc. NA. *Potamotherium* U. Olig.-L. Mioc. Eu., ?L. Plioc. NA. *Promartes* L. Mioc. NA. *Promeles* [*?Polgardia*] L. Plioc. Eu. *Promellivora* L. Pleist. As. *Promephitis* [*?Nannomephitis*] ?Mioc., L. Plioc.-L. Pleist. Eu., L.-U. Plioc. As. *Proputorius* Mioc., ?L. Pleist. Eu., L. Plioc. As. *Pseudictis* U. Mioc. Eu. *Rhabdogale* Pleist. Eu. *Sabadellictis* L. Plioc. Eu. *Semantor* L. Plioc. WAs. *Sinictis* L. Plioc.-Pleist. As., L. Plioc. Eu. *Sivalictis* L. Plioc. Eu. As. *Sivaonyx* L. Plioc. Eu. NAf., U. Plioc. As. *Spilogale* L. Pleist.-R. NA. *Sthenictis* M. Mioc.-L. Plioc. NA. *Stipanicicia* Pleist. SA. *Stromeriella* L. Mioc. Eu. *Taxidea* U. Plioc.-R. NA. *Taxodon* M. Mioc. Eu. *Trigonictis* L. Pleist. NA. *Trocharion* M. Mioc. Eu. *Trochictis* M. Mioc.-L. Plioc. Eu. *Trochotherium* M. Mioc. Eu. *Vishnuonyx* L. Plioc. As. *Vormela* U. Plioc.-R. As., Pleist.-R. Eu. **Canidae**, *Absonodaphoenus* L. Mioc. NA. *Actiocyon* Mioc. NA. *Aelurodon* M. Mioc.-L. Plioc. NA. *Afrocyon* Mioc. NAf. *Agnotherium* [*Tomocyon*] M. Mioc.-L. Plioc. Eu. *Aletocyon* L. Mioc. NA. *Alopecocyon* [*Alopecodon ?Galecynus Viretius*] M. Mioc. Eu. *Alopex* [*?Xenalopex*] Pleist.-R. Eu. As. *Amphicynodon* [*Cynodon Paracynodon Plesiocyon*] U. Eoc.-L. Olig. Eu. *Amphicyon* [*Amphicyonops Ictiocyon*] M. Olig.-L. Plioc. Eu., L.-U. Mioc. NA., ?L. Mioc. Af., U. Mioc.-L. Plioc. As. *Amphicyonopsis* M. Mioc. Eu. *Arctamphicynodon* U. Plioc. As. *Borocyon* L. Mioc. NA. *Borophagus* [*Hyaenognathus*] U. Plioc.-Pleist. NA. *Brachyrhynchocyon* [*Brachycion*] M. Olig. NA. *Campylocynodon* L. Olig. NA. *Canis* [*Aenocyon Dinocynops Theriodictis Thos*] L. Plioc.-R. Eu. As. Af. Aus., Pleist. SA., L. Pleist.-R. NA. *Chrysocyon* Pleist.-R. SA. *Cuon* [*Crassicuon Cyon Semicuon Sinocuon Xenocyon*] L. Pleist. Eu., Pleist.-R. As., R. EInd. *Cynarctoides* L. Mioc. NA. *Cynarctus* M. Mioc.-L. Plioc. NA. *Cynodesmus* L.-M. Mioc. NA. *Cynodictis* Eoc. EAs., U. Eoc.-L. Olig. Eu. *Daphoenocyon* L. Olig. NA. *Daphoenodon* L. Mioc. NA. *Daphoenus* L. Olig.-L. Mioc. NA. *Dusicyon* Pleist.-R. SA. *Enhydrocyon* L. Mioc. NA. *Euoplocyon* U. Mioc. NA. *Gobicyon* U. Mioc. As. Eu. *Hadrocyon* L. Plioc. NA. *Haplocyon* U. Olig.-L. Mioc. Eu. *Haplocyonoides* L. Mioc.-L. Plioc. Eu. *Hesperocyon* [*Pseudocynodictis*] L. Olig.-L. Mioc. NA. *Ischyrocyon Leptocyon* U. Mioc.-L. Plioc. NA. *Lycaon* Pleist.-R. Af. *Mammaocyon* L. Mioc. NA. *Mesocyon* U. Olig.-L. Mioc. NA. *Neocynodesmus* L. Mioc. NA. *Nothocyon* L. Olig.-L. Mioc. NA. *Nyctereutes* L. Plioc.-R. As., L. Pleist. Eu. *Osteoborus* U. Mioc.-U. Plioc. NA. *Otocyon* [*Prototocyon*] Pleist.-R. Af. *Oxetocyon* U. Olig. NA. *Pachycynodon* Olig. Eu., L. Olig. As. *Paradaphaenus* L. Mioc. NA. *Parictis* L. Olig.-L. Mioc. NA. *Pericyon Philotrox* L. Mioc. NA. *Phlaocyon* L.-M. Mioc. NA. *Pliocyon* U. Mioc.-L. Plioc. NA. *Pliogulo* L. Plioc. NA. *Proamphicyon* L.-M. Olig. NA. *Procynodictis* U. Eoc. NA. *Protemnocyon* M. Olig. NA. *Protocyon* [*Palaeocyon Palaeospeothos*] Pleist. SA. *Pseudamphicyon* U. Eoc.-L. Olig. Eu. *Pseudarctos* M.-U. Mioc. Eu. *Pseudocyon* U. Olig.-M. Mioc. Eu. *Simamphicyon* U. Eoc. Eu. *Simocyon* [*Araeocyon Metarctos*] L. Plioc. Eu. NA. *Sivacyon* Pleist. As. *Speothos* [*Icticyon*] Pleist.-R. SA. *Sunkahetanka* L. Mioc. NA. *Temnocyon* U. Olig.-L. Mioc. NA. *Tephrocyon* M. Mioc.-L. Plioc. NA. *Thaumastocyon* M. Mioc. Eu. *Tomarctus* L. Mioc.-L. Plioc. NA. *Urocyon* Pleist.-R. NA., R. SA. *Vishnucyon* L. Plioc. As. *Vulpes* ?U. Mioc., L. Plioc.-R. NA., U. Plioc.-R. Eu. Af., Plioc.-R. As. **Procyonidae** R. As. NA. SA., *Ailurus* ?Pleist. Eu., R. As. *Allocyon Bassariscops* L. Mioc. NA. *Bassariscus* U. Mioc.-R. NA. *Brachynasua* Pleist. SA. *Cyonasua* [*Amphinasua ?Chapalmalania Pacynasua*] U. Mioc.-U. Plioc. SA. *Edaphocyon ?Kolponomos* M. Mioc. NA. *Nasua* Pleist.-R. SA., R. CA. *Parailurus* L.-U. Plioc. Eu. *Plesictis* [*Mustelavus*] ?U. Eoc., L. Olig.-L. Mioc. Eu., L. Olig. NA. *Procyon* L. Pleist.-R. NA., Pleist.-R. SA. *Sivanasua* [*Ailuravus Schlossericyon*] L.-U. Mioc. Eu., L. Plioc. As. *Zodiolestes* L. Mioc. NA. **Ursidae** R. Eu. As. NA. SA., *Agriotherium* [*Agriarctos Hyaenarctos Lydekkerion*] ?U. Mioc., L. Plioc.-Pleist. Eu., L.-U. Plioc. NA., U. Plioc.-Pleist. As. *Ailuropoda*

[*Aelureidopus Ailuropus*] L. Pleist.-R. As., Pleist. EInd. *Arctodus* [*Arctoidotherium Arctotherium Dinarctotherium Pararctotherium Proarctotherium Pseudarctotherium Tremarctotherium*] ?U. Plioc., Pleist. SA., Pleist. NA. *Cephalogale* M. Olig.-L. Plioc. Eu., L. Mioc. As. *Dinocyon* U. Mioc.-L. Plioc. Eu., U. Mioc. As. *Helarctos* U. Plioc. Eu., Pleist.-R. As. *Hemicyon* [*Harpaleocyon Phoberocyon Plithocyon*] L.-U. Mioc. Eu., U. Mioc. As., U. Mioc.-L. Plioc. NA. *Indarctos* L. Plioc. Eu., L.-U. Plioc. As. NA. *Melursus* Pleist.-R. As. *Plionarctos* L. Plioc. NA. *Tremarctos* Pleist. CA., R. SA. *Ursavus* L. Mioc.-L. Plioc. Eu., ?M. Mioc. NA. *Ursavus* M. Mioc.-L. Plioc. Eu. *Ursus* [*Euarctos Selenarctos Thalarctos Ursulus*] L. Plioc.-R. Eu., Pleist. NAf., Pleist.-R. As. NA.

SUBORDER PINNIPEDIA. **Otariidae** R. Pac., *Allodesmus* L. Mioc. WNA, Plioc. EAs. *Arctocephalus* Plioc. SA., Pleist. Aus. SAf. NZ., R. SOc. *Atopotarus* Mioc. Eu., Mioc.-Pleist. EAs., U. Mioc. NA. *Desmatophoca* U. Mioc. WNA. *Dusignathus* U. Plioc. WNA. *Eumetopias* Plioc.-Pleist. WNA. ESA., R. NPac. *Neotherium* L. Mioc. WNA. *Otaria* Pleist. NA. SA., R. SPac. SAtl. *Pithanotaria* U. Mioc.-Plioc. WNA. *Pliopedia* U. Plioc. Pac. NA. *Pontolis* [*Pontoleo Pontoleon*] U. Mioc.-L. Pleist. WNA. *Zalophus* ?U. Plioc.-Pleist. WNA., Pleist. Aus. NZ. EAs., R. Pac. **Odobenidae** R. Arc. NAtl., *Alachtherium* U. Plioc. Eu. *Odobenus* [*Rosmarus Trichechus*] Pleist. ENA. Eu. EAs., R. Arc. *Prorosmarus* U. Mioc. ENA. *Trichecodon* Pleist. Eu. ENA. *Valenictis* L. Plioc. Pac. NA. **Phocidae,** *Callophoca* U. Plioc. Eu. *Cystophora* Pleist.-R. NA. *Gryphoca* U. Plioc. Eu. *Leptophoca* M. Mioc. ENA. *Mesotaria* U. Plioc. Eu. *Monachus* [*Pliophoca*] U. Plioc. Eu., R. Oc. *Monotherium* U. Mioc. Eu. *Palaeophoca* U. Plioc. Eu. *Phoca* [*Monachopsis ?Pontophoca ?Praepusa*] M. Mioc.-R. Eu. NA., Plioc.-R. NAs., R. NAtl. NPac. *Phocanella Platyphoca* U. Plioc. Eu. *Pristiphoca* [*Miophoca*] M. Mioc.-L. Pleist. Eu., L. Pleist. NAf. *Prophoca* U. Mioc. Eu.

PINNIPEDIA INCERTAE SEDIS: *Necromites* Mioc. WAs.

ORDER CONDYLARTHRA. **Arctocyonidae (Oxyclaenidae),** *Anacodon* ?U. Paleoc., L. Eoc. NA. *Arctocyon* [*Arctotherium Heteroborus Hyodectes*] *Arctocyonides* [*Creodapis Procynictis*] U. Paleoc. Eu. *Baioconodon Carcinodon* L. Paleoc. NA. *Chriacus* [*Lipodectes*] L. Paleoc.-L. Eoc. NA. *Claenodon* [*Neoclaenodon*] M.-U. Paleoc. NA. *Colpoclaenus* U. Paleoc. NA. *Deltatherium Deuterogonodon* M. Paleoc. NA. *Elpidophorus* M.-U. Paleoc. NA. *Eoconodon* L. Paleoc. NA. *Goniacodon* M. Paleoc. NA. *Loxolophus* L. Paleoc. NA. *Mentoclaenodon* U. Paleoc. NA. *Metachriacus Mimotricentes* M. Paleoc. NA. *Oxyclaenus* L.-M. Paleoc. NA. *Paradoxodonta* [*Paradoxodon*] Paleoc. NA. *Paratriisodon* U. Eoc. EAs. *Prothryptacodon* M. Paleoc. NA. *Protogonodon* L. Paleoc. NA. *Protungulatum* U. Cret.-L. Paleoc. NA. *Spanoxyodon* M. Paleoc. NA. *Thryptacodon* U. Paleoc.-L. Eoc. NA. *Tricentes* M.-U. Paleoc. NA. *Triisodon* M. Paleoc. NA. **Mesonychidae,** *Andrewsarchus* U. Eoc. EAs. *? Apterodon* L. Olig., ?L. Mioc. Af., ?L. Olig. Eu. As. *Dissacus* [*Hyaenodictis*] M. Paleoc.-L. Eoc. NA., U. Paleoc.-M. Eoc. Eu. *Gandakasia* M. Eoc. SAs. *Hapalodectes* L. Eoc. NA., ?U. Eoc. As. *Harpagolestes* M.-U. Eoc. NA., U. Eoc. EAs. *Hessolestes* U. Eoc. NA. *Ichthyolestes* M. Eoc. SAs. *Mesonyx* M.-U. Eoc. NA., U. Eoc.-Olig. EAs. *Microclaenodon* M. Paleoc. NA. *?Olsenia* U. Eoc. EAs. *Pachyaena* L. Eoc. Eu. NA. *Synoplotherium* [*Dromocyon*] M. Eoc. NA. **Hyopsodontidae,** *? Adapisorex Apheliscus Asmithwoodwardia* U. Paleoc.-L. Eoc. SA. *Choeroclaenus* L. Paleoc. NA. *Dracoclaenus* M. Paleoc. NA. *Epapheliscus* L. Olig. Eu. *Haplaletes* M.-U. Paleoc. NA. *Haplomylus* U. Paleoc.-L. Eoc. NA. *Hyopsodus* L.-U. Eoc. NA. *Jepsenia* M. Paleoc. NA. *Kopidodon* M. Eoc. Eu. *Litaletes* M. Paleoc. NA. *Litolestes* U. Paleoc. NA. *Litomylus* M.-U. Paleoc. NA. *Louisina* U. Paleoc. Eu. *Mioclaenus* M. Paleoc. NA. *Oxyacodon* L. Paleoc. NA. *Oxytomodon* M. Paleoc. NA. *Paratricuspiodon Promioclaenus* [*Ellipsodon*] ?L., M.-U. Paleoc. NA., M. Eoc. SAs. *Protoselene* M., ?U. Paleoc. NA. *Tiznatzinia* L. Paleoc. NA. *Tricuspiodon* [*Conaspidotherium Plesiphenacodus*] U. Paleoc. Eu. **Meniscotheriidae,** *Meniscotherium* [*Hyracops*] U. Paleoc.-L. Eoc. NA. *Orthaspidotherium Pleuraspidotherium* U. Paleoc. Eu. **Periptychidae,** *Anisonchus* L.-U. Paleoc. NA. *Carsioptychus* [*Plagioptychus*] L.-M. Paleoc. NA. *Conacodon* L. Paleoc. NA. *Ectoconus* L. Paleoc. NA. *Haploconus* M. Paleoc. NA. *Hemithlaeus* L. Paleoc. NA. *Periptychus* M.-U. Paleoc. NA. **Phenacodontidae,** *Almogaver* M. Eoc. Eu. *Desmatoclaenus* L.-M. Paleoc. NA. *Ectocion* [*Gidleyina*] M. Paleoc.-L. Eoc. NA. *Phenacodus* U. Paleoc.-L. Eoc. NA., L.-M. Eoc. Eu. *Tetraclaenodon* [*Euprotogonia*] M.-U. Paleoc. NA. **Didolodontidae,** *Archaeohyracotherium Argyrolambda* L. Eoc. SA. *Didolodus* [*Cephanodus Lonchoconus Nephacodus*] L. Eoc., ?M. Eoc. SA. *Enneoconus* L. Eoc. SA. *Ernestokokenia* [*Notoprogonia Progonia*] U. Paleoc.-L. Eoc. SA. *Lamegoia* U. Paleoc. SA. *Lophiodolodus* L. Olig. SA. *Megadolodus* U. Mioc. SA. *Paulogervaisia Proectocion* L. Eoc. SA.

ORDER AMBLYPODA

SUBORDER PANTODONTA. **Pantolambdidae,** *Caenolambda Pantolambda* M.-U. Paleoc. NA. **Barylambdidae,** *Barylambda* U. Paleoc. NA. *Haplolambda* [*Archaeolambda*] U. Paleoc. NA. EAs. *Ignatiolambda Leptolambda* U. Paleoc. NA. **Titanoideidae,** *Titanoides* [*Sparactolambda*] M.-U. Paleoc. NA. **Coryphodontidae,** *Coryphodon* [*Letalophodon Loxolophodon*] U. Paleoc.-L. Eoc. NA., L. Eoc. Eu. EAs. *Eudinoceras* U. Eoc. EAs. *Hypercoryphodon* M. Olig. EAs. *Procoryphodon* L. Eoc. EAs. **?Pantolambdodontidae,** *Pantolambdodon* U. Eoc. As.

SUBORDER DINOCERATA. **Prodinocerotidae,** *Bathyopsoides* U. Paleoc. NA. *Mongolotherium* L. Eoc. EAs. *Probathyopsis* [*Prouintatherium*] U. Paleoc.-L. Eoc. NA., ?L. Eoc. EAs. *Prodinoceras* U. Paleoc. EAs. **Uintatheriidae,** *Bathyopsis* L.-M. Eoc. NA. *Eobasileus* [*Uintacolotherium*] U. Eoc. NA. *Tetheopsis* M.-U. Eoc. NA. *Uintatherium* [*Dinoceras Elachoceras Ditetrodon Laoceras Octotomus Paroceras Platoceras Tinoceras Uintamastix*] M. Eoc., ?U. Eoc. NA. **Gobiatheriidae,** *Gobiatherium* U. Eoc. EAs.

SUBORDER XENUNGULATA. **Carodniidae,** *Carodnia* [*Ctalecarodnia*] U. Paleoc. SA.

SUBORDER PYROTHERIA. **Pyrotheriidae,** *? Carolozittelia* L. Eoc. SA. *Griphodon* ?Olig. SA. *Propyrotherium* U. Eoc. SA. *Pyrotherium* L. Olig. SA.

ORDER PROBOSCIDEA

SUBORDER MOERITHERIOIDEA. **Moeritheriidae,** *Moeritherium* U. Eoc.-L. Olig. Af.

SUBORDER EUELEPHANTOIDEA. **Gomphotheriidae (Trilophodontidae),** *Amebelodon* U. Plioc. NA. *Anancus* [*Di-*

bunodon Pentalophodon] L. Plioc.-Pleist. Eu. As., Pleist. Af. *Cuvieronius* [*Cordillerion Teleobunomastodon*] L.-U. Pleist. NA., Pleist. SA. *Eubelodon* L. Plioc. NA. *Gnathabelodon* U. Plioc. NA. *Gomphotherium* [*Bunolophodon Choerolophodon ?Geisotodon Genomastodon ?Hemilophodon Megabelodon ?Protanancus ?Stegotetrabelodon Tatabelodon Tetrabelodon Trilophodon*] L. Mioc.-L. Plioc. Eu., L. Mioc.-L. Pleist. As., L.-U. Mioc., ?Plioc., ?Pleist. Af., U. Mioc.-L. Plioc., ?L. Pleist. NA. *Notiomastodon* Pleist. SA. *Palaeomastodon* L. Olig. NAf. *Phiomia* L. Olig., ?U. Olig. NAf. *Platybelodon* [*Torynobelodon*] U. Mioc. As., L.-U. Plioc. NA. *Rhynchotherium* [*Aybelodon Blickotherium Dibelodon*] L. Mioc. Af., U. Mioc.-L. Pleist. NA., L.-U. Plioc. As. *Serridentinus* [*Hemimastodon Ocalientinus Serbelodon Serridanancus Tetrabelodon*] L. Mioc.-L. Plioc. As., L. Mioc.-U. Plioc. Eu., M. Mioc.-U. Plioc., ?L. Pleist. NA. *Stegomastodon* [*Aleamastodon Haplomastodon Rhabdobunus*] U. Plioc.-Pleist. NA., Pleist. SA. *Synconolophus* U. Mioc.-U. Plioc. As., L.-U. Plioc. Eu. *Tetralophodon* [*Lydekkeria Morrillia*] U. Mioc.-L. Plioc. As., L.-U. Plioc. Eu., L.-U. Plioc., ?Pleist. NA. **Mastodontidae (Mammutidae)**, *Mastodon* [*Mammut Mastolophodon Miomastodon Pliomastodon Turicius Zygolophodon*] L. Mioc.-L. Pleist. Eu. As. Af., M. Mioc.-Pleist. NA. **Elephantidae,** *Elephas* [*Hypselephas Platelephas*] L. Pleist.-R. As. *Loxodonta* [*Hesperoloxodon Omoloxodon Palaeoloxodon Phanagoroloxodon Pilgrimia Sivalikia*] L.-U. Pleist. Eu. As. EInd., L. Pleist.-R. Af. *Mammuthus* [*Archidiskodon Dicyclotherium Metarchidiskodon Parelephas Stegoloxodon*] L.-U. Pleist. Eu. As. EInd. Af., Pleist. NA. *Stegodon* [*Parastegodon Platystegodon Sulcicephalus*] U. Plioc.-Pleist. As., Pleist. EInd. Af. *Stegolophodon* [*?Eostegodon*] L. Mioc.-Pleist. Af., U. Mioc.-Pleist. As., U. Plioc. Eu., L.-U. Pleist. EInd.

SUBORDER DEINOTHERIOIDEA. **Deinotheriidae,** *Deinotherium* [*Prodeinotherium*] L. Mioc.-U. Plioc. Eu., L. Mioc.-M. Pleist. Af., M. Mioc.-U. Plioc. As.

?SUBORDER BARYTHERIOIDEA. **Barytheriidae,** *Barytherium* U. Eoc. NAf.

ORDER SIRENIA. **Dugongidae (Halicoridae)** R. WPac. Ind. Oc. Red Sea, *Anomotherium* U. Olig. EAs., *Caribosiren* M. Olig. WInd. *Eotheroides* [*Archaeosiren Eosiren Eotherium*] M.-U. Eoc. NAf. *Felsinotherium* [*Cheirotherium Halysiren*] L. Plioc. NAf. NA., L.-U. Plioc. Eu. *Halianassa* [*?Dioplotherium ?Haplosiren Masrisiren Metaxytherium*] L.-U. Mioc. Eu., U. Mioc., ?L. Plioc. WNA. *Halitherium* [*Manatherium*] L. Olig.-L. Mioc. Eu., Olig. Mad., ?Mioc. NA. *Hesperosiren* M. Mioc. NA. *Indosiren* Mioc. EInd. *Lophiodolus* Olig. SA. *Miosiren* L. Plioc. Eu. *Prorastomus* Eoc. WInd. *Protosiren* M. Eoc. Eu. NAf. *Prototherium* [*Mesosiren Paraliosiren*] U. Eoc. Eu. *Rytina* [*Hydrodamatis Rhytina*] SubR. NPac. *Rytiodus* [*Rhytiodus*] U. Olig. Eu. *?Sirenavus Thalattosiren* M. Eoc. Eu. *Prohalicore* M. Mioc. Eu. **Manatidae (Trichechidae),** *Manatus* [*Trichechus*] Pleist.-R. ENA. WInd., R. WAf. *Potamosiren* L. Mioc. SA. *Ribodon* M. Plioc. SA.

ORDER DESMOSTYLIA. **Desmostylidae,** *Cornwallius* L. Mioc. WNA. EAs. *?Cryptomastodon* Pleist. EInd. *Desmostylus* [*Desmostylella Kronokotherium*] L. Mioc.-L. Plioc. WNA. EAs. *Paleoparadoxia* L. Mioc.-L. Plioc. EAs., M. Mioc.-L. Plioc. WNA. *Vanderhoofius* M. Mioc. WNA.

ORDER HYRACOIDEA. **Geniohyidae,** *Bunohyrax Geniohyus Megalohyrax* [*Mixohyrax*] *Titanohyrax* L. Olig. NAf. **Hyracidae (Procaviidae),** R. Af. SWAs., *Hyrax* [*Procavia Prohyrax*] L. Mioc.-R. Af., R. SWAs. *Meroehyrax* L. Mioc. EAf. *Pachyhyrax* L. Olig. NAf. L. Mioc. EAf. *Saghatherium* L. Olig. NAf.

ORDER EMBRITHOPODA. **Arsinoitheriidae,** *Arsinoitherium* L. Olig. NAf.

ORDER NOTOUNGULATA

SUBORDER NOTIOPROGONIA. **Arctostylopidae,** *Arctostylops* L. Eoc. NA. *Palaeostylops* U. Paleoc. EAs. **Henricosborniidae,** *Henricosbornia* [*Hemistylops Microstylops Monolophodon Pantostylops Polystylops Prohyracotherium Selenoconus*] U. Paleoc. M. Eoc. SA. *Othnielmarshia* [*Postpithecus*] *Peripantostylops* ?U. Paleoc., L. Eoc. SA. **Notostylopidae,** *Edvardotrouessartia* L. Eoc. SA. *Homalostylops* [*Acrostylops*] U. Paleoc.-L. Eoc. SA. *Notostylops* [*Anastylops Catastylops Entelostylops Eostylops Isostylops Pliostylops*] L., ?M. Eoc. SA. *Otronia* M. Eoc. SA. *?Seudenius* U. Paleoc. SA.

SUBORDER TOXODONTIA. **Oldfieldthomasiidae (Acoelodidae),** *?Acoelodus* L. Eoc. SA. *?Allalmeia* L. Olig. SA. *?Brachystephanus* U. Eoc. SA. *Colbertia Kibenikhoria* U. Paleoc. SA. *Maxschlosseria* [*Paracoelodus*] *Oldfieldthomasia* L. Eoc. SA. *Tsamnichoria* M. Eoc. SA. *Ultrapithecus* L. Eoc. SA. *?Xenostephanus* L. Olig. SA. **Archaeopithecidae,** *Acropithecus Archaeopithecus* L. Eoc. SA. **Isotemnidae,** *?Brandmayria* U. Paleoc. SA. *?Calodontotherium ?Distylophorus* M. Eoc. SA. *Eochalicotherium* [*Amphitemnus Dimerostephanus*] L. Eoc. SA. *?Grypolophodon* M. Eoc. SA. *Isotemnus* [*Prostylops*] U. Paleoc.-L. Eoc. SA. *?Lafkenia* M. Eoc. SA. *?Lophocoelus* L. Olig. SA. *Periphragnis* [*Proasmodeus*] M. Eoc. SA. *Pleurocoelodon* L. Olig. SA. *Pleurostylodon* [*Pleurotemnus*] L. Eoc. SA. *Rhyphodon* M. Eoc. SA. *Thomashuxleya* L. Eoc. SA. *Trimerostephanos* M. Eoc.-L. Olig. SA. **Homalodotheriidae,** *Asmodeus* ?M. Eoc., L. Olig. SA. *Chasicotherium* [*Puntanotherium*] L. Plioc. SA. *Diorotherium* U. Olig. SA. *Homalodotherium* L.-U. Mioc. SA. *Prochalicotherium* U. Olig. SA. **Leontiniidae,** *Ancylocoelus* L. Olig. SA. *Colpodon* U. Olig. SA. *Henricofilholia Leontinia Scarrittia* L. Olig. SA. **Notohippidae,** *Argyrohippus* U. Olig. SA. *Eomorphippus* M. Eoc. SA. *Eurygenium* L. Olig. SA. *Interhippus* M. Eoc.-L. Olig. SA. *Morphippus Nesohippus* L. Olig. SA. *Notohippus* L. Mioc. SA. *Perhippidium* U. Olig. SA. *Pseudostylops* M. Eoc. SA. *Rhynchippus* L. Olig. SA. *Stilhippus* U. Olig. SA. **Toxodontidae,** *Abothrodon* [*?Neotrigodon*] ?Plioc. SA. *Adinotherium* L.-M. Mioc. SA. *Alitoxodon Chapalmalalodon* U. Plioc. SA. *Dinotoxodon* [*?Palaeotoxodon*] *Eutomodus* ?M. Plioc. SA. *?Gallardodon* ?Mioc. SA. *Gyrinodon* ?L. Plioc. SA. *Haplodontherium* M. Plioc. SA. *Hemitoxodon* L. Plioc. SA. *Hyperoxotodon* U. Mioc. SA. *Mixotoxodon* Pleis. SA. CA. *Neoadinotherium* L. Plioc. SA. *Nesodon* L.-U. Mioc. SA. *Nesodonopsis Nonotherium* M. Plioc. SA. *Ocnerotherium* L. Plioc. SA. *Palyeidodon* M. Mioc. SA. *Paratrigodon* U. Mioc. SA. *Phobereotherium* Mioc. SA. *Proadinotherium* L.-U. Olig. SA. *Prototrigodon* M.-U. Mioc. SA. *Stenotephanos* ?M. Plioc. SA. *Stereotoxodon* U. Mioc. SA. *Toxodon* [*?Posnanskytherium*] U. Plioc.-Pleist. SA. *Toxodontherium* ?M. Plioc. SA. *Trigodon* [*Eutrigodon*] ?M., U. Plioc. SA. *Trigodonops* ?Plioc. SA. *Xotodon* M.-U. Plioc. SA.

SUBORDER TYPOTHERIA

SUPERFAMILY TYPOTHEROIDEA. **Interatheriidae,** *Archaeophylus* [*Progaleopithecus*] L. Olig. SA. ? *Caenophilus* U. Mioc. SA. *Cochilius* L.-U. Olig. SA. *Epipatriarchus* U. Mioc. SA. *Guilielmoscottia* M. Eoc. SA. *Interatherium* [*Icochilus*] L., ?M. Mioc. SA. *Miocochilius* U. Mioc. SA. *Notopithecus* [? *Pseudadiantus*] ?U. Paleoc., L.-?M. Eoc. SA. *Paracochilius* U. Olig. SA. *Plagiarthrus* [*Argyrohyrax*] L. Olig. SA. *Protypotherium* [*Patriarchus*] U. Olig.-L. Plioc. SA. *Transpithecus* ?U. Paleoc., L. Eoc. SA. **Typotheriidae (Mesotheriidae),** *Eutypotherium* [*Tachytypotherium Typothericulus*] U. Mioc. SA. *Proedium* [*Isoproedrium Proedrium*] L. Olig. SA. *Pseudotypotherium* M.-U. Plioc. SA. *Trachytherus* [*Eutrachytherus*] L. Olig. SA. *Typotherium* [*Bravardia Mesotherium Typotheridion*] M. Plioc.-Pleist. SA. *Typotheriopsis* [? *Acrotypotherium*] L.-M. Plioc. SA.

SUPERFAMILY HEGETOTHEROIDEA. **Archaeohyracidae,** *Acoelohyrax* L. Eoc. SA. *Archaeohyrax* ?U. Eoc., L. Olig. SA. *Degonia* M. Eoc. SA. *Eohyrax* ?U. Paleoc., L. Eoc. SA. **Hegetotheriidae,** *Eohegetotherium Eopachyrucus* M. Eoc. SA. *Ethegotherium* L. Olig. SA. *Hegetotherium* [*Selatherium*] U. Olig.-M. Mioc. SA. *Hemihegetotherium* M. Plioc. SA. *Munizia* ?U. Mioc. SA. *Pachyrukhos* [*Pachyrucos*] U. Olig.-U. Mioc., L. Plioc. SA. *Paedotherium* L. Plioc.-L. Pleist. SA. *Prohegetotherium Propachyrucos Prosotherium* [*Medistylus Phanophilus*] L. Olig. SA. *Pseudohegetotherium* L. Plioc. SA. *Tremacyllus* L. Plioc.-Pleist. SA.

NOTOUNGULATA INCERTAE SEDIS: *Acamana* U. Eoc. SA.

ORDER ASTRAPOTHERIA. **Trigonostylopidae,** ? *Albertogaudrya* [*Scabellia*] L. Eoc. SA. ?*Shecenia* U. Paleoc. SA. *Trigonostylops* [*Chiodon, Staurodon*] U. Paleoc.-L. Eoc. SA. **Astrapotheriidae,** *Astraponotus* [*Notamynus*] U. Eoc. SA. *Astrapothericulus* L. Mioc. SA. *Astrapotherium* U. Olig.-U. Mioc. SA. *Parastrapotherium Proastrapotherium* L. Olig. SA. *Scaglia* L. Eoc. SA. ? *Uruguaytherium* ?Mioc. SA. *Xenastrapotherium* U. Mioc. SA.

ORDER LITOPTERNA. **Proterotheriidae,** *Anisolambda* [*Eulambda Josepholeidya*] Eoc. SA. *Brachytherium* M.-U. Plioc. SA. *Deuterotherium* L. Olig. SA. *Diadiaphorus* L. Mioc-L. Plioc. SA. *Diplasiotherium* U. Plioc. SA. ? *Eolicaphrium* L. Eoc. SA. *Eoproterotherium* L. Olig. SA. *Guileilmofloweria* Eoc. SA. *Heteroglyphis* U. Eoc. SA. *Licaphrium* L., ?M. Mioc. SA. *Licaphrops* U. Olig.-L. Mioc. SA. *Neolicaphrium* Pleist. SA. ? *Phoradiadius* L. Olig. SA. *Polyacrodon* [*Decaconus Periacrodon Oroacrodon*] *Polymorphis* [*Megacrodon*] U. Eoc. SA. *Prolicaphrium* U. Olig. SA. *Proterotherium* L. Mioc.-L. Plioc. SA. *Protheosodon* L. Olig. SA. *Prothoatherium* U. Olig. SA. *Ricardolydekkeria* [*Heterolambda Lopholambda*] U. Paleoc.-L. Eoc. SA. ? *Rhoradiadus* U. Eoc. SA. *Thoatherium* L. Mioc. SA. *Wainka* U. Paleoc. SA. *Xesmodon* [*Glyphodon*] U. Eoc. SA. **Macraucheniidae,** ? *Amilnedwardsia Coniopternium* [*Caliphrium Notodiaphorus*] L. Olig. SA. *Cramauchenia* U. Olig. SA. *Cullinia* L. Plioc. SA. ? *Ernestohaeckelia* L. Eoc. SA. *Macrauchenia Macraucheniopsis* U. Plioc., Pleist. SA. *Oxyodontherium* ?M. Plioc. SA. *Paramacrauchenia* U. Olig. SA. *Paranauchenia* ?M. Plioc. SA. *Promacrauchenia* M.-U. Plioc. SA. ? *Ruetimeyeria* L. Eoc. SA. *Scalabrinitherium* ?M. Plioc. SA. *Theosodon* U. Olig.-L. Plioc. SA. *Victorlemoinea* U. Paleoc.-L. Eoc. SA. *Windhausenia* ?U. Plioc., L. Pleist. SA. **Adianthidae,** *Adianthus* [*Adiantus*] U. Olig.-L. Mioc. SA. *Adiantoides Proadiantus* L. Olig. SA. *Proheptoconus* U. Olig. SA.

ORDER PERISSODACTYLA

SUBORDER HIPPOMORPHA

SUPERFAMILY EQUOIDEA. **Equidae** R. As. Af., *Anchilophus* M.-U. Eoc. Eu. *Anchitherium* [*Kalobatippus Paranchitherium Sinohippus*] L.-U. Mioc. NA., M. Mioc. CA., M. Mioc.-L. Plioc. Eu., M. Mioc.-L. Plioc. As. *Archaeohippus* L.-U. Mioc. NA. M. Mioc. CA. *Calippus* U. Mioc.-U. Plioc. NA. *Epihippus* [*Duchesnehippus*] U. Eoc. NA. *Equus* [*Allohippus Allozebra Amerhippus Asinus Dolichohippus Hemionus Hesperohippus Hippotigris Kolpohippus Kraterohippus Neohippus Onager Plesippus* etc.] L.-U. Pleist. NA. Eu., L. Pleist.-R. As. Af., Pleist. SA. *Haplohippus* Olig. NA. *Hipparion* [*Hemihipparion Hippotherium Notohipparion Proboscidohipparion Stylohipparion*] U. Mioc.-L. Pleist. Eu. Af., L. Plioc.-L. Pleist. As. NA. *Hippidion* [*Hippidium Plagiohippus Stereohippus*] Pleist. SA. *Hypohippus* [*Megahippus*] M. Mioc.-L. Plioc. NA., L. Plioc. As. *Hyracotherium* [*Eohippus Protorohippus*] L. Eoc. Eu. NA. *Lophiotherium* M.-U. Eoc. Eu. *Merychippus* [*Protohippus*] M. Mioc.-L. Plioc. NA. *Mesohippus* [*Pediohippus*] L.-M. Olig. NA. *Miohippus* M. Olig.-L. Mioc. NA. *Nannippus Neohipparion* L. Plioc.-L. Pleist. NA. *Onohippidium* [*Hyperhippus*] Pleist. SA. *Orohippus* M. Eoc. NA. *Pachynolophus* M.-U. Eoc. Eu., ?L. Olig. As. *Parahipparion* [*Hyperhippidium*] Pleist. SA. *Parahippus* L.-U. Mioc. NA. *Pliohippus* [*Astrohippus*] L.-U. Plioc. NA. *Propachynolophus* L. Eoc. Eu. *Propalaeotherium* M. Eoc.-L. Olig. Eu. ?L. Eoc., M. Eoc.-L. Olig. As. **Palaeotheriidae,** *Palaeotherium* U. Eoc.-L. Olig. Eu. *Plagiolophus* [*Paloplotherium*] M.Eoc.-L.Olig.Eu.

SUPERFAMILY BRONTOTHERIOIDEA. **Brontotheriidae (Titanotheriidae)** *Brachydiastematherium* U. Eoc. or L. Olig. EEu. *Brontops* [*Diploclonus Megacerops*] L. Olig. NA. ?Eu. *Brontotherium* L. Olig. NA. *Desmatotitan* U. Eoc. As. *Diplacodon* U. Eoc. NA. *Dolichochinoides* U. Eoc. As. *Dolichorhinus* U. Eoc. NA. *Embolotherium* L.-M. Olig. As. *Eometarhinus* M. Eoc. NA. *Eotitanops* L. Eoc. NA., ?M. Eoc. SAs. *Eotitanotherium* U. Eoc. NA. *Epimanteoceras Gnathotitan* U. Eoc. As. *Hyotitan* M. Olig. As. *Lambdotherium* L. Eoc. NA. *Limnohyops* M. Eoc. NA. *Manteoceras* M.-U. Eoc. NA. *Megacerops* [*Symborodon*] L. Olig. NA. *Menodus* [*Allops Titanotherium*] L. Olig. NA., ?Eu. *Mesatirhinus Metarhinus* M.-U. Eoc. NA. *Metatelmatherium* U. Eoc. As. NA. *Metatitan* L.-M. Olig. As. *Microtitan* U. Eoc. As. *Notiotitanops* U. Eoc. NA. *Pachytitan* U. Eoc. EAs. *Palaeosyops* L.-M. Eoc. NA. *Parabrontops* ?U. Eoc., L. Olig. EAs. *Protembolotherium* U. Eoc.-L. Olig. EAs. *Protitan* U. Eoc. As. *Protitanops* L. Olig. NA. *Protitanotherium* U. Eoc. NA. EAs. *Rhadinorhinus* U. Eoc. NA. *Rhinotitan* U. Eoc.-L. Olig. EAs. *Sivatitanops* U. Eoc. As. *Sphenocoelus Sthenodectes* U. Eoc. NA. *Teleodus* U. Eoc.-L. Olig. NA. *Telmatherium* M. Eoc. NA. *Titanodectes* U. Eoc.-L. Olig. EAs.

SUBORDER ANCYLOPODA. **Eomoropidae,** *Eomoropus* U. Eoc. NA. EAs. *Grangeria* U. Eoc. EAs., ?NA. *Litolophus* U. Eoc. EAs. *Lophiaspis* L.-M. Eoc. Eu. *Lunania* U. Eoc. EAs. *Paleomoropus* L. Eoc. NA. **Chalicotheriidae,** *Ancylotherium* [*Circotherium Nestoritherium*] L. Plioc. Eu., Plioc.-Pleist. EAs. L. Pleist. Af. *Borissiakia* U. Olig. WAs. *Chalicotherium* [*Macrotherium*] L. Mioc.-L. Plioc.

As., L. Mioc. EAf., M. Mioc.-L. Plioc., Eu., M. Mioc. NA., ?Pleist. EInd. *Moropus* L.-M. Mioc. NA., L. Mioc. As. *Oreinotherium* L. Olig. NA. ?*Pernatherium* U. Eoc. Eu. *Phyllotillon* [*Metaschizotherium*] U. Olig.-L. Mioc. As., U. Olig. ?L. Plioc. Eu., ?L. Mioc., L. Pleist. Af. *Postschizotherium* U. Plioc.-Pleist. As. *Schizotherium* Olig. Eu., L. Olig.-L. Mioc. As. *Neoschizotherium* Plioc. Eu.

ANCYLOPODA INCERTAE SEDIS: *Schizotheroides* U. Eoc. NA.

SUBORDER CERATOMORPHA

SUPERFAMILY TAPIROIDEA. **Isectolophidae,** *Homogalax* ["*Systemodon*"] L. Eoc. NA., L. (?)Eoc. EAs. *Isectolophus* [*Parisectolophus Schizolophodon*] M.-U. Eoc. NA. **Helaletidae (Hyrachidae),** *Colodon* [*Paracolodon*] U. Eoc., ?L. Olig. NA., U. Eoc.-U. Olig. EAs. *Colonoceras* M. Eoc. NA. *Dilophodon* [*Heteraletes*] M.-U. Eoc. NA. *Ephyrachyus* M. Eoc. NA. *Helaletes* [*Chasmotheroides Desmatotherium Veragromovia*] M.-U. Eoc. NA., U. Eoc. EAs. *Heptodon* L. Eoc. EAs. NA. *Hyrachyus* L.-U. Eoc. NA., U. Eoc. EAs. *Metahyrachyus* M. Eoc. NA. **Lophialetidae,** *Breviodon Lophaletes* ?*Pataecus* ?*Rhodopagus* U. Eoc. EAs. *Schlosseria* M. or U. Eoc. EAs. **Deperetellidae,** *Deperetella* [*Cristidentinus Diplolophodon*] U. Eoc. EAs. *Teleolophus* ?M., U. Eoc.-L. Olig. EAs. **Lophiodontidae,** *Atalonodon* M. Eoc. Eu. *Chasmotherium* L.-U. Eoc. Eu. *Lophiodochoerus* L. Eoc. Eu. *Lophiodon* L.-U. Eoc. Eu. **Tapiridae** R. SAs. SA., *Miotapirus* L. Mioc. NA. *Palaeotapirus* [*Paratapirus*] L. Mioc. Eu., ?Mioc. As. *Protapirus* L. Olig. Eu., M. Olig.-L. Mioc. NA., ?L. Mioc. As. *Tapiravus* M., ?U. Mioc., ?L. Plioc. NA. *Tapirus* [*Megatapirus Tapiriscus*] ?Olig., ?U. Mioc., Plioc.-Pleist. Eu., L. Plioc.-Pleist. NA. L. Plioc.-R. As., Pleist. EInd. Af., Pleist.-R. SA., R. CA.

TAPIROIDEA INCERTAE SEDIS: *Indolophus* U. Eoc. As.

SUPERFAMILY RHINOCEROTOIDEA. **Hyracodontidae,** *Ardynia* [*Ergilia*] U. Eoc.-L. Olig. EAs. *Caenolophus* U. Eoc. EAs. *Hyracodon* L. Olig.-L. Mioc. NA. *Parahyracodon* L. Olig. As. *Prothyracodon* U. Eoc. NA. *Teilhardia* M. Eoc. EAs. *Triplopus* M.-U. Eoc. NA. **Amynodontidae,** *Amynodon* [*Sharamynodon*] U. Eoc. NA. EAs. *Amynodontopsis* U. Eoc. NA. *Cadurcodon* L.-U. Olig., ?Mioc. As., M. Olig. Eu. *Gigantamynodon Hypsamynodon* L. Olig. EAs. *Lushiamynodon* U. Eoc. EAs. *Megalamynodon Mesamynodon* U. Eoc. NA. *Metamynodon* [*Cadurcopsis*] L.-M. Olig. NA., ?U. Eoc., L. Olig. EAs. *Orthogonodon* U. Eoc. NA. *Paramynodon Procadurcodon* U. Eoc. As. *Sianodon* U. Eoc. EAs. **Rhinocerotidae,** *Aceratherium* [*Acerorhinus Turkanatherium*] M. Olig.-L. Plioc. Eu., U. Olig.-L. Plioc., ?Pleist. As., Mioc. Af., U. Plioc. EInd. *Amphicaenopus* L.-U. Olig. NA. *Aphelops* M. Mioc.-U. Plioc. NA. *Baluchitherium* U. Olig.-L. Mioc. As. *Benaritherium* U. Olig. Eu. *Brachypotherium* [*Indotherium Thaumastotherium*] U. Olig.-L. Plioc. As., L. Mioc. Af., L. Mioc.-L. Plioc. Eu., *Caenopus* L.-U. Olig. NA. *Chilotherium* L. Mioc.-L. Pleist. As., U. Mioc.-U. Plioc. Eu. *Coelodonta* [*Tichorhinus*] Pleist. Eu. As. NAf. *Diceratherium* [*Metacaenopus Menoceras*] U. Olig. Eu., U. Olig.-L. Mioc. NA., M. Mioc. CA. *Dicerorhinus* [*Ceratorhinus*] U. Olig.-R. As., L. Mioc.-Pleist. Eu., "U. Mioc." NAf. *Diceros* [*Atelodus Ceratotherium Opsiceros Pliodiceros* ?*Serenageticeros*] U. Plioc. Eu., U. Plioc.-R. Af. *Dromaceratherium* L. Mioc. Eu. *Eggysodon* M.-U. Olig. Eu. *Elasmotherium* Pleist. Eu., L.-U. Pleist. NAs. *Eotrigonias* U. Eoc. NA. EAs. *Epiaceratherium* [*Allocerops*] L.-U. Olig. Eu., M.-U. Olig. As. *Epitriplophus* U. Eoc. NA. *Floridaceras* L. Mioc. NA. *Forstercooperia* [*Cooperia*] U. Eoc. As. *Gaindatherium* L.-U. Plioc. As. *Gobitherium* L. Mioc.-L. Plioc. EAs. *Hispanotherium* M. Mioc. Eu. *Ilianodon* U. Eoc., ?L. Olig. EAs. *Indricotherium* L.-U. Olig. As. *Iranotherium* L. Plioc. As. *Juxia* U. Eoc. EAs. *Meninatherium* U. Olig. Eu. *Orthogonoceros* L. Pleist. Eu. *Pappaceras* U. Eoc. EAs. *Paraceratherium* [*Aralotherium*] U. Olig. As. *Peraceras* U. Mioc.-L. Plioc. NA. *Plesiaceratherium* M.-U. Mioc. EAs. *Pleuroceros* U. Olig.-?L. Plioc. Eu., L. Mioc.-?L. Plioc. As. *Preaceratherium* Olig. Eu. *Prohyracodon* M. Eoc. Eu., U. Eoc.-L. Olig. EAs. *Protaceratherium* M.-U. Olig. Eu. *Rhinoceros* [*Procerorhinus*] L. Pleist.-R. As. EInd. *Ronzotherium* [*Paracaenopus*] L.-U. Olig. Eu. *Sinotherium* [*Parelasmotherium*] L. Plioc. As. *Subhyracodon* L.-M. Olig. NA. *Symphysorrhachis* Olig. EAs. *Teleoceras* [*Aprotodon*] U. Mioc.-U. Plioc. NA., Plioc. EAs. *Tongriceros* Olig. Eu. *Trigonias* L. Olig. NA. *Urtinotherium* L. Olig. EAs.

CERATOMORPHA INCERTAE SEDIS: *Toxotherium* L. Olig. NA.

PERISSODACTYLA INCERTAE SEDIS: *Caucasotherium* Mioc. EEu.

ORDER ARTIODACTYLA

SUBORDER PALAEODONTA. **Diacodectidae,** ?*Bunophorus Diacodexis* [*Trigonolestes*] L. Eoc. NA. ?*Eohyus* ?L. Eoc. NA. *Wasatchia* L. Eoc. NA. **Leptochoeridae,** *Leptochoerus* [*Laopithecus*] M.-U. Olig. NA. *Nanochoerus* Olig. NA. *Stibarus* L. Olig. NA. **Homacodontidae,** *Antiacodon* [*Sarcolemur*] M. Eoc. NA. ? *Auxontodon* U. Eoc. NA. *Bunomeryx* U. Eoc. NA. *Hexacodus* L. Eoc. NA. *Homacodon* [*Nanomeryx*] M. Eoc. NA. *Hylomeryx* [?*Sphenomeryx*] *Mesomeryx* U. Eoc. NA. *Microsus* M. Eoc. NA. *Mytonomeryx Pentacemylus* ?*Tapochoerus* U. Eoc. NA. **Dichobunidae,** *Dichobune* [?*Thylacomorphus*] M. Eoc.-L. Olig. Eu. *Haqueina* M. Eoc. SAs. *Hyperdichobune* U. Eoc.-L. Olig. Eu. ?*Kirtharia* M. Eoc. SAs. *Meniscodon* M. Eoc. Eu. *Metriotherium* M. Olig. Eu. *Mouillacitherium* M. Eoc.-L. Olig. Eu. *Pilgrimella* M. Eoc. SAs. *Protodichobune* L. Eoc. Eu., U. Eoc. EAs. *Synaphodus* M. Olig. Eu. **Achaenodontidae (Helohyidae),** *Achaenodon* [*Protelotherium*] *Apriculus* U. Eoc. NA. *Helohyus Lophiohyus* M. Eoc. NA. *Parahyus* U. Eoc. NA.

SUBORDER SUINA

SUPERFAMILY ENTELODONTOIDEA. **Choeropotamidae,** ?*Brachyhyops* U. Eoc.-L. Olig. NA. *Choeropotamus* U. Eoc.-M. Olig. Eu. *Gobiohyus* U. Eoc. EAs. **Cebochoeridae,** *Cebochoerus* [*Acotherulum Leptacotherulum*] M. Eoc.-L. Olig. Eu. *Choeromorus* M.-U. Eoc., L. Olig. Eu., ?Eoc. EInd. *Mixtotherium* M.-U. Eoc. Eu. L. Olig. NAf. **Entelodontidae (Elotheridae),** *Archaeotherium* [*Choerodon Megachoerus Pelonax Scaptohyus*] L.-U. Olig. NA., U. Olig. EAs. *Dinohyus* [? *Ammodon* ?*Boochoerus* ?*Daeodon*] L. Mioc. NA. ?*Dyscritochoerus* U. Eoc. NA. *Entelodon* [*Elodon Elotherium*] L.-M. Olig. Eu., ?NA. *Eoentelodon* U. Eoc. EAs. *Ergilobia* [*Brachyodon*] L.-M. Olig. EAs.

SUPERFAMILY SUOIDEA. **Suidae,** *Babyrousa* [*Babirussa*] Pleist.-R. EInd. *Celebochoerus* Pleist. EInd. *Chleuastochoerus* L. Plioc. As. *Conohyus* L. Mioc.-U. Plioc. As., ?L. Mioc. NAf., M. Mioc.-Plioc. Eu., *Diamantohyus* L. Mioc. SWAf. *Dicoryphochoerus Hippohyus* [*Hyosus*] L. Plioc.-Pleist. As. *Hylochoerus* Pleist.-R. Af. *Hyotherium* M.-U. Mioc. Eu. *Kubanochoerus* M. Mioc. WAs. *Libycochoerus* L. Mioc. NAf. *Listriodon* L.-M. Mioc. Af., L. Mioc.-L. Plioc. Eu., U. Mioc.-U. Plioc. As. *Lophochoerus* L. Plioc. As. *Nyanzachoerus* Pleist. Af. *Omochoerus* [*Mesochoerus*] L.-U. Pleist. Af. *Orthostonyx* Pleist. EAf. *Palaeochoerus* U. Olig.-M. Mioc. Eu., L. Mioc. Af., U. Mioc.-U. Plioc. As. *Paradoxodonides* [*Paradoxodon*] U. Eoc. or Olig. Eu. *Phacochoerus* [*Afrochoerus Gerontochoerus Kolpochoerus Metridiochoerus Notochoerus Potamochoeroides Potamochoerops Prontochoerus Stylochoerus Synaptochoerus Tapinochoerus*] L. Pleist.-R. Af., Pleist. WAs. *Potamochoerus* [*Choiropotamus Eostopotamochoerus Koiropotamus Postpotamochoerus Propotamochoerus*] ?L. Mioc., L. Pleist.-R. Af., L. Plioc. Eu., L. Plioc.-Pleist. As. *Propalaeochoerus* L.-U. Olig. Eu. *Sanitherium Schizochoerus* L. Plioc. Eu. As. *Sivachoerus* U. Plioc.-Pleist. SAs., L. Pleist. NAf. *Sivahyus* U. Plioc. As. *Sus* [*Microstonyx*] U. Mioc.-R. Eu. As., ?U. Mioc., Plioc.-R. Af., Pleist.-R. EInd. *Tetraconodon* L. Plioc.-Pleist. As. *Xenochoerus* U. Mioc. Eu. **Tayassuidae (Dicotylidae),** *Catagonus* Pleist. SA. *Choeromorus* [*Choerotherium Taucanamo*] M.-U. Mioc. Eu. *Doliochoerus* M. Olig. Eu. *Dyseohyus* U. Mioc. NA. *Floridachoerus* L. Mioc. NA. *Hesperhys* [*Desmathyus Pediohyus*] L.-U. Mioc. NA. *Leptotherium* Pleist. SA. *Mylohyus* Pleist. NA., ?SA. *Pecarichoerus* L. Plioc. SAs. *Perchoerus* [*Bothrolabis Thinohyus*] L. Olig.-L. Mioc. NA. *Platygonus* [*Parachoerus*] L.-U. Pleist. NA., U. Plioc.-Pleist. SA. *Prosthenops* U. Mioc.-U. Plioc. NA., ?Pleist. SA. *Selenogonus* Plioc. SA. *Tayassu* [*Dicotyles Pecari*] ?U. Plioc., Pleist.-R. SA., Pleist.-R. NA.

SUPERFAMILY HIPPOPOTAMOIDEA. **Anthracotheriidae,** ? *Anthracobune* M. Eoc. SAs. *Anthracochoerus* L. Olig. Eu. *Anthracohyus* U. Eoc. EAs. *Anthracokeryx Anthracosenex Anthracothema* U. Eoc. As. *Anthracotherium* U. Eoc.-L. Mioc. Eu., L. Mioc., ?L. Plioc. As. *Arretotherium* L. Mioc. NA. *Bothriodon* [*Aepinacodon Ancodon Ancodus Hyopotamus*] U. Eoc.-M. Olig. EAs., L. Olig. Eu., L.-M. Olig. NA., Olig.-Mioc. Af. *Bothriogenys* Olig. Af. *Brachyodus* ?U. Eoc. L. Olig.-L. Mioc. Af., L. Olig.-M. Mioc. Eu., As. *Bunobrachyodus* Olig. Eu. *Choeromeryx* L.-U. Plioc. As. *Elomeryx* L. Olig.-L. Mioc. Eu., U. Olig. NA. *Galasmodon* Olig. As. *Gobiotypus* U. Eoc. EAs. *Haplobunodon* M.-U. Eoc. Eu. *Hemimeryx* M. Olig.-L. Plioc. As. *Heptacodon* L.-U. Olig. NA. *Hyoboöps* [*Merycops*] U. Olig.-L. Plioc. As., L. Mioc. Af. *Kukusepasutauka* Mioc. NA. *Lophiobunodon* M. Eoc. Eu. *Merycopotamus* L. Plioc. NAf., U. Plioc.-Pleist. As., L.-U. Pleist. EInd. *Microbunodon* [*Microselenodon*] M. Olig. Eu. *Octacodon* U. Olig. NA. *Parabrachyodus* L. Mioc. SAs. *Probrachyodus* U. Eoc. EAs. *Prominatherium* Eoc. Eu. *Rhagatherium* [*Amphirhagatherium*] M. Eoc.-L. Olig. Eu., L. Olig. NAf., ?L. Plioc. As. *Telmatodon* [*Gonotelma*] L. Mioc.-L. Plioc. As. *Thaumastognathus* U. Eoc. Eu. **Hippopotamidae,** *Choeropsis* Pleist. Mad. NAf., R. WAf. *Hippopotamus* [*Hexaprotodon Hippoleakius Prochoeropsis Tetraprotodon*] U. Plioc.-SubR. As., U. Plioc.-R. Af., Pleist. Eu. EInd.

SUBORDER RUMINANTIA

INFRAORDER TYLOPODA

SUPERFAMILY CAINOTHEROIDEA. **Cainotheriidae (Caenotheriidae),** *Caenomeryx* L.-M. Olig. Eu. *Cainotherium* [*Caenotherium Procaenotherium*] M. Olig.-M. Mioc. Eu. *Oxacron* U. Eoc. Eu. *Paroxacron* U. Eoc.-L. Olig. Eu. *Plesiomeryx* M.-U. Olig. Eu.

SUPERFAMILY ANOPLOTHEROIDEA. **Anoplotheriidae,** *Anoplotherium* U. Eoc.-L. Olig. Eu. *Catodontherium* U. Eoc. Eu. *Dacrytherium* M.-U. Eoc. Eu. *Diplobune* U. Eoc.-M. Olig. Eu. *Ephelcomenus* [*Hyracodontherium*] M. Olig. Eu. *Leptotheridium* M.-U. Eoc. Eu. *Tapirulus* M. Eoc.-L. Olig. Eu. **Xiphodontidae,** *Dichodon* [*Tetraselenodon*] M. Eoc.-L. Olig. Eu. *Haplomeryx* M.-U. Eoc. Eu. *Xiphodon* U. Eoc.-L. Olig. Eu. **Amphimerycidae,** *Amphimeryx* U. Eoc.-L. Olig. Eu. *Pseudamphimeryx* M. Eoc.-L. Olig. Eu.

SUPERFAMILY MERYCOIDODONTOIDEA. **Agriochoeridae,** *Agriochoerus* L. Olig.-L. Mioc. NA. *Diplobunops Protoreodon* [*Agriotherium Chorotherium Eomeryx Hyomeryx Mesagriochoerus Protagriochoerus*] U. Eoc. NA. **Merycoidodontidae (Oreodontidae),** *Bathygenys* L. Olig. NA. *Brachycrus* [*Pronomotherium*] M. Mioc. NA. *Cyclopidius* L.-U. Mioc. NA. *Desmatochoerus* [*Hypselochoerus Paradesmatochoerus*] L.-M. Olig. NA. *Eporeodon* [? *Eucrotaphus*] M. Olig.-L. Mioc. NA. *Hypsilops* L. Mioc. NA. *Leptauchenia* U. Olig.-L. Mioc. NA. *Limnenetes* L. Olig. NA. *Mediochoerus* M. Mioc. NA. *Megoreodon* L. Mioc. NA. *Merychyus* [*Metoreodon*] L. Mioc.-U. Plioc. NA. *Merycochoerus* M. Mioc. NA. CA. *Merycoides* L. Mioc. NA. *Merycoidodon* [*Oreodon*] L.-M. Olig. NA. *Mesoreodon* L. Mioc. NA. *Miniochoerus* [*Paraminiochoerus*] M.-U. Olig. NA. *Oreodontoides* [*Paroreodon*] L. Mioc. NA. *Oreontes* L. Olig. NA. *Paramerychyus* L. Mioc. NA. *Parastenopsochoerus* M. Olig. NA. *Phenacocoelus* L.-M. Mioc. NA. *Platyochoerus* M.-U. Olig. NA. *Prodesmatochoerus* L.-M. Olig. NA. *Promerycochoerus* [*Paracotylops Paramerycochoerus Parapromerycochoerus Pseudopromerycochoerus*] U. Olig.-L. Mioc. NA. *Promesoreodon* U. Olig. NA. *Pseudodesmatochoerus Pseudomesoreodon* L. Mioc. NA. *Stenopsochoerus* [*Pseudostenopsochoerus*] M.-U. Olig. NA. *Subdesmatochoerus* M.-U. Olig. NA. *Submerycochoerus* L. Mioc. NA. *Superdesmatochoerus* L. Mioc. NA . *Ticholeptus* [*Poatrephes*] M. Mioc. NA. *Trigenicus* L. Olig. NA. *Ustatochoerus* U. Mioc.-L. Plioc. NA.

SUPERFAMILY CAMELOIDEA. **Oromerycidae,** *Camelodon* U. Eoc. NA. *Eotylopus* L. Olig. NA. *Malaquiferus Oromeryx Protylopus* U. Eoc. NA. **Camelidae,** *Aepycamelus* L. Plioc. NA. *Alticamelus* M.-L. Mioc. Plioc. NA. *Camelops* Pleist. NA. *Camelus* [*Cameliscus*] U. Plioc.-Pleist. EEu., L. Pleist.-R. As., Pleist. NA., Pleist.-R. NAf. *Eschatius* Pleist. NA. *Gentilicamelus* Mioc. NA. *Gigantocamelus* L.-U. Pleist. NA. *Hesperocamelus* L. Plioc. NA. *Lama* [*Auchenia, Hemiauchenia, Palaeolama, Vicuyna*] U. Plioc.-R. SA. *Leptotylopus* Pleist. NA. *Megacamelus* L. Pleist. NA. *Miolabis* M. Mioc. NA. *Nothokemas Oxydactylus* L. Mioc. NA. *Paracamelus* [*Megatylopus, Neoparacamelus*] L.-U. Plioc. NA., U. Plioc.-Pleist. As. EEu. *Paratylopus* M.

Olig.-L. Mioc. NA. *Pliauchenia* L. Plioc.-L. Pleist. NA. *Poebrodon* U. Eoc. NA. *Poebrotherium* M. Olig. NA. *Procamelus* U. Mioc.-L. Plioc. NA., ?L. Plioc. Eu. *Protauchenia* Pleist. SA. *Protolabis* M. Mioc.-L. Plioc. NA. *Protomeryx* [*Dyseotylopus Gomphotherium*] U. Olig.-L. Mioc. NA. *?Pseudoceras* ?Plioc. NA. *Pseudolabis* U. Olig. NA. *Rakomylus* L. Plioc. NA. *Stenomylus* L. Mioc. NA. *Tanupolama* U. Plioc.-Pleist. NA. *Titanotylopus* Pleist. NA.

INFRAORDER PECORA

SUPERFAMILY TRAGULOIDEA. **Hypertragulidae,** *Archaeomeryx* U. Eoc. As. *Bachitherium* L.-M. Olig. Eu. *Floridatragulus* L. Mioc. NA. *Gobiomeryx* L.-M. Olig. EAs. *Hypertragulus* [*Allomeryx*] M. Olig.-L. Mioc. NA. *Hypermekops* L. Mioc. NA. *Hypisodus* M. Olig. NA. *Indomeryx* U. Eoc. SAs. *Leptomeryx* L. Olig.-L. Mioc. NA. *Leptoreodon* [*Camelomeryx Hesperomeryx Merycodesmus*] *Leptotragulus* [*Parameryx*] U. Eoc. NA. *Miomeryx* U. Eoc.-L. Olig. As. *Nanotragulus* U. Olig.-L. Mioc. NA. *Poabromylus Simimeryx* U. Eoc. NA. **Protoceratidae,** *?Heteromeryx* L. Olig. NA. *Paratoceras* ?L. Plioc. NA. *Protoceras* [*Calops Pseudoproceras*] L.-U. Olig. NA. *Syndyoceras* L. Mioc. NA. *Synthetoceras* [*Prosynthetoceras*] L. Mioc.-L. Plioc. NA. **Gelocidae,** *Cryptomeryx* U. Eoc.-L. Olig. Eu. *Gelocus* [*Paragelocus Pseudogelocus*] U. Eoc.-M. Olig. Eu. *Lophiomeryx* L.-M. Olig. Eu., L.-U. Olig. As. *Phaneromeryx* U. Eoc. Eu. *Prodremotherium* L.-M. Olig. Eu., M.-U. Olig. WAs., L. Mioc. Af. **Tragulidae,** R. As. Af., *Dorcabune* L.-U. Plioc. As. *Dorcatherium* L. Mioc. Af., M. Mioc.-L. Plioc. Eu., U. Mioc.-L. Pleist. As. *Hyemoschus* [*Hyaemoschus*] Pleist.-R. Af. *Tragulus* ?U. Plioc. Pleist. R. As., Pleist. EInd.

SUPERFAMILY CERVOIDEA. **Palaeomerycidae,** *Aletomeryx* [*Dyseomeryx Sinclairomeryx*] L.-M. Mioc. NA. *Amphitragulus* U. Olig.-L. Mioc. Eu., L. Mioc. As. Af. *Barbouromeryx* L.-M. Mioc. NA. *Blastomeryx* [*Parablastomeryx Problastomeryx Pseudoblastomeryx Pseudoparablastomeryx*] L. Mioc.-U. Plioc. NA. *Climacoceras* L. Mioc. EAf. *Cranioceras* M. Mioc.-L. Plioc. NA. *Dremotherium* U. Olig.-L. Mioc. Eu. *Drepanomeryx* M. Mioc. NA. *Dromomeryx* M.-U. Mioc. NA. *Eumeryx* L.-U. Olig. EAs. *Lagomeryx* [*Heterocemas*] L.-U. Mioc. Eu., U. Mioc.-L. Plioc. EAs. *Machaeomeryx* L. Mioc. NA. *Palaeomeryx* L. Mioc. Af. As., M. Mioc.-L. Plioc. Eu. *Procervulus* L.-U. Mioc. Eu. *Rakomeryx* M.-U. Mioc. NA. *Yumaceras* U. Plioc.-L. Pleist. NA. **Cervidae,** *Agalmaceros* Pleist. SA. *Alce* [*Alces Libralces*] Pleist.-R. Eu. As. NA. *Antifer* [*Paraceros*] U. Plioc.-Pleist. SA. *Axis* L. Plioc.-R. As., Pleist. Eu. EInd. *Blastocerus* Pleist.-R. SA. *Capreolus* U. Plioc.-R. Eu. As. *Cervalces* Pleist. NA. *Cervaviscus Cervavitulus* Mioc. Eu. *Cervocerus* [*Cervavitus Damacerus Procervus*] L. Plioc. Eu., L.-U. Plioc. As. *Cervus* [*Cervodama Deperetia, Elaphus Epirusa Euctenoceros Nipponicervus ?Procoileus Pseudaxis Rucervus Rusa Sika*] U. Plioc.-R. Eu. As., Pleist. NAf. EInd., Pleist.-R. NA. *Charitoceros* Pleist. SA. *Ctenocerus* [*Pliocervus*] L. Plioc.-L. Pleist. Eu. *Dama* U. Plioc.-R. As., Pleist. NAf., Pleist.-R. Eu., *Dicrocerus* [*Euprox Heteroprox*] L. Mioc.-U. Plioc. Eu., U. Mioc.-L. Plioc. As. *Elaphodus* Pleist.-R. As. *Elaphurus* L. Pleist.-R. EAs. *Eostylocerus* Plioc.-Pleist. EAs. *Eucladocerus* [*Polycladus*] U. Plioc.-Pleist. Eu. As. *Eustylocerus* L. Plioc. Eu., L. Plioc.-Pleist. As. *Georgiomeryfi* L. Plioc. SWAs. *Hippocamelus* Pleist.-R. SA. *Hydropotes* ?Pleist., R. EAs. *Kosmelaphus Longirostromeryx* M. Mioc.-L. Plioc. NA. *Mazama* Pleist.-R. SA. *Megaloceros* [*Dolichodoryceros Megaceros Megaceroides Orthogonoceros Sinomegaceroides Sinomegacerus*] L.-U. Pleist. Eu., Pleist. NAf. As. *Metacervulus* [*Paracervulus*] U. Plioc.-L. Pleist. EAs., L. Pleist. Eu. *Metaplatyceros* Plioc. Japan *Micromeryx* [*Orygotherium*] M. Mioc.-L. Plioc. Eu. *Morenelaphus* [*Habromeryx Pampaeocervus*] Pleist. SA. *Moschus* ?L. Plioc., Pleist.-R. As. *Muntiacus* [*Cervulus*] L. Plioc.-R. As., U. Plioc. EEu., Pleist. EInd. *Muva* Pleist. SAs. *Odocoileus* [*Palaeoödocoileus Protomazama*] L. Pleist.-R. NA. SA. *Ozotoceras* Pleist.-R. SA. *Palaeoplatycerus* M. Mioc.-L. Plioc. Eu. *Paradicrocerus* Mioc. EEu. *?Pediomeryx* Plioc. NA. *Platycemas* U. Plioc. EAs. *Procapreolus* L. Plioc. Eu. As. *Pronodens* L. Mioc. NA. *Pseudalces* U. Plioc. EEu. *Pudu* Pleist.-R. SA. *Rangifer* Pleist.-R. Eu. As. NA. *Rohania* Pleist. SA. *Stephanocemas* M.-U. Mioc. Eu., M. Mioc.-L. Plioc. As. *Strongyloceros* Pleist. Eu. *Tamanalces* L. Pleist. EEu. *Walangania* L. Mioc. EAf. **Giraffidae,** *Birgerbohlinia* L. Plioc. Eu. *Bohlinia* [*Orasius*] U. Mioc.-L. Plioc. Eu. *Bramatherium* U. Plioc. SAs. *Csakvarotherium* L.-U. Plioc. Eu. *Decennatherium* U. Mioc.-L. Plioc. Eu. *Giraffa* [*Camelopardalis*] L. Plioc. Eu., L. Plioc.-Pleist. As., L. Pleist.-R. Af. *Giraffokeryx* U. Mioc.-L. Plioc. Eu. SAs. *Helladotherium* [*Panotherium*] L. Plioc. EEu. SWAs., ?Plioc. Af. *Honanotherium* U. Plioc. EAs. *Hydaspitherium* U. Plioc. SAs. *Libytherium* Plioc.-L. Pleist. Af. *Okapia* Pleist.-R. Af. *Palaeotragus* [*Achtiaria*] U. Mioc.-L. Plioc. EEu., U. Mioc.-Pleist. As., L. Plioc. NAf. *Prolibytherium* L. Mioc. NAf. *?Progiraffa* L. Mioc. As. *?Propalaeomeryx* L. Plioc. SAs. *Samotherium* [*Akicephalus Cheronotherium Shansitherium*] U. Mioc.-L. Plioc. Eu. As., L. Plioc. Af. *Sivatherium* [*Griquatherium Indratherium ?Orangiatherium*] Pleist. SAs., L.-U. Pleist. Af. *?Triceromeryx* [*Hispanocervus*] L. Mioc. Eu. As. *Vishnutherium* U. Plioc. SAs.

SUPERFAMILY BOVOIDEA. **Antilocapridae,** *Antilocapra* [*Neomeryx*] Pleist.-R. NA. *Capromeryx* [*Breameryx Dorcameryx*] Pleist. NA. *Ceratomeryx* L. Pleist. NA. *Hexobelomeryx* [*Hexameryx*] L. Plioc. NA. *Ilingoceros* U. Plioc.-Pleist. NA. *Meryceros* [*Submeryceros*] ?L. Plioc. NA. *Merycodus* [*Cosoryx Paracosoryx Subcosoryx Subparacosoryx*] M. Mioc.-U. Plioc. NA. *Osbornoceros* ?U. Plioc. NA. *Proantilocapra* L. Plioc. NA. *Ramoceros* [*Merriamoceros Paramoceros*] U. Mioc.-L. Plioc NA. *Sphenophalos* [*Plioceros*] Plioc.-L. Pleist. NA. *Tetrameryx* [*Hayoceros Stockoceros*] Pleist. NA. *Texoceros* U. Plioc.-L. Pleist. NA. **Bovidae,** *Addax* Pleist.-R. Af., R. SWAs. *Adenota* Pleist.-R. Af. *Aeotragus* Pleist. Af. *Aepyceros* Pleist.-R. Af. *Alcelaphus* [*Bubalis Pelorocerus*] ?Pleist. SWAs., L. Pleist.-R. Af. *Anoa* [*?Probubalis*] Pleist.-R. As. EInd. *Antilope* L. Pleist.-R. As., Pleist. EInd. *Antilospira* L.-U. Plioc., ?L. Pleist. EAs. *Bathyleptodon* Pleist. EAf. *Beatragus* L. Pleist.-R. Af. *Bibos* ?Plioc., Pleist.-R. As., Pleist. Eu., Pleist.-R. EInd. *Bison* [*Gigantobison Parabison Platycerobison Simobison Superbison Stelabison*] L.-U. Pleist. As., Pleist.-R. Eu. NA. *Boöpsis* L. Pleist. EAs. *Boötherium* Pleist. NA. *Bos* [*Poephagus*] Pleist.-R. Eu. As. Af. Alaska *Boselaphus* Pleist.-R. As. *Bubalus* L. Pleist. Eu., L.-U. Pleist. Af., L. Pleist.-R. As. EInd. *Bu-*

capra L. Pleist. SAs. *Budorcas* U. Plioc.-R. As. *Bularchus* Pleist. Af. *Cambayella* L. Pleist. SAs. *Capra* [*Aegoceros Ibex*] U. Plioc.-R. Eu., ?Pleist. NA., Pleist.-R. As. NAf. *Capricornis* Pleist.-R. EAs., R. EInd. *Cephalophus* Pleist.-R. Af. *Connochaetes* ?Plioc. Eu., Pleist.-R. Af., *Criotherium* L. Plioc. EEu. WAs. *Damalavus* L. Plioc. NAf. *Damaliscus* L. Pleist.-R. Af. *Damalops* L. Pleist. EAs. SAs. *Deperetella* L. Pleist. Eu. *Dorcadoryx Dorcadoxa* L. Plioc. As. *Duboisia* Pleist. SAs. EInd. *Eotragus* [*Eocerus Murphelaphus*] Mioc. Af. As., M.-U. Mioc. Eu. *Euceratherium* [*Aftonius Preptoceras*] Pleist. NA. *Fenhoryx* Plioc. EAs. *Gangicobus* L. Pleist. SAs. *Gazella* [*Antidorcas Gazelloportax Procapra*] U. Mioc.-L. Pleist. Eu., L. Plioc.-R. As. Af. *Gazellospira* L. Pleist. Eu. EAs. *Gidleya* Pleist. NA. *Gobiocerus* M. Mioc. EAs. *Gorgon* L. Pleist.-R. Af. *Helicoportax* L. Plioc. SAs. *Helicotragus* [*Helicoceras Helicophora*] L. Plioc. EEu. SWAs. *Hemibos* [*Amphibos Peribos ?Probubalis*] L. Pleist. As. *Hemistrepsiceros* L. Plioc. EEu. *Hemitragus* ?Plioc.-Pleist. As., Pleist. Eu., R. As. *Hesperoceras* L. Pleist. Eu. *Hippotragus* U. Plioc. Eu., U. Plioc.-L. Pleist. As., L. Pleist.-R. Af. *Hippotragoides Homoioceras* Pleist. Af. *Hydaspicobus* U. Plioc.-L. Pleist. SAs. *Hypsodontus* M. Mioc. EEu. *Indoredunca* L. Pleist. SAs. *Kobus* [*Cobus*] Pleist. As., Pleist.-R. Af. *Kobikeryx* Plioc. SAs. *Leptobison* L. Pleist. EAs. *Leptobos* [*Epileptobos*] L.-U. Pleist. Eu. As., U. Pleist. EInd. *Leptotragus* L. Plioc. EEu. SWAs. *Lunatoceras* Pleist. Af. *Lyrocerus* U. Plioc.-L. Pleist. EAs. *Makapania Megalotragus* Pleist. Af. *Megalovis* U. Plioc.-L. Pleist. EAs., L. Pleist. Eu. *Menelikia* L. Pleist. Af. *Microtragus* L. Plioc. EEu. SWAs. *Miotragocerus* [*Dystychoceras*] U. Mioc.-L. Plioc. Eu. *Myotragus* Pleist. Med. *Naemorhaedus* L. Pleist. Eu., Pleist.-R. As. *Neotragocerus* Pleist. NA. *Nesotragus* Pleist.-R. Af. *?Nothobos* Pleist. SA. *Numidocapra* L. Pleist. NAf. *Oioceros* L. Plioc. Eu. SWAs. *Olonbulukia* L. Plioc. EAs. *Onotragus* Pleist.-R. Af. *Oreamnos* Pleist.-R. NA. *Oreotragus* Pleist.-R. Af. *Oryx* L. Pleist.-R. Af., R. SWAs. *Ovibos* [*Parovibos, Praeovibos*] Pleist. Eu. As., Pleist.-R. NA. *Ovis* [*Ammotragus, Caprovis, Pachyceros*] L. Plioc.-R. As., L. Pleist.-R. Eu. NA. NAf. *Pachygazella* L. Plioc. EAs. *Pachyportax* L. Plioc.-L. Pleist. SAs. *Pachytragus* L. Plioc. EEu. SWAs., Plioc. NAf. *Palaeohypsodontus* M. Olig. EAs. *Palaeoreas* U. Mioc.-L. Plioc. Eu., L. Plioc. As. NAf. *Palaeoryx* L. Plioc. As., L. Plioc.-L. Pleist. Eu. Af. *Palaeotragiscus* Pleist. Af. *Pantholops* Pleist.-R. As. *Parabos* L.-U. Plioc. Eu. Af. *Paraboselaphus* Plioc. Eu. As. *Parabubalis* Pleist. As. *Paraprotoryx* L. Plioc. ?Eu. EAs. *Parapseudotragus* U. Mioc. Eu. *Paratragocerus* M. Mioc. EAs. *Parestigorgon Parmularius* Pleist. Af. *Parurmiatherium* L. Plioc. SWAs. EEu. *Pelea* Pleist.-R. Af. *Pelorovis* L. Pleist. Af. *Perimia* L. Plioc. SAs. *Phaleroceros Phenacotragus* Pleist. Af. *Philantomba* Pleist.-R. Af. *Platycerabos* [*Parabos*] Pleist. NA. *Platybos* L. Pleist. SAs. *Plesiaddax* L. Plioc. EAs. *Pliotragus* [*Deperetia*] L. Pleist. Eu. *Praedamalis Praemadoqua* Pleist. Af. *Proamphibos* U. Plioc.-L. Pleist. SAs. *Proboselaphus* L.-U. Pleist. EAs. *Procamptoceras* L. Pleist. Eu. *Procobus* L. Plioc. EEu. *Prodamaliscus* L. Plioc. EEu. SWAs. *Proleptobos* L. Plioc. SAs. *Pronetragus* L. Plioc. EEu. *Propalaeoryx* L. Mioc. Af. *Prosinotragus* L. Plioc. EAs. *Prostrepsiceros* L. Plioc. EEu. SWAs. *Protetraceros* Mioc. EAs. *Protoryx* L. Plioc. EEu., L. Plioc.-L. Pleist. As. *Protragelaphus* L. Plioc. EEu. SWAs. *Protragocerus* [*Paratragoceros*] L.-M. Mioc. Eu. *Pseudobos* Plioc. As. *Pseudotragus* L. Plioc. EEu. SWAs. *Pultiphagonides* L. Plioc. As., L. Pleist. Af. *Qurliqnoria* L. Plioc. EAs. *Rabaticerus* Pleist. NAf. *Raphicerus Redunca* [*Cervicapra*] Pleist.-R. Af. *Rhynotragus* Pleist. Af. *Rupicapra* L. Pleist.-R. Eu. SWAs. *Rueticeros* M. Plioc. As. *Saiga* Pleist. Alaska, Pleist.-R. As. EEu. *Samotragus* L. Plioc. SWAs. *Selenoportax* L.-U. Plioc. SAs. *Simatherium* L. Pleist. Af. *Sinoreas* U. Plioc. EAs. *Sinoryx* L. Plioc. EAs. *Sinotragus* L. Plioc. EAs. *Sivacapra* L. Pleist. SAs. *Sivaceros* L.-U. Plioc. SAs. *Sivacobus Sivadenota* L. Pleist. SAs. *Sivaportax* L. Plioc. SAs. *Sivatragus* U. Plioc.-L. Pleist. As. *Sivoreas* L. Plioc. SAs. *Sivoryx* L. Pleist. SAs. *Soergelia* Pleist. Eu. *Spirocerus* L.-U. Pleist. As. *Strepsiceros* [*Tragelaphus*] L. Pleist. EEu., L. Pleist.-R. Af. *Strepsiportax* L. Plioc. As. *?Strogulognathus* ?L. Mioc. Af., U. Mioc. Eu. *Sylvicapra* Pleist.-R. Af. *Symbos* [*Scaphoceros*] Pleist. NA. *Syncerus* Pleist.-R. Af. *Taurotragus* [*Oreas*] ?Plioc., L. Pleist.-R. Af., Plioc.-Pleist. As. *Tetracerus* Pleist.-R. SAs. *Thaleroceros* L. Pleist. EAf. *Tossunnoria* L. Plioc. EAs. *Tragocerus* [*Austroportax Dystichoceras Graecoryx Indotragus Pikermicerus Pontoportax Tragoportax*] U. Mioc. Pleist. Eu., L. Plioc. NAf., L.-U. Plioc. As., *Tragoreas* U. Mioc.-L. Plioc. EEu. WAs. *Tragospira* Pleist. Eu. *Tsaidamotherium* L. Plioc. EAs. *Urmiabos* L. Plioc. WAs. *Urmiatherium* L. Plioc. As. *Vishnucobus Vishnumeryx* L. Pleist. SAs. *Xenocephalus* L. Pleist. Af. *Yakopsis* Pleist. Eu.

ORDER EDENTATA

SUBORDER PALAEANODONTA. **Metacheiromyidae,** *Metacheiromys* M. Eoc. NA. *Palaeanodon* U. Paleoc.-L. Eoc. NA. **Epoicotheriidae,** *Epoicotherium* [*Xenotherium*] L. Olig. NA. *Pentapassalus* L. Eoc. NA. *Tetrapassalus* M. Eoc. NA. *Xenocranium* L. Olig. NA.

SUBORDER XENARTHRA

INFRAORDER LORICATA (CINGULATA)

SUPERFAMILY DASYPODOIDEA. **Dasypodidae,** R. SNA. SA., *Astegotherium* L. Eoc. SA. *Cabassous* [*Xenurus*] Pleist.-R. SA. *Chaetophractus* U. Plioc.-R. SA. *Chlamyphorus* [*Chlamydophorus*] Pleist. SA. *Chorobates* U. Plioc. SA. *Dasypus* [*Praopus Tatu Tatusia*] ?U. Plioc., Pleist.-R. SA., Pleist.-R. NA. *Doellotatus* [*Eutatopsis*] ?L., M.-U. Plioc. SA. *Euphractus* [*Scleropleura*] Pleist.-R. SA. *Eutatus Hoffstetteria* Pleist. SA. *Kraglievichia* U. Mioc.-U. Plioc. SA. *?Machlydotherium* L.-U. Eoc. SA. *Macroeuphractus* L.-U. Plioc. SA. *Meteutatus* [*Sadypus*] L. Eoc.-L. Olig. SA. *Paleuphractus* L. Plioc. SA. *Pampatherium* [*Chlamytherium Chlamydotherium Holmesina*] ?U. Plioc., Pleist. SA., Pleist. NA. *Plaina* M.-U. Plioc. SA. *Proeuphractus* L.-U. Plioc. SA. *Proeutatus* U. Olig.-U. Mioc. SA. *Propraopus* Pleist. SA. NA. *Prostegotherium* L. Eoc. SA. *Prozaedyus* U. Olig.-U. Mioc. SA. *Pseudeutatus* [*Anutaetus Isutaetus Pachyzedyus*] M. Eoc., ?L. Olig. SA. *Pseudostegotherium* ?L. Eoc., ?U. Olig. SA. *Ringueletia* U. Plioc. SA. *Stegotheriopsis* U. Olig. SA. *Stegotherium* L. Mioc. SA. *Stenotatus* [*Prodasypus*] U. Olig.-M. Mioc. SA. *Tolypeutes* U. Plioc.-R. SA. *Utaetus* [*Anteutatus Orthutaetus Para-*

taetus Posteutatus ? Coelutaetus] L.-M. Eoc. SA. *Vassallia* M. Plioc. SA. *Vetelia* L. Mioc. SA. *Zaedyus* U. Plioc.-R. SA. **Peltephilidae,** *Anantiosodon* L. Mioc. SA. *Epipeltephilus* U. Mioc., ?L. Plioc. SA. *Parapeltocoelus* U. Olig. SA. *Peltephilus* ?U. Olig., L. Mioc. SA. *Peltocoelus* U. Olig.-L. Mioc. SA. **Pseudorophodontidae,** *Pseudorophodon* U. Olig. SA.

SUPERFAMILY PALAEOPELTOIDEA. **Palaeopeltidae,** *Palaeopeltis* U. Eoc.-M. Olig. SA.

SUPERFAMILY GLYPTODONTOIDEA. **Glyptodontidae (Hoplophoridae),** *Asterostemma* L.-U. Mioc., Pleist. SA. *Berthawyleria* M. Plioc. SA. *Brachyostracon Castellanosia Chlamyphractus* Pleist. SA. *Chlamydotherium* [*Boreostracon*] Pleist. NA. SA. *Cochlops* L. Mioc. SA. *Comaphorus* ?M. Plioc. SA. *Daedicuroides* U. Plioc., ?L. Pleist. SA. *Daedicurus* Pleist. SA. *Eleutherocercus* M.-U. Plioc. SA. *Eosclerocalyptus* M., ?U. Plioc. SA. *Eosclerophorus* M. Plioc. SA. *Eucinepeltus* L. Mioc. SA. *Glyptatelus* M. Eoc.-L. Olig. SA. *Glyptodon* [*Glyptocoileus Glyptostracon Glyptopedius Stromatherium Xenoglyptodon*] ?U. Plioc., Pleist. SA., Pleist. NA. *Glyptodontidium* M. Plioc. SA. *Glyptotherium* L. Pleist. NA. *Hoplophorus* [*Sclerocalyptus*] Pleist. SA. *Hoplophractus* M. Plioc. SA. *Isolinia* U. Mioc. SA. *Lomaphorelus* M. Eoc. SA. *Lomaphorops* M. Plioc. SA. *Lomaphorus* ?Plioc., Pleist. SA. *?Metopotoxus* L. Mioc. SA. *Neosclerocalyptus Neothoracophorus Neuryurus* Pleist. SA. *Nopachtus* U. Plioc. SA. *Palaehoplophorus* M. Mioc.-M. Plioc., ?U. Plioc. SA. *Palaeodaedicurus* U. Plioc. SA. *Panochthus* Pleist. SA. *Paraglyptodon* U. Plioc. SA. *Parahoplophorus* ?M. Plioc. SA. *Peiranoa Phlyctaenopyga* M. Plioc. SA. *Plaxhaplous* Pleist. SA. *Plohophoroides* U. Plioc. SA. *Plohophorus* M. Plioc.-Pleist. SA. *Prodaedicurus* M.-U. Plioc. SA. *Propalaeohoplophorus* U. Olig.-U. Mioc. SA. *Propanochthus* U. Plioc. SA. *Protoglyptodon* ?M. Plioc. SA. *Pseudothoracophorus Pseudoeuryurus* M. Plioc. SA. *Stromaphoropsis* Mioc.-Plioc., ?Pleist. SA. *Stromaphorus* M. Plioc. SA. *Trachycalyptus Urotherium* ?M. Plioc., U. Plioc. SA. *Xiphuroides* ?U. Plioc., L. Pleist. SA.

INFRAORDER PILOSA

SUPERFAMILY MEGALONYCHOIDEA. **Megalonychidae,** *Acratocnus* [*Miocnus*] Pleist. WInd. *Amphiocnus* ?M. Plioc. SA. *Megalocnus* [*Miomorphus*] Pleist. WInd. *Megalonychops* ?L. Plioc., U. Plioc., ?Pleist. SA. *Megalonyx* [*Onychotherium*] L. Plioc., L.-U. Pleist. NA. *?Menilaus* ?M. Plioc. SA., *Mesocnus* [*Parocnus*] *Microcnus* Pleist. WInd. *Ocnopus* [*Parascelidodon*] Pleist. SA. *Ortotherium* ?M. Plioc. SA. *Paulocnus* Pleist. Curaçao *Pliomorphus Protomegalonyx* ?M. Plioc. SA. *?Sinclairia* U. Mioc. NA. **Megatheriidae,** *Analcimorphus* L. Mioc. SA. *Diheterocnus* [*Heterocnus*] U. Plioc., ?L. Pleist. SA. *Eremotherium* [*Pseuderemotherium Schaubia Schaubitherium*] Pleist. NA. SA. *Eucholoeops* L. Mioc. SA. *Hapaloides* U. Olig. SA. *Hapalops* [*Parhapalops Pseudhapalops Xyophorus*] L.-M. Mioc., ?L. Plioc. SA. *Hyperleptus* L. Mioc. SA. *Lophiodolodus* Olig. SA. *?Megalonychotherium* L. Mioc., ?Plioc. SA. *Megathericulus* ?L., U. Mioc. SA. *Megatheridium* [*Pliomegatherium*] L. Plioc. SA. *Megatherium* [*Paramegatherium*] ?U. Plioc., Pleist. SA., U. Plioc.-Pleist. NA. *Neohapalops* ?M. Plioc. SA. *Neoracanthus* Pleist. SA. *Nothropus* L. Pleist. SA. *Nothrotherium* [*Coelodon Nothrotheriops*] Pleist. NA. SA. *Pelecyodon* L. Mioc. SA. *Planops* [*Prepotherium*] L.-U. Mioc. SA. *Plesiomegatherium* L., ?M. Plioc. SA. *Promegatherium* [*Eomegatherium*] M.-U. Mioc. SA. *Pronothrotherium* [*Senetia*] U. Mioc.-U. Plioc. SA., ?U. Plioc. NA. *Proschizmotherium* U. Olig. SA. *? Protobradys* L. Eoc. SA. *? Pseudoprepotherium* Mioc. SA. *?Pyramiodontherium* [*Megatheriops*] M. Plioc. SA. *Schizmotherium* L. Mioc. SA. *Synhapalops* U. Plioc. or L. Pleist. SA. *? Valgipes* Pleist. SA. **Bradypodidae** R. SA. CA.

SUPERFAMILY MYLODONTOIDEA. **Mylodontidae,** *Analcitherium* L., ?M. Mioc. SA. *Chubutherium* U. Olig. SA. *Diodomus* Mioc. SA. *Elassotherium* M. Plioc. SA. *Glossotheridium* U. Plioc. SA. *Glossotherium* [*Eumylodon "Mylodon" Oreomylodon Pseudolestodon*] ?U. Plioc., Pleist. SA. *Inthodon* Mioc. SA. *Laniodon* Plioc. SA. *Lestodon* [*Prolestodon*] *Mylodon* [*"Glossotherium" Grypotherium Neomylodon*] Pleist. SA. *Nematherium* L. Mioc., ?M. Mioc. SA. *Neonematherium* U. Mioc. SA. *Nephotherium* ?M. Plioc. SA. *Octodontotherium* L. Olig. SA. *Octomylodon* ?M. Plioc. SA. *Orophodon* L. or M. Olig. SA. *Paramylodon* [*"Mylodon"*] Pleist. NA. *Platyonyx* Pleist. SA. *Pleurolestodon* M. Plioc. SA. *Promylodon* ?M. Plioc. SA. *Proscelidodon* U. Plioc. SA. *Ranculcus* M. Plioc. SA. *Scelidodon* Pleist. SA. *Scelidotherium* [*Catonyx*] ?U. Plioc., Pleist. SA. *Sphenotherus* M. Plioc. SA. *Stenodontherium* L. Mioc. SA. *Strabosodon* ?M. Plioc. SA. **?Entelopsidae,** *Entelops* L. Mioc. SA.

INFRAORDER VERMILINGUA. **Myrmecophagidae** R. SA. CA., *Myrmecophaga* [*Neotamandua*] M. Plioc.-R. SA., R. CA. *Nunezia* ?M., U. Plioc. SA. *Palaeomyrmedon* M. Plioc. SA. *Promyrmephagus* [*Protamandua*] L. Mioc. SA. *Tamandua* Pleist.-R. SA., R. CA.

?EDENTATA INCERTAE SEDIS: *Chungchienia* U. Eoc. EAs.

ORDER PHOLIDOTA. **Manidae,** *? Galliaetatus* Mioc. Eu. *? Leptomanis* Olig. Eu. *Manis* Pleist. ?Eu. EInd., Pleist.-R. As., R. Af. *? Necromanis* Olig. Eu. *? Teutomanis* Olig., Mioc. Eu.

ORDER TUBULIDENTATA. **Orycteropodidae,** *Myoricteropus* L. Mioc. EAf. *Orycteropus* U. Mioc.-R. Af., L. Plioc. EEu. SWAs. *? Palaeorycteropus* Eoc.-Olig. Eu. *Plesiorycteropus* Pleist. Mad. *? Tubulodon* L. Eoc. NA.

ORDER CETACEA

SUBORDER ARCHAEOCETI. **Protocetidae,** *? Anglocetus* L. Eoc. Eu. *Eocetus* [*Mesocetus*] U. Eoc. NAf. *? Pappocetus* M. Eoc. WAf. *Protocetus* M. Eoc. NAf., ?NA. **Dorudontidae,** *Dorudon* U. Eoc. NAf. NA. *Kekenodon* L. Mioc. NZ. *Phococetus* L. Mioc. Eu. *Zygorhiza* U. Eoc. Eu. NA. **Basilosauridae (Zeuglodontidae),** *Basilosaurus* [*Hydrargos Zeuglodon*] U. Eoc. NA. NAf. *Mammalodon* ?Eoc. Aus. *Platyosphys* L. Olig. Eu. *? Pontobasileus* ?Tert. NA. *? Pontogeneus* U. Eoc. NA. *Prozeuglodon* M.-U. Eoc. NAf.

SUBORDER ODONTOCETI. **Agorophiidae,** *Agorophius Xenorophus* U. Eoc. NA. **Squalodontidae,** *Colophonodon* U. Mioc. NA. *Metasqualodon* L. Mioc. Aus. *Microcetus* U. Olig. Eu. *Microsqualodon ?Microzeuglodon Neosqualodon* L. Mioc. Eu. *Parasqualodon* L. Mioc. Aus. *Phoberodon* L. Mioc. SA. *Phocogenius* Mioc. NA. *Prionodelphis* L. Plioc. SA. *Prosqualodon* Olig. Aus., L. Mioc. SA. NZ. *Rhytisodon* L. Mioc. Eu. *Saurocetus* U. Mioc. NA. *Squalodon* [*Phocodon*] L.-U. Mioc. Eu., M.-U. Mioc. NA., Mioc. NZ. *Tangarasaurus* L. Mioc. NZ. *Trirhizodon*

L.-M. Mioc. Eu. **Platanisidae** R. As. SA., *Anisodelphis* Plioc. SA. *Goniodelphis Hesperocetus* L. Plioc. NA. *Hesperoinia* Mioc. Eu. *Ischyrorhynchus* Plioc. SA. *?Pachyacanthus* U. Mioc. Eu. *Plicodontinia* ?Pleist. SA. *Proinia* L. Mioc. SA. *Saurodelphis* [*Pontoplanodes Saurocetes*] Plioc. SA. *Zarhachis* M. Mioc. NA. **Ziphiidae** R. Oc., *Anoplonassa* U. Mioc. NA. *Belemnoziphius* U. Mioc. Eu. NA. *Berardiopsis* U. Plioc. Eu. *Cetorhynchus* M.-U. Mioc. Eu. *Choneziphius* U. Mioc.-U. Plioc. Eu., U. Mioc. NA. *Eboroziphius* U. Mioc. NA. *Incacetus* Mioc. SA. *Mesoplodon* U. Mioc.-U. Plioc. Eu., U. Mioc. NA., Plioc. Aus., R. Oc. *Notocetus* [*Diochotichus*] L. Mioc. SA. *Palaeoziphius* U. Mioc. Eu. *?Pelycorhamphus* Mioc. NA. *Proroziphius* U. Mioc. NA. *Squalodelphis* L. Mioc. Eu. *Ziphioides* M. Mioc. Eu. *Ziphirostrum* [*Mioziphius*] U. Mioc. Eu. *Ziphius* ?L. Plioc. Eu., R. Oc. **Delphinidae** R. Oc. SA., *Agabelus* M. Mioc. NA. *Allodelphis Araeodelphis* Mioc. NA. *Belosphys Ceterhinops* M. Mioc. NA. *Delphinavus* L. Mioc. NA. *Delphinodon* M. Mioc. NA. *Delphinopsis* U. Mioc. Eu. *Delphinus* L. Plioc.-Pleist. Eu., R. Oc. *Doliodelphis* Mioc. NA. *Globicephala* Pleist. NA., R. Oc. *Grypolithax* M. Mioc. NA. *Iniopsis* L. Mioc. Eu. *Ixacanthus Kentriodon Lamprolithax* M. Mioc. NA. *Leptodelphis* U. Mioc. Eu. *Liolithax* M. Mioc. NA. *Lonchodelphis* U. Plioc. NA. *Loxolithax* M. Mioc. NA. *Macrochirifer* U. Mioc. Eu. *Macrodelphinus* Mioc. NA. *Megalodelphis* M. Mioc. NA. *Miodelphis* Mioc. NA., Plioc. Eu. *Nannolithax Oedolithax* M. Mioc. NA. *Orcinus* U. Plioc.-Pleist. Eu. EAs., R. Oc. *Pelodelphis* M. Mioc. NA. *Pithanodelphis* U. Mioc. Eu. *Platylithax* M. Mioc. NA. *Pontistes Pontivaga* Plioc. SA. *Priscodelphinus* Mioc. NA. *Protodelphinus* L. Mioc. Eu. *Pseudorca* L. Pleist. Eu. EAs., R. Oc. *Sinanodelphinus* Mioc. EAs. *Sormatodelphis* U. Mioc. Eu. *Steno* L.-U. Plioc. Eu., R. Oc. *Stenodelphis* [*Pontoparia*] Pleist. NA., R. SA. *Stereodelphis* M. Mioc. Eu. *Tretosphrys* M. Mioc. NA. *Tursiops* [*Tursio*] L. Pleist. Eu. NA., R. Oc. **Eurhinodelphidae,** *Argyrocetus* L. Mioc. SA. *Eurhinodelphis* M. Mioc. NA., U. Mioc. Eu. EAs. *Ziphiodelphis* L. Mioc. Eu. **Hemisyntrachelidae,** *Hemisyntrachelus* L. Plioc. Eu. *Lophocetus* M. Mioc. NA. **Acrodelphidae,** *Acrodelphis* L.-U. Mioc. Eu., M. Mioc. NA. *Champsodelphis* M.-U. Mioc. Eu. *Eoplatanista* L. Mioc. Eu. *Heterodelphis* M.-U. Mioc. Eu. *Pomatodelphis* M. Mioc. Eu., L. Plioc. NA. *Schizodelphis* [*Cyrtodelphis*] L. Mioc. Af., L. Mioc.-L. Plioc. Eu., M. Mioc., L. Plioc. NA. **Monodontidae** (**Delphinapteridae**) R. Oc., *Delphinapterus* Pleist. Eu. NA., R. NOc. *Monodon* Pleist. Eu. NA., R. NOc. **Phocaenidae** R. Oc., *Palaeophocaena* U. Mioc. Eu. *Phocaena* [*Phocaenoides*] Pleist. EAs., R. Oc. *Phocaenopsis* Pleist. NZ. *Protophocaena* U. Mioc. Eu. **Physeteridae** R. Oc., *Apenophyseter* L. Mioc. SA. *Aulophyseter* M. Mioc. NA. *Balaenodon* U. Plioc. Eu. *Diaphorocetus* L. Mioc. SA. *Dinoziphius* U. Mioc. NA. *?Graphiodon* Mioc. NA. *Hoplocetus* M. Mioc. Eu., ?L. Plioc. NA. *Idiophyseter* M. Mioc. NA. *Idiorophus* L. Mioc. SA. *Kogia* Plioc. EAs., R. Oc. *Kogiopsis* L. Plioc. NA. *Ontocetus* M. Mioc. EAs., M.-U. Mioc. NA. *Orycterocetus* M. Mioc. NA. *Physeter* L. Mioc.-Pleist. Eu., U. Mioc.-Pleist. NA., R. Oc. *Physeterula* U. Mioc. Eu. *Physetodon* L. Plioc. Aus. *Placoziphius* U. Mioc. Eu. *Priscophyseter* U. Plioc. Eu. *Prophyseter* U. Mioc. Eu. *Scaldicetus* ?L. Mioc., U. Mioc. Eu., ?L. Plioc. Aus. *Scaptodon* Mioc. Aus. *Thalassocetus* U. Mioc. Eu.

SUBORDER MYSTICETI. **Patriocetidae,** *? Agriocetus* U. Olig. Eu. *Archaeodelphis* U. Eoc. NA. *Patriocetus* U. Olig. Eu. **Cetotheriidae,** *Aglaocetus* L. Mioc. SA. *Amphicetus* L. Plioc. Eu. *Aulocetus* Mioc. Eu. *Cephalotropis* U. Mioc. NA. *Cetotheriomorphis* U. Mioc. Eu. *Cetotheriopsis* U. Olig. Eu., L. Mioc. SA. *Cetotherium* U. Mioc.-L. Plioc. Eu. *Cophocetus* M. Mioc. NA. *Eucetotherium* U. Mioc. Eu. *Herpetocetus* U. Mioc. Eu. *Heterocetus* L. Plioc. Eu. *Isocetus* ?M. Mioc. NA., U. Mioc. Eu. *Mauicetus* L. Mioc. NZ. *Mesocetus* ?M. Mioc. NA., M.-U. Mioc. Eu. *Metopocetus* U. Mioc. NA. Eu. *Mixocetus* U. Mioc. NA. *Nannocetus* Plioc. NA. *Pachycetus* M. Olig. Eu. *Palaeobalaena* Mioc. SA. *Parietobalaena* ?M.-U. Mioc. NA. *Periplocetus* M. Mioc. NA. *Plesiocetopsis* [*Plesiocetus*] Mioc. SA., U. Mioc. NA., L. Plioc. Eu. *Rhegnopsis Siphonocetus* U. Mioc. NA. *Tiphyocetus* M. Mioc. NA. *Tretulias Ulias* U. Mioc. NA. **Eschrichtiidae** (**Rhachianectidae**) R. NPac., *Eschrichtius* [*Rhachianectes*] ?Pleist. Eu., R. NPac. **Balaenopteridae** R. Oc., *Balaenoptera* [*Cetotheriophanes*] U. Mioc.-Pleist. Eu., L. Plioc.-Pleist. NA., Pleist. As., R. Oc. *Burtinopsis* U. Plioc. Eu. *Idiocetus* Mioc. EAs., U. Plioc. Eu. *Megaptera* L. Plioc., Pleist. NA., U. Plioc. Eu., R. Oc. *Megapteropsis* L. Plioc. Eu. *Mesoteras* U. Mioc. NA. *Notiocetus* Plioc. SA. *Palaeocetus* U. Plioc. Eu. **Balaenidae** R. Oc., *Balaena* Plioc. Aus., L. Plioc.-Pleist. Eu., R. Arctic Oc. *Balaenotus Balaenula* U. Plioc. Eu. *Eubalaena* Pleist. Eu. SA., R. Oc. *Morenocetus* L. Mioc. SA. *Protobalaena* L. Plioc. Eu.

ORDER RODENTIA

SUBORDER SCIUROMORPHA (PROTROGOMORPHA)

SUPERFAMILY ISCHYROMYOIDEA. **Paramyidae,** *Ailuravus* [*Megachiromyoides*] M. Eoc. Eu. *Cedromus* L.-M. Olig. NA. *Decticadapis* L. Eoc. Eu. *Franimys* L. Eoc. NA. *Ischyrotomus* M.-U. Eoc. NA. *Leptotomus* L.-U. Eoc. NA. *Lophiparamys* L. Eoc. NA. *Manitscha* L.-M. Olig. NA. *Maurimontia* U. Eoc. Eu. *Microparamys* L. Eoc. Eu., L.-U. Eoc. NA. *Mytonomys* U. Eoc. NA. *Paramys* U. Paleoc.-M. Eoc. NA., L.-M. Eoc. Eu. *Pelycomys* L.-M. Olig. NA. *Plesiarctomys* M.-U. Eoc. Eu. *Plesiospermophilus* M. Olig. Eu., U. Olig. As. *Prosciurus* L. Olig.-L. Mioc. NA., U. Olig. EAs. *Pseudotomus* L.-M. Eoc. NA. *Rapamys* U. Eoc. NA. *Reithroparomys* L.-U. Eoc. NA. *Tapomys* U. Eoc. NA. *Thisbemys* L.-U. Eoc. NA. *Uriscus* U. Eoc. NA. **Sciuravidae,** *? Advenimus* U. Eoc. EAs. *Dawsonomys* L. Eoc. NA. *Floresomys* Eoc. NA. *Knightomys* L. Eoc. NA. *Mysops* L.-M. Eoc. NA. *Pauromys* M. Eoc. NA. *Sciuravus* L.-U. Eoc. NA., U. Eoc. EAs. *Taxymys* M.-U. Eoc. NA. *Tillomys* M. Eoc. NA. *Tsinlingomys* U. Eoc. EAs. **Cylindrodontidae,** *Ardynomys* L. Olig. EAs. NA. *Cyclomylus* U. Olig. As. *Cylindrodon* L. Olig. NA. *Pareumys Presbemys* U. Eoc. NA. *Pseudocylindrodon* L. Olig. NA. EAs. *Pseudotsaganomys* L. Olig. As. *Sespemys* U. Olig., ?L. Mioc. NA. *Tsaganomys* U. Olig. As. **Protoptychidae,** *Protoptychus* M.-U. Eoc. NA. **Ischyromyidae,** *Ischyromys* L.-U. Olig. NA. *Titanotheriomys* L. Olig. NA.

SUPERFAMILY APLODONTOIDEA. **Aplodontidae** R. NAs. NA., *Allomys* U. Olig. Eu., L. Mioc. NA. *Ameniscomys* M. Mioc. Eu. *Aplodontia* Pleist.-R. NA. *Eohaplomys* U. Eoc. NA. *Haplomys* L. Mioc. NA. *Liodontia* M. Mioc.-Plioc. NA. *Meniscomys* L. Mioc. NA. *Niglarodon* U. Olig.-L. Mioc. NA. *Pseudaplodon* L. Plioc. As. NA. *Sciurodon* M. Mioc. Eu. *Sewellelodon* L. Mioc. NA. *Tardontia* L. Plioc. NA. **Mylagaulidae** *Ceratogaulus* U. Mioc. NA. *Epigaulus* L. Plioc. NA. *Meso-*

gaulus L.-U. Mioc. NA. *Mylagaulodon* L. Mioc. NA. *Mylagaulus* U. Mioc.-U. Plioc. NA. *Promylagaulus* L. Mioc. NA.

SUPERFAMILY SCIUROIDEA. **Sciuridae** R. Eu. As. Af. NA. SA., *Ammospermophilus* L. Plioc.-R. NA. *Arctomyoides* U. Mioc. NA. *Burosor* Pleist. NA. *Citellus* [*Otospermophilus Pliocitellus Spermophilus*] M. Mioc.-R. NA., U. Plioc.-R. Eu., ?U. Tert., Pleist.-R. As. *Csakvaromys* L. Plioc. Eu. *Cynomys* ?U. Mioc., L. Plioc.-R. NA., R. As. *Eutamias* U. Mioc.-R. As., Pleist.-R. NA. L. Plioc.-R. Eu. *Getuloxerus* U. Mioc. NAf. *Glaucomys* Pleist.-R. NA. *Heteroxerus* U. Mioc. Eu. *Marmota* [*Arctomys Stereodectes*] L. Plioc.-R. NA., ?U. Tert., Pleist.-R. As., Pleist.-R. Eu. *Miosciurus* L. Mioc. NA. *Miospermophilus* L.-M. Mioc. NA. *Paenemarmota* U. Plioc.-L. Pleist. NA. *Palaeoarctomys* L.-U. Mioc. NA. *Petauria* Pleist. Eu. *Petaurista* Pleist.-R. As. EInd. *Plesispermophilus* U. Eoc. or L. Olig. Eu. *Pliopetes* ?L. Plioc., U. Plioc. Eu. *Pliosciuropterus* L. Plioc. Eu. ?*Protosciurus* L. Olig.-L. Mioc. NA. *Protospermophilus* L. Mioc.-L. Plioc. NA. *Pteromys* [*Sciuropterus*] U. Mioc.-L. Plioc. NA., U. Mioc.-R. Eu., Pleist.-R. As. *Ratufa* ?L. Mioc., Pleist.-R. EInd., R. SAs. *Sciurotamias* Pleist.-R. As. *Sciurus* U. Olig.-R. Eu., M. Mioc.-R. NA., ?Olig., Pleist.-R. As. SA., R. NAf. *Tamias* [*Neotamias*] L. Mioc.-R. NA., Pleist. EAs. *Tamiasciurus* Pleist.-R. NA. *Xerus* ?Mioc. Eu., Pleist.-R. Af.

SUBORDER CAVIOMORPHA

SUPERFAMILY OCTODONTOIDEA. **Octodontidae** R. SA., *Acaremys* U. Olig.-L. Mioc. SA. *Alterodon* Pleist. WInd. *Palaeoctodon* ?M. Plioc. SA. *Phthoramys* L. Plioc. SA. *Pithanotomys* U. Plioc.-Pleist. SA. *Plagiodonta* SubR.-R. WInd. *Plataeomys* Pleist. SA. *Platypittamys* L. Olig. SA. *Pseudoplataeomys* Plioc. SA. *Sciamys* L. Mioc. SA. *Strophostephanus* ?M. Plioc. SA. **Echimyidae,** R. SA. CA., *Adelphomys* L. Mioc. SA. *Aphaetreus* Pleist. WInd. *Boromys* SubR. CA. *Brotomys* SubR. CA. *Carterodon* Pleist.-R. SA. *Cercomys* Pleist.-R. SA. *Deseadomys* L. Olig. SA. *Dicolpomys* Pleist. SA. *Echimys* [*Echinomys Loncheres*] Pleist.-R. SA., R. CA. *Eumysops* [*Proaguti Proatherura*] ?L., U. Plioc.-L. Pleist. SA. *Euryzygomatomys* Pleist.-R. SA. *Haplostropha* ?M. Plioc. SA. *Heteropsomys Homopsomys Isolobodon* Pleist. WInd. *Isothrix* Pleist.-R. SA. *Ithydontia* Pleist. WInd. *Kannabateomys Mesomys* Pleist.-R. SA. *Palaeoechimys* U. Plioc. SA. *Proëchimys* Pleist. WInd., Pleist.-R. SA., R. CA. *Prospaniomys* U. Olig. SA. *Protacaremys* [*Archaeocardia Eoctodon* "*Palaeocardia*"] *Protadelphomys* U. Olig. SA. *Spaniomys* [*Graphimys Gyrignophus*] *Stichomys* L. Mioc. SA. **Ctenomyidae** R. SA., *Actenomys* [*Dicoelophorus*] U.. Plioc. SA. *Ctenomys* [*Paractenomys*] U. Plioc.-R. SA. *Eocoelophorus* L. Pleist. SA. *Eucoelophorus* U. Plioc. SA. *Megactenomys* L. Pleist. SA. *Xenodontomys* M. Plioc.-Pleist. SA. **Abrocomidae** R. SA., *Abrocoma* M. Plioc.-R. SA. *Protabrocoma* M. Plioc. SA. **Capromyidae** R. SA. CA. WInd., *Capromys* Pleist.-R. WInd. *Isomyopotamus* U. Plioc. SA. *Macrocapromys* Pleist. WInd. *Myocastor* [*Myopotamus*] U. Plioc.-R. SA. *Paramyocastor* M. Plioc. SA. *Quemisia* Pleist. WInd. *Tramycastor* L. Pleist. SA.

SUPERFAMILY CHINCHILLOIDEA. **Chinchillidae** R. SA., *Euphilus* Plioc. SA. ?*Eusigmomys* Mioc. SA. *Lagidium* Pleist.-R. SA. *Lagostomopsis* U. Plioc. SA. *Lagostomus* U. Plioc.-R. SA. *Perimys* U. Olig.-L. Mioc. SA. *Pliolagostomus Prolagostomus Scotaeumys* L. Mioc. SA. *Scotamys* L. Olig. SA. *Sphaeromys* Olig.-Mioc. SA. *Sphodromys* L. Mioc. SA. **Dasyproctidae** R. SA. CA., *Cephalomys* [*Orchiomys*] L. Olig. SA. *Cuniculus* [*Agouti Coelogenus Coelogenys Stichtomys*] Pleist.-R. SA. *Dasyprocta* Pleist.-R. SA. WInd., R. CA. ?*Homocentrus* L. Mioc. SA. *Litodontomys* L. Olig. SA. *Neoreomys* L.-U. Mioc. SA. *Olenopsis* L. Mioc.-Plioc., ?Pleist. SA., *Scleromys* [*Lomomys*] ?U. Olig., L., ?U. Mioc. SA. **Dinomyidae** R. SA. *Carlesia* ?M. Plioc. SA. *Clidomys* Pleist. WInd. ?*Colpostemma Dabbenea Diaphoromys* ?M. Plioc. SA. *Dinomys* ?L. Mioc. R. SA. *Doellomys* ?M. Plioc. SA. *Eumegamys* [*Megamys*] *Eumegamysops* ?M. Plioc. SA. *Gyriabrus* L.-M. Plioc. SA. *Isostylomys Morenia Neoepiblema Pentastylodon Pentastylomys* ?M. Plioc. SA. *Phoberomys* U. Mioc.-Plioc. SA. *Potamarchus* [*Discolomys*] L.-M. Plioc. SA. *Protomegomys Pseudosigmomys* ?M. Plioc. SA. *Rusconia* ?M. Plioc. SA. *Simplimus* U. Mioc. SA. *Speoxenus* Pleist. WInd. *Spirodontomys* Pleist. WInd. *Telicomys* U. Plioc. SA. *Telodontomys* ?M. Plioc. SA. *Tetrastylomys* ?M. Plioc. SA. *Tetrastylopsis* M. Plioc. SA. *Tetrastylus* Plioc. SA. **Elasmodontomyidae,** *Amblyrhiza Elasmodontomys* [*Heptaxodon*] Pleist. WInd.

SUPERFAMILY CAVIOIDEA. **Eocardiidae,** *Asteromys* L. Olig. SA. *Chubutomys* L. Olig. SA. *Eocardia* [*Dicardia* ?*Hedymys Tricardia*] L. Mioc. SA. *Luantus* [*Luanthus*] *Phanomys* U. Olig.-L. Mioc. SA. *Schistomys* [*Procardia*] L. Mioc. SA. **Caviidae** R. SA. , *Allocavia* U. Plioc. SA. *Cardiomys* [*Caviodon Diocartherium Lelongia Neoprocavia Pseudocerdiomys*] L.-U. Plioc. SA. *Cavia* Pleist.-R. SA. *Caviops Dolicavia* U. Plioc. SA. *Dolichotis* [*Paradolichotis*] U. Plioc.-R. SA. *Galea* Pleist.-R. SA. *Macrocavia* U. Plioc. SA. *Microcavia* [*Nanocavia*] ?U. Plioc., Pleist.-R. SA. *Neocavia* M. Plioc. SA. *Orthomyctera* Plioc. SA. *Palaeocavia* U. Plioc. SA. *Parodimys* ?M. Plioc. SA. *Pascualia Pliodolichotis* U. Plioc. SA. *Procardiomys* L. Plioc. SA. *Prodolichotis* U. Mioc.-Plioc. SA. *Propediolagus* ?U. Plioc. SA. *Xenomicavia* Plioc. SA. **Hydrochoeridae** R. SA. CA., *Anchimys* ?M. Plioc. SA. *Anchimysops* ?M., U. Plioc. SA. *Cardiatherium* [*Cardiotherium*] M.-?U. Plioc. SA. *Eucardiodon* [*Cardiodon*] ?M. Plioc. SA. *Hydrochoeropsis* U. Plioc. SA. *Hydrochoerus* Pleist. NA. WInd., Pleist.-R. SA. *Neoanchimys* U. Plioc. SA. *Neochoerus* [*Palaeohydrochoerus Pliohydrochoerus Prohydrochoerus Protohydrochoerus*] ?U. Plioc.-Pleist. SA., Pleist. NA. CA. *Nothydrochoerus* U. Plioc., L. Pleist. SA. ?*Phagatherium* Plioc. SA. *Plexoechoerus* M., ?U. Plioc. SA. *Procardiatherium* ?M. Plioc. SA. *Xenocardium* L. Plioc. SA. *Xenohydrochoerus* ?U. Plioc., L. Pleist. SA.

SUPERFAMILY ERETHIZONTOIDEA. **Erethizontidae** R. NA. SA., *Chaetomys* Pleist.-R. SA. *Coendu* Pleist.-R. SA., R. CA. *Eosteriomys* U. Olig. SA. *Erethizon* L. Pleist.-R. NA. *Hypsosteiromys* U. Olig. SA. *Neosteiromys* M. Plioc. SA. *Protosteiromys* L. Olig. SA. *Steiromys* [*Parasteiromys*] U. Olig.-Plioc. SA.

CAVIAMORPHA INCERTAE SEDIS: *Drytomomys* Pleist. SA. *Lomodelphomys* Mioc. SA. *Ronchophorus* Pleist. SA.

SUPERFAMILY MUROIDEA. **Cricetidae** R. Eu. As. Af. NA. SA., *Akodon* Pleist.-R. SA. *Allocricetus* L. Pleist. Eu. WAs. *Allophaiomys* L. Pleist. Eu. *Alticola* Pleist.-R. As. *Anatolomys* L. Plioc. EAs. *Anomalomys* U. Mioc.-U. Plioc. Eu. *Aralomys Argyromys* Olig. WAs. *Arvicola* Pleist.-R. Eu. As. *Baiomys* U. Plioc.-R. NA. *Bensonomys* Pleist. NA. *Blarinomys* Pleist.-R. SA. *Bothriomys* L. Pleist. SA. *Cimmaronomys* U. Plioc.-Pleist. NA. *Cleithrionomys* [*Evotomys*] Pleist.-R. Eu. As. NA. *Copemys* U. Mioc.-L. Plioc. NA. *Cotimus* U. Mioc. Eu. NA. *Cricetinus* Pleist. EAs. *Cricetodon* L. Olig. EAs., L. Olig.-L. Plioc. Eu., M. Olig. NA., U. Mioc. NAf. *Cricetops* U. Olig. As. *Cricetulus* L. Plioc.-R. As., Pleist.-R. Eu. *Cricetus* L. Plioc.-R. Eu., ?Plioc., L. Pleist.-R. As. *Cudahyomys* Pleist. NA. *Democricetodon* [*Megacricetodon*] L.-U. Mioc. Eu. *Dicrostonyx* Pleist.-R. Eu. As., R. NA. *Dinaromys* Pleist. Eu. *Dolomys* [*Pliomys Apistomys*] ?L. Plioc., U. Plioc.-R. Eu. *Eligmodontia* L.-U. Pleist. NA., R. SA. *Ellobius* Pleist.-R. Eu. As. NAf. *Eothenomys* Pleist.-R. As. *Epicricetodon* L. Plioc. EAs. *Eumys* L. Olig.-L. Mioc. NA., U. Olig. As. *Eumysodon* Olig. WAs. *Gerbillus* L. Plioc.-R. As., R. Af. *Golunda* Pleist.-R. As. *Hesperomys* Pleist.-R. SA. *Heterocricetodon* L. Olig. Eu. *Holochilus* Pleist.-R. SA. *?Horatiomys* U. Mioc. NA. *Hypogeomys* Pleist.-R. Mad. *Kanisamys* L.-U. Plioc. EAs. *Kislangia Lagurodon* Pleist. Eu. *Lagurus* L. Pleist. Eu., R. As. NA. *Laugaritiomys* L. Pleist. Eu. *Leidymys* M. Olig.-L. Mioc. NA. *Lemmus* [*Myodes*] Pleist.-R. Eu. As. NA. *Lophocricetus* L. Plioc. EAs. *Macrotarsomys* Pleist.-R. Mad. *Majoria* Pleist. Mad. *Megalomys* Pleist.-R. WInd. *Melissiodon* L. Olig.-M. Mioc. Eu. *Meriones* Pleist.-R. NAf., R. EEu. As. *Mesocricetus* Pleist.-R. As., R. Eu. *Microtodon* [*Baranomys Prosomys*] L. Plioc. As., U. Plioc. NA., U. Plioc.-Pleist. Eu. *Microtoscoptes* [*Goniodontomys*] U. Plioc.-L. Pleist. EAs. NA. *Microtus* [*Chionomys Pedomys Phaiomys*] U. Plioc.-R. Eu., L. Pleist.-R. NA., Pleist. As., Pleist.-R. NAf. *Mimomys* [*Cosomys*] U. Plioc.-Pleist. Eu. NA., Pleist. As. *Miochomys* U. Mioc. NA. *Miromus* Pleist. Eu. *Myocricetodon* U. Mioc. NAf. *Myospalax* [*Miotalpa Siphneus*] U. Plioc.-R. As. *Mystromys* Pleist.-R. Af. *Nannocricetus* L. Plioc. EAs. *Nebraskomys* Pleist. NA. *Necromys* Pleist. SA. *?Neocometes* M. Mioc. Eu. *Neocricetodon* Pleist. Eu. *Neofiber Neotoma* L. Pleist.-R. NA. *Ogmodontomys* U. Plioc.-L. Pleist. NA. *Ondatra* [*Anaptogonia Sycium*] L. Pleist.-R. NA. *Onychomys* U. Plioc.-R. NA. *Oryzomys* Pleist.-R. NA. SA. *Oxymycterus* Pleist.-R. SA. *Paciculus* L.-M. Mioc. NA. *Palaeocricetus* M. Mioc. Eu. *Paracricetodon* L. Olig. Eu. *Paracricetulus* L. Plioc. EAs. *Parahodomys* L. Pleist.-R. NA. *Peromyscus* [*Haplomylomys*] U. Plioc.-R. NA. *Phenacomys* L. Pleist.-R. NA. *Phodopus* Pleist. Eu., R. As. *Phyllotris* Pleist.-R. SA., *Pitymys* Pleist.-R. Eu. As. NA. *Plesiodipus* [*Plesiocricetodon*] L. Plioc. EAs. *Pliolemmus* L. Pleist. NA. *Pliophenacomys* Plioc.-Pleist. NA. *Pliopotamys* L.-U. Pleist. NA. *Pliotomodon* U. Plioc. NA. *Poamys* U. Mioc. NA. *Promimomys* U. Plioc. Eu. *Proreithrodon* U. Plioc. SA. *Prosiphneus* U. Mioc.-L. Pleist. EAs. *Pseudomeriones* U. Mioc. or L. Plioc. EAs. *Ptyssophorus* Pleist. SA. *Reithrodon* Pleist.-R. SA. *Reithrodontomys* L. Pleist.-R. NA., R. SA. *Rhipidomys* Pleist.-R. SA., R. CA. *Ruscinomys* U. Mioc.-L. Plioc. Eu. *Scafteromys* Pleist.-R. SA. *Scottimus* M. Olig.-L. Mioc. NA. *Selenomys* U. Olig. EAs. *Sigmodon* U. Plioc.-R. SA., L. Pleist.-R. NA. *Sinocricetus* L. Plioc. EAs. *Stachomys* [*Leukaristomys*] U. Plioc.-L. Pleist. Eu. *Symmetrodontomys* L. Pleist. NA. *Synaptomys* [*Microtomys Mictomys*] L. Pleist.-R. NA. *Tatera* Pleist.-R. Af. As. *Thomasomys* Pleist.-R. WInd., R. SA. *Tretomys* Pleist. SA. *Trilophomys* L.-U. Plioc. Eu. *Tyrrhenicola* Pleist. Eu. *Ungaromys* [*Germanomys*] ?L. Plioc., U. Plioc.-Pleist. Eu. *Villanyia* Pleist. Eu. *Zygodontomys* Pleist.-R. SA., R. CA. **Muridae** R. Eu. As. Aus., *Acomys* Plioc.-R. Eu., Pleist.-R. Af., R. SWAs. *Anthracomys* L. Plioc. Eu. ?As. *Apodemus* U. Plioc.-R. Eu., Pleist.-R. As. NAf. *Arvicanthus* Pleist.-R. Af. WAs. *Coryphomys* Pleist. EInd. *Colomys Dasymys Dendromys Grammomys* Pleist.-R. Af. *Gyomys* Pleist.-R. Aus. *Lemniscomys* ?Pleist., R. Af. *Lenomys* Pleist.-R. EInd. *Lophuromys Malacothryx* Pleist.-R. Af. *Mastacomys* Pleist.-R. Aus. *Micromys* Pleist.-R. As., R. Eu. *Mus* [*Leggada*] ?Plioc., Pleist.-R. Eu. As. Af., R. Cos. *Nesokia* Pleist.-R. SAs., R. NAf. *Notomys* Pleist.-R. Aus. *Otomys* [*Myotomys*] Pleist.-R. Af. *Palaeotomys* Pleist. SAf. *?Palustrimus* L. Mioc. NA. *Papagomys* Pleist.-R EInd. *Parapodemus* U. Mioc.-Pleist. Eu., ?L. Plioc. As. *Pelomys* Pleist.-R. Af. *Progonomys* L. Plioc. Eu., ?SWAs. *Prototomys* Pleist. Af. *Pseudomys* Pleist.-R. Aus. *Rattus* [*Aethomys Epimys Mastomys Thallomys*] ?Plioc., Pleist.-R. As., Pleist.-R. Eu. Aus. Af. EInd., R. Cos. *Rhabdomys* Pleist.-R. Af. *Rhagamys* Pleist. Eu. *Saccostomus* Pleist.-R. Af. *Spelaeomys* Pleist. EInd. *Stecitomys* Pleist.-R. Af. *Stephanomys* U. Plioc. Eu., ?Plioc., Pleist. As. *Zelotomys* Pleist.-R. Af.

SUPERFAMILY DIPODOIDEA. **Dipodidae** R. Eu. As. Af., *Allactaga* L. Pleist.-R. As., Pleist.-R. Eu., R. NAf. *Allactagulus* Plioc. Eu. *Brachyscirtetes* Plioc. As. *Dipus* Pleist.-R. As. EEu. *Jaculus* Pleist.-R. Af. *Paradipoides* Pleist. NA. *Paralactaga* U. Mioc.-L. Plioc. EAs. EEu. *Protalactaga* U. Olig.-L. Plioc. As. *Sminthoides* L. Plioc. As. **Zapodidae** R. Eu. As. NA., *Heterosminthus* L. Plioc. As. *Macrognathomys* L. Plioc. NA. *Napaeozapus* Pleist.-R. NA. *Plesiosminthus* [*Parasminthus Schaubenemys*] U. Olig. Eu. As., L.-M. Mioc. NA. *Pliozapus* U. Plioc. NA. *Sicista* U. Plioc.-R. Eu. As. *?Simimys* U. Eoc. NA. *Sminthozapus* L. Pleist. Eu. *Zapus* L. Pleist.-R. NA., R. AS.

SUPERFAMILY GEOMYOIDEA. **Eomyidae,** *Adjidaumo* [*Gymnoptychus*] L.-M. Olig. NA. *Aulolithomys Centimanomys* L. Olig. NA. *Eomys* [*Paradjidaumo*] U. Eoc.-M. Olig. Eu., L.-M. Olig. NA. *Kansasimys* U. Plioc. NA. *Leptodontomys* U. Plioc. NA. *Ligerimys* L. Mioc. Eu. *Meteomys* Pleist. Eu. *Namatomys* L. Olig. NA. *Omegodus* L. Mioc. Eu. *Protadjidaumo* U. Eoc. NA. *Pseudotheridomys* M. Olig.-L. Mioc. Eu., L. Mioc. NA. *Rhodanomys* M. Olig.-L. Mioc. Eu. *Ritteneria* L. Mioc. Eu. *Yoderimys* L. Olig. NA. **Geomyidae** R. NA., *Cratogeomys* L. Pleist.-R. NA. *Dikkomys* L. Mioc. NA. *Diplolophus* [*Gidleumys*] M. Olig. NA. *Entoptychus* L. Mioc. NA. *Geomys* [*Nerterogeomys Parageomys*] Plioc.-R. NA. *Grangerimus* L. Mioc. NA. *Gregorymys* L. Mioc. NA. *?Griphomys* U. Eoc. NA. *Heterogeomys* Pleist.-R. NA. *Plesiosaccomys* Plioc. NA. *Plesiothomomys* Pleist. NA. *Pleurolicus* L. Mioc. NA. *Pliogeomys*

U. Plioc. NA. *Pliosaccomys* L.-U. Plioc. NA. *Thomomys* L. Mioc.-R. NA. **Heteromyidae** R. NA. NSA., *Akmaiomys Apletotomeus* M. Olig. NA. *Cupidinimus* L. Plioc.-L. Pleist. NA. *Dipodomys* L. Pleist.-R. NA. *Diprionomys* L.-U. Plioc. NA. *Etadonomys* Pleist. NA. *Florentiamys* L. Mioc. NA. *Heliscomys* L. Olig.-L. Mioc. NA. *Hitonkala* L. Mioc. NA. *Liomys* L. Pleist.-R. NA. *Mookomys* L.-M. Mioc. NA. *Peridiomys* M.-U. Mioc. NA. *Perognathoides* U. Mioc.-L. Plioc. NA. *Perognathus* U. Mioc.-R. NA. *Prodipodomys* ?U. Mioc., U. Plioc.-L. Pleist. NA. *Proheteromys* L.-U. Mioc. NA. *Trogomys* L. Mioc. NA.

SUPERFAMILY GLIROIDEA. **Gliridae (Myoxidae)** R. Eu. As. Af., *Amphidroyomys* Plioc. or Pleist. Eu. *Brachymys* Mioc. Eu. *Caenomys* L. Mioc. Eu. *Dryomys* [*Dyromys*] U. Olig.-R. Eu., R. As. *Eliomys* Mioc.-R. Eu., R. NAf. WAs. *Gliravus* U. Eoc.-L. Mioc. Eu. *Glirulus* L. Mioc. Eu. *Glis* [*Myoxus*] L. Mioc.-R. Eu., R. WAs. *Heteromyoxus* L.-M. Mioc. Eu. *Hypnomys* Pleist. Balearic Is. *Leithia* Pleist. SEu. Malta *Muscardinus* ?L. Plioc., U. Plioc.-R. Eu., R. WAs. *Myodryomys* Pleist. Eu. *Myomimus* [*Dryomimus Philistomys*] Pleist. Eu., Pleist.-R. WAs. *?Neocometes Pentaglis* M. Mioc. Eu. *Peridyromys* U. Olig.-L. Mioc. Eu. **Seleviniidae** R. As., *Plioselevinia* L. Plioc.-L. Pleist. Eu.

SUPERFAMILY SPALACOIDEA. **Spalacidae** R. Eu. WAs. Af., *Pliospalax* L. Plioc. Eu. *Prospalax* U. Plioc.-L. Pleist. Eu. *Spalax* ?Plioc., Pleist.-R. Eu., Pleist.-R. SWAs. NAf. **Rhizomyidae** R. As. Af., *Brachyrhizomys* L. Pleist. EAs. *Bramus* Pleist. NAf. *Pararhizomys* L. Plioc. EAs. *Protachyoryctes* L. Pleist. SAs. *Rhizomys* U. Mioc.-R. As. *Rhizospalax* M.-U. Olig. Eu. *Tachyoryctes* Pleist.-R. Af. *Tachyoryctoides* L. Mioc. EAs.

SUPERFAMILIES NOT ASSIGNED TO SUBORDERS

SUPERFAMILY CASTOROIDEA. **Castoridae** R. Eu. As. NA., *Agnotocastor* L.-U. Olig. NA. *Anchitheriomys* [*Amblycastor*] L.-U. Mioc. NA., M. Mioc. Eu., M. Mioc.-Plioc. As. *Capacikala Capatanka* L. Mioc. NA. *Castor* L. Plioc.-R. NA. Eu. As. *Castoroides* Pleist. NA. *Dipoides* Plioc. Eu. As., L. Plioc.-L. Pleist. NA. *Eucastor* [*Sigmogomphius*] L.-U. Plioc. NA., Plioc. As. *Eucastoroides* Pleist. NA. *Euhapsis* L. Mioc. NA. *Hystricops* U. Mioc. or L. Plioc. NA. *Monosaulax* L. Mioc.-L. Plioc. NA. Eu., Mioc. EAs. *Palaeocastor* U. Olig. WAs., L. Mioc. NA. *Palaeomys* [*Chalicomys*] U. Mioc.-L. Plioc. Eu. *Paradipoides* Pleist. NA. *Pipestoneomys* L. Olig. NA. *Procastoroides* L. Pleist. NA. *Sinocastor* Plioc. EAs. *Steneofiber* [*Steneotherium*] U. Olig.-L. Plioc. Eu. *Trogontherium* [*Conodontes*] L.-U. Pleist. Eu., Pleist. As. **Eutypomyidae,** *Eutypomys* L. Olig.-L. Mioc. NA.

SUPERFAMILY THERIDOMYOIDEA. **Pseudosciuridae,** *Adelomys* [*Sciuroides*] U. Eoc.-L. Olig. Eu. *Masillamys* U. Eoc. Eu. *Pseudosciurus* U. Eoc.-L. Olig. Eu. *Suevosciurus* U. Eoc.-M. Olig. Eu. **Theridomyidae,** *Altinomys* Olig. Eu. *Archaeomys Issiodoromys Nesokerodon* M. Olig. Eu. *Pararchaeomys* Olig. Eu. *Protechimys* ?L., M. Olig. Eu. *Pseudoltinomys* L. Olig. Eu. *?Pseudosciuromys ?Sciuromys* U. Eoc. or L. Olig. Eu. *Taeniodus* M. Olig. Eu. *Theridomys* [*Oltinomys Trechomys*] U. Eoc.-M. Olig. Eu.

SUPERFAMILY THRYONOMYOIDEA. **Phiomyidae,** *Apodector* L. Mioc. EAf. *Bathyergoides Diamantomys* L. Mioc. SAf. *Gaudeamus Metaphiomys* L. Olig. NAf. *Neosciuromys Paraphiomys* L. Mioc. Af. *Phiomyoides* L. Mioc. EAf. *Phiomys* L. Olig. NAf. *Phthynilla* L. Mioc. EAf. *Pomonomys* L. Mioc. SAf. *Pseudospalax* L. Mioc. EAf., ?NAf. **Thryonomyidae** R. Af., *Paraulacodus* L. Plioc. As. *Thryonomys* [*Aulacodus*] L. Mioc.-R. Af. **Petromuridae** R. Af., *Palaeopetromys* Pleist. Af. *Petromus* [*Petromys*] Pleist.-R. Af. ?**Bathyergidae** R. Af., *Cryptomys* Pleist.-R. Af. *Georychus* ?Pleist., R. Af. *Gypsorhychus* Pleist. Af. *Heterocephalus* Pleist.-R. Af.

FAMILIES NOT ASSIGNED TO SUPERFAMILIES OR SUBORDERS. **Hystricidae** R. Eu. As. Af., *Acanthion* Pleist. As., Pleist.-R. EInd. *Atherurus* Pleist.-R. Af., R. As. *Hystrix* [*Xenohystrix*] ?Olig., M. Mioc.-R. Eu., L. Plioc.-R. Af., U. Plioc.-R. As., Pleist. EInd. *Sivacanthion* L. Plioc. As. **Ctenodactylidae** R. Af., *Africanomys ?Dubiomys* Mioc. NAf. *Karakoromys ?Pellegrina* Pleist. Eu. *Leptotataromys* U. Olig. EAs. *Pectinator* ?L. Plioc. As., R. Af. *Sayimys* ?Olig., L. Plioc. As. *Tataromys Yindirtemys* U. Olig. EAs. **Anomaluridae** R. Af., *Anomalurus* ?L. Mioc., R. Af. **Pedetidae** R. Af., *Megapedetes* L. Mioc. EAf. *Parapedetes* L. Mioc. SAf. *Pedetes* Pleist.-R. Af.

ORDER LAGOMORPHA. **Eurymylidae,** *Eurymylus* U. Paleoc. EAs. *?Mimolagus* ?Olig. EAs. **Ochotonidae** R. As. NA., *Alloptox* [*Metochotona*] U. Mioc.-L. Plioc. EAs. *Amphilagus* U. Olig.-U. Mioc. Eu. *Austrolagomys* L. Mioc. Af. *Bellatona* U. Mioc. EAs. *Desmatolagus* M.-U. Olig. EAs., M.-U. Olig., ?M. Mioc. NA. *Hesperolagomys* L. Plioc. NA. *Heterolagus* U. Mioc. Eu. *Kenyalagomys* L. Mioc. EAf. *Marcuinomys* U. Olig. Eu. *Ochotona* [*Lagomys Lagotona Pliochotona Proochotona*] L. Plioc.-Pleist. Eu., L. Plioc.-R. As., U. Plioc.-R. NA. *Ochotonoides* ?L., U. Plioc.-Pleist. As. *Ochotonolagus* U. Olig. EAs. *Opsolagus* [*Lagopsis*] L.-U. Mioc. Eu. *Oreolagus* L.-M., ?U. Mioc. NA. *Paludotona* L. Plioc. Eu. *Piezodus* U. Olig.-M. Mioc. Eu. *Procaprolagus* Olig. EAs. *Prolagus* M. Mioc.-Pleist. Eu. *Sinolagomys* U. Olig. EAs. *Titanomys* U. Olig.-M. Mioc. Eu. **Leporidae,** *Alilepus* L.-U. Plioc. Eu. As. ?NA., ?Pleist. NA. *Agispelagus* U. Olig. EAs. *Archaeolagus* L. Mioc. NA. *Caprolagus* Plioc. Eu., Plioc.-R. As. *Gobiolagus* U. Eoc.-L. Olig., ?U. Olig. EAs. *Hypolagus* ?L., M. Mioc.-Pleist. NA. ?L. Plioc., U. Plioc.-Pleist. Eu. As. *Lepus* L. Plioc.-R. Eu., Pleist.-R. As. Af. NA. *Litolagus* Olig. NA. *Lushilagus* U. Eoc. EAs. *Megalagus* L. Olig.-L. Mioc. NA. *Mytonolagus* U. Eoc. NA. *Nekrolagus* L. Pleist. NA. *Notolagus* [*Dicea*] Plioc.-Pleist. NA. *Oryctolagus* L. Pleist.-R. Eu. NAf. *Palaeolagus* L. Olig.-L. Mioc. NA. *Panolax* Plioc. NA. *Pliolagus* Pleist. Eu. *Pratilepus* L. Pleist. NA. *Procaprolagus* U. Olig. EAs. *Pronolagus* Pleist.-R. Af. *Serengetilagus* Pleist. Af. *Shamolagus* U. Eoc. EAs. *Sylvilagus* [*Palaeotapeti*] L. Plioc.-R. NA., Pleist.-R. SA.

BIBLIOGRAPHY

The books and papers listed below include but a small proportion of even the more fundamental and important works dealing with fossil vertebrates. Special emphasis has been given to comprehensive monographs, bibliographic sources, and well-illustrated descriptions of typical forms. Some notes regarding workers in the field have been included.

Vertebrate paleontology was essentially founded by Cuvier somewhat over a century ago. As a comparative anatomist he studied petrifactions in the light of his knowledge of modern types. Much of his work is summarized in his *Ossemens fossiles* (No. 15). Sir Richard Owen, working somewhat later in the nineteenth century, was, again, a comparative anatomist and the first English paleontologist of importance. Von Meyer, his contemporary, published many fine works on German fossils, and Agassiz's important work on fossil fishes (No. 28) also dates from the first half of the century. Neither Cuvier, Owen, nor Agassiz believed in evolution. Darwin did no original work in paleontology, but the acceptance of his theories gave a new point to the study of fossils and led to a wide expansion of the field.

In America little work was done until exploration had opened up the great western fossil country. Joseph Leidy, the anatomist, whose active career in the field of paleontology was mainly between 1850 and 1872, was the first American worker of importance. In the late sixties, Cope of Philadelphia and Marsh of New Haven entered the field and dominated it until the closing years of the century.

In the late 1800's and the early part of the present century, research and publication was centered in western Europe—France, England, and most notably, Germany—and North America. Prominent figures of this period include, for example, Smith Woodward in England, Boulé in France, Broili and Schroeder in Germany, Osborn, Scott, Williston, Matthew, and Gregory in North America, Ameghino in Argentina, Broom in South Africa. In recent decades, study of vertebrates has spread to nearly every region, with major increase in work in Russia and China.

BIBLIOGRAPHIES

Zoological Record. 1864–, London.

A year-by-year bibliography of all zoological papers, arranged taxonomically. The literature on fossils is well covered in the more recent volumes, less so in the earlier issues.

Biological Abstracts. 1926–, Philadelphia.

Covers all biological literature, including paleontology.

Neues Jahrbuch für Mineralogie, Geologie und Paläontologie. Referate, 1820–1943. Continued as *Zentralblatt für Geologie und Paläontologie*, 1943–.

Contains abstracts of most of the literature.

Hay, O. P. 1902. Bibliography and Catalogue of the Fossil Vertebrata of North America. *Bull. U.S. Geol. Surv.*, No. 179. Pp. 868.

———. 1929. Second Bibliography. . . . 2 vols. *Publ. Carnegie Inst. Washington*, No. 390. Pp. 2003.

These works list every citation, to the year 1927, of every American fossil form, together with the general literature on every group ever present in North America.

Romer, A. S. and Others. Bibliography of Fossil Vertebrates Exclusive of North America, 1509–1927. *Mem. Geol. Soc. Amer.*, No. 87. 2 vols. Pp. 1544.

A bibliography of all works mentioning vertebrate fossils outside of North America; this with the two Hay bibliographies list the entire literature of the subject to 1927. It is hoped that a catalogue of the contents will be prepared.

Camp, C. L. and Others. 1940–. Bibliography of Fossil Vertebrates. *Geol. Soc. Amer., Spec. Papers and Memoirs.*
Covers the world literature, in continuation of Hay and Romer from 1928 on. Six volumes have already appeared.

Society of Vertebrate Paleontology, Bibliography, 1947–.
Annual bibliographies, including all major works and a fraction of the pertinent minor papers.

JOURNALS

Most of the literature is scattered through a great variety of biological and geological journals and publications of museums and learned societies. Many are included in the paleontological journals listed below.

Palaeontographica, 1846–. Stuttgart.
This and the next five journals are primarily devoted to extensive monographs (quarto).
Palaeontographical Society, 1847–. London.
Société Paléontologique Suisse, Mémoires, 1874–. Basel.
Société Géologique de France, Mémoires, 1833–. Paris.
Geologische und Paläontologische Abhandlungen, 1882–. Jena.
Annales de Paléontologique, 1906–. Paris.
Akademiia Nauk S.S.S.R., Paleontologicheskii Institut, Trudy, 1932–. Leningrad.
Journal of Paleontology, 1927–. Tulsa.
Paläontologische Zeitschrift, 1914–. Berlin.
Palaeontology, 1957–. London.
Vertebrata Palasiatica, 1957–. Peking.
Palaeontologia Africana, 1953–. Johannesburg.
Ameghiniana, 1957–. Buenos Aires.

GENERAL WORKS

1. Piveteau, J. (ed.). 1952–. *Traité de Paléontologie.* 7 vols. Paris.
Written mainly by French paleontologists. Vols. 4–7 give a comprehensive up-to-date account of fossil vertebrates.
2. Orlov, J. A. (ed.). 1959–64. *Osnovy Paleontologii [Fundamentals of Paleontology].* Moscow.
A comprehensive work, comparable to the last. Three volumes cover the vertebrates.
3. Grassé, P. P. (ed.). 1948–. *Traité de Zoologie: Anatomie, Systématique, Biologique.* Paris.
Primarily devoted to Recent forms but fossils are discussed as well.
3a. Grassé, P. P. and Devillers, C. 1965. *Zoologie. 2. Vertébrés.* Précis de Sciences Biologiques. Paris. Pp. 1129.
4. Colbert, E. H. 1955. *Evolution of the Vertebrates.* New York. Pp. 479.
An excellent introductory account.
5. Gregory, W. K. 1951. *Evolution Emerging.* 2 vols. New York.
Includes a full volume of useful illustrations.
6. Romer, A. S. 1959. *The Vertebrate Story.* 4th ed. Chicago. Pp. 437.
A semipopular account of vertebrates and their evolution.
7. Lehman, J.-P. 1959. *L'Évolution des Vertébrés Inférieurs.* Paris. Pp. 188.
This volume and the next emphasize work and theories of the Stockholm group.
8. Jarvik, E. 1960. *Théories de l'Évolution des Vertébrés.* Paris. Pp. 104.
9. Kuhn-Schnyder, E. 1953. *Geschichte der Wirbeltiere.* Basel. Pp. 156.
Includes historical notes.
10. Abel, O. 1912. *Grundzüge der Palaeobiologie der Wirbeltiere.* Stuttgart. Pp. 703.
A stimulating study of adaptations in living and fossil vertebrates.
11. ———. 1927. *Lebensbilder aus der Tierwelt der Vorzeit.* 2d ed. Jena. Pp. 714.
Interesting discussion of a number of representative faunas from Permian to Pleistocene.
12. ———. 1935. *Vorzeitliche Lebensspuren.* Jena. Pp. 644.
Discussion of nonskeletal remains—footprints, eggs, coprolites, etc.
13. Watson, D. M. S. 1951. *Paleontology and Modern Biology.* New Haven. Pp. 216.
Discusses various evolutionary problems among lower vertebrates.
14. Westoll, T. S. (ed.). 1958. *Studies on Fossil Vertebrates.* London. Pp. 263.
Essays on various paleontological problems dealing with lower vertebrates.
15. Cuvier, G. 1834–36. *Recherches sur les ossemens fossiles, où l'on rétablit les caractères de plusieurs animaux dont les révolutions du globe ont détruit les espèces.* 4th ed. 10 vols.+2 vols. pls. Paris.
The first great paleontological work. First issued in 1812.
16. Zittel, K. A. von. 1887–93. *Handbuch der Palaeontologie*, Vols. 3, 4: *Vertebrata.* Munich and Leipzig. Pp. 1699.
A rather full account of the older material. Also published in a French edition.
17. ———. 1923. *Grundzüge der Paläontologie.* Munich and Berlin.
Originally an abbreviation of No. 16. The last German edition of Vol. 2 (*Vertebrata*), revised by Broili and Schlosser, was published in 1923. An English translation, revised by Sir Arthur Smith Woodward, was published in two volumes, that on lower vertebrates in 1932, that on mammals in 1925.

VERTEBRATE STRUCTURE

18. Bolk, L. Göppert, E., Kallius, E., and Lubosch, W. (eds.) 1931–38. *Handbuch der vergleichenden Anatomie der Wirbeltiere.* 6 vols.+index. Berlin.
A comprehensive work on vertebrate anatomy by a large number of specialists.
19. Goodrich, E. S. 1930. *Studies on the Structure and Development of Vertebrates.* London. Pp. 907. Reprinted 1958. 2 vols. New York and London.
Excellent discussion of many problems dealing with the skull, backbone, fins, etc. in lower vertebrates.
20. Romer, A. S. 1962. *The Vertebrate Body.* 3d ed. Philadelphia. Pp. 627.
21. Young, J. Z. 1962. *The Life of Vertebrates.* 2d ed. Oxford. Pp. 820.
22. Reynolds, S. J. 1913. *The Vertebrate Skeleton.* 2d ed. Cambridge. Pp. 535.
A standard work, prepared from the paleontologist's point of view, although dealing primarily with living forms.
23. Edinger, T. 1929. Die fossilen Gehirne. *Ergeb. Anat. Entwickl.* Vol. 28. Pp. 249.
Endocranial structure in all vertebrate classes. A supplement in *Fortschr. d. Pal.* 1:234–51 (1937). An annotated bibliography, bringing the work up to date, is in preparation.

24. Dechaseaux, C. 1962. *Cerveaux d'Animaux Disparus.* Paris. Pp. 148.
25. Edmund, A. G. 1960. *Tooth Replacement Phenomena in the Lower Vertebrates.* Life Sciences Division, Royal Ontario Museum, Contr. No. 52. Toronto. Pp. 190.
26. Enlow, D. H. and Brown, S. O. 1956–58. A comparative histological study of fossil and Recent bone tissue. Parts 1–3. *Texas Jour. Sci.* 8:405–43; 9:186–214; 10:187–230.

LOWER CHORDATES

27. Grassé, P. P. (See No. 3). Vol. 11 (1948) gives an excellent account.

FISHES, GENERAL

The first comprehensive study of fossil fishes was that of Louis Agassiz, nearly a century ago (No. 28). A more recent thorough survey of the field is that of Smith Woodward (No. 29). With these two workers should be ranked Traquair, who published many important papers on Scottish Paleozoic fishes during the latter decades of the nineteenth century; and no mention of fossil fishes could be made without reference to Hugh Miller, the Scottish stonemason, who collected much of the Old Red Sandstone material studied by Agassiz and published several volumes compounded of paleontology, theology, and homely philosophy. Among the comparatively few early American workers who devoted themselves mainly to fishes may be mentioned Newberry, Dean, and Eastman. Among European workers of the nineteenth century and the early part of the twentieth may be mentioned Pander and Rohon, older Russian workers on Paleozoic material; Leriche, Priem, and Sauvage in France; Hennig, Broili, and the erratic but brilliant Jaekel in Germany; Kiaer in Norway. A considerable number of workers are now active or have been active in recent decades in paleoichthyology. A partial list includes (in Europe) Aldinger, Arambourg, Brough, Casier, Danilchenko, Devillers, Griffiths, Gardiner, Gross, A. Heintz, N. Heintz, Jarvik, Kulczyski, Lehman, Miles, Nielsen, Ørvig, Obruchev, Patterson, Rayner, Ritchie, Saint-Seine, Stensiö, Tarlo, Vorobeva, Watson, Westoll, and White and (in North America) Appleby, Bardack, Denison, Dineley, Dunkle, Nursall, Schaeffer, and Thomson.

28. Agassiz, L. 1833–44. *Recherches sur les poissons fossiles.* 5 vols. Pp. 1420+369 pls. With supplement, *Monographie des poissons fossiles du vieux grès rouge ou système Dévonien (Old Red Sandstone) des Iles Britanniques et de Russie.* Neuchatel. Pp. 171+43 pls.
An account of all forms known at that time. While the text is antiquated, the numerous colored plates are still very valuable.
29. Woodward, A. Smith. 1889–1901. *Catalogue of the Fossil Fishes in the British Museum.* 4 vols. London. Pp. 2493+70 pls.
Not merely a catalogue; in reality a summary of our knowledge of fossil fishes to the date of publication. It is still extremely useful, especially for Mesozoic and Tertiary forms.
30. Dean, B. 1916–23. *A Bibliography of Fishes.* 3 vols. New York. Pp. 2160.
A complete bibliography of all works on fishes, Recent and fossil, to the date of publication. An author list with comprehensive indexes.
31. Goodrich, E. S. 1909. *A Treatise on Zoölogy* (ed. E. Ray Lankester), Part 9: *Vertebrata craniata.* Fasc. 1: Cyclostomes and fishes. London. Pp. 534.
An excellent account of fishes, living and fossil, with particular attention to morphology. Still good except for Paleozoic groups.
32. Berg, L. S. 1958. *System der rezenten und fossilen Fischartigen und Fische.* Berlin. Pp. 310.
Translated from the 2d Russian edition.
33. Norman, J. R. 1947. *A History of Fishes.* 3d ed. London. Pp. 463.
34. Moy-Thomas, J. A. 1939. *Palaeozoic Fishes.* London and New York. Pp. 149.
35. Romer, A. S. 1946. The early evolution of fishes. *Quart. Rev. Biol.* 21:33–69.
36. Westoll, T. S. 1960. Recent advances in the palaeontology of fishes. *Liverpool and Manchester Geol. Jour.* 2: 568–96.
37. Romer, A. S. 1955. Fish origins—fresh or salt water? *Deep-Sea Res.* 3 (suppl.): 261–80.
This and the following are among a series of papers discussing early fish environments.
38. Denison, R. H. 1956. A review of the habitat of the earliest vertebrates. *Fieldiana: Geology* 11:359–457.

FISH FAUNAS

A fraction of the faunal literature, arranged approximately in stratigraphic order.

39. Ørvig, T. 1957. Notes on some Paleozoic lower vertebrates from Spitsbergen and North America. *Norsk Geol. Tidsskr.* 37:285–353.
40. Gross, W. 1933–37. Die Wirbeltiere des rheinischen Devons. *Abhandl. preuss. geol. Landesanst.*, new ser. No. 154: 1–83; No. 176: 1–83.
Early marine faunas.
41. Ørvig, T. 1957. Remarks on the vertebrate fauna of the lower Upper Devonian of Escuminac Bay, P.Q., Canada, with special reference to the porolepiform crossopterygians. *Arkiv Zool.* ser. 2, 10(6):367–426.
42. Davis, J. W. 1883. On the fossil fishes of the Carboniferous limestone series of Great Britain. *Trans. Roy. Dublin Soc.* ser. 2, 1:327–548.
Teeth of many pavement-toothed sharks are figured.
43. Aldinger, H. 1937. Permische Ganoidfische aus Östgrönland. *Meddel. om Grønland* 102 (3):1–392.
44. Piveteau, J. 1934. Paléontologie de Madagascar. No. 21. Les poissons du Trias inférieur. *Ann. Paléont.* 23:81–180.
45. Nielsen, E. 1942, 1949. Studies on Triassic fishes from East Greenland. *Palaeozool. Groenl.* 1:1–403; *Meddel. om Grønland* 146:1–309.
46. Lehman, J. P. 1952. Étude complémentaire des poissons de l'Eotrias de Madagascar. *Handl. K. Svenska Vetenskapsakad.* 2(6):1–201.
47. Lehman, J. P., Chateau, C., Laurain, M. and Nauche, M. 1959. Paléontologie de Madagascar. 28. Les poissons de la Sakamena Moyenne. *Ann. Paléont.* 45:3–45.
48. Dartevelle, E. and Casier, E. 1943, 1949, 1959. Les poissons fossiles du Bas-Congo et des régions voisines. *Ann. Mus. Roy. Congo Belge* ser. A, 2:1–586.
49. Woodward, A. Smith. 1916–19. The fossil fishes of the English Wealden and Purbeck formations, *Monogr. Palaeontogr. Soc. London.* Pp. 148.
Mainly holosteans, also *Leptolepis* and *Hybodus*.
50. Saint-Seine, P. 1949. Les poissons des calcaires lithographiques de Cerin (Ain). *Nouv. Arch. Mus. Hist. Nat. Lyon.* Fasc. 2:1–357.
51. ———. 1955, 1962. Poissons fossiles de l'étage de Stanleyville (Congo Belge). *Ann. Mus. Roy. Congo Belge,* Sér. in 8°, Sci. Géol. 14:1–126 (1955); *Ann. Mus. Roy. Afrique Centrale,* 8°, 44:1–52. (1962) (with Casier).

52. Woodward, A. Smith, 1902–12. The fishes of the English chalk. *Monogr. Palaeontogr. Soc. London.* Pp. 264.
Mainly early teleosts, with many good figures.
53. Arambourg, C. and Signeux, J. 1952. Les vertébrés fossiles des gisements de phosphates (Maroc-Algérie-Tunisie). *Notes Mém. Serv. Géol. Maroc.* No. 92. Pp. 374.
54. Arambourg, C. 1954. Les poissons Crétacés du Jebel Tselfat (Maroc). *Notes Mém. Serv. Géol. Maroc.* No. 118. Pp. 185.
55. Casier, E. 1946. La faune ichthyologique de l'Yprésien de la Belgique. *Mém. Mus. Hist. Nat. Belg.* 104:3–267.
56. ———. 1958. Contribution à l'étude des poissons fossiles des Antilles. *Mém. Suisses Paléont.* 74(3):1–96.
57. Leriche, M. 1951. Les poissons Tertiaires de la Belgique (suppl.). *Mém. Inst. Roy. Sci. Nat. Belg.* 118: 474–600.
58. Eastman, C. R. 1911. Catalog of fossil fishes in the Carnegie museum. I. Fishes from the Upper Eocene of Monte Bolca. *Mem. Carnegie Mus.*, 4(7):349–91. Also a supplement in the same series, 6(5):315–48 (1914).
59. David, L. R. 1943. Miocene fishes of southern California. *Geol. Soc. Amer., Spec. Papers.* No. 43. Pp. 193.
60. Arambourg, C. 1925. Revision des poissons de Licata (Sicilie). *Ann. Paléont.* 14:1–96.
61. ———. 1927. Les poissons fossiles d'Oran. *Matér. Carte Géol. Algérie.* ser. 1. No. 6. Pp. 218.

AGNATHA

62. Stensiö, E. A. 1927. The Downtonian and Devonian vertebrates of Spitzbergen. I. Family Cephalaspidae. *Skr. om Svalbard og Nordishavet.* No. 12. 2 vols. Pp. 391+112 pls.
An exhaustive study of the cranial structure of the Osteostraci, with an account of previous work and a discussion of the structure and relationships of all the agnathous types. The next is a companion volume. Cf. Nos. 64, 65, 69.
63. ———. 1932. *The Cephalaspids of Great Britain.* London. Pp. 220+66 pls.
63a. ———. 1964. Les cyclostomes fossiles ou ostracodermes. *Traité de Paléontologie* 4(1): 96–382.
64. Heintz, A. 1939. Cephalaspida from Downtonian of Norway. *Skr. norsk. vidensk.-Akad. i Oslo, mat.-nat. Kl.* No. 5. Pp. 119.
65. Wängsjö, G. 1952. The Downtonian and Devonian vertebrates of Spitsbergen. 9. Morphologic and systematic studies of the Spitsbergen cephalaspids. *Norsk Polarinst. Skrift.* No. 97, Pp. 615.
66. Denison, R. H. 1951. Evolution and classification of the Osteostraci. The exoskeleton of early Osteostraci. *Fieldiana: Geology* 11:155–218.
67. Westoll, T. S. 1945. A new cephalaspid fish from the Downtonian of Scotland, with notes on the structure and classification of ostracoderms. *Trans. Roy. Soc. Edinburgh* 61:341–57.
68. Denison, R. H. 1952. Early Devonian fishes from Utah. Part 1. Osteostraci. *Fieldiana: Geology* 11: 263–87.
69. Watson, D. M. S. 1954. A consideration of ostracoderms. *Phil. Trans. Roy. Soc. London B*238:1–25.
70. Kiaer, J. 1924. The Downtonian fauna of Norway. 1. Anaspida. *Skr. vidensk. selsk. Kristiania, mat.-nat. Kl.* Vol. 1. No. 6. Pp. 139.
First discovered by Traquair in the late nineties, the anaspids became adequately known only as a result of Kiaer's work. A thorough description of the morphology and evolutionary significance of typical anaspids.
71. Stensiö, E. A. 1939. A new anaspid from the Upper Devonian of Scaumenac Bay in Canada, with remarks on the other anaspids. *Handl. K. Svenska Vetenskapsakad.* ser. 3, 18:1–25.
72. Parrington, F. R. 1958. *On the Nature of the Anaspida. In: Studies on Fossil Vertebrates* (ed. T. S. Westoll). London. Pp. 108–28.
73. Heintz, A. 1958. *The Head of the Anaspid Birkenia elegans, Traq. In: Studies on Fossil Vertebrates* (ed. T. S. Westoll). London. Pp. 71–85.
74. Ritchie, A. 1960. A new interpretation of *Jamoytius kerwoodi* White. *Nature* (London) 188:647–49.
75. ———. 1964. New light on the morphology of the Norwegian Anaspida. *Skr. norsk. vidensk.-Akad. i. Oslo. Mat.-nat. Kl.* new ser. No. 14:1–35.
76. Kiaer, J. 1932. The Downtonian and Devonian vertebrates of Spitsbergen. 4. Suborder Cyathaspida. *Skr. om Svalbard og Ishavet.* No. 52. Pp. 26.
77. Ørvig, T. 1958. *Pycnaspis splendens,* new genus, new species, a new ostracoderm from the Upper Ordovician of North America. *Proc. U.S. Nat. Mus.* 108:1–23.
78. Heintz, A. 1962. Les organes olfactifs des Heterostraci. *In:* Problèmes Actuels de Paléontologie (Evolution des Vertébrés). *Coll. Internat. Centre Nat. Recher. Sci.* No. 104. Pp. 13–29.
79. Denison, R. H. 1953. Early Devonian fishes from Utah. Part 2. Heterostraci. *Fieldiana: Geology* 11(7):299–355.
80. Gross, W. 1963. *Drepanaspis gemuendenensis* Schlüter Neuuntersuchung. *Palaeontographica* 121*A*:133–55.
81. Denison, R. H. 1964. The Cyathaspidae, a family of Silurian and Devonian jawless fishes. *Fieldiana: Geology* 13: 307–473.
82. Tarlo, L. B. 1962. The classification and evolution of the Heterostraci. *Acta Palaeont. Polonica* 7:249–90.
83. White, E. I. 1935. The ostracoderm *Pteraspis* Kner and the relationships of the agnathous vertebrates. *Phil. Trans. Roy. Soc. London* B225:381–457.
84. Denison, R. H. 1963. New Silurian Heterostraci from southeastern Yukon. *Fieldiana: Geology* 14:105–41.
85. Kiaer, J. 1932. New coelolepids from the Upper Silurian on Oesel (Esthonia). *Arch. Naturk. Estlands* ser. 1, 10: 169–74.

ARTHRODIRES

86. Ørvig, T. 1951. Histologic studies of placoderms and fossil elasmobranchs. 1. The endoskeleton, with remarks on the hard tissues of lower vertebrates in general. *Ark. Zool.* ser. 2, 2:321–454.
87. Bystrow, A. P. 1957. The microstructure of dermal bones in arthrodires. *Acta Zool.* 38:239–75.
88. Westoll, T. S. 1948. The paired fins of placoderms. *Trans. Roy. Soc. Edinburgh* 61:381–98.
89. Heintz, A. 1931. Untersuchungen über den Bau der Arthrodira. *Acta Zool.* 12:225–39.
90. ———. 1929. Die downtonischen und devonischen Vertebraten von Spitsbergen. 2. Acanthaspida. *Skr. om Svalbard og Ishavet.* No. 22. Pp. 81. *Nachtrag.* Pp. 20.
91. Stensiö, E. 1945. On the heads of certain arthrodires. 2. On the cranium and cervical joint of the Dolichothoraci (Acanthaspida). *Handl. K. Svenska Vetenskapsakad.* 22(1):1–70.
92. Denison, R. H. 1958. Early Devonian fishes from Utah. Part 3. Arthrodira. *Fieldiana: Geology* 11:459–551.
93. White, E. I. 1952. Australian arthrodires. *Bull. Brit. Mus.* (*Nat. Hist.*), *Geol.* 1(9):249–304.

94. Heintz, A. 1931. Revision of the structure of *Coccosteus decipiens* Ag. *Norsk Geol. Tidsskr.* 12:291–313.
95. Stensiö, E. 1934. On the heads of certain arthrodires. 1. *Pholidosteus, Leiosteus,* and acanthaspids. *Handl. K. Svenska Vetenskapsakad.* 13(5):1–79.
96. Heintz, A. 1932. *The Structure of Dinichthys: A Contribution to Our Knowledge of the Arthrodira.* Bashford Dean Memorial Volume. New York. Pp. 115–224.
97. Stensiö, E. 1959. On the pectoral fin and shoulder girdle of the arthrodires. *Handl. K. Svenska Vetenskapsakad.* 8(1):5–229.
98. ———. 1963. Anatomical studies on the arthrodiran head. Part 1. Preface, geological and geographical distribution, the organization of the arthrodires, the anatomy of the head in the Dolichothoraci, Coccosteomorphi and Pachyosteomorphi. Taxonomic Appendix. *Handl. K. Svenska Vetenskapsakad.* 9(2):1–419.
99. Lehman, J. P. 1956. Les arthrodires du Dévonien Supérieur du Tafilalet (Sud Marocain). *Notes Mém. Serv. Géol. Empire Chérifien (Rabat).* No. 129:1–70.
100. Heintz, A. 1934. Revision of the Estonian Arthrodira. 1. Family Homostiidae Jaekel. *Publ. Geol. Inst. Univ. Tartu.* No. 38. Pp. 114.

PETALICHTHYIDS

101. Stensiö, E. 1925. On the head of the macropetalichthyids, with certain remarks on the head of the other arthrodires. *Field Mus. Nat. Hist., Geol. Ser.* 4(4):87–197.
102. Gross, W. 1961. *Lunaspis broilii* und *Lunaspis heroldi* aus dem Hunsrückschiefer (Unterdevon, Rheinland). *Notizbl. Hess. Landesamt. Bodenforsch.* 89:17–43.
103. ———. 1962. Neuuntersuchung der *Stensiöellida* (Arthrodira, Unterdevon). *Notizbl. Hess. Landesamt. Bodenforsch.* 90:48–86.

RHENANIDS

104. Gross, W. 1963. *Gemuendina stuertzi* Traquair. *Notizbl. Hess. Landesamt. Bodenforsch.* 91:36–73.
105. ———. 1959. Arthrodiren aus dem Obersilur der Prager Mulde. *Palaeontographica* 113:1–35.
Actually earliest Devonian, not Silurian.

PTYCTODONTS

106. Watson, D. M. S. 1938. On *Rhamphodopsis*, a ptyctodont from the Middle Old Red Sandstone of Scotland. *Trans. Roy. Soc. Edinburgh* 59:397–410.
107. Ørvig, T. 1962. Y a-t-il une relation directe entre les arthrodires ptyctodontides et les holocéphales? *In:* Problèmes Actuels de Paléontologie (Évolution des Vertébrés). *Coll. Internat. Centre Nat. Recher. Sci.* No. 104. Pp. 49–61.
108. Westoll, T. S. 1962. Ptyctodontid fishes and the ancestry of Holocephali. *Nature* (London) 194:949–52.
A different theory of placodont-holocephalian relationships is advocated by C. Patterson (122a).

ANTIARCHS

109. Gross, W. 1931. *Asterolepis ornata* Eichw. und das Antiarchi-Problem. *Palaeontographica* 75:1–62.
110. Denison, R. H. 1941. The soft anatomy of *Bothriolepis. Jour. Paleont.* 15:553–61.
111. Stensiö, E. 1948. On the Placodermi of the Upper Devonian of East Greenland. 2. Antiarchi: Subfamily Bothriolepinae. *Palaeozool. Groenland.* 2:1–622.

PALAEOSPONDYLUS

112. Moy-Thomas, J. A. 1940. The Devonian fish *Palaeospondylus gunni* Traquair. *Phil. Trans. Roy. Soc. London B*230:391–413.
The nature of this little fish has been much debated.

ELASMOBRANCHS

113. Moy-Thomas, J. A. 1939. The early evolution and relationships of the elasmobranchs. *Biol. Rev.* 14:1–26.
An excellent review.
114. Casier, E. 1954. Essai de paléobiogéographie des Euselachii. Vol. Jubilaire Victor van Straelen 1:575–640.
115. Bigelow, H. B. and Schroeder, W. C. 1948–53. Fishes of the Western North Atlantic. No. 1, Part 1. Cyclostómes and sharks; No. 1, Part 2, Sawfishes, guitarfishes, skates, rays. *Mem. Sears Found. Mar. Res.* Pp. 29–546; 1–514.
A comprehensive illustrated account of modern shark and skate genera; many known as fossils.
116. Dean, B. 1909. Studies on fossil fishes (sharks, chimaeroids, and arthrodires). *Mem. Amer. Mus. Nat. Hist.* 9: 211–87.
Good illustrations of *Cladoselache* and late Devonian arthrodires.
117. Moy-Thomas, J. A. 1936. The structure and affinities of the fossil elasmobranch fishes from the Lower Carboniferous rocks of Glencartholm, Eskdale. *Proc. Zool. Soc. London* 1936: 761–88.
118. Romer, A. S. 1964. The braincase of the Paleozoic elasmobranch *Tamiobatis. Bull. Mus. Comp. Zool.* 131:87–105.
119. Casier, E. 1953. Origine des Ptychodontes. *Mém. Inst. Roy. Sci. Nat. Belg.* ser. 2, 49: 1–51.
120. Fowler, H. W. 1911. A description of the fossil fish remains of the Cretaceous, Eocene, and Miocene formations of New Jersey. *Bull. Geol. Surv. New Jersey* 4:22–182.
One of numerous illustrated accounts of shark teeth.

CHIMAERAS

121. De Beer, G. R. and Moy-Thomas, J. A. 1935. On the skull of Holocephali. *Phil. Trans. Roy. Soc. London B*224: 287–312.
122. Moy-Thomas, J. A. 1936. On the structure and affinities of the Carboniferous cochliodont *Helodus simplex. Geol. Mag.* 73:488–503.
122a. Patterson, C. 1965. The phylogeny of the chimaeroids. *Phil. Trans. Roy. Soc. London B*249: 101–219.

ACANTHODIANS

123. Watson, D. M. S. 1937. The acanthodian fishes. *Phil. Trans. Roy. Soc. London B*228:49–146.
124. Miles, R. S. 1965. Some features in the cranial morphology of acanthodians and the relationships of the Acanthodii. *Acta Zool.* 46:233–55.

ACTINOPTERYGIANS

Many of the faunal papers listed earlier (Nos. 43–61) treat of actinopterygians.

125. Schaeffer, B. 1956. Evolution in the subholostean fishes. *Evolution* 10:201–12.
126. Lehman, J. P. 1947. Description de quelques exemplaires de *Cheirolepis canadensis* (Whiteaves). *Handl. K. Svenska Vetenskapsakad.* 24(4):1–40.

127. Goodrich, E. S. 1927. *Polypterus a palaeoniscid? Palaeobiologica* 1:87–92.
128. White, E. I. 1927. The fish fauna of the Cement Stones of Foulden, Berwickshire. *Trans. Roy. Soc. Edinburgh* 55: 255–86.
129. Moy-Thomas, J. A. and Dyne, M. B. 1938. The actinopterygian fishes from the Lower Carboniferous of Glencartholm, Eskdale, Dumfriesshire. *Trans. Roy. Soc. Edinburgh* 59:437–80.
Numerous palaeoniscoids; good figures.
130. White, E. I. 1939. A new type of palaeoniscoid fish, with remarks on the evolution of the actinopterygian pectoral fins. *Proc. Zool. Soc. London* B109:41–61.
131. Westoll, T. S. 1944. The Haplolepidae, a new family of late Carboniferous bony fishes. *Bull. Amer. Mus. Nat. Hist.* 83:1–121.
With discussion of general aspects of early actinopterygian evolution.
132. Gardiner, B. G. 1963. Certain palaeoniscoid fishes and the evolution of the snout in actinopterygians. *Bull. Brit. Mus. (Nat. Hist.), Geol.* 8:255–325.
133. Rayner, D. H. 1948. The structure of certain Jurassic holostean fishes with a special reference to their neurocrania. *Phil. Trans. Roy. Soc. London* B233:287–345.
134. Brough, J. 1936. On the evolution of bony fishes during the Triassic period. *Biol. Rev.* 11:385–405.
135. Westoll, T. S. 1937. On the cheek bones in teleostome fishes. *Jour. Anat.* 71:362–82.
136. Brough, J. 1939. *The Triassic Fishes of Besano, Lombardy.* British Museum (Nat. Hist.). London. Pp. 117.
137. Rayner, D. H. 1941. The structure and evolution of the holostean fishes. *Biol. Rev.* 16:218–37.
138. Gardiner, B. G. 1960. A revision of certain actinopterygian and coelacanth fishes, chiefly from the Lower Lias. *Bull. Brit. Mus. (Nat. Hist.), Geol.* 4(7):239–384.
139. Stensiö, E. 1935. *Sinamia zdanskyi*, a new amiid from the Lower Cretaceous of Shantung, China. *Paleont. Sinica* ser. C. Fasc. 1. Pp. 48.
140. Schaeffer, B. and Rosen, D. E. 1961. Major adaptive levels in the evolution of the actinopterygian feeding mechanism. *Amer. Zool.* 1(2):187–204.
141. Griffith, J. and Patterson, C. 1963. The structure and relationships of the Jurassic fish *Ichthyokentema purbeckensis. Bull. Brit. Mus. (Nat. Hist.), Geol.* 8:1–43.
142. Greenwood, P. H., Rosen, D. E., Weitzman, S. H., and Myers, G. S. 1966. Phyletic studies of teleostean fishes with a provisional classification of living forms. *Bull. Amer. Mus. Nat. Hist.* In press.
A modern attack on the problems of teleost evolution and classification.
143. Gregory, W. K. 1933. Fish skulls: a study of the evolution of natural mechanisms. *Trans. Amer. Phil. Soc.* 23: 75–481.
144. Rayner, D. H. 1937. On *Leptolepis bronni* Agassiz. *Ann. Mag. Nat. Hist.* ser. 10, 19:46–74.
145. Dunkle, D. H. 1940. The cranial osteology of *Notelops brama* (Agassiz), an elopid fish from the Cretaceous of Brazil. *Lloydia* 3:157–90.
146. Patterson, C. 1965. A review of Mesozoic acanthopterygian fishes, with special reference to those of the English chalk. *Phil. Trans. Roy. Soc. London* B247:213–482.

CROSSOPTERYGIANS

147. Jarvik, E. 1948. On the morphology and taxonomy of the Middle Devonian osteolepid fishes of Scotland. *Handl. K. Svenska Vetenskapsakad.* 25:1–301.
148. Romer, A. S. 1937. The braincase of the Carboniferous crossopterygian *Megalichthys nitidus. Bull. Mus. Comp. Zool.* 82:1–73.
See also *Jour. Morph.* 69:141–60 (1941).
149. Gross, W. 1956. Ueber Crossopterygier und Dipnoer aus dem baltischen Überdevon im Zusammenhang einer vergleichenden Untersuchung des Porenkanalsystems paläozoischer Agnathen und Fische. *Handl. K. Svenska Vetenskapsakad.* 5(6):1–140.
150. Jarvik, E. 1942. On the structure of the snout of crossopterygians and lower gnathostomes in general. *Zool. Bidrag Uppsala* 21:235–675.
In this and later papers Jarvik advocates a theory of dual origin of tetrapods from crossopterygians.
151. ———. 1944. On the dermal bones, sensory canals and pit-lines of the skull in *Eusthenopteron foordi* Whiteaves, with some remarks on *E. säve-söderberghi* Jarvik. *Handl. K. Svenska Vetenskapsakad.* 21(3):3–48.
152. ———. 1954. On the visceral skeleton in *Eusthenopteron* with a discussion of the parasphenoid and palatoquadrate in fishes. *Handl. K. Svenska Vetenskapsakad.* 5(1):1–103.
153. ———. 1962. Les Porolépiformes et l'origine des Urodèles. *In:* Problèmes Actuels de Paléontologie (Évolution des Vertébrés). *Coll. Internat. Centre Nat. Recher. Sci.* No. 104:87–101.
154. ———. 1963. The composition of the intermandibular division of the head in fish and tetrapods and the diphyletic origin of the tetrapod tongue. *Handl. K. Svenska Vetenskapsakad.* 9(1):1–74.
155. Kulczycki, J. 1960. *Porolepis* (Crossopterygii) from the Lower Devonian of the Holy Cross mountains. *Acta Palaeont. Polonica* 5:65–106.
156. Thomson, K. S. 1962. Rhipidistian classification in relation to the origin of the tetrapods. *Breviora, Mus. Comp. Zool.* No. 166:1–12.
157. ———. 1964. The comparative anatomy of the snout in rhipidistian fishes. *Bull. Mus. Comp. Zool.* 131:315–57.
158. Millot, J. and Anthony, J. 1958. *Anatomie de Latimeria chalumnae. I. Squelette, Muscles et Formations de Soutien.* Centre Nat. Recher. Sci. Paris. Pp. 122.
159. ———. 1965. *Anatomie de Latimeria chalumnae. II. Système nerveux et organes des sens.* Centre Nat. Recher. Sci. Paris. Pp. 130.
160. Schaeffer, B. 1952. The Triassic coelacanth fish *Diplurus*, with observations on the evolution of the Coelacanthini. *Bull. Amer. Mus. Nat. Hist.* 99:25–78.
161. Stensiö, E. 1921. *Triassic Fishes from Spitzbergen.* Part 1. Vienna. Pp. 307.
Especially coelacanths.
162. ———. 1937. On the Devonian coelacanthids of Germany with special reference to the dermal skeleton. *Handl. K. Svenska Vetenskapsakad.* ser. 3, 16:1–56.
163. Moy-Thomas, J. A. 1937. The Carboniferous coelacanth fishes of Great Britain and Ireland. *Proc. Zool. Soc. London* B107:383–415.

DIPNOI

164. Dollo, L. 1895. Sur la phylogénie des Dipneustes. *Bull. Soc. Belge Géol.* 9:79–128.
A brilliant essay in which the evolutionary trend of the lungfish series was first clearly demonstrated.
165. Westoll, T. S. 1949. On the evolution of the Dipnoi. *In:*

Genetics, Paleontology, and Evolution (eds. G. L. Jepsen, E. Mayr, and G. G. Simpson). Part 3. Pp. 121–84.

166. Lehmann, W. and Westoll, T. S. 1952. A primitive dipnoan fish from the Lower Devonian of Germany. *Proc. Roy. Soc. London B*140:402–21.
167. Säve-Söderbergh, G. 1952. On the skull of *Chirodipterus wildungensis* Gross, an Upper Devonian dipnoan from Wildungen. *Handl. K. Svenska Vetenskapsakad.* 3(4): 1–28.
168. White, E. I. 1965. The head of *Dipterus valenciennesi. Bull. Brit. Mus.* (*Nat. Hist.*), *Geol.* 11(1):1–45.
169. Graham-Smith, W. and Westoll, T. S. 1937. On a new long-headed dipnoan fish from the Upper Devonian of Scaumenac Bay, P.Q., Canada. *Trans. Roy. Soc. Edinburgh* 59:241–66.
170. Lehman, J. P. 1959. Les Dipneustes du Dévonien Supérieur du Groenland. *Meddel. om Grønland* 160(4):3–58.

AMPHIBIA–GENERAL AND LABYRINTHODONTS

The early finds of amphibians were principally described by Huxley (British material), Fritsch and Credner (Continental Europe), and Cope (America). Among later and current workers on the class may be cited Augusta, Broili, Broom, Brough, Bystrow, Efremov, Heyler, Huene, Lehman, Nilsson, Panchen, Säve-Söderbergh, Steen, Spinar, and Watson in Europe; Baird, Carroll, Case, Chase, Colbert, Eaton, Hotton, Langston, Olson, Romer, Sawin, Seltin, Welles, and Williston in North America; Cosgriff in Australia.

General accounts are given in Piveteau (No. 1), Orlov (No. 2), and Huene (No. 226).

171. Schaeffer, B. 1965. The rhipidistian-amphibian transition. *Amer. Zoologist* 5:267–76.
172. Fritsch, A. 1879–1901. *Fauna der Gaskohle und der Kalksteine der Permformation Böhmens.* 4 vols. Prague. Pp. 492.
    Many colored plates of amphibians, particularly "branchiosaurs" and lepospondyls.
173. Romer, A. S. 1947. Review of the Labyrinthodontia. *Bull. Mus. Comp. Zool.* 99:1–368.
174. ———. 1964. Problems in early amphibian history. *Jour. Anim. Morph. Physiol.* 2(1):1–20.
175. Williams, E. E. 1959. Gadow's arcualia and the development of tetrapod vertebrae. *Quart. Rev. Biol.* 34(1):1–32.
176. Westoll, T. S. 1943. The origin of the tetrapods. *Biol. Rev.* 18:78–98.
177. Jarvik, E. 1952. On the fish-like tail in the ichthyostegid stegocephalians. *Meddel. om Grønland* 114(12):1–90.
178. ———. 1955. The oldest tetrapods and their forerunners. *Scient. Monthly* 80(3):141–54.
179. Carroll, R. 1964. Early evolution of the dissorophid amphibians. *Bull. Mus. Comp. Zool.* 131(7):161–250.
180. Watson, D. M. S. 1919. The structure, evolution and origin of the Amphibia—the orders Rachitomi and Stereospondyli. *Phil. Trans. Roy. Soc. London B*209:1–73.
181. ———. 1962. The evolution of the labyrinthodonts. *Phil. Trans. Roy. Soc. London B*245:219–65.
    On Permian and Triassic temnospondyls.
182. ———. 1956. The brachyopid labyrinthodonts. *Bull. Brit. Mus.* (*Nat. Hist.*), *Geol.* 2(8):315–92.
183. Bystrow, A. P. 1935. Morphologische Untersuchungen der Deckknochen des Schädels der Wirbeltiere. 1. Mitteilungen. Schädel der Stegocephalen. *Acta Zool.* 16: 65–141.
    Many good figures of labyrinthodonts and diplocaulids.
184. Romer, A. S. and Witter, R. V. 1942. *Edops*, a primitive rhachitomous amphibian from the Texas Red Beds. *Jour. Geol.* 50:925–60.
185. Sawin, H. J. 1941. The cranial anatomy of *Eryops megacephalus. Bull. Mus. Comp. Zool.* 88:407–63.
186. Credner, H. 1881–93. Die Stegocephalen und Saurier aus dem Rothliegenden des Plauen'schen Grundes bei Dresden. 1–10. *Zeitschr. Deutsch. Geol. Gesell.* Vols. 33–35, 37–38, 42. Pp. 345.
    "Branchiosaurs," etc.
187. Langston, W. 1953. Permian amphibians from New Mexico. *Univ. Calif. Publ. Geol. Sci.* 29(7):349–416.
188. Romer, A. S. 1939. Notes on branchiosaurs. *Amer. Jour. Sci.* 237:748–61.
    Branchiosaurs are larval labyrinthodonts. Cf. Watson.
189. Watson, D. M. S. 1963. On growth stages in branchiosaurs. *Palaeontology* 6(3):540–53.
    Agrees that branchiosaurs are larval labyrinthodonts.
190. Bystrow, A. P. and Efremov, J. A. 1940. *Benthosuchus sushkini* Efr.—a labyrinthodont from the Eotriassic of Sharjenga River. *Trav. Inst. Paleont. Acad. Sci. U.R.S.S.* 10:1–152.
    Structure of a primitive stereospondyl.
191. Nilsson, T. 1946. On the genus *Peltostega* Wiman and the classification of the Triassic stegocephalians. *Handl. K. Svenska Vetenskapsakad.* ser. 3, 23:3–55.
192. Welles, S. P. and Cosgriff, J. 1965. A revision of the labyrinthodont family Capitosauridae. *Univ. Calif. Publ. Geol. Sci.* 54:1–148.
    Stereospondyls.
193. Watson, D. M. S. 1958. A new labyrinthodont (*Paracyclotosaurus*) from the Upper Trias of New South Wales. *Bull. Brit. Mus.* (*Nat. Hist.*), *Geol.* 3: 235–63.
194. Fraas, E. 1889. Die Labyrinthodonten der schwäbischen Trias. *Palaeontographica* 36:1–158.
    A classic paper on stereospondyls.
195. Sawin, H. J. 1945. Amphibians from the Dockum Triassic of Howard County, Texas. *Univ. Texas Publ.* No. 4401:361–99.
    Anatomy of "*Buettneria*" (*Metoposaurus*).
196. Colbert, E. H. and Imbrie, J. 1956. Triassic metoposaurid amphibians. *Bull. Amer. Mus. Nat. Hist.* 110(6): 399–452.
197. Panchen, A. L. 1959. A new armoured amphibian from the Upper Permian of East Africa. *Phil. Trans. Roy. Soc. London B*242:207–81.
198. Nilsson, T. 1946. A new find of *Gerrothorax rhaeticus* Nilsson, a plagiosaurid from the Rhaetic of Scania. *Acta Univ. Lund* 42:1–42.
199. Watson, D. M. S. 1926. The evolution and origin of the Amphibia. *Phil. Trans. Roy. Soc. London B*214:189–257.
200. ———. 1930. The Carboniferous Amphibia of Scotland. *Paleont. Hungarica* 1:221–52.
    Skulls of some of the oldest and most primitive tetrapods.
201. Romer, A. S. 1964. The skeleton of the Lower Carboniferous labyrinthodont *Pholidogaster pisciformis. Bull. Mus. Comp. Zool.* 131(6):129–59.
202. ———. 1963. The larger embolomerous amphibians of the American Carboniferous. *Bull. Mus. Comp. Zool.* 128(9):415–54.
203. ———. 1957. The appendicular skeleton of the Permian embolomerous amphibian *Archeria. Contr. Mus. Paleont. Univ. Michigan* 13(5):103–59.

204. White, T. E. 1939. Osteology of *Seymouria baylorensis*. *Bull. Mus. Comp. Zool.* 85:325–402.
205. Bystrow, A. P. 1944. *Kotlassia prima* Amalitzky. *Bull. Geol. Soc. Amer.* 55:379–416.
206. Špinar, Z. V. 1952. Revise některých moravských Diskosauriscidů. [Revision of some Moravian Discosauriscidae.] *Roz. Ustred. Ustav. Geol.* 15:1–159.
207. Romer, A. S. 1964. *Diadectes* an amphibian? *Copeia* 1964(4):718–19.
208. Olson, E. C. 1965. Relationships of *Seymouria, Diadectes,* and Chelonia. *Amer. Zoologist* 5:295–307.

PALEOZOIC LEPOSPONDYLS

209. Romer, A. S. 1950. The nature and relationships of the Paleozoic microsaurs. *Amer. Jour. Sci.* 248(9):628–54.
210. Gregory, J. T., Peabody, F. E., and Price, L. I. 1956. Revision of the Gymnarthridae, American Permian microsaurs. *Bull. Peabody Mus. Nat. Hist.* 10:1–77.
211. Sollas, W. J. 1920. On the structure of *Lysorophus* as exposed by serial sections. *Phil. Trans. Roy. Soc. London B*209:481–527.
212. Gregory, J. T. 1965. Microsaurs and the origin of captorhinomorph reptiles. *Amer. Zoologist* 5:277–86.
213. Steen, M. 1938. On the fossil Amphibia from the Gas Coal of Nýřany and other deposits in Czechoslovakia. *Proc. Zool. Soc. London B*108:205–83.
Lepospondyls and labyrinthodonts.
214. Baird, D. 1965. Paleozoic lepospondyl amphibians. *Amer. Zoologist* 5:287–94.
215. Beerbower, J. R. 1963. Morphology, paleoecology, and phylogeny of the Permo-Pennsylvanian amphibian *Diploceraspis*. *Bull. Mus. Comp. Zool.* 130(2):31–108.

MODERN AMPHIBIAN ORDERS

216. Parsons, T. S. and Williams, E. E. 1963. The relationships of the modern Amphibia: a re-examination. *Quart. Rev. Biol.* 38:26–53.
217. Piveteau, J. 1937. Un amphibien du Trias inférieur; essai sur l'origine et l'évolution des amphibiens anoures. *Ann. Paléont.* 26:135–76.
218. Hecht, M. K. 1962–63. A re-evaluation of the early history of the frogs. *Syst. Zool.* 11(1):39–44; 12(1):20–35.
219. Reig, O. A. 1957. Los anuros del Matildense. *Acta Geol. Lilloana* 1:231–97.
220. Casamiquela, R. M. 1961. Nuevos materiales de *Notobatrachus degiustoi* Reig. *Rev. Mus. La Plata,* new ser., Sec. Pal. 4:35–69.
221. Schaeffer, B. 1949. Anurans from the early Tertiary of Patagonia. *Bull. Amer. Mus. Nat. Hist.* 93:45–68.
222. Tihen, J. A. 1965. Evolutionary trends in frogs. *Amer. Zoologist* 5:309–18.
223. Estes, R. 1965. Fossil salamanders and salamander origins. *Amer. Zoologist* 5:319–34.
224. Westphal, F. 1958. Die Tertiären und Rezenten Eurasiatischen Riesensalamander (Genus *Andrias,* Urodela, Amphibia). *Palaeontographica* A110:20–92.

REPTILES—GENERAL AND FAUNAL

Owen and von Meyer in Europe and Cope and, to a lesser extent, Marsh in America were among early workers on fossil reptiles. To a later generation belong such figures as Seeley, Lydekker, Nopcsa, Broili, Schroeder, Wiman, and Peyer in Europe; Williston, Case, and Gregory in America; and Broom in South Africa. Among living researchers, Huene in Germany and Watson in England are important senior workers. Other current or recently active reptilian paleontologists are (in Europe) Appleby, Attridge, Augusta, Bergounioux, Bystrow, Charig, Chudinov, Cox, Devilliers, Efremov, Heyler, Hoffstetter, Janensch, Könjukova, Kühne, Kuhn, Kuhn-Schnyder, Lapparent, Lehman, Orlov, Panchen, Parrington, Persson, Piveteau, Robinson, Rozhdestvenskii, Špinar, Steen, Tarlo, Tatarinov, Viushkov, Walker, and Westphal; (in America) Auffenberg, Baird, Beerbower, Brattstrom, Camp, Carroll, Chase, Colbert, Crompton, Eaton, Estes, J. T. Gregory, Hotton, Langston, Mook, Olson, Ostrom, Parsons, Peabody, Romer, Sawin, Seltin, Sternberg, Swinton, Tihen, Vaughn, Welles, and Zangerl. Workers in Asia include Haas (Israel) and Liu and Young (China); in Africa, Brink, Boonstra, Ewer, Kitching, and Toerien; in South America, Bonaparte, Casamiquela, Price, and Reig.

General accounts in Piveteau (No. 1) and Orlov (No. 2).

225. Romer, A. S. 1956. *Osteology of the Reptiles.* Chicago. Pp. 772.
A comprehensive account of the skeleton and classification of reptiles, fossil and recent.
226. Huene, F. 1956. *Paläontologie und Phylogenie der niederen Tetrapoden.* Jena. Pp. 716. Nachträge und Ergänzungen. Jena. 1959. Pp. 58.
227. Nopcsa, F. 1926. *Osteologia Reptilium Fossilium et Recentium.* Fossilium Catalogus, Pars 27. Berlin. Pp. 391. Also *Supplement.* Fossilium Catalogus, Pars 50 (1931). Pp. 62.
Lists (with comments as to content) every paper of importance dealing with fossil reptiles and amphibians to 1930.
228. Colbert, E. H. 1965. *The Age of Reptiles.* New York and London. Pp. 228.
229. Bellairs, A. d'A. 1957. *Reptiles.* London. Pp. 195.
230. Swinton, W. E. 1962. *Fossil Amphibians and Reptiles.* 3d ed. British Museum (Natural History). London. Pp. 118.
231. Owen, R. 1849–84. *A History of British Fossil Reptiles.* Reprinted from publications of the Palaeontographical Society. London.
232. Williston, S. W. 1914. *Water Reptiles of the Past and Present.* Chicago. Pp. 251.
Semipopular account of aquatic groups.
233. Kuhn, O. 1958. *Die Fährten der vorzeitlichen Amphibien und Reptilien.* Bamberg. Pp. 64.
234. Olson, E. C. 1951–58. Fauna of the Vale and Choza: 1–14. *Fieldiana: Geology* 10 (partim).
235. Romer, A. S. 1958. The Texas Permian redbeds and their vertebrate fauna. *In: Studies on Fossil Vertebrates. Essays Presented to D. M. S. Watson* (ed. T. S. Westoll). London. Pp. 157–79.
236. Peyer, B. 1944. Die Reptilien vom Monte San Giorgio. *Neujahrsbl. Naturf. Ges. Zürich* 146:5–95.
237. Huene, F. 1940. Die Saurier der Karroo-, Gondwana- und verwandten Ablagerungen in faunistischer, biologischer und phylogenetischer Hinsicht. *Neues Jahrb. Min. Geol. Paläont.,* Beil.-Bd., Abt. B 83:246–347.
Lists and comments on many of the older reptilian faunas.
238. ———. 1944. *Die fossilen Reptilien des südamerikanischen Gondwanalandes.* Ergebnisse der Sauriergrabungen in Südbrasilien, 1928–29. Munich. Pp. 332.

COTYLOSAURIA

239. Carroll, R. L. 1964. The earliest reptiles. *Jour. Linn. Soc. London (Zool.)* 45:61–83.
240. Romer, A. S. 1946. The primitive reptile *Limnoscelis* restudied. *Amer. Jour. Sci.* 244:149–88.
241. Watson, D. M. S. 1954. On *Bolosaurus* and the origin and classification of reptiles. *Bull. Mus. Comp. Zool.* 111:297–449.
242. ———. 1957. On *Millerosaurus* and the early history of the sauropsid reptiles. *Phil. Trans. Roy. Soc. London* *B*240:325–400.
243. Efremov, J. A. 1940. Die Mesen-Fauna der permischen Reptilien. *Neues Jahrb. Min. Geol. Paläont.*, Beil.-Bd., Abt. B 84:379–466.
244. Watson, D. M. S. 1914. *Procolophon trigoniceps*, a cotylosaurian reptile from South Africa. *Proc. Zool. Soc. London* 1914:735–47.
   See also Broili and Schroeder (No. 389). Part 21.
245. Colbert, E. H. 1946. *Hypsognathus*, a Triassic reptile from New Jersey. *Bull. Amer. Mus. Nat. Hist.* 86(5): 225–74.
246. ———. 1960. A new Triassic procolophonid from Pennsylvania. *Amer. Mus. Novit.* No. 2022:1–19.
247. Huene, F. 1912. Die Cotylosaurier der Trias. *Palaeontographica* 59:69–102.
248. Haughton, S. H. and Boonstra, L. D. 1929–34. Pareiasaurian studies. 1–11. *Ann. South African Mus.* Vols. 28, 31.

CHELONIA

249. Hay, O. P. 1908. The fossil turtles of North America. *Publ. Carnegie Inst. Washington* No. 75:1–568.
250. Jaekel, O. 1914–16. Die Wirbeltierfunde aus dem Keuper von Halberstadt. *Paleont. Zeitschr.* 1:155–215; 2:88–214.
   Describes *Proganochelys* [*Triassochelys*], a primitive Triassic form.
251. Zangerl, R. 1953. The vertebrate fauna of the Selma Formation of Alabama. Part 3. The turtles of the family Protostegidae. Part 4. The turtles of the family Toxochelyidae. *Fieldiana: Geology Mem.* 3(3, 4):61–277.
252. Zangerl, R. and Sloan, R. E. 1960. A new specimen of *Desmatochelys lowi* Williston, a primitive cheloniid sea turtle from the Cretaceous of South Dakota. *Fieldiana: Geology* 14(2):7–40.
253. Simpson, G. G. 1938. *Crossochelys*, Eocene horned turtle from Patagonia. *Bull. Amer. Mus. Nat. Hist.* 74:221–54.
254. Zangerl, R. 1958. Die oligozänen Meerschildkröten von Glarus. *Schweiz. Paläont. Abhandl.* 73:5–56.
255. Watson, D. M. S. 1914. *Eunotosaurus africanus* Seeley and the ancestry of the Chelonia. *Proc. Zool. Soc. London* 1914:1011–20.

EOSUCHIA

256. Broom, R. 1914. A new thecodont reptile. *Proc. Zool. Soc. London* 1914:1072–77; see also *Bull. Amer. Mus. Nat. Hist.* 51:67–76 (1924).
   On *Youngina.*
257. ———. 1925. On the origin of lizards. *Proc. Zool. Soc. London* 1925(1):1–16.
258. Parrington, F. R. 1935. On *Prolacerta broomi*, gen. et sp. n., and the origin of lizards. *Ann. Mag. Nat. Hist.* ser. 10, 16:197–205.
259. Camp, C. L. 1945. *Prolacerta* and the protorosaurian reptiles. *Amer. Jour. Sci.* 243:18–32, 84–101.
260. Merriam, J. C. 1905. The Thalattosauria, a group of marine reptiles from the Triassic of California. *Mem. California Acad. Sci.* 5(1):1–52.
261. Kuhn-Schnyder, E. 1952. *Askeptosaurus italicus* Nopcsa. *Schweiz. Paläont. Abhandl.* 69(2):1–73.
262. Parks, W. A. 1927. *Champsosaurus albertensis*, a new species of rhynchocephalian from the Edmonton Formation of Alberta. *Univ. Toronto Studies, Geol. Ser.* 23:1–40.
263. Russell, L. S. 1956. The Cretaceous reptile *Champsosaurus natator* Parks, *Bull. Nat. Mus. Canada* 145:1–51.
   See also Peyer (Nos. 236, 356).

SQUAMATA

264. Robinson, P. L. 1962. Gliding lizards from the Upper Keuper of Great Britain. *Proc. Geol. Soc. London.* No. 1601:137–46.
   A full account is in preparation (1966).
265. Gilmore, C. W. 1928. Fossil lizards of North America. *Mem. Nat. Acad. Sci.* 22:1–201.
266. ———. 1943. Fossil lizards of Mongolia. *Bull. Amer. Mus. Nat. Hist.* 81:361–84.
267. ———. 1943. Osteology of *Polyglyphanodon*, an Upper Cretaceous lizard from Utah. *Proc. U.S. Nat. Mus.* 92: 229–65.
268. Fejérváry, A. M. 1923. Beiträge zur einer Monographie der fossilen Ophisaurier. *Paläont. Hungarica* 1:123–220.
269. Estes, R. 1964. Fossil vertebrates from the late Cretaceous Lance Formation, eastern Wyoming. *Univ. Calif. Publ., Geol. Sci.* 49:1–187.
   Principally lizards.
270. Williston, S. W. 1898. Mosasaurs. *Univ. Geol. Surv. Kansas* 4:83–221.
271. Osborn, H. F. 1899. A complete mosasaur skeleton, osseous and cartilaginous. *Mem. Amer. Mus. Nat. Hist.* 1:167–88.
272. Camp, C. L. 1942. California mosasaurs. *Mem. Univ. Calif.* 13:1–68.
273. Bellairs, A. d'A. 1951. The origin of snakes. *Biol. Rev.* 26:193–237.
274. Hoffstetter, R. 1962. Revue des récentes acquisitions concernant l'histoire et la systématique des squamates. *In:* Problèmes Actuels de Paléontologie (Évolution des Vertébrés). *Coll. Internat. Centre Nat. Recher. Sci.* No. 104:243–79.
275. Gilmore, C. W. 1938. Fossil snakes of North America. *Geol. Soc. Amer. Spec. Papers.* No. 9:1–96.
276. Simpson, G. G. 1933. A new fossil snake from the *Notostylops* beds of Patagonia. *Bull. Amer. Mus. Nat. Hist.* 67:1–22.
277. Auffenberg, W. 1963. The fossil snakes of Florida. *Tulane Studies in Zool.* 10(3):131–216.

RHYNCHOCEPHALIA

278. Huene, F. 1939. Die Verwandtschaftsgeschichte der Rhynchosauriden. *Physis* 14:499–522.
   See also Huene (No. 238).
279. ———. 1938. *Stenaulorhynchus*, ein Rhynchosauride der ostafrikanischen Obertrias. *Nova Acta Leopoldina.* new ser., 6:83–121.
280. ———. 1939. Die Lebensweise der Rhynchosauriden. *Paläont. Zeitschr.* 21:232–38.

THECODONTIA

281. Hughes, B. 1963. The earliest archosaurian reptiles. *South African Jour. Sci.* 59:221–41.

282. Ewer, R. F. 1965. The anatomy of the thecodont reptile *Euparkeria*. *Phil. Trans. Roy. Soc. London B*248:379–435.
283. Walker, A. D. 1961. Triassic reptiles from the Elgin area: *Stagonolepis*, *Dasygnathus* and their allies. *Phil. Trans. Roy. Soc. London B*244: 103–204.
284. Colbert, E. H. 1952. A pseudosuchian reptile from Arizona. *Bull. Amer. Mus. Nat. Hist.* 99:561–92.
285. Huene, F. 1936. The constitution of the Thecodontia. *Amer. Jour. Sci.* ser. 5, 32:207–17.
286. ———. 1921. Neue Pseudosuchier und Coelurosaurier aus dem württembergischen Keuper. *Acta Zool.* 2:329–403.
287. Sawin, H. J. 1947. The pseudosuchian reptile *Typothorax meadei*, new species. *Jour. Paleont.* 21(3):201–38.
288. Camp, C. L. 1930. A study of the phytosaurs with description of new material from western North America. *Mem. Univ. Calif.* 10:1–161.
289. Colbert, E. H. 1947. Studies of the phytosaurs *Machaeroprosopus* and *Rutiodon*. *Bull. Amer. Mus. Nat. Hist.* 88: 53–96.
290. Gregory, J. T. 1962. The genera of phytosaurs. *Amer. Jour. Sci.* 260:652–90.
291. Swinton, W. E. 1960. The history of *Cheirotherium*. *Liverpool Manchester Geol. Jour.* 2(3):443–73.
See also Huene (No. 238).

CROCODILIA

292. Haughton, S. H. 1924. The fauna and stratigraphy of the Stormberg series. *Ann. South African Mus.* 12:323–497.
Particularly *Sphenosuchus* and *Notochampsa*.
293. Mook, C. C. 1934. The evolution and classification of the Crocodilia. *Jour. Geol.* 42:295–304.
294. ———. 1925. A revision of the Mesozoic Crocodilia of North America. *Bull. Amer. Mus. Nat. Hist.* 51:319–432.
295. Colbert, E. H. and Mook, C. C. 1951. The ancestral crocodilian *Protosuchus*. *Bull. Amer. Mus. Nat. Hist.* 97(3):143–82.
296. ———. 1946. *Sebecus*, representative of a peculiar suborder of fossil Crocodilia from Patagonia. *Bull. Amer. Mus. Nat. Hist.* 87(4):217–70.
297. Fraas, E. 1902. Die Meer-Krokodilier (Thalattosuchia) des oberen Jura unter specieller Berücksichtigung von *Dacosaurus* und *Geosaurus*. *Palaeontographica* 49:1–72.

PTEROSAURIA

Seeley in England and Plieninger in Germany were among the more important workers on European Jurassic pterosaurs. The American Cretaceous forms were described mainly by Marsh and Williston.

298. Seeley, H. G. 1901. *Dragons of the Air*. London. Pp. 240.
A good summary of older work, but illustrations none too good.
299. Plieninger, F. 1901. Beiträge zur Kenntniss der Flugsaurier. *Palaeontographica* 48:65–90.
Pterodactylus, etc.
300. ———. 1907. Die Pterosaurier der Juraformation Schwabens. *Palaeontographica* 53: 209–314.
301. Eaton, G. F. 1910. Osteology of *Pteranodon*. *Mem. Connecticut Acad. Sci.* 2:1–38.
A good account of this large form.
302. Broili, F. 1936. Weitere Beobachtungen an *Ctenochasma*. *Sitzungsber. Bayer. Akad. Wiss. München* 1936:137–56.
303. ———. 1938. Beobachtungen an *Pterodactylus*. *Sitzungsber. Bayer. Akad. Wiss. München* 1938: 139–54.
304. Young, C. C. 1964. On a new pterosaurian from Sinkiang, China. *Vert. Palasiatica* 8:239–55.

DINOSAURS—GENERAL

Most older European remains were described by Owen. In America, apart from the famous footprints from the Connecticut Valley described by Edward Hitchcock, Marsh was the principal early student of dinosaurs. Later or current workers on dinosaurs include, notably, Huene, Janensch, and Lapparent in Europe; Brown, Colbert, Gilmore, Lambe, Langston, Lull, Ostrom, Parks, Sternberg, and Swinton in North America; and Young in China.

305. Colbert, E. H. 1961. *Dinosaurs—Their Discovery and Their World*. New York. Pp. 300.
306. Swinton, W. E. 1934. *The Dinosaurs: A Short History of a Great Group of Extinct Reptiles*. London. Pp. 233.
307. Marsh, O. C. 1896. The dinosaurs of North America. *Sixteenth Ann. Rept. U.S. Geol. Surv.* Pp. 133–244.
Summarizes all of American types known to that date. Still valuable.
308. Sternberg, C. M. 1946. Canadian dinosaurs. *Bull. Nat. Mus. Canada* 103:1–20.
309. Colbert, E. H. 1962. The weight of dinosaurs. *Amer. Mus. Novit.* No. 2076:1–16.
310. Huene, F. 1929. Los saurisquios y ornitisquios del Cretáceo Argentino. *An. Mus. La Plata* ser. 2a, 3:1–196. Atlas of 44 pls.
311. Lapparent, A. F. de. 1960. Les dinosauriens du "Continental Intercalaire" du Sahara Central. *Mém. Soc. Géol. France*, new ser., 88:1–56.
312. Young, C. C. 1958. The dinosaurian remains of Laiyang, Shantung. *Palaeont. Sinica*. new ser. C, 16:53–138.
313. Bergounioux, F. M. 1956. Les reptiles fossiles des dépôts phosphatés Sud Tunisiens. *Ann. Min. Géol. Roy. Tunis.* No. 15:1–105.
314. Rozhdestvenskii, A. K. 1958. *Auf Dinosaurierjagd in der Gobi*. Leipzig. Pp. 240.

SAURISCHIA—BIPEDS

315. Colbert, E. H. 1964. Relationships of the saurischian dinosaurs. *Amer. Mus. Novit.* No. 2181:1–24.
315a. Charig, A. J., Attridge, J., and Crompton, A. W. 1965. On the origin of the sauropods and the classification of the Saurischia. *Proc. Linn. Soc. London* 176(2): 197–221.
316. Huene, F. 1932. Die fossile Reptil-Ordnung Saurischia, ihre Entwicklung und Geschichte. *Monogr. Geol. Pal.*, 1st ser. 4:1–368.
A modern summary of our knowledge of saurischian dinosaurs, with special reference to Triassic forms.
317. Lull, R. S. 1953. Triassic life of the Connecticut Valley. Rev. ed. *Bull. Connecticut Geol. Nat. Hist. Surv.* 81:1–331.
Dinosaurs, thecodonts, footprints.
318. Huene, F. 1926. The carnivorous Saurischia in the Jura and Cretaceous formations, principally in Europe. *Rev. Mus. La Plata* 29:35–167.
Review of knowledge of theropods (in broad sense) to 1921; particularly discusses "*Megalosaurus*" of European Jurassic and Lower Cretaceous, and *Compsognathus*.
319. Gilmore, C. W. 1920. Osteology of the carnivorous Dinosauria in the United States National Museum, with special reference to the genera *Antrodemus* (*Allosaurus*) and *Ceratosaurus*. *Bull. U.S. Nat. Mus.* 110:1–159.
320. Osborn, H. F. 1917. Skeletal adaptations of *Ornitholestes*,

Struthiomimus, Tyrannosaurus. Bull. Amer. Mus. Nat. Hist. 35:733–71.
A well-illustrated account of these forms.
321. Lambe, L. M. 1917. The Cretaceous theropodous dinosaur *Gorgosaurus. Mem. Geol. Surv. Canada* 100:1–84.
Osteology of a large Cretaceous carnivore.
322. Walker, A. D. 1964. Triassic reptiles from the Elgin area: *Ornithosuchus* and the origin of carnosaurs. *Phil. Trans. Roy. Soc. London B*248:53–134.
323. Huene, F. 1926. Vollständige Osteologie eines Plateosauriden aus dem Schwäbischen Keuper. *Geol. Paläont. Abhandl. Jena.* new ser., 15:1–43.
A prosauropod.
324. Janensch, W. 1925. Die Coelurosaurier und Theropoden der Tendaguru-Schichten Deutsch-Ostafrikas. *Palaeontographica* Suppl. 7:1–99.
325. ———. 1929. Ein aufgestelltes und rekonstruiertes Skelett von *Elaphrosaurus bambergi. Palaeontographica* Suppl. 7:279–86.
326. Young, C. C. 1941. A complete osteology of *Lufengosaurus huenei* Young (gen. et sp. nov.) from Lufeng, Yunnan, China. *Palaeont. Sinica.* new ser. C. 7: 1–53.
See also Haughton (No. 292).

SAUROPODS

327. Wiman, C. 1929. Die Kreide-Dinosaurien aus Shantung. *Palaeont. Sinica.* new ser. C. 6:1–67.
328. Hatcher, J. B. 1901. *Diplodocus* (Marsh), its osteology, taxonomy and probable habits, with a restoration of the skeleton. *Mem. Carnegie Mus.* 1:1–63.
See also Holland, *Mem. Carnegie Mus.* 2:225–64 (1906) and 9:379–403 (1924).
329. Gilmore, C. W. 1925. A nearly complete articulated skeleton of *Camarasaurus,* a saurischian dinosaur from the Dinosaur National Monument, Utah. *Mem. Carnegie Mus.* 10:347–84.
330. ———. 1936. Osteology of *Apatosaurus* with special reference to specimens in the Carnegie Museum. *Mem. Carnegie Mus.* 11:175–300.
331. Janensch, W. 1935. Die Schädel der Sauropoden *Brachiosaurus, Barosaurus* und *Dicraeosaurus* aus den Tendaguru-Schichten Deutsch-Ostafrikas. *Palaeontographica* Suppl. 7:147–248.
332. ———. 1950. Die Wirbelsäule von *Brachiosaurus brancai.* Die Skelettrekonstruktion von *Brachiosaurus brancai. Palaeontographica* Suppl. 7:27–103.
333. Young, C. C. 1958. New sauropods from China. *Vert. Palasiatica* 2(1):1–29.
See also Huene (Nos. 310, 316), Marsh (No. 307).

ORNITHOPODS

334. Crompton, A. W. and Charig, A. J. 1962. A new ornithischian from the Upper Triassic of South Africa. *Nature* (London) 196:1074–77.
335. Swinton, W. E. 1936. Notes on the osteology of *Hypsilophodon* and on the family Hypsilophodontidae. *Proc. Zool. Soc. London* 1936:555–78.
336. Gilmore, C. W. 1915. Osteology of *Thescelosaurus,* an orthopodous dinosaur from the Lance Formation of Wyoming. *Proc. U.S. Nat. Mus.* 49:591–616.
337. ———. 1909. Osteology of the Jurassic reptile *Camptosaurus. Proc. U.S. Nat. Mus.* 36:197–332.
338. Hooley, R. W. 1925. On the skeleton of *Iguanodon atherfieldensis.* sp. nov. *Quart. Jour. Geol. Soc. London* 81:1–60.
339. Casier, E. 1960. *Les iguanodons de Bernissart.* Brussels. Pp. 134.

340. Lull, R. S. and Wright, N. E. 1942. Hadrosaurian dinosaurs of North America. *Geol. Soc. Amer., Spec. Papers.* No. 40. Pp. 242.
341. Ostrom, J. H. 1961. Cranial morphology of the hadrosaurian dinosaurs of North America. *Bull. Amer. Mus. Nat. Hist.* 122(2):33–186.
342. Brown, B. and Schlaikjer, E. M. 1943. A study of the troödont dinosaurs with the description of a new genus and four new species. *Bull. Amer. Mus. Nat. Hist.* 82: 121–49.
343. Osborn, H. F. 1924. *Psittacosaurus* and *Protiguanodon:* two Lower Cretaceous iguanodonts from Mongolia. *Amer. Mus. Novit.* No. 144:1–12.

STEGOSAURS

344. Gilmore, C. W. 1914. Osteology of the armored Dinosauria in the United States National Museum, with special reference to the genus *Stegosaurus. Bull. U.S. Nat. Mus.* 89:1–143.
345. Hennig, E. 1925. *Kentrurosaurus aethiopicus:* Die Stegosaurier-Funde vom Tendaguru, Deutsch-Ostafrika. *Palaeontographica* Suppl. 7:101–254. See also restoration in the same work, pages 255–76 (Janensch).

ANKYLOSAURS

No comprehensive descriptions have been made of members of this interesting group.

346. Lull, R. S. 1921. The Cretaceous armored dinosaur *Nodosaurus textilis* Marsh. *Amer. Jour. Sci.* ser. 5, 1:97–126.
347. Sternberg, C. M. 1928. A new armored dinosaur from the Edmonton Formation of Alberta. *Trans. Roy. Soc. Canada.* ser. 3, 22:93–106.
348. Nopcsa, F. 1929. Dinosaurierreste aus Siebenbürgen. *Geol. Hungarica. ser. Paläont.* 4:1–72.
349. Russell, L. S. 1940. *Edmontia rugosidens* (Gilmore), an armoured dinosaur from the Belly River series of Alberta. *Univ. Toronto Studies, Geol. Ser.* 43:3–28.

HORNED DINOSAURS

350. Colbert, E. H. 1948. Evolution of the horned dinosaurs. *Evolution* 2:145–63.
351. Hatcher, J. B., Marsh, O. C., and Lull, R. S. 1907. The Ceratopsia. *Monogr. U.S. Geol. Surv.* 49:1–157.
Includes all the earlier work.
352. Lull, R. S. 1933. A revision of the Ceratopsia or horned dinosaurs. *Mem. Peabody Mus. Nat. Hist.* 3(3):1–135.
353. Brown, B. and Schlaikjer, E. M. 1940. The structure and relationships of *Protoceratops. Ann. New York Acad. Sci.* 40:133–266.
354. Sternberg, C. M. 1950. *Pachyrhinosaurus canadensis,* representing a new family of Ceratopsia from southern Alberta. *Bull. Ann. Rept. Nat. Mus. Canada* 118:109–20.

EURYAPSIDS

355. Vaughn, P. P. 1955. The Permian reptile *Araeoscelis* restudied. *Bull. Mus. Comp. Zool.* 113(5):303–467.
356. Peyer, B. 1931–37. Die Triasfauna der Tessiner Kalkalpen. 1–12. *Abhandl. Schweiz. Paläont. Ges.* Vols. 50–59.
Including *Tanystropheus,* placodonts, and nothosaurs.
357. Gregory, J. T. 1945. Osteology and relationships of *Trilophosaurus. Univ. Texas Publ.* 4401:273–359.

358. Edinger, T. 1935. *Pistosaurus. Neues Jahrb. Min. Geol. Paläont.*, Beil-Bd. 74:321–59.
359. Andrews, C. W. 1910–13. *A Descriptive Catalogue of the Marine Reptiles of the Oxford Clay.* 2 vols. London. Pp. 411.
360. Welles, S. P. 1952. A review of the North American Cretaceous elasmosaurs. *Univ. Calif. Publ. Geol. Sci.* 29(3): 47–144.
361. ———. 1943. Elasmosaurid plesiosaurs with description of new material from California and Colorado. *Mem. Univ. California* 13:125–254.
362. Fraas, E. 1910. Plesiosaurier aus dem oberen Lias von Holzmaden. *Palaeontographica* 57:105–40.
Figures of two fine skeletons.
363. Williston, S. W. 1903. North American plesiosaurs. 1. *Publ. Field Mus. Dept. Geol.* 2:1–77 (continued in *Amer. Jour. Sci.* ser. 4, 21:221–36 [1906]; *Proc. U.S. Nat. Mus.* 32:477–89 [1907]; and *Jour. Geol.* 16:715–36 [1908]).
364. Persson, P. O. 1963. A revision of the classification of the Plesiosauria with a synopsis of the stratigraphical and geological distribution of the group. *Lunds Univ. Arssk.* N.F. Adv. 2. 59:1–60.
365. Kuhn-Schnyder, E. 1960. Ueber Placodontier. *Palaeont. Zeitschr.* 34(1):91–102.
366. Peyer, B. 1955. Die Triasfauna der Tessiner Kalkalpen. XVIII. *Helveticosaurus zollingeri* n.g. n.sp. *Schweiz. Paläont. Abhandl.* 72:1–50.
367. Broili, F. 1912. Zur Osteologie des Schädels von *Placodus. Palaeontographica* 59:147–155.
Good figures.
368. Drevermann, F. 1933. Das Skelett von *Placodus gigas* Agassiz im Senckenberg Museum. *Abhandl. Senck. Naturf. Ges.* 38:319–64.
Followed by a discussion by Huene, pp. 365–82.
369. Huene, F. 1936. *Henodus chelyops*, ein neuer Placodontier. *Palaeontographica*, Abt. A 84:99–148.
Further description in *ibid.*, Abt. A. 89:105–14 (1938).
See also Peyer (No. 356), Parts 3 and 8.
See also Peyer (No. 236).

MESOSAURS

370. McGregor, J. H. 1908. *On Mesosaurus brasiliensis, nov. sp. from the Permian of Brazil.* Commisão de Estudos das Minas de Carvão de Pedra do Brazil. Pp. 302–36.
371. Huene, F. 1941. Osteologie und systematische Stellung von *Mesosaurus. Palaeontographica* 92*A:* 45–58.

ICHTHYOSAURS

372. Huene, F. 1922. *Die Ichthyosaurier des Lias und ihre Zusammenhänge.* Berlin. Pp. 114.
With a general discussion of ichthyosaur evolution and a bibliography.
373. Gilmore, C. W. 1905. Osteology of *Baptanodon* Marsh. *Mem. Carnegie Mus.* 2:77–129.
A good account of the same form (*Ophthalmosaurus*) in Andrews (No. 359).
374. Sollas, W. J. 1916. The skull of *Ichthyosaurus*, studied in serial section. *Phil. Trans. Roy. Soc. London* *B*208:63–126.
375. Appleby, R. M. 1956. The osteology and taxonomy of the fossil reptile *Ophthalmosaurus. Proc. Zool. Soc. London* 126:403–47.
376. Merriam, J. C. 1908. Triassic Ichthyosauria with special reference to the American forms. *Mem. Univ. California* 1:1–196.
377. Wiman, C. 1910. Ichthyosaurier aus der Trias Spitzbergens. *Bull. Geol. Inst. Univ. Upsala* 10:124–48.
See also *ibid.* 11:230–41 (1912).
378. Huene, F. 1949. Ein Schädel von *Mixosaurus* und die Verwandtschaft der Ichthyosaurier. *Neues Jahrb. Min. Geol. Palaeont.*, Monatsh. Abt. B. Pp. 88–95.

PELYCOSAURS

The earlier work on these forms was done principally by Cope, Case, and Williston; Broom, Watson, Huene, and Olson have also contributed.

379. Romer, A. S. and Price, L. I. 1940. Review of the Pelycosauria. *Geol. Soc. Amer., Spec. Papers.* No. 28:1–538.
380. Romer, A. S. 1948. Relative growth in pelycosaurian reptiles. *Roy. Soc. South Africa, Spec. Publ., Robt. Broom Comm. Vol.* Pp. 45–55.
See also Romer (Nos. 235, 382), Olson (Nos. 234, 383).

THERAPSIDA

381. Watson, D. M. S. and Romer, A. S. 1956. A classification of therapsid reptiles. *Bull. Mus. Comp. Zool.* 114:37–89.
382. Romer, A. S. 1961. *Synapsid evolution and dentition.* Internat. Colloq. on the Evolution of Mammals. Kon. Vlaamse Acad. Wetensch. Lett. Sch. Kunsten België, Brussels 1961. Part 1:9–56.
383. Olson, E. C. 1962. Late Permian terrestrial vertebrates, U.S.A. and U.S.S.R. *Trans. Amer. Phil. Soc.* new ser. 52(2):1–224.
384. Orlov, J. A. 1960. Les dinocéphales rapaces de la faune d'Isheevo (Titanosuchia). *Trad., Bur. Recher. Geol. Min., Paris* No. 2244.
French translation of Russian article appearing in *Trudy Paleont. Inst. Akad. Nauk S.S.S.R.* 72:1–114 (1958). Excellent account of some early therapsids.
385. Efremov, J. A. 1940. Preliminary description of the new Permian and Triassic Tetrapoda from U.S.S.R. *Trav. Inst. Paleont. Acad. Sci. U.R.S.S.* 10(2):1–156.
Particularly primitive dinocephalians.
386. ———. 1940. *Ulemosaurus svijagensis* Riab.—ein Dinocephale aus den Ablagerungen des Perm der UdSSR. *Nova Acta Leopold.* new ser. 9:155–205.
387. Boonstra, L. D. 1963. Early dichotomies in the therapsids. *South African Jour. Sci.* 59:176–95.
388. Broom, R. 1932. *The Mammal-like Reptiles of South Africa and the Origin of Mammals.* London. Pp. 376.
389. Broili, F. and Schroeder, J. 1934–37. Beobachtungen an Wirbeltieren der Karroo-formation. 1–28. *Sitzungsber. Bayer. Akad. Wiss. München* 1934–37.
Good description of *Cynognathus* and other cynodonts, gorgonopsians, therocephalians, bauriamorphs, dicynodonts, and dinocephalians; also *Chasmatosaurus, Procolophon, Tritylodon*, and labyrinthodonts.
390. Olson, E. C. 1944. Origin of mammals based upon cranial morphology of the therapsid suborders. *Geol. Soc. Amer., Spec. Papers.* No. 55. Pp. 136.
391. Broom, R. 1930. On the structure of the mammal-like reptiles of the sub-order Gorgonopsia. *Phil. Trans. Roy. Soc. London* *B*218:345–71.
392. Boonstra, L. D. 1934. A contribution to the morphology of the Gorgonopsia; additions to our knowledge of the South African Gorgonopsia, preserved in the British Museum (Natural History); a contribution to the mor-

phology of the mammal-like reptiles of the sub-order Therocephalia. *Ann. South African Mus.* 31:137–267.

393. Colbert, E. H. 1948. The mammal-like reptile *Lycaenops*. *Bull. Amer. Mus. Nat. Hist.* 89(6):353–404.
394. Broom, R. 1948. A contribution to our knowledge of the vertebrates of the Karroo Beds of South Africa. *Trans. Roy. Soc. Edinburgh* 61(2):577–629.
395. Brink, A. S. 1957. Speculations on some advanced mammalian characteristics in the higher mammal-like reptiles. *Palaeont. Afr.* 4:77–96.
396. ———. 1955. A study on the skeleton of *Diademodon*. *Palaeont. Afr.* 3:3–39.
397. Broom, R. 1936. On the structure of the skull in the mammal-like reptiles of the suborder Therocephalia. *Phil. Trans. Roy. Soc. London B*226:1–42.
398. Attridge, J. 1956. The morphology and relationships of a complete therocephalian skeleton from the Cistecephalus zone of South Africa. *Proc. Roy. Soc. Edinburgh B*66(4): 59–93.
399. Brink, A. S. 1957. On *Aneugomphius ictidoceps* Broom and Robinson. *Palaeont. Afr.* 4:97–115.
400. Cox, C. B. 1965. New Triassic dicynodonts from South America, their origins and relationships. *Phil. Trans. Roy. Soc. London B*248:457–516.
401. Watson, D. M. S. 1948. *Dicynodon* and its allies. *Proc. Zool. Soc. London* 118:823–77.
402. ———. 1960. The anomodont skeleton. *Trans. Zool. Soc. London* 29(3):131–208.
403. Kühne, W. G. 1956. *The Liassic therapsid Oligokyphus*. British Museum (Natural History). London. Pp. 149.
404. Hopson, J. A. 1964. The braincase of the advanced mammal-like reptile *Bienotherium*. *Postilla, Peabody Mus. Nat. Hist.* No. 87:1–30.
405. Crompton, A. W. 1958. The cranial morphology of a new genus and species of ictidosauran. *Proc. Zool. Soc. London* 130(2):183–216.
406. ———. 1963. On the lower jaw of *Diarthrognathus* and the origin of the mammalian lower jaw. *Proc. Zool. Soc. London* 140:697–753.

See also Huene (No. 238).

BIRDS

The literature on fossil birds is widely scattered. A. Wetmore, Loye and Alden H. Miller, H. Howard, and P. Brodkorb in America and the late K. Lambrecht of Budapest have been the more recent active workers.

407. Brodkorb, P. 1963. Catalogue of fossil birds. Part 1. Archaeopterygiformes through Ardeiformes. *Bull. Florida State Mus., Biol. Sci.* 7(4): 179–293. Part 2. Anseriformes through Galliformes. *Ibid.* 8(3): 195–335, 1964.
408. Lambrecht, K. 1933. *Handbuch der Palaeornithologie*. Berlin. Pp. 1024.

   A well-documented account.
409. Swinton, W. E. 1958. *Fossil Birds*. British Museum (Natural History). London. Pp. 63.
410. Wetmore, A. 1951. Presidential address. Recent additions to our knowledge of prehistoric birds, 1933–49. *Proc. 10th Internat. Ornithol. Congr. Uppsala*, 1950. Pp. 51–74.
411. Howard, H. 1955. Fossil birds. *Los Angeles County Museum, Sci. Ser.* No. 10. Pp. 40.
412. Pycraft, W. P. 1910. *A History of Birds*. London. Pp. 489.

   A general account, with some mention of fossil forms.
413. Heilmann, G. 1926. *The Origin of Birds*. London. Pp. 208.

   A well-illustrated and readable discussion of bird origins, with an account of Jurassic and Cretaceous birds, the reconstruction of a hypothetical "Proavis," and a résumé of the archosaurian reptile groups.
414. Holmgren, N. 1955. Studies on the phylogeny of birds. *Acta Zool.* 36(3):243–328.
415. Beer, G. R. de 1954. *Archaeopteryx lithographica*. British Museum (Natural History). London. Pp. 68.
416. Heller, F. 1960. Der dritte *Archaeopteryx*-Fund aus den Solnhofer Plattenkalken des oberen Malm Frankens. *Jour. Ornith. (Berlin)* 101:7–28.
417. Marsh, O. C. 1880. Odontornithes: a monograph on the extinct toothed birds of North America. *Rept. Geol. Expl. 40th Parallel*. Pp. 201.
418. Gregory, J. T. 1951. Convergent evolution: the jaws of *Hesperornis* and the mosasaurs. *Evolution* 5(4):345–54.
419. Simpson, G. G. 1946. Fossil penguins. *Bull. Amer. Mus. Nat. Hist.* 87:1–99.
420. ———. 1957. Australian fossil penguins, with remarks on penguin evolution and distribution. *Rec. S. Austr. Mus.* 13:51–70.
421. Matthew, W. D. and Granger, W. 1917. The skeleton of *Diatryma*, a gigantic bird from the Lower Eocene of Wyoming. *Bull. Amer. Mus. Nat. Hist.* 37:307–26.
422. Patterson, B. and Kraglievich, J. L. 1960. Sistemática y nomenclatura de las Aves fororracoideas del Plioceno Argentino. *Publ. Mus. Mun. Cien. Nat. Trad. Mar del Plata* 1(1):1–49.
423. Milne-Edwards, A. 1867–71. Récherches Anatomiques et Paléontologiques pour servir à l'Histoire des Oiseaux Fossiles de la France. 2 vols. text; 2 vols. pls. Paris.

   Tertiary fossil types and numerous illustrations of skeletons of modern forms as well.
424. Howard, H. 1927. A review of the fossil bird, *Parapavo californicus* (Miller), from the Pleistocene asphalt beds of Rancho La Brea. *Univ. California Publ., Bull. Dept. Geol. Sci.* 17:1–62.
425. ———. 1932. Eagles and eagle-like vultures of the Pleistocene of Rancho La Brea. *Publ. Carnegie Inst. Washington*. No. 429:1–82.
426. ———. 1964. Fossil Anseriformes. *In: The Waterfowl of the World*. Vol. 4. Pp. 233–326.
427. Owen, R. 1879. *Memoirs on the Extinct Wingless Birds of New Zealand*. London. Pp. 513.

   Moas, etc. Mainly a reprint of the author's previous separate papers on moas.

MAMMALS—GENERAL

Cuvier and Owen were early European students of fossil mammals, followed later by such workers as Gaudry, Gervais, and Filhol in France; Forsyth Major and Lydekker in England; Kowalevsky and Rütimeyer in central and eastern Europe. Among later generations of European students of mammals may be cited Abel, Andrews, Bate, Berckhemer, Borissiak, Boule, Dal Piaz, Dépéret, D'Erasmo, Forster-Cooper, Helbing, Heller, Hopwood, Pilgrim, Revilliod, Reynolds, Roman, Schlosser, Sickenberg, Simionescu, Soergel, Stehlin, Stromer, Teilhard de Chardin, Weitzel, and Zdansky. There are numerous active or recently active European workers on fossil mammals, including Adam, Astre, Arambourg, Beliaeva, Bohlin, Butler, Crusafont-Pairo, Dechaseaux, Dehm, Dietrich, Ehrenberg, Erdbrink, Flerov, Friant, Gabuniia, Ginsburg, Gromov, Gromova, Hoffstetter, Hooijer, Hürzeler, Kälin, Kahlke, Kermack, Klebanova, Koenigswald, Kowalski, Kretzoi, Kurtén, Lavocat, LeGros Clark, Leonardi, MacInnes, Musset, Oettingen-Spielberg, Piveteau, Rakovec, Reig, Russell, Saban, Savage, Schaub, Sutcliffe, Thaler, Thenius, Tobien, Trofinov, Vallois, Viret, Weigelt, and Zapfe.

Indian material was described in earlier times by Falconer and Lydekker, more recently by Pilgrim, Forster-Cooper, Matthew, Colbert, Sahni, and Dehm. Much Chinese mammal material was described by a series of writers in *Palaeontologia Sinica* on the basis of material collected by Swedish expeditions. Mongolian mammals were collected by the American Museum (New York), and numerous papers have been published by Matthew and Granger. In recent years native writers such as M. M. Chow, W. C. Pei, and C. C. Young have begun important contributions to the literature of paleontology in China. Japanese workers include Shikama and Takai.

The important Fayûm, Egypt, fauna has been described mainly by Andrews and Schlosser. South and East African mammals (including Australopithecines) have been mainly described by Broom, Cooke, Dart, Ewing, Robinson, and Singer. In Australia, Pleistocene remains were early described by Owen; Ride is a current active worker. Stirton and others have worked in recent years to push the Australian record back into the Tertiary.

In America, all of the three principal early workers—Leidy, Cope, Marsh—published voluminously on mammals. Leidy's and Cope's major works are cited below, while Cope published numerous smaller papers in the *American Naturalist* and most of Marsh's many publications are scattered through the *American Journal of Science*. Osborn, Scott, Wortman, Matthew, Barbour, Gidley, Granger, Gregory, Loomis, Merriam, Peterson, and Sinclair were among later entrants into the field. In North America, as in Europe, a large number of vertebrate paleontologists are included among those who are now working, or have recently been active, in the field of mammals. Among them: Alf, Bader, Black, Churcher, Clemens, Dalquest, Dawson, Dorr, Downs, Edinger, Edmund, Fields, Galbreath, Gazin, Green, Guilday, Hibbard, Hopkins, Hough, Jepsen, Kellogg, Lundelius, Macdonald, McGrew, MacIntyre, McKenna, Olsen, Orr, Patterson, Quinn, Radinsky, Ray, Repenning, J. T. Robinson, P. Robinson, L. Russell, Savage, Schultz, Shotwell, Simons, Simpson, Slaughter, Stirton, Tanner, Tedford, Turnbull, Van Valen, T. E. White, Whitmore, J. A. Wilson, R. E. Wilson, A. E. Wood, and H. E. Wood.

Much of the important South American fossil mammal material was described by Florentino Ameghino on the basis of material collected by his brother Carlos; this work, however, was to some degree obscured by the erection by Ameghino of impossible phylogenetic hypotheses in which, patriotically, South America was made a great center of general mammalian evolution. A number of types, particularly from the Pleistocene, were described by Owen, Burmeister, Gaudry, Moreno, Roth, Lydekker, and other early writers. The Santa Cruz (Miocene) forms were monographed by Scott and Sinclair on the basis of material collected for Princeton by Hatcher; Loomis revised the fauna from the *Pyrotherium* beds. Early Tertiary material has been recently described by Simpson, and later Tertiary collections by Riggs have been described by him and by Patterson and Scott. Recent South American writers include Bordas, Cabrera, Castellanos, Kraglievich, Pascual, Paulo Couto, Reig, and Rusconi.

For general accounts see Piveteau (No. 1) and Orlov (No. 2).

428. Thenius, E. and Hofer, H. 1960. *Stammesgeschichte der Säugetiere. Eine Uebersicht über Tatsachen und Probleme der Evolution der Säugetiere.* Berlin. Pp. 322.

429. Young, J. Z. 1957. *The Life of Mammals.* Oxford. Pp. 820.

430. Weber, M., Burlet, H. M. de, and Abel, O. 1928. *Die Säugetiere.* 2d ed. 2 vols. Jena.

A standard work on mammalian anatomy and classification. The treatment of fossils is uneven.

431. Simpson, G. G. 1945. The principles of classification and a classification of mammals. *Bull. Amer. Mus. Nat. Hist.* 85:1–450.

432. Flower, W. H. and Lydekker, R. 1891. *An Introduction to the Study of Mammals, Living and Extinct.* London. Pp. 763.

Still a good account of living forms, although much of the paleontology is out of date.

433. Flower, W. H. 1885. *An Introduction to the Osteology of the Mammalia.* 3d ed. London. Pp. 383.

An old, but very useful, little book.

434. Blainville, H. M. D. 1839–64. *Osteographie des Mammifères.* 4 vols. Paris.

An early comprehensive illustrated account of fossil mammals. Cuvier's still earlier work (No. 15) is mainly devoted to mammals also.

435. Matthew, W. D. 1915. Climate and evolution. *Ann. New York Acad. Sci.* 24:171–318.

Important on geographic history of mammals. A new edition with notes was published in 1939.

436. Osborn, H. F. 1910. *The Age of Mammals in Europe, Asia and North America.* New York. Pp. 652.

A faunal treatment, out of date but still useful.

437. Scott, W. B. 1937. *A History of Land Mammals in the Western Hemisphere.* 2d ed. New York. Pp. 786.

A good semitechnical account of the more prominent groups of American fossils.

438. Patterson, B. 1957. Mammalian phylogeny. *Publ. Internat. Union Biol. Sci., Paris*, Ser. B. No. 32:15–49.

439. Leidy, J. 1854. The ancient fauna of Nebraska. *Smithson. Contr. Knowl.* 6(7):1–126.

This and the following contain the original descriptions of many American fossil types.

440. ———. 1869. The extinct mammalian fauna of Dakota and Nebraska, including an account of some allied forms from other localities, together with a synopsis of the mammalian remains of North America. *Jour. Acad. Nat. Sci. Philadelphia.* ser. 2, 7:1–472.

441. Cope, E. D. 1884. The Vertebrata of the Tertiary formations of the West. *Rept. U.S. Geol. Surv. Territ.* Vol. 3. Pp. 1044.

"Cope's Bible," a huge tome, which figures and describes a great number of Tertiary mammals. A basic work of importance, although many of the forms included have been described later from more complete material.

FAUNAL AND REGIONAL PAPERS

Following more general accounts, Nos. 451 to 494 are arranged in approximately stratigraphic sequence.

442. Ameghino, F. 1889. Contribución al conocimiento de los mamíferos fósiles de la República Argentina. *Actas Acad. Nac. Cienc. Córdoba* 6:1–1028.

Reprinted as Vols. 6–9 of *Obras Completas y Correspondencia Científica de F. Ameghino* (La Plata). This and the following are two of Ameghino's basal contributions to the paleontology of South America.

443. ———. 1906. Les formations sédimentaires du Crétacé supérieur et du Tertiaire de Patagonie. *An. Mus. Nac. Hist. Nat. Buenos Aires*. ser. 3, 8:1–568.
Reprinted as Vol. 16 of *Obras completas*.

444. Borissiak, A. A. 1962. A survey of fossil sites of Tertiary land mammals in the U.S.S.R. *Internat. Geol. Review* 4: 845–67.

445. Andrews, R. C. 1932. *The New Conquest of Central Asia: A Narrative of the Central Asiatic Expeditions in Mongolia and China*, 1921–1930. New York. Pp. 678.
Includes data on vertebrate fossils.

446. Wood, H. E., Chaney, R. W., Clark, J., Colbert, E. H., Jepsen, G. L., Reeside, J. B., Jr., and Stock, C. 1941. Nomenclature and correlation of the North American continental Tertiary. *Bull. Geol. Soc. Amer.* 52:1–48.
A revision, edited by Wood, Patterson, and Simpson is in preparation.

447. Olsen, S. J. 1959. Fossil mammals of Florida. *Florida Geol. Surv., Spec. Publ.* No. 6. Pp. 74.

448. Simpson, G. G. 1940. Review of the mammal-bearing Tertiary of South America. *Proc. Amer. Phil. Soc.* 83: 649–709.
See also *Amer. Scientist* 38:361–89 (1950).

449. ———. 1942. Early Cenozoic mammals of South America. *Proc. 8th Amer. Sci. Congr.* 4:303–32.

450. Scott, W. B. 1942. The later Cenozoic mammalian faunas of South America. *Proc. 8th Amer. Sci. Congr.* 4:333–57.

451. Sloan, R. E. and Van Valen, L. 1965. Cretaceous mammals from Montana. *Science* 148:220–27.

452. Clemens, W. A., Jr. 1964. Fossil mammals of the type Lance Formation, Wyoming. Part 1. Introduction and Multituberculata. *Univ. California Publ. Geol. Sci.* 48: 1–105.

453. Simpson, G. G. 1937. The beginning of the age of mammals. *Biol. Rev.* 12:1–47.

454. Matthew, W. D. 1937. Paleocene faunas of the San Juan Basin, New Mexico. *Trans. Amer. Phil. Soc.* new ser. 30:1–510. (eds. W. Granger, W. K. Gregory, and E. H. Colbert.)

455. Simpson, G. G. 1937. The Fort Union of the Crazy Mountain field, Montana, and its mammalian faunas. *Bull. U.S. Nat. Mus.* 169:1–287.

456. Gazin, C. L. 1941. The mammalian faunas of the Paleocene of central Utah, with notes on the geology. *Proc. U.S. Nat. Mus.* 91:1–53.

457. ———. 1956. Paleocene mammalian faunas of the Bison Basin in south-central Wyoming. *Smithson. Misc. Coll.* 131(6):1–57.

458. Van Houten, F. 1945. Review of latest Paleocene and early Eocene mammalian faunas. *Jour. Paleont.* 19(5): 421–61.

459. Simpson, G. G. 1948. The beginning of the age of mammals in South America. *Bull. Amer. Mus. Nat. Hist.* 91(1):1–232.

460. Paula Couto, C. de 1952. Fossil mammals from the beginning of the Cenozoic in Brazil. Condylarthra, Litopterna, Xenungulata, and Astrapotheria. *Bull. Amer. Mus. Nat. Hist.* 99(6):355–94.

461. Russell, D. 1964. Les mammifères Paléocènes d'Europe. *Mém. Mus. Nat. Hist. Natur.* new ser. C. 13:1–321.

462. Stehlin, H. G. 1903–16. Die Säugetiere des schweizerischen Eocaens: critischer Catalog der Materialien. *Abhandl. Schweiz. Paläont. Ges.* 30:1–153; 31:155–455; 32:447–595; 33:597–690; 35:691–837; 36:839–1164; 38: 1165–1298; 41:1297–1552.

463. Matthew, W. D. and Granger, W. 1915–18. A revision of the Lower Eocene Wasatch and Wind River faunas. 1–5. *Bull. Amer. Mus. Nat. Hist.* 34:1–103, 311–28, 329–61; 429–83; 38:565–657.

464. McKenna, M. C. 1960. Fossil Mammalia from the early Wasatchian Four Mile fauna, Eocene of northwest Colorado. *Univ. California Publ. Geol. Sci.* 37(1):1–130.

465. Gazin, C. L. 1962. A further study of the Lower Eocene mammalian faunas of southwestern Wyoming. *Smithson. Misc. Coll.* 144(1):1–98.

466. ———. 1952. The Lower Eocene Knight Formation of western Wyoming and its mammalian faunas. *Smithson. Misc. Coll.* 117(18):1–82.

467. Dehm, R., and Oettingen-Spielberg, T. 1958. Paläontologische und geologische Untersuchungen im Tertiär von Pakistan. 2. Die mitteleocänen Säugetiere von Ganda Kas bei Basal in Nordwest-Pakistan. *Abhandl. Bayer. Akad. Wiss.* N.F. 91:1–54.

468. Bohlin, B. 1942–46. The fossil mammals from the Tertiary deposit of Tabenbuluk, western Kansu. 1. Insectivora and Lagomorpha. *Palaeont. Sinica*. Ser. C. No. 8*a*:1–113 (1942). 2. Simplicidentata, Carnivora, Artiodactyla, Perissodactyla, and Primates. *Palaeont. Sinica*. Ser. C. No. 8*b*:1–259 (1946).

469. Scott, W. B., Jepsen, G. L., and Wood, A. E. 1941. The mammalian fauna of the White River Oligocene. *Trans. Amer. Phil. Soc.* N.S. 28:1–980.
An excellent and well illustrated discussion of American Oligocene mammals.

470. Galbreath, E. C. 1953. A contribution to the Tertiary geology and paleontology of northeastern Colorado. *Univ. Kansas Publ., Paleont. Contr., Vertebrata*. Art. 4: 1–120.

471. Lavocat, R. 1951. *Révision de la faune des Mammifères Oligocènes d'Auvergne et du Velay*. Sciences et Avenir, Paris, Pp. 153.

472. Filhol, H. 1876–77. Recherches sur les phosphorites du Quercy: étude des fossiles qu'on y rencontre et spécialement des mammifères. *Ann. Sci. Géol.* 7:1–220; 8:1–340. Also published as a separate work.

473. ———. 1882. Étude des mammifères fossiles de Ronzon (Haute-Loire). *Ann. Sci. Géol.* 12:1–271.
Oligocene European mammals.

474. ———. 1881. Étude des mammifères fossiles de Saint-Gérand le Puy (Allier). *Ann. Sci. Géol.* 11:1–86.
Lower Miocene mammals.

475. ———. 1891. Étude sur les mammifères fossiles de Sansan. *Ibid.* 21:1–319.
Middle Miocene.

476. Matthew, W. D. 1924. Third contribution to the Snake Creek fauna. *Bull. Amer. Mus. Nat. Hist.* 50:59–210.

477. Whitmore, F. C., Jr. and Stewart, R. H. 1965. Miocene mammals and Central American seaways. *Science* 148: 180–85.

478. Wilson, R. W. 1960. Early Miocene rodents and insectivores from northeastern Colorado. *Univ. Kansas Publ., Paleont. Contr., Vertebrata*. Art. 7:1–92.

479. Thenius, E. 1952. Die Säugetierfauna aus dem Torton von Neudorf an der March (ČSR). *Neues Jahrb. Min. Geol. Paläont.* 96(1):27–136.

480. Downs, T. 1956. The Mascall fauna from the Miocene of Oregon. *Univ. California Publ. Geol. Sci.* 31(5):199–354.

481. Colbert, E. H. 1935. Siwalik mammals in the American Museum of Natural History. *Trans. Amer. Phil. Soc.* 26:1–401.
482. Gaudry, A. 1867. *Animaux Fossiles et Géologie de l'Attique.* 2 vols. Paris. Pp. 363.
A classic account of a Pontian (Pliocene) fauna.
483. Savage, D. E. 1951. Late Cenozoic vertebrates of the San Francisco Bay region. *Univ. California Publ. Geol. Sci.* 28(10):215–314.
484. Viret, J. 1954. Le loess à bancs durcis de Saint-Vallier (Drôme) et sa faune Mammifères villafranchiens. *Nouv. Arch. Mus. Hist. Nat. Lyon* 4:1–200.
485. Romer, A. S. 1933. Pleistocene vertebrates and their bearing on the problem of human antiquity in North America. *In: The American Aborigines* (ed. D. Jenness). Toronto. Pp. 49–83.
486. Kurtén, B. 1960. Chronology and faunal evolution of the earlier European glaciations. *Soc. Scient. Fennica Comment. Biol.* 21(5):3–62.
487. Reynolds, S. H. 1902–39. A monograph of the British Pleistocene Mammalia. *Monogr. Palaeont. Soc. London.* Vols. 2, 3.
488. Hibbard, C. W. 1958. Summary of North American Pleistocene mammalian local faunas. *Papers Michigan Acad. Sci.* 43:3–32.
489. Cooke, H. B. S. 1963. Pleistocene mammal faunas of Africa, with particular reference to southern Africa. *In: African Ecology and Human Evolution* (eds. Howell, F. C. and Bourlière, F.) Chicago. Pp. 65–116.
490. Colbert, E. H. and Hooijer, D. A. 1953. Pleistocene mammals from the limestone fissures of Szechwan, China. *Bull. Amer. Mus. Nat. Hist.* 102(1):1–134.
491. Hay, O. P. 1923–27. The Pleistocene of North America and its vertebrated animals. *Publ. Carnegie Inst. Washington.* Nos. 322, 322*A*, 322*B*. Pp. 1230.
492. Hibbard, C. W. and Taylor, D. W. 1960. Two late Pleistocene faunas from southwestern Kansas. *Contr. Mus. Paleont. Univ. Michigan* 16(1):3–223.
493. Stock, C. 1930. Rancho La Brea: a record of Pleistocene life in California. *Publ. Los Angeles Mus.* No. 1:1–82.
494. Leakey, L. S. B. 1965. *Olduvai Gorge* 1951–1961. *Fauna and Background.* Cambridge, England. Pp. 109.

THE MESOZOIC GROUPS

495. Simpson, G. G. 1928. *A Catalogue of the Mesozoic Mammalia in the Geological Department of the British Museum.* London. Pp. 225.
This and the following give a complete account of all early finds of pre-Tertiary mammals.
496. ———. 1929. American Mesozoic Mammalia. *Mem. Peabody Mus. Yale Univ.* 3(1):1–235.
497. ———. 1962. *Evolution of Mesozoic Mammals.* International Colloquium on Evolution of Mammals. Brussels. Part 1. Pp. 57–95.
498. Kermack, K. A. 1965. The origin of mammals. *Science Journal* (Sept. 1965): 66–72.
499. Kermack, K. A. 1963. The cranial structure of the triconodonts. *Phil. Trans. Roy. Soc. London B*246:83–103.
500. Simpson, G. G. 1937. Skull structure of the Multituberculata. *Bull. Amer. Mus. Nat. Hist.* 73:727–63.
On multituberculates see also Simpson (No. 455. Pp. 80–104), Matthew (No. 454. Pp. 277–96), Sloan and Van Valen (No. 451), Clemens (No. 452), Russell (No. 461. Pp. 22–41), McKenna (No. 464. Pp. 33–41).
501. Jepsen, G. L. 1940. Paleocene faunas of the Polecat Bench Formation, Park County, Wyoming. 1. *Proc. Amer. Phil. Soc.* 83:217–340.
Multituberculates.
502. Patterson, B. 1956. Early Cretaceous mammals and the evolution of mammalian molar teeth. *Fieldiana: Geology* 13(1):1–105.

MARSUPIALS

503. Simpson, G. G. 1930. Postmesozoic Marsupialia. *Fossilium catalogus*, 1. Animalia. Pars 47. Berlin. Pp. 87.
Literature and a general discussion of marsupial evolution.
504. Hofer, H. 1952. Ueber das gegenwärtige Bild der Evolution der Beuteltiere. *Zool. Jahrb.*, Abt. Anat. 72:365–437.
505. Simpson, G. G. 1940. The development of marsupials in South America. *Physis* 14:373–98.
506. Sinclair, W. J. 1906. Mammalia of the Santa Cruz beds: Marsupialia. *Rept. Princeton Univ. Exped. Patagonia* 4: 333–460.
507. Patterson, B. 1958. Affinities of the Patagonian fossil mammal *Necrolestes. Breviora, Mus. Comp. Zool.* No. 94: 1–14.
508. Owen, R. 1878. *Researches on the Fossil Remains of the Extinct Mammals of Australia.* London.
Reprints of papers previously published separately in *Phil. Trans. Roy. Soc. London* (1859–72); *Diprotodon, Thylacoleo*, and other Pleistocene forms.
509. Stirton, R. A. 1955. Late Tertiary marsupials from South Australia. *Rec. South Australian Mus.* 11(3):247–68.
510. Stirton, R. A., Tedford, R. H., and Miller, A. H. 1961. Cenozoic stratigraphy and vertebrate paleontology of the Tirari Desert, South Australia. *Rec. South Australian Mus.* 14:19–61.
511. Ride, W. D. L. 1959. On the evolution of Australian marsupials. *In: The Evolution of Living Organisms, a Symposium of the Royal Society of Victoria.* Melbourne. Pp. 281–306.
See also Simpson (No. 459. Pp. 32–69), Filhol (No. 473. Pp. 51–68), Scott (No. 469. Pp. 960–63).

INSECTIVORA

512. Gregory, W. K. and Simpson, G. G. 1926. Cretaceous mammal skulls from Mongolia. *Amer. Mus. Novit.* No. 225:1–20.
513. Simpson, G. G. 1951. American Cretaceous insectivores. *Amer. Mus. Novit.* No. 1541:1–19.
514. Butler, P. M. 1956. The skull of *Ictops* and the classification of the Insectivora. *Proc. Zool. Soc. London* 126(3): 453–81.
515. McKenna, M. C. 1963. Primitive Paleocene and Eocene Apatemyidae (Mammalia, Insectivora) and the primate-insectivore boundary. *Amer. Mus. Novit.* No. 2160:1–39.
516. Butler, P. M. 1948. On the evolution of the skull and teeth in the Erinaceidae, with special reference to fossil material in the British Museum. *Proc. Zool. Soc. London* 118(2):446–500.
517. ———. 1956. Erinaceidae from the Miocene of East Africa: *Brit. Mus.* (*Nat. Hist.*), *Fossil Mamm. Afr.* No. 11:1–75.
518. Patterson, B. 1965. The fossil elephant shrews (family Macroscelididae). *Bull. Mus. Comp. Zool.* 133(6):295–335.

BATS

519. Revilliod, P. 1917–22. Contribution à l'étude des Chiroptères des terrains Tertiaires. *Mém. Soc. Pal. Suisse* 43:1–57; 44:63–128; 45:133–95.

An excellent Lower Eocene bat is being described by Jepsen.

TAENIODONTS

520. Patterson, B. 1949. Rates of evolution in taeniodonts. *In: Genetics, Paleontology, and Evolution* (eds. G. L. Jepsen, E. Mayr, and G. G. Simpson). Princeton. Pp. 243–78.

See also Matthew (No. 454. Pp. 228–77).

TILLODONTS

521. Gazin, C. L. 1953. The Tillodontia: an early Tertiary order of mammals. *Smithson. Misc. Coll.* 121(10):1–110.

522. Van Valen, L. 1963. The origin and status of the mammalian order Tillodontia. *Jour. Mamm.* 44:364–73.

PRIMATES

523. LeGros Clark, W. E. 1960. *The Antecedents of Man: An Introduction to the Evolution of the Primates.* Chicago. Pp. 374.

524. ———. 1955. *The Fossil Evidence for Human Evolution.* Chicago. Pp. 181.

525. Simons, E. L. 1963. A critical reappraisal of Tertiary primates. *In: Evolutionary and Genetic Biology of Primates* (ed. J. Buettner-Janusch). New York. Pp. 65–129.

526. Simpson, G. G. 1962. Primate taxonomy and recent studies of nonhuman primates. *Ann. New York Acad. Sci.* 102(2):497–514.

527. ———. 1940. Studies on the earliest primates. *Bull. Amer. Mus. Nat. Hist.* 77:185–212.

On early primates see also Simpson (No. 455. Pp. 141–169), Matthew (No. 454. Pp. 103–62), Matthew and Granger (No. 463. Part 4), Filhol (No. 472. Pp. 275–326), Scott, Jepsen, and Wood (No. 469. Pp. 26–29), Gazin (No. 465. Pp. 27–42. No. 457. Pp. 19–25), McKenna (No. 464. Pp. 63–81), Russell (No. 461. Pp. 80–123).

528. Gregory, W. K. 1920. On the structure and relations of *Notharctus*, an American Eocene primate. *Mem. Amer. Mus. Nat. Hist.* N.S. 3:49–243.

Includes much general material on primate evolution.

529. Piveteau, J. 1948. Recherches anatomiques sur les lémuriens disparus: le genre *Archaeolemur. Ann. Paléont.* 34: 125–72.

530. Simons, E. L. 1962. A new Eocene primate genus, *Cantius*, and a revision of some allied European lemuroids. *Bull. Brit. Mus. (Nat. Hist.), Geol.* 7(1):1–36.

531. ———. 1961. Notes on Eocene tarsioids and a revision of some Necrolemurinae. *Bull. Brit. Mus. (Nat. Hist.), Geol.* 5(3):43–69.

532. Gazin, C. L. 1958. A review of the Middle and Upper Eocene primates of North America. *Smithson. Misc. Coll.* 136(1):1–112.

533. Straus, W. L., Jr. 1963. The classification of *Oreopithecus*. *In: Classification and Human Evolution* (ed. S. L. Washburn). Chicago. Pp. 146–77.

534. Ferembach, D. 1959. Les limnopithèques du Kenya. *Ann. Paléont.* 44:149–249.

535. Simons, E. L. and Pilbeam, D. R. 1965. Preliminary revision of the Dryopithecinae. *Folia Primatologica* 3:81–152.

536. LeGros Clark, W. E. and Leakey, L. S. B. 1951. The Miocene Hominoidea of East Africa. *Brit. Mus. (Nat. Hist.), Fossil Mamm. Afr.* No. 1:1–117.

537. Howells, W. 1962. *Ideas on Human Evolution.* Cambridge, Mass. Pp. 555.

538. Simons, E. L. 1964. The early relatives of man. *Scient. Amer.* 211(1): 50–62.

539. Leakey, L. S. B. 1961. *The Progress and Evolution of Man in Africa.* London. Pp. 50.

540. Broom, R. and Schepers, G. W. H. 1946. The South African fossil ape-men, the Australopithecinae. *Mem. Transvaal Mus.* 2:1–272.

541. Robinson, J. T. 1962. Australopithecines and the origin of man. *Ann. Rept. U.S. Nat. Mus.* for 1961. Pp. 479–500.

542. Boule, M. and Vallois, H. V. *Fossil Men. A Text-Book of Human Palaeontology.* 5th French ed. Revised and translated by M. Bullock. London. Pp. 535.

CREODONTS

543. Van Valen, L. 1966. Deltatheridia: a new order of mammals. *Bull. Amer. Mus. Nat. Hist.* (In press.)

544. Wortman, J. L. 1901–2. Studies of Eocene Mammalia in the Marsh collection, Peabody Museum. I. Carnivora. *Amer. Jour. Sci.* ser. 4, 11:333–48, 437–50; ser. 4, 12: 143–54, 196–206, 281–96, 377–82, 421–32; ser 4, 13: 39–46, 115–28, 197–206, 433–48; and ser. 4, 14: 17–23. (Only the last four parts apply to Creodonta as defined here.)

545. Matthew, W. D. 1909. The Carnivora and Insectivora of the Bridger Basin, Middle Eocene. *Mem. Amer. Mus. Nat. Hist.* 9: 291–567. (Creodonts, pp. 466–84.)

546. Denison, R. H. 1938. The broad-skulled Pseudocreodi. *Ann. New York Acad. Sci.* 37:163–256.

547. Gazin, C. L. 1957. A skull of the Bridger Middle Eocene creodont, *Patriofelis ulta* Leidy. *Smithson. Misc. Coll.* 134(8):1–20.

See also Matthew and Granger (No. 463, Part 1. Pp. 5–13, 42–103), Scott, Jepsen, and Wood (No. 469. Pp. 30–53), Filhol (No. 472. Pp. 169–250), and Colbert (No. 481. Pp. 75–82).

CARNIVORES

548. Butler, P. M. 1948. The evolution of carnassial dentitions in the Mammalia. *Proc. Zool. Soc. London* 116(2):198–220.

549. Ginsberg, L. 1961. La faune des carnivores de Sansan (Gers). *Mém. Mus. Nat. Hist. Natur.* N.S. C 9:1–187.

550. Dehm, R. 1950. Die Raubtiere aus dem Mittel-Miocän (Burdigalium) von Wintershof-West bei Eichstätt in Bayern. *Abhandl. Bayer. Akad. Wiss., Math.-Nat. Abt.* 58:1–141.

551. Hough, J. R. 1948. The auditory region in some members of the Procyonidae, Canidae, and Ursidae. *Bull. Amer. Mus. Nat. Hist.* 92(2):67–118.

See also Viret (No. 484. Pp. 31–91), Thenius (No. 479. Pp. 41–67), Colbert and Hooijer (No. 490. Pp. 41–71), Galbreath (No. 470. Pp. 75–80), Lavocat (No. 471. Pp. 89–106), Bohlin (No. 468, Pt. 1), Downs (No. 480. Pp. 231–39).

MIACIDS AND VIVERRIDS

552. Gregory, W. K. and Hellman, M. 1939. On the evolution and major classification of the civets (Viverridae) and

allied fossil and Recent Carnivora: a phylogenetic study of the skull and dentition. *Proc. Amer. Phil. Soc.* 81: 309–92.

On miacids see Wortman (No. 544, ser. 4, vol. 11, pp. 338–450; ser. 4, vol. 12, pp. 143–54), Matthew and Granger (No. 463. Part 1. Pp. 16–42), Simpson (No. 455. Pp. 207–15), Matthew (No. 545. Pp. 339–406). (No. 454. Pp. 100–103), Gazin (No. 466. Pp. 54–59. No. 465. Pp. 56–61).

On viverrids see Colbert (No. 481. Pp. 102–4), Pilgrim (No. 555. Pp. 96–109), Filhol (No. 475. Pp. 118–26).

FELIDS

553. Matthew, W. D. 1910. The phylogeny of the Felidae. *Bull. Amer. Mus. Nat. Hist.* 28:289–316.

554. Merriam, J. C. and Stock, C. 1932. The Felidae of Rancho La Brea. *Publ. Carnegie Inst. Washington.* No. 422. Pp. 231.

555. Pilgrim, G. E. 1932. The fossil Carnivora of India. *Palaeont. Indica* N.S. 18:1–232.

See also Scott, Jepsen, and Wood (No. 469. Pp. 109–49), Filhol (No. 472. Pp. 152–69. No. 475. Pp. 47–94), Matthew (No. 476. Pp. 146–49), Colbert (No. 481. Pp. 116–25), Gaudry (No. 482. Pp. 105–21).

MUSTELIDS

556. Teilhard de Chardin, P. and Leroy, P. 1945. Les mustélidés de Chine. *Publ. Inst. Geobiol.* 12:1–56.

See also Matthew (No. 476. Pp. 128–46), Filhol (No. 475. Pp. 94–117), Pilgrim (No. 555. Pp. 59–96), Colbert (No. 481. Pp. 94–102).

CANIDS

557. Matthew, W. D. 1930. The phylogeny of dogs. *Jour. Mammal.* 11:117–38.

558. Hatcher, J. B. 1902. Oligocene Canidae. *Mem. Carnegie Mus.* 1:65–108.

Particularly *Daphoenus.*

559. Peterson, O. A. 1910. Description of new carnivores from the Miocene of western Nebraska. *Mem. Carnegie Mus.* 4:205–78.

Particularly *Daphoenodon.*

560. Hough, J. R. 1948. A systematic revision of *Daphoenus* and some allied genera. *Jour. Paleont.* 22:573–600.

See also Scott, Jepsen, and Wood (No. 469. Pp. 55–106), Filhol (No. 472. Pp. 53–144. No. 473. Pp. 19–39. No. 475. Pp. 126–68), Pilgrim (No. 555. Pp. 13–36), Matthew (No. 476. Pp. 88–128), Colbert (No. 481. Pp. 82–88), Reynolds (No. 487. Vol. 2. Part 3).

PROCYONIDS

561. McGrew, P. O. 1938. Dental morphology of the Procyonidae with a description of *Cynarctoides*, gen. nov. *Publ. Field Mus. Nat. Hist., Geol. Ser.* 6:323–39.

BEARS

562. Erdbrink, D. P. 1953. A review of fossil and Recent bears of the Old World. Deventer. Pp. 597.

563. Frick, C. 1926. The Hemicyoninae, and an American Tertiary bear. *Bull. Amer. Mus. Nat. Hist.* 56:1–119.

564. Merriam, J. C. and Stock, C. 1925. Relationships and structure of the short-faced bear *Arctotherium*, from the Pleistocene of California. *Publ. Carnegie Inst. Washington* No. 347:1–35.

565. Abel, O. and Kyrle, G. 1931. *Die Drächenhohle bei Mixnitz.* Speläol. Monogr. Nos. 7–9. Pp. 953.

Cave bears.

566. Kurtén, B. 1958. Life and death of the Pleistocene cave bear. *Acta Zool. Fennica* 95:1–59.

567. Davis, D. 1964. The giant panda. A study of evolutionary mechanisms. *Fieldiana. Zoology, Mem.* 3:1339.

See also Colbert (No. 481. Pp. 89–93), Pilgrim (No. 555. Pp. 36–52), Reynolds (No. 487. Vol. 2. Part 2).

PINNIPEDS

568. Kellogg, R. 1922. Pinnipeds from Miocene and Pleistocene deposits of California . . . and a résumé of current theories regarding the origin of the Pinnipedia. *Univ. California Publ. Bull. Dept. Geol. Sci.* 13:23–132.

Includes a summary of all pinnipeds then known.

CONDYLARTHS

Cope was the describer of *Phenacodus*, central type of the order, and responsible for emphasizing its primitive nature. See Matthew (No. 454. Pp. 1–100, 103–62, 185–209), Simpson (No. 455. Pp. 170–207, 216–69), Gazin (No. 456. Pp. 24–52), Matthew and Granger (No. 463. Pp. 13–16, 311–61), Matthew (No. 545. Pp. 485–502, 508–22), Wortman (No. 544. Vol. 12. Pp. 285–432; Vol. 13. Pp. 39–46), Russell (No. 461. Pp. 133–268), Gazin (No. 466. Pp. 50, 59–61. No. 457. Pp. 19–47), Paula Couto (No. 460. Pp. 361–63), Sloan and Van Valen (No. 451. P. 226), Simpson (No. 459. Pp. 94–142), Gazin (No. 465. Pp. 50–52, 61–69), McKenna (No. 464. Pp. 89–91, 97–114), Dehm and Oettingen-Spielberg (No. 467. Pp. 11–21).

TUBULIDENTATA

569. MacInnes, D. G. 1956. Fossil Tubulidentata from East Africa. *Brit. Mus. (Nat. Hist.), Fossil Mamm. Afr.* No. 10:1–38.

570. Colbert, E. H. 1941. A study of *Orycteropus gaudryi* from the island of Samos. *Bull. Amer. Mus. Nat. Hist.* 78: 305–51.

AMBLYPODS

571. Marsh, O. C. 1884. Dinocerata, a monograph of an extinct order of gigantic mammals. *Monogr. U.S. Geol. Surv.* 10:1–255.

572. Simpson, G. G. 1929. A new Paleocene uintathere and molar evolution in the Amblypoda. *Amer. Mus. Novit.* No. 387:1–9.

573. Flerov, C. C. 1957. A new coryphodont from Mongolia, and on evolution and distribution of the Pantodonta. *Vert. Palasiatica* 1:73–81.

574. Patterson, B. 1939. New Pantodonta and Dinocerata from the Upper Paleocene of western Colorado. *Field Mus. Nat. Hist., Geol. Ser.* 6:351–84.

Earlier accounts of *Barylambda* appeared in *Proc. Amer. Phil. Soc.*, 73:71–101 (1934); 75:143–62 (1935).

575. Wheeler, W. H. 1961. Revision of the uintatheres. *Bull. Peabody Mus. Nat. Hist.* 14:1–93.

576. Simons, E. L. 1960. The Paleocene Pantodonta. *Trans. Amer. Phil. Soc.* 50(6):3–81.

See also Loomis (No. 589), Paula Couto (No. 460. Pp. 370–86), Matthew (No. 454. Pp. 162–84).

SUBUNGULATES

577. Andrews, C. W. 1906. *A Descriptive Catalogue of the Tertiary Vertebrata of the Fayûm, Egypt*. London. Pp. 324.
Describes the earliest known subungulate faunas—*Moeritherium*, *Palaeomastodon* (and *Phiomia*), *Arsinoitherium*, early hyracoids, and sirenians.
578. Schlosser, M. 1911. Beiträge zur Kenntnis der oligozänen Landsäugetiere aus dem Fayüm, Ägypten. *Beitr. Pal. Geol. Oesterr.-Ungarns* 24:51–167.
579. Matsumoto, H. 1926. Contribution to the knowledge of the fossil Hyracoidea of the Fayûm, Egypt. *Bull. Amer. Mus. Nat. Hist.* 56:253–350.
580. Osborn, H. F. 1936–42. *Proboscidea: A Monograph of the Discovery, Evolution, Migration, and Extinction of the Mastodonts and Elephants of the World*. Vol. 1. *Moeritherioidea, Deinotherioidea, Mastodontoidea*. Vol. 2. *Stegodontoidea, Elephantoidea*. New York. Pp. 1675.
581. Watson, D. M. S. 1946. The evolution of the Proboscidea. *Biol. Rev.* 21:15–29.
582. Reinhart, R. H. 1959. A review of the Sirenia and Desmostylia. *Univ. California Publ. Geol. Sci.* 36:1–146.
583. Depéret, C. and Roman, F. 1920. Le *Felsinotherium serresi* des sables pliocènes de Montpellier et les rameaux phylétiques des siréniens fossiles de l'ancien monde. *Arch. Mus. Hist. Nat. Lyon* 12:1–55.
584. Sickenberg, O. 1934. Beiträge zur Kenntnis tertiärer Sirenen. *Mém. Mus. Hist. Nat. Belgique* 63:1–352.
A brief account in *Palaeobiologica* 4:405–44.
585. Simpson, G. G. 1932. Fossil Sirenia of Florida and the evolution of the Sirenia. *Bull. Amer. Mus. Nat. Hist.* 59: 419–503.

SOUTH AMERICAN UNGULATES

586. Scott, W. B. 1910. Litopterna of the Santa Cruz Beds. *Rept. Princeton Univ. Exped. Patagonia* 6:287–300.
587. Sefve, I. 1923. *Macrauchenia patagonica*. *Bull. Geol. Inst. Univ. Upsala* 19:1–21.
Osteology, restoration, habits.
588. Simpson, G. G., Minoprio, J. L., and Patterson, B. 1962. The mammalian fauna of the Divisadero Largo Formation, Mendoza, Argentina. *Bull. Mus. Comp. Zool.* 127: 239–93.
589. Loomis, F. B. 1914. *The Deseado Formation of Patagonia*. Amherst. Pp. 237. Oligocene forms.
590. Sinclair, W. J. 1909. Mammalia of the Santa Cruz beds: Typotheria. *Rept. Princeton Univ. Exped. Patagonia* 6: 1–110.
591. Scott, W. B. 1912. Mammalia of the Santa Cruz beds. 3. Entelonychia. *Ibid.* 6:239–86.
592. ———. 1912. Mammalia of the Santa Cruz beds: Order Toxodontia. *Ibid.* 6:287–300.
593. ———. 1928. Astrapotheria of the Santa Cruz beds. *Ibid.* 6(4):301–42.
Much of the early work in Ameghino (Nos. 442 and 443); see various later papers by Simpson and Patterson. See also Matthew and Granger (No. 463. Pp. 429–33), Paula Couto (No. 460. Pp. 367–70, 387–94), Simpson (No. 459. Pp. 142–222).

PERISSODACTYLS

See Viret (No. 484. Pp. 135–63), Thenius (No. 479. Pp. 100–109), Colbert and Hooijer (No. 490. Pp. 81–102), Lavocat (No. 471. Pp. 107–20), Wilson (No. 478. Pp. 51–87), Dehm and Oettingen-Spielberg (No. 467. Pp. 21–26).

HORSES

594. Simpson, G. G. 1951. *Horses*. New York. Pp. 247.
595. Edinger, T. 1948. Evolution of the horse brain. *Mem. Geol. Soc. Amer.* 25:1–177.
596. Stirton, R. A. 1940. Phylogeny of North American Equidae. *Univ. California Publ., Bull. Dept. Geol. Sci.* 25(4): 165–98.
597. Filhol, H. 1888. Études sur les vertébrés fossiles d'Issel (Aude). *Mém. Soc. Géol. France* (3)5:1–188.
Palaeotheres.
See also Filhol (No. 472. Vol. 8:158–71. No. 473. Pp. 68–74), Stehlin (No. 462. Parts 2 and 3).
598. Kitts, D. B. 1956. American *Hyracotherium* (Perissodactyla, Equidae). *Bull. Amer. Mus. Nat. Hist.* 110:1–60.
599. Osborn, H. F. 1918. Equidae of the Oligocene, Miocene, and Pliocene of North America, iconographic type revision. *Mem. Amer. Mus. Nat. Hist.* N.S. 2(1):1–330.
See also Scott, Jepsen, and Wood (No. 469. Pp. 911–53), Matthew (No. 476. Pp. 153–75), Filhol (No. 475. Pp. 109–94).
600. Sefve, I. 1927. Die Hipparionen Nord-Chinas. *Palaeont. Sinica*. Ser. C. Vol. 4. Fasc. 2. Pp. 91.
See also Colbert (No. 481. Pp. 129–62) and Gaudry (No. 482).

TITANOTHERES

601. Osborn, H. F. 1929. The titanotheres of ancient Wyoming, Dakota and Nebraska. *Monogr. U.S. Geol. Surv.* No. 55. 2 vols. Pp. 953.
Complete account of all known forms. On American Oligocene forms see also Scott, Jepsen, and Wood (No. 469. Pp. 871–910).
602. Granger, W. and Gregory, W. K. 1943. A revision of the Mongolian titanotheres. *Bull. Amer. Mus. Nat. Hist.* 80: 349–89.

CHALICOTHERES

603. Radinsky, L. B. 1964. *Paleomoropus*, a new early Eocene chalicothere (Mammalia, Perissodactyla), and a revision of Eocene chalicotheres. *Amer. Mus. Novit.* No. 2179:1–28.
604. Holland, W. J. and Peterson, O. A. 1913. The osteology of the Chalicotheroidea, with special reference to a mounted skeleton of *Moropus elatus* Marsh, now installed in the Carnegie Museum. *Mem. Carnegie Mus.* 3:189–406.
See also Colbert (No. 481. Pp. 162–76).

TAPIROIDS

605. Radinsky, L. B. 1963. Origin and early evolution of North American Tapiroidea. *Bull. Peabody Mus. Nat. Hist.* 17:1–106.
606. ———. 1965. Early Tertiary Tapiroidea of Asia. *Bull. Amer. Mus. Nat. Hist.* 129:183–263.
607. Schaub, S. 1928. Der Tapirschädel von Haslen: ein Beitrag zur Revision der oligocänen Tapiriden Europas. *Abhandl. Schweiz. Paläont. Ges.* 47:1–28.
See Stehlin (No. 462. Part 1).

608. Radinsky, L. B. 1965. Evolution of the tapiroid skeleton from *Heptodon* to *Tapirus*. *Bull. Mus. Comp. Zool.* 134(3):69–103.

On *Protapirus* see Scott, Jepsen, and Wood (No. 469. Pp. 749–62).

609. Wood, H. E. 1934. Revision of the Hyrachyidae. *Bull. Amer. Mus. Nat. Hist.* 67:181–295.

RHINOCEROSES

610. Osborn, H. F. 1923. The extinct giant rhinoceros *Baluchitherium* of western and central Asia. *Nat. Hist.* 23: 209–28.

611. Matthew, W. D. 1932. A review of the rhinoceroses with a description of *Aphelops* material from the Pliocene of Texas. *Univ. California Publ., Dept. Geol. Sci.* 20:411–80.

612. Wood, H. E. 1941. Trends in rhinoceros evolution. *Trans. New York Acad. Sci.* ser. 2, 3:83–96.

613. ———. 1949. Evolutionary rates and trends in rhinoceroses. *In: Genetics, Paleontology, and Evolution* (eds. G. L. Jepsen, E. Mayr, and G. G. Simpson). Princeton. Pp. 185–89.

614. Peterson, O. A. 1920. The American diceratheres. *Mem. Carnegie Mus.* 7:399–456.

615. Ringstrom, T. 1924. Nashörner der *Hipparion*-Fauna Nordchinas. *Paläont. Sinica.* Ser. C. Vol. 1. Part 4. Pp. 150.

616. Zeuner, F. E. 1945. New reconstructions of the woolly rhinoceros and Merck's rhinoceros. *Proc. Linn. Soc. London* 156:183–95.

ARTIODACTYLS—GENERAL

The literature on artiodactyls is vast, and there are few comprehensive treatments.

617. Pilgrim, G. E. 1941. The dispersal of the Artiodactyla. *Biol. Rev.* 16:134–63.

618. Matthew, W. D. 1929. Reclassification of the artiodactyl families. *Bull. Geol. Soc. Amer.* 40:403–8.

Based mainly on foot structure. For classifications on other bases see Pearson (No. 622), Stehlin (No. 462).

619. Schaeffer, B. 1948. The origin of a mammalian ordinal character. *Evolution* 2:164–75.

See also Gazin (No. 466. Pp. 70–77. No. 465. Pp. 80–86), Savage (No. 483. Pp. 252–79), Dehm and Oettingen-Spielberg (No. 467. Pp. 26–38), Downs (No. 480. Pp. 299–318), Lavocat (No. 471. Pp. 121–41), Galbreath (No. 470. Pp. 83–91), Colbert and Hooijer (No. 490. Pp. 102–30), Thenius (No. 479. Pp. 70–99), Viret (No. 484. Pp. 105–33).

PALAEODONTA

620. Sinclair, W. J. 1914. A revision of the bunodont Artiodactyla of the Middle and Lower Eocene of North America. *Bull. Amer. Mus. Nat. Hist.* 33:267–95.

On European Eocene artiodactyls see Stehlin (No. 462. Part 4); see also Scott, Jepsen, and Wood (No. 469. Pp. 365–78).

621. Macdonald, J. R. 1955. The Leptochoeridae. *Jour. Paleont.* 29:439–59.

SUINA

622. Pearson, H. S. 1927. On the skulls of early Tertiary Suidae, together with an account of the otic region in some other primitive Artiodactyla. *Phil. Trans. Roy. Soc. London B*215:389–460.

The braincase of primitive European artiodactyls, with suggestions as to classification based on the presence or absence of the mastoid.

623. Peterson, O. A. 1909. A revision of the Entelodontidae. *Mem. Carnegie Mus.* 4:42–158.

See also Scott, Jepsen, and Wood (No. 469. Pp. 378–441).

624. Colbert, E. H. 1938. *Brachyhyops*, a new bunodont artiodactyl from Beaver Divide, Wyoming. *Ann. Carnegie Mus.* 27:87–108.

625. Kowalevsky, W. 1873. Monographie der Gattung *Anthracotherium* Cuv. und Versuch einer natürlichen Classification der fossilen Hufthiere. *Palaeontographica* 22: 131–346.

A classic work on ungulate classification and evolution.

626. Colbert, E. H. 1938. Fossil mammals from Burma in the American Museum of Natural History. *Bull. Amer. Mus. Nat. Hist.* 74:255–436.

Particularly primitive anthracotheres; cf. Colbert (No. 481. Pp. 266–79); Dehm and Oettingen-Spielberg (No. 467).

627. Macdonald, J. R. 1956. The North American anthracotheres. *Jour. Paleont.* 30(3):615–45.

628. Pearson, H. S. 1928. Chinese fossil Suidae. *Palaeont. Sinica.* Ser. C. Vol. 5. Part 5. Pp. 75.

629. Gidley, J. W. 1921. Pleistocene peccaries from the Cumberland Cave deposit. *Proc. U.S. Nat. Mus.* 57:651–78.

630. Lundelius, E. L. 1960. *Mylohyus nasutus.* Long-nosed peccary of the Texas Pleistocene. *Bull. Texas Mem. Mus.* No. 1:9–40.

See also Stehlin (No. 462. Part 5), Filhol (No. 473. Pp. 85–240. No. 474. Pp. 6–40. No. 475. Pp. 205–32), Scott, Jepsen, and Wood (No. 469. Pp. 441–93), Gaudry (No. 482. Pp. 218–44), Colbert (No. 481. Pp. 214–94).

TYLOPODA

631. Gazin, C. L. 1955. A review of the Upper Eocene Artiodactyla of North America. *Smithson. Misc. Coll.* 128(8): 1–96.

632. Schultz, C. B. and Falkenbach, C. H. 1940–56. Contribution to the revision of the oreodonts (Merycoidodontidae). Nos. 1–7. *Bull. Amer. Mus. Nat. Hist.* 77:213–306 (1940); 79:1–105 (1941); 88:157–286 (1947); 93:69–198 (1949); 95:87–150 (1950); 105:143–256 (1954); 109: 373–482 (1956).

633. Hürzeler, J. 1936. Osteologie und Odontologie der Caenotheriden. *Abhandl. Schweiz. Paläont. Ges.* 58:1–88; 59: 91–112.

634. Peterson, O. A. 1904. Osteology of *Oxydactylus*. *Ann. Carnegie Mus.* 2:434–76.

635. Webb, S. D. 1965. The osteology of *Camelops*. *Bull. Los Angeles County Mus., Sci.* No. 1: 1–54.

See also Stehlin (No. 462. Part 6), Filhol (No. 472. Pp. 355–76, 404–45. No. 473. Pp. 79–85. No. 475. Pp. 20–36), Scott, Jepsen, and Wood (No. 469. Pp. 602–45).

TRAGULOIDS

636. Colbert, E. H. 1941. The osteology and relationships of *Archaeomeryx*, an ancestral ruminant. *Amer. Mus. Novit.* No. 1135:1–24.

See also Filhol (No. 472. Pp. 445–67. No. 473. Pp. 240–54. No. 475. Pp. 232–47), Scott, Jepsen, and Wood (No. 469. Pp. 506–603).

PALAEOMERYCIDS

637. Stirton, R. A. 1944. Comments on the relationships of the cervoid family Palaeomerycidae. *Amer. Jour. Sci.* 242: 633–55.

638. Frick, C. 1937. Horned ruminants of North America. *Bull. Amer. Mus. Nat. Hist.* 69:1–669.

Essentially an illustrated catalogue of a vast collection of American material, mainly palaeomerycids and antilocaprids.

639. Whitworth, T. 1958. Miocene ruminants of East Africa. *Brit. Mus. (Nat. Hist.), Fossil Mamm. Afr.* No. 15:1–48.

See also Filhol (No. 474. Pp. 40–65. No. 475. Pp. 247–64), Matthew (No. 476. Pp. 193–98).

CERVIDS

640. Thenius, E. 1950. Die tertiären Lagomeryciden und Cerviden der Steiermark. *Sitzungsber. Oesterr. Akad. Wiss., Math.-Naturw. Kl.* 159:219–54.

641. Zdansky, O. 1925. Fossile Hirsche Chinas. *Palaeont. Sinica.* Ser. C. Vol. 2. Fasc. 3. Pp. 90.

642. Colbert, E. H. 1936. Tertiary deer discovered by the American Museum Asiatic expeditions. *Amer. Mus. Novit.* No. 854:1–21.

See additional notes in *Amer. Mus. Novit.* No. 1062 (1940).

See also Filhol (No. 475. Pp. 268–93), Gaudry (No. 482. Pp. 304–8); Colbert (No. 481. Pp. 314–23).

GIRAFFIDS

643. Singer, R. and Boné, E. L. 1960. Modern giraffes and the fossil giraffids of Africa. *Ann. South African Mus.* 45: 375–548.

644. Bohlin, B. 1926. Die Familie Giraffidae mit besonderer Berücksichtigung der fossilen Formen aus China. *Palaeont. Sinica.* Ser. C. Vol. 4. No. 1. Pp. 179.

645. Crusafont Pairó, M. 1952. Los jiráfidos fósiles de España. *Mem. Comm. Inst. Geol. Barcelona* 8:1–239.

See also Colbert (No. 481. Pp. 323–75), Gaudry (No. 482. Pp. 245–71).

ANTILOCAPRIDS

646. Colbert, E. H. and Chaffee, R. G. 1939. A study of *Tetrameryx* and associated fossils from Papago Spring Cave, Sonoita, Arizona. *Amer. Mus. Novit.* No. 1034:1–21.

With notes on antilocaprid phylogeny and taxonomy.

See also Frick (No. 638), Matthew (No. 476. Pp. 198–206).

BOVIDS

647. Pilgrim, G. E. 1947. The evolution of the buffaloes, oxen, sheep, and goats. *Jour. Linn. Soc. London* 279:272–86.

648. Pilgrim, G. E. and Hopwood, A. T. 1928. *Catalogue of the Pontian Bovidae of Europe in the Department of Geology of the British Museum.* London. Pp. 144.

649. Pilgrim, G. E. 1939. The fossil Bovidae of India. *Palaeont. Indica.* N.S. Vol. 26. No. 1. Pp. 356.

See also Gaudry (No. 482. Pp. 271–304).

EDENTATES

650. Simpson, G. G. 1931. *Metacheiromys* and the Edentata. *Bull. Amer. Mus. Nat. Hist.* 59:295–381.

See also Matthew and Granger (No. 463. Part 5), Scott (No. 469. Pp. 955–59).

651. Hoffstetter, R. 1956. Caractères ancestraux et phylogénie des édentés xénarthres. *In:* Problèmes Actuels de Paléontologie. *Coll. Internat. Centre Nat. Recher. Sci.* No. 60:87–98.

652. Burmeister, H. 1874. Monografía de los glyptodontes en el Museo Público de Buenos Aires. *An. Mus. Nac. Buenos Aires* 2:1–412.

653. Scott, W. B. 1903–4. Mammalia of the Santa Cruz Beds: Edentata. *Rept. Princeton Univ. Exped. Patagonia.* 5: 1–364.

654. Stock, C. 1925. Cenozoic gravigrade edentates of western North America. *Publ. Carnegie Inst. Washington.* No. 331. Pp. 206.

Detailed description of *Nothrotherium* and *Mylodon* and general discussion of occurrence and relationships of ground sloths.

655. Matthew, W. D. and Paula Couto, C. 1959. The Cuban edentates. *Bull. Amer. Mus. Nat. Hist.* 117(1):1–56.

See also Simpson (No. 459. Pp. 70–94).

WHALES

656. Kellogg, R. 1928. The history of whales: their adaptation to life in the water. *Quart. Rev. Biol.* 3:29–76, 174–208.

657. Slijper, E. J. 1936. Die Cetaceen: Vergleichend-anatomisch und systematisch. *Capita Zool.* 7:1–590.

Both fossil and Recent forms.

658. ———. 1962. *Whales.* London. Pp. 475.

659. Kellogg, R. 1936. A review of the Archaeoceti. *Publ. Carnegie Inst. Washington.* No. 482:1–366.

660. Beneden, P. J. Van and Gervais, P. 1880. *Ostéographie des cétacés vivants et fossiles.* Paris. Pp. 642.

RODENTS

661. Ellerman, J. R. 1940–49. *The Families and Genera of Living Rodents.* With a list of named forms (1758–1936) by R. W. Hayman and G. W. C. Holt. Vol. 1. Rodents other than Muridae. Pp. 689 (1940). Vol. 2. Muridae. Pp. 690 (1941). Vol. 3. Part 1. Additions and corrections to vols. 1 and 2, with appendices 1 and 2. Pp. 210 (1949). London. British Museum (Natural History).

662. Wood, A. E. 1959. Eocene radiation and phylogeny of the rodents. *Evolution* 13(3):354–61.

663. Lavocat, R. 1962. Réflexions sur l'origine et la structure du groupe des rongeurs. *Coll. Internat. Centre Nat. Recher. Sci.* No. 104:287–99.

664. Wood, A. E. 1962. The early Tertiary rodents of the family Paramyidae. *Trans. Amer. Phil. Soc.* N.S. 52(1): 1–261.

665. Stehlin, H. G. and Schaub, S. 1951. Die Trigonodontie der Simplicidentaten Nager. *Schweiz. Paläont. Abhandl.* 67:1–384.
666. Wilson, R. W. 1949. Early Tertiary rodents of North America. *Publ. Carnegie Inst. Washington.* No. 584:67–164.
667. Shotwell, J. A. 1958. Evolution and biogeography of the aplodontid and mylagaulid rodents. *Evolution* 12(4): 451–84.
668. Black, C. C. 1963. A review of the North American Tertiary Sciuridae. *Bull. Mus. Comp. Zool.* 130(3):109–248.
669. Stirton, R. A. 1935. A review of the Tertiary beavers. *Univ. California Publ., Bull. Dept. Geol. Sci.* 23:391–458.
670. Wood, A. E. 1935. Evolution and relationships of the heteromyid rodents with new forms from the Tertiary of western North America. *Ann. Carnegie Mus.* 24:73–262.
671. ———. 1965. Grades and clades among rodents. *Evolution* 19(1):115–30.
672. ———. 1950. Porcupines, paleogeography and parallelism. *Evolution* 4(1):87–98.
673. Wood, A. E. and Patterson, B. 1959. The rodents of the Deseadan Oligocene of Patagonia and the beginnings of South American rodent evolution. *Bull. Mus. Comp. Zool.* 120(3):282–428.
See also Viret (No. 484. Pp. 93–104), Colbert and Hooijer (No. 490. Pp. 30–41), Galbreath (No. 470. Pp. 51–74), Lavocat (No. 471. Pp. 38–85), Bohlin (No. 468. 2. Pp. 13–197), Downs (No. 480. Pp. 215–31).

LAGOMORPHS

674. Wood, A. E. 1957. What, if anything, is a rabbit? *Evolution* 11:417–25.
675. Dawson, M. R. 1958. Later Tertiary Leporidae of North America. *Univ. Kansas Paleont. Contr., Vertebrata.* No. 6:1–75.

INDEX

The index is mainly taxonomic. It includes references to all genera and groups of higher rank contained in the classification as well as in the text, and to the more common popular names.

A few of the more general accounts of major skeletal features have been indexed, and references are given to the place where various structures are first described or defined. For the details of anatomy in the various types, consult the name of the genus or group. Various geologic and geographic terms are included.

[References to classification in boldface; to figures in italics. **R** and **L** in references to classification indicate right and left columns on the page.]